A GENERAL HISTORY OF HOROLOGY

Gaston Bogaert, *Le Passé*.

An architect, theatre designer, and graphic artist who taught at the l'Ecole supérieure des techniques de publicité, Brussels, the Franco-Belgian Bogaert (1918–2008) worked in the surrealist tradition of Magritte and Delvaux. Philosopher, he published several works: *Procès d'une métaphysique* in 1980, *Propylées* in 1988. In 1989, he accompanied his exhibition at the Galerie 2016 with an essay entitled *L'Enigme du temps*. Time is one of his major sources of inspiration and reflection. Refined and poetic, constantly renewed, the compositions of Bogaert display his ideas on the fall of civilizations, the enduring time of nature, and the ephemeral time of humanity. Mechanical clocks often appear in the 'figuration spiritualisée' of his conception of the world. In *Le Passé*, he evokes the ruins of a lost civilization which are opposed to the eternal force of nature represented by the snowy summits of the mountains. In the centre, a young woman, allegorical of human temporality, indicates a complex monumental skeleton clock in the form of a gateway. Bogaert himself said of *Le Passé* that the clock carries a message of hope, not one of ineluctable destruction.

Oil on board, 54 × 65 cm, 1989, private collection.

Title Page Illustration

The seal of an 'Horlogiarius', 'Robert the clockmaker from (?)Yarmouth', *c.*1300. © York Archaeological Trust, GB. Reproduced by permission.

A GENERAL HISTORY OF HOROLOGY

EDITOR

ANTHONY TURNER

ADVISORY EDITORS

JAMES NYE AND JONATHAN BETTS

OXFORD
UNIVERSITY PRESS

OXFORD
UNIVERSITY PRESS

Great Clarendon Street, Oxford, OX2 6DP,
United Kingdom

Oxford University Press is a department of the University of Oxford.
It furthers the University's objective of excellence in research, scholarship, and education by
publishing worldwide. Oxford is a registered trade mark of Oxford University Press in the UK
and in certain other countries

Supported by Sotheby's

© Oxford University Press 2022

The moral rights of the authors have been asserted

Impression: 1

All rights reserved. No part of this publication may be reproduced, stored in
a retrieval system, or transmitted, in any form or by any means, without the
prior permission in writing of Oxford University Press, or as expressly permitted
by law, by licence or under terms agreed with the appropriate reprographics
rights organization. Enquiries concerning reproduction outside the scope of the
above should be sent to the Rights Department, Oxford University Press, at the
address above

You must not circulate this work in any other form and you must impose this same condition
on any acquirer

British Library Cataloguing in Publication Data

Data available

Library of Congress Control Number: 2021936263

Data available

ISBN 978-0-19-886391-5

DOI: 10.1093/oso/9780198863915.001.0001

Printed in Great Britain by
Bell & Bain Ltd., Glasgow

Links to third party websites are provided by Oxford in good faith and for information only.
Oxford disclaims any responsibility for the materials contained in any third-party website
referenced in this work.

To write down all I contain at this moment
I would pour the desert through an hour-glass,
The sea through a water-clock,
Grain by grain and drop by drop
Let in the trackless, measureless, mutable seas and sands.

Kathleen Raine, 'The Moment', 1946.

Today most professional historians 'specialise'. They choose a period, sometimes a very brief period, and within that period they strive, in desperate competition with ever-expanding evidence, to know all the facts. [. . .] Theirs is a static world. They have a self-contained economy, a Maginot Line, and large reserves which they seldom use; but they have no philosophy. For a historical philosophy is incompatible with such narrow frontiers. It must apply to humanity in any period. To test it, a historian must dare to travel abroad, even in hostile country; to express it he must be ready to write essays even on subjects on which he may be ill-qualified to write books.

Hugh Trevor-Roper, *Historical Essays*, 1958, Foreword.

. . . time that brings all things to ruin, perfects also everything.

Thomas Browne, *Pseudoxia epidemica* (1648), 1672, 302.

CONTENTS

List of Contributors		x
Introduction		xii
1.	Time Measurement in Antiquity *Jérôme Bonnin*	1
2.	India and the Far East	27
	Section One: Horology in India *S. R. Sarma*	27
	Section Two: China to 1900 *David Chang*	41
	Section Three: Modern China *Ron Good and Jon Ward*	53
	Section Four: Japan *Katsuhiro Sasaki*	59
3.	Late Antiquity and the Middle Ages	77
	Section One: Sundials and Water Clocks in Byzantium and Islam *Anthony Turner*	77
	Section Two: Time Reckoning in the Medieval Latin World *Mario Arnaldi*	99
	Section Three: Water Clocks in Christian Europe and Early Escapements *Sebastian Whitestone*	121
	Section Four: Sand-Clocks, Sand-Glasses, Fire Clocks *Anthony Turner*	133
4.	Public Clocks from the Thirteenth to the Eighteenth Centuries *Marisa Addomine*	137
5.	The Domestic Clock in Europe	153
	Section One: From the Thirteenth Century to the Invention of the Pendulum *Dietrich Matthes*	153
	Section Two: From Huygens to the End of the Eighteenth Century *Wim van Klaveren*	171
6.	Watches 1500–1800 *David Thompson*	185

CONTENTS

7.	The Structures of Horological Manufacture and Trade: Sixteenth to Eighteenth Centuries *Anthony Turner*	215
8.	The Development of Sundials: Fourteenth to Twentieth Centuries *Denis Savoie*	231
9.	Clocks as Astronomical Models	253
	Section One: 'The Heavens Daily in View': Planetary Clocks in Europe, Fourteenth to Sixteenth Centuries *Karsten Gaulke, Michael Korey, and Samuel Gessner*	253
	Section Two: Nineteenth and Twentieth Centuries *Denis Roegel*	273
10.	Musical and Automaton Clocks and Watches: Sound and Motion in Time-Telling Devices *Sharon Kerman*	289
11.	The Quest for Precision: Astronomy and Navigation *Jonathan Betts*	311
12.	Decimal Time *Anthony Turner*	341
13.	Clock- and Watch Making from the Nineteenth to Twenty-First Centuries Industrial Manufacture and Worldwide Trade	347
	Section One: The Mixed Fortunes of Britain *James Nye*	347
	Section Two: American Horology and its Global Reach *Michael Edidin*	355
	Section Three: The Horological Endeavour in France *Joëlle Mauerhan*	363
	Section Four: The Challenge of the Swiss and their Competitors *Johann Boillat*	370
	Section Five: Developing the German Industry *Sibylle Gluch*	378
	Section Six: The Pendule de Paris, From the Workshop to the Factory, 1800–1910 *Françoise Collanges*	391
14.	Precision Attained: The Nineteenth and Twentieth Centuries *Jonathan Betts*	403
15.	Responding to Customer Demand: The Decoration of Clocks and Watches from the Renaissance to Recent Times *Catherine Cardinal*	421
16.	Eighteenth-Century Clock Exports from Britain to the East Indies *Roger Smith*	443
17.	Public Clocks in the Nineteenth and Twentieth Centuries *Marisa Addomine*	463
18.	Wristwatches from their Origins to the Twenty-First Century *David Boettcher*	473
19.	Electricity, Horology, and Networked Time *James Nye and David Rooney*	495
20.	Women in Horology *Joëlle Mauerhan*	531

CONTENTS

21. Keeping Clocks and Watches: Maintenance, Repair, and Restoration — 541
    Jonathan Betts
22. Accessories in Horology — 555
    Estelle Fallet
23. Applications of Clockwork — 569
    - Section One: Planetary Models — 570
      Jim Bennett and Anthony Turner
    - Section Two: Timing and Driving Systems — 575
      Paolo Brenni
    - Section Three: Metronomes — 582
      Anthony Turner
    - Section Four: Car Clocks — 586
      James Nye
    - Section Five: The Noctuary or Watchman's Clock — 591
      Jonathan Betts
    - Section Six: Roasting-jacks — 594
      Anthony Turner
24. Horology Verbalized, Horology Visualized — 599
    Christina J. Faraday
25. The Literature of Horology — 621
    Bernhard Huber
26. Collecting and Writing the History of Horology — 643
    Anthony Turner

Glossary — 653

Bibliography & Abbreviations — 661

Index — 730

LIST OF CONTRIBUTORS

Marisa Addomine holds a doctorate in engineering, is an independent horological researcher, and President of the Italian turret clock register.

Mario Arnaldi is an artist and sundial designer who researches and publishes widely on the history and development of Medieval and Renaissance dials.

Jim Bennett FSA is Emeritus Director of the Museum of the History of Science, Oxford.

Jonathan Betts MBE, FSA is Curator Emeritus of the Royal Museums, Greenwich and Horological Adviser to several heritage bodies, including the National Trust. He is Vice-Chairman and Honorary Librarian of the Antiquarian Horological Society.

David Boettcher is a Chartered Engineer and Fellow of the British Horological Institute who has researched the early history of wristwatches.

Johann Boillat teaches at the Haute Ecole de Suisse Occidentale HES-20, Neuchâtel, Switzerland.

Jérôme Bonnin holds a doctorate in Roman archaeology. He is mainly interested in ancient timekeeping, sources, artefacts, representations, and social needs.

Paolo Brenni[†] was President of the Scientific Instrument Commission of the International Union of History and Philosophy of Science and President of the Scientific Instrument Society.

Catherine Cardinal is Emeritus Professor of the History of Art at the University of Clermont-Auvergne and former Scientific Director and Curator of the Musée International d'Horlogerie, La Chaux-de-Fonds, Switzerland.

David Chang is Vice-President of the Macau Horology Association and Member of the Chinese Society of Cultural Relics.

Françoise Collanges is a consultant in conservation to public museums and heritage sites. Trained as a conservator in horology and related objects at West Dean College, UK, she is also a keen historian of French clockmaking.

Michael Edidin is Professor Emeritus of biology, Johns Hopkins University. He has published extensively on immunology and biophysics, and continues to study horology.

Estelle Fallet is Chief Curator of horology, enamels, jewels, and miniatures in the Musée d'Art et d'Histoire, Geneva, and a university lecturer. She is the author of numerous publications concerning Swiss horology and related subjects.

Christina Faraday is a research fellow of Gonville and Caius College and affiliated lecturer in the history of art in the University of Cambridge.

Karsten Gaulk is Head of Astronomisch-Physikalisches Kabinett in Kassel.

Samuel Gessner is a historian of mathematical cultures, and is affiliated with the Observatoire de Paris and CIUHCT, the inter-university centre for the history of science and technology in Lisbon.

Sibylle Gluch, PHD in German Studies, curated the exhibition 'Simple and perfect: Saxony's Path into the world of international watchmaking', Mathematisch-Physikalischer Salon, Dresden, 2015, and subsequently engaged in a research project on the beginnings of German precision horology funded by the Gerda-Henkel-Stiftung. She is currently leading research, funded by the Deutsche Forschungsgemeinschaft (DFG), on the development of precision standards for astronomical timepieces in the 18th century.

Ron Good is Curator of the primarily virtual Alberta Museum of Chinese Horology in Peace River, and a grateful student of his many teachers in China.

Bernhard Huber has been responsible since 2003 for the library of the German horological society, Deutsche Gesellschaft für Chronometrie, Nuremberg, which is probably the largest horological library in Europe.

Sharon Kerman worked in the fields of music, horology, and mechanical musical instruments for many years. A particular focus of her research is singing birds and related automata.

LIST OF CONTRIBUTORS

Michael Korey is a mathematician and senior curator at the Mathematisch-Physikalischer Salon in Dresden.

Dietrich Matthes is a quantum physicist by training and researches Gothic and Renaissance horology in its technical and cultural context.

Joëlle Mauerhan, an independent scholar, was the founding director of the Musée du Temps, Besançon.

James Nye FSA, is the Antiquarian Horological Society chairman; writer and lecturer in horology; principal sponsor, the Clockworks Museum, London; Master, the Clockmakers Company (2022).

Denis Roegel is an associate professor at the University of Lorraine, France, and conducts research in the history of computing and on astronomical and tower clocks of which he has examined nearly a thousand.

David Rooney is a writer, curator, former Keeper of Technology and Engineering at the Science Museum, London, and Curator of Timekeeping at the Royal Museums Greenwich; currently a Research Associate at Royal Holloway, University of London.

S. R. Sarma is retired Professor of Sanskrit, Aligarh Muslim University, India.

Denis Savoie is a historian of Science at Universcience and the Observatoire de Paris, and a specialist in the history and practice of gnomonics.

Katsuhiro Sasaki is an Honorary Member of the National Museum of Nature and Science, Tokyo where he ended his career as Director of the Department of Science and Engineering 2001–6. He continues his research on Japanese clocks, time measurement, and astronomical clocks.

Roger Smith FSA studies international aspects of the organization of the horological trade in the eighteenth century, including exports and the movement of components and skilled workers.

David Thompson FSA was formerly Curator of Horology at the British Museum, London.

Anthony Turner is an independent scholar and consultant for the history of horology and scientific instruments.

Wim van Klaveren, initially an engineer, studied and taught Medieval English language and literature. In the early 1980s, he trained in clock restoration, at the same time becoming a horological text editor.

Jon Ward is a collector of vintage Chinese mechanical watches in Saskatchewan, Canada.

Sebastian Whitestone FSA is a dealer in, and student of, antiquarian horology in the Early Modern period.

INTRODUCTION

Horology in modern usage designates the entire range of time-finding, time-keeping, and time-telling instruments with the exception of calendrical instruments. Since 1899, when F. J. Britten published his *Old Clocks and watches and their makers*, a work that, despite its defects, remained in print for nearly a century and is still consulted, there have been few attempts at a complete survey of the development of horology apart from Willis I. Milham, *Time and timekeepers . . .* , 1923 (reprint 1975) and David S. Landes, *Revolution in time, clocks and the making of the modern World*, 1983 (revised French edition 1987; extra-illustrated edition of the French translation without revision of the text, 2017). Apart from these, the only substantial work to have appeared, but restricted to Europe, is that by Giuseppe Brusa, *L'Arte dell' orologeria in Europa, sette secoli di orologi meccanici*, Milan, 1978. A new general survey of the history of horology is therefore desirable.

Such a survey is the more needed because a very considerable amount of new research on the development of horology has been carried out in the last five decades. Most of this, however, has been published in the journals of national societies for horology and its history and unites collectors, museum curators, historians, and practising clock- and watchmakers worldwide. It is now therefore appropriate to attempt to synthesize this new knowledge from several languages and sources of limited circulation into a single, large-scale volume offering a general survey of the whole subject, thus enabling it to be considered in a long perspective. We are conscious that this is not complete. Australia and New Zealand, for example, have a minimal presence and horological trade from Europe to the Americas is also, like the role of publicity and advertising in horological marketing, only briefly mentioned. All are casualties to the exigencies of space.

HOROLOGY, THE WORD

'The term horology is at present more particularly confined to the principles upon which the art of making clocks and watches is established'.[1] Such is the earliest instance that the *Oxford English Dictionary* offers in support of its definition of horology as 'the art or science of measuring time; the construction of clocks and watches'.[2] The formulation by the authors of the *Pantologia*, however, suggests that the term had already been in use for some time but had changed its meaning. This is certainly the case. 'Clock, in *Horology*' begins the article of that name in Rees' *Cyclopedia*.[3] Horology is here used as a collective noun that embraces all that is defined in *Pantologia* or the *OED*. Despite the date of publication (1819–20) affixed to the completed work, Rees' *Cyclopedia* was actually issued in parts between 1805 and 1813 with many of the articles on clock- and watch work, all written by William Pearson F.R.S. (1767–1847), appearing in 1807.[4] 'Horology' as a collective noun, therefore, was clearly employed in the early 1800s, perhaps even before, coming into use shortly after a corpus of systematic writing on the subject was developed by such writers as Alexander Cumming (1733–1814), Thomas Hatton (*fl*. Pre-1757–74), and Ferdinand Berthoud (1727–1807) in the third and fourth quarters of the eighteenth century.[5]

The word, however, did not arise from a void. As a singular noun, 'horology' has a far longer presence in English. 'An horology', or its many variant forms,[6] from the fourteenth century onwards had the general meaning of 'an instrument for telling the hour'. This is clearly expressed by Thomas Blundeville in 1594 when explaining that 'the most part of horologies or clockes in the east countries . . . ' marked twenty-four hours.[7] The word

---

1 Good, Gregory & Bosworth 1819, v.
2 *OED* 1972, i, 1332: 391.
3 Rees 1819–20, viii, sig. 373ʳ. The opening of the article 'Chronometer' is similar 'Chronometer, . . . , is a term in *Horology*,', Sig. B1ʳ.
4 Harte 1973, 92*ff*.
5 For their works, see Bibliography.
6 Listed in OED 1972, i, 1332: 390; see also Robey & Linnard 2017, 193.
7 Blundeveille 1594, f. 172v. A similar use is *Othello* II. 3: 'He'll watch the horologe a double set/If drink rock not his cradle'.

could be applied to sundials, as by John Wycliffe in his commentary on Isaiah xxxviii: 8 (1382) 'the shadowe of lynes bi the whiche it hadde go doun in the oriloge'[8] or, a few years later, by Chaucer as a synonym for a clock when vaunting Chauntecleer, whose crowing was more reliable than 'a clokke or an abbey orlogge'.[9]

Any kind of time-measuring instrument then could be described as an horology, although Thomas Powell restricted the term to devices 'which by the motion of several Wheels, and Springs, and Weights, and couterpoizes should give an account of the time, without Sun or Stars.'[10] Nevertheless, Sir Thomas Browne (1605–82) tells us, 'Before the daies of *Jerom* there were Horologies, and several accounts of time; for they measured the hours not only by drops of water in glasses, called Clepsydræ, but also by sand in glasses, called Clepsammia'.[11] The word could also take an adjectival form. In the earliest book in English devoted to sundials, Thomas Fale (*fl*.1586–1604) explained that he had omitted the 'Horological Cylinder';[12] when he eventually published the universal equinoctial ring dial, William Oughtred (1575–1660) described it as *the generall horologicall ring*;[13] and the earliest book in English devoted exclusively to clocks and watches appeared under the title of *Horological Dialogues*.[14] As an adjective, even in the seventeenth century, 'horological' seems to have held a more general meaning than the noun, and this would eventually lead to a generalization of the sense of 'horology' to describe the entire subject.

It is this generality of meaning that 'horology' has acquired since the early nineteenth century that explains and justifies the inclusion, in a general history of the subject, of sundials, fire-clocks, pneumatic clocks, and sand-glasses, which all depend upon different principles from the water-weight, solid-weight-, and spring-driven devices that constitute the greater part of the matter to be treated here. The origin of the term is the Greek *horologion*, a noun compounding *hora*, hour (and by extension time) and *logion*, indicator, or shower. In old English and French generally written in some such form as *orlogge*, *orloge*, *orologge*, or *oriloge*, mediation through the Latin *horologium* led the aspirate 'h' of the Greek to become a written consonant, and to the use of a medial 'o' as in modern English 'horologe', although the latter failed to maintain itself in modern French *horloge*. This, although its etymology has never been in doubt, had serious gender problems, but the generality of the term is displayed by Dominique Jacquinot who, explaining how to find the difference of longitude between Paris and Lyon by comparing the time shown by his clock or watch (which had been set by his astrolabe on departure), with that found using his astrolabe on arriving, describes the former as his 'monstre d'horloge' to make the distinction between the two instruments clear.[15] Modern Italian *orologio* kept the medial 'o' but without the consonantal 'h', while Spanish *reloj* (via old Catalan *relotje* and *orollotje*) decapitated the earlier forms, as also occurred in France in the regions of Berry and Burgundy, where *reloje* was used. In general, the shift from the singular name of the object to a collective name for the making of the objects took place in the mid- to later eighteenth century: 'horology is the art of making machines which, by means of wheelwork, measure time by dividing it into equal parts, and indicating this division by intelligible signals.'[16] Fuller than that offered by the *Dictionnaire* of the Académie Française,[17] such a definition, as Jaubert makes clear in his following paragraphs, reflects a new perception of the clockmaker's craft as an art based on scientific principles.[18]

HOUR SYSTEMS

Hour systems have varied widely across both time and space. The natural time division upon which they all depend is the day defined as the total period of daylight and darkness that elapses between two sunsets, two sunrises, or two other definable moments. This period can be treated as a single unit—the **nychthemeron**—and uniformly divided up, or as two distinct units—that of daylight and that of darkness—each of which may be separately and uniformly divided. Because of the change in solar declination throughout the year, however, the period of daylight and the period of darkness are equal to each other only at the spring and autumn equinoxes, these being the mid-points of the half-year from the winter solstice to the summer solstice (during which the daylight period lengthens) and the half-year from the summer solstice to that of winter (during which the daylight period declines). In consequence, a uniform division of the daylight period will not be equal to a similar uniform division of the dark period, nor will the lengths of the divisions of either remain equal from day to day. Hours obtained in this way are therefore doubly unequal and are thus designated *unequal* hours. Uniform

---

8 Cited from OED (n. 2).
9 Geoffrey Chaucer (*c*.1368), 'The Nun's Priest's Tale' (line 34), cited from Robinson 1957, 199. It is of course possible that Chaucer was here using the words as alternatives, not synonyms in which case the distinction must be between a clock sounding the hours, and a monastic alarm. See, however, the suggestion by Robey & Linnard 2017, 193 that in other contexts the distinction made is between the clock movement and the dial. Barrington 1778, 422 thought a bell and a clock were in question.
10 Powell 1661, 6.
11 Brown 1672, ch. V. xvii, 301.
12 Fale 1593, aiii[v].
13 Oughtred 1652.
14 Smith 1675.

15 For the gender of *horloge* see Havard 1887–90, ii, 1292–3; Maddison 1994. For the longitude, see Jacquinot 1545 f. 52[v].
16 Jaubert 1773, ii, 401.
17 'The art of making clocks, pendulum clocks and watches', *Dictionnaire* 1772, i, 611.
18 Specific terms in horology also repay investigation. On the term 'foliot', for example, see Bradley 2015, Linnard 2015, and Robey 2015.

division of the total period of light and dark (the solar day), however, provides intervals that are invariable. These are designated *equal* hours. Other terminology has been used. *Seasonal, variable,* or *temporal* may be found used for *unequal hours*, and *equinoctial* or *invariable* for *equal* hours.

Both the number of hours contained in a day (or a day and night period), and the point from which the count is begun, are arbitrary. The origins of the double twelve-hour count familiar in much of Europe seem to lie in third millennium BCE Egypt.[19] Before this, however, the Egyptians probably distinguished four periods in the solar day–night: two periods of twilight and daylight, and darkness. These were subdivided to twelve, daylight to ten and the periods of twilight to one each. In Rome, two different systems were used:

1. a 'natural day' of twelve hours counted from sunrise to sunset in unequal hours, and an equivalent 'natural night' counted from sunset to sunrise, with the hours gathered into groups of three hours each, the 'vigils', generally used for everyday communal life.
2. a 'civil day', in which the night was considered as an integral part of the day with a count of twenty-four hours starting from midnight, generally used for civil and legal purposes.

In China,[20] three main systems have existed. One system divided the **nycthemeron** from midnight to midnight into 100 *ke* (notches or graduations). A second system divided it into twelve *shi* (double hours), the first of which was divided by midnight. Each *shi* was given the name of one of the signs of the Chinese zodiac. The sequence therefore was

11pm–1am	*zi*	rat	11am–1pm	*wu*	horse
1am–3am	*chou*	ox	1pm–3pm	*wei*	sheep
3am–5am	*yin*	tiger	3pm–5pm	*shen*	monkey
5am–7am	*mao*	hare	5pm–7pm	*you*	cock
7am–9am	*chen*	dragon	7pm–9pm	*xu*	dog
9am–11am	*si*	snake	9pm–11pm	*hai*	boar

A third system divided the night from sunset to sunrise into five equal parts called *geng*:

Rigu	sunset
Hun	dusk
Chugeng	10 *ke* after dusk
Diadem	period of waiting for dawn
Xiao	dawn

The first two of these are **equal hour** systems, the third, an **unequal hour** count.

In Japan, the so-called Edo hours were determined by dividing the astronomical day in two: day and night based on dawn and dusk (not on sunrise and sunset) and dividing each into six equal parts. Therefore, an Edo hour was equivalent to two hours on average, and people called it *toki* (時). A *toki* is roughly equivalent to a double hour, similarly half a *toki* (*han-toki*, 半時) is equivalent to a single hour. Edo hour names start with 9 (*kokonotsu*, 九つ) at 12:00 midnight, then subtract successively by 1 from 9, which gives 8 (*yatsu*, 八つ), 7 (*nanatsu*, 七つ), dawn 6 (*ake mutsu*, 明け六つ), 5 (*itsutsu*, 五つ), and 4 (*yotsu*, 四つ). The count then starts again with 9 at 12 noon and proceeds as before 8, 7, dusk 6 (*kure mutsu*, 暮れ六つ), 5, and 4. The hour names originated from the number of times the temple bell was struck in ancient Japan, as *Engishiki* showed. This curious system of declining hour numbers follows from the fact that 9 is a significant number in *Onmyo* thought. By multiplying the number from 1 to 6 by 9, numbers 9, 18, 27, 36, 45, 54 are obtained. Subtracting ten places from each number gives the number of the reverse order. These hours were tolled by the time bells, *tokinokane* (時の鐘).[21]

In India, the *Taittirīya-Brāhmaṇa* and the *Satapatha-Brāhmaṇa* (c.700–600 BCE), two texts of the Vedic corpus, divide the civil day into thirty *muhūrtas*. The *Vedāṅga-jyotiṣa* (c.400 BCE), also a part of the Vedic corpus, divides the *muhūrta* into two *nāḍikās*. Accordingly, the civil day is divided into sixty equal units of *nāḍikās* (later called *ghaṭīs*). This became the standard unit of time measurement and remained so until the end of the nineteenth century.

The *Vedāṅga-jyotiṣa* subdivides the *nāḍikā* into ten 1/20 *kalās*, a *kalā* into four *pādas*, a *pāda* into thirty-one *kāṣṭhās*, and a *kāṣṭā* into five *akṣaras*. Other texts contain different subdivisions. In the early sixth-century Āryabhaṭa (BCE 476) standardized these into a sexagesimal system, parallel to the sexagesimal division of the circle into minutes, seconds, and so on:[22]

Nychthemeron	=	sixty *nāḍikās* (each of twenty-four minutes)
one *nāḍikā*	=	sixty *vināḍikās* (each of twenty-four seconds)
one *vināḍikā*	=	sixty *guru-akṣaras* (time to utter one long Sanskrit syllable, approx. 0.4 second)

In this system, the sixty *nāḍikās* are an unequal hour count. While the duration of the *nāḍikās* remains constant, the number of *nāḍikās* from sunrise to sunset or from sunset to sunrise varies according to local latitude and the seasons.

Most societies have a variety of terms to designate divisions of the day, but these being subjective are neglected here although they could give rise to quantified systems. These, in early societies, tended to be used only in religious and dynastic contexts, while the use of equal hours was virtually exclusive to astronomers. In Antiquity and the Middle Ages, the hours of everyday life throughout Europe, the Near East, and India were the unequal hours. In Europe these were gradually abandoned from the fourteenth century. In Islamic regions, where the unequal hours were

---

19 Neugebauer & Parker 1960, 120.
20 For details see Bedini 1994, 14–15.

21 Urai 2014; Robertson 1931, 198–203.
22 Aryabhaṭa 1976, 85–6.

intimately linked with prayer times, and in Japan, unequal hours remained customary.

However, conversion to equal hour measurement in Europe did not lead to the harmonizing of hour counts:

- Italian (or Bohemian, Czech, Silesian, Polish, or Welsh) are hours counted 1–24 from sunset or a little after. Gradually abandoned from the seventeenth century onwards, they nonetheless maintained themselves in Italian usage until well into the nineteenth century.[23]
- Babylonian or Greek hours counted 1–24 from sunrise.[24]
- Nuremberg hours were counted from sunrise and sunset, each point being considered 0 as in the unequal hour system. The count however was an equal hour count of the unequal periods of day and night. Thus an early summer day would count from sunrise to sunset up to 13 hours 10 minutes. At sunset, the count would begin again although the night would only extend to 10 hours 50 minutes.

Variants could be found in other German and Central European regions. Basel, for example, employed a double twelve-hour count but began it at 11 p.m. and 11 a.m. Astronomers everywhere used the double twelve-hour count beginning the day at noon, while the same count, but begun at midnight (known also as 'common hours') was typical of France, Britain, and Northwest Europe. In all cases, however, the time employed was, at least until the end of the eighteenth century, local time, so even within the confines of a single country, the time of day varied with difference of longitude.

CONVENTIONS

On the first mention of a person in the text (this can be ascertained from the index), his/her dates are given in full. If exact dates are not known they are given in the form *c.*1921–*c.*42 where there can be a presumption that the actual date(s) though unknown was/were within five to ten years of the date(s) offered. Where there is greater uncertainty but at least one definite date is known, *fl.* 1921 is used. Where no more than a period can be indicated, this is indicated as 2nd half 8th century. Biographical details of makers are generally not given but can be readily ascertained from the several national dictionaries and other lists that are available.[25] The appearance of a technical term in **bold** type in the text indicates that it is explained in the glossary. A few terms are included in the glossary but appear very frequently throughout the book. These have therefore been set in **bold** only on their first appearance in each chapter. References in the notes are given by author's name, date of publication, and page; these key to the bibliography. Discussions of works in the text may, however, have a fuller citation.

Chapters 1, 8, 13, Section 3, 13, Section 4, 15, 20, and 23 were originally written in French and have been translated by Anthony Turner.

ACKNOWLEDGEMENTS

The editors' first gratitude must go to the authors for accepting arduous assignments and patiently tolerating our exigencies; equally to the staff of the Oxford University Press for undertaking the work and giving it physical form, in particular to the editorial team of Sonke Adlung, Francesca McMahon, and Sharmila Radha who have been helpful and supportive throughout. Our thanks also to Sotheby's London, whose sponsorship has enabled the work to be colour-printed and for a professional index to be compiled. Well beyond the call of duty, Roger Smith has helpfully commented on several chapters besides writing his own. The following have offered us advice, encouragement, and answered questions: Katy Barrett, Alun Davies, Peter de Clercq, John Davis, Jennifer Speake, and Tony Weston. The debts of individual authors are acknowledged in their several chapters.

---

23 Arnaldi 2007; Dohrn-van Rossum 1996, 114, and Catamo 2008. From at least the sixteenth century onwards in Italy the hour count was begun half an hour after sunset so as to coincide with the ringing of the bell for the *Angelus* prayer. The origins of this custom are in doubt. Tailliez n.d. suggests distinguishing hours counted from sunset itself (however defined) as italic hours, and those counted from thirty minutes after sunset as *Italian* hours. The distinction, however, does not work in Italian. See Arnaldi 2006, 2007; Dohrn-van Rossum 1996, 114; Catamo 2008; and Schneider 2017. The last known use of Italian hours was for the meridian line constructed in 1891 in the church of San Giorgio, Modica, Sicily.

24 *Nuovo Almanacco per l'anno bisestile 1776 arricchito di notizie utili e dilettevoli*, Venezia [1775/76], 52, notes that 'the Greeks nowaday are the only ones, who begin the day at sunrise.'

25 Abeler 1977; Basanta Campos 1972; Chenakal 1972; Fraiture 2009; Loomes 2006; Morpugo 1972; Patrizzi 1998; Pipping, Sidenbladh & Elfström 1995; Pritchard 1997; Sposato 1983; Tardy 1972; Turicchia 2018.

CHAPTER ONE

TIME MEASUREMENT IN ANTIQUITY

Jérôme Bonnin

To write the history of horology before Antiquity is next to impossible. It is only in 'historical' periods—those from which textual evidence has survived—that the historian can find traces of the material organization of time in simple or complex systems that permit the synchronization of human activity. Many elements remain unknown. How time was thought of in Antiquity has no answer, for an exact study of the history of time, and theories and concepts about it, is still needed. In this volume it is more appropriate to deal with the reality of time as experienced in everyday life, rather than to engage in speculation about the concept of timekeeping, which, even if it reveals an approach (essentially that of the elite), is rather far removed from the actual principles and practice of horology. The notion of time, however, is important in literature.[1] More than the Greeks, the Romans attributed a material reality to time, and linked it to their everyday life and to the success of their projects. For the Romans, it was a philosophical rather than a divine reality—a natural entity that did not unfold by chance. Time seemed to be controlled not by a divinity, but by natural laws.[2] The Romans had an essentially materialist attitude that linked the idea of time to the purely material matters of daily activity, work, and business of all kinds. It is to the Romans that we owe ideas, like Horace's 'Beware of seeking what tomorrow will bring; profit from the day whatever destiny may bring you', that relate time with both eternity and enjoyment of the moment.[3] Even so, reflection about the measurement of time largely precedes Græco-Roman Antiquity.

THE ORIENTAL ORIGINS OF TIME MEASUREMENT

The origin of time measurement is an insoluble question. The origins and forms of the first time-measuring instruments are also unknowable. It would be illusory to examine all the geographical regions and civilizations known from Antiquity seeking to disinter traces of such instruments. In discussing time measurement, two cultures demand attention—Egypt and Babylon—although only a brief survey of what is known can be given here.

The Egyptians divided the day from dawn to sunset into twelve equal parts, each part having a specific name.[4] The oldest witness to a division of day and night into twenty-four parts occurs in a twelfth century BCE papyrus preserved in Cairo.[5] The oldest known time-measuring instruments also come from Egypt, as do the oldest writings about them; it was there also that the first idea of 'divine' time developed with a need of instruments to measure it.[6] Time was a major element of civilization, although this aspect is often not recognized. The obelisks, for example, were not parts of **sundials** or meridians in Ancient Egypt.[7] Schematizing, it can be said that time measurement in Egypt has two major elements: the creation of several sundial-type instruments that nonetheless had no later influence, and the perfecting of effective hydraulic instruments

---

1 For a complete study of the subject see Baran 1976, 2–20.
2 Seneca, *Letters* 101, 'Time unfolds according to strict but impenetrable laws'.
3 Horace, *Odes*, I, 9, 10.

4 For Egyptian hour divisions, see above, 'Introduction'. For hour names, see Maddison & Turner 1999, 126–129.
5 Museum of Egyptian Antiquities, Cairo: Inv. no 86637. Concerning this text, the division of time and Egyptian astronomical concepts more generally, see Clagett 1995, 98ff.; Neugebauer & Parker 1960, 114.
6 See Symons 1999, whose section on 'Shadow Clocks and Sloping Sundials', 127–51, is currently the definitive treatment of the subject; see also Symons 2002.
7 Symons 1999, 128, n. 130.

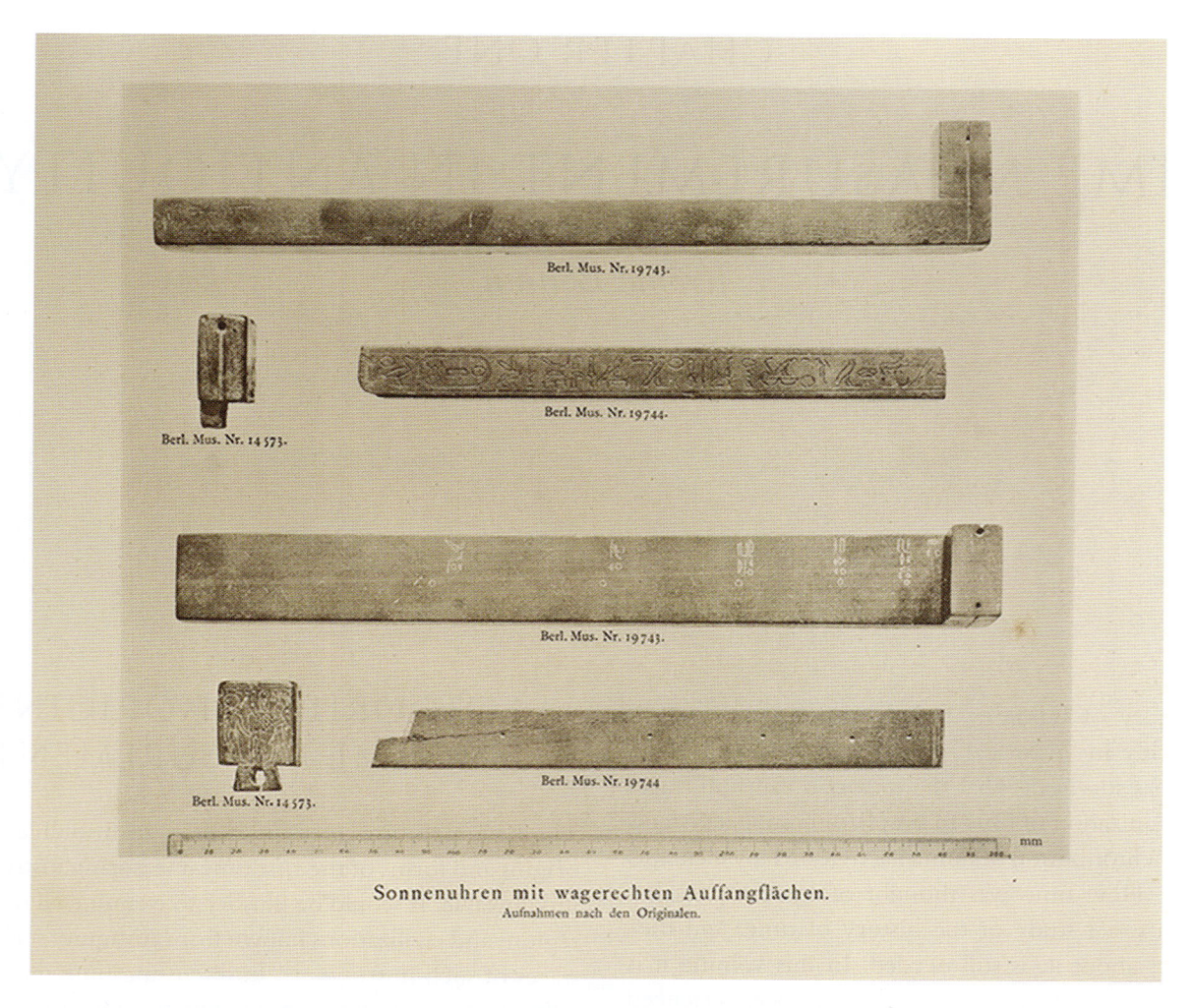

Figure 1 Ruler-type Egyptian shadow clocks, Berlin Museum after Borchardt 1920, pl. 12. Photo: Jean-Baptiste Buffetaud.

of a kind that can still be found used at Rome in the first century CE and linked with religious ceremonies. In Egypt, solar instruments were of greater number, and with a specific way of operating, time being indicated by the length of a shadow rather than by its direction. It is this system that later became habitual.[8]

The oldest known instrument indicating time by the length of the **gnomon** shadow dates from the time of Thutmose III (first half fifteenth century BCE)[9] and several examples are known (Figure 1). The device is formed of a rule with a rectangular block at one end. A small hole and a guideline worked in this block serve to fix and allow the use of a plumb line to level the instrument. The upper surface of the rule is pierced with five small holes with the corresponding name of the hour sometimes marked beside them, beginning at sunrise. Once levelled, the instrument is pointed towards the Sun so that the shadow falls exactly on the rule, the hour being read from the position of its tip. The device is not especially accurate as the hour positions were determined empirically, not by calculation. This is confirmed by examining the relation between the height of the gnomon-block and the spacing of the hour holes, which seem to be standardized on the known examples. Further examination of these dials also shows that they were only really usable in Egypt from the end of spring to the beginning of autumn. Therefore, their use (quite different from **water clocks**, which deployed **unequal hours**) seems to have been reserved for specific tasks, mainly religious, the instruments taking on standard forms that lasted throughout the Egyptian period.

This 'ruler-type' dial was probably the precursor of the inclined plane dial, apparently developed between the seventh and sixth centuries BCE. Used as offertory objects in the sanctuaries, several examples have survived, most of them complete.[10] They are composed of a stone block carrying at one end a rectangular gnomon-block and facing it a similar prismatic block, but with the upper face cut at an angle to form an inclined plane. The gnomon-block has a rectangular cavity with a vertical line incised beneath it as a guide for the plummet. Abridgements of the Greek names of the Egyptian months are marked on the top of the

---

8 For a recent attempt to classify Egyptian Sundials, see Symons & Khurana 2016.

9 Berlin, Ägyptisches Museum, Inv. no 1974. Such an instrument is depicted and described in a pictorial inscription in the Cenotaph de Seti I, Abydos (c. 1291–78 BCE). Frankfort 1933, ch. viii.

10 See Bosticco 1957, 33–49.

Figure 2 Dial from the Valley of the Kings after Bickel & Gautschy 2014. Photo: © University of Basel, Kings' Valley Project, M. Kacicnik.

angled block and seven lines corresponding with them are incised on the inclined face of the instrument. Also set along each of these seven lines are six points placed at unequal intervals, their distance diminishing from the top to the bottom. They are used in a way similar to that of the first group, except that the gnomon shadow falls on the inclined plane and not on a horizontal rule. As before, once levelled they are pointed towards the Sun, with the time being read on the line corresponding with the month. Despite a more refined conception and a smaller size (an average length of 120mm), they pose problems of accuracy and use like those of the first group, particularly as no account appears to have been taken of latitude in their construction.

Beside these two relatively well-documented types of dial, which even supplied the graphism for some hieroglyphs signifying 'sundial', there may have existed a further, more problematic, group, of which four examples are known.[11] These appear to use the *direction* of the gnomon shadow, not its *length*, i.e. they act as vertical **direction dials**. One from Gaza and one discovered by a University of Basel excavation team (Figure 2) are the oldest, dating from the thirteenth century BCE. The two others are Græco-Roman and could have been influenced by the vertical plane dials of this period. Discussion of these objects today concerns their function and the nature of the lines and hour system that they show—if they are indeed sundials. A semicircular face is divided by lines that converge towards the centre into twelves sectors of 15 degrees each. This design raises alternative hypotheses. Firstly they are dials showing **equal hours**, which implies that the Egyptians of the thirteenth century BCE knew of the **polar gnomon**, although there is no other attested historical evidence. Secondly, the objects show a moment of time by means of the shadow of a straight gnomon set at right angles to the graph on the surface. A further difficulty is the positioning of these dials, as there is no means of attaching them to a support, and the user has to orient them, which presupposes knowledge of the north–south direction; this was uncommon in Antiquity and required a lengthy period of measurement to determine it, so the utility of a portable dial is lost. In sum, it seems best to consider these objects simply as solar pectorals and to exclude them from the corpus of Egyptian sundials.

Just as the oldest solar instruments derive from Egypt, so do the oldest **outflow water clocks**. The earliest archaeological referent for an outflow water clock dates from the reign of Amenhotep

---

[11] The first was found at Gaza in the early twentieth century but its present whereabouts is unknown. The second, found at Luxor, is preserved in Berlin (Inv. no 20 322), the third in Brussels (Inv. no E 7330). The last was found in 2013 in the Valley of the Kings by Swiss archaeologists. See Gautschy & Bickel 2014.

III (c. 1415–1380 BCE).[12] It is an alabaster vase in the shape of a truncated cone graduated on the inside wall with hour marks placed as functions of the twelve months. At daybreak or sunset, the vase was filled with water, which flowed out slowly though a small orifice worked in the base. The time was read on the inner wall using the month scale appropriate for the time of the year. This was not entirely straightforward in low light, and the meniscus could hinder locating the exact level of the water in the restricted space of the vase. A much-later papyrus (third century BCE) found at Oxyrhynchus gives some indications about the construction of such instruments.[13] The document is written in Greek by a Greek or Hellenistic bureaucrat, probably summarizing an Egyptian technique transmitted down to the Ptolemaic period. This kind of instrument would still be found many centuries after their creation in some temples of Isis during the Roman period.

What was the use of these solar and hydraulic instruments? The most plausible hypothesis is that they had religious rather than civil use. No surviving text about them refers to their use in everyday life. They were most probably employed in the cults to order the hours of prayer and the services to be rendered to the Gods. The fact that all the known examples of both classes of sundial are portable suggests that they were not permanent but could be used for religious ceremonies inside or outside the temples. Moreover, many of these instruments have been found in funerary contexts together with other cult or everyday objects intended to serve the dead person in afterlife.

All this raises an important point concerning time-measuring instruments in the Græco-Roman period. The Greeks seem to have inherited little from Egyptian civilization. Egyptian instruments, apart from the water clocks, are unique and have hardly any equivalents in later periods; this may be due to their unreliable nature, a function of their essentially ritual, religious role, which made them inappropriate for civil time measurement. An omnipresent religion, personified in the Pharaoh, furnished explanations for the order of the universe and its movement. Thus, time instruments were not linked technically with the natural imperatives of unequal hours, latitude, and seasonal variations, but with religious ritual. This, however, was not the case for the civilization of Babylon (second millennium BCE—mid-first century BCE)—another culture that made far-reaching explorations into time measurement.

The Babylonians would influence all the peoples of the Mediterranean basin, and in particular, the Greeks. They would produce exact documents about the movements of the planets, the Moon, and the stars. Complex ephemerides would be drawn up in order to predict events (notably eclipses), to establish the religious calendar, or to construct horoscopes. It is well known that these 'Babylonians' attained a high level of scientific knowledge—notably in mathematics and astronomy—from the eighth century BCE onwards. That they should have made time-measuring instruments is well-founded and corroborated by Greek and Roman testimony. By contrast with Egyptian and later periods, the soil of Babylon has yielded neither sundials nor water clocks, with the exception of one small **sinking bowl** vessel.[14] Is this the result of chance operating in archaeological investigation, or were such instruments generally created from perishable materials like wood? Moreover, the documents about time keeping that exist from the Babylonian empire are extremely succinct (unlike for mathematics and astronomy) and fall into two groups: those concerning the measurement of a quantity of liquid, and those concerning the measurement of a shadow. The first group consists of two kinds of text: tablets containing mathematical problems and exercises (such texts actually mention an instrument, the *maltaktum* or *didbibdu* in Akkadian, and give its height and diameter, but not the timescale); and tablets that specifically discuss the linked subjects of astronomy and astrology—principally the first of the *Mul-Apin* tablets.[15] The latter give information about the volume (expressed as weight) of a liquid escaping from a container without any details as to the form or name of the container. These were identified as water clocks.[16] Specialists suggest that, known in Mesopotamia from the eighth century BCE onwards, they were prismatic or cylindrical in shape and operated on the **outflow** principle. Whatever the case, the instrument seems to have been definitive for all astronomical measurements that included a lapse of time. The frequency with which they are mentioned in astronomical and mathematical texts throughout the first millennium BCE shows to what degree they were considered indispensable.

The second group of instruments is still more problematic than the first. It would be simplest to say that no text exists with a description of an instrument that could be interpreted as a sundial or a gnomon and that we therefore know nothing about them. However, sources do exist, although only a few cuneiform tablets mention the measurement of a shadow—the second clay tablet of the *Mul-Apin* series dated between the tenth and the eighth century BCE.[17] This tablet tells us that, to determine the relationship between day and night, the length of the day should be

---

12 That some, less developed, outflow clocks of cylindrical or prismatic form existed in second-millennium Babylon is clear from texts concerning them, which are interpreted by Michel-Nozières, 2000.

13 P. Oxy. III no 470. Borchardt 1920, 10–14 and plates 7a, 7b interprets and reproduces this text.

14 This was found at Nimrud by David Brown (Brown 2000, 103–22). It is a pierced bowl weighing 250g preserved in the British Museum, London (inv 91238). In form it recalls other sinking-bowls known from other cultures notably India (see Chapter 2, Section 1 and Sarma 2008, ch. 5), the Far East (e.g. two bowls, one from Colombo and one from Sri Lanka also in the British Museum, inv As 1946.0708.17 135gr; As1898.0703.291.a) and the Maghreb (see Glick 1969), and BM inv Af 1952.23.1 from Tolga, an irrigation timer that sinks in between four and five minutes). The Babylon bowl is a short duration timer (Fermor & Steele 2000, 214–15 and 220–1), and remains isolated, a *unicum*.

15 *Mul-Apin* II. i, 10, 24, first published in King 1912.

16 Brown, Fermor, & Walker 2000, 130–48; Hoyrup 1998, 192–4.

17 *Mul-Apin* II. ii, 21–42, first published by Weidner 1924.

measured by a 'shadow instrument'; but that this is a sundial is only a hypothesis. The measure of a shadow is noted, but no specific instrument is mentioned. This problem of nomenclature is a constant in Babylonian writings on time. The second, and last, evidence must also be treated with caution. It is found in two highly fragmented tablets,[18] which include in the introduction a phrase translated as 'if you wish to construct a gnomon ...'. So a term supposed to mean 'gnomon' exists in Sumerian u_4-sakar, which gives uskaru or askara in Akkadian. However, these tablets are of a very late date, possibly even from the Hellenistic period, and even if the term originates in an earlier period, it is no proof of the age of the instrument.

If these direct sources are of doubtful interpretation and little can be derived from them, the secondary sources are not any clearer. The oldest is Greek and is a well-known passage in Herodotus (480–25 BCE):[19]

> For the Greeks learned the *polos* πόλοσ, and the *gnomon* (γνώμονα), and the twelve parts of the day from the Babylonians.

These two lines have caused much ink to be spilt. The testimony is more complex than it looks, for it is the first Greek text that clearly mentions instruments for the measure of time. Setting aside the problems associated with the terms *polos* and *gnomon*, what is important here is the Babylonian origin of the one or two instruments mentioned, as well as the reference to the 'twelve parts of the day'. There is no need to doubt Herodotus concerning the Greek debt to Babylonian knowledge for mathematics, astronomy, and astrology. That Herodotus mentions it in the fifth century BCE implies that such knowledge was known and used in Greece from the sixth century BCE or even earlier. It was on the basis of this inheritance that the Greeks would develop their own idea of astronomy and gnomonics using all the geometrical skills we know they possessed.

GREEK TIME MEASUREMENT: SUNDIALS

The origin of time-measuring instruments in Greece is much debated. The fundamental instrument was the gnomon, the first astronomical instrument used, and from which sundials derived. Herodotus first mentions them in the fifth century BCE, which poses serious difficulties of interpretation, notably for the terms *polos* and *gnomon*. These words relate to no known archaeological vestiges and produce only hypotheses. Given the context and the mention of the division of the day into twelve hours, it seems logical to suppose that the *polos* and the *gnomon* are two different instruments used to divide up the day. However, other sources suggest that originally the 'gnomon instrument' was used for more than determining the hour. Anaximander of Miletus (d. post-546 BCE), one of the great names of ancient astronomy, is associated with this device. Unfortunately, all sources about him were written long after his period.[20] Diogenes Laertius (first half of the third century CE) claims Anaximander as the inventor of the gnomon, but we must consider this statement with caution. If a gnomon is simply a stake thrust into the ground in order to measure variations in shadow length, no one in particular has a claim to its invention. If we consider it as a more complex device incorporating a gnomonic scale that allows the day to be divided, the attribution to Anaximander falls foul of Herodotus' evidence, as well as that the Babylonians had divided up the day with the aid of a complex instrument probably similar to a gnomon well before Anaximander. Often, in consequence, the text of Diogenes is explained by the supposition that Anaximander did not invent the gnomon but introduced it into Greece from Babylonia. However, all the sources concerning Anaximander are unanimous that he used the gnomon to determine the seasons, the solstices, and the equinoxes.

The oldest use of the gnomon was indeed probably to indicate the solstices. It is for that reason moreover that it was called a heliotrope (ἡλιοτρόπιον), 'index of the variations of the Sun'. Perhaps its first mention is in Homer's *Odyssey*, although in a rather obscure form.[21] Later sources attest to the existence of large instruments of this kind. The legendary Pherecydes installed one on the Island of Syros;[22] others mentioned are those of Thebes,[23] and of Syracuse dated to the fourth century BCE cited by Plutarch (46–125 AD).[24] It is far more likely that these early instruments served to determine the dates of the solstices and the equinoxes

---

18 Sachs 1955, No xxxiv.
19 Herodotus, *Histories*, ii 109.

20 Anaximander's work is known firstly from a passage in Pliny (*Natural History*, ii. 78), According to him the gnomonic art can be traced back to Anaximenes (fl. 546–25), Anaximander's pupil, who should have been the first to erect a gnomon in Lacedaemonia. However, this differs from the evidence of Diogenes Laertius, *Lives and Opinions of the Philosophers* (probably written second century CE), according to whom it was Anaximander, Anaximenes' teacher, who first discovered that the length of a gnomon shadow varied with latitude and that it was he who invented the gnomon. Placing the construction of such an instrument in Lacedaemonia agrees with Pliny. Eusebius in the fourth century CE further mentions Anaximander, and the last source concerning him is the *Suda* (tenth century CE).
21 *Odyssey*, xv, 403–4. Eumaeus speaks of his country, the Island of Syra (Syros), where the 'variations of the Sun' are found. Szabo & Maula 1986 discuss the possible interpretations of this passage and conclude that it refers to a heliotrope.
22 Diogenes Laertius, i, 119: 'His heliotrope is also preserved on the isle of Syros'.
23 Polybius, *Histories*, v 99.8. The description of preparations for the siege of Thebes by Philip mentions the place where a third of the army was stationed, a place called 'heliotrope'.
24 Plutarch, *Life of Dion*, 29. 'At the foot of the fortress and the Pentepyles, there was an immense heliotrope visible from afar, which had

than to give the hour. Anaximander in the sixth century BCE was perhaps not the inventor of the gnomon, but the first to use it to calculate correctly the date of the equinoxes.

Herodotus, however, does not only speak of the gnomon, but also of the *polos*—a highly uncertain term. If the gnomon he mentions is only a simple pole planted vertically in the ground, then the term designates only that which stands above the soil. Could not the *polos* correspond with the plane receiving the shadow cast by the gnomon, perhaps a surface other than the ground—a plate of wood or stone, fixed or not, engraved—a carrier of the indications? *Polos* has various meanings: 'terrestrial pole', 'celestial pole', 'the heavens which turn around these poles', or even 'sundial'.[25] On this basis, the *polos* is traditionally, and still generally today, considered to be the ancestor of the spherical dial, the *scaphe*. The meaning, 'celestial vault', and the resemblance to this of the spherical sundial are the principal reasons for the association. Only a few scholars reject the idea.[26] This rejection is reinforced by a passage from Athenaeus (c.170–c.223 CE). It is difficult to conceive that a dial called a *polos*, made 'in imitation of the heliotrope of Achradine', could have been a spherical type of dial, notably because up to this point the heliotropes are always characterized by their monumental aspect and, following from this, are generally horizontal. Moreover, only a rather limited horizontal dial could imitate a heliotrope and not a spherical dial.

The last point mentioned by Herodotus, the division of the day into twelve parts, seems almost an anecdote. Certainly, the day was divided into twelve since the Egyptians and the Babylonians; the idea is not new. However, examining the phrase more carefully raises a problem. Does Herodotus speak of a division into hours? If so, what kind of hours? The idea of the hour applied to the twelve divisions of the day is known with certitude among the Greeks from the end of the fourth century BCE and seems to have become an everyday term about the same time. Earlier than this, it was a technical term specific to astronomy. Moreover, the astronomical hour was not the same as the hour of everyday life. The Greeks made a clear distinction between **equal hours** and **unequal hours**. The first were used essentially for astronomical observations and calculations. They appear in Greece shortly before the fourth century BCE and were probably used by Eudoxus (408–355 BCE). The second refers to the **unequal hours** given by ordinary sundials. These the Greeks used in everyday life. When it is stated in a text that an event occurred at the sixth hour, that does

Table I Concordance of the unequal hours of Antiquity with modern equal hours

Latitude 42° Obliquity 24°	Summer solstice	Equinoxes	Winter solstice
Sunrise	4 h 25 m	6 h	7 h 35 m
1st hour	5 h 41 m	7 h	8 h 19 m
2nd hour	6 h 57 m	8 h	9 h 03 m
3rd hour	8 h 13 m	9 h	9 h 47 m
4th hour	9 h 28 m	10 h	10 h 32 m
5th hour	10 h 44 m	11 h	11 h 16 m
6th hour	12 h 00 m	12 h	12 h 00 m
7th hour	13 h 16 m	13 h	12 h 44 m
8th hour	14 h 31 m	14 h	13 h 28 m
9th hour	15 h 47 m	15 h	14 h 13 m
10th hour	17 h 03 m	16 h	14 h 57 m
11th hour	18 h 19 m	17 h	15 h 41 m
12th hour	19 h 34 m	18 h	16 h 25 m

not mean 6 o'clock in the morning or the evening, but midday since, at least among the Romans, the hours were counted from daybreak to sunset. The night period was also divided into twelve parts and was counted from sunset to sunrise. Since unequal hours are not easily assimilable to modern hours, Table I shows the correspondences between modern hours and those for the latitude of Rome (most of the texts to be cited in what follows derive from the Italian peninsula) and is calculated using 24 degrees as the value of the obliquity of the **ecliptic**.[27]

Herodotus spoke of the division of the day into twelve hours in the context of astronomical instruments, not that of everyday life. It follows therefore that he was probably referring to **equal hours**. These could have been known to Greek astronomers even earlier than the fourth century BCE, given that the Babylonians already used them. Therefore, the text should probably be read as 'the division of the day into twelve equal parts'. That Herodotus does not use the word 'hour' is certainly because in the fifth century BCE the term was normally used to designate 'season' and would not have been understood as 'hour' by his readers.

GREEK TIME MEASUREMENT: WATER CLOCKS

The Babylonians and the Egyptians not only developed methods and instruments for finding and measuring time by the Sun, but also probably transmitted the first instruments for measuring time using water. The clepsydra is perhaps one of the oldest means. Etymologically the word κλεψύδραν signifies 'water thief'. It derives either from the name of a rather insignificant rivulet in Athens or from a quite specific utensil. In the sixth century BCE, experiments on various pneumatic and hydraulic phenomena

---

been installed by Denys'. Denys the Old lived 431–367 BCE. It was probably the work of his son.

25 Diels 1920, 157. Some uses: Euripides, *Orestes*, i, 1685 (the size and the geometric characteristics fifth century BCE) in the sense of 'celestial vault'; Aristophanes, *Geryatedes* (fr. 8739,11 N), in the sense of 'sundial', although in the *Detalien* its use is equivalent to 'heliotrope'; Moschion as reported by Athenaeus (v. 207e) mentions a structure placed on the ship of Hieron II of Alexandria, a ship supposed to have been constructed under the direction of Archimedes: 'On the roof a sundial (*polos*) is found in imitation of the heliotrope of Achradine'.

26 Diels 1920, 157, n. 1; Hannah 2009, 71–2.

27 Table calculated by Denis Savoie.

provoked the creation of a device that is today known as a 'toddy lifter': a pierced hollow body fitted with exit tubes that can be stopped with a finger and, as with a pipette, raise water, wine, or oil from a container. The possibility of slowing the outflow, or observation of the slow immersion of such a body in a liquid could have led to adaptation of the device for determining a lapse of time.[28] This latter interpretation is the most probable, as it was only in the fourth century BCE that the Athens rivulet received the name of 'clepsydra'. If the instrument could serve scientific, astronomical, and military purposes from the fourth century BCE and medical from the third century BCE, it was primarily in the judicial realm that it would develop. It is certain that in the fifth century BCE this device was used in the courts to measure (and limit) the length of orations. Speeches were made as a function of 'water', not as a function of 'time'.

> Is it decent thus to ruin a blameless old man before the clepsydra (κλεψύδραν), a comrade who has laboured much, who so many times has been in a hot and glorious sweat, a veteran who fought at Marathon for the Republic?[29]

Aristophanes' plays date from the late fifth century BCE. They contain the earliest references to the judiciary use of the clepsydra, yet already the term serves as a metonym for the court—the instrument has become customary, habitual. Its use may date from well before the beginning of the fifth century BCE with the development of democracy. Use of the clepsydra as a way of limiting the length of speeches so as to ensure an equitable distribution of time to all is linked with the world of politics and civic life.

A fortunate discovery allows us to know the form and the characteristics of one such clepsydra dated to the fourth century BCE from the archaeological context in which it was found.[30] It takes the form of a terracotta vase with a thickened rim and two handles (Figure 3).

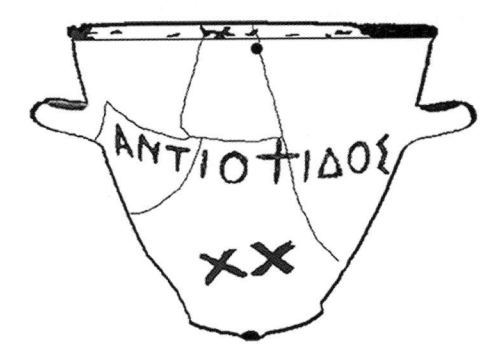

Figure 3 Clepsydra from the Athens *Tholos*. Drawing by J. Bonnin after Young 1939.

An upper orifice placed just below the rim serves as an overflow to evacuate excess water. An orifice at the base, with an inset tube of bronze, allows water to escape into a second vessel. The interior of the vase is covered with an impermeable glaze, and the exterior carries the inscription 'Antiochidos' followed by 'XX'. Antiochidos is the name of an Athenian tribe, and 'XX' represents the capacity of the vase, which is two *choes*, the equivalent of 6.4 litres. This quantity allowed the orator to speak uninterruptedly for six minutes. Evidently, there would have been different clepsydras in the courts, their capacity related to the importance of the process. Like weights and measures, the standards for these clepsydras would have been kept in the *Tholos* or *Skias*, which served as the 'depot for the standards'. It was obligatory to place the instrument where all could see it.

If the clepsydra sufficed for judicial needs, it did not give the hour, even if some adjustments (before the invention of large water clocks) could have allowed the transformation of these 'timers' into instruments to show time. It was by means of clepsydras that military watches were regulated (devalued in periods of time on the model of a 'timer', not of a timekeeping clock). Such a system did not show an hour but only measured the elapse of a period of time. It was perhaps to allow all citizens to know the time of public assemblies that large, hour-showing clocks were developed, such as that installed on the Agora at Athens and close to the most important tribunals. Constructed on the same model as a clepsydra, it had instead a capacity of 1,000 litres, although its flow rate is indeterminable.

This clock was discovered in 1953 in the southwest corner of the Agora, although it was not immediately identified as such. Only the unusual layout of the construction, which had no resemblance to that of public fountains or other water-using systems, led the archaeologist Vanderpool to deduce that it was a water clock, a hypothesis confirmed twenty years later by Armstrong and McCamp.[31] Ceramic matter found in the infill of the foundations and around the clock suggest that it was erected towards the end of the fourth century BCE and underwent many subsequent repairs and adaptations. It probably ceased to be used during the second century BCE. A second, large-capacity clock existed at Oropos where the fourth century BCE site has been conserved and well analysed.[32] Two clocks in the Agora at Samos are mentioned in a decree from the second century BCE.[33] From these gnomonic and hydraulic appliances the Romans would derive their techniques and instruments.

TIME PIECES IN ROME: AN UNPRECEDENTED DIFFUSION

If the Romans invented only a few instruments, they took up Greek techniques, employed Greek artisans, and developed time

---

28 Dohrn-van Rossum 1996, 22–3.
29 Aristophanes, *The Archanians*, 694.
30 Young 1939, 274–84.

31 Armstrong & McCamp 1977; for technical details, see 147–61.
32 Theodossiou, Katsiotis, Manimanis, & Mantarakis 2010, 159–67.
33 Thompson & Wycherley 1972. For the decree itself see *SEG*, 41, 711.

measurement in cities in a unique way, thus turning it into an indispensable element of civilization. The first mention of clocks in Roman history is by Pliny the Elder (23–79 CE)[34] who, in a long passage, presents 'the third agreement between peoples', i.e. a scientific notation of the hours. He relates that this arrived only late in Rome, from the fourth century BCE onwards, and in his *Natural History* he tells us that the first dial appeared in Rome in 293 BCE, an unproblematic date, since dials were known well before in Greece. But Pliny also recounts Varro's version that the first dial appeared in 263 BCE, at the beginning of the Punic War, when a Greek dial and its accompanying hours arrived in Rome as war booty from the Greek Sicilian city of Catania. Finally, in 164 BCE the censor Quintus Marcius Philippus had the first dial calculated for the latitude of Rome installed, contrasting with that drawn for Catania, which the Romans had nonetheless used for ninety-nine years. Varro (116–27 BCE) also tells us that a water clock was installed in Rome at the latest by 159 BCE:

> *Solarium* designates the dial on which the hours are seen from the Sun, or the clock that Scipio Nasica placed in the shadow of the Basilica of Emily & Fulvia.[35]

This remark proves that a need to know a fixed hour, even during the night, must rapidly have developed. It is difficult to conjecture the form of the instrument, but its location suggests a large, sophisticated device.

The story of the dial from Catania is frequently cited in modern studies of Roman timekeeping. Reading the text of Pliny and other later writers leaves an impression that the Romans blindly obeyed an 'imprecise' sundial for ninety-nine years without knowing it. But, some corrections should be made. Firstly, it is probable that the Romans noted the error but accorded little importance to it given that they rarely had need of exact hours. What is important is that the entire community refers to the same hour regardless of its precision. Moreover, the error in question was perhaps not as upsetting as might be thought.[36] The anomaly evoked by Pliny is always explained by the removal of the dial from Catania to Rome.[37] If, as according to Denis Savoie, it would be impossible to recognize that the dial indicated an incorrect hour, it was possible, by contrast, to realize that the dial was not adapted for use in Rome. As Savoie notes

> if one removes a horizontal dial from Catania to Rome, the shadow of the gnomon at Noon will always be longer than in Catania whatever the date. Since the Greeks terminated the hour lines at the hyperbolic arc for the winter (and summer) solstice, it follows that an hour-reading would be impossible around the winter solstice since the shadow would go well beyond the winter arc.[38]

This leads him to suppose that it was by the inappropriate length of the gnomon as well as the shadow that, given that it was correctly oriented, the error of the dial could be seen. This proposition is convincing and leads us to question Pliny's insistence on this unimportant error. He seems to stigmatize excessively the lack of scientific knowledge among the Romans. Is this an example of what Paul Veyne calls the superiority/inferiority complex of the Romans in relation to the Hellenes?[39] Once the dial of Marcus Philippus was erected, the Romans had an instrument calculated for the latitude of the leading city and no longer needed to depend on a foreign one to control political and daily life.[40]

Roman literature is not lacking in early references to time, nor about the earliest dials and their role. Their introduction was rapid and left profound traces. A passage in *The Beotien* is interesting in this respect:[41]

> May the gods damn the man who first invented hours, and especially he who first installed a sundial here: to my distress he has cut up my day in slices. When I was a boy my belly, the best and most exact of clocks, was my dial. In any place, it told me when to eat, except when there wasn't anything. Now, even if there is, one eats only by permission of the Sun so much the city is full of dials. Already most of the population languish, dried out by hunger.

The author of this comedy takes up here a commonplace conceit of the Greeks—the hanger-on complaining about the introduction of timepieces into private life. Critics generally accept that this piece by Plautus (c.250–c.184 BCE) dates from the late third century BCE. The outburst certainly has comic effect, but the piece also gibes at the recent introduction of dials in Rome. According to this fragment, the city was already *opleta solariis*, full of sundials. However exaggerated, to make its effect, the remark has to be grounded in reality. When Plautus addresses his audience, he expects them all to know what a sundial is, what it is used for, and where it is to be found. All this presupposes a well-established assimilation of the instrument, effective at least from the third quarter of the third century BCE, which more or less coincides with the date of 293 BCE advanced by Pliny.

---

34 *Hist. Nat.*, vii, 212–15.
35 Varro, *The Latin Language*, vi. 4.
36 Savoie 2007, 1170–5; Savoie 2014a, 21–7.
37 Catania has a latitude of 37° 30′; Rome 41° 54′.
38 Savoie 2007, 1172.
39 Veyne 2005, 196.
40 This story of the dial seems risible compared with other, far more embarrassing, Roman errors. For example, the civil calendar was in advance of the astronomical data by some four months at the end of 190 BCE, March falling in October. This should have been an evident problem, not the imperceptible displacement of a few minutes on a sundial.
41 Aulus Gellius, *The Attic Nights*, iii, 3. 4–5.

Figure 4 Inscription and dial. Image created by Jérôme Bonnin based on the dial from Bevagna (Museo Archeologico Nazionale, Perugia, inv. 50,028.

Archaeological sources do not go back so far. The oldest known sundial comes not from Rome, but from Bevagna, a town in modern Umbria.[42] It is a spherical dial, without declination curves, mortised into its base (Figure 4). This suggests that it was mounted high, probably on a column. It measures 43cm in height by 40cm by 26cm. The base carries an inscription in Umbrian written in Etruscan characters recording that it was the gift of Norinus and Iantus Aufidius *quaestores fararii* (food magistrates). Following this inscription the dial has been dated to between 263 and 190 BCE[43] using as criteria the disappearance of Etruscan characters for Umbrian after 190 BCE. This in no way contradicts the texts of Pliny or Vitruvius, nor any known archaeological findings. The Romanization of Umbria began from the 290s BCE onwards. Therefore, it is likely to find a dial at the end of the third century BCE. Dials were already present in Rome at this period, and they would spread more and more quickly in the cities that came under Roman control.

It was indeed exactly at this time that timepieces became a social necessity and were no longer the province of a few savants. Briefly put, the Greeks, inheriting Babylonian knowledge in astronomy, and rather less from the Egyptians, developed increasingly complex instruments. Used first strictly in a scientific context in the sixth and fifth centuries BCE for studying the sky and its apparent movements, they slowly left this restricted area and became used by political authorities. In the fourth century BCE dials and water clocks expanded into the major cities, becoming indispensable for some people, although still objects of curiosity for others. This technology was introduced into Rome towards the middle of the third century BCE and little by little spread through the region around the city. By contrast with what occurred in Greece, the sundial had a hitherto unprecedented development in the Roman world—opposite to that of the water clock. To understand this fully, and to correct some persistent errors about Roman timepieces, it is first necessary, after having defined the term *horologium* precisely, to present the characteristics of the archaeological pieces in detail.

HOROLOGIUM: A DECEPTIVE TERM

As David Landes (1924–2013) noted as regards the Middle Ages,[44] the word *horologium* is ambiguous, and does not allow us to know

---

42 Concerning this dial, see Ciotti 191, 81–5 and Bonnin 2015, A-46.
43 Filippetti 2000, 64.

44 Landes 1983/2000: ' . . . it is one of the misfortunes of scholarship that there was only one word for clock in the western Europe of the Middle Ages: *(h)orologium*. This generic term referred to every kind of timekeeper, from sundial to clepsydra to fire clock to mechanical clock. So, when in the late thirteenth century, we get an unprecedented spate

exactly what kind of object is referred to in an ancient text. The Romans themselves did not fail to note this ambiguous, generalizing aspect of the term. However, by contrast, there did not exist in Antiquity a single term in either Latin or in Greek.

In Greek a water clock, like a sundial, is called ὡρολόγιον. In literary texts this can be replaced by ὕδριον ὡροσκοπεῖον.[45] Generally it is completed by a composite term using the word 'water', either by the addition of a qualifier or by a periphrasis. This linguistic procedure is found in Lucian (120–180 CE)[46] when he mentions the clock of Hippias, which had two devices for showing the time: 'one by means of water and roaring, the other displaying it by means of the Sun': references to the Sun and water are classic for designating sundials and water clocks. 'Roaring' refers to a compressed air system that indicated the passing hours. In literary texts a somewhat perturbing use of 'clepsydra' is found. But the difference between the water clock and the clepsydra is distinct: two instruments of different usage are concerned. Even so, since the clepsydra probably lies at the origin of water clocks, some authors in Antiquity occasionally used the term 'clepsydra' in a loose or ambiguous way. It should not, however, be thought that 'clepsydra', at least before Late Antiquity, could be a synonym for water clock. Ancient authors were far more aware of the difference than modern authors,[47] who still, unfortunately, use 'clepsydra' for 'water clock', and *vice versa*. Different generic terms also exist in Greek to name the various types of sundial. As for the water clock, the commonest is ὡρολόγιον or ὡρολογεῖον. These are followed by an equally widely used term, ὡροσκόπιον or ὡροσκοπεῖον (literally 'watcher of the hours'), which always indicates a sundial in general of no particular type.[48] The terms ἡλιοτρόπιον or σκιαθηρικόν are much rarer and apparently designate particular instruments of the oldest kind (discussed earlier). They are found almost exclusively in the early texts and are very rarely used for sundials.

In Latin, *horologium*, calqued on the Greek, is also generic. It is the term most commonly used in epigraphs (78% of recorded inscriptions) and in literary works. As in Greek it serves indifferently for sundials and water clocks. Vitruvius (*c*.90–*c*.20 BCE)[49] has to add *ex aqua* to specify the kind of instrument of which he treats (*horologiis ex aqua*), although only at the beginning of the chapter, the context sufficing thereafter to make his meaning clear. In the sixth century CE, Cassiodorus (*c*.485/90–*c*.580/5 CE) still uses qualifiers to differentiate the two kinds of timepiece.[50] Other terms however were used such as *solarium*, a term specific to the Romans. Initially it designated a sundial but, little by little became applied to water clocks when they showed the day hours. Censorinus (3rd century CE), among other authors, uses *solarium* for the sundial, and *horarium* for the water clock, while explaining that the latter takes the name of *solarium* when it shows the hours of the day:

> P. Cornelius Nasica made a time-piece (*horologium*) functioning with water, which is also called a *solarium* from the name of the Sun, which makes the hours known.[51]

Finally, therefore, by a displacement of sense, *solarium* (a word initially reserved for sun-based timepieces) came to mean either sundial or water clock. Censorinus' explanation is the more interesting since water clocks would also be calculated and regulated by sundials.

So what is to be understood, in terms of objects, when speaking of water clocks or sundials in Graeco-Roman Antiquity? The former are presented first as they constitute only three per cent of the archaeological corpus but, unlike sundials, have a very substantial bibliography because they have often been studied during the nineteenth and mid-twentieth centuries.[52]

Water clocks

In the first place, the *clepsydra* is an object that cannot be included in the class of instruments allowing time to be read (*horologia*). Although no ancient author confused the two instruments, many modern commentators use the terms interchangeably, as if they were synonyms, which is not the case. In modern terms the clepsydra differs from a water clock only by its use. It does not

---

of references to clocks, we cannot be sure *prima facie* what kind of device our sources are talking about'.
45 For example, in Hero of Alexandria *Pneumatica*, i, 1: 'We have moreover been led to write about this subject since we found it a natural extension of our treatise in four books about water clocks'.
46 *Hippias or the Baths*, viii.
47 It was certainly modern authors who introduced errors and found problems that did not exist. For example, in his 1684 translation Claude Perrault was astonished that 'Vitruvius, who so affects bringing in Greek names to signify things that have Latin ones, employs here a Latin circumlocution instead of using the term clepsydra, use of which was very common among the Romans'. In fact, Vitruvius knew very well that neither in Greek nor Latin did clepsydra designate water clocks.
48 This term is found in scientific literature notably in Geminos (*Introduction to the Phénomena*, ii, 35 and xvi, 13) and Strabo (*Géography*, ii, 5, 14: τα ὡροσκοπεια). It is also used by Diogenes Laertius (*On the Lives and Opinions of the Philosophers*, ii, 1) and in epigraphs showing clearly that the word refers to sundials.

49 *De Architectura*, ix, viii.
50 Cassiodorus *Institutiones*, i. 30.5. 'This is why I have prepared for you a timepiece that goes by the light of the Sun; and another that indicates, day and night without stopping, the number of the hours'.
51 Censorinus, *Of the Natal Day*, xxiii.
52 For technical and mechanical details, proposed reconstructions, and the automata and other ornaments mentioned directly or indirectly in the texts, reference can be made to older but fundamental works with numerous references to the writers of Antiquity. Blümmer 1875–88; Beck 1899; Neuburger 1919. Diels 1917, although its reconstitutions of water clocks contain many errors, remains valuable. Essential information is also presented in Kubitscek 1928; Meerwaldt 1921; Beaujeu 1948; Drachmann 1948; Price 1975; Turner 1984b, 1–9; Nordon 1991, 83–91; Turner 1994, ch. 1; Lewis 2009; Turner 2000; Hannah 2008. For automata, see Hill 1976; Hammerstein 1986.

indicate the hour but measures a discrete interval of time. It is a 'timer', and cannot function continually, but, like a sand-glass, is set in motion at a specific moment, stops after the predetermined period built into it, and has to be restarted when needed. In Ancient times, this system of stop and restart became an integral part of law court procedure, and the expression 'stop the water' is very familiar in Greek texts. For example:

> After having delayed paying the sum adjudicated, he at last paid after an accommodation with the adversary. I can produce witnesses to this effect. Clerk **stop the water** and bring in the witnesses. ... Before he could know about this arrangement he fled from here for fear of Aristodikos and settled in Thebes. Now, you know, I think that, had he been a Plataean, it is probable that he would have settled anywhere else other than in this town. I shall produce witnesses to show that he had lived in Thebes for a long time. Clerk **stop the water** and bring in the witnesses.[53]

Examples of this kind abound. This use continued in Rome and has left its trace in Latin texts. The 'timer' function was essential and in writings of the Roman period *clepsydra* is well attested, most of the time being used in its correct limited sense.

It is rather difficult to confuse a clepsydra with a large water clock. The former, according to a passage in Athenaeus,[54] can be traced as far back as Plato (428–348 BCE), and their development is marked by other great names such as Ctesibios of Alexandria (3rd century BCE), Archimedes (287–12 BCE), Philo of Byzantium (3rd century BCE), and later, after Vitruvius, Hero of Alexandria (1st century CE). Vitruvius, our main source concerning Roman water clocks, offers a synthesis of the different types of structures known, 'the same authors also sought a way to realise timepieces through the use of water'.[55]

Vitruvius gives copious technical information. Firstly, by his training, he was often in contact with such machines, which had become essential in the military world. He knew them well, could have used them, and have helped in their construction, or simply had the time to examine them and understand the way they worked. Thereafter, 'he sought equally to make them known and understood, in a word, to educate his reader'.[56] The first device described should be classed, according to Vitruvius, in the category 'amusements'. This was the water clock invented or, more probably, improved by Ctesibios of Alexandria. His first innovation was to form the outlet hole (at the base of the clock) in a piece of gold or a gemstone. Choosing gold, perhaps for its precious and inalterable nature, was not without drawbacks, given its soft, deformable character. Moreover, despite Vitruvius' belief to the contrary, it did not hinder the deposit of impurities. The second innovation was to replace the internal hour scale, which was difficult to read, with a more convenient external one. This was accomplished by the addition of a *scaphium inversum*: a cork, wood, or ceramic float in the form of an inverted bowl filled with air[57] and placed in the lower vessel. In its simplest form the instrument functioned as follows: the float carried a vertical rod surmounted by a figure that indicated the hours with a pointer on a structure (a pillar or column) placed beside the clock carrying an hour scale that was changed each day.

Following this, Vitruvius describes two different methods for reading the hour. The first is a clock with a fixed graduation in which the length of the unequal hour is simulated by varying the outflow rate. The explanations of this are abundant but difficult to follow, and at times he seems to doubt the efficacy of these regulating systems. The second method employs a constant water outflow, where the graduated pillar is replaced by a column on which the hours are shown by curves drawn 'after the analemma', a graphic realization of the variation of the unequal hours throughout the year. The column needs to be turned each day so that the pointer is set on the line for the day and month of use. No mechanism for doing this is indicated, but the column can easily be manually turned. Finally, Vitruvius tells us, other mechanisms gave aural indications and moved automata, all this in the great tradition of Alexandria, which would continue until at least the sixth CE and which Arabic engineers would pursue still further. But had Vitruvius actually seen these automata, or did he take the description of them from the *Commentarii* of Ctesibios? He only mentions them, without explanations, despite the fact that they are complex devices apt to astound.

Finally, Vitruvius describes an instrument that he seems to have seen working and for which archaeological evidence is available to help reconstruct it. This is the anaphoric clock, also known as the *horologia hiberna* (winter clock).[58] In winter, human activity continues after sunset, which justifies the use of such devices. It is generally agreed by contemporary authors that such clocks showed the rise and set of stars, a hypothesis supported by the Greek term which, for astronomers signified 'the rising of a star'. Realizing such a clock, however, is complex, as is the description given by Vitruvius.

The dial of an anaphoric clock is composed of a metal grid (*virgulis aeneis*) through which a mobile disc placed behind it can be seen. The stationary grid is composed, according to Vitruvius, of twenty-four separate hour lines curving from the centre to the circumference and of seven concentric circles representing the months grouped in pairs except for June and December. Behind the grid, the vertical disc, carried on a horizontal arbor turned by a counterweight, shows the constellations, the ecliptic, and the signs of the zodiac. In such clocks it is the position of the degree of the ecliptic corresponding with the date that indicates the hour as it moves behind the fixed metal grid. In

---

53 Lysias, *Against Panelion*, xxiii, 14–15.
54 *Deipnosophists*, iv. 174b.
55 *De Architectura*, ix, 8–2.
56 Fleury 1994, 205

57 Bilfinger 1886, 42; Ardaillon 1900, 262. The Greek synonym given by Vitruvius, *phellos*, meaning 'oak cork' suggests that the first floats were worked in cork, to be replaced later by others in wood or ceramics.
58 On this type of clock, see especially Turner 2000.

Figure 5 Reconstruction of the dial of the Salzburg anaphoric clock. By Jérôme Bonnin, after Kuenzel 2000, 548.

addition, the ecliptic circle is pierced with small holes, each one corresponding with a day of the year. A bright indicator was set manually in the hole for the appropriate date and was sometimes furnished with an index. The latter offered the advantage that it allowed the indicator to be more easily distinguished behind the hour grid and facilitated manipulating it. It is for this reason that the remnant of such a clock, found at Salzburg (Figure 5), carries the names of the months on the reverse of the disc to guide its keeper in resetting it each day. It was this that supplied confirmation of the nature of this fragment as part of an anaphoric clock.[59]

Found in 1901, the Salzburg clock is composed of part of a bronze plaque originally of about 1.2m in diameter and two millimetres thick. On one side a fracture follows the circular path of a series of small holes which can be identified without difficulty as the *cava*, which Vitruvius tells us were set in the ecliptic. No doubt a public clock, it would have been particularly impressive and prized. The mechanism and the decoration around would together have taken up a considerable amount of space; it was perhaps housed in its own protective shelter. Knowledge of it allows the accuracy of Vitruvius' account to be gauged.

A second instrument of the same type was found in 1886 at Grand (Vosges, France), a well-known Roman thermal station.[60] It consists of a fragment of bronze similar to that of the fragment from Strasbourg except that it is smaller and has no zodiacal indications. A piece of bronze that would originally have been part of a disc of 35cm diameter recovered in 2008 on the military site of Vindolanda (on Hadrian's Wall, England), was initially thought to be part of an anaphoric clock or a parapegma.[61] It has perforations and shows the names of the months (only September survives), abbreviations for the Kalends, Nones, and Ides, and indications for the equinox. It comes from a ring of bronze that supplied the outer, upper, edge of a hemispherical outflow water clock, the holes being probably to receive a marker to indicate which scales on the interior of the vessel were to be used for the time of year. It is similar to the rim of a complete clock held by the Frankfurt Archaeological Museum since 2000.[62] If the Vindolanda find-site suggests a military purpose for the instrument, serving perhaps to fix the times of the watches or other administrative formalities such as the arrival and the despatch of messages, other evidence, including a further such fragment found on the rural site of Hambledon, Hampshire (England), suggest that the use of simple water clocks was widespread. That is, their use was not confined to any specific task despite some strong evidence that enables them to be associated with the control of hours for male and female frequentation of public baths.[63]

Sundials

It is not usually difficult to recognize an ancient sundial. They adhere to a general schema that is fairly simple whether in terms of the material used, their shape, or their constitutive elements. Rather few materials are used: only local or imported stone and the metal pieces needed to make them function. Less essential, decoration is sometimes added, such as stucco or coloured pigments. Marble, followed by limestone or local stone, is the most common material. Portable dials are generally made of bronze or a similar metal alloy, but examples in bone also occur. The gnomon was usually made of bronze but there exist some rare examples in iron. Whatever the case, it was always anchored in place with lead, and many dials retain traces of this. Occasionally the negative imprint of a prismatic gnomon is still clearly visible, as are traces of the tool used to drive out the air and allow the gnomon to be perfectly adjusted to the lead and the stone.[64] Additional material can enhance (or camouflage if poorly executed) the quality of the instrument. This is notably the case of stucco used to cover stone of little aesthetic appeal, such as tuff and some calcareous stones. Its absence may be the result of erosion or of pieces being abandoned. Finally, some rare survivals show that the hour and declination lines could be highlighted with black or red pigment; this was probably generally the case, not an exception. But dials made of tuff, covered with stucco, and painted are fragile. It is difficult to imagine that they were placed in the open air. Perhaps they were placed on the edge of the peristyles of the *domus* or more modest dwellings, or they were protected from rain by a light structure that did not hinder the passage of sunrays when reading

---

59 Kuenzel 2000, 548; Rehm & Weiss 1903, 33–41 & 41–9, figs. 18–22.
60 Maxe-Werly 1887, 170–8; Nordon 1990, 27–42; & fig. 8, Turner 2000, 540–2. It is conserved in the Musée des Antiquités Nationales, St Germain-en-Laye.
61 Lewis 2009, 13–17.

62 Stutzinger 2001; Meyer 2019.
63 Meyer 2019, 194–9; for the hours and the baths, see Bonnin 2015, 229–34.
64 So the gnomon of a conical dial preserved at Athens in the reserves of the Museum of the Stoa of Attalos (inv ST. 147). *Cf.* Gibbs 1976, 230.

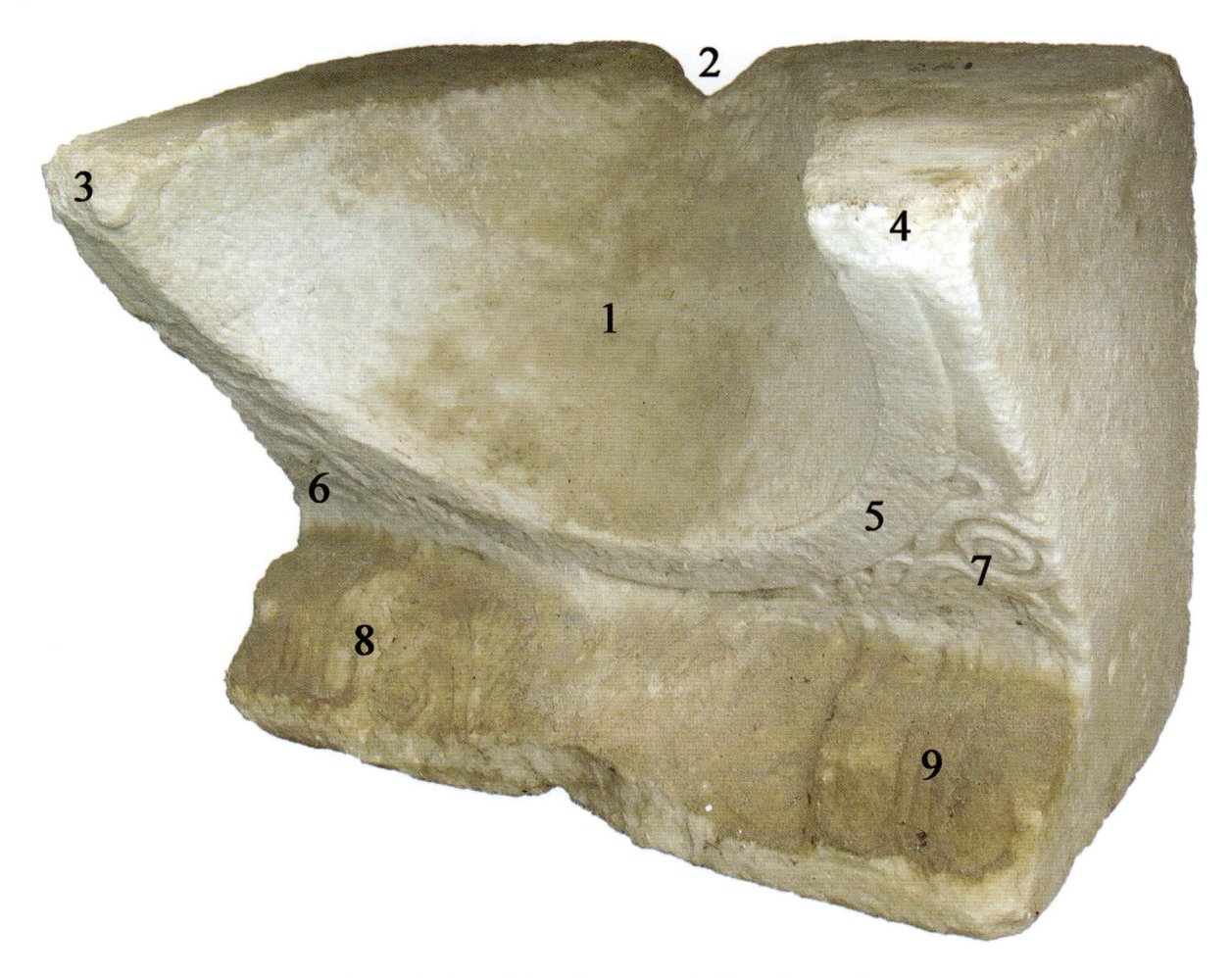

Figure 6 Constituent parts of a standard sundial (Delos, inv 261). Photo J. Bonnin.

the hour. Many vertical dials today are placed beneath the gutters or eaves of buildings, thus protecting them without hindrance to their functioning. This remark applies specifically to private dials; public ones were generally of a far better quality. For this reason many of them could be mounted on columns or in high positions, easily visible for all. It was probably the same for dials placed in the necropols, where vertical plane **dihedrons** and spherical dials with **aperture-gnomons** preponderate. These forms rendered the stone less liable to water damage.

Sundials formed from a hollowed block (conical, spherical, spherical with **aperture gnomon**, cylindrical, or multiple) are composed from various typical elements that are found in the majority of examples. This is also true, though in a minor degree, of dials drawn on plane surfaces. In the majority of cases they have two distinct sections—the base, and the receiving surface—and usually seven fixed parts (Figure 6).

The part that receives light is composed of (1) the receiving surface itself, (2) the **gnomon** or its position when the **gnomon** itself has disappeared, (3) the left extremity, (4) the right extremity. This upper part of the dial is sometimes separated from the base by (5) a moulding. Between the base and this moulding shaped corner-pieces (6) and (7) carry figurative or non-figurative decoration. Finally, the base is usually flanked with two feet of variable form (8) and (9).

The hour lines, usually eleven in number, divide the receiving surface into twelve equal parts.[65] They are frequently complemented by **declination** curves, usually three in number. In this case the equinoctial line is always at the centre of the instrument. The curve representing the summer solstice line is found farther away from the base of the gnomon on surfaces cut into a block or on plane vertical dials, and closer in plane horizontal dials. The opposite is true for the winter solstice line.

Up to now, archaeological investigations have brought over 650 different dials to light. The difficulty for the historian is to find a correspondence between them and the typology derived from the ancient sources—Vitruvius in particular. The details of ancient dials and how a particular form can be attributed to a particular

---

65 Six dials show either too few (9 or 10), or too many (12 or 13) hour lines. These are late dials no longer using the classic hour system or, for those with thirteen lines, a desire to represent the first and the twelfth hours by separate lines. For dials with too many lines, see Gibbs 1976, 168 (1054G from Durostorum); Bonnin 2012, fiche A_449 (unpublished sundial from Masada). For dials with too few lines, see Gibbs 1976, 359 (5018 from Akrai); Bonnin 2015 395, A_337 (dial from Richborough); Bonnin 2012, fiche A_511 (dial from Ville-Pommeroeul); Schaldach 2006, 126 (dial from Isthmia).

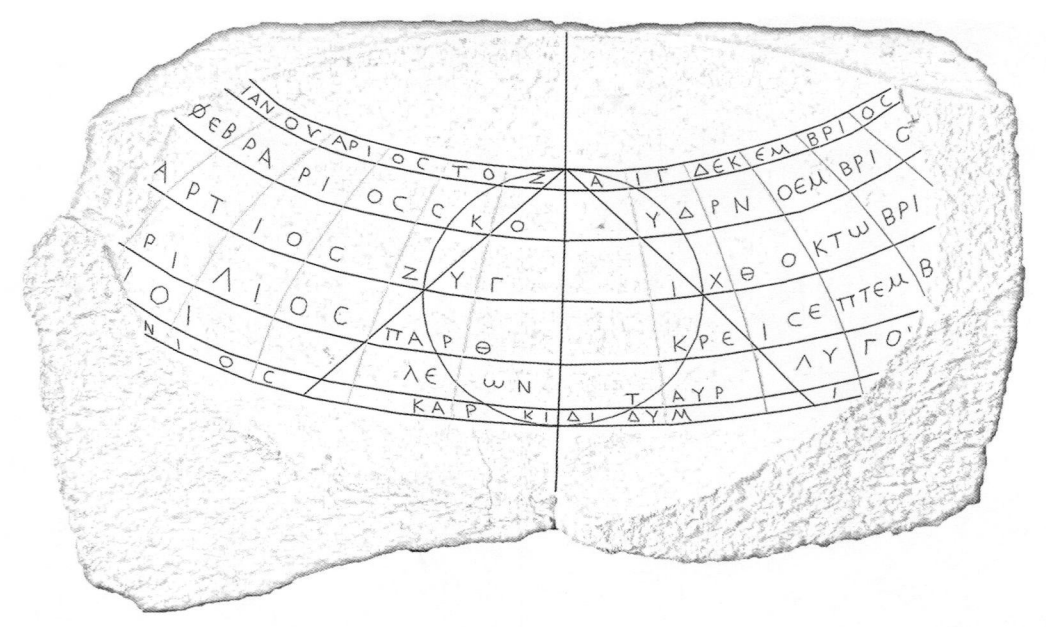

Diagram 1 Hemispherical dial.

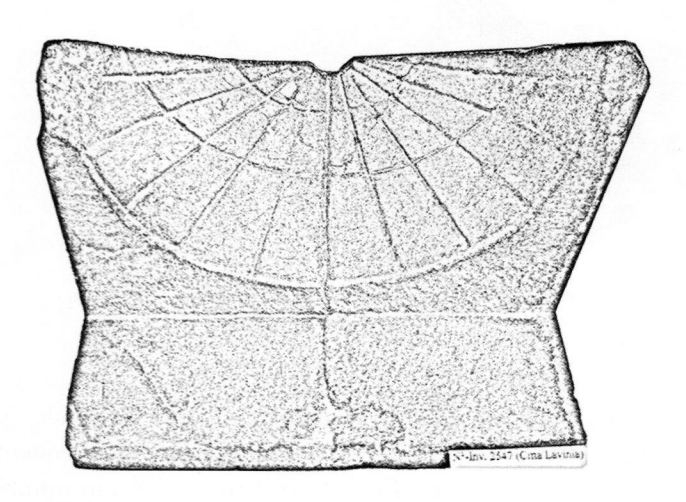

Diagram 2 Quarter-spherical dial.

Diagram 3 Spherical dial with aperture gnomon.

name are discussed elsewhere.[66] Briefly, however, fifteen types of dials are known from archaeological sources.

Hemispherical dials As the name indicates these dials (Diagram 1) have a hemispherical receiving surface (not a quarter sphere). If not *the* oldest form, it is certainly one of the earliest to have been developed.

Quarter-spherical dials Spherical dials have a receiving surface that forms a quarter sphere (Diagram 2). Although one of the commoner types, few of the many surviving specimens are accurately executed.

Spherical dials with aperture gnomon This type is characterized by an inclined, half-spherical receiving surface that corresponds with the latitude of the place where it is erected (Diagram 3). It is fitted with a zenital aperture gnomon. The hour grid is complex and difficult to execute. When it has survived, the upper part of the instrument may still retain traces of the seat of the circular pierced bronze plaque that formed the **aperture gnomon** fitted with metal pins anchored in lead.

Conical dials With its dial surface formed from the trunk of a cone (Diagram 4), the lines on these dials are shallow curves and thus easier to draw than those for a spherical dial. As a result this is the commonest of surviving dials.

Conical dial with aperture gnomon Only two examples of this type are known. Made in the same way as a spherical dial with aperture gnomon but with a conical receiving surface, the gnomon is replaced by an orifice placed at the supposed extremity of a bar (Diagram 5). Technically complex, it is unlikely that many dials of this sort were ever constructed.

---

66 See Bonnin 2015, 98–126.

Diagram 4 Conical dial.

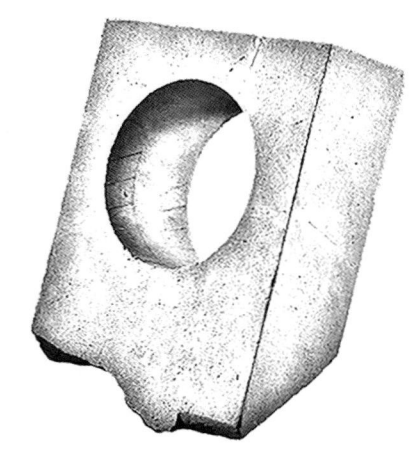

Diagram 6 Inclined cylindrical dial.

Diagram 5 Conical dial with aperture gnomon.

Diagram 7 Spherical dial.

Inclined cylindrical dial Although only one example of this category is known, it is indisputably a separate type. The sole example comes from Aï Khanoun, a city founded by the Greeks in the fourth century BCE. It is composed of a parallel-sided block with a large circular opening at the centre (Diagram 6). The bloc is inclined appropriately for the latitude. The hour lines are drawn on the inner, lower surface of the cylinder formed by the central opening. The **gnomon**, in the form of an inverted 'T', is set in the top of the cylinder and shows the hour on the two lower edges of the cylinder according to the season. An illustration of the high degree of mathematical and astronomical knowledge attained by the Greeks, it was probably abandoned for simpler instruments. It should be noted, however, that this unique example, although correctly inclined for the latitude of Aï Khanoum, does not have hour lines correctly drawn for this location.[67]

Spherical dials Two dials only, both indisputable technical masterpieces, are known of this kind. The general form is very simple. The dial is drawn on a solid sphere (Diagram 7). The hour is read thanks to a shadow line (a terminator), created by the sphere on itself.

Plain horizontal dials The dial face is a horizontal slab on which the hour line grid takes the form of a butterfly with outspread wings or a double headed axe (Diagram 8). What matters in this group is not the nature of the support, the size, or the geometric characteristics, but rather the form of the receiving surface.

Vertical plane dials The dial face is set in the vertical plane, usually south facing, and carries an hour grid composed of twelve sectors (Diagram 9). It is these 'protractor' dials[68] that seem to be the origin of the similar dials and 'mass' dials found throughout Europe during the Middle Ages. They are neither very elaborate nor particularly well made. Some seem more to be preliminary layouts for a dial rather than true scientific or everyday instruments. The only real public dial of this class is found on the Tower of the Winds in Athens, but this deploys a quite different hour diagram.[69]

---

67 See Savoie 2007.

68 Term from Gibbs 1976, 46.
69 For a presentation of this emblematic monument of ancient horology see Figure 7.

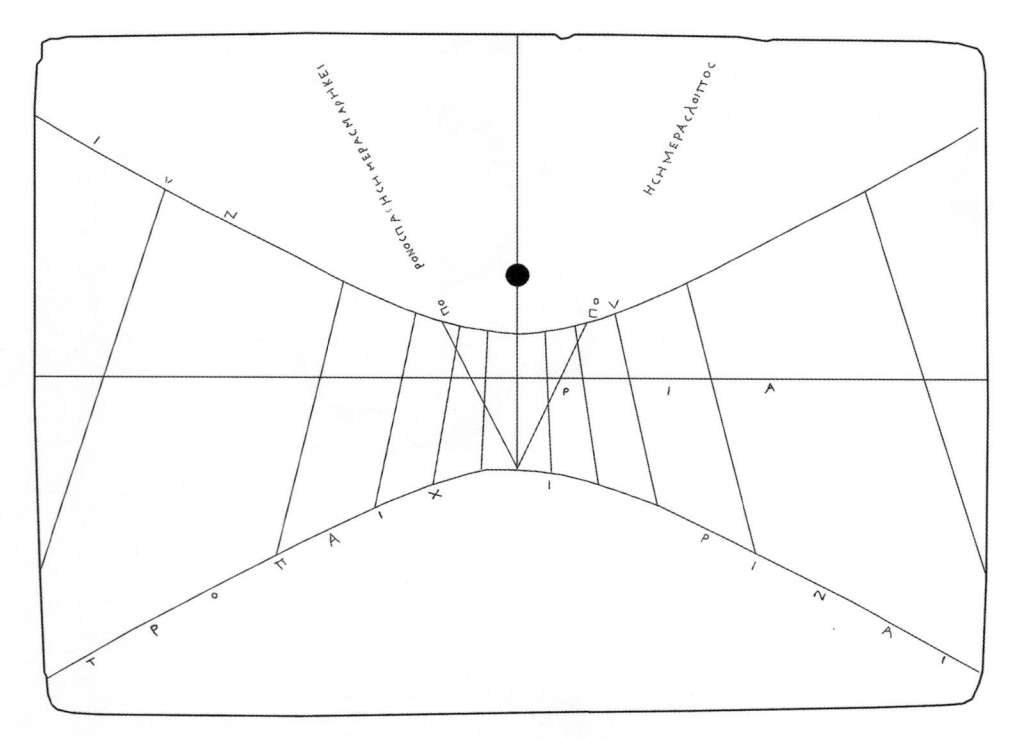

Diagram 8 Plain horizontal dial.

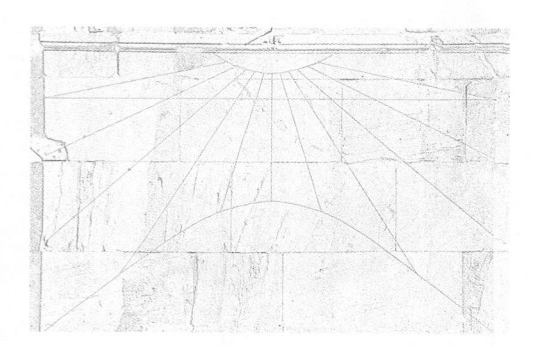

Diagram 9 Vertical plane dial.

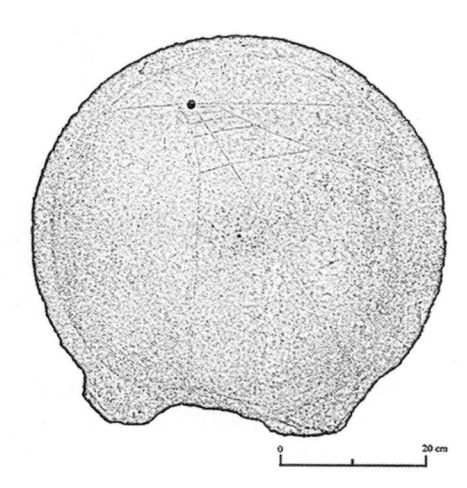

Diagram 10 Circular plane vertical declining dial.

Declining plane vertical dials This group of dials are known as 'declining' because they do not face due south. Rather, they 'decline' away from it towards the east or west and the form of the hour diagram changes although this is always composed of six sectors.

Circular plane vertical declining dial Known from only two examples, dials of this group are formed of a circular stone disc with parallel faces set vertically on a base (Diagram 10). Each face carries a direct east- (morning) or west- (afternoon) facing hour grid. The edge of the disc may also carry a **meridian** line with its own **gnomon** since such dials do not allow a time-reading close to midday.

Plane dial, inclined Once again, only two examples of this kind of dial are known. It consists of a narrow stone plate with two dial faces, one for summer, the other for winter. The entire instrument is set in position inclined at an angle corresponding with the latitude of the place of use. The form of the hour grid resembles that of plane vertical dials.

Multiple face dials Dials in this group have no specific characteristics except in possessing several separate receiving surfaces (Diagram 11). The twenty-two known examples, all different, are among the most astonishing products of ancient dialling.

Plane vertical dihedral dials These dials are formed of two narrow stone plates usually set at right angles to each other (Diagram 12) on a base (in most cases now missing). The gnomon, of prismatic or extended cylindrical form, is found at the intersection of the two plates, fixed with several mortises and sealed in with lead.

Portable dials Easily transportable because of their small size (Diagram 13), and rapidly prepared for use, portable dials fit uneasily into a single category since many of them are *sui generis*.

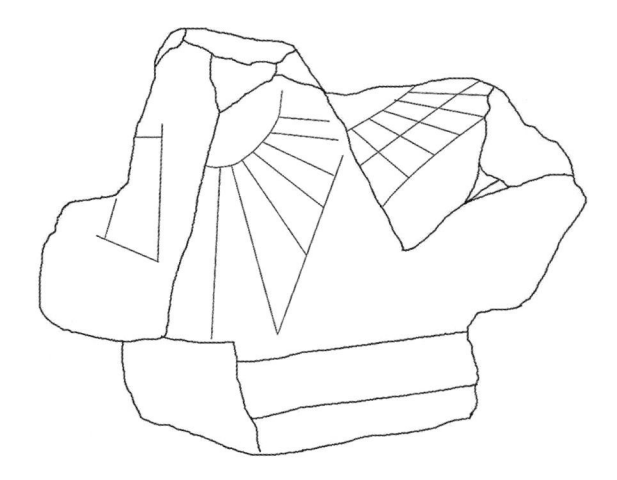

Diagram 11 Multiple face dial.

They are currently the subject of much revision, with an increasing number of studies of them and with new discoveries (or rediscoveries) enlarging the corpus.[70]

Precision

The diversity of forms and technical possibilities sketched above is, in itself, evidence of the interest shown by Greeks and Romans in time measurement and its instruments. This was such that *horologia*, sundials, or water clocks, became omnipresent in civic life, this gradually effacing their earlier scientific, astronomical, use. In the third century CE, and for almost everyone after Cetius Faventinus (3rd century CE), the sundial is no more than a means 'to know time in the quickest way'.[71] From this affirmation it can be understood that often the accuracy of the dial mattered little if it allowed this need to be satisfied. This impression can be corroborated from archaeological sources. The study of the 'precision' of ancient dials reveals much about the role assigned to them, or at least about those that are the most frequently encountered.[72] However, it is necessary to understand and to accept the limits. Many modern studies have pushed the results available too far, mathematically overinterpreting them.

The Greeks, and even more the Romans, did not share our modern demand for precision to the last minute. Indeed, such a conception is completely anachronistic. An error of five to ten minutes, in modern terms, was of little importance; approximate time was the norm. These five or ten minutes of difference from 'real time' could only be perceived with difficulty since no standards existed by which to compare measurements. Moreover, if craftsmen, often of Greek origin, were perfectly capable of producing reliable, well-calculated instruments, perhaps few people used the declination curves on dials to locate themselves in the year. To rely on a dial for calendrical ends would have been the source of many errors. But since dials in origin were prestigious astronomical instruments that offered complicated information to those who knew how to use them, they retained these attributes for a long time. Their signification was rather symbolic than astronomical, or even pedagogic. Eventually, the dial surface of everyday instruments came to be used uniquely for reading the hour. In the light of this, we can better understand why Cetius Faventinus cuts short his explanations of the most complex dials, insinuating that they interested no one, particularly not the wealthy land owners to whom his manual was addressed. It is also easier to understand why Vitruvius cites extremely rapidly the kinds of instruments known and their characteristics. A sundial should serve to locate oneself in the day and all 'furniture' was to be excluded.[73] Finally, it should always be remembered that the notion of an exact and constant hour did not exist; when someone asked the time, he was reassured to have several replies,[74] as Seneca implies:

> I think you will understand better if I tell you that we were in the month of October, and at the third day of the Ides of October. I cannot tell you the hour exactly. Philosophers can be made to agree more easily than timepieces. Even so, it was between the sixth and the seventh.

This passage does not prove that all the timepieces at Rome were of poor quality. It simply attests to their variable quality. One instrument necessarily differed from another as they were individual artisanal works lacking uniformity or a single conceptual approach. And even on two identical instruments, the reading would be different for different observers because, unless the gnomon-shadow fell exactly on an hour line, the reading was always approximate, there being no intermediary graduations.[75] This can become habitual, as with modern watches without numerals, although the reading is less precise. But it was of little importance. A Greek or Roman having a meeting at the third hour did not regard the dial as we do. He simply organized himself not to arrive too early or too late.

MAKERS AND METHODS

There are few elements that allow us to reconstruct how an Ancient sundial was created. If some hypotheses can be advanced for spherical, conical, or plane dials, for others, notably the portable dials, it is difficult. Ancient authors furnish almost no really useful information on the subject. The text of Cetius Faventinus is symptomatic of the problem. He does not tell us who realized the devices, nor what was the manufacturing chain.

---

70 See Arnaldi & Schaldach 1997; Savoie & Goutaudier 2012; Hoët-Van Cauwenberghe 2010b; Talbert 2017; Talbert 2019.
71 Cetius Faventinus, xxix.
72 On this subject, see in particular Savoie 2014a, 17–32.

73 By contrast, Vitruvius devotes several pages to water clocks the reading of which was more complex and which, probably because of their cost, combined several uses in order to justify this to their users.
74 Carcopino 1939, 189.
75 Except on some extremely rare dials, where the half hour is marked.

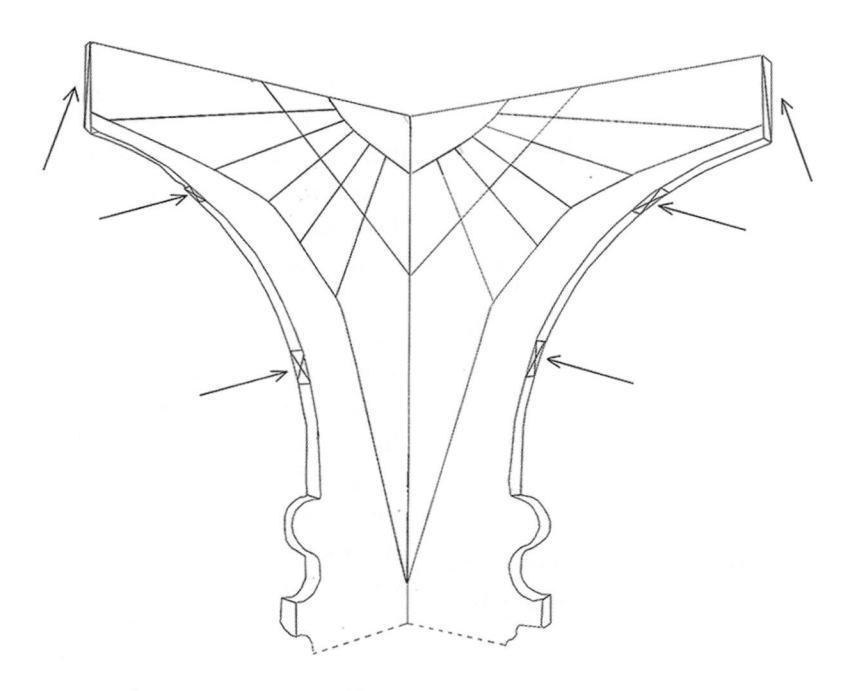

Diagram 12 Plane vertical dihedral dial.

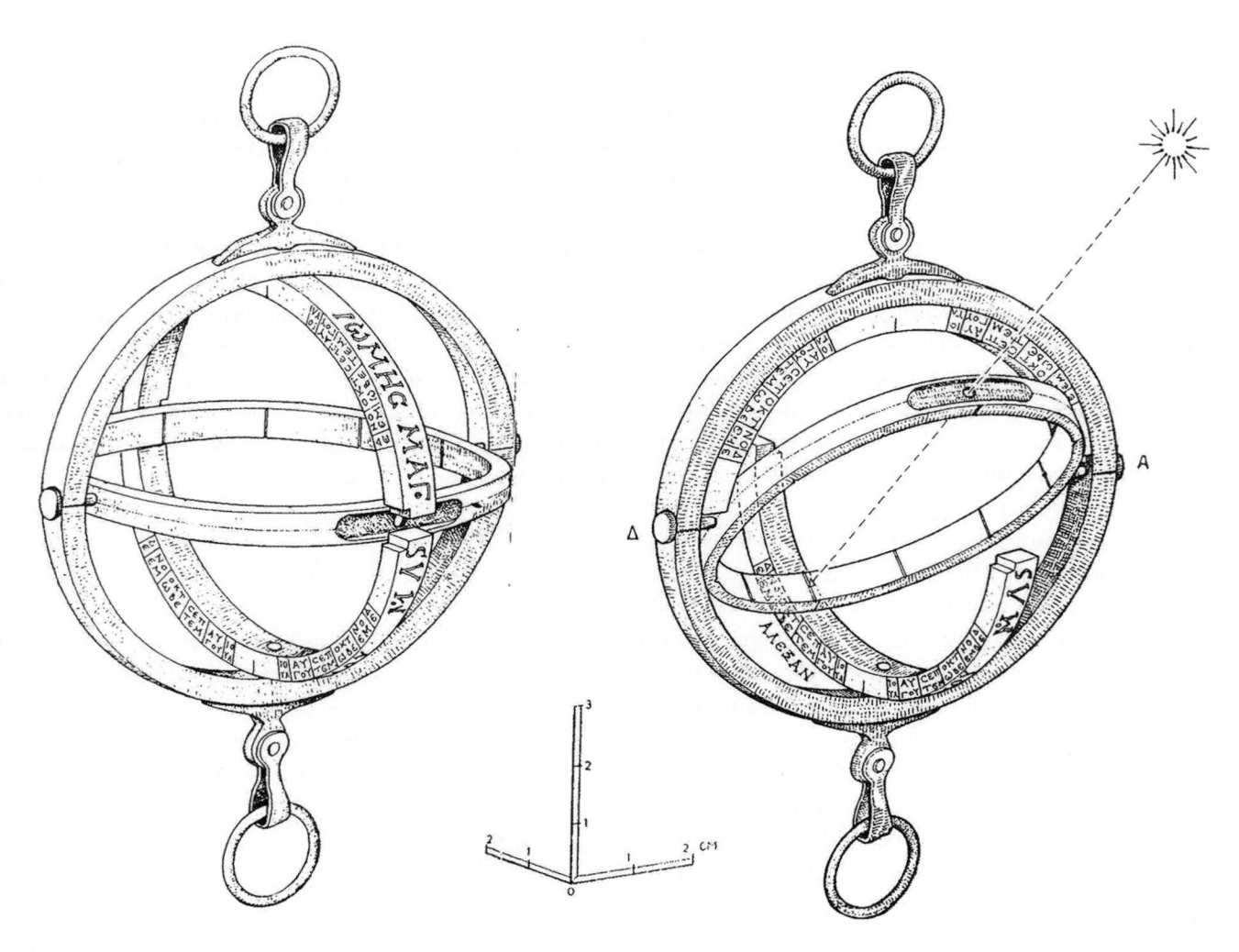

Diagram 13 Portable ring dial.

For the *pelecinium* dial, albeit in simplistic terms, he attempts to show the progression from 'two plaques of stone, larger at the top, narrower at the bottom' to a finished dial. But what he describes cannot lead to a finished dial and his 'method' seems rather imaginary than otherwise. If the main steps are exact (shaping of the plaques, drawing the lines, installation of the gnomon), details are either non-existent or wrong. When Cetius Faventinus discusses the spherical dial with aperture gnomon he gives no indication of how to draw the declination curves. It is the same for the hour lines 'drawn as eleven, equally spaced, straight lines to represent the hours'. In reality it was far more complicated, in particular for an aperture gnomon spherical dial. He ends by indicating how the upper emplacement for the pierced bronze plaque is made but without exactly describing how its position is determined. Generally his language is descriptive, not technical. 'In effect, we mark, we place, we insert, we arrange, we lay out, we set out, we align, we lay off ...'. There is never question of a geometrical, mathematical method, of verification, or of an astronomical analysis. Vitruvius' text, notably that concerning the analemma,[76] offers other difficulties.[77]

Unfinished dials or those with construction marks still visible are more reliable sources than the brief literary indications. However, only two unfinished dials exist. The first, conserved at Bucharest in the National Archaeological Museum,[78] was discovered at Constanta. Unfortunately, it has never been completely published and no image of it is known. That it is incomplete can be deduced from the lack of visible hour lines and that the position of the gnomon is not even marked. The second, more interesting, instrument comes from Delos.[79] It is characterized by a roughened area divided by a vertical line and bounded by a semicircle marking the area of the receiving surface. A vertical band left intact at the centre is difficult to explain as it has no technical use. Once the whole was drawn out it would be possible to hollow out the interior. Only at this moment could the hour grid be drawn and cut. This stage can be partially reconstituted thanks to construction marks surviving on some instruments. This is the case on a second dial at Delos: a conical dial with an eroded dial surface.[80] On the upper surface there are three, non-equidistant parallel lines, which seem to correspond with the internal declination lines, at least with those for the summer solstice and the equinoxes. However, the first line does not correspond with the winter solstice but rather with the position of the base of the gnomon or that of the axis for the generation of the cone.

Workshops and their personnel are just as difficult to perceive as the construction details. Nevertheless, the distribution and characteristics of archaeological finds allow two probable centres of production to be identified: Delos and Aquileia. The island of Delos contains twenty-seven dials of different types. In itself this number does not identify the island as a manufacturing centre. Pompei, for example, has furnished forty-three instruments, Rome twenty-three, and they are not considered to be centres of production. What distinguishes Delos is the presence of unfinished dials. Since it is unlikely that they would have been brought unfinished to the site, it may be supposed that a dialling workshop existed there. It is difficult to say whether this workshop contented itself with suppling the (limited) needs of the island, or whether it exported its products to Greece and Asia Minor. The case of Aquileia is different. Here no unfinished dials have been found, but a concentration of aperture gnomon spherical dials suggest that the city specialized in the making of this type of instrument. Of fourteen 'fixed' instruments coming from the city, ten are of this kind, while it is perhaps not an accident that two portable dials functioning with aperture gnomons have also been found there. Such a typological concentration has been found nowhere else in the Empire.

Generally speaking, however, apart from some specially reputed centre, there were probably not defined places of production. Many instruments would have been realized close to their future place of use, a fact confirmed by the use of local stone particularly in the western provinces of the Empire, and by the regional nature of their decoration. Thus a dial discovered at Bettwiller in Alsace was made of grey Vosges sandstone, while the supporting sculptural work comes from a local workshop. Errors in the delineation of the lines also show that the piece was not imported from a great centre of production. The craftsman had probably used existing patterns without really understanding the work of a dial maker. This is but one of many examples. Moreover the artisans employed were probably neither particularly competent, nor even trained, to produce such items. Stone-cutting centres would have responded to these slightly special orders without having specialists to complete them.

Only the designer of the object, at a superior level, had some importance. Charged with calculating and realizing the working diagram, he would need to be someone sufficiently competent to respond to the varied requests from cities and private individuals. An inscription, probably from Nicaea and dating to the first century CE, mentions one of them, a certain Aelianus Asclepiodotus:

> For good fortune, Aelianus Asclepiodotus, specialist in gnomonics, has dedicated the statue to the Goddess Nemesis.[81]

The term used, γνωμονηκὸς, could be a synonym for the Latin *gnomonicus*, signifying 'he who occupies himself with gnomonics', but Aelianus Asclepiodotus in this inscription certainly specifies his profession, not his hobby. The task of a professional gnomonist is, among other things, to create sundials or at least the design for these, and not to cut the stone, a task that would be consigned to others. The information summarized here is all that is available

---

76 Vitruve, *De architectura*, ix, vii. 1–7.
77 On which see Gibbs 1976, 105–17.
78 Gibbs 1976, 179, no 1065.
79 Gibbs 1976, 179, no 1064.
80 Gibbs 1976, 243, no 3025.

81 *SEG* XXXVI, 1153: αϕγαθῆ/τύχη/θεὰς τὰς Νεμέσεις/Αἰλι ανὸς ᾿Ασκληπιόδοτος/γνωμονηκὸς ἀνέθηκε.

about dial makers. The case of water clock makers is even more obscure.

MEASURING TIME WITHOUT TIMEPIECES: SOME PARALLEL TECHNIQUES

For the majority of the population of the Ancient World, who were rural and more or less literate, the idea of time was fundamentally that of a measurable period. If across the centuries methods of time measurement are specific to particular societies, then Græco-Roman society had a relatively straightforward time need. Often it was no more than the simple succession of days, nights, and seasons—in all the movements of natural phenomena. What phenomena are more evident than the stars? What changes more obvious than the changes of the seasons? What proof of the movement of the Sun throughout the day is more tangible than the movement of the shadow of a man or of an instrument conceived to show it? But even before instruments were devised, other means of time recording existed that were probably known to and used by all.

Parapegmata (from the Greek verb meaning to drive in, to fix) in their simplest form are calendars showing the position of the Sun in the zodiac, stellar phases, and, sometimes, meteorological predictions. These indications were often cut in stone or drawn on walls with a series of holes around them into which a wood day-marker, or a round headed pin (*bulla* for the Romans), was pushed.[82] This pin, which had to be moved every day, is moreover close to the indicator pin used in anaphoric water clocks, as described by Vitruvius. Used at first simply as a way of following the calendar throughout the year, the *parapegmata* ended giving astronomical, astrological, and meteorological information. Those surviving do not all give the same information, nor have the same form, whence arise the difficulties in analysing them or proposing their reconstruction if they are defective.[83] Making them presupposes a certain amount of accurate astronomical and calendrical knowledge, as well as popular demand for them. To be able to predict not only the weather, but also the time of the city or the stars, without being a necessity, must have been attractive for many city dwellers. In the countryside, the natural rhythm of the seasons would have been sufficient. Popular Roman traditions transmitted by Ovid, Columella, and Pliny the Elder mark the beginnings of the seasons: for summer with the morning rise of the Pleiades; for autumn with the morning setting of Lyra; for winter with the morning setting of the Pleiades; and for spring by a meteorological, rather than an astronomical phenomenon: the first west wind, Zephyr or Favonius. This astronomical and meteorological knowledge was sufficiently widely spread that rural dwellers had no need of more sophisticated ways of for locating themselves in the year using *parapegmata* or a *menologium*. For positioning oneself in the day, it was not exactly the same, and for this, natural methods would gradually be accompanied, even supplanted, by more complex, artificial means.

The first way of dividing the day is still evident—observing the sky with the succession, free or imagined, of natural phenomena in space. The second method is less familiar—consulting a set of shadow tables to find one's position in the day—a method certainly older than that of using timepieces. The shadow table in its developed form consists of a series of columns showing the length of a shadow as a function of the time of year indicated either by a month or by the position of the Sun in the zodiac. The shadow could be given by a fixed casting element or be that cast by the observer himself. Simple, even primitive, it is known historically from Egypt whence a table has survived from the Middle Empire (2033–1786 BCE).[84] It can be found in Greece perhaps from the early fourth century BCE. Several comedies from this period mention measuring the length of one's own shadow to arrive at a dinner at the proper time. If the length of the shadow noted does not yet correspond with a particular numerical hour, the procedure is straightforward and should have sufficed to amuse the spectators of the comedies.

Blepyros But who will cultivate the fields?
Praxagora The slaves. You will have no more concern than, when the shadow is ten feet long, to go to supper fat and shiny with oil.[85]

In the play, the parasite Cherephon has already been mentioned. Now Menander, in his piece called *Kekryphales*, recalls him, and in another entitled *Anger* says:

This man differs in nothing whatever from Cherephon. Having been invited to a supper when the shadow should be twelve feet long, he rushed in the morning to see the shadow of the Moon and arrived just as day was breaking, excusing himself for being a little late.[86]

From the Hellenistic period onwards, shadow tables were compiled as a function of the seasons, and the idea of numbered hours appeared, each shadow-length being linked to an hour of the day changing according to the season of the year. This is primordial

---

82 Petronius, *The Satyricon*, xxx, 4. 'Two tablets were fixed to the door jambs? One was inscribed, in so far as I remember, "the IIIrd of, and the day before, the Calends of January, Gaius our master will dine in town". The other showed the course of the Moon, the seven planets, the propitious and inpropitious days, marked by pins of different colours.'
83 On *parapegmata* in general, see Pritchett 1963; Lehoux 2007.

---

84 Borchardt 1920, 27.
85 Aristophanes, *The Assembly of Women*, 651–2.
86 Athenaeus, *The Deipnosophists*, vi, 243a.

for understanding the diffusion of timepieces, since the hour given by the shadow tables should be the same as that given by a sundial. What is more, the appearance of hours in the shadow tables seems to have coincided with the development of dials in Greek cities. These tables would be reproduced by the Romans and continue to be used into the Medieval period.[87] A 'user's manual' of one such set of tables is supplied in a Byzantine source in which an 'hour master' dedicates a shadow table to a 'King Philip'. There is no better explanation than that provided by this text:[88]

> Whatever the moment when you wish to know what time it is, I will tell you here how to proceed. At the place where you are walking you should measure your own shadow; and once you have found the shadow of your head, mark the place and walk from the place where you are, step by step. Count how many paces you take, look at the work on the wall [i.e. the shadow table] on which the months are marked. Thus you find the hour of the day for each month.

The text continues with an enumeration of the nature of each hour—fundamental evidence allowing the use of the table to be better understood.

EMBLEMATIC INSTRUMENTS: THE TOWER OF THE WINDS

Alongside the everyday instruments already presented, there also existed in Antiquity some exceptional instruments that commanded the admiration of all and have left traces in literature as well as in the urban landscape. Here we present two of the most remarkable: The Tower of the Winds in Athens and the monumental Meridian of Augustus.

The Tower of the Winds (Figure 7), built under the direction of Andronikos Cyrrestes in the second or first century BCE, is set in the centre of the city immediately beside the Roman Agora. It is a primordial building for understanding the place of time measurement in the ancient city. It is exceptional—one of the rare ancient structures to be preserved almost intact and that has been used for almost the whole of its existence.[89] However, after two and half centuries of research, many problems still surround it. Mystery surrounds the origins of its design, its function in the urbanism of the city, Andronikos' role in its conception, and the date of its construction.[90]

Nevertheless, the building is relatively well documented even from Antiquity since it is mentioned by Varro.[91] Vitruvius also mentions the Athens tower but in a wider manner:

> ... Andronicus of Syrrhos who as proof [that there are eight winds] built the marble octagonal tower in Athens. On the several sides of the octagon he executed reliefs representing the several winds, each facing the point from which it blows; and on top of the tower he set a conical shaped piece of marble and on this a bronze Triton with a rod outstretched in its right hand. It was so contrived as to go round with the wind, always stopping to face the breeze and holding its rod as a pointer directly over the representation of the wind that was blowing.[92]

Since Vitruvius describes the building in the context of a discussion of winds, it is not surprising that he fails to mention the dials that appear on each face of the building. He would discuss dials further on in his treatise, so perhaps he had not seen the building and did not know that it carried plane vertical dials. Interestingly, it was also the winds, and not the dials, that interested M. Cetius Faventinus in the late third century CE,[93] but it is the dial that are of interest here.

They form the outstanding group of ancient planar dials.[94] Eight are engraved on the walls and a ninth on the circular annex is interpreted as the cistern of a water clock. These dials, which appear to resemble modern dials, are extremely rare in Antiquity. They are plane vertical dials calculated and engraved according to the orientation of the face on which they are placed, from direct south to direct north with other dials declining from the west to the northeast. It is a display of technical mastery showing the intellectual capacities of Andronikos, his mastery of gnomonical science, and his knowledge of astronomical phenomena.[95]

The interior of the building seems to have contained a complex water system interpreted as a water clock[96] or, more recently, as a form of planetarium.[97] Evidently nothing remains of the moving parts, but traces in the soil (of the ducts, mortises, fixing holes, etc.) together with the presence of a cistern outside the building are clear indicators of an important hydraulic system within. Above all, the building, and its collection of measuring instruments, allows us to assess the important place occupied by time

---

87 On shadow tables in general, see Neugebauer 1975, ii, 736–46; Lippincott 1999, 107 reproduces a shadow table from a manuscript of Abo of Fleury (late tenth century). For Muslim shadow tables, see King 1997, 207.
88 *CCAG*, 7. 188.4–189-2. This explanation, although drawn from a Byzantine source, can be applied to any earlier set of tables, the underlying principles being the same.
89 It is this that explains its excellent external state of preservation. See Webb 2017.

90 The most recent assessments place it *c*.140 BCE. See Kienast 2014; Webb 2017.
91 Varro, *Rural economy*, iii, 5, 17.
92 Vitruvius I. 6. 4. Translation from Morgan 1960, 26.
93 Cetius Faventinus II. 2.
94 For a detailed studies of them, see Alberi 2006; Schaldach 2006.
95 For details of Andronikos, see Bonnin 2015, 290–1.
96 Noble and Price 1968.
97 Kienast 2014; Webb 2017.

Figure 7 The Tower of the Winds, Athens, today. Photo: J. Bonnin.

measurement at the turn of the second/first century BCE. The building was ostentatious, symbolic, and uncommon in Antiquity and, as such, provoked some Latin authors to mention it. In a practical sense, it paired the natural phenomena of winds with astronomical phenomena like the sundials (and perhaps the interior hydraulic structure, which could have included an 'anaphoric' presentation). The entire assembly is a miniature cosmos: a weathervane at the summit moves as the wind changes, the shadows of the gnomons fall on the dials throughout the day, and the interior mechanism's many mobile parts (whatever they may have represented) are always in motion.

The site was chosen as a function of these imperatives: in a high place, open at the time, and close to a water source. But who had access to this structure? Was it reserved for an elite group, and regarded as a 'curiosity' to which access was strictly controlled, even closed off at certain moments? What was its purpose? Who created it? We do not know whether the commission was public or (more probably) private.

For how long did it function? From the time when the area to the west of the tower was developed, the working of the west facing and perhaps the northwest and southwest dials could have been limited, if not entirely blocked. Similarly, once the rectangular building today known as the *pseudo-agoranomion* was constructed, it certainly prevented the meridian on the circular annex from functioning. In addition, the adjacent Roman developments upset Andronikos' entire intended presentation. It was no longer possible to walk around the building to see the time on all the dials. Furthermore, the construction of the *pseudo-agoranomion* set immediately against the reservoir may have interrupted the water supply that maintained the hydraulic system. It is unlikely that the internal works still existed in the Roman period as they would have been pillaged by the troops of Sulla (138–78 BCE) after the capture and sack of Athens in 87–6 BCE. This event may explain why Varro and Vitruvius fail to describe the interior.

If the Tower of the Winds lost its scientific function quite quickly, the building nonetheless remained intact and was used for other purposes. The desire remained to provide the hour–for strictly utilitarian purposes–on the dials that were still visible. In this, we see an element constant throughout the Roman use of time measuring instruments: utility above all, with knowledge and experimentation coming second.

EMBLEMATIC INSTRUMENTS: THE MERIDIAN OF AUGUSTUS

The meridian of Augustus Caesar (63 BCE–14 CE) is one of the major monuments of Imperial Rome and of ancient gnomonics. Apart from the Tower of the Winds, no other relic of time measurement has provoked so many studies, entrenched positions, and passionate explanations. Since its rediscovery in the years 1975–1980, it has generally been referred to as the *Horologium Augusti* or the *Solarium Augusti*. More recent investigators[98] designate it the 'Meridian of Augustus', which alone is coherent with the literary and archaeological evidence.

The monument can be reconstructed thanks to Pliny the Elder who described Egyptian obelisks in Rome.[99] After surveying several, he paused over that erected in the *Campus Martius*, the first to be imported into the city, the use of which for him was particularly remarkable (*mirabilem usum*). It was not that it marked shadows and determined the length of day and night that was special, but rather that an obelisk had replaced the gnomon habitually used for this purpose, albeit on a lesser scale. Historically, pharaonic gnomons were not used for this purpose; this was the first time they were so employed. Pliny describes use of the meridian in relatively simple astronomical terms. The only novelty in what he recounts is that the progression of the shadow through the seasons is marked by bronze rules, an innovation attributed to Facundus Novius, about whom nothing else is known. Pliny's description corresponds exactly with that of a meridian.

Mostly forgotten since it fell, or was destroyed, in the eighth century, the obelisk was rediscovered at the end of the fifteenth century, and in 1748 its five broken parts were transported for re-erection in the Piazza di Monte Citorio opposite the Italian parliament. Thanks to the inscription on its base,[100] it is possible to date the original erection of the obelisk to between the months of July year 10 and June year 9 BCE. The information given there is not extensive, but the inscription mentions the Sun, Egypt, and the function of Augustus as *pontifex maximus*, three important elements concerning the meridian.

Thanks to archaeological investigations carried out between 1979 and 1981 underneath house 48 in the Campo Marzio, the meridian line mentioned by Pliny was rediscovered. The investigators uncovered a stone pavement with a band at its centre oriented exactly north–south, several inscriptions, and bronze rules.

Also found were some zodiacal signs with calendrical inscriptions. Finally, the archaeological exploration showed that during the period of Hadrian the meridian line was displaced by a reservoir and that the obelisk had also been thrust two to three metres deeper into the soil.

While currently no monument is comparable with that discovered in the Campo Marzio, nor with that described by Pliny, meridians were nonetheless familiar in Antiquity[101] and, as already noted, had a name—*heliotrope*. They were large instruments and visible from afar, as was that of Augustus. They served calendrical purposes. In Græco-Roman gnomonics, they were incorporated in many dials.

---

98 Heslin 2007 & 2011; Alberi 2011; Frischer 2017.
99 Pliny the elder, *Natural History*, xxxvi. 2.
100 *CIL* vi, 702.
101 See, for example, Jones 2014, 175–88.

Today, the Meridian of Augustus is as much a monument of historiography as it is of gnomonics. No discussion of it is complete without also covering the various hypotheses about it—some proved, others not—and Buchner's is worth recounting here.

In 1976 Edmund Buchner (1923–2011) advanced a theory, based on Pliny's text and the (supposed) original height of the gnomon, to restore the form and function of the monument.[102] He supposed, in the first place, that the instrument was a gigantic sundial serving simultaneously as timepiece, calendar, and windrose, the obelisk itself acting as the gnomon. On the basis of the hypothetical reconstruction (made before the investigations of 1979 and 1981), he attempted to explain its purpose: the three monuments erected by Augustus on the Campo Marzio—the mausoleum, the Ara Pacis, and the *horologium*—were linked in a complex mathematical relationship. The position of the Ara Pacis, its relation to the mausoleum and the obelisk, its orientation, and its size were determined by two chief elements of the *horologium*: the equinoctial line and the circle drawn around the winter solstice. In the proposed restitution the equinoctial line is directed towards the Ara Pacis. At the Autumn equinox, which was also the birthday of Augustus, the gigantic shadow of the obelisk would indicate the Ara Pacis. At the winter solstice, the first day of Augustus' zodiacal sign, Capricorn, the circle is drawn around this point and is tangent to the south of the Ara Pacis—a link between the two monuments. The *horologium* and the Ara Pacis would thus have functioned as a birthday group, as, according to Buchner, the two monuments were consecrated in the same year, if not on the same day. The ideological implications that follow from this are obvious: the group should be understood as a public demonstration that Augustus was destined to bring back peace and assure a prestigious future to the Roman people by establishing an enduring political system.

This hypothetical reconstruction was rapidly questioned[103] because the excavations had revealed only the meridian line. The prevailing consensus today is that nothing suggests a timepiece, while everything suggests a meridian. In 45 BCE, Julius Caesar (100 BCE–44 BCE) reformed the Roman calendar so that it would correspond with the seasons. However, the following pontiffs did not understand this reform and established a leap year every three years instead of four. This meant that there was once more a lack of correspondence for thirty-six years. Then, in 9 BCE, when Augustus became *pontifex maximus* on the death of Lepidus in 13/12 BCE, he officially declared this calendar to be erroneous, re-established Caesar's work, and realigned the Roman calendar with the seasons. It was exactly at this moment that he erected the obelisk and the meridian, and his message was clear. The meridian demonstrated to everyone that it was not an ideological manipulation. At the equinoxes and the solstices the shadow touched exactly the rules representing the equinoxes and the solstices, as it did accurately for the other days of the year. The purpose of the monument was therefore calendrical and religious. Furthermore, the term *pontifex maximus* is emphasized in the inscription by not being abbreviated, which is rather rare in Latin epigraphy; this title is often abridged and is not usually deployed on a whole line.

Whatever may be its origins and its links, real or imaginary, with an ideological programme close to modern propaganda, the meridian of Augustus should be considered as an important element of the landscape of Ancient Rome. If it was not a monumental timepiece, it is nonetheless linked with time measurement—since it identifies a position in the year and shows exactly the moment of noon each day—and more importantly, carries a special, calendrical symbolism.

CONCLUSION

The trajectory of ancient timepieces is well worthy of attention. In origin the instruments were complex and prestigious. They seem to have first appeared in Asia Minor, and then in Greece well before the Romans concerned themselves with the question of time. Greek influence led to time-keeping instruments becoming a part of everyday Roman life—first in Rome itself from the third century BCE, and thereafter, following the Roman conquests, throughout the Italian peninsula. From the first century BCE there was not a city in Italy without a timepiece, and these spread to all the conquered provinces without exception. For six centuries from the second BCE to the fourth century CE timepieces would be a habitual part of 'civilized' life in both West and East. This challenges the modern presupposition that timepieces never had a great presence in Rome or much success in the Roman world. Both historical texts and archaeological remains show that it was exactly during the Roman period that time measurement became mentally embedded,[104] and that timepieces themselves became a measure of a man's civility. They were diffused throughout a vast territory and had a hitherto unknown temporal continuity. Lewis Mumford wrote that 'The clock is not just a way of following the march of the hours, it is also a means for synchronising human activity',[105] and this synchronization began in Antiquity.

The concept of a time—not uniform, but at least available everywhere for everyone—constitutes a first example of temporal control. However, one Roman characteristic should not be denied—the disappearance of all research concerning time measurement. For if the Romans inherited Greek knowledge in this matter, they also inherited the use of timepieces in daily life. The essentially scientific aspect of the first instruments in classical Greece, conserved to a lesser degree during the Hellenistic

---

102 Buchner 1976; Buchner 1982.
103 By Heslin, 2007, 2011; Haselberger 2011; Frischer 2017.

104 Bonnin 2013.
105 Mumford 1934.

period, gave way to the utilitarian aspect. One went to the forum during certain hours, to the baths or the gymnasium at others. Markets opened only at specific times, as did libraries and public centres in general. Examples of the use of the hour and the timepiece are numerous. In general the common denominator of these public uses was regulation. It was obligatory to institute hours so that civic life could unfold correctly: heat the baths at the right time, avoid promiscuity, avoid overloading on the roads, summon the citizen body when needed, coordinate complex activities, or even travel with the empire in one's hand.[106]

---

106 To paraphrase the subtitle of Talbert 2017.

CHAPTER TWO

INDIA AND THE FAR EAST

SECTION ONE: HOROLOGY IN INDIA

S. R. Sarma

UNITS OF TIME MEASUREMENT

The history of horology in India has to be reconstructed from brief literary references and extant time-measuring instruments. From the earliest times, some kind of astronomical activity must have taken place for determining the seasons, the equinoxes, and the solstices for agricultural and religious purposes. The *Ṛgveda*, the earliest available text, dateable to *c.*1500–1200 BCE, contains references to seasons, to the length of the year, and so on. In an allegoric passage, it speaks of a wheel with twelve spokes that turns without stopping (alluding to the year of twelve months that repeats itself constantly) and of pairs of suns whose total number is 720 (i.e. 360 days and 360 nights).[1] The *Taittirīya-Brāhmaṇa* and the *Śatapatha-Brāhmaṇa* (*c.*700–600 BCE), two later texts of the Vedic corpus, divide the civil day into 30 *muhūrtas*. The *Vedāṅga-jyotiṣa* (*c.*400 BCE), also a part of the Vedic corpus, divides the *muhūrta* into two *nāḍikās*. Accordingly, the civil day is divided into 60 equal units of *nāḍikās* (later called **ghaṭīs**). This became the standard unit of time measurement and remained so until the end of the nineteenth century. The instruments that were used to measure this unit are the gnomon and the water clock.

GNOMON

The earliest mention of the **gnomon** occurs in the *Arthaśāstra* of Kauṭilya, which was composed and redacted between the second century BCE and the third century CE. This text lays down that the king should divide his day and the night separately into eight parts, each by means of the **water clock** (*nālikā*) or by the length of the gnomon-shadow (*chāyā-pramāṇa*), and devote each part for a specific administrative or personal task:

> He should divide the day into eight parts as also the night by means of *nālikās,* or by the measure of the shadow [of the gnomon]. [A shadow measuring] three *pauruṣas*, one *pauruṣa*, [and] four *aṅgulas* and midday when the shadow disappears are the four eighth parts of the day. By them are explained the later [four].[2]

Here Kauṭilya shows how to divide the period from sunrise to midday (*n*) by means of the gnomon-shadow. *Pauruṣa* (literally 'of the man') is used here in the sense of gnomon height (*g*). The gnomon can be of any height, but it is divided into 12 parts called *aṅgulas* (*g*/12). At sunrise, the shadow length is three times a man's height (*pauruṣa*), which actually means thrice the gnomon's height (3*g*). When the shadow length is 1*g*, it is ¼ *n* from sunrise; when it is four *aṅgulas* (4/12 = 1/3*g*), it is ¾*n* from the sunrise. When there is no shadow it is midday. The same shadow lengths in the reverse order indicate the four parts of the time between midday and sunset. But these shadow lengths are correct only on equinoctial days. For other times, an increase of two *aṅgulas* per month in shadow lengths is prescribed.[3]

---

1 Griffith 1889, 1.164.11.

2 Kauṭilya 2010, 46

3 Jacobi 1920, 253–4.

Thereafter, no information is available until Brahmagupta composed the *Brāhma-sphuṭa-siddhānta* in CE 628. The third chapter of this text is designated 'the chapter on three questions or topics' and discusses the determination of the cardinal direction (*diś*), terrestrial latitude (*deśa*), and time (*kāla*) by means of the gnomon (*śaṅku*). As in the *Arthaśāstra*, a straight staff is set up temporarily on a level ground on which lines are drawn for the determination of these three parameters and others, such as the times of sunrise and sunset, rising of the zodiac signs, the ascendant, and so on. This practice of devoting the third chapter to gnomonics is emulated by nearly all the Sanskrit texts of the genre *Siddhānta*.

The *Brāhma-sphuṭa-siddhānta* describes some other instruments of permanent nature and so do several subsequent texts.[4] Although many types of instruments are described in these texts for measuring time by the sunlight, those that are represented by extant examples are just three: the ring dial (*Cūḍā-yantra*), the column dial (*Kaśā-yantra*), and the horizontal dial with a triangular gnomon (*Palabhā-yantra*).

Ring dial (*Cūḍā*-yantra)

The earliest of these is the ring dial.[5] Here, a small hole in the breadth of the ring allows sunlight to fall upon the inner concave surface of the opposite side, which is graduated in degrees to measure the solar altitude. Local time can also be measured directly if the inner surface is provided with separate scales for each solar month.

The ring dial was first described by Āryabhaṭa about the beginning of the sixth century CE in his *Āryabhaṭa-siddhānta*[6] and by Varāhamihira in his *Pañcasiddhāntikā* in the middle of the sixth century CE.[7] In his *Yantraprakāśa* (1427), Rāmacandra Vājapeyin describes three varieties that work on the same principle but differ in size. The *Valaya-yantra* measures a **cubit** in diameter, the *Cūḍā-yantra* a **span** or less, and the *Mudrikā-yantra* is much smaller. The inner concave surface is graduated in *ghaṭīs* for measuring time and the rim in 360 degrees for measuring the solar altitude.[8]

Of these, the *Cūḍā-yantra* appears to have been more popular; it is represented in several Mughal miniatures, where astronomers are depicted holding a small ring dial to measure the solar altitude, and not the astrolabe as might be expected.[9] Because of its frequent occurrence in paintings depicting the Mughal court, one is apt to think that it may have been of Islamic origin. But the ring dial was not known to the Islamic world and must have been developed in India itself.

---

4 Cf. Sarma 1986–1987a.
5 Sarma 2019a, Section O.
6 Shukla 1967, 93, 98; Ôhashi 1994, 236–8.
7 Varāhamihira 1968, 80; Ôhashi 1994, 238–9.
8 Rāmacandra, 1886–92, 61–2.
9 Sarma 1992, 249–52.

Sawai Jai Singh of Jaipur (1688–1743), before he designed the huge masonry instruments for his observatories at Delhi, Jaipur, and three other places, caused the compilation of the *Yantraprakāra* between *c.*1716 and 1724, which deals with the construction and use of several portable instruments, culled from diverse sources. This text contains a detailed description of the *Cūḍā-yantra*, together with an elaborate set of tables to be used in conjunction with it. There are nineteen separate tables, all for the latitude of Delhi (28°39´). The first and last of these tables are prepared for the first **decans** of Capricorn and Cancer respectively; of the remaining seventeen tables, each one is meant for a pair of opposite decans.[10]

Compared with horizontal and vertical sundials, the ring dial is more difficult to construct and also more difficult to use. It is probably for this reason that only three examples are extant. Two well-crafted specimens are preserved in Jai Singh's Observatory at Jaipur (Figure 8). These must have been made for Jai Singh himself in the first quarter of the eighteenth century at Jaipur. The third, of an inferior make, is in the Museum of Indology at Jaipur.

Column dial (*Kaśā-yantra*)

The column dial, also known as cylinder dial, consists of a straight wooden staff, with a circular or prismatic cross-section, divided lengthwise into several columns, each carrying a separate scale of *ghaṭīs* for measuring time in a particular solar month.[11] The *ghaṭīs* are numbered serially from the top to the bottom according to the length of the half-day from sunrise to midday in that particular season. In the upper part, just above the scale, is a hole in each column to receive the horizontal gnomon.

For measuring time, the horizontal gnomon is inserted into the hole above the scale meant for the current solar month and the staff turned slowly towards the Sun so that the gnomon throws its shadow exactly on the scale below. Where the end of the shadow touches the numbered scale, the number indicates in the forenoon the *ghaṭīs* that have elapsed since the sunrise, and in the afternoon, the number of *ghaṭīs* that are to elapse up to sunset.

The column dial is described in four Sanskrit texts that were composed between 1428 and 1439.[12] In these texts, the column dial is called *Cābuka-yantra*, *Kaśā-yantra*, or *Pratoda-yantra*. *Cābuka* is a loan word from the Persian and denotes a horsewhip; *Kaśā* and *Pratoda* are Sanskrit renderings of the Persian term. This suggests that the instrument was borrowed from the Islamic world.

The earliest column dial surviving from the Islamic world was made by Abî al Farāj 'Īsā in AH 559 (1163/64 CE). It has twelve columns and the hours on these are divided by continuous curves.[13] In the thirteenth century, al-Marrākushī also describes

---

10 Sarma 1986–1987b, 27–8, 79–82, 105–14.
11 Sarma 2019a, Section P.
12 Ôhashi 1998.
13 Casanova 1923; Turner 2018, 187–9.

Figure 8 Ring dial from the Jai Singh Observatory, Jaipur. Photo: S. R. Sarma.Column Dial (*Kaśā-yantra*).

the column dial in the same manner.[14] But neither source contains any reference to a horsewhip.

The column dial has a continuous history in Europe since Late Antiquity.[15] In Europe there exist in museums several Renaissance and later specimens, but only a few Islamic examples have survived.[16] Both the Islamic and the European specimens differ in construction from the extant Indian examples. They are much smaller, with lengths of about 200mm. The twelve scales on their columns are divided by continuous curves that flow from one to another. The top with the gnomon can be rotated so that the gnomon rests on the desired month.

Compared with these, the Indian versions are much longer, ranging between 1100mm to 1550mm. Instead of twelve separate scales for the twelve solar months, they usually have eight scales, employing one common scale for two solar months that are at equal distance from the equinoxes. The scales are not divided by continuous curves, but by straight lines unconnected with those on the adjacent column. In other words, these are cruder imitations of those produced in the Islamic world and in Europe.

It is clear that the idea of the column dial came from the Islamic world—we cannot identify the exact process of transmission—but the Sanskrit authors merely borrowed the name, not the principal feature, namely, marking the hours on the different scales by continuous curves.

The extant specimens of Indian column dials are of three types: those made of metal or ivory, those made of wood on which scales are painted, and those made of timber on which the scales are carved. There exist just two specimens of the first group. The first is an exquisitely crafted steel column dial in the museum of the History of Science, Oxford, 95.6cm, with all the scale lines, numbers and decorative patterns inlaid in gold (Figure 9). It is topped with an ornate finial and the other end terminates in a sharply polished blade. It must have been created for some prince in Rajasthan. The other, made of ivory, is also of excellent workmanship, with a beautifully carved finial at the top and an ornate end at the bottom; it was in the now-defunct Time Museum, Rockford, Illinois, USA.[17]

In the second category are wooden column dials on which the scales and numbers are painted. There exist three specimens of no great merit.[18]

---

14 Ōhashi 1998, 195–196.
15 This volume Chapter 3. ii; Zinner 1979, 50–1.
16 See Chapter 3, Part One and Part Two. Ottoman pillar dials are held in the Institut du Monde Arabe, Paris (Naffah 1989), the Adler Planetarium, Chicago (Pingree 2009, 222), and two in the Kandili Observatory, Istanbul (Danisan 2020).

17 Published in Sotheby's 2004, 194, no. 717.
18 One of these is in Jai Singh's Observatory, Jaipur, the others in the Shri Sanjay Sharma Museum & Research Institute, Jaipur.

Figure 9 Column dial of damascened steel, Museum of the History of Science, Oxford, Inv. no. 50,041. Photo: S. R. Sarma.

The third group is the largest with some eighteen specimens, their lengths varying from 1034mm to 1541mm.[19] On these the numbers and letters are carved in relief. Designed for a latitude of about 27°, they were produced in the Himalayan foothills in the region of Darjeeling (27° 3′ N, 88° 16′ E) and Kalimpong (27° 3′ 36″ N, 88° 28′ 12″ E). Nearly all these column dials are equipped with an iron spike at the bottom, which has to be set in the ground so that the column stands upright and the scale for the current month faces the Sun. Two of them carry dates that correspond with 1869 and 1884 CE; the others must have also been made about the same time in the later nineteenth century. All the extant specimens are in museums and private collections outside India.

Horizontal dial (*Palabhā-yantra*)

In the *Palabhā-yantra*, the horizontal sundial is equipped with a triangular gnomon, the hypotenuse of which points to the north celestial pole.[20] Like the column dial, this one is also borrowed from the Islamic world. The earliest known specimen was made by Ibn al-Shāṭir in the second half of the fourteenth century for the Umayyad Mosque, Damascus.[21] From the Islamic world, it was transmitted westwards to Europe and eastwards to India. Because of the simplicity of its construction, it was frequently produced in Europe and North America.

In India it was absorbed into the Sanskrit repertoire of instruments under the title of *Palabhā-yantra*. *Palabhā* denotes the noon equinoctial shadow, i.e. the length of the shadow thrown by a gnomon at midday on the days of the equinox. A *palabhā-kṣetra* then is a right-angle triangle where the length of the shadow forms the base and the height of the gnomon the vertical. The hypotenuse of this triangle subtends an angle that is equal to the terrestrial latitude of the place. For measuring time, the instrument is so set up that the gnomon rests on the north–south line upon a horizontal dial that is divided into *ghaṭīs*. The angles separating the *ghaṭī* lines will be unequal. Since the gnomon is fashioned according to the terrestrial latitude of a specific locality, it can be used only at that latitude.

The *Palabhā-yantra* is described for the first time in the *Yantraprakāra* in the early eighteenth century.[22] This text teaches how to calculate the angles between the hour lines on a horizontal dial for the latitude of Delhi at 28° 39′ and for that of Amber at 27°. Thereafter, the *Yantraprakāra* provides a table showing the angular distances between the lines indicating one *ghaṭī* up to fifteen *ghaṭīs* for the latitudes of Delhi and Amber. There are several other eighteenth-century texts that describe it.

The instrument appears to have been popular in Rajasthan. In Sawai Jai Singh's Observatory at Jaipur, a *Palabhā-yantra* is incorporated at the top of the *Nāḍīvalaya-yantra*. The Observatory also owns a *Palabhā-yantra* made by Gokula Nātha Śarmā in 1882 (Figure 10). There must have been several others erected in public places. Some have been removed and preserved in museums, many are lost.

WATER CLOCKS

An obvious disadvantage with sundials is that they can be used only in daytime and only at the latitude for which they were designed. In contrast, **water clocks** can be used both in the day as well as in the night, and at all latitudes. In India **outflow water clocks** and **sinking bowl clocks** were used, not simultaneously but one after the other. Both types were designed to measure the same unit of time, namely, one-sixtieth part of the nychthemeron. In the texts where the outflow water clock is mentioned, this time unit is mentioned variously as *nālikā*, *nālī*, *nāḍikā*, or *nāḍī*. For convenience, we shall use the form *nāḍikā*. But the instrument itself is not mentioned by any name; it will be called *Nāḍikā-yantra*. The sinking bowl type of water clock is mentioned as *ghaṭikā-yantra*, *ghaṭī-yantra*, or *jalaghaṭī-yantra*, and the unit of time measured by it as *ghaṭikā* or *ghaṭī*. Here also, for the sake of consistency, the instrument will be referred to as *ghaṭikā-yantra* and the time unit as *ghaṭī*.

Outflow water clock

The outflow water clock (Figure 11) is described in four texts: the *Vedāṅga-jyotiṣa*, *Arthaśāstra*, *Śārdūlakarṇāvadāna*, and *Jyotiṣkaraṇḍaka*.[23] The dates of the first two texts are given above;

---

19 Winter 1964 describes an undated specimen at the John Gershom Parkington Memorial Collection, Bury St Edmunds, Suffolk, UK.
20 Sarma 2019a, Section N.
21 Berggren 2001, 12.

22 Sarma 1986–1987b, 26, 76–7.
23 The first three sources are discussed in Fleet 1915.

Figure 10 *Palabhā-yantra* made by Gokula Nātha Śarmā in 1882, Jai Singh Observatory, Jaipur. Photo: S. R. Sarma.

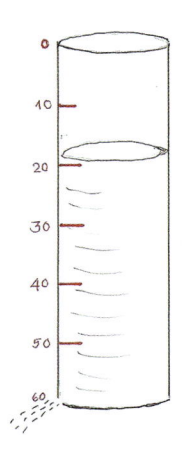

Figure 11 Outflow water clock (*Nāḍikā-yantra*).

the last two texts belong roughly to the early centuries of the Christian era. The descriptions they give are extremely brief and not very coherent. In the absence of actual specimens, it is difficult to interpret these sources properly.

The outflow water clock there described consists of a vessel with a very small hole at the bottom through which the water in the vessel flows out and indicates time. The shape and size of the vessel are not mentioned. The names of the time unit *nālikā* or *nālī* are diminutive forms of *nala*, which denotes, among other things, a reed or a tube, or a hollow cylinder. Accordingly, the vessel must have been of cylindrical shape. A cylindrical vessel has the advantage that its height can easily be divided to show the water level at various subdivisions of a *nāḍikā*.[24]

On the amount of water with which the vessel was to be filled, the texts make contradictory statements. However, since the length of the day varies according to seasons, the *Vedāṅga-jyotiṣa* prescribes that a *prastha* of water should be added every day when the length of the daylight or of the night increases and the same amount of water is removed when the length decreases.[25]

The most remarkable feature is the prescription regarding the size of the hole at the bottom of the vessel. The *Arthaśāstra* and the *Jyotiṣkaraṇḍaka* prescribe that the hole should be so large that a gold wire, four *māṣas* in weight and four *aṅgulas* long, should just fit into it. The *Śārdūla-karṇāvadāna* retains the length of four *aṅgulas*, but prescribes one *suvarṇa* weight. As pure gold is malleable, a gold wire of a uniform diameter can be drawn with a given weight of gold. The hole being a minute one, there is no better way of defining it. The Babylonian and Chinese records on outflow water

---

24 The cylindrical shape is confirmed by Bīrūnī, 1910, i, 334, who visited India in the first quarter of the eleventh century.
25 Lagadha 1985, 44.

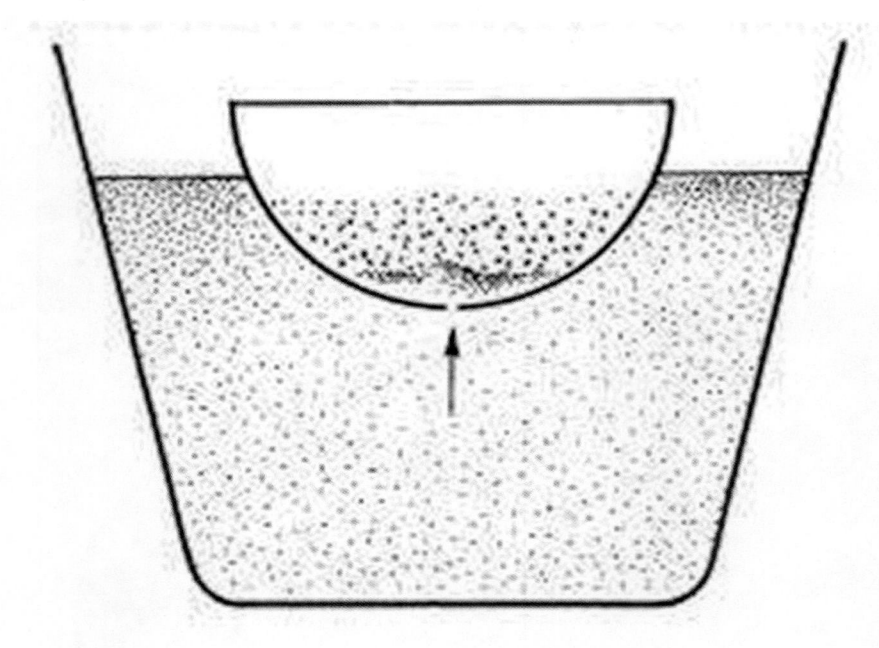

Figure 12 Sinking bowl water-clock (*Ghaṭikā-yantra*).

clocks do not appear to be familiar with this method. Therefore, this method of micro-measurement may have developed in India. Even when the outflow water clock was replaced by the sinking bowl type, the dimension of the hole in the bowl was defined in a similar manner, as will be shown later.

Jacobi argues that, in a cylindrical vessel, as the water level goes down the pressure decreases, and so does the speed of discharge through the hole at the bottom. Consequently, the time units indicated by this device will not be of a uniform duration; they will be shorter at first and then become longer and longer. Therefore, the outflow water clock was gradually replaced by the sinking bowl type of water clock.[26]

Sinking bowl water clock

The sinking bowl water clock (Figure 12) consists of a hemispherical bowl made of a thin sheet of copper with a fine hole at the centre of the bottom. When this bowl is floated on water in a larger basin, water percolates into the bowl through the hole and fills it, and the bowl sinks. The bowl is then lifted up, emptied, and set up once again on the surface of the water. The hole is so made that the bowl sinks sixty times across a day and a night, i.e. the bowl takes 24 minutes to fill and sink. Since the bowl is called *ghaṭikā* or *ghaṭī* (diminutive form of *ghaṭa*, 'pot'), the time unit measured by this instrument also came to be called *ghaṭikā* or *ghaṭī*.

Āryabhaṭa is the first author to describe the *Ghaṭikā-yantra* in his *Āryabhaṭasiddhānta*:[27]

> Get a round (i.e. hemispherical) bowl made with ten *palas* of copper. Let its height be six *aṅgulas* and the diameter at the mouth twelve *aṅgulas*. Get a hole bored at its bottom [so that a gold wire] one *pala* [in weight] and eight *aṅgulas* [in length can pass through it].

After Āryabhaṭa, this instrument is described in several astronomical texts.[28] Lalla (eighth or ninth century), in his *Śiṣyadhīvṛddhida-tantra*, while retaining the dimensions of the bowl, changed the size of the perforation—from the gold needle of one *pala* weight and eight *aṅgulas*' length to a gold needle three and one-third *māṣas* in weight and four *aṅgulas* in length.[29] Since sixty-four *māṣas* make one *pala*,[30] the perforation prescribed by Lalla would be 5/48th of that prescribed by Āryabhaṭa. While the size and weight of the bowl remain the same, such stark reduction in the size of the perforation will greatly increase the duration of the time needed for the bowl to fill and sink, and consequently the duration of the *ghaṭī* as well. One is therefore led to suspect that these specifications are fictitious and have no connection with actual practice. Therefore, Bhāskarācārya in his *Siddhāntaśiromaṇi* (1150) dismisses these prescriptions as illogical and difficult to implement.[31]

In spite of this confusion in the textual prescriptions, countless specimens must have been made throughout the centuries and these must have kept reasonably correct time of one *ghaṭī* of twenty-four minutes. But it is doubtful whether any artisan has ever produced a bowl precisely according to the textual prescriptions. In the few specimens that survive in modern collections, the bowls rarely have the exact shape of a hemisphere (Figure 13). The sizes and weights also vary considerably. The holes were obviously

---

26 Jacobi 1920, 251.
27 Shukla 1967.
28 Some of these descriptions are translated and discussed in Sarma 2004, 303–10.
29 Lalla 1981, part i, 246.
30 Śrīdhara, text, 5; translation, 3.
31 Sarma 2019b, 332.

Figure 13 Water clocks in the Pitt Rivers Museum of Ethnology, Oxford. Upper row, left to right: from Mirzapur, Myanmar, unknown source; Second row: from Rampet, unknown source, Sri Lanka; Bottom row: made of coconut shell from Malabar. Photo: S. R. Sarma.

made by trial and error, comparing the new bowl with one that showed the correct time.

HOROLOGICAL VOCABULARY

Even after the *Nāḍikā-yantra* was completely replaced by the *Ghaṭikā-yantra*, the terms related to the older device—*nālikā*, *nāḍikā*, *nāḍī*—continued to be used along with the new terms *ghaṭikā* and *ghaṭī* to denote the basic unit of time of twenty-four minutes. From these two sets of terms, two other sets of names are derived to designate the sixtieth part of a *ghaṭī* (i.e. twenty-four seconds), for example, *vināḍikā*, *vināḍī*, and *vighaṭikā*, *vighaṭī*. But this unit is more frequently called *pala* in northern India.

The influence of the *Ghaṭikā-yantra* on the horological vocabulary is all-pervasive in India. In many modern Indian languages, timekeeping devices, however sophisticated they may be, are still called *ghaḍī* (from Sanskrit *ghaṭī*). The term *ghaṭikālaya*, originally the designation of the timekeeping establishment, engendered the names of the timekeeping devices in Telugu (*ghaḍiyāramu*, *gaḍiyāramu*) and Malayalam (*gaḍigāram*); it also gave rise to the term *ghaḍiyāl*, which denoted the 'gong' in North India and clock in Gujarati. Finally, in many Indian languages, when it is, for example, four o'clock, one says 'it is striking four' or something similar; such an expression is not derived from the chimes of the European clocks but from the old practice of striking the *ghaṭīs* on a gong.

Measuring the fractions of a *ghaṭī*

Sanskrit sources suggest an ingenious method for measuring the fractions of a *ghaṭī*. It was mentioned above that Āryabhaṭa divides the *nāḍikā* into sixty *vināḍikās* and each *vināḍikā* into sixty long syllables; that is to say, a *vināḍikā* is the time taken to utter sixty long syllables. Therefore, concludes Varāhamihira, a *nāḍikā* or *ghaṭī* is the time taken to recite sixty times a verse made up of sixty long syllables.[32] Hence, if one wishes to measure four *ghaṭīs* and twenty-one *palas* since sunrise, one sets the perforated bowl upon the water when half the solar orb rises above the horizon. After four immersions of the bowl, one recites twenty-one times a verse consisting of sixty long syllables.

Another possibility is to mark the divisions of the *ghaṭī* on the inner side of the vessel, but it is very difficult to graduate geometrically the inner wall of the small bowl into sixty

---

32 Varāhamihira, 14.32; interestingly, this verse itself is made up of sixty long syllables! For three other verses of this nature, see Sarma 2004, 320–2.

vighaṭīs. But it should be possible to empirically divide the bowl, if not into sixty parts, at least into ten parts (= 2 min 24 sec) each. Gilchrist reports in 1795 that in some specimens he saw in Bengal, the bowl was marked with divisions.[33] Mrs Meer Hasan Ali[34] and Thurston[35] also speak of the marks made on the bowl to indicate the subdivisions. However, during my survey I did not come across a single specimen with such marks of graduation.

Prahara and yāma

While time was measured in equal *ghaṭīs* for astronomical and astrological purposes, for common people broader segments were sufficient, such as the fourth part of the day and of the night. The fourth part of the day is called *prahara* and that of the night *yāma*, although *prahara* is frequently used for both day and night. These time units were announced by strokes on a drum at a central place and hence were called *prahara* (literally 'stroke'). The duration of this *prahara/yāma* is variable according to the geographical latitude and the season. Astronomical texts hardly mention *prahara* or *yāma*, but they occur in other texts.

This raises the question how the variable *prahara* was measured with water clocks, which measured only the fixed unit of one *ghaṭī*? In other words, how were the variable units of *prahara* reconciled with the fixed units of *ghaṭīs*? On this, there is no evidence before the mid-sixteenth century, but then it appears that, although the *prahara* is one-fourth of the daylight, in practice, however, the *praharas* are so arranged that they always consist of an integral number of *ghaṭīs*.

John Gilchrist explains how this was done with the help of an elaborate 'Hindoostanee Horal Diagram.' According to him, in a day all the *praharas* were not of equal length but each *prahara* consisted of an integral number of *ghaṭīs*. Thus, on equinoctial days, the first and fourth *prahara* of the day consisted of eight *ghaṭīs* each, whereas the second and third contained seven each, making a total of thirty *ghaṭīs*. This would mean that the second *prahara* ends at noon, thus dividing the day into two equal parts. Secondly, the length of the *prahara* is not adjusted every day, but only when the day or the night becomes longer or shorter by two *ghaṭīs* (forty-eight minutes).[36]

Ghaṭikā-yantra in inscriptions and literary texts

Institutions for timekeeping are attested, from the seventh century onwards, at Buddhist monasteries, royal palaces, town squares, and the like, where time was measured constantly with the water clock and the passage of each *ghaṭī* and completion of each quarter of the day (*prahara*) or of the night (*yāma*) was broadcast. There are references to endowments made for the maintenance of the attendants. Ideally, there should always be two attendants, one to lift the bowl and empty it when it has sunk, the other to announce the completion of the *ghaṭī*. There should be at least four such pairs who take turns: two in the day and two at night. Of course, to lift the bowl, empty it and replace it carefully on the water surface takes time; at a minimum of fifteen seconds for the process, it will make a difference of fifteen minutes, when this is done sixty times in a day and night.

The Chinese traveller I-Tsing (modern spelling Yi Jing), who spent some ten years (c.675–85) at the Buddhist monastery of Nālanda, gave a detailed account of the timekeeping establishment there; time was measured by means of sinking bowl water clocks and announced by means of drums and conch shells.[37] He adds that these water clocks, 'together with some boys are the gifts from kings of many generations, for the purpose of announcing hours to monks.' At the beginning of the eleventh century, Al-Bīrūnī reports about a timekeeping establishment in Peshawar:

> The Hindus have a popular kind of division of the nychthemeron into eight *prahara*, i.e. changes of the watch, and in some parts of their country they have clepsydrae regulated according to the *ghaṭī*, by which the times of the eight watches are determined. After a watch which lasts seven and half *ghaṭī* has elapsed, they beat the drum and blow a winding shell called *śaṅkha*, in Persian *sped-muhra*. I have seen this in the town *Purshūr* [Peshawar]. Pious people have bequeathed for these clepsydrae, and for their administration, legacies and fixed incomes.[38]

The buildings which house the water clocks are called *ghaṭikālaya*, *ghaṭikāgṛha*, or *ghaṭīgṛha* in inscriptions and other documents from Gujarat belonging to the thirteenth and fourteenth centuries. For example, the *Lekha-paddhati*, a collection of model drafts for official and private documents, compiled in the thirteenth century, lists *Ghaṭikāgṛha-karaṇa* among thirty-two administrative departments (*karaṇa*); this would then be the department which maintains and supervises timekeeping establishments in different cities in the kingdom.[39]

The water clock house is pictorially depicted on the wooden cover of a palm-leaf manuscript (*c.* thirteenth century).[40] Both the sides of this cover are painted with events related to the historic debate between Vādi Devasūri of the Svetāmbara Jain sect and the monk Kumuda Candra of the Digambara Jain sect, which took place at the court of Siddharāja Jayasiṃha at Patan in 1124 and in which Devasūri emerged victorious.

---

33 Gilchrist 1795, 87.
34 Ali 1974, 55–6.
35 Thurston 1907, 565–6.
36 Gilchrist 1795, 83.

37 I-Tsing 1896/1966, 144–6. For a critique of his account, see Sarma 2019a, 3779–83.
38 Bīrūnī 1910, i, 337–8.
39 Strauch 2002, 240–1. For other sources, see Sarma 2019a, 3784.
40 Chandra 1949, 59–62, Figures 193–8; Sarma 2019a, 3785–7.

The front side of the cover is devoted to scenes at Āśāpalli (modern Ahmedabad) where Devasūri was challenged for a debate by Kumuda Candra. Here, next to the shrine of Neminātha, is a structure identified by the label above as the *ghaṭikāgṛha*. No details of the water clock or of the attendants are shown, except the half-open door between two pillars. Above the pillars is the roof with a parapet consisting of a series of roundish battlements.

The reverse side contains scenes from the city of Pāṭan where the debate was to be held at the court of King Jayasiṃha. Here, the *ghaṭikāgṛha* is depicted next to the royal harem. One sees just the back of the house. It is, however, noteworthy that in both cities the *ghaṭikā-gṛhas* are located at central places; at Āśāpallī next to the shrine of Neminātha and at Pāṭan next to the royal harem. This would suggest that the *ghaṭikā-gṛha* was an important feature in the urban landscape of twelfth-century Gujarat.

Three inscriptions (dated respectively 1228, 1403, and 1404) from the Telugu-speaking region on the east coast refer to the endowments made for timekeeping with water clocks (*ghaḍiyāramu*) at Hindu temples.[41] Besides these inscriptions, several Telugu literary works also contain references to water clocks.[42]

WATER CLOCKS AT THE COURTS OF THE TUGHLUQS AND MUGHALS

In the first half of the eleventh century, al-Bīrūnī described the announcement of time by drum and conch shell. During the next centuries, these sound-producing instruments appear to have been replaced by the brass or bronze gong (*ghaḍiyāl*) on which a series of rapid blows separated the strokes for the *ghaṭīs* and those for the *praharas*. The Muslim rulers of Delhi adopted the water clock and the gong.

Shams-i Sirāz 'Afīf, a contemporary and chronicler of Fīrūz Shāh Tughluq, the Sultan of Delhi (1351–88), narrates that the Sultan installed the *ṭās-i ghaḍiyāl* at his palace.[43] Afīf does not provide any description of the device. But the Persian word *ṭās* means 'cup' or 'bowl' and therefore the expression *ṭās-i ghaḍiyāl* stands for the ensemble of the sinking bowl and the gong.

Bābur (r. 1526–30), the first Mughal emperor, was so impressed by this system of timekeeping that he adopted it and introduced improvements in the mode of announcement.[44] The same system continued in Akbar's reign (1556–1605) as reported by his chief chronicler Abū al-Faḍl.[45] More importantly, the water clock is depicted very accurately in two miniature paintings executed in Akbar's atelier.[46] Indeed, these are the only pictorial depictions of the device. The first painting relates to the birth of Akbar.[47] It shows Akbar's father Humāyūn seated on a throne, with astronomers in attendance. They have measured the time of birth by means of a water clock and a ring dial and drawn up the horoscope. The water clock, with the bowl floating in a large basin, is drawn very clearly. The second miniature is related to the birth of Akbar's son Jahāngīr. It depicts the Hindu and Muslim astronomers seated together, measuring the time of birth with a water clock, the Sun's altitude with a ring dial, and drawing up the horoscope.[48]

The sinking bowl type of water clock was in use throughout India until the end of the nineteenth century. Powell reports in 1872 that '[t]his article is in common use, and by it all police guards, &c, keep the time, striking their gong as each hour comes round.'[49] One would therefore expect scores of water clocks to be surviving in every part of India. But in India unused copper or brass vessels are generally recycled and therefore there are few specimens extant. Of late, some former Maharajas have established museums inside their palaces and display artefacts of historical interest. Thus, in the palace museums at Bharatpur, Bundi, Kota, Udaipur (all in Rajasthan), and Ramnagar (in Uttar Pradesh), the time apparatus of perforated bowl, water basin, and gong is displayed.

The largest collection of water clocks—of different shapes, sizes, and materials—is preserved in the Pitt Rivers Museum of Ethnology, Oxford (Figure 13). While some of these were made to measure the traditional Indian unit of *ghaṭī* or its fractions, others were made to measure the hour of sixty minutes, introduced by the British colonial administration, or its fractions. Yet there is one more specimen which was not intended to measure time as such, but to measure the duration of irrigation water supply into different fields.

ORIGIN AND DIFFUSION OF THE SINKING BOWL WATER CLOCK

The sinking bowl type of water clock was not confined to India, but rather was used in many countries from Iran to Spain in the west and from Myanmar to Indonesia in the east. It was used not only for regular timekeeping, but also to measure segments of time for the distribution of irrigation water in the west and to measure segments of time in cock fights in Southeast Asia. But where did it originate?

The earliest mention of the sinking bowl occurs in the Pali commentary of the *Majjhimanikāya*, written by Buddhaghoṣa in

---

41 Sarma 2019a, 3788–92.
42 Sarma 1948, 324–7.
43 Elliot & Dowson 1871, III, 338.
44 Bābur 2006, 516–17.
45 Abū al-Faḍl, 17–18.

46 Sarma 1992, 241–3.
47 British Library, Ms Or. 12,988, f. 20b. Sen 1984, 130–1, Pl. 57.
48 Museum of Fine Arts, Boston, 17.3112. Welch 1978, 70–1, Pl. 16.
49 Powell 1872, 200.

the first half of the fifth century CE in Sri Lanka.⁵⁰ This commentary narrates the use of a sinking bowl that measured one *yāma* at a Buddhist monastery. At the low geographical latitude of Sri Lanka, *yāma* may have been treated as equal to the constant value of 3¾ *ghaṭīs*. Bowls that measured one *ghaṭī* must be still older. It is therefore reasonable to suppose that the sinking bowl type of water clock was developed some time in the fourth century in the Indian Ocean region, either in Sri Lanka, or on the southern coast of India. It is also likely that the original inspiration for the sinking bowl came from the coconut shell, which is naturally endowed with a hole; water clocks made of well-scrubbed halves of the coconut shell survive in museums.

There is, however, one problem with this scenario. The use of the sinking bowl in Iran was mainly associated with the distribution of irrigation water from the underground water channels called *qanāt*. While this *qanāt* system is said to be very ancient,⁵¹ it is not known when the distribution of irrigation water to individual farmers began to be regulated by means of the sinking bowl. Therefore, it is difficult to say whether the sinking bowl used with the *qanats* reached India and from there spread to Southeast Asia, or whether it originated either in Sri Lanka or India, and thence spread westwards and eastwards.

The sinking bowl as an irrigation clock

The *qanāt* system spread from Iran to other areas in the Middle East and then to North Africa and Spain. In some of these places, distribution of water to individual cultivators was regulated by sinking bowls. For their use in Iran, Wulff offers a very detailed account,⁵² as does Glick for North Africa and Spain.⁵³

The sinking bowl may have been used as an irrigation clock in India also, but there do not seem to be any published records of this. The only tangible evidence is a copper bowl from Rampet, Pallar Valley, in Tamilnadu, in the Pitt Rivers Museum, Oxford (Figure 13).

Nepal For Nepal, Thurston reports as follows:⁵⁴

> In Nepāl the measurement of time is regulated in the same manner [as in India]. Each time the vessel sinks, a gong is struck, in progressive numbers from dawn to noon. After noon, the first ghari struck indicates the number of gharis which remain of the day till sunset. Day is considered to begin when the tiles on a house can be counted, or when the hairs on the back of a man's hand can be discerned against the sky.

In Kathmandu, at Hanumandhoka Palace, and also at Gorkha Palace, it is said that time was measured with a water clock and announced regularly up to the beginning of the twentieth century. Recently, Olivia Aubriot studied the irrigation system in a village in central Nepal and reported that the sinking bowl was used there as the irrigation clock.⁵⁵

Sri Lanka The large sinking bowl that measured one *yāma* in the fifth century CE at a Buddhist monastery in Sri Lanka has already been mentioned. It is not known how time was measured in subsequent centuries in Buddhist monasteries. In the seventeenth century, however, according to Robert Knox, measuring time with the sinking bowl had become the exclusive privilege of the king:⁵⁶

> They have no *Clocks*, *Hour-glasses*, or *Sun-Dials*, but keep their time by guess. The King indeed hath a kind of Instrument to measure time. It is a *Copper Dish* holding about a Pint, with a very small hole in the bottom. This Dish they set a swimming in an Earthen Pot of water, the water leaking in at the bottom till the Dish be full, it sinks. And then they take it out, and set it empty on the water again, and that makes one *Pay*. Few or none use this but the King, who keeps a man on purpose to watch it continually.

But in later times, sinking bowls appear to have been used even outside the royal palace. The Pitt Rivers Museum, Oxford, owns a specimen from Sri Lanka that must have been made at the end of the nineteenth or the beginning of the twentieth century (Figure 13).

Myanmar For Myanmar (Burma), fortunately, a sinking bowl exists which was used at the emperor's palace in the nineteenth century (Figure 13) as well as a description by John Nisbet:⁵⁷

> Each day was under Burmese rule divided into sixty hours (*Nayi*), and subdivided into eight watches, each of about three hours, which varied in length at different seasons of the year according as the days and nights were relatively longer or shorter. The *Nayi* or 'time measurer' was a copper cup having a tiny perforation at the base, which, being inserted in water, sank to a particular mark within a given time.... As each *Nayi* was thus measured off a gong was beaten, and at every third hour the great drum-shaped gong was sounded from the *Pahózin* or time-keeper's tower within the inner precincts of the royal palace at the eastern gate. One beat of the drum denoted nine o'clock in the morning or evening, two beats twelve o'clock, three beats three o'clock, and four

---

50 Hinüber 1978, 224–5.
51 On *Qanāts*, see Wulff 1966, 249–56; Wulff 1968.
52 Wulff 1966, 254–6.
53 Glick 1969, 425–6.
54 Thurston 1907, 563–4.

55 Aubriot 2004, 175–82. I owe this information to Dr Jérôme Petit, Paris.
56 Knox 1681/1995, 111.
57 Nisbet 1901, II, 288–289.

beats six o'clock. From Pahó the beats were repeated on large bells by all the guards throughout the palace.... Now, under British rule, wherever there are jails, police stations, treasury guards, and so forth, the hours are marked off by beat of a gong. Hence, in towns, the word *Nayi* has now come to mean both the hour, measured by the European method, and the clock or watch by which it is measured.

The sinking bowl in Southeast Asia

There can be no doubt that the sinking bowl spread eastward from India, in some cases with Indian terminology. In Indonesia, there is a considerable influence of Sanskrit language and Hindu religion in the islands of Java and Bali. Consequently, many Indian units of time measurement have been absorbed there, as reported by Lewis Pyenson:[58]

> Indian astronomy was certainly present in Old Javanese texts of Brahman inspiration dating from the ninth and tenth centuries. The Indian time units of day and night—divasa for 24 hours; muhurta, ksana for 48 minutes; ghati, ghatika, nadi, nadika for 24 minutes; kala for 48 seconds—all are present in the language. In Old Javanese poetry composed in Indian meter, however, the natural day is divided into equal parts of 8 hours, calculated from sunrise to sunset. The word for 'hour' is also that for 'stroke' or 'fall', suggesting hours being signalled by a striking device.

Aside from time measurement, the sinking bowl was also used in Bali for timing cockfights, which formed part of the ritual of exorcising evil spirits.[59] In Thailand, by contrast, cockfights were held to commemorate the victory of the Siamese prince Naresuan over a Burmese prince in a legendary cockfight in the second half of the sixteenth century. The duration of this fight was also determined by the sinking bowl water clock.

In China, although the prevailing water clock was the inflow type with a series of water tanks, according to Needham, the sinking bowl type was also known:[60]

> The Chinese also knew another archaic device, the inverse variant of the outflow clepsydra, a floating bowl with a hole in its bottom so adjusted that it took a specific time to sink.... A Chinese example is afforded by the work of the Thang monk Hui-Yuan, who arranged a series of lotus-shaped bowls to sink one after another during the twelve double-hours.

THE INTRODUCTION OF EUROPEAN HOURS

It has already been shown that from the fourteenth century onwards the Muslim rulers in India adopted the local system of measuring the equal intervals of *ghaṭī* and *pala* and also the variable intervals of *prahara* with the water clock, broadcasting these intervals with the gong called *ghaḍiyāl*. At the same time, they also continued their traditional practice of using **unequal hours** (*al-sā'āt al-zamāniya* for their prayer times, and **equal hours** (*al-sā'āt al-I'tidāl*) for astronomical purposes. But these hours, temporal or equal, did not spread beyond the Muslim community until the advent of the Europeans.

Realizing that the Portuguese and the Dutch were making huge profits from the trade with the 'East Indies', some English merchants formed the East India Company to trade with India and received a charter for this purpose from Queen Elizabeth in 1600. They sent embassies to the Mughal court, seeking license to establish trading posts in India. The presents these envoys carried to the Mughal court included mechanical clocks. Thus, Sir Thomas Roe, the Company's envoy to the Mughal Court at Agra 1615–19, presented a clock to Emperor Jahāngīr. Sir Robert Shirly is said to have presented a silver clock to him in 1616. In the same year, Shāh Abbās of Iran sent an embassy to Jahāngīr with many gifts, which included five European clocks. However, Jahāngīr, although he was interested in astronomy and technical innovations, did not mention the gift of European clocks in his memoirs. Other Europeans also made gifts of clocks to different members of the Mughal nobility. Apparently, these mechanical devices did not arouse the interest either of Jahāngīr or of his courtiers. One of the reasons for the lack of interest may be that these clocks showed time according to the European style of 12×12 hours, starting from midnight and midday, and not according to the Indian style of 60 *ghaṭīs*.[61]

The English trading posts, 'factories', must have possessed clocks and made use of them, but these had to be imported from England. Using the local water clock and the gong was more convenient. In his *Geographical Account of Countries Round the Bay of Bengal, 1669 to 1679,* Thomas Bowrey narrates that many wealthy Muslims set up water clocks in their front porches with two servants attending all the time, one to lift the bowl and set it up again, the other to strike the gong, and that this custom was emulated by the English and Dutch in their factories.[62] It is quite likely that in these factories, the hole of the bowl was adjusted to measure half an hour or a full hour and that European hours were announced with the gong.

In the course of the next two centuries, the East India Company, supported by a large army, gradually expanded its trade and political power so that by the middle of the nineteenth century it ruled large parts of the Indian subcontinent. In 1858, the British

---

58 Pyenson 1998.
59 Eiseman 1990, 240–50.
60 Needham 1959, 315 and note h.

61 Qaisar 1982, 64–9.
62 Qaisar 1982, 68.

Crown assumed direct control of the subcontinent, and continued the Company's practice of measuring time with the water clock and announcing the passage of European hours from all their offices and institutions. In the late nineteenth century, water clocks were progressively replaced by pendulum clocks, but announcement of the hours continued.[63] Thus by the beginning of the twentieth century, the traditional system of *ghaṭīs* of twenty-four minutes was completely replaced by the European hours of sixty minutes.

Public clocks also contributed to the dissemination of European hours, at least in some of the larger towns. The earliest public clocks were set up in churches. As early as 1516, a public clock was attached to the Church of St Francis at Cochin in Kerala. In 1660, the Basilica of the Holy Rosary was constructed along with a clock tower in the Portuguese settlement at Bandel in Bengal. Later, clock towers were also set up by administrative and secular institutions, either incorporated in their main buildings or as free-standing structures.[64]

Some sundials were also set up, with horizontal dials and triangular gnomons, resembling the *Palabhā-yantras* in design, but with lines to measure hours which are numbered either in Roman numerals or in modern international numerals. But these played only a marginal role in timekeeping and the dissemination of European hours. With increasing use of European hours, however, trade in mechanical clocks increased. British and Swiss companies began to export public clock movements, wall clocks for offices and private homes, alarm time pieces, and wrist watches. They also manufactured clocks and wrist watches specially for use in India, with the emblems of the maharajas or with pictures of tigers or of elephants, or which showed time in *ghaṭīs* and *palas*.

The Benares College founded by Jonathan Duncan of the East India Company in 1791 (now Sampurnanand Sanskrit University) owns a clock which shows time in European hours as well as in Indian *ghaṭīs* (Figure 14). It was manufactured by the Synchronome Company, London, in 1951. The dial carries four graduated scales. The innermost scale is divided into twenty-four hours and every second hour is numbered in Roman numerals from II to XXIV, with the label HOUR written below. The next scale is divided in sixty *ghaṭīs* and every second *ghaṭī* is numbered in 'English numerals' from 2 to 60, with label GHATI below. The third scale is also divided into sixty parts, but not numbered. The label below reads PALA, but *palas* (= twenty-four seconds) cannot be measured on this scale. The outermost scale is also divided in sixty parts; here every second unit is numbered in red in English numerals. The label below reads VIPALA, but again the units of *vipalas* (= twenty-four thirds) cannot be measured on this scale. In other words, all the three outer scales are divided into sixty units and can measure only the *ghaṭīs*. The labels PALA and VIPALA may have been added for didactic reasons. Surprisingly, there is no scale to measure the minutes.

The dial is equipped with three hands. The shortest one, which is reticulated, reaches up to the scale labelled GHATI; the next one, lance-shaped, reached up to the scale marked PALA; the third and longest one, painted red, reaches up to the outermost scale named VIPALA. What these three hands actually show is not known as when I saw the clock in 1991, it was not functioning.

Clock-making on a commercial scale does not appear to have commenced in India until the middle of the twentieth century, but there were individual clockmakers who repaired foreign clocks and watches and sometimes even produced innovative pieces. One such person is B. Mulchand who created a large astronomical clock for the Maharaja of Benares in 1872. This clock shows the hours and minutes according to the European fashion and *ghaṭīs* and *palas* in Indian style, but English months and dates. It strikes hours, half hours, and quarter hours. There is a separate dial to indicate the times of sunrise and sunset in hours and minutes and the Sun's position in the zodiac in signs and degrees. Another dial displays the phases of the Moon and the lunar days (*tithis*). Yet another dial shows the name of the weekday and an image of the planetary deity after whom the day is named. It also shows the ascendant and other astrological parameters.[65] This clock is now displayed in the museum of the Maharaja's palace. When Albert Edward, Prince of Wales (the future Edward VII), visited Benares on 5 January 1876, the Maharaja presented him with a smaller version of this astronomical clock, which is preserved at his seat at Sandringham.[66] One other astrological clock made in Gujarat in 1850–1 is known; a set of seventeen detached parts of it have survived and these are in a private collection.

INDIAN STANDARD TIME

Even though the European hours were adopted by the beginning of the twentieth century, there remained a major problem. Every city followed its own local time based on the mean sunrise and sunset. John Goldingham, the first official astronomer of the Madras Observatory, determined, on the basis of his observations of the eclipses of the satellites of Jupiter made between 1794 and 1802, the longitude of Madras at 80° 18′ 30″, corresponding to 5 hours 21 minutes and 14 seconds ahead of Greenwich Mean Time.[67] This came to be known as 'Madras Time'. By the late 1860s, railways and telegraphic communications expanded rapidly, and telegraphs began to use 'Madras Time' uniformly

---

63 This practice continues even now. Between 1974 and 1981, the author lived in a house in a north Indian town where he could hear the regular announcement of the hours from the district jail as well as from the office of the district collector.

64 In his forthcoming work on the clocktowers of India, Debasish Das describes some 120 clock towers erected during the colonial period.

65 Sen 2015, 170–1; Sarma 2019a, 4081–5.

66 Sarma 2014. Our thanks to Rufus Bird, Keeper of the Royal Collections, for locating the clock (RCIN 7810).

67 Anon 1809.

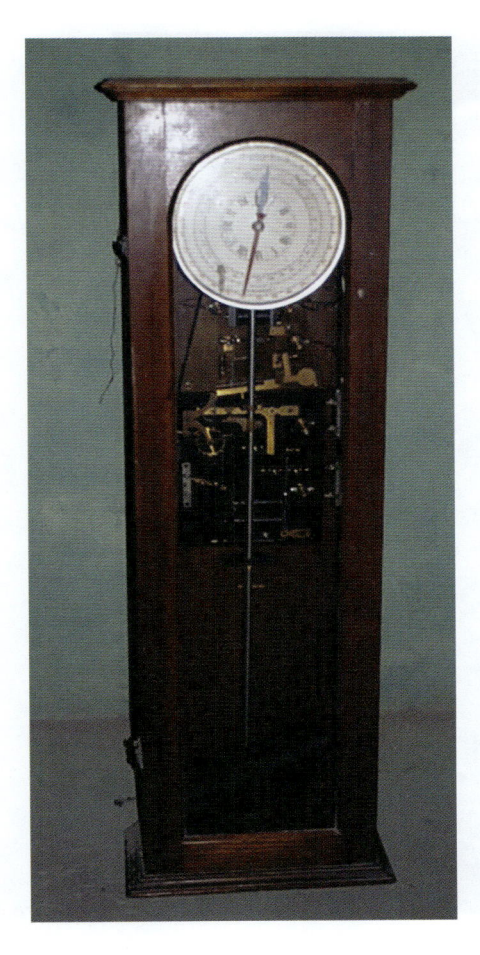

Figure 14 Synchronome clock with a dial for hours and *ghaṭīs*. Photos: S. R. Sarma.

in all their dispatches.[68] The Railways adopted 'Madras Time' as their standard time in 1870 and began publishing their timetables accordingly, with the result 'Madras Time' began to be called 'Railway Time'.

In 1904–5, the Royal Society recommended to the colonial government to divide the country into two time zones, fixing the time at six hours ahead of the GMT in the east and five hours in advance of the GMT in the west, but this recommendation did not find favour with the government, who desired a single time zone for the entire subcontinent.[69] The government replaced Madras Time of 5 hours, 21 minutes, and 14 seconds by a more convenient period of 5 hours and 30 minutes in advance of the GMT and declared the meridian passing through the longitude 82° 30′ E as the central meridian for India. This Indian Standard Time came into effect at midnight 1 July 1905.[70] It was immediately adopted by all the major railway companies and all government offices. However, there was strong resistance in Calcutta, which was proud of being the capital of British India, and also in Bombay, the industrial and commercial metropolis, and these two cities retained their local time. After India attained independence from British rule in 1947, the government established Indian Standard Time as the official time, but Calcutta and Bombay followed their local times until 1948 and 1950, respectively.

WATER CLOCKS AT PLACES OF WORSHIP

Even after the *Ghaṭikā-yantra* was replaced by European clocks, the former continued to be used for ritual purposes in the places of worship of all major faiths in the Indian subcontinent. On the day of *Janmāṣṭamī*, the birth of the Hindu god Kṛṣṇa is celebrated at all the temples of Mathura (27° 30′ N, 77° 40′ E) in Uttar Pradesh. The festivities commence at midnight when Kṛṣṇa is supposed to have been born. At the Dwarakadhish temple, the precise time of midnight is said to be determined by means of the water clock. At the Jain temple dedicated to Tirthankar Shantinath at Jhalawar (24° 35′ N, 76° 10′ E) in Rajasthan, the *Ghaṭikā-yantra* is still used to determine the time of various rituals.[71] In the town of Sehwan in Sindh, Pakistan, at the mausoleum of Qalandar Shahbaz, the water clock was in use as late as 1973 for determining

---

68 Krishnan 2013, 47
69 Krishnan 2013, 33.
70 Krishnan 2013, 78.

71 Reported in *The Hindu*, New Delhi edition, 24 April 1994; cf. Sarma 1994; this water clock was shown in the BBC TWO documentary series *What the Ancients Did for Us*, Episode 5: 'The Indians', broadcast on 5 March 2008.

Figure 14 detail.

the times of prayer and the time of the dance of the dervishes (*dhammal*).[72]

SUNDIALS AS URBAN SCULPTURE

Sundials also have not been completely forgotten. Following European fashion, a few large sundials have been set up as urban sculptures in public places. A notable specimen is a sundial in the shape of the *Palabhā-yantra* set up recently near the Barapullah Flyover in New Delhi by the Delhi Development Authority. The brass gnomon has a height of 12.7m and a length of 24.5m. The hour lines are laid out in coloured marble and numbered in large Devanagari numerals.[73]

---

72 Baloch 1979.

73 Information from Debasish Das.

SECTION TWO: CHINA TO 1900
David Chang

'Respecting the heavens and giving time' was a core idea of ancient Chinese civilization.[74] Firstly, astronomy is the major attribute of Chinese imperial rule—observation and forecasting of astronomical phenomena was the first task of the Emperor. Secondly, in order to maintain power, the dynasty had to promulgate time rules with the calendar at their centre. The system of ancient Chinese astronomy had been established before CE 220.[75] Since the primary interest of this discipline was political, research was accorded official support. Making various astronomical instruments was one important task, and this included the manufacture of time measurers, timekeepers, and mechanical astronomical timepieces to assist observations. The time frame of the calendar determined the units of measurement. The year, month, and day are all time units from nature. Their periods can be determined through astronomical observations. As the 'day' has the shortest period and can be intuitively perceived, it was the basic time unit in ancient China.[76] Faced with astronomical changes, the Chinese formed a unique concept of cyclical time. The Chinese characters, represented by ten heavenly stems and twelve earthly branches, cooperate with each other and are shown by a pair of stem branches every day, with the sixtieth day as a cycle, which constantly repeats.[77] In order to divide the **nycthemeron**, the Chinese adopted three main systems (Figure 15): twelve 'double hours', 100 'quarters', and five 'night watches'.[78]

CHINESE TRADITIONAL TIMEPIECES

The Chinese official Xue Jixuan (CE 1134–73) mentioned four types of timepieces in his book: the **sundial**, the incense stick, the rolling ball, and the **water clock**.[79] Among these typical ancient Chinese timekeeping devices, only the sundial is a time finder.[80] Records are preserved in Chinese literature of work with gnomons concerning the Sun's azimuthal position as a function of time from CE 102 and CE 594.[81] Later, it was realized that for the scale to be evenly divided, the dial needs to be set parallel to the celestial equator, with the gnomon directed towards the celestial pole. This is the equinoctial sundial, the type most commonly used in ancient China. It appeared no later than CE 851.[82] The territory of the Southern Song Dynasty (1127–1279) was dominant south of the Yangtze River, at which time the equatorial sundial appeared in various cities and became popular.[83] Until the Qing Dynasty (1644–1911), this kind of sundial was placed outdoors in government offices and palaces. Although the sundial can only be used during the day, when there are no clouds or rain, of the four types of timepieces it was the most popular. Around CE 1612, the Spanish missionary Diego de Pantoja (1571–1618) and Sun Yuanhua (1581–1632) collaborated in writing a book in Chinese, recording various Western sundial manufacturing methods.[84] Since then, the Chinese have not stopped studying Western sundials, and the use of portable dials is also very common.

The incense stick is the most common timekeeper in popular use because of its low cost and simplicity. The solid incense baton includes straight lines, circles, and Chinese characters. Time is displayed by the burning position. An inscription from CE 1073 records that there were not only incense sticks that showed the twelve 'double hours' and 100 'quarters', but also sticks dedicated to the night showing the five 'night watches'.[85] The incense powder was formed into various patterns using a mould, and the patterns could also be timed after the incense trail was ignited. The date and origin of these methods are unknown, but they were probably known from the eighth century CE, becoming popular

---

74 Ruan 1935, 2. In 1799, Ruan Yuan (1764–1849) completed a biography of a Chinese scientist in which he noted that the most important task for the emperor was to effect astronomical observations.
75 Jiang & Niu 1998, 19–21.
76 There are twelve months in a year and twenty-nine or thirty days in a month since the ancient Chinese calendar combined the solar year and the lunar month. In order to coordinate the number of days between the two methods, in every determined period two consecutive months will be thirty days, and there will be thirteen months in a given year, see Zhang 2019, 82–6.
77 Ten heavenly stems: 甲乙丙丁戊己庚辛壬癸. Twelve earthly branches: 子丑寅卯辰巳午未申酉戌亥.
78 See this volume, 'Introduction'; Needham, Wang, & Price 1986, 199–205.

79 Xue 2003, 446. *The New York Times* 18 July 1875 recorded that the Chinese were still then using the sundial, the incense stick, and the water clock; see Zheng 2001, 88.
80 For dials in general, see Chapter 1, Chapter 3, Section 1, and Chapter 8. In ancient China, the equinoctial direction dial used the twelve earthly branches as the scale of the dial, and the gnomon passed vertically through the centre of the dial. The top face of the dial was used for the summer period from the spring equinox to the autumn equinox, the lower face for the winter period from the autumn equinox to the spring equinox.
81 Needham 1959, 306; Quan 2013, 379.
82 Wang 1986, 8.
83 Quan 2013, 379–80.
84 Fung, 2004, 348–9. The book title is *Sundial Diagram* 日晷圖法.
85 Wang 1989, 260–1.

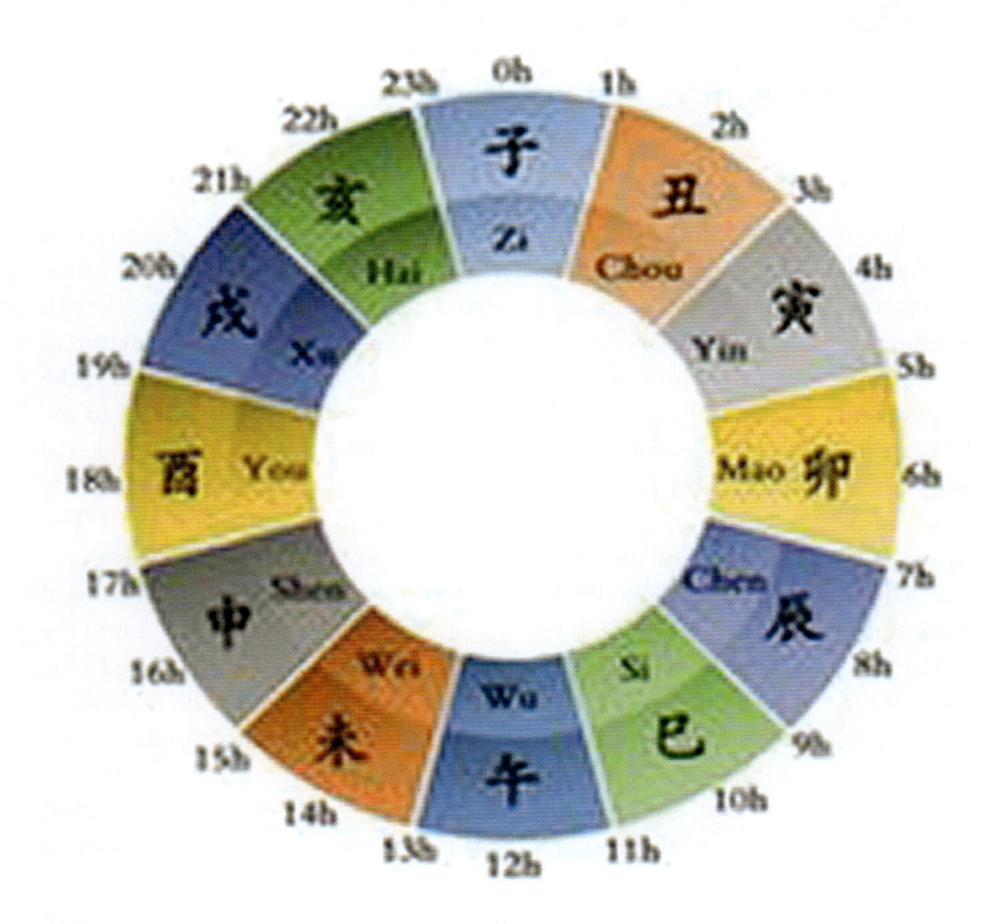

Figure 15 Correspondence between the twelve 'double hours' and twenty-four hours.

during the Northern Song Dynasty (960–1127). The most useful form was perhaps the dragon boat clock (Figure 16) in which an incense stick was laid lengthways along the body of a holder in the form of a dragon boat. Cords, with small bronze weights at each end were hung across it at appropriate positions so that as the stick burned along its length, they would fall, giving an aural indication of the hour, on a metal platter below. By placing only one weighted cord in a predetermined position the device could also act as an 'alarm clock'.[86]

The rolling ball was invented by a monk in the Tang Dynasty (618–907). This is a very simple timepiece that uses a bevelled track made of bamboo or wood. A ball is placed on it to roll down under the action of gravity. It records time from the accumulated number of the descents which occupy equal intervals of time. As this relies on manual control, it cannot guarantee timing accuracy. Because it can be used at any time, and anywhere, rolling balls were generally used when travelling and marching, at least until the Yuan Dynasty (1271–1368).[87] Obviously, neither this, nor the incense timepiece, was used in astronomy.

Figure 16 A dragon-boat incense timepiece. Photo: Courtesy of Macau Timepiece Museum.

Among the four types of timepieces, the water clock is the oldest and the most widely disseminated. It was used for astronomical timekeeping during the Western Han Dynasty (206 BCE– CE 8), using the change in water level caused by its flow into or out of

---

[86] Needham 1959, 330. A far more detailed account is given by Bedini 1994a.

[87] Dai 1988, 268–9. Guo 1988, 182–3. Needham's understanding of the rolling ball timer as a spring device is wrong; he was misled by Ruan Yuan's writings. See Needham 1965, 527.

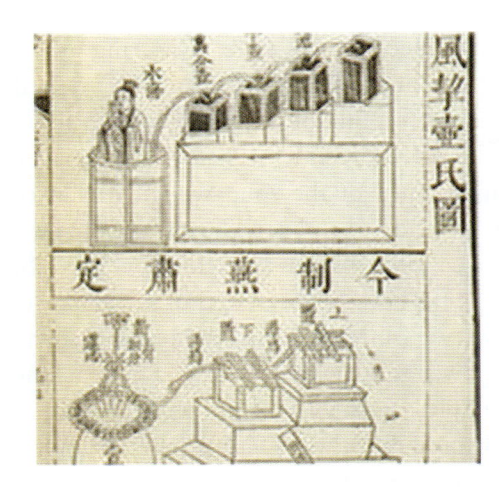

Figure 17 Lu Cai's inflow type clepsydra with are three compensation tanks. (top) Yan Su's inflow type clepsydra, has an overflow tank on the ground. (bottom) Illustration from Liu Jing Tu Kao 六經圖考 of Yang Jia (c.1110–84), reprinted in 1722, Book vi 37.

a container to display the time.[88] To maintain a constant speed of flow for inflow type clepsydras overflow tanks were introduced to ensure a constant head of pressure. A famous example in this regard is the work of official Lu Cai (b. ?–CE 55) (Figure 17, top). The official, Yan Su (CE 961–1040), designed the overflow tank in CE 1030 (see Figure 17, bottom). Shen Kuo (CE 1031–95) then improved this design and the accuracy of the instrument.[89]

The advantage of the water clock is that it can be used during both day and night, and it shows subdivisions using the 100 'quarters' system. In addition to its applications in astronomical work, different types of the instrument were in popular use. The most accurate was the steelyard **clepsydra**. It displayed time using the principle of equalizing the weight of dripping water per unit time,[90] but it is an intermittent timepiece that cannot display the time continuously and has uses similar to those of a 'stopwatch'. This timepiece was invented by Li Lan (fl.450) in the fifth century CE. At first it was a small device, then a large one was made for the emperor by the officials Yu Wenkai (CE 555–612) and Geng Xun (CE ?–618) in about CE 605. From the Sui Dynasty (581–618) to the Tang and the Northern Song Dynasty, the steelyard clepsydra was the main astronomical timepiece of the imperial family.[91]

ASTRONOMICAL TIMEPIECES FROM THE SECOND TO THE THIRTEENTH CENTURY

The history of Chinese mechanical timekeeping begins with astronomical timepieces. About CE 130, the first astronomical timepiece was made by the official Zhang Heng (78–139): a celestial globe driven by water. In addition to the celestial globe running once a day, the mechanical transmission could also display the number of days in the lunar month (similar to a calendar mechanism).[92] But there is no detailed description of the power and control devices in the literature concerning this timepiece.[93] About an astronomical timepiece by Geng Xun, the text clearly states that it was not only rotated by water, but also 'required no manpower'. In order to achieve such independence, some kind of control device was certainly used.[94] Although a clear answer cannot be given as to what it was, water power was clearly traditional in Chinese astronomical timepieces. These prototypes of the mechanical timekeeping that appeared in China in the second century CE laid the foundation for the development of later Chinese astronomical timepieces.

In the 600 years after Zhang Heng, these became increasingly complex.[95] The astronomical demonstration is more intuitive and the timing function is more prominent, which meant that Chinese mechanical timekeeping had matured. The astronomical timepiece made by the scholar Yi Xing (683–727) and the official Liang Lingzan (fl.721) in CE 725 was a turning point. It added sun and moon motions to the celestial globe. The transmission was further complicated: as there were two wooden jacks, one striking a bell to indicate the hours, and the other striking a drum to indicate the quarters.[96] Henceforward audible time-announcement became standard in Chinese mechanical timepieces. However, the power and control devices designed by Yi Xing are still not clearly described.

A somewhat more detailed record appears in the introduction to astronomical timepieces by Zhang Sixun (fl.976).[97] This astronomical timepiece completed in CE 980 indicated the quarters

---

88 Chen 2016, 1247–50. Hua 1991, 22–4. water clocks are mainly divided into two types: **outflow** and **inflow**. The **outflow** type is the earlier, but the use of the **inflow** type was more popular.
89 Hua 1991, 61–4, 84–94. The scales used by Shen Kuo's clock could also display the five 'night watches'; see Guo 1988, 96.
90 The water weight unit is converted into a time scale, which is displayed directly on the weighing rod. The siphon is used to stabilize the water flow, ensuring the equal weight of the water flow in equal time.
91 Hua 1991, 65–7.

92 Mechanical transmission relies on gears. Before Zhang Heng, gears had already been used in China, see Needham 1965, 85–8. For research on calendar mechanisms, see Liu 1962, 116–19.
93 Scholars have conjectured; see Liu 1962, 99, Figure 115. Li 2014, 81, figs. 1–45. Needham, Wang & Price 1986, 112, fig. 3.3.
94 This astronomical timepiece was made around CE 590, see Needham 1965, 482. Considering Geng Xun's familiarity with the steelyard clepsydra, it is likely that he used the principle of leverage as a control device.
95 For the development of astronomical timepieces from Zhang Heng to Geng Xun, see Needham 1965, 482–4; Forte 1988.
96 Needham 1965, 473–4.
97 Needham 1965, 470–1. The instrument uses mercury instead of water as the power source, thereby eliminating the problem of water freezing. However, the physical properties of mercury pose their own problems, see Guo 2011, 475–6.

audibly (seven jacks strike different instruments), and the hours visually (twelve jacks hold hour signs). Obviously, the twelve 'double hours' and the 100 'quarters' are expressed in two different ways, visual and aural, respectively. This was the first large astronomical timepiece in the Northern Song Dynasty, which, by adopting a more complex gear system and control device, allowed Chinese mechanical timekeeping to reach its summit. In CE 1092, led by the officials Su Song (1020–1101), Han Gonglian (fl.1088), and others, a great astronomical timepiece, called Shui Yun Yi Xiang Tai 水運儀象台, was completed.[98] In addition to the devices for driving the armillary sphere and the celestial globe, the mechanical design is outstanding. It includes a driving water wheel with a weighing-rod timing device, linkwork control devices, gear transmission devices, and time display. A new mechanism also appeared in this instrument. This was a water wheel and steelyard clepsydra mechanism, which consisted of a timing steelyard clepsydra and a water wheel lever. The former was used to generate uniform and periodic motion, and the latter to generate periodic vibration to check and release the intermittent motion of the water wheel.[99]

The time display device is located in a wooden pavilion of five stories with different wooden jacks on different levels. These jacks can display all three Chinese horary systems: the twelve 'double hours', the 100 'quarters', and the five 'night watches'. They also indicate the quarters of each hour and the night watches by striking different instruments.[100] The entire astronomical timepiece is located in a building about 12m high (see Figure 18). The armillary sphere is placed on the upper floor, the celestial globe is placed on the middle floor, and the wooden pavilion is placed on the lower floor. Chinese mechanical timekeeping had reached a new stage of development. Following a change of dynasty, this timepiece was moved to Beijing in CE 1127, but after a few years only the device for observing astronomical phenomena survived. It was damaged by lightning in CE 1195. After the Mongols overran Beijing in CE 1214, the remaining parts of the astronomical device disappeared.[101]

Su Song's astronomical timepiece is the most complicated mechanical timekeeping device known from ancient China. It combines the three major functions of observation, demonstration, and timekeeping. The 'escapement' that appeared in the timekeeping machine has been claimed as 'an intermediate stage or "missing link" between the time-measuring properties of liquid flow and those of mechanical oscillation'.[102] Whether this had any direct influence on European timepieces remains uncertain,[103] but, in the history of world mechanical timepieces, especially before the appearance of the solid weight-driven timepiece, it does provide a vivid example of the fusion of the motor, distribution and oscillator systems In fact, the value of Shui Yun Xi Xiang Tai goes far beyond the scope of a timepiece. From this huge machine, we can deduce many concepts such as tracking observation, power gear, and *remontoir*.[104] This machine occupies a very important place in the world history of precision technology.

If the mechanical clock was a 'fallen angel' from the astronomical world,[105] then several 'angels' appeared in the Yuan Dynasty (1271–1368). These were rather more independent of astronomy. They are still driven by water power but are more focused on automata display.[106] In CE 1276, the official Guo Shoujing (1231–1316) designed a mechanical timepiece for the Imperial Palace and used a hydraulic display device for the first time to show the quarters on a rotating ring against the finger of a jack.[107] The last emperor of the Yuan Dynasty (r. 1333–68) had a similar mechanical timepiece made in CE 1354, but the structure was more complicated. It employed many kinds of automata, jacks, and animals, which performed at the specific times required.[108]

After the establishment of the Ming Dynasty (1368–1644), officialdom no longer attached importance to the construction of mechanical timepieces, but civil research and innovation continued. The most famous example is the work of Zhan Xiyuan (fl.1370), recorded by the official Song Lian (1310–81).[109] His timepiece is different from those made previously and has become a simple timekeeper. It has two main features. Firstly, dial and hand are used to display the time.[110] Secondly, the driving wheel is moved by sand and power is transmitted through five

---

98 In CE 1096, Su Song completed *New Design for an Armillary and Globe* 新儀象法要, which was the first and most detailed monograph on astronomical instruments in China, and a scientific and technological work of world significance. The astronomical is introduced and illustrated in this book.
99 Yan 2007, 176–82.
100 Needham, Wang & Price 1986, 34.
101 Needham, Wang & Price 1986, 132–3; Dai 2010, 370. Many scholars participated in the research and restoration of this timepiece, see Zhang & Zhang 2019, 47–55.

102 Needham, Wang, & Price 1986, 59. From the analysis of words and images, this timepiece certainly had a mature control system although this is different from that employed in European weight-driven clocks. Needham believed that water wheel linkwork mechanisms were first used by Yi Xing, see Needham 1965, 474, note c. Dai believed it was by Zhang Sixun, see Dai 1988, 52–3. Wang believed it was by Su Song; see Wang 1989, 273.
103 Landes 2000, 18–20.
104 More information about this machine in Needham, Wang, & Price, 1986.
105 Price 1975, 369. Lu, however, denies mechanical clocks to be by-products of astronomical instruments, and thinks their development paths were different, see Lu 2017, 65–70.
106 Since Yi Xing's astronomical timepiece, the tradition of using jacks has not changed, and it has continued in Chinese clockmaking.
107 Guo 2011, 524–5. Needham, Wang & Price 1986, 135–6.
108 Needham 1965, 507.
109 Song 2014, 2193. Needham, Wang & Price 1986, 158–9.
110 Needham 1965, 512, fig. 668. This is the same display as the European mechanical clock of the same period. This timepiece still uses the jacks to tell time.

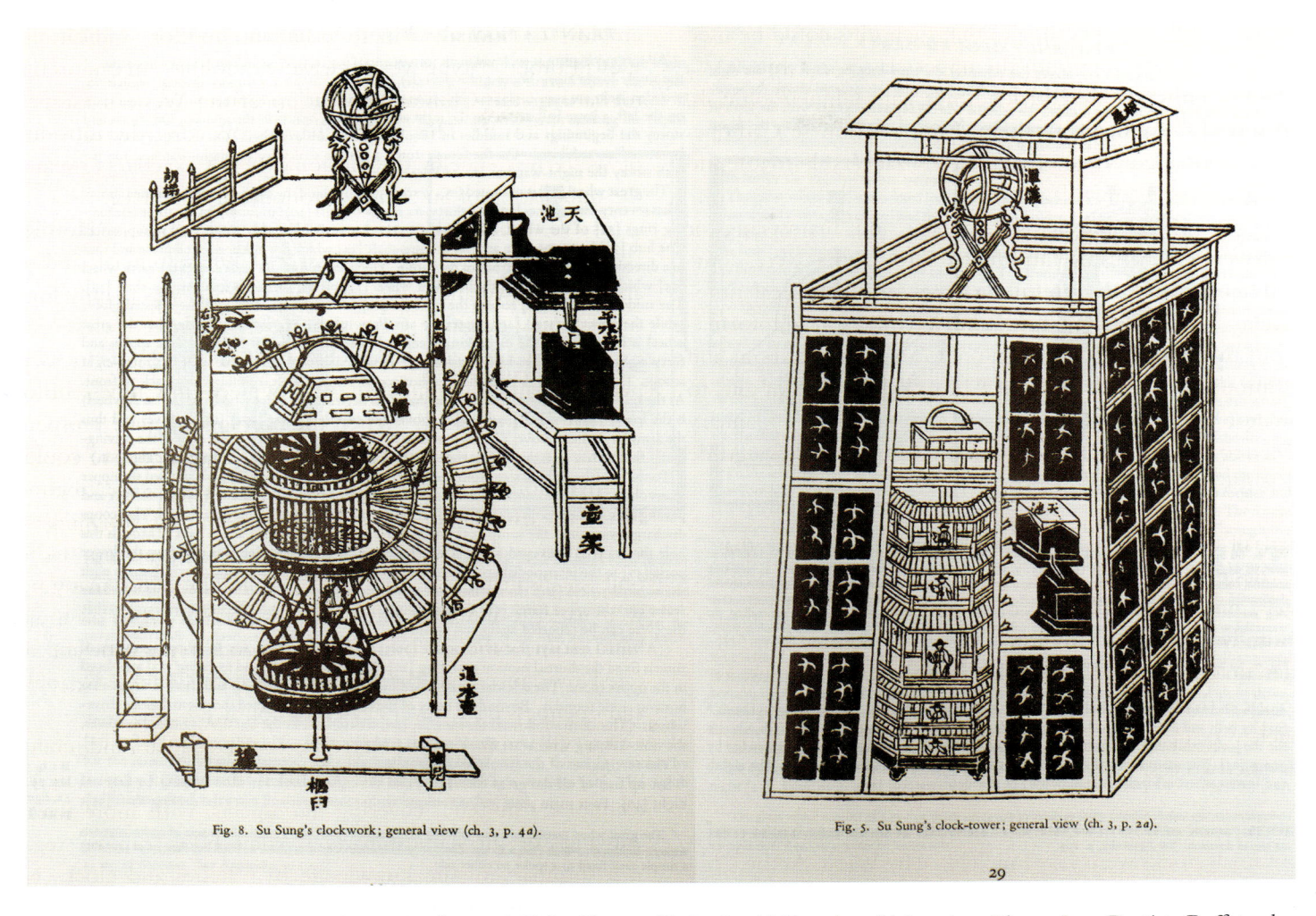

Figure 18 Su Song, *New Design for an Armillary and Globe*, Chapter iii, 85, 89. (a) Exterior; (b) interior. Photo: Jean-Baptiste Buffetaud.

gears. However, compared with water flow, the controllability of sand flow is not ideal. Two hundred years later Zhou Shuxue (mid-sixteenth century) made a series of improvements to it.[111] From the water clock to this piece, after more than 1,000 years of development, Chinese timekeepers seemed to be entering a more modern phase. At the same period, however, the history of independent timepieces in China was reaching an end. European missionaries brought weight- and spring-driven clocks to China and the development of Chinese timepieces became involved with that of Western clocks.

THE INTRODUCTION OF EUROPEAN CLOCKS

From the middle of the sixteenth century, Jesuits priests preached in the Far East, introducing weight- and spring-driven clocks to Japan and China as concomitant to their endeavour.[112] We cannot know how many clocks the missionaries donated or made in China in the twenty years from 1581 to 1601, beginning with Michele Ruggieri (1543–1607). But it is certain that Macao was the starting point for missionary 'horological diplomacy' in China. Luxury clocks brought by Jesuits or shipped from Europe would be collected in Macao and would play an important role.[113] When the missionaries wished to live in Zhaoqing, in mainland China, the clock was a gift to Chinese officials that eased the way, as they were curious about it and liked it. In September 1583, the missionary Matteo Ricci (1552–1610) arrived in Zhaoqing, and from there began a seventeen-year attempt to reach the Ming Dynasty capital in Beijing. The following year Ruggieri sent him the best clockmaker in Macao: 'a Canary Islander who had come from India, black-skinned' to make a clock for the Zhaoqing prefect Wang Pan (fl.1580).[114] Ricci carried clocks to Shaozhou (1589), Nanchang (1595), and Nanjing (1598), displaying or giving them to local officials. Knowledge of European clocks was thus spread.[115] Ricci had realized that, for the Chinese, Western clocks

---

111 Bai & Li 1984, 140.
112 Bedini 1975, 454–8. In CE 1581, Ruggieri gave a clock to an official in Guangzhou, see Bernard 1936, 190. For a comparative perspective, see Hiraoka 2020.
113 Chang 2020. 'Horological diplomacy' is a very vivid term, see Jin & Wu 2007, 538.
114 Spence 1984, 184.
115 Tang 2012, 291–2.

and watches were very special, and that such gifts could be used to build close relationships with the upper classes in China. As a result, he not only established friendships with Chinese officials, but also spread Catholicism to them. Famous Chinese Catholics included Xu Guangqi (1562–1633), Li Zhizao (1571–1630), and Yang Tingyun (1562–1627).[116]

The crown wheel escapement, foliot, and mainspring or weight-driven mechanisms introduced into China were completely different from the prevailing types of mechanical timepieces there. This made the Chinese curious and spread information about them; they also led some scholars to think about and accept European scientific thought.[117] In a description of gifts to the Emperor Wanli (Zhu Yijun, reigned 1573–1620) on 27 January 1601, Matteo Ricci explicitly recorded two clocks.[118] Diego de Pantoja wrote:

> When all things were put in order, particularly those that were designed for the king, there were two clocks with wheels, one large and made of iron, in a very large case, beautifully made, with multiple carved representations of golden dragons, that are the emblem of the Emperor, just as the eagle is that of the King [of Portugal and Spain]. Another, smaller, very beautiful one, one palm high, inside a golden case. The two cases were made in China.[119]

Western clocks not only entered the Chinese court for the first time, but small clocks were very much liked by Wanli. As the Chinese did not understand the operation of such clocks, the emperor sent eunuchs to learn clock technology from Ricci, whom he also needed to visit the palace regularly to maintain his clocks. It was thus that Ricci was able to live in Beijing for a long time.[120] In order to facilitate the eunuchs' education, Ricci also translated the names of clock parts into Chinese. It is worth noting that Ricci had translated or edited this work, because in the Chinese book on Western mechanics prepared by the missionary Johann Terrenz Schreck (1576–1630) in 1627, a Chinese work called *Explanations of the Clock* 自鳴鐘說 was cited as a reference.[121] Wang Zheng (1571–1644), who worked with Schreck, published a book on mechanical engineering, which included descriptions, with diagrams, of his own work. Among them was a timepiece that employs traditional Chinese visual and auditory indications but uses a European-type weight drive with foliot—an important step forward in the Chinese application of the Western escapement and oscillator.[122] This major change following the introduction of Western science by the missionaries was exactly 530 years after the publication of Su Song's astronomical timepiece.

As Western watches and clocks entered China, a 'Horological Road' formed from West to East.[123] Chinese officials and scholars both praised and criticized clocks. Praise was mainly for their timekeeping qualities. Critics thought they were useless—simply expensive. Concerning the question of the origin of clocks, some people regarded the mechanical timepieces in Chinese history as the origin of clocks.[124] When the horological trade prospered in China, they emphasized that China is a vast land, and could create inventions on its own without having to buy foreign goods.[125] From the middle of the nineteenth century, when Westerners penetrated into Qing society, clocks became contradictory objects, inspiring love and hate. On the one hand, many people were eager to purchase foreign goods as fashionable; on the other hand, some officials found foreign goods repugnant and advocated resistance. On the Horology Road, Chinese xenophobia and openness have never ceased to conflict.

Xu Chaojun (1752-1823) was both scholar and craftsman in Shanghai. He published China's first book on horological technology in 1809, covering ten aspects, with more than fifty drawings of mechanical parts.[126] This book can be regarded as a summary of the knowledge of watches and clocks before the nineteenth century. Among other topics it gave a brief description of the gearing and showed how to disassemble a mechanism. At this time, there was already a clockmaking industry in China, especially in Shanghai, and Xu's work may not have been widely distributed or read.[127] Another scholar, Zou Boqi (1819–69), also studied clocks in Guangzhou. He understood the principle of clock pendulums and realized the fundamental difference between a clock and a sundial. The clock displays mean solar time while the sundial displays true solar time, which are not the same.[128] Zou's scientific spirit

---

116 Xu and Li were particularly interested in Euclid and Clavius' *Sphaera* on which they worked with Ricci. Spence 1984, 152

117 Li 2012, 18–19. Of course, this attraction still existed in the Qing Dynasty.

118 This document dated to the twenty-fourth day of the twelfth month, and the twenty-eighth year of Wanli, see Zhu 2001, 232–3.

119 Ye 2019, 408. Thanks to Fernando Correia de Oliveira for the translation from Spanish to English.

120 Bedini 1975, 462. For a discussion of why Ricci lived in Beijing, see Jin & Wu 2007, 536–67.

121 Korean scholar Li Guiying (1788–?) records *Explanations of the Clock* as written by Ricci; see Tang 2017, 549. This book has not been seen in China. It remains to be researched by scholars.

122 Needham 1965, 513.

123 In the history of Sino-Western cultural exchanges, the Silk Road has always occupied an important position, but the road of timepieces from Europe to China also deserves attention as much for the light it can cast on religion, trade, and diplomacy as on technology dissemination and ideology.

124 Li 2012, 58–61.

125 Li 2012, 112–13.

126 Chen 1987, 43.

127 Chinese artisans mainly used experience to teach, rather than systematic theoretical learning. Although focusing on practice enhanced craftsmanship, it stuck at the level of imitation and was not innovative.

128 Zou 2009, 172. Thanks to Mr Ye Yongjian for the documentation provided.

helped the development of Guangzhou and Chinese clockmaking even if watchmaking did not develop on a large scale. On the one hand, it lacked a theoretical basis of mathematical and mechanical knowledge; on the other, it relied too much on manual manufacturing and lacked professional machinery and equipment.

MANUFACTURE OF THE IMPERIAL CLOCKS

Xu Guangqi, the friend of Ricci, not only accepted Western learning but also applied the results of Western science. When he presided over the calendar reform of the late Ming Dynasty in 1629, he suggested to the emperor that three clocks should be made.[129] This did not occur before the fall of Beijing to the Manchus and the establishment of the Qing Dynasty (1644–1911), so the ideas of Ming Dynasty officials were realized by the Qing emperors. Initially the court simply maintained its collection of watches and clocks, but clocks soon were made. In 1689, the Emperor Kangxi (r. 1662–1722) established a scientific and artistic institution in the Forbidden City. The main purpose of this was to sustain the imperial manufacture, and to facilitate Western academic activities.[130] There is no doubt that there were many workshops in this institution, including a clock workshop, which not only manufactured new clocks and modified existing ones, but also maintained, repaired, and installed those that were on display.[131]

Just as Matteo Ricci served the Wanli of the Ming Dynasty, so the principal makers of clocks for the Emperors of the Qing Dynasty were Christian missionaries. For over a century, more than ten of them—from Gabriel de Magalhães (1609–77) to Charles Paris (1738–1804)—were successively court clockmakers.[132] The Swiss Frantz Stadlin (1658–1740) served in the clock workshop from 1707 and was the first to make a pocket watch for Kangxi. In his later years, Kangxi talked about how the missionaries improved the accuracy of clocks.[133] Once technical problems were solved at court, it became possible to produce pocket watches.[134] Combining Western technology with the Chinese horary system had also always been a main focus of the workshop. On 16 October 1736, the French Valentin Chalier (1697–1747) described in a letter a clock showing the five 'night watches' that he had made. It took him four months to complete, and it could strike the unequal night watches.[135] Another French maker, Jean-Mathieu de Ventavon (1733–87), was good at creating automata. At the end of 1785, the Emperor Qianlong (r. 1736–95) ordered him to develop a figure that could write Manchu.[136] Under Qianlong, the clock workshop reached a peak, but unlike his grandfather Kangxi, Qianlong preferred to treat the clock as a mechanical toy with Chinese elements. Indeed, while European clockmakers developed highly decorative and highly precise timepieces during the eighteenth century, the Chinese imperial clocks only became increasingly artistic and adorned with automata. It was their most distinctive feature.

The imperial clock was a fusion of Chinese art and European technology, produced by Western clockmakers and Chinese craftsmen working together. Craftsmen from other workshops in such crafts as enamel, wood carving, and gilding supported the clockmakers.[137] Imperial clocks are not simply replicas of Western clocks. They have an entirely Chinese style of decoration while incorporating various types of automatic musical devices that continue the tradition of medieval Chinese timepieces. The abundance of people or animals in the clocks led to the aural and visual indication of time being neglected. The usefulness of the imperial clock as a timekeeper was not emphasized.[138] The impact of these imperial clocks was not only material, but also spiritual. Firstly, in the field of literature, members of the imperial family or central officials wrote poems or essays to introduce clocks, and expressed in these works their perception of time.[139] Secondly, clocks appeared in court paintings, shown as display items in the living room or the playthings of nobility.[140] Illustrations and explanatory texts about clocks and pocket watches appear in the imperial book of ordinances and artefacts completed in 1759.[141] This shows that, in the consciousness of the emperor, clocks and watches were not simply timepieces, but also

---

129 Xu 1963, 336.
130 Kangxi's interest in western learning also gave the missionaries a certain position in the court, and they served as the emperor's teachers. See Yan 2015, 312–28.
131 About the clock workshop at the court, see Pagani 2001, 37–9.
132 Pagani 2001, 46–7. Biography of the missionary, see Guo 2011, 197–234.
133 Kangxi 1994, 134–5.
134 For some pocket watches made in the Kangxi period, see Chapuis 1919, 42–3. Patrizzi 1980a, 67

135 Pelliot 1920, 64–5. The letter stated there were more than 4,000 scientific supplies such as clocks and watches from Paris and London at the Chinese court.
136 Hou 2009, 105. Cf. this volume Chapter 16, n. 59 After 1750, the production of automata figures was the main task of missionaries.
137 In addition to the clock workshop, there were dozens of workshops for making other artefacts.
138 Cipolla 1967/2003, 87. Cipolla believes that the Chinese only regarded the mechanical clock as a toy, but he only saw one aspect. In the eighteenth century, the timekeeping function of clocks and watches was also valued, but not solely, and at that time people who could own a clock were a minority.
139 Li 2012, 24–8.
140 Guo 2013, 303–11. For images of some clocks that appeared in the Qing Dynasty, see Li 2014 206–15.
141 Pagani 2001, 62.

Figure 19 Wood and enamel clock in the form of a two-storey building. H. 88cm, W. 51cm, D. 40cm. Royal clock workshop. Qianlong period. Palace Museum, Beijing. Photo: Qi Haonan.

represented a ritual instrument equivalent to those of astronomy. Was not the bell ringing every day in the court a symbol of the order of the emperor? In imperial clock manufacturing, the emperor firstly proposed his own idea, then repeatedly revised it, before the plan was finally passed to a clockmaker to be made. The function of the clock was mainly display. For some special occasions, such as the birthday of the emperor or his mother, special clocks or 'blessing automata' were made.[142] But, some were made and used for timing, such as the large striking clocks (see Figure 19), and some portable clocks fitted with an alarm function.

In the early nineteenth century, following an official ban on missionary activities, there were no longer any Europeans in the clock workshop to serve the imperial family. As the watch trade brought Swiss pocket watches to the court, these smaller, more convenient devices became popular with the imperial family. However, the clock workshop lacked watchmaking talents, and craftsman could not fulfil imperial requirements for pocket watch manufacture.[143] More importantly, the new emperor, Jiaqing (r. 1796–1820), the son of Qianlong, lacked his father's interest. He did not like decorative clocks and automata, only practical timepieces. The imperial clock workshop was bound to decline.[144]

TRADE INPUT AND THE IMPACT OF POCKET WATCHES

For China and Europe the eighteenth century was an era of trade development.[145] In 1757, Qianlong decided that all Western trade would be concentrated in one port, Guangzhou. Missionaries had used clocks to open China's doors to themselves 175 years earlier, and now clocks were one of the several goods that European merchants used to reduce the amount of silver bullion needed to pay for the tea and other Chinese goods desired in Europe that stimulated trade during the Qing Dynasty.[146] By the middle of the eighteenth century, the English East India Company brought mechanical clocks worth more than £20,000 from London to Guangzhou each year.[147] In 1791, before the British ambassador George Macartney (1737–1806) visited China, the number of 'Sing-songs' recorded by the Guangdong Customs reached 1025 pieces.[148] After visiting the Chinese palace in 1793, Macartney

---

142 Pagani 2001, 40–1. Hou 2009, 108–9.
143 Chapuis 1919, 40.
144 Guo 2011, 171.

145 In the late Ming Dynasty, the Portuguese had begun trade with China. Later, the Dutch, the British, the French, the Danish, the Austrian, and the Swedish all participated; see Zhang 2019, 473–9.
146 European missions with commercial purposes also gave clocks to the imperial family as gifts, but most clocks entered China through trade routes.
147 Braga 1967, 69. The English East India Company (1600–1874) had focused on trade in Guangzhou since 1716, and Britain had become a major trading country with China by the 1750s at latest.
148 Guan 2000, 88.

thought that the clocks and watches there were all made in England.[149] The British were indeed at the forefront of the clock trade, benefiting from an advanced clockmaking industry, and an advantageous trade position with China.[150]

In the Sino-British horology trade, the works of two British families, Cox and Ilbery, are representative. James Cox (1723–1800) entered the Chinese market in the mid-1760s. His goods were characterized by an extravagant appearance and incorporated dynamic automata and musical functions in a Chinese style born of Western imagination, although this did not weaken Chinese enthusiasm for them,[151] and English clocks, to a certain extent, formed a technical interaction with the manufacture of imperial clocks. Following Cox's bankruptcy in 1778, partly caused by payment failure in China, his son John Henry Cox (c.1750–91), went to Guangzhou in 1781 to recover debts and continue to trade in watches, clocks, and other goods.[152]

By the 1780s, the watchmaker John Ilbery (1750–1808) had a well-established business into which he later incorporated two of his sons: William Ilbery (1772–1852) was in charge of the business in London, and James Ilbery (1784–1839) was in charge in Guangzhou.[153] They developed a typical 'Chinese Market Watch', in association with Swiss makers.[154] They used the French Lépine calibre (Figure 88) to reduce the thickness of the movement. The cases were decorated with pearls and enamel (Figure 165).[155] This type of watch was introduced at the end of the Qianlong period, and at first was not standardized.[156] Some of them had a calendar and date display. Perhaps because the dials were too complicated, Qianlong stated in 1782 that such watches were not needed, but only chiming enamel watches.[157] This demand by the largest Chinese buyer of Western watches underlines the importance of the fine enamel painting on the case.

In the 1780s, the Swiss clock and watchmaking firm of Jaquet-Droz provided Cox in Guangzhou with pocket watches.[158] Other Swiss makers soon sought to exploit the market. Functionality in these watches took second place to artistry in the case decoration and ingenuity in the mechanism, which was frequently fitted with repeating work and/or music. The introduction of the **hanging barrel** allowed the back plate to be suppressed so that the highly decorated movement could be seen. Although Swiss watches frequently used a cylinder or standard duplex escapement, a characteristic form of the latter with forked locking teeth became popular for these 'Chinese' watches. A centre seconds hand, which showed at a glance that the watch was active, was obligatory, and since the watch was for show it was worn on a belt.[159] Although these models were expensive, they were welcomed by wealthy Chinese customers particularly because the enamelling of the case was completely different from Chinese painting techniques. Realistic figures or flowers aroused Chinese attention. Musical and striking functions were also always a Chinese preference. Such watches were often sold in pairs, a traditional Chinese concept signifying harmony (Figure 114).[160] Even court display clocks may be in pairs.[161]

Until the early nineteenth century, British traders in Guangzhou were predominant in the market. One such was Charles Magniac (1776–1824) who arrived in Guangzhou in 1801. His father Francis Magniac (1751–1823) was a clockmaker in London. Their company had been selling clocks and watches in China, and even after the 'clock crisis' of the 1810s, their clock business continued, ceasing only in 1824.[162]

From the 1820s, Switzerland increasingly dominated the pocket watch trade with China.[163] In 1818, Edouard Bovet (1797–1849) arrived in Guangzhou. Responding to the current state of the Chinese market, he made several partnerships for manufacturing and selling pocket watches.[164] These models gradually became typical. In Chinese, they had a specific name, the Chinese calibre watch 大八件, and were described as such in a work published in 1832—they were the final version of the Chinese market watch.[165] Characterized by the use of a going barrel and suspended movement, Chinese calibre watches divide into two groups: the luxury version and the simple version. The former has a case usually

---

149 Cranmer-Byng 1962, 261, but this is probably an exaggeration.
150 Chen 2014a, 120. About British watch and clockmaking in the eighteenth century, see Stirling-Middleton 2018, 37–41.
151 Chen 2014b, 116–17; Smith 2013, 25; this volume Chapter 16. For technical details about British clocks in the Palace Museum, see Wang & Qi 2017.
152 Ye 2008, 132. J. H. Cox's company is the predecessor of Jardine Matheson & Co. For Cox see White 2012, 94–150 and 158–207 for his clocks. For J. H. Cox, see White 2012, 151–6.
153 White 2019, 340–1. My thanks to Ian White for supplying the birth and death dates.
154 Vaucher 2003. The Ilbery family enjoyed strong commercial and technical cooperation with the Swiss watchmaking world. And James' son, James William Henry Ilbery (1811–96), was an apprentice in Fleurier.
155 Tellier 2010, 17–18. Such watches manufactured by Ilbery entered the Chinese market in the early nineteenth century.
156 Qing 2016, 245.
157 Guo 2013, 135.
158 For some original archival records, see Chapuis 1919, 62–3.

---

159 A good survey of Swiss watchmakers working for the Chinese market is given by Chapuis 1919.
160 Chapuis 1944, 137. The resulting mirror enamel pair is also a feature of the Chinese market watch, see Didier 2010, 27–9.
161 Chang 2016, 44–5. The most recent survey of Chinese watches is White 2019.
162 Greenberg 1951, 27–8, 86–7.
163 Bonnant 1964, 41–5. By the early nineteenth century, the British paid more attention to the profits brought by the export of cotton and subsequently opium from India, and the watch trade was gradually abandoned.
164 Chang 2016, 24–8. At first Edouard Bovet also came to Guangzhou as an employee of a British watch company.
165 Deng 1832, Vol. 6, 3.

Figure 20 Bovet pocket watch, simple version of the 'Chinese calibre watch' with an enamelled movement, using the shape of a Chinese dragon. D: 60mm. *c*.1860. Private collection. Photo: Tao Yanhe.

made of gold with enamel and pearl decoration, while the latter was usually silver cased without any exterior decoration.[166] Although externally different, the structure of the movement is similar but made in two materials: one of brass with engraved patterns, the other in polished steel. Special movement designs used the Chinese dragon and phoenix shape, sometimes with enamel decoration (Figure 20). Although all this increased the difficulty of manufacturing, they vividly reflect Chinese taste, indicating that the Swiss manufacturers had closely studied Chinese preferences.[167]

With the opening of trade in Shanghai and other cities in 1842, the Guangzhou trade monopoly was destroyed, and brands other than the Chinese calibre watch entered the mainland.[168] Specialized watch stores appeared in Shanghai, and opened branches in different cities. Laidrich & Vrard (later renamed L. Vrard & Co.), founded in 1860, was the most famous. The firm was founded by the Swiss, Edouard Laidrich (?–1869), and the French Ludovic Vrard (1833–1916). The northern branch was independently operated by Pierre Loup (1840–99) from 1881.[169] The sale of Chinese calibre watch continued until the end of the nineteenth century. The reason they achieved such unprecedented success was not only promotion by the sales network, but also, more importantly, the integration of this type of watch into Chinese culture. Firstly, almost all brands used Chinese patterns and characters as trademarks. Secondly, some models used the Chinese twelve 'double hours' method to comply with the Chinese traditional horary system.[170] Of course, there were also visual elements, such as centre seconds, the transparent cuvette, and the engraved movement, all of which were favoured in China and which differed from contemporary European pocket watches.[171] After 1850, the enamel painting on watchcases used more Chinese elements, such as landscapes, flowers, birds, and portraits.[172] This customized concept showed that the Chinese were not passively buying watches, but were actively proposing different models.

The position of Chinese calibre watch in the Chinese market was well summarized by Kiu Tai Yu (1946–2020):[173]

> This looks characteristic of the Chinese calibre pocket watch (Figure 20), the type and shape of the escapement in the movement are various. The materials, layout design, manufacturing process, and craftsmanship of the movement differ. Therefore, each of the Chinese calibre watches has its own characteristics, and no one is exactly the same.... Chinese calibre watches have the longest monopoly in the huge Chinese market and have the largest number of Chinese brand names. With its quality, beauty, variety, quantity, and coverage, it has created the biggest miracle in watches![174]

How did pocket watches affect the Chinese? In 1670, the traditional 100 'quarters' system was changed to the ninety-six 'quarters', a phenomenon of the Qing Dynasty using the Western calendar, which essentially accorded with the Western time system.[175] The Emperor Yongzheng (r. 1723–35) inherited the calendar promulgated by his father Kangxi. The Imperial Astronomical Agency used timepieces to check the time when observing astronomical phenomena.[176] In the history of the Qing Dynasty, Yongzheng was known for his diligence and dedication, and the eunuch in charge of pocket watches provided his timekeeping. He asked the imperial workshop to make some cases for pocket watches in 1726, perhaps for the emperor so that he could use the pocket watch at any time to keep track of time.[177] The clocks and watches used by Qianlong not only appeared in the court but were also used while travelling. For example, during the southern tours of 1756 and 1761, twenty small clocks and watches

---

166 Chapuis 1919, 162–3.
167 Chang 2016, 72–9.
168 The main brands are Vaucher (富硕), Dimier (點耶), Juvet (有喊), etc., all from Fleurier. For an introduction, see Patrizzi 1980b, 107.
169 Niklès van Osselt 2013, 81; White 2019, 6–7.
170 Taking Bovet as an example, it adopted Chinese (播喊) as a trademark in the 1830s. In 1878, Bovet launched a twelve 'double hours' pocket watch, See Chang 2016, 43, 89.

171 Obviously, the Chinese market watch tends to be more artistic. Chinese buyers tended to appreciate pocket watch design more than the utility of timekeeping. At that time, the Chinese preferred the centre seconds design, so that it could have visual dynamics.
172 Chapuis 1919, 186–7. For an overview of the development of pocket watch enamel technology, see Fallet 2015, 260–3.
173 Kiu Tai Yu is a well-known watch collector, watchmaker, and horology historian in China.
174 Kiu 2006, 63–4.
175 Zhan 2010, 133. A day is divided into ninety-six quarters, and each 'double hour' is exactly eight quarters, which is equivalent to four quarters in an hour. Through the use of clocks and watches, people in the Qing Dynasty gradually accepted the timekeeping method of Western.
176 Zhu et al. 1999.
177 Guo 2013, 112–14.

were required for each. Obviously, these were prepared for timing.[178] Although the frugal Jiaqing publicly stated that he did not like luxury items from the West, he wrote a poem specifically about pocket watches,[179] in which he described their accuracy. Although the emperor's clock had an amusement function, we cannot deny the importance of timing to him, and the importance of the pocket watch.

In the Qianlong period, pocket watch users were not only members of the imperial family, but also were senior officials in the court. To facilitate the scheduling of government affairs, these officials became accustomed to using watches.[180] There is documentation that shows how ministers fixed small watches in their belt buckles to facilitate using them. When Yu Minzhong (1714–80) wrote an official document, he would place his pocket watch next to the inkwell to prevent delay in submitting it. When the clock in Jiaotai Hall was ringing at noon, he instructed his colleagues to wind up the watch. Similarly the attendants of Fu Heng (1722–70) all had pocket watches that they checked with each other.[181] But until the middle of the nineteenth century, pocket watches were generally owned only by officials and the wealthy. It was during the Emperor Guangxu's reign (1875–1908) that increasing numbers of men and women at more popular levels of society began to use pocket watches, with consequent effects on their concept of time, especially for those living in cities.[182] It should be noted, however, that although pocket watches were becoming popular during the Qing Dynasty, no watchmaking industry had yet been established in China. Superficially the large quantity of imported watches would seem to have inhibited the development of domestic watches, but the root cause was the lack of knowledge and the technology for watchmaking in China.

POPULAR CHINESE CLOCKMAKING

In the first half of the seventeenth century, as missionaries preached throughout China, clocks became familiar to increasing numbers of people, and curiosity led to their imitation. The imitators were mainly to be found in Nanjing, Shanghai, and Zhangzhou.[183] The missionary Álvaro Semedo (1585–1658) mentioned in his book that Chinese craftsmen could make table clocks,[184] indicating that the Chinese already used mainsprings by this time. In the second half of the seventeenth century, in addition to the above-mentioned areas, records of clocks in the Guangzhou, Suzhou, and Hangzhou regions also appeared. A Russian mission met with Kangxi in 1676, and a member of the mission noted that Guangzhou artisans had full ability to copy and make large clocks.[185] The most detailed record, however, is of the work of Ji Tanran (fl.1655). He had seen Westerners make clocks and imitated one that he called the 'Heavenly Pagoda' (通天塔), using brass gears, and fitting a striking device and automata.[186] It was closer to the Western clocks than Wang Zheng's design. Generally, this was a period of apprenticeship and imitation, as an original craft industry had not yet formed.

The development of the Chinese clock industry came in the eighteenth century. With the increase of family workshops in different regions, local clocks took on their own design characteristics.[187] The scholar Qian Yong (1759–1844) noted that artisans in Guangzhou, Nanjing and Suzhou could all make clocks.[188] Although they were not the only places where clocks were made, by the mid-nineteenth century, Guangzhou and Nanjing were the two major clock-making centres, producing different styles of clocks: the Guang clock (廣鐘) and the Su clock (蘇鐘).

European clocks coming through the Guangzhou trade were mainly supplied to the Emperor through the tribute system or acquired by wealthy merchants The Guang clock developed in this market.[189] In the 1780s, the Guang clock entered a peak period

---

178 Guan 2011, 140.
179 Interestingly, he was the first emperor to write poems on the theme of pocket watches, indicating that his focus on pocket watches focused on practicality.
180 In 1816, Clarke Abel (1780–1826), a member of the Amherst Mission, discovered that Chinese officials were curious about pocket watches. He could not determine which aspect aroused this curiosity—as timer or ornament? His doubts were justified because pocket watches were mainly owned by the Imperial family and ministers at that time, and local officials had not generally begun to use them.
181 Chang & Bai 2009, 87, 111.
182 Zhan 2010, 127–8. With the advent of the public clock tower, the individual's emphasis on the accuracy of the timer will become more obvious.

---

183 Zhou Hui (1546–c.1627) documented that Huang Fuchu (fl.1611) produced a clock in Nanjing, Chen Xinfu was a craftsman from Zhangzhou, see Li 2012, 44–5. When missionary Lazzaro Cattaneo (1560–1640) was in Shanghai, locals were able to make oversized clocks; see Tang 2017, 457.
184 Semedo 1655, 27.
185 Zhu 2018, 46. In 1676, a craftsman named Cham in Suzhou was able to repair clocks for missionaries, see Golvers 1999, 548–9. Huang Luzhuang (1656–?) and Zhang Shuochen (fl.1662) in Hangzhou also made clocks; see Li 2012, 47.
186 Needham 1965, 515. This clock had a Chinese design style, but the quality was relatively low and it could not run for a long time, indicating that the manufacturing of the mainspring was immature. Entering the eighteenth century, Sun Ruli of Zhangzhou was able to manufacture 10cm clocks, and the mainspring technology was relatively mature.
187 The market and demand will determine the development of an industry. In the Qing Dynasty, the clock was regarded as a status symbol. This consumer psychology affected the style of the local clock industry, without influencing that of Western clocks made for the Chinese market.
188 Qian 1997, 321. The clock-making technology also interacted between Fujian and Guangzhou. Leng 2012, 51.
189 Records of Guang clocks already appear in the court archives in 1723. At first, they were only simple imitation of Western clocks,

characterized by three elements: firstly, a resemblance to Chinese architecture or goods; secondly, the use of enamel decoration in blue, green and yellow; thirdly, the provision of many automata and musical devices.[190] In addition to gorgeous and mechanically complex clocks, the Guang clock was also produced in simpler market versions. Production reached a peak of more than 2,000 sales per annum in the 1870s, and local horological guilds also appeared at that time.[191]

Nanjing was the centre of a clockmaking industry, and clockmakers from there took their skills to surrounding cities. The famous Suzhou clock-making centre had a close relationship with Nanjing. The inscriptions recorded in the cemetery of the Suzhou horology guild in 1816 not only indicate that some local clockmakers came from Nanjing, but also that Suzhou clock-making was already of a good size.[192] In 1853, there were two clock-making workshops in Yangzhou. The workshop owners came to Yangzhou from Nanjing in order to avoid the war of the Taiping Heavenly Kingdom.[193] The clock-making methods in these cities were based on family workshops, using manual manufacturing and a division of labour.[194] Compared with Guang clocks, Su clocks were mainly aimed at the mass market, and did not emphasize gorgeous appearance. They can be divided into three categories: striking clocks for the 'night watches', clocks with automatic jacks, and 'screen clocks' (插屏鐘), the main style, popular in the last 25 years of the nineteenth century. Clock cases were usually made of wood, with a round dial mounted on a square, copper plate engraved with various patterns. The movements were mostly spring driven, with a fusee, a crown wheel escapement, and a bob pendulum.

On 4 July 1851, Daniel J. Macgowan (1814–1893), the American missionary in Ningbo, wrote about the scale of the clockmaking industry in Nanjing, Suzhou, Hangzhou, and Ningbo. He believed that the annual output of these four places was about 1,000 pieces, their prices at 7–100 dollars, with an average price of about twenty-five dollars.[195] These data reflected the fact that the Chinese clock industry still relied on manual manufacturing and could not mass produce. Although the appearance and style of Chinese clocks had been formed, they lacked innovation in movement technology.[196] After the Opium War, China's commercial centre moved from Guangzhou to Shanghai, and the clock-making industry also developed there on the basis of available business capital.[197] In 1876, Ningbo native Sun Tingyuan (fl.1906) founded the May War Lee (美華利) company in Shanghai, mainly dealing in watches and clocks from Europe. Later, his son Sun Meitang (1884–1959) introduced mechanized production, which meant that China's clock-making industry entered the era of industrialization.[198]

---

and were not recognized by the emperor because of quality problems. Beginning in the 1760s, they gradually absorbed the creativity of British clocks in automata, and in appearance became less dependent on Western elements, a more local Chinese design being used, see Huang 2013, 18–19.

190 Huang 2013, 25–7. At the end of the eighteenth century and the beginning of the nineteenth century, Guang clocks became cheaper, at only one-third the price of similar British goods; see Cipolla 2003, 109.

191 Li 2012, 50. There were many simple versions that used a weight-drive, such as wall clocks.

192 In the late Qing Dynasty, there were more than twenty workshops, with fewer than 100 employees in Nanjing; nearly twenty workshops, with more than fifty employees in Suzhou. The movements, wooden cases, and engraved copper plates were made locally in Nanjing, but some accessories needed to be purchased from other places, such as dials from Guangzhou and bells from Suzhou and Yangzhou. Song 1960, 18–21; Pagani 2001, 79–80.

193 In the late Qing Dynasty, there were twenty workshops, with nearly 100 employees in Yangzhou. Most of the fusee chains produced in Yangzhou were made by rural women and supplied to Suzhou and Shanghai. Yangzhou was famous for making small Screen Clock whose movement height was within 10cm. Wu 1984, 105–8.

194 Chen 1981, 90.

---

195 Macgowan 1852, 336. This is an article about the Chinese clock market that he wrote for the US Patent Office. He recorded the number of clockmaking workshops: forty in Nanjing, thirty in Suzhou, seventeen in Hangzhou, and seven in Ningbo.

196 The movement still used the crown wheel escapement, and costs could not be reduced. As a result, Chinese clocks could not compete with German and Japanese imported clocks in the 1890s, and they gradually withdrew from the market.

197 In the late Qing Dynasty, there were nearly ten workshops, with fewer than forty employees in Shanghai; see Wu 1984, 115.

198 Sun 1925, 1. May War Lee established a clock factory in Ningbo in 1906, relocated to Shanghai in 1912, and used machine manufacturing in 1915.

SECTION THREE: MODERN CHINA

Ron Good and Jon Ward

At the turn of the twentieth century, domestic Chinese production was limited to what could be produced by skilled craftspeople in small and independent workshops dispersed throughout China. Although these smaller shops used then-current tools, they were still centred around manual production of individual pieces. There were no large-scale manufacturers, nor was there any assembly-line production of clocks or watches. Moreover, these few domestic clock makers had to compete against imports that dominated the Chinese market. By the 1920s, however, assembly-line production was firmly in place, largely spearheaded by the vision of two entrepreneurs, Sun Meitang, founder of the Meihuali Watch and Clock Co. Ltd, with its beginnings in 1905, and Li Dongshan, founder of the Baoshi Clock Factory (or Yantai Baoshi Clock Factory) in 1915. Sun Meitang was the son of the owner of the Meihuali Watch Shop, founded in Shanghai in 1875 as a franchise seller of imported watches and clocks. Convinced that China should and could manufacture clocks to compete with outside manufacturers, Sun Meitang opened the Meihuali Watch and Clock Co., Ltd in Ningbo, Zhejiang Province, just south of Shanghai, in 1905, and moved its operations to China's first modern assembly-line style clock factory in the Yangshupu district of Shanghai, in 1913. By 1915, Meihuali was producing clocks for domestic and foreign markets and even winning awards at a Panama Pacific World Exhibition in San Francisco.[199] The year 1915 also saw the very significant birth of the Yantai Baoshi Clock Factory, later known widely as Yantai Polaris, which is presently China's longest existing watch and clock company. Li Dongshan, the owner of a Yantai hardware retail company, the Shunde Xing Hardware Co. Ltd, had also been inspired by foreign clocks carried in his store. He realized, like Sun Meitang in Shanghai, that this represented an opportunity for new domestic industry. Leaving Yantai in 1913, Li travelled to Germany to learn about clockmaking. He returned in 1915, and after a further six months in Japan, opened his own factory, which produced its first finished clocks in 1918 (Figure 21):

> Most of these 'Nanjing clocks', [so named because the style was born in Nanjing during the Jiaqing period of the Qing Dynasty], have the shape of a square box, and the clock stands are carved from rosewood or mahogany, generally divided into three layers. The chassis is slightly larger with four tiger feet supporting it. The middle layer is a screen inserting frame, with groove inserts on both sides; the upper movement box can be inserted into the middle layer frame imitating the traditional screen inserting process, so it is called the screen inserting clock. The upper movement glass frame is mostly inlaid with mahogany snails. The lower and middle layers are carved and embossed, which is exquisite, antique, and elegant. Even more exciting is the clock face, which has a variety of white porcelain, enamel and bronze carving styles, as well as the copper base plate around the round clock face, decorated with eight immortals and auspicious patterns of opera, etc., to add beauty and interest.[200]

In the years that followed Baoshi's founding, and as its production of wall, alarm, and pendulum clocks became successful and were exported broadly through Southeast Asia,[201] over thirty more clock factories were established in other coastal cities, mostly by Yantai entrepreneurs.[202] The most notable of them today, Tak On Clock Factory in Shanghai, opened in 1932. After suffering extensive damage from bombing in 1937, it was rebuilt the next year and renamed Jin Xing Industrial Association, producing alarm clocks with the *Zuanshi* (Diamond) brand name.[203] In the ensuing years, it became one of China's most prominent horological enterprises. By 1925, the Meihuali Watch and Clock Co. Ltd had 643 employees and twenty-five retail branches in China, but it suffered a different fate. Meihuali's facilities were fully destroyed by the wartime bombing that had only disrupted Tak On, causing it to cease operations altogether.[204] Other, newer, Shanghai companies had better fortunes. The China Clock Factory, for example, introduced a '555' brand fifteen-day clock in 1940. Although production stalled during the Japanese occupation, by the early 1950s '555' clocks were again becoming very popular, and the company also developed highly accurate astronomical clocks.[205]

Domestic watchmaking in China, up to the mid-1950s, was almost exclusively limited to workshops established between 1926 and 1943 (including the most prolific, Huacheng Watch Case Factory) that only cased foreign movements. Domestic movement makers did not exist.[206] That changed when work began in both Shanghai and Tianjin to create the first Chinese-made wristwatches. A small workshop in Tianjin was the first to produce a prototype on 24 March 1955.[207] Named *Wuxing* (Five Stars),

---

199 Zhang, 2016.

200 Lan 2010.
201 Li 2015.
202 China Daily 2015.
203 Chan 2007c.
204 Zhang 2016.
205 Zhang 2014.
206 OSC 2003.
207 Jin 2013.

Figure 21 Nanjing Clock.
Photo: David Chang.

Figure 22 Tianjin 51 model wristwatch. Photo: Jon Ward.

Figure 23 Tianjin Sea-Gull ST5 movement. Photo: Ron Good/AMCHPR.

it used a virtually handmade copy of a Sindaco five-jewel pin-lever movement. Shanghai's first prototype, modelled after an A. Schild 1187, was finished in September 1955. Soon, Tianjin and Shanghai were producing watches for retail sale: Tianjin with its '51'-branded models (Figure 22), and Shanghai with the 581 series. Other popular series from both companies soon followed, along with steady improvements in durability (anti-shock modules, for example) and adaptations to facilitate more efficient mass production.

The third of the original major Chinese watch factories, Beijing Watch Factory, was founded on 19 June 1958, producing under 4,000 of its original BS1 model by 1961. That year, Beijing purchased an entire Swiss assembly line and began production of its anti-shock BS2 model, and by 1968 had produced 166,861 BS2 watches.

Other watch factories were also established in 1958, in an additional five cities: Guangzhou, Nanjing, Qingdao, Jilin, and Andong (now Dandong). At the same time, four Shanghai watch-case companies were merged with Huacheng to manufacture cases for the factories in both Shanghai and Tianjin. In 1959, Shanghai's Jin Xing (formerly Tak On Clock Factory) added stopwatch production, initially mostly for the military.[208] These stopwatches later found a wider market and proved so successful that the factory's name was changed to Shanghai Stopwatch Factory seven years later. Also in 1959, wristwatch production began at the renamed Yantai Clock Factory (formerly Yantai Baoshi). They were given the brand name *Beijixing* (Polaris), a name later given to all of the factory's products. Yantai had greatly expanded its range of clocks, including developing China's first marine chronometer in 1957, so watchmaking remained a minor activity. The factory became a state-owned company, Yantai Clock and Watch Factory, in 1962.[209]

Tianjin began work on China's first wrist chronograph project in 1961 using equipment purchased from Switzerland's Venus Watch Company.[210] Shanghai introduced the first Chinese calendar watch in 1962, the A623, which Premier Zhou Enlai famously wore for the rest of his life.[211] Tianjin unveiled the first entirely Chinese-designed and manufactured watch movement in 1966. Branded *Dongfeng* (East Wind), reportedly named by Mao Zedong himself,[212] it sported the ST-5 (Figure 23), a reliable and accurate movement and the first watch movement in China that was not highly derivative of a foreign calibre. Shanghai's SS1 movement, a dramatic redesign of its A-581, was an improvement introduced in the same year. Beijing's home-grown improvement of its Swiss-inspired BS2 movement, the SB-5, followed in 1968.

By this time, Shanghai, Tianjin, and Beijing Watch Factories were the 'Big Three' of the Chinese watch industry and captured most of the domestic market for wristwatches, although factories in other cities survived trading to their immediate regional markets.

China's Ministry of Light Industry opened the Clock and Watch Research Institute in Xi'an in 1967, charged with formulating and revising national horological standards, product quality testing, and vocational training. Its research projects led to technological advances in timepieces used for military and civilian purposes, including the development of an aviation clock, mechanical and tuning fork timing devices for satellites, and a quartz marine astronomical clock.[213] The number of watches manufactured in China increased from 500,000 in 1960 to 3.5 million in 1970.[214] A new central government plan to further expand watch production was initiated in 1969. Major investments were made opening watch factories in almost every province and developing a standardized men's watch movement, ordered to be manufactured by almost every watch factory in China. The movement

---

208 Chan 2007c.
209 Li 2015.
210 Adelstein 2014.
211 Dong & Wang 2009.
212 Adam 2010.
213 ZBYJS a,b.
214 Byrd & Tidrick 1992, 60.

Figure 24 Peacock Brand Tongji movement. Photo: Ron Good/AMCHPR.

(given the name *tongyi jixin*: unified movement, normally abbreviated to *tongji*) allowed factories to be more easily constructed all over the country, while interchangeable parts made repairs easier and cheaper. Most important, it allowed for an increase in production, making watches accessible to a greater number of people.[215] *Tongji* design (Figure 24) drew expertise and resources from the eight leading watch manufacturers, the Clock and Watch Research Institute, and Tianjin University. Blueprints for the new seventeen-jewel manual winding movement were finalized in 1971. Over the next few years almost all Chinese watch factories used it.[216] Visually, the resulting movement resembled the Enicar 1010, to a degree that suggests at least some influence, but with a seventeen-jewel count compared to the Enicar's usual 21.

The first mass-produced *tongji* watches were released to the public in 1972 by Shanghai No. 2 Watch Factory with the brand name *Baoshihua* (Gemstone Flower).[217] Beijing produced its first *tongji* the same year. Other factories soon followed. Even though movement design was standardized, some factories developed their own variations containing parts compatible with other factories' movements, adding higher jewel counts or simple complications. The Shanghai Watch Factory marketed its first *tongji* watches in 1974. These nineteen-jewel variants were of particularly good quality, and quickly became the most desired domestically manufactured watches in China.[218] The English language text which appears on some of them suggests that they were intended for export as well as domestic use, although Chinese watches were still exported in only very small numbers at this time.[219] The *tongji* was a remarkable domestic success. Within a decade, thirty-eight enterprises were manufacturing complete watches—more than in any other country at the time.[220] Some of them also delivered movements to one or more of the dozens of additional factories which just cased them.

The Tianjin Watch Factory had been exempted, along with two other companies, Shanghai Stopwatch and Nanjing, from the mandatory switch to the *tongji*. Consequently, Tianjin continued to manufacture its ST-5 *Dongfeng*, which became China's first export watch the year after receiving official approval in 1973. The export version was given a new brand name: the now well-known Tianjin Sea-Gull.[221] Many millions of people were served as production by state-owned factories increased year after year into the 1980s. In addition, the opening of the market after the end of the Cultural Revolution had a formidable effect on exports. By 1981, China had become the world's fifth-largest producer of watches and exported about ten per cent of the overall 28.7 million it produced that year. China also manufactured 27.7 million clocks in 1981, exporting thirty-five per cent of them. Shanghai's factories, amalgamated under the umbrella corporation Shanghai Clock and Watch Company, dominated the domestic industry, producing one-third of the country's watches.[222]

While quartz timepieces were becoming dominant in much of the rest of the world, China's industry was still overwhelmingly based on mechanical movements. This was due to domestic consumer demand; quartz was more difficult to have repaired, and components were harder to obtain. Digital watches were widely shunned by a population that regarded a wristwatch as a treasured possession. While a few factories did manufacture complete quartz timepieces, they could not be produced as economically as their domestic mechanical and foreign quartz counterparts.[223] Demand for some clocks remained high, for example the 555 desk clock,[224] but by this time China's clock market had generally become saturated, and overall domestic production dramatically decreased. In an effort to avoid the same fate for the watch industry, production limits were placed on the factories. No such limits were placed on clock production, which was controlled at the provincial level. However, these limits were not strictly enforced. In addition, some unauthorized factories were founded in a number of smaller cities.[225] But it did not take long for market conditions and consumer tastes to undergo dramatic transformations.

As China's economy opened under the leadership of Deng Xiaoping, a number of Special Economic Zones were established in the southern provinces of Guangdong and Fujian, spurring the formation of new watch companies with joint Chinese and foreign ownership. Lacking the experience of the established

---

215 Teng 2015.
216 Chan 2008b.
217 Chan 2008b.
218 Teng 2015.
219 Byrd & Tidrick 1992, 60.

220 Byrd & Tidrick 1992, 63.
221 Adam 2010.
222 Byrd & Tidrick 1992, 59–60.
223 Byrd & Tidrick 1992, 63–4.
224 Yang 2007.
225 Byrd & Tidrick 1992, 60–1.

Figure 25 Yantai Polaris 100th Anniversary Clock 2015. Photo: Ron Good/AMCHPR.

Figure 26 Modern Chinese Jintuofei Brand tourbillon. Photo: Ron Good/AMCHPR.

factories, they installed inexpensively made foreign quartz movements, mostly from Japan and British Hong Kong, in locally manufactured cases. Soon a huge number of inexpensive, thinner, more accurate watches would flood the domestic market.[226] The Shanghai Clock and Watch Company quickly worked to combat the threat these new competitors posed. A new mechanical movement, designated SBS, was developed to power thinner watches which were coming into fashion. It was a joint project by Shanghai and Shanghai No. 4 (formerly Shanghai Stopwatch) Watch Factories.[227] Within a few years it was the movement used in both factories' flagship wrist watches. A few other factories, including Beijing and Nanjing, designed new, less bulky movements of their own, but despite these efforts, the thinner manual winding movements ultimately failed to achieve the factories' goals.

There was nevertheless a place for mechanical watches in some markets, as long as they were automatic and inexpensive. Guangzhou Watch Factory, hit particularly hard by its nearby southern China competitors, developed an automatic version of its ladies' watch movement. Similarly, Tianjin used its women's ST-6 as a base movement for men's and women's automatic watches. Both factories sold these inexpensive-to-produce movements to be cased by other enterprises in China and Hong Kong. Meanwhile the new enterprises in Guangdong thrived. By 1987 more watches were manufactured in Guangdong than in Shanghai, which had dominated the industry for decades.

The turn of the century saw a new 'Big Three' emerge in China's fashion watch market: Rossini in Zhuhai, and Fiyta and EverBright (now Ebohr), both in Shenzhen. They primarily specialized in the production of quartz watches to meet the demand of a domestic population with ever-increasing incomes.[228] The year 1997 saw the beginning of an official government policy to privatize almost all state-owned enterprises.[229] Within the next few years, a massive reorganization of the Chinese watch industry occurred. Most of the *tongji*-era factories closed, and their assets were sold to private investors. The number of mechanical watch movements manufactured in China plummeted.[230] However, also in 1997, Tianjin Sea-Gull introduced a new (ST16) movement, based on Miyota's workhorse 8200 series.[231] The mechanical watch renaissance occurring outside of China became a welcome opportunity for renewed growth. Several other factories, including Guangzhou and Beijing, also developed their own Miyota-inspired movements. Some of the surviving original Chinese watch factories continued to make complete watches, but a large part of their activities consisted of the manufacture of reliable movements to be used in watches cased elsewhere—a complete reversal of the industry's business model in the first half of the twentieth century.

Sea-Gull's resurrection of its 1960s mechanical PLAAF (Chinese air force) chronograph in 2003 was immediately successful. Beijing released its first tourbillon watches in 2004. In the years following, Sea-Gull, Shanghai, Hangzhou, and, more recently, Guangzhou and Dandong's Peacock Watch released their own tourbillons.[232] A shortage of ETA-style movements in the early 2,000s presented a further opportunity for Chinese movement manufacturers. Sea-Gull and Hangzhou, in particular, took advantage of this, producing their own alternatives. Other companies, like Shandong's Liaocheng Zhongtai Watch Co., Ltd found a profitable niche producing a wide range of inexpensive but decorative watch movements for its own watches as well as for major southern Chinese manufacturers. Dial-side open heart movements are a common example, found world-wide. By 2007, Sea-Gull alone manufactured a quarter of the world's mechanical watch movements, more than any other single company at the time.[233]

More than two decades after the industry's 1997 upheaval, the descendants of six of the eight 1958 watch factories remained in operation. A century after its first assembly-line clock factory was founded (Figure 25), China manufactured nearly ninety per cent of the world's clocks, exporting about eighty per cent of them; 540 million clocks were produced in 2014.[234] In 2009, China accounted for seventy per cent of the world's watch output (although this was only ten per cent of world market value[235]), Sea-Gull and other Chinese companies continue to innovate, producing (besides classic three-handers) quarter and minute repeaters, alarm movements, micro-rotor automatics, tourbillons of varied complexity (Figure 26), and perpetual calendar watches.

---

226 Dong & Wang 2009.
227 Chan 2007c.
228 Teng 2015.
229 Chen et al. 2018.
230 Teng 2015.
231 Jin 2013.
232 Adam 2010.
233 Adam 2010.
234 China Daily 2015.
235 Dong & Wang 2009.

SECTION FOUR: JAPAN

Katsuhiro Sasaki

THE ANCIENT *KEIHYO* SUN-DIAL AND TRAVEL SUNDIAL

One of the earliest mentions of **sundials**, *guibiao* (圭表), in East Asia, appears in the official history books, *Former Han Book,* compiled in 206–8 BCE. Sundials were called *guibiau*(圭表) in ancient China and Korea.[236] The word *guibiau* is made by combining the characters *gui* (圭), a time-scale placed on the ground to measure the length of the shadow, and *biao* (表), a gnomon to make the shadow of the Sun.[237] There is also the word *tugui* (土圭), from the characters *tu* and *gui* (圭) pronounced 'tokei'. Here *biao* (表) is a gnomon that stands on the ground to cast a shadow from the sun, and *gui* (圭) is the scale that is placed (or drawn) on the ground to measure the length of that shadow (*biao*).[236] The Chinese character *guibiao* made by combining the two characters *gui* and *biao* is the word which means sundial in China, Korea,[237] and Japan. A Chinese character *tugui*, having the same meaning as *gui*, is pronounced as 'tokei'. Thus, the Japanese word 'tokei' for clocks originated from the word for sundials.

The first Japanese sundial, a *keihyo* (圭表), was a *guibiao* imported from China. *Keihyo*s were brought with Buddhism and many artefacts by Japanese missions, the *kenzuishi* (遣隋使) and the *kentoshi* (遣唐使), who were sent to China during the Sui Dynasty (581–618) and the Tang Dynasty (618–907), during the Asuka period (538–710) of Japan. Throughout the Nara period (710–94) and Heian period (794–1192), *keihyo*s were used in the bureau of divination, the *Onmyoryo* (陰陽寮), which performed astronomical observations and compiled calendars. When Harumi Shibukawa (1639–1715) compiled and completed the *Jokyoreki* (貞享暦) calendar in 1684, a *keihyo* was one of his main observation instruments.

Many portable sundials are found during the Edo period (1600–1868). Many of them have a bowl-shaped depression 2–3cms in diameter, which carries the hour lines and a gnomon, and they are fitted with a compass. They divide the period between sunrise and sunset into six parts, an **unequal hour** system used throughout the Edo period. Some portable sundials were designed as *netsuke* (根付け) for *inro* clocks. In addition, portable sundials with accessories such as a compass, writing brush, brush and ink case, ear pick, knife, tweezers, and abacus were also made (Figure 27). Japanese people in the Edo period are noted as being among the world's leading travel lovers. The post towns on each highway were always crowded with visitors, such as *daimyos* en route for the capital, Edo, people visiting the Ise Shrine or the eighty-eight temples in Shikoku, and many others. Portable sundials were a necessity for travellers, and paper sundials were also prepared.[238]

THE EMPEROR TENJI'S WATER CLOCK: *ROKOKU*

The *rokoku* (漏刻) is an **inflow water clock** time being shown by the rising water level in the receiving tank. The precision of water clocks, insufficient in early days, was greatly improved during the Tang Dynasty by inserting multiple tanks to equalize the flow rate between the upper tank and the lower one. In particular, the Tang bureaucrat Ryosai (呂才, 606–65), who excelled in astronomy, the calendar, medicine, and music, is known for making the five water-tank type *rokoku*.[239] The *rokoku* is mentioned in the *Nihon-shoki* (日本書紀), which was compiled in the Nara period and recounts the history of ancient Japan. In part 27, in the article on the Emperor Tenji (天智天皇, 626–72) is the statement:

> On the 'April *Kanoto* (辛) *U* (卯)' in summer, the Emperor installed the *rokoku* in a new building and told the time with a bell and a drum for the first time. This *rokoku* was made by himself when he was Emperor Prince. [author's translation]

This suggests that Emperor Tenji started the work of timekeeping and time-telling in 671. 'April *Kanoto U*', 25 April in the lunisolar calendar, corresponds to 10 June in the Gregorian calendar. It is commemorated by the 'Time Day', *Tokino-kinenbi* (時の記念日), in Japan, inaugurated on 10 June 1920 at the Tokyo Educational Museum (now the National Museum of Nature and Science, Tokyo), where a centenary 'Anniversary of Time' exhibition was held in June 2020.

The *Nihon-shoki* also mentions the Emperor Prince's *rokoku* in the article (part 26), on the Emperor Saimei (594–661). 'And Emperor Prince made the first *rokoku* and told the time to the people'. This statement is confirmed by one for 671 stating that the Emperor Prince, Naka-no-Oe-no Oji (中大兄皇子), had already made a *rokoku* eleven years earlier. In December 1981, the Nara National Cultural Properties Research Institution carried out archaeological excavations at Asuka in Nara. It was concluded that lacquer pieces discovered in the excavations must be parts of the *rokoku* made by Naka-no-Oe-no Oji, and of great significance, and it became a big topic.[240]

---

236 For sundials and water clocks in Korea, not otherwise treated in this volume, see Jeon 1974, 42–71; Needham, Gwei-Djien, Combridge, & Major 1986; Hovey 1986.
237 Needham 1959, 285, fig. 110.
238 Tsunoyama 1984, 63–4; Chuko-Shinsho 715.
239 Needham 1959, 324, caption to figure.
240 Asuka 1983.

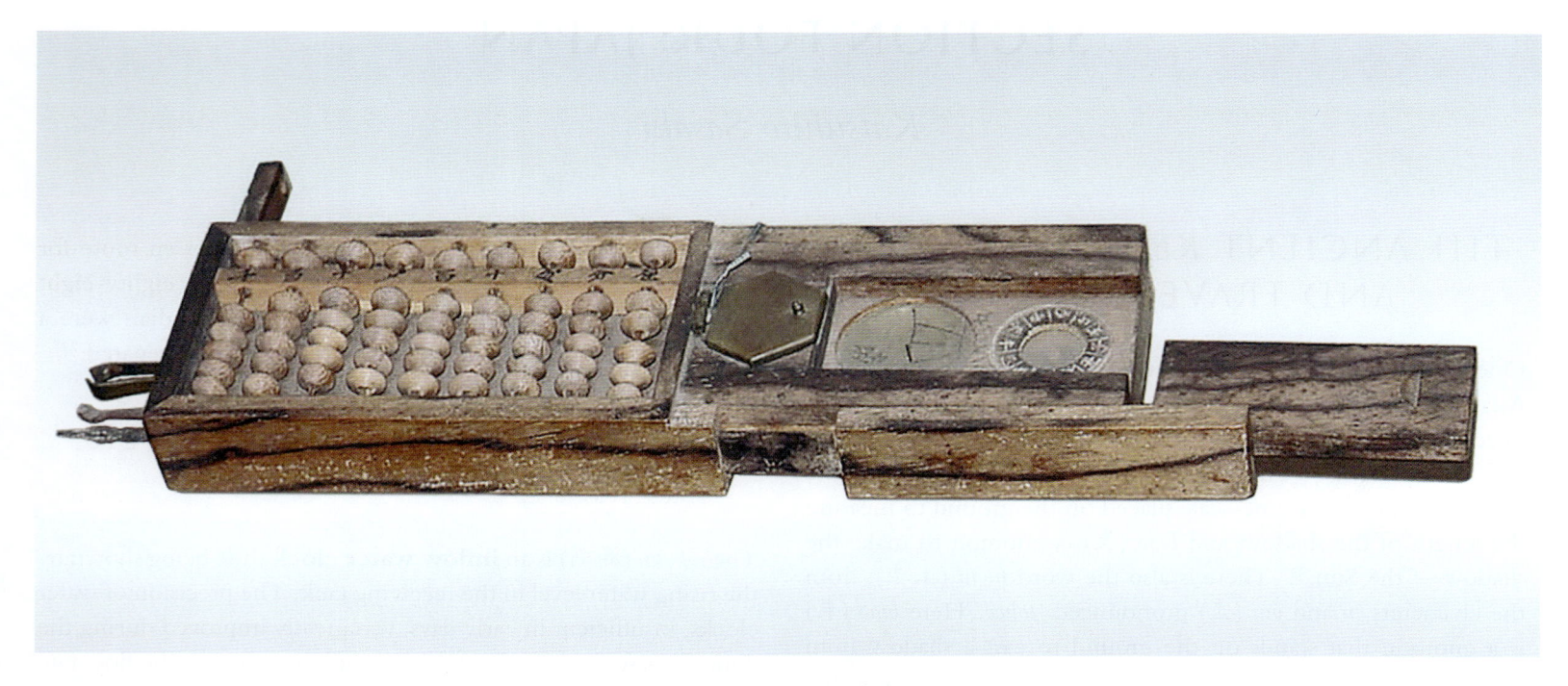

Figure 27 A portable sundial with seven accessories. Takabayashi collection, NMNS Tokyo.

Emperor Tenji's *rokoku* has not survived, so it is impossible to know exactly what it was like. However, it is illustrated in the *Rokoku-setsu* (漏刻説) by Yosen Sakurai (桜井養仙) of Chikugo, Yanagawa, in 1732 (Figure 28). This shows it to have consisted of four water tanks, named from the top *yatenchi* (夜天池), *nittenchi* (日天池), *heiko* (平壺), and *suikai* (水海). A dragon-shaped outlet was attached to the bottom of each tank from the first to the third. The fourth tank had a float with a time-scaled arrow, *sen* (箭), the rising level of water against this indicating the time. *Engishiki* (延喜式), a book compiled in the Heian period (794–1185), describes in detail the enforcement regulations for the *Ritsuryo* legal system, and their basis. In the article on the *Onmyoryo* (the divination bureau) the *Engishiki* gives additional details for using the *rokoku*: to indicate the time of sunrise and sunset, the opening time of the main gate of the Palace, the opening and closing times of other gates, and the leaving time for officials.[241] The times given in the description show that time was expressed in the capital by *toki* (辰刻) and *koku* (刻). Twelve *toki*s made up one day and four *koku*s made up a *toki*.

One of the most famous works of classical literature in the Heian period was the *Makurano-soushi* (枕草子) by Sei-shonagon (清少納言), a court lady serving the Empress Teishi (中宮定子), consort of the Emperor Ichijo (一条天皇, 980–1011). A poetess, she was a leading writer of *tanka* based on close observation of daily life and the nature of the four seasons in a delicate feminine way. Time is treated in section 250 of her essay:[242]

It is very interesting how the time officer performs his work. On a very cold night, I hear the sound of dragging wooden shoes like '*ko ho, ko ho*', the sound of flicking the string of the amulet bow, the elegant voice with which the time officer says his name and tell the time like '*ushi mitsu* (three *koku* of bull *toki*, 2:00 AM)' or '*ne yotsu* (four *koku* of rat *toki*, 0:30 AM)', etc., and also the sound of inserting a wooden plate with the time name to the stand. All of them are very interesting. [author's translation]

The Department of State, *Dajokan* (太政官), the highest organ of the judiciary, administration, and legislation under the *Ritsuryo* system, which Sei-shonagon recorded, was on the east side of the council hall, *Daigokuden* (大極殿), in the centre of the Imperial Palace, while the *Onmyoryo* was on the north side of the *Dajokan*.[243] Seishonagon, who served near the *Onmyoryo*, wrote vividly about various sounds such as the dragging of shoes and even the insertion of the time plate.

FIRE CLOCK: *JIKOBAN* OF NIGATSUDO HALL AND THE GEISHA'S INCENSE-STICK CLOCK

A fire clock utilizes the constant burning speed of candles or incenses. Particularly important, was the *jokoban* (常香盤) (Figure 29) commonly used in Buddhist temples to keep the sacred fire burning and to purify uncleanliness. This was used in the following way.

A trail, 'powder line', for the incense is drawn in a levelled bed of ashes. The incense is lit and its burnt positions mark dawn, dusk,

---

241 Engishiki 1929, i 587–602, O-okayama Shoten, keyword: 延喜式.
242 Kaneko & Tachibana 1955, 250th step, 675, 'Meiji Shoin'.

243 Saito 1995, Figure IX–2.

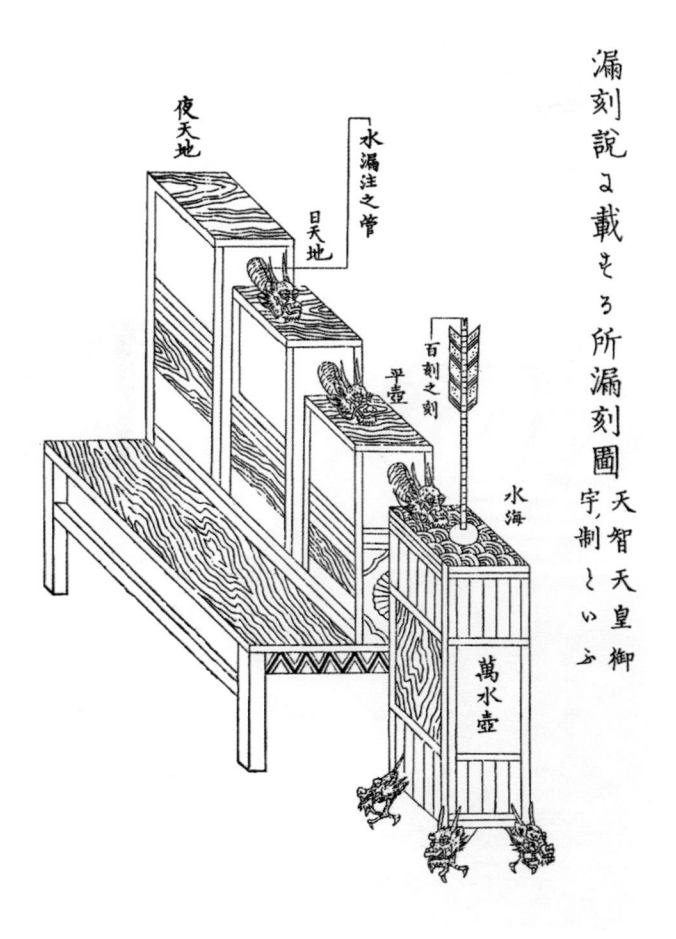

Figure 28 Emperor Tenchi's *Rokoku* from the book *Rokoku-setsu*.

and the following dawn. This determines the burned length of the incense powder line during the day and night, respectively. Using the measured burning length, the dawn-to-dusk or dusk-to-dawn positions on the incense line are marked, each length divided into six equal parts. A bamboo or copper time tag is attached to each position. By reading the burnt position of the incense line with the tag, one can know the hour. *Jokobans* were so simple and easy to use that they were widely used in not only in Buddhist temples but also the houses of the leading local farmers.

It is known, for example, that a *jokoban* has been used at the Water-Drawing Festival, *Omizutori* (お水取り), held in the Nigetsudo Hall of Todaiji Temple in Nara every year, where the *jokoban* was called '*jikoban* (時香盤)'. This festival, in which monks pray for peace, and the order of society, while dedicating sacred water to the 11-faced Kannon in Nigatsudo Hall, has been held for over 1,200 years. It comes to a climax with a powerful fire display using large torches, on 12 March. The *jikoban* is still used for time management at the festival, and is therefore a valuable cultural link between the Nara period (710–794) and the present day, together with various tools, customs, and words related to the *Omizutori*.

Another example of fire clocks is the unique incense-stick clock, *snkodokei* (線香時計), used in geisha houses. The incense-stick clock exhibited in the National Museum of Nature and Science in Tokyo consists of a stand and an abacus (Figure 30). The stand has thin cylinders to hold incense-sticks, and small sprigs or posts on which wooden tags can be hung—the tags bearing the geisha's name.

Using modern time notation, the use of the incense-stick clock is as follows: assume the geisha's working hours are from 6pm to midnight, and that each incense stick burns for approximately 30 minutes. Each geisha has twelve incense sticks. She sells the incense sticks one by one according to the customer's requests, and when they are sold out, her time is finished. This method can be expected to prevent troubles such as double booking, the *senkodokei* can be said to be a very useful work planner for a geisha.[244] As we know that Japanese clocks often appeared in ukiyoes (woodblock prints) painted by artists such as Harunobu Suzuki (鈴木春信, 1725–70) and Utamaro Kitagawa (北川歌麿, 1753–1806), a geisha house was a place where people had a strong interest in time. Unlike artisans, who were paid on completion of their work regardless of how long it had taken, geishas were paid for their time almost literally converted into money.

THE ENCOUNTER WITH EUROPEAN MECHANICAL CLOCKS: FROM XAVIER TO IEYASU

The oldest-known record of the importation of a mechanical clock to Japan is that of the clock given to Yoshitaka Ouchi (大内義隆, 1507–51), the war lord of Suo (current Yamaguchi prefecture), by a missionary of the Society of Jesus, Francis Xavier (1506–52), in 1551. A description which seems to refer to it is found in the geographical work, *Ouchi-Yoshitaka-ki* (大内義隆記).[245]

> Many ships come from Tang, India, and Korea. One of the gifts from India, in particular, which controls the length of the twelve hours equally and makes the bell sound, performs the pentatonic and twelve scales without the playing of a 13-stringed harp by a person. [author's translation]

From this, the gift presented by Xavier can be understood as a musical clock. On the other hand, another description that may refer to the same clock is given by Jean Classet (1681–92).[246] However, he describes the clock given by Xavier as 'a small striking clock', which does not match with the clock described in the *Ouchi-Yoshitaka-ki*. Unfortunately, Yoshitaka's clock was burnt along with the palace during a revolt by his vassal, Takahusa Sue (陶隆房, 1521–55), so its type can no longer be confirmed.

---

244 Sasaki 1996; Bedini 1994a, 181–4.
245 Hanawa 1793–1819 Hoki-ichi Hanawa, 'O-uchi Yoshitaka-ki', *Gunsho Ruiju*, vol. 394, Kassen-no-bu, ch. 26. *Cf.* the translation, discussion and account of this period in Hiraoka 2020.
246 Classet 1880 i, 176–7.

Figure 29 An incense clock, *jokoban*. Old Hirai collection, NMNS Tokyo.

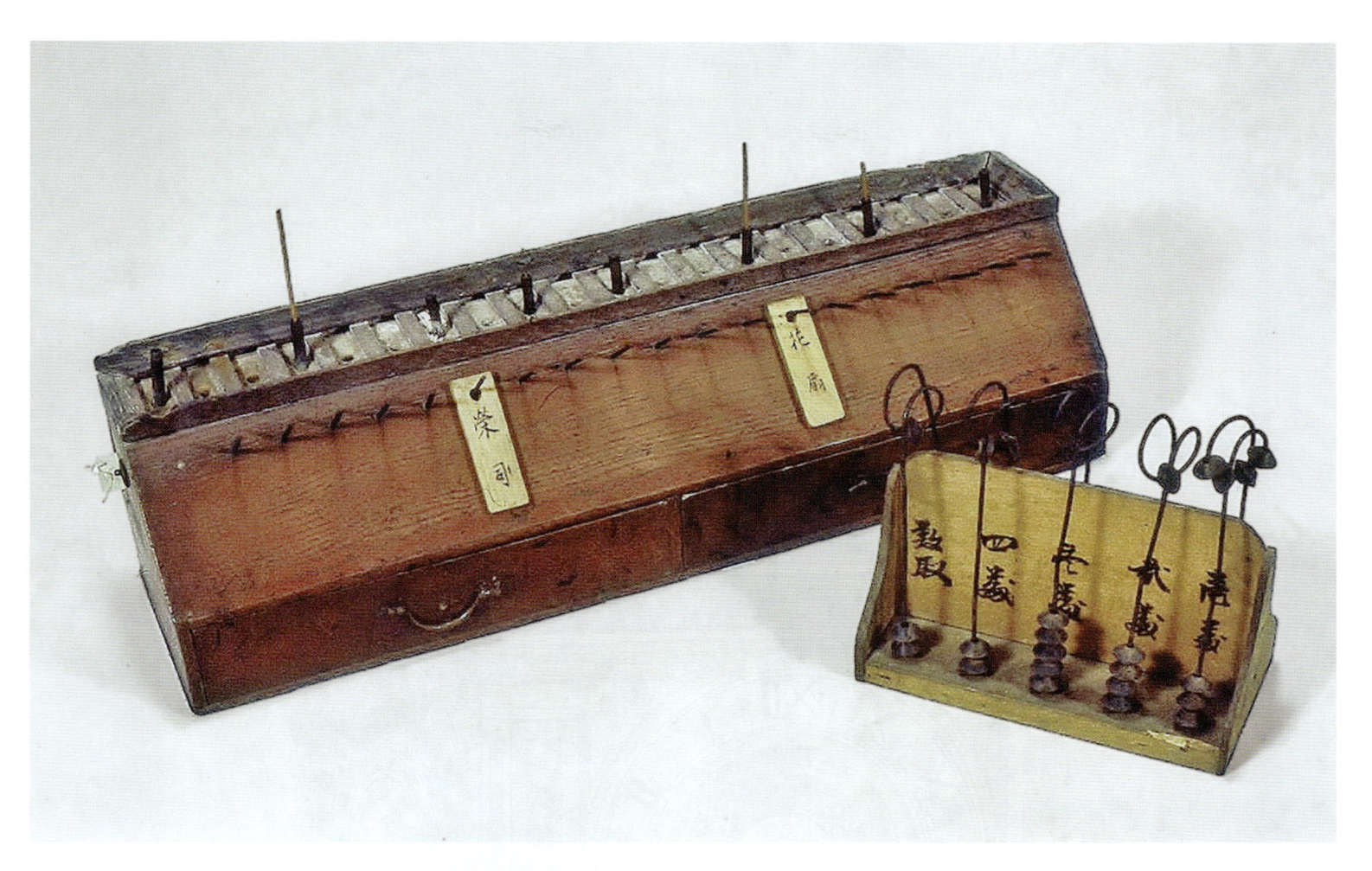

Figure 30 A Geisha's incense-stick clock, *senko-dokei*. Takabayashi collection, NMNS Tokyo.

Half a century after Xavier, in the confusions of the warring states period, the war lord Nobunaga Oda (織田信長, 1534–82) and the first minister Hideyoshi Toyotomi (豊臣秀吉, 1537–98) were eager to unify the country. They also encountered European mechanical clocks. A description of clocks appears in *Japanese History* written by the missionary Luís Fróis (1532–79), who observed Japan first-hand for over thirty years. A clock is mentioned by Fróis at the time of his visit to Nobunaga at Nijojo castle in Kyoto in 1569 to express his gratitude for permission to propagate Christianity.[247] Nobunaga was very pleased when Fróis brought him a small alarm clock that he had requested, but did not accept it, saying that it was difficult to use, and he would break it.

After Nobunaga's death, Hideyoshi—who achieved the unification of the country—was also greatly interested in clocks. The circuit inspector Alessandro Vallignano (1539–1606) and four boys of the Tensho Boy Mission to Europe, which had returned to Japan in the previous year, were called by Hideyoshi to the *Jurakudai* (聚楽第) residence where they had an audience with him in March 1591. Fróis also wrote about gifts to him, including a clock commissioned by the Indian Archbishop and four carriage clocks presented by Prince Mantua. The day after the audience, Hideyoshi called back one of the mission boys, Mantio Ito (伊藤マンショ, 1569–1612), and a Japanese interpreter João Rodrigues (1561–1633) to ask them how to adjust the clocks.[248]

The European table clock kept in the Kunozan Toshogu Shrine is the oldest existing clock in Japan (Figure 31). This clock is spring-driven, given in 1611 to Ieyasu Tokugawa (徳川家康, 1543–1616), the first Shogun of Edo, by the former governor of the Philippines, Don Rodrigo de Vivero y Valesco (1564–1636) to express his gratitude for the rescue of more than 300 sailors from his ship, the *San Francisco*, which was wrecked off the coast of Iwawada in Chiba in 1609.[249]

The signature plate on the front of the clock reads 'HANS DE EVALO ME FECIT EN MADRID A 1581'. In his report on dismantling the clock in 2012, at the request of Toshogu Shrine, David Thompson, curator of horology at the British Museum, revealed some new and interesting facts.[250] Ieyasu's clock is extremely valuable for the following reasons: Flemish clocks from the late sixteenth century are few. Hans de Evalo (c.1533–98), clockmaker to Philip II of Spain, is one of the few Flemish clockmakers known, so any clock by him is important.

---

247 Frois 2000 ii, Chapter 35, 153–4.

248 Frois 2000 v, ch. 28, 121.
249 Hoeve & Thompson 2014, 1063.
250 Thompson & Sasaki 2012.

Figure 31 Ieyasu's table clock. Kunozan Toshogu Museum, Shizuoka.

Ieyasu's clock has had but few modifications, having been kept as a treasure for over 400 years, and is therefore in substantially its original state. Thompson also noted that Flemish clockmakers usually engraved their signature directly onto the movement or the case and did not commonly use a signature plate. He suggested that the plate was added later, hiding another signature.[251]

Following his report, a radiographic examination requested by Toshogu Shrine from the Institute of Electronics, Shizuoka University, revealed that a completely different signature lay under the plate on the bottom of the clock—that is, 'NICOLAVS DE TROESTENBERCH ME FECIT ANNO DNI 1573 BRVXELENCIS'. Nicolaus de Troestenberch (mid-sixteenth century), was the clockmaker to the Holy Roman Emperor Charles V (1500–58).[252] Here, a new mystery emerged as to why Evalo's plate was fixed above Troestenberch's signature.[253]

ACQUISITION OF CLOCKMAKING TECHNIQUES BY JAPANESE CLOCKMAKERS

One source showing how Japanese clockmakers learned clockmaking techniques is the local geographical book, *Owari-shi* (尾張誌), published by Masatsugu Fukada (深田正韶, 1773–1850) in 1844. It states that when a clock presented from Korea to Ieyasu was damaged, Sukezaemon Tsuda (津田助左衛門) of Kyoto not only repaired the clock but also made another one using it as a model and presented to Ieyasu. With this achievement, Sukezaemon was employed as clockmaker to the Owari Tokugawa family in 1598.[254] For these reasons, *Owari-shi* praises him as the first clockmaker in Japan. However, recent research has called the reliability of this statement into question.[255]

In fact, the acquisition of clockmaking techniques by Japanese craftsmen was closely linked with the arrival of Christianity in the second half of the sixteenth century. According to Schilling, seminaries founded in places such as Azuchi, Arima, or Nagasaki had vocational schools attached to them where subjects like oil and watercolour painting, copperplate engraving, printing, organ production, clockmaking, and astronomical instrument making were taught.[256] A priest taught the making of geared clocks at the seminary in Nagasaki.[257] The students not only made clocks with simple mechanisms, but also made astronomical clocks showing the motion of the Sun and the Moon which were presented to the major *daimyos*. One of the astronomical clocks was delivered to Ieyasu at Fushimi castle in 1606 by João Rodrigues, as a gift from the Jesuit vice-provincial of Japan.[258] It is said that some of the students made clocks to make a living. Such practical education in geared clockmaking ended in 1612 when Christianity was banned.

It is certain that Japanese clockmakers learned skills themselves through Christianity, but some, like Tsuda Sukezaemon, acquired their skills themselves. This was particularly so for those involved in gun making. In 1543, a European gun was brought to Japan by the Portuguese and within twenty years domestic gun production began. The many battles fought by feudal loads in the warring states lead to a surge in the demand for guns and Kunitomo in Omi, and Sakai in Izumi prospered as gun making centres. At the end of the sixteenth century, Japan was the largest possessor of guns in the world.[259] Excellent craftsmen gathered in the production centres and advanced techniques such as metallurgy, forging, and metalworking were acquired.

When the Edo period began and peace came, the demand for guns declined and their production almost stopped. Gun making has similarities with mechanical clockmaking, and it is thought that some of the gunsmiths without work turned to clockmaking, since it would be quite easy for a highly skilled blacksmith to restore or make a clock, as did Sukezaemon. Regardless of whether the Sukezaemon story is true or false, Japanese clockmakers may have learned the techniques of mechanical clockmaking by the end of the sixteenth century or the beginning of the seventeenth century either through Christianity or gunsmithing, or both.

The making of mechanical clocks steadily took root, clockmakers were active in major Japanese cities, including Kyoto, Edo, Nagoya, Sendai, and Nagasaki in the second half of the seventeenth century. In the famous guide book *Kyo-habutae* (京羽二重) published in 1685, the names of Musashi Hirayama (平山武蔵), Hokyo Motoza (法橋元佐), Katsuji Miyake (三宅勝次), and others are given as active clockmakers in Kyoto. In the geographical booklet *Zoho-so-Edo-kanoko Meisho-taizen* (増補惣江戸鹿子名所大全) published in 1690, the names of Yumi-cho Tokeiya Riemon (弓町 時計屋理右衛門), Kajibashi Kawa-gishi Oumi-no-kami Motonobu (鍛冶橋川岸 近江守元信), Yumi-cho Ichibei Tanaka (弓町 田中市兵衛), and Kanda Norimono-cho Kita-Yoko-cho Masatsugu Fujiwara (神田乗物町北横丁 藤原正次) are noted as clockmakers in Edo. And in the clockmaker section of the artisan encyclopaedia *Jinrinkunmo-zui* (人倫訓蒙図彙), the names of Kyo Goko-machi Hachiman-cho Agaru-cho Musashi Hirayama (京御幸町八幡丁上ル丁 平山武蔵), Horikawabe Nakatateuri-cho Agaru-cho Motoza (堀川辺中立売丁上ル丁 元佐), Edo Yumi-cho Riemon (江戸弓町 理右衛門), Kajibashi Motonobu (鍛冶橋 元信), and Norimono-cho Masatsugu (乗物町 正次)

---

251 Hoeve & Thompson 2014.
252 Fraiture & van Rompay 2011, 35–6.
253 Sasaki & Saito 2016.
254 Hukada 1832, *Owari-shi*, vol. 4.
255 Kawamoto 2013. A recent survey of this period to 1612 is given by Hiraoka 2020, 212–17.
256 Schilling 1943, 251.
257 Matsuda 1988, 85.

---

258 Classet 1880, ii, 342.
259 Perrin 1991.

are found. This shows that clockmaking was not only established as a profession, but that clockmakers were rather star artisans.

THE UNEQUAL HOUR SYSTEM AND THE IMPROVEMENT OF MECHANICAL CLOCKS

In the early Edo period when mechanical clockmaking began, people used an **unequal hour** system.[260] In Europe, invention of mechanical clocks seems to have accelerated use of the **equal** hour system; in Japan, by contrast, clockmakers adapted the clockwork to indicate unequal hours.

At first, Japanese clockmakers changed from a European dial to a Japanese one and adapted the European count wheel to one appropriate to Japanese bell-striking. Following this, they devised mechanisms to count unequal hours, and installed them in clocks. The mechanism is a double-foliot one, *nichoutenpu-kiko* (二挺天符機構), or a split-piece dial, *warikomashiki-mojiban* (割駒式文字盤) (Figure 32).

The double-foliot mechanism has two foliots: an upper one for daytime, a lower one for night. They switch automatically from one to the other at *ake mutsu* (dawn) and *kure mutsu* (dusk).[261] By adopting this mechanism, it was no longer necessary to adjust the cursor weights on the foliot twice a day, but only to adjust for the seasonal hour changes every fifteen days, namely at the transition between each one of the twenty-four seasons—the *nijushi-sekki* (二十四節気). Just who invented the double-foliot mechanism is uncertain. However, it is most likely to be Musashi Hirayama because of his signature on the double-foliot lantern clock preserved at the museum in Kitakyushu.

The split-piece dial is another mechanism for displaying the unequal hour system. This is a kind of variable dial, which is adjusted to the unequal hour system by metal pieces engraved with the hour name characters sliding in a circumferential groove on the dial.[262] As a metal piece resembles a shogi piece, *koma* (駒) in Japanese, the split-piece is called *warikoma* (割駒) in Japanese. There is little information about who invented the split-piece dial, nor when. The accepted view is that it appeared in the second half of the eighteenth century. This is because split-piece dials tend to be used for lantern clocks with four-legged stands and bracket clocks, and most of them were made of brass and had circular balances and balance-springs, they could thus be considered a century later than iron lantern clocks with double foliots. In the Edo period, such mechanical clocks were called '*tokei* (土圭, 斗鶏, 時計)', '*jimeisho* (自鳴鐘)', and '*jimeiban* (時鳴盤)'. They have been called '*wadokei* (和時計)' since the 1930s, when basic research on Japanese clocks began.

TYPES OF JAPANESE CLOCKS AND SOME NOTABLE EXAMPLES

Japanese clocks can be classified formally as follows: hanging lantern clocks, *kake-dokei* (掛時計), lantern clocks with pyramidal stands, *yagura-dokei* (櫓時計), lantern clocks with four-legged stands, *dai-dokei* (台時計), bracket clocks, *makura-dokei* (枕時計), pillar clocks, *shaku-dokei* (尺時計), and others (Figure 33).

The first three types of lantern clocks seem to differ only in how they are placed, but there are also differences in the mechanism and materials used. In the hanging type, the number of wheels in the train is one less than in the others, which manifests itself in the difference in the drop distance of the driving weight. For this reason, the hanging type needs to be hung high on the wall, whereas in the pyramid type and the four-legged type, their stands are only some dozen centimetres high. In the hanging type and the pyramid type the movements are made of forged iron, while in the four-legged type the movements are made of rather brass.

In a bracket clock the gilt-brass and spring-driven movement are in a rosewood case. The clock is placed it in an alcove or on the display shelf of a Japanese-style room. Most of the movements have a balance wheel with a balance-spring and a split-piece dial. They are also called *daimyo* clocks because only *daimyos* could afford them.

A pillar clock is designed to take advantage of the constant rate of descent of the weight to which a pointer for the hour scale is attached. The clocks have three types of hour scale: the linear split-piece type, *warikoma-shiki* (割駒式), the multi-scale type, drawn for thirteen kinds of hours for the twenty-four seasons on the front and back of seven plates each, *setsuita-shiki* (節板式), and the graph-type, with the hours for all twenty-four seasons drawn graphically on one plate, *namiita-shiki* (波板式). The linear hour scales of the pillar clock have no precedent in European clocks and are unique to Japan.

Other Japanese clocks that should be mentioned include *inro* clocks which placed the movement in a small case of *inro*, *inro-dokei* (印籠時計); table clocks that have a dial on the top, *takujo-dokei* (卓上時計); clocks that served also as a paper weight, *kesan-dokei* (掛算時計); a kind of astronomical clock, *suiyokyu-gi* (垂揺球儀); and a kind of orrery, *shumisen-gi* (須弥山儀).

A notable example of Japanese clocks is the single-foliot lantern clock with an astronomical display made by Musashi Hirayama. This clock is a pyramid-type lantern clock with a total height of 98cm, width and depth of 33cm, conserved in a private collection in Tokyo (Figure 34). On the right side of the front frame pillar of the movement, the signature 'Kyo Gokomach-ju Hirayama Musashi-jo Naganori (京御幸町住 平山武蔵掾長憲)'

---

260 Edo hours are described in the Introduction under 'Hour systems'.
261 Sasaki 2010, fig. 1. Tsukada 1960, 122–2.
262 Edwardes 1996, 173 and Plate VI/6–7; Tsukada 1960, 107–10.

Figure 32 (a) Double-foliot mechanism, (b) split-piece dial.

is engraved. The astronomical display has a Sun hand and a Moon-phase window as a pointer. The Sun hand indicates the hour and the moon pointer indicates the day of the lunar calendar; they also indicate the positional relationship between them. The mechanism is well executed and employs a differential arrangement between wheels of fifty-seven and fifty-nine teeth to indicate the moon phase.[263] Musashi-jo (武蔵掾), which appears in the signature, is a kind of official rank name which Hirayama received on 14 December 1661, and which helps to estimate the date of the clock.

Next, a double-foliot pyramid-type lantern clock, attributed to Sukezaemon, is stored in the Seiko Museum in Tokyo, should be mentioned. The clock is about 36cm in height, 11.5cm in width, and 21.5cm in depth, and engraved with the date 'Jokyo 5 nen, tatsu no 12 gatsu (貞享五年辰ノ十二月)', December 1688.

Throughout the Edo period (over 250 years) from Sukezaemon Tsuda the first, Masayuki (政之) to Sukezaemon Tsuda the tenth, Yoshitomo (良晁), only five clocks are certainly known to have been made by them. These include a single-foliot pyramid-type lantern clock made in 1678, preserved in the British Museum.[264] Although the official work of the Owari Tokugawa family was clockmaking, their actual job was blacksmithing, supervising the making of nails and clamps for ship building. This probably explains why only a few clocks were made. About this lantern clock, from the production date of December 1688, the gold and silver inlay decorations on the front of the case, and a heart-shaped cam used instead of a hoop-wheel to control striking, it appears almost certain that the maker of the clock was Sukezaemon Tuda the third, Nobutsura (信貫,—c.1700).[265]

Other special examples of Japanese clocks include astronomical clocks, *suiyougyu-gi* (垂揺球儀). That described here is one of two *suiyokyu-gi*s in the Ino Tadataka Museum in Sawara, Chiba Prefecture, and was made in 1796 by Tozaburo Toda (戸田東三郎),

---

263 Sasaki et al. 2015.

264 Robertson 1931, 216.

265 Sasaki & Kondo 2008.

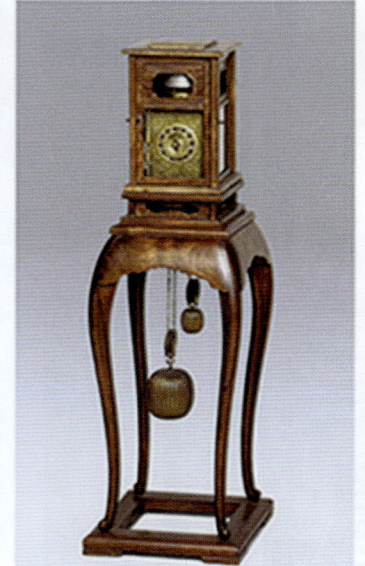

Figure 33 Types of Japanese clocks: (a) hanging clock, *kake-dokei*; (b) pyramid stand clock, *yagura-dokei*; (c) four-leg stand clock, *dai-dokei*; (d) bracket clock, *makura-dokei*, and (e) pillar clock, *shaku-dokei*.

a well-known metal worker at Shijo-Karasuma in Kyoto. It has three decimal dials and at the bottom two small windows displaying the two higher numbers. Together, they constitute a five-digit decimal counter that can count up to one million pendulum swings.

'Suiyokyu (垂搖球)' is the Japanese word for a pendulum; therefore, the word *suiyokyu-gi* means a time measuring device using a pendulum. It was developed by Goryu Asada (浅田剛立, 1734–99), who opened a private school, *Senjikan* (先事館), to study European Astronomy in Osaka during the 1770s, together with his two excellent students, Yoshitoki Takahashi (高橋至時, 1764–1804) and Shigetomi Hazama (間重富, 1756–1816). Goryu and his students initially measured the time of astronomical phenomena such as an eclipse or occultation by counting the oscillations of a hand-held pendulum, following instructions given in the Chinese translation, *Reitai-gishou-shi* (靈台儀象誌), of a work by Ferdinand Verbiest (1623–88). The task of keeping the pendulum swinging for more than a day was arduous, so in order to reduce the work they designed a device that automatically counted the swings of the pendulum, and had it made by Tozaburo Toda.

In March 1795, the Shogunate ordered Yoshitoki and Shigetomi as already well-known, excellent astronomers to move to Edo and work on the compilation of the calendar. They performed astronomical observations using the *suiyokyu–gi* at the Asakusa Observatory founded in 1782, completing the first calendar based on European astronomy in Japan, the *Kansei–reki* (寛政暦). This was enforced from 1798.

Another special example of Japanese clocks is the large pillar clock for astronomical observation, *shojiban* (正時版). The *shojiban* has a large lacquered case 185.4cm high, 36.5cm wide, and 18.2cm deep, but unfortunately the movement is missing (Figure 35). The clock has two scale plates, on which the **unequal** hours are drawn as a graph. It should be noted that *han-toki* (半時), half a *toki* in length (about one hour), called '*yo* (餘)' is inserted before *mutsu* (六つ) of dawn and dusk each, and here *yo* means the time of surplus. This is the unique unequal hour system in which day and night are divided into thirteen equal parts each.[266]

The clock was exhibited as a precision timepiece, *seimitsu-shaku-dokei* (精密尺時計). In 1989, the author and colleagues investigated a *suiyokyu-gi* which was known as the *shojiban-hutenki* (正時版符天機) stored in the Koju-bunko (高樹文庫— current Shinminato Museum) in Imizu-city in Toyama. The possibility immediately arose that it was the missing movement of the *seimitsu-shaku-dokei*. The daily descent distance of the weight of the *suiyokyu-gi* matched the ruler length for a day of the *seimitsu-shaku-dokei*, and documentation in the Koju-bunko showed that

---

266 Sasaki et al. 1989.

Figure 34 The single-foliot weight-driven lantern clock with astronomical display made by Musashi Hirayam. The Katsuyuki Kondo private collection, Tokyo.

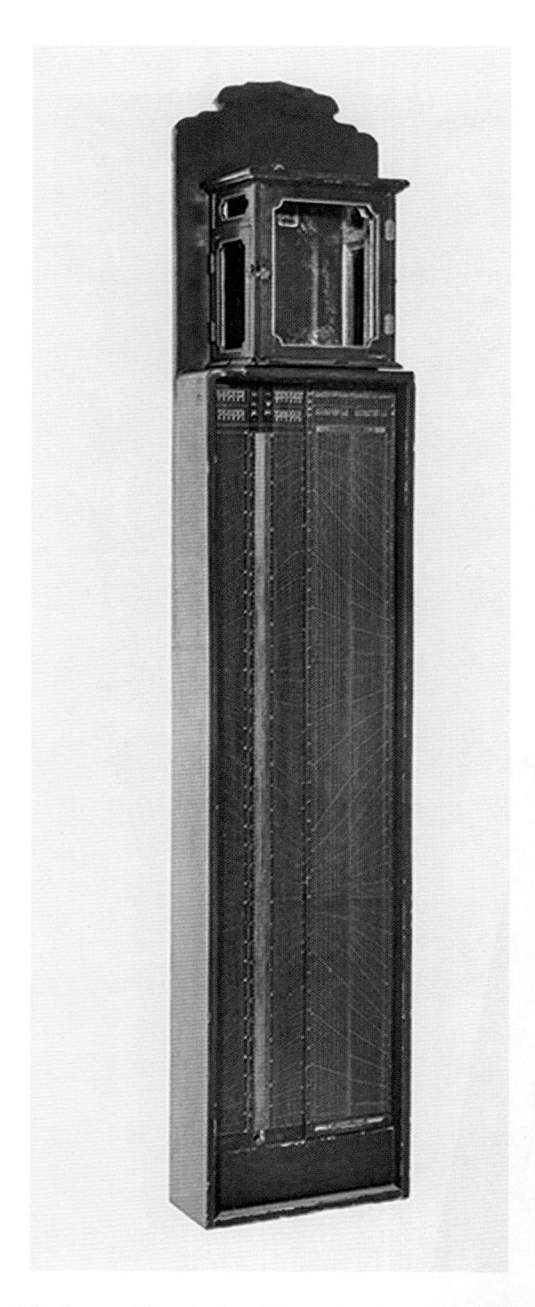

Figure 35 The large pillar clock with precise time scale boards: *shojiban*, with the door-clasp carrying the *umebachi-mon* design on the top case. Takabayashi collection, NMNS Tokyo.

Figure 36 The *suiyokyu-gi*, which must be the movement of the *shojiban*: '*shojiban-hutenki*'. Shinminato Museum, Toyama.

the *suiyoukyu-gi* was used in the revision of the hour system carried out by the Kaga clan. Near the end of the investigation, the Kaga clan emblem *umebachi-mon* (梅鉢紋) was discovered as a door clasp design on the case (Figure 35). This led to the conclusion that the *seimitsu-shaku-dokei* was the *shojiban* used in the Kanazawa castle in 1820s and the missing movement was the *suiyokyu-gi* of the Kouju-bunko.

Originally, the Kaga province used the inaccurate 'thirteen-division' unequal hour system including the *yo* hour. To correct this, the Kaga feudal retainer Takanori Endo (遠藤高環) was ordered to revise the hour system. The revision was done twice. The first revision was to an exact 'twelve-equal-division' unequal hour system in 1823; the second was to a 'thirteen-equal-division' unequal hour system in 1825. The first revision had aroused strong public opposition with the deletion of the *yo* hour—a surplus free hour which people could use as they wished. This is why the revision had to be made again. In the first revision, observational instruments such as the *shojiban* were used, and methods for determining time were established. Because of this, in the second revision, Takanori and his colleagues had to adopt the exact 'thirteen-equal-division' instead of the original obscure 'thirteen-division'. This unique hour system was used in Kaga until 1872, when the calendar was revised by the Meiji government.[267]

AUTOMATIC DISPLAY MECHANISMS FOR THE UNEQUAL HOUR SYSTEM

In the seventeenth and eighteenth centuries, Japanese clockmakers were still trying to improve mechanisms for showing the unequal hour system. An automatic display mechanism for these was invented in the late Edo period, of which three examples are here described: two of them have an automatic split-piece dial and one has an automatic extending and contracting clock hand on a round graph-type dial.

A pendulum hanging clock made by Tadayuki Iwano (岩野忠之), and owned by the Takekawa family in Matsuzaka-city in Mie (Figure 37), has a split-piece drive mechanism on the back of the rotating dial. The split-pieces are attached to twenty-two radial arms and an elliptical disc with twelve slits drives these arms (Figure 38a, b). The elliptical disk is driven by a seventy-two teeth annual gear with an eccentric circular cam. The product of the number of teeth of the annual gear (seventy-two) and the gear ratio of five in the motion work on the back of the dial means that the dial rotates 360 times while the split-pieces go and return

---

267 Sasaki & Watanabe 1996.

Figure 37 The automatic split-piece dial pendulum lantern clock made by Tadayuki Iwano. Private collection in Mie.

Figure 38 The automatic split-piece dial mechanism of Iwano's clock.

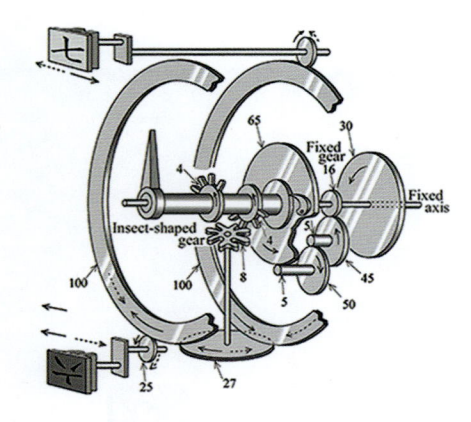

Figure 39 The automatic split-piece dial mechanism of the Myriad Year Clock.

once, namely in this motion work the 365.2422 days of a solar year are approximated to 360 days.[268]

From the signature 'Kishu-ju Iwano Tadayuki (紀州住岩野忠之)', the clockmaker's name, Tadayuki, and the production place, Kishu (current Wakayama prefecture), can be known, but the date is unknown. The gilt-brass movement and the use of a pendulum suggest that the clock was probably made around 1800.

The Myriad year clock, *Mannendokei* (万年時計), on display at the National Museum of Nature and Science deposited by Toshiba (Figure on outer back dust wrapper and Figure 39), is a calendar clock, completed in 1851 by Hisashige Tanaka (田中久重, 1799–1881), with lacquered, mother-of-pearl, and cloisonné enamel decoration. On the upper part of the clock, there are six dials, namely, an automatic split-piece dial for unequal hours, a twenty-four-season *Nijushi-Sekki* dial, a ten-stem and twelve-zodiac *Jukkan-Junishi* (十干十二支) dial, a week *Shichi-Yo* (七曜) dial, a lunar day dial with a moon-phase display, and a European equal hour dial. There is also an orrery with the Sun and the Moon on the top of the clock. The internal mechanism of the split-piece dial, which is unique to Hisashige, is as follows: each split-piece is driven by connecting it to the tip of a crank and the inversion of each crank gear every half a year by a special insect-shaped gear (Figure 39). The dial rotates 365.625 times while the split-piece goes and returns once. This shows that one solar year is approximately 365.625 days.[269]

In 1875, Hisashige Tanaka established a telegraph factory, the Tanaka-seizoujo (田中製造所), in Tokyo. After his death, the factory developed into the Shibaura Seisakujo Co., Ltd, and then merged with Tokyo Electric Co., Ltd to become the major global company Toshiba. It is said that the foundations of the successful modernization of Japanese science and technology during the Meiji period (1868–1912) were laid during the Edo period. Hisashige Tanaka and his Myriad year clock are emblematic of this process. In 'The Myriad Year Clock Restoration–Reproduction Project'[270] conducted by the National Museum of Nature and Science in Tokyo in 2004, the clock was disassembled, examined in detail, and a replica made for exhibition use.[271]

A further example of the automatic unequal hour mechanism is provided by a hanging clock (private collection, Tokyo), made by Arimasa Tatsunosuke Iyo (伊豫辰之助在政), the clockmaker from Aki, in Hiroshima in 1835. This clock (43.3cm high, 15.8cm wide, 10.9cm deep), is designed to indicate the unequal hour on the round graph-type dial by an hour hand that automatically extends and contracts according to the season (Figure 40). The mechanism for the hand is provided by a differential arrangement between wheels of seventy-two teeth and seventy-three teeth. The product of seventy-three and the motion work ratio of five means that the hand rotates 365 times while the hand expands and contracts once. In the mechanism, one solar year is approximated by 365 days.[272]

THE MEIJI CALENDAR CHANGE AND A NEW WATCH AND CLOCK INDUSTRY

When the Meiji era began, exchanges with foreign countries became more frequent and the modernization of currency, weights and measures, the calendar, and the hour system became necessary. The Japanese government changed to the Gregorian calendar with 3 December 1872 as 1 January 1873 and adopted

---

268 Sasaki et al. 2005.
269 Sasaki et al. 2005.
270 *Reports of the Myriad Year Clock Restoration–Reproduction Project*, Edono-monodukuri Research Team, 2005.
271 As at the Aichi Expo, *Ai Chikyuhaku* (愛.地球博), held in Nagoya in 2005, and in the Japanese Embassy, London, 29 October to 1 December 2014, see *AH*, xxxv, 1062. The Myriad Year Clock was designated as a Nationally Important Cultural Property of Japan in 2006 and was certified as Mechanical Heritage No. 22 in 2007.
272 Sasaki & Kondo 2009.

Figure 40 The round graph-type dial clock by Arimasa Iyo. The hand on the scale on the outermost circle indicates the summer solstice. Thereafter it contracts into the central disk until the winter solstice. It then extends outwards once more. NMNS Tokyo, deposited by Katsuyuki Kondo.

the equal hour system. As a result, Japanese clocks which displayed the time in the unequal hour system could no longer be used. Some Japanese clocks were simply modified by using only one of the two foliots and replacing the dial with a European-style one. However, with the calendar revision and the import of high-precision wall clocks, the demand for traditional Japanese clocks was lost and the history of traditional Japanese horology ended with the lantern clock made in 1887 by Risuke Katsu (勝利助, 1818–92) in Nasubigawa in Ena, Gifu prefecture.[273]

When Yokohama port was opened in 1859, many foreign trading-houses set up in Kannai, Yokohama. Over a dozen of them were European and American watch and clock trading companies, including Favre-Brandt & Co. of Switzerland. Imported pocket watches and wall clocks were sold to the public through newly founded Japanese watch and clock retailers. Successful dealers accumulated capital, and within twenty years local clock production began. One such was the clockmaking factory Seikosha, established by Kintaro Hattori (服部金太郎, 1860–1934) in Tokyo in 1892.[274] Throughout the 1890s, watchmaking and clockmaking factories were established one after another in Nagoya, Kyoto, and Osaka, including Seikosha in Tokyo. Typical factories are those of Jiseisha by Ichibe Hayashi (林市兵衛), c.1890, the Aichi Clock & Co. Ltd. by Ihei Mizuno

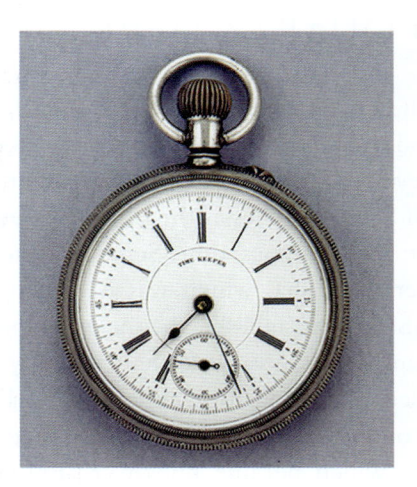

Figure 41 The first pocket-watch produced by Seikosha in 1890, 'Time Keeper'. NMNS Tokyo.

273 Sasaki 2002.
274 Uchida 2002, pt.1, ch. 1, 321–49.

Figure 42 'Cristal Chronometer QC-951' completed in 1963: (a), 'Seiko Quartz Astron 35SQ' (b), released in 1969. Note that the size has reduced from 20cm to 3.6cm in six years. Seiko Museum, Tokyo.

(水野伊兵衛) and others in 1892, and Osaka Watch Manufacturing, Inc. by Masayasu Tsuchio (土生正泰) and others in 1889.[275] However, the price war that resulted from the establishment of so many factories was fierce, and wall clock prices slumped during the recession of 1900. As a result, most factories were shut down; only the production area in Nagoya and Seikosha in Tokyo survived.[276] The Nagoya production area survived the recession by adopting a low-cost, low-quality strategy, using low-cost parts manufactured by domestic subcontractors and compensating for the low quality by careful adjustment. During the Taisho era (1912-1926), Nagoya wall clock production developed into a regional industry that not only dominated the domestic market but also exported sixty per cent of its production.

Seikosha, by contrast, adopted a strategy of maintaining high quality and high prices, with integrated production from parts to the final product, careful planning of production quantities, and selling their products at relatively high prices. Seikosha did not succumb to the momentum of the Nagoya industry thanks to their development of pocket watches, already begun around 1890. The first pocket watch manufactured by Seikosha, imitating a Swiss watch, was a twenty-ligne model, named 'Timekeeper', with a cylinder escapement (Figure 41). In the decades that followed, Seikosha developed into a leading Japanese company.

THE JAPANESE WATCH AND CLOCK INDUSTRY AFTER THE SECOND WORLD WAR: FROM POST-WAR RECONSTRUCTION TO SEIKO QUARTZ DEVELOPMENT

Overcoming the chaos after the Second World War, Seikosha entered a period of high growth. One of the factors that allowed Seikosha to develop into a global watchmaking company was the adoption of Seiko watches as the official timekeepers for the Tokyo Olympics. In the 1960s, three Seikosha factories competed with each other: Seikosha, which was K. Hattori's production division; Daini Seikosha established by the watch production division of K. Hattori; and Suwa Seikosha (currently Seiko Epson), which was based on the factory evacuated by Daini Seikosha during the Second World War. Suwa Seikosha early established a research team which became known as 'Project 59A', since it began in 1959 and worked on the development of quartz timekeeping.

---

275 Uchida 1985, 187.
276 Uchida 1985, 197.

The main goal for the Seikosha Group was to be official timekeepers for the Tokyo Olympics to be held in 1964. Seikosha was in charge of the display board and the large clock in the stand, Daini Seikosha was responsible for stopwatches, and Suwa Seikosha for quartz crystal chronometers. Suwa Seikosha was also aiming at a battery-powered, portable quartz watch. Overcoming many problems, including power saving for battery operation, impact resistance for portability, and many other, the Crystal Chronometer QC-951—20cm long, 16cm wide, and 7cm high, 3 kg in weight, and with a 0.2 seconds daily rate—was completed in the autumn of the year before the games. The name of Seiko became known around the world with the success of the Tokyo Olympics and this experience led to the development of the later quartz wristwatch.

In 1963, Suwa Seikosha participated in the chronometer competition of the Neuchatel Observatory in Switzerland with the crystal chronometer QC-951, being ranked eleventh and thirteenth as first-time foreign makers in the marine chronometer category (Figure 42a). When Daini Seikosha and Suwa Seikosha participated with mechanical wristwatches in 1964, the results were not so good, 154th, and 144th place, respectively. However, the grade improved year by year, and in 1967, Daini Seikosha achieved good results with fourth, eighth, and thirteenth places, while Suwa Seikosha achieved twelfth, twentieth, and twenty-fifth places. These factories developed to a level at which they could compete with leading European makers in just four years. The technology created for the competition was used in the top brands of Seiko wristwatches: Grand-Seiko and King-Seiko.

Five years after the Tokyo Olympics, development of a quartz wristwatch by Suwa Seikosha had also rapidly proceeded. There were various problems, such as miniaturization of the crystal unit, development of a flip-flop circuit, and development of a micro motor to drive the second hand. In response to these, the size of the crystal unit was reduced from the 10cm of the QC-951 to 2cm, the hybrid integrated circuit was developed—a dedicated flip-flop circuit which reduces the oscillation frequency of 8192 Hz to 1 Hz—and a power-saving ultra-small stepper motor was developed with a distributed arrangement for the rotor and stator. On 25 December 1969, at the watch and jewellery shop, K. Hattori & Co. Ltd, Ginza, Tokyo, the world's first quartz wristwatch 'the Seiko Quartz Astron 35SQ', in a gold case with an outer diameter of 36mm (movement 30mm), a thickness of 11mm, (movement 5.3mm), and a daily rate of 0.2 sec, developed by Suwa Seikosha was announced and released for public sale. This was unusual in the trade and made strong worldwide impact. The Astron has changed not only our image of the watch in terms of mechanism and precision, but also our concept of time (Figure 42b).

CHAPTER THREE

LATE ANTIQUITY AND THE MIDDLE AGES

SECTION ONE: SUNDIALS AND WATER CLOCKS IN BYZANTIUM AND ISLAM

Anthony Turner

LATE-ANTIQUITY AND BYZANTINE SUNDIALS

Late Antiquity and the early Middle Ages (thought of here as the fourth to eighth centuries of the Christian era) inherited from the Græco-Roman classical and Hellenistic worlds a substantial body of horological instruments in material form, which was probably accompanied by a corpus of texts concerning their construction. Vitruvius (first century BCE) is quite clear that such a literature existed alongside kinds of dials that he had not described.[1] Ptolemy's discussion of ways of finding celestial angles useful to diallists in his *Peri analemas*,[2] and Cetius Faventinus' text,[3] show that it continued to be augmented. That some of it was known to later diallists is suggested by a description of the determination of a meridian by three shadow lengths measured in the course of a single day given by Diodorus Siculus (first century CE) in his lost *Analemma*, on which Pappus (fl. c.320) wrote a commentary (also lost) in the fourth century, mentioned by Ibrāhīm ibn Sinān (AH 295–334 [CE 907/8–945/6]) and by al-Bīrūnī (AH 362–440 [CE 972/3–1048/9]).[4] The anonymous text (eighth to tenth century) describing a portable vertical **sundial** included in the group of writings that became ascribed to Bede under the name of *Libellus de mensura horologiis* is also likely to be a survivor from the literature mentioned by Vitruvius whose own work remained known in Latin speaking areas of Late Antiquity although virtually unknown in the East.[5]

Both embodied and recorded knowledge of dials was then available throughout the regions of the Late Roman Empire. As this disintegrated from the fourth/fifth centuries onwards into the Eastern Byzantine Empire, the nominally controlled West, and later the Arab–Islamic domains, sundials developed variously, although on sometimes parallel courses. Certainly, there was continuity, and some regional preferences may have been perpetuated. While spherical dials have been found throughout the Mediterranean basin and Italy, conical dials concentrate in the eastern Mediterranean, although a few have been found in Gaul, where the majority of the rather small number of planar dials have also been found. All these dials occur both in public and private environments—marketplaces, temples, baths, and theatres—while open spaces near them were popular sites for erecting public dials. Others have been found in the precincts of private dwellings.[6]

In Late Antiquity we may posit that the sundial, a product of urban civilized life, was an integral part of that life and was incorporated wherever a new town, or even a large urban-style country villa, was built. Because it was accepted, the making of

---

1 *De Architectura*, IX. vii i. 1 Soubiran 1969, 30–1.
2 Ptolemy's *Analemma* was first published with a commentary and an appendix explicitly applying it to dialling by Federico Commandino in 1562. It was edited by Heiberg 1907 (187–223), from the seventh-century MS Ambrosiana L99, Milan and the thirteenth-century Latin translation by William of Moerbeke, Vatican Library MS Ottobono Lat 1850. Modern discussions are to be found in Delambre 1814, Drecker 1925, Luckey 1927, Neugebauer 1975, 839ff, and Gibbs 1976, 109–17. Savoie 2014a; 2014b, ch. 4.
3 Discussed by Pattenden 1979a; 1979b; see also Plommer 1973.
4 Berggren 1985; Kennedy 1959.
5 See Arnaldi 2011a; 2011b. ODLA 2018, ii 1573.
6 On the positioning of dials, *cf.* the remarks by Olszawski 2012, 6–17.

Figure 43 Fragment of a Byzantine geared portable dial. The British Museum, London, Inv. 1997.0303.1

sundials was reduced to rules and precepts and had largely departed from the world of intellectual discourse, whereas in the third and second centuries BCE, it had posed interesting mathematical problems for a small group of Greek intellectuals. It is these two characteristics of dialling that conditioned its very different fortunes in East and West.

Two characteristics of the Byzantine Empire were fundamental for the development of dialling (Figure 43). The first was the continuity it enjoyed with earlier Greek learning. The second was the new organization of Christian life that took place there. The continuity of the Eastern Empire implied a similar continuity for sundials. An accepted object of civic life, there are few literary references to dials and the lack of any technical literature about them means that either this has all disappeared or that a relatively small demand was met by craftsmen in stone and metal whose empirical knowledge was orally transmitted. We do know that sundials were made; eight of the known portable dials surviving from Late Antiquity are inscribed in Greek and include Constantinople in their latitude lists. They therefore post-date the founding of the Eastern Empire.[7]

Where or why they were made is not known, although a military usage has been suggested.[8] The construction of Santa Sophia (532–8) naturally included time measurers. The *Horologion* placed in the south wall towards the west corner was probably a sundial. It was set above the door leading into the chamber of the Holy Well and so into the *Augustaion*, the entrance normally used by the Emperor.[9] Close to the church, but in an independent structure, Justinian caused a large water clock to be built. This, the *horologion* of the *milium*,[10] was also provided with sundials; however, no details of them seem to have been preserved.[11] Other churches were also adorned with dials for one on the church of SS. Sergius & Bacchus is mentioned by Constantine Porphyrogenitos.[12] Of

---

[7] Talbert 2017, ch. 2, numbers 3, 8–12, 15, and 16.

[8] Field 1990, 127.

[9] Paspates 1893, 118–19. Anderson 2014, 24, however, takes it to be a water clock and implies that it was identical with the *horologion* of the *milium*.

[10] The *milium* was the golden marker from which all distances in the Empire were measured. Swift 1940, 187. Anderson 2014, 24, suggests that like the *milium*, the clock which John of Lydos described as the 'clock of the city' had 'official status in the measurement of time'.

[11] Swift 1940, 180. Associating clock and dials would have allowed for twenty-four-hour time telling.

[12] Paspates 1893, 119, n. 4. This basilica had been constructed on the orders of Justinian between *c*.527 and *c*.536.

the form of such dials we know nothing, but it is not unlikely that, like so many of the surviving Greek and Roman dials, they were three-dimensional concave dials with sculpted ornament and/or supporters.

Concave dials were also probably used in free-standing positions mounted on pillars or other kinds of plinth. A few manuscript illustrations show such set ups, and one of them, a sixth-century miniature, derives stylistically either from Constantinople itself, or more probably from the Syro-Palestinian region.[13] If dials were commonly presented in this way in the cities and gardens of the Byzantine Empire, then the lack of surviving examples becomes understandable and suggests that the *horologion* erected by Justin II and his wife Sophia (565–78), mentioned in the Palatine Anthology as having been stolen but recovered by Julian at some time between its erection and the ninth century, was a sundial, rather than a water clock.[14] Free-standing dials of this kind would have been especially susceptible to damage and destruction, gradually by time, but also by siege and pillage in the invasions that resulted in Syria, Palestine, Iraq, and Egypt passing under Arab–Islamic sway in the mid-seventh century.[15] The shocks of this conflict and defeat were little conducive to learning, and it is not until the ninth century, particularly during the reign of Leo the Wise, that we again find traces of activity in the exact studies of mathematics and astronomy. That dialling profited from this revival is suggested by the fact that, in his work on surveying the anonymous author, known as Hero of Byzantium in the mid-tenth century,[16] he states that not only had he laid out a meridian line in the green room of the 'high place' (? = an observing room) of the imperial Boukoleon Palace,[17] but also he has explained how to do this in his treatise of sundials. Unfortunately, this treatise has not survived, but that it existed shows that a literature concerning sundials was available in Constantinople. Because it was available in the mid-thirteenth century, Nicephoras Blemmydes could use the fact that sundials may be drawn anywhere on the sphere of the Earth as a natural illustration in his explanation that the Earth is placed at the centre of the Sun's sphere.[18] By the time that Nicephoras wrote, a startling and remarkable free-standing dial had only recently been destroyed at Constantinople. This had occurred when, during the fourth Crusade, the city was seized by Latin troops in April 1204. According to Nicetas:[19]

> In the Hippodrome was placed the brazen eagle the work of Apollonius of Tyana who, when visiting Byzantium had been asked for a charm against the venomous bites of the serpents which infested the place. For this purpose, he employed all his natural skill, with the devil as his coadjutor, and elevated upon a column a brazen eagle. The wings of this bird were extended for flight; but a serpent in his talons twining round him, seemed to impede his soaring. The head of the reptile seemed to be approaching the wings to inflict a deadly bite, but the crooked points of the talons kept him harmless; . . . The eagle was looking proudly, and almost crowing out 'victory'. . . . But the figure . . . was more admirable still, for it served as a dial, the horary divisions of the day were marked by lines inscribed on its wings; these showed the hours of the day to those who knew how to read these characters when the Sun's rays were not interrupted by clouds.

The ascription of the dial in this passage to Apollonius of Tyana (first century CE) is unlikely, for not only is there no evidence in Philostratus' *Life*[20] that Apollonius ever visited Constantinople, but also the survival of an old monument in that new city is improbable, especially in view of the antagonism of early Christianity to Apollonius 'the holy man', around whose memory an anti-Christian cult had developed in Asia, and especially Syria, during the late third century. What is more likely is that the sundial eagle was erected by a Byzantine Apollonius who was mistakenly identified with Apollonius of Tyana. 'Apollonius the Carpenter, the geometrician', for example, seems to have been the Byzantine author of a brief treatise on a musical automaton associated with extant manuscripts of the *Kitāb Arshimīdas fī 'amal al-binkāmāt*.[21] He, or someone like him, working in collaboration with a sculptor and a brass-founder, could have constructed the sundial eagle.

Early Christianity had relatively little concern for the measurement of time. If the business of man on Earth was there to perfect himself spiritually in preparation for an ultimate union with the divine in the ever-present time of *aion*, absolute time, eternity, then the divisions that would be of interest to him were those of rites and sacraments, not those of the secular world, and still less the arbitrary subdivisions of the day produced by water clocks and sundials. For Tertullian (c.160–c.222), 'there are three hours that are notable in human affairs, that divide up the day, that punctuate business and that are publicly sounded',[22] these being the third, sixth, and ninth hours. These hours (with an obvious reference to the passion),[23] were designated for prayer, but they were not the only occasions for saying prayers. All the other

---

13 Weitzmann 1997, 84–5. See also illustrations 5, 7–9 in Bonin 2011.
14 See Anderson 2014, 25 for references for this episode.
15 On which see Butler 1978.
16 Dain 1933, 16.
17 Vincent 1858, 239.
18 Epitome Isagogicæ, liber II Epitome physicæ, xxviii, 5 in Migne 1863, 142, col. 1272b.
19 Translated from Buchon 1828, 333.
20 Edited and translated by Conybeare 1969. Apollonius was, however, associated with time in Byzantine tradition. See Maddison and Turner 1999, 135, where a list of hour names ascribed to Apollonius is given.
21 Hill 1976, 5, 9.
22 *De Jejeunis*, cap X in Migne, ii, 966. *Cf. de Oratione*, xxiii–xxv.
23 Mark xv, 25, 33, 34. Arnaldi 2000a, 64–8. *Cf.* the mosaic reproduced in Van Cauwenberghe 2012a; 2012b; 2012c, 3, which uses a column sundial to illustrate that 'the ninth hour has passed'. Although annexed by Christianity, these significant hours derive from Roman judicial practice. See Bastien 2012, 4.

prayer-times, however, were fixed by recurrent events, such as before each meal, at sunrise, or sunset.[24] They did not therefore require an artificial means for their determination. In the Christian day, although some such device as a sundial would be needed to indicate the third, sixth, and ninth hours, the twelve-fold division of the Græco-Roman dial was superfluous.

Tertullian's formulation was one of the earliest presentations of the method of organizing the day into periods of prayer that would gradually be adopted as a basic structural element of monasticism. This first appeared in eastern Christianity in the third century, and shortly afterwards in the West. At first anarchic, a degree of order was introduced, in the East by Basil of Cappadocia, Pachomius, and Jerome, and in the West by Augustine and Cassian between about 380 and 425. The rules then laid down were elaborated during the fifth and sixth centuries until, in the rule known as 'of the master' in the first third of the sixth century, and that of St Benedict (530–60), the fullest codification was produced.

An examination of the various rules[25] reveals not only how minutely time-structured were monastic days and nights, but also by how much they varied. In most monasteries, responsibility for time-measurement was given to one or more specified members of the community. That it was the abbot himself who was so charged in the rule of St Benedict[26] reveals the high significance ascribed to it. That all monks should do the same thing at the same time was part of their obedience and loss of personality in the discipline that led them to God. It also gave greater force and weight to their prayers. More important still, lack of punctuality was disrespect to God, and so intolerable. The responsibility of the monk who sounded the time for the offices was therefore heavy. Although two references to use of the *horologium* in the Rule of the Master cannot be taken to apply either to a water clock or to a sundial,[27] and determining time by star positions[28] was probably the most common method used until the eleventh and twelfth centuries, it was not exclusive. Use of sundials in monasteries is not only attested by Cassiodorus' gift of one, together with a water clock, to the monastery that he founded at Vivarium (modern Saletti, near Squillace, Calabria),[29] but dials have themselves survived from religious sites. A conical dial marked with a cross was found in the ruins of the Kastellion monastery founded by St Sabbas in 492 in Judæa. Since the monastery had disappeared by the mid-seventh century, the dial is fairly closely datable to the late fifth or sixth centuries.[30] A similar Syrian dial was found at Bir al-Saba,[31] and others without crosses have also been found in Palestine.[32] A vertical dial inscribed in Syriac and found at Tel-Brisé has been ascribed to the fifth century.[33] From a somewhat later date are two vertical dials from Skripu and from Thebes, and both can be associated with ecclesiastical buildings; the former is dated to c.873/4 and the latter to 876/7.[34] The example from Thebes is similar to yet another dial found in the ruins of the monastery of Arnaut Kevi (Arvanitochari), but for this no date indications are available.[35] Archaeological work in Israel has also revealed dials carved with Christian crosses from the Byzantine period and, in one case, associated with an ecclesiastical site. All are concave dials and were perhaps mounted on free-standing pillars or supports.[36]

The existence of this small group of ecclesiastical dials illustrates that sundials were used in early near-Eastern monastic communities and provides some further evidence for the use of dials in the Byzantine Empire. Barbarian invasions (as they seemed to those who suffered them), would reorient their use. Dials such as that at the Kastellion Monastery, overrun either by Bedouin tribesmen or during the Arab conquests, before the middle of the seventh century, displayed the usefulness of sundials in religion for determining the times of prayers. It was precisely this use that was to be emphasized in the Arab–Islamic lands.[37] The possibility that the example of the Christian use of the sundial for prayer-time indication influenced its adoption for such purpose in Islam should not be discounted, particularly given the correspondence that can be seen between the prayer times of the Syrian Christians and those of Islam.[38]

---

24 Rambaux 1984, 304 who notes that Tertullian's formulation should be related to those of Clement of Alexandria and Hippolytus of Rome. *Cf.* Bouvaert 1984, who conveniently brings these passages together with discussion.
25 Biarne 1984, 99–101.
26 'The indicating of the hour for the Works of God by day and by night shall be the business of the abbot. Let him either do it himself or entrust the duty to such a careful brother that everything may be fulfilled at its proper time.' *Rule*, ch. 47, cited from McCann 1976. On the communitarian totalitarianism of monastic life and its need of time structuring, see d'Haenens, 1980.
27 It was probably a table of the position of the Sun, or of stars, permitting the hour to be estimated. Biarne 1984, 117, or less probably the schedule of psalms or lessons to be said or chanted such as that described in Mateos.
28 For which, see McCluskey 1990.
29 *Institutiones*, XXXV, iv & v; Jones 1946, 135. Water clocks are difficult to trace in monasteries before the late tenth/eleventh century, but Hildemar of Corbie in his mid-ninth century *Expositio regulae*, ch. 8, noted that to find the times of the night offices rationally a water clock was needed. Text and translation available online at *The Hildemar Project*.
30 Marder 1929, 123.
31 Illustrated in Abel 1903, 430.
32 Marder 1929, 124–5.
33 R. D. 1928, 80.
34 Strzygowski 1894. For a good illustration of the Skripu dial, see Zervos 1935, pl. 326. South-facing vertical dials are also to be found in Armenia whence the earliest, that from the temple of Zvartnots, dates from the seventh century; see Tumanian 1974, 97.
35 Mendel 1914, 567.
36 Ben-Layish 1969–71. Two fragments of concave dials found during excavations of the Wailing Wall in strata dating from the second Temple period (first century CE), may, however, be presumed to be late Roman
37 The fullest available survey of Arab–Islamic dialling is that by Ferrari 2012, which covers all the ground sketched out in the present chapter and in far greater detail.
38 King 2004–5, 600–1.

ARAB–ISLAMIC DIALS

In what may be the earliest reference to dials in an Islamic context, it is the use of an instrument to indicate prayer-times that is emphasized:

> ... 'Umar ibn 'Abd al-'Azīz set up [an instrument for finding] hours (*nasaba sā'āt*) in order to know the times of prayer and the remainder of the day. His muezzin used to make the call for the *ẓuhr* prayer if six hours had passed and the beginning of the seventh had been reached, which is when the Sun starts to decline, and [*the muezzin*] used to make the call for the *'aṣr* prayer when the nine hours had passed and the beginning of the tenth hour had been reached. Then he would give the order for the prayer to be performed ... [39]

'Umar ibn 'Abd al-'Azīz (AH 63–101 [CE 682/3–719/20]), reigned at Damascus as Caliph 'Umar II from AH 99–101 (CE 717/18–719/20),[40] although nothing indicates that this story, which derives from his contemporary al-Awzā'ī, relates specifically to the brief period of his caliphate. What it does show is that, as early as *c.*700 CE, an instrument was being used for prayer-time determination, and that since an unusual term *nasaba sā'āt*, is used,[41] it may have been relatively unfamiliar. What the instrument might have been is impossible to determine, but since the times of Islamic prayers are fixed by the position of the Sun, that it was some form of shadow-casting instrument, perhaps even a true sundial, is not improbable.[42] Whether in this case the instrument was an existing Græco-Roman dial that was recycled, or was a new instrument ordered from local craftsmen, must remain an open question. But the story may accurately indicate the moment when the Arab ruling classes first detected the social utility of instruments for regulating the times of prayer. Some slight confirmation of this is found in the remark of an anonymous commentator on a different, later, version of the story given by the eleventh century Andalusian historian and legal scholar Ibn 'Abd al-Barr that there was some resistance to using instruments for regulating prayer-times in his (the anonymous commentator's) day.[43] It should further be noted that what is expected to be indicated by the dial is how much of the day has elapsed, i.e. it is the completion of a period that is indicated, not a specific numerical time. This interest in portions of time as proportions of the day emerges even more clearly from a briefer version of the story also recorded by al-Aṣbāḥī:

> 'Umar ibn 'Abd al-'Azīz set up [an instrument for finding] hours in order to know from them how much of his day had passed on him and how much of it remained and these hours are light [i.e. shorter] in winter and heavy [i.e. longer] in summer.[44]

It has been argued elsewhere[45] that what was required from sundials by their early Christian users was not an exact hour measurement, but an indication that a certain proportion of the daylight period had elapsed. The same expectation seems to have been held of Islamic dials, although the Islamic prayer-times became more specifically defined than their Christian counterparts. Islamic prayer-times consist of bands of time during which the prescribed prayer must be said, the limits of the bands being determined in terms of the apparent position of the Sun relative to the local horizon. This being so it was natural to discuss the prayer periods in terms of shadows, and the sundial was an equally natural instrument to use to indicate them. The study of shadows, explained al-Bīrūnī, 'pertains to the types of mathematical science which [facilitate the solution] of problems of everyone who resorts to religion, depending on the way of excellent truth ... '.[46] To the study of this *'ilm al-mīqāt*, the sciences of timekeeping for religious purposes, al-Bīrūnī devoted an extensive volume, *The Exhaustive Treatise on Shadows*, which summarized the developments of the previous two centuries. In it, although he had a great deal to say about the **gnomon**, he had rather little to say about sundials, despite being aware of other timekeeping devices such as sand- and water clocks. This could mean that even in his time (first half of the eleventh century), dials were not widely used in Eastern Islam,[47] even though a great deal was known to scholars about their theory. Rather the simple gnomon and shadow-length method was employed. Al-Bīrūnī, however, invested it with a symbolic, religious, significance that almost made the method specific to Islam and which must also have coloured the Arab–Islamic attitude to true sundials.

> ... the shadow of the gnomon extends horizontally on the ground, like the kneeler placing his head on the ground, throwing dust on his face, with his shadow moving from one side to another, being carried (by the Sun) from one place to another, and from one side to another, indicating its cause, which is the motion of the Sun from sunrise to sunset. It is among the most mighty of indicators and the most clear (indicator) of them of the prime mover who moves.[48]

Although the ultimate injunction to prayer comes from the *Qur'ān*:

---

39 Al-Aṣbāḥī, *al-Yawāqīt*, cited from King 2004–5, 581.
40 On him, see ODLA, ii, 1540.
41 King 2004–5, n. 39, and private discussions.
42 It could perhaps have been a simple gnomon, but if so, it would not have allowed 'Umar to know either the prayer-times or 'the remainder of the day' without calculation, which is presumably what he wished to avoid.
43 King 2004–5, 582.

44 King 2004–5, 581.
45 Turner 2004, 32–3.
46 Al-Bīrūnī 1976, ii. i.
47 That is, the heartlands of Islam, from Arabia to Egypt and Syria and in Persia, as opposed to Western Islam, composed of al-Andalus (Islamic Spain) and North Africa (the Maghreb).
48 Bīrūnī 1976, i, 42.

> Perform the prayer
> at the sinking of the Sun to the darkening of the
> night
> and the recital of dawn;
> surely of dawn is witnessed

or

> So, glory be to God
> both in the evening hour
> and in the morning hour.
> his is the praise
> in the heavens and the Earth
> alike at the setting Sun
> and in your noontide hour.[49]

The indications there given are not specific. It was only in the course of the second (eighth) century as the Hadith (the traditions of the Islamic community, which combine with the *Qur'ān* to supply its political, legal, and doctrinal basis) were systematized that the prayer times were defined.[50]

The Muslim day begins at sunset once the solar disc has entirely disappeared beneath the horizon. The first prayer therefore is the *maghrib* which should be said during the period of dusk or twilight between sunset and true nightfall. The second prayer, *'ishā*, is to be said after nightfall and before a third or a half of the night has passed. The third prayer, the *fajr*, or morning prayer, is to be said between daybreak and the rising of the Sun. The two day-prayers were both said in the afternoon. The *zuhr* prayer was said in the period running from immediately after the moment of Noon (i.e. from the moment when the Sun had crossed the meridian and shadows were beginning to lengthen) until the shadow of an object equalled its own height. The *'asr* prayer is said during the period between the last moment of the *zuhr* until sunset or until the shadow of an object was twice its own length.[51]

It was of course unequal hours that were shown on the dials of Græco-Roman antiquity, and which the Arabs encountered in the Hellenized lands of the Near East. The story of Caliph 'Umar II suggests a desire to adapt such remnants from a displaced but still admired civilization to the purposes of that, more-righteous-in-the-eyes-of-God society, which had supplanted it. Emulation of the Greeks and absorption of their knowledge and skills of learning was indeed a hallmark of Islamic scholars from the mid-second (eighth century) onwards. Sundials and the theory of sundials partook of this general movement. Nonetheless, for nearly a century little is known about what use was made of sundials surviving *in situ*, nor of what Greek writings on the subject came to the attention of Arab scholars. Although Ptolemy's treatise on the analemma seems never to have been translated into Arabic, it could have been known to Greek-speaking non-Arabs like Thābit ibn Qurra, and an earlier analemma, that of Diodorus (first century BCE), is mentioned by Thābit's grandson Ibrāhīm ibn Sinān and by al-Bīrūnī.[52] Berggren argues that it was the investigation of analemma methods for transforming coordinates in the context of the vertical meridian dial with an east-facing gnomon that led to the development of one method for calculating the azimuth of the qibla.[53] Interest in meridians and gnomons is further indicated by the title of a treatise on this subject, *A Gnomon for the Determination of Noon*, recorded by Ibn al-Nadīm as having been written by al-Fazārī.[54] Ibrāhīm ibn Ḥabīb al-Fazārī (second half of the eighth century) was an astrologer/astronomer who assisted in laying out the plan of the new city of Baghdad for al-Manṣūr,[55] compiled a *zīj* (a collection of numerical tables for astronomical use), based on Indian sources, and according to Ibn al-Nadīm, was one of the first men in Islam to make and write about the astrolabe.[56] If the *mīzān al-Fazārī* mentioned by al-Marrākushī in the late seventh/thirteenth century is to be associated with him he may even have gone further in the investigation of gnomonics.

Although little material has survived from this early period, from it one negative conclusion seems possible. The encounter with Græco-Roman dials does not seem to have led to any great interest in spherical, hemispherical, or conical dials. Instruments surviving from later periods and the treatises confirm that most known Arab–Islamic dials are planar dials. The main exception to this, the cylinder dial, can be considered as a special case of the vertical plate dial. The typical horizontal dial of Islam does, however, strongly resemble the *pelecinium* type among the Græco-Roman dials mentioned by Vitruvius, of which several examples have survived.[57] Therefore, it is possible that this form of dial was consciously taken over by Islamic scholars exploiting mathematical methods for its construction derived from Greek texts. However, since it employs curves that are not easy to construct geometrically, and since the makers of dials in Late Antiquity and early Islam were trying to construct the same thing—a horizontal dial with a vertical gnomon showing the unequal hours—empirical constructions of the dial would necessarily have produced similar looking results. In the course of the day, the rays from the Sun passing over a vertical gnomon form a double cone that is cut

---

49 Arberry 1972, 412, xxx, 16. Ali 1973, 777, xxi 30: 2, 17–18.
50 On this, see Burton 1994.
51 For the prayer times see Wensinck 1913–38; Kennedy 1985; King 2004–5, part IV; King 1980 gives the Turkish forms for the prayer time names. For four different definitions of the *'asr*, see Ryckmans & Moreau 1926.
52 Berggren 1980, 12; Kennedy 1959.
53 Berggren 1980; Berggren 1985. That the problem of the *qibla* (the direction of Mecca from any given place, in which direction one must face when praying), was mathematically analogous with some aspects of gnomonic theory can only have intensified interest in dials for mathematicians and confirmed their respectability as their usefulness to religion was thus underlined.
54 Dodge 1970, 679.
55 Dunlop 1971, 215.
56 Dodge 1970, 649.
57 Gibbs 1976 lists fifteen examples; *cf.* Bonin 2015, 162–3; Jones 2017a, 82.

by the plane of the horizon and must result in a hyperbolic for the solstice lines drawn on a horizontal surface. It is just as likely that empirically drawn dials in early Islam prompted mathematical, theoretical research into their properties as the reverse.

The earliest known treatises on dialling in Arabic date from the early third (ninth) century, and some of them have survived. In *al-Fihrist*, Ibn al-Nadīm listed the following scholars as having written about sundials:

- Muḥammad ibn ʿUmar (Ibn Ḥafṣ) ibn al-Farrukhān, Abū Bakr al-Ṭabarī (early ninth century),[58] who wrote a work called *The Gnomon*.
- Muḥammad ibn Mūsā al-Khwārizmī (d. 850).[59]
- Ḥabash ibn ʿAbd Allāh al-Marwazī al-Ḥāib (ninth century).[60]
- Muḥammad ibn al-Ṣabbāḥ (ninth century), who wrote an epistle *On the Construction of Sundials*.[61]
- Abū ʿAlī al-Ḥusayn ibn Muḥammad al-Ādamī, a scholar working probably in the late ninth century, who wrote a work called *Techniques, Walls and Making Sundials*, which may imply that it was specifically concerned with mural dials.[62]
- Aḥmad ibn Muḥammad ibn Kathīr al-Farghānī (Alfraganus).[63]
- Abū ʿAbd Allāh Muḥammad ibn Ḥasan ibn Akhrī Hishām al-Shaṭawī, who wrote on *Making Oblique Sundials*, which probably treated of declining dials.[64]
- Abū Yūsuf Yaʿqūb ibn Isḥāq al-Kindī.[65]

Even from this incomplete list (Thābit ibn Qurra's dialling works, for example, are not mentioned), it is clear that there was considerable interest, at least in the theory of dialling, throughout the third/ninth century. Some of this interest can be localized since four of the savants mentioned by Ibn al-Nadīm (al-Ḥāsib, al-Farghānī, al-Khwārizmī, and Ḥabash) were associated with the court of al-Maʾmūn (AH 198–218: CE 813–33), and with the *Bayt al-Ḥikma* (House of Wisdom), that remarkable library (*khazīna*) and centre of translation and research, established in the ʿAbbāsid capital. Astronomical observations were performed there, so craftsmen skilled in instrument-making were probably not lacking. It seems reasonable to suppose that some sundials were made. It is perhaps some confirmation of this that of the treatises surviving from this period one—that by al-Khwārizmī—is a practical treatise on construction that supplies tables—for eleven different latitudes—of the coordinates needed to define the points of intersection of the lines for a fixed latitude horizontal dial. A second table supplies a more detailed set of coordinates for the latitude of Baghdad (33°). A third table offers shadow lengths for the unequal hours at the equinoxes and solstices for a polar dial for latitude zero.[66] Such a dial is a polar equinoctial dial, but it can only be presumed that al-Khwārizmī was aware that it could have been rendered universal by a hinge.

Investigation of al-Khwārizmī's tables by David King has shown that some of the trigonometric formulae that underly their construction derive from Indian sources, and he has further identified a copy of al-Khwārizmī's tables in the work on dialling and astrolabe-making by the tenth-century astronomer al-Sijzī. The tables seem therefore to have had some influence, but that al-Khwārizmī calculated them himself or in collaboration with his colleagues in the *Bayt al-Ḥikma*, borrowed them from other scholars (as the attribution of the horizontal dial tables to Ḥabash may suggest), or derived them from an earlier source remains to be determined. That no such tables are extant in Greek mathematical literature is without significance, given that no Greek texts on dialling have survived. Such texts, however, *did* exist and at least one of them was known to Arab mathematicians. This geometrical method has no relation with the computational dialling of al-Khwārizmī. However, both approaches to the subject were known at an early stage in the Islamic world even if methods of geometrical construction were rather quickly abandoned, at least among mathematicians, in favour of computational methods.[67]

Two treatises on dialling by a late contemporary of al-Khwārizmī, Thābit ibn Qurra (?AH 209–88: CE 824/5–900/1), and a fragment of a third have survived.[68] Thābit, a Sabæan from Ḥarrān, and thus native to a city with a tradition of precision instrument-making centred around astrolabes and balances, was a protégé of the Banū Mūsā (early to mid-ninth century) in Baghdad. There he was formed in the traditions of the *Bayt al-Ḥikma* following al-Khwārizmī, al-Farghānī, Ḥabash, and the Banū Mūsā themselves, to become the leading translator (from Greek and Syriac), and astronomer-mathematician of his generation. Dialling, it should be noted, occupies a very minor place in Thābit's output—two, perhaps three, treatises out of forty-one or forty-two devoted to astronomy. The interest of dialling for Thābit, as his most recent interpreter emphasizes, was in the mathematical problems that it poses.[69]

Thābit's treatises are masterpieces of mathematical writing that deal with every possible variant of the problem considered. Thābit's aim, which would be that of many writers on the theory of dialling for centuries after him, was to discover general solutions of the greatest rigour for a problem proposed. The point of

---

58 Dodge 1970, 650 and 1059.
59 Dodge 1970, 652–4 and 1033.
60 Dodge 1970, 653–4 and 991. Al-Nadīm states that he lived for over a hundred years. For a summary of what is known of him and his work (although much concerning sundials seems to be lost), see King 1999, 40–1 and 351–8.
61 Dodge 1970, 633 and 944
62 Dodge 1970, 663 and 944.
63 Dodge 1970, 660 and 985.
64 Dodge 1970, 665 and 1054.
65 Dodge 1970, 619 and 1033.

---

66 King 1983.
67 'No treatises are known which give instructions for making sundials by geometrical construction'. King 1987, ch. I, 10.
68 There is some doubt about Thābit's dates. For a discussion, see Morelon 1987.
69 Morelon 1987, xxxi–xxxiii.

such exercises was supposed to be to supply a standard for practical gnomonics, but one may doubt that the makers of dials could have learned a great deal from Thābit's high-level mathematics. Practical utility was a rhetorical ploy used not infrequently by intellectuals in early Islam to justify their investigations and should not be taken too seriously. The surviving fragment of Thābit's third tract does, however, seem more practical. This, devoted to the construction of single latitude fixed dials (horizontal, vertical, or in the plane of the meridian and inclining or declining in respect to any one of these planes) offers step-by-step instructions to the reader as to how to proceed.[70]

Thābit is not generous with indications of his sources. Although it is clear that in the 'Description with figures ...', he makes considerable use of Apollonius, he mentions him only once,[71] and in the 'Book on instruments ...' he offers no clue to his sources or procedures, although a general debt to Ptolemy's *Analemma* seems indisputable.[72] Basic to Thābit's work is the case of a dial projected on the plane of the horizon, and it is worth noting that, also probably because he was an astronomer, Thābit deals indifferently with drawing lines for equal or unequal hours. This even leads him to discuss a possible case that would result in an equinoctial dial. Because of their later importance, it is worth noting that knowledge of equal hour and equinoctial dials was available in the ninth century; however, it should be stressed that at the time this was purely theoretical knowledge. No one, except an astronomer, would have been interested in equal-hour dials in the ninth or tenth centuries, and that Thābit describes them tends to underline the non-utilitarian, theoretical nature of his work. Equal-hour dials had no function in ʿAbbāsid society.

The treatises of al-Khwārizmī and Thābit ibn Qurra are unusual among early treatises on dialling in that they have been the subject of modern study. Other treatises have been lost, while others remain unpublished and unexamined. Of those mentioned by Ibn al-Nadīm, the treatise by al-Farghānī (early ninth century), which is no longer extant, could have been of particular importance if it had anything like the same scope as his (surviving) treatise on astrolabes, and King has suggested that al-Farghānī may have been the author of some of the dialling tables found in al-Khwārizmī's treatise.[73] The treatises by al-Sabbagh, al-Kindī, and perhaps al-Ādamī are extant but unstudied.[74] A number of other early treatises are also known, such as a treatise on the construction of the unequal hour lines for domes by al-Faḍl ibn Ḥātim al-Nayrīzī (late ninth century, Baghdad), and chapters on sundial calculation in more general works by such well-known astronomers as al-Battānī (AH 244–317: CE 858/9–929/10), al-Sijzī (AH 362–c. 410: CE 973–c.1020), and Ibn Yūnus (d. AH 400: CE 1009).[75]

This early corpus of material in Arabic concerning sundials derives, like the equivalent materials concerning astrolabes, from Eastern Islam. Unfortunately, and unlike astrolabes, there are no early examples of sundials themselves to set against it. Surviving Arab–Islamic dials are extremely rare, but the nine earliest examples known all derive from Western Islam. Not until AH 696 (CE 1296/7), is there any information about a dial from the Eastern Islam, and even this has not survived.[76] The fact that at least a dozen treatises were written on sundials must mean that dials existed to excite the curiosity of mathematicians. Even if this curiosity was initially aroused by Græco-Roman dials surviving *in situ*, it is difficult to believe that investigation of them did not lead to some construction. Instrument-makers were available, spreading out from Ḥarrān, in the later ninth and tenth centuries, and Ibrāhīm ibn Sinān (H 296–339: CE 908/9–950/51) established a general geometrical theory of dials in his *Fī ālāt al-azlā* to replace the ways in which the ancients and their successors had each used their own individual methods to draw a dial on a specific plane. At the same time. manuals intended for artisans were specifically written by the likes of Abū ʿAlī al-Ḥasan ibn al-Ḥasan ibn al-Haytham (H 354–431: CE 965–1039/40, Alhazen in the Latin tradition) and Abū al-Wafāʾ al-Buzjānī (H 328–998: CE 940–998).[77] Possibly, as suggested above, dials were immediately annexed for religious purpose to indicate prayer-times and declined in interest for the regulation of secular, social life. But in a society in which sacred and secular were inextricably mingled, and in which prayer-times structured both religious and secular life, such a distinction is forced. Sundials in mosques were just as useful socially as if they had been located elsewhere. Mosques, however, were frequently rebuilt and dials may have disappeared or been replaced in the course of such reconstruction.

The earliest sundial to have survived from any region of Islam comes from Western Islam—from Al-Andalus. Ascribable to c.1000 CE, it is close in date to the earliest astrolabe extant from Western Islam. The dial, only half of which survives, is signed by Aḥmad ibn al-Ṣawwār or al-Ṣaffār.[78] That he was aware of any

---

70 Morelon 1987, 134.
71 Morelon 1987, ccxxiii.
72 For the relation with Ptolemy, see Luckey 1937–38, who concludes that Thābit's treatise offered a very sophisticated development of the Ptolemaic material. Savoie 2014a, 91, has also noted a similarity with the treatment of the gnomon in the *Surya-Siddhanta* (fourth–fifth century).
73 Implicitly in King 1983, and explicitly in a private communication to the author.
74 In order, Sezgin 1974; 1978. V, 253 & VI, 148; VI, 154; VI, 216–7.

75 If al-Nayrīzī's treatise (Sezgin 1974, vol. 6, 192) treats, as its title suggests, of concave or convex dials of the Græco-Romano type, it would be of particular interest as there are few discussions of this form of dial in the Arabic literature (see Charette 2003, 4.4.1). Al-Battānī's discussion of dials makes up Chapter 56 of his *Zīj*. See Nallino 1899–1907. For Sijzī, see King 1983, 18 and 21, and for Ibn Yūnus, Schoy 1923, F86ff.
76 The dial made for the Ibn Tulum mosque in Caro of which Marcel recorded the fragments; see Sédillot 1844, 56–8.
77 Rashed 1996–2008, v, 685, 687.
78 The first reading of the name is that of Cabanelas 1958. King 1978, 360, n. 10 prefers the reading Ibn al-Ṣaffār and identifies the maker with the astronomer Aḥmad ibn al-Ṣaffār, the brother of the maker of the first

text on sundial construction is unknown, as indeed he is, if his name was really al-Ṣawwār. Whatever the case, his dial is not of the highest quality. That the latitude of 39° 30' calculated from it by King[79] is not too far removed from the standard medieval value for Cordoba, 38° 30' may or may not be significant, but the equinoctial line on the instrument is bent, as are some of the hour lines. Whether these errors arose during the design or the construction of the dial it is impossible to tell, just as it is impossible to know whether al-Ṣawwār/Ṣaffār cut the dial himself or had it carved following his instructions by a stonemason. Whatever the case, although dial making was clearly present in al-Andalus, as presumably elsewhere in the Islamic world at this date, none of the surviving examples is particularly accurate.[80] There seems to have been a considerable gap between theory as expressed in the abstract treatises of the astronomers and mathematicians and practise, as shown by the earliest dials extant.

Accurate, inaccurate, well sculpted, or not, the Andalusian dials display some, or all, of the prayer-lines, as well as the hour lines. One even goes further to include an approximate *qibla* indication.[81] The primacy of the religious function in these dials, which is borne out by their origin in mosques, is thus underlined. All the dials are plane horizontal dials as is also a curiously primitive circular, or semicircular, dial,[82] of which a description is ascribed to Ibn al-Ṣaffār (d. AH 426: CE 1034/5),[83] although this was probably preceded by one by the fourth/tenth century Cordovan scholar, Qāsim ibn Mutarfiff al-Qaṭṭān. The same kind of dial is mentioned in the *Tarbula Jehen* of Ibn Muʿadh (late fourth/tenth to early fifth/eleventh century),[84] and a summary description was given by Maimonides (1135 or 1138–1204), in his commentary on the *Misna, Eben has aʿot* (mid–late twelfth century). As translated by King, this runs:

> A piece of marble (*rukhāma*) is fixed on the ground and straight lines are drawn [as radii] with the names of the hours written on them [to form] a circle. In the centre of that circle there is a nail standing perpendicular [to the plane of the circle], and whenever the shadow of that nail is in the same direction as one of those lines, it is known how many hours of daylight have passed. The name of this instrument which is used by the astronomers, is the *balata*.

This primitive form of azimuth dial may have been developed by analogy with the simple vertical 'protractor' dials, which derive from Late Antiquity, and became the typical Christian dial of the early Middle Ages.[85] Examples of these could have been known in Spain, but the instrument would only have given very approximate results. It might however, as Samso suggests, be an example of Latin influence on the Arab–Islamic tradition.[86]

In the same period (late tenth/eleventh century, as the earliest dials have survived) two linked treatises were written by the well-known scholar Ibn al-Haytham.[87] The first of them was an instruction manual for craftsmen to make horizontal dials 'to know the hour' and to know 'at each moment the portion of the day that has elapsed and what remains'.[88] In the second, placing himself in the tradition of what had already been written about sundials, in particular by Ibn Sinān, al-Haytham attempts a universal theory of dials.

In the absence of surviving examples and with many texts still to be studied, it is impossible to delineate in any detail developments during the following two centuries. That it was not negligible however is shown by the range of dials known to a well-informed *mīqātī* in thirteenth century Cairo. They are described in the enormous encyclopaedia of astronomy, *Jāmiʿ al-mabādī wa-l-ghāyāt fī ʿilm al-mīqāt* (*All the Beginnings* [sc. principles] *and Ends* [sc. results] *concerning astronomical timekeeping for the appointed* [prayer-] *times*), compiled by Abū ʿAlī Ḥasan ibn ʿAlī al-Marrākushī, in Cairo (AH 675–80: CE 1275–1281/2).[89]

Little is known of al-Marrākushī except that he was of Maghribī, perhaps Moroccan, origin, and lived in Cairo for most of his life, where he died before AH 725 (CE 1324/5) and was

---

astrolabe to have survived from Western Islam. Although King's reasons in his n. 10 are unconvincing, he tells us in the 'Addenda & Corrigenda 7 to King XIV that the reading has been confirmed by Julio Samso after inspection of the dial. Mann 1992, 243 also reads the name as al-Saffār, but thinks that this 'does not necessarily indicate that the sundial was made during al-Saffār's lifetime, but that the instrument was copied from one of his [making], or that his name was added as a sign of the excellence of the instrument'. The last suggestion has to be taken as ironic in the light of King's analysis of the shortcomings of the dial. For Aḥmad ibn al-Saffār's life see Saʿīd 1991, 66; Goldstein in *EI* 2, iii, 924.

79 King & Millburn 1978, 362; King 1987, 95.

80 See the detailed analysis of eight dials in King 1992 and discussions in King & Millburn 1978; Barcelo & Labarta; Cabaneles 1958; Carandell 1989; Orus.

81 Found in Cordova and now in the Museo de la Alhambra, Granada. King 1978, 364–7.

82 That described above in note 79.

83 The description is contained in *Kitāb al-asrār fī natāʾij al-afkār* (*The Book of secrets about the result of thoughts*), MS Medicea-Laurenziana Or. 152 ff. 47r-47vn a twelfth-century Andalusian treatise on mechanical devices by Khalad al-Murādī, who has yet to be identified with certainty. The dialling text almost coincides with that of ch. 22 of the *Kitāb al-Hayʾa* (*The Book of Astronomy*) of Qāsim. For a detailed discussion, see Casulleras 1993.

84 Samso 1994 6.

85 Gibbs 1976, 45–6; 350–68; Turner 2004 and Chapter 3, Section 2 in this volume.

86 Samso 1994, 6. In this context it is worth remembering other evidence for the transmission of Latin works in Islamic Spain, the most notable being the translation during the tenth century of Orosius' *Histories* (fifth century CE). See Dunlop 1971, 20–1.

87 For his life see Sabra in DSB and Rashed 1996–2008, vol. 2, 1–18.

88 Rashed 1996–2008, vol. 5, 820, who, at 683–850 of the same volume edits and translates the two texts into French with introductions and commentaries.

89 Charrette 2003, 9.

probably a more than usually well-informed *mīqātī*.[90] His treatise, which describes the whole range of instruments used in astronomy, includes a substantial section on gnomonics. It is entirely practical supplying the reader with instructions how to draw dials using geometrical constructions, which are presented without demonstrations, or by using tables in the *Analemma* tradition.[91] Al-Marrākushī does not expand on the theory of dials, nor on his sources. He mentions few earlier writers by name, although he claims to have corrected and completed them. But if his treatise is of little help in understanding the development of Islamic dialling, it does supply a copious and valuable synthesis of the subject in the late thirteenth century.

The dials mentioned by al-Marrākushī are:

1. The *ḥāfir* (hoof), a fixed latitude horizontal dial with radiating zodiac lines traversed by hour arcs, time being indicated by the tip of the shadow of the vertical gnomon.[92]
2. A universal horizontal dial drawn in the form of a spiral with a single turn somewhat resembling a snail.[93]
3. Universal- and single-latitude cylinder dials, which may be made in hardwood or in brass (see Figure 44).[94]
4. A vertical plate dial known as *al-sāq al-jarāda* (leg of the locust), for single latitudes or universal.[95]
5. Universal- or single-latitude conical dials.[96]
6. A composite instrument called *al-mīzān al-Fazārī* (the Fazārī balance), which is a nomographical device for carrying out conversions concerning shadows and solving problems relating to the Sun and incorporating sundials.[97]

From this first group of instruments al-Marrākushī distinguishes a second group, which are considered more convenient because the cardinal points and the azimuth of the *qibla* are marked on them by straight lines. Dials in this group may be drawn on planar surfaces, horizontal or vertical, in cylindrical or conical forms, or even on hemispheres, concave or convex.[98] Among the variants that he specifies are several worthy of remark, particularly a vertical 'protractor' dial for latitude 0°, a dial drawn on adjacent plane surfaces having a common axis, and equal hour dials of various types. Of one of these al-Marrākushī remarks that 'it makes part of the things not in use that we give in this work as a result of our meditations and reflections'.[99] Clearly al-Marrākushī thought that at least a part of his work was original.

For much however, he draws on his predecessors. Conical and hemispherical dials, like those drawn on plane surfaces, were familiar to Arab–Islamic mathematicians from the earliest times, having been investigated in the ninth century. Thābit ibn Qurra, for example, in a fragment appended to surviving copies of his *Book of the Instruments that Indicate Hours, Called Sundials*, but which is probably independent of that treatise, describes the construction of a dial drawn on adjacent planes set at an angle to each other. Although Thābit's example is drawn for about 34° (anywhere between Baghdad and Samara), and al-Marrākushī's for Cairo, the two dials are basically similar.[100] Although not restricted simply to the business of time-finding, the horizontal instrument, *al-musātara*, or *al-musātira*, which al-Marrākushī describes at the beginning of his section on projections and the astrolabe,[101] should also be mentioned here. The ancestor of the double horizontal dial, current in late sixteenth and seventeenth-century Europe,[102] the instrument employs stereographic projection onto the plane of the local horizon. Al-Marrākushī's is the earliest description of it, but his sources are unknown.

The planar version of the pillar dial, the 'locust's leg',[103] was also extant before al-Marrākushī described two forms of it. A portable version of this dial, signed by Abī al-Faraj ʿĪsā in AH 554 (CE 1159/60) is both the oldest portable dial to have survived from Islam, and the oldest dial of any sort surviving from Eastern Islam. It is inscribed that it was made for Nūr al-Dīn Maḥmūd ibn Zangī 'for knowing the hours of time and the moments of prayer'. He was Sultan of Syria AH 5431–69 (CE 1146/7), a famous anti-crusader, and an exceptional and innovative ruler.[104] The survival of this dial perhaps tells us something about the social class for whom personal dials were made. Clearly, although it has left little trace, there was dialling activity in Eastern Islam. Equally as clear, al-Marrākushī was in part describing dials that were actually made.

The cylindrical version of the vertical plate dial, the cylinder or pillar dial, is described by both al-Marrākushī and Najm al-Dīn, who present it as a standard device.[105] Since Roman

---

90 On him see King in EI 2, vi 582–3. Charette 2003, 1.2. (9–12). King ascribes to al-Marrâkushî an astrolabe signed from Cairo in AH 621 [CE 1282/3] by Hasan ibn ʿAlî, now in the Museum of the History of Science, Oxford (IC N° 107). See Gunther 1932, i 239; Mayer 1956, 46; Brieux & Maddison 2021, HSN ALI 1.
91 Savoie 2014a; 2014b, 93–5.
92 Sédillot 1835, ii 422–30; Charette 2003, 146–8.
93 Sédillot 1835, ii, 430–1.
94 Sédillot 1835, ii, 433–40.
95 Sédillot 1835, ii, 440–50. Al-Marrākushī is perfectly aware of the equivalence of this dial with the preceding one. He states that he regards the dial on the plane surface as 'a development of the cylinder'.
96 Sédillot 1835, ii, 451–7.
97 Sédillot 1835, ii, 458. Sédillot 1844, 46–54, transcribes and translates the headings for fifty uses of the instrument; Charette 2003, 6.2.
98 Sédillot 1835, ii, 475ff.

99 In a gloss, however, Sédillot fils 1844 makes this remark refer to the whole class of equal hour dials, not to one particular type among them.
100 Morelon 1987, 294; Sédillot 1835, ii 540ff.
101 Sédillot 1844, 34, 151–2 and see the discussion in Charette 2003, 2.5 who discusses in detail Najm al-Dīn's description of the instruments and supplies a list of later Arabic texts concerning it.
102 Turner 1982a; 1982b & 1985; Davis & Lowne 2009.
103 On which appellation, see Charette 2003, 3.2.1.
104 Casanova 1923; Mayer 1956, 52–3; Brieux & Maddison 2021; King 1994, 436–7; Turner 2019. For Nūr al-Dīn see Grunebaum 1970, 165–8; Runciman 1951–54, ii 398.
105 Charette 2003, 141–2.

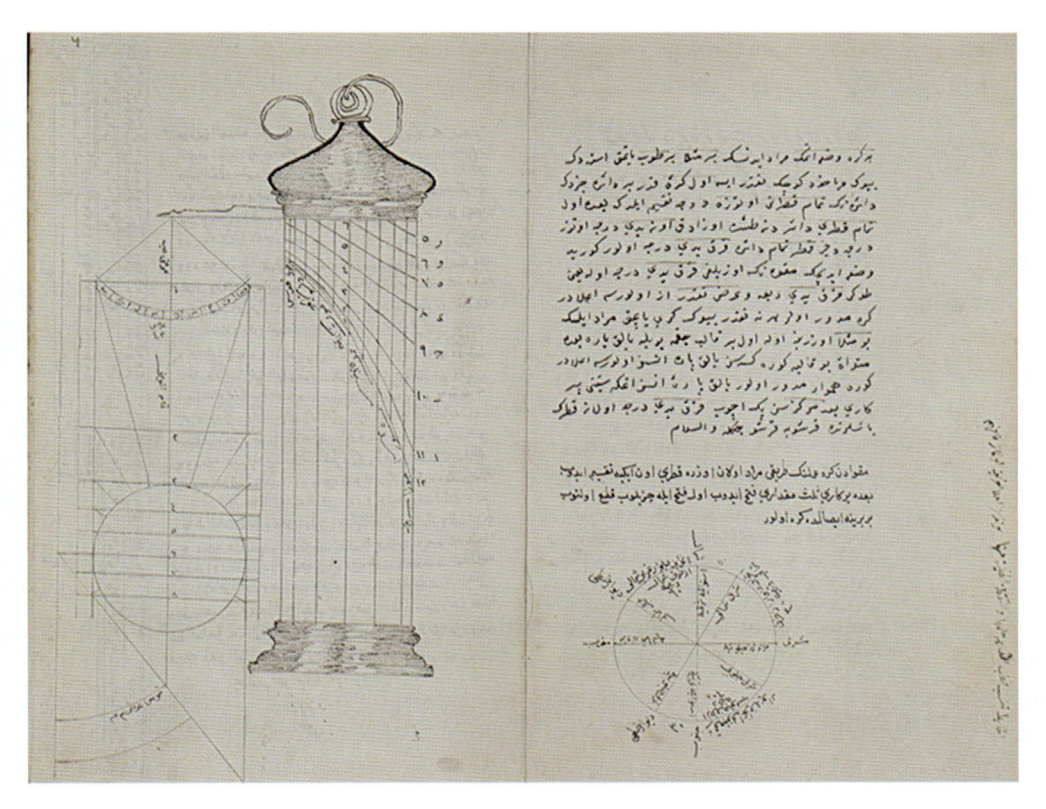

Figure 44 Drawing of an Ottoman pillar dial from an anonymous and untitled collection of notes on instruments. Kandili Observatory ms 39, copied in AH 1057 (CE 1647). Photo: courtesy of the library of KOERI, Boğaziçi University, Istanbul.

examples from the first and third centuries CE are extant,[106] that they continued to be made throughout Late Antiquity into the early Islamic period seems probable.

In AH 747 (CE 1346/7), Aḥmad ibn Muḥammad al-Lamṭī drew a form of a fixed version of the dial reduced to the meridian (noon) line, named prayer lines, and the solstice and equinoctial lines on an onyx column in the Palace of the Victories at Mansura (5km west of Tlemcen). This column was subsequently moved to the Sīdī al-Ḥalwī mosque in Tlemcen, where it became non-functional.[107] Something, however, is known from textual sources of a related dial, the *mukhula*, which is a conical dial similar in construction, though not in calculation, to the classic pillar dial. The body of the *mukhula*, which is a portable dial, is a truncated cone of which the top is covered by a hemisphere with a long lip. This top fitted tightly into the body of the instrument and could be rotated. A suspension ring was set in the top, and there was a metal gnomon that could be folded into the head. The dial is described in two manuscripts—one by ʿAbbās al-Saʿīd, of whom nothing is known, and the other by the seventh/thirteenth-century scholar Ibn Yaḥyā al-Siqillī.[108] Between the two texts, both of which appear to be incomplete, there is an as-yet unelucidated relationship, but that the dial was more widely known is suggested by the fact that a single-latitude version of it is described in an anonymous ninth-century treatise,[109] that it is mentioned in al-Khwārizmī's encyclopædia, *Mafātīḥ al-ʿUlūm* (tenth century),[110] and by al-Bīrūnī in the following century.[111] Al-Siqillī explicitly states that the dial is made of box wood and noted that it could be used in different latitudes if the gnomon length were adjusted. To render the instrument 'universal' he equipped it with three gnomons of different lengths.[112] The device also retained the attention of al-Marrākushī and Najm al-Dīn.[113]

Of the dials mentioned by al-Marrākushī, at least six (the vertical plate and the pillar dial; conical dials; the various kinds of horizontal and vertical planar dials, and their variant drawn on two adjoining surfaces) and perhaps a seventh (if the *mīzān al-Fazārī* is correctly ascribed to the eighth-century astronomer and is a dial) can be traced back in texts to the ninth and even the eighth century. In Western Islam, the evidence of surviving examples shows that dials were made there from at least the fourth/eleventh century onwards and although nothing is yet known to survive from earlier than the mid-sixth/twelfth century, that there was no production of dials in Eastern Islam seems unlikely. That they were drawn directly by craftsmen using traditional step-by-step geometrical methods or were calculated by mathematicians leaving only the cutting of the stone and the calligraphy to masons is not yet determinable, but probably varied in different times and places, as also according to what kind of dial was being made. The relationship between scholars and craftsmen in Islam remains to be delineated, but in a society where learning was

---

106 Hoet-van Cauwenberghe 2012a; 2012b; 2012c, 40–1. This volume Chapter 3, Section 2.
107 Marçais & Marçais 1903, 290–4.
108 Livingston 1992, 299.
109 Istanbul, Aya Sophia MS 4832. Information from David King.
110 Schoy 1923, 72.
111 Livingston 1992, 300.
112 Livingston 1992, 301, 307.
113 Charrette 2003, 142–5.

validated by its utility and conformability with religion, close involvement of a learned man in the making of an instrument, or even the making of it himself, would have been no derogation from his dignity or status. Indeed, as is shown by the case of Ibn Bāṣo (d. AH 709/CE 1309/10), *muwaqqit* of the great mosque of Granada, it could enhance his position.[114]

Muwaqqits were astronomers and mosque officials responsible for determining the times of prayer and qibla directions. They seem to have developed as a specialized, professional group in the course of the thirteenth century and to have originated in Egypt. Similar to them, but usually without an institutional attachment, were the *mīqātī*—professional teachers and specialists in spherical astronomy and astronomical timekeeping.[115] The groups overlap, and both occupied themselves with instruments. Ibn Bāṣo's contemporary, Ibn al-Raqqām (d. AH 75: CE 1315/16), wrote a typical *mīqātī* work—the *Risāla fī 'ilm al-zihāl*—which described graphical methods based on the *analemma* for drawing the hour and prayer curves on dials destined for use in religious settings.[116] The work of al-Marrākushī, *mīqātī* in Cairo, has already been mentioned. Ibn al-Shāṭir (AH 15 Sha'bān 785—Rabī' I 777: CE 1 March 1306–August 1375), an outstanding astronomer, responsible for two of the most remarkable sundials to have survived from the Middle Ages, was a *muwaqqit*,[117] as was his elder contemporary Muḥammad ibn Aḥmad al-Mizzī.[118]

Sundials in Islam had always been closely connected with prayer-time determination. It was this that gave them social relevance. Al-Siqillî considered the chief purpose of the *mukhula* to be finding the correct times for prayer. The fact that dials exist that carry only lines for prayer-times[119] illustrates the lack of importance of the twelve/twenty-four-hour system in social life. With the institutionalization of local time services in the form of *muwaqqits* in the mosques, dials were also consecrated, absorbed totally into that knowledge of natural phenomena employed for religious purposes that makes up the quintessence of Islamic science. Dials are rarely found outside religious buildings and in this they differ from water clocks and automata, which are most frequently encountered in court contexts. Given that the indication of prayer-times was more important than the indication of numbered hours, it becomes clear that mosques and madrasas were their natural habitat. Increasing use of mosques in the thirteenth and fourteenth centuries as places of social intercourse 'for loafing and gossip', which was condemned by *Sunni* writers (traditionalists dedicated to identifying and condemning 'innovations' in Muslim life), gave dials a social role. Indeed this 'socialization' of the mosque, combined with the institutionalization of *muwaqqits*, may have been a factor encouraging the erection of dials. Here, an exceptional dial may be described, constructed exactly during this period, in the Ibn Tulum mosque, Cairo, founded in AH 259 (CE 872–3).

The dial, which no longer survives, was found in fragments by J. J. Marcel (1776–1854), epigrapher to the team of *savants* sent by Bonaparte to survey and describe Egypt in 1798. Assembling the fragments, Marcel was able to reconstruct the dial almost entirely and to take a typographic impression of it. His intention to remove the fragments to safety on the following day was however thwarted by their disappearance overnight. The dial, which was drawn on a stone slab 69cm × 53cm, is a horizontal dial that offers the particularity of having its eastern and western halves superimposed on each other with the north–south line doubled at each edge of the diagram, while the east and west edges form a sort of pyramid at the centre with an inverted 'V' top. Such a procedure allows the width of the dial to be considerably reduced besides giving rise to a pleasing geometrical pattern.

Since the second-largest piece missing from the dal when found fell precisely in the middle of the inscription, the maker's name (if it were given), is lost, although the date AH 696 (CE 1296/7) is clearly marked. The dial carries lines for the signs of the zodiac, unequal hours, and for the time of the *'aṣr* prayer, all bounded in the usual way by the solstice arcs and traversed by the straight line followed by the equinox shadow. Exploration of this dial[120] has revealed errors in the drawing of the line of *'aṣr* (although with some effort to correct them) and in that of the summer solstice. Apart from these, however, it is well executed, employing a particularly elegant *qarmartian* kufic script for the inscriptions.[121]

---

114 On Baso, see Turner (2019) and references there given.
115 King (role), in King 2004–2005, i, part V, 623–77; Sabra 1996, 668–9. For astronomical timekeeping in general, see King 2004, part VII.
116 Carandell 1984.
117 For Ibn al-Shāṭir, see below; King in DSB; Kennedy & Ghanem 1976; Charrette 2003, 1.3.5.
118 Mayer 1956, 61–2; see also Charrette 2003, 1.3.3. Al-Mizzî's instruments were highly prized and highly priced. Five quadrants by him are known to have survived.
119 For example, those in the Sīdī al-Ḥalwī mosque in Tlemcen and that by al-Shaddād in Tunis.

120 By Janin & King 1978, 337–8 and 351.
121 Sédillot 1844, 56, n. 4 whose characterization was literally repeated by Janin and King 1978, 335. Of this script Safadi 1978, 12–13, notes that the name has not been satisfactorily explained:

'Two possible answers may be offered. One is that its name related to *al-Qarāmitah*, a rebellious Muslim movement which extended to many parts of the Islamic empire, including Khurasan in Eastern Persia, where the Qarmatian script was often used for copying the *Qur'ân* and other important religious works. A more likely explanation is purely linguistic: *qramata* is a verb which forms part of an Arabic idiomatic phrase including the word *khatt* (calligraphy), and reading *qarmata fī-khatt*, which means to make the letters finer and write the ligatures closer together. Close ligatures are indeed a feature of the Qarmatian script as compared with standard Kufic. Although extant specimens of Qarmatian Kufic are relatively rare, they are among the most splendid examples of Arabic calligraphy'.

For the Qarmatian sect see Grunebaum 1970, 111–13. Safadi's linguistic explanation seems probable and is strengthened by the evidence of this inscription from a leading orthodox mosque.

Beyond its intrinsic interest, the dial is notable for the originality of its design and for being the oldest monumental dial known from Egypt. One other dial of this type, that by Khalīl ibn Ramtash AH 726 (CE 1325/6),[122] has survived. But that the design continued to be known is shown by its occurrence in a treatise on dialling by Ibn al-Muhallabī in AH 829 (CE 1425/6). Ibn al-Muhallabī was also a *muwaqqit* in Egypt, and his dial was more strictly related to prayer-time determination than that of his predecessors.[123]

The integration of dials into the organized time service of the mosque created ideal conditions for their development. The most extensive treatment of their mathematical construction was given in the early fourteenth century by Najm al-Dīn al-Miṣrī.[124] The position of *muwaqqit* became an attractive one for mathematicians and astronomers, and dials like other instruments offered scope for intellectual curiosity and invention. The eighth and ninth (fourteenth and fifteenth), centuries were a highly productive period for Arab–Islamic dialling. Just as the office of the *muwaqqit* seems to have developed first in Egypt, so Egypt—in particular Cairo—seems to have been at least a local centre for religious timekeeping and dialling. At the same time as al-Marrākushī was writing his exhaustive compendium, Shihāb al-Dīn al-Maqsī, his contemporary in Cairo, produced accurate tables for time-telling and wrote a treatise on gnomonics that included tables for marking out the curves for horizontal dials over a range of latitudes but for vertical dials only in the latitude of Cairo.[125] Since al-Maqsī's tables remained in use until the thirteenth (nineteenth) century, his dialling treatise may also have had some influence—it was, for example, praised by Ibn al-Mahallabī. It may not, however, have reached as far afield as did the work of al-Marrākushī, to whom Ibn al-Mahallabī also referred his reader,[126] whose treatise influenced scholars in Egypt, in Syria, in Rasulid Yemen, and in Ottoman Turkey.[127]

From Syria astronomers travelled to Egypt in search of instruction and al-Mizzī and Ibn al-Shāṭir are said to have studied with the physician and encyclopaedist Ibn al-Akfānī (d. AH 749: CE 1348/9). From him they would have received a broad, general education. Men such as al-Mizzī and Ibn al-Shāṭir should be seen as general scholars with a particular interest in astronomy and mathematics. They were quite clearly not professional instrument-makers or metalworkers in the way that the Andalusian astrolabist Muḥammad ibn Fattūḥ al-Khamā'irī (*fl.* AH 604–634: CE 1207/8–1236/7), or the Persian maker Muḥammad ibn Ḥāmid ibn Maḥmūd al-Iṣfahānī (late sixth/twelfth) century, were. This may explain why only a few examples of work associated with them are known and why they were so esteemed. Generally, it remains mysterious as to whether the works carrying their names were actually made by them, or whether they were produced by a craftsman following their designs and perhaps their verbal instructions.

Ibn al-Shāṭir (AH 705–77; CE 1306–75) claims a place in any history of dials by virtue of the remarkably full, complete, and elegant dial that is associated with him in the Umayyad mosque, Damascus, and for his portable multiple instruments incorporating a dial, the *Sandûq al-yawâqît*. Orphaned at an early age, he was taught inlaying in ivory, wood, and mother-of-pearl by an uncle, and mathematics and astronomy by another. Thereafter he probably studied in Cairo and Alexandria. His earliest known work, an original form of universal astrolabe plate is dated AH 733 (CE 1317/18). At an unknown date, probably no earlier than AH 750 (CE 1349/50),[128] he became *muwaqqit* of the Umayyad mosque. He enjoyed high prestige and was sufficiently wealthy to inhabit one of the finest houses in Damascus.

Ibn al-Shāṭir's primary studies were in astronomy, in particular in planetary theory, but he is associated with two sundials. One of these, a large, fixed dial for the Umayyad mosque, has been described as 'the most sophisticated of all known dials from before the European Renaissance'.[129] The instrument, which according to the inscription, he cut himself ' ... by the hand of he who designed it', was indeed special. Composed of a slab of marble 2.06m × 1.01m, it was set up on a south-facing passage way at the foot of the tower of the *al-'Arūs* (the bride) minaret, and below the balcony used by the muezzins from which it could be observed, as it could also be from a window overlooking it. On the surface of the slab, a huge central dial was engraved, flanked by two subsidiary dials on the north and south sides. The signature inscription was placed along the west edge. The dial showed both equal and unequal hours and, although equal hours were certainly known earlier, Ibn al-Shāṭir's instrument is the oldest known surviving dial to mark them using a polar gnomon so to do.

Ibn al-Shāṭir's dial was a commission from Sultan al-Ashraf Sha'bān (AH 764–78; CE 1362/3–1376/7), not something spontaneously produced by Ibn al-Shāṭir. But by designing such an object he was seen by his contemporaries to have done something special. According to the *Shazanat al-Zahab*, 'Damascus celebrated

---

122 Now in the Victoria and Albert Museum, London. It is drawn for the latitude of Cairo, 30°. See Eden & Lloyd 1900, 180–1; Janin & King 1978, pl. 7.
123 Janin & King 1978, 351–2.
124 Edited with extensive commentary in Charrette 2003.
125 King 1993a, 8.
126 Janin & King 1977, 351.
127 Some indication of the interest evoked by al-Marrākushī's text is given by the note of the copyist of BNF MS or. 147–8, that this was the seventh time he had copied the work. This Paris copy, which was the basis of the Sédillots' work, was based on an autograph copy by al-Marrākushī himself. Given to the Umayyad mosque in Damascus in AH 813 (CE 1410/11), the copy was damaged by fire at an early date but survived to reach the hands of the famous Turkish astronomer Taqī al-Dīn in AH 971 (CE 1563/4). Sédillot 1835, 14.

128 This being the death date of Muḥammad al-Mizzī, whom Ibn al-Shāṭir probably succeeded.
129 King 1994, 439. *Cf.* Janin 1972a; 1972b.

the day of its inauguration'.[130] This was perhaps because dials were unusual, little known, and because the instrument ingeniously combined the equal hours of interest to an astronomer with diagrams allowing the times of prayers to be conveniently read off without calculation. Elegant both in conception and execution, it marks a high point in Arab–Islamic dialling. Nonetheless, it was a very traditional dial, its filiation with Arab–Islamic and Greek forebears apparent.[131] It was with an instrument produced a few years earlier that Ibn al-Shāṭir may have been, as he claimed in the inscription on it, original.

The *Ṣandūq al-yawāqīt* (chest of precious stones) is a small (120mm × 120mm × 3mm) portable instrument which, according to one of its two inscriptions, offers 'the means of knowing the hours of prayer made in an original way [or made and designed] by 'Alī ibn Ibrāhīm Ibn al-Shāṭir ... '. According to the second inscription, it was made for the Royal Library at the request of Sayf al-Dīn, governor-general Munḳalī-Bughā al-Ashraf al-Shamsī, a notable governor of Damascus from AH 764–68 (CE 1362/3–1366/7). Although the instrument is now incomplete, it is clear that originally the box contained a magnetic compass needle used to orient the instrument when it was used to find *qibla* directions or equal hours, via a sighting apparatus (now missing) that rotated over the equal hour scale (expressed in degrees), engraved on the top of the lid which could be inclined to the angle appropriate to the latitude of the place where it was used. For six major cities listed on the sides of the box this setting could be made directly through an apparatus now missing; for other places, a quadrantal degree arc, also now missing, was used. The compass-needle was held in place by a plate slid into the box, engraved on its northern part with the hour diagram for a pin-gnomon polar dial and the position of the *'aṣr* line. On its southern part, the plate carried a semicircular scale for the azimuth of the *qibla* for ten named places. Within the circular scale of hours, marked in groups of fifteen degrees subdivided to three, on the cover there is, apart from one of the signature inscriptions, a series of arcs drawn in stereographic projection labelled 'horizons for all places', and a separate single horizon for Damascus used with some form of *rete* to simulate local phenomena and resolve certain time-related astronomical questions such as the determination of the ascending point of the ecliptic for a given date, or finding the height of the Sun above the horizon.[132]

The *Ṣandūq al-yawāqīt*, AH 767 (CE 1365/6), is the second oldest known Arab–Islamic portable dial, and the oldest known direction dial from perhaps any part of the world. Of the origins of such dials we know nothing, although, since they depend for their functioning on the use of a magnetic compass, they cannot predate the appearance of that device. This, however, still means that the dials could go back to the seventh (thirteenth) century, or even a little earlier.[133] The earliest evidence for dry-mounted magnetic needles occurs in discussions of *qibla* indications and prayer-time measurement, so the context in which the direction dial originated in Islam is fairly clear. Exactly when or where this occurred, however, remains conjectural and the possibility cannot be ruled out that the originality which Ibn al-Shāṭir claimed for his instrument was exactly that of using the compass to produce a portable instrument for finding both time and direction.

Ibn al-Shāṭir was a well-known scholar and not without influence. If his great dial for the Umayyad mosque is traditional, his portable dial seems to stand much closer to the beginning of a tradition. That tradition—of small portable direction dials—developed at about the same time as, and in parallel with, similar developments in Latin Christendom, but was never to attain the same importance. In Europe, time was already in the fourteenth century becoming secularized. Need was beginning to be felt in mercantile and civic life for a system of day and night time-markers more complete than that offered by the traditional indications of the church. The availability of portable timepieces, whether mechanical or gnomonical, enabled time to become personal and secular. As a result, different types and styles proliferated. In Islam such was not the case. Here, the unity of religious and secular life, the integration of each in the other, meant that the religious divisions of the day and night remained primordial and supplied a structure strong enough to serve both its own religious purposes, and secular needs. The concept of personal time, with its concomitant personal timekeepers, had less meaning in Islam than in Christendom and portable dials were correspondingly less numerous. Such as there were, moreover, retained religious functions, a fact reflected in the characteristic graduation of 'hour scales' in degrees of hour-azimuth, rather than in numbered hours. Periods of time were important for the location of the moments for prayer, arbitrarily numbered grids of hours imposed on day and night, were not. That Arab–Islamic dials almost routinely include *qibla* indications is also a reflection of this situation.

Ibn al-Shāṭir wrote a treatise on the *Ṣandūq al-yawāqīt*, parts of which have survived.[134] This was known and used by the late ninth/fifteenth century Egyptian astronomer Shams al-Dīn Muḥammad ibn Abī al-Fatḥ al-Ṣūfī.[135] Shortly before al-Ṣūfī wrote, however, 'Abd al-'Azīz ibn Muḥammad al-Wafā'ī al-Mīqātī (d. AH 876. CE 1471/2), also working in Egypt, made himself a (slightly adapted) version of the instrument.[136]

Al-Wafā'ī, a freelance specialist in astronomical time-telling (hence his *nisba* of *al-Mīqātī*), may have written on a form of

---

130 Reich & Wiet 1939–40, 70.

131 Simplified versions of it were made in later centuries. See Ferrari 2009, 4.

132 For more detailed descriptions of the instrument see Reich & Wiet 1939–40; Janin & King 1977.

133 Turner 2019.

134 Published and translated by David King in Janin & King 1977, 191–6 and 243–47.

135 Published and translated by David King in Janin & King 1977, 196–8 and 248–50.

136 Part of it survives in the History of Science Museum of the Kandili Observatory, now part of the Boğaziçi University. On al-Wafā'ī, see Charrette 2003, 1.3.9.

inclining dial,¹³⁷ and he developed two forms of equinoctial dial the *muqawwar* and the *dā'irat al-mu'addil* (equinoctial semicircle). Primarily intended as a universal instrument for measuring hour-angles; the latter was to remain in use into the twelfth/nineteenth century. It consists of a semicircular ring graduated either in hour-degrees or directly in hours, hinged to a circular or semicircular base, with which it has virtually the same diameter, carrying an inset compass. The hour ring may be set against a latitude quadrant mounted on the meridian line of the instrument and an alidade with a slotted arched upper sight is pivoted at the centre. Usually the base carries *miḥrāb* indications radiating from the compass.

A practical instrument with no detached parts to get lost, the *dā'irat al-mu'addil* was described by al-Wafā'ī himself¹³⁸ and in the mid-sixteenth century by the Turkish admiral Seydi Ali Reis (d. 1562).¹³⁹ Other unpublished descriptions survive, as do several specimens of the instrument itself. Although not ubiquitous as painted on wood horary quadrants would become in the eighteenth and nineteenth centuries, the *dā'irat*, with the pillar dial, was one of the more widely diffused gnomonic instruments of the Ottoman Empire.¹⁴⁰ Large fixed dials continued to be made, such as the notable vertical dial made by 'Alā al-Dīn 'Alī ibn Muḥammad al-Qūshjī (d. AH 879; CE 1474) for the *kulliye* (place of higher learning) of the Fatih Mosque, Istanbul, in AH 878 (CE 1473–4).¹⁴¹ Most of the (rather few) fixed dials that have been recorded are not signed and are found in mosques or madrasas, only rarely in other public spaces.¹⁴² Rather little is known about their production perhaps because in the Ottoman Empire dialling held a recognized, but unobtrusive, place. It was not particularly emphasized in teaching, and daily life, but neither was it discouraged.

This being so, as diplomatic and commercial contacts developed in the Early Modern period between Europe, the Ottoman Empire, and Persia, so Islamic dialling responded to new models. The most important of these was the adaptation of the European compass dial, which conveniently indicated both time and direction, to the specific needs of Islam. This, a cylindrical box, usually of less than 10cm diameter, that opens into two, almost equal parts, contains a compass with a pivoted needle in the lower half over which an open plate engraved either with an hour scale or an azimuthal, hour-angle, degree scale and carrying a folding gnomon is placed. Tables giving the angular distance of Mecca from different towns are engraved on the inner and/or outer surfaces of the lid. Such dials apparently originated in Persia whence there was considerable intercourse with Europe during the Safavid period, and from there spread to other parts of the Islamic world where they would continue in use until the mid-nineteenth century. During this period other European dials would also be introduced.¹⁴³

Although dialling activity under the Ottomans hardly compares in fecundity with that of earlier periods, it should not be dismissed as without interest. Quite complex fixed planar dials continued to be erected—a notable polar dial was constructed at this time by Ibrāhīm al-Faradī al-Kurdī in AH 1201 (CE 1786/7) in the mosque of St John of Acre.¹⁴⁴ Dials carrying only the lines of prayer are known¹⁴⁵ and there was also activity among portable dials. Many more pillar dials (Figure 44) than have survived were probably in circulation and cartographic *qibla* indicators, sometimes incorporating a dial, were also produced. These instruments, painted on lacquered wood—a technique characteristic of the Ottoman period—are usually protected by a thin wood or leather box. If to these one adds the various forms of *qiblanumā*, the *dā'irat*, and the ubiquitous horary quadrant, it becomes clear that portable dials, when required, were not lacking in the Ottoman Islamic world, and were probably commercially produced.

Quadrants punctuate a thousand years of Islamic time-finding. Although they are not sundials as defined at the beginning of this chapter, they challenge a place here as widespread horological instruments. Four groups have been distinguished:

1. Sine quadrants.
2. Horary quadrants either universal or for use in a single latitude.
3. Astrolabe or almucantar quadrants.
4. The *shakkāziya* quadrant.

The first of these are trigonometric devices that solve graphically problems relating to spherical trigonometry and so *inter alia*, those concerning time. The second and the third find time more directly by positioning a bead-marker, mounted on a plumb line, according to the position of the Sun along the ecliptic, and allowing the plumb line to fall freely so that the position of the bead among the hour lines gives a time reading.

Like the sine quadrant, horary quadrants probably derive from ninth-century Baghdad. The known treatises describing them are all anonymous, although one, describing the single latitude quadrant, has been associated with the circle around al-Khwārizmī.¹⁴⁶

---

137 For this, see Janin and King 1977, 215.
138 The description is edited with an English translation by Tekeli 1962.
139 See Brice, Imber, and Lorch 1976, and the review of this work by King 1979. See also Dizer 1977.
140 For the pillar dial, see Naffah 1989; Danisan 2020. For the *dā'irat*, Maddison 1997, 277–8; Rohr 1988.
141 For which, see Turner 2019, and references there given.
142 For an initial survey, see Unver 1954. Of the forty-three dials he lists, only eleven are signed. Dials on mosques in Istanbul are surveyed by Meyer 1980. According to Ferrari 2012, 10, some ninety-four public dials from throughout Turkey are now known, the two oldest dating from 1409 (Konya) and 1470 (Fatih Mosque, Istanbul).

143 For these exchanges see Turner 2019.
144 Michel & Ben Eli 1965. The dial was specially designed for the mausoleum of Hajj Ahmad Pasha al-Jazzār who had commissioned it. For examples of eighteenth and nineteenth fixed planar dials see Savoie 2014a; 2014b, 102, 105, 107, 110–13, 116–17.
145 Ferrari 2012, 7–9.
146 King 1983, 30–1; Charrette 2003, 108. See also Lorch 1981.

Several variant forms of horary dials are described in the treatises of al-Marrākushī and Najm al-Dīn in the thirteenth and fourteenth centuries.[147]

While the *shakkāziya* quadrant, developed in fourteenth-century Syria, was perhaps not broadly used, the third device—the astrolabe- or almucantar-quadrant—however, was to become one of the most widely used time-finding instruments of the Ottoman Empire.[148] Derived from the astrolabe, the earliest known description of it is contained in a twelfth-century Cairo manuscript in which the author does not claim invention of the device. Thereafter descriptions of it are known from fourteenth-century Syria, as are five fine examples in brass by al-Mizzī.

However, these are exceptional. The typical astrolabe–quadrant (Figure 45) from the Ottoman period was made of wood, the scales and inscriptions being drawn in black and red ink on an orange ground, either on paper, or directly onto close-grained wood, while the edges of the instrument were usually coloured red and could carry instructions for use or other inscriptions (dedications, ownership, or a maker's formula). The faces of the instrument were lacquered. Attractive, and less expensive than a full astrolabe in brass, such instruments normally carried a sine-quadrant on one side, and the astrolabe-quadrant on the other. There was, however, some variation, while an unequal hour diagram was frequently drawn in the apex of the quadrant.

Widespread fascination with dials leading to amateur activity in their design and construction was less common in Islam than in Europe. Only very occasionally does one encounter untrammelled enthusiasm for dialling. An example is provided by Aḥmad Pasha, who was appointed governor of Egypt in AH 1161 (CE 1748). Fascinated by mathematics, he sought out in Cairo the noted astronomer and man of letters Ḥasan ibn Ibrāhīm ibn Ḥasan al-Zaylaʿī al-Jabartī (AH 1110–88; CE 1698/9–1774). With him he studied the art of dialling in which Ḥasan was well skilled his knowledge deriving from earlier studies—he owned, for example, half of Najm al-Dīn al-Miṣrī's enormous set of 'tables for the hour-arc' (c. AH 700–30; CE 1310–39).[149] He made several monumental dials in stone and marble that were set up in mosques throughout the city although none of them, apparently, has survived. Delighted by his lessons, Aḥmad Pasha made gifts of some value to his teacher as well as himself designing and cutting a number of dials. Of these three are known.[150]

If the most notable, Aḥmad Pasha, was not al-Jabartī's only pupil. The names of two others—Muḥammad ibn Ismāʿīl Nifarawī and Maḥmūd ibn al-Ḥasan al-Nishī, a *muwaqqit*—are known. Even so, with the disappearance of al-Jabartī's own dials, only a few astronomical treatises being left of his learned labours, it is unlikely that anything much would be known of this gnomonic activity in the mid-eighteenth century were it not that al-Jabartī's son, ʿAbd al-Rhaman ibn Ḥasan al-Jabartī, became a historian and wrote at length about his father in his biographical memoir of his own times. Probably it was knowledge of his father's interest in dialling that led the younger al-Jabartī to note the dialling activities of one of the French astronomers during the occupation of Cairo 1798–1801:

> The astronomer Tūt [? = Nouet] occupied the house of Ḥasan Kāshif Jarkis. In the upper court of the house he drew on the entire paved surface, the usual lines to indicate the degrees of the hours up to midday. Instead of a rod gnomon, he placed on the front of the house itself, a circular plate pierced with several holes in such a way that the rays of the Sun, passing through fell on the lines and their divisions in degrees. So, one can know how much time remains before noon, what are the signs of the zodiac, of which the signs drawn make the position of the Sun exactly known each month.
>
> He also drew, on the upper wall of the ground-level courtyard between the two houses, a sundial with a bar-gnomon to show the hours before or after the middle of the day. But this dial is not like ours which indicate the time of *ʿaṣr*, the degrees of the hours until the setting of the Sun, the arc corresponding with dusk and dawn, the direction of the *qibla*, the divisions into degrees etc., in order to determine the times of prayers. Since the French are not concerned with these indications, they do not bother with them.
>
> The same astronomer also drew several lines on the surface of a square plate of brass. He set this plate on a column a little less tall than a man in the middle of the garden. The bar-gnomon is replaced by a triangle of iron, of which the shadow of the point is projected on the lines divided in degrees. It is a very fine piece with all the indications necessary around the edge. The maker's name is written in Arabic of fine and beautiful script, engraved like the brass and decorated with silver inlay.[151]

Al-Jabartī's account points up neatly some of the major differences between Islamic and European dialling. Traditional dialling did not disappear from the Arab–Islamic world in the twelfth/nineteenth century. Instruments were made and used, prayer-times respected. A science that is fully integrated into its community, functions, continues, serves, but has no apparent

---

147 Charrette 2003, Chapter 3.
148 On astrolabe-quadrants see King 2004–5, ii, 77–80; Charrette 2003, sect. 2.4.
149 Charette 2003, n. 114. Hasan owned Charette's MS A.
150 Mayer 1956, 38.
151 Translated from the French version by Cuoq 1979, 134–5. The vertical dial that al-Jabartī mentions is very well shown in a watercolour (preserved in a private collection) of the inner courtyard of the house, by Nicolas-Joseph Conti (1755–1805), one of the artists attached to the expedition. It was briefly visible in 1998 in the context of the exhibition at the Muséum nationale d'Histoire Naturelle, Paris, *Il y a 200 ans, les savants en Égypte*. See also Bret 2019, 218–19.

Figure 45 An Ottoman astrolabe-quadrant, face, with its case. Photo: Jean-Baptiste Buffetaud.

history. Even so, the old tension between astronomers and legal scholars as to how *al-mīqāt* should be conceived could still surface. A fine marble dial in the al-Daqqāq mosque in Damascus carries an inscription that tells us that it was specifically calculated by ʿAbd al-Qādir ibn Muḥammad al-Ṭanṭāwī 'on the visual horizon' in order to satisfy those scholars who objected to its being drawn on the astronomers' virtual horizon.[152]

---

152 Brieux & Maddison 2021, ABD QUDR TNTAWI. The dial was completed at the beginning of *Rabīʿ I*, 1305 [CE early November 1887].

In Islam, due to their specific usefulness in mosques, dials were not displaced as time tellers as early as they were in Europe. They continued to function and to be made. They even found incarnation in printed form when printing developed in Islamic society towards the end of the nineteenth century. In AH 1303 [CE 1885/8], Aḥmad Mukhtar (Pasha's) *Riyāḍ al-Mukhtar mīrʾāt al-mīqāt wa-l-adwār majmūʿāt al-ashkāl* was published at Bulāq (Cairo), although it had been written in Istanbul. Primarily concerned with practice, the work also shows some traces of antiquarian interest in dialling, a reflection perhaps of the discovery of the 'Middle Ages' by Ottoman historians and of the intrinsic interest of the past.[153] A late dial is that realized by Ahmet Ziya Akbulut for the Bayazīt *madrasa*, Istanbul in 1916.[154] Thereafter, there is little to record of Arab–Islamic sundials except their gradual abandon and neglect for most of the fourteenth (twentieth century). Their fundamental prayer-time indicating functions being now usurped by printed year-diaries and wall-calendars, times for individual days being also printed in daily newspapers, the obsolete object, totally so in the secular, modernizing society of post-1923 Turkey, did not even challenge a place as an antiquity and was ignored.[155]

However, it was not totally forgotten. Just as in the West dialling as an intellectual recreation and as a pedagogical tool enjoyed a renaissance in the closing decades of the fourteenth/twentieth century, so, at least in Iran, a new interest has developed, in part inspired by study of the history of Islamic and Iranian science, and in part as an offshoot of the development of astronomy to which dialling is a natural adjunct, and which serves as a pleasant way of teaching basic mathematical techniques and astronomical concepts. As a result, dialling in Iran has enjoyed a renewal based on the long tradition recorded in early manuscripts and modern studies of them, and on contemporary investigations of dialling in Europe and North America. A general treatise on the subject in Persian was published in 1985 by M. A. Ahya'I, and a sundial kit has been prepared by Muhammad Bagheri who, more recently, designed an analemmatic dial with a human gnomon in the National Park, Bustan-e Mellat, at Rasht in the province of Gilān. Here the Thāqeb Astronomical Society planned a 'sundial park', for the effecting of which on the 27 September 2002 a 'sundial group' was inaugurated within the society.[156]

The long tradition of dialling in Islam has been revived and adapted.

WATER CLOCKS IN THE BYZANTINE WORLD

Like sundials, techniques for the construction of water clocks derived from Hellenistic Antiquity. Unlike sundials, some of the literature embodying these techniques survived in writings by Ctesibios (*fl.* 270–250 BCE), Archimedes (*c.*287–213 BCE), Philon of Byzantium (*fl. c.* 230 BCE), and Hero of Alexandria (first century CE),[157] and was accompanied by such tangible remains as the Tower of the Winds in Athens (late second/first half first century CE), a *horologion* in the forum at Antioch, or the monumental automaton clock displaying the labours of Hercules at Gaza that would be described in the early sixth century by Procopius. Such devices required maintenance, probably supplied by mechanicians such as Pappus (late third/early fourth century) evoked:

> Now the mechanicians of Hero's school tell us that that the science of mechanics consists of a theoretical and a practical part. The theoretical part includes geometry, arithmetic, astronomy and physics, while the practical part consists of metalworking, architecture, carpentry, painting, and the manual activities connected with these arts. One who has had instruction from boyhood in the aforesaid theoretical branches, and has attained skill in the practical arts mentioned, and possesses a quick intelligence, will be they say, the ablest inventor of mechanical devices and the most competent master-builder.[158]

Universality, however, is only rarely possible. It is better to specialize. Among the specialists, Pappus includes those

> ... who contrive marvellous devices [...] Sometimes they employ air-pressure as Hero in his *Pneumatica*; sometimes ropes and cables to simulate the motion of living things, e.g. Hero in his works on *Automata* and *Balances*; and sometimes objects floating on water, e.g. Archimedes in his work *On Floating Bodies*, or water clocks, e.g. Hero in his treatise on

---

153 Strauss 2012.
154 Hitzel 2012, 17.
155 For an illustration of an abandoned dial, see Savoie 2014a; 2014b, 109. *Cf.* King & Millburn 1978, who reports a sundial gnomon twisted to carry a telephone wire in Istanbul and another in Damascus obscured by a drainpipe, and Meyer 1980, 194 who noted of the three sundials on a mosque built (1501–6) by the son of Sultan Bayazit that 'One of these cannot be recognised anymore but I remember having seen its polos which had an eye for better reading the time'. For literary laments and reflections on the demise of the old hour system in Turkey, see Hasim 2012, written in 1921, and the novel by Ahmet Hamdi Tanpınar, *The Time Regulation Institute* (1962).

156 Bagheri 1998 and 2002.
157 For whom see the entries in *ODLA* 2018 and Drachmann 1948; Gille 1980, ch. 6.
158 Downey 1947.

that subject, which is evidently connected with the theory of the Sun dial.[159]

Clockmakers existed in the early Byzantine empire among mechanics deploying inherited knowledge. They constructed sundials, water clocks, and automata, together or separately, and were associated, as they would continue to be in Islam, with the whole range of Heronic devices. For none of them however has archaeological evidence survived and distinguishing between dials and water clocks under the general appellation of *horologion* employed in Byzantine literary sources is an uneasy matter.[160] Gregory of Nyassa seems to refer to a true water clock in the later fourth century,[161] and an epigram in the Greek *Anthology*, compiled between the fifth and tenth centuries, describes one 'thrice giving voice ... when the water is compressed towards the narrow mouth and the air sends forth a far-reaching blast'.[162] However, whether the *horologion* of the Augustaion was a dial or a water clock remains uncertain, as must the question whether it was moved to the Chalke Gate—the principal ceremonial entrance into the palace—in 538. It is more certain that a clock was placed there, as described in a seventh-century Chinese description of Constantinople. The clock is said to be outside the palace and to incorporate 'golden' (*sc.* gilt) spheres, one of which dropped each hour to give an aural indication of passing time. That this is the same clock as one 'artfully crafted in copper' that fell, broke, and was repaired by Hypatios, as recorded in a tenth-century account, also remains uncertain but the story at least attests to the existence of a ball-dropping, monumental clock in the eighth or ninth century.[163]

The nature of the *horologion* of Hagia Sophia is much clearer. This water clock was probably erected in the reign of Theophilus (829–42) above the notable bronze doors of the south entrance completed in *c.*840. The clock was described by an Arab prisoner in Constantinople, Hārūn ibn Yaḥyā, *c.*881. He records that there were twenty-four small doors, one of which opened and closed at each passing hour and that it was created by Apollonius.[164] Shortly before, the rule of Theodore of Stoudios (759–826) prescribed a small alarm water clock as necessary for the monk charged with waking his brethren for the night services.[165] But activity continued around 'marvellous devices'. Already during the reign of Theophilus, the father of the patriarch Antony Cassimatas (d. 837) had constructed automata that included singing birds and roaring lions to adorn the imperial throne, possibly in emulation of automata at the Abbasid palace in Baghdad.[166] At about the same time (*c.*840), Leo the philosopher suggested a method for relaying messages from Loulon on the Cilician frontier to the imperial palace in Constantinople by synchronized water clocks.[167] Theophilus' automata and an elaborate water clock were among items melted down by Michael II to pay his army, but new automata created for the Solomon throne of Constantine VII (905–59) caused wonderment to Liutprand of Cremona during his embassy of 946.[168] Heronic technology had not been forgotten, at least in Constantinople. Among the rich gifts that Basil II made to the cathedral treasury on the Parthenon in Athens to celebrate his victory over the Bulgars in 1014 was a gilt dove that flapped its wings and a self-feeding lamp.[169] An eleventh-century letter from Michael Psellos (1020–*c.*1110) to the Patriarch of Constantinople mentions artificial songbirds and other mechanical devices.[170] By this time, however, Islamic scholars had long absorbed the fruits of their predecessor's labours.

ARAB–ISLAMIC WATER CLOCKS

When Gaza succumbed to Muslim arms (632–4), the new rulers were confronted by the free-standing Hercules clock erected by an unknown craftsman early in the sixth century and described by Procopius;[171] he mentions combined automata with hour openings and that this clock offered a precedent for the clock of Hagia Sophia. Arab engineers now began to seek out the principles of such constructions. Small water clocks were built using existing artisanal knowledge and the **sinking-bowl water clock** may have been transmitted from India. Although seldom mentioned in literary texts, it was widely used for the timing of water distribution in the irrigation systems of Spain, the Caspian Mountains, the Yemen, North Africa, and Iran. It was also employed as a component in al-Jazarī's more complex mechanisms.[172] Jāḥiẓ (d. AH 255: CE 868/9) mentions in his *Kitāb al-Ḥayawān* the use

---

159 *Mathematical Collection* viii, 1. Translation from Cohen & Drabkin 1966, 183–4. *Cf.* Downey 1947.

160 This is compounded in ecclesiastical sources by use of the word *horologion* to designate the book containing the extracts to be recited in monasteries by the psalmist, lector, or choir at fixed intervals throughout the day. For one example, see Mateos 1964.

161 *Patrologia Graece*, xlv, 969B, cited in Talbot 1991.

162 Cited from Gow & Page 1965, i, 102–3. For the *Anthology*, see ODLA 2018, I, 680–1.

163 Anderson 2014, 24 and references therein.

164 Anderson 2014, 26–8. For Apollonius, see nn. 20 and 21.

165 *Patrologia Graece*, xcix, 1704C, cited in Talbot 1991.

166 Brett 1954.

167 Aschoff 1980; Pattenden 1983; Anderson 2014, 31.

168 For Liutprand's description and the ceremonial context of such display, see Walker 2012, 102.

169 Miller 1908, 16.

170 Thorndike 1964, 9.

171 For details of it see Diels 1917; Hill (1981), 13.

172 For details, see Turner 1984a; 1984b, 9–11; Glick 1969; Turner 2002a, 211–13. An interesting description of its use (translated from Chardin 1711, ii 73a–b) in Persia is given by the late seventeenth-century traveller John Chardin:

'For what concerns the distribution of water from rivers and springs, it is effected weekly or monthly as necessary in the following way: a round copper bowl is placed on the channel that leads water into the

of domestic water clocks by monarchs and scholars to find the time at night.[173] Also in the ninth century, Qusṭā ibn Lūqā translated Hero's *Mechanics* into Arabic. Ibn al-Nadīm tells us that the astronomer Muḥammad ibn al-Ḥasan ibn Akhrī Hishām al-Shatawī, apart from writing on different types of sundials, also described 'the technique of balls', that is of water clocks.[174] Ibn al-Nadīm also recorded that a book on weight-dropping water clocks was extant among the works of Archimedes. A work under his name was certainly known in the Arabic tradition. The surviving text, however, following Hill's analysis,[175] is composite, made up of Hellenistic, Byzantine and/or Persian, and Arabic elements. These were probably intended to be used individually, following the craftsman's whim or the instructions he had been given, and not to be combined as a single clock. The date of compilation of the work is unknown although one section of it is perhaps no earlier than the mid-twelfth century.

From the eleventh to thirteenth centuries, four important treatises have survived, the last of them being the magisterial work by Ibn al-Razzāz al-Jazarī, which illustrates that the tradition of ingenious mechanical devices was still a unitary one (Figure 46). The six water clocks and the two candle clocks there described are accompanied by a range of automata, drinking vessels, mechanical phlebotomy devices, and combination locks. Since the descriptions of water clocks given by Pseudo-Archimedes, Ibn Khalaf al-Murādī (fifth/eleventh century), al-Khāzinī (written AH 515; CE 1121–2), Riḍwān (written AH 600; CE 1203–4), and al-Jazarī (written AH 602; CE 1205–6), have been fully analysed and explained by Donald Hill, it is unnecessary to repeat his details.[176] Suffice it to say that the tradition of mechanical devices in Islam was practical. The authors of the texts fulfilled the prescription of mechanicians in the Heronic tradition enunciated by Pappus so many centuries earlier and presented both theory and practice. The devices recorded were, and still can be, made following the descriptions given.

Scattered evidence across the centuries confirms their currency. Water clocks are mentioned in the tenth-century encyclopaedia *Mafātīḥ al-ʿulūm* by Abū ʿAbdallāh al-Khwārizmī, and there are stray references to them (without details) in works by al-Bīrūnī.[177] Two large water clocks showing the phases of the Moon, erected on the banks of the Tagus by Ibn al-Zarqellu (d. AH 493; CE 1100), were discovered by Christian forces on the fall of Toledo in 1085 and one of them continued in service until 1133.[178] An anonymous description, *'Amal al-sandūq li-l-sāʿāt* (Operation of the Hour Box), describing a ball-dropping water clock with a zodiacal disk, was copied into a collection of philosophical and scientific tracts in the mid-twelfth century in Baghdad.[179] What may have been a water clock with an astrolabe dial was installed in the dwelling of Ibn al-Shāṭir the noted instrument-maker,[180] and the *muwaqqit* and instrument-maker al-Mizzī (d. AH 750 [CE 1349]) also constructed mechanical devices.[181] The remains of two combination locks signed by an astrolabist as maker, similar to and contemporary with those described by al-Jazarī, have survived.[182] So, too, have several copies of al-Jazarī's treatise, with dates ranging from AH 602 (CE 1205/6) to the seventeenth century and with a Persian translation copied as late as AH 1291 (CE 1874).[183] A very late nineteenth-/early twentieth-century copy can now be added to their group.[184] All attest to the continuing prestige and interest of al-Jazarī's work in the Muslim world.

Clocks in the tradition continued to be built. In AH 633 (CE 1235/6), a monumental water clock by ʿAlī ibn Taghlib ibn Abī al-Ḍiyā al-Sāʿātī (AH 601–84: CE 1204/5–1284/5), that announced both the hours of prayer and the time by day and night, was erected in the entrance hall to the Mustanṣiyya college in Baghdad. In form it seems to have been similar to the Castle water clock of al-Jazarī.[185] In AH 685 (CE 1286/7), a clock was built by Muḥammad ibn al-Ḥabbāq al-Tilimsānī at the Qarawiyyin mosque in Fez. It was replaced early in the following century by Muḥammad al-Sinhāghī and this in turn was reconstructed in AH 763 (CE 1361/2) by Muḥammad al-ʿArabī. Parts of this clock,

---

field, water enters it little by little and when the bowl sinks the measure is full and the cycle restarted until the agreed amount of water is in the field. Normally the bowl sinks after two or three hours. This contrivance also serves to measure time in the East. It is the only dial and clock in many parts of India, above all in fortresses and in the houses of great men where a guard is mounted.

173 Siddiqi 1927, 246.
174 Dodge 1970, 663–6.
175 Hill 1976, 7–9; Hill 1981, 17.
176 See his works listed in the bibliography. For a contextual account of his studies, see Turner 2002a; 2002b.

177 Al-Khwārizmī 1895, 235. For al-Bīrūnī, see Kennedy 1985, i, 12, 150, 229.
178 Millas-Vallicrosa 1950, 6–9.
179 Zāhiriyya, Damascus, MS 4871. See Ragep & Kennedy, 1981, 97, N° 15.
180 It was seen there by the historian al-Safadī in AH 743 (CE 1342–3), although what he says of the astrolabe as reported by al-Nuʿaymi is paradoxical: ' . . . it turned unceasingly day and night without the help of machinery'. Rihaoui 1961–2, 210. The interpretation by King 1975b, 362, however, is more enlightening.
181 Charrette 2003, 13.
182 Maddison 1987, 154, notes that 'A Hellenistic/Byzantine source, practical, perhaps, rather than literary, for the dial combination locks would be consistent with some of the ultimate sources of al-Jazarī's more elaborate devices'.
183 They are listed in Hasan 1977, 60–2; Hill 1981, 89–91.
184 It was sold in Paris, 3 December 2018 by Millon auctioneers, as lot 288 in their sale of 'Arts d'orient & orientalisme'.
185 Le Strange 1900, 267; Jawad ?1960, 31–4.

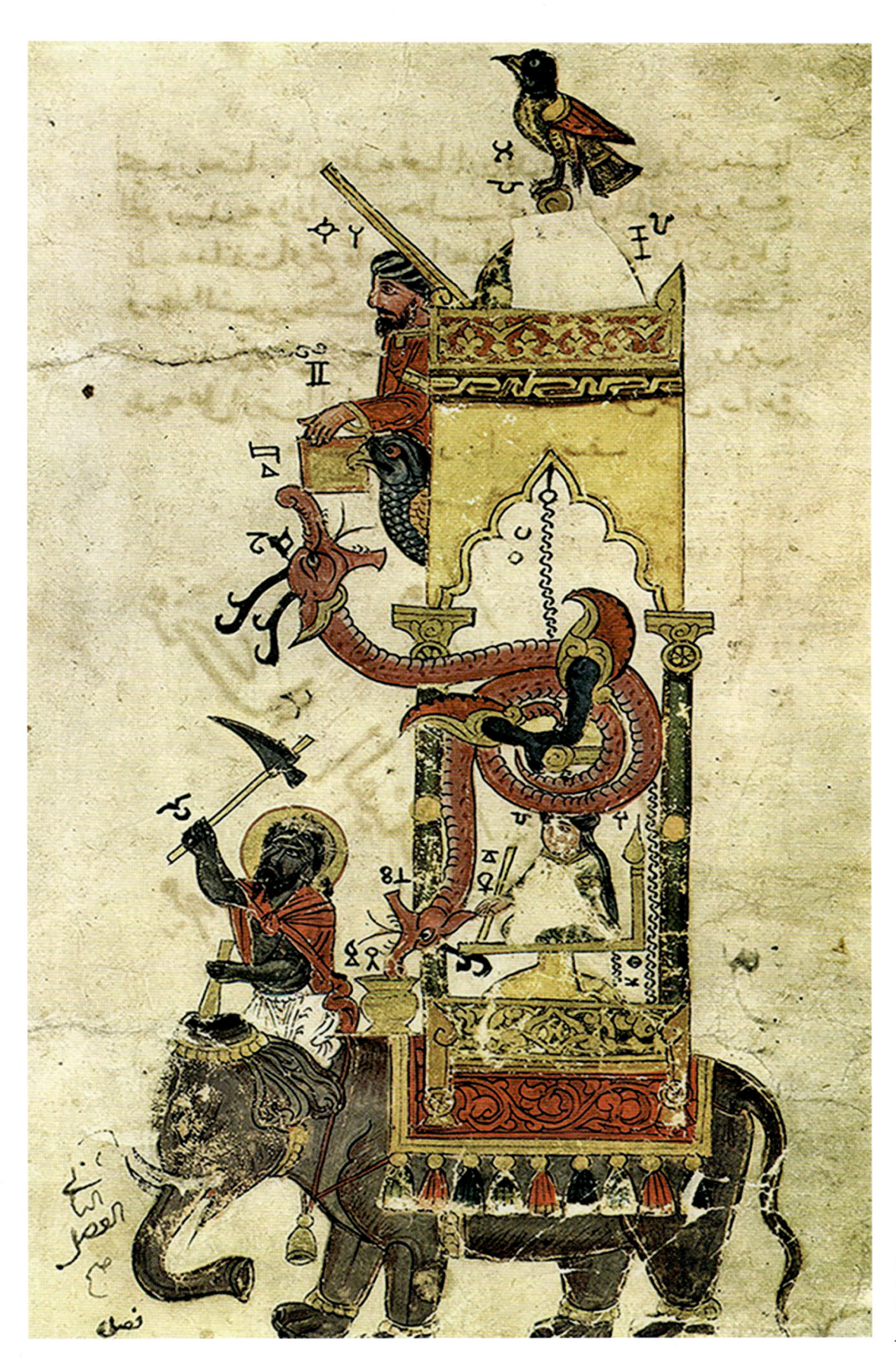

Figure 46 Al-Jazarî, the elephant water clock from ms Graves 27, the Bodleian Library, Oxford. Photo: Jean-Baptiste Buffetaud.

including a revolving astrolabe dial, survive,[186] as do the facade, gongs, windows, and ball-channels of another clock in Fez on the Bū'anāniyya mosque, which was completed on AH 14 Jumādā I 758 (CE 6 May 1357) by Abu al-Ḥasan ibn Aḥmad al-Tilimsānī.[187] Clearly a professional, he was also the maker of an automaton clepsydra kept in the royal palace of Abū Ḥammū (AH 760–91: CE 1358/9–1388/9) and was displayed at the feast of *Mawlid* (the prophet Muḥammad's birthday), one of the rare occasions when the king showed himself in public. Apart from the standard row of doors for the hours, two eagles to drop balls into bowls, and a moon moving through a semicircle to mark the time at night, there were automata: a bird and two chicks in a rosebush, underneath which lurked a serpent, which, at each hour, snaked up the bush to attack the chicks, whose parent made defensive noises. From the hour door a slave girl came forth holding in one hand a tablet inscribed with verses and raising the other to her mouth in salutation. The device continued in use until at least AH 814 (CE 1411/12).[188]

Although references to them are sparse, water clocks continued to command attention in the fifteenth and sixteenth centuries. In the last decades of the fifteenth century, the library of the Vizir at Herat contained a clock on which a figure held a stick with which it sounded the hours on a drum.[189] The astronomer Ibn Abī al-Fatḥ al-Ṣūfī, who completed a copy of al-Jazarī's text in AH 891 (CE 1486), himself wrote a short tract on water clocks.[190] Soon, however, the tradition would begin to weaken, rivalled by the tinkling attractions of European weight- and spring-driven timepieces adorned with automata arriving in the Ottoman Empire as diplomatic gifts and items of trade.[191] By the mid-sixteenth century, Taqī al-Dīn, although he discussed different types of water clocks in another work, thought that 'there is no profit [*sc.* use] in [the water clock] [. . .] The difficulty of its construction is more than its profit, generally, it is not possible to carry them [sic] from one place to another'.[192]

---

186 Price, 1962a; 1962b. There is a photograph of the dial by Wim Swaam in Landau 1967, 119.
187 Price, 1962a; 1962b; Mayer 1956, 40. It is illustrated in Landau 1967, 91 and has been the subject of many early twentieth-century postcards.
188 Bargès 1859, 368–76.

189 Details kindly communicated by Souren Milikian.
190 King 1975a; 1975b, 288, n. 8. The very accurate copy is now Bodleian Library MS Graves 27.
191 On which, see Kurz 1975, chs. 1 and 2; Mraz 1980.
192 Tekeli 1966, 143.

SECTION TWO: TIME RECKONING IN THE MEDIEVAL LATIN WORLD

Mario Arnaldi

The Middle Ages inherited the day division and hour systems (**equal hours** for astronomical computation; unequal **hours** for daily life) of Antiquity practically unaltered; hours were counted with ordinal numbers from the first to the twelfth, the sixth always marking the middle of the day, while the night was divided by the unchanged military watches. Each hour line on a sundial showed the end of an hour, not the beginning of it.[193] As had late Roman scholars, medieval writers compiled works that explained time reckoning, its many parts, and the name of each part. Isidore of Seville (c.560/4–636) summarizes the new Christian idea of the day:

> It is called 'day' the time taken by the Sun to reach sunset from the moment of its rise. It is used to define the 'day' in two ways: 'effective', from the rising of the Sun until it returns to the point of the subsequent rise, 'improper' from the rising of the Sun until its sunset. The spaces of the day are two: diurnal and nocturnal. Each contains 12 hours and therefore the 'effective' day consists of 24 hours.[194]

Compared with antiquity there is little that differs: only a few definitions and names change. What Censorinus (fl. 268) called a 'natural day', measured from the rising of the Sun until sunset, was defined by Isidore as 'improper',[195] by Bede (c.673–735) as 'common' or 'ordinary',[196] by William of Conches (1090–1154) as 'usual' or 'usable';[197] frequently, it was called 'artificial'. The twenty-four-hour Roman civil day, changed its name to 'legitimate' or 'effective' day in Isidore, and to 'legitimate' and 'natural' in Bede, and also in Rabanus Maurus (780–896), abbot of Fulda. Other authors, for example, Byrhtferth of Ramsey (970–1020),[198] Honorius of Autun (1080–1151),[199] William of Conches,[200] and Vincent of Beauvais (1186–1264), called it the 'natural' day.[201]

The two parts of the **nychthemeron** were not very different from those of the civil day of ancient Rome (Table II). The diurnal portion was divided into three parts: *mane* (morning)—from the rising of the Sun until the end of the fourth hour, *meridies* (midday)—from the end of the fourth to the end of the eighth hour, and *suprema*—from the end of the eighth hour to sunset. The nocturnal part was divided into seven portions of variable length: *vesper* (at sunset), *crepusculum* (twilight immediately after sunset lasting about an hour), *conticinium* (at the end of the first eve, that is at the end of the third hour of the night), *intempestum* (the central hours of the night), *gallicinium* (at the end of the third **vigil**, that is, at the end of the ninth hour of the night), *matutinum* (shortly before dawn), and *diluculum* (at dawn).[202]

As we have seen from Isidore, and can also read in his *Etymologiae*, the computation of the *nychthemeron* or 'legitimate' day however changed substantially, the sequence of twenty-four hours running from sunrise to sunrise.[203] This system became established until at least the tenth century.

At the turn of the eleventh and twelfth centuries, the Justinian *Digest* of Roman civil law once again became of interest to jurists. Revisions of them were produced by the Bologna lawyer Irnerio (1050–1125). A passage from the *Decretals* (1235) of Gregory IX (1145–1241) reissued by Innocent IV (1195–1254) testifies that the Church also accepted that the legal and ecclesiastical day should start at midnight according to ancient Roman usage.[204] St Thomas Aquinas (1225–74) reaffirmed in his writings the same ecclesiastical custom, and William Durand (1230–96) justified its use on

---

193 Beda Venerabilis, 'De ratione computi', ch. 2, (*MPL*, xc, col. 579). For medieval timekeeping in general see Lejbowicz 1992, passim.
194 Isidorus Hispaliensis, 'De natura rerum', 1,1, (*MPL*, lxxxiii, col. 963).
195 Isidorus Hispaliensis, 'Etymologiarum libri xx', v, 30, 1, (*MPL*, lxxxii, col. 215B; and 'De natura rerum', 1,1, (*MPL*, lxxxiii, col. 963).
196 Beda Venerabilis, 'De temporibus liber', ch. 2, (*MPL*, xc, col. 279); and 'De temporum ratione', ch. 5, (*MPL*, xc, col. 309).
197 William of Conches in Maccagnolo 1980, iv, 332; William of Conches (*alias* Beda Venerabilis), (Elementorum philosophiae), ii, (*MPL*, xc, col. 1153); William of Conches (*alias* Honorius Augustodunensis), 'De philosophia mundi', ii, (*MPL*, clxxii, col. 71).
198 Bridefertus Ramesiensis, 'Glossae' in Beda 'De temporibus', ch. 3, (*MPL*, xc, col. 303).

199 Honorius Augustodunensis, 'De imagine mundi libri tres', ii, 12, (*MPL*, clxxii, col. 147D).
200 William of Conches in Maccagnolo 1980, iv, 331.
201 Vincentius Bellovacensis, *Speculum naturale Vincentii*, iii, 76, 1494.
202 Isidorus Hispaliensis, 'Etymologiarum libri xx', v, 30, 13–16 and 31, 4–14, (*MPL*, lxxxii, coll. 217–218).
203 Isidorus Hispaliensis, 'Etymologiarum libri xx', v, 30, 1, (*MPL*, lxxxii, col. 215); and 'De natura rerum', 1,1, (*MPL*, lxxxiii, col. 963); Rabanus Maurus, 'Liber de computo', ch. 20 (*MPL*, cvii, col. 679); Bridfertus Ramesiensis monachus, 'glossae', (*MPL*, xc, col. 307); Honorius Augustodunensis, 'De solis affectibus seu affectionibus liber', ch. 16, (*MPL*, clxxii, col. 106).
204 Gregory IX, *Decretales*, i, 29, 24; Innocent IV, *Apparatus super quinque libros Decretalium*, Venetia, 1495, fol. giiiir.

Table II Comparative table of the parts of the day in Ancient Rome and in the Middle Ages

Hours		XII tables (V cent. BCE)	Censorinus (III cent. CE)	Macrobius (V cent. CE)	Isidore of Seville (VII century)			Hours
1			Mane	Mane				1
2			(2ᵐ Diluculum)		Mane			2
3		Ante meridiem						3
4			Ad meridiem	Ad meridiem				4
5	Day						Day	5
6					Meridies			6
7			Meridies	Tempus occiduum				7
8		Post meridiem	De meridie					8
9								9
10			Suprema		Suprema			10
11				Suprema tempestas				11
12		Suprema tempestas	Vesper	Vesper	Vesper			12
1		1th vigil	Crepusculum	Prima fax	Crepusculum	1th vigil		1
2			Prima face					2
3			Concubium	Concubia	Conticinium			3
4		2nd vigil	Nox (?)	Ad mediam noctem	Intempesta (nox)	Intempestum (tempus)	2nd vigil		4
5								5
6	Night		Intempesta	Media nox			Night	6
7		3rd vigil		De media nocte	Mediae noctis inclinatio			7
8								8
9								9
10		4th vigil	Nox (?)	Gallicinium	Gallicinium	Gallicinium	4th vigil	10
11			Conticinium	Conticinium	Matutinum			11
12			Ante lucem Diluculum	Diluculum	Diluculum			12

purely theological grounds.²⁰⁵ The system remained in place until the Renaissance and beyond. After six centuries of oblivion, the Roman civil day returned to being counted from midnight, as it still is today.²⁰⁶

The Romans divided each of the twelve portions of the day into at least two parts, commonly called *semis* (half) or *semihorae* (half hours). They often identified intermediate times with expressions such as 'between the seventh and eighth hours of the day' or 'between the tenth and eleventh hours of the day'.²⁰⁷ The hour

---

205 Thomas Aquinas, *Summa Theologiae*, iii, 80, 8.5; Gulielmus Durandus, *Rationale divinorum officiorum*, vii, 1.
206 On this subject, see Quinlan-McGrath 1995, 57–9.

207 Pliny, *Nat. Hist.*, II, 72, 180.

Table III The medieval hour and its subdivisions

	1 Hour	1 Point	1 Minute	1 Part	1 Moment	1 Ostent	1 Athom
POINTS	4 (5 with the moon)						
MINUTES	10	2,5					
PARTS	15	3,75	1,5				
MOMENTS	40	10	4	2,6			
OSTENTS	60	15	6	4	1,5		
ATHOMS	22.560	5.640	2.256	1.504	564	376	Indivisible

was subdivided into even smaller portions (the 'scrupuli'),[208] but in common life these were generally ignored. Only astronomers and astrologers needed small subdivisions. Ptolemy divided an hour into halves, thirds, quarters, fifths, and sixths,[209] but only the half hours—and that rarely—were easily inserted among the lines of a sundial.[210] Nevertheless, the date and time of death, in the Christian funerary epitaphs of the fourth and fifth centuries, were often minutely noted. In addition to the name and a commemorative phrase, chronological data were carefully transcribed and the length of life—especially for children—was noted with almost manic attention, as on the tomb of Innocentius who lived only one year, nine days, three hours, and a half. However, although recordings of hourly fractions were at the time considered to be fairly accurate if indicated with a simple *plus* or *minus*,[211] in some cases they could be exaggerated in their precision, as in the inscription for Silvana, who lived twenty-one years, three months, four hours, and six scrupuli.

In the Medieval **Computus**, the hour had six different subdivisions (Table III): 'Points' (a quarter of an hour), 'Minutes' (a tenth of an hour), 'Parts' (a fifteenth of an hour), 'Moments' (a fortieth of an hour), 'Ostents' (a sixtieth of an hour), and 'Atoms' (a 22,560th of an hour). Table III shows the full scheme of hour fractions in the Middle Ages.

The Computus developed notably in Ireland of the High Middle Ages, where important monastic schools were founded.[212] The fundamental texts throughout the Middle Ages were those of Isidore of Seville and Bede and were the basis of study in all the monastic schools in Europe. This early computation was later developed by scholars such as Rabanus Maurus, Helpericus of Auxerre (ninth century), Pacificus of Verona (c.776–844), and later Honorius of Autun. With them, the *Computus* reached its highest development. Time was divided into very small parts, used almost exclusively for intellectual, astrological, and 'scientific' purposes to solve problems of chronology, astronomy, or cosmology.

MONASTIC TIME-TELLING

A first example of an *horarium* (timetable) can be recognized in an epigram by Martial, who distinguishes the various phases of the daily life of the rich Roman.[213] It was only with the establishment of organized monasticism in the early Middle Ages that time discipline became necessary and was imposed. A monk's day was totally controlled by the Rule of his order. There were precise times for: praying, working, reading, eating, sleeping, washing, shaving, and even for attending to bodily needs. The *Ora et labora* left no room for laziness, and the concept previously unthinkable of 'being late' developed.[214] The monk who came after the *Gloria* of the first psalm to the prayer of the Divine Office had been sung, for example, was severely punished and, if he repeated the offense, could be expelled from the community. For St Benedict of Norcia (c.480–547), whose codified rules for the life of the monastic community were widely accepted, punctuality with God was so important that he urged the monks to leave whatever they were doing as soon as they heard the first sound of the call to prayer and, to be sure that the timetable was followed scrupulously, he ordered the abbot himself, or a reliable and zealous monk, to mark the passage of time. Time perception changed Christianity. Time was no longer public and private, as in ancient Rome, but the property of God. Everyone subjected himself to His rhythms and the sound signals that marked the liturgical day of the monks soon became signals to which the lay community also paid attention.[215] As a consequence, even sundials underwent a process of 'Christianization'.[216] The sign of the cross was frequently engraved near the hour grid (this is the case with many Irish sundials),[217] but

---

208 'Scrupulo' was the twenty-fourth part of an hour. See Forcellini 1871, *Lexicon Latinum*, 'scrupulus', 394.
209 Ptolemy 1998, 7.
210 It was, however, a very rare thing; in fact, only two sundials with a similar subdivision are known: see Gibbs 1976, 226, n. 3007 and 239, n. 3020. For the splitting of hours, see Arnaldi 1999.
211 Leclerq & Cabrol 1907–53, s.v. *Heure*, col. 2370.
212 Ó Cróinín 1983.
213 Martial 1990, *Epig.*, iv, 8.
214 Zerubavel 1981, 66.
215 Le Goff 1977, 26; 'The unit of work time in the medieval West is the day ... defined by the mutable reference to natural time, from rising to the setting of the Sun, and approximately emphasized by the religious time, that of the *horae canonicae*, derived from the Roman antiquity'.
216 Turner 2004.
217 Arnaldi 2000b.

we find many also in old Palestine) or, as in the famous sundial of Bewcastle, England, it was the cross itself that housed the sundial;[218] sometimes the Christian sign was placed within the very design of the hour lines, as at Santa Maria della Strada at Taurisano, Italy.

The monastic day was measured in unequal hours, which in consequence structured the times of the offices, the 'canonical hours'. These were: 'Vigil' or 'Nocturne' (at midnight), 'Matins' (shortly before twilight, at the beginning of the fourth Vigil), 'Prime' (at the beginning of the first hour of the day, sunrise), 'Terce' (at the third hour of the day, i.e. at mid-morning), 'Sext' (at the sixth hour of the day, i.e. near noon), 'Nones' (at the ninth hour of the day, about mid-afternoon), 'Vespers' (at the twelfth hour, i.e. at sunset), and Compline (at the beginning of the night or at the end of twilight).[219] Before being standardized by Benedict, canonical hours had their own, overlapping, development. Briefly, this proceeded from only two with Pachomius (292–396 CE),[220] four with Augustine (354–430), five with St Cyprian (d. 258), and Cassian (c.360–435),[221] six in the *Apostolic Constitutions* of St Jerome (347–420), and finally, in the *Regula Magistri*, eight–the same number as would be enshrined in St Benedict's Rule. These, constituted by seven diurnal and one nocturnal hour, would eventually become standard in the European monastic world. Obviously, those shown by a vertical fixed medieval sundial, such as is found traced on numerous Romanesque and Gothic churches, are only those that can be seen in daylight from the shadow of a gnomon (practically only six), that is: Prime, Terce, Sext, Nones, Vespers, and sometimes, Compline. As a tool for this purpose only the lines that determined the times of the canonical hours were strictly necessary. A good number of medieval sundials, therefore, had a quadripartite division of the space included in the 180 degrees that makes up the semicircle within which the hour lines were usually enclosed. A line every 45 degrees—that is, every three hours—marked the end of the third, of the sixth, and the ninth hour, to which the Divine Offices of the Terce, Sext, and Nones were formerly combined. The two horizontal lines, which defined geometrically the diameter of the semicircle, referred to the times of the recitation of Prime, which was recited at *solis ortu* (sunrise), and of the Vesper, at *solis occasu* (sunset), when the shadow of the stylus was practically horizontal.

Canonical hours are not a real hour system but a sequence of daily times initially bound to the unequal hours. The three main daytime Offices (Terce, Sext, and Nones) took their name from the hour at which they were originally celebrated, exactly at the end of the third, sixth, and ninth unequal hours. The difference between the hours intended as fractions of the day (unequal hours) and the canonical hours, intended as liturgical time, was well defined by Dohrn-van Rossum with the two clear expressions: *hora quoad tempus* (hours of time) and *hora quoad officium* (hours of prayer).[222] The canonical hours of Terce, Sext, Nones, Vespers, and Compline underwent considerable temporal displacement over the centuries. While their names remained unchanged, they ended up being sung at other hours of the day, sometimes very different from the original ones. Since the sixth century they began to change their position on the dial, following the times of the liturgy and the calendar. It is therefore not easy to trace a single and unequivocal diagram, because the times of prayers, so changeable even among the various monastic traditions, cannot be set in a constant time frame. But apart from the irregularity of the temporal hours, it was perhaps also the medieval use of sundials with equidistant radial lines that was partly responsible for the displacement of the times of the offices.

At latitudes very close to the equator, the tracing of a sundial 'with equidistant radial lines' overlaps with the tracing of a sundial showing unequal hours with good approximation. However, as the latitude increases, the two paths diverge more and more dramatically until they no longer agree even to the equinoxes. At 50° latitude, the errors become unacceptable, as in winter differences of up to two hours can be reached between the two tracks.

MEDIEVAL HOUR SYSTEMS

The dislocation of the times of the canonical hours, which had been carried over from monastic communities to churches serving the secular world, was probably a strong contributory cause for the development of new systems of time division. Regularly divided, medieval European fixed dials can be variously grouped. Best known are those with the semicircular surface divided into twelve sectors, but others with four, six, eight, and eleven equal spaces are also widespread. In addition, there are others that are more difficult to place in an historical-cultural framework, which present ten, thirteen, and even sixteen or eighteen divisions of equal size.[223]

Twelve-sector dials

In this group, all the hour lines that converge towards the common centre have a regular spacing of 15°; each space represents a whole hour of the daylight period and the shadow of a gnomon at right angles to the plane moves from the first to the twelfth showing a 'presumed' division of the day into classical unequal hours.

This kind of sundial is, in practice, the matrix of all the other models that we describe. Its purpose was to mark the hours of the day, not those of the offices. However, in some twelve-division

---

218 See Aked 1973a; 1973b, 1995a, 1995b, but the fullest account is now Orton, Wood, & Lees 2007.
219 The time periods of the hours are given here in generic form; for more information on the correct placement of the canonical hours during the daily season, see Arnaldi 2005 and Arnaldi 2011a, part I, chs. 4–5.
220 Verheul 1981, 229–22.
221 Cassianus, *De diurnis orat.*, chs. 1 and 2.

222 Dohrn-van Rossum 1996, 30–1.
223 Rau 2000.

dials, the canonical hours are highlighted in some way. Only in these cases can we speak of a double use (temporal and canonical) of the same dial. In Greece and in all the geographical areas where Byzantine culture had influence, dials are generally divided into twelve sectors and this is also present in Italy, especially along the Adriatic coast.

Four-sector dials

One of the most widespread models in all medieval Europe was that divided into only four equal sectors. Many scholars have considered that this division of the artificial day was of North European origin, particularly in the British islands, perhaps stemming from Iceland or the Scandinavian Peninsula. In the ancient Viking and Icelandic language there was a term, *eykt* (eighth), still used today to identify each of the eight temporal portions into which the *nychthemeron* and the horizon can be divided.[224] Perhaps in the British Isles, the Saxon word *Tide* or *Tid* was borrowed with the sense of the Nordic language as an eighth portions of the natural day, and so identified the period of time equivalent to an *eykt*; in other words, a *tid* defined a quarter of the daylight period. Daniel Haigh, along with many other British authors after him, argued that one *tide* lasted three hours.[225] Turner, however, rebutted this by showing that representative Anglo-Saxon texts used the word as a synonym for the Latin *hora* (hour) or even in the generic sense of 'a period of time'.[226]

The fourfold division of the day had been known in Antiquity. A few Græco-Roman sundials carry, in addition to the classic twelve hours, markings for these four fundamental moments of civil and religious life.[227] What canonical time, which approximates to the quaternary division, is indicated by the lines of a medieval sundial is not easy to say, as in most cases there are no indications apart from some sign such as crosses, points, forks, and similar. Fortunately, among the many sundials that have come down to us, some survive that carry the initial letters of the canonical hour on the appropriate lines. Thanks to these dials, and some literary evidence, we can confirm that even in the case of a daily partition as simple as the quadripartite division, sundials were not always in agreement with each other concerning the time of the offices.

In Italy, the sundial of the abbey of San Tommaso in Acquanegra sul Chiese, in the province of Mantua, is quadripartite and near the line that determines the end of the first quarter we read the letter T (Terce) and near the line which determines the end of the third part the letter N (Nones) is engraved; on the vertical line of midday the letter S (Sext) should have been placed, but it was not engraved. In this case, the canonical hours of Terce and Nones were sung exactly in the times of the hours originally intended for them (the third and the ninth). The sundials on the Benedictine abbey of Santa Reparata (Figure 47) in Marradi (province of Florence), that on the cathedral of Ascoli Piceno, and that on the Benedictine abbey of Santa Maria a Piè di Chienti near Montecosaro—province of Macerata—are divided into twelve equal sectors. All of them have initial letters for the canonical functions engraved at the end of some of the hour lines. Thus, in Marradi, on the hour lines showing the end of the third, sixth, and ninth hours we read the letters T, S, and N, and also in this case the canonical hours govern exactly the times of the hours originally destined for them. The same occurs on the sundial of the cathedral of Ascoli Piceno: the position of the functions of the Terce, Sext, and Nones corresponds to the original hours, but to these are added the initials of the hours of Prime (P) and Vespers (V) which, departing from tradition, are placed at the end of the first and the beginning of the twelfth hour.

On the Montecosaro sundial, on the other hand, the letters T, S, and N are all advanced by one hour, suggesting that Terce was recited at the end of the second hour, Sext at the end of the fifth hour, and Nones at the end of the eighth hour.

In the fourteenth century, at least in Florence, canonical hours were placed on the sundial in a different way. It is Dante who explains this to us in his *Convivio*, speaking of the division of time into four parts:

> And these parts are made similarly in the year, with spring, summer, autumn and winter; and in the day, this is up to Terce, and then up to Nones (leaving Sext, in the middle of this part, for the reason that is discerned), and then up to Vespers, and from Vespers on.[228]

Dante divides the day using the terminology of the canonical hours (Figure 48), which at that time had also become common among the laity, and therefore tells us that the first part went from the sunrise to Terce, the second from Terce to Nones (midday in thirteenth-century Italy), leaving Sext in the middle of this second part to better discern it, so that it did not overlap with Nones at noon. The third part went from noon (Nones) to Vespers (the bell of the Vespers sounded at the end of the ninth hour), and the fourth from there until sunset.

Eight-sector dials

The division of the 'artificial' day into eight parts was a direct consequence of the quadripartite scheme; this diurnal partition

---

224 Vilhjálmsson 1991 and Vilhjálmsson 1997. See also Wikander 2010.
225 Haigh 1879.
226 Turner 1984a; 1984b.
227 The Greek dial from Herakleia Latmia, Turkey, second quarter of the third century (BCE 227), the Roman sundial of Tor Paterno, near Ostia, second century CE, both in Evans 2017, Figure VI, 10.

---

228 Dante Alighieri, *Convivio*, IV, 23, 14. Author's translation.

Figure 47 Marradi (Florence): Abbey of Santa Reparata. The medieval sundial with letters for the canonical Hours and a probable Mass sign at the end of the second hour.

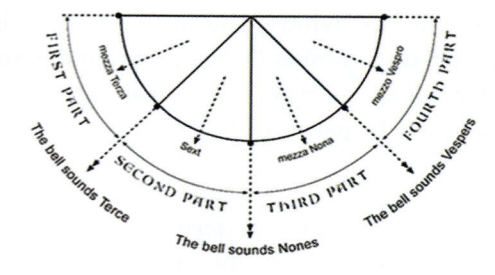

Figure 48 The hour divisions described by Dante Alighieri in the Convivio.

probably arose from the need to have greater precision. This second temporal subdivision, known in Anglo-Saxon England,[229] was adopted in Italy at the turn of the thirteenth and fourteenth centuries and soon became the most readily accepted time division by both the people and the Church in many regions of Italy. It was a mixed time system that can be called 'canonical-secular', known in Italy through popular terms widely used in the literature of the fourteenth and fifteenth centuries. In practice, the four-hours were joined by their halves: 'mezza Terza' (half Terce), 'mezza Nona' (half Nones), and 'mezzo Vespro' (half Vespers).[230]

The precise location of these moments is again revealed by Dante in his *Convivio*:

> And yet the office of the first part of the day, that is Terce, is said at the end of that part, and that of the third (Nones) and the fourth part (Vespers) is said at the beginning of them. And so it is said half-Terce before it sounds for that part, and a half-Nones after the bell sung for that part; and so half-Vespers. And yet, let each one know that the bell for the right Nones must always sound at the beginning of the seventh hour of the day (noon): and this is enough for the present digression.[231]

---

229 Among the best known we can list the sundial of Kirkdale in North Yorkshire, that of North Stoke, and that of Great Edstone; see many examples in Scott & Cowham 2010.

230 Quotations related to this medieval time system are scattered almost everywhere in the Italian literature of those centuries, for example in Dante Alighieri (*Inf.* xxxiv), or (*Convivio*, iv, 23), or in Giovanni Boccaccio (1313–1375), in the novels of his *Decameron* (i.e. introduction to the third day, introduction to the eighth day, or conclusion of the seventh day).

231 Dante Alighieri, *Convivio*, IV, 23, 16. Author's translation.

Thus 'half Terce' is placed in the middle of the part called 'Terce' (the first part of the day), 'half Nones' in the middle of the part called 'Nones' (the third one) and 'half Vespers' in the middle of the part called 'Vespers' (the fourth and last part of the day). No mention is made of the second part of the day, the one in which Sext should be celebrated, but Dante had already spoken about it, albeit briefly, in paragraph 14 of the same chapter (Diagram 14). In fourteenth-century England, the hours thus conceived were called 'vulgar' or 'common' hours, at least by Robert Stikford (*fl.*1396–1401), fourth prior of St Albans monastery, in a manuscript on dialling discovered in 2005 in the Ambrosiana Library, Milan.[232]

The eight-part time division was also known in Germany and France. In medieval French texts (Figure 49) the eight moments of the day were designated:

1. *Prime* = Prime, sunrise
2. *Haute Prime* = between Prime and Terce (It. mezza terza)
3. *Tierce* = third hour
4. *Haute tierce* = between Terce and midday
5. *Haute nonne* = midday, Nones, sixth hour, *Midi*
6. *Basse nonne* = past Nones (It. mezza nona)
7. *Vespre* or *haut vespre* = ninth hour
8. *Bas vespre* = past Vesper (It. mezzo vespro).[233]

Six-sector dials

Around the twelfth century, some European medieval sundials began to divide the day into six parts. A similar time system had already been used by the Babylonians,[234] and this double hour, considered as a single unit, continued to be used by astrologers, as Marcus Manilius (first century BCE), confirms in the third book of his astronomical poem,[235] and perhaps persisted into late antiquity.[236] However, no six-divided sundial from Babylon or Late Antiquity has survived, although Bede in chapter 39 of his *De temporum ratione* seems to imply that there were astronomers who used such a division.[237] At present it is not possible to affirm a practical use of double hours in northern Europe during the seventh or eighth centuries; rather, it seems to be a purely computational system. The several six-divided monastic sundials that have come down to us are probably the result of the cultural revival of the twelfth century.[238] We find them widely spread in France and in Germany, with others in Austria, in the former Yugoslavia, and in Italy. They remained one of the last ecclesiastical systems of time reckoning before the definitive disappearance of the temporal hours.[239]

THE COMPUTUS OF PHILIPPE DE THAON

The place of the canonical hours in a six-divided sundial is shown clearly in the poetic treatise on the computus by de Thaon, written between 1113 and 1119.[240] In his rhymes (lines 247–64 of the MS), he describes the popular–religious use to group the twelve temporal fractions of the bright day in six groups of two hours each. I reproduce below the original text of the verses in the old French idiom and in translation:

Nepurquant par demures,	Nevertheless by means of demures (time intervals),[241]
Que nus apelum 'hures',[242]	That we call 'hours',
En est division	We get a division
Par itele raisun,	According to the rule,
Char prime apelent le une,	That 'Prime' is called the first one,
Tierce, midi e nune,	[Then follow] 'Terce', 'midday', and 'Nones'
La quinte, remuntee,	The fifth [portion is called] 'awakening',
E la siste, vespree.	And the sixth, 'Vespers'.
Encore entre chascune	And then within each [of these demures]
En i laissent il une,	We omit one,[243]
Ço est pur le cunter	This is done just to count them
E pur tost remembrer.	And remember them more easily.
Mais ki dreit volt numbrer,	But who wants to number them correctly,
Duze en i pot truver;	It can count twelve;
E quant eles sunt passees,	And when they passed,
Tutes sunt renuvelees,	All are renewed,
En ordre lur curs	According to their course
Tenent tuz a estrus	Without fail.

---

232 Robert Stikford, *De umbris versis et extensis*, MS Ambrosiana & 201 bis sup.; see Bellettini 2007 and Davis 2011.
233 Bilfinger 1892, 27.
234 Bilfinger 1888.
235 Manilius, *Astronomicon*, lib. 3. 537–559; see Goold 1977, 206–9.
236 Bilfinger 1888.
237 Beda Venerabilis, '*De temporum ratione*', 39, MPL. xc, coll. 469A–469B.
238 English 'Anglo-Saxon' specimens seem generally to be no older than the twelfth century. See Scott & Cowham 2010.

239 Many Franciscan convents in Italy and the former Yugoslavia dating back to the sixteenth century have sundials divided into six sectors. See Tadić 1988, 1997.
240 de Thaon 1984; *cf.* de Thaon 1873.

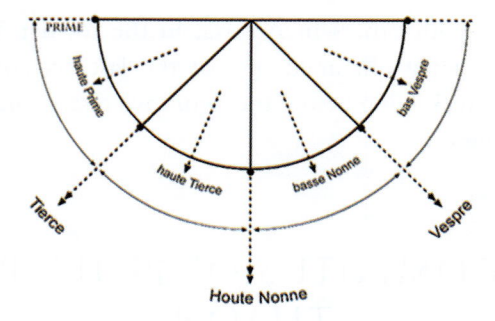

Figure 49 The hour division in 8 sectors described in medieval French texts.

Philippe's text is particularly clear and is one of the best testimonials about dividing daylight into six parts. The author defines the various hourly subdivisions of the day with the term *demures*. In the Anglo-Norman language dictionaries, *demure* generally appears with a meaning of 'delay' or 'protraction'; in Philippe's text we must understand demure as 'a certain amount of time', like the Anglo-Saxon *tide*.[244]

For de Thaon, the *demure* were, therefore, six; the first one was called 'Prime', the second one 'Terce', the third 'noon', the fourth 'Nones', the fifth 'awakening', and the sixth 'Vespers' or 'evening'. It is a mixed set of definitions, partly ecclesiastical, partly popular, and certainly not indicating canonical hours, since the *midi* and *remuntee* hours do not refer to a time of prayer. The hour of *remontée* or *remontière*, otherwise known as *ravaler* or *relevée*, which the author places after Nones, was the time of awakening in the afternoon from a nap after the midday meal, or it was the end of the noon rest period of both monks and workmen. This moment was between Nones and Vespers. More accurately it was between the *basse nonne* and the *vespre*, that is, at the moment also called haut vesper (see Figure 49).[245]

The same sequence—except for the hour of *remuntee*, which never appears—can be found on numerous French six-divided dials. On the abbey of St Pierre in Uzerche, for example, the letters [P], T, M, N, and V can be read at the end of the lines (Figure 50).

The same order is also found on the sundial of the Collegiate church of Moustier in St-Yrieix-le-Perche (Vienne). In this case, the indication of the canonical hours is provided by the first two letters: PR, TE, ME, NO, and VE.[246]

On the facade of the transept to the south of the church of Notre-Dame-de-Porporières in Mérindol-Les-Oliviers is another dial with six divisions, with the letters P, T, M, N, and V. The letters are not placed at the end of each line, but at the top, in a sort of abbreviation (PTMNV), immediately below the eloquent inscription OROLOGIUM. Considering that the letter M refers unequivocally to the vertical central line (midi, meridies), I consider the placing of the other letters at the end of each line to the right and left of M, as corresponding with the same hour lines as those of Uzerche and St-Yrieix-le-Perche (see Figure 51).

As already noted in Italy, it was not easy to equate quadripartite sundials with the offices; it is even more difficult for six-part dials. We have only three documents: a commentary on a passage in Dante's *Divine Comedy* by Francesco di Bartolo da Buti (1324–1406), an Italian manuscript, and the sundial of Santa Maria della Strada in Taurisano, in the province of Lecce. In the first tercet of canto fifteen of the Purgatorio,[247] the Sun is said to be as many degrees away from the point of sunset as there are from the point of rising up to the point where the third hour is ending.[248] It was, therefore, Vespers (three hours to sunset). In his commentary, however, Francesco di Bartolo gave an astronomical explanation of the division of the Dantean day into six parts.[249] This is based on the circle of the ecliptic which with its motion lets rise and set six signs of the Zodiac every day.[250]

The design of the six-divided medieval sundial on the fol. 26v of the codex LJS 497, a small manuscript composed 1501–2 for a latitude of 45° North (more or less all of northern Italy), shows a different sequence. The five lines below the horizontal that represent the rising and setting of the Sun are expressly named: *Tertia*, *Sexta*, *Nona*, *Vespero*, and *Completorio*, corresponding to the second, fourth, sixth, eighth, and tenth medieval temporal hours (Figure 52).[251]

Finally, there is a dial on the church of Santa Maria della Strada in Taurisano (province of Lecce), inhabited by one of the many Greek-speaking communities of ancient Puglia. The dial of Santa Maria is the only Italian example that has the hour lines marked by Greek letters. The hours, therefore, are identified with: 'Π' (Prime) on the first horizontal line, then 'T' (Terce), 'C' (Sext), 'N' (Nones), 'B' (Vespers), and 'K' (Compline). The midday line has no canonical identification (Figure 53).[252]

---

241 *Demure* is a generic term that is used to mean 'a certain amount of time' or, as de Thaon explains, also every hour of the day (in this case one double hour).
242 de Thaon 1984, 2056, 28: '*E qui cunte les hures / Quë apelum demures*' (And who counts the hours, that we call *demures*).
243 That is, jumping two hours in two hours.
244 Turner 1984a; 1984b.
245 Bilfinger 1892, 30–1.
246 The same position of the canonical hours is present on one of the two medieval sundials engraved on the Elisabeth-Kapelle of Hameln, Germany.

247 Dante Alighieri, *Divina Commedia*, Purg. 15, 1–5.
248 'How much [time] remains between the end of the third hour and the beginning of the day … so much seemed in the evening to be left to the Sun at the end of its course.'
249 In reality, Dante had never mentioned it.
250 Arnaldi 1998.
251 I thank Nicola Severino for pointing out the manuscript.
252 See Jacob 1985; Arnaldi 2013.

Figure 50 Sundial on the abbey of St Pierre, Uzerche (Corrèze): six divided with letters for the canonical hours. Photo: Bernard Rouxel.

Eleven-sector dials

Particular attention must be paid to medieval dials divided into eleven-hour sectors (see Figure 54). This way of dividing up the daily space was common to all the areas of Greek–Byzantine culture, including France, although this has not yet been closely studied.

Six of the nine Greek medieval sundials illustrated by Schaldach have eleven divisions.[253] Similar dials can be found in Armenia (Aghjots Vank, Ereruk', Haghartsin, Makaravank, Saghmosavank, Zvarnots),[254] in Israel (on the south wall of the Armenian church in Jerusalem),[255] in France (abbey of Sainte Marguerite in Bouilland, Côte d'Or, Saints Cosma and Damian in Gigondas, in the Vaucluse, the Collegiale of Saint Geniez in Thiers),[256] and in Italy (Conversano, Valenzano, Piacenza).[257] The consequence of this kind of division was the disappearance of the meridian line at the sixth, or noon, hour.

How were the twelve hours of the day read on such a sundial? Indeed, were there always twelve hours, or do the eleven division dials represent a totally different time system? Schaldach found that Byzantine documents always deal with a twelve-hour day. The question therefore is: if there were twelve hours in Byzantine culture areas, how could an eleven-sector dial show them? Why choose a division of this type? There are no safe answers, but a possible solution is suggested by a twelfth-century dial surviving on the monastery of Saint Benedict in Conversano, (province of Bari); the hour lines and the containment semicircle are materialized in small porphyry-coloured mosaic squares. At the end of each hour line the initial letters of the hours are highlighted with turquoise squares:

P[rima], S[ecunda], T[ertia], Q[uarta], Q[uinta], S[exta], S[eptima], O[ctava], N[ona], D[ecima], U[ndecima], e D[uodecima]).

This sundial, the only one of the three Italian examples to have hour line numbering, has the initial letters of the hours placed at the end of each hour line. If we consider even the morning and evening twilight in what we call 'day' then each hour line would represent the centre of each of the twelve hours. Some Greek-Byzantine models seem to confirm this. The sundials of Aghia Triada in Merbaka (thirteenth century) and that of Amfissa number their lines, from the first horizontal, consecutively with the letters of the Greek alphabet A, B, Γ, Δ, E, S, Z, H, Θ, I, IA,

---

253 Schaldach 2006, 44–52.
254 *Cf.* mainly Cuneo 1988, ii, 812–13; Lush 2011.
255 Adam 2001.
256 Cowham 2007, and Schneider 2007, 81–3.
257 All three Italian sundials have a relationship with Byzantine culture. See Arnaldi 2003a and 2003b, Azzarita 2005.

Figure 51 Sundial at Mérinol-Les-Oliviers, Notre-Dame-de-Porporières: six divided with the letters of the canonical hours. Photo: Jean Louis Labaye.

IB, just like in Conversano, and the same arrangement is found in the Zvarnots sundial in Armenia.[258]

Other sundials, however, such as that at Orchomenos (873–874) in Boeotia, central Greece (see Figure 54), and Oshakan in Armenia, start numbering at the first oblique line below the horizontal to the left, indicating, in fact, the end of every hour passed. In this case, therefore, the last hour line (the horizontal to the right) establishes the end of the eleventh hour of the day. Did the twelfth pass with the Sun under the horizon, or was it not counted at all? The same question arises for the sundial of Aghios Laurentios (tenth century) in Greece and for the Armenian dials of Aghjots Vank (thirteenth century) and of Haghartsin (thirteenth century), which, by contrast, have the numbering between the two lines that limit the space of each of the eleven hours. Someone saw in this division into eleven sectors a possible link with the declination of the wall. Schaldach points out that all the Greek sundials still *in situ* and eleven-divided (Orchomenos, Aghia Triada, and the two sundials on the Church of the Saviour in Amfissa) are located on east-declining walls of 10°, 20°, and 15°.[259] This east declination, however, does not, in itself, seem to be crucial for the design of the lines, because the same layout is also found on sundials drawn on walls declining towards the west.[260] Computer simulations have shown that the temporal hours shown by the shadow of an horizontal stylus on an eleven divided dial at the latitude of Orchomenos (38°: 30′ North, dec. 10° East), and with the same declination, will only correspond to the lines traced on the dial with an acceptable tolerance at the equinoxes. Obviously, the same conclusions apply in a mirror image for equal declinations towards the west.[261]

Ten-, thirteen-, fourteen-, sixteen-sector dials, and other anomalies

A small group of medieval sundials with subdivisions into ten, thirteen, fourteen, and sixteen equal parts pose a further problem

---

258 On the sequence number A, B, Γ, Δ, E, S, Z, H, Θ, I, IA, IB, and its characteristics, see Schaldach 2006, 52, 221.

259 Löschner 1906; Schaldach 2006, 47–9.

260 The sundial of Conversano declines about 15°, 8′ towards West and that of Valenzano is located on a wall declining towards west of 14°, 23′, that is, the exact opposite of the Greek ones.

261 The best match between the shadow and the hour lines is obtained only with a wall declination equal to 12° East, or West, at a latitude of 40° North and a solar declination of 0°.

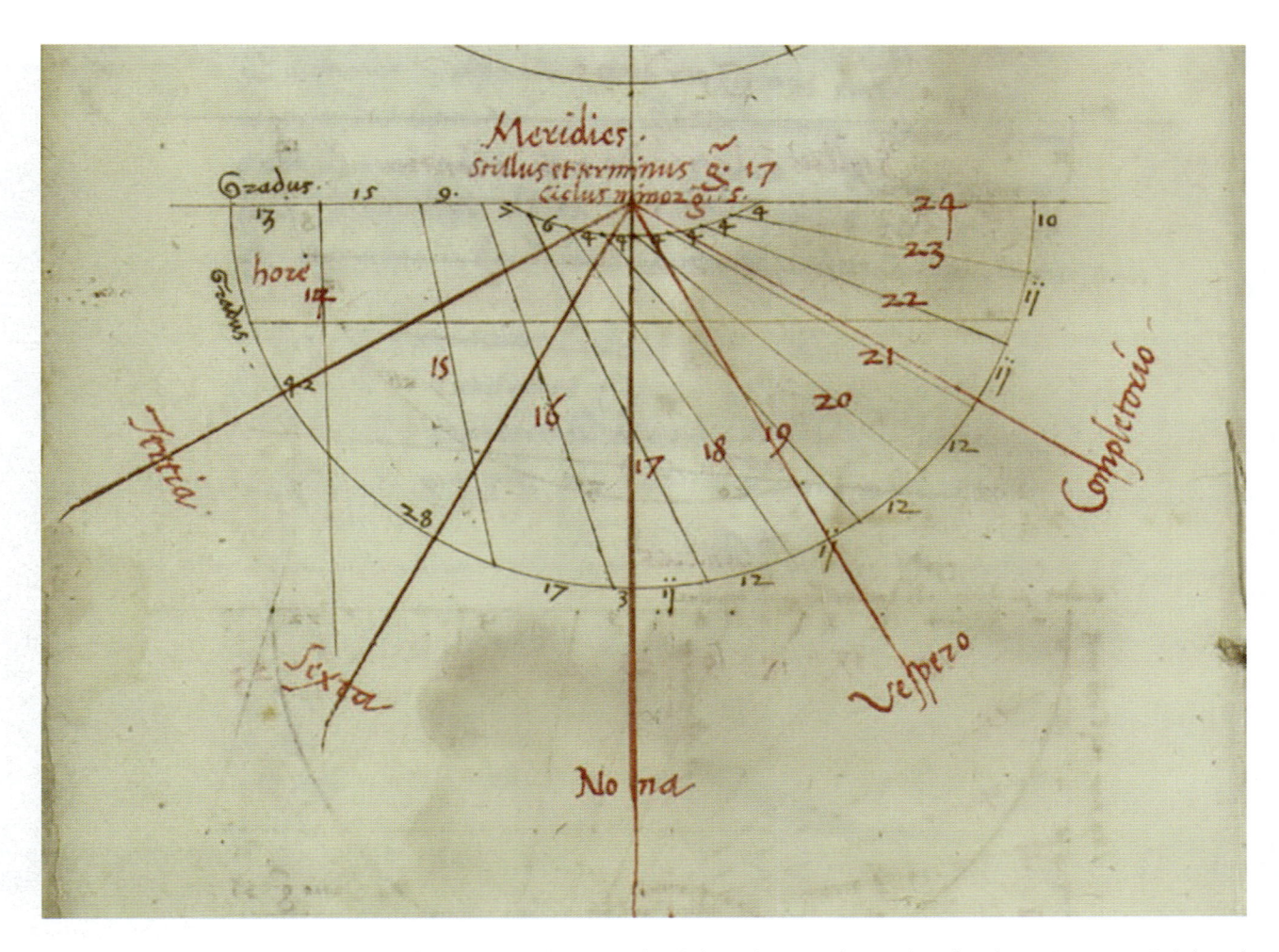

Figure 52 A six-divided sundial with the names of the canonical hours. The dial overlaps another with Italian lines. Lawrence J. Schoenberg Collection of Manuscripts: The Kislak Centre, University of Pennsylvania, Ms LJS 497, fol. 26v (early sixteenth century).

of interpretation and only inferences and likely probabilities can be presented. Ten equal-part dials are uncommon but to be found in Italy and elsewhere. Not many have survived, and they were probably few even in the Middle Ages. Figure 55 shows the sundial on the church of Telč, in Bohemia.

For thirteen-part dials a minimal explanation is suggested by a diagram—almost a doodle—drawn at the end of a martyrology dated 1066 preserved in the library of the Abbey of St Gall.[262] The drawing depicts a sundial divided into thirteen sectors, titled: *Discite crisocomi motus distinguere phebi*.[263] A somewhat poetic title, but what interests us is that across some of the fourteen hour lines depicted in the manuscript we read the initial letters of some diurnal canonical hours, and this is the key to understanding this anomalous division. On the second line, we read the letter P, on the fourth line T, on the seventh S, on the tenth N, and on the twelfth V.[264] There are at least two variants of this system: the first consists in dividing the semicircle of the hours into fourteen equal parts (as seen in the sundial of the church of Cochstedt, or in that of Großenwieden in Germany); the second, also present on the Schönebeck sundial, Germany, shows the division of the fourteen sectors by 15° each in an arc of 210°, even if the two sectors traced above the horizontal line can never receive the shadow of the gnomon. They can however be considered as symbolizing the periods of morning and evening twilight.

The division into sixteen sectors can be considered a further variation of the division of the *nychthemeron* into eighths. It can, therefore, be seen as an improvement of the division into eight spaces, which in turn had improved on that into four sectors. An example of this division was illustrated by Haigh and recognized by Hall at Sinnington and Lockton.[265] Other anomalous divisions are those in fifteen, seventeen, and more equal sectors in a semicircle of 180°, but for these very rare cases no acceptable explanation has yet been found.[266]

---

262 MS Saint Gall, 450, 46
263 Learn how to divide the motion of the Sun; '*crisocomi phebi*', literally 'Phoebus with golden hair', that is, the Sun god.
264 I am grateful to Nicola Severino for pointing out the manuscript page.

265 Haigh 1879; Hall 1997, 3.
266 The only person who has tried to explain these anomalous subdivisions, especially the latter, seems to be Herbert Rau (see Rau 2000 and his other writings).

Figure 53 Sundial with Greek letters for the canonical hours. Santa Maria della Strada, Taurisano (Lecce). Photo: Antonio Ciurlia.

EARLY MEDIEVAL SUNDIALS

As in antiquity, so in the Middle Ages were different kinds of sundials used. They could be fixed or portable, common or unusual. Fixed dials were usually vertical and engraved on south-facing walls, while horizontal stone slabs on the ground are less common. Portable dials took different forms of which the horary quadrant is probably the most important. Such instruments coexisted with the fixed sundials traced on the walls of Romanesque churches. Evidence for them comes from manuscripts and from surviving examples.

Fixed vertical sundials

The shape of almost all medieval fixed dials was very simple: a semicircle engraved on a flat surface and divided into as many segments as there were hours or times to show, as discussed above. All the hour lines converged on the centre of the horizontal diameter where a metal gnomon (*virgula ferrea*) was placed perpendicular to the wall.[267] It is not difficult to prove that a design thus conceived did not have the scientific rigour present in most Græco-Roman sundials. But, however incorrect, it was sufficient for the needs of early medieval societies. A medieval sundial was, above all, a symbol even before it was an instrument for measuring time. Even its divisions could have different allegorical meanings: twelve for the apostles, four for the elements, the ages of the world, and of man, six was twice the perfect number three, and so on. Sometimes dials placed in cemeteries automatically assumed the explicit meaning of a *memento mori* or of the *rota temporis* as a perennial warning of the transience of human life. In many cases, the sundial even took on a magical-thaumaturgical character.[268]

Azimuth sundials

The Greeks and the Romans fully understood the course of the Sun and the various set points on the horizon. The application of this knowledge to the wind roses described by Vitruvius[269] and by Pliny the Elder[270] was not therefore difficult. Diagrams began to appear in some manuscripts written at the turn of the eighth and ninth centuries—images certainly recovered from an older tradition—where the course of the Sun is illustrated at the three primary moments of the year: the two solstices and the

---

267 For the origin of this fan-shaped sundial, see the discussion by Bonnin, in Chapter 1 of this volume.

268 See Arnaldi 2000a.
269 M. Vitruvius Pollio, *De Architectura libri x*, i, 6, 6–7.
270 Pliny, *Nat. Hist.*, xviii, 326–333.

Figure 54 Sundial with eleven sectors showing letters for ten hours of the day. Orchomenos (Boeotia), central Greece. Photo: Karlheinz Schaldach.

equinoxes with three diurnal arcs.²⁷¹ Later, a circular diagram was diffused, where the diurnal arcs measured in equal hours were often combined with a precise sequence of unequal hours. The most important and significant example is one traced in a manuscript compiled in the early eighth century at Echternach that belonged to St Willibrord (658–739).²⁷² This is the diagram of a 'dial' which had an extraordinary diffusion in the Carolingian period (see Figure 56).

In layout this dial is contained in a circle whose radius is divided into five equal portions by four concentric circles. The first circle, the central one, represents the *mundus*, the second, going towards the outside, describes the diurnal arc at the winter solstice, the third reproduces the day arc at the equinoxes and the fourth represents the diurnal arc at the solstice in summertime. The diurnal arcs are measured by eight rays placed at forty-five degrees to each other, emerging from the central circle. From the second minor circle (winter solstice) four other segments are traced exactly in the middle of the four quarters of the upper semicircle which represents the diurnal part. These segments show the third and ninth hours in the two solstices. Figure 56 faithfully reconstructs the diagram of Willibrord's *horologium* in MS Paris, BnF lat. 10,837, and omits only the superfluous explanations for its gnomonic use. The solstice diurnal arcs begin and end on the lines inclined at forty-five degrees with respect to the cross of the principal cardinal points. This exact location refers to the latitude or 'climate' that has a longest day equal to eighteen equinoctial hours and six for the shortest one. In the Middle Ages these are the values for Ireland or northern Britain—regions and places familiar to Willibrord.²⁷³

More than just a sundial with a gnomon, this was a geometric astronomical scheme, which allowed an approximate reading of the hours of the day from the centre of the diagram by visually evaluating the direction of the Sun above the horizon. The hour lines are all oriented according to a fan that goes from east, through south, to west; it is therefore a matter of solar

---

271 Obrist 1997, 57–9.
272 St Willibrord was one of the first evangelizers of Transrenian Germany. At the age of twenty he was sent to study in Ireland at the abbey of Rath Melsigi, now identified with the abbey of Mellifont (County Louth) near Monasterboice. From there he was then sent along with twelve other companions to evangelize the Germanic tribes of Friesland. He travelled a great deal and came several times to Italy. He died at the age of eighty-one in Echternach (Luxembourg) in the small monastery he founded.

273 For a complete description, and for the original image of the MS, see Obrist 2000.

Figure 55 A ten-divided dial from Telč church (Bohemia). Photo: Reto Ambrosini.

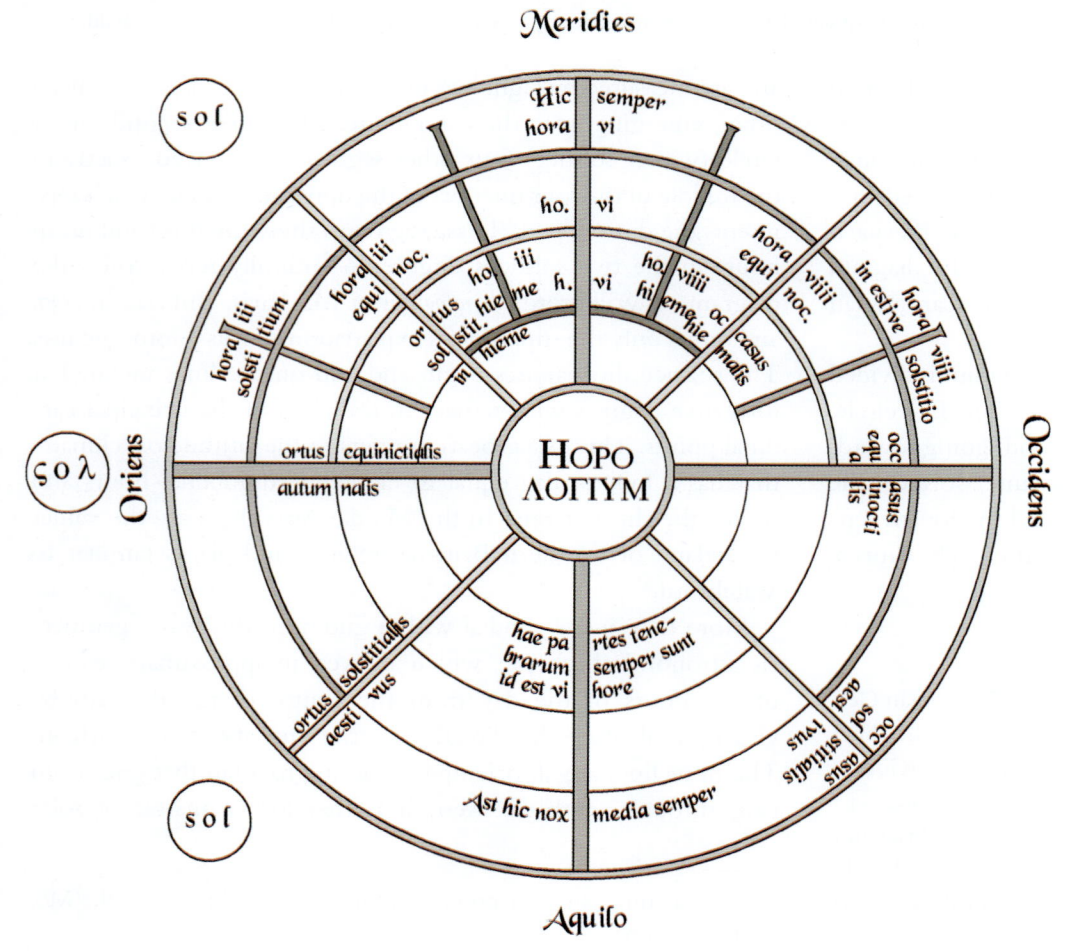

Figure 56 The St Willibrord Horologium, redrawn from BnF Paris, MS lat. 10,837, fol. 42r.

Table IV The shadow scheme of Rutilius Palladius. The values below the month columns show the length of the shadow measured in 'feet' at every unequal hour of the day

hours	Ian & Dec	Feb & Nov	Mar & Oct	Apr & Sep	Mai & Aug	Iun & Iul
1 & 11	29	27	25	24	23	22
2 & 10	19	17	15	14	13	12
3 & 9	15	13	11	10	9	8
4 & 8	12	10	8	7	6	5
5 & 7	10	8	6	5	4	3
6	9 Feet (*pedes*)	7	5	4	3	2

directions and not lines crossed by the shadow of a gnomon. Following this scheme we find other, later, manuscripts[274] that show the same diagram conceived as a real horizontal sundial engraved on a flat stone.[275] It is, therefore, easy to recognize this path in several medieval sundials, in particular in some Anglo-Saxon models such as those at Coates, Aldborough, and Darlington.[276]

Shadow-length schemes

Finding time from the length of the shadow of a gnomon or a person was known in Babylon and was adopted in the fourth century BCE by the Greeks.[277] Monks of the early Middle Ages also knew such 'shadow schemes' and adopted them as a real sundial for travellers. By placing themselves with their shoulders facing the Sun they measured the length of their shadow using the length of their foot as a unit. The time was given by the agreement between the length of the shadow and the numerical value provided by a particular mnemonic scheme drawn up for each hour of the day and for all twelve months of the year. This procedure was known under different names as: *horologium viatorum*, *orologium viarum*, *oralogium horarum*, *horologio solare*, or more simply, horologium. The best-known Latin shadow scheme is that drawn up by Rutilius Palladius (*c.*375–*c.*450 CE) in his work *De re rustica* (Table IV).

The 'shadow scheme' developed widely during the Middle Ages. It was widespread in both Islamic and Christian lands. The earliest medieval Latin tables that we know date from the eighth century. They were written to complement astronomical calculation manuals, missals, and calendars, or they appeared in notebooks together with other gnomonic subjects.[278] Starting from the shorter length (sixth hour from June to July), the ratio between the various shadow lengths of each row and of each column follows an almost regular and constant increase, which is easy to memorize. However, depending on the tradition followed, the schemes differed from each other in the sequence of numerical values of the shadows. Four main models can thus be distinguished: Aachen, Flavigny, the Irish or St Gall model, and Mozarabic. The tables were reproduced in various forms: tabular, textual, and graphic. Many of those proposed graphically were presented with the numerical values of the 'feet' entered in the spaces obtained from the partition of a semicircle whose final design very much resembled a fixed medieval sundial.[279]

THE DEVELOPMENT OF PORTABLE ALTITUDE DIALS

During the entire Middle Ages, time could be measured with portable devices useful to those who, travelling, did not find themselves near a church with a dial. Portable sundials were called *viatoria pensilia* by Vitruvius, precisely because they were small objects easily transportable in the pocket, and *pensilia* because to be able to read the hours it was necessary to keep them suspended by means of a cord or a string. Vitruvius avoided describing them because, according to him, those interested could easily consult the numerous books (now lost) that were in circulation at that time.[280] He limited himself to citing two models: one for a given latitude (climate) and one for several latitudes.

Several ancient portable dials found in recent archaeological excavation are of different geometric, sometimes bizarre,

---

274 Ten manuscripts are known. Eight of them are investigated in Arnaldi 2012b, and two more recently found by the writer are not yet published.
275 For a complete explanation of these kinds of sundial and the manuscripts known, see Arnaldi 2012b.
276 Scott & Cowham 2010.
277 See the account in Bonnin, this volume Chapter 1.

278 For shadow schemes in Antiquity see Bilfinger 1886, 55ss; Neugebauer 1975, 736–74; Hannah 2009, 75–80; for Islamic examples, see King 1990.
279 For the medieval schemes see Valdés Carracedo 1997; Borst 1998, 476; Schaldach 2008; Arnaldi 2020.
280 Vitruvius, *De architectura*, ix, 8.

shapes (like the famous 'ham dial' discovered at Herculaneum),[281] for example, plates, cylinders, fixed and rotating disks,[282] or of rings.[283] A first classification was published in 1997,[284] but today it can be updated from more recent publications.[285]

This Roman tradition was not completely lost. The Middle Ages inherited ancient precepts for the construction of portable dials, probably through simple, popular manuals now known only through small and scattered textual fragments. A few rare examples of actual dials have also survived. Judging from some manuscript illustrations, small pocket rules showing the values of the shadow schemes discussed above may also have existed. These schemes were based on the height of the Sun above the horizon: the higher the Sun, the shorter the shadow cast by the gnomon on a horizontal plane, the longer on a vertical plane. This was basically the principle behind many Roman portable dials like the Portici 'ham', or the several examples with interchangeable discs held in a round box with a gnomon hole in the side.[286]

'Primitive' portable sundials

Three very simple sundials, perhaps of Nordic origin, lacking in mathematical sophistication, and typically early medieval, but so far known only in the British Isles, are known only from a drawing published by Way and by Gatty,[287] a description by Gatty, and a third kept in the Museum of the History of Science in Oxford (Figure 57).

There are doubts about the first two objects being sundials but they seem to be confirmed by the surviving artefact[288] that is clearly a portable early medieval sundial. It is pear shaped (83 × 110 × 23mm) and made from a small sandstone block. It has been ascribed only an unspecified 'medieval' age, but can likely be attributed to the Anglo-Saxon period. The execution is good: the block is divided into an upper part with a circle divided into eight equal portions by straight lines passing through a central hole, and a lower part with a typical semicircular vertical sundial of Anglo-Saxon design with four equal sectors. Each space of the eight included in the upper circumference is further divided

Figure 57 Medieval portable sundial. Photo courtesy of the Museum of the History of Science, Oxford.

into three by two equidistant points, thus generating a sequence of twenty-four equal parts. It is unclear how such a dial was used, but the upper circle could be used as a compass to find south, and once found, the lower dial was placed vertically pointing south.[289] Such portable sundials were probably used during the rest periods that punctuated the travels of itinerant monks, in places where there was no fixed sundial nearby.[290]

The cylinder or pillar dial

One of the best known and most widespread portable timekeepers of the Middle Ages, and which would continue in use until to Renaissance and beyond, was the cylinder dial. This dial descended from the first century CE almost unaltered and in medieval times, like other travellers' dials, was known as an

---

281 National Museum, Naples. It is a small portable metal dial in the shape of a leg of pork, the gnomon being provided by the tail; see Anon., *Le pitture antiche di Ercolano e contorni incise con qualche spiegazione*, T. iii, prefazione, Naples mdcclxii, v–xvii; cf. Severino 1997; an excellent study of this sundial is in Ferrari 2009.
282 See Baldini 1754; Tölle 1969; De Solla Price 1962b; Field 1990; Wright 2000a; 2000b.
283 Gounaris 1980.
284 Arnaldi & Schaldach 1997.
285 Hoët-van Cauwenberghe 2012a; 2012b; 2012c, 99; Schaldach 2017, 91.
286 Because of their fixed gnomon, in these latter the azimuthal angle also came into play: strictly therefore they were alt-azimuthal dials.
287 Way 1868, 221–3; Eden & Lloyd 1900, 86–7.
288 Museum of the History of Science, Oxford, inv. No. 38841.

289 See Arnaldi 2011a, 221.
290 In the Middle Ages, the reference to the *viator* who carried a *horologium viatorum* did not refer to travellers in general, but to a particular type of itinerant monk. The movements of this type of monk are characterized by more or less long stops in one place; see Morini 1999.

horologium viatorum.²⁹¹ Today it is best referred to simply as a 'cylinder dial' (see Figure 44). It was usually made of boxwood or some other hard material suitable for turning. It was composed of a rotatable cap and a hollow cylindrical body, on the outside of which the hour lines were traced. One or two metal gnomons were pivoted in the turned cap, and they could be folded down and enclosed within the cylindrical body of the dial itself when not in use. A suspension ring was mounted on the cap.

The hour lines on the exterior of the dial were curved in the horizontal and intersected a series of vertical lines (six or twelve) representing the months of the year. Use was very simple: the gnomon was erected and the cap rotated until the gnomon was placed above the month line for the date on which a reading was to be made. The dial was then suspended and pointed so that the gnomon tip was in the same plane as the Sun, and its shadow, exactly vertical, fell on the hour lines. The operation of the cylinder was based on the angular measure of the height of the Sun on the horizon in every period of the year and in every hour of the day; the higher the Sun, the longer the shadow of the gnomon on the vertical cylinder scales (see Figure 44).

Existing Arabic manuscripts describe a very similar instrument, the *Mukḥula*, which has been traced back to before the tenth century, when it is already present in a list of astronomical instruments compiled by Abu Abdallah al-Khwarizmi.²⁹² The Mukḥula, however, was not cylindrical, but conical. The oldest description of a Mukḥula, at this time probably cylindrical, dates back to the ninth century, and is illustrated in a manuscript found some years ago by David A. King.²⁹³ The oldest Latin document describing the construction of such dials was compiled at Reichenau by the Benedictine monk Herman, called 'the Lame' or 'the Cripple' (1013–54).²⁹⁴ The late Anglo-Saxon dial known as 'the Canterbury pendant', although flat, belongs to the same Roman gnomonic tradition; its construction follows the same rules as the portable cylinder sundial (see Diagram 14).²⁹⁵ As an alternative to the fixed gnomon mounted on a rotating body, the 'pendant' had a mobile gnomon, separated from the body of the object, which was inserted in a housing hole at the top of each monthly column.

A model similar to this, but of more rigorous construction, survives in the twelfth-century Syrian dial by Hamiyyat Allah al-Astūrlabī,²⁹⁶ while a dial similar to the Canterbury dial but made of bog oak and bone, and of an indeterminable late medieval or early Renaissance date, is also preserved.²⁹⁷

---

291 Thorndike 1929; Krenn 1977.
292 Wiedemann & Würschmidt 1916; see also Livingston 1992.
293 See Jordan & King 1988, 26; *cf.* Arnaldi & Schaldach 1997, 115; King 2004, part iv, ch. 7, par. 7.4, 585–6.
294 Hermannus Contractus, *Demonstratio componendi cum convertibili sciothero horologeci viatorum instrumenti*, MPL, cxliii, coll. 405–8. See Krenn 1977.
295 Turner 1990 94, N° 168 and references there given. Arnaldi 2011b, c and Arnaldi 2012a.
296 Casanova 1923; Turner 2019, 187–289.
297 Collections of the Adler Planetarium and Astronomical Museum. Illustrated and briefly described in Turner 1990, 95, N° 169.

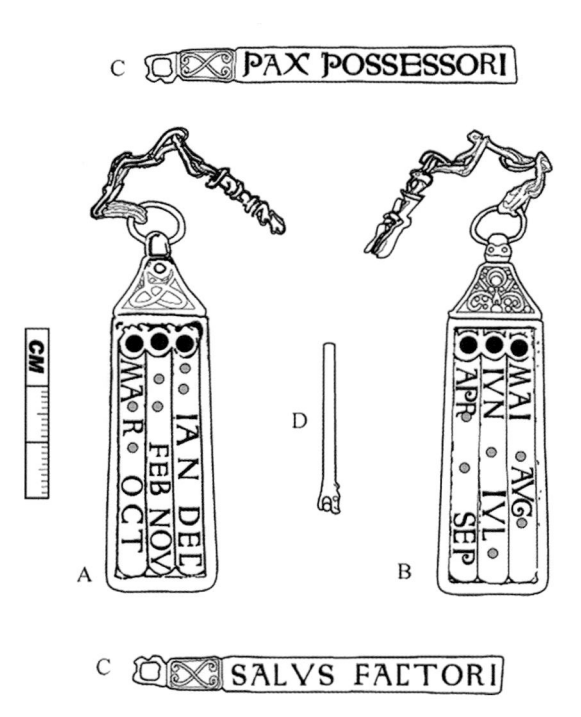

Diagram 14 The portable sundial found at Canterbury in 1938. A: winter face; B: summer face; C: right and left sides; D: the gnomon.

Rectilinear dials

In the first half of the fourteenth century, a particular type of multi-latitude, portable, high-altitude dial began to spread in Europe—the rectilinear dial. Those that developed within the medieval era, specifically no later than the first half of the sixteenth century, are the *navicula de venetiis*, the *organum Ptolomei*, the *quadratum horarium*, the *parallelogrammum*, and the *horoscopion*. Sundials of this type are equipped with two pierced pinnules for sighting the Sun, and a plumb line equipped with a sliding bead, which shows the hours of the day on the intersection between the bead and the hour lines. The main feature of medieval straight-line dials was that their hour lines are all parallel to each other and show **equal hours**.

The *Navicula de Venetiis* and the *Organum Ptolomei*

Together with the common *horary quadrant* discussed further on, the universal sundial called *Navicula de venetiis* (Little Ship of the Venetians) is one of the most interesting sundials of the Middle Ages and at the same time one of the very few created by calculation. It belongs to the family of universal dials—those able to read time at any latitude. Its name derives from its shape similar to that of a small commercial Venetian cargo ship (Figure 94). The mathematics and geometry behind the construction of this dial, and of rectilinear sundials in general, have been fully described.²⁹⁸ However, even today, the mental procedures that led gnomonically inclined mathematicians to formulate such a construction are unknown to us.

---

298 Archinard 1988; Archinard 1995; Fantoni 1988, 390–415; King 2003; Eagleton 2010.

The operation of the *navicula*, compared with its complex construction, is not difficult. The cursor which slides on the rod is first positioned for the local latitude marked along the inclinable shaft of the instrument (a scale of latitudes is generally engraved on the mainmast from 0° to 66. 5°). Next, the mast is tilted to the right or left making sure that its index at the bottom positions itself on the date of the day on the calendar scale. The plumb line is then stretched towards the lateral calendar and it moves the mobile bead on the date of the day. At this point, the Sun is sighted through the pinnules at the base of the two small bars and when the plumb line is dropped, the hour is read on the intersection between the hour line and the plumb line (see Figure 95).

The *Navicula de Venetiis* is quite rare. Today only nine are known: five English examples from the fifteenth century, and four of various origins made in the sixteenth century.[299] Recently David King has argued that universal rectilinear dials (though not specifically the navicula) originated in ninth-century Baghdad. However, the *navicula* as we know it seems to have originated in late medieval England.[300] In the fifteenth century, a transmission of English manuscripts to the continent somehow took place, in particular towards Germany; more properly in the Viennese area. A second model, *organum Ptolomei*,[301] developed with a simplified geometrical construction. It does not seem that this variant was very successful outside the area of the Viennese school, from whence come most manuscripts relating to it. So far, no specimen precisely corresponding with this model has been found.

The universal rectilinear dial of Johannes Müller of Königsberg (Regiomontanus) In Nuremberg in 1474, Johannes Müller from Königsberg (1436–76), commonly known as Regiomontanus, published *Kalendarium*, at the end of which he inserted a drawing of a universal rectilinear sundial that he called *quadratum horarium generale*. This particularly ingenious dial was equipped with an articulated metal arm on which the plumb line was attached. The instrument acquired Regiomontanus' name, even though he acknowledged his debt to an unidentified *antiquus compositor*. In effect, the graph is very similar to the *Navicula*, but the movable limb has been replaced by a grid of latitudes that also incorporates into itself the calendar scale marked on the lower part of the *Navicula*. To read the time, the articulated arm on the grid has to be placed at the exact point where the local date and latitude coincide. The model would remain in use for nearly three centuries.

The *parallelogram* and the universal quadrant of Peter Apian (Bienewitz) The *parallelogram* is form of rectilinear dial that has affinities with those just described, but of which the origins are unknown. A drawing and a description of it were first published in 1531 by Sebastian Münster (1488–1552) in his *Compositio Horologiorum*, and again in 1533 in *Horologiographia*,[302] and claims a place here as Münster relates that it had been circulating for some time.[303] The instrument, which today is better known as a 'Capuchin' or single-latitude rectilinear dial, worked only for the latitude for which it was built. It no longer had a movable arm, but a cursor to be positioned along a calendar scale. Peter Apian (or Bienewitz, 1495–1552) made the *parallelogram* universal in 1532 when he published a drawing with a description in his work *Quadrans Apiani Astronomucus*. Münster's instrument, the astronomical quadrant, like that of Regiomontanus, extended the date scale into the latitude grid. It was therefore again necessary to have a jointed arm to place the point of attachment of the plumb line on the correct date and latitude. In the same year, Apian published a second work, in which he proposed another version of the instrument, calling it *Horoscopion*. In the following year development ended with the 'poplar leaf', a graphically enriched instrument, simplified with respect to the *horoscopion*, designed specifically to please Lord John William of Loubemberg, whose coat of arms featured three diagonally placed poplar leaves.[304]

Many authors have tried to reconstruct the evolution of this type of sundial, but it remains disputed. Stebbins believed that the first models of the *Navicula de venetiis* dated from the thirteenth century and possibly even earlier.[305] Fantoni was convinced that the *Navicula* was an 'evolutionary variant' of the *quadratum horarum* of Regiomontanus, 'superbly decorated', but 'technically imperfect'.[306] For him, priority rested with the Capuchin dial. Archinard and Severino have also advanced hypotheses for the priority of the Capuchin (at least in its primitive form).[307] The most recent, and perhaps most convincing discussion is that of Yvon Massé.[308] It should, however, be noted that the use of the orthographic projection in these dials provides interesting evidence for its use before the publication of Ptolemy's *Analemma* by humanist scholars.

---

299 The five fifteenth-century instruments are described in Eagleton 2010, ch. 2, the most detailed discussion of the instrument. Three of the sixteenth-century dials are also described there in Eagleton 2010, ch. 8. For the fourth, constructed of boxwood, see Delalande, Delalande, & Rocca 2020, 573–85.
300 King 2003.
301 For this development, see Eagleton 2010, ch. 7.

302 Münster 1531; Münster 1533.
303 Münster 1533, cap. xli, 250.
304 Apian 1532; on that instrument both equal hours, and unequal hours could be read.
305 Stebbins 1961.
306 Fantoni 1988, 374–415.
307 Archinard 1988; Severino 1995, 31–40.
308 Massé 2008. Although the chronological reconstruction of Massé is based on factual data as well as on logical assumptions, it still has some errors. It confuses the *Organum Ptolomaei* with the *Navicula* and so postpones the date of the latter instrument. Massé proposes a hypothetical Arabic instrument, the 'carca-quibla' as a possible origin for rectilinear dials.

Origin and development of the nocturnal as a complement to sundials

Because of the Earth's rotation, the fixed stars seem to rotate constantly around the celestial poles and the constellations rise and set. Already monks in the sixth century used this phenomenon to know the hours of the night.[309] The *significator horarum*, the monk in charge of indicating the time for the offices, was advised to observe the stars.[310] The practice was already known to John Cassian,[311] was familiar in the sixth century to Gregory of Tours,[312] and known in the eleventh century to Peter Damian. Detailed night-time tables were compiled in order to facilitate the task of the *significator horarum* who, positioning himself in a particular place in the cloister, observed the passage of the constellations over some architectural point of the monastic building.[313] Some monasteries were equipped with **water clocks** to measure night-time, but even these had to be regulated either by the stars or by the daytime shadow of the gnomon on a sundial.[314]

This cosmic mechanism suggested the possibility of determining the night hours simply by evaluating the position of a given star with respect to the pole star. Thus c.800, Pacificus, the archdeacon of Verona, may have invented[315] a method which was to become the basis of all later Renaissance nocturnals.[316] The method, described in a poem entitled *Argumentum Horologii*, consisted essentially in dividing the northern sky into four quadrants by drawing an invisible cross, in whose centre the axis of the celestial pole rotated. By considering the rotation of a particular bright star, called *computatrix* (that is, the one that allows us to count the hours), close to the North celestial Pole, it was possible to determine the hour.[317]

The diagram described by the archdeacon is similar to a circle divided into four quarters by a cross where the north celestial Pole resides.

Details of this device are lacking, but a good idea can be drawn from later manuscript illustrations accompanying the poem, and by the texts accompanying the diagrams related to 'shadow

---

Spera caeli quater senis horis dum revolvitur,	While the celestial sphere rotates in four times six hours
omnes stellae fixae caeli, que cum ea ambiunt,	all the fixed stars of the sky that rotate with it travel
Circa axem breviores circulos efficiunt.	through smaller circles around the axis [of the poles].
Illa igitur, quae polo apparet vicinior,	The star, therefore, that appears closer to the pole,
inter omnes tamen ei splendor est praecipuus,	and among all is the brightest, is called the 'counter'
ipsa noccium horarum computatrix dicitur.	(computatrix) of the night hours.
Argumentum en inventum: cardini oppositum recta linea si	Here is the rule found: if from the pole the gaze is kept
serves luminum intuitu,	along a straight line, the time of the night can be known
horas noctis nosse potes galli sine vocibus.	without the need for the crowing of the cock.
O quam pulchrum stema tenet clavorum positio;	Oh what a beautiful diadem forms the position of the
crucis Christi rotae fixi hoc in horologio,	spokes of the wheel of this watch; the cross on which
in quia ipse carne pendens pro salute hominum.	the body of Christ was affixed for the salvation of men.
Dextra, leva et profunda, quae tendit ad aethera,	To the right, to the left, to the bottom and up to the
serva semper computatrix per distincta tempora,	sky always serves the star that counts [the hours] in the
aequinoctia designans atque solistitia.	different seasons that designate the Equinoxes and the Solstices.
Ante axem siquis volvens curiosus steterit,	If the observer is in front of the pole he will see the
aequinoctium vernale a sinistra noverit,	spring equinox to his left and to his right he will notice
cernere ad dextram sui autumnale poterit.	the autumnal one.
Solistitia duobus indita temporibus:	The Solstices are linked to two seasons: summer is called
aestivalis, qui erectus ad superna ducitur,	the rising ray, winter is said to fall on the descending ray.
radius ad ima mersus hiemalis dicitur.	

---

309 See Sheridan 1896; Poole 1915; Constable 1975.
310 McCluskey 1990.
311 Joannes Cassianus, *De cenobiorum institutis*, ii, 17 (MPL, xlix, col. 109A).
312 Gregory of Tours in MGH I, pt 2, 854–72.
313 See Poole 1915 and Constable 1975.
314 In the second half of the thirteenth century, for example, the sacristan of the abbey of Villers (Brussels), who performed *significator horarum* duties, possessed a detailed handbook on the correct way to set the clock that was inside the church. The clock of the abbey was undoubtedly a water clock and the sacristan had to follow scrupulous operations to set it accurately, carefully following the movements of the Sun through the church windows by day and, at night, that of stars; see Sheridan 1896.

315 The epitaph dedicated to Pacificus which is found in the Verona cathedral reads as follows: 'Horologium nocturnum nullus ante viderat./En invenit argumentum, et primus, fundaverat./Horologioque carmen sperae caeli optimum,/Plura alia grafiaque prudens inveniet'.
316 The nocturnal as an astronomical and horological instrument was developed already in the early Renaissance (several examples were made by the makers Della Volpaia in Florence); they consisted of a set of toothed wheels centred in the centre by an eyelet through which the polar star was aimed. The circumpolar stars that served to allow the hourly reading of the night were identified by a long mobile target also rotating around the centre.
317 The conviction of some that the archdeacon Pacificus had used a rose window in the same church where he resided (Saint Zeno in Verona) as a tool to read the night hours, is not, in my opinion, a thesis supported by convincing arguments.

schemes'. Some manuscripts accompany the text with a figure that illustrates its use.[318] The drawings show a simple instrument that consists of a stand upon which an optical tube (*fistula*) is fixed and through which the Pole star is observed.[319] The position of the *computatrix* was probably evaluated with the other free eye on the edge of a wheel positioned around the tube.[320] The *fistula* was thus equipped with a *circumcisa rotella* (a gear wheel),[321] but it is not clear where this was placed. The *circumcisa rotella* is shown in the drawing in MS Chartres 173, but even more clearly in MS BM Old Royal 15 B IX, fol 78v. Other manuscripts, with particular reference to Gerbert of Aurillac, add short texts to the images.[322] In these cases, however, the *fistula* does not appear to have been mounted, but instead was held freely in the hand and introduced at the centre of a hemicycle whose nature is unclear.

It must be premised that the system works only if we consider a subdivision of time in sidereal equal hours; in this case the *computatrix* is located on each arm of the cross, in the position corresponding to that indicated in the poem at 6 a.m. (modern) or at 6 p.m.; in the first case the equinoctial arms are inverted, in the second the solstitial arms. Which is chosen is indifferent. As the purpose of the instrument was to find the hours of the night at a given date, the twelve hours of the day were omitted. The time count started on one arm; midnight, or the sixth unequal hour of the night, was found when the *computatrix* was at the end of the eighteenth sector. From there it was easy to find the hours of the vigil by knowing the number of equinoctial hours that night. This was by more or less concise tables, already widespread in antiquity. For most Italian latitudes, the 'climate' of the Hellespont was used, at the summer solstice the day was fifteen hours long, and the night nine. Counting on the right and left of the midnight point symmetrically as many hours as made up the night in question, the points of sunset and rising could be obtained. The position of the Vigil is easily found halfway between the midnight point and the newly found extreme points. By placing a ball of wax on all the points

---

318 MS Vat. Lat. 644, fol. 76r, tenth–eleventh century; MS Chartres 173, destroyed by bombing in 1944, publ. by Michel 1954; Venezia Biblioteca Nazionale Marciana MS lat. viii 22, fol. 1r, twelfth century; MS St Gallen Stiftsbibliothek MS 18, 43, tenth century (unfortunately this manuscript is a palimpsest; its original text has been totally scraped away to make way for a liturgical text not related to the image); the MS Avranches 235, fol. 32v, twelfth century, the MS B.N.Paris lat. 7412, fol. 15 and the MS London Roy. 15.B.IX, fol. 76v (these last three also mentioned by Poulle 1985). See also Maddison 1969, 30–5.
319 Michel 1954.
320 MS Avranches 235.
321 MS Avranches 235, fol. 32v.
322 MSs. Avranches 235 and Chartres 173. Thietmar of Merseburg wrote that Gerbert of Aurillac built a dial in Magdeburg that *recte illud constituens secundum quandam stellam, nautare ducem, quam consideravit per fistulam miro modo*. It is obvious that we are not referring to an astrolabe, but to a dial similar to a nocturnal or, more precisely, similar to the optical tube discussed here.

Figure 58 Italian Nocturnal with an altitude sundial on the reverse by Girolamo della Volpaia, 1575. Photo courtesy of the Istituzione Biblioteca Classense, Ravenna.

detected in the wheel (*rotella*) in order to be able to feel it in the dark, the user could prepare the apparatus in time before darkness arrived.

The need to know night-time hours led to a subsequent development of portable instruments, called nocturnals, sometimes combined with an altitude sundial engraved on the rear (see Figure 58). This instrument made its appearance in Europe towards the end of the fifteenth century and then spread widely in the sixteenth and seventeenth. It was usually composed of various toothed disks pivoted in the centre with an eyelet through which the polar star could be observed. One or more movable arms enabled following circumpolar stars thanks to which the required time could be known.

Development of time reckoning with fixed dials

In fourteenth-century Europe, with reliable time measurement increasingly necessary, the technical innovation of the weight-driven clock transformed prevailing hour systems and so sundials; also contributing to this transformation was the combination of the growing secularization of social life and mercantile pressure for a more structured time division than that offered by canonical and unequal hours.[323] In the course of the century, unequal hours were largely displaced in general usage by their equal hour equivalents. The hour systems then adopted were largely modernizations of Ancient schemes: Roman civil hours, counted from midnight; the hours ab occasu (now called '**Italian**'), as had been counted by the Athenians from sunset, and the hours ab ortu (now called '**Babylonian**'), as counted by the Umbrians starting from sunrise.[324] While unequal hours were abandoned, three elements were maintained: the separation of the hours of the night and day by the so-called **Nuremberg hours**, the mobile beginning of the

---

323 On this subject see Le Goff 1977, chs. 2 & 3.
324 Gellius, *Noct. Att.*, iii, 2, 4–6

hour count by Italian and Babylonian hours, and the two distinct twelve-hour counts by common hours which kept a separation between two series of twelve hours.[325]

Equal hours Despite traces of equal hours in Antiquity (dials from Olympia and Oropos)[326] and even in the Middle Ages, how it was possible in the mid-fourteenth century to conceive the construction of a sundial subdivided in equal hours? It is difficult, if not impossible, to answer this question at the moment. Certainly, equal-hour dials were described by Islamic mathematicians as early as the ninth century, but no such dials are known from these regions before the second half of the fourteenth century.[327] That construction methods for equal-hour dials came to Europe from Islam is uncertain. There could easily have been an independent 'discovery' in Latin Europe when the need arose. What is certain is that in England, at the end of the fourteenth century, Robert Stikford wrote a treatise on gnomonics in which he described the construction of equal hours (designated 'natural hours').[328] His treatise, of which a unique fifteenth-century copy survives, does not use Arabic methods, nor the typical terminology of such texts. Whence the technique derived remains unknown, but it is perhaps significant that Stickford's tract emanates from St Albans abbey, where Richard of Wallingford (1292–1336) had worked.[329]

Equal hours and the 'Erfurt rule' The advent of mechanical clocks showing equal hours did not cause the definitive disappearance of the unequal, or seasonal, hours.[330] Two sundials built in the sixteenth century are testimony to the survival of unequal hours. The sundial on the Chamaret cathedral in Provence, for example, is divided into twelve equal sectors, but is dated 1548,[331] while the similar sundial in the Franciscan convent of Slano in Dalmatia dates from 1586.[332]

A rule was written in Erfurt and copied into numerous manuscripts between the mid-fourteenth century until the second half of the fifteenth century.[333] It represents one of the first attempts to modernize the tracing of the hour lines of sundials.

---

325 Bilfinger 1892, 185–95.
326 Maybe these two were used for astronomical or didactical meanings, as they were unpractical in common daily life. See Schaldach 2004; Schaldach 2017.
327 Ferrari 2012, 117–22.
328 Davis 2011.
329 Richard of Wallingford (1292–1336), was abbot of the St Albans monastery and discovered the Albion and also made a famous astronomical clock.
330 See Poulle 1999.
331 Opizzo 1998, 21, 27 (fig. 1.1)
332 Tadić 1988, 9–10 (photo 10–11); Tadić 2002, 76–7.
333 The oldest draft of this rule was made in Erfurt, Germany, in 1364. It is certainly not the original copy, because the rule was known in Germany since at least 1334; see Schaldach 2002.

Table V Angular openings for hour lines following the Erfurt rule

TEMP. HOURS a.m.	I	II	III	IV	V	VI
DEGREE	10	11	13	16	18	22
PARTS	5	5.5	6.5	8	9	11
TEMP. HOURS p.m.	XII	XI	X	IX	VIII	VII

It was not a question of adapting the medieval sundials to show a new time system, but rather of improving the reading of the unequal hours on a sundial built on a perfectly south-facing wall at a latitude of 48° (Paris), simply opening the fan of the lines with different angles. In all probability, the 'rule' was devised by Johannes Danck, alias John of Saxony (fl.1322–55), who was considered 'the most learned master of astronomical science who lived in that era in the German countries'.[334]

The rule consisted of some construction indications that roughly mirrored the simplicity of a medieval sundial. One had to first draw a semicircle, marking the centre where a horizontal gnomon was to be placed, after which it was necessary to follow, literally, a numerical table corresponding to the different angular openings of the hour lines. To find the right meridian direction, the system of '**Indian**' or '**Hindu circles**' was recommended.[335] The table divided the semicircle into 90 parts, equivalent to 180°. The first hour was assigned five of these parts (= 10°), the second five and a half (= 11°), the third six and a half (= 13°), the fourth eight (= 16°), the fifth nine (= 18°), and the sixth hour eleven parts (= 22°). The other hours were mirror images of the first six (Table V). The angular values expressed in the text of the Erfurt rule, however, were supported neither by calculation nor by geometric constructions; they were listed one after the other and had to be accepted with confidence by the reader. The resulting figure corresponded to a direct south-facing dial with unusual openings between the hour lines: wider in the central hours of the day and increasingly narrower as the lines became closer at the extreme hours.

The angular values provided by the rule referred to an opening of the lines in a very precise position, which corresponded to the equinoctial shadow angles, at the end of each unequal hour, at the

---

334 And this, as Schaldach points out, says a lot about the degree of teaching of gnomonics in the European universities of that time.
335 Independently, and known by a different name, this system for detecting the meridian was used since ancient times (see Hyginus Gromaticus in the second century) and consisted of a vertical gnomon whose base was the centre of a series of concentric circles drawn on a horizontal plane. The meeting points were marked with the apex of the shadow on one of the circles at a certain hour in the morning and at a relative post-meridian hour. Joining these two places with a line and finding the middle point, a straight line was traced through the base of the stylus and the midpoint found—this line represented the north–south direction.

Table VI Calculated angular values (in degrees) to the Equinoxes at a different latitude, compared with those of the Erfurt rule

Latitude	Calculated angular values (in degrees) to the Equinoxes at a different latitude, compared with those of the Erfurt rule					
	I–XII	II–XI	III–X	IV–IX	V–VIII	VI–VII
40°	11.6	12.3	13.6	15.5	17.7	19.3
45°	10.7	11.5	13.1	15.5	18.5	20.8
48°	10.2	11.0	12.7	15.4	19.0	21.8
50°	9.8	10.6	12.4	15.3	19.3	22.6
Rule data	10	11	13	16	18	22

latitude of 48° (see Table VI).[336] Thus, at this latitude, the error in reading hours on a common, twelve-divided, medieval sundial, decreased.[337]

But the suggested method, besides having a poor gnomonic value, lacked any proof, either geometric and mathematical, and it did not provide other tables for adaptation to a different latitude, nor did it state for which latitude it was composed; it acquired a spurious universality. Despite these obvious shortcomings, the Erfurt rule was successful. Born in the Parisian university environment, it was copied in many codices and gained widespread popularity in Germany, France, and, marginally, even in Italy.

It took nearly a century for its deficiencies to be realized.[338] Commentators tried, therefore, to repair the errors, without rewriting it. Even then, a true method based on correct gnomonic principles was not formulated. The first change to the Erfurt rule was the adaptation of the hour lines to read equal hours, simply giving a new meaning to the hour numbers, effectively inverting the angular values. By chance, the inversion of the angular values of the original rule generated an almost perfect diagram for an equal-hours sundial for the same latitude, 48°, provided, however (even if in the text is never specified), that the gnomon was parallel to the terrestrial axis.

The rule thus modified could also be applied to other latitudes without incurring obvious reading errors; it could be used with errors negligible for that time, from 44° (Florence) up to 51° (Erfurt),[339] effectively becoming a simple 'semi-universal' method for building simple sundials.

From the rule of Erfurt many others derived, each more or less similar to the next. Ernst Zinner (1886–1970) divided the series of manuscripts known to him—over seventy—into four fundamental groups ranging from the original rule to variants for horizontal dials with sixteen hours to those with different numerical series.[340]

The modified Erfurt rule arrived in Italy through Fra' Giocondo of Verona,[341] who learned it in France and inserted it in one of his manuscript volumes, now preserved in the Laurentian Library of Florence (MS Plut. 29.43 fol. 65r).[342] But in Italy this construction method seems never to have taken root; **Italian hours** were established. A general rule for a different time system, however simple, was of no interest. Shortly however, the spread of equal hours would provoke the adoption, or rediscovery, of the polar gnomon.

---

336 Ferrari 2000.
337 It means a semicircle divided into twelve equal sectors of 15° each.
338 The two sundials engraved on the cathedral of Braunschweig, in Germany, have been dated 1334 and 1346. The latter was probably traced in replacement of the first one. The design of the first is based on the original rule, while that of the second follows the directives imposed by the 'reformed Erfurt rule'. See Roslund 2005.
339 These two latitudes must be understood as extreme limits, beyond which, at certain times of the day, the error is no longer tolerable.
340 Zinner 1939, 102–5. A rule similar in simplicity to the Erfurt rule, but obtained with ruler and compass, was proposed by G. B. Vimercato in the mid-sixteenth century in his work *Dialogo de gli horologi solari*, and highlighted by Gunella 2006.
341 Brother Giocondo (Monsignori Giovanni, 1433–1515) was a Dominican friar, a scholar, a mathematician, and an architect. He made the drawings of the Loggia del Consiglio and the door of the bishop palace in Verona and the Fondaco dei Tedeschi in Venice, and in 1509 of the fortifications of Treviso. Vasari wrote of him that he was 'a very rare and universal man in all the most praised faculties'. In the years 1496–9 he was invited as *architecte royal* to Paris by the king of France. In Paris he collaborated in the construction of the bridge of Notre Dame. In Paris he discovered eleven letters of Pliny, which were later published in 1508 by Manutius. In 1511, he edited the first illustrated edition of Vitruvius' *De Architectura*.
342 On the original Erfurt rule, see Schaldach 2002; on the reformed Erfurt rule, see Arnaldi 2009.

SECTION THREE: WATER CLOCKS IN CHRISTIAN EUROPE AND EARLY ESCAPEMENTS

Sebastian Whitestone

WATER CLOCKS

The principal stellar-independent method of timekeeping in Late Antiquity and the Middle Ages was the **water clock** or **clepsydra**.[343] Two possible early sinking bowl specimens, where a small perforation in the bottom of the bowl causes it to sink at a measured rate in a reservoir of water, have been unearthed in England. One, a shallow bowl of four inches diameter inside a clay container, was unearthed in an Anglo-Saxon burial ground at Market Overton in Rutland. The time required for this bowl to sink was found by testing to be approximately one hour. The other bowl, now in the History of Science Museum at Oxford, was discovered at Lakenheath Fen in Norfolk and is nine inches in diameter and of indeterminate date.[344] Unfortunately, their context of use is indeterminable.

For keeping track of time for Christian devotions,[345] two aspects of monastic life together made water clocks useful: the necessity to rise during hours of darkness, and the adherence to **unequal** hours, which disturbed man's inner clock by varying the allotted sleeping period. The problem caused by unequal hours is clear from the admonition to monks to use clepsydrae, written around 845 in the *Commentaries of St. Benedict* by the French monk Hildemar de Corbie (*c*.821–50). In explaining what Benedict meant when he said that monks should rise at the eighth hour of the night, Hildemar drew a circular chart of the temporal night hours entitled '*horologium innocte*' to illustrate exactly where this point occurs in the hours of darkness. He then writes, 'For he who realistically wishes to do this, a water clock is necessary'.[346] It is hard to judge the extent to which this recommendation was followed and how widespread the monastic use of water clocks became. However, it is likely that the early references in monastic rules and consuetudes to *horologia* refer mainly to simple alarm devices. Mechanical devices for measuring time were variously reported from the beginning of the Middle Ages. The word *horologium* is frequently found in early Latin texts and could refer to any device that indicated time or even to written time tables. Reference to a mechanical device is usually indicated by a qualifying adjective such as *aquae, aquatile, aquare, nocturnum*, or *automatum*. Even where there is no mention of water it would seem safe to suppose that, before the advent of the oscillator clock, any automated function performed by a horologium was likely to involve hydraulics. However, caution is required. The '*horologium nocturnum*' that 'no-one had seen before' invented by Archdeacon Pacificus of Verona (*c*.776–844), and mentioned in an epitaph in Verona Cathedral, was long thought to be either the earliest reference to a weight-driven clock with wheels or, at the very least, a water clock. However, rediscovery of the text and the illustration in the '*argumentum*' or instructions that accompanied the device show it to have been a sort of nocturnal with a sighting tube.[347] Similarly, the *horologium nocturnalis* devised *c*.1060 by William, Abbot of Hirschau (1030–91) that was 'agreeable by nature to the celestial hemisphere' probably refers to the stone hemisphere in the monastery of St Emeran at Regensburg and had no mechanical components.[348]

The late-Roman senator Cassiodorus (*c*.485–*c*.580) records that the Emperor Theodoric the Great (*c*.454–526) sent a water clock together with a **sundial** to the Burgundian King Gundobad (?–516) in 507. Gundobad had heard of such clocks from his ambassadors in Rome and wished to have one. Cassiodorus wrote the letter that Theodoric sent to the philosopher Boethius (*c*.480–*c*.524) ordering him to arrange the supply of these instruments. When they were sent, they were accompanied by technicians, as requested by Gundobad.[349]

Cassiodorus provided his monks in the monastic-style centre of learning he established at 'Vivarium' (Staletti, near Squillace in Calabria) with a water clock, noting that this *horologium aquatile* showed the correct hour of the night and of the day when a clear Sun was absent.[350] In 756, King Pepin of the Franks (*c*.718–68) received a *horologium nocturnum* from Pope Paul I (700–67).[351] Reports of the mechanical marvels in the Islamic Near East reached the West. The clepsydra sent as a diplomatic gift to Charlemagne (748–814) in 807 by the Abbasid Caliph Harūn al-Rashīd (786–809) caused a sensation at the Frankish court, according to an account by the chronicler Einhard.[352] The clock was 'not like our ones' and had twelve automata horsemen that appeared and then disappeared through twelve windows, and the

---

343 See discussions of these in Chapters 1 and Chapter 3, Section 1
344 Turner 1984b, 10. For sinking-bowl clocks in India, see Chapter 2, Section 1.
345 For the time structuring of Christian life, especially in monasteries in the Early Middle Ages, see Chapter 3, Section 2.
346 Hildemar de Corbie, *Exposito in Regulam Sancti Benedicti*, ch. 8, 1 quoted in Martene 1736, iv, 5.
347 Maddison 1969, 3–33; Turner 1994, 115; Dohrn-Van Rossum 1996, 54.
348 Beckmann 1817, i, 428; Dohrn-Van Rossum 1996, 55.
349 Barnish 1992, 20–4. For Cassiodorus, Gundobad, Theodoric, and Boethius see the entries concerning them in ODLB.
350 'Institutiones', I, xxxv, iv, & v., trans. in Jones 1946, 80–96.
351 Beckmann 1817, I, 425.
352 Calmet 1728, I, 582; Dohrn-Van Rossum 1996, 72.

hour was struck on a small bell. Several foreign ambassadors invited to the *majlis* of the Imperial Palace of Magnaura in Constantinople wrote of a device with automata that presumably operated hydraulically. Of these ambassadors, the papal envoy Bishop Liutprand of Cremona (*c*.920–972) in the middle of the tenth century, gave the following account of it:

> Before the emperor's seat stood a tree, made of bronze gilded over, whose
> branches were filled with birds, also made of gilded bronze, which uttered
> different cries, each according to its varying species. The throne itself was so
> marvellously fashioned that at one moment is seemed a low structure, and at
> another it rose high into the air. It was of immense size and guided by lions,
> made either of bronze or wood covered over with gold, who beat on the
> ground with their tails and give a dreadful roar with open mouth and quivering
> tongue.[353]

A similar device existed in Baghdad, where in 917 an Arab chronicler reported that visitors were amazed at the caliph's silver tree with its waving branches and birds singing various songs. Knowledge of the Islamic hydraulics that produced these complicated automata devices reached Europe via the Iberian Peninsula and Sicily, as well as by returning travellers and crusaders from the East. Apart from the studies sponsored by Alfonso X of Castile, Roger II King of Sicily (*c*.1095–1154) erected a timepiece that may have been a clepsydra at his Palace in Palermo in 1142, but no specificity is offered by the inscription in Greek, Latin, and Arabic that recorded the event and still survives. This gives no details of the timepiece but, if a clepsydra, it may have been similar to one in Malta that had a moving figure who successively entered twelve arches and dropped the number of balls that corresponded to the hour, onto a metal dish below.[354]

A returning crusader utilized Islamic hydraulics in ludic diversions. Robert II Count of Artois, returning in 1270 to Picardy by way of Palermo (and probably inspired by the Saracenic gardens there as well as by what he encountered in Syria) created famous water features at the Palace of Hesdin.[355] One room reproduced the effects of a thunderstorm and another area concealed 'eight pipes for wetting ladies from below and three pipes by which, when people stop in front of them, they are whitened and covered in flour'.

In spite of the availability of sophisticated hydraulics, few descriptions survive of clepsydrae with automata or astronomical indications in medieval Europe. Gerbert of Aurillac describes the clepsydra as useful in determining the equinoxes and is credited with having built one at Ravenna.[356] By the beginning of the tenth century the word clepsydra is replaced in monastic texts by *horologium*, often qualified by words such as *aquae*, *aquatile*, or *aquare*, its last recorded use is in the older consuetudes of Fleury *c*.1000.[357] Why should the Greek word disappear? One theory is that its desuetude reflected the eclipse of the water feature of simple outflow models by various, more memorable features that came with mechanical elaboration.[358] One of the many manuscripts from the Benedictine monastery of Santa Maria de Ripoll, probably dating from the eleventh century and thought to come from Lotharingia, describes a simple alarm clepsydra that would have prompted the sacristan to sound the bell for the offices.[359] The description mentions weight drive, rotating wheels with pins that struck bells in an adjustable order, water vessels, and oil lubrication. The vocabulary and style are completely independent of Vitruvius' descriptions and also unrelated to Arab models.[360] Two dials are mentioned: one which divides day and night into quarters and the other an unknown division into hundredths. An analysis of the manuscript is given by Maddison, Scott, and Kent[361] and a reconstruction based on the surviving section of the manuscript was made by Edouard Farré-Olivé.[362]

The importance of clepsydrae in monastic life is shown in the consuetudes of the Abbey of Fruttuaria near Turin, written around 1100 for the Benedictine Abbey of Göttweig, in Lower Austria:

> The sacristan shall rise at night as the horologium falls, and, if the sky is clear, examine the stars. And if it is the time of rising ... he shall proceed to the horologium, pour the water from the small vessel into the larger one, pull up the rope and lead, and after this, strike the bell.[363]

It would appear from this extract that the 'falling' of the clock was some sort of acoustic alarm that woke the sacristan who, in turn, had to sound a *scilla* to wake his brother monks. The *scilla* could be a bell similar to the squilla, listed by the French liturgist and theologian Jean Beleth (*fl*.1135–82) for use in refectories. He also speaks of *nolula* and *duplula* bells for a *horologium*, but mention of

---

353 Eamon 1983, 70 & 175.
354 So Ungerer 1931, 28, who affirms 'according to contemporary records' without citing or even locating them.
355 Maguire, 1997, 175 (with translation), also see Kieckhefer, 1989, 101.

356 Turner 1984a; 1984b, 28; Dohrn-Van Rossum 1996, 55.
357 Dohrn-Van Rossum 1996, 60.
358 Dohrn-Van Rossum 1996, 60.
359 Ripoll MS no. 225. Archivo de la Corona de Aragón, Barcelona.
360 Dohrn-Van Rossum 1996, 64–5.
361 Maddison, Scott, & Kent 1962.
362 Farré-Oliva 1989.
363 Spätling & Dinter 1985, ch. 4, n. 40. Also Dohrn-Van Rossum 1996, 60.

these bells is not found elsewhere in connection with clocks.³⁶⁴ In the Cistercian Rules of the time there are several references to aural rather than visual signals, such as *ad sonitum horolgii* and *audito horologio*. It would appear from these Rules that by the twelfth century the horologium had been acoustically amplified to reach more congregants and also extended to sound at regular intervals. The Cluniac Rules mention *horologium sonens* in a passage on winter risings.³⁶⁵ The phrase '*inspecta hora in horologio*' suggests that devices now possessed dials that displayed the time.³⁶⁶ In the Chronicle of the Benedictine monk Jocelyn de Brakelond (*fl. c.*1160–*c.*1205), there is mention of the 'horologium' of Bury St Edmunds Abbey in 1198. According to this account the clock had just sounded the call to matins (*cecidit horologium ante horas matutinas*) and this raised the master of the vestry, who discovered that a fire had taken hold of the shrine of St Edmund. The fire was extinguished with water from the well and also with water from the clock, '... *nostri propter aquam currentes, quidem ad puteum quidem ad horologium*'.³⁶⁷ Whether the additional water was from the clock itself or from a supply adjacent to the clock has been questioned. Nevertheless, in view of the coincidence, there is a strong probability that the clock referred to was indeed a clepsydra.³⁶⁸

In one of the only two known medieval depictions of a water clock, bells are clearly visible. The clepsydra is shown in an illumination in a French moralized bible of the late thirteenth century.³⁶⁹

The Latin text, which appears on the left of the illustration, translates as follows: 'King Hezekiah was sick unto death and prayed to the Lord and the Lord added fifteen years to his days and gave him a sign by the prophet Isaiah that the Sun would be moved back 10 degrees in the clock'. This refers to an incident that appears both in 2 Kings XX, 5–11 and Isaiah XXXVIII 8. In the illustration the Isaiah is shown pointing to a clepsydra whereas in the scriptures, the sign of divine authority is a shadow of the Sun moving backwards. Isaiah offers the King a choice as to whether this shadow should move backwards or forwards and the King chooses the former and the shadow duly moved backwards. Various hypotheses on the workings of the clepsydra depicted in the Hezekiah illustration (Figure 59) have been advanced.³⁷⁰ It is generally agreed that the wolf's head above the bowl at the bottom represents the mouth of an outflow tank and that, as the large vertical wheel rotates, teeth at its circumference cause the five bells to be sounded. One explanation for the number of teeth on the wheel is that they represent the fifteen divisions of the monastic day practised in northern France from whence the bible originated, although evidence of such practice dates only from the following century.³⁷¹ A structure that could have housed a similar clepsydra is depicted in Villard de Honnecourt's mid-thirteenth century sketchbook, also from northern France. The sketch is entitled 'housing for a clock' but there is no information on the clock inside or its size. No clock dial is evident, suggesting that the clock gave only aural indication.³⁷²

The second known depiction of a clepsydra is a sketch on the verso of the last folio of a legislative compilation drawn up during the reign of King Jaume II of Catalonia (1291–1327) and written in Catalan. The drawing is clearly not connected with the text and could therefore be of later date. A musical score at the top of the page has been analysed, along with details in the sketch. This analysis suggests the device was designed at around the time of the introduction of the oscillator clock, or sometime between 1291 and 1430 and operated as follows.

The water clock in the lower part of the structure appears to be an outflow clepsydra calibrated in equal hours, which triggers the playing of the carillon above. This carillon consists of a circle of ten bells of varying tones, five of which are not shown. It is thought that the musical score at the top of the page was written for the sketched device, not only because of its proximity to the sketch but also because the semibreve tempo notation is contemporary with it, being in use from the thirteenth century until about 1430. In not ending on the same note as the first, the descendent scale is a feature of mechanical rather than vocal music.³⁷³

In the two depictions that we have of medieval clepsydrae, no dial is evident. Nevertheless there is evidence of time indication in the early twelfth century from the Cistercian rules, with the phrase 'look at the hour of the horologium'.³⁷⁴ From the same source we also find that the sacristan shall set the clock *horologium temporare*.³⁷⁵ From the Cistercian Abbey of Villars, near Brussels, founded in the twelfth century, there remain fragments of slates that are dated from their surroundings to 1267/8 and are inscribed with instructions for the sacristan on how to set the abbey clepsydra. The clock is to be set for each day of the year with reference to the correct periods of daylight and the correct length of service for the current festival:

> Always set the clock, however long you may delay
> on 'A', afterwards you
> shall pour water from the little pot that is there, into
> the reservoir until it
> reaches the prescribed level, and you must do the

---

364 Dohrn-Van Rossum 1996, 62.
365 Hallinger, 1983, vii 2, 136.
366 Dohrn-Van Rossum 1996, 62.
367 Butler 1949, 106.
368 Drover 1954, 55.
369 MS Bodl. 270b fo. 183v. Bodleian Library, Oxford.
370 Simoni 1965, 16.

371 Sleeswyck 1979, 488–94. See also letters in *AH* 1955, 84 and Drover 1980, revisiting the illustration.
372 Paris, Bibliothèque Nationale de France, MS F19093. Reproduced with commentary in Bechmann 1991, 112–17 who notes that the title 'C[es]t li masons don orologe' is written in a different ink and in a later hand from the rest of text and who interprets the drawing rather as a belfry than a clock housing.
373 Farré-Olivé 1989.
374 De Cangey 1517, Liber Usum Cisterciensis, Chapter 96.
375 De Cangey 1517, Liber Usum Cisterciensis, Chapter 114

Figure 59 A water clock illustrating the story of Hezekiah from MS Bodl. 270b fo. 183v. Photo: Jean-Baptiste Buffetaud.

same when you set (the

clock) after complines (dusk) so that you may sleep soundly.[376]

The first medieval clockmaker whose name is recorded is Father Hermann Joseph of the Norbatine Abbey of Steinfeld near Cologne. A life of Father Hermann written by his prior around 1230 tells us that he was asked to make clocks for monasteries not thus provided.[377] He was canonized in 1958.

The water clock in the *Libros del Saber* (Figure 60) was one of three timekeepers described that were intended to work indoors. The design is in the tradition of Græco–Arab technology. As with the descriptions of the other devices, the *Book of the Water Clock* is divided into two sections: how to make the clock and how to use it.[378] The introduction points to deficiencies in other designs and outlines the following principles for optimizing performance: outflow tanks should not be drilled at the bottom as this produces a variable outflow according to the head of water in the tank; lids should be fitted to tanks to prevent dust from entering; only clean water should be used, which should be strained through linen and further purified using ground alum.

In the illustration the reservoir (on the right) is higher than the receiving tank on the left. It is specified that the reservoir should be twice as high as it is wide. The volume of water is calculated by timing the flow from a full bucket perforated with a hole. Dividing that time into the desired duration of the clock will give the number of bucketsful required. Interestingly, the instruction states that an astrolabe should be used to measure this period of efflux. The reservoir is fitted with a series of internal pipes that maintain a constant head of water at the small rectangular outflow basin between the two main containers, called the *pila*. This works as follows: the curved pipe acts as a cistern and disgorges the water into the *pila*. The straight pipe is an air-intake pipe, which also opens inside the *pila*. As the level of the water rises inside the *pila* it shuts out the air-intake opening thus stopping the flow of water until it can breathe once more as the level descends. The outflow pipe from the *pila* is of the same dimension as the pipe feeding it. The water thus flows via the *pila*, where it is maintained at a constant level, into the lower tank on the left. This contains a float to which a ladder-like frame is attached, with a board nailed to it. This board is fitted with plaques showing signs of the zodiac in such a way that the constellations that are ascending, culminating, and descending at any one time are shown in a line of three vertical columns rising out of the top of the receiving vessel. These constellation plaques would need to be changed daily. The back of the rising board shows the position of the Sun throughout the year in both temporal and equinoctial hours. That this clepsydra

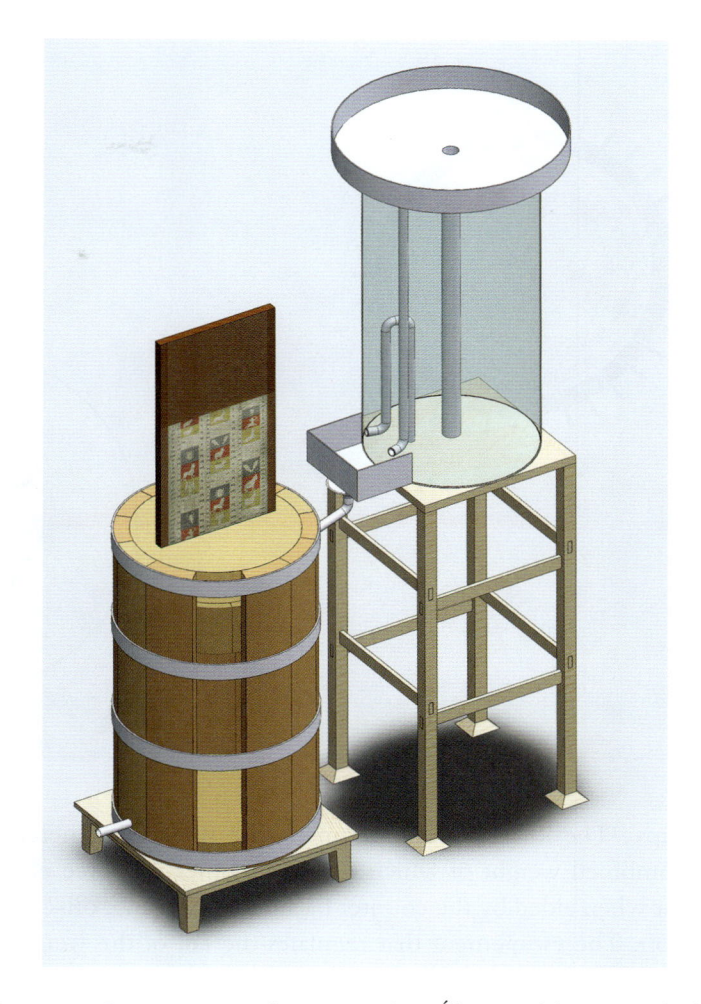

Figure 60 Reconstruction by Victor Pérez Álvarez of the water clock described in the Libros del Saber, 1276. Photo: V. Pérez Álvarez.

was intended to be mysterious is shown by the following admonition to those who may wish to build one: ' . . . and cover the whole mechanism and vessels in order that no part is visible except the panel that rises. When the mechanism is hidden the clock will seem more beautiful.'

A full description of the water clock with reference to a virtual reconstruction was published in 2012 by Víctor Pérez Álvarez.[379] With the exception of an oscillator, the mercury clock (Figure 61) described in the manuscripts has all the features of the mechanical clock of modern connotation.[380] It has a driving weight, gears, **escapement**, chiming mechanism, and even a dial with an indicator that rotates clockwise in twenty-four hours. The problem (faced by all pre-oscillator devices using gravity) was how to prevent the drive from gathering momentum so that a useful period of elapsed time could be evenly measured. In the mercury clock, the solution lies in the drum that has to be rotated by the falling weight attached to it by a chord that is wrapped around it. The drum is divided into twelve internal compartments, each with a small hole in both dividing walls. The instructions state that each compartment should be 'caulked well with wax and resin, as they

---

376 Drover 1954, 56. Sheridan 1869, 203–15. 405–51.
377 Dohrn Van Rossum 1996, 63.
378 Rico y Sinobas 1863–7, iv, 24–64.

379 Pérez Álvarez 2012, 196–207. See also Whitestone 2019.
380 Rico y Sinobas 1863–7, iv, 65–76.

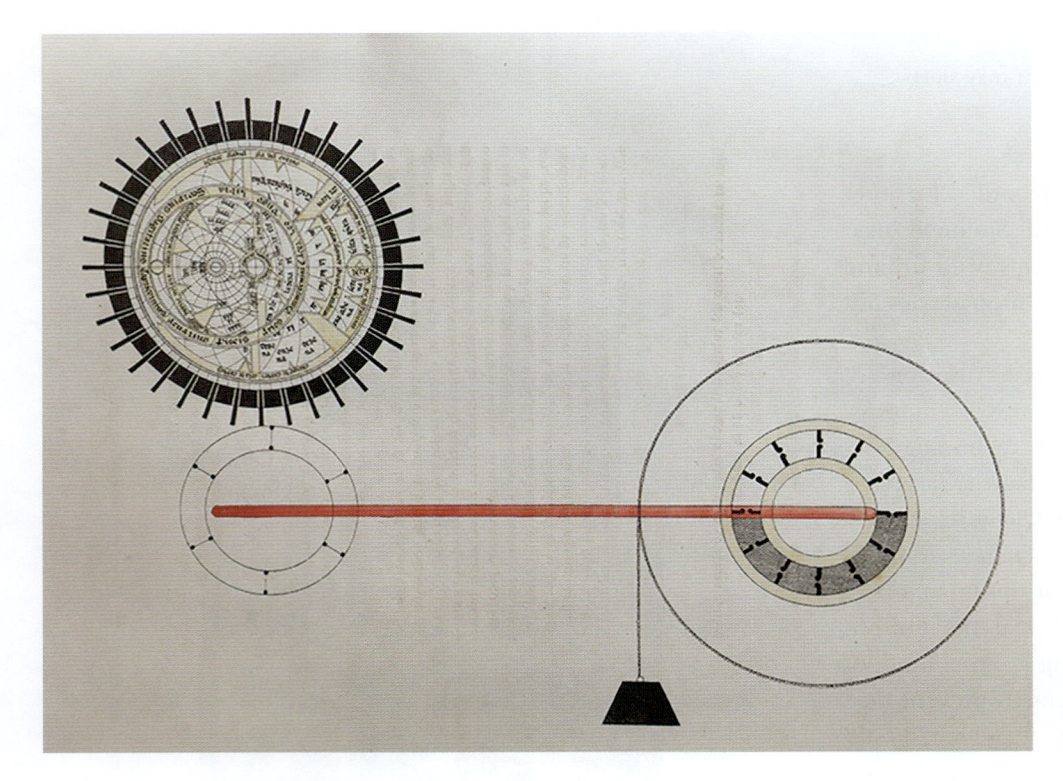

Figure 61 Mercury clock from the Libros del Saber, 1276 as redrawn by Rico y Sinobas, 1866. Photo: S. Whitestone.

do ships'. The lower six compartments are filled with mercury, the combined weight of which only overcomes the force of the weight when aided by the counter balance of its rising, off-centre position. The escapement that regulates the fall of the weight is the gradual flow of mercury from higher to lower compartments, through the holes in the dividing walls. The drum is carried on an arbor (drawn in red), which carries a stub pinion of six leaves that turns once in four hours.

It is interesting to note how the text describes this very early use of a pinion: ' ... and make a wheel similar to the wheel that they call in Arabic *açenna* ... '. The word 'açenna' exists today in Spanish in the form 'aceña', meaning water mill. Thus, having no word for the new device, a term derived from the Arabic for a water wheel is used to describe a pinion. This pinion moves a wheel of thirty-six teeth one revolution in twenty-four hours. This wheel carries a rete that rotates clockwise on the astrolabe dial. In many respects this clock would seem to be the forerunner of the earliest church clocks with astrolabic dials that were produced a century later. An interesting difference is that the mercury clock uses a latitude plate with a south polar projection, just as does a normal astrolabe. However, almost all the earliest church clocks have astrolabe plates using a north polar projection. These include the first Strasbourg Cathedral clock of 1354; the Lund Cathedral clock of c.1380; the clock in the church of St Nicholas in Stralsund made in 1394; the clock of the Church of St Mary, Lübeck, made in 1405; the Prague Town Hall clock of 1490; and the Chartres Cathedral clock made c.1520. The north pole projection is better suited to clock dials as it shows the celestial latitudes in the correct order for viewing the Sun's passage over the horizon, i.e. with the outermost of the three concentric circles being that of the tropic of Cancer. This outer circle thus gives a higher arc above the horizon, with the inferior circles of the equator and Capricorn giving progressively shallower arcs.[381] That the rete could, in theory, be replaced by a simple hour hand, turning clockwise to be read against a twenty-four-hour dial, shows that a tradition of time indication, shortly to be adopted by the mechanical clock, was already in place. With regard to the construction of the clock, there is the very useful evidence of A. A. Mills, who built and tested a replica in 1988.[382] His study does not consider the ancillary chiming mechanism. He concluded that the translator, and possibly also the author of the original Arabic manuscript, had never seen the instrument they attempt to describe. He found also that the range of possible periods of rotation of the mercury-filled drum was, at best, a little over half the four hours claimed in the description and of poor reproducibility (±10 per cent). Alan Lloyd also described the clock,[383] presumably drawing on an article in the *Deutsche Uhrmacher Zeitung* by F. M. Feldhaus,[384] as he repeats that author's error that the recommended wood for the construction of the wheel is oak, whereas walnut and jujube wood are specified. On the original source of the design for the mercury clock, the text states that Alfonso ordered that 'Rabiçag's book on how the clock is made, should be done in the manner of the book of "Iran" the philosopher which spoke of how to lift heavy things, and instructs that it should be made in this way'.[385] This is

---

381 Wåhlin 1932, 'Astrolabe Clocks' in Gunther 1932, ii, 539–49.
382 Mills 1988, 329–44. For a study of compartmented water clocks, see Bedini 1962.
383 Lloyd 1958, 4.
384 Feldhaus 1930, 608–12.
385 Rico y Sinobas 1863–6.

a reference to Hero of Alexandria, whose treatise *Mechanica* deals with the lifting of heavy objects.

The advent of the oscillator clock ended the age of the water clock. The above-mentioned inscriptions on slate at Villers Abbey in 1266/7 are the last written evidence of the use of monastic water clocks in Europe, although the Hezekiah and Catalan illustrations post-date them. The rich heritage of Hero and Vitruvius would continue to echo in the water clocks of Renaissance savants, most notably that of Oronce Fine (*c.*1494–1555), who illustrates a device where a model of a ship acts as the floating counterpoise to the driving weight that rotates a drum attached to an hour hand. A siphon channels the efflux of water up through the mast and then down to a reservoir below, thus maintaining a constant level of water upon which the vessel floats.[386] Scientific use of water clocks, however, would take longer to disappear. The German philosopher Nicholas of Cusa (*c.*1405–64) proposed the use of a clepsydra for assessing pulse rates and respiration. His method, like that used by Galileo, described below, involved proportional comparison of volumes of water rather than any measure against a standard time.[387] Tycho Brahe (1546–1601) resorted to the use of a clepsydra filled with mercury in spite of commissioning four oscillator clocks for his observatory on the island of Hven. One of these oscillator clocks had a wheel with 1200 teeth. However, none seem to have functioned satisfactorily.[388]

It is remarkable, given the high level of workmanship displayed in northern European horology at this date, that a clepsydra could have any practical place among precision instruments. For it is certain that flowing water is not a natural agent of accurate timekeeping. Changes in the rate of efflux caused by varying heads of water, defined by Evangelista Torricelli (1608–47) as $V=\sqrt{2gh}$ (where V = the velocity of efflux, g = gravitational acceleration, and h = the head of water), were avoided by simple overflow devices or tapering scales. Nevertheless, variations in viscosity caused by changes in temperature, and other complex properties of fluid dynamics such as surface tension, could have a considerable effect on accuracy.[389] Nevertheless scientists such as Da Vinci, Galileo, Viviani, Newton in his youth, and Robert Hooke would continue to design and use water timers. Da Vinci's drawing only shows the mechanism from above where twenty-four pipes of varying diameter but uniform height are successively filled, one at each hour. It is not clear how the mechanism functions nor whether the figure striking a bell is an automaton or a human.

Galileo (1564–1642) used a water timer in his observations, which proved crucial in developing his theories of motion and especially of falling objects. He describes its use thus:

> As to the measure of time we had a large pail filled with water and fastened from above, which had a slender tube affixed to its bottom through which a narrow thread of water ran; this was received in a little beaker during the entire time that the ball descended along the channel or parts of it. The little amounts of water collected in this way were weighed from time to time on a delicate balance. The differences and ratios of the weights giving us the differences and ratios of the times, and with such precision that, ..., these operations repeated time and again never differed by any notable amount.[390]

Several folios of Galileo's worksheets show this process in action. In codex 72 in the National Central Library of Florence, folio 189v shows a column of numbers that represent different weights assembled to balance the weight of a volume of expelled water. That volume represented a period of time taken to complete a certain number of swings of a pendulum. A quarter period of a pendulum, returning to the perpendicular, is comparable to the circular descent of an object. This could then be compared to vertical descent and inclined descent, where a ball rolls down an inclined plane. All such motions were timed by Galileo's water clock to obtain their relative ratios without direct reference to any time standard.[391]

To what extent clepsydrae were constructed in medieval times with astronomical indications, similar to those on early mechanical tower clocks, is hard to judge. When considering Renaissance clepsydrae such as those illustrated by John Bate[392] it would seem likely that the main system used for rotational drive was that employing two weights attached to a rope that is wound around a drum. The heavier weight, resting on a float, prevents the drum from rotating, but as the float rises it permits the lighter weight on the other side to rotate the drum at the rate of the rising water. This is one explanation for the mention of 'supplying clock ropes' (*cordas ad ortelogium*) in a list of duties of the treasurer of York Minster in the Statutes of 1221. The alternative explanation is that the mechanical clock with oscillator escapement occurred far earlier than previously thought.[393]

EARLY OSCILLATOR ESCAPEMENTS

The invention of the oscillating escapement that spawned the mechanical clock is the subject of continuing speculation. In this escapement, the interaction of **pallets** attached to an oscillating balance slows an escape wheel to an even, usable rate of rotation. Owing to the wear involved, these new machines were made of metal. They were driven by a descending weight, attached by cord

---

386 Fine 1560, 73–4. *Cf.* Turner 2009, 191–6.
387 Von Kues 1982, iii, 616; Dohrn-van Rossum, 1996, 284.
388 Hevelius 1673, 143.
389 Mills 1982.

390 Galileo 1638, *Discorsi . . . intorno à due nuoue scienze . . .*, Leiden, trans. Drake 1974, 170.
391 Hill, 1994, Whitestone, 2017 and for modern testing of the method, see Settle 1961.
392 Bate 1634, 39–43.
393 Statutes of Lincoln Cathedral in Wordsworth 1897, ii, 100.

to a drum on the arbor of a wheel that provided rotation to the train of wheels geared from it. The invention passed uncelebrated, and the early life of the machine is hidden within the generic *horologium*, where even the additional mention of a weight fails to separate it from water clocks that also used them. This linguistic problem is encapsulated in a key text of 1271. A Dominican priest known as Robert the Englishman or Robertus Anglicus (*fl.*1270), who taught at the University of Montpelier and possibly also at the University of Paris, wrote a commentary on Johannes de Sacrabosco's treatise on astronomy, *De Sphera Mundi*, which includes the following passage:

> Nor is it possible for any clock to follow the judgment of astronomy with
>
> complete accuracy. Yet clockmakers (*artifices horologiorum*) are trying to make
>
> a wheel which will make one complete revolution for every one of the equinoctial circle, but they cannot quite perfect their work. But if they could, it would be a really accurate clock and worth more than the astrolabe or other astronomical instruments for reckoning the hours, if one knew how to do this according to the method aforesaid. The method of making such a clock would be this, that a man make a disc of uniform weight in every part so far as could be possibly done. Then a lead weight would be hung from the axis of that wheel and this weight would move that wheel so that it would make one complete revolution from sunrise to sunrise, minus as much time as about one degree rises, according to an approximately correct estimate. For from sunrise to sunrise the whole equinoctial rises and about one degree more, through which degree the Sun moves against the motion of the firmament in the course of the natural day.
>
> Moreover, this could be done more accurately if an astrolabe were constructed with a network (rete) on which the entire equinoctial circle was divided up. Then the degree at which the Sun was at rising would be noted, then the point of time or degree on the equinoctial which touched the horizon. And similarly on the day following would be noted when the Sun touched the horizon and likewise the degree of the equinoctial touching the horizon and it would be about one degree more than the revolution of the entire equinoctial. Then let the time corresponding to one degree of the equinoctial be subtracted from the entire time from sunrise to sunrise; then having the said wheel complete its revolution in that time and let it be divided into twenty-four equal parts; then the situation of each part would show the hour in the sky.[394]

This passage is taken by historians as evidence that the oscillator clock was unknown to its author at the time of writing and thus yet to emerge.[395] However, the opposite could well be the case, that it is the earliest testament to its existence. The first sentence, which states: 'Nor is it possible for any clock to follow the judgment of astronomy with complete accuracy', is most likely not a reflection on the incapacity of contemporary clockwork, but rather on the incapacity of clock time, i.e. mean time, to portray in equal beats of solar time the motion of the heavens '*secundum veritatem*', 'in accordance with the truth'. This is precisely what an astronomer might say in response to vaunted improvement in clock accuracy. Upon hearing the news, in 1658, that the new pendulum clocks 'fail not a minute in 6 moneth', Sir Robert Moray (1608/9–73) cavilled, 'this you will believe as little as I do, for I can demonstrate that it must go wrong to keep foot with the Sun'.[396]

Historians may also have been misled by Robert's emphasis on the wheel and its uniform weight. This emphasis conjures an image of rotation balanced with drive, inviting comparison with its contemporaries, the wheel of the Mercury clock in the *Libros del Saber*, and with the rotating magnetic sphere of Pierre de Maricourt.[397] However, it may denote nothing more than the importance of evenness in a large, multi-toothed wheel and of the challenge its manufacture would present. Were Robert to have had in mind some form of hydraulic/weight mechanism, he would hardly have specified lead-weight drive without mentioning the fluid with which it would interact, any more than the word 'mercury' would be omitted from the title of the *Libros del saber* clock or the word 'magnet' from de Maricourt's globe. Oscillator control, on the other hand, is notably lacking in early descriptions of clocks that had them, no doubt owing, in part, to the lack of established nomenclature but also, as argued below, because of the lack of a noteworthy feature to distinguish this device from existing striking mechanisms. Thus Robert's peculiar omission of any indication as to how the motion of this clockmakers' wheel might be controlled is because we are to assume this would be done by (their) clockwork. And that they 'cannot quite perfect their work' suggests that completion hinges on something less fundamental than an unknown principle of such control. Although certainty eludes us on this point, the description below on how the oscillator escapement may have evolved is supported by the assumption that Robert was indeed reporting on the prospect of detailed astronomical display being delivered by a *horologium verax valde*, 'a really accurate clock', with the caveat

---

394 Thorndike 1949, 180–230.

395 Thorndike 1949, 230. Drover 1954, 57. Dohrn-Van Rossum 1996, 89–90.

396 Letter from Sir Robert Moray to the Earl of Kincardine, 13/23 April 1658, Stevenson 1988, 190.

397 As mentioned earlier, Mills 1988 concluded that neither the original Arab author nor the translator writing in the *Libros del Saber* had seen the clock they described. For the perpetually rotating sphere of Pierre de Maricourt, see Dohrn-van Rossum 1987, 89, who quotes the 'Epistola de magnete' from the edition in Hellman 1898, 7f.

of divine limitation to such ambition, the *aequatio dierum* or **equation of time**. This interpretation claims a close conformity with Robert's own words, the implausibility of foreknowledge of an 'impending invention' implied by the traditional interpretation,[398] and the coincidence of a subsequent spate of horological activity.

In 1273/4 the rolls of Norwich Cathedral priory record a payment for a clock as '*pro horologio*'. As is so often the case, no further detail is given and there is nothing to distinguish this entry from previous monastic accounts referring to water clocks or sundials. Its repair was recorded in 1290/1 as '*in emendacione horologii*'. However, at the time of constructing an astronomical clock for another site in the cathedral in 1321, which certainly had an oscillator movement, the same rolls record *Ecclesia. Item 1 corda ad antiquum horlogium*, suggesting not only that this earlier clock was weight-driven but also, in being distinguished only by the word *antiquum*, was similar to the one being built.[399] One of the strongest clues to the existence of the oscillator clock in the thirteenth century suggests that it was already well established by 1284. In that year, the French bishop William Durand of Mende (1250–96) wrote his *Rationale divinorum officiorumn*, a widely disseminated treatise on church rites.[400] He proposed the inclusion in all church naves of 'a clock where the hours are measured (gathered) and read'.[401] Such a general recommendation implies a widely available, standard form of mechanized time indication which, together with the designated location in the nave, is a departure from previous references to water clocks. An unlikely siting for a water clock is that above the rood screen (*pulpit*) where, according to the annals of Dunstable Priory, a *horlogium* was erected in 1283.[402] A succession of references to clocks follows, either *orologium* or *horlogium*, in English churches and colleges.[403] In 1284, Exeter Cathedral made a grant to Roger de Ropford, a bell founder, for duties that included the maintenance of the clock. Old St Paul's Cathedral in London granted rations to an *orologviarius* in 1286. The records of Merton College, Oxford, note *c.*1288 of '*expense orolgii*'. At Ely Abbey in 1291 there are the first records of quarterly payments *pro custodia orolgii*. From Christ Church Cathedral, Canterbury, thirty pounds were paid for a *novum horologium magnum in Ecclesia*. The earliest description that specifically points to an oscillator clock appears in the chronicle of the Dominican monastery of St Eustorgio, Milan, in 1306, where the acquisition of an 'iron' clock is recorded.[404] Iron is also mentioned in work for a clock at the Palace of Hesdin in 1300.[405] In 1301, quantities of iron, metal, and charcoal were delivered to the armourer Gilbert de Louvre for use in the construction of a clock that Philip the Fair (1286–1314) wished to give to the Abbey of Poissy.[406]

The mention by Robertus of *artifices horologiorum* is one of several references in the last quarter of the thirteenth century that suggest a new group of craftsmen. The ledgers of Beaulieu Abbey record an *orologiarius* in 1269/70.[407] However, that a guild and street of clockmakers existed in Cologne from 1183 onwards has been shown by Dohrn-van Rossum to be founded on a mistranslation of *Gewerbe* as 'guild' instead of 'trade' or 'craft'. The street designation is perhaps owing to the coincidence of a resident's name, 'Urlong', being confused with the German for clock 'uhr'.[408]

Clocks also begin to appear in literary sources around the time of Robert Anglicus' commentary. The French poet Ruteboef (1230–85) mentions a striking or alarm clock in his poem of *c.*1260 *De secrestain et de la fame au chevalier*:

> All rose eagerly from their sleeping
> To hear what was most out of keeping,
> And in something close to shock
> Heard ring not bell, nor campanile nor clock,
> Now all most willingly that night describe
> When sacristans were drunk
> So much did they imbibe.[409]

In the *Roman de la Rose*, Jean de Meung (1240–1305) also mentions chiming clocks.[410] His *orloges* with wheels in continual motion (*pardurable*) give an image of clocks with striking or alarm mechanisms as part of domestic furniture, and strengthens the notion that this is a reference to the new solid weight-driven clocks. But the earliest literary reference implying oscillator clocks is generally taken to be that in Dante's *Paradiso c.*1318:

> And as the wheels in the mechanism of a
> clock turn so that, to one who watches, the first
> seems to be motionless and the last to fly.[411]

The last wheel here mentioned is probably the foliot wheel which, as in the *astrarium* of Giovanni Dondi (1318–88), could well have been a balance wheel as a bar. A second passage translates as:

> Then, like a clock that calls us in the hour
> when the bride of God rises to sing a dawn

---

398 See Thorndike, 1949, 180, 230. Also Dohrn-van Rossum, 1996, 90.
399 North 1976, ii, 316.
400 It was one of the first books to be printed in Mainz in 1459, and ran to forty editions before 1500.
401 Brusa 1990, 486.
402 Beeson 1971, 13.
403 Beeson 1971.
404 Galvano Fiamma, *Cronica ordinis praedicatorum* (ed. B. M. Eichert), Rome 1897, 107, cited from Brusa 1990, 488; Dohrn-van Rossum 1996, 22.

405 Dohrn-van Rossum 1996, 92.
406 Dohrn-van Rossum 1996, 92.
407 Hockey 1975, 235. Cf. the seal of an orologiarius c. 1300 reproduced on the title page of the present work.
408 Dohrn-van Rossum 1996, 96.
409 Jubinal 1874, 315.
410 Cited in this volume, Chapter 7, Section 1.
411 Dante 2011, *Paradiso*, x, 139–45, trans. R. M. Durling, iii, 387.

song to the Bridegroom, that he may love her,
Whose one part pulls and other pushes
Sound the *Tin* so sweet a note
That a well-disposed spirit swells with love:
So I saw the glorious wheel turning,
Voice answering voice, with tempering sweetness
That cannot be known except there
Where rejoicing forever is.[412]

The oscillator clock had probably emerged in its simplest form as a timepiece some time before Robert's commentary of *c*.1271, and quickly acquired alarm and striking mechanisms. Independence from hydraulic drive appears to have been little appreciated, perhaps because water clocks were easier to construct. As described below, the earliest descriptions are of clocks with complicated astronomical indications that took many years to complete and presumed a familiarity with the mechanisms. The country of origin could be anywhere in Europe, with England, France, and Italy as the strongest contenders.

The first oscillator clock described in any detail is the famous astronomical clock made under the direction of Richard of Wallingford (1292–1336), abbot of St Albans Abbey and described in his *Tractatus horologii astronomici* of 1327.[413] Here attention is confined to those aspects of the clock that are relevant to the emergence of the oscillator movement. The St Albans clock was undoubtedly the most complicated clock of its day, but was not the first of its type. The rolls of Norwich Cathedral show that work on a large astronomical clock was begun there in 1321. The Norwich rolls mention a 'Laurencius orologiarius', and a 'Magister Rogerus orologiarius', who are probably the same two craftsmen, one being Laurence of Stoke, as worked on the St Albans clock.[414] Both clocks were probably originally placed in the transepts of the churches. Although the St Albans clock was begun fifty years after Robertus Anglicus' commentary, several aspects of it resonate with his words and support the interpretation of them given above.

The description of the clock says relatively little about the escapement or the going train, as if both were already well known at the time of writing and would be readily understood. There is nothing in any of the manuscripts to suggest the going train and escapement were a novelty. The clock has an astrolabe on the dial and a wheel turning in sidereal time just as Robert proposed. Further it had an ovaloid, or kidney-shaped, **contrate** wheel of three hundred and thirty-one teeth to display the equatorial speed of the Sun, as if addressing Robert's caveat about the equation of time. That the clock was not finished in the eight years between Wallingford's description and his death from leprosy in 1335 provides a model for the sort of project, albeit much earlier, to which Robert might have referred. When Edward III (1312–77) visited St Albans he rebuked Richard for embarking on such a costly and time-consuming undertaking when the fabric of the abbey was in disrepair. Richard is said that to have replied that while plenty of abbots would succeed him capable of repairing the building, none could complete the clock.

Indeed, Richard was uniquely qualified for such an ambitious construction. He was the son of a blacksmith, and a mathematician and astronomer of the highest order. His monks would have had experience of similar technology in the renovation of the abbey flour mills.[415] His iron movement is estimated to have been approximately ten feet wide and to have struck the hours. The immensely complicated astronomical indications included the position of the Earth, Sun, Moon, and stars and also perhaps the planets. The display not only showed the age and phase of the Moon, but also predicted lunar eclipses. From the description in the *Tractatus*, John North reconstructed the form of the escapement now commonly known, following Wallingford's terminology, as the *strob* (Figure 62).

It is clear that, in common with early **crown-wheel** escapements, the clock had a **verge** or vertical shaft, forming a 'T' with a horizontal beam or **foliot**. Weights were suspended from each arm of this and placed equidistant from the centre. The further removed these weights were from the centre, the greater the moment of inertia about the verge axis and the slower the clock would go. The verge had one pallet, rather than two, engaging with the **escape wheel**, the rear edge of which forms a semicircle. On the flat side of this semicircle are two smaller concave semicircles that alternately engage with the teeth of two coaxial escape wheels rotating on either side of the verge. The teeth of these escape wheels are formed as pins. The bottom of the verge turns in a pivot hole and the top may also have done, or it may have been suspended by a cord. Such a suspension provides minimal torsion and does not give any restoring force to the balance, but it does reduce friction. In fact, in common with early crown-wheel escapements, the foliot balance of the strob escapement is not in itself an oscillator, but relies on the force of the succeeding pallet face for its return and thus its oscillatory motion. The escapement is a **recoil escapement** requiring considerable force to give the necessary impulse, thus causing severe wear on the pallet face and the pins. There were fifteen pins on each escape wheel producing an amplitude of vibration of approximately 84°. As interpreted by North, this escapement has a tight pallet-face angle that could lead to banking, where an escape pin hits the back of the pallet face. A solution to this problem could lie in setting back the centre of the semicircle from the centre of rotation. The text of the *Tractatus* uses the term *semicirculu* for the double pallet, the escape wheel

---

412 Dante 2011, *Paradiso*, xxiv, 13–15, trans. R. M. Durling, iii, 859.
413 Bodleian Library, MS Ashmole 1796 (mid-fourteenth century); Gonville & Caius College Library, MS 230/116 (late fifteenth century); Bibliothèque Royale, Brussels (late fifteenth/early sixteenth century). The Ashmole manuscript alone provides some 80% of the details about the clock. North 1976, ii, 309.
414 North 1976, iii, 271.

415 Whyte.

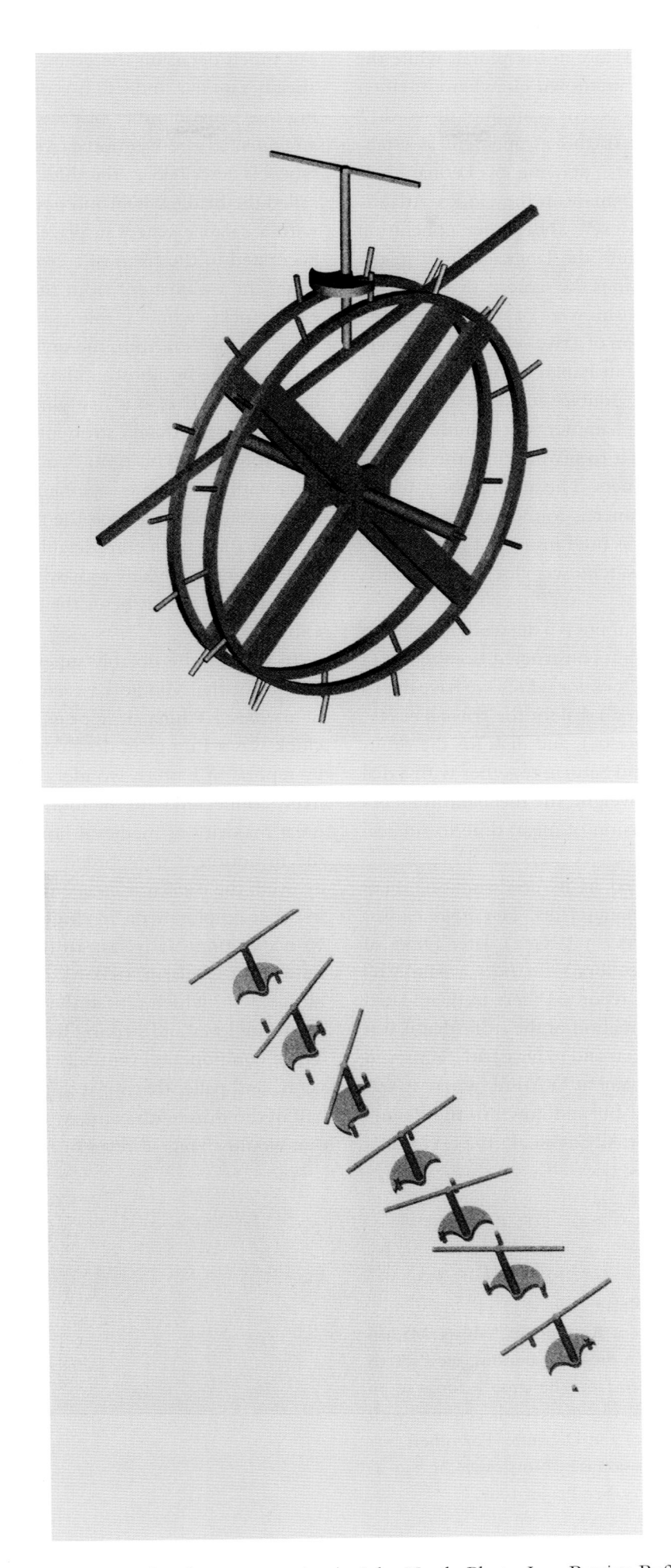

Figure 62 The 'strob' escapement and its action after the reconstruction by John North. Photo: Jean-Baptiste Buffetaud.

is termed *rota strob*, appearing invariably in the plural, while the verge is called *hasta strob*. 'Strob' is unknown elsewhere in either Latin or English and is translated by North as an English word.[416]

The general form and working of this escapement as established by North are confirmed by a drawing by Leonardo da Vinci (1452–1519) of a similar escapement in the Codex Atlanticus *c.*1495, and in the Madrid codex.[417] The other conclusion arrived at by North was that because the language used in the *Tractatus* to describe the striking train is the same as that referring to the escapement, the two mechanisms were of similar character. This is strengthened by the illustration of a two-wheeled hammer-striking mechanism in, once again, an Italian manuscript.[418] The striking mechanism has the same double wheel as the escape wheel, but with fewer pins, and the hammer post is similar to the verge, with a double pallet causing it to pivot, while the arms of the foliot act as alternate bell hammers. The arrangement where a foliot-type bar acts as a bell hammer is found on several early clocks where a single crown wheel is used instead of the double-wheel strob as, for example, in the watchtower clock of St Sebald's Nuremberg.[419]

The next earliest clock for which there is a detailed description is the *astrarium* of Giovanni de' Dondi constructed between 1348 and 1364. It is the earliest oscillator clock for which we have drawings. The heptagonal frame contained a dial for the *primum mobile*, which is an astrolabe dial (see Figure 98), and dials for the orbits of Mercury, Venus, Mars, Jupiter, and Saturn, together with a dial for the Moon. Powered by a single weight, the mechanism incorporates the earliest recorded conversion of mean time to sidereal time.

Only copper and brass were used in its construction, there being no mention of 'iron' in the manuscripts. The *astrarium* was acquired by Gian Galeazzo Visconti (1351–1402), later Duke of Milan, and installed in 1381 in the library at the Castello Visconteo in Pavia. The clock was last recorded there in dilapidated state in 1529/30. There is no confirmation of the story that it was offered to Emperor Charles V—who is said to have had it removed to his retirement home in the Convent at St Yuste, where it was presumed lost after the convent was burned down in the Peninsular wars in 1809.[420] As with the St Albans clock, the text leaves out much detail on the going train and escapement as if it were already familiar; indeed, it is described as a 'common clock' giving the 'usual number of beats'. Unlike the St Albans clock, the *astrarium* has a conventional crown-wheel escapement. The escape wheel is described as 'the third wheel which in common clocks is called the bridle wheel and has teeth like a large saw on the side', with a balance wheel in place of a foliot. The balance wheel is decorated in the form of a crown. It is the one feature that separates the escapement from those to be found in all subsequent early clocks, where a foliot with adjustable weights is standard. Thus, in the astrarium, the only way of regulating the going train appears to be by adjusting the driving weight. With its depiction, however, we have our first glimpse of an escapement that would eclipse the St Alban's strob and be, in some form, used for the next 500 years. This is no doubt owing to its robustness. In common with the strob escapement it is a recoil escapement and the angle subtended between the pallet surfaces is normally between 90° and 100°, with considerable diversity of opinion as to its ideal proportions. It is likely to have been far less prone to banking and other problems that beset the strob and led to its relatively swift extinction.

In spite of the notably superior hydraulic and automata technology of the Middle East and of the great astronomical clocks of medieval China, it appears almost certain that the oscillator clock was a European development, albeit one that was inspired by reports of Eastern wonders. Furthermore, its origins are most likely connected with the monastery, for it is in this realm that the great majority of medieval horological activity is reported. That the invention is unrecorded is most likely owing to the way it evolved: the oscillator escapement passed unnoticed for centuries in other employment, in ringing bells, lifting arms, and flapping wooden wings. It is easy to overlook, when considering going trains, that all alarm, striking, and automata mechanisms have an escapement, which caused the action they performed. And every returning lever and striking hammer is an oscillating component. These components eventually took the form of the crown-wheel, verge, and pallet that we recognize today and that, in the second half of the thirteenth century, were put to the more serious work that would change forever our daily lives.

---

416 See his discussion of the term in North 1976, i, 375, n. 6.
417 Codex Atlanticus f. 348v. Biblioteca Ambrosiana, Milan. Madrid Codex, i ff. 7r, 27r. For commentary on these illustrations, see Pedretti 1957, 103–4. See also Reti 1974, 249 & 256. The strob belongs to an escapement family that was used until the nineteenth century, with examples along the way by Volet (1742) and Debaufre. For others, see Gros 1913, 68, 80, 110. The fullest discussion is by North 1976, ii, 330–4.
418 British Museum, MS 34, 113 f. 187v. See Prager 1968.
419 Robertson 1931, 22–4.
420 Bedini & Maddison, 1966, 39.

SECTION FOUR: SAND-CLOCKS, SAND-GLASSES, FIRE CLOCKS

Anthony Turner

While water is perhaps the most obvious flowing substance to employ for time-measurement, it has its disadvantages and other such substances, whether liquid or granular, have been used. Mercury was employed by al-Murādī (late-fifth/eleventh century) in tilting balance arms in some of the clocks he described and it was also deployed in the compartmented cylinder that controlled the chiming, astrolabe, clock of Isaac b. Sīd described in the *Libro del Saber de Astrologia* (1276, Figure 61).[421] That the largest European source of mercury is also found in Spain at Almadén may explain this local use of the substance. Considerably later, mercury would be proposed by Leonardo da Vinci for use in a **time-glass**,[422] and was employed as the motive force of a compartmented cylindrical clepsydra developed by the brothers Matteo, Pier Tommaso, and Giuseppe Campani in 1655/66 while in search of a silent **escapement** for night clocks. Later, *c.*1815, Henry Constantine Jennings (1731–1819) would use it in his new form of log-glass.[423]

SAND-CLOCKS AND SAND-GLASSES

Sand, however, if suitably prepared, like similar flowing granular substances such as millet-seed or mustard-seed, could also be used as a motive force. Hero of Alexandria's book on automata is the earliest known source (Figure 63). There he describes the mechanism for a self-moving automata theatre powered by the fall of millet-seed from the upper compartment of a reservoir, through a hole of appropriate size, under pressure from a lead weight. A string attached to the top of this was then held in tension as it passed over pulleys to join the axle of a traction-wheel, which, turning, drew the theatre towards the spectators.[424]

Such a mechanism could be adapted to drive any number of devices, including clocks and astronomical models, as Abū 'Abdallah al-Khwārizmī explained, nearly a millennium later, in his *Mafātīh al-'Ulūm*:

> As for the movements that arise from ... those that work by sand and ... with mustard-seeds or millet: an instrument is made in the elongated form of the tube. Its lower part is pierced with a small hole, and its top is open. Then it is filled with sand or mustard-seed or something similar. A piece of lead is placed on top of it. The lead draws a thread or a cord tight: to the thread is attached what is necessary for the motion. Then the tube is placed in a vertical position, so that the sand or other [material] can come out of the hole at the bottom of it. As the sand gradually diminishes, the weight is moved downwards and it moves what is connected to it. In this way wonderful motions of various types are set up.

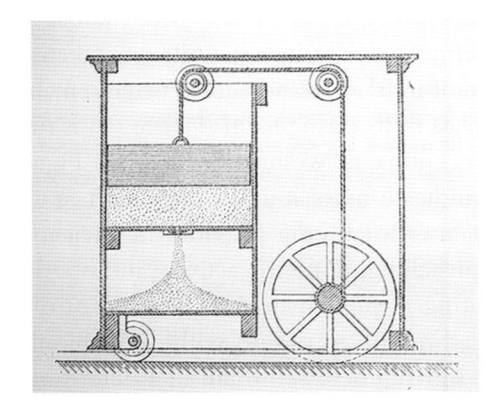

Figure 63 The mechanism of Hero's mechanical theatre. Diagram from Albert de Rochas, Les Origines de la science et ses premières applications, Paris, 1884, 155. Private collection. Photo: Jean-Baptiste Buffetaud.

Shortly afterwards, *c.* AH 391 (CE 1000/01), al-Bīrūnī mentioned water, sand, and seed clocks as equivalents when noting that 'equal motions have become the measuring units of time'.[425] A century later (pre-CE 1115), al-Khāzinī would employ machinery similar to al-Khwārizmī's to turn a celestial sphere continually across 24 hours.

Lengthy empiricism is described: 'I prepared many holes and tested them repeatedly, until I discovered the narrowest hole through which sand comes out without getting clogged'.[426] The sand itself was also carefully prepared, as he describes in his later work, *The Book of the Balance of Wisdom* (*c* AH 515: CE 1121–2). 'Sluggish sand, it is repeatedly washed free from dirt and earth, and dried. Then it is sieved in two sieves of different meshes, to discard large stones and small pieces, so that evenly flowing sand

---

421 The classic, though not entirely reliable, edition of this (Codex 156, Complutense University, Madrid), is that by Rico y Sinobas, 1863–66. The most recent discussion is Whitestone 2019, with references to earlier studies.

422 Reti 1974, 241. On Almadén in the early eighteenth century, see Jussieu 1719.

423 Bedini 1962, 127–30; Bryden 2018, 20.

424 De Rochas 1884, 152–3; Gille 1980, 142–3 & 125 for Hero's works.

425 Al-Bīrūnī 1976, I, 24, presumably thinking only of astronomical time-measurement.

426 Lorch 1980, 306.

remains. It is protected from small particles falling into it by a sieve placed over its container'.[427]

If the lead weight pressing the sand through the orifice in these devices is removed and the walls of the reservoir are inclined at an appropriate angle to the orifice, then the weight of the sand itself will cause it to flow into the lower section of the reservoir. The principle of the sand- or time-glass is implicit in the Hellenistic and Muslim mechanisms, but there is no evidence that the crucial step of inclining the container walls was taken either in the Ancient World or in Islam. Where it occurred remains unknown, but the sand-glass in its classic form (Figure 64) is first attested in the fourteenth century in Western Europe (see Figure 211), with some possible indications for it in the thirteenth century.[428] It had a multitude of uses for timing activities in navigation, teaching (Figure 214), preaching, domestic activity, industrial processes,[429] and the executing of piece-work. Its basic form of two conical ampoules placed mouth to mouth with a suitable flowing granular substance moving between them in a determined period of time varied little from the fourteenth to the twentieth centuries, although the stand in which it was mounted offered considerable scope for decorative ingenuity and was executed in materials ranging from the humblest woods to amber, silver, or gold, sometimes with enamelled decoration.[430]

Glasses were also mounted in batteries of two to four, each individual instrument running for a different time, and they were also sometimes fitted with a manual dial on which a record of time elapsed could be maintained.[431] Sand-glass making was, however, perhaps not much esteemed as it 'requires very slender Parts to become Master of; he is partly a Turner and buys his glass from the Glass-House; there are not very many of them, nor much to be made by those who are employed'.[432]

Although ubiquitous, the double-ampoule sand-glass, did not displace other forms of sand-clock. Late Renaissance writers, exploring the classical and medieval legacy, imagined a number

Figure 64 The classic sand-glass, French or German, early eighteenth century. Private collection. Photo: Jean-Baptiste Buffetaud.

of forms;[433] some, such as that of Radi, followed by Martinelli, exploited the principle of the compartmented cylinder while offering a battery of twelve one-hour sand-glasses held within a cylinder as an alternative.[434] Semi-automatic and automatic turning of glasses was developed,[435] while attempts were made to apply forms of the classic glass in astronomy, navigation,[436] the household, and for timing pulse-beats in medicine which, with the navigational log-glass, and the egg-timer, was perhaps the most successful.[437] Sand as a source of power of constant force for

---

427 Hill 1981, 49.
428 Two Arabic treatises, both in the British Library, London, one probably thirteenth century, the other fourteenth century, describing sand-run timepieces await investigation. The sand-glass, because it measures equal divisions of time, could only have secular use in Islamic societies. Even so, Evliya Celebi noted fifteen sand-glass workshops employing twenty craftsmen in seventeenth-century Istanbul. Hitzel 2012, 19–20.
429 For some examples see Dohrn-van Rossum 1996, 302. Cf. the interesting description in a late-fifteenth-century Welsh poem, Linnard & Owen 2012, 632–3.
430 For the origins of sand-glasses and their development, see Turner 1982a;1982b; Turner 1984a; 1984b, 73–84, for the 'sand', Drover et al 1960, Delalande 2015, 114–19; for their variety Turner 1984a; 1984b, 85–114; Attali 1997 (with a good selection of paintings depicting them), Delalande 2015, 127–393.
431 An example is shown in Aked 1979, fig. 1.4, where other forms of glass are also illustrated.
432 Campbell 1747, 323.

433 E.g. Schott 1664.
434 Radi 1665; Martinelli 1669, the latter with French translation in Ozanam 1694.
435 Turner 1984a; 1984b, 83–4; Baillie 1951, 111; Turner 2016a.
436 Such investigation was stimulated by the Paris Académie des Sciences which, in 1725, offered a prize for finding the best way to preserve the stability of sand-glasses and water clocks at sea. The prize was obtained by Daniel Bernoulli, but was entirely theoretical. See Engelbert 1994, 183ff.
437 Turner 1984a; 1984b, 81–4; for the pulse-glass Blaufox & Constable 2008; Turner 1977, 88, N° 158.

geared time-mechanisms, however, continued to be investigated in later centuries.[438] Sand-glasses were also adapted to new uses; one, which had some success, was the 'sablier-compteur' for use in photography, patented by the Daguerrotypiste and alchemist Théodore Tiffereau (1819–1909) in 1852.[439] Another was the incorporation of a small, log-type glass into the 'Fluidimètre' developed by the Etablissements Lefranc in the early twentieth century, to measure the viscosity of varnishes and enamels.[440]

FIRE CLOCKS

Even less is known of the origins of fire clocks than of sand-glasses. Since substances such as oil, tallow, wax, or incense burn at a regular rate of consumption this can be used to measure equal intervals of time, although this was a disadvantage before **equal hours** came into general use from the late Middle Ages onwards. Three types may be distinguished: candle clocks, lamp-clocks, and incense-clocks, of which only the first two are discussed here.[441] Clay oil lamps, which could be organized in function of their body size, type of wick, and type of oil to burn for a set period, were used as timers in Antiquity in such diverse activities as mine-drainage and the conjuring of spirits.[442] Absorbed into the Hellenistic tradition of mechanical devices, self-feeding lamps were developed by Hero. Working in the same tradition, in the mid-ninth century CE, the three Banū (sons of) Mūsā ibn Shākir in their *Kitāb al-Hiyal* (Book of Ingenious Devices) described four lamps that derive from Hero's work. The third of them, with an automatic oil feed and an automatic wick supply, combined the functions of the first two. Of it they noted:

> ... it is possible with these arrangements to make a lamp that shows the hours and whenever an hour elapses a ball drops ... [dropping of a ball per day was also possible]. The people of the religions require this lamp—they who see it believe that it is a perpetual lamp ... They are the magians. And in churches, the Christians place the column and the reservoir for the oil in the wall, and [everything] is hidden except the lamp, [which is] more beautiful for the viewers of the lamp.[443]

Figure 65 A late-eighteenth century oil lamp timekeeper. Private collection. Photo: Jean-Baptiste Buffetaud.

The candle clock was more typical of Christian Europe, where oil was less readily obtainable, and was more primitive. 'In England, Alfred the Great, wishing to devote half his time 'to God from the service of his body and his mind', could not 'equally distinguish the lengths of the hours by night, on account of the darkness, and ofttimes of the day, on account of the storms and clouds'. After reflection he had six candles of given weight and equal length made each marked with twelve divisions:

> ... those six candles burned for twenty-four hours, a night and day, without fail ... but sometimes when they would not continue burning a whole day and night, till the same hour that they were lighted the preceding evening, from the violence of the wind, which blew day and night without intermission through the doors and windows of the churches ... the king therefore considered by what means he might shut out the wind, and so by a useful and cunning invention, he ordered a lantern to be beautifully constructed of wood and white ox-horn, which, when skilfully planed till it is thin, is no less transparent than a vessel of glass.... By this contrivance, then, six candles, lighted in succession, lasted

---

438 A well-made late-eighteenth-century sand-clock movement is held in the collections of the Musée des Arts et Métiers, Paris, inv 10,645. A device that used the weight of sand fallen to determine time was described in Prosper 1727.
439 Three examples are held in the Musée des Arts et Métiers, Paris, inv. 14,009; for Tiffereau, see Principe 2014, 24–53.
440 Lefranc 1925, 247.
441 For incense clocks, and other fire clocks in China and Japan, see Chapter 2, Section 2 and Chapter 2, Section 4.
442 Hannah 2009, 96–7.
443 Hill 1979, 237. For the Hellenistic origins of this technology *cf.* 21, 'Many of the Banū Mūsā's devices are elaborations of basic ideas contained in the works of Philo or Hero, or both'.

four and twenty hours, neither more nor less, and, when these were extinguished, others were lighted.[444]

Whether the *Life of Alfred* was written by Asser (d. 909) in 893 as traditionally supposed, or is a later compilation by Byrthferth,[445] need not cast doubt that candle clocks were used in the religious institutions and entourage of Alfred. However, since they would burn at a constant rate their weight would have had to be adjusted throughout the year for them to correspond with the changing length of the unequal hours.[446]

The sophisticated Islamic tradition and the simpler Christian tradition would continue in parallel development throughout the Middle Ages. The chandelier clock of Yūnus al-Husayn al-Asturlābī (mid-sixth/twelfth century) functions in similar manner to Alfred's candles but employs oil-lamps burning a weight of oil different for each hour so as to measure unequal hours.[447] Another of Yūnus' timepieces employed an automatically positioned candle. This was examined by al-Jazarī who dismissed it ' … the design was useless, its failure being due to the overflow of the wax.'[448] Jazarī himself describes four variant forms of automatically setting, and self-trimming candle clocks which combine utility with playful ingenuity and decorative automaton figures.[449]

The latter however are not employed in the weight and pulley candle clock, of which the motive force is the weight of the candle itself, described by Samuel el-Levi in the *Libro del Saber de Astrologia*.[450] A rather basic Persian ball-dropping candle lamp ('the best'), is described in a late-fourteenth or mid-fifteenth century Byzantine text.[451] In Christian Europe meanwhile several candles, or a single large one, graduated up to twenty-four hours, were used to measure devotional periods by Louis XI (1214–70) and Charles V (1337–80) of France, and no doubt by others.[452] Certainly, c.1305, the deployment of candles and oil-clocks by monks of St Albans Abbey in their cells was disapproved of by the abbot.[453]

Long known in the Near East, the fountain feed lamp seems not have appeared in Europe much before the sixteenth century when Jerome Cardan's description of it in his *de Subtilitate* (1550) perhaps marks the moment when it began to be employed. Investigation of it for time-measurement quickly followed. Many forms were developed, some of them more than once,[454] and the candle-timekeepers were largely displaced. The interest of fire as the motive force and illuminant for a silent night clock, a crucial field of investigation in mid-seventeenth century Italy, was explored.[455] The longest lasting form of lamp-clock, however, was that perhaps first described by the Jesuit Pierre Bobynet (1593–1668) in a book primarily devoted to sundials.[456] This consisted of a vertical glass column closed at the top and screwed into an oil reservoir carrying the wick (Figure 65). This in turn was mounted on a high, turned base. Two straps ran down opposite sides of the glass tube, one graduated for hours, the other carrying a handle. Simple, safe, and portable, the model remained in use throughout Europe for over two centuries.

---

444 Smyth 2002, 40–1.
445 As argued by Smyth 2002. For a succinct account of the work and defence of its authenticity, see Wormald.
446 See Birth 2018 for details of the time context of Alfred's candles.
447 Kennedy & Ukashah 1969, and dissent in Turner 1995.
448 Hill 1974, 87.
449 Hill 1974.
450 Rico y Sinobas 1863–6; Whitestone 2019.

451 Transcribed and translated with commentary in Tihon 2000.
452 Havard n.d., 59–60; Solente 1936, i, ch. 36.
453 *Gesta abbatum monasterium S. Albani*, 1302–8, cited from Brockhaus 1955–60, ii, 94, N° 5490.
454 Bedini 1965; Turner 1984a; 1984b, 120–1.
455 Martinelli 1669, best consulted in the French translation by du Bosc, published by Jacques Ozanam 1694, who added extensive critical commentaries.
456 Bobynet 1663, 303–8 and pl. 21.

CHAPTER FOUR

PUBLIC CLOCKS FROM THE THIRTEENTH TO THE EIGHTEENTH CENTURIES

Marisa Addomine

DEFINING A PUBLIC CLOCK

'Public clock' can be a misleading term, as every time-teller not strictly conceived for private use can be so considered. Here a 'public clock' will be defined as a large, mechanical, weight-driven clock designed for a community, even if sometimes privately owned. This definition is far from being obvious, especially looking back to a past when Latin was the language of educated people and the word *horologium* and its variants indicated all sorts of time-telling devices.

To some extent, even the concept of *community* was different from that used today: it is common to identify a public clock as something related to a church (whether isolated or in an urban setting), or to an important building in a town or city, especially when we think of the beautiful clocks which can still be admired on town halls in those areas historically under German cultural influence. Nevertheless, in the late Middle Ages an essential role was still played by monastic communities: not only as groups of people sharing a common faith and living according to Christian principles, but also as cultural centres, often—as in the case of the Cistercian Order—as real concentrations of technical culture, especially in hydraulics, metallurgy, and mechanics. A large monastic complex encompassed much more than the monastery and its church: it included mills, forges, looms, and workshops; the majority of those activities would become the main source of income for artisans who would later move to towns and cities, when the general political and social situation evolved, especially in the twelfth and thirteenth centuries. In a large monastic community, monks and laymen, peasants, craftsmen, and their families lived in a well-organized structure, where religious *officia* were regularly held at fixed moments, but were variable throughout the seasons according to the pattern of **unequal hours** and different latitudes. Every activity was organized around a complex time schedule.[1]

Sundials were largely employed as public timekeepers, but they had intrinsic limitations, being unable to drive a bell, and working only during the day and in favourable weather conditions. Nevertheless, sundials remained indispensable companions to public mechanical clocks, as they offered the necessary reference to local noon to allow for clock adjustment, after stoppages resulting from breakdown, missed winding, repairs, or other interventions. But the information sundials provided was necessarily available only to a user standing in front of them. **Water clocks**[2] represented a technological leap and their roots can be traced back to Hellenistic times. As Vitruvius tells us, some incorporated gearing. Acoustic signals were present in the water clock that Gundobad, king of the Burgundians, received as a gift from Theoderic the Great, king of the Visigoths. Described by Cassiodorus in his *Variae*, it was a clock intended for public use, of the same type that Cassiodorus, writing on behalf of Theoderic, described as commonly present in the urban environment of Roman tradition.[3]

These clocks, conceived as public devices, had gears, a dial (perhaps more than one), and were capable of offering time-reckoning to distant listeners thanks to an aural system. They were also of large size: this feature compelled the mathematicians who were in charge of calculating the ratios of toothed wheels, to face

---

1 For the latter, see this volume, Chapter 3, Section 2. For the scope of monkish craftsmanship in the twelfth century as revealed by Theophilus, Dodwell 1961.
2 See Chapter 1, Chapter 3, Section 1, and Chapter 3, Section 4, and for a more general survey of public time indication, Turner 2014c.
3 Cassiodorus in Barnish 1992, 20–4. *Cf.* Chapter 3, Section 4, n. 7.

problems that they would not normally encounter when dealing with small clocks. Since Antiquity, machines existed that had no mathematical ambitions or precision requirements, but were capable of exerting substantial forces, and withstanding heavy duty conditions: mills. This is not to suggest that clocks were invented by mill makers. However, a primary factor in the design of large clocks was a familiarity with the technology necessary to master sizeable weights, large masses, and environmental challenges much more significant than those faced by a table clock in an Italian palace *studiolo*. A public clock, therefore, is *a clock designed to tell time to a community; it is mechanical, of large size, and weight driven*.

Designing a real, working large mechanical device, as modern mechanical engineers know very well, is not simply a factor of scale. Just to mention the most intuitive obstacle, linear dimensions grow by a factor of two, but volumes—and masses—grow by a cube law. Large, heavy wheels are less sensitive, or entirely insensitive, to minor problems, like the proverbial speck of dust that can stop the best watch. On the other hand, in a large clock, friction, wear and tear, corrosion, environmental challenges, the action of wind, rain, snow—even birds on the hands—and the need to ring large bells, as well as general mechanical stress, are problems that have to be overcome using completely different criteria from those which apply when dimensions are smaller. A public clock is *not* simply a common clock, made larger.

TIME OF MONKS, TIME OF MERCHANTS

In a well-known essay, the *Annales*-school historian Jacques Le Goff (1924–2014) analysed the change of time perception in Western Europe during the period between the late Middle Ages and the Renaissance.[4] He discussed the evolution from timekeeping in a rural society, where every activity was necessarily bound to sunlight, nothing required to be evaluated on a precision basis, and timekeeping depended on canonical hours, to a new concept of time dictated by different social conditions and a new attitude towards work, salaries, production, and trade. He compared peasants and monks, sharing days and nights of variable length, across the seasons as a function of latitude, based on the church bell, with what he defined as the *time of the merchant*—a new urban time—that began to appear around 1200.

People needed to manage their working day in a different way: a new lifestyle, involving travel and more encounters, appeared in cities and towns and the need for a shared time, rather than one tied to the labour in fields, sunlight, and prayer, accompanied it. An individual might be involved in many events on the same day, but in different places: knowing the time and sharing it became vital in a new societal precursor of the Modern Era. The production of goods for trade implied the evaluation of costs related to manpower, where employee wages started to be calculated on an hourly basis. The modern concept of productivity was born. Public clocks offered an easy-to-understand, uniform signal, thus becoming the beating heart of a new economy—and so of a new world.

EARLIEST DOCUMENTS

Very few examples of early public clocks survive. By contrast, generally speaking, few timekeepers are as richly documented as those that do, and for good reason: a public clock was an important asset. Especially in the first period, from the thirteenth to sixteenth centuries, it was often the only reasonably reliable timekeeper available for the community. But it was expensive. Being installed by the community, all expenses, from acquisition to maintenance (whether regular or unscheduled), had to be documented, as they were paid from the common purse. The clock required daily winding, not to mention periodical time adjustment and repairs, both major and minor, all of which required recorded payments.[5]

Horological researchers working on late Medieval papers in Latin are often troubled when faced with accounting notes or contracts dealing with facts or expenses related to a *horologium*. Was it a mechanical, weight-driven one? Luckily, sometimes a detailed list of interventions or parts is present, leading to an identification beyond doubt. Thus, wherever possible, archival documentation should be explored when dealing with public clocks, although this may be impossible, for example, if a clock has been moved from its original location and its vital provenance data is missing—especially when there is no maker's signature. Some methodological warnings are also necessary: the absence of information about a clock does not mean that a clock was not present. The presence of a clock in a given place, and the presence of an ancient document mentioning a clock in that very place, do not automatically relate to each other.

Public clocks were subject to repairs, but also to replacement: sometimes, in the same building, four or five clocks have been installed successively across the centuries. Sometimes repair was more expensive than replacement: sometimes a later clockmaker was unable to repair a complicated clock, so the appeal of reduced cost was the excuse for hiding his inability to intervene on something he did not understand.

England appears to be the country whose archives contain the oldest records dealing with early public clocks. However, there is debate between those who think that clocks were born small and simple, and later grew to bigger dimensions, and those who argue that big clocks, even mechanically complex examples, appear in the earliest records. Here we stick to present archival knowledge. The ecclesiastical rolls of Dunstable Priory, Bedfordshire

---

4 Le Goff 1960, reprinted in Le Goff 1977; English translation in Le Goff 1980.

5 On maintenance and repair see Chapter 21.

(1283) and Exeter, Devonshire (1284) mention expenditure related to something described only as an *horologium* placed inside the church, while at St. Paul's Cathedral in London, payments are recorded in 1286 to *Bartholomo Orologiario*. Expenditure for the installation and maintenance of a clock appears in the books of Merton College, Oxford between 1289 and 1290. Still, in the late thirteenth century, another clock and its expenses are documented at Norwich Cathedral Priory, Norfolk (1290); similar entries are found in the books of the Benedictine Abbey of Ely, Cambridgeshire (1291), and in Christchurch Cathedral, Canterbury, Kent (1292).[6]

The Sacrists' Rolls of Norwich Cathedral from 1322–5 are extremely interesting. First, they record the construction of an astronomical public clock and mention Master Roger de Stoke (a very skilled clockmaker who will later be one of the craftsmen involved in making Richard of Wallingford's masterpiece at St Alban's). Roger's intervention was requested because unexpected troubles occurred during the construction of the clock, and his skill and experience were needed to solve problems that the contracted maker could not overcome.[7] The list could be extended.[8]

The demand for public clocks in Europe spread quickly as the circulation of knowledge, both theoretical and practical, was much greater than might be thought. In the Middle Ages, large monastic orders had yearly meetings, during which priors and abbots exchanged information and managed the movement of monks with special skills from one site to another, helped by their shared knowledge of Latin. Artisans and craftsmen, not bound to the countryside as peasants were, moved from place to place—and from court to court—if their products were in demand. With public clocks, itinerant makers were less common, essentially because wrought iron, the most commonly used metal in construction, implied a forge. In some cases, as in the construction of the Perpignan clock,[9] a forge was built on the spot, but generally most clocks were made in the workshops of makers, tested, dismantled, transported, and reassembled on site, frequently being installed in rooms high above ground level, accessible only by means of narrow ladders.

To list the some 450 datable clocks known from before 1500 is not possible here.[10] Table VII illustrates the spread of public timekeepers in Western Europe before the fifteenth century.

EARLIEST SURVIVORS

Time devours all, but as regards public clocks, in common with many other artefacts, man often destroyed more than time. A large

---

6 Beeson 1971, 13–15.
7 Beeson 1971, 16; North 2005, 141–3.
8 As it is in Beeson 1971, 18–24.
9 Beeson, 1982
10 Dohrn-van Rossum 1987 offers a statistical survey of this diffusion.

Table VII Some early European Clocks

City	Country	Year
Milan St Eustorgius	Italy	1309
Milan Holy Virgin	Italy	1335
Padua	Italy	1344
Monza	Italy	1347
Vicenza	Italy	1349
Trieste	Italy	1352
Genoa	Italy	1353
Florence	Italy	1353
Bologna	Italy	1356
Pisa	Italy	1356
Siena	Italy	1360
Avignon	France	1353
Strasbourg	France	1354
Perpignan	France	1356
Vincennes	France	1359
Paris	France	1370
Bourges	France	1372
Prague	Czech Republic	1354
Exeter	England	1284
London	England	1286
Oxford	England	1289
Norwich	England	1290
Ely Abbey	England	1291
Canterbury	England	1292
Lincoln	England	1324
St Albans	England	1327
Windsor Castle	England	1351
Toledo	Spain	1366
Salamanca	Spain	1378
Seville	Spain	1380
Burgos	Spain	1386
Regensburg	Germany	1358
Rostock	Germany	1379
Bad Doberan	Germany	1390
Stralsund	Germany	1394

clock consisted of a large amount of valuable metal—high-quality iron for the frame and wheels, sometimes a large copper dial in the case of astronomical examples. It was common practice, evidenced by the earliest documents, when commissioning a new clock, to offer the old one to the clockmaker, obtaining a price discount on the new installation based on scrap weight. This was the sad end of extraordinary clocks all over Europe. Nevertheless, some survived, even if at present no clocks are known dating back to the earliest period.

Dating a public clock requires close attention. Clocks were replaced over the years and great care must be applied when relating archival papers to existing objects. At present, two clocks contest the title for earliest surviving clock in the world: both of them are public clocks. Documents for both the Salisbury Cathedral clock and for the Chioggia clock indicate their existence already by 1386. The earliest Salisbury document and the earliest Chioggia text are both dated and are only forty days apart. Both are quite basic clocks, using a large birdcage iron frame, with time and striking trains, driving a single hand. Their construction is very different, and the English clock is much more refined when compared to the coarse, though robust, Italian one. The universally known Salisbury clock, re-discovered in 1928 by T. R. Robinson (1905–83), has been the subject of several studies. Keith Scobie-Youngs' (the present official keeper of the historical clock) recent examination[11] starts from the similarity between this timekeeper and the non-astronomical section of the Wells clock—something already noted in the existing literature—as well as the proposition that both clocks were constructed at the wish of Bishop Erghum, who had links with both sites. Scobie-Youngs then highlights strong similarities with a third English clock, that at Rye Church, Sussex. The three mechanisms are astonishingly similar, including the style of forging of the iron bars, dimensional values, and other details. All three clocks were modified over the centuries, so interrogation must focus essentially on frame, layout, general design, construction choices, and the maker's personal style.

Dating the Salisbury and Wells clocks to the fourteenth century relies on the existence in the Salisbury Archives of a document dated 1386: a contract for the lease of a house and a shop by the Dean and chapter to Reginald Glover and his wife Alice, on condition that they maintain the clock in the bell tower. This was considered sufficient evidence to conclude the clock already existed in 1386, while that of Wells is recorded from 1392. That both clocks were to be associated with Ralph Erghum (?–1400), who held the sees of Salisbury and Wells in succession, was advanced in the early literature but has been vigorously denied in recent investigation.[12] The known documents testify to the existence of a public clock needing a keeper in Salisbury in 1386, but we cannot conclude that the clock on display inside the cathedral is necessarily the one to which the parchment refers. Scobie-Youngs poses the basic question: given that all three clocks appear to share a provenance in the same workshop, should one date them all to fourteenth century, or should one date them all to the Tudor period?

As recently as 2004, a public clock was discovered in the ancient tower of St Andrew's Church in Chioggia, Italy, not far from Venice.[13] A serendipitous conjunction of facts, people, and events led to an extensive exploration of the exceptionally rich local archive and to the discovery of a document that mentions a public clock already in existence on 26 February 1386. This discovery was followed by a series of several hundred other documents and notes that allow the story of the clock to be traced to the present day, including its removal from its original location and reinstallation in the nineteenth century in the tower where it remains, thus excluding the possibility of any replacement.

The most striking evidence is a note about a particular repair on the main wheel in 1423 that was defined as 'unusual' and subject to a very well-detailed contract between the town and the restorer from Venice. The intervention is still visible and is a quite rare substitution of four broken teeth through a dovetail insertion of a new metal part. This evidence confirms that the clock in the tower is the one mentioned in the chronicles and in the accounts.

The Chioggia clock design is a good one from a mechanical point of view, but at the same time economical, using thin iron bars and pillars. It is more coarsely wrought than the Salisbury clock and also smaller. Despite this, it worked satisfactorily until the 1970s, when the lack of a volunteer for daily winding meant it was retired and replaced by an electronic timekeeper.

MAKERS AND KEEPERS

Making a public clock out of wrought iron involved skills different from those normally required from makers of small timekeepers. But being a good blacksmith did not qualify anyone, in itself, as a potential clockmaker, as the computation of gear ratios required at least some elementary knowledge of arithmetic.

Precision in the movement of the hour hand had to be obtained by a mechanical contrivance that was simultaneously capable of exerting non-negligible forces, using large size metal parts, susceptible to dimensional variation owing to temperature change, and subject to heavy mechanical stress. Temperature and humidity, whose values could vary widely, affected the organic lubricants that every maker used. Before modern metallurgy allowed the making of large, high-quality toothed **pinions**, all pinions were of the cage type: their construction was simple and reliable, and the substitution of a broken element could be performed simply, minimizing the cost and effort of the repair.

This technology was already widely used by mill makers and by engineers designing large building-site machines. These normally employed wood and cage pinions, with cylindrical elements orthogonal to the facing plates and parallel to the rotation axis in which the grain of the wood lay longitudinally, matching good mechanical resistance with low cost and ease of making.

The possible origin of large weight-driven clocks in monasteries is an acceptable hypothesis: in a monastery, the average cultural level was much higher than in late medieval lay communities; people with different skills cooperated and in-house workshops, including forges, were often present. The case of Richard of Wallingford, abbot of Saint Albans, demonstrates that astronomical clocks, implying difficult calculations for gear ratios and the design of non-standard wheel configurations, were the result of teamwork. Such a clock needed cooperation between the designer, who had the mathematical and astronomical capabilities, and the fabricators (who also knew about horology, even if not at the highest levels), who had the necessary skills to forge bars, wheels, and pinions, an activity requiring both long training and physical strength. Richard of Wallingford relied on two members of the Stokes family, both of them attached to Saint Albans,

---

11 Scobie-Youngs 2018 citing the earlier literature. For the Rye clock, see Tyler 1976.
12 McKay 2016 in *AH*, 37, 135–6.
13 Addomine 2016.

the younger probably a monk; both were already skilled clockmakers before Richard started to design his famous astronomical clock. According to contemporary chronicles, their contribution was required in other cities, and sometimes they were consulted when a clockmaker encountered major troubles in constructing a clock.[14]

The distinction between designer and maker is also well evident in almost all astronomical public clocks, which did not normally follow a repetitive design pattern. Jacopo Dondi (1293–1359), father of Giovanni (c.1330–88), maker of the Astrarium, designed the astronomical clock of Padua but certainly did not forge its wheels personally; Conrad Dasypodius (1532–1600), the celebrated mathematician, designed the second Strasbourg clock but did not physically make it, and so forth.

Early documents show that a public clock maker in the late Middle Ages and throughout the sixteenth century was often a man familiar with technology in a broad sense. He could be known for his ability to make church organs, as in the case of Wauter Lorgoner (1344) in London, or Giovanni degli Organi in Modena, who built the first city clock in Genoa in 1354 following the orders of the ruler of Milan Bernabò Visconti (c.1321–85). Or, he could sometimes be a maker of firearms—a totally new technology at the time. The German maker Bodo of Hardessen was referred to as 'organista et orologista' and Jacobus de Novaria, from Novara in northern Italy, worked as turret clock maker about 1410 in Caffa, Crimea, a then-Genoese colony on the Black Sea, and was known as 'magister orologii et organorum'. The Della Volpaia family, working in Florence between 1470 and 1600 across three generations, were skilled engineers, designers, and suppliers of domestic as well as turret clocks, but also of scientific instruments, mills, and water lifting machines. The Rye clock was supplied in 1561–2 by Lewis Billiard, who is recorded in the city accounts as a clock- and crossbow-maker.[15]

Sixteenth-century documents in the Venetian State Archives show that the city guild of blacksmiths in charge of making large-sized iron objects, distinct from the small items commonly considered as locksmith's work, was invariably presided over by a member who was a clockmaker. Reading autograph texts by these makers reveals that they were far from being humble, almost illiterate men, as they competently managed complex supply contracts and trade agreements and wrote in a proper style showing the fluent and neat handwriting typical of those who were well acquainted with reading and writing.

A public clock was always an expensive item: being supplied to a community, it was normally subject to a contract specifying in detail its features, delivery parameters, the maker's obligations, and terms of payment. If the maker lived close by, he could also be assigned the role of clock-keeper; otherwise, he was often obliged to guarantee proper training for a local person for such daily duties and for routine maintenance. In many cases, the acceptance of the clock, necessary for final payment to be made, involved examination and testing by other makers.

The first description of a clock-keeper's duties, though probably referring to mechanical water-driven clocks, are to be found in the twelfth-century rules of the Cistercian Order.[16] As a sundial requires no daily supervision, the obligations of the keeper, with punishments for those who did not attend the clock properly, are evidence for the presence of a mechanical clock—water- or solid-weight driven. They also suggest—as the rules applied in all Cistercian monasteries—that *all* monasteries should have had such a clock. The Cistercian Order represents a very interesting milieu for early clockmaking. Firstly, its members were admitted according to a strict selection, being subject to a test of their cultural qualifications, and whether they showed a tendency towards the development of engineering and technology skills. Secondly, there was what could be defined as a Cistercian obsession with synchronicity: *officia*, songs, and prayers should theoretically be performed at the same moment in all Cistercian communities, as their simultaneity would reinforce their appeal to God—a sort of collective resonance effect. In order to perform something at a given time, time had to be known precisely, so the clock really was essential in Cistercian communication.

A keeper's duties were of two kinds: daily, and simple periodical activities. Each day the clock required winding, which could also be performed by another local person, adjusting the clock by comparison to a local sundial, often close to a public clock. Local noon was the preferred reference point, and a good sundial could guarantee a precision of the order of a few minutes—more than acceptable in a world where time had a new role but was not yet required to be accurate to our modern standards. Periodical activities involved what we consider to be preventive maintenance: cleaning, lubricating, and the occasional substitution of simple parts subject to wear and tear, such as ropes, all without the dismantling of the clock.

Major maintenance—required whenever the clock stopped owing to breakdown of some mechanical part—necessarily required competent intervention. Looking at public clocks, very often damage was caused by incompetent keepers, unable to understand especially complex and non-standard designs, like those of astronomical clocks. This dearth of knowledge sometimes led to irreversible alterations and to the elimination of parts that a new keeper was unable to fix.

Whenever a clock had a major breakdown, the main choice lay between repairs or replacement. When commissioning a new clock, a common strategy was to secure a discount from the maker by offering him the metal parts of the old one, i.e. the movement and (sometimes) the old dial, valuing them on the basis of their

---

14 North 2005, 141–3.
15 Dohrn-van Rossum 1996, 99 and additionally for Giovanni, Turicchia 2018, 144; for the della Volpaia, Maccagni 1969–71; Pagliara 789–92; for Billiard, Loomes 1981a, 97.

---

16 Choisselet & Vernet 1989.

scrap metal weight. Such was the inglorious end of many extraordinary astronomical clocks, including the fifteenth-century clock from St. Mark's Square in Venice, replaced in 1757 by Bartolomeo Ferracina (1692–1777)[17] with nothing of the original timekeeper left, save the Moors striking the bells. It is easy to see that the supplier of the new clock might deliberately suggest repair was not the right option, even if the movement were not in bad condition. There was more to be gained by the sale of a new clock and then receiving the old one at scrap value, which could then be repaired and resold as a second-hand movement. In other cases, however, when the clock owner was an organized institution or a wealthy patron, old working clocks were moved to secondary residences or other locations, and new ones were installed in more important places in the buildings.

In most cases, the maker was the first keeper and sometimes the supply contract also covered several years of full maintenance. Sometimes, being a clock-keeper did not simply require horological competence. In the German-speaking areas, i.e. in a wide territory from the Baltic Sea to part of Northern Italy, there are examples of special tasks included in the duty list for a candidate clock-keeper. For example, in the now-Italian Vipiteno (Sterzing)—which was, until the end of the First World War, part of the Austro-Hungarian Empire—the public clock was installed in the tower at the city entrance gate built in the fifteenth century. This building also housed the city gaol. Several chronicles demonstrate the difficulties encountered by the community in finding a clock-keeper as he also had to be, by contract, gaoler, in charge of feeding the prisoners, managing them, preventing their possible escape, and taking on himself responsibility for their lives, including the repression of possible riots.

The clock-keeper's task was often transmitted from father to son or kept as a family activity: it guaranteed a fixed income and, in some cases, important benefits. For example, in Venice, the keeper of the St Mark's Square clock also had an apartment inside the tower housing the clock where he and his family could live, a tradition dating back to the sixteenth century. This was discontinued only a few years ago, after the latest restoration of the clock.

TYPICAL MECHANICAL FEATURES OF EARLY PUBLIC CLOCKS

A full description of the movements of all types of public clocks and of their components and variants would require a volume to itself: only a brief survey can be given here drawn from descriptions taken from the earliest known sources (Figure 218), up to the end of eighteenth century. Firstly, the trains—each clock section had its own weight and a set of wheels. The basic, minimum, and indispensable train is the wheelwork providing the time; its main and central element is the **escapement**. This **going train** is powered by a weight connected to a rope wound around a wood or metal drum. The winding operation, originally always performed manually, was possible thanks to capstan-like spokes on the main wheel, or to a crank. The winding of public clocks with large weights relied on the elementary lever effect, using handles or other elements placed at quite a large distance from the centre of the relevant wheel.

The immediate evolution for a basic clock was the addition of a second train for driving the geared section related to hour (and eventually half-hour) striking. This second **striking train** was mechanically linked to the going train but was powered by a separate weight and consequently had its own drum. In the oldest clocks the two drums were placed with their wheels parallel to each other, with their arbors in-line, and the barrels end-to-end, as this configuration is generally called.

This design, with a few exceptions, was common across Europe. From the late seventeenth century onwards, however, several innovations were made. Immediately on Huygens' publication of the pendulum in 1658 it was applied by Coster to a clock at Scheveningen near The Hague.[18] A pendulum clock was proposed to the canons of Rouen in 1664, and at Fécamp in 1667,[19] while in England the clocks of Church Hanborough (Oxfordshire) in c.1670, Hanwell (Oxfordshire) in 1671, and Hornchurch (Essex) in 1674 were all fitted with horizontal verge and pendulum control.[20] However, by this time the anchor escapement, which reduces the inconveniently wide swing required of the pendulum by the horizontal verge, was coming into use. The earliest known example is that fitted to the clock of Wadham College, Oxford, by Joseph Knibb in 1669/70, with several others known immediately afterwards.[21]

A further fundamental change was introduced in the early eighteenth century when Julien Le Roy conceived laying out a clock movement not in a vertical 'bird-cage' frame with the arbors above each other, but in a horizontal frame—a 'flatbed'. This, he claimed, would be more economical to construct, and therefore less expensive, and would function better as friction would be reduced.[22]

This new style was not immediately adopted everywhere. Even in France, where it was most popular during the eighteenth century, hybrid forms appeared. Like the birdcage form, it could easily accommodate the addition of a third train for quarter-striking, but some rare clocks obtain quarter-striking using only two trains thanks to an extremely complicated system of notches on the **count wheel**. A further train could be associated with special chiming features, such as a carillon drum.

---

17 For whom, see Turicchia 2018, 114. For some further examples, see Chapter 7.

18 OCCH, ii, 126.

19 Ungerer 1931, 96; Lenôtre 1986, 49–54.

20 Beeson 1971, 52–3; Tyler 1991, 618.

21 Beeson & Maddison 1969, 78–9.

22 Le Roy 1737, 338–50.

Northern European three-train public clocks sometimes adopted a T-type configuration for the drums, with two of them placed end-to-end and the third orthogonally. The presence of a third train for striking quarters was relatively common in northern Europe. For example, this T-type or cross-type configuration is encountered in the Chiaravalle Milanese astronomical clock in Italy. As this clock was described as being brought to Chiaravalle from far away, a foreign provenance cannot be excluded—the Cistercians had hundreds of monasteries across Europe and one of the best organized technological networks of their time.

As a general rule, even if subject to a high number of variants introduced by the ingenuity of the clockmaker, the astronomical section, driving further gearwork, or the section driving automata and/or other contrivances, was separate from the main clock and external to it. This was in spite of the fact that it required a mechanical link in order to have its action synchronized with the going train.

After considering the number and layout of trains, a public clock must be analysed according to its *frame* features, the shape, and the material of its construction. The earliest clocks had all their toothed wheels and drums arranged inside a wrought iron square or rectangular section frame—the 'bird cage'. The corner pillars and vertical structures are held together by horizontal iron bands and kept in place and consolidated by a system of wedges. Nuts and bolts would appear only towards the end of the eighteenth century, although later restoration sometimes placed them in older structures. Sometimes elements were soldered or riveted, but this made disassembly more complicated. Toothed wheels, pinions, and drums are mounted on arbors inserted in the vertical elements and turning inside bearings often made of brass or bronze.

In a 'bird cage' frame, all the elements are easily visible, and ordinary, almost daily, maintenance can easily be performed. Nevertheless, should a part need changing or be subject to a major modification, the whole clock had to be disassembled and reassembled afterwards, with all the risks this entailed. Surviving early clocks from northern Europe from the fourteenth to the sixteenth centuries and their pictorial representations show their frame design reflecting the Gothic style. The corner pillars are often at 135 degrees with respect to the planes of the sides they are connecting, and with a well decorated profile reminiscent of Gothic architectural elements, such as buttresses. This Gothic pattern, however, seems not to have influenced the design of contemporary Italian public clock frames—these were much simpler and essentially based on flat, plain iron strips, without any embellishment.

The only known exception to this is the so-called Brunelleschi clock in Scarperia, not far from Florence. Unfortunately, it has been heavily repaired and modified across the centuries. The local archives reveal that the clock was bought for the civic offices around 1440. It was supplied by the celebrated architect Filippo Brunelleschi (1377–1446), who was also a skilled designer of machinery for building sites and, during his youth, a maker of clocks. The Scarperia timekeeper is considered to have been made at the end of his life, since final payment was made, after his death, to his son. This small public clock retains the original fifteenth-century frame but is strangely reminiscent of the German style in the disposition of its structural elements and in the elementary decoration chiselled on the pillars, as well as their Gothic profiles. Since Florence was a rich city and attracted skilled craftsmen from all over Europe, the hypothesis that Brunelleschi simply bought a clock from a German maker working locally and resold it to the city cannot be excluded, but evidence is wanting.

An interesting and rare turret clock frame shape appears in northern Italy in the area between the cities of Reggio Emilia and Modena. The design is very compact and develops vertically on two levels, to some extent recalling some lantern timekeepers, with one peculiar feature—at the lower level, the structure has a square section, which at the top ends in a round crowning. The frame is wrought iron, as usual. Unfortunately, only a few examples survive. Despite their archaic look, according to supporting documents they were made in the sixteenth and early seventeenth centuries.

Some rare examples of wooden-frame bird cage turret clocks exist, and some of them also have at least some of their wheels made in wood. Examples were made until the nineteenth century in German-speaking countries, following the local tradition of wooden clockmaking. They are rare, but triangular-shaped wooden-frame clocks do exist in England, for example, the seventeenth-century Bletsoe clock, now at Bedford Museum.

Two other completely different frame structures were used in a small number of public clocks in England.[23] They can be classified into two categories, according to their mechanical structure. They share a quite impressive general feature—the fact that they are essentially flat and two-dimensional, as opposed to the substantially massive, three-dimensional structures of bird cage frames. The first group is designated as having a 'field-gate' frame: they have end-to-end trains, a rectangular wooden or iron frame, and a verge and foliot escapement in the earlier clocks, sometimes an anchor in the later ones (or when clocks have been renovated). Examples of this rare frame style are the clocks in the St John's Chapel of Hartford, Devon, and in Porlock, Somerset. Both of them, despite their archaic appearance, date only to the seventeenth century. The second, much older, group is represented by the so-called 'door frame' clocks. These clocks have two trains one above the other, with the striking train generally above the going. They are among the oldest of surviving public clocks, as some of them date back to between the late fifteenth and early sixteenth centuries. The wheelwork is distributed in a very simple way along a vertical rod, either iron or wood. In some cases, as in the extremely interesting clock at Cotehele House, Saltash, Cornwall, the clock has no dial, and it simply tells the time by striking a bell. Other door frame-type clocks can be found in a

---

23 Special thanks to Andy Burdon for information on this type of frame.

portion of the astronomical clock in Exeter Cathedral at Marston Magna, Somerset, and the faceless clock at St Andrew's Church at Castle Combe, Wiltshire.[24]

Discussion of early clock escapements normally concerns the well-known verge and foliot design. Very few early examples survived conversion to pendulum. A parallel escapement related to a public clock, the *strob*, can be derived from Richard of Wallingford's *Tractatus Horologii Astronomici* (pre-1336).[25] Very little is known about its fortunes, although it appears in some clock drawings by the Tuscan artist Mariano di Jacopo, nicknamed Taccola ('jackdaw'), around 1440, unaccompanied by any note,[26] and in Leonardo da Vinci's notebooks.[27] By then, however, it had generally been displaced by the verge and foliot, even if the construction of this is more complicated and implied the use of a sawtooth wheel, not a simple pin wheel.

The regulation of oscillation could be obtained by the **foliot** bar or by a circular **balance**, sometimes referred to as *corona freni*, i.e. the crown brake, in contemporary Latin sources. The analysis of the wheel trains of early public clocks shows that the standard oscillation period was in the order of seconds, much longer than the values encountered in domestic examples. The verge and foliot configuration was the preferred solution for public clock escapements until the introduction of the pendulum in the later seventeenth century.

The striking train of early monumental clocks, if present, necessarily relied on count wheel notches that determine the number and the sequence of strikes, engaging or stopping the action of the system on the hammers. Count wheels may have notches cut off from the inner rim of the wheel, often associated with a vertical locking system, or more commonly—and in more recent times—notches along the outer rim, selected by a locking arm. A more modern system is **rack striking**. The substitution of a count wheel of the first type with one of the second type over time was not infrequent, and sometimes also associated with the choice of a different sequence of strikes, probably following the request of the clock owners. The speed of the action of the striking system on the hammers was regulated by means of a **fly**, acting as an air brake. The speed of the fly could normally be regulated by changing the effective surface of its blades. The larger the effective surface, the higher the air resistance and the slower the frequency of strikes.

The power source for a turret clock was in the great majority of cases one or more stones, roughly carved in the earliest clocks and transformed into modular groups of calibrated weights in later times. As soon as electricity became available, clockmakers tried to replace the clock-keeper's daily duty of rewinding by effecting it with an electric motor, although clocks could be damaged when this was improperly introduced. An earlier idea was that incorporated in an extraordinary turret clock made around 1780 by Giambattista Rodella (1749–1834) for Villa Cavalli in Bresseo, Italy. This showed and struck both the Italian and French hour systems, counting from sunrise and sunset. It was rewound by a wind vane.[28]

HOUSING A PUBLIC CLOCK

For English speakers, 'turret clock' and 'tower clock' are synonymous with public clocks, suggesting that the collective unconscious associates public clocks with towers or belfries, topped by one or more bells and having one or more dials. German speakers call them *Turmuhren*, again linking the idea of clock (*Uhr*) to the idea of tower (*Turm*). Italians generally speak about *Orologi da Torre*, like the Germans, or *Orologi da Campanile*, specifying that the tower is a *campanile*, a bell tower usually associated with a church. The French talk about *Horloges d'Edifice*, i.e. clocks for buildings, whereas Spanish speakers simply name them *Relojes de Torre*, i.e. turret clocks.

Surveying the history of public clocks, however, it is clear that these, especially non-standard ones, could be installed inside a large building (usually a church or cathedral), but still have a communal use. This was especially the case for clocks enriched with ancillary functions as automata or complex astronomical dials; these ensured the necessary shelter against adverse weather conditions or avian visitors. Such clocks, always extremely expensive, became attractions for pilgrims and educated travellers, who frequently described them in their travel journals, remarking on their presence as a sign of distinction, wealth, and of the technical power of the community that supported their construction. A fine clock was also a symbol of a well-regulated society, and thus, indirectly, testament to the skill of the ruling class.

The wealthy Hanseatic League cities on the Baltic Sea still preserve some high-quality monumental astronomical clocks made between the fourteenth and sixteenth centuries.[29] All these clocks, now in Germany (with the exception of Lund and Gdansk), were placed inside cathedrals. In England, the first fully documented public clock by Richard of Wallingford (1292–1336) was placed in the upper part of the transept of St Albans Cathedral.

Countries enjoying more favourable climates— Southern France, Spain, or Italy—had clocks mainly placed on the façades of religious and civil buildings. It is interesting to note that the most important early Italian public clocks are more frequently

---

24 Beeson 1971, 81–93.

25 Described in Chapter 3, Section 3. See also North 1976, ii 330–4; North 2005, 175–82.

26 Anonymous drawings attributed to Taccola and Francesco di Giorgio Martini, MS Palatino 767, Biblioteca Centrale Nazionale Firenze, 43.

27 Codex Atlanticus f348ᵛ reproduced in North 1976, iii, 64, with commentary at ii, 331.

---

28 My thanks to the late Luigi Pippa for information about this clock. On Rodella, see Turicchia 2018, 242–3.

29 Schukowski 2006; 2009.

found on town halls or civic towers, and not on churches, especially in the territory of the Republic of Venice. This deliberate placement is emblematic of how civil authority wanted to emphasize its control over citizens' lives, as well as display city pride, until the end of the Renaissance.

LOOKING AT A PUBLIC CLOCK, LISTENING TO A PUBLIC CLOCK

The English word *clock* is related to the late medieval Latin term *clocca*, indicating a bell—clocks are intimately linked to their activity of striking the hours, and public clocks are no exception. Most common activities from the city to the village began with the tolling of the local bell. Broadly after the year 1,000, the opening and closing of city gates, the beginning of daily labour for peasants and its end at dusk, or the opening and closure of markets were all marked by the sound of a bell, tolled by a man, often a city guard. An interesting manuscript in the Salisbury archives describes very well the relationship between the ringing of the city bell and the activities in the community related to the monastery, including the surrounding countryside. This applies as well to foreign merchants, pilgrims, and visitors. In this document, dated 1306, the bell is rung manually, and its sound is heard everywhere. No sales were allowed in the market before the opening bell, nor after the evening one.

The first important role for a public clock was as an automatic bell ringer: the human element was no longer required, and by both night and day the population was duly informed about the hour. Expert opinion is divided, as some believe that the earliest public clocks had no dial and limited their function to striking, while others believe that the first clocks may have had dials but no bells. As evidence for the latter view, the late medieval chronicles of Milan—one of the wealthiest and richest cities in Europe and famous for the skill of its metal workers—report that in 1309 a weight-driven clock was present in the Basilica church of St Eustorgio. However, only a few years later the local historian Galvanus Fiamma (1283–1344) records that in 1335 the iron clock in the belfry of the Holy Virgin, a church now dedicated to St Gottardo in Corte, was *extended* and it struck once at 1 o'clock, two at 2 o'clock, and so on, up to the final twenty-four strokes at the end of the day; according to the tradition of *Hora Italica*, about sunset. The novelty of this feature is remarked on as being extremely useful, giving the reader the feeling that the other city clocks either were not striking or were simply marking each new hour by a single strike. These clocks did not survive.

The striking pattern of public clocks followed local traditions and styles. In Italy, it was based on the twenty-four-hour *Hora Italica* but could be modified to 2 × 12 to make it more intelligible, as counting a long sequence of strikes could lead to errors. The transformation of most dials and striking systems into the standard twelve-hour system became almost universal in Italy after the Napoleonic military campaigns and consequent French domination. Many churches adopted a mixed style, striking two different methods, at an interval of about one minute, in order to acknowledge—in a very typical Italian way—both the Pope and the dominant political power. Chiefly in central and southern Italy, additional strike sequences allowed a clock to mark meaningful moments in the day. Some of these had a civil meaning: the start of school for students, or of work at a factory, or labour in the field for manual workers, and correspondingly the end of such daily duties. Some had a religious significance, such as the *Angelus*, introduced by the Catholic Church as the prayer at sunset.

Sometimes in central and southern Italy, long sequences of strikes reminded the population of ancient curfew times—when all citizens had to be inside the city walls, behind locked gates. In some towns, the tradition is maintained of having the local clock strike one hundred times at midnight, to signal the end of the day. In other locations, at Easter, the clock might strike thirty-three at midnight, as a remembrance of Christ's age at his death.

In a uniquely Italian setting, the Papal States, which represented most of central Italy (excluding Tuscany) until the arrival of the Kingdom of Italy, dials and striking systems split the twenty-four hours of a day into four six-hour segments. Even if this derived from an ancient time-reckoning system, peculiar to the Church and to earlier divisions of the day, such six-hour mechanical clocks only appeared in the sixteenth century. Several public clocks, in territories where the Catholic Church's fortunes waxed and waned in terms of local control, employed a mixed economy, striking the hours under a twelve-hour system, and then under a six-hour system, a minute later.

Dials in Italy were generally laid out in six-hour or twelve-hour designs, dependent on local choices and the degree of freedom that local citizens had when the dial was made or renewed. A six-hour dial offered two advantages: looking at a dial, each hour intersected a sector of sixty degrees, and a higher spatial resolution allowed a better identification, at a glance, of the hour fraction; the total number of hammer blows in a day was eighty-four instead of the 156 demanded by a twelve-hour dial or, even more, the 300 blows needed for twenty-four-hour dial. Fewer hammer blows allowed a shorter run for the striking-train weight, less rewinding, and facilitated the placement of the movement in low towers.

Chiming was a feature even in early public clocks: hour blows were either replaced by a short, polytonal tune, or special tunes were played at given times and activated by the clock. Such carillon clocks were especially popular in the Low Countries. It is natural today to associate the sound of hour striking with the end of the corresponding hour: when a clock strikes one, this means that the first hour after midday, marked by the number one, has elapsed. A clock dial showing 1:20 means that the first hour after noon has been *completed* and twenty minutes of the second hour have elapsed. But this way of measuring time, *hora Completa* as it was called, is a matter of convention. Another possibility was

Hora Incipiens, i.e. naming the hour from the very moment when it begins: such a choice, consequently, would have one struck at noon, with the hour hand showing I, as in that moment the first hour begins.

This style was only rarely used, although it was famously the norm in Basle, Switzerland, being abandoned only in 1798 against fierce resistance by the citizens.[30] The large clock in the middle of the engraving by Jan van der Straet (Johannes Stradanus, 1523–1605) in the *Nova Reperta* collection (Figure 92) may show this system. The numeral I appears where XII would be expected, with the other numbers correspondingly shifted.

When clock dials were placed outdoors, situating them in a high position made them easily visible even to people far from the immediate vicinity of the tower or wall housing the timekeeper. At the same time, a high position offered a good drop for the weights, without the need of mechanical interventions such as pulleys, to shorten their path. Bells were normally situated in an even higher position: acoustic waves benefited from the lack of obstacles and a wider expansion cone and could therefore be heard at higher distances from the belfry.

Multiple hour dials, placed on different sides of a tower or on opposite walls of a building housing important clocks, as in Venice or Brescia, were already present in the sixteenth century. These secondary dials did not show the same detailed information offered by the main dial, but only displayed the basic current time. A very old secondary dial can be found in the cathedral church of Chartres, France: the main dial is on a wall of the choir and dates back to the fifteenth century, with astronomical indications and a rich set of data (Figure 66). The external dial, driven by the same movement, by contrast simply offers the current time to passers-by outdoors, thanks to a single hand on a twenty-four-hour dial.

Public clock dials follow local styles and traditions. A dial hanging indoors allowed a choice of decoration completely different from one placed outdoors, exposed to sun, rain, snow, and birds. Public clock dials can be in wood, metal, painted like frescoes on masonry, covered by majolica tiles, or carved stone. Early dials had only one hand, an hour hand on a twelve-hour dial in Northern Europe countries, or on a twenty-four-hour dial, where the *Hora Italica* system was used, or wherever an astronomical dial was required.

The original single hand, showing hours, generally gained a companion showing minutes from the end of the seventeenth century. The increase in precision obtained from the introduction of the pendulum was a common reason for upgrading older timekeepers and a marketing plus in the design of new ones. The standard practice of having the hour hand shorter and the minute one longer was not the initial choice in German countries, where older public clocks sometimes show the opposite convention, having a shorter, inner index for minutes and a longer one for hours. A well-known example is the dial of the tower in Graz, Austria, locally known as the *Grazer Uhrturm*, modified in the eighteenth century to add a minute hand to the existing clock. This dial is representative of German-style dials, with the inner minute circle divided into four sectors, numbered I—II—III—IV, and typical clover-like gilt hands. It is worth remarking on the great importance attributed by the German school to quarters of hours, echoing to some extent the Roman *puncta*, retained in the *computus* tradition. Bede, for example states that the day had twenty-four hours and that one *punctum* was equivalent to one fourth of a full hour. In northwest European countries, quarters were not only evidenced on dials, but also in striking, and their use on public clocks was far more common in that cultural milieu than elsewhere in Europe.

The earliest documented presence of a minute hand on a public clock is that on the Cistercian Monastery of Chiaravalle Milanese in the suburbs of Milan, Italy. A sixteenth-century manuscript chronicle describes the structure of the clock, which is described as 'already very old' and which was documented at the end of fifteenth century in a wonderful drawing by Leonardo da Vinci in his *Codex Atlanticus*, and again in some drawings in his *Codex Madrid I*. Both cover details of the extended drive assembly (remarkably articulated through ninety degrees) used in Chiaravalle for the motion of the Moon, a real example of mechanical ingenuity.[31]

In the Chiaravalle clock two separate dials showed hours and minutes, whereas a major dial, above them, showed the position of the Sun in the Zodiac, the phase of the Moon, and the day of the year. Leonardo's drawings show clearly the two separate arbors for the hour hand and for the minute hand and are identified by captions that remove any doubt about their function. These are perfectly coincident with the rich verbal description provided by Benedetto de Blachis, the layman of the monastery in charge of writing the Chronicle, who was evidently informed about the clock and its features by someone knowledgeable about it—probably the monk who was also the clock-keeper.

The hour hand did not necessarily rotate clockwise in early public clocks. A famous example is the clock in the Santa Maria del Fiore church of Florence, 1443, of which the dial was painted by Paolo Uccello (1397–1475). It also has an unusual method of presenting the hour numerals. On typical dials, the numerals are displaced around the periphery, and the base of each numeral points towards the centre—and the centre line of each numeral defines the beginning of the hour. In Uccello's dial, a deep form of peripheral chapter ring is made up of twenty-four painted segments, each carrying a numeral, but the numerals are orthogonal to the standard placement, and instead run along the inner edge of each segment, towards the centre—a stylistic feature common to the earliest Italian dials.

---

30 Ackermann 1986, 33.

31 Addomine 2007.

Figure 66 Dial of the clock of Chartres cathedral. https://commons.wikimedia.org/wiki/File:Chartres_-_Horloge_astro_03.jpg. Creative Commons Attribution-Share Alike 3.0 Unported license. Image in public domain.

TERRITORIAL EXPANSION OF PUBLIC CLOCKS

Skilled clockmakers, capable of making turret clocks, were highly qualified craftsmen and they moved from court to court, from town to town, crossing borders like other artisans. Chronicles reveal that King Edward III of England (1327–77) had Lombard as well as Dutch clockmakers at his court, and in the same century we find German clockmakers settled in Florence and in Venice.

The expansion of public clocks became an irreversible phenomenon—all towns, all parishes, all communities wanted one. In larger cities, more clocks appeared on towers. Clocks were no longer confined to monasteries and town halls: from the fifteenth to the seventeenth centuries, more and more civil buildings and churches wanted a timekeeper, and a good second-hand market of affordable items blossomed. Even in the Western Ottoman empire some public clocks were available, although these were widely thought to be detrimental to the authority of the *muezzins*.[32] In Europe, however, by the end of the eighteenth century most town churches had a clock. The production of turret clocks was still limited to the traditional methods of the forge where wrought iron was transformed into wheels, pinions, and frames: productivity was constrained, and parts were all hand-made, one by one. Production levels increased when the cast iron process was introduced in the nineteenth century, with a consequent reduction in costs.

FROM AUTOMATA TO ASTRONOMICAL CLOCKS

One of the most striking aspects encountered in the records of public clocks is the high degree of complexity, which appears even in the earliest accounts. Like Minerva springing fully armed from the brow of Zeus, public clocks emerge in the earliest medieval chronicles complete with automata, decorated dials, angels, trumpets, and multiple bells.

Public clocks are eminently *public*: they represent a community and its wealth—they are a status symbol *ante litteram*. If they are inside a cathedral, they will become an attraction for humble pilgrims as well as for educated visitors, who will spread their fame: the monumental clocks of the late Middle Ages, like that at Strasbourg (perhaps the most famous), were more for theatrical effect than timekeeping: they were unforgettable. Automata could be very simple: jacks striking their bells, for example, or the Three

---

32 Kreiser 2012; Turner 2019, 111

Kings paying homage to the Virgin and Baby Jesus—or could be more complex representations, coordinating mechanical and musical effects.

Astronomical clocks, on the other hand, required much higher skills, a thorough knowledge of astronomy, mathematics, and kinematics, and a real mastery of the craft in order to transform drawings and calculations into a working system. A real-time representation of the position of Sun, Moon, and possibly of the then-known planets in the sky and of their reciprocal position was essential in order to determine the astrological aspects, i.e. the reciprocal angular positions of the celestial bodies. Starting from the astrological aspects, and from the position of Sun, Moon, and planets in the different Zodiac signs, the astrologer could determine the degree of opportunity, in each moment of a day, for taking an action of whatever kind.

Automata and astronomical sections were at the same time the great strength and yet also the Achilles' heel of many extraordinary clocks: their presence witnessed the wealth of their communities as well as the skill of their designers and of their makers, but when the first makers and their direct descendants were no longer available to take care of them, their complexity became a nightmare for following generations.

An adequate discussion of astronomical clocks is outside the scope of this work,[33] but some main points may be adduced. Astronomical clocks were real simulators of the sky: they offered the observer the celestial configuration in real time, at any moment, whatever the weather conditions, night or day, and at a glance. Without them, even a skilled astronomer would have needed to use astronomical tables and make long and complex calculations to know where planets were at a given date and time. Moreover, almost everybody was interested in this sort of information as, increasingly in the later Middle Ages, astrology pervaded daily life: every moment of the day could be influenced, positively or negatively, by the stars.

Different types of public astronomical dials can be characterized as follows:

All astronomical dials represent twenty-four hours, sometimes in two lots of twelve: their presentation can be different, as they reflect totally different design styles. The first astronomical Italian clock with concentric dials was designed by Jacopo Dondi, father of Giovanni, the maker of the *Astrarium*, and is that in Padua (Figure 67), made in 1344. The present clock was installed about eighty years later, as the original one was lost in a fire. According to historians, it is a faithful replica of the original one.

Italian astronomical dials showed moon phases, the age and date of the Moon, and its phase and the day of the lunar synodic month, composed of 29½ days. They are based on a system of concentric discs, showing calendric information, the zodiacal position of the Sun and Moon, and (only in one case, in Cremona) the Head

---

33 But see Chapter 9.

Figure 67 Dial of the Padua cathedral clock. https://commons.wikimedia.org/wiki/File:Celestial_clock.jpg. Creative Commons Attribution-Share Alike 4.0 International license.

and Tail of the Dragon to represent eclipses. Only one example in Italy is recorded of an early public clock with an astrolabe dial, although this has not survived.

Nothing survives of what would have been the extraordinary production of the celebrated Raineri family, working in northern Italy during the fifteenth and sixteenth centuries. The Raineri made the first clock in St Mark's Square, Venice; the clock was later removed by Bartolomeo Ferracina as scrap (including the dials) when he installed a wholly different clock in the eighteenth century. The Raineri specialized in astronomical and planetary clocks but did not refrain from adorning them with automata of all sorts; only descriptions are left.

In Germanic countries astronomical clocks had a moon dial, often a three-dimensional representation of a sphere—the so-called *Mondkugel*, or moon ball—but also an astrolabe dial showing a representation of the sky. The methodology adopted was the one used for planispheric astrolabes, i.e. a stereographic projection, using the South Pole as the centre of projection. A second, large dial appears below that of the main astrolabe: it is finely traced and carries the names of the saints and of the feasts throughout the year. This is the so-called calendar wheel, and its position, closer to the ground, allows the observer to check among the seemingly infinite list of names and indications. In one case, in Münster, Westfalia, the large and richly decorated astrolabe dial is flanked by movable sequences of metal boards carrying the names of the planets that in astrology were thought to 'control', i.e. influence actions, in a given hour of the day or night. The finest examples among these extraordinary astronomical timekeepers in the German style are the wonderful clocks of the Hansa region, in the north—they are masterworks of ingenuity and strong declarations of the power of the communities who could afford their construction in the late Middle Ages.

In France astronomical clocks have astrolabe dials, for example the Saint-Omer clock dating back to 1558, the wonderful masterpiece in the cathedral of Saint Jean in Lyon, or the remarkable clock in the choir of Chartres cathedral. Designed by Jean Fusoris

Figure 68 Dial of the Wells Cathedral clock. https://commons.wikimedia.org/wiki/File:Wells_clock.jpg. Creative Commons Attribution-Share Alike 4.0 International license.

(1365–1436), the recently restored astronomical clock at Bourges cathedral is a true masterwork both for its high antiquity and its extremely high quality. The Strasbourg clock in Alsace belongs to a region that has alternated between French and German control. The city cathedral housed three extraordinary clocks across six centuries.[34] Very little is known about the first one, constructed 1352–4: it had an imposing height of about 18m, showed several automata, including the then-fashionable representation of the Three Kings and a carillon, and had a classical Germanic double-dial structure. This exceptional clock was dismantled in the sixteenth century, replaced by one much more complex and designed by the mathematician Conrad Dasypodius, and implemented by a team including the Swiss clockmakers Isaac and Josiah Habrecht. This second clock was delivered in 1574. It was replaced in 1843 by a third astronomical clock by Schwilgué.[35]

English astronomical clocks show peculiarities distinguishing them from their continental counterparts. Some of their dials, for example, share a peculiar style, with the labels carrying the twenty-four hour indicators spreading out from the outer circle and concentric discs. Unfortunately, the lack of early representations of these dials does not allow certainty about their original appearance, which has no counterpart in domestic clocks of the same age and provenance. Some of these clocks have their companion jacks, often wearing military dress. Among the English astronomical clocks, the Wells double dial (Figure 68) is the most interesting for its complex structure, topped by jousting automata, for its rich decoration, including the winged figures in the spandrels, and the twin jacks tolling the bells. Should it eventually be confirmed to be the original 1390 dial, it would cast new light on the subject of early astronomical dials.

The astronomical clock at Hampton Court, heavily modified over the centuries, reflects with its multiple concentric discs a design closer to the Italian dial style and is unique in Britain. Made by Nicolas Oursian, clockmaker to the King, his initials 'N O' are stamped on an iron bar on the inside of the dial, one of the few original parts still surviving. Replacement movements, including examples by Langley Bradley and Vulliamy among others, have been installed over a long period, the most recent being a new clock by Gillett & Bland in 1878–9.[36] There are very few astronomical public clocks in England, and the only early examples still surviving are those at Wells, Exeter, Wimborne Minster, Ottery St Mary, Durham, and Hampton Court.

The celebrated astronomical clock of Prague has a long and complex history. In style it is Germanic, with the double dial so

---

34 Oestmann 2020; Bach *et al.* 1992.
35 Described in Chapter 9, Section 2.
36 Hellyer & Hellyer 1973.

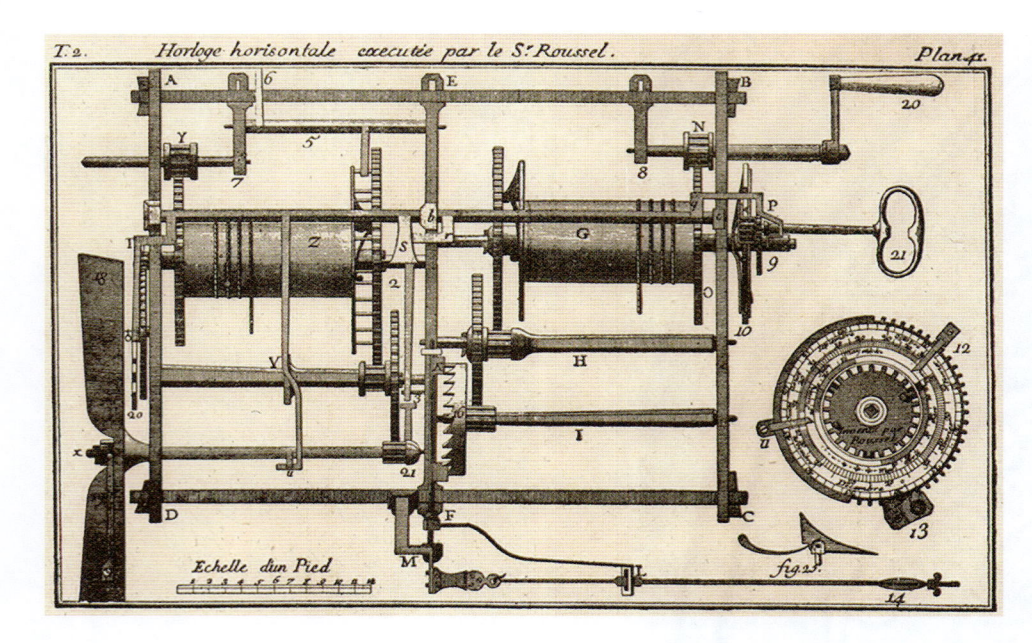

Figure 69 Flatbed tower clock seen from above, as depicted in Thiout 1741, pl. 41.

frequently encountered in the Baltic clocks: the upper dial for the astrolabe and the lower for the calendar table. The present clock, recently fully restored, is a combination of elements, some of them belonging to the very first clock, dating back to the early fifteenth century, and some of them more recent. While the clock has undergone several modifications and restorations, some owing to war damage, it undoubtedly remains one of the most intriguing of timekeepers.

MUSICAL PUBLIC CLOCKS

Clocks and bells have been associated since time immemorial: but in some periods and in some areas, public clocks and music have become even more intimately related. Since the late Middle Age, a strong tradition developed in the Low Countries for automated bell systems, the *carillons*, originally played by hand by means of special keyboards but later activated by public clocks. When a chiming train was present, it was formed of an additional train, including a drum where movable pegs were placed in slots and which acted in turn on rods or other lifting levers, thus ringing different bells in a desired sequence. One of the most interesting aspects of these large carillon drums from a technologically historical point of view is that they were among the first examples of programmable units, as they were intrinsically reconfigurable by the simple replacement and relocation of their mobile elements—the pegs, or pins—whose position and sequence determine the tune played.

A carillon (in German, *Glockenspiel*) is by definition a set of at least twenty-three bells—a large one can involve more than a hundred!—and the complexity of the melody depends on the drive-system adopted. The oldest surviving carillons date back five centuries. Turret clocks could act as the time base for an automated start of chimes at given intervals: either fixed short tunes, for example, at the quarters, followed by a longer melody at the hour; or perhaps a special tune at a given moment of the day or on specific occasions, according to tradition or local choice.

THE CENTURY OF REVOLUTION

The eighteenth century was the real century of transformation for public clocks. If the introduction of the pendulum must strictly speaking be considered as pertaining to the seventeenth century, it was the century of the Enlightenment that generally saw public clocks benefit from it and the anchor escapement. In most cases, pendulums were simply applied to existing clocks, and intervention was minimized, essentially for reasons of cost: this implied the retention of the crown wheel and the verge and pallets, but rotating the crown wheel from a horizontal axis to vertical, so that the pendulum rod could directly drive the verge. Better results would be obtained when the anchor escapement was introduced. The improvement was dramatic and as a result practically all turret clocks were modified and only a very few survived that retained the original foliot or balance. This greater precision offered the possibility of a more refined time display, and thus minute hands started appearing on public clocks. New clocks, of course, were made according to the new demands of the market.

Another discovery in this century was the escapement that would prove extraordinarily successful in public clocks. In 1741 the French clockmaker Jean-Louis Amant presented the first version of his pinwheel, dead-beat escapement, which would be developed in innumerable versions. Thanks to its design and its features, it proved extremely suited to the harsh environmental conditions of many turret clocks. As noted above, a revolutionary innovation had been made with the flatbed turret clock (Figure 69). The flatbed structure is not simply another type of frame of different design: it represents a totally new approach. In a birdcage frame, every intervention requires the clockmaker to dismantle the whole clock and to reassemble all the parts, whereas in a flatbed the intervention can be limited and confined to the part requiring action, dramatically reducing both time spent, and the risk of damaging parts. Moreover, a flatbed requires less room for its installation and less time to be assembled. It heralds the beginning of modern turret clock making. Originally, the frame was constructed of wrought iron bars, but cast iron was soon used. A new era had begun.

CHAPTER FIVE

THE DOMESTIC CLOCK IN EUROPE

SECTION ONE: FROM THE THIRTEENTH CENTURY TO THE INVENTION OF THE PENDULUM

Dietrich Matthes

THE DOMESTICATION OF GEARED CLOCKS

The first specific reference to a geared clock that is clearly domestic comes from c.1275 in the *Roman de la Rose*: the author Jean de Meun(g) (c.1240–c.1305) lets the hero Pygmalion 'make his clocks sound/In his rooms and his houses/With very fine wheels/in their steady movements'.[1] The reference to domestic geared clocks in the most successful and influential work of medieval French literature means that, at that time, clocks were widespread enough to be known to his (civic rather than courtly) audience. Small clocks were or soon became a favourite item with high-ranking nobility as shown by a series of archive records. A silver clock listed in the inventory of King Charles V of France had been made between 1268 and 1314.[2] The chamberlain Jean Engilbert purchased a clock for the papal chamber, Avignon, in 1336 for eleven gold florins.[3]

Other early domestic clocks are the guard-tower clocks—small weight-driven clocks placed inside the bell towers of large cities and were used to keep local time and alert the guards to strike (manually) a bell at the predefined times of public time-announcement. Two of these are preserved with a clear medieval provenance and a location for their use (Figure 70).[4] They show the earliest construction for domestic clocks: a flat vertical frame of iron or brass contains the bearings for the drum on which the rope is wound and the **great wheel** and those of the **second (escape) wheel** vertically above. The teeth of the **escape wheel** mesh with the **pallets** of the **verge**, which has a straight **foliot** above suspended by a thread. The foliot has notches on both sides and small weights may be moved along them to adjust the rate of the clock to show unequal hours throughout the year.[5] The great wheel drives the **hour wheel** with the hand by means of a **pinion** at the front. Winding (these clocks have only one weight with the rope wound on a drum and need frequent winding), is achieved by a capstan mounted on the great wheel axis that allows for manual rewinding of the rope on the wheel. An alarm with a bell is set off at a predefined time by inserting a pin in a series of

---

1 Pomel 2012, 23. I am grateful to Anthony Turner for pointing out this information to me. The quoted lines from Strubel 1992, verse 21, 037–43
2 Labarte 1878: 'Item, a clock all entirely [made] of silver without iron, which was of King Philip the Fair [1268–1314], with two counterweights of silver filled with lead'. The goldsmith Pierre Pipelard received payments in 1299 and 1300 'for some clocks he made for the king'; see Tardy 1981, 24.
3 Zinner 1954, 29; Schäfer 1914, 54; Dohrn van-Rossum 1998, 14.
4 One is in the Germanisches Nationalmuseum, Nuremberg (WI 999) and was found in the northern tower of Sebalduskirche in Nuremberg. It is typically dated to the fourteenth century (later changes). See Maurice 1976, n. 34. The second one is in the Historisches Museum Hanover (VM 030 126) and was originally located in the tower of the market church, Sts George and James, for which the acquisition of a tower guardian clock is recorded for 1392 (see Figure 70), the possible date for this clock (later changes). Similar clocks, known from Germany, Burgundy and Italy, are usually dated to the fourteenth to sixteenth centuries—Maurice 1976, nn. 34, 35, 36, 37, Krombholz 1984, ill. 56 & 68, and Edwardes 1965, pl. 2–3, pl. 4–5, Zeller 1969 pl. 1.
5 Dohrn-van Rossum 1998, 49–50.

holes on the hour wheel. The alarm has a separate small weight mounted on the side of the clock—once unlocked, it will sound until the weight has dropped the full length of its rope. The clocks are hung on a wall and have a horizontal bar with two backward-projecting support spikes at the lower end of the frame to avoiding tilting, as the weight(s) are asymmetrically mounted on the clock.

This, the simplest construction of a weight-driven domestic wall clock, was soon improved. Most importantly, rather than mounting the trains in a flat vertical frame, a three-dimensional cage was used. This allowed for mounting the **going train** and the separate **striking train** end to end.[6] It furthermore avoided any tilting of the clock, since the two main weights could be placed on opposite sides—therefore, this meant that the two great wheels were turning in opposite directions.[7] With simpler (e.g. non-striking) clocks, it also provided a support for new elements added throughout the fourteenth and fifteenth centuries, such as striking work with **count wheel** and a bell. The count wheel was typically mounted at the back or the side of the cage,[8] the latter on top of the (often four-legged) frame in a bell canopy. The entire clock could either be hung on the wall by means of lugs on its rear side or placed standing on a bracket. The front of the clock carried a dial that was open in the earliest known clocks but was increasingly painted on a metal plate covering the entire front. The cage-like construction allowed clock movements to be enclosed with sheet metal sides and tops for protection from dust and other contamination. The transition from open movements to closed clocks happened throughout the fifteenth century.[9]

At that point, clock design had reached a point that was to prevail for almost 300 years in the shape of what is usually called '(gothic) iron clock'(Figure 75, left). The construction with two trains in a cage-like frame of iron was simple enough to be created by lock- or blacksmiths without much formal training in clockmaking. This is why—even when technological advances had occurred—there was both a demand for and supply of these simple iron clocks. More refined and advanced clocks would be created by specialized clockmakers, usually in urban centres. Given that the simple iron clocks were not usually produced under guild regulation, most are unsigned, and they also avoid elaborate decoration (such as engraving and gilding), having only painted dials. One noteworthy exception to the anonymity of their makers is the Liechti family in Winterthur who created iron clocks, both domestic and public, for several generations.[10] Iron clock production spread all over Europe, from Italy to Scandinavia and from Spain to Poland.

SPRING-DRIVEN CLOCKS

The major innovation in clockmaking around 1400 was the complementing of gravity-driven weight clocks by those driven by a coiled spring. Their origin is not certain. Indirectly, the earliest mention of springs in the context of horology is by Antonio Manetti *c.*1480[11] in his life of Filippo Brunelleschi.[12] The specific mention is to '*generazione di mole*',[13] which literally means 'production' or 'generation of springs'. This could refer either to a mainspring—which would have been made from a series of pieces of elastic wire or sheet metal—or could mean that Brunelleschi introduced several (flat) springs in clocks. A number of flat springs are used in clocks[14] to improve the mechanical properties and it is not entirely clear whether Manetti refers to their introduction as a driving force. The first image of a (probably) spring-driven clock occurs in a manuscript by Henri Arnault de Zwolle (*c.*1400–66),[15]

---

6 The trains were mounted 'behind' each other (as seen from the dial). The going train was placed at the front and the striking train at the back. An alternative way of positioning the frame of such a clock is to have one train on the left and one train on the right at right angles to the dial (Simoni 1968). This 'right-angle' or 'orthogonal' positioning of the frame requires an orthogonal transmission of the driving power—which is usually obtained by a right-angled spur or 'star' wheel. The right-angle positioning of the frame of such clocks seems to have been the 'standard' layout in Burgundy but was also used in early horology in Italy. The advantage of the right-angle positioning is that the axes for winding are easily accessible from the sides of the movement even with a dial (Leopold 1971, 13). Therefore it constitutes a significant advantage for operation before winding keys were in regular use. The first known domestic clock of three-dimensional cage construction is shown in a French miniature from 1400-4 illustrating *L'Épître d'Othea* (BNF, MS Fr. 12778, f. 108v).
7 Leopold 1971, 13.
8 In a few cases it was mounted between the two trains, see, e.g. a French clock from the fifteenth century in Matthes 2019, 109.
9 The oldest depiction of a clock enclosed on all sides is a manuscript illumination from 1406 (L'Orloge de Sapience, BNF, MS Fr. 926, f. 118). Archival evidence of (possibly spring-driven) clocks at the ducal court of Burgundy also mentions clocks in closed cases as early as 1423 (see n. 26 (2))

10 Schenk 2006.
11 *Vita di Filippo Ser Brunellesco*, Toesca 1927, 19, following manuscript MS II.II.325, Biblioteca Nazionale, Florence, fol. 299v.
12 Simoni 1954; Prager 1968. Filippo Brunelleschi (1377–1446), a trained goldsmith, is most famous as an architect and mathematical practitioner who excelled in engineering.
13 A direct transcription from the manuscript gives '*varie ediuerse generatoni dimole*', which Prager 1968, reads as '*varie e diverse generationi di mole*'.
14 E.g. for the hammer striking the bell and for the unlocking assembly of the striking train properly engaging with the hoop- and count-wheels. Their introduction allows for a significant improvement of the otherwise unreliable functioning of going and striking trains.
15 Paris, Bibliothèque Nationale, MS 7295, fol. 76.

Figure 70 Tower-guardian clock from the market church of St. Georgii and Jacobi Hanover, made in or before 1392. Historisches Museum Hanover, N. 030126. Photo: Reinhard Gottschalk (HMH).

Figure 71 Spring-driven table clock in the shape of a double-spire gothic façade, bearing the arms of Philip the Good of Burgundy and the fire-striker of the order of the golden fleece. This is the oldest preserved spring-driven clock. Burgundy, c.1430–5. Germanisches Nationalmuseum, Nürnberg, HG 9771. Photo: Georg Janssen (GNM).

a pupil of Jean Fusoris (1365–1436),[16] c.1420–25.[17] The oldest preserved spring-driven clock (c.1430–35) is that associated with Philip the Good of Burgundy (Figure 71).[18]

This clock can be considered the starting point of spring-driven clockmaking, although it was certainly not the first spring-driven clock—many of its elements are already rather refined. It consists of an openwork, twin tower, gothic case of gilt brass. The closed rectangular base rests on four animal feet. An open architectural structure forms the main section containing the movements for the going and striking trains in a right-angle setup. This part is crowned by a balcony of pierced tracery panels supporting two spires and a small empty canopy between them. Two lions with the coat of arms of the Duke of Burgundy and a fire-striker are placed on top of the spires.[19] The front of the main section has a circular dial with numerals I–XII on a blue-enamelled background and an hour hand. The going (four-wheel) and striking (three-wheel) trains are mounted vertically with substage mainspring barrels located in the closed base, the **fusees** are mounted in small architectural houses with the great wheels outside and the further (brass) wheels on horizontal iron arbors with iron pinions vertically above them, with the bearings set in hollow brass columns.[20] The verges of the circular **balances** are located inside the hollow brass columns and the balances themselves inside the pierced tracery-panel balcony (originally suspended by threads,

---

16 Poulle 1963, 27
17 The drawing of the movement of the clock does not specifically show the driving force. Yet the pins for unlocking the striking train on the great wheel of the going train show that the great wheel made one revolution every three hours. It has been noted (Leopold 1971, 15) that the great wheels of weight-driven clocks in the fourteenth and fifteenth centuries all made one full turn per hour. Spring-driven clocks used slower great wheels with one revolution every three to five hours. Since the fusee had a limited number of turns to achieve a satisfactory overall duration of the clock, the great wheel revolution speed had to be reduced. The inference from this is that the clock shown was spring driven. Arnault's name is variously spelt Arnaud or Arnault. We follow here the form employed by Wickersheimer 1979.
18 Germanisches Nationalmuseum, Nuremberg, HG 9771. The authenticity of this clock had been disputed in the past (see Lloyd 1959 and references therein), but a detailed examination in the 1970s has confirmed its authenticity (Maurice 1976, 86). This has been further corroborated by new archival material about clockmaking in the first half of the fifteenth century which confirms that some of the previously disputed aspects were indeed standard practice in fifteenth-century Burgundy. This pertains to the use of screws (criticized by Lloyd 1959 and solution by Pérez Álvarez 2013, 501) and the unique use of the striking train great wheel as the count wheel (Maurice 1976, 86; Poulle 1963, 27). Zech 2019 provides a historiographical study of the clock.

---

19 Both lions are attached to the spires by screws.
20 One of the two escape wheels (that of the going train) shows a punch mark 'J' in a gothic font. Since the two escape wheels are different and the use of a gothic font for punch marks is not standard, its authenticity is not certain.

now with later **balance cocks**). Both the going and the striking train are controlled by verge escapements—a construction rarely seen after the fifteenth century. The hexagonal canopy between the two spires (and two further hollow vertical hexagonal 'chimneys' in the middle of the movement) are thought to have contained automata.[21] Given that it can be closely dated, the clock is a cornerstone of horological history. It is not yet held in a cage-like frame but integrally within a complex (open) case that has to be completely dismantled in order to dismount the trains. The **arbors** of the trains are set horizontally between the vertical columns with winding from the sides via a disc with holes mounted on the great wheel arbor using a Y- or H-shaped key. The great wheel of the striking train has slots in the rim and serves as the count wheel at the same time. These slots control a complex lever system for **unlocking**[22] and striking the hours on a bell contained in the base between the two mainspring drums.

Although the origins of spring-driven clocks are not certain, the court of Philip the Good was the earliest centre of spring-driven horology. A remarkably large number of spring-driven clocks can be traced to it[23] at a time when there are few or no traces elsewhere. Other, later, fifteenth-century spring-driven clocks survive from Burgundy,[24] together with some illustrations[25] and detailed archive notes.[26] Soon afterwards, a series of detailed descriptions of spring-driven clocks are available from many regions of Europe—notably Italy,[27] France,[28] and Germany.[29]

FORCE EQUALIZATION IN MAINSPRING-DRIVEN CLOCKS

The key elements in the construction of spring-driven clocks are:

1. A **mainspring**, which supplies the going force, and
2. Compensation for the change in the force exerted by the spring as it unwinds.[30]

---

21 Bassermann-Jordan 1927, 20. An unpublished reconstruction of the mechanism has been undertaken by Bernhard Huber.
22 The unlocking assembly is released by five unlocking pins on the great wheel of the going train.
23 Independent until the succession war (1477–93) and the death of Mary of Burgundy in 1482, by the treaty of Senlis, 1493, the Duchy of Burgundy centred on Dijon became part of France, while the County of Burgundy and the Burgundian Netherlands became part of the Holy Roman Empire. All three areas are here described as 'Burgundy'. To attribute clocks to the different regions is not possible.
24 (2) A hexagonal table clock that was originally spring-driven (later converted to weight-drive) from around 1450 (British Museum, London—on loan from Victoria & Albert Museum, London, Reg. M11–1940); (3) A (probably Burgundian) small clock or pouch-watch from around 1460–80 (Figure 72)
25 (1) The manuscript by Henri Arnault de Zwolle (n. 15), contains a spring-driven clock made by Fusoris for Philip the Good; (2) a painting (Antwerp, Musée des Beaux Arts) ascribed to Rogier van der Weyden shows Jehan Lefèvre, Chancellor of Philip the Good with a small clock in the shape of an open tower hanging from a nail on the wall by a chain. The clock is constructed on a square base with substage barrels, open sides, and a bronze or brass bell contained in a canopy on top (see Matthes 2018, 8 for a detailed illustration). The clock is spring driven and has the motto of Philip the Good (upon his third marriage to Isabel of Portugal in 1430) on the dial (*Tant que je vive—autre N'Auray*); (3) a miniature in a manuscript (Bibliothèque Royale, Brussels, MS IV, iii, f. 13v; see Figure 212) shows a spring-driven hexagonal table clock movement open on a table. This, while of uncertain origin, is the first depiction of a spring-driven clock with vertical arbors and is an early depiction of the fusee.
26 (1) A '... small square clock, the outside gilt, with a white enameled zodiac, with a bell for the hours on top ...' in the inventory of Philip the Good, (2) A '... small gilt clock where the sides and the dial are made of gilt silver ...' (both Cardinal 1985, 13); (3) A spring-driven table clock made of gilt silver that is described in detail in the inventory of Isabel of Castilia in 1504 and that can be traced back to the Burgundian court of Philip the Good or Philip the Bold (Pérez Álvarez 2013; Pérez Álvarez 2015a; 2015b).
27 (1) A manuscript drawing of two spring-driven clocks with *helical* springs (British Library, Manuscript 34,113, fol. 74r and 156r, Italy, 1445–75) that have been linked to the archival notes of Filippo Brunelleschi's clocks (Prager 1968); (2) A letter by Comino da Pontevico to the Margrave of Mantua from 1482 detailing ' ... the clock has a steel spring, which is contained in a spring drum of brass, around which the gut is wound. You connect the gut with the spring drum so that you transmit the force to the screw (fusee), which connects to the wheeltrain ... This is how clocks work that are made by masters who make clocks without weights'; (3) The Almanus Manuscript (*Ars horologica*, Augsburg, Staats- und Stadtbibliothek 2° Cod. 209 (Leopold 1971) written around 1477–85 by brother Paulus Almanus from Augsburg in Rome. This manuscript is the most valuable source on early domestic horology; (4) The famous clock/watch surprise from 1488 described in a letter by Jacopo Trotti from Milano to the Duke of Ferrara: three portable clocks/watches had been sewn into dresses or garments handed to two guests (and the host) of a party. The two in the garments of the guests rang throughout the festivities to the surprise of everybody (Morpurgo 1954, 72). Further archival evidence in Italy points to 'small clocks'; see Brusa 1990.
28 Next to the manuscript (n. 15) which describes a spring-driven clock made possibly in Burgundy or in France, there are notes (2) from 1459 about clocks that Jehan de Lyckbourg delivered to King Charles VII and of which ' ... the fifth is gilt and without weights'; (3) from 1480 that Louis XI pays Jehan of Paris for a ' ... clock with a dial, which strikes the hours [...] The named Sir has bought this clock so that he can take it wherever he goes ... ' (both Cardinal 1985, 14).
29 The earliest references concerning spring-driven clocks are (1) the Almanus manuscript (n. 27(3)). Most of the eight spring-driven clocks in this manuscript are Italian (Leopold 2003), but one of them (clock 16 in Leopold 1971, 156) is probably German as is (2) a spring-driven clock with substage barrels in the Germanisches Nationalmuseum, Nuremberg (WI 163) which is datable to 1480–1520 and showing the 'large **Nuremberg Hours**'; see Maurice 1976 N. 84; Eser 2014, cat. 42, 178 and references therein.
30 Later formulated by Robert Hooke that the force needed to extend or compress a spring by some distance varies linearly with respect to that distance. Hooke first stated this as a law in 1676 in a Latin anagram

The actual making of mainsprings in the context of clockmaking was first described in 1722 as far as is known.[31] Mainsprings had to be hammered from long, sheet-steel plates, then drawn and filed to shape. After that, they needed to be heat-treated and formed to a spiral[32] shape. For this step and tempering,[33] the spring was inserted (wound) into a cylindrical metal wire basket of 5–6cm diameter. This description is more than three hundred years later and can only be expected to give a faint indication of mainspring making in the fifteenth century. However, the detailed steps required and undertaken earlier were passed on from master to apprentice verbally and are hence unknown today. It is worth noting that, during the early centuries of spring-driven horology, not all springs were made of steel—some were of brass.[34] It has been speculated (Prager 1968) that springs in the early fifteenth century were made by successively hammering and welding small sections of wire onto each other since metal in the required shape and size was not available. This is what could have been meant by *generazione di mole* in the text on Filippo Brunelleschi's creations.

The second challenge in spring-driven clockmaking was compensation for decreasing force during the uncoiling of the spring. Four solutions for this were found between the fifteenth and seventeenth centuries:

1. The **fusee** is considered to be the earliest regulating device since it is known to have been used already in the early fifteenth century.[35] The oldest preserved specimens are those in the Philip the Good clock (Figure 71). Theoretically, it produces the best results. The principle of the fusee is to insert a lever of varying length between the spring and the movement. This lever exerts a torsional moment on the great wheel. By adapting the lever length, it is possible to compensate the changing force from the spring as it unwinds—the lever being shortest when the spring is fully wound and exerts most force, and longest when the spring is mostly unwound and exerts least force. The shape of the fusee needs to be carefully adapted to the properties of the spring.[36] However there are several drawbacks to the fusee:
   a. It occupies a lot of space both in length and in width—being separate from the spring barrel. This became a problem only when miniaturization of timekeepers started in the late fifteenth century.
   b. It requires a connecting link. This was usually made with a gut line. Gut, being organic, changes in length with changes in temperature and humidity. Apart from the risk of breakage, this adversely affects the going of the timekeeper. A convenient alternative is the use of a linked **chain**. This is reported as first used in horology in 1527 by Peter Henlein of Nuremberg (c.1480–1542),[37] but has the significant drawback of being very difficult to make. Another alternative is to use the principle of the fusee within the gearing. This had already been considered by Leonardo da Vinci in codex Madrid I[38] and was taken up again by Jean de Hautefeuille (1647–1724) in 1718.[39]

   Fusees come in two very distinct shapes: firstly with few turns and so a rather coarse spiral; secondly a very slender version with many turns and a very fine spiral. The former is typical of sixteenth-century German clocks, the latter of sixteenth-century French clocks.[40]

2. The vertical **stackfreed**: The vertical stackfreed is an equalizer consisting of a special spring that exerts a decreasing pressure upon the mainspring barrel. The spring works upon a solid helical drum (in surviving examples) or a helical plane covering

---

(*ceiiinosssttuu*) and published the solution in 1678 as '*ut tensio, sic vis*' (as the extension, so the force); see Petroski 1996, 11 and references therein.
31 Leutmann 1722 pl. 4, figs. 2–5; Leonardo da Vinci gives a short account of spring-making for clocks in Codex Madrid I. f.14v (Biblioteca Nacional de España, MS 9937).
32 One early illustration shows a helical spring being used in a clock (see n. 27(1)). This was speculatively described as an experimental step towards the construction of mainsprings for clocks (Prager 1968). All other known sources where this detail can be ascertained deal with spiral springs.
33 For metallurgical analysis of the optimization of material properties for springs in early clockmaking, see Wayman 2000, 29 ff. (sixteenth century), 53 ff. (seventeenth century and later).
34 This is the case for the well-known cylindrical table-clock made by J. Zech in Prague in 1525 (British Museum, London on loan from the Society of Antiquaries, London).
35 See Brusa 1978, 2nd edn., 1982, 32 and relevant addendum. One early example, Brusa 1978, Figure. 22, is an illustration of a fusee in *Bellifortis* written by the Eichstätt engineer Konrad Kyeser before 1402 (Niedersäsische Staats- und Universitätsbibliothek, Göttingen, Cod. Phil 63 fol. 76v.).

36 Honig 1980. This careful adaption is the reason why some fusees in Renaissance clocks were made of wood for easier fine-tuning of the shape. The theoretical shape of a fusee was calculated by Pierre de Varignon in 1702 based on Leibniz's differential calculus (Gowing 1997). The theoretical leverage laws for force of the spring and associated lever (diameter of the fusee) was formulated by Leibniz in 1718 (Gerland 1906, 133).
37 Matthes 2018, 263, with the letters published in (Eser 2016) and (Starsy 1985): 'Thus the chain that pulls the balance does not wind properly and falls from the rills of the fusee and gets on top of itself . . . ' (letter by the Chancellor from Schwerin to Peter Henlein in 1527). Linked chains were well-known by that time and had been drawn by Leonardo da Vinci (n. 38).
38 Leonardo da Vinci, Codex Madrid I (1490–9) and II (1503–5), Biblioteca Nacional de España, Madrid, Inv. 8936 & 8937, f. 4r, f. 16r, & f. 45r.
39 Hautefeuille 1718; he proposed the geared fusee in order to dispense with a cord or chain, quoted after Brusa 1990.
40 See (Leopold 2003) in relation to the fusee in a clock in the Almanus manuscript which had the most turns and was described as 'slender' (*subtilis*).

the barrel (in the drawing by da Vinci[41]). The equalization is achieved because exerting the force increases friction[42]—and hence reduces the power applied by the mainspring in the wound state—all the way to a small incremental amount of friction when the mainspring is almost unwound. This use of friction for force equalization is a very inefficient way to equalize the mainspring force as it loses much of the energy contained in the spring after winding. The vertical stackfreed is one possible alternative to a fusee. Never widely used,[43] it has however been employed for clockmaking since the late fifteenth century. Evidence for this is varied—it ranges from its first mention in drawings by da Vinci via a description by Girolamo Cardano[44] to a picture by Tāqī al-Dīn.[45] It is thought that it was invented in the late fifteenth century.[46]

3. The horizontal stackfreed is a variant of the vertical stackfreed using the same principle: a horizontal spring exerts decreasing pressure upon the side of a **cam** that is mounted horizontally upon the great wheel. The stackfreed spring force hence decreases as the mainspring unwinds. Similarly to the vertical , its effect is mostly brought about by additional friction in the movement—leading to a loss of energy from the spring. This loss of energy is shown by the significant amount of wear on stackfreed cams. The first drawing of a horizontal stackfreed is by da Vinci in the late fifteenth century.[47] Since da Vinci was often inspired for his drawings by existing devices, he had probably seen an example. The oldest preserved horizontal stackfreed is dated 1533, and its first known use in a watch is 1548.[48]

4. The Wheel-Lock Chain is known from one extant clock, dated 1545.[49] It may be a survivor of alternative early force equalization experiments. It has force equalization that is similar to that of contemporary wheel-lock pistols: a flat spring exerts a force on a wheel with a concentric lever arm setup. The connection between spring and wheel is by means of a small, linked chain that helps a strong flat spring to exert force on the wheel. It has frequently been noted that the use of springs in clocks seems to start at around the same time as the development of spring-driven wheel-locks in pistols—leading to a hypothesized derivation of the clock spring-drive from hand-held guns. However, given the dates of several early sixteenth-century clocks, if any link existed, it was rather the reverse with horology influencing weapon technology.[50]

Of these four force equalization techniques, only the fusee and the horizontal stackfreed were widely employed in the Early Modern period. The fusee became the standard method while the horizontal stackfreed saw brief employment in Germany during the later sixteenth and earlier seventeenth centuries. Given the advantage it offered for making 'slim' pendant watches, it became standard in German Renaissance watchmaking before being superseded by the fusee even in Germany in the mid-seventeenth century.

These technologies allowed clocks to be accurate enough for everyday purposes, but not accurate enough for scientific measurements. The fusee shape could not be sufficiently adjusted to individual mainsprings, which were inhomogeneous. A weight-driven clock was easier to produce than a spring-driven one—yet it also suffered from inaccuracies since the driving force (the mass of the weight) increases over time because the weight of the unwound rope has to be added.[51] The challenge of achieving a constant driving force in mainspring-driven clocks was solved by Jost Bürgi (1552–1632)[52] during his time as court clockmaker to Wilhelm IV of Hessen in Kassel.

Bürgi started from the fact that the change in the driving force of the mainspring happens across a lengthy period: that of the complete unwinding of the mainspring. His solution was to introduce a second spring. This second, much weaker, spring (or a small weight) was fitted to a wheel further up the going train and was rewound at short frequent intervals, controlled by the going movement itself, usually every fifteen minutes when the quarter striking train is released. The second spring can either be wound by the going train mainspring or even the spring of the quarter striking train itself, removing the need for the going train mainspring altogether. This device, which is known as a **remontoire**, operates always at the same overall tension for the short interval between rewindings.[53] The resulting

---

41 Codex Madrid I, f. 13v (n. 38).

42 It also exerts an additional driving force on the movement—yet this is small in most cases.

43 Only two clocks are known with vertical stackfreeds: an Italian clock of the early sixteenth century now in the Istituto e Museo di Storia della Scienza in Florence and a small table clock (case: 86×61×51mm) in the British Museum (Reg. No. MLA 1874, 7–18,3). For a discussion of both, see Brusa & Leopold 1999.

44 Cardano 1557, 362–8: a barrel similar to a snail set around the mainspring.

45 Tekeli 1966, 295.

46 Brusa & Leopold 1999.

47 Manuscript M, Biblioteca Ambrosiana Milano, in the Institut de France, Paris since its removal from Milan by Napoleon, f. 81r.

48 Matthes 2015. The pendant watch in a cylindrical case marked 'C' and 'W' (? Caspar Werner, Nuremberg) dated '1548', now in the Patek Philippe Museum in Geneva.

49 A table clock dated 1545 (MDXLV). Height 47.6cm, width 22.4cm (squared). Kunsthistorisches Museum, Vienna, Inv. Nr. Kunstkammer 856. See also Maurice 1976, ii, Figure 91, and Maurice & Mayr 1980 cat. N. 48.

50 Maurice 1976, 84, and references therein.

51 This had already been pointed out by Tycho Brahe (1546–1601).

52 For Bürgi see Maurice 1980a; Matthes & Sánchez-Barrios 2019, Oestmann 2019.

53 In some of Bürgi's clocks, the going train is weight-driven and the remontoire spring lifts the weight very frequently so that the clock itself

drive is almost constant, which significantly improves going train accuracy.

MINIATURIZATION OF SPRING-DRIVEN CLOCKS

The key advantage spring-driven clocks have over weight-driven clocks is their portability. During medieval times there was no social need to be *on time*—yet horology became a matter of prestige. Already in 1377, the King of France owned a 'portable clock' and from 1387 onwards he employed a 'clock carrier' at his court.[54] This was almost certainly a weight-driven clock. In 1481, Marin Guerier was paid fifty sous for carrying Louis XI's clock on his horse for ten days,[55] which may refer to a spring-driven clock. Miniaturization and portability brought about two challenges for clockmakers:

1. Reducing the movement size to fit a small space.
2. Ensuring that *all* elements of the clock work independently of gravity.[56]

Reducing the movement size is a matter of practical construction but in its early stages suitably small tools were not available. Hence, the miniaturization of clocks proceeded in tandem with the development of small hand tools. It is therefore not surprising to see miniature horology flourish from the late fifteenth century onwards when such tools became readily available throughout Europe—and vice versa, the small tools were produced when demand increased.[57]

The second challenge involves two elements: the driving force and the escapement. The driving force found a satisfactory solution with the introduction of the mainspring around 1400.[58]

In the **escapements** of early clocks (both weight- and spring-driven), the verge with the foliot or balance is suspended on a thread from a gallows above the movement. This construction serves two purposes:

1. The suspension reduces friction since the weight of the verge does not need to be carried on one of the two bearings in which it runs, but hangs, with a consequent reduction in friction.
2. The twisting of the double thread (made of organic material) exerts a torsional moment on the verge. This torsional moment implicitly regulates the movement—although this was probably unknown to medieval clockmakers and the effect is small.

The significant disadvantage of the construction is obvious: the suspension depends on gravity and makes the clock sensitive to movements even when spring driven. Furthermore, the physical properties of the organic suspension thread and the associated torsional moment it exerts depend significantly on local humidity and temperature—next to the risk of breakage. These disadvantages can be alleviated by replacing the entire suspension with a balance cock on the upper side, and a hardened and hard-wearing endplate under both pivots to support the verge axially.[59] It is this specific construction element, plus the use of a mainspring, that enabled miniaturization and portability of clocks.

The origin of the balance cock and endplates cannot be traced with certainty. It has been pointed out[60] that it was one of the key innovations in the small cylindrical table clocks/watches that became fashionable in the early sixteenth century. It can also be observed or inferred during the fifteenth century in several instances. A drawing by Henri Arnault de Zwolle[61] in his *c.*1420 manuscript shows a falling-ball-clock. In this case, the clock with going and striking train[62] is contained in a cylindrical case that is suspended by a rope from the ceiling and the clock is driven by its own weight. The clock has horizontal going train arbors with a vertical escape wheel arbor and a horizontal escape wheel meshing with the pallets of a horizontal verge. This means that the verge required a balance cock as the outside bearing and endplates at either end. Similarly, a miniature from around 1455[63] shows an octagonal spring-driven clock movement on a table. The escapement is not shown, but the setup of the clock as a small table clock without bell canopy makes it likely that it had a balance cock.

---

is portable because the weight is contained inside the case. An example of this construction is the 'Wiener Planetenuhr' (Kunsthistorisches Museum, Wien, Kunstkammer 846, Cat. 53; see Maurice 1980a). An example of the construction with a remontoire spring that winds the mainspring of the movement is the 'Wiener Kristalluhr' (Kunsthistorisches Museum, Wien, Kunstkammer 1116), the only signed clock by Bürgi (Maurice 1980a with detailed pictures of the movement, 95–8).

54 Eser 2014, 18.

55 Paris, Archives nationales, KK 64, Fol. 138, cited from Cardinal 1985, 227.

56 Gravity affects many elements of early watches with an undesired impact on timekeeping accuracy. The smaller impacts were only alleviated with contrivances such as the Tourbillon in 1795 (invention)/1801 (patent). What we are concerned with here is a very coarse requirement towards practical usability only, rather than specific and accurate timekeeping beyond an accuracy of up to ten minutes per day.

57 Dick 1925, 23–31.

58 For the debate about when and where watches were 'developed', see Eser 2014 and Oestmann 2018 and references there given.

---

59 Flores & Mundschau 2017.

60 Flores & Mundschau 2017.

61 See manuscript in n. 15, fol. 59–60.

62 This is the striking train that is similar to the Philip the Good clock in that it uses the great wheel of the striking train as the locking plate, see n. 18. For a movement reconstruction see Matthes 2020.

63 See n. 25 (3). It is discussed in (Michel 1962) and has a large reproduction in Matthes 2018, Figure 5. This volume Figure 195.

Figure 72 Small portable clock with iron movements, originally suspended from a chain. This is the oldest preserved miniaturized clock. Presumed Burgundy, c.1460–80. The illustration shows a virtual cut through the clock with the striking train left and the going train right. Private collection, Switzerland. Illustration: 3D computer tomography: Fraunhofer IIS—EZRT, Fürth.

The earliest known balance cock is in a small Burgundian clock from around 1450–80[64] (Figure 72). This small clock seems expressly made to be portable since not only does it employ a balance cock, but also the movement is compressed: the bearings of the horizontal arbors are not set vertically above each other but are moved to the sides to make better use of space. Additionally, the trains are 'folded' to the sides in a right angle at the bottom such that the movement bars have the shape of an 'L' or an upside-down 'Y'. This construction was not only more compact, but also—together with the five-wheel trains—allowed for miniaturizing the clock without the need to make finer teeth on the wheels: the insertion of a fifth wheel, compared to the normal three- or four-wheel movements allows for the same overall gearing ratio without the smaller teeth and pinions. Thence, it was a small step to construct yet smaller clocks. The way was open for the development of the watch.

IMPROVING CLOCK TECHNOLOGY IN THE SIXTEENTH AND SEVENTEENTH CENTURIES: THE ESCAPEMENT

Once a reliable driving force and portability had been achieved during the fifteenth century, the next area for investigation was the time displayed and so the behaviour of the escapement. This needed to be reliable whatever external condition (temperature, humidity, angle of the entire system when moved) was in play. The motivation for this was twofold: firstly, during the third quarter of the sixteenth century, astronomers realized that clocks could be used for timing observations, but that the existing escapements were not precise enough for this task; secondly, the discoveries of Christopher Columbus (1451–1506), the treaty of Tordesillas,[65] and generally the massive increase in long-distance nautical activity made accurate navigation of key interest. Use of a portable clock for longitude determination had been suggested by Gemma Frisius (1508–55) in 1530[66] and was taken up by Alonso de Santa Cruz (1505–67) on the order of Philip II of Spain in 1554,[67] although initial attempts had quickly uncovered the many shortcomings of the available instruments.

The first significant innovation during the fifteenth and sixteenth centuries in this area was the regulation of the escapement. For centuries, the basic crown wheel/verge arrangement was used without modification. Experiment had been made only with the adjustable weights on the foliot.[68] During the swinging of the escapement, several phases have to be distinguished:

1. Acceleration phase—the escape wheel tooth is in contact with a pallet and pushes it in one direction.
2. Free motion—after the pallet leaves engagement with the escape wheel tooth, it starts a motion where there is no

---

64 See n. 24 (3).

65 On the relevance of the treaty of Tordesillas in relation to horology, see Matthes & Sánchez-Barrìos 2017.

66 In *De principiis astronomiae et cosmographiae*, Amsterdam, 1530, ch. XVIII fol. D3r with addendum in ch. XIX in the edition of 1553; see Pogo 1935; Oestmann 2013.

67 *Libra de las longitudines y manera de hasta agora se ha tenido en el arte de navegar*, Biblioteca Nacional de España, Raros y manuscritos, 9941.

68 The following analysis rests on and develops that by von Bertele 1953.

acceleration of the foliot/balance, but frictional forces act on it.
3. Recoil phase—begins once the second pallet engages with a tooth of the escape wheel and continues with a braking of the foliot until it stops.
4. This second pallet is accelerated as the first one was before.

The shortcomings of this system are manifold:

1. It is an aperiodic balance system, i.e. the system has no natural frequency. This means that any distortion leads to non-**isochronic** behaviour.
2. In each period, some of the angular momentum and energy of the rotating system is lost: the recoil means that some of the kinetic energy of the foliot system is converted into friction which causes variations. So performance must be irregular to the same degree as friction in the various parts is irregular.
3. When the pallets hit the tooth, the braking (and rebound) are not elastic but lead to a vibration of the entire system that further distorts **isochronism**—however, this effect is much less significant than the first two.

The second and third shortcomings improved during the fifteenth and sixteenth centuries via escapement regulation, i.e. the limitation of the movement of the foliot (and hence of the pallets) by means of a braking or stopping device known as the bristle. It reduces the recoil with the associated vibration- and friction-induced loss of energy and isochronism.

The first bristles employed were rigid and made of thin metal wire as in the small Burgundian clock of c.1450–80 (Figure 71), and the 1525 clock by Jakob Zech. The next important step was the use of flexible bristles. On 15 January 1524,[69] Peter Henlein made the first mention of hog's bristles: answering a letter from the Duke of Mecklenburg's chancellor, he mentions that 'he gives the messenger two bristles and a small bundle of strings'. The use of elastic bristles is a significant improvement in accuracy. The elastic bristle acts as a harmonizing element to the escapement. Bristle time standards can perform quite satisfactorily, 'even', Von Bertele claims, 'comparably with hairspring balance clockwork for a few days and for one particular setting',[70] although, being organic, they are still subject to climatic variation.

A further improvement in precision occurred with the introduction into horology of the cross-beat escapement by Jost Bürgi.[71] He replaced the pallet-carrying verge of the verge and foliot escapement by two arbors connected by a toothed wheel, each carrying one pallet. The ends of the two arbors carry one overlapping foliot each. The overlapping cross-shaped movement that they execute when the clock is in motion named the escapement. Both foliot and cross-beat are recoil escapements—hence, the latter has no technological advantage over the former. It is mostly the exceedingly high quality of craftsmanship in Bürgi's clocks that led to their superior performance in scientific applications. The cross-beat escapement is rather difficult to build with a high accuracy and therefore never achieved a wide dissemination.

In 1593 or 1594 an escapement was invented by Christoph Margraf (1565?–1604),[72] the imperial court clockmaker to Rudolph II. Margraf obtained an 'imperial privilege', a fifteen-year monopoly patent, in 1595 for making clocks with a rolling-ball escapement.[73] His idea was to use a different time standard that was seen as constant.[74] In the late sixteenth century it became clear that bodies, even of different mass, always fall the same distance during the same time interval. This insight was developed by Galileo Galilei (1564–1642) during 1589–92, but before him two Dutch physicists Simon Stevin (1548–1620) and Jan Cornets de Groot (1554–1640), had already conducted falling-ball experiments from the tower of the Nieuwe Kerk in Delft. They published them in 1586,[75] and it might be this publication that inspired Margraf.

Margraf's escapement uses balls rolling on an inclined plane. Although they suffer from friction, and were never capable of accurate timekeeping, their behaviour is otherwise the same as in the falling-ball experiment. Balls roll down an inclined plane consecutively such that when one ball finishes its trajectory at the bottom of the plane, a second follows the same path from the top. A counting device registers the number of trajectories and shows it on a dial. Since the time required by a ball to complete its trajectory is always theoretically the same, the counting index acts directly as a clock hand. The clock movement has two tasks: to release a new ball whenever one has finished its trajectory and to transport the balls from the end of the trajectory back into a ball reservoir at the start. A tight metal wire is stretched in zigzag lines across an inclined glass plate which shows a painting underneath. The rolling ball (made of ivory, metal, or rock crystal) crosses and passes through a semi-circular channel on one end to cross again

---

69 Eser 2016.
70 Von Bertele 1953, 806
71 Von Bertele 1955. Bürgi did not 'invent' the cross-beat escapement. It was known in the mining industry for pumping water. Maurice 1980a, Figure 18 reproduces an engraving from Jacques Besson's *Theatrum Instrumentorum* (1569/1570) of the mechanism employed in a water pump.

---

72 Christoph Margraf succeeded Georg Emmoser (d.1584) as imperial clockmaker and was the predecessor of Bürgi, appointed in 1604.
73 Vienna, Haus- Hof- und Staatsarchiv, Reichsregister Rudolf II., vol. 17, 8 May 1595; transcribed in Von Bertele & Neumann 1963, 94–6.
74 Neumann 1958.
75 In their *Behinselen der Weeghconst* (The Principles of Weighing), 1586, see Dijksterhuis 1955, i, 509–11: 'Let us take ... two balls of lead, the one ten times bigger and heavier than the other, and let them drop together from 30 feet high, and it will show, that the lightest ball is not ten times longer under way than the heaviest, but they fall together at the same time on the ground ... This proves that Aristotle is wrong'.

in the other direction. The entire trajectory takes between fifteen seconds and one minute.

The rolling ball escapement enjoyed some popularity in the seventeenth century. Grollier de Serviere (1596–1689), who had lived in Prague, had seven different models made for his private mechanical museum on his return to France.[76] Giuseppe Campani (1635–1715) also worked with the escapement; two rolling ball clocks by him are known,[77] as are examples by other makers.

CLOCKS AS ITEMS OF LUXURY, POWER, AND KNOWLEDGE

How and why did the outstanding place that clocks enjoyed in western society in the sixteenth and seventeenth centuries come about? Most of the earliest clocks were made and acquired for dynastic representation and the demonstration of wealth, power, knowledge, and, importantly, as indicators of temperance and an ordered life. These aspects become apparent in the succession of clocks shown in portraits—starting with Philip the Good of Burgundy via Jehan Lefèvre, Philip's chancellor in 1455, with his clock hanging behind him, Louis XI (1423–83) of France with a (travelling?) clock next to him on a table,[78] the portrait from 1532[79] of the German merchant Georg Gisze (1497–1562) with a cylindrical drum watch on the table, the portrait from 1538[80] of Eleonora Gonzaga della Rovere, Duchess of Urbino (1493–1570), with a small table clock next to her, the portrait of around 1550 of Henry II (1515–59) of France picking up a drum watch from a table,[81] and the portrait from 1555[82] of Mary Neville Baroness Dacre with a drum watch next to her on the table. Members of the high nobility throughout Europe chose to be portrayed in the presence of a clock.[83]

The Renaissance saw an unprecedented re-evaluation of 'nature' and with it, throughout Europe, the creation of *studioli/cabinets de curiosités/Wunderkammern*. These *Wunderkammern* were not stereotyped but represented their creators' personal interests.[84] An initial focus on *naturalia* and antique sculpture widened throughout the century and collecting was seen as a promethean activity since it dealt with the apprehension of nature—Prometheus had given the fire of reason to humanity and was considered as the founding father of technology. *Wunderkammern* therefore could be expanded to include technical objects. The *Studiolo* of Francesco I de Medici (1541–87) in the Palazzo Vecchio in Florence contained 'precious stones, medals, cut stones, crystal and mechanical inventions',[85] and the *Wunderkammer* in the castle of Ambras 1576 contained '*naturalia*, ... musical instruments, ... instruments, ... clocks and automata' and further mechanical machines.[86] Joachim Rheticus (1514–74) in his 'First Report' on the Copernican heliocentric system asked rhetorically in 1540 'why should we not consider God, the creator of nature, as equally skilled as ordinary clockmakers—who refrain from adding a little wheel to the movement that is either superfluous or the purpose of which could better be performed in another way'.[87]

This view was general in sixteenth-century Europe.[88] The imperial astronomer Johannes Kepler (1571–1630) noted his admiration for a series of clocks and automata in 1598. They inspired him with the intention 'to show that the heavenly machine is not like a divine being but as a clock'.[89] This was a culmination of a mechanist tradition that derived its authority from the statement that God had created and ordered 'everything according to measure, number and weight'.[90] Clocks were therefore included in *Wunderkammern* all over Europe.

Clocks also had political meaning during the seventeenth century as a symbol for absolutist power. Diego de Saavedra Fajardo (1584–1648) best summarized this in 1640 when he compared the state to the movement of a clock and 'this is why the ruler should not only be the hand of the clockwork but also the escapement/balance which gives the pulse to all other wheels'.[91] In consequence, the acquisition and display of clocks became a passion with European nobility and of all affluent members of society (Figure 213). The fact that many are not primarily utilitarian tools for measuring time can easily be observed in their construction and shape. Rather, they are made of precious materials, show the movements of the sky while not being useful as scientific instruments, or contain automata that imitate life. For two centuries the fact that they were unreliable timekeepers was irrelevant for one

---

76 On which, and Grollier, see Turner 2016a.
77 Florence, Museo di Storia delle Scienza: the rolling ball in this clock is thrown back to the start by a catapult, and the clock has been converted from spring drive to weight drive; Kassel, Physikalisch-Mathematisches Kabinett (destroyed in 1943, only fragments are preserved).
78 The painting was destroyed by fire in Madrid. A copy of it has been made in 1836 by Jean Lugardon (based on an engraving that had been created prior to destruction) and can be found in Versailles.
79 By Hans Holbein, the elder, Berlin, Gemäldegalerie.
80 By Tiziano Vecelli (called Titian), Florence, Uffizi.
81 Ascribed to Francesco Primaticcio, Chantilly, Musée Condé.
82 By Hans Eworth, Ottawa, National Gallery of Canada.
83 This choice is certainly helped by the fact that clocks were an item of ultimate luxury due to their price. See also chapter 24 below.
84 Bredekamp 2000, 33 ff.

85 Schaefer 1976.
86 Scheicher 1985.
87 Rheticus/Zeller 1540/1944, 56.
88 Further examples are Philippe de Mornay (1549–1623): 'Sure the sky is as the great wheel of a clock ... ' (Philip of Mornay, *A Work Concerning the truenesse of Christian Religion*, London, 1617, 95) or John Norden (1547(?)–1625): 'This moving world, may well resembled be, T'a Jacke, or Watch, or Clock, or to all three ... ' (John Norden, *The Labyrinth of Man's Life*, London, 1614, D-2 v.). See further in Mayr 1980.
89 Kepler/Frisch 1858–1871, ii, 84.
90 The statement appears in the apocryphal *Wisdom of Salomon*, 11.21.
91 Saavedra 1640, 441.

of their key uses. This Renaissance view is encapsulated in a small astronomical table clock certainly made for one of these *Wunderkammern* (Figure 73). It shows Hercules holding up the sky for Atlas while the latter fetches the apples of the Hesperides. There is an implied meaning that would have been evident to Renaissance men. Hercules tricked Atlas thanks to Prometheus's gratitude for Hercules freeing him from the rock in the Caucasus. Hence, this work also embodies Prometheus's fire of reason and technology given to mankind and epitomized by the understanding of the celestial movements shown on the clock.

REGIONAL CLOCKMAKING SCHOOLS DURING THE SIXTEENTH AND SEVENTEENTH CENTURIES

From its beginnings to the sixteenth century, clockmaking was an international craft, with innovations travelling quickly through Europe. This is testified by the rapid spread of any innovation—so rapidly in many cases (e.g. the introduction of the mainspring or the miniaturization of clocks) that it is difficult today to identify their origin. This international transfer has been examined in detail through the specific example of Janello Torriani.[92] Technical knowledge moved through individual craftsmen—itinerant masters who made tower clocks in the fourteenth and fifteenth centuries, those who followed their aristocratic clientele, and wandering apprentices who had to leave their home cities and travel for two to three years to learn from masters elsewhere before returning to construct their masterpiece.[93] However, with the significant increase in the number of clockmakers in the main European centres,[94] local horological schools developed from the late fifteenth century onwards.

Burgundy

Burgundy was the earliest region to create spring-driven clocks in some number. Burgundian domestic clocks (both weight- and spring-driven) show a characteristic construction. They are typically of a vertical layout with the bearings for the arbors arranged above each other in bars or plates and the arbors are horizontal. The trains of the movements are typically arranged at right angles to the dial (also usually the locking plate—although this is not invariable) and the bell is struck by a hammer moving in a vertical plain from the inside. Although the movements are usually constructed from iron, already in the fifteenth century they are accompanied by movements mostly made of brass. Given that Burgundy ceased to exist as an independent state in 1493, a proper Burgundian tradition of horology cannot be followed any further, but some sixteenth-century Flemish clocks still show several of the features mentioned previously. Since Flanders, which had been part of Burgundy, had a weak association with its new Imperial rulers, this might have helped maintain a specific clockmaking tradition there.

France

Horology in France was largely based on demand from the court and the leading aristocracy—a situation markedly different from that in Italy and Germany, although demand rose in urban centres such as Paris and Lyon during the sixteenth century. Spring-driven clocks had been adopted quickly in France on a significant scale shortly after their development in Burgundy (shown by the several mentions of portable clocks for French kings in the fifteenth century and their depiction in portraits). It has been suggested that influence from Italy led to the development of fine clockmaking.[95] While some such influence could have occurred, a strong manufacture existed before sustained contact with Italy. Technological exchange with Burgundy, which became part of France in 1493, is likely to have been a more important influence.[96] The pattern of royal and aristocratic demand led to the development of horological centres close to the court in Blois and Paris.[97] Until 1530, Blois was the main royal residence. Already by 1504, Julien Coudray was established in Blois as royal clockmaker and, up to his death in 1530, his was the only workshop known in Blois. However, the main commercial centre was Paris, which led to a significant concentration of the horological craft in that area, followed by the establishment of the first Paris clockmaker corporation by royal decree in 1544.[98] Paris alone accounted for a total of thirty-two makers when the entire country of France had only seventy-six between 1500–50.[99]

Iron chamber clocks with end-to-end movements were the typical French standard. The hammer struck externally on the bell, attached to the top finial in the middle, operated by a wire

---

92 Zanetti 2014 & 2017.
93 Groiss 1980a, 66–9 and this volume Chapter 7.
94 See Turner 2014a; 2014b, 9–11 for France, Groiss 1980a; 1980b, 83 for Augsburg, and Matthes & Sanchez-Barrios 2019, 222 & 225 for Southern Germany.

95 Fourrier 2001–2005, ii, 32.
96 It is worth noting that Jean Fusoris, one of the first to make spring-driven clocks, was active both in France (Paris) and at the court of Burgundy (see nn. 14–16) already in the 1420s. This might be the 'bifurcation point' of the two clockmaking schools.
97 The early distribution of clockmakers in France (with a focus on the makers of 'small clocks', i.e. the more difficult to make spring-driven clocks) has been thoroughly analysed in Turner 2014c, whom we follow here.
98 Cardinal 2020.
99 Turner 2014a; 2014b, 9 ff.

Figure 73 Miniature globe table clock carried by Hercules, signed 'Hanns Reinholdt Augspurg' (i.e. Johann Reinhold the Younger (1587–1639), Augsburg, around 1620. Private collection, Germany.

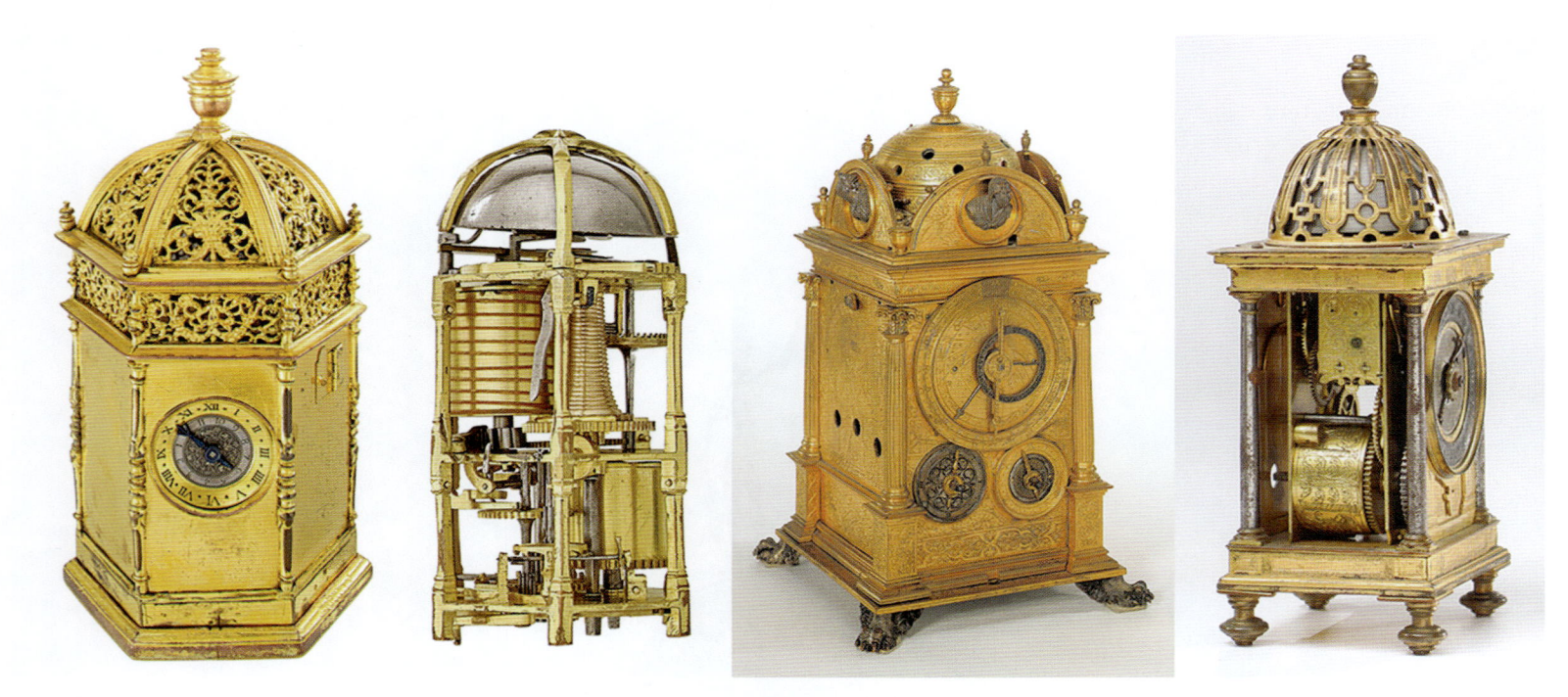

Figure 74 From left to right: Hexagonal French table clock. Aix-en-Provence c.1530, and movement. The going train is on the upper level, the striking train on the lower. Patek Philippe Museum, Geneva. Lavishly decorated German table clock on a square base of brass and silver with astronomical indications and an all-iron etched movement. Hans Gruber, Nuremberg c.1560. Private collection, Italy. Italian (or Flemish) table clock signed 'P. Filassieri' c.1580 with a combined brass-iron movement and fixed barrels. Private collection, Italy.

or thread. The domestic horological tradition of France from the fifteenth to the seventeenth centuries shows one peculiarity that distinguishes French clockmaking even from that of Burgundy: French spring-driven clocks are almost exclusively constructed with vertical arbors standing in bushes between horizontal plates at the top and bottom, which in turn are joined by vertical columns. This is the 'natural' way to construct small timepieces like horizontal table clocks or the small cylindrical drum watches that became fashionable in the early sixteenth century. However, it is a unique construction compared with spring- and weight-driven domestic horology in all other European countries where, in most examples, the arbors of 'vertical clocks' are horizontal between vertical bars or plates. The earliest 'French style' construction can be seen in the miniature in the Brussels manuscript (Figure 212) and it later developed into the standard French tower-shaped table clock design achieved by placing the going and striking trains above each other, rather than next to each other, end-to-end, or orthogonally, as in all other horological schools.[100] The layout can be either square or characteristically hexagonal with a bell canopy on top (Figure 74).

The fact that horology was, to a significant extent, a courtly art is also testified by a number of other aspects. French spring-driven timepieces into the seventeenth century are of outstandingly fine craftsmanship when compared with the average German production of the time and are significantly rarer today. Many of the preserved pieces have additional complications, such as astrolabes, although this was not as standard a feature as in German clocks of the second half of the sixteenth century. However, an early French specialty was clockwork-driven armillary spheres and miniature globes. Some of these carry a mark for Blois and date already from the first half of the sixteenth century,[101] making a link to the royal court likely in these cases as well. Compared with other European regions the material composition of French clocks is advanced. While domestic gothic clocks in France (as in most parts of Europe) were made mostly of iron, the makers of spring-driven clocks switched to an advantageous iron–brass combination very early. The first clocks using a significant number of brass parts in the movements appeared in the 1530s,[102] supplanting the iron movements that had also previously been employed in France. One driving force of this development was the fact that several early clockmakers in France were also makers of scientific instruments—which were typically worked in brass.

---

100 The first extant examples of these 'vertical arbor' trains is a group of four travelling timepieces of which the earliest two date to around 1500–10 and the others soon after (Matthes 2018, Figures 24–35). For an overview of typical French timepieces, see Tardy 1981, i, 43–116.

101 This is the case, e.g. for a globe clock with armillary sphere resting on four columns on a triangular base. It is stamped 'Blois' and '1533' together with a fleur-de-lys (McCord Museum, Montreal) Cardinal 1986.

102 Even the earliest known spring-driven clocks from France from around 1500 employ combination-metal escape wheels where the brass teeth engage with iron pallets—significantly reducing friction in the place where it is most crucial.

Italy

Markedly different from France and Germany, Italy was a country without a strong central power. In Italy, a land of small duchies and principalities, clockmaking was essentially a local and regional court activity serving aristocratic clients. Permanent workshops reaching out to a mercantile and bourgeois market were few.[103] This background provided ample breeding ground for the making and spreading of public clocks at a rather early stage. Furthermore, it served as the foundation for a refined horological craft scene early on. Next to the reference of Brunelleschi's early spring clocks, Italian archives mention 'jewel-like' miniature timepieces as court art early and numerously throughout the second half of the fifteenth century. Highly refined clocks and in particular the earliest known planetary clocks testify to a clockmaking craft scene of the highest level. Additionally, Rome acted as a magnet to would-be fashionable men from all over Europe (e.g. Paulus Almanus from Augsburg and the advanced clocks from a geographically diverse background that he saw in Rome[104]) who brought knowledge of the latest technological developments from a variety of countries to the city and the surrounding region. However, the demand profile was for refined (and expensive) timepieces in small numbers, a situation similar to that in France. Turner has examined the numbers of 'fine' clockmakers that can be traced.[105] The result indicates a decline in Italian fine clockmaking

> ... probably to be related to the instability of the region through internecine warfare and foreign invasion in the early 16th century, Italy being Europe's battlefield up to the Peace of Cambrai in 1529. General horological manufactures did not become established in even the largest of Italian towns, a failure palliated by importations, and by the activities of itinerant craftsmen.'[106]

The number of clockmakers in Italy per decade fell from fifteen in the 1480s to a low point of four in the 1510s—slightly recovering to eight in the 1540s. This might explain why early Italian domestic clocks are of exceeding rarity—even more so than pieces from France.[107]

Successful local schools of clockmakers developed in Italy during the sixteenth century which are traceable not only in archives[108] but also in surviving examples (Figure 74).[109] Italy continued to be characterized by a mixture of settled and itinerant craftsmen as in the centuries before, since clockmakers working for courts had to be highly mobile. The outstanding Italian clockmaker of the sixteenth century, Giovanni Torriani/Juanelo Turriano (1500(?)–1585) spent much of his working life abroad for his employer, the emperor Charles V.[110]

Germany

From a horological perspective, the Holy Roman Empire had a significant advantage compared with France and Italy. The political structure was balanced between a central power (the emperor and his court) and the local aristocracy with more than 100 duchies and principalities of varying independence, power, and resources. The free or imperial cities—fifty large autonomous cities subject only to central imperial authority and taxation—were a third force. This led to a trichotomy of power and to a wealth distribution that created a large domestic market. Local merchant elites in the imperial cities had a representation aspiration that was as pronounced as that of the aristocracy. It was also backed by a large concentration of wealth,[111] although, on average, this was lower than that of the aristocracy. Hence, precious metals and stones were replaced by something equally desirable and luxurious: clocks and automata. The resulting domestic customer base for clockmakers in Germany was an order of magnitude higher than in other European countries and led to an explosion of clockmaking as of other technical activity from around 1500 onwards as a sedentary craft.[112]

Southern Germany provided a perfect breeding ground for clockmaking: not only was there an established trade network connecting economic and political centres from Italy (Milan and Venice) and France (Lyon and Paris) to Eastern Central Europe (Cracow and Budapest), but also Germany had an established

---

103 Turner 2014a; 2014b, 4.
104 Leopold 2003, 671.
105 Although this approach cannot cover local blacksmiths who made tower clocks and, in some instances, other clocks, this restriction holds true for all countries and so does not necessarily influence the results.
106 Turner 2014a; 2014b, 5.
107 Few domestic clocks of probable Italian manufacture are known to predate 1550. Next to several weight-driven open-frame clocks of 'alarum' type (see n. 4), and gothic clocks with a cage-like frame made of brass and probably from the late fifteenth century (e.g. Leopold 2003, 668 & 669, & private collection, Italy), those are a spring-driven table clock from around 1490 (British Museum, Waddesdon Bequest,

WB.222) (Matthes 2020) and a hexagonal clock movement with vertical arbors on two storeys (Florence, Museo Galileo) Brusa & Leopold 1999, 161–71; Matthes 2020. A few further clocks could be identified as Italian. See also Cherbel 1971.
108 Brusa 1990, 498–511; Leopold 2003, 671.
109 Leopold 2003, figs. 3a & 3b, Cherbel 1971, figs. 1–5.
110 Zanetti 2014 & Zanetti 2017.
111 The Augsburg merchant Jakob Fugger (1459–1525) is considered by some the richest man who ever lived.
112 Turner 2014a; 2014b, 7; Matthes & Sanchez-Barrios 2019, 222 & 225. Early non-aristocratic clock owners can be identified in a few cases from paintings (see, e.g. Eser 2014, Figures 57, 63, & 69–70; cat. 22) and by inscriptions on clocks as, e.g. the watches owned by Nuremberg merchant families from the 1550s and 1590s; Eser 2014 cat. 48 & 49, or a Metzger-type clock with the printer's mark of the early Mainz printing office of Franz Behem around 1540 (private collection).

metalworking focus with supplies of metals in all shapes and the respective skilled craftsmen. 'As Italian small clock- and watch-making, always court-centred and small scale, faltered in the decades 1500 to 1530 and failed to establish permanent manufacturing structures, there was little competition for the south German products'.[113] By the middle of the sixteenth century, the number of clockmakers in all of France was less than five times that of Augsburg alone.

The products were also similar, although more numerous, to those of Italian and French clockmakers. Besides a few complex planetary clocks for the elite, the majority of the output was horizontal table clocks or vertical tower-shaped *Türmchenuhren* (Figure 74). These were characterized by end-to-end movements with **'nag's head'** striking train release by a star-wheel behind the dial. The hammer moves in a horizontal plain inside the bell, sitting on a vertical arbor in the spring-driven specimens, and strikes from the outside, reaching up from the movement in the weight-driven specimens. The movements were made almost entirely of iron until 1560–70 (with some brass bushes earlier) while cases were made mostly of brass.[114] Usually, the indication was simply hours. A peculiarity of German clockmaking was the so-called Metzger-style clocks after Jeremias Metzger/Metzker (c.1530–99) of Augsburg, who made not the earliest, but most famous, examples.[115] These were highly complex astronomical machines with calendrical and astronomical indications. The two striking trains are arranged at right angles on either side of the going train.

A rigid guild system, which was established from around the 1550s for clockmaking, with exactly prescribed types of clocks to be made, allowed for standardization and a consistently high quality of products despite a very diverse range of artisans for their production. On the other hand, it hindered innovation both of types of products and of their accuracy. Once demand patterns changed in the early seventeenth century and innovation was actively discouraged by the guilds,[116] the clockmaking profession in Germany went into a decades-long decline. Together with the effects of the Thirty Years' War (1618–48), it led to an almost complete eradication of Germany as a horological centre by the late seventeenth century.

England

There is little physical evidence for English makers of domestic clocks before the late sixteenth century. However, domestic clocks were produced locally as testified by archive entries starting with William More, Prior of Worcester who, having previously owned a domestic clock that had to be sent to London for repairs, bought a new one in 1526: 'At Worcester—Item to Walter Smyth for ye makyng of a new lytull clock to convey to our manere … 41s, ye case 4d'.[117] Perhaps, at least for standard domestic weight-driven clocks, a local production already existed in the late fifteenth century. Like most other European countries, there was a significant royal interest in horology. According to the inventory of his goods, Henry VIII (1491–1547) owned some 150 timepieces—although they were not all geared clocks, and most of the latter were possibly of foreign make.[118]

What may be the earliest surviving domestic clock made in England is a small iron clock made by James Porrvis in 1567 for John Webbe of Salisbury. It shows considerable similarity with French and Flemish styles in the gothic corner pillars set at forty-five degrees to the frame and the overall movement structure, but it also has some features that are not found on continental gothic iron clocks, such as (among others) the brass front plate.[119] Early British-made spring-driven clocks also survive from the second half of the sixteenth century, made by Bartholomew Newsam (?–1593), clockmaker to Queen Elizabeth I,[120] closely following French horological traditions. But it is lantern clocks that are considered the first genuinely British development in horology.[121] When compared with French and German gothic clocks, English lantern clocks are quite distinct.[122] However, they do share some characteristics with their continental spring-driven counterparts, and more with Flemish and German examples than with French ones. This could indicate that the technology was imported by foreign clockmakers into Britain during the late sixteenth/early seventeenth centuries.[123][124] The earliest fully developed British

---

113 Turner 2014a; 2014b, 6.
114 Usually, the cases were made of engraved sheet brass, yet cast brass cases with relief decoration occurred for the more luxurious production. Coole & Neumann 1972, 22–9 & 88–94. The predominant use of iron for the movements of early German clocks is explained by the fact that clockmaking was part of the locksmith corporations.
115 Neumann 1967, 31.
116 The innovations of escapements described here were mostly initiated before the establishment of the guild system (e.g. by Peter Henlein in the 1520s) and later in a court context.

117 Beeson 1963, 8; Robey & Gillibrand 2013, 517.
118 Turner forthcoming.
119 Robey & Gillibrand, 2013, *passim*.
120 Thompson 2004, 38.
121 Following the most recent overview by Robey 2016 and Robey 2017.
122 Differences include the use mostly of brass for the construction rather than iron (except for arbors and pinions as well as a number of other parts—see Robey 2016), the use of screws, the use of warned striking, the construction of hammer arbors of lantern clocks pivoted between cruciform extensions of the front and rear movement bars, the location of the hammer inside the bell, and the presence of movable hands for easier setting of the clock; see Robey 2016.
123 To a significant extent by Flemish clockmakers who sought religious freedom after the Union of Arras (1579) led to stronger Spanish (Catholic) influence in Flanders.
124 Numerous of their characteristics can be found in spring-driven continental clocks such as the cruciform bars in Burgundian examples,

Figure 75 Left: South German weight-driven gothic iron clock, 2nd half 15th century. End-to-end movement with the alarm train visible on the upper left. Uhrenmuseum Winterthur, Kellenberger collection. Photo: Steffi Tremp. Right: English lantern clock signed and dated 'William Bowyer fecit 1617', the earliest known dated lantern clock. The typical rear cruciform bar is visible through the side. Private collection. Photo: Bonhams, London.

lantern clocks appeared around 1610 and are linked to makers such as Robert Harvey (1580/83–1615), Henry Stevens (around 1577–1638), and a little later William Bowyer (around 1590–c.1653) (Figure 75).

It has been suggested[125] that England's lack of a developed horological manufacture in the late sixteenth and early seventeenth centuries became an advantage. Innovation stagnated in Europe, notably in Germany, partly because of the conservatism of guild members. Accuracy, in everyday clocks, was also perceived as being low. So Shakespeare:

> A woman, that is like a German clock,
> Still a-repairing, ever out of frame,
> And never going aright, being a watch,
> But being watch'd that it may still go right.[126]

The typical British lantern clock, with a modern design, simple construction, devoid of unnecessary functions, and easy to make by clockmakers not accustomed to forging iron, was a response to this situation. This simplicity of construction and decoration was to become their key advantage: the use of a brass (rather than filed iron) frame meant that its parts could be serially produced in foundries—rather than forging and filing (often requiring several hands and a high accuracy to fit rigidly) in the clockmaker's workshop—allowing to do away with the large forge, as well as necessity for assistants, plus cooperation by engravers and gilders—all of which added considerably to the costs. Use of the characteristic cruciform front and rear movement bars that carry the hammer and striking-work arbors furthermore allowed hinging the side doors directly between the top and bottom movement plates so avoiding a separate case as found on continental clocks. The highly reliable yet cheap and simple method of producing clocks led to a significantly widened market amongst the rapidly expanding merchant class in towns and cities and became the standard clock of the seventeenth and well into the eighteenth century in England and beyond.

Spain

In terms of socio-economic structure, Spain lay between Italy and France. There were a small number of independent kingdoms, such as Castile and Leon, Navarre, and Aragon in permanent war with Granada up until 1492. Consequently, domestic horology was largely a royal and courtly interest. Early indications of domestic horology exist in Spain from 1376,[127] although it is

---

the conveniently simple way of setting the hands, and the combination of iron arbors with brass wheels.
125 Robey 2017.
126 William Shakespeare, *Love's Labour's Lost*, (1594–5), 3.1.183–6.

127 In a letter dated 6 February 1376 and written by King Pedro IV of Aragon (1319–87) to his daughter, he describes a clock he sent: a small alarm clock with three bells and an astrolabe dial showing the zodiac,

unclear whether the very refined clock described was of local production. Several clocks are described in the death inventory of Isabel of Castile (1451–1504). Next to a domestic weight-driven clock, there were two spring-driven ones—both in elaborate and part silver cases. One had Isabel's coat of arms on it, so might have been locally made (unless the coat of arms was added later). The other one is known to be of Burgundian origin, and Isabel's clockmaker is referred to as coming from France.[128] This international exchange is also reflected in pictorial evidence: fifteenth-century Spanish paintings show domestic gothic iron clocks with features of French horology (the hammer attached to the top finial, striking externally on the bell operated by a wire or thread).[129] The importation of clocks and clockmakers is a continuing feature of early domestic horology at the courts of Spain: that the Kingdom of Spain was governed together with the Holy Empire for almost half a century by Charles V (1500–58) led to an influx both of clocks and clockmakers encouraged by Charles's keen interest in horology. Influential figures came from Italy (Juanelo Torriani) and the Flemish Habsburg domains (Hans de Evalo and others). They dominated the local production of fine clocks.

CONCLUSION

Early users of domestic clocks thought of them as musical instruments—but soon began to regard them as both practical instruments and symbolic representations of the state, the universe, or man's relationship with God. The development of domestic horology reflects these several approaches. Only in the seventeenth century did 'efficient and accurate timekeeping' prevail as the principal use of clocks.

The fascination that contemporaries felt for the intricate mechanisms of clocks made them attractive to potentates and intellectuals alike—which significantly aided the rapid international dissemination of horological knowledge. Given domestic clocks were a 'high-value-easy-transport' product, they were always a primary item for import/export activities. This led to a competitive environment that furthered rapid innovation at an international level. Once larger domestic markets had developed, different and local schools of clockmaking appeared and increasingly supplied broader segments of the population—until clocks had become an everyday product in the second half of the seventeenth century.

Sun, Moon, and fixed stars and with plates adjustable for any latitude in the Kingdom of Aragon (quoted after Pérez-Álvarez 2013, 493.
128 Pérez-Álvarez 2013, 497.
129 E.g. Juan de Flandes, *The Birth and Naming of John the Baptist*, around 1496, painted for the monastery of Miraflores (Burgos), The Cleveland Museum of Art, John L. Severance Fund 1975.3 and Juan de Flandes, *Cena en Casa de Simon* (part of the Poliptico de Isabel de Castilia), late 15th century, Palacio Real, Madrid.

SECTION TWO: FROM HUYGENS TO THE END OF THE EIGHTEENTH CENTURY

Wim van Klaveren

INTRODUCTION

One of the most significant breakthroughs in the history of clockmaking was undoubtedly the application of the pendulum to clock movements. It effected a momentous advance in the accuracy of timekeeping, from about fifteen minutes to about fifteen seconds per day, and provoked a boom in clockmaking across Europe. Galileo Galilei (1564–1642) experimented with the idea after discovering the regularity of an oscillating pendulum, but it was the Dutch scientist Christiaan Huygens (1629–95) who successfully applied this principle to clockwork. In the course of 1656, he developed a pendulum clock. The side view in his *Horologium Oscillatorium* (1673) incorporates a print of 1657 that is probably Huygens' first model and the missing design shown to the States General in June of that year for a patent.[130]

Through the Republic of Letters, his ideas spread quickly across Europe, finding fertile ground in the clockmaking traditions of the various European countries. Although each nation developed its own style of clockmaking, close trade and the exchange of ideas resulted in shared influences. Initially, existing clocks were often converted from **balance-wheel escapement** to a horizontal **verge** escapement combined with a pendulum to improve their timekeeping, and later, after *c.*1670, with an **anchor** escapement. At the same time, clockmakers also began to produce new clocks specifically designed for regulation by a pendulum.

In the Netherlands, pendulum regulation led to the so-called Hague clocks, which were made in large numbers, not only in The Hague but also in other parts of the country. In France, its counterpart was the *pendule religieuse*, also made in great numbers, mainly in Paris but also in the larger provincial centres. In England, both spring clocks (whether for placing on a table or on a wall bracket) and, a little later, longcase clocks were produced. In Italy, pendulum clocks were also manufactured,[131] a specific model for night use being more widespread there than elsewhere in Europe. In the German lands, which had been the champions of Gothic and Renaissance clocks, development was slower, partly because of the Thirty Years' War (1618–48).[132] Many of these early pendulum clocks were relatively austere (Figure 76). They also betray a common origin, and evidence of the close communications between various parts of Europe.[133]

With the technical mastery of pendulum clockmaking the emergence of distinct styles is noted, usually going hand in hand with the styles of furniture and, to some extent, architecture. In most countries there was not only a mainstream evolution centred on the capital, but also local development in regional centres of clockmaking, each with its own characteristic features. Development shifted from The Netherlands, initially the leading centre, towards England and France, which dominated most of the eighteenth century. These countries were the driving influences on clockmaking in Europe, but where possible, this section also discusses clock development in other countries.

EARLY DEVELOPMENTS

Christiaan Huygens was a mathematician interested in both astronomy and in solving the longitude problem at sea. This was his primary motive for developing accurate clocks.[134] However, being a theoretician, he needed craftsmen to put his ideas into practice.[135] Initially, he worked with the clockmaker Salomon Coster (*c.*1620?–1659),[136] who made the first commercial Hague clock in 1657. As Huygens was in The Hague in the years preceding the appearance of this clock, he may have worked with Coster to experiment with clock movements. Indeed, it is known that Huygens and Coster later carried out experiments converting the turret clock of a church in Scheveningen (near The Hague) to an extremely long pendulum (*c.*7m), as well as with the clock of Utrecht cathedral, shortly after the production of the first commercial pendulum clocks in the Netherlands.

The exchange of ideas and scientific developments was rapid, and news of Huygens' breakthrough in accurate time measurement quickly spread. However, Huygens also had a business interest in clockmaking—selling Coster clocks in France. For this reason, Claude Pascal (before 1635–74), who emigrated from the

---

130 Huygens 1658. Recent revisions of pendulum pre-history are Whitestone 2008; Whitestone 2012a; Whitestone 2012b; Whitestone 2017; Whitestone 2020, Whitestone 2022. See also Bedini 1991.

131 Ferdinand II de Medici came into the possession of a Hague clock within a year of the first being was made Plomp 1979, 15.

132 On this see Rabb 1962, Munck 2005, 29–31.

133 A good comparative study of early pendulum clocks is Van den Ende et al. 2004.

134 Huygens was the son of a distinguished diplomat and had an extensive network of highly placed correspondents. See the letters published in *OCCH*, vols. i–x, xxiii.

135 For them, see Leopold 1979

136 But see Whitestone 2008 arguing that Huygens' earliest clock was made in Paris by Isaac Thuret. After Coster's sudden death in 1659, Severijn Oosterwijck (*c.*1637–*c.*1694) worked for Huygens.

Figure 76 Three early pendulum clocks. From left to right: A Hague clock by Salomon Coster, 1657. A wall clock by Ahasuerus Fromanteel, c.1658. A *pendule religieuse* by Louis Brulefer, 1660–5. Left: Photo of the Rijksmuseum Boerhaave, Leiden. Centre: Courtesy of Carter & Wright. Right: Photo of Nico v.d. Assem, Netherlands.

Geneva area to Holland around 1654, worked as an intermediary between Huygens and his French customers. Nicolas Hanet (?–1723), another French clockmaker, went to The Hague to work with Coster for some months before officially importing his clocks into Paris.[137] At least eleven Coster clocks, probably more, were exported to France in this way.[138] Isaac Thuret (c.1630–1706), clockmaker to Louis XIV, was also in contact with Huygens from his time in France and there is a theory that Thuret made pendulum clocks with a long pendulum[139] that resemble Huygens' original diagrams more closely than the Coster clocks and probably precede his clock of 1657. However, much depends on the precise dating of the existing Thuret clocks.[140]

When the pendulum clock was introduced in England, the most common domestic clock was the lantern clock, fitted with a balance. There were many clockmakers who produced this type of clock, such as Peter Closon (c.1595–1660/61) and William Bowyer (1590s–1653 Figure 75, right), as well as Ahasuerus Fromanteel (1607–93,), the latter being instrumental in the widespread introduction and development of the pendulum clock. He was born in Norwich of Flemish descent and probably spoke Dutch. In 1657, after the news of the pendulum clock had spread, he sent his eldest son John (1638–1689) to the Netherlands to work in Coster's workshop. This event is recorded in a rather cryptic contract between John Fromanteel and Salomon Coster, the details of which are still widely debated.[141] A year later, the elder Fromanteel published an advertisement in two journals:

Mercurius Politicus of 17 October and the *Commonwealth Mercury* of 25 November:

> There is lately a way found out for making clocks that go exact and keep equaller time than any now made without this Regulator (examined and proved before his Highness the Lord Protector by such Doctors whose knowledge and learning is without exception), and are not subject to alter by change of weather, as others are, and may be made to go for a week, or a month or a year, with once winding up, as well as those that are wound up every day, and keep time as well, and is very excellent for all House-Clocks that go either with Springs or Weights. And also Steeple-Clocks that are most subject to differ by change of weather. Made by Ahasuerus Fromanteel who made the first that were in England: you may have them at his house on the Bank-side in Mosses Alley, Southwark, and at the sign of the Maremaid, in Lothbury near Bartholomew Lane end London.

The advertisement coincided with a dynamic development in clockmaking—a period of great activity and innovation.[142] Pendulum clocks almost immediately drifted away from the prototype, the box-shaped wall/table clock, and rapidly advanced both technically and stylistically. An early form in England was the 'diamond'-shaped wall clock.[143] Many early longcase and spring clocks were made with a so-called architectural top—a shallow triangular pediment—although rectangular spring clock cases

---

137 On Hanet, see Hordijk 2018 and review by Whitestone in *AH*, xxxix, 2018, 549–50; Hordijk 2020.
138 Plomp 1979, 19 following Huygens' correspondence.
139 For an example, see Vehmeyer 2004, 810–11.
140 Whitestone 2008.
141 Leopold 2005; Van Kersen 2005; Taylor 2010. That the cradle of the pendulum was in England rather than The Hague is a speculation advanced in recent years and most recently set out in Garnier and Hollis 2018. However, a severe critique of this position was earlier offered by Whitestone 2012b.
142 An excellent and lavishly illustrated introduction is Garnier & Carter 2015.
143 See Betts 2018, who mentions nineteen examples, and Pettifer 2020; see also Robey 2020, who add three more examples. Vehmeyer 2004, ii, 572–3.

continued to be made, veneered in either ebony or walnut. During the 1660s, apart from Fromanteel, famous clockmakers such as Edward East (1602–96), Samuel Knibb (1625–70), his cousin and apprentice Joseph Knibb (1650–1711), William Clement (before 1622–1709), and Henry Jones (1634–95), an East apprentice, among others, contributed to the development of clock production. It was in this context that the innovative maker and businessman Thomas Tompion (1639–1713) worked in London (from 1671), as did Daniel Quare (1647/48–1724), both of whom rose to prominence producing many works in different styles. Tompion's expertise is amply illustrated by the clocks he made for the newly built observatory in Greenwich, two of them with a two-seconds pendulum (length about 4m).[144]

THE LONGCASE CLOCK

Early pendulum clocks were not only table clocks but also wall clocks. The latter developed into so-called hooded clocks (Figure 77), elegantly housed in cases in the architectural style with movements of eight-day duration. However, they were heavy. The longcase clock likely resulted from a combination of technical necessities and changing furniture tastes. Technical necessities included the need to support the heavy weights that powered eight-day **striking** clocks in a space (the trunk) that also protected them from interference. Skills in cabinetmaking and veneering could also be displayed in the woodwork of the longcase.

Early longcase clocks, usually ebony veneered and sometimes panelled, were small and elegant, the square dials being only eight inches across, but these dimensions increased and by the end of the century the dial size was ten to twelve inches. The early case designs featured predominantly architectural elements, such as the aforementioned triangular pediment, and originally had no door in the (rising) hood. To access the dial for winding, the hood had to be lifted and was held in position by a special sprung catch on the backboard of the case. This somewhat cumbersome system gave way to a hinged-door hood, which could be moved forwards for access to the movement.

It was not only the case design but also the mechanisms that changed. In the early years, lantern clocks—which continued to be made both in London and in other parts of England throughout the seventeenth and the first half of the eighteenth centuries—were sometimes also installed in longcases. However, in general movements specifically designed for longcase clocks were developed. Often the changes in the movement and the case responded to each other, with changes in one leading to changes in the other, and vice versa. For example, not long after the beginnings of the longcase clock, the value of longer pendulums, which ensured better timekeeping, was appreciated, but this proved to be problematic as the amplitude of a pendulum in combination with a verge was too wide for the narrow cases. So, in the late 1660s, several clockmakers, such as East, Fromanteel, Clement, and Joseph Knibb, began to experiment with alternative escapements to the verge. This resulted in the emergence of the anchor escapement, an innovation which was to be applied throughout the following centuries. Who invented the new escapement is uncertain, but William Clement (1646/7–1709) and Joseph Knibb (1640–1711) have the strongest claims.[145] It led to the application of a long pendulum to domestic clocks, usually a seconds pendulum (sometimes referred to as a Royal pendulum), or, more rarely, a 1¼-seconds pendulum.

As soon as the pendulum was introduced into clockwork, more complex clocks were made, such as month-going and even year-going clocks, as well as quarter-striking clocks. A particular feature of the English longcase clock movement is the so-called bolt-and-shutter maintaining power (Figure 78). In this design, metal shutters cover the winding holes during normal operation. Upon opening the shutters, power to the **going train** is maintained via a sprung bolt to the centre **arbor**, thereby preventing the movement from running backwards when winding, or even stopping, as might otherwise occur. A less-expensive version of the longcase clock was one with a thirty-hour movement, usually driven by one weight via an endless rope or chain (the so-called Huygens system), which intrinsically provided maintaining power. This

Figure 77 Hooded clock by Ahasuerus Fromanteel, c.1665. Author's photograph from Dawson et al. 1982, 164, fig. 214.

---

144 Tompion's life and work is discussed in depth in Evans, Carter, & Wright 2013. For Quare, see Kenney 1979; Kenney 2016.

145 A claim by Robert Hooke (1635–1703) to have invented the anchor escapement as early as 1657 may be discounted. Joseph Knibb's turret clock for Wadham College, Oxford was installed and paid for before Michaelmas 1670. It had been ordered on 10 July 1669. Beeson 1971, 'Invention', and Beeson 1989, 64–5. That the clock installed by Clement in Middle Temple Hall before Michaelmas 1667 had an anchor escapement is an argument that depends entirely on its relatively high price. Bates 2018, 268–72.

type of clock was widely made, particularly in provincial towns all over the country, until well into the nineteenth century.

The striking train of early clocks was controlled by an external, and later an internal, **count wheel**. The former was mounted on the outside of the backplate, while the latter was positioned between the plates and fixed on the **great wheel** of the striking train. The disadvantage of this system is that the strike can get out of synchronization with the time indication. Rack-and-snail striking, invented probably in the mid-1670s, overcomes this problem,[146] as it always indicates the right number of strokes corresponding to the position of the hour hand. In addition, it allowed clocks to have trip repeat, making it possible for the previous hour to be sounded at any particular moment by pulling a cord. Around this time, an aperture showing the date and a subsidiary ring with a small hand indicating the seconds began to appear more often and would become a standard feature of longcase clocks.

In the period 1665–1700, the shape of English longcase clocks underwent a gradual change, with the architectural top disappearing and the square hood with mouldings emerging, flanked by columns, either plain or barley twist. Cases were mostly walnut veneered and usually rested on bun feet. They became taller and were often embellished by marquetry (originally from Italy, this style became popular first in Holland and thereafter in England), green-stained bone inlays, or exquisite veneers, such as olive oyster shells or cocobolo wood. The trunk doors could have clear glass lenticles allowing observers to view the movement of the pendulum bob; in the case of a 1¼ seconds pendulum, the lenticle was in the base. In the 1680s, crestings on the hood became more common and towards the end of the century caddies—first shallow, later more pronounced—began to appear, often flanked and/or surmounted by shaped wooden or metal finials, foreshadowing the more imposing longcase clocks of the eighteenth century.

From England, longcase clocks spread to other areas of Europe. They were probably introduced in the Low Countries by the English maker Joseph Norris, who settled in Amsterdam in or shortly after 1675—one of many clockmakers who sought their fortune in Holland. By then, the longcase in England was produced on a large scale and Norris could continue the work he had been doing in Abingdon, Oxfordshire where he was born and where he had been apprenticed to his brother Edward. Norris was not alone. The Fromanteels were regular visitors to the Netherlands and around 1680 set up a business producing longcase clocks, first in The Hague and later in Amsterdam. Not long after, Christopher Clarke (c.1668–1735) also came to Holland, and became the partner of his father-in-law, Ahasuerus II Fromanteel, to form the company Fromanteel & Clarke. Other English clockmakers who emigrated to the Netherlands included Edward Brookes and Stephen Tracy. They not only made clocks there, but also often imported parts or sometimes whole movements from England. This influenced Dutch clockmaking, even that of Hague clocks, as seen in the application of the **fusee** and corner spandrels. Early Dutch makers include Pieter Klock (1665–1754), Jacob Hasius (1682–c.1725), and Steven Huygens (c.1653–1720).

The early Low Country longcase clocks showed their English origins, with ten-inch dials and similar designs. They sometimes had marquetry cases with intricate inlays while ebony-veneered or ebonized cases also occur, but the majority were walnut veneered. By the time eleven-inch dials came into fashion, Dutch cases began to be embellished by wooden carvings, still sharing similarities with the English clocks of the time. These carvings could sometimes be quite elaborate, particularly on top of the square hood. However, around the beginning of the eighteenth century, the Dutch longcase clocks began to diverge from their English models and develop a style of their own, particularly when arched dials gradually replaced square dials.

English influence extended from the Low Countries into the neighbouring German lands, and beyond.[147] Clockmaking in the Holy Roman Empire had been reviving since the 1650s and both longcase and spring-driven clocks were welcomed. Clockmaking in the English style spread much farther, not only to Scandinavia, but also to Austria and Northern Italy. Later, English influence could also be seen in the Iberian Peninsula.[148]

French longcase clocks were quite different from their English counterparts. They, too, first appeared in the 1670s by makers such as Samuel Panier (?–1700) and Antoine Gaudron (?1650–c.1707).[149] It seems that case makers first positioned a *pendule religieuse* on a shaped pedestal. This design was the basis of the clocks that followed and was maintained and further developed well into the eighteenth century, with the cases becoming more and more ornate (Figure 156). Longcase versions were often waisted and shaped, had a lenticle in the trunk door, and were embellished with ormolu mounts. Shortly after 1700, longcase clocks were being made all over Europe, the styles varying from very English-looking clocks to more ornate examples following French styles. Nevertheless, while there were the two broad English and French styles by the turn of the century, longcase clocks were made with features unique to their region as clockmakers adapted to local taste and fashion.

---

146 The Lancashire clergyman and watchmaker Edward Barlow (1639–1719) was long held to be the inventor of rack striking, but Robey 2005 shows this to be incorrect.

147 For examples, see Mühe & Vogel 1976, 66–7.
148 Montañes, 1968.
149 For a fine example by Gaudron in the J. Paul Getty Museum, California with a case attributed to André-Charles Boulle; see Wilson et al. 1996, 2–9.

Figure 78 Longcase clock with rising hood and bolt and shutters by James Clowes, *c.*1665. Author's photograph; his former collection.

LONGCASE CLOCKS IN THE EIGHTEENTH CENTURY

In the seventeenth and the first half of the eighteenth centuries, most clocks were still costly objects.[150] Makers such as Tompion and Quare in England made the most exquisite clocks, for instance, those produced for the courts of William and Mary and Queen Anne, and for wealthy clients at home and abroad. French clocks were becoming increasingly 'royal', with very expensive, richly adorned cases, changing in style across time, while new styles came into vogue during the Baroque and Rococo periods. It was French styles that were to dominate furniture design on the Continent, and hence clock case design as well.[151] The pendulum was introduced into France during Louis XIV's reign. During this period, lavishly decorated furniture was made, becoming more delicate towards the end of the period. Celebrated French designers and cabinet makers included André-Charles Boulle (1642–1732) and Gilles-Marie Oppenordt (1672–1742), the latter of Dutch descent. During the Regency (1715–1723), when the future Louis XV was still too young to govern by himself, a style transitional between Baroque and Rococo was characterized by slightly more delicate and sinuous shapes, curved lines, and ornamentation. The Louis XV period (1723–1774) is defined by the highly symmetrical designs of previous eras gradually making way for the Rococo style, characterized by asymmetric features and flowing lines. Ormolu scroll mounts embellished all sorts of furniture, including clocks, which were made throughout the century in various degrees of richness, coming to a peak around the middle of the Louis XV period (Figure 90). From after c.1750 there is a return to more austere designs, mostly because of a renewed interest in Greek and Roman antiquity, foreshadowing the neoclassical style that was to follow. The Louis XVI style itself saw a trend towards delicacy, which lasted until the end of the century, merging into that of the **Directoire** (1795–9).[152] In this period, clocks were made with dials having Arabic rather than Roman hour numerals. There was also some colour, for instance, a maker's signature marked in red or blue, a trend first seen in the late Louis XVI period. The importance of French decorative styles was such that they were followed all over the Continent, most notably in the Netherlands and Germany, but occasionally in England as well.

While French clocks developed largely under the influence of wider changes in furniture fashion, in England technical progress remained an important driver of changes in appearance.[153] In the early eighteenth century, c.1710, the first arched dials began to appear in England, initially quite shallow, but by 1725, a semicircle. One of the reasons for this was the increasing complexity of the dial, incorporating indications of the phases and age of the Moon, sometimes combined with the time of high water in a particular port. The arch could also be used for a strike/silent lever or for date indication. More rarely, it offered space for automata. Longcase clocks were becoming larger. Towards the end of the seventeenth century, clocks became taller because of the application of caddies, sometimes double caddies, and this trend continued well into the eighteenth century. Paralleling developments in furniture design, cases were sometimes decorated with an imitation of oriental lacquer work—japanning also referred to as *Chinoiserie*—a fashion which lasted until the 1760s, not only in England but also on the continent, for instance, in Sweden.[154] Mahogany cases largely replaced walnut, the export of which to England was restricted by the French government.[155] Cases became taller and more monumental, the tops often embellished with brass finials, culminating in superb mahogany-veneered pagoda tops around the middle of the century.

At the end of the eighteenth century some aspects of English clockmaking became industrialized, increasing specialization, and new production techniques emerged. For example, around 1770, the painted dial was introduced. Until then, clocks had brass dials with applied, usually silvered, chapter rings and gilt corner spandrels (though by the end of the 1760s, dials were also made without applied chapter rings and were entirely silvered). These painted dials were made by specialists, based particularly in the Birmingham area, and were cheaper than their brass counterparts. Initially, they followed the layout of their predecessors, with painted spandrels in the corners, later followed by other motifs, as flowers, symbolic representations of the four seasons, or the four continents. Images of birds on dials were also very popular. To facilitate matching a painted dial to a movement, an intermediate plate was developed, the so-called 'false plate', which allowed the clockmaker to finish the movement and fit the purchased dial afterwards.[156] This development was hardly ever seen outside Britain.

In the first quarter of the eighteenth century, Dutch longcase clocks began to drift away from their English models. The Amsterdam longcase clock became a status symbol, made with an expensive oak case, which was almost always burr walnut veneered. Starting from the 1720s, these clocks were truly imposing, reaching heights of up to three metres or more at their peak, and had elaborate mechanisms. For example, they often had dials showing a full (sometimes perpetual) calendar indicating the day,

---

150 However, simpler and less expensive country clocks were increasingly available and brought timekeeping to a wider public; see Loomes 1976.
151 See Chapter 15.
152 The *Directoire* ended the period of the Terror and saw the return of high society to France.
153 On English clockmaking, see Darken & Hooper 1997; Darken 2006.

154 See Falck *et al* 1996, 74–3.
155 Timber in France had become so scarce that its export was banned, causing England to turn to the American colonies, first for walnut and, from the 1720s, for mahogany.
156 Loomes 1974; Loomes 1994; Tennant 2009; McEvoy 2011; Robey 2019.

the month, and the date. There was almost always a moon dial, together with high water indication. This was an important feature for wealthy merchants, enabling them to know at what time ships could enter port and trade begin. Clocks invariably had **Dutch striking**, which is also, but rarely, to be found on English clocks, sometimes in combination with one stroke on the quarter hours. In addition, they often had a weight-driven alarm, not a common feature on English clocks. More prestigious clocks had a musical train, with a choice of up to twelve tunes to be played on the hour through a nest of up to sixteen bells and a shorter tune on the half hour. Other prestigious clocks had automata, often rocking ships against the background of a city, but also other scenes, such as a blacksmith hammering iron on an anvil or a garden party near a country house with a girl on a swing.

Dutch makers such as Paulus Bramer (?–1770) produced astronomical clocks that showed a whole range of astronomical indications in the arch, for instance, the time of sunrise or sunset, or the position of the Sun and Moon in the Zodiac. The cases became more monumental throughout the century, not only in height but also in their bases. The initial rectangular base was fortified with buttresses on the front corners and after 1760 became *bombé*. At the same time, the automata moved from the top of the dials to the bottom, for which even larger hoods became necessary, adding to the monumental impression created by the clocks. Most longcase clocks were typically surmounted by a figure of Atlas on the elaborate hood, which had pierced sound frets, flanked by two trumpet-blowing angels. Some cases clearly show the influence of French furniture making.[157]

Early eighteenth-century longcase clocks in Scandinavia were closely modelled on English examples. Sometimes cases were imported from either England or the Netherlands. In Denmark, longcase clocks were made in Copenhagen, but also on the isle of Bornholm in the Baltic Sea. Tradition has it that the rise of the industry was the result of a Dutch ship carrying English longcase clocks running aground there. Soon affordable clocks based on the shipwrecked clocks were made on the island, particularly in the main town of Rönne, and these found their way to other parts of Denmark.[158] Painted cases (Figure 113) were popular in Scandinavia, especially in Denmark and Sweden, where by the middle of the century specially shaped clocks with painted cases and circular dials were made, notably in Mora, hence the term Mora clocks.

The situation in German-speaking countries was different. As mentioned, the first longcase clocks were modelled on either English or French examples and this trend can be seen well into the eighteenth century, resulting in austere cases side by side with extremely ornate examples. A special kind of clock was made in the area around Aachen, with a panelled solid oak case, often embellished by carvings. This type of clock was also seen in the areas around Liège, Luxembourg, and Alsace, and even in Switzerland. In the second half of the eighteenth century among the most striking German types of clock were those intended to be impressive pieces of furniture, some of them made by the cabinetmaker David Roentgen (1743–1807), who collaborated with the notable clockmaker Peter Kinzing (1745–1816). They became famous for producing *Flötenuhren* (organ clocks) and other clocks with complicated musical mechanisms, housed in magnificent mahogany-veneered cases, often embellished with marquetry.

In France, oak cases were also used, particularly to accommodate Comtoise clocks, although other wood was also used such as chestnut and cherry. These clocks, made in the Franche Comté, the area in the east of France bordering Switzerland, began to appear at the beginning of the eighteenth century.[159] Sturdy, iron-framed clocks were made in this area and were distributed by itinerant clock sellers. Though intended as wall clocks, they could also be installed in a case, which was usually made by a local joiner at the place of destination. Comtoise clocks have verge escapements with a long pendulum, consisting of pieces of wire and a lead, pear-shaped bob, the long verge pallets limiting the amplitude. The striking train usually indicates the hours fully, repeated after two minutes, and sounds the half hours by one stroke.[160] They were functional clocks, waking the household of a farm with their weight-driven alarms. Early varieties have pierced brass front frets and brass chapter rings, the time usually indicated by a single hand, while around 1750 full enamel dials and cast brass frontons begin to appear, initially depicting a cockerel looking over its shoulder, and later, after 1790, eagles or sunburst motifs. The cockerel often surmounted a medallion with three fleurs-de-lys, the royal emblem. During the Revolutionary period, these fleurs-de-lys were sometimes filed away to prevent the owner being accused of being a royalist.[161] Across the border in the Swiss Jura, there were also clockmaking centres, notably La Chaud-de-Fonds, Le Locle, and Neuchâtel. Though not famous for producing longcase clocks, some were made, particularly in the later years of the eighteenth century, by makers such as Pierre Jaquet-Droz (1721–90). The clocks are quite austere and show specific characteristics for the area, which are discussed in the section on table clocks.

---

157 The longcase collection of the Rijksmuseum Amsterdam shows the great variety of styles that followed, from Boulle marquetry cases (*c*.1720) to rich Rococo cases (*c*.1765).
158 Thorsten n.d.
159 They are sometimes referred to as Morbier or Morez clocks, after the clockmaking centres that produced them, or occasionally *Mayet* clocks after the Mayet family, prominent early makers.
160 The earliest clocks do not have this two-minute repeat striking system but sound the hours fully only on the hour.
161 For general surveys see Nemrava 1975; Schmidt 1983; Maitzner & Moreau 1985; Caudine 1992.

PRECISION LONGCASE CLOCKS OR REGULATORS

Huygens applied the pendulum to a clock in his quest for greater precision, but the desire for ever-increasing accuracy continued in the eighteenth century. The introduction of the dead-beat escapement was a notable technical advance. The anchor escapement, by then common, had one disadvantage: the recoil of the escape-wheel (and the rest of the train). This can be observed on most longcase clocks, with the seconds hand bouncing back after every beat. George Graham (1673–1751), who worked with Tompion for nearly two decades, is acknowledged as the first to implement successfully the dead-beat escapement, which eliminated the recoil. Although, a form of the escapement was initially developed around 1675 by Richard Towneley (1629–1707) working with Tompion,[162] Graham's version was applied in precision clocks, or regulators. These clocks were used for scientific purposes and were made in England by such makers as John Shelton (fl.1719–1777), Thomas Mudge (1715–1794), and his partner, William Dutton (c.1720–1794), both Graham apprentices. Regulators typically have dials, showing the minutes on the outer track and the seconds very clearly on a subsidiary chapter ring with the hours less prominently displayed, often in an aperture.

To enhance the accuracy of regulators they had compensation pendulums, which maintained the pendulum centre of mass at the same distance from the point of suspension independent of fluctuations in temperature. Several systems were developed,[163] but the most common was the gridiron pendulum, which was composed of brass and steel bars which extend or shrink in opposite directions. It was invented by John Harrison (1693–1776). The other compensation pendulum that became popular was one with a mercury-filled glass jar; the mercury in the glass rising when the pendulum rod extended. In France, regulators were also made but looked quite different, less technical, and more like normal longcase clocks. Robert Robin (1741–99) produced many, as did his contemporary, the Swiss-born clockmaker and prolific author Ferdinand Berthoud (1727–1807).[164] Among other subjects, he discussed the **equation of time**, which is of prime importance for astronomers, and was an immediate problem when the first pendulum clocks registered the difference. Purchasers wanted a clock that would 'go by the Sun'. Huygens was the first to produce a correction table. He was followed by clockmakers, the English ones using the tables of Flamsteed.[165] Occasionally, a copy of these tables can be seen on the inside of a longcase clock. Initially, clocks were made with a manual equation—a solar hand having to be adjusted continually to indicate the difference of the day. Towards the end of the seventeenth, the so-called 'equation kidney' was developed, which controlled a solar hand (often with a sun emblem) allowing both mean and solar time to be displayed on the chapter ring.

THE SPRING-DRIVEN TABLE CLOCK

Early spring-driven pendulum clocks were wall-mounted on a bracket or placed on furniture or a mantelpiece. Initially austere, the clocks rapidly developed into more ornate styles. In England, spring pendulum clocks were usually of eight-day duration, contrasting with the early Hague clocks, which only ran for a couple of days after winding. Another notable difference was that Hague clocks, like many early French clocks, had a pendulum suspended on a silk thread playing between cycloidal cheeks[166] and engaging with a crutch, whereas in English spring clocks the pendulum was firmly attached to the verge arbor with a knife-edge suspension at the back. As the cases developed, they became glazed on all sides. This meant that the backplate, which initially only bore the maker's signature, was visible and became a decorative site.[167]

The commonest case model to appear in the last quarter of the seventeenth century and the early years of the eighteenth century was the so-called domed top (see Figure 79), which was quite often adorned with pierced repoussé mounts, with or without turned (gilt) brass finials on the corners. Around 1680, so-called basket tops began to appear: instead of the wooden dome a pierced gilt brass basket-shaped (or sometimes even a double basket top) surmounted the case, giving the clock a richer look. In rare cases, the mounts and the basket were made of silver, as in examples by Tompion and Knibb. The clocks had a carrying handle so that they could be taken from one room to another, and since they were often equipped with a strike/silent device, bedrooms were not excluded. If the owner wanted to know the time in the night, he could activate the pull repeat, sounding the last hour struck and also, usually, the last quarter, struck on two or more bells, so that the time could be ascertained audibly to within a quarter hour. Another common feature of the table clock was a false pendulum aperture showing from the front that the clock was running. For convenience, once steel ribbon suspension was used in table clocks towards the end of the century, 'rise-and-fall' regulation was introduced—a subsidiary dial in one top corner allowing the clock to be regulated from the front. Symmetrically opposite there

---

162 Howse 1970; Howse 1971; Evans et al. 2013, 31 ff.; Turner 1972 documents Graham's introduction of the dead-beat anchor.
163 See Chapter 10.
164 His workshop is conserved in the Musée de Arts et Métiers, Paris.
165 Turner 2015b.

166 Huygens had theoretically established that the cycloidal arc of the pendulum bob resulted in better timekeeping than the circular arc. The cheeks in combination with the silk suspension realized such a trajectory. However, in practice it made no significant difference. Cf. the opinion of a contemporary, Raillard, cited in Chapter 26.
167 Although some later Dutch and Austrian table clocks followed the English model, engraved backplates were largely confined to England. For an in-depth study of the engraving on English clocks, see Dzik 2019.

Figure 79 Domed spring clock by Jonathan Lowndes, c.1695. Photo: Alexander George, Fine Antiques.

usually was a strike/silent dial in the other top corner, with engraving in between. This resulted in spring clocks growing taller and the dial becoming rectangular rather than square, with the area above the chapter ring often used for the maker's signature.

This was also the period of the increasing influence of Huguenot makers. Having fled religious oppression in the period following the revocation of the Edict of Nantes in 1685, protestant French clockmakers such as Simon De Charmes, Claudius Duchesne, and Nicholas Massey continued making high-quality clocks in England. With workmanship that displayed unique case designs, embellishments, and engraving styles, the Huguenots made their mark not only on English clockmaking, but also on horology in the Netherlands.

Around 1710 the first **inverted bell-top** case designs appeared in England, followed by **bell-tops** from c.1740. Over the eighteenth century, cases might be embellished by caryatids on the corners. More complicated clocks, which were usually larger and more imposing, had quarter striking, a musical train or automata, and (in rare cases) a combination of two or three of these properties.[168] Although mainstream clock cases remained quite austere, around 1760–70 a French decorative influence is noticeable in certain clocks, which have doors shaped in Rococo style with correspondingly rich gilt brass mounts.

While clocks for the English market were influenced by decorative trends from France and elsewhere, enterprising clockmakers took up the opportunities of Britain's growing world trade and adapted their clocks for overseas markets. Some, such as the Markwick Markham business (active c.1729–c.1742) and George Prior (1735–1814), concentrated on the Ottoman market and produced clocks with 'Turkish' numerals on the dial.[169] Clocks were also in demand in the Orient and Russia, where magnificent English clocks were popular. These were usually exquisite items with automata and/or musical mechanisms, housed in very richly executed cases, for instance, those produced by the entrepreneur James Cox (1723–1800).[170] English makers also settled in Russia. Although the clocks produced there looked English, they also displayed characteristics which were typically seen in Russian furniture. Examples include clocks made by Robert Hynam

---

168 There is a most impressive four-train table clock by Stephen Rimbault in the Ashmolean Museum, Oxford.

169 Kurz 1975, especially ch. 3; White 2012, chs. 3, 9, and appendices B, C, & F.

170 On whom see Smith 2000; White 2012, chs. 5–7. This volume Chapter 16.

(1737–1817), who served the Imperial Court in St Petersburg from c.1776.[171]

Meanwhile, in the Dutch Republic, the making of Hague clocks with their typical velvet-covered dials had spread to other cities. Over time, the cases developed into more mature forms than the early box-shape type and were given arched pediments, architectural tops, and—most commonly—broken arch pediments (see Figure 80).

The single-barrel driven movements made way for twin-barrel clocks of eight-day duration, but only in rare cases with fusees, in contrast to their English counterparts. The chapter ring had minute divisions, indicated by Arabic numerals (early French clocks followed this habit, too). Well-known makers include Pieter Visbach (c.1643–1722),[172] Severijn Oosterwijck (c.1637–c.92), and Johannes van Ceulen (c.1657–1715).[173] Hague clocks were made until c.1710, when they were superseded by English-style spring clocks, which, from the end of the seventeenth century, were also made in German-speaking countries, although the movements had characteristics, such as spring barrels screwed to the front plate, which belonged to German clockmaking traditions. Some cases were different from the English models and more reminiscent of late Renaissance altar clocks. A clock unique to the German lands was the so-called *Zappler* ('nervous ticker'). These clocks, often of only one day duration, had short front pendulums with a verge escapement, hence their name. The often richly shaped and sometimes pierced dial plate was made in relief and could be of gilt brass, silver, silvered copper, or brass. It hid the movement completely, which could be a timepiece, striking or even quarter striking, protected by a metal case. The earliest date from around 1720, and they were made throughout the eighteenth century.

When the first pendulum clocks arrived in Italy, they found fertile soil. During his stay at the court in Florence, Johann Treffler (1625–98), an immigrant clockmaker from Augsburg, worked with Vincenzo Viviani (1622–1703). Several clocks by Treffler's hand are known, mainly night clocks. The concept of the night clock had been developed by the brothers Campani (Giuseppi, Pietro Tomasso, and Matteo) in the 1650s.[174] They had dials with pierced numerals with a light behind the dial so that the time could be told in the dark. Because these clocks were meant for the bedroom, the Campani brothers also developed a silent escapement, a verge in combination with a crank mechanism and a pendulum. That news travelled fast in the late seventeenth century is once more demonstrated by the appearance of night clocks in England around 1665, some being made by Edward East and Joseph Knibb. Night clocks were also made elsewhere on the continent, for instance, in the Netherlands by Pieter Visbach. Others made in the second quarter of the eighteenth century in Switzerland were significantly smaller than, and quite different from, their Italian predecessors, the height being only some 30cm with a gilt brass tripod carrying the movement and a rotating dial with pierced numerals and half hour markers.

In France, the *pendule religieuse* developed along similar lines as the Hague clock. However, the cases rapidly became much more ornate than their Dutch counterparts, closely following the fashion in furniture. This resulted in tortoiseshell veneered examples with string inlays, either in pewter or in brass, as in Boulle-marquetry clocks, produced by makers such as Isaac Thuret (1630–1706), Jean Lemaire (c.1635–?), Antoine I Gaudron (c.1640–1714), Matthieu Marguerite (?–after 1702), and others. Early clocks were also often embellished by pierced cast brass pediments and/or mounts. While English spring clocks began to supersede Hague clocks in the Netherlands in the late seventeenth and early eighteenth centuries, table clocks in France developed virtually independently. The *tête-de-poupée* ('doll's head') is a good example—a French clock with a uniquely shaped case which

Figure 80 Hague clock by Pieter Visbach, c.1680. Author's collection.

---

171 For clockmaking in Russia and English makers there, see Chenakal 1972.
172 After Coster had suddenly died in 1659, Visbach was asked to take over the workshop. It was stipulated that Christiaan Reijnaert, Coster's apprentice at the time, would be able to stay on and finish his apprenticeship.
173 Plomp 1979.

174 Bedini 1983, Introduction.

appeared in the last quarter of the seventeenth century, often veneered with Boulle-type marquetry. As in other countries, clocks became larger, their shapes varying widely. During the Regency, a particular type of clock emerged: the *cartel*. This was a clock that was either placed on a surface or, if it had a matching bracket, could be mounted on the wall (see Figure 65). They were made throughout the century, though in later years the term was applied to any wall clock, even the *oeil-de-boeuf* (bull's eye). Early *cartels* have richly embellished, shaped wooden cases, which could be tortoiseshell veneered or, in later examples, covered in horn (*corne verte*, or 'green horn') or lacquered (*vernis Martin*). These clocks could also have cast brass or bronze cases, often in typical Rococo shapes with asymmetrical elements and flowing lines, always chased and fire-gilt. The movements had an almost standard construction, with count wheel striking and a short silk-suspended pendulum, often with regulation from the front with a watch key. Quarter-striking clocks in this style occur, too. This type of clock was also being produced in the Netherlands, Sweden, and Switzerland, where they are called *Neuchâteloise* after the main centre of production: Neuchâtel in the Swiss Jura.[175]

The chapter ring on French clocks could be solid, but frequently consisted of enamel cartouches for each hour numeral embedded in a brass dial cast in relief. In the first quarter of the eighteenth century a centre plaque was added, and sometimes plaques for the minutes, to produce thirteen or twenty-five piece dials. Around 1740, enamelling techniques developed to allow the making of single piece enamel dials. These came to be used on most clocks, which continued to become more ornate. The extravagance of the *style rocaille*, an alternative name for the Louis XV style, is evident on clock cases embellished by porcelain images often depicting a pastoral scene or an abundance of flowers. Movements could also be carried by cast bronze animals, horses, rhinos, camels, or elephants (Figure 140), while with the most exclusive clocks the animal was positioned on an elaborately adorned base containing a musical mechanism. In the Louis XVI period, mantel clocks often depicting symbolic or classical representations began to appear.[176] As France had become the leading nation in the field of furniture making, French styles were followed in many places across Europe. Hence, clocks based on French models were seen in the Netherlands, the German-speaking lands, Scandinavia, Italy, Russia, and other countries, either produced by local makers or imported with the name of a retailer on them. This meant that these types of clock were made in a great number of countries alongside English-style clocks.

Apart from stylistic changes, horological developments in the eighteenth century are characterized by improvements seeking greater precision. The development of the dead-beat escapement was a first step, followed by the compensation pendulum. In France, a first refinement was the adjustable twin-pallet escapement devised by Marie-Henri de Béthune, Chevalier de l'ordre de St Louis (c.1670–1744) before 1727. This design had been employed by London clockmakers, such as Joseph Knibb, in the late 1660s seeking an escapement to work with a long pendulum. Béthune's version was used and described by Thiout.[177] The pin-wheel escapement, invented by Louis Amant (?–before 1753), was later modified and improved by Jean-André Lepaute (1720–89). This escapement was often used in table regulators, which appeared in various forms in the later eighteenth century. To make it possible for the seconds to be indicated on the dial of clocks with ½-seconds pendulums, the *coup perdu* 'lost beat' escapement was developed. This allows the clock to show true seconds on a subsidiary seconds ring as the clock impulses on one pallet only, the other one being jointed and acting as a dummy. Table regulators could be housed in rectangular glazed gilt brass or austere mahogany cases or could be made as skeleton clocks. The latter show the quality craftsmanship and outstanding skills of leading makers like Antide Janvier (1751–1835), who often made these clocks with all sorts of complications, such as the equation of time or astronomical indications. Other makers include Robert Robin (1741–99) and Ferdinand Berthoud (1727–1807). Skeleton table regulators were also made by Hubert Sarton (1748–1828) and his followers in the area around Liège.[178] Most of the skeleton clocks produced in this period were spring-driven but there are also weight-driven table regulators, sometimes with a remontoir mechanism, such as that devised by Robin,[179] enhancing their accuracy. Another type of table clock developed in the Louis XVI period was the *cercles tournantes*—these globe- or vase-shaped clocks had a fixed hand with two horizontal rotating chapter rings, the top one indicating the minutes, the lower one the hours. Cases could also take the form of a temple, with impressive columns, or the movement might be carried by a statue representing the Three Graces.

At the end of the century there was a great interest in nature under the influence of the philosopher Jean-Jacques Rousseau (1712–78). This resulted in an idealization—the noble savage—a native living in the natural world without the influence of modern society.[180] In the field of horology, notably in the *Directoire* period, this interest was expressed in *pendules nègres*, the cases of

---

175 On French clocks, see Plomp 2017; Augarde 1996; Kjellberg 1997; Niehüser 1999; Tardy 1981 for the *Neuchâteloise*, Chapuis 1917; Chapuis 1931.
176 Niehüser, 1999.
177 Thiout 1741, 100–1.
178 Nève 2016.
179 It remained in use for some decades. For a description, see Le Normand 1830, i, 166–8.
180 For Rousseau's ideas see *Émile ou de L'éducation*, 1762, and *Du contrat social ...*, 1762. A further important influence was *Paul et Virginie*, Bernardin de Saint-Pierre 1789.

which were dominated by patinated cast bronze figures like native Americans with head dresses or figures of African origin.[181]

The French Revolution substantially changed French art and culture. Several clockmakers fled Paris, among them the Swiss-born Abraham-Louis Breguet (1747–1823), one of the most inventive and creative clockmakers of the time. When the political climate stabilized, he was able to return to Paris and continue his work.[182] Clockmakers who remained in Paris were required to make timepieces according to the newly instituted revolutionary or decimal time reckoning system.[183] This system, which was extremely short-lived,[184] resulted in clocks with decimal dials, sometimes with the old and the new system side by side.

TRAVELLING CLOCKS

Travelling clocks can be regarded as a subcategory of table clocks, though there are also travelling wall clocks, for instance, small lantern clocks in wooden travelling cases and so-called sedan clocks, small wall clocks with a watch movement usually of thirty-hours' duration. The first travelling table clocks after the pendulum clock appeared in France. They were facilitated by the application of springs to the balance—a subject independently investigated by Robert Hooke (1635–1703)[185] and by Huygens. The latter's design was the more practical and was quickly adopted throughout Europe. One of the earliest travelling clock designs were miniature *têtes-de-poupées*, which had a verge escapement with both a balance wheel in combination with a coiled balance spring as well as a pendulum. Thuret, for example, made such clocks in the last quarter of the seventeenth century, as did Mathieu Marguerite, Pierre Margotin (1626?–c.1700), and others.[186] In England, spring clocks with a wooden travelling case are known, as are some examples of dual control clocks by Tompion.[187]

Travelling clocks appeared in greater numbers in various countries, notably France, Switzerland, and Austria,[188] in the second half of the eighteenth century. These Louis XVI-style clocks usually have a day-going movement housed in an ornate cast, gilt brass case, surmounted by a carrying handle, often in the shape of an *ouroboros*, a snake eating its own tail, symbol of eternity. Except in miniature models, they almost invariably have an alarm and a striking train that can be set to *grande sonnerie*, *petite sonnerie*, and *silence* with trip repeat. In France, this type of clock was rivalled by the *capucine*. The cases of these officer's clocks are made of brass plate and are surmounted by a bell with carrying handle—initially with a tear-drop shaped handle and, later with a handle shaped in the form of a flat oval. They often have the double Comtoise-type of striking, the hour strike being repeated after two minutes.[189]

In England, some exquisite travelling clocks were made by Thomas Mudge, inventor of the **lever escapement**, which would become the standard escapement for carriage clocks in the nineteenth century. A famous example, now in the British Museum, is the Polwarth clock, a costly drum-shaped clock with a wooden travelling case. Finally, the Russian maker Ivan Petrovich Kulibin (1735–1818), active in St Petersburg, and later in Nizhny Novgorod, where he was born, made an egg-shaped pierced and chased striking musical automaton clock. This unique object was presented to Catherine the Great in 1769, playing a cantata at noon, which the maker had composed himself. Upon opening the egg, an automaton is revealed enacting the resurrection.[190] This egg, which is now in the Hermitage State museum, predates the famous Fabergé eggs by some 130 years.

THE WALL CLOCK

As mentioned, the most common pre-pendulum clock was the lantern clock, a nineteenth-century name for a domestic or chamber clock made of brass and steel, which first appeared at the beginning of the seventeenth century and which was produced well into the eighteenth century, not only in England, but also in France, Italy, and the Netherlands.[191] Early lantern clocks almost invariably had a single hand only. Around the time of the introduction of the pendulum, the typical English lantern clock had an engraved foliate front fret depicting two 'dolphins' (fish-like sea creatures would be a better description) with twisted tails, although frets with a coat of arms or an arched gallery also occur. Their duration is about twelve hours. This all changed after lantern clocks were made with a verge escapement in combination with a pendulum, driven by a single weight via a Huygens endless rope or chain, which extended their duration. There were also, but less commonly, lantern clocks with a seconds pendulum and anchor escapement. In the last quarter of the seventeenth century lantern clocks were made with 'wings' in the doors and a central anchor-shaped pendulum visible when swinging. Miniature lantern clocks were also made and could be alarm timepieces, although some are also striking clocks. These smaller versions of lantern clocks were also made in France, both in Paris and in the

---

181 Fine examples are housed in the François Duesberg Museum in Mons, Belgium.
182 Chapuis 1953; Daniels 1975.
183 It was introduced on 5 October 1793. See this volume Chapter 12, and Shaw 2011.
184 Its mandatory use was ended on 18 Germinal of Year III (7 April 1795).
185 Hooke's work on balance springs occurred in the context of his attempt to produce a marine timekeeper. See Wright 1989.
186 Vehmeyer 2004, ii,836–7.
187 Evans, Carter, & Wright 2013, *s.v.* index.
188 For these, see Fritsch 2010.

---

189 Allix & Bonnert 1974; Roberts 1993.
190 See Dallas 2020; Chenakal 1972, 34–5.
191 White 1989; Loomes 2008.

Figure 81 Two lantern clocks side by side. Left: An English lantern clock by John Ebsworth London, c.1660; Right: A French lantern clock by *Normand à Paris*, c.1730. Left: Photo of Mario Crijns, Oosterhout. Right: Photo of Peter Zwaanenburg, Dordrecht.

provinces, notably Normandy, Brittany, and the Lyonnais.[192] In appearance they were very much like the ones made in England, although the frets were always different, with the front one usually depicting an engraved satyr head (Figure 81).

Some French lantern clocks made in the provinces look quite different from the 'standard' model and could have a cast brass dial with enamel cartouches or a full enamel dial on later examples. In Italy lantern clocks were made throughout the country, sometimes with six-hour dials and six-hour striking. Early lantern clocks had narrow chapter rings, which become wider on later models. By the beginning of the eighteenth century, they were extending beyond the case to such an extent that they nearly covered the whole of the front ('sheep's head clocks'). Soon after, lantern clocks were produced with full arched dials, some of them with a wooden case and referred to as hooded clocks.

Although lantern clocks were also made in the Low Countries, the more common wall clock was the so-called *stoel* clock, made in the country away from the larger trading cities. Initially they had balance-controlled escapements, of which only a few survive, but they continued to be made after the pendulum was introduced in, for instance, the north of the province of Holland (Zaandam clocks), in Friesland, and in the east of the country, each with their own specific shape. This type of clock consisted of a solid, heavy, shaped backboard with brackets, on which the metal-cased short-duration movement sat (hence the name, *stoel*, meaning seat or chair in Dutch), driven by a single weight. The free-hanging pendulum, suspended from a metal knife-edged hook in the backboard, was impulsed by a driver, a stiff metal wire attached to the vertical verge, the shaped pendulum bob showing on each side of the movement housing. As an example of the regional variation, the Zaandam clock had a hollow backboard in which the pendulum was accommodated, and the pendulum bob (often in the shape of a rider on horseback) appeared with each swing in a shaped window. Hooded clocks were also made in the Low Countries in the second half of the eighteenth century. These grew into the *staartklok* ('tail clock'). This was a kind of drop dial with a long trunk (the tail) accommodating the pendulum and was a less expensive version of the Amsterdam longcase clock, usually with thirty-hour movements with Dutch striking. The tail clock became extremely popular later in the century, notably in Friesland. They usually had oak

---

192 Bollen 1978; Lerouxel 1981; Sabrier 1988; Vehmeyer 2004, ii, 838–43.

Figure 82 *Cartel d'alcove* by Charles Baltazar Paris, *c.*1750. Photo: Sunny Dzik, Boston, MA.

cases with arched hoods surmounted by three gilt wooden finials and painted dials.

In the German-speaking lands there was also a tradition of making wall clocks, notably clocks with movements largely made of wood. The Black Forest was probably the most famous area for this type of clock. Early ones are balance or foliot-controlled and as soon as the pendulum appeared, it was incorporated into their construction. They had painted wooden dials and were weight driven. Some clocks even had a musical train, playing their tunes on glass bells.[193] In the middle of the eighteenth century, the first cuckoo clocks made their appearance. Wooden wall clocks were known in Switzerland as early as the second half of the seventeenth century. Clocks developed in quite different ways in different areas, in the often-isolated communities of these mountainous areas. In the area around Winterthur, generations of the Liechti family made Gothic iron clocks, continuing until the last maker of the family, Hans Jakob, died in 1741. In the area around Bern, wall clocks were made with austere cases, whilst in the Jura, all sorts of clocks were produced, the earlier Neuchâtel clocks being severe ebonized clocks, while later, more ornate clocks appeared, gradually leading up to elegantly waisted clocks by the end of the century, which continue to be made up to the present day.

A wall clock related to the *Zappler* made its first appearance at the end of the seventeenth century: the *Telleruhr* ('dish clock'). This clock, which also often had a front pendulum, was suspended from a support by a suspension eye above the dish, which was usually circular, later shaped, and richly embellished. Clocks with repoussé dishes became the most common. *Telleruhr* clocks are in sharp contrast to the large English wooden wall clock—the tavern clock. These clocks date from the first quarter of the eighteenth century and were housed in austere cases that were often japanned against a black background. They were designed as timepieces and developed into circular wall clocks referred to as dial clocks, which continued to be made well into the twentieth century.[194] Early ones have fusee verge escapements and silvered dials, sometimes with a false pendulum aperture, and are of eight-day duration. The plain cases are usually made of mahogany, with a so-called salt box behind the dial containing the movement. The French counterpart at this time was the *cartel* clock, by now a richly shaped ormolu brass wall clock, initially in Louis XV style. They could be striking clocks or timepieces. The latter type was produced in a special bedroom version, the *cartel d'alcove* (Figure 82), which had pull quarter repeat on one or two bells, so that an approximation of the time could be ascertained in the dark.

Louis XVI *cartels* are more austere and symmetrical in shape. In later years, round versions with bronzed or gilt brass cases appeared, with ormolu floral or leaf embellishments and pearl strings, sometimes referred to as *oeil-de-boeuf*. *Cartel* clocks, often with gilt wooden cases, were made in the Netherlands, Germany, Sweden, and even England. Finally, a special eighteenth-century clock made in Germany and France should be mentioned: the rack clock. This could be a table clock as well as a wall clock, the rack being mounted on a board. The weight of the clock itself, often increased by an internal piece of lead, supplies its motive force. The clock runs down the rack which engages with the movement. Some of these clocks worked in reverse: they were spring-driven and would climb up the rack, being wound by pushing it back down again.

CONCLUSION

Having reached the end of the eighteenth century, a quite different picture emerges from that seen at the beginning of this chapter. Starting with the pendulum, scientific research had required and delivered increasingly accurate clocks. Meanwhile, the rise of the bourgeoisie, combined with new production techniques and methods, led to the clock becoming more widespread, and as a result, timekeeping was no longer only the preserve of the very wealthy. Consequently, a wide variety of styles developed, the most important of which have been discussed in this chapter. With the Industrial Revolution, around 1800, time had become an ever-more important feature of daily life, which was soon to touch every European household.

---

193 Schaaf 1988, pp. 99–101.

194 All thoroughly discussed in Rose 1994. See also Gatto 2010.

CHAPTER SIX

WATCHES 1500–1800

David Thompson

Watches are portable timekeepers that can be worn or carried on the person. They were originally carried in a pouch or purse, suspended from the neck or a belt, and later carried in a pocket in clothing. In essence, watches are directly related to clocks. Each has a driving force, usually in the form of a coiled spring contained in a barrel to restrict its expansion, with a series of gear wheels to transfer the force of the **mainspring** to the **escapement**. Then, by means of an oscillating device, the rate at which the watch runs can be controlled. In early watches, this took the form of a bar-balance, later a circular **balance**. A dial indicated the time, initially with a single hour hand, later with concentric hour and minute hands, and finally with a seconds hand. Complications such as striking mechanisms, alarms, calendar dials, and, later, repeating functions were also added. These movements were contained in cases ranging from a simple plain case to elaborately decorated examples according to the purpose of the watch and/or the owner's taste. Watches from their beginnings until today were functional means of telling the time, symbols of wealth and status, or both.

The development of the watch is closely related to the changing fortunes of different countries in Europe over the centuries, the prosperity and security of which ebbed and flowed with religious persecution, war, and economic fluctuations. It is against this background that watches evolved from their origins as inaccurate and unreliable timekeepers at the beginning of the sixteenth century to the precision machines made by eminent and skilled makers at the end of the eighteenth century.

ORIGINS OF THE WATCH

The watch could not have been developed without the introduction of a coiled spring as the driving force in clocks in the mid-fifteenth century. There are two surviving examples of clocks from this period that are known without doubt to have had spring-driven mechanisms[1] From such early developments it took very little time for them to proliferate in the clockmaking centres of Europe: northern Italy, southern Germany, France, and Flanders. By the latter part of the fifteenth century such clocks had become relatively common.

Exactly where and when the watch originated, in northern Italy or southern Germany, remains controversial,[2] but a probable source is the small drum-shaped horizontal table clock. Miniaturizing the machine to reduce weight, adding a cover for the front to protect the hand from disturbance, and adding a loop so that the device could be suspended brought the truly portable device now called a watch into being.

It then developed in specific, often differing ways, across Europe, and while there was cross-fertilization in ideas, methods of construction, and styles, considering its developments in the context of different geographic areas brings both clarity and distinctiveness.

The German States

One of the earliest references to the existence of a watch comes from Johannes Cocleus' *Cosmographia Pomponii Melae* of 1512, where it describes a young man named Peter Hele[3] who made objects that astonished the most learned mathematicians; 'out of very little iron he assembled timepieces with many wheels, which, without any weights and in any position indicated and ran for

---

[1] See this volume, Chapter 5.1.
[2] Morpurgo 1954; Defossez 1956, 175–80; Oestmann 2018.
[3] Peter Hele or Henlein of Nuremberg has often been described as the inventor of the watch. While early references to him making watches show that he was an early exponent of the art, there is no evidence to suggest that he actually made the first watch.

forty hours even if contained in a purse or pouch'.[4] In Nuremberg archives from 1524 there is a record of a payment of 50 florins to a Mr Henlein for a gilt musk-ball with a timepiece. A slightly later reference is an account written in Nuremberg by Johann Neudorfer, published in 1546, with mention of Andreas Henlein 'one of the first who has contrived these small timepieces in musk-balls'.[5] Of the watches that survive from this period, the earliest datable example is a spherical-cased watch, made in 1530 for Philip Melanchthon.[6] Some other spherical-cased watches survive, in particular a small unsigned watch now in the Ashmolean Museum in Oxford.[7] An early characteristic of the watch is that the hinged halves of the case are closed by a latch. From the number of surviving watches made in the second quarter of the sixteenth century, it becomes clear that the most prolific area of production was South Germany, in particular Nuremberg, Augsburg, and Munich, and that the most common form was the drum-shaped or tambour-cased watch derived from the small, horizontal spring-driven table clocks of the period.[8] Extant numbers are considerable and, given the fact that the survival rate must be very small, it is clear that they were probably produced in thousands. One early dated example, albeit with a questionable dial and hand, has the punched inscription CW 1548 inside the case and is attributed to the Nuremberg maker Caspar Werner.[9]

These watches were highly treasured objects owned by few as items of status worn at banquets and other court, family, and social occasions, their function being primarily decorative. As such, watches were frequently used as accoutrements in portraiture, as a sign of wealth, or to show that the owner, in possessing such a machine, was a person of learning. They may also offer a metaphorical allusion to the fleeting nature of life. A portrait of Viet Konrad Schwartz from 1558 shows the typical fashion for wearing the watch by suspending it on a cord, ribbon, or chain around the neck, a prominent place for a fine possession.[10] The fact that the watch at this time was worn more to impress the onlooker than to provide the owner with the time might account for the fact that, with only one or two exceptions, the twelve numeral is at the top as seen by the onlooker. This arrangement was impractical for the owner, wearing the watch, to see the time, since when lifted up the indications would be upside down.

The German Watch Movement

In the early period all German watch movements were made from a crude form of ferrous alloy, commonly referred to as iron. It has a degree of carbon content, closely allied to the metal used by armourers, locksmiths, and blacksmiths, but it is not steel in the modern sense.[11] Mainsprings in the early watches were not contained in a barrel but were restricted in their movement by studs riveted into the movement plate. The pillars separating the plates were normally riveted to the back rather than the front plate, the practice in other watchmaking countries. The simple gear train of five wheels[12] transferred the motion to the **escapement**, which was, without exception, a **verge** escapement controlled by an oscillating balance either in bar form, in the early period, or of wheel form later in the sixteenth century. To even out the unequal torque produced by the mainspring as it unwound, a device known as a **stackfreed** was used.

The main cause of inaccuracy in early watches was the great variation in the power output of the **mainspring** as it unwound to drive the watch. A coiled spring produces its greatest energy when fully wound, and its energy output lessens progressively as the spring uncoils. Because the oscillating balance in the watch has no natural period, its rate of oscillation is determined by its mass and mean radius (its **moment of inertia**) and the amount of energy imparted to it by the escapement. This means that, without some means of evening out the force exerted by the mainspring via the train of gears and the escapement, the watch runs fast when the mainspring is fully wound, and progressively more slowly as it unwinds. To ensure reasonable accuracy, the German watchmakers introduced the **stackfreed**.[13] This device was almost exclusively German apart from a small number made by craftsmen on the borders of German states (Figure 83).

The geared stackfreed, which is mounted on the outside of the movement frame, allowed makers to use a four-wheel gear train, which enabled watches to run for more than a day at a single winding. For fine adjustment of the going rate of a watch, German makers favoured a device known as a **'hog's bristle regulator'**.[14] Towards the end of the sixteenth century this regulator was often geared and operated by a key.

---

4 Translation from Tait & Coole 1987, 10.
5 Peter Henlein and his early watches are discussed in Eser 2014, and Matthes 2015, 183–94, although the item in the British Museum (1888, 1201.105) is clearly a table clock. It has no ring for suspension, there is no cover to protect the dial, and its size would make it unwearable.
6 Walters Art Gallery, Baltimore. See Gahtan and Thomas, 2001, 377–88.
7 Ashmolean Museum (WA1947.191.1)
8 For which see Matthes 2018.
9 Wuppertaler Uhrenmuseum.
10 There is no doubt that the item being worn is a watch as the text on the portrait clearly states, 'so I have a small striking watch hanging from the neck, the gift of my dear father in 1557, while I was then in Venice'. *Das Trachtenbuch des Viet Konrad Schwartz*, (II, 27, June 1558) Herzog-Anton- Ulrich-Museum, Brunswick.

11 On which see Wayman 2000.
12 This consists of a great wheel, second wheel, third wheel, contrate wheel, and escape wheel.
13 See Glossary and discussion in Chapter 5, Section 1.
14 See Glossary and discussion in Chapter 5, Section 1.

Figure 83 Stackfreed watch movement, maker 'C K', Southern Germany, c.1575. Photo: Courtesy of Sotheby's, London.

The German WatchCase

The earliest surviving examples show hinged covers and backs closed using latches, with the hinge at three o'clock and the loop at twelve. A typical example is a watchcase in the British Museum dating from c.1540,[15] where the case is also furnished with a **volvelle** showing the age and phase of the Moon and a sundial designed for use at latitude c.45° north. By about 1560, the system had changed to a friction-fit back and a cover with the hinge at twelve o'clock. The components of the case were of cast brass, which was pierced, chased, and fire-gilt.[16]

During the second half of the sixteenth century German watch shapes moved away from a rigid tambour case towards a more rounded form, with domed cover and back and with a *bombé* form for the case band. In non-striking watches the back would be solid cast and the cover would be pierced to allow the position of the hand to be seen. **Striking** and alarm watches, however, would have a pierced back to allow the sound of the bell to escape. The German dial usually had a decorated centre within the integral chapter ring, and the hours and half-hours would be marked with an outer circle I–XII in Roman numerals and an inner circle of Arabic numerals 13–24. The single hand might also indicate against a further circle calibrated to quarter-hour marks.

Throughout the sixteenth century, case components were cast, fire-gilt, and commonly decorated with arabesque and foliate motifs or hunting scenes in mannerist style. Evidence from surviving watches strongly suggests that in South German centres such as Augsburg, Munich, and Nuremberg there were craftsmen who specialized in case components supplying different makers. According to town guild regulations, the master watchmaker had to mark his work, and this was normally a shield containing his initials and occasionally a town or city mark—a pine cone for Augsburg, a right-facing monk's head for Munich, and crossed arrows for Nuremberg—although these town marks are uncommon. Towards the end of the sixteenth century and into the seventeenth century, practice changed to the maker writing his name in full and the place where the watch was made. So a watch made by Hans Koch in Munich c.1580 has a shield with his initials HK and the town mark of Munich,[17] whereas a watch by

---

15 Tait 1987, cat. no.2.
16 Fire, or mercury, gilding involved cleaning the metal to be gilt, mixing gold powder with mercury, painting the metal with the resultant paste, and then burning the mercury off as a vapour to leave the gold layer permanently on the base metal below. In watches this was nearly always brass, which performed well in the process.

---

17 Ashmolean Museum (WA1947.191.7). See Thompson 2007, 8–11.

Niklaus Rugendas, Augsburg, in the early seventeenth century has his name and, sometimes, the place written out in full.[18]

By the 1580s a change took place, and watches began to appear with gilt-brass front and back plates for the movement, but still with steel wheels. However, by the early seventeenth century, movement wheels were also being made from brass, although springs and other components of back plate furniture were made from steel. From the outset, German striking alarm watches were round and for the most part remained so. In the early seventeenth century, however, smaller oval watches appeared. Examples that struck the hours had a small oval bell secured in the case. The Thirty Years' War, however, had a devastating effect on German watchmaking, causing its eclipse as an international producer for at least half a century.

France

As in Germany, watchmaking in France began in the early sixteenth century. Among the earliest references is an order dated 31 December 1518 by Francis I for Jean Sapin to pay Julien Coudray in Blois the sum of 200 *écus* for two daggers with 'horloges' in the pommels.[19] Following this early reference, watches are often mentioned in inventories, although production was very small, and nothing has survived from before the mid-century (Figure 152). The trade in watches was set back considerably by the wars of religion from the 1560s onwards. The St Bartholomew's Day Massacre on 23–24 August 1572 was followed by an intense persecution of Huguenots that led many, including watchmakers and watchcase makers, to emigrate. From Paris and the towns in western and northern France they usually went to London where they tended to congregate in Blackfriars, just outside the jurisdiction of the city livery companies. From cities more to the east, e.g. Lyon, they went to Geneva where they made an important contribution to early watchmaking in Switzerland. The Edict of Nantes in 1598, however, restored freedom to worship and trade to the Huguenots. In the first half of the seventeenth century production of watches was high, with watchmakers in nearly every major town and city in the country, even if production was concentrated in Paris, Blois, Rouen, and Lyon. During the second half of the seventeenth century the watch trade remained both local and national. A watch attributable to Noël I Cusin of Autun[20] c.1585 for example has a plain gilt brass case, originally with a full metal lid, but at some later date this was opened up and glazed.[21] The Revocation of the Edict of Nantes in 1685 caused a further considerable exodus of watchmakers from the traditional centres of the craft.[22]

From the later sixteenth century, Blois was among the leading centres of the art. Concentrated around the royal court, still frequently present in the city, and later around that of Gaston d'Orléans (1608–60), goldsmiths, silversmiths, and engravers supplied cases for the movement-makers to produce high-quality watches for a wide range of customers.[23] The Cuper family in particular had a long history. The founder of this dynasty was Paul I Cuper who, born in Liège, was established in Blois by 1553. Thence, the family descends in unbroken succession until the 1840s. The Gribelin family was also significant in Blois. The founder of the dynasty, Simon Gribelin, was working in Blois from 1588 until his death in 1633. Probably the best known is Abraham (1589–1671), son of Simon Gribelin. He was established as a watchmaker in his own right in 1614 and succeeded his father as Clockmaker to the King. He fathered fifteen children between 1620 and 1638 with his wife Judith Festeau. Numerous surviving watches bear his signature; all are of very good quality.[24] A particularly fine example with a calendar dial and a silver case shows Orpheus charming the animals.[25] Among Abraham's sons were Isaac, a watchcase enameller, Jacob, an engraver, and Nicolas a clock and watchmaker. In the seventeenth century many other watchmakers, designers, and engravers excelled.[26]

From the beginning, French watches differed fundamentally from the German model. In France, a **fusee** was used to equalize the power output from the mainspring, and it was connected to the mainspring barrel using a gut line, so the watches had a short duration of about fifteen hours, determined by the weakness of the gut line, which allowed only for a four-wheel gear train. For regulation, a ratchet and click system was generally used. Nonetheless, French watch movements during the first half of the seventeenth century did not differ greatly from those found in other watch-producing countries. One significant difference is that, generally speaking, the movement and dial were not hinged to the case but were held in by two catches at three and nine o'clock with a common U-shaped spring under the dial.

Decoration using champlevé enamel dates back in the Limoges region to the twelfth century, to which painting on enamel was added in the later fifteenth century.[27] However, this does not appear on watchcases until the early seventeenth century and then

---

18 Tait & Coole 1987, 68
19 Archives Nationales de Paris, KK289, fol.444.
20 Noël Cusin lived and worked in Autun about 80km from Dijon, between 1539 and 1585. There in 1559 he is recorded as keeper of the cathedral clock; Charmasse 1888. His family played a notable role in the establishment of watchmaking in Geneva.
21 See Thompson 2008, 18–19.

22 For detailed accounts of watchmaking in France in the sixteenth and seventeenth centuries, see Cardinal 1989, Chapiro 1991; for biographies of makers, Tardy 1972.
23 Detailed studies are Develle 1917; Fourrier 2000, Fourrier 2001–2005.
24 Over ten Gribelin watches are recorded.
25 Ashmolean Museum (WA1947.191.31). See Thompson 2007, 20–3.
26 For watchcase decoration, see this volume Chapter 15.
27 See Speel 2008, ch. 2, for an account of Limoges enamelling and ch. 10 for the nineteenth-century revival of the art and a discussion of imitations.

probably with little success since only a small group have survived and they are mainly fitted with later movements.[28] Some may be nineteenth-century amalgamations.

By the 1630s a different form of enamelling began to appear in which the body of the case was enamelled in a plain base colour and then overlaid with a pierced gold frame which itself was enamelled in translucent colours. A fine example of this work, if somewhat damaged, exists in a watch in the British Museum containing a movement by Louis Vautier, Blois c.1635.[29] It is without doubt that during this period of the 1630s Blois, Chateaudun, and Paris were the leading exponents of the art of enamelling watchcases and took it to a high level of sophistication. The art of painting on enamel was developed by Jean I Toutin (1578–1644) and his sons Henri and Jean. Together with Isaac Gribelin in Blois they perfected better metal oxide colours. Their techniques were soon adopted by others, and a number of workshops in both Blois and Paris supplied an ever-growing market for the finest enamel watchcases and dials (see Figure 154 and Figure 155). By the middle of the century, the French pictorial enamel watchcase represented perhaps the highest level of luxury. The only cases that would cost more would be those with pictorial enamel and precious stones, of which a number survive. One spectacular watch, contained in what is unquestionably a French case, is by the London maker David Bouguet,[30] a Huguenot who left Paris for London in the early part of the seventeenth century. The watch has a standard movement for c.1650 when it was made, but the gold case has a stunning array of realistic flowers in high-relief colours over a body of black enamel. As if that were not enough, the cover is adorned with ninety-two diamonds. The dial and counter-enamel are painted with pastoral scenes. Clearly Bouguet did not sever his links with the case makers of France and ordered this special commission to which he added a movement before selling the watch to a very wealthy customer.[31]

Pictorial polychrome enamel cases were extensively used by Jacques Goullons in mid-seventeenth-century Paris.[32] A typical example in the Metropolitan Museum of Art in New York has a somewhat old-fashioned movement with ratchet and click set-up for the mainspring, rather than the more modern tangent screw system. However, the attraction of the watch lies not in its movement but in the superbly enamelled case. It depicts the Virgin and Child visited by an angel on the case back and Joseph awakened by an angel on the cover, both painted following engravings by Pierre Daret (1605–78) and Michel Dorigny (1617–65), after the original paintings by Vouet.[33]

Lyon was another flourishing centre of watchmaking.[34] An outstanding watch combines a movement by Jean Vallier with an oval case beautifully engraved with depictions of Apollo and the Nine Muses. The calendar dial is of the highest quality with fine engraved detail around the various indications. There are depictions of Diana holding a crescent moon (The Moon) and Apollo with his Lyre (The Sun). At the top, a small rectangular aperture shows the date, above a subsidiary dial indicating the age and phase of the Moon. To the left, another dial shows the days of the week, with the ruling deity of each day shown in a sector. On the right, an engraved silver chapter ring indicates the quarters. The bottom dial is for the hours with an alarm-setting disc at its centre. Two sectors in the upper right-hand area show the seasons and the months, with the number of days in each. Since the dial inscriptions are in Italian it was presumably commissioned by an Italian customer. The finely made movement has gear trains for the time, hour striking, and an alarm, making it a very sophisticated watch for its time.[35] Lyon makers had good trade contacts with Geneva, ordering watchcases, among other items, from there. Thus, in 1623, Pierre Louteau in Lyon ordered three rock-crystal cases for watches from the lapidary Paul Tillier.[36] Alongside the domestic market in France, there were also export opportunities. An example is the calendar watch made in 1613 by Pierre Combret in Lyon and acquired by Alethea, Countess of Arundel.[37]

The Netherlands and Flanders

Little is known of the origins of watchmaking in the Netherlands and Flanders, and no early work has survived. Two watches from the second half of the sixteenth century, however, are important. Both show German influence in their construction, but at the same time come from a different tradition, taking some technical aspects from the small horizontal table clocks being made in the region at the time. One, by a maker with the initials 'WA',[38] is engraved on the dial with an armorial shield and the date 1571. The hybrid nature of the watch is demonstrated by the fact that it has a fusee, rather than a stackfreed, but employs a hog's bristle regulator. The watch is also early in using gilt brass for the movement plates, albeit with iron wheels and escapement. It seems

---

28 Discussed by Vincent 2007.

29 For Vautier, see Develle 1917, 345. For the British Museum (1892,0523.1), see Thompson 2008, 36–7.

30 The name of this maker is also quoted as Bouquet in many sources. While the watches show a signature suggesting Bouguet, two clocks, one in the Victoria & Albert Museum and the other in the Science Museum, London, show the name as Bouquet.

31 https://www.bmimages.com/preview.asp?image=00035777001.

32 Cardinal 2018a; 2018b, 185–201. For enamelling techniques, see Speel 2008.

33 Vincent, Leopold, & Sullivan 2015, 66–69; Cardinal 2018a; 2018b, 196, Corpus 2

34 See Vial & Côte 1927.

35 British Museum (1888.12-1177), Vial and Côte 1927, 118 Thompson, 2008, 34–5. Another similar watch, this time mounted on a statue of Atlas, is in the Rijksmuseum, Amsterdam (BK-17,012).

36 Jaquet & Chapuis 1970, 22.

37 Illustrated and described in Williamson 1912, 24–6; Vial and Côte 1927, 105.

38 Vincent, Leopold, & Sullivan 2015, 26–31; https://www.metmuseum.org/art/collection/search/194139.

1607 watch is also a very early example of one with a glazed cover to reveal the dial, a practice not normally found until the 1640s, except for watches with rock-crystal covers; in those, the intention was not to make the dial visible, but to add to the opulence of the watch.

There were clearly only a small number of watchmakers in Flanders and the southern Netherlands in the last quarter of the sixteenth century.[40] It is also perplexing to find that there are apparently no examples of watches by Ghylis van Gheele, Michael Nouwen (*fl.* 1582), or his relative Francis Nouwen (*fl.* 1580–93), made in their native country before they emigrated to London and began making watches there in the 1580s.[41] It seems however, that they arrived already trained in the skills needed to make watches and it comes as no surprise that those they made were in the style of their homeland as at that time there was no existing 'English' style.

Despite the emigration of some leading watchmakers, particularly to London, watchmaking continued to thrive in the Netherlands even during war and occupation. Stylistically they parallel closely what could be seen in London, which might be expected given the origins of the makers there in the late sixteenth century. By the early years of the seventeenth century timepiece watches had become smaller and were either of an oval or elongated octagonal shape. Decoration concentrated on biblical, mythological, or historical scenes. From watches surviving from the period in the Netherlands, France, and England it is clear that the wearing of a watch to demonstrate religious sentiments was popular. A good example dating from *c.*1600 is a watch by Wybe Wybrandi of Leeuwarden, finely engraved with biblical scenes. On the inside of the lid is a blank shield with a crowned helmet and, around the outside, a table of months with the number of days in each.[42]

From the early seventeenth century the tulip became a celebrated flower in Europe, with a massive market in the Netherlands. This 'tulip mania' spilled over into clockmaking and watchmaking. In a new fashion for **'form watches'** the tulip was the commonest of those made in the shape of flowers. Dutch examples, however, are strangely rare, when compared to the numbers made in Geneva. It is not astonishing that when watches in flower form became fashionable, they soon gained popularity in London. What is difficult to establish is which location had precedence in setting the fashion, although when it comes to exotic flowers such as the snake's head fritillary, it is tempting to say that the fashion must have begun in the Netherlands. Soon, tulip-form

Figure 84 A calendar watch by Jan Janssen Bockelts the Elder, inv. WA1947.191.13. Photo: © Ashmolean Museum, University of Oxford.

partly Flemish or Dutch in character, partly Germanic. Vincent and Leopold firmly establish an Antwerp origin for this watch with some of the evidence coming from the engraved decoration the case. It is an early example of an oval-cased watch when compared with the drum-shaped watches still being made in Augsburg and Munich at the time.

The second watch, also in the Metropolitan Museum of Art, is by a known maker, Jan Janssen Bockelts the elder (*fl.* 1590–1626).[39] This very fine calendar-dial watch with sundial can be dated to about 1605 and is unusual in that the case decoration is also signed 'IAN IANSEN BOCKELS INVE ET SCVLP', showing that Bockelts was not only a watchmaker but also an artist engraver. It is rare that cases were signed by their makers, but here Bockelts shows himself to be multi-skilled. The watch compares very closely with another now in the Ashmolean Museum (Figure 84), signed *Jan Janss Bockelts* with an inscription dedicated to Abraham Ampe and the date 1607. By this time Bockelts had moved from Aachen to live in Harlem where he was a friend of Ampe, who was 21 years old in 1607.

Bockelts has here a similar concept for his watch in that it has both a fusee and a hog's bristle regulator for the balance. The

---

39 Vincent, Leopold, & Sullivan 2015, 44–8.

40 For makers in the northern Netherlands, see Morpurgo, 1970; for the southern Netherlands, see Fraiture 2009.
41 The numerous variants of the name (Nawe, Nauwe, Noway, Neuvers, Nowe) are rationalized here to that which appears on watches by Michael. Fraiture 2009, 533 & 535 however, enters them under Nawe and Nowan. For a clock watch by Michael made for the 7th Earl of Shrewsbury in 1599, see Beeson 1965.
42 Ashmolean Museum (WA 1974.214); Morpurgo 1970, 144.

watches were being made in Geneva, London, and to a lesser extent Germany. The fritillary was a less-common form, but one of this type was made in Amsterdam by Daniel van Pilcom and has a deeply engraved silver case producing a chequered pattern.[43]

During the middle years of the seventeenth century, probably as a result of the continuing conflict in the southern Netherlands and Flanders, watchmaking appears to have declined, with only a small number of makers in the major cities and towns. An oval silver pair-cased watch by Matthys Bockelt of Haarlem, made in about 1620, is one of a very small number of surviving examples of his work.[44] In that period, Dutch makers tended to produce watches similar to those made in other European centres. A watch made by Nicholas Goediej differs little from a rock-crystal-cased watch made in Geneva at the same time.[45]

England

There is so far no evidence to suggest that there were any watchmakers in England before about 1570. Any watches known to have been in England before then were almost certainly imported, particularly from the German watchmakers of Nuremberg and Augsburg, or perhaps from France. The probate inventory of Henry VIII's goods contains a small number of ambiguous entries that could indicate a watch,[46] but there are no clear references to watches in royal possession until the reign of Elizabeth I. Important information, however, appears in inventories of Elizabeth I's jewels made in the 1580s. These provide fascinating descriptions of the highly decorated watches that were worn by the Queen herself and perhaps by other members of the court. Examples taken from *A booke of soche Jewells - delivered to the charge and custodie of Mistress Mary Radclyffe* give an idea of the sumptuous nature of the watches concerned—'A watch of Aggat made like an egg garnished wth gold. Item. a litle watche of christall slightlie gar. wth golde wth her ma^{ts} picture in it. Item. a litle watche of gold enameled wth sundry colors on both sides alike'.[47] Although no surviving portrait of Queen Elizabeth depicts her wearing a watch, the dresses she wore give an insight into just how magnificent these watches must have been. A probable reason for the lack of surviving watches may be that, becoming quickly obsolete in the changing world of court fashion, they were broken down and their jewels either reused or sold to realize cash to fund more urgent requirements. Nor is there evidence to suggest who might have made these sumptuous pieces.

It is clear from the records of clock and watchmakers in London in the latter part of the sixteenth century that there were a number of immigrants living and working there by the end of the 1560s.[48] Many came from the southern Netherlands and Flanders. A typical example is Ghylis van Gheele, mentioned in the 'Returns for the Aliens', a register of foreign immigrants living in London in the 1580s. Referred to as 'Van galand', he is described as a 'Clockmaker born under the obeyance of the King of Spaine, payeth tribute to no companye and is of the Dutche church'. He probably came to London from Geel in Flanders before 1582. He was married in London in that year and had at least two children, baptized in 1585 and 1590. A measure of the output of his workshop is the survival of eight watches and a sundial, all of which bear his name.[49] Other immigrant watchmakers in this period include Nicholas Vallin, Michael and Francis Nouwen, and Jacques Bulcke.

It was not long before native British makers began to ply their trade in London. One of the earliest of these was Randolf Bull, a member of the Goldsmiths' Company in the city of London. An early example of his work is a watch in the British Museum, a plain oval gilt brass cased watch dated 1590.[50] In this early period watchmakers within the city of London tended to be members of the Goldsmiths' Company, founded by royal charter in 1327 or in the Blacksmiths' Company established in 1571. The makers from the Low Countries congregated in the parish of Austin Friars where they would be among fellow Dutch-speaking immigrants and where the 'Dutch Church' gave them a place to worship. Another significant development was an influx of protestant Huguenots from France who were watchmakers, goldsmiths, and silversmiths. Watchmaking styles were bound to be international.

THE FUSEE WATCH MOVEMENT

In its early years, the London watch movement followed the style prevalent in the Low Countries (Figure 85). Generally, timepiece watches were oval, while striking and alarm watches were

---

43 https://research.britishmuseum.org/research/collection_online/collection_object_details/collection_image_gallery.aspx?assetId=316,435,001&objectId=58,133&partId=1 (accessed January 2020).
44 Morpurgo 1970, 12. The watch is in the Museum of London (34.181/26).
45 British Museum (1888, 1201.195).
46 Discussed in Turner forthcoming.
47 British Library (MS Royal Append 68).

---

48 Loomes 1981a, 1981b, gives information on several watchmakers in London in the decades before 1600.
49 Museum of London, dated 1591 (reg.no.34.181.90), Webster Collection, sold Sotheby's 27 May 1954, lot.85, Webster Collection sold Sotheby's 19 October 1954, lot 132. Webster Collection sold Sotheby's 19 October 1954, lot 132. Ernst Sarasin-Vonder Muhll collection sold Basel, Galerie Fischer, 18 November 1948 (possibly not genuine), Ashmolean Museum formerly owned by Percy Webster, sold Christie's in 1941, and presented to the Ashmolean by Francis Mallett in 1947, Musée international de l'horlogerie, La Chaux-de-Fonds (Inv.no.1190). The Germanic style of the case suggests that it is unlikely to be original to the movement and dial, Christie's London December 2006, lot 74. Sundial dated 1592 in the Victoria & Albert Museum.
50 The case is badly worn and the movement somewhat distressed, but an important survival. British Museum (1874,0718.11).

Figure 85 Movement of an early fusee watch by Robert Grinkin senior *c.*1610. Photo: Courtesy of Sotheby's, London.

round so as to house a round bell in the back of the case. The movement consisted of two separate plates held apart by four pillars, usually of baluster form. In the early period, the back-plate would be engraved with a foliate border, a form almost exclusively English.[51]

Between the plates was the **mainspring** barrel connected by a gut line to the **fusee**. The mainspring commonly had about twelve turns and was retained in the barrel by a dovetailed cap. Using the combination of the ratchet set up and the stop-work on the fusee, only the middle turns of the spring were used where the output of force from the unwinding spring was most constant. Setting up the spring with more turns already wound would make the watch go faster and with fewer turns slower. The profile of the fusee then compensated for the reducing force of the mainspring as it unwound during the going of the watch. This setting up was only intended to be done by a watchmaker to adjust the rate of the watch before the owner had access to it. The ratchet and

click set up mechanism was gradually superseded from about 1640 by a tangent-screw mechanism to which the owner could make adjustments using a crank-key. This would have two squares: one larger for winding the watch and one smaller for adjusting the set-up.

Until the introduction of the balance spring in 1675, the performance of watches was erratic and unpredictable and could vary by as much as half an hour per day. Because of the weakness of the gut line only a gear train of four wheels could be used, which meant that the duration of the watch was normally about fifteen or sixteen hours, necessitating the winding of the watch twice a day. The resulting erratic performance mattered little; watches were probably not part of everyday wear but reserved for special occasions.

The **great wheel** was mounted on the fusee. It drove the second and **contrate** wheels, each with a **pinion** to drive the next in the gear train.[52] The contrate wheel then drove the escape wheel, which operated with the verge and which had two pallets engaging alternately with the teeth of the escape wheel. An hour wheel, mounted under the dial, was driven by a pinion on the fusee **arbor** and the hour wheel carried the hour hand, which rotated once in twelve hours. The single hand, usually of iron or blued steel with a short, pointed tail, was secured friction tight to the hour wheel arbor so that the time could be adjusted as necessary. On the back plate there was a **balance cock** mounted on a rectangular stud secured by a transverse pin. The balance cock provided the upper pivot hole for the balance staff. Riveted to the underside of the back plate was a potence (from this period in England the plate generally became referred to by makers as the **potence-plate** and the front plate to which the pillars were riveted became known as the **pillar-plate**), which provided the lower pivot hole for the balance staff and the front pivot for the escape wheel. The balance was made from steel and had a single diametral cross arm riveted at the centre to the verge staff. The balance would oscillate quickly, typically anywhere between 12,000 and 18,000 oscillations per hour. Screwed to the inside of the back plate was the counter-potence, which had a separate stud in which was a bearing hole for the rear escape wheel pivot. The ratchet and click system for setting up the mainspring was also on the upper surface of the back plate. Riveted to the inside of the back plate was a block with a spring-loaded **stop-work finger**

---

51 Exceptions to this are watches by Johannes Bock in Frankfurt and Joost Silleman of Leeuwarden. The former is in the Sandberg collection, sold by Antiquorum, 31/3/2001, lot 252, two by the latter are in the Louvre, Cardinal 2000, N° 131, 132.

52 Two typical gear trains of the time would have a great wheel with sixty or fifty-five teeth driving a pinion of five leaves on the second wheel. This wheel would then have fifty-five or forty-five teeth driving a pinion of five leaves on the contrate wheel which would have forty-five or forty teeth, driving a pinion on the escape wheel of five leaves, and the escape wheel would then have seventeen or fifteen teeth. However, there were no set rules for this, and other counts are commonly found. Care has to be taken as it is not uncommon to find watches of this period where wheels and pinions have been replaced, creating a misleading result.

that stopped the rotation of the fusee when the watch was fully wound. The whole movement was then hinged into the case at twelve o'clock and retained by a sprung catch at six o'clock.

STRIKING AND ALARM MECHANISMS

A typical striking train in a pre-balance spring verge **clock watch** consisted of a fixed mainspring in a barrel screwed to the underside of the back plate, with pierced and engraved foliate scrolls on the visible side. The great wheel attached to the mainspring arbor had a winding ratchet and drove a further five wheels, the last being a simple pinion fitted with a turned inertial weight. The second wheel had the pins for operating the hammer and the third wheel had a stud for locking and warning. There would be two spring-loaded steel 'gates': one with release and warning detents, the other with locking and count-wheel detents. The striking was controlled by a count-wheel on the outside of the back plate, driven by a pinion of report on the extended second-wheel arbor. To retain the count-wheel there was a pierced and engraved foliate cock. The barrel had geared stop-work mounted on the back plate beneath a decoratively pierced cover. Beneath the dial the strike-release lever was lifted once per hour by a twelve-point star wheel.

The alarm mechanism consisted of a fixed mainspring barrel screwed inside the movement. There were up to three wheels in the gear train: the great wheel, mounted on the barrel arbor, drove the second and contrate wheels. The contrate wheel drove a verge escape wheel running in a riveted potence. When the alarm was released, the escape wheel oscillated a verge with two pallets on which was mounted a semi-circular hammer that struck on the inside of the bell. A pin in one hammer arm acted in a slot cut in the movement plate to limit the traverse of the hammer. The alarm release consisted of a pivoted lever with a return spring riveted to the plate. The inner end of the lever acted on a cam mounted on the hour wheel. At the other end, the lever caught a pin on the hammer arm. The cam had a slot cut in it, and the release cam had a squared arbor that carried the alarm-setting disc. When the lever dropped into the slot, the whole train was released to sound the alarm.[53]

Further sophistication could be in the form of a calendar dial indicating day, date, month, and age and phase of the Moon. They might even be truly astronomical, showing the position of the Sun and Moon in the zodiac throughout the year, although such highly sophisticated watches were rare.[54] The duration of some of these large-sized watches could be about thirty hours, and thus needing winding only once a day, but these were much less common than those needing two winds per day. In the pre-1600 period chapter rings were normally integral with the dial plate, with Roman hour numerals and half-hour marks between them. There would also be a circle of touch pins at the hours with a higher more pointed one at XII, by which the time could be ascertained by touch in darkness, poor light, or for those with failing eyesight. These, however, were not universally applied.

By the second decade of the seventeenth century there was a trend towards much smaller watches as wearing them moved from neck to waist. There are many portraits in which the sitter is depicted wearing a watch in this way. Typical of pre-1600 watches, is one shown in the frontispiece of Sir John Harrington's translation of Ludovico Ariosto's *Orlando Furioso*.[55] That watch design in London embraced a stronger French influence may be the result firstly of the considerable numbers of Huguenot makers who had come to London and set up in the areas of Blackfriars and Covent Garden where they were outside the jurisdiction of the city companies, making watches with finely engraved silver and gilt-brass cases. It arose secondly from King James' appointment in 1613 of David Ramsay as Page of the Bedchamber in which role he also took care of clocks and watches in the royal household.[56] Ramsay had been living and working in France and brought with him French taste and access to French engravers, either those living in France or those who had emigrated to London. All of this fitted well with the king's liking for French fashion. There can be little doubt that what the king favoured spread quickly through the court and thence into wider society. Watches made for the king were very much in contemporary French manner and have cases decorated with portraits of the king himself and with biblical scenes.[57] These watches also present rare examples of cases signed by the artists involved in their production. One is signed by Gérard de Heck who, having left Blois, lived and worked in London. All three of his oval watches have a sophisticated calendar dial thought

---

53 Exact details of the making of a watch in London in the second quarter of the seventeenth century are given a document recently transcribed from the papers of John Evelyn and currently being prepared for publication by the present author. For an interim announcement, see Thompson 2019.

54 A watch in the British Museum with an astronomical dial signed H. Roberts is discussed and illustrated in Thompson 2008, 24–5. Dials such as this were already in existence in table clocks.

55 Ludovico Ariosto, *Orlando Furioso*, epic poem first printed 1512, trans. by Sir John Harrington, published in 1590. The frontispiece has a miniature portrait of Sir John with an oval watch on the table in front of him. His arms are engraved on the inside of the open lid.

56 For a biography of David Ramsay, see Finch, Finch, & Finch 2019, 177–99.

57 An example of this group of watches is illustrated and described in Camerer Cuss 2009, 34–5. The watch is in the Victoria & Albert Museum. A second example is illustrated in Finch, Finch, & Finch 2019, 186. A third example is in the Royal Museums of Scotland. A further watch with a case decorated by de Heck is a star-shaped watch in the collections of the Worshipful Company of Clockmakers, see Camerer Cuss 2009, 36.

to have been of Ramsay's design and was adopted by other makers later in the century. In the royal accounts there is reference to 'watches, three bought of Mr. Ramsey the Clockmaker lxi [61]', giving a unit price of just over twenty pounds each.[58]

By 1610 the number of watchmakers in London had increased considerably. In 1622 sixteen makers signed a petition to King James I for the incorporation of clock and watchmakers. To this number, forty-five foreign watchmakers can be added, including their apprentices, who were accused of making poor-quality movements even though their outward appearance was commonly beautiful.[59] Subsequent petitions culminated in the incorporation of a Company of Clockmakers by Charles I in 1631. The first Master was David Ramsay, although he rarely appeared at the meetings and his duties were carried out by the senior warden, Henry Archer. It is interesting to note that in spite of the petitions, some of the 'foreign' watchmakers soon became Free Brothers in the company, prominent among them being David Bouguet.[60]

Timepiece watches could be round but were more commonly oval or of the new elongated octagonal shape. It was during this period that silver became more prominent, used for separate chapter rings for the hours and half-hours as well as for applied decorated case bands, covers, and backs. A typical example is a watch by Edmund Bull of Fleet Street, London, made in about 1615, where the cover and back are of silver and decorated with scenes from the bible. The movement, while smaller in size, retains exactly the same components as its sixteenth-century predecessors.[61] Another maker in London with a prolific output was Edward East, his productivity partly explained by his long life. Born in Bedfordshire in 1602, he was apprenticed in the Goldsmiths' Company in London and went on to be a founder member of the Clockmakers' Company and its Master in 1645. He died probably in late 1695, his will being proved on 23 February 1696.[62] An interesting aspect of East's workshop was its concentration on watch production until the introduction of the pendulum to clocks in 1657, after which he seems to have concentrated on clocks. There are very few East watches with a balance spring. A good example is a silver cased clock watch with alarm, c.1645, the back of the case pierced and engraved with realistic flowers and the movement signed 'Eduardus East Londini'.[63]

In the same period the **'form watch'** began to appear in London. It is difficult to judge exactly where this originated, but by about 1635 the new fashion was proliferating in Geneva (Figure 153) and London and to a lesser extent in France, Germany, and the Netherlands. There are numerous surviving examples from London between 1635 and 1645. One typical example in a sea-urchin case is by John Charlton.[64] At least two other examples survive with this form of case: one by François Thomas of Nantes and the other by Thomas Chamberlaine of Chelmsford.[65] Clearly, if there was a need for engravers, metal casters, and so on, then local specialist craftsmen were at hand to supply cases and dials to the resident watchmakers. This is borne out by two watches with similar shell-shaped cases. The first houses a watch by Richard Masterson and the second is by Robert Grinkin junior. Both cases have cast silver cases of similar design[66]

Perhaps the finest decorative watches by London makers date from the middle of the seventeenth century with good quality movements housed in exquisite enamel cases. A good example is a gold and enamel-cased watch c.1640, signed by Edward East. It has an exquisite blue floral enamel case and dial with pastoral scenes on the dial and inside the case back. The case was almost certainly made in Blois.[67]

In stark contrast to the highly decorative form watches and the fine enamel-cased watches are those watches now described as 'puritan' watches on account of their totally plain cases, supposedly reflecting the simplicity required by Puritans. The earliest examples date from c.1635–40 and appear both in the Netherlands and in London. They are typically made from silver, although there are some in gilt brass and at least one in gold, possibly owned by Oliver Cromwell.[68] Silver-cased versions of this style exist in numerous examples. A typical one is by Richard Crayle of Fleet Street. It is housed in two cases: the inner of plain silver and the outer of leather-covered brass with silver piqué pin decoration. The cover is glazed to reveal the time. The watch dates from about 1650 and with its glazed cover and outer leather case it was clearly intended to be worn as a means of knowing the time on a daily basis.[69]

---

58 Atkins & Overall 1881, 171.

59 The 1622 petition, National Archives, SP14/127 reel 178. Finch, Finch, & Finch 2019, 343–64, 478–90.

60 https://www.bmimages.com/preview.asp?image=00035777001

61 British Museum (1992, 0514.1).

62 Finch, Finch, & Finch 2017, 487. Watches are in the British Museum and the Victoria & Albert Museum. Examples in private hands can be seen in Camerer Cuss 2009, pl. 17, 19, 34–36, 47.

63 V & A (M.64:1, 2-1952) http://collections.vam.ac.uk/item/O78625/clock-watch-east-edward/. An example of a puritan-style watch is in the Metropolitan Museum of Art, https://www.metmuseum.org/art/collection/search/194060.

---

64 John Charlton, King Street, London, from about 1630, member of the Blacksmiths' Company. British Museum watch (1958, 1201.2341).

65 Both watches are in the Ashmolean Museum: first a silver cased verge watch by François Thomas, Nantes, made c.1640 (WA 1947.191.60), Second, a silver-gilt cased verge watch with calendar dial by Thomas Chamberlaine, Chelmsford, Essex, c.1650 (WA 1947.191.61).

66 The Masterson watch is in the British Museum (1958, 1201.2301) that by Grinkin in the Bowes Museum, Barnard Castle, see Tuck 1989, 185.

67 Victoria and Albert Museum: http://collections.vam.ac.uk/item/O112705/watch-east-edward/.

68 A plain gold double-cased watch; see Thompson 2008, 38–9.

69 Illustrated in Camerer Cuss 2009, 75.

For many centuries, a person would keep belongings and valuables in a purse, or pocket, suspended from a belt around the waist and concealed under outer garments. The pocket was then accessed through a fitchet, a slit in the outer garment. By the middle of the seventeenth century the pocket was transformed into a sewn-in addition, particularly in the new garment, the waistcoat. The fashion for integral pockets, which began in the French court, soon became popular. It is therefore not surprising that once the watch could be hidden away in everyday clothing, it could be worn all day. As well as this, the application of a pendulum to clocks increased their accuracy, in good examples to less than a minute a day. As a result, time consciousness developed in urban society, and in London and elsewhere by about 1660 a new form of watch became commonplace. The new watches were larger than form watches and the small oval pair-cased puritan watches and the introduction of a fifth wheel into the gear train to produce a running time of about thirty hours—allowing the watch to be wound just once per day—necessitated a stronger mainspring. This in turn commonly caused problems with breaking gut lines between mainspring-barrel and fusee. The solution to this was to utilize a fusee chain, which very quickly superseded the gut line in watches. These new-style watches normally had a plain gilt-brass or silver inner case and an outer case made from brass covered with leather and with piqué pin-work decoration to match the metal of the inner case. Now the watch was not reserved for special occasions but was worn on a daily basis. In urban areas, the owner would put the watch right by his own clock, which was rated and set against a sundial or he might use the local church clock as his reference. It was now possible to arrange meeting times, albeit somewhat vaguely, but 'meet me in the Swan and Anchor at half past six' was a viable proposition. As well as this, the valuable watch could be kept out of the view of the eyes of pickpockets and footpads. Throughout the seventeenth century watches with striking and/or alarms were not unusual, but from the middle of the century they began to go out of fashion.

SWITZERLAND

The practice of watchmaking began in northern Switzerland in the early years of the seventeenth century. It is logical that the watches made there followed the pattern of their close neighbours in south Germany. In Zug there were whole families involved in watchmaking, notably those of Hans Jacob Zurlauben, Paulus Bengg, and Jean Baptiste Letter. Some makers, such as Zurlauben, used the stackfreed to equalize the power output of the mainspring, while others a little later favoured the fusee. A typical watch from this area by Paulus Bengg made in about 1640 has a fusee, but the silver and enamel dial clearly shows a Germanic ancestry. The heart-shaped case with a rock-crystal band, back, and cover is unusual.[70] Another watch from the region by Zurlauben dates from the first decade of the seventeenth century. It is one of a small number of watches provided with a detachable baluster stand so that it could stand on a desk or be worn.[71]

In contrast to the modest number of watchmakers active in rural towns, Geneva was very different, with more than twenty-six makers and their descendants producing watches in fine cases, particularly in rock-crystal and enamel. As a result of the persecution of Protestants in France, many watchmakers, particularly from Lyon, moved to the independent Calvinist city state of Geneva in the closing years of the sixteenth century. Although the by-laws of Geneva forbade the wearing of jewellery and watches,[72] makers tapped into a flourishing export market and some, such as the Sermand family, traded in the Ottoman Empire.

During the mid-seventeenth century, the popularity of watches in rock-crystal cases increased, and they proliferated in Switzerland. The cases took many forms, but principally had octagonal and lobed oval shapes. The case body was cut from a single piece of rock-crystal and had a rim of gold, silver, or gilt brass around its upper edge with a hinge element at XII. There would then be a full cover of similar shape, also with a rim and a hinge element. The movement and dial were latched into the case. While the majority of these watches were of small size, there is one that stands out. It is signed 'J. Sermand' and dates from c.1660, a product of the workshop of Jacques the nephew. The mechanism of the watch is of a common sort but finished to a high standard. The case, on the other hand, is anything but standard, consisting of a single piece of facetted rock-crystal onto which are mounted gold rims enamelled in colours with a beautiful array of realistic flowers. The dial sets the watch apart from many of its contemporaries. Laid on to a gold base is a white enamel chapter ring with the usual Roman hour numerals. It surrounds a beautiful pictorial enamel, rustic scene with a man seated in the foreground looking towards what appears to be a fortified bridge. It is a truly spectacular watch.[73]

Another specific design for the rock-crystal case was the cruciform shape, and many examples survive today. One such is a watch by Jacques Sermand's brother-in-law, Jean Rousseau in Geneva. The case body is made from a single piece of rock-crystal, the space for the movement cut out using a small grinding wheel.

---

70 Illustrated in Jaquet & Chapuis 1970, pl. 21.
71 While the surname of the maker has been removed, it is still possible to establish that is was 'Zurlauben'; British Museum (1888, 1201.167), Tait & Coole 1987, cat. no.49.
72 In the *Sumptuary Ordinances and Laws concerning Clothing, furnishings and other excesses of similar kinds*, reviewed by the Little Council and the Grand Council of the Republic of Geneva in February and June 1668, is the following sentence: '*Item: Women and girls are forbidden all gold needles and pins and all things made with precious stones on their dress and accoutrements . . . all watchmakers' watches and mirrors on their girdles*'.
73 British Museum (1958, 1201.2341).

There are gold rims at the top of the case and around the cover. The dial is engraved with the four ages of man: childhood at the bottom, youth at the top, manhood to the left, and old age to the right. The figure of childhood sits significantly on a skull, the symbol of mortality. This decoration is unusual on a crucifix watch in being of a secular nature. The subjects chosen for these watches designed to express a religious sentiment are almost without exception either scenes from the Life of Christ or the Instruments of the Passion. A caveat is required in considering rock-crystal cases, for there are a number of watches in existence today that might have a seventeenth-century movement and dial but have a crystal, or even a glass case, that dates from the nineteenth century (Figure 159). In some instances, it may have been a matter of fitting a new case to replace a broken one, but there are nevertheless numerous examples with fanciful cases made to house watches that originally had metal cases.

In common with all the other watchmaking countries, Switzerland produced form watches. One is of a style encountered almost nowhere else: watches with cases in the form of miniature creatures. Jean Baptiste Duboule made a watch with a small oval movement of modest proportions. The one-piece silver dial simply shows the hours around a charming townscape. The case is in the form of a miniature silver dolphin with tiny ruby eyes. It was made in about 1635 and is now in the Patek Philippe Museum. Another watch also in the form of a dolphin is by Jacques Sermand.[74] A miniature lion with a movement by Jean Baptiste Duboule is in the Ashmolean Museum, and the British Museum has a watch in the form of a little dog containing a movement by Jacques Joly. Lastly, the Musée de l'horlogerie et de l'émaillerie, Geneva had a shell by Jean Rousseau and a rabbit by Pierre II Duhamel.[75] It would seem very likely that there was just one small workshop where these cases were made, but none of them is signed by a maker (Figure 153).[76]

Another type of watch that became popular from the late sixteenth century onwards was the coach watch, particularly in countries other than England.[77] These watches of large size were intended to be used when travelling—hence the name. They usually had striking and alarm functions as well as a calendar dial with the usual date and lunar indications. An early example in the British Museum was made by Hans Gruber in Nuremberg in about 1575.[78] In this instance the watch only has hour-striking and no alarm. A later Geneva watch of this type is by Jean Baptiste Duboule, housed in a finely decorated silver case with an outer protective leather case with silver rims and hinge. The watch is heavy and with a diameter of 10cm clearly not intended to be worn. In addition to hour-striking and alarm functions, there is an elaborate calendar dial with sector apertures showing months and the signs of the zodiac, as well as the four seasons. To the right a large dial shows the age and phase of the Moon, and at the bottom is a gilt chapter ring and blued-steel pointer to show the hours, with a pierced and engraved dial at the centre to indicate the alarm time. There should be over this a blued-steel hand for setting the alarm. To the lower left a sector aperture shows the periods of the day. Above that, in the middle left a gilt chapter ring has the date, while an aperture within it shows the days of the week, each named in Latin and illustrated by its ruling planet.[79]

THE BALANCE SPRING AND BEYOND

Precedence for the invention of the spiral balance spring for watches claimed by Christiaan Huygens in 1675 was challenged by Robert Hooke, who asserted that he had conceived the idea many years before. In defence of his claim, Hooke commissioned a watch from the leading London watchmaker, Thomas Tompion, and had it engraved with the inscription, *R, HOOK invenit an.1658 T. Tompion fecit 1675*.[80] The dispute about Huygens' or Hooke's invention continues to this day, but it is generally agreed that Christiaan Huygens' spiral balance spring produced the practical solution.[81] The balance spring enabled a watch to keep time to within less than a minute per day. The rate of oscillation of the balance in watches without a spring was determined by the mass and radius of the balance and the amount of impulse given to it by the escapement from the mainspring and the gear train. This was bad for accurate timekeeping as the oscillator was heavily influenced by variations in power imparted to the balance. With a spiral balance spring attached at one end to the watch plate and at the other to the balance staff the varying deformation of the spring caused by variations in the impulse would result in greater or lesser speed of rotation to return the balance to its neutral position and so produce relative **isochronism**—within certain limits. There were, of course, still other factors which affected the accuracy of the watch. Temperature changes had a direct influence on the elasticity of the balance spring. The degeneration of oils available at the time and indeed, the position of the watch itself, dial up and down or pendant up, all affected timekeeping. However, the resultant time variations in the balance spring watch were as nothing compared to what went before. The result was

---

74 Patek Philippe Museum. A similar example was sold at Sotheby's London auction, 11 November 2019, lot 5. https://www.sothebys.com/fr/auctions/ecatalogue/2019/masterworks-time-adolf-lange-golden-era-glashutte-ge1924/lot.5.9Y9V6.html (accessed January 2020).
75 Menz & Sturm 2003, 62–3. Stolen, present whereabouts unknown.
76 Jaquet 1943, *passim*.
77 For an overview of these watches, see Stolberg 1993; Ackermann 1983.
78 Tait & Coole 1987, cat. no. 62.

---

79 British Museum (1888, 1201.229), Thompson 2008, 44–5.
80 The whereabouts of this watch are unknown.
81 For the beginnings of the balance spring applied to watches, see Wright 1989; Ball 2006.

a transformation in the timekeeping properties of the watch, and the new invention was rapidly adopted. By 1680, the majority of watches made in Europe were fitted with a balance spring, and countless numbers of older watches were upgraded to improve their accuracy and make them useable in the new age.

The introduction of the balance spring in London brought about a change in the form of the watch. In the period immediately preceding its introduction London makers had already realized that a movement with a greater distance between the plates would allow for a higher barrel with a higher spring producing more force. This would be better suited to the longer duration, thirty-hour, movement, which needed more power to drive it.[82] As well as this, the introduction of the fusee chain c.1660 had provided greater reliability with the stronger power source. At the same time as the introduction of the balance spring, watchmakers began fitting larger balances, which provided better timekeeping stability but also required more power from the mainspring to give the balance sufficient amplitude to achieve good timekeeping qualities. For these reasons post-balance spring watches generally had a greater distance between the movement plates. Another development in the appearance of the watch was a change in the position of the mainspring set-up work. The rate of going of a watch was now determined by the effective length of the balance spring, and a new system was introduced. It consisted of a geared sector on which were mounted two 'curb' pins through which the outer end of the balance spring passed. The geared sector was moved by rotating a pinion with a squared arbor using a key. Turning the key would move the pins one way or the other to adjust the point where the balance spring passed through, thus changing the effective length of the spring. Also on the regulator pinion was a calibrated disc, usually of silver, which provided a reference for adjustment. This new regulator became known as the 'Tompion' regulator, but there is no evidence to support the suggestion that it was invented by him. The balance spring regulator needed a retaining cover, which in English watches usually took the form of a pierced and engraved plate. This meant that there was no longer room for the mainspring set-up mechanism previously mounted on the back plate. So in the new design, watches had the mainspring set-up worm gear and brackets mounted on the underside of the front plate. A good example of an early balance spring watch (between 1682 and 1683), is one by Jonathan Puller (c.1662–c.1707) in which the balance cock is fully open with both table and foot pierced and engraved, the tri-form design of the table covering a balance spring of few turns.[83]

Fitted with a balance spring, the ordinary watch was now worthy of a minute hand, and it was not long before there were also watches with a seconds hand, although these remained relatively rare. The under-dial wheel work for the new hour and minute hands also saw a change from the drive to the single hand being taken from a pinion mounted on the fusee arbor to the watch having an extended centre-wheel arbor on which was a pipe, or cannon pinion, carrying the minute hand and rotating once per hour. The cannon pinion then drove a minute wheel mounted on the movement front plate. On this was a pinion, which then drove the hour wheel and pipe that carried the hour hand, so that the two hands were concentric. The introduction of two hands with the necessary motion wheels beneath the dial for the differential movement of the hour and minute hands meant that there was no room for an alarm-release mechanism. From this time onwards, the alarm watch became very scarce.

The introduction of the balance spring quickly made a new form of watch fashionable in England. At the same time, the pair-cased watch also became the norm, with two cases, one inside the other. The inner case was, at the time, usually referred to as the box and the outer case, the 'case' or 'pair-case'. The outer pair-case had a solid back and a hinged open bezel at the front. The inner case or 'box' had a glazed front bezel and a hole in the back for the winding key. More expensive watches had gold pair cases, lesser ones, silver, the cheapest gilt-brass cases. A watch could also have an outer case made from silver or brass covered with **tortoise shell**, horn, leather, or **shagreen**. These were produced in specialist workshops, one such belonging to the Brisset family (probably Obrisset) from Dieppe. In the case of these organic materials there was often a design of piqué pins decoratively placed, which kept the material more stable and gave partial protection from rubbing. The use of these robust cases further enabled the watch to be worn on a daily basis, and London newspapers of the time contained countless advertisements for lost or stolen watches to be handed in to a specified person in exchange for a reward. Such was the level of losses that the London clock and watchmaker Daniel Delander placed the following advertisement in the *Post-Man* between 19 and 22 January 1706:

> Whereas the ladies often lose their watch-cases from their sides, occasioned by the wearing of the spring, This is to give notice, that there is a new invention of a spring to fasten the cases in such a manner that will infallibly prevent them from being lost, by the aforesaid Daniel Delander, watchmaker in Devereux Court.

HUGUENOTS: THE SECOND WAVE

The Revocation of the Edict of Nantes in France in 1685 provoked a second wave of Huguenot refugees in London. By 1700 the Huguenot population is thought to have increased by about 25,000 among them a considerable number of watchmakers. Among the most prominent were David Lestourgeon, Simon Decharmes, Peter Garon, and Henry Massy. Generally they worked in Soho and Covent Garden, and some were welcomed

---

82 An example is a pre-balance spring watch made in Thomas Tompion's workshop in about 1672, now in the collections of the Worshipful Company of Clockmakers and illustrated in Evans, Carter, & Wright 2013, 564

83 Camerer Cuss 2009, 114.

into the Clockmakers' Company as Free Brothers. They were not allowed to take apprentices in the Company but nevertheless ran flourishing workshops. Competition with the established London makers was stiff, and one way in which the newcomers tried to attract business was by making watches with unusual dials—wandering hours, Sun, and Moon, jumping hours, and so on. A very fine example of one of these watches is by Peter Fardoil, freeman in Blois in 1684. The watch has a sophisticated fly-back hour hand which traverses a semi-circular aperture in the dial from six o'clock in the morning to six o'clock in the evening, at which point it automatically flies back.[84] It is perhaps significant that the established craftsmen with a large and rich clientele rarely attempted to compete with the newcomers in this style. Another form of sales practice was that of David Lestourgeon, who made a number of watches that could be termed 'allegiance watches', marking significant events such as the death of William III or the accession of Queen Anne. While Huguenot watchmakers' names died out to some extent during the eighteenth century, they continued through generations as watchcase makers.[85]

In London, watchcase makers did not mark their products until the end of the seventeenth century. The earliest maker's mark recorded is on a watch by Robert Seignior made in about 1675, with case marked with the mark ND, that of Nathaniel Delander. From this time on it became common, although far from universal, for case makers using gold or silver to mark their work according to new regulations from the Goldsmiths' Company and the Assay Office.[86] Usually, the inner case (box) and the outer case were made by the same maker, but for a short period at the beginning of the eighteenth century, there were watches with the two cases made in different workshops and consequently bearing different makers' marks. An example of this can be seen in a watch with a movement by the immigrant German maker John Bushman.[87] In this example, the inner silver case was made by William Jacques in sterling silver using the mark 'WI', while the outer case was made by John Willoughby, using Britannia silver of higher purity and using his Britannia mark 'Wi'.[88] Another feature worth noting is the case design of the period around 1700 and up until about 1720 with a square-ended hinge on the outer 'pair' cases. After this, the hinge ends were chamfered and rounded.[89]

REPEATING WATCHES

Following the invention by Edward Barlow (1639–1719) of rack striking for clocks, he commissioned Tompion to make a **repeating** watch in 1687 to secure a patent for it. Daniel Quare contested this with just such a watch which he submitted to the Privy Council on 2 March 1687 in opposition to the Barlow–Tompion model. According to Derham:

> The King, upon tryal of each of them, was pleased to give the preference to Mr *Quare's*: of which, notice was given soon after in the Gazette. The difference between these two Inventions was, Mr *Barlow's* was made to Repeat by pushing in two pieces on each side of the Watch-box: one of which Repeated the Hour, the other the Quarter. Mr *Quare's* was made to Repeat, by a Pin that stuck out near the Pendant which being thrust in (as now 'tis done by thrusting in the Pendant) did repeat both the Hour, and Quarter, with the same thrust.[90]

The Ashmolean Museum holds just such a watch by Daniel Quare, London, no. 611,[91] for which there is no doubt that it was made for King James and as such is an important documentary piece. From this time on in London, a gold pair-cased quarter-repeating watch represented a high point in excellence.[92] In the period up to about 1725 there were many different versions of Quare's original design, and by 1725 the system had become standardized. One man who specialized in repeating work, which he made for London watchmakers, was Matthew Stogden (pre-1717–70). He introduced his specific version of the mechanism in about 1728 when he was working for George Graham. In the following years, repeating work became more sophisticated, with watches sounding half-quarters, then five minutes, and finally minutes. There were also watches known as dumb-repeaters, which had no bell in the case-back but had hammers that simply struck the case-back to produce a series of taps.

On present evidence the minute-repeating watch appears to have originated in Friedberg during the first quarter of the eighteenth century. Examples signed *Marqüch, London* and *Lekceh,*

---

84 See Thompson 2008, 60–1, and https://www.bmimages.com/results.asp?txtkeys1=fardoil.
85 For a brief survey of Huguenot watchmakers in London, see Thompson 1996, 417–30. See also Murdoch 1985.
86 For a detailed account of English watchcase making, see Priestley 2000.
87 This watch is illustrated in Thompson 2008, 58–9 and at https://research.britishmuseum.org/research/collection_online/collection_object_details.aspx?objectId=57955&page=1&partId=1&searchText=wandering%20hour.
88 Following the Britannia Act of 1697, the law stated that all silver goods were to be made from Britannia silver alloy 98.3% pure and that all makers should use the first two initials of their surname as their mark registered at the Goldsmiths' Hall, London. The law was not always obeyed: see watch by William Jacques with a sterling silver case 92.5% pure, marked WI—the initials of both his names.

89 In the English watchmaking trade the hinge between the two parts of the case became commonly known as the 'joint'.
90 Derham 1696.
91 Ashmolean Museum (WA1947.191.85). Thompson 2007, 46–9.
92 For the development and function of the repeating watch see Wadsworth 1965, 364–7 and Wadsworth 19, 24–6 and 48–52.

London are known.[93] The name Marquch is highly unusual and is perhaps a foreign interpretation of the name of the London maker Markwick, which, if transmitted verbally, could result in a bizarre spelling. The name Lekceh is clearly the reverse of Heckel, a Friedberg maker. Because of the complexity of making reliable mechanisms to repeat hours, quarters, and minutes, these watches are quite rare, and these early examples did not mark the beginning of a trend for making them in large numbers.

The earliest minute-repeating watch made in England so far recorded is signed 'Jno Ellicott London 1961' and is in a private collection. While the inner case of this watch has no hallmarks, the Ellicott firm's serial number gives it a date of about 1738.[94] Who actually made the Ellicott watch is open to question, but it was not likely to have been Ellicott himself. Thomas Mudge also made watches with a minute-repeat function. A surviving example is one in the British Museum that may relate to a request to Thomas Mudge from Miguel Smith, Ferdinand VI of Spain's horological representative, who wrote to Mudge from Madrid on 26 December 1757:

> Sr. my last to you was ye. 19 current wch time then would not permit me to Give you the orders that his Majesty desired me vizt me that you would put in hand Directly a large Gold strikeing quarter minit Repeating watch I mean that it must have all the performances that the crook-head has that you have made, that it must strike the Houres quarters and minits single blows as far as 14 minits & the Double Blows at ye quarters; in Regard to the sise of it, Let it be as nigh as you can Gess to the sise of a Large quarter clock watch that you made about 4 years a Gon by my order which ye King wares & allows to be the best watch for performance that he ever had.[95]

So far just four minute-repeating watches by Mudge are known.[96] The very small number suggests that they were difficult and time consuming to make, the cost outweighing the impact they made in the public domain.

DIALS IN ENGLAND

From the 1660s, a standard form of dial began to emerge that probably had its origins in the clock dials of the time shedding decoration in the interest of accurate time indication. The one-piece engraved watch dial of the mid-seventeenth century was superseded by the champlevé dial with a raised chapter ring for the hours I–XII with half hour marks often in the form of a raised lozenge, over a matted ground. Around the outside the minute circle was numbered 0–60 at five-minute intervals. In the last quarter of the century there could also be a circle inside the hour numerals divided to either quarter or half hours. In London work the dials had a separate centre charged with the name and place of the maker on two decorative cartouches. This may have been done to facilitate the making of dials for any number of makers, the only difference being the separate name disc. Where the dial was of one piece and had no name or place, the central area of the dial could be engraved with a decorative floral pattern. These champlevé dials continued in use until the advent of the one-piece white enamel dials in the 1720s, although they were not completely superseded, particularly for watches made for the Turkish market, where the champlevé dial was used until quite late in the century. Also in this period hour and minute hands were usually made from blued steel, the favoured metal for hands for the coming century.

DUST CAPS

Another development in watch design in London was the introduction of the movement ring or cap to prevent the ingress of dust into a watch in everyday use.[97] One of the earliest makers to realize that dust was a major enemy of good running was Francis Nouwen as early as the 1580s. His watch movement and dial are fitted snugly into the case, and a post on the back of the movement passes through a small hole in the back of the case, locked off by a rotating disc. As well as this, the hole for applying the key for winding the watch has a shell-shaped cover. Nouwen clearly did not want his watch to be opened in order to wind it.[98] The two devices of ring and cap appear at much the same time. The dust ring is simply a gilt-brass ring that fits down over the movement back plate so that it encloses the open side of the movement. The dust-cap[99] is a gilt brass cover that fits over the whole movement with openings for the hammers in repeating watches. It appears in London in about 1705, and while its originator is not known, there is no doubt that Daniel Quare was one of the earliest to fit them to his watches. In the early days there were variations in design and the way in which the cap was secured to the watch, but it did not take long for a standard form to evolve with a full cap in gilt brass or occasionally silver, which covered the whole movement back and sides and was retained on

---

93 Whitestone 1993, 145–57. See also https://www.barnebys.com/realized-prices/lot/the-earliest-known-minute-repeating-watch-marquech-london-produced-in-friedberg-germany-circa-1710-extremely-rare-and-important-silver-pair-cased-minute-repeating-watch-xe8AgiV5lR. Whitestone 2010.
94 Thompson 1997, 306–21, 429–42.
95 Turner & Crisford 1977, 581.
96 The fourth of these is illustrated in Camerer Cuss 2009, 226.

97 Sully, 1737, 173, is adamant that a watch should only be opened if absolutely necessary to adjust the balance spring.
98 See Thompson 2007, 12–13.
99 In English watchmakers' terminology this is normally referred to simply as 'the cap'.

two posts on the movement back plate by a blued-steel curved locking-slide with raised pin in the middle to aid moving it from locked to unlocked.[100] Throughout the eighteenth century a cap was standard for high-quality watches.

JEWELLING

By the turn of the eighteenth century, people were expecting rather optimistic performances from their watches, and watchmakers had to cope with constant wear to components in the movements. Friction had for long been seen as the clock and watchmaker's main enemy. Oils were of poor quality with a relatively short life, being made from either rendered-down cattle bone, 'neat's foot', or refined whale oil. It would be common for a watch to need cleaning and re-oiling every two years or so. One step towards better performance came in 1704, when the Swiss mathematician Nicolas Fatio de Duillier (1664–1753), in collaboration with Pierre and Jacob Debaufre, obtained a patent for a new use of precious or more common stones 'as internall and usefull part of the [watch] work or engine itself'. On applying for an extension to the patent, however, on 6 December it was opposed by the Clockmakers' Company and refused.[101] It was nevertheless recognized that the combination of a hardened and polished steel pivot running in a jewelled hole with oil for lubrication was the optimum arrangement. Nonetheless, the art of piercing jewels remained in the hands of a small number of experts. It was a time-consuming (and therefore expensive) process, which meant that it was not applied to watches in general for many years. In the early part of the eighteenth century the best-quality London watches had jewelled bearings and end-stones for the balance pivots only; in other countries it was a rarity right up to the later eighteenth century. Daniel Quare in London was one of the earliest to incorporate the new technology in his watches. It was soon adopted as normal practice for all the leading London makers in their best quality work.

TOMPION, GRAHAM, AND ELLICOTT

London became the leading watchmaking centre of Europe from about 1680 onwards. From then on, there was a succession of makers who excelled, beginning with Thomas Tompion (1639–1713), who arrived in London and became a Free Brother in the Clockmakers' Company in 1671. The basis of his success was threefold. Firstly, he excelled as a watch and clockmaker using his ingenuity to develop watches of the highest quality and produce them in large quantities (Figure 86). Secondly, he organized his workshop to use components supplied only by the best makers in London at the time and employed a large group of apprentices and highly skilled craftsmen. Thirdly, establishing a high reputation, he was able to sell watches to crowned heads, aristocracy, and wealthy classes throughout Europe. From 1671 until his death in 1713, his workshops produced watches of consistently high quality. Allowing for watches made before he began serial numbering, Tompion's workshops produced about 5,100 watches.[102]

At his death, Tompion's hugely successful business was continued by his former partner and nephew by marriage, George Graham (1673–1751). In 1713, an advertisement in the *London Gazette* explained the situation concisely:

> George Graham, Nephew of the late Mr. Thomas Tompion, who lived with him upwards of seventeen years and managed his trade for several years past, whose name was joined with Mr. Tompion's for some time before his death, and to whom he left all his stock and work, finished and unfinished, continues to carry on the said trade at the late Dwelling House of said Mr. Tompion at the sign of the Dial and Three Crowns, at the corner of Water Lane, in Fleet Street, London, where all persons may be accommodated as formerly.

Graham went on to become the most eminent clock and watchmaker in London and a Fellow of the Royal Society. Apart from his innovations in the clock world, he was responsible for devising the **cylinder** escapement in about 1726, a significant advance in performance for watches in the eighteenth century. This new escapement was based on an earlier idea of Tompion's c.1695, and such faith did Graham have in it that he used it exclusively in preference to the usual verge escapement. Other makers quickly adopted it, particularly John Ellicott junior. The advantage of the escapement was that it was effectively a **dead-beat** escapement with no recoil in its operation. Frictional interference on the oscillating balance by the escapement was thus reduced. It was nevertheless a **frictional rest** escapement with timekeeping errors caused by the rubbing of the escape wheel teeth on the cylinder mounted on the balance staff.[103] By the time of Graham's death in 1751, the centre-seconds cylinder escapement watch with stop-lever was the most accurate to be had, and even with its shortcomings, continued in use throughout Europe until

---

100 Illustrated in https://www.invaluable.com/auction-lot/strixner-london-pump-quarter-repeating-verge-fu-49-c-f3088a011, although here the cap is inscribed *London*, it is a Friedberg watch made by Jacob Strixner, see Thompson 2007, 54–5.
101 Patent no. 371, May 1704, granted for fourteen years. Atkins & Overall 1881, 245–6; Chamberlain 1941, 310–12.

---

102 For a comprehensive account of Tompion, his life and work, see Evans, Carter, & Wright 2013.
103 For a technical discussion see, Gazeley 1956, 131–53.

Figure 86a, b Backplate and side view showing the fusee and 'tulip' pillars of a watch by Thomas Tompion hallmarked 1698. Photo: Courtesy of Sotheby's, London.

the end of the nineteenth century (Figure 159).[104] Graham's workshops produced watches of all kinds—from simple timepieces to quarter-repeating watches and centre-seconds stop-watches. A gold pair-cased, quarter-repeating cylinder watch and an outer case decorated in repoussé by Henry Manley was perhaps the finest that could be purchased at the time.[105]

Others would also have flourishing businesses in London. Particularly successful was the Ellicott family. Their business was begun in 1693 by John Ellicott senior. He formed a partnership with his son John Ellicott junior in the 1720s and the business continued under Ellicott junior's son Edward. By the time it closed in the 1840s, 11,000 watches had been sold.[106]

The Lever Escapement

Thomas Mudge introduced a new form of escapement for portable timekeepers, which he first used in a table clock primarily designed to show the latest and most accurate lunar indication using a series of epicyclic gears.[107] The clock was completed in about 1755 and contained within it was Mudge's new escapement partly based on Graham's dead-beat escapement for clocks. The escapement incorporated a standard dead-beat escape wheel and pallets, but these were arranged to impulse a large-diameter slow-beating balance. The crutch (which in a regulator would have impulsed a pendulum) was arranged to have an open-ended fork, which engaged with a piece projecting from the balance staff. In its later developed form, the crutch in this escapement would form what became known as *the lever*. The balance, in moving the lever to and fro, would facilitate locking and unlocking while at the same time the lever would impart impulse to the balance, just as with the dead-beat escapement, except that for some of the balance swing, it was not connected to the fork of the lever. From his writings, however, he did not see the introduction of the fork on the end of the lever as a means of achieving a balance free of the escapement as it oscillated—a detached function. Indeed, Mudge felt that the escapement should at all times be engaged with the escapement throughout its oscillation. In his *Thoughts on the Means of Improving Watches* (1765), he stated: ' . . . the less space the balance moves beyond the arc that is necessary to its scaping, the greater quantity of motion can be procured by the same force from the mainspring . . . These are the reasons why it appears to me necessary that the arc it scapes should be large . . . '.[108]

It was to be after 1765, having witnessed the disclosure of Harrison's H4 timekeeper with its highly successful large watch balance (2.25 inches in diameter) running at high frequency and high amplitude, that he followed the lunar clock with a small portable clock, known today as the Polwarth clock.[109] In this clock he employed a modified version of the escapement with a specification of balance identical to that of H4, with a much larger amplitude and frequency. The clock went extraordinarily well and inspired him, in 1769, to create a pocket watch with a smaller version of the same escapement and balance; it was made for King George III, who presented it to Queen Charlotte in 1770.[110] This was to be the forerunner of the modern detached-lever escapement watch.

---

104 For a general account and manufacture in the French Jura, see Belmont 1975/1989.
105 For Graham watches with repoussé outer cases, Edgcumbe 2000.
106 Thompson 1997, 306–21 and 429–42.
107 British Museum 1958,1201.2118, Thompson 2004, 118–119.
108 Betts 2017, 38–9.
109 British Museum 1995, 0207.1, Thompson 2004, 120–1.
110 Royal Collection (RCIN 63,759) https://www.rct.uk/collection/63759/queen-Charlottes-lever-watch-and-pedestal. An informative animation of the function of the escapement can be found at https://vimeo.com/99220922.

Mudge's original concept, however, was complicated, difficult, and expensive to replicate. It would not be until the 1770s that the escapement was developed by others. Josiah Emery was the maker who took the lever escapement to the next level, making it a viable escapement for everyday watches. In producing it, Emery employed Richard Pendleton, a highly skilled escapement maker, and the result of their combined efforts was a series of pocket watches incorporating Emery's lever escapement. Emery was apprenticed in London in 1749–56 and established himself as a watchmaker in Warwick Street 'Opposite Hedge Lane'. His reputation as a watchmaker quickly grew, and in consequence he gained the attention of Mudge's patron, John Maurice, Count von Brühl, who encouraged Emery to develop Mudge's escapement. As Emery said:

> About seventeen or eighteen years ago Count Brühl repeatedly urged me to make him a watch upon an escapement Mr. Mudge had invented. I for a long time successively answered that Mr Mudge was the properist person for such an undertaking; for to own the truth, I doubted whether it would be possible to ever make a common sized pocket watch with an escapement on so large a scale. But the Count not contented with my repeated refusal, at last prevailed upon me. It was then he brought me a large frame, like a clock escapement, but at the same time he gave me no rules concerning it, nor any of the smallest hint of the construction of Mr Mudge's watch; nor did I ever see the said watch till many years after (for the Queen had it) and his Majesty himself did me the honor to shew it to me.

Between 1782 and 1794 Emery produced a series of lever escapement watches—at least thirty-eight were made. A good example is in the British Museum, signed *Josiah Emery, Charing Cross, London 1089* on the movement. The eighteen-carat gold case has the hallmarks for London 1786/7 and bears the case maker's mark 'VW', the mark of Valentine Walker. The white enamel dial is of the so-called **regulator** type. The movement is characteristic of Emery's watches, with finely gilt plates and a decoratively engraved bridge on the back plate for the balance. The balance is of a type designed by John Arnold and commonly known as the 'double S' referring to the two 'S' shaped bi-metallic compensation devices attached to the balance arms, intended to obviate timekeeping errors caused by temperature change.[111] Following Josiah Emery's pioneering work, others joined in the quest for a truly detached lever escapement, among them John Leroux and John Grant in London and Robert Robin and Abraham-Louis Breguet in Paris.

Shortly after Emery's development of the lever escapement in London, another English watchmaker, Peter Litherland (1756–1804), developed a different concept in which the watch balance was linked to the escapement by a rack and pinion, the escapement now referred to as the 'rack lever'. Litherland obtained two patents for his new escapement (Figure 87).[112] In partnership with Thomas Whiteside in Liverpool he produced large , the patented escapement.[113] Between 1792 and 1820 more than 10,000 watches were made. While the rack lever escapement had inherent friction problems, it was nevertheless robust in use and cheap and easy to make in comparison with Emery's escapement. As well as watches made and retailed under the Litherland & Whiteside name, which continued after Peter Litherland's death in 1804, countless others were made bearing the names of retail watchmakers throughout Britain.

Edward Massey (1768–1852) worked as a watch finisher from about 1790 and is known to have had business in Hanley and Burslem between 1795 and 1804. He was later based in Ironmonger Row, Coventry 1813–c.1821 before moving to London in the 1830s. In his early career, he developed a series of ship's logs and sounding devices from 1802 onwards. He was, however, primarily an escapement maker, providing escapements for others to incorporate in their watches. In 1812 he patented a new form of lever escapement for watches, with a further patent in 1814 for a revised version.[114] In essence, Massey's new escapement consisted in its first form of a solid steel roller, mounted on the balance staff, which engages in a notch in the end of the pivoted lever. As the balance oscillates, the single 'pallet' on the roller moves the lever back and forth so that the two pallets at the other end of the lever alternately lock and unlock the teeth of the escape wheel. As it rotates, the escape wheel teeth push the lever to and fro and in so doing the lever gives impulse via the pallet on the balance staff to keep the balance oscillating. This new escapement was subsequently modified by Massey so that eventually there were five different versions.[115] A rare and interesting example by Massey, which combines the use of his lever escapement and a function specifically designed for maritime use, is a watch he designed for marine surveying.[116] The main dial is numbered 1–4 and has hour and minute hands, although the minute hand is a plainly fashioned replacement. The minute hand revolves once in twenty minutes and the hour hand once in four

---

111 For examples of watches with the lever escapement, see Randall & Good 1990, especially 130–2 for Emery no.1089. For the life and work of Josiah Emery, see Betts 1996–1997, 26–44, 134–50, 216–30, and Betts 1996, 394–401, 510–23.

112 British patents N° 1830 14 October 1791, and N° 1889, 12 June 1792.

113 For an account of Litherland and his watches, see Mercer 1962, 316–23, Evans 2010, 93–6.

114 Patent no. 3559, 30 April 1812, 'Chronometers' and Patent no. 3854, 17 November 1814, 'Chronometers and pocket watches'.

115 For Massey see Treherne 1977; Kemp 1982, 558–64.

116 For details of the watch, see Thompson 2008, 108–9, or https://research.britishmuseum.org/research/collection_online/collection_object_details.aspx?objectId=56857&partId=1&searchText=Edward+Massey&page=1

Figure 87 Movement of a watch by Peter Litherland hallmarked 1798. Photo: Courtesy of Sotheby's, London.

hours. Each hour division on the dial is divided into fifteen—i.e. four-minute intervals, and there are further divisions within the chapter-ring at twenty-minute intervals. At the bottom are two subsidiary dials, the one on the right has a hand revolving once in twenty seconds, its circle divided into 2½ second intervals, and the dial on the left is divided into 2½ seconds and divided to ½ second intervals. Thus, in operation, this seconds hand revolves around the dial in 2½ seconds—no mean feat for a mechanical device.

The purpose of this timer was for surveying at sea, based on the difference between the speed of light and the speed of sound. A small boat, or a number of boats, would be despatched some distance from a mother ship but remain in sight of it. A cannon would be fired, and the time between seeing the gun flash and hearing the report would be taken. Sound travels at 720mph at sea level and so the distance can be calculated from the time difference. The system could equally be used on land, and Massey, himself, in a letter written to the Board of Longitude on 17 June 1823, describes a similar watch:

A watch with decimal seconds for the purpose of ascertaining with accuracy Distance by Sound and also for ascertaining minute divisions of time in taking observations: I applied myself to produce this result at the suggestion of Captain Owen. This watch also answers very well for the pocket.[117]

The Massey lever escapement was adopted and used by a number of Liverpool makers, particularly Robert Roskell, well into the nineteenth century, until eventually the table roller lever escapement gained preference, either in the 'English' pattern, generally referred to as the 'ratchet-tooth lever' or the 'English lever', or in the Swiss form, now referred to as the 'club-tooth lever escapement'. The difference between the two is the shape of the escape wheel teeth; the former has teeth tapering to a point, whereas the latter has a shaped end dividing the lift of the escapement between both tooth and pallet and aiding oil retention.[118]

The Duplex escapement

In parallel with the development of the lever escapement, another escapement appears in English watches towards the end of the eighteenth century—the duplex. This had its origins

---

117 The specific maritime use of Massey watches like these is described in detail by Belcher 1835.
118 For a technical analysis of the later lever escapements, see Gazeley 1956, 168–220.

in an escapement designed by Daniel Delander (c.1678–1733) of London in about 1715 and fitted to a series of longcase clocks.[119] It was introduced for watches by the Paris maker Jean-Baptiste Dutertre (1715–42). In essence, the escapement has two escape wheels: one for locking and unlocking and one for impulse.[120] A refined form was developed by Pierre Le Roy in 1750. In 1782 in England, Thomas Tyrer was granted a patent (N° 1311), for his version. At the outset, the escapement relied on two separate wheels, but this changed to a single escape wheel with radial teeth for locking and unlocking and a series of raised teeth on the upper surface of the wheel rim for impulse. While this escapement was less used in other countries, it became one incorporated in good-quality English watches from the later eighteenth century; however, it was more commonly used in the early part of the nineteenth. A later form known as the Chinese duplex was extensively used in Swiss watches made for export to China, particularly by makers such as Bovet in Fleurier.[121] The escapement, chosen for its ability to produce the appearance of a jump seconds hand, was popular with the Chinese.

MUDGE, ARNOLD, AND EARNSHAW

The improvements made to watches in London in the second half of the eighteenth century owe an enormous debt to the work of John Harrison (1693–1776) in his quest to make a timekeeper capable of maintaining a sufficiently accurate rate to establish longitude at sea.[122] As part of this quest, the production in 1752–3 of a watch designed by Harrison, now referred to as the 'Jefferys watch', and later the making of Harrison's marine timekeeper H4, completed in 1759, have great significance for the subsequent development of precision timekeeping for domestic use. Harrison demonstrated the great advantage of providing jewelled bearings for the watch going train as well as for the balance. An interesting aspect of this was the apparent secrecy surrounding the use of jewelled bearings in English watches and chronometers. Jewelled bearings for pivots were still not being produced in France fifty years after Nicholas Fatio and the Debaufre brothers had developed them. Berthoud is quoted as saying, 'If some parts of the watch would be difficult to make, there are others which could not be done at all in France. I mean the pierced rubies carrying the pivotal staffs', and in 1771 Pierre Le Roy reiterated the point saying 'Whereas in Harrison's watch and indeed in all good watches in England, the balance staff and the last wheels of the train are set and move in pierced rubies, we in France have not the secret of making these rubies'. The practice of including jewelled bearings for the escapement in high-quality London-made watches was common, but it was not until the second half of the eighteenth century that jewelled bearings were more frequently provided for the train wheels. This then became standard for precision timekeepers.

Thomas Mudge (1715–94) was without doubt one of the most accomplished and ingenious watchmakers working in eighteenth-century London. Apprenticed to George Graham, he finished his apprenticeship in 1738, became a Freeman of the Clockmakers' Company, and set himself up in Fleet Street not far from his former master. There he made watches for other makers, in particular, John Ellicott junior. In 1750 he went into partnership with William Dutton, also a former apprentice of Graham. Watches signed *Mudge & Dutton* represent the 'bread and butter' of their production, but Mudge continued to make his own watches as well as special commissions for other makers. Of particular note are two perpetual calendar watches made in the 1760s, the earliest to be made in England. One is in private ownership and a slightly later one, hallmarked 1764, in the British Museum.[123] These extraordinary watches not only show the date with longer and shorter months changing automatically, but also adjust for leap years.[124]

In 1752, the clockmaker Michael (Miguel) Smith, Ferdinand VI of Spain's agent, was in London and visited Mudge, already well aware that he was making watches showing the equation of time for John Ellicott junior. On 3 July of that year Smith wrote to Mudge and made mention of a watch with equation of time indication that he had seen in Mudge's workshop and explained that he, Smith, had recommended Mudge to the king and court:

> Since my return to this Court wch was ye 20 of May I have Being Asked By ye King, ye Prime minister and severall of ye Nobility who was the most Capable person now Mr Graham is dead and Believing Mr Mudge worthey of my Recommendation I Gave you the preference.[125]

Mudge & Dutton watches are of the high quality expected of London watchmakers at the time and, continuing in the tradition of Graham, they have cylinder escapements. There are centre-seconds timepieces as well as quarter-repeating watches in fine

---

119 For these see Evans & McBroom 2020, 525–6.
120 It is described in Thiout 1741, 102. See also Chapiro 1991, 322–3.
121 Described in White 2019, 98–9.
122 Discussed in Chapter 11.

---

123 See Camerer Cuss 2009, 219, pl. 129 for watch no. 525 and http://www.sothebys.com/en/auctions/ecatalogue/2016/john-harrison-enduring-discovery-l16055/lot.28.html. For the British Museum watch, no. 574, see Good 1981, 181.
124 For the life and work of Thomas Mudge, see Daniels 1981, 150–73, Allix 1981, 627–34, Cliborne 1981, 144–5, and Cheetham 2002, 90–3.
125 Turner & Crisford, 1977, 580–2.

gold or silver cases, many of which bear the case maker's punch mark, PM, for Peter Mounier.[126]

John Arnold (1736–99) was born in Bodmin, Cornwall, son of Richard Arnold, to whom he was apprenticed in c.1754. By c.1763 he had moved to London and set up business in Devereux Court, The Strand, London. He was a member of the Clockmakers' Company from 1783. His son John Roger Arnold (1769–1843) was apprenticed to his father and was then in partnership with him until his father's death.[127] Arnold became one of the foremost watchmakers and chronometer makers of his time, contributing more to the development of the successful marine chronometer, post Harrison, than any other maker. Following the death of his father, John Roger continued the business and was in partnership with Edward John Dent from 1830 to 1840.

On the domestic front John Arnold proved himself to be an extraordinary watchmaker, not least by making two miniature watches for King George III. The first, in 1764, was a tiny watch set into a finger ring, which had a quarter-repeating mechanism and a cylinder escapement. In a description of the watch published in the *Gentleman's Magazine*, the whole movement was described as weighing only 2 dwts, 2 grs, and an eighth of a grain. A second finger-ring watch was presented to the King in 1770 with a ruby cylinder escapement, the first of its kind to be made. In a description of the ruby cylinder in this watch it was said to be only 1/54 part of an inch in diameter and only 1/47 part of an inch long. Following John Arnold's pioneering work on the development of the detached spring detent escapement for marine chronometers and temperature compensation in portable timekeepers, the Arnold business produced pocket chronometers made for personal use that incorporated these improvements. A typical Arnold pocket chronometer had jewelled bearings for the gear train and escapement, maintaining power to keep the watch going during winding, a spring detent escapement, and a temperature compensated balance, making the watch a true precision timekeeper for the pocket.[128] A typical example is a pocket chronometer signed *John Arnold, London. Inv et Fecit. No.21/68*.[129] Here the watch, of large size, incorporates the latest refinements in precision timekeeping. Interestingly, Arnold's chronometer no. 1/36 was the first to be termed a 'chronometer'.

Thomas Earnshaw (1749–1829) also had a profound influence on the development of precision timekeeping. Like John Arnold, he was to become a central character in the story of precision timekeeping and marine chronometry. Earnshaw was born in Ashton-under-Lyne and is said by Alexander Dalrymple to have been apprenticed to William Hughes in Holborn, London.[130] In contrast to John Arnold, Earnshaw's life was in some ways more troubled. He began his career as a watch finisher and escapement maker, and even after a spell in debtor's prison, his dogged determination saw him become eminently successful.[131] During his career, Earnshaw not only made watches that he retailed under his own name, but also made movements and escapements for other watch 'makers'.[132] His production included watches with either cylinder, or duplex, escapements, and spring-detent escapements in his pocket chronometers. He also developed his own temperature-compensated balance, which became standard for nineteenth-century precision timekeepers. Earnshaw's watch serial numbers reflect his business production; he used two numbers, one for the individual number of the watch and the second for his overall production, which included those which he made for other retailers. A typical example of an Earnshaw pocket chronometer is no. 506/2849 in the British Museum. It has a plain brass movement, 'sugar tongs' temperature compensation, and spring-detent escapement. It is housed in gold pair cases made by Thomas Carpenter and hallmarked London 1800.[133] After his death, his son carried on business in High Holborn, London.

While this was a period of exceptional development in precision timekeeping, the enormous number of more everyday watches with verge and cylinder escapements should not be overlooked. A measure of the volume of watches being made in the eighteenth century is provided by the business of James Upjohn. As the century progressed there were an increasingly large number of watch 'makers' outside the main production centres of London and Liverpool, affixing their names to watches they purchased. Upjohn, a watchmaker with business in London and Exeter, had a very considerable workshop production. He explains in his autobiography that:

> I did not quit them [retailers] for I had never refused their orders, but I had so much business at home and abroad I could not wait on them in person. I had done a great deal of business all over England and had made 2,500 watches a year for many years successively for the country trade only.

In furthering his business, Upjohn travelled to towns and cities in the Netherlands and Flanders, as well as Germany and France. In addition he sent his sons Edward, James, and Francis to Amsterdam

---

126 Mark entered at Goldsmiths' Hall in 1761, address given as Frith Street Soho. Still working in 1773.
127 For Arnold's life and work, see Mercer 1972 and Staeger 1997. *Cf.*, as for Mudge and Earnshaw, this volume Chapter 11. Also Betts 2017.
128 For a detailed analysis of marine and pocket chronometers by John Arnold and his successors, see Randall & Good 1990, Staeger, 1997
129 See Randall & Good 1990, cat. no. 8, and Thompson 2008, 96–7.
130 In *Longitude, a full answer . . .*, 1806, but no confirmation of this is available.
131 For Earnshaw's career, see Betts, 2017
132 Randall 1988, 367–71 lists known Earnshaw watches and chronometers.
133 See Randall & Good, 1990, cat. No. 61, and Thompson 2008, 104–405, also https://research.britishmuseum.org/research/collection_online/collection_object_details.aspx?objectId=57637&partId=1&searchText=Thomas+Earnshaw&page=1 (accessed 2020).

to further the trade. In Upjohn's account, there is also information concerning his prices. In 1763 he supplied James Cox with two skeletonized-movement watches for thirty-six guineas and charged £300 for two gold and enamel-cased repeating watches set with diamonds.[134]

THE GERMAN STATES AND THE AUSTRO-HUNGARIAN EMPIRE

Outside England in this period, developments were slower and less significant. German watchmaking was badly affected during the Thirty Years' War and took a very long time to recover. Like France and Switzerland, it was still underdeveloped at the beginning of the eighteenth century, with a backward-looking view of the art prevailing in such older centres as Augsburg in contrast with the forward-looking English makers or indeed with those of Friedberg. As the eighteenth century progressed a series of disputes between the fragmented German states and those in the Austro-Hungarian empire also took their toll. With cultural life decentralized and provincial, it was not until the mid-eighteenth century that there was a revival of the economy and culture.

Evidence of the efforts that led to such revival was the publication of Johann Georg Leutmann's treatise on clocks and watches, one of the first general treatments published.[135] It displays a desire to disseminate precise information about the intricate details of the machines. Slowly revival occurred. One German maker who stands out is Philipp Matthäus Hahn (1739–1790), born in Scharnhausen auf den Fildern. Two of his sons, Christof Matthäus and Christian Gottfried Hahn also became watchmakers. Hahn was minister of the parishes of Onstmettingen, Kornwestheim, and Echterdingen and while following his career as a clergyman, he also trained as a watchmaker and set up a flourishing workshop where members of his family worked. Hahn produced a number of good-quality watches, many of which had calendar and lunar indications.[136] A place where watchmaking thrived was Friedberg, near Augsburg. Georg Bayr and Leonhardt Engelschalk were working there in the 1660s making fashionable watches; examples of their work survive to this day.[137] Later, Friedberg watches met considerable success and were widely exported around Europe. In the mid-eighteenth century, one maker stands out. He had an interesting line in what were effectively forgeries in terms of how he signed his work. He was Joseph Spiegel, and he seems to have specialized in making coach-watches of large size, and while the signatures are misleading, the watches are of good quality. A number of examples are known today, some of which are signed *Legeips London*, 'Legeips' being his name backwards, his name Spiegel meaning 'mirror'. There are also coach watches by him which are signed *Miroir London* and *Miroir à Paris*.[138] Clearly, he was adding the prestigious place name in a spurious manner, but in case he was challenged, spelled his name backwards as a cover. This practice was also carried out by at least one maker in Augsburg, a member of the Graupner family who signed his watches *Renpuarg London*.[139] A further example is a watch by Johann Heckel where the movement plate is signed *Lekceh, London* and the silver cap is decorated with the arms of Queen Anne.[140] Watches made in Friedberg during the eighteenth century often followed the English pattern with a balance cock, rather than the bridge widely used throughout Europe at the time. But as elsewhere in Europe, German watchmaking was threatened by a growing preponderance of Swiss imports adversely affecting local trade.

FRANCE

The emigration of Huguenot watchmakers from France after 1685 contributed to a significant decline in watchmaking. By 1700 it was clear that watch production was seriously affected. At the same time there was a marked increase in the number of London watches coming into France.[141] Catherine Cardinal makes the point that, in London, watchmakers themselves were involved in the theoretical science of the machine, whereas in France research was in the hands of non-practising *savants*. This, however, is not to say that there were not makers of fine watches in France. The Martinot family, Balthazar Martinot in particular (1636–1712), received royal patronage enabling his workshop to produce quantities of fine watches, many of which survive to this day.[142] In about 1690 a style of watch emerged known today as an 'oignon', the name derived from its bulbous shape.[143]

---

134 Leopold & Smith 2016, 127 ff.
135 Leutmann 1722. See the discussion of it in Chapter 26 this volume.
136 See Abeler 1968, 248–50. Hahn watches are described and illustrated in Väterlein 1989, I, 467–516. An interesting watch with twenty-four-hour dial and calendar indications can be seen at: https://www.uhrenmuser.de/en/49002/hahn-a-echterdingen-philipp-matthaeus-hahn-pocket-watch (accessed Feb 2020).
137 For a survey of watch and clock making in Friedberg, see ; Riolini-Unger, Friess, & Hügin 1993.
   See also, Arnold-Becker 2014, 663–82 and 783–95; Whitestone 1993; this volume Chapter 7, Section 1.

138 There is an example in the British Museum, see Thompson 2008, 72–3, British Museum (1888, 1201.256).
139 The watch is signed 'Renpuarg London no. 4', with a dust cap engraved with the British Royal Arms. As part of the Belin Collection, it was sold at Sotheby's, Geneva, 17 November 1997, lot 175.
140 Museum im Wittelsbacher Schloss Friedberg, Inv. Nr. 1996/234.
141 See Cardinal 1983, ch. 3.
142 Balthazar Martinot (1636–1714), *valet de chambre* to Louis XIV. A typical example of a Martinot movement can be seen at: https://watch-wiki.org/index.php?title=Datei:Balthasar_Martinot_%C3%A0_Paris_Spindeltaschenuhr_(7).jpg.
143 A very typical 'oignon' watch by Gauthier of Paris is described and illustrated in Thompson 2008, 64–5. For detailed discussion of the different types see Chapiro 1991, 35–100.

In the same way that London watchmakers moved to a watch with more space between the plates to allow for a higher and consequently stronger mainspring, so too did French makers, but their distance between the plates was much greater. The need for an even stronger mainspring was brought about by the use of a much larger, and therefore much heavier, balance than that used by English makers. The French craftsmen believed that a large-diameter heavy balance would produce a better performance from the watch.

The typical 'oignon' had a single case with a glazed bezel, and for the most part was made either from silver or from gilt brass with repoussé or engraved decoration on the back, band, and bezel (Figure 88); gold-cased examples are unknown. Early examples of such watches had only one hand indicating time against a chapter ring with separate enamel plaques for the hours and an enamel minute circle. Winding the watch was carried out using a key applied to a square at the centre of the dial. In comparison with London watches with fine gold cases and concentric hour and minute hands, these 'oignons' were somewhat old-fashioned in their appearance, albeit robust in their performance. The movements had a verge escapement and a bridge for the large-diameter oscillating balance and doubtless served their task well. By about 1710, a one-piece enamel dial had been introduced, with either a single hand or concentric hour and minute hands. Given the number of French watchmakers living and working in Geneva, it is not surprising that the 'oignon' was also produced there as indeed elsewhere.[144]

Julien and Pierre Le Roy, father and son, are eminent in French watchmaking.[145] Both held the office of *horloger du roi*, and their workshops produced large numbers of good-quality verge watches as well as some incorporating newly invented watch escapements.[146] Julien Le Roy published a short guide to regulating a watch,[147] in which he says that a watch should be able to keep time to within a minute a day and that watches should be adjusted according to a sundial or good pendulum clock. Local church clocks, he says, should not be trusted. It is estimated that the business of Julien Le Roy, followed by his son Pierre, made about 5,000 watches between 1720 and 1779.[148] Pierre Le Roy was also active in improving the timekeeping of portable machines, but made no progress with his work on the temperature compensation balance for watches, and continued to use bi-metallic curbs for compensation in these smaller timekeepers. It was during the early years of Julien Le Roy that the ordinary French watch became smaller and more compact allowing for easier use in the pocket than the 'oignon', had been. By this time in France, the business of making a watch had developed into a diverse one. Cardinal gives an excellent account of the way in which Julien Le Roy organized his workshop and the watchmakers who specialized in different aspects of producing the completed and working watch.[149]

Ferdinand Berthoud was one of the most celebrated watchmakers in France during the second half of the century. Born 18 March 1727 in Plancemont in the canton of Neuchâtel, Switzerland, he was apprenticed at the age of fourteen to his brother Jean-Henri Berthoud to learn the art of clock and watchmaking. In 1747 he left for Paris where in 1753 he became a master watch and clockmaker and was to spend the rest of his life until his death in 1807. He was a member of the Institut de France, a Fellow of the Royal Society in London, and was awarded the medal of the *Légion d'honneur*. He enjoyed a royal pension and was General Inspector of marine timepieces. In addition to Berthoud's work on marine timekeepers, his workshops were prolific in their output of good-quality verge watches housed in cases with enamel decoration or in cases of multicoloured gold, in line with the popular style of the second half of the eighteenth century. Besides being an accomplished maker, Berthoud was also one of the rare horologists who wrote on the subject, and did so extensively following in a line of practising Paris watch and clockmakers who wrote about their art.[150]

THE VIRGULE ESCAPEMENT

In 1752 a new escapement was introduced in France by Pierre Auguste Caron, later known as Beaumarchais (1732–99). While it has similarities with the cylinder escapement, it differs in having an escape wheel with teeth in the form of raised pins on each side of the wheel that interact with two hook-shaped pallets. The cylinder on the balance staff is cut away to provide locking and unlocking, and the hook-shaped pallets provide the impulse in both directions of the balance oscillation. At about the same time, Jean André Lepaute (1720–89) designed a similar escapement, also with two pallets. Jean-Antoine Lépine (1720–1814), who married one of Caron's sisters and worked in partnership with his father-in-law André Charles Caron (1697–1775), produced a simpler version of Lepaute's escapement. This version had raised teeth in the plane of the wheel with upturned ends, which acted with a single virgule providing impulse in one direction only. The virgule escapement was not widely used, even in France.[151]

---

144 Examples in Chapiro 1991, 96–7, 100.
145 For Julien Le Roy see Cardinal & Sabrier 1987; for Pierre Le Roy, Brusa & Allix 2006, 645–62 and 775–89. Pierre, however, has to be distinguished carefully from his uncle, also Pierre, on whom, see St-Louis 2020.
146 A good, standard Le Roy watch in a gold and bloodstone case, movement signed *Jul^n. Le Roy A Paris,* in the Victoria and Albert Museum: http://collections.vam.ac.uk/item/O113171/watch-le-roy-julien/.
147 Le Roy 1719. It was reprinted in 1741.
148 See Chapiro 1991, 127.

149 See Cardinal 1989, 52–3.
150 Cardinal 1984; this volume Chapter 26.
151 Gazeley 1956, ch. 10. Also, British Museum (1958, 1201.1429) signed *L'Epine horloger du Roy A Paris N.1349.* See https://research.british

Figure 88 Movement of a French verge watch (*oignon*) by P. Dvchesne, Paris; back showing the large elaborately decorated bridge and the edges of the shagreen case with silver picqué decoration. Photo: Jean–Baptiste Buffetaud, courtesy of Chayette & Cheval auctioneers, Paris.

Until the 1760s, a watch movement usually consisted of two full plates between which the wheels and escapement were mounted. On the back was a separate cock or bridge for the top pivot of the balance. Jean-Antoine Lépine introduced a new arrangement, now generally referred to as the 'Lépine' calibre (Figure 89). Here, the watch is built up on a single plate to which separate cocks and bridges are screwed to provide bearings for the various wheels and pinions. In addition, the fusee was abandoned and power was provided by a hanging going-barrel, with the first gear wheel forming an integral part of the barrel wall.

The aim of this new construction was to facilitate assembly and simpler access for maintenance as well as to allow the watch to be made thinner. The new *montres classiques* began to appear in the 1762–72 period but it was between 1772 and 1792 that the main production took place.[152]

ABRAHAM-LOUIS BREGUET

If there is one maker who eclipses all others in the history of French watchmaking, it is Abraham-Louis Breguet. During his lifetime he built a reputation second to none; his watches were owned and prized by the crowned heads and aristocracy of Europe and particularly favoured by the Tsar and the imperial court of Russia. The Duke of Wellington owned several pieces by Breguet.[153] Born in Neuchâtel, Switzerland, at the age of 15 he was sent to France, first to Paris and then to Versailles, where he studied with the watchmaker Etienne Gide. He married Cécile L'Huillier in 1775 and set up in business at Quai de l'Horloge in Paris in partnership with Xavier Gide, the brother of his former master. Before the Revolution he built a high reputation in Paris with close court contacts. This imperilled his security during the years of the French Revolution, so he took refuge in his birthplace during the worst excesses. On his return to Paris in 1795, his business was virtually in ruins. However, during his stay in Switzerland, he had been planning for his return, formulating new ideas and new calibres of watches. One of his new projects was the *souscription* watch, which was to be simple to produce, but reliable in function. The watches cost 600 *livres* in silver, 800 *livres* in gold, with the customer paying a quarter in advance and the

---

museum.org/research/collection_online/collection_object_details. aspx?objectId=57168&partId=1&searchText=Virgule&page=1.
152 Chapiro 1988.

153 Breguet 1997, 200–1. Apart from this standard biography see Ruellet Daniels 1975.

Figure 89 Backplate layout of a watch by Jean Antoine Lépine c.1780. Photo: Courtesy of Sotheby's, London.

remainder on completion. The first record of such a watch concerns N° 96 begun in 1794 and completed two years later. In 1797, Breguet published a printed brochure soliciting subscriptions.[154]

The *souscription* was a great success, resulting in a simple, elegant watch of quite large diameter with a single hand to show the time on a white enamel dial. The movement had no refinements. The mainspring barrel was mounted centrally and was wound by inserting a key through a hole in the centre of the dial and hand. A simple gear train terminated in a cylinder escapement. Here, however, in place of the normal steel cylinder, the Breguet *souscription* watches have a cylinder made from a single piece of ruby, fashioned into a hollow tube with part of the side cut away very precisely to act with the revolving escape wheel, controlling its rotation and, at the same time, giving impulse to the balance. Another of Breguet's refinements was a bi-metallic compensation device attached to the balance cock. Operating on the end of the balance spring, it changes its position as the temperature changes and thus compensates for these changes, which have a profound effect on the elasticity of the spring and consequently on timekeeping. Breguet designed his own version of this bi-metallic steel and brass compensation device.[155]

Breguet applied himself to the improvement of watches and was a leader in the introduction of new technology. He was an early maker of the watch with automatic winding, which he called the *perpetuelle*. He devised a resilient top pivot bearing for the balance which is known as the '**parachute**' suspension. He devised his '**tourbillon**' watch, in which the escapement and balance are mounted in a carriage which is constantly rotating in an attempt to minimize the timekeeping errors caused by a watch being in different positions.[156] Following the invention of the terminal curve by John Arnold in 1780, he also introduced a system where the terminal outer coil of the balance spring was curved inwards so that its fixing point was as near to the centre of oscillation as was possible so that the balance spring would expand and contract

---

154 Reprinted with commentary in Rigot 2017, 14, now the fullest study of *souscription* watches.

155 A typical example of a Breguet *souscription* watch can be found in the British Museum; see https://research.britishmuseum.org/research/collection_online/collection_object_details.aspx?objectId=58298&page=1&partId=1&peoA=77126-3-17&people=77126, and is described and illustrated in Thompson 2008, 102–3. See also the many examples in Rigot 2017.

156 For an account of the tourbillon watch, see Meis 1986.

equally in all directions, giving a better performance and thus improving timekeeping.

Breguet was able to achieve a consistently high quality for his watches (Figure 58) by employing the best craftsmen in his workshops. Whereas many would employ cheap labour in the form of apprentices to carry out the basic work, Breguet employed men highly qualified in all aspects of the work and, when satisfied with their work, paid them generously. An example of this can be seen in the watch that Breguet modified in memory of his friend, the London chronometer maker John Arnold. Arnold's son John Roger Arnold gave him one of his father's pocket chronometers, no. 11. Breguet modified it by designing a tourbillon escapement for it and re-casing it as a desk watch. In all, there were fifteen different, highly skilled craftsmen involved in the work, ranging from Renevier, who made the escapement, to Tavernier, who made the new silver case.[157]

From the post-revolutionary period, the Breguet workshops continued to produce relatively large numbers of watches of all types (Figure 158), all of the same high quality, whether simple *souscription* watches or highly complex pieces made on special commission. Without doubt, one of the most celebrated watches to come from the Breguet workshops was the incredibly complicated watch commissioned as a gift for Marie Antoinette. She did not live to see it completed; nor did Abraham-Louis himself, for it was not finally finished by a team of Breguet craftsmen under the direction of Antoine-Louis Breguet until 1827.[158] At the time it was the most complicated watch ever made, with an automatic winding system, minute-repeating mechanism, perpetual calendar, and equation of time indication as well as an independent centre-seconds function. In addition, there were dials to show the state of wind of the watch and a thermometer.[159]

In France in 1792, following the Revolution, a new calendar and division of time was introduced.[160] There are a small number of surviving watches using the decimal system, but few are outstanding. Robert Robin (1742–99)[161] made a small group of watches with decimal indications, one of which is in the Palazzo Falson, Mdina, Malta. As well as having a ten-hour dial, this watch contains Robin's own version of the lever escapement. The watch is signed 'Robin a Paris No 2, Echappement Libre inventé en 1791'.[162]

THE NETHERLANDS

The introduction of the balance spring to watches in 1675 saw a resurgence in watchmaking in the Netherlands, and a new generation began to establish a more robust trade. Particularly active were the workshops of Fromanteel & Clarke in Amsterdam. One of the sons (or both) of the London-based clockmaker Ahasuerus I Fromanteel set up a business in Amsterdam with Ahasuerus I's son-in-law Christopher Clarke.[163] Although not universally the case, some watches inscribed with their name were clearly made in London.[164] Dutch movements are typified by large balance cocks and extended, pierced regulator plates, which leave little room for a signature and consequently no space for a place name to be inscribed, often leading to the misconception that they were indeed London made. The Dutch dial in the later part of the seventeenth century was in many ways a parallel to those found on watches made in London, in gold, silver, or gilt metal to match the case metal and of the champlevé type with raised Roman hour numerals and half-hour marks for the chapter ring with a numbered minute circle around the outside. On the inside of the chapter ring was a circle normally calibrated to quarter-hours. Not long into the eighteenth century a new design appeared in which the circumferential minute circle was arcaded. Later, in the 1730s, when white enamel dials became the fashion, this design continued to be used and today it is commonly referred to as the 'Dutch' dial. Another form of dial popular in the Netherlands at the turn of the eighteenth century was the sun-and-moon dial. Here, the dial has an outer circle for minutes indicated by a single hand. The upper half of the dial has a semi-circular aperture surrounded by hour numerals beginning at VI on the left, with midday XII at top centre and VI to the right. Within the aperture is a disc with sun and moon images that rotate once in 24 hours. Below the centre a depiction of *Chronos* is a common feature. An example can be seen in a watch by Jacob Hasluck of Amsterdam.[165]

---

157 For a detailed description, Randall & Good 1990, 201–4, N° 176, and Thompson 2008, 84–5. See also https://research.britishmuseum.org/research/collection_online/collection_object_details.aspx?objectId=58522&partId=1&searchText=Breguet&page=1.

158 https://www.breguet.com/en/house-breguet/manufacture/marie-antoinette-pocket-watch.

159 The watch was bequeathed to the L. A. Mayer Institute for Islamic Art, Jerusalem, by Sir David Salomons in 1925, along with 56 other fine-quality Breguet watches. See Daniels & Markarian 1980, 34–5.

160 For details, see Shaw 2011, also this volume Chapter 12.

161 Fima 2010, *passim*.

162 For the Robin watch see http://www.palazzofalson.com/collection_detail.aspx?id=74003. For a watch by Chambon of Paris in the British Museum with dual time showing decimal and duodecimal hours and minutes see https://research.britishmuseum.org/research/collection_online/collection_object_details.aspx?objectId=57888&partId=1. See also this volume, chapter 12.

163 For the Fromanteel family see Loomes 1981a; 1981b, 236–8 and references there given.

164 A good example is a watch sold by David Penney's Antique Watch Store in 2020. Fromanteel & Clarke no.1728, *c.*1700, housed in silver pair-cases made by the London case maker William Jaques. https://www.antiquewatchstore.com/archive/2360-fromanteel-clarke-no-1728.html.

165 Jacob Hasluck, Amsterdam (working 1695, died 1747). The watch is in the Metropolitan Museum of Art, New York (17.190.1499a, b).

In common with London and Geneva, the Netherlands saw an influx of French Huguenots from 1685, for example, Estienne II Hubert of Amsterdam, originally based in Rouen.[166] However, the numbers involved were small and do not appear to have exercised any influence on the style of the Dutch watch; there are no French style 'oignons' to be found with Dutch names on the movements.

In Rotterdam there was a flourishing watchmaking industry with makers such as William Gib, father and son, at the forefront of the business. Interestingly, authentic William I Gib watches have the movement number punched on the shoulder of the inner case next to the pendant stem. Those that do not have this feature appear to be Swiss. In terms of movement design, there was a period at the end of the seventeenth century and into the eighteenth where the balance oscillates beneath a balance bridge which has an aperture, which allows visibility of a small disc set on the balance, rather like a pendulum swinging in a clock. Indeed, these watches are described as having a 'mock pendulum', long after the pendulum was introduced to clocks but reflecting a belief that the balance spring was its equivalent. They were popular in the Netherlands from about 1690–1730. Another aspect of Dutch watches in the second quarter of the eighteenth century is the extensive use of repoussé silver cases decorated with classical scenes. It seems likely that these outer cases may have been imported from Switzerland where they were being made in large quantities. A typical example of a Dutch watch from about 1730 is a silver pair-cased watch by William II Gib which has a mock-pendulum balance beneath a balance bridge with an applied silver decorative piece with cupid reading a book. The outer silver repoussé case is decorated with a depiction of The Abduction of Helen of Troy, signed by Daniel Cochin in a reserve at the bottom of the design.[167]

Judging from the number of surviving watches, production in the Netherlands was not huge, and while there were watchmakers in Amsterdam and all the other major urban centres, they were relatively few compared with London, Paris, or Geneva. Also significant was the level of watch imports from Switzerland sold in the Netherlands at prices considerably lower than those locally made. Evidence for trade links with Geneva comes from the Daniel Cochin outer case on the William II Gib watch, but other evidence exists in the numerous watches with movements by Dutch makers contained in fine Geneva enamelled cases. One such watch exists with a movement dating from *c.*1700 signed *Pieter Klock Amsterdam*. The movement is in English style with a balance cock, although the open pierced movement pillars are stylistically associated with continental makers. The movement is housed in a finely enamelled case by the enamellers Jean-Pierre and Amy Huaud in Geneva.[168] Like others, although the watch has a modern movement, the dial is outdated, having only a single hand. Probably Swiss dominance caused Dutch watchmaking to decline towards the end of the eighteenth century, never to recover.[169]

SWITZERLAND IN THE EIGHTEENTH CENTURY

From the end of the seventeenth century, the fortunes of the Swiss watchmaking industry were somewhat in decline, and the great names from that century such as Sermand, Martin, and Ester were no longer in business. There was, however, a corpus of makers who continued to work, particularly in the St Gervais district of Geneva, and in the early part of the eighteenth-century watchmaking began to develop in Neuchâtel, the Jura, and the Vallée de Joux. An Italian refugee estimated that in 1686 there were 100 master watchmakers and 300 journeymen in Geneva.[170] The numbers increased steadily, and it is estimated that in 1790 the city was exporting something in the region of 14,000 gold-cased watches and 45,000 silver-cased watches. The same story pertains in Le Locle, where in 1750 there were forty-one watchmakers and forty case makers, while La Chaux-de-Fonds had sixty-one watchmakers and thirty-eight specialists. In 1764 it is estimated that together they produced about 15,000 watches. Given the high percentage exported, there was evidently a marked impact on the trades of other watchmaking countries, with ready markets in regions such as Scandinavia, Spain, and Portugal, where local watchmaking was negligible.[171]

An aspect that aided the Geneva makers in the production of very large quantities in the eighteenth century was the structure of their trade. They set up a system in which different parts of the watch were made in different workshops, all the various components then being brought together in the business of the maker whose name would appear on the watch in its finished form, the *établisseur*. He then sold the watches directly or passed them on to merchants who would distribute them to customers widely dispersed around Europe, especially to those countries where there were no native makers, such as Russia and Scandinavia. Swiss traders also made inroads into the French market,

---

https://www.metmuseum.org/art/collection/search/120008803?rpp= 60&pg=4&ao=on&ft=Pocket+watch&pos=225 (accessed Feb 2020).
166 Estienne II Hubert, member of a celebrated family descended from Robert Hubert, recorded in Amsterdam from 1702.
167 The inscription 'D. Cochin' appears on numerous examples of silver repoussé watchcases; see Patrizzi 1998, 139.

168 Pieter Klock, clock and watchmaker of Amsterdam, 1665–1754. The watch is now in the Metropolitan Museum of Art (17.190.1436a, b). https://www.metmuseum.org/art/collection/search/194029 (accessed 2020).
169 Jaquet & Chapuis 1970, 122–3, has a short note of Swiss trade with Holland and Belgium. For Dutch watches in general see Beringen 2012.
170 Babel 1938.
171 See Jaquet & Chapuis 1970, 88–9.

the Swiss watch being cheaper than the homemade commodity. The making of the multitude of different components was in the hands of *cabinotiers*, including movement makers, case makers, spring-makers, or enamellers, working in their own small workshops where they produced the finished components from the raw materials.[172] By 1788, the number of people involved in the watchmaking trades in Geneva numbered 6,423, comprising watchmakers, case makers, engravers, spring-makers, jewellers, goldsmiths, lapidaries, and enamellers, as well as others in allied trades. Across the country a subtle change began to take place in the way the watch was made. It was here that the **ébauche** was developed, the making of a completed rough movement which would then be supplied to an engraver and a gilder. It would then pass on to an escapement maker and watch finishers, from whence it would be ready for a dial, hands, and case.[173]

In contrast to this was the emergence in Geneva of the making of magnificent pictorial enamel watchcases. The leading family in Geneva for such work was that of Huaud. Jean Huaud was a goldsmith in Châtellerault in France, but his son Pierre left for Geneva where he finished his apprenticeship and became a highly accomplished goldsmith and painter on enamel. Further members of the Huaud family were also involved in this highly skilled art, and there are many watches bearing the name, discreetly placed in the enamel decoration. That their business was international can be seen from the number of Huaud cases with movements by Dutch makers. From this time on, Geneva makers excelled in the making of exceptionally fine gold and enamel watchcases, with verge movements which, though well made, were of less significance than the cases that housed them.

During the second half of the eighteenth-century, makers of considerable skill and industry worked in Geneva and the towns around Lake Neuchâtel, Le Locle, La Chaux-de-Fonds, and Neuchâtel itself. It was here that Daniel JeanRichard introduced the system of the *établisseur* bringing together the work being carried out in small workshops.[174] The Swiss system produced massive numbers of watches, a considerable portion of them being exported, especially to France. Julien Le Roy commented on the Geneva watches flooding into the south of the country and that they were of poor quality. As well as this, the Swiss makers were selling large numbers of cheap watches to Parisian traders. Le Roy observed that the Swiss concentrated far more on the outward appearance of their watches than they did the quality of the mechanism held within—'by this means they succeed very well in making the public fall into the trap that they hide under the deceptive outside appearance of beauty and a bargain price, while most of their watches could not even run three months without breaking down'.[175] A further advantage that the Swiss had over both the French and English makers was that there was no real control over the purity of metals used for cases. In this respect, the Swiss were free to use gold of lower purity and silver alloys, another way of lowering the production costs.[176] They also made much thinner cases.

Jean Moîse Pouzait (1743–93) was director of the watchmaking school in Geneva. In *c.*1786 he produced a model of a new escapement that he had devised, a drawing of which survives in the University of Geneva. The escapement is a form of lever in which the lever and pallets operate in conjunction with a large-diameter balance oscillating once per second, allowing a seconds hand to proceed at second intervals—a dead-second indication.[177]

THE EQUATION OF TIME

With more accurate time measurement, the phenomenon of the **equation of time** became evident to the general public. Largely undetectable before the introduction of the pendulum and the balance spring, it provoked an important movement of popular science education and the ingenuity of clock and watchmakers to produce watches which automatically indicated the equation of time. One such was made by Ferdinand Berthoud in Paris in 1760. It has automatic indication of the equation of time in which a central white enamel disc, calibrated in minutes and numbered 0–60, is controlled by a cam beneath the dial, which automatically rotates back and forth so that true solar time is indicated by the tail of the mean-time minute hand. That Berthoud was proud of this watch is demonstrated by his describing it in his *Essai sur l'horlogierie . . .*, where in the introduction, he says, 'In chapter XVIII I give all the details of the construction of the equation watches I have made. I here explain first one which has a dumb repeat and seconds and goes for 8 days and which marks the months of the year . . . '. He refers to two more watches, stating of the second, 'I also present the calibre of an equation watch, jump centre-seconds, repeating, going for a month without re-winding, which marks the months of the year'.[178]

An example from an English maker is one of a group of four made for John Ellicott in London in 1747 by Thomas Mudge. In

---

172 See Jaquet & Chapuis 1970, 76. Babel 1916; Babel 1938.
173 Jaquet & Chapuis 1970, 79–80.
174 For an extended account see Jaquet & Chapuis 1970, ch. 2; see also Favre 1991.

175 'Mémoire contenant les moyens d'augmenter le commerce et la perfection des ouvrages d'horlogerie'. Manuscrits de Julien et Pierre Le Roy, Musée national des techniques, Paris, Archives 4°, 126. Cardinal 1985, 55–7.
176 See Cardinal 1989, 56.
177 See Chamberlain 1941, 62–7. For an example signed, 'Ls George & Ce. Hr du Roy à Berlin', *c.*1820 in the British Museum (1958, 1201.1176), https://research.britishmuseum.org/research/collection_online/collection_object_details.aspx?objectId=57461&partId=1&searchText=Pouzait&page=1.
178 Berthoud 1763, 1786. For the Berthoud equation watch in the British Museum (1988, 1104,1), see Thompson 2008, 76–7.

these watches, the white enamel dial on the front shows hours and mean solar minutes as well as true solar minutes. Thus, in conjunction with the outer rotating enamel ring showing months and days as well as degrees and signs of the zodiac, it shows the equation of time throughout the year. The equation hand is controlled by a kidney-shaped cam mounted beneath the dial. On the back is a small white enamel dial showing seconds.[179]

THE TURKISH MARKET

From as early as the mid-seventeenth century, trading in watches in the Ottoman empire was recognized as a lucrative business. Establishing such business was achieved through the services of specialist merchants who had permission to carry on their trade in specific locations.[180] Genevan makers were the first to profit from this trade in Istanbul, but their commerce declined towards the end of the century. Thereafter, it was London makers who exploited the opportunity presented by this profitable market. There are names of London makers who are synonymous with this trade, particularly the businesses of Markwick, Markham, & Perigal, as well as that of George Prior. The so-called Turkish Market watch differed in some respects from watches intended for the home market. It is rare to see such a watch with a cylinder escapement, the verge being almost universally preferred, probably for its robust operation. While there are some watches with champlevé silver or gilt-brass dials, with the advent of the more easily read white enamel dial, it became the preferred option. The numerals on the dial are idiosyncratic. Their origin is obscure, but they seem to have been a western version of Arabic numerals mixed with a Roman concept of symbols for the units and the fives. It is neither Arabic nor European, and not intrinsic to either culture. Cases were usually of gilt-brass or silver, but rarely, if ever, gold. In the latter part of the eighteenth century it was common for a 'Turkish Market' watch to have triple case, presumably to provide further protection from rough usage, dust, and sand. A common material for outer cases was **shagreen** and a further, fourth conically shaped case in chased silver or silvered metal was not infrequently locally added.

THE CHINESE MARKET

From as early as the late seventeenth century there had been a market in clocks made in London and elsewhere for export to China. Following the popularity of these curious and luxury items, firstly with the emperors and then spreading down the wealth scale to members of the imperial court and to high-ranking officials, the market proliferated. The export of watches also took place but in far fewer numbers. James Cox (1723–1800), who excelled in exporting clocks to China, is also noted for a small group of watches among the amazing variety of clocks and automata produced in his workshops to designs by Joseph Merlin and other gifted goldsmiths, jewellers, and automata makers.[181] William Ilbery was also active in this field, producing a far greater number of watches, mostly made in Fleurier, Switzerland. Towards the end of the eighteenth century, the popularity of European clocks diminished drastically, leaving makers such as James Cox in serious financial difficulties. However, at the same time there was a growing popularity in watches, and it was the Swiss makers who exploited it, developing a lucrative trade in high-quality watches and bijouterie, many with ingenious automata displays. The makers Pierre Jaquet-Droz and his successors stand out in this trade, producing exquisite watches and superb automata.[182]

FORGERIES

When London became the watchmaking capital of Europe, it took no time at all for less scrupulous makers to involve themselves in an illicit trade, both at home and abroad, as reported: '1657, October 1st, Mr John Smith was fined and paid 10s for putting Esme Hubert's name on a watch'.[183] In July 1704, it was reported by the Master of the Clockmakers' Company that certain persons in Amsterdam were in the habit of putting on their works the names of Tompion, Windmills, Quare, Cabrier, Lamb, and other well-known London makers, and selling them as English.[184] This problem persisted throughout the eighteenth century and later. There were instances where London watchmakers were illicitly exporting cases, dials, and dial-plates abroad to have poor-quality cheap movements fitted, but equally examples of makers importing poor-quality movements to carry out the duplicity in London.

One maker with a high reputation overseas was Daniel Quare, who seems to have been the most frequently copied of all London makers during his lifetime. When Quare died, the *Daily Post* of Thursday, 26 March, reported: 'Last week dy'd Mr Daniel Quare, watchmaker in Exchange Alley, who was famous both here and at foreign courts for the great improvements he made in that art'. There are many watches surviving today signed *Quare London* but without a serial number and clearly not from his workshop in

---

179 For one of the Ellicott equation watches, see https://research.britishmuseum.org/research/collection_online/collection_object_details.aspx?objectId=52227&partId=1&searchText=Ellicott&page=2 (accessed Feb 2020). Also, Daniels 1981, 153, and Thompson 1997, 319–20.
180 For a description of the trading arrangements, see White 2012, ch. 3.
181 See White 2012. In addition to an extensive account of the trade in clocks, there is mention of the watches. Cox's watches are described on 180–2. Two examples survive, one in the British Museum and the other in the Museum of London. *Cf.* this volume Chapter 16.
182 See Chapuis 1919. Many fine examples can be seen in White 2019.
183 Clockmakers' Company Journal, Vol. 1.
184 Atkins & Overall 1881, 258.

Exchange Alley. A good example of a Quare forgery can found in the Metropolitan Museum of Art.[185] Here the watch has a wandering hour dial, a type not generally made by the established London city makers of the time. Looking at the movement, it is of poor quality, and the back plate is so crowded with a very large regulator index that there is almost no room for the signature. Similarly, Thomas Tompion's name and reputation spread far and wide, and there are a number of surviving contemporary forgeries. A very obvious forgery is a watch in the British Museum, where the watch is clearly a French 'oignon' with alarm but is signed *Tompion London*.[186] That extensive forgery was practised at the end of the seventeenth century is evident.

If things were bad at the beginning of the eighteenth century, they deteriorated as it progressed. In the mid-eighteenth century there were forgeries, described at one time as 'Dutch Forgeries'. The name seems to be based on the fact that they commonly had an arcaded minute circle on the dial, similar to that popular on Dutch clocks and watches at the time. More recently they have been termed 'Geneva Forgeries', it being thought the watches were made there. They were inscribed with fictitious makers' names and with the place 'London'. Common names on the watches were 'Tarts' and 'Wilter' and later in the century 'Sampson'. The cases of these watches are generally of silver with cast or repoussé classical scenes. Some of the best-quality cases bear the inscription *Cochin f* in a reserve in the decoration. The watch movements are of poor quality and have a bridge for the balance, a practice very rarely found in London work. It is likely that these watches were destined to be sold in Europe generally where the place 'London' on a watch would add kudos and buyers would not easily recognize the inferior quality of the workmanship. Today the vast majority of these watches are seen to have a continental European provenance, rather than an English one. The high surviving numbers of what were essentially poor-quality pieces suggest that they were produced in large quantities, albeit perhaps in only a few workshops.[187]

The nature of forgery changed during the nineteenth century, when a dramatic and lasting change took place in the world of collecting from the 1820s and 1830s onwards. One of the pioneers in that change was Debruge Dumesnil in France, a pioneer of collecting what became known as industrial art—that of objects used in everyday life.[188] In terms of horology, collectors were mainly interested in acquiring watches from the early period and, until later in the nineteenth century, were seldom interested in watches with a balance spring. In the early part of the century, there was little to be acquired. Fakers and forgers filled the gap, sometimes with out-and-out forgeries, otherwise making iron movements for tambour-cased stackfreed watches or taking mundane movements and placing them in exotic rock-crystal cases. A particularly interesting example in the British Museum is a large gold and enamel-cased oval watch. It has no maker's name on the back plate or dial and at first glance looks to be a sumptuous piece from about 1620. Further investigation shows otherwise, and it is now generally agreed that it is a sophisticated piece of nineteenth-century work.[189]

A maker who suffered extensively at the hands of forgers was Abraham-Louis Breguet.[190] The problem was such an issue for this eminent maker that he devised a system to confound them. He had a large master signature engraved on a plate which was then transferred using a pantograph to the dial beneath or either side of the numeral 12 where it would be exceedingly small. These signatures are referred to as 'secret', although they can be seen with a good eye.[191] An example of the signature can be seen on a Breguet *souscription* watch in the British Museum.[192] However, the secret signature was only a temporary solution. It was not long before the fakers were doing exactly the same thing using their own pantographs. The business of faking has continued. Even today there are those who set about confounding the unsuspecting buyer. There is no doubt that there are watches which have had the back plate thinned down to remove a pedestrian maker's name, the name Thomas Tompion added, and then the plates re-gilt. *Caveat emptor*.

A precise calculation of the number of watches produced between 1500 and 1800 is impossible. Apart from the fact that many were adapted long after their production, many were simply destroyed or lost, and some may lurk unknown somewhere. The number that has survived is perhaps more remarkable, preserved in private and public collections. Many are in major permanent exhibitions, some single items or smaller collections are displayed in museums, and others exist behind the scenes, but may often be viewed upon application. The subject merits much further study and documentation.

---

185 https://www.metmuseum.org/art/collection/search/194089 (accessed February 2020).
186 https://research.britishmuseum.org/research/collection_online/collection_object_details.aspx?objectId=56,470&partId=1&searchText=Thomas+Tompion&page=1 (accessed February 2020).
187 There are numerous examples of these watches in museum collections. For a typical watch by 'Wilter' see Thompson 2008, 80–1.
188 Labarte 1847, also this volume Chapter 26.
189 The watch was assessed in Tait 1986, 247–57. http://wb.british museum.org/MCN5072#1543832001. See also the historicist watch in Figure 159.
190 A typical Breguet forgery can be seen in the British Museum. https://research.britishmuseum.org/research/collection_online/collection_object_details.aspx?searchText=Breguet&ILINK|34484,|assetId=789221001&objectId=58293&partId=1 (accessed Feb 2020).
191 The Breguet pantograph was sold by Sotheby's, in the George Daniels collection London 6 November 2012, lot 68. http://www.sothebys.com/en/auctions/ecatalogue/2012/george-daniels-so-l12313/lot.68.html (accessed Feb 2020).
192 See Thompson 2008, 102–3; Rigot 2017, 146–59.

CHAPTER SEVEN

THE STRUCTURES OF HOROLOGICAL MANUFACTURE AND TRADE
SIXTEENTH TO EIGHTEENTH CENTURIES

Anthony Turner

Clocks and watches emerged in a hierarchical society.[1] From their beginning until at least the eighteenth century, and to some extent even today, they were usually luxury objects to be coveted and displayed. The structure of the trade that produced them reflects both these elements. Neither unitary nor static in geographical space and time, this structure nonetheless displays similarities that allow us to establish a general schema. Basic to horological production was the workshop. In an urban setting, this was, directly or indirectly, generally subject to civic control, which was exercised via the guilds.[2] However, workshops dependent on royal, princely, or noble patronage were free of such regulation,[3] as were those located in the **liberties**, which were therefore particularly attractive to immigrant craftsmen. Both in towns and rural or semi-rural areas where the trade was generally free, workshops were usually domestic, housed in the craftsman's dwelling or with a shop close by. They could be entirely independent but increasingly, as specialization developed during the latter part of this period, they became absorbed into the **putting out** system controlled by retail masters or wholesalers, who were primarily merchants.

GUILDS AND REGULATION

Because clockmaking and watchmaking depend fundamentally on the working and decoration of metal, it was among the smiths that it first developed: blacksmiths for heavy work; locksmiths for fine work; gold and by silversmiths for decoration chasing, incrustation, piercing, enamelling, and the like. From the beginning, some ancillary trades were also essential, such as bell-founders and the makers of the tooled leather cases that protected precious portable timepieces.[4] Throughout Europe such craftsmen were organized in guilds—which either controlled their trade directly (trade guilds), or they assimilated to one that controlled several related crafts and with which the trade had a natural association (envelope guilds). Clockmakers in Liège, for example, were

---

My thanks to Jonathan Betts and Roger Smith for helpful comment on this essay.

1 Exceptions include small domestic alarms, watchtower guard clocks (see Figure 192), and the like.

2 The fullest account of European guilds is now Ogilvie 2019, with its extensive bibliography, although, to some extent, Ogilvie's picture is vitiated by being framed in terms of the debate about whether the guilds helped or hindered economic growth—a question that would have been incomprehensible to those controlled by or controlling them. For a counterpoise, see the informative essays collected in Epstein & Prak 2008. The discussion of corporatism in Brockliss & Jones 1997, part I. 3, is also helpful for understanding the system.

3 For example, in Paris artisans in the Galleries du Louvre, and the Hôtel des Gobelins, painters and sculptors who were members of the Académie, those maintained in houses of princes of the blood, those who held the title 'marchand privilégié suivant la cour' (for whom see Brockliss & Jones 1997, 242–3), and those in the university colleges were largely exempt from the authority of the corporations. Gabourd 1885, i, 515; Saint-Leon (1922), 399–404, 550–1; Augarde 1996, 40–5. For further examples, i.e. some for London, see in the Glossary. For **liberties** more generally in Europe, see Ogilvie 2019, 470–3.

4 The latter could themselves be subdivided, as they were in Paris, between the *gainiers*, who made the cases that the *doreurs sur cuir* embellished with blind or gold tooling; see Saint-Léon 1922, 474.

included among the Smiths, who embraced all metalworking activities.[5] Because it was a new trade, separate horological guilds (Table VIII) appeared late in the cities of Europe,[6] but craftsmen in earlier periods were still subject to guild, or other, control.[7] So, in Nuremberg, where there were no guilds as such, artisans were either members of a restricted, closed group, the *Geschwornen Handwerke,* or of a free craft *Freien Kunste*. However, both groups were strictly controlled by the central city government, and the structure of masters, journeymen, and apprentices prevailed with provision for **masterpieces** and restricted entry to the trade. Sundial making in Nuremberg, by origin a free craft, had also effectively become a closed, controlled occupation by the end of the seventeenth century.[8] In the eyes of those who governed hierarchical society, in which every person was supposed to fit into a preordained structure, a free, unregulated tradesman was a dangerous element that could lead to anarchy.[9]

The lives of those who conformed were, in principle, controlled by their betters from youth to the grave. The career of Eustache François Houblin (1722–pre-1789) offers a paradigmatic case (Figure 90). He was the son of François Houblin, laundryman, and Marie Madeleine Blanchard, living in the parish of St Etienne du Mont, Paris, where Houblin was baptised on 9 March 1722. On 6 April 1739, aged seventeen, he was apprenticed to Nicolas Lenoir for six years at a fee of 300 livres. On 11 June 1744, he was turned over, with Lenoir's agreement, to Auguste Fortin;[10] on 11 October 1745, he was accepted into the corporation of Paris clockmakers as a free master by apprenticeship and the presentation of a masterpiece. This he had probably made in the workshop of Lenoir since it was from this address (Rue Pallu,

Table VIII Dates of some horological guilds in Europe

1544	Paris	1564	Augsburg
1565	Nuremberg	1600	Blois
1601	Geneva[a]	1616	Rouen
1631	London	1659	Lyons[b]
1688	The Hague[c]	c.1690	Nuremberg[d]
		1695	Stockholm[e]
1705	Prague[f]	1755	Copenhagen
1775	Danish Clockmakers' Company[g]	1781	Liège

[a]Babel 1916, 56. [b]Vial & Cote 1927, 23 [c]Spierdijk 1965, 98, 233. In Rotterdam and Groningen, clockmakers were included in the general metal-working guilds of St Eloy. [d]sand-glass makers' guild, Lunardi 1974, 78; von Bertele & Neumann 1963, 340. [e]Pipping 2000, 24. Originally included in the Blacksmith's guild, in the 1650s the clockmakers and the toolmakers established a separate guild from which the clockmakers separated. [f]Folta 1997, 407. Previously with locksmiths in the Smiths guild. [g]Previously in the Blacksmith's Guild. Knap 1984, 618.

Place Maubert) that it was presented. Two years later he booked his first apprentice, who, before 1775, was followed by five others. He remained in the parish of St Etienne du Mont all his life, maintained a workshop of some size, and fulfilled his duties in the Corporation, being *garde-visiteur* in 1764–66 and 1770–72 and *Député* in 1778 and 1789. However, in the latter year, his wife was listed as a widow in the list of Paris clockmakers.[11]

As a dutiful guildsman and citizen, Houblin accepted control of his working hours and of the number of apprentices he could have; usually, this was only one, although a second could be booked as the first neared the end of his time, and other exceptions could be made especially in periods of heavy demand. He also accepted control of the quality and number of the goods that he produced, the places where he could sell them, the ways in which he could employ and dismiss his workmen, and the wages that he could pay them. Apprentices and journeymen were similarly hedged round with regulations, although the insistent repetition of regulations in the various issues of corporation statutes and the plethora of conflicts at all periods suggests that guild controls were regularly ignored or circumvented. Guild courts had authority to settle internal disputes and to impose their regulations, but this authority ultimately rested on that of the city, town, seigneurial, or regal power that had established them. Internal and inter-guild disputes were therefore frequently heard in public courts.[12]

The multitude of guild regulations existed supposedly to ensure social harmony but could easily be manipulated in the interests of the masters just as the guilds themselves were exploited by the state. They were therefore in themselves a source of conflict.

---

5 Simon-Peret 2020, 37–8.
6 For the lack of foundation for the long-standing claim that a guild of clockmakers existed in thirteenth-century Cologne, see Dohrn-van Rossum 1996, 96–8. NB: with increasing demand for domestic clocks, small clockmakers tended to split away from 'great clockmakers' as had occurred by 1623 in Zurich (Fallet 1948, 21), and in 1647 in Basel (Ackermann 1986, 35).
7 Clockmakers and watchmakers were found, for example, among the smiths in Augsburg, Aberdeen, Dublin, Edinburgh, Glasgow, Haddington, London, and Rotterdam; in Annenberg and Nuremberg they were domiciled among the locksmiths. For further examples and discussion, see Dohrn van-Rossum 1996, 194. Generally, with time, clockmakers broke away from the envelope guilds, but a counterexample is offered by Orléans, where in 1776 the clockmakers were joined to the goldsmiths and jewellers, perhaps because they were very few—three in 1787; see Cuissard 1897, 199–200.
8 Gouk 1988, 45–6, 63. For the masterpieces prescribed for the small clockmakers in 1565, see Baillie 1951, 17, and 27 for those in nearby Ansbach.
9 *Cf.* Willan 1976, 50–6, in the context of wholesale and retail trade.
10 The reasons for an apprentice being 'turned over' to another master are rarely explained but they could include legal fiction (see Loomes 1983a, 38–9 for this in London), antipathy between master and apprentice, death of the master, or enabling the apprentice to acquire specific, specialized skills that his first master could not impart.

11 Account based on the Fiches Brateau; Augarde 1986, 334. Had Houblin been apprenticed twenty years later he would probably have been bound for eight years (Saint-Léon 1922, 554).
12 For a quite detailed account of the day-to-day activity of a guild, see Bates 2018, chs. 2 & 3; for internal and inter-guild squabbles in a provincial town, see Develle 1917. In London, Loomes 1976, 22–31.

Figure 90 Cartel with hunting decoration by Eugène François Houblin. Photo: Jean-Baptiste Buffetaud, courtesy of Chayette & Cheval, Paris.

Nonetheless, throughout Europe,[13] the guilds were a basic element of social organization accepted perforce even by those who gained least from them. Urban clockmaking and watchmaking developed within this unavoidable context. Only in rural areas (and towns like Coventry, where horology was a late implantation) could any alternative structure develop.

SHOPS, WORKSHOPS, AND TOOLS

In late medieval and sixteenth-century Europe, the shop that a free master could maintain was both workplace and selling place. The 1544 Paris regulations prescribed that the clockmaker was to work in public view.[14] Contemporary illustrations from Italy and Germany show him doing so at a bench that juts out into the street, and which is without much protection from the elements. Here, his work could be critically appraised and disputed by would-be purchasers as is shown in Jost Amman's well-known woodcut representing such a scene perhaps in Frankfurt.[15] To prevent fraud, work was to be done in only one place in the master's house, by preference the open shop. An Augsburg regulation laid down in January 1562 further stated: 'Nor shall anyone allow any journeyman to work outside the house; but if a man has a *permanent sales shop*, he shall have work done not in the house and the shop, but in the shop alone or else in the house, and only in the one place, whichever he wishes'.[16] How far clockmakers and watchmakers worked to order and how far for stock is difficult to determine. Certainly in trades such as shoemaking, saddlery, pewter, and even goldsmithing, some craftsmen retailers in sixteenth-century provincial England held stock against future orders.[17] Clocks and watches, for which demand was more limited, were different. In rural areas, clocks were generally made to order; in urban centres some stock could be held for passing trade, or components for rapid assembly when an order was received, as was apparently the practice in Thomas Tompion's shop.[18] In eighteenth-century London, urgent orders might also be met by drawing on the stock of fellow clockmakers, but practice certainly varied throughout Europe. In Augsburg watches were already being made for stock in the later sixteenth century. They could be sold in markets and fairs, but door-to-door hawking was prohibited, as it also was in England.[19]

The counters of the early shops were often made to be raised, converting into a lower shutter for night security, that met an upper one used, during the day when swung up, as a protective roof. However, working in the open during daylight hours was neither safe nor agreeable. At the end of the sixteenth century, and on through the seventeenth, workbenches moved fully inside the shop and were usually set beneath large windows for light, and fairly close to the furnace, which thus simultaneously heated the metals and those who worked them. Although much work was, necessarily, still bespoke, shop stock increased, and the need to display it led to the development in large towns during the later seventeenth century of the agreeably appointed, fully enclosed (and so more secure) selling spaces that would entice fashionable customers (Figure 91) throughout the eighteenth century during the 'season' (October to June).[20]

According to the 1741 rules of the Paris corporation, shops were to be open from 8 a.m. to 6 p.m. in winter and from 6 a.m. to 8 p.m. in summer (1 April to 1 October).[21] Purely retail shops in provincial towns and the country were less likely to be specialized in the goods they sold, and manufacturing tradesmen could operate from a shop or their own homes. Even so, shops—whether solely retail or providing both work and selling space—would include living accommodation above, which was occupied either by the owner himself and his family or, increasingly in the later seventeenth and eighteenth centuries, by his apprentice or journeyman, or simply let out.[22]

In both town and country, shop hours were long: from 6 a.m. to 9 p.m. in summer, from sunrise to sunset in winter. Keeping the shop could also be exhausting: the fixed-price shop only appeared at the very end of the eighteenth century and, even then, was seen as an aberration. Haggling was customary for most commodities, the more so for luxury goods, and even when agreement was reached, credit, often for up to a year, was expected to be offered

---

13 Ogilvie 2019, 239 estimates that between 78,000 and 98,000 guilds (all trades) existed in France, Spain, the two Netherlands, Italy, and Germany during the long eighteenth century to 1810.
14 'The said masters may not work at the said trade unless they maintain a shop and work openly in compliance on the public way'. Translated from Franklin 1888b, 182.
15 Rifkin 1973, 75 and, for a similar, but late-fifteenth-century Italian shop, Ward 1980, 172. Both are conveniently reproduced in Matthes 2018, 229.
16 Cited from Groiss 73, n. 104. However, in Nuremberg, the workshop of Peter Henlein was apparently placed on the third floor of his house; see Matthes 2017, 326–7.
17 Willan 1976, 56–7.
18 Evans, Carter, & Wright, 2013, 140, 44, 147, 149.

19 For the prohibition of pedlars in towns, see Willan 1976, 54.
20 The internalization of the shop is further illustrated in the depiction by Luyken 1694 of a clockmaker's shop showing the bench within but still with the shop open to the street. *Cf.* Davis 1966, 101–2, 181. However, open shops maintained themselves in various trades until well into the eighteenth century; see Davis 1966, plate 9 and, for the 'season', 122. For a wider view of the role of fashion and the development of shopping in the eighteenth century, see Berg 2005, especially ch. 7, and Coquery 2011. For a view of high-level shopping in the related trade of instrument-making, see Bennett 2002; for Sophie von La Roche's visit to Vulliamy's shop in 1786, see Williams 1933, 100; Hutchinson 1992.
21 Raillard 1752, 39.
22 For example, Jeremy Dehind, brother of the clockmaker Peter Dehind, lived in a shop leased by his brother; see Evans 1999, 551; Germain Pecquet (*c.*1653–1731), 'horloger à Paris', however, lived above his shop, which gave on to the street, in the Rue Froidmanteau. AN Y 11661, 2 August 1731, cited from the Fiches Brateau.

Figure 91 A clockmaker's shop and workroom, anonymous, second quarter, eighteenth century. Musée du Temps, Besançon, inv. 1896.1.167. Photo: Pierre Guénat.

in both wholesale and retail trading and payment could also sometimes be in kind.[23] This would normally only be decided by the master, but his employees otherwise had considerable responsibilities. Inevitably it was the apprentice who was charged with taking up the shutters at daybreak.

In general four or five people comprised the manufacturing and trading unit: the master, his wife, a journeyman, and one or two apprentices—progression through the hierarchy was that of from youth to age. Usually, apprenticeship, which supposedly was obligatory, began at or close to fourteen years of age—at Augsburg the minimum was twelve years.[24] It could last from three to ten years—in Paris the stipulated period was eight years, in Britain normally seven but could vary from thirteen months to ten years, and in Geneva from two to six years.[25] The length and cost of apprenticeship might vary with fluctuations in trade, as in the Neuchâtelois mountains during the eighteenth century.[26] Apprentices, like journeymen, usually received board and lodging and sometimes bedding, washing, and some clothing. Although exceptions occurred, as in London,[27] all had to produce a masterpiece in order to be accepted into the freedom of the company. While awaiting permission to construct his masterpiece, the qualified, but unfree, craftsman could work as an employee for free masters either in the city of his apprenticeship or elsewhere. In the Jura and Geneva, he could acquire advanced skills through a further, usually short (one to four years), remunerated period of work under a master, as a *réassujeté*.[28] Post-apprenticeship was also the period when he might wander through Europe, sometimes to fix himself elsewhere,[29] or court his master's daughter or

---

23 Davis 1966, 155, 183–4; Mraz 1980 45; see also Willan 1976, who offers examples from different trades. See esp. 93–4 where its use and risks in the setting up of a business are illustrated. It was a two-way affair. At his death in 1624, Thomas Aspinwall owed £180 and £165 was owed to him for items supplied; see Bailey & Barker 1969, 3. For payment in kind, and the length and scale of credit, see Willan 1970, 26–7.
24 Groiss 1980a, 59. Twelve or thirteen years was also a common age for apprenticeship in Paris, although, as the example of Houblin shows, it could be considerably later. An exceptional group were the *alloués* (those agreed), trainees who could be taken on by masters outside the guild structure. They could begin at any age and were usually instructed for between five and six years. However, they could not proceed to the freedom; see Dequidt 2014, 159–60.

25 Augarde 1996, 20; Moore 2003, 4; Babel 1916, 31, 58–9.
26 Cortat 1999, 9–11. For England, see Minns & Wallis 2013.
27 See, for example, the case of Josef Weidenheimer in Mainz, described in Friess 1999, 524–5
28 Fallet & Cortat 2001, 31–2; Mottu-Weber 1970, 334–5; Babel 1916, 430–3 & 468–70.
29 See the fictional, but realistic, account, of such travel given by Camus described in Turner 2015a. Liegois clockmakers, it was said in 1781, perfected themselves in England, France, and Switzerland (Simon-Perret 2020, 38). For the perambulations of a London-trained, Yorkshire clockmaker in northeast England, see Bates 2018, 172–3. For discussion in a wider guild and chronological context, see Ogilvie 2019, 448–56, who

Figure 92 A late-sixteenth-century clockmaker's workshop by Philip Galle after Jan van der Straet (Stradanus). Plate 5 from the *Nova Reperta*, c.1600. Photo: Jean-Bapiste Buffetaud.

widow, so obtaining preferential terms by which to mount his own workshop; however, the requirement of the masterpiece was rarely waived. In theory this was to be built in his master's, or another master's, workshop to ensure that the aspirant really did construct it all himself.[30] This hurdle overcome, the costs of acceding to the freedom were nonetheless high, a fact favouring the entry of masters' sons, who paid a lower fee, and the marriage route. The average age of sons of masters taking their freedom in Augsburg from the mid-sixteenth to mid-seventeenth century was twenty-eight, and that of newcomers to the trade thirty-one to thirty-two. This was probably general to Europe but there could be wild individual fluctuations. Since there seem always to have been too few journeymen available, as many institutional obstacles as possible were set in their way to their freedom.

Much of what has been described in the foregoing applied equally to other manufacturing trades, particularly to the makers of portable compass sundials. Such dials had been made since, at the least, the late fifteenth century in Paris and probably Rouen, and from the late sixteenth century onwards, the trade was also developing in Dieppe, which would become the dominant centre by the mid-seventeenth century. The dials were produced by specialists among the *tablettiers*—craftsmen in inlay work employing ivory and precious woods. Their output was not inconsiderable. In February and September 1631, Jean de la Roche exported bundles of four and six dozen such dials, respectively, from Rouen to the Iberian Peninsula. More were sent the following year,[31] as they were no doubt to other destinations.

Regulations, and efforts to circumvent them, were endemic to Early Modern crafts. More distinctive of individual trades were the materials and the tools employed. In the sixteenth century, clockmaking, and even more, watchmaking, stood out because they were new trades. At first assimilated to smithing, horology shared materials (brass, iron, and steel) with it and many tools (several shown in Figure 92), such as hammers, small anvils, chisels,

---

notes that travelling was more prevalent in German-speaking Central Europe where it was sometimes mandatory, but that there was considerable regional variation. Success was not always attained: Thomas Hagon, despite a seven-year apprenticeship with Thomas French in Norwich for a fee of £10, died in the Bell Inn at Kesgrace Suffolk in 1796 'a poor travelling watchmaker', victim of a 'Visitation of God'; see Loomes 2006, 335.

30 See the remarks by Julien Le Roy on laxity in examining masterpieces in mid-eighteenth-century France, cited in Augarde 1996, 16.

---

31 De Beaurepaire 1905, 432, 433, 446.

simple lathes and drills, saws, gouges, files, rasps, and punches.³² However, cutting well-divided gear wheels with equally profiled teeth by hand was lengthy, difficult, and tedious. Ways to facilitate their division and cutting were therefore actively and successfully sought. If the dividing plate and the cutting engine seem to have developed first as an aid to producing the very large wheels of astronomical clocks in the mid-sixteenth century,³³ they were quickly applied to more everyday clockmaking. Already in February 1608/9, the London clockmaker Peter Dehind could bequeath two 'instruments to cut wheels' that he had presumably used for some decades earlier.³⁴ Dividing plates (*plateformes*) are found listed in the probate inventories of Blois watchmakers,³⁵ a dividing plate with a ruler was part of an order for tools disputed between Samuel Aspinwall and Ellis Bradshawe in 1656,³⁶ and in December 1670 there is report of a watch-wheel cutter contrived by Thomas Mossoke in County Durham.³⁷ In October 1673 in Blois, 'a tool to cut wheels' was listed in an inventory of the goods of Charlemagne Viet,³⁸ just after a misleading mention of the device in Robert Hooke's diary.³⁹ By this time the instrument had long been in use and in London dividing plates and cutting engines were already made by specialist maker Benjamin Bell.⁴⁰

By the end of the seventeenth century, a range of horological tools to effect specific tasks in the manufacturing process existed and affirmed the difference between clockmakers and other trades. Viet's inventory included, besides a generous array of standard tools, one large *plateforme* and two smaller ones, presses for riveting, a lathe for balances, two tools for 'equalizing' the **escape wheel**, and a machine for cutting **fusees**.⁴¹ In London, Bell had a **profile cutter**,⁴² Tompion used a special instrument to adjust the fusee to the force of the spring,⁴³ and had devised his own form of wheel cutter—a large machine built into his shop.⁴⁴

Throughout the eighteenth century, as the individual stages of clockmaking and watchmaking became specialist trades, specialist tools were developed. Already in 1741, the description of them given by Thiout occupies some sixty pages and thirty-eight plates in his treatise on horology.⁴⁵ By the end of the century specialist toolmakers would respond to a constantly growing demand by issuing printed catalogues of their goods.⁴⁶

SUPPLIERS

A recurrent theme in the history of horology is the growth of specialization. 'Of late years the Watch-Maker…scarce makes anything belonging to a Watch; he only employs the different tradesmen among whom the Art is divided,… '.⁴⁷ Campbell's remark, which he claims was also true of clockmaking, was applicable to major horology centres throughout Europe. While he itemized ten separate craftsmen as contributing to the watch, thirty years later Pierre Jaubert listed twenty-seven,⁴⁸ and by the end of the century William Pearson could enumerate thirty-five.⁴⁹ The advantages of such subdivision were early recognized.

> A Watch is a work of great variety, and 'tis possible for one Artist to make all the several Parts, and at last to join them altogether; but if the Demand of Watches shou'd become so very great as to find constant imployment for as many Persons as there are Parts in a Watch, if to every one shall be assign'd his proper and constant work, if one shall have nothing else to make but Cases, another Wheels, another Pins, another Screws, and several others their proper Parts; and lastly, if it shall be the constant and only imployment of one to join these several Parts together, this Man must needs be more skilful and expeditious in the composition of these several Parts, than the same Man cou'd be if he were also to be imploy'd in the Manufacture of all these Parts. And so the Maker of the Pins, or Wheels, or Screws, or other Parts, must needs be more perfect and expeditious at his proper work, if he shall have nothing else to pulse and confound his skill, than if he is also to be imploy'd in all the variety of a Watch.⁵⁰

Even if Henry Martyn cast this description in the conditional tense, such specialization had occurred early in and developed

---

32 For some late-sixteenth-century/early-seventeenth-century examples that may have belonged to Peter Henlein, see Matthes 2017, 326–7, with a fuller description in Mathes 2018, 241–50. For some account of file making in northern England, see Jars 1774, i, 228–31; 233–4; 259ff. See also letters from Dietrich Matthes and John Griffiths in *AH* xli, 2020, 572–6.
33 The earliest known is that by Janello Torriani (*c.*1500–85) conceived for his construction of a planetary clock for Charles V. See Woodbury 1958, 45–6; Zanetti 2016, 120–1, who relates it to the Milan manufacturing context with possible influence from Leonardo.
34 Evans 1999, 551
35 Berthier 1996.
36 Griffiths 2002, 171.
37 Peter Nelson to Henry Oldenburg 15 December 1670. Hall & Hall 1970, 326.
38 Fourrier 2001–5, 139, 141.
39 Robinson & Adams 1935, 5.
40 These references are given fully, with discussion, in Turner 2019. For the first (?) engine in North America, see Crom 1987, 350.
41 Fourrier 2001–2005, 140–1.
42 Robinson & Adams 1935, 35.
43 De Hautefeuille 1694, 7–8.
44 John Locke to Nicolas Toinard, 16 September 1680 in Ollion 1912, 72.

45 Thiout 1741, volume i. A summary of recent machines is offered by Jaubert 1773, ii 423–4
46 Roberts 1976, 1–20; Smith 1978; Crom 1980, 119–64.
47 Campbell 1747, 250.
48 Jaubert ii, 417–19. In a survey of eighteenth- and early-nineteenth-century (pre-1858), London wills Ponsford 2008 recorded twenty-one separate trades for both non-clockmaking and clockmaking horological practitioners. Crespe 1804, 29–31, in Geneva listed twenty-nine.
49 'Watch' in Rees 1819–20, although this was written *c.*1807, see Harte 1973.
50 Martyn 1701, 42–3

throughout the seventeenth century[51] in parallel with—perhaps in part enabled by—the development of specialist tools. These in the sixteenth and seventeenth centuries were often designed and made by the clockmaker or watchmaker himself. 'When I was an apprentice', John Carte tells us, 'I have made several Tooles for my Master [Samuel Watson] which have not been common ... he being a most curious Inventor of <his owne> Tooles'.[52] According to Leupold, the *kleinuhrmacher* (small clockmaker), Johann Mathias Willebrand (*c.*1658–1726), best known for his elegant pocket sundials, also made wheel cutting engines.[53] However, during the first half of the eighteenth century, toolmaking became increasingly the province of tool specialists who made horological tools as one element in a wider production for different trades before emerging as an independent activity from the mid-century onwards. At least such occurred in England, where John Wyke established a toolmaking business in the late 1740s, and in Switzerland, where the clockmaker Abram Borel-Jaquet (1731–1815)—perhaps guided by Ferdinand Berthoud—similarly specialized in toolmaking from (probably) the 1750s.[54] At Montécheroux (Doubs), horological toolmaking was grafted onto the existing cutlery-making manufacture *c.*1780 and gradually supplanted it.[55]

If horological craftsmen could thus, from the mid-eighteenth century, obtain their tools ready-made from specialist suppliers, both of them required raw material—primarily brass, steel, and iron—on which to work. Of these three, the first used in clockmaking, in the earliest watches and largely in toolmaking, was iron. Although generally available throughout Europe, local conditions could vary. In England, iron had been produced since Roman times and production was widely spread throughout the country, the geographical dispersion being determined by the availability of small wood for conversion into the charcoal essential to refining and forging. This voracious demand during the late sixteenth and seventeenth centuries incurred government disapproval and also caused a steep rise in price.[56] As a result, increasing quantities of raw iron were imported, attention turning from iron *production* to iron *working*. In 1728 Daniel Defoe wrote: 'No particular manufacture can be named which has increased like this of the hardware.... The best knife blades, scissors, surgeon's instruments, watches, clocks, jacks [presumably roasting jacks], and locks that are in the world, and especially toys and gay things, are made in England'.[57] National hubris aside, this remark neatly situates clocks and watches in the manufacturing world. Crude steel was supplied by iron merchants.[58] The development of its manufacture depended on the spread of cementation furnaces from the mid-seventeenth century[59] for producing blister steel, which would be essential to crucible steel when this was developed between 1740 and 1751 by the clockmaker Benjamin Huntsman (1704–66), who sought a better steel for clock-springs. Primarily, the iron used came from Sweden, especially suitable for steelmaking, shipped to Hull, and then transported on to Sheffield. Sweden was also a primary supplier for the Low Countries, cut off by the great revolt from its traditional sources in Spain via the port of Bilbao, as were iron supplies in Germany and Central Europe, disrupted by the Thirty Years' War. Not until the end of the seventeenth century would Swedish iron be challenged by that produced in the Alps, Poland, and, increasingly, Russia.[60] The last would be a potent competitor throughout the eighteenth century.

The raw materials of clock and watchmaking could come from afar and were subject to variation in both price and availability. Much the same was true of copper and brass. That almost no brass watchmaking existed in England until the closing decades of the sixteenth century was, in part, because there was almost no brass manufacture in the country until after the finding of calamine in Somerset in the 1560s.[61] Even after the establishment of the Mineral and Battery works in 1568, brassworking had a chequered development in seventeenth-century England, while the quality of the brass produced (despite new works built in the region of Bristol and in Buckinghamshire) remained doubtful. John Carte records that 'The cheife Master-Watch makers in England buy

---

51 For specialization in London clockmaking in the late seventeenth century and the probability of the wholesaling of movements to different retailers by a small number of anonymous workshops, see Dawson, Drover, & Parkes 1982, 323ff, whose argument is developed in Parker 2019. For the putting out of engraving to a small number of craftsmen working for several maker–retailers, see Dzik 2019. For a hypothetical reconstruction of internal specialization and putting out in Tompion's workshop see Evans, Carter, & Wright 140–7.
52 Turner 2014a, 56.
53 Leupold 1724, 53; Crom 1980, 1. For Willebrand's dials, pedometers and other devices see Bobinger 1966, 379–98. The variety of his output is a useful reminder of the diversity of horological instruments that a skilled practitioner could produce.
54 Smith 1978, 4–5 who offers an edition of Wyke's catalogue; Cavin 1983, 317 with a useful brief survey of the main tools. Context is provided by Cummins & Ò Gràda 2019, 7–16. For what may be the first wheel cutting engine in America, see Crom 1987, 350–1; for an American-made wheel cutting engine *c.*1776, see Crom 1987 368–9, and for one of *c.*1779, see Hummel 169–70. The slightly later, and very successful, file maker and toolmaker Peter Stubbs is studied by Ashton 1939; see also Dane 1973.
55 Bonnet 2017.

56 High prices and supply difficulties no doubt caused many clockmakers to use substandard raw materials, as did Thomas Boteler, from whom low-quality Flemish iron was seized during a Blacksmiths' Company search in January 1613–14. See Evans 2000, 388–9.
57 *A Plan of the English Commerce*, 290–1, quoted from Lipson 1948, ii 156.
58 So 'le Livre commode', 1692. Fournier 1878, ii 48. *Cf.* Bouthier 2011.
59 For this process see Smith 1968, ch. 6; see also Bouthier 2011.
60 For the situation in England, see Lipson 1948, ii 155ff; Evans 2000, 388–90. For Swedish iron see the survey in the collective work HMGSIA. See also the remarks in Wright 2000b, 529–30, Berg & Nettell, and the important paper by Barraclough 1990.
61 For traces of very small-scale brass making in England in the later Middle Ages see Blair, Blair, & Brownsword 1986, 87. For pertinent remarks on brass in the English context of sundial making, see Davis & Lowne 2009, 100–3.

their Brass & Steel themselves, having made it their Study to know the best from the slighter sort,[62] and such was doubtless the case throughout Europe, purchases being made either directly from a nearby foundry, a local depot, or from travelling agents in the form of rods, sheets, or castings. The best brass for horological use came from Germany, usually via Antwerp, the main market for copper and brass in Northwest Europe, whence German brass and Polish calamine were re-exported to England, France, Portugal, and Spain. This state of affairs prevailed from the mid-sixteenth century to late into the eighteenth century, when the Toulouse clockmaker Pierre Vigniaux[63] stated categorically that the best brass came from the Low Countries, that it should be worked slowly and gently, and that unworked cast brass should never be used. The difficulty of obtaining good brass was therefore an incentive to use it economically and to recycle old brass (shruff).[64] Iron was also recycled. Dealing in scrap iron was one of the many ways in which John Willan, curate of St Peter's, Lincoln, augmented his stipend,[65] and numbers of contracts for the replacing of public clocks state that the iron of the old machine became the property of the contracting clockmaker.[66]

Good raw materials having been obtained, they were not always immediately worked on by the clockmaker or watchmaker. As noted earlier, specialized subdivision early established itself in the manufacture. Early clockmakers and watchmakers may have inherited an ideal that the fully competent craftsmen should be capable of making every part of the machine. However, this was largely a function of isolation or restricted markets. Many of the skilled clockmakers of the late Middle Ages worked in enclosed, monastic communities for which self-sufficiency was a declared objective, especially if a house was situated in a remote area. A high degree of self-sufficiency in metalworking skills was also essential to the itinerant craftsman. Until well into the eighteenth century, horological craftsmen in isolated communities, for example, Samuel Roberts of Llanfaircaereinion, Montgomeryshire (and the many craftsmen like him scattered through rural Wales as through rural and small-town Europe generally), would routinely make most parts of his clocks, experimenting from one to the next, a practice made evident by the variety encountered in his movements.[67] Similar self-sufficiency is shown by the Leicestershire clockmaker Samuel Deacon (1746–1816)[68] and many other provincial makers, even though the making of the clock case was usually contracted out. Self-sufficiency in clockmaking was also evident in the American colonies, Daniel Burnap (1759–1838) being a notable example.[69]

However, the growing horological trade of the guild-controlled towns was different. Here in the hierarchically organized workshop, tasks were already divided up between its various members—a practice that increased efficiency in both production and in training.[70] Despite Campbell's belief that 'At the first appearance of watches ... they were begun and ended by one Man,'[71] even in sixteenth-century Germany and France, sub-contracting of such items as bells, protective leather pouches, engraving and other forms of decoration for cases, and even dials occurred. An early subdivision was the making of steel springs.[72] Although the training of a clockmaker was intended to make him a master of all parts of the craft (which, at least in the sixteenth and early seventeenth centuries he probably was), division of labour advanced rapidly. The small parts of domestic clocks (and even more, watches) were light and easily transportable. Therefore, they could be made in either town or country as a full- or a part-time activity. Labour in the countryside was less costly than in the cities where retailing took place, and specialization also helped diminish demarcation disputes between guilds.

Subcontracting and division of labour need to be distinguished, although in practice this is not always straightforward, as is also the case with subcontracting and putting out.[73] To some extent, the guild system encouraged the first since it implies an agreement

---

62 Turner 2014a, 55. For brass and copper generally in England, see Hamilton 1926. For remarks about the quality of the brass used by German musical instrument makers, see Hachenberg 1991, 240. For advice on how to assess iron and steel, see Smith 1968, chs. 4 and 5.

63 Vigniaux 1788, 8. For English brass workers' protests against the duties levied on imported brass, see Hamilton 1926, 114–15.

64 For the parsimony with which Samuel Roberts, a mid-eighteenth-century Montgomeryshire clockmaker, used brass, see Pryce & Davies 1985, 66–7, and, for an example of poor-quality brass used for the front plate of one of Roberts' clocks, 72. Cf. the remarks on brass use in Loomes 1976, 77–8. Davis & Lowne 2009, 102, note the recycling of medieval tomb brasses in the century after the Reformation in England and cite the relatively high prices paid by Robert Hooke for brass in the 1670s. An old dial ring was re-used as a support ring in the case of Jakob Zech's Berlin clock 1528 (Matthes 2018, 204 ill. 6.4 and 470 lower left.

65 Hart 1962, xiii.

66 For example, the worn out parts of Westminster Palace clock abandoned to Wauter Lorgoner in 1344 (see Chapter 21 at n. 7; Madden 1855, 3; further remarks in Chapter 4). The 'iron, brass and materials' of the old clock of the Hôtel de Ville in Paris were contractually abandoned to Lepaute when he agreed to replace it in 1780. This practice was lacerated by Law 1891 i, discussing Vulliamy's removal of the sixteenth-century movement of the Hampton Court clock.

67 Davies 1985; Pryce & Davies 1985, 98.

68 Daniell 1975, 9–12. Deacon's papers, which have been studied in Thornton 2000a, Thornton 2000b, Thornton 2001, and Thornton 2002, are a mine of information on the supply and cost of raw materials, tools, and outwork in the later eighteenth century.

69 Hohmann, 2009, 31.

70 Landes 1983/2000, 207.

71 Campbell 1747, 250.

72 The gold dial with enamel numbers and decoration of a clock watch ordered by Gilbert Talbot, 7th Earl of Shrewsbury from Michael Nouwen in 1599 was separately made, and separately paid for, from a goldsmith; see Beeson 1965, 372. In his will made in September 1665, the London clockmaker Isaac Pluvier acknowledged debts of twenty shillings 'to a graver' and another twenty shillings 'to one Horne, a Casemaker'; see Stevens 1962, 18. For spring-making, see Landes 1983/2000, 205, and Evans 2002.

73 For the fluidity of the latter categories, see Lis & Soly 2008, 81–95.

between manufacturers in separately constituted trades using different materials. It was also a guarantee of quality, as a specialist leather worker might be expected to make a better protective case for a watch or portable clock than a brass worker. The casting of decorative panels, corner pieces, finials, and the like by founders can be added to the subcontracting trades listed earlier,[74] as can the working of glass or rock crystal for side panels, the occasional case, and, in the eighteenth century, watch-glasses, while joiners were regularly commissioned to make clock cases, as were, particularly in eighteenth century France, sculptural bronze founders. However, case making for watches was often a bone of contention between the watchmakers themselves, goldsmiths, and jewellers.[75]

Division of labour occurs when several craftsmen are employed to make individual components of a single product. As an economic phenomenon it was not unique to clockmaking and watchmaking; rather, it was ubiquitous in the textile and metalworking trades. The classic example, following Adam Smith, is that of pin-making—essential to the clothing trades—although he offered many others.[76] Campbell in 1747 explained that the movement-maker forged his wheels to the appropriate size, but sent them to another workman to be cut, he simply finishing and **profiling** them on their return. Steel pinions were prepared at the mill, springs and chains were made by other workers, the latter frequently by women, while caps and studs were made by separate makers, as were cases and dial-plates. Gilding was a specialized occupation, as was assembling all the parts, which was the task of the finisher. The final specialist was the master who had commissioned all this work and now had to sell it.[77]

Jaubert added detail to this description, which must perforce be applied retrospectively to watchmaking from the previous century. In the first place he specifies that it was the master who established the **calibre** to be followed. He details the functions of the finishers who dealt with both teeth and pivots and notes that watch-finishers were divided into those that dealt with plain watches and those who adjusted repeating watches, although both 'finished the pivots, the wheels, and gears: they balanced the **fusee** with its spring, made the ordinary escapements, and adjusted the movement in its case'.[78] Wheels for repeating watches were made by someone who did nothing else, as was also the case for the **cadrature**; wheel-cutting was a specialized occupation, as was spring making. Pendulum bobs, hands, brass weights, and the silvered dials of clocks were all made by the same artisan, although engraving the hands was created by another, who also engraved the cocks and bridges for watches. However, engravers of the dials of seconds-showing clocks were separate. Polishing the parts was performed by specialists, while the makers of enamelled watch dials did not make clock dials. Different people were responsible for making the hollow shank of clock or watch keys and its mount. Alongside all these workmen were the chasers, gilders, joiners, and founders who worked on the case. The latter were not the same as the founders who cast the movement parts, nor were case gilders the same as those (mainly women) who gilded the movement. To all should be added specialists making **fusees**, **escapements**, the rough cast watchcase without its joint (added by the workman who adjusted the movement within it), the various case decorators, and the makers of watchcase chains and châtelelaines.

Most of this work was achieved by independent craftsmen, working in their own homes or in small, often rented, workshops.[79] They were independent in the sense that they were not directly employed on a permanent basis, nor obliged to work fixed hours in a specific overseen space. The making of pieces for clocks and watches was carried out in both town and country. In both, it could be executed as a full-time job or be a supplement to another occupation. As a full-time occupation, executed by chambermasters, it was probably more frequent in towns, although even here it could be combined with other activities. In London, watchmaker Peter Cuff worked in his own premises as a finisher for Daniel Quare for some twenty years before finally establishing his own business in 1718. This he combined with partnership in a brewing venture.[80] In Nuremberg, Albrecht Karner and his three sons, Melchior, Georg, and Hans, combined the family business of making ivory diptych sundials with careers as professional musicians,[81] and the London maker Davis Mell combined music with clockmaking.[82] In rural areas the widespread cottage or domestic industries were usually combined with agriculture, although it should be noted that this occurred exclusively in regions of pastoral farming, which, being less labour intensive than arable farming, left time free for manufacture, and also offered space to expand and attract further workers if the manufacture prospered.[83]

---

74 Patterns in wood for pillars, or in lead for wheels and other parts could be sent to Le Roux, founder Rue Beauvoisine, Rouen, who used good brass not brittle and liable to break; see Beuriot 1719, 66–7.
75 For disputes between goldsmiths and watchmakers in Blois, see Develle 1917, 85–92. For case making in Britain, see Priestley 2018.
76 Thirsk 1978, 149 and 78–83 for the development of pin-making, which, by the mid-eighteenth century, involved nineteen distinct operations. See also 145–9 for a description of how the division and multiplication of labour became seen as valuable in contemporary economic thought.
77 Campbell 1747, 250.
78 Jaubert 1773, ii 418.

---

79 Smith 2006, 353, who offers an important survey of some of Gray & Vulliamy's outworkers and suppliers c.1760.
80 Holland 2019, 2–3. Brewing, however, was not a success and by 1728 the partners were bankrupt. Two of Quare's other outworking finishers were Daniel Grignion (?1684–c.1748) and his son Thomas (b.1713). Loomes 1981a, 267–8.
81 Gouk 1988, 53.
82 Fernandez & Fernandez 1987.
83 Thirsk 1978, 110. On rural industries in general, see Thirsk 1961. For specific examples in Wales concerning Samuel Roberts and Henry Williams, see Davies 1985; Pryce & Davies 1985 112 & 320: '... farming was the more reliable means of supporting the whole family'; Cloutman

General smithing or foundry work were also occupations that accompanied clockmaking and watchmaking as, not infrequently, did locksmithing and cannon casting. Innkeeping, school teaching, potting, and even harp making are among the many other associated occupations that can be found, although some of these may have been conducted by clockmakers' wives or other family members.[84]

The description of the part work involved in producing a watch or clock given earlier is specific to the eighteenth century. However, it had certainly started in the early to mid-seventeenth century, if not earlier. A clock watch ordered in 1599 by Gilbert, 7th Earl of Shrewsbury from Michael Nouwen had a three-colour enamel dial made and paid for separately from the watch.[85] Watches by David Ramsay (c.1580–1660) have French-style cases, two of them being signed 'de Heck sculp',[86] and the supply of painted-on-enamel watchcases from Blois and Paris in the mid-seventeenth century and later is relatively well documented.[87] Many of these were exported for use elsewhere in Europe, and centres of piece production could also be at some distance from the finishing and retail outlets. Paris clockmakers c.1700 purchased their springs from makers based in either Geneva or London, and springs had earlier been exported from Augsburg.[88] By the mid-seventeenth century the making of watches, parts of watches, and tools was already well established in Lancashire, especially in the area of Toxteth Park near Liverpool and Prescot, the products mainly serving London makers.[89] It would endure, a fruitful and profitable trade, until the twentieth century and was described in 1795:

> The tool and watch-movement makers are numerously scattered over the country from Prescott [sic] to Liverpool, occupying small farms in conjunction with their manufacturing business, in which circumstance they resemble the weavers about Manchester. All Europe is more or less supplied with the articles above mentioned made in this neighbourhood.[90]

'All Europe' exaggerates somewhat, and the makers 'numerously scattered over the country' could be organized by local entrepreneurs. Richard Wright, who regularly supplied leading London retailers with watch movements, was himself ' ... the son and heir of a well established land-owning Lancashire family at Cronton ... a few miles south of the watch-movement making centre of Prescot, who augmented his income by employing artisan craftsmen in the watch-making business while at the same time attending to his family estates'.[91] Wright's craftsmen, however, retained their independence. Royal initiatives in France at Bourg-en-Bresse in 1765 and in Paris in 1785, and others such as that of Voltaire at Ferney in 1770, aimed at a more closely integrated manufacture staffed by dependent, immigrant workmen.[92] Artificial implantations had only limited success and only a brief existence.

By the end of the seventeenth century another manufacture, that in Friedberg, was set for a fruitful future. Situated only a few kilometres from the free Imperial city of Augsburg and more or less opposite it on the Bavarian bank of the River Lech, it existed mainly within that city's economic orbit. Nonetheless, from the late sixteenth century onwards, horology flourished in Friedberg, and despite devastation during the Thirty Years' War (in 1632 and 1646), in 1688, 6.5 per cent of the free citizens were watchmakers or small clockmakers. It would not cease to grow as Friedberg products came increasingly to supply the horological masters of Augsburg, who now became primarily retailers. Professionalization, and an increasing division of labour, marked the entire town, ' ... We find the names of goldsmiths together with case makers, engravers, key makers, spring makers, producers of cock saws and files as well as of watch chains and shagreen cases'.[93] An important role in this wide-ranging horological production was played by women—wives, widows, and daughters of the master craftsmen—who produced shagreen cases, cocks and cock-saws, files, springs, hands, and bridges. If a high proportion of Friedberg production passed to Augsburg, there to be adopted under an Augsburg master's name, a similarly high proportion was exported, particularly to London, there to be sold under false or bogus names as London work and distributed elsewhere in Europe. However, unlike other imitation English work stemming mainly from Switzerland,[94] Friedberg products were

---

& Linnard 2003, ch. 1, who cite (10) the epitaph for Williams as 'a Clock and Watchmaker and a great farmer'.

84 Loomes 1976, 72–3.

85 Beeson 1965, 372.

86 Clutton & Daniels 1975, 8, No 7. De Heck was Gérard de Heck (1585–c.1629), an engraver from Cleves, Gelderland, who had worked in Blois (Develle 1917, 137) before moving to England c.1618–c.1623. His cases for Ramsay may therefore have been produced in London. For Ramsay see Smith 1921, 306–8; Thompson 2008, 32–3; *ODNB* 1975, 1738 (682–3); Finch, Finch, & Finch 2019 who conjecture, 187–8, that Ramsay maintained close contacts with French suppliers from whom he may have received parts, or part-made movements, as well as cases.

87 Cardinal 2018a; Cardinal 2018b. For the making of bells between Augsburg and Lyon, see Groiss 1980a, 72.

88 Blakey 1780, viii. Groiss 1980a, 69 n. 65. The French and Genevan industries were also largely dependent on England for pinion wire, files (but there was also an export trade of these from Germany), and other tools. Crom 1980, 17–20; Griffiths 1994.

89 Bailey & Barker 1969; Griffiths 2002, esp. 176. Cummins & Ó Gráda 2019, 3–7.

90 Aiken 1795, 310.

91 Smith 1985, 608.

92 Beillard 1895, 98–9; Chapiro 1991, 142–3.

93 Arnold-Becker 2014, 678.

94 By the mid-eighteenth century the revived French trade was also suffering from cheap Swiss imitation goods. See the complaint of the Paris clockmaker François II Beliard (1723–post-1789), who describes the introduction 'of an immense quantity of vile foreign horological

both well-made and innovative. They played a major role in the burgeoning European and nascent worldwide trade.[95]

In the mountains of the Jura a similar combination of division of labour, specialized tools, and subcontracting also led to the development of a flourishing horological industry. While this would only reach its full extent in the nineteenth century, the foundations were laid earlier. In the Morez–Morbier region in the heart of the French Jura, clockmaking developed out of a pre-existent ironworking manufacture making scythes, sickles, reaping hooks, and, in particular, nails.[96] In the course of the eighteenth century, although it remained a subsidiary industry, horology there, with something over 4,000 clocks being produced in 1789, became firmly established as an independent rural industry. Horological apprenticeships developed lasting from two to five years and ancillary techniques, such as the making of white enamel dials and, slightly later, of files, were introduced and domesticated.[97] Elsewhere in the French Jura, at Septmoncel and Les Bouchoux, watch movements were produced in quantity and sold for ten francs each in Geneva, where the *fabrique* became increasingly concentrated on finishing, casing, and retailing.[98] By the second half of the eighteenth century, watch part-making was also well established in Savoy. From Araches a regular courier carried movements to Geneva every two weeks, returning with fresh supplies of materials and necessary tools.[99] In the Swiss Jura, division of labour was established among the mountain workers.[100] The brothers Jacot-Guillarmod, horological traders based in La Cibourg, exported the greater part of the production they ordered to Lisbon. Their papers record the names of over one hundred and thirty suppliers, besides many others who remain anonymous, throughout La Chaux-de-Fonds, Le Locle, Geneva, Bienne, and Renan. Although making took place in the Neuchâtel mountains, casing was carried out in Geneva. Charles David Jacot-Guillarmod wrote to his brother on 16 January 1798 that, 'I much prefer that the cases are made in Geneva and to suffer some delay and expense than that they should be made by our bally mountain porkers'.[101] But the international trade of the brothers was subject to disruption by the Revolutionary and Imperial wars of the turn of the century. They palliated its consequences by maintaining, even enlarging, the agricultural holdings that assured them a basic income.[102]

In England, France, Switzerland, the Black Forest region of southern Bavaria (where a manufacture of clocks in which wood was the principal component based on a pre-existing woodcarving trade developed in the later seventeenth century),[103] and elsewhere in Europe, clockmaking took its place among the many secondary, rural industries that enabled smallholding families to make a living that exceeded mere subsistence. Division of labour between horology, cultivation, and dairy production occurred within the household where specific processes could be learned quickly and the familial structure allowed costs to be minimized. With abundant labour at low cost, which was moreover disciplined and could be trained, when aptitude was shown, to execute more skilled matters such as springing, finishing, and adjusting, the cottage industry in horology, especially in the Jura, could fully furnish urban centres, thus pushing them further along the path to finishing casing and retailing.[104] However, cottage industry depended on merchant middlemen.

DISTRIBUTION

The distribution of goods was essential both as input and output in horological manufacture. In town and country, masters, journeymen, and outworkers all depended on the distribution by metal producers of their primary material. This once obtained and used, they depended equally on the good distribution of their products, to markets near or distant, for their remuneration. In the initial supply of raw materials a distinction has to be drawn between an 'artisan system', in which small masters themselves purchased their raw materials, and the 'putting out' system,[105] in which retailing masters or agents acting for them supplied, or put out, raw materials to be worked up by craftsmen for a wage paid by the piece. Inevitably there was a blurred frontier between the two and a tendency with time for successful merchants to extend the putting out system to exert increasing control over craftsmen. But

---

goods, as much from Geneva as from Switzerland assembled (*établie*), or ready to be assembled'; see Beliard 1767, 3.

95 For Friedberg, see Riolini-Unger 1993; Whitestone 1993; Arnold-Becker 2014.

96 Olivier 2000; Olivier 2002, 21–7; 50–6; Buffard 2017, 33. *Cf.* Thirsk 1978, 112, 'the centres of manufacture for new projects was never chosen in a completely haphazard or arbitrary way. Projects tended to build upon existing structures however rudimentary. They positioned themselves in places where an earlier tradition of manufactures, or some associated occupation, existed already'.

97 For the similar flexible apprenticeship of two to five years developed in the Neuchâtel mountains, see Cortat 1999; Fallet 1999; Fallet & Cortat 2001.

98 Babel 1916, ch. xviii, and for efforts to develop retail outlets for the *fabrique* in Paris and Madrid by Jacques-François Deluc in the mid-eighteenth century. Eisler 2014/15, 26; Eisler 2016/18.

99 Perrin 1902, 12.

100 For the origins of the Neuchâteloise industry in myth and reality, see Favre 1991; Favre 1992; the *loci classici* for its development are Chapuis 1917; Chapuis 1931; Fallet & Cortat 2001, ch. 1.

---

101 Translated from Scheurer 1995b, 161.

102 Scheurer 1995b, 165.

103 For which see Tyler 1977, ch. 1; also Jüttemann 1991.

104 For a critical survey of the concept of the 'peasant clockmaker/watchmaker' which suggests qualifications to refine the notion, see Scheurer 1995a. For a breezy survey of the development of the Neuchâtel industry, see Landes 1983/2000, 257–61.

105 The terms are those of Coleman 1960, 3–4, whose brief pamphlet, although focused on the textile industry, gives a lucid description of the domestic system.

in the Early Modern period a standardized product was not yet entirely viable. Customer taste, changing fashions, and bespoke ordering still played a central role. Preparing for a commercial voyage in search of markets, Charles David Jacot-Guillarmod told his brother David (who would supply the watches to be taken) that he had sufficient plain polished gold and silver watches but would like four in engine-turned cases and two with calendar work. Others could have their cases engraved in multicolour gold so that he would have a good variety. Even so, a degree of standardization can here be discerned; in the sixteenth and seventeenth centuries customer desire was more direct, particularly on the decoration of cases.

This, however, is to anticipate. Brass was available in the form of ingots, sheets, plates, wire, rods, and bars, and iron available in bars or small plates. That bars could be worked to a specific size to order is not unlikely. Since land carriage was slow, dangerous, and expensive whenever possible both brass and iron were shipped from the furnaces to the centres of its use by river and, for countries blessed with coasts, by sea. Because many of the artisans working the metals, particularly clockmakers and watchmakers, required only small quantities or were craftsmen of limited means, stocks of the metals quickly became the preserve of capitalist merchants who tried to control both input and output. From the mid-seventeenth century onwards, warehouses were maintained in major cities and these, with agents to run them, gradually extended through the regions in the course of the eighteenth century.[106] Independent factors and provincial wholesale merchants could also supply the raw material as other clock parts, as did Abraham Dent who, exceptionally, on 21 January 1756, purchased for an unidentified customer 60lb weight of clockwork, 6lb of bells, three sets of pinions for 30-hour clocks, and one set for an eight-day clock from John Latham in Wigan.[107] From the mid-eighteenth century toolmakers also needed to set up distribution networks, as did specialist makers of semi-standard parts who, from at least the later seventeenth century onwards, were supplying bells, cast brass wheels and other components, engraved dials, and fittings to whoever needed them. The physical distribution of such items throughout Europe was effected by travelling agents, who could be either the manufacturer himself, a member of his family, or an independent chapman. Thus, the London clock engraver Mandeville Somersall in the mid-eighteenth century both travelled himself and advertised in provincial newspapers that 'Country Chapmen may be furnished with all sorts of Clock Dial-Plates compleately fitted up: As also all Sorts of Tools and Materials for Clock and Watch-making at the lowest prices'.[108] Delivery of such goods could also be effected by carriers or, for small packets of parts or watches, entrusting them to known persons using the coaching services, as did Richard Wright sending goods to London, or Dieudonné Sarton on the Liège–Sedan–Paris route.[109]

If national distribution was important to the maker of clock and watch parts whose local market supplied the foundation of his trade,[110] both national and international distribution was crucial to the master-craftsman–retailer. Advertising from the mid-seventeenth century onwards was an important aspect of this, at first through handbills such as the equation of time tables that leading makers began sometimes to issue, if like Tompion they had an international trade, in Latin and in French.[111] Subsequently, larger format broadsheets intended to be posted in public places, and finally advertisements appeared in the new and burgeoning newspapers and periodicals.[112] Commercial travelling was also essential. The Paris-based master-maker–retailer Noël Héroy (fl. 3rd quarter 18th century) travelled to his clients in north and northwest France,[113] as occurred throughout Europe. James Upjohn travelled through England and on the continent. Even travelling to Exeter to introduce his new wife to his family, 'As I went along I called on all the Watchmakers whom I thought it proper to do Business with: and in my way down I pick'd up a few Orders'. During a stay in Exeter of some weeks he travelled the countryside looking for business, as he did again while returning to London: 'I called on all the Trade as I came along, and got some few Orders, upon the whole as many as I wanted...'. Later he would travel through France, Belgium, Germany, and the Low Countries.[114] In 1756, Pierre Jaquet-Droz with his father-in-law and Jacques Gavril carried an assortment of clocks to Spain to sell, successfully, at the royal court.[115] Early in the next century systematic travelling would be essential to the success of the Vacheron Constantin business.[116]

---

106 For this development in England, see Hamilton 1926, 296–304.
107 Willan 1970, 37. As this was a first order Latham allowed no credit and stipulated that the wrapping paper and cord were to be returned. He added that the 'Pinnions ... are made by a Toolmaker of the Best of Steel and cut down on an engine and everyone who has of them like them'. Following a second order from Dent in April, Latham informed him that he also had rope or gut line for clocks, drill bows, and lacquer available. For Latham, see Loomes 2006, 465.
108 Robey 2012a 485; see also Robey 2012b. For Birmingham suppliers of dials and false plates in the late-eighteenth and early-nineteenth century, see Osborne 1966. For the work of a dial painter, William Whitaker of Halifax (1718–1800), see Robey 2019, 361–3.
109 Smith 1985, 611–13; Simon-Perret 2020, 41–2.
110 For example, see the entries for work done locally in Richard Wright's notebook, Smith 1985, 606–7. For a case study of demand in a single English village in the Early Modern periods, see Williams 2005, passim.
111 For English and Latin examples, see Evans, Carter, & Wright 2013, 47, 300, 301.
112 McCleod & Millburn 1998–1999.
113 Dequidt 2008 offers an excellent picture of the functioning of retail trade at this period.
114 Leopold & Smith 2016.
115 Tissot 1982.
116 See Jean-Claude Sabrier, 'The Saga of Vacheron Constantin', Antiquorum, 1994, 28–68.

CENTRES OF HOROLOGICAL TRADE

The manufacturing processes of clockmaking and watchmaking have the particularity that they could be effected in both town and country, and the unit of production ranged from the domestic hearth to a factory of hundreds of workmen. However, development was not linear, from smaller to larger or from country to town. Rather, a series of overlapping units were created in response to social and entrepreneurial pressures.

While the making of public sundials found its natural home among stonemasons and stonecutters, water-driven clocks and the solid weight-driven clocks that accompanied them were domiciled among metalworkers—smiths of all sorts, blacksmiths in particular, for the large and heavy iron structures that were the movements of late Medieval and Early Modern public clocks. As demand for them increased, specialized smiths emerged to make them. They were often itinerant, working on site,[117] for the travel costs of one or two men were considerably less than those for transporting heavy iron. Itinerancy was constant throughout the period, but it could take different forms: travelling to fulfil orders on site either locally or in a wider national—even international—area; travelling to the order of a patron;[118] voluntary travelling in search of experience or commissions or both;[119] forced travelling as a result of religious or ethnic persecution or of economic pressure; travelling as the showman of an exceptional horological creation;[120] or travelling for the advancement of commerce.

The mobility of craftsmen was a major factor in the spread of horological knowledge and the implantation of horological manufactures throughout Europe. Local blacksmith–clockmakers multiplied and retained their place. Only in the second half of the eighteenth century would easier transport on improved roads and the newly constructed canals make fabrication at a distance possible for public clocks. Domestic clocks, travelling clocks, and watches did not suffer from this restriction and centres of production developed in towns or regions where skills, raw materials, trade routes, commercial networks, and local demand conveniently came together. Small clockmaking and watchmaking developed among locksmiths, who were skilled in working steel and iron; watchmaking only freed itself from this association in the closing decades of the sixteenth century as brass became the favoured material for the product.

The main urban centres of horology in the sixteenth century were Nuremberg, Augsburg, Paris, Milan, other North Italian cities, and perhaps Florence. These were the chief exporting centres. Alongside them however, mainly in the latter part of the century, a smaller scale, local, manufacture developed in cities such as Lyons, London, Antwerp, and Blois. Some of these would develop more fully in the following century; others would remain local. In parallel with both, privileged, protected craftsmen worked in, or were attached to, royal and princely courts throughout Europe.[121] Between the three groups there was, inevitably, much overlap, in part a result of the itinerancy mentioned earlier. Nuremberg in the middle decades of the century exported both watches and their makers;[122] the development of the horological craft in Paris and Blois was considerably stimulated by the presence of the royal court in both cities for long periods of the year. Court workshops, however, especially those in minor courts, were ephemeral their existence dependent on that of the ruler who had created them. Essential to the establishment of an enduring horological trade was the combined presence of court and courtiers, permanent state administrators, craftsmen skilled in both construction and commerce, and good international trade links. Even this might not be enough. In the later sixteenth century, Nuremberg, perhaps because slow to switch from iron to brass, lost ground to Augsburg, and the Italian cities failed to develop a sustained international trade.

Even though watchmaking developed rapidly in Blois and in Rouen, Paris maintained its position. In the first half of the seventeenth century, the three cities seem to have worked more in symbiosis than in competition. Men, materials, and products were exchanged between the three, although with the departure of the court from Blois and the increasing difficulties experienced by the mainly Huguenot watchmakers in the face of government bigotry, manufacture there had largely disappeared by the end of the century. The trade of the Paris–Rouen axis was also confronted during the later decades of the century by the surge forwards of fine-quality watchmaking and clockmaking in London, and the steady growth of more everyday production in Geneva. Therefore, by 1700, French watchmaking had declined into a semi-local trade, only recovering an international presence in the second half of the eighteenth century thanks to the work of a small number of highly talented craftsmen.

For most of the eighteenth century the international horological trade was dominated by London and Geneva, with Paris claiming an increasingly important part in the second half of the century.[123] In the first two cities watches, like domestic clocks,

---

117 As, for example, Roger and Laurence de Stokes between St Albans and Norwich in the mid-fourteenth century, North 1976, ii 315–17, or Petrus de Metz, a clockmaker from Romans who worked in Grenoble, for whom see Maignien 1887.
118 I.e. Jost Burgi, travelling between Kassel and Prague, Maurice 1980a, 88–9.
119 As in the case of Martin Altman. See Pérez Álvarez 2020.
120 Pastre 1721.

121 *Cf.* Brockliss & Jones 1997, 335, who, in the context of medicine, note 'a Europe-wide tendency for the court to provide protection and succour for innovative minds, threatened by entrenched corporative institutions'.
122 Matthes & Sanchez-Barrios 2019, 230.
123 Dequidt 2008; Dequidt 2013; Dequidt 2014. *Cf.* Beliard 1767, 2, 'French horology has acquired such a reputation throughout Europe,

were finished, cased, and retailed. The success of both depended on their manufacturing hinterlands. For London, this was southwest Lancashire, and for Geneva the Swiss—and in part, the French—Jura. A similar interdependence occurred elsewhere, though on a smaller scale. The most notable example is that of Friedberg supplying complex movements to Augsburg, although, as some Friedberg work was supplied directly to London, as was some Augsburg-cased Friedberg work, both cities can be seen as a part of the supply chain to that world dominant luxury entrepôt.

European horology in the Early Modern period displays a complex pattern of large and small enterprises, local manufacture for local markets, and quantity production for international markets. Some of its structures—guilds, workshops, itinerancy—were general to Europe as a whole, while others, such as the domestic system, were specific to particular regions. While horology was a trade of major economic importance, it came also to hold a place in the calculations of statesmen as astronomers and mathematicians revealed the crucial role that it could play in precise measurement and navigation. This, as a motive for technical improvement, worked in parallel with the commercial imperative for advance. Precision and elegant presentation combined became increasingly important during the eighteenth century for commercial success, their marriage being perhaps consummated in the works of Breguet from the 1780s onwards. However, response to market demand was not only to novelty and new fashions. The burgeoning bourgeois market led to the development of production in quantity of raw clock movements, as in St Nicolas d'Aliermont (Normandy), supplied to the Paris finishers, and raw watch movements made in semi-industrial quantity by Frédéric Japy (1749–1818) at Beaucourt.[124] By contrast, English makers maintained a near monopoly of the Ottoman market until well into the mid-nineteenth century because they were willing to continue producing old-fashioned watches and clocks that suited conservative Turkish taste. Technically proficient and commercially astute, Early Modern European horology offered a potent springboard for future development.

---

that most nations have hardly any other watches than French watches, or those made in imitation of them'. It may have been the latter development that led the corporations of Paris (see Raillard 1752), Lyons, and Marseilles in the mid-eighteenth century to reaffirm their statutes and seek reinforcing legislation against non-guild members. Thus in 1763, Guillaume Zacharie and Nicolas-Jacques Lenoir, the wardens of the Corporation of clockmakers of Lyon, petitioned the city for powers to proceed against 'an infinite number of false workmen, commonly called "Chamber-masters", who work for themselves in secret and without authorisation'. Reglements 1743 [*recte* 1763], 25.

124 There is no study of clockmaking in St Nicolas during the eighteenth century, only what can be gleaned from notices of Charles-Antoine Croutte (1690–1768), reputedly the first clockmaker there; a brief chapter in Cournarie 2011, 20–6, who indicates eight workshops in 1750 and twenty-seven in 1789; Lombardi & Gebus 2015–16; and local histories. For Japy, see Lamard 1985; Lamard 1984/1988; Lamard 1999; Bergeron 1995.

CHAPTER EIGHT

THE DEVELOPMENT OF SUNDIALS
FOURTEENTH TO TWENTIETH CENTURIES

Denis Savoie

Novelty of form and technical advances in construction characterize Sundials produced in the Early Modern and modern periods, although their variety makes it difficult fully to describe them. In this field of instruments, we must examine three sources: manuscripts, printed works, and the dials themselves.[1] We can subdivide the final category into:

1. fixed external dials: horizontal, vertical, inclining/declining.
2. table dials: these are portable but not in a pocket. They were intended for use on windowsills, and to be transported between locations.
3. portable or 'pocket' dials.

The **first** have an evident social function—to indicate the hour and, from the late seventeenth century onwards, act as a standard for the setting of clocks. Pocket and table dials, however, combine these roles with that of ostentation, precision being perhaps of lesser importance. Masterpieces of decoration and engraving they are often of a complexity that only an élite intellectual, social class could contemplate and enjoy. Such pieces are exemplified by the **chalice-dials** of Georg Brentel (*fl.*1573–1619), or Marcus Purmann (*fl.*1588–1619), the exuberant **altitude dials** of Wenzel Jamnitzer (1508–86), or the **polyhedral dials** of Nicolas Kratzer (1487–1550). A similar abundance of lines and astronomical indications is also to be found on some fixed external dials (or internal in the case of **reflex** dials), which had perhaps less importance in themselves than in the image that they projected of their owner. In both cases however, a decorative function was sought, often augmented by a religious device or a sententious reflection on the passage of time.

During the Renaissance, dials were once again incorporated into buildings. Architects like Danielo Barbaro (1514–70), Jean Bullant (1520–78) Salomon de Caus (1576–1635), or Girard Desargues (1593–1662) all followed Vitruvius, who had affirmed that a complete treatise on architecture must discuss buildings, dials, and machines. Sundials appear frequently in architectural writings, even, in the eighteenth and nineteenth centuries, in treatises on stereometry (the study of shadows). Dials had a civic role. But the subject has a triple parentage for it was also considered to be a subsection of astronomy and as an area for research in mathematics. A great many of the tracts on Sundials from the sixteenth and seventeenth centuries were written by mathematicians or astronomers. The importance of dialling for mathematical research should be underlined, especially in the fields of plane and spherical trigonometry and for all that concerned conic sections. Some problems indeed were not solved until the early twentieth century.

Essentially the Sundial indicates the solar hour, or more precisely, the hour angle of the Sun. This can be measured directly by dials with **polar gnomons**, or by means of the height of the Sun above the horizon (**altitude dials**), or by its **azimuth** (**direction**, **azimuthal**, or **analemmatic dials**). In all three cases the latitude of the place of observation is the fundamental parameter.[2] A Sundial is drawn for a specific latitude. Known since Antiquity, altitude dials have the advantage of not requiring orientation along the meridian, a requirement that

---

[1] The bibliography of dialling is extensive. A good basis is provided is given by Aked & Severino 1997; a survey of the history of dials by Turner 1989a.

[2] Only in 1978 would a dial independent of latitude be devised by J. G. Freeman, although this is rather a curiosity than a useful instrument.

long–handicapped the construction of vertical dials.³ The inclusion of small compasses in portable direction dials partly enabled the problem of orientation to be overcome. Even so, use of both altitude and direction dials requires knowledge of the date. In fourteenth-century Western Europe a major transformation took place with the introduction of the polar gnomon which would lead to a generalization of the use of **equal** hours and radically modify the methods of gnomonic calculation. Throughout the Early Modern period, a variety of hour systems—astronomical, **Italian, Babylonian**—remained in use, although **unequal** hours largely disappeared during the fourteenth century.⁴

PORTABLE ALTITUDE DIALS

The principle of altitude dials (Figure 93) is straightforward both for the change in the height of the Sun during the day and throughout the year. The three basic parameters are the hour, the date, and the place. The Sun is higher at noon in summer (maximum at the summer solstice), lower in winter (minimum at the winter solstice). Solar height also depends upon latitude. The Sun is higher in the sky on approaching the equator. For example, at Athens (latitude 38°), solar altitude at the summer solstice is 75° 26′, whereas at Rome (latitude 42°) it is 71° 26′. Finally, over a day the height of the Sun varies from its rising to setting, with a maximum at solar noon. Therefore, finding the time by the height of the Sun is achieved as follows: two elements, the latitude and the date, are given, and variation in the third (altitude) is used to determine the fourth (the hour). All can be represented on a plane or three-dimensional surface by a graphical construction that allows a direct reading of the hour and avoids trigonometrical calculations. Variation in the height of the Sun is the most natural way of finding the time and can be traced back to Antiquity, but it is, above all, the late Middle Ages and the Early Renaissance that were the golden age of altitude dials, their makers competing to produce ingenious, elegant, and sometime complex designs.

The best known, apart from the cylinder dial (Figure 44), are the several forms of universal rectilinear dials—the *Navicula*, the *Capucine*, that described by Regiomontanus, and the best of them, the astronomical ring dial. This was known in Antiquity, although only one example has survived.⁵ The use of the cylinder dial, also known in Antiquity, was extended through the Middle Ages to the Renaissance. It offered the advantage to Renaissance diallists that it could show unequal or equal hours, something impossible with any other type.⁶

This entire, highly varied, family of dials has advantages and disadvantages. Their basic advantage is that they are all portable, usually being of a small size and easy to carry about. Strictly, however, the cylinder dial is usable at only one latitude, although the *navicula* and the Regiomontanus dial can be adapted to a range of latitudes. Their small size means that their precision is poor, and they are difficult to use towards solar noon when the altitude variation is minimal. Since the Sun passes twice a day over the same altitude it is also needful to know if it be morning or afternoon.

Cylinder dials could be executed in wood, bone, ivory, or metal and carried a set of hour curves usually doubly numbered 4 to 12 from the top towards the bottom, although they could also be drawn for **Italian** hours, as per Clavius.⁷ At top or bottom there is a calendar giving the date by month or sign of the zodiac, or both. The whole instrument is topped by a removable cover, to which the gnomon (which can be folded down inside the body of the instrument when not in use) is attached. Having set the gnomon above the appropriate date line, the cylinder is held vertically towards the Sun so that the shadow of the gnomon falls on the hour grid. The tip of the shadow indicates the time. Most surviving cylinder dials are of small dimensions (less than 10 cm tall); they were intended for use in only a limited geographical zone. It is evident that a difference of ± 1° of latitude will have little effect on their timekeeping.⁸ From the *Compositio horologiorum* of Sebastian Münster, 1531, right up to the end of the twentieth century, most dialling books discuss the cylinder dial, thus making it the most

---

3 What is today called the *gnomonic declination* of a wall, that is the azimuth of its vertical from geographical south is usually called the *inclination* or the *declination* in earlier texts.
4 But see Poulle 1999. Ginzel, 88–97.
5 Gounaris 1980. Talbert 2017, 76–81.

6 For ancient portable dials, see Chapter 1 this volume, and Bonnin 2015, 106–109. Three examples of cylinder dials from Antiquity are currently known. For one (fourth-century CE), see Arnaldi & Schaldach 1997; for that excavated at Amiens (third-century CE) Hoët-Van Cauwenberghe 2012b; for both Talbert 2017 10–13. For the third, Hoët-Van Cauwenberghe 2012c. The ancient universal portable dials are typical examples of dials that show only unequal hours. See Savoie 2014a, 33–51.
7 Clavius 1581, 645.
8 For example, imagine a cylinder 10 cm high fitted with a gnomon $a = 3$ cm functioning at 42° latitude: the length y of the vertical shadow is given by $y = a\ tg\ h$ where h is the height of the Sun. At solar noon at the summer solstice the shadow measures $y = 8.9$ cm ($h = 90° — 42° + \delta$, with $\delta = 23°$, 433). If the latitude is changed by 1° (43°), the shadow then measures 8.4 cm, and the 5mm of difference are visible to the naked eye. However, before 9h and from 15h onwards, the difference decreases to 1mm and so becomes undetectable. Around the winter solstice the difference is 1mm at solar noon and not detectable before 11h and after 13h. To notice a 2-mm difference in winter with the naked eye it is necessary to move through 3° or 4° of latitude (i.e. about 333km north or south). Moreover, altitude dials are affected by refraction changing the height of the Sun, and therefore the hour read on the dial, and this the more so as the Sun is lower above the horizon. For example, with H as the hour angle, an altitude dial drawn for latitude 43° showing H $= 95°$ (18h 20m) with $\delta = 5°$ has an error of about −3 minutes.

Figure 93 A highly decorated altitude dial by Wentzel Jamnitzer (1508–85), Nuremberg 1578, carrying indications for equal and unequal hours. Collection and photo: Observatoire de Paris.

frequently described dial in the literature, especially as there are many medieval tracts devoted to it.[9] The appellation, 'shepherd's dial', is an unfounded term invented in the mid-nineteenth century.[10]

---

9 Thorndike 1929; Zinner 1930. On Fusoris' construction method, Poulle 1963, 67–9. For early, (twelfth-century) mentions in the West, see Krenn 1977; Jacquemard, Desbordes, & Hairie, 2007.

---

10 See Savoie 2012b. The description was given by a journalist in 1857 and has since been copied repeatedly.

Drawing such a dial requires a solar altitude table established trigonometrically as a function of date and latitude.[11] These are so common in astronomy that numerous variants have been devised to allow drawing a grid either on a plane surface or on a more elaborate one, such as a ring. Indeed, ring dials can function in different latitudes on pointing them towards the Sun when a small opening allows a spot of sunlight to fall on the hour lines drawn on the inner face of the ring as a function of date. Already known in fourteen-century Syria,[12] the precision of this dial is limited because of its small dimensions, but it continued to be described in some gnomonical works up the late eighteenth century[13] and to be made by craftsmen-jewellers even in the twentieth century.

There are too many variants of altitude dials drawn on a plane surface to be described here, particularly as some were produced in minute quantities and others never at all. Already in the sixteenth century, the first gnomonic works describe plane portable dials with curved hour lines, often no bigger than a playing-card, and fitted with a weighted thread on which a sliding bead indicates the hour. All such dials derive from the medieval *quadrans vetus*, *quadrant* being the generic word used during the Middle Ages for all instruments constructed on a quarter of a circle irrespective of their scales. Of Arabic or Persian origin,[14] one of the earliest mentions of the *quadrans* is by Hermann the Lame (1013–54),[15] but it was largely spread in Europe by Sacrobosco (late twelfth–early thirteenth century),[16] and Robertus Anglicus (*fl* 3rd quarter 13th century).[17]

In effect, the *quadrans vetus* is an adaptation of the unequal hour diagram found on the back of astrolabes completed by a sighting system and a weighted thread. It is a universal altitude dial giving the unequal hour independently of latitude and date. It is composed of a quadrant with a sighting system, usually pinnules, set on one of its straight sides, a plumb line mounted at the apex of the hour diagram, and a 90° scale marked on the curved limb. Six arcs (the hour lines) run from the apex to the limb, which they divide into six sectors each of 15°. During the twelfth century a **cursor**, moving in a slot in the rim and marked with a zodiacal calendar, was added to give the meridian height of the Sun as a function of the date. It can be considered as a sort of declination table for the Sun.

Adapted to show equal hours, the horary quadrant became latitude dependent and lost its universal quality.[18] The hour lines, arcs of circles, no longer converge at the apex of the quadrant.[19] In 1513, Johannes Stöffler (1452–1531) proposed a variant in which he folded the solstices one onto the other. A first circle represented the equator, a second the solstices. Then, on these circles the height of the Sun was indicated at different times and linked by a straight line. There are many variants of such horary quadrants where the constructor has simply modified the scales. It should be noted that what Fine calls **astrolabe quadrants**[20] relate more closely to **astrolabes** than to Sundials. Some, such as that of the English mathematician Edmund Gunter (1581–1626), are particularly complex and employ **stereographic projection** to give the quadrant astronomical uses.[21] There are many fine examples of this sophisticated instrument and it was made for over a century. Other instrument-makers, like Wenzel Jamnitzer in Nüremberg, or the della Volpaia dynasty in Florence, produced portable dials functioning with gnomons perpendicular to the Sun, adapting a conventionalized hour grid.[22] The hour was read from the shadow projected tangentially by the gnomon onto the plane of the dial at the intersection of the curve representing the hour as a function of solar height and the date line.

The problem for all these dials was to read the apparently infinite shadow on the surface of the dial. The hour grid could be composed of parallel straight lines. In this case the dial has no fixed gnomon, time being determined from the position of a bead mobile on a plumb line when the dial is oriented to receive radial light from the Sun. Commonly designated as 'rectilinear dials' they are not easy to use.[23] Many portable dials

---

11 If ϕ is the latitude of place, δ the declination of the Sun, H the **hour angle**, the height h is calculated by $\sin h = \sin \phi \sin \delta + \cos \phi \cos \delta \cos H$. The use of an astrolabe to determine h is attested.

12 Charette 2003, 142–4.

13 See for example in Oddi 1638, 222–5, or Ozanam 1741, ii, 120–7 (1778 edition, iii), where the construction is acknowledged to Deschales. Bion 1716, 359–32 describes the construction of such a dial, which is analysed in Drecker 1925, 90–1.

14 King 2002; Charette 2003, 211–15.

15 Study of the texts of this period, including that attributed to Hermann the Lame allowed J.–M. Millas-Vallicrosa (1897–1970) to show that two forms of *quadrans* existed: the *vetustissimus*, and later, the *vetus*. See Millas-Vallicrosa 1932. See also the pertinent remarks in Poulle 1983, 9–13, and Poulle 1972, 36–8.

16 The fundamental study is Knorr 1997.

17 The Greek and Latin text is edited in Tannery 1897, see however the remarks by Knorr 1997.

18 On the construction and theory of the *quadrans vetus*, see Savoie 2014b. Strictly the *quadrans vetus* is only accurate at the equinoxes. Clavius 1581, 645, describes its adaptation to Italian hours.

19 Illustrations of the hour diagram of the *quadrans vetus* for both unequal and equal hours are given in Fine 1560, 143, 151. Both also appear in Fine 1532, f. 189^r; f. 191^r.

20 Fine 1534, which only gives the uses without any illustration, although this can be found in the *Protomathesis*, 1532, f. 204^r. On medieval quadrants, see Poulle 1964.

21 Gunter 1623; Gunter 1624. Gunter's gnomonical works are collected in Gunter 1673. The third book is an application of trigonometry to the calculation of various forms of dials using stereographic diagrams. On Gunter's quadrant, see Higton 1996, 92–108, with a briefer version in Higton 2001, 70–6. The remarkable collection of horary quadrants in the Royal Museums of Greenwich can be consulted in Higton et al. 2002, 337–58. On the theory of astrolabe quadrants, see D'Hollander 1999, 213–33.

22 See Savoie 2017b, and Savoie 2012c.

23 Archinard 1988, although here the quadrant of Apian is not discussed, while the Saint Rigaud quadrant, despite what is said, is correct. See also

were first described by Münster,[24] but there can be no doubt that he described existing instruments for which written instructions were in circulation. The pedagogic functions of such dials partly explain their longevity. Dials are concrete applications of geometrical and trigonometric constructions that can be tested via the Sun, that is, portable altitude dials allow comprehension of the way in which the hour can be obtained from solar motion in relation to two parameters: the latitude and the declination. They do not rival externally placed fixed dials, since they have different functions. Some, such as the *Navicula de venetiis* and the universal rectilinear dial of Regiomontanus, have largely monopolized the attention of historians, despite the rather small number of surviving specimens, because of their sophistication—a Renaissance sophistication which apparently left little to be done thereafter. However, the origins of many altitude dials remain in doubt.

The *Navicula de Venetiis*, little ship of the Venetians— (Figure 94) is a portable, ship-shaped dial about the size of a hand,[25] which gives the hour as a function of solar altitude for all latitudes. It comprises a latitude scale on the 'mast' of the ship of a zodiacal calendar and a mobile thread with bead attached to a cursor. Once this has been set for latitude, the mast is inclined so that its index corresponds with the date in the zodiacal calendar. The thread attached to the cursor is then drawn out to cross the date. The bead slides on the midday line. When the dial is oriented in the vertical plane of the Sun, the bead shows local solar time in the grid of parallel lines. Strictly speaking, the *Navicula* is exact only at the solstices and the equinoxes, but its makers skilfully distorted certain graduations to render this error almost imperceptible. Long miSunderstood, this apparently simple dial is mathematically subtle.[26]

Only six medieval examples and four sixteenth-century examples are known of this instrument. All the medieval survivors are probably fifteenth century and made in England.[27] It was only thanks to the printed word that it became known elsewhere.

The spearhead work was the *Protomathesis* of Oronce Fine,[28] whose dial is considered to be a variant form of the universal rectilinear dial of Regiomontanus. Some writers, such as Clavius or Kircher, would follow Fine's description without really understanding how the dial functioned.[29] Kircher offers a Baroque illustration[30] with the hour scale drawn on a dove. The origins of the Navicula remain debatable.[31]

Notable astronomer and mathematician Regiomontanus (Johannes Müller von Königsberg, 1436–76) founded a printing press to encourage the advance of mathematical subjects and was interested in the production of related instruments. His name is attached to a universal altitude dial that he discussed at the end of his *Calendarium*,[32] although there is no certainty that he conceived it. Usually rectangular, it is composed of a grid of straight, parallel hour lines in the lower half of the instrument, and above this a **trigon** with a 47° opening. This contains a lateral, straight line with graduations for latitudes up to the Polar circle; the summit is placed on the central, or VI, line in the hour scale. To right and left, a second trigon of 47° is drawn perpendicular to the first, and this permits adjustment for date against a zodiacal scale. When the instrument is held vertically, the top of the instrument is fitted with two pinnules for sighting the Sun. Previously, however, its use required setting the tip to the appropriate latitude on the trigon of a **brachiolus**, to the extremity of which a weighted thread, with a sliding bead, was attached. The thread is then pulled taut and the bead placed over the date on the second lateral trigon. When held in the Sun the shadow of the bead falls on the hour grid to mark time.

The best-known source for the construction of this rectilinear dial is Münster's treatise,[33] although his teacher, Johannes Stöffler (1452–1531), had already described it.[34] After Fine, other noted authors, for example, Andreas Schöner (1477–1547) or Clavius, described ways of constructing it, but a full geometrical demonstration was only provided by Claude François Milliet Deschales (1621–78) in 1674.[35] This was taken over into the posthumous

---

the survey and discussion of possible connections between altitude dials in Massé 2009.

24 Münster 1531.

25 The fundamental work on this dial is Eagleton 2010. The name remains mysterious (despite the fanciful discussion in King 2004–5, ii, 288–90), but the dial could derive either from the *organum* of Ptolemy, which is drawn using orthographic projection (illustrated in Apian), or from the universal rectilinear dial associated with Regiomontanus. For discussion, see King 2004–5, ii, 267 ff.; see also Eagleton 2009.

26 Kragten 1989.

27 Two remain in England, a third is lost, the fourth is in the Musée d'Histoire des Sciences de Genève, and the fifth in the Museo Galileo, Florence. All are discussed in Eagleton 2010, ch. 2. A sixth in Edinburgh, mentioned by Delalande & Rocca 2020, 581, is unpublished. See Archinard 1995; see also Brusa 1980, who discusses some later examples, Turner 2007, 60–3, and Delehar 1993, 186–8. Three sixteenth-century examples are discussed in Eagleton 2010, ch. 8, and the fourth is discussed in Delalande & Rocca 2020, 573–85.

28 Fine 1532, f. 199^{r-v} and 200^{r-v}. An example of the *Navicula* in ivory, signed by Fine and dated 1524, is in the Poldi Pezzoli Museum, Milan (Figure 94). See Brusa 1980. A crown and a salamander stamped on the mast link it to the court of François I, and perhaps to the king himself; see Eagleton 2009, 84

29 For confusions in the European manuscript tradition before and after Fine, see Eagleton 2009, 86.

30 Kircher 1646, 506–7, and Eagleton 2010, 141–4.

31 Eagleton 2006, 42–59. See also the long analysis in King 2004–5, 267–336.

32 Published in Nuremberg in 1474.

33 Münster 1531, ch. 36, 'that is what Johannes Regiomontanus taught in his *Calendarium* but using a quite different layout of the lines, as we have explained in Chapter 5 . . . '.

34 Stöffler 1518.

35 Deschales 1674, iii, 257–63. The geometrical demonstration was also realized in a thesis maintained at Jena by Johann Friedrich Schuman. The

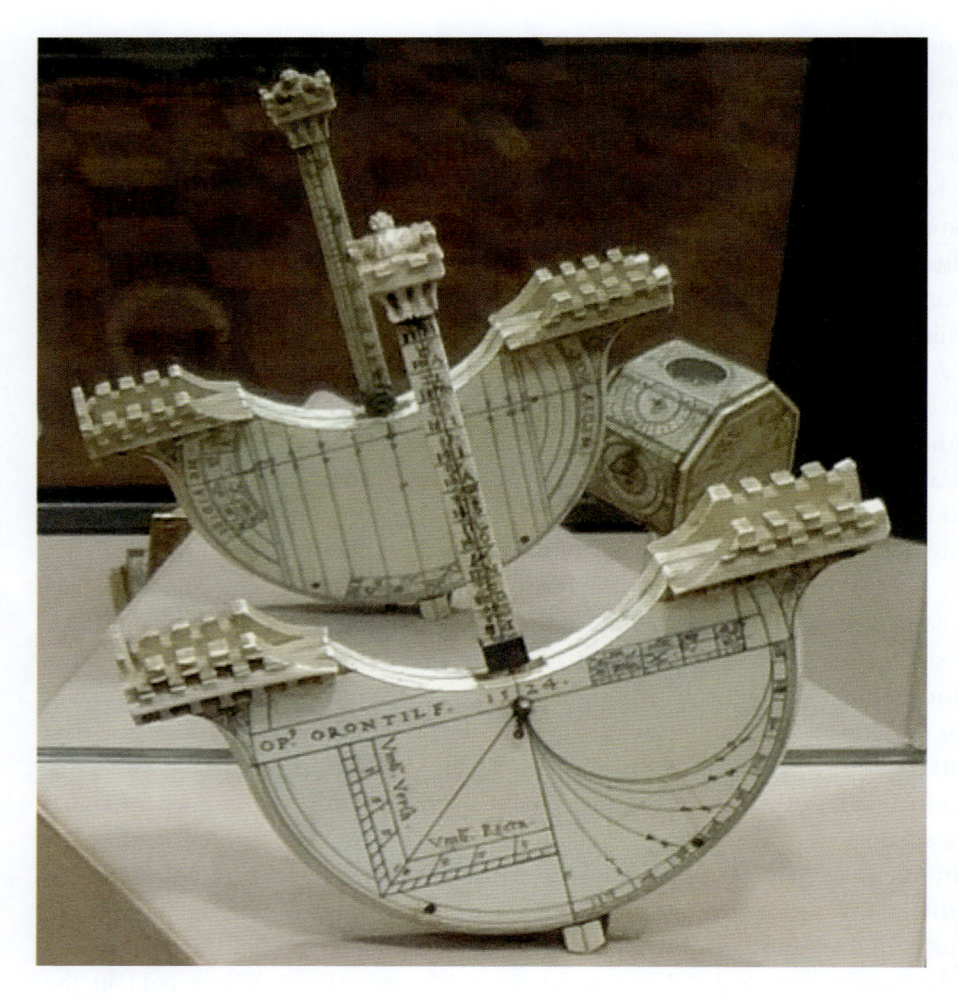

Figure 94 A 'navicula de Venetiis' in ivory by Oronce Fine 1524 (reverse and obverse, with a small ivory polyhedral dial in the background). Museo Poldi Pezzoli, Milan. Photo: Anthony Turner.

editions of Ozanam's *Récréations mathématiques*,[36] in which the dial is described as 'universal rectilinear'; the name 'universal Regiomontanus dial' seems to derive from Delambre.[37] The dial was also adapted to show different kinds of hours, in particular by the German map-maker Johannes Stabius (1450–1522) who, in 1512, published in a broadsheet a 'Regiomontanus' dial showing Italian and unequal hours.[38]

The origin of this dial remains uncertain. Some authors derive it from the *Navicula*, while other disagree. Ingenious, aesthetically appealing, and universal, it is less frequently described in the literature of dialling than the cylinder dial, although it is always mentioned in twentieth-century writing on the subject.[39] However, the fixed latitude version of the universal rectilinear dial may underlie another type of straight line altitude dial—the Capuchin, so named from the shape of the diagram that resembles in the upper part of the dial a Capuchin's cowl, although the dial itself has no link with the Capuchin order; the name was probably given by Ozanam. Its earliest description is once more found in Münster, and solar time is read by the position of a bead, placed according to the date, by radial sunlight when the dial is set in the plane of the Sun. For some authors, the Capuchin dial could be the origin of universal rectilinear dials,[40] but the relations between them and another version of the dial—that devised by François de Saint-Rigaud (1606–1673)[41]—were notably confused during the later eighteenth century by Montucla. Like the other rectilinear dials, the Capuchin has once more become popular in the twentieth century.[42]

---

aim of these works was to prove that the dial materializes the equation: $\sin h = \sin \phi \sin \delta + \cos \phi \cos \delta \cos H$. See the modern proof in Drecker 1925, 95.

36 Ozanam 1685, Ozanam 1723, Ozanam 1741, ii, 99–120. The dial does not appear, in the edition prepared by Montucla in 1787.

37 Delambre 1814–27, (1819), 323–4, who gives a translation of Regiomontanus' description.

38 *Horoscopion omni generaliter congruens climati*, 1512. For Stabius, see Kremer 2016. For the making of rectilinear dials and their theory, see de Vries 1998.

39 See for example the excellent account in Fantoni 1988, 390–407.

40 Severino 2000, 38–44. The Musée d'Histoire des Sciences in Geneva holds two examples of this dial, one of them dated 1573. On the relation of the Capuchin with Apian's universal dial, see de Vries 1999, 4–8. It can be shown that the Capuchin is a fixed latitude version of Apian's universal dial.

41 The text *Analemma novum* in which he described it remains unknown. For Saint-Rigaud see Sommervogvel 1890–1911 (1896, vii, 439–40). Cf. Massé 2009, 41. Note that Ozanam affirms that Saint-Rigaud's instrument can also be used as an horizontal altitude dial, which is rare for this class of dial

42 See the description in Cousins 1969, 168–74, based on Stebbins 1961, 49–56.

The German astronomer Peter Apian (1495–1552) made some important modifications to the Regiomontanus universal dial.[43] He drew the zodiac and latitude scales in a single diagram, the first being vertical, the second inclined. He also proposed using the dial to find the time at night by using bright stars—a unique suggestion.[44] Although Apian's dials would later be discussed by Andreas Schöner,[45] they seem never to have been actually built—only described in his books.

The *astronomical ring dial* is the most elaborate of altitude dials. Universal and a derivative of the armillary sphere, its origins are lost in Antiquity.[46] It is composed of three open rings, one within the next and the third movable. It is uncertain whether the device was transmitted from Antiquity by routes now lost or was reinvented during the Renaissance. Of the various types, usually in metal but occasionally wood, the most common are formed of two or three rings. The outer circle is the meridian fitted with a suspension ring and a system to adjust the dial for latitude; the middle ring represents the equinoctial and carries the hour scale; the third ring carries the sighting apparatus. The dial requires little calculation for its construction, it is simple to use, its orientation does not change throughout the day, and it can be used in all latitudes. The disadvantage of the three-ring model is that it cannot be used around midday (the meridian ring putting the third circle into shadow), nor near the equinoxes, when the thickness of the equinoctial circle hinders the passage of the Sun's rays. Perhaps for this reason relatively few examples of the three-ring dial have survived; those that have are associated with the instrument shops of Louvain.[47]

An innovation in the 1630s attributed to the English mathematician William Oughtred (1574–1660) would return the instrument to popular favour. Oughtred replaced the third ring, on the polar axis of the instrument, by a bridge fitted with a sliding sight and engraved with a zodiacal calendar.[48] Once set for date, the sight allowed a spot of sunlight to fall onto the hour scale engraved on the equinoctial ring. Although the dial suffered from the same disadvantages as the three-ring model, it was lighter in construction. Later eighteenth-century models would be fitted with alidades mounted in a mobile ring in order to read the time throughout the day. Recommended as an instrument for navigators in some texts,[49] it was esteemed highly and widely used throughout the eighteenth century, despite requiring considerable skill in its construction.[50]

PORTABLE DIRECTION DIALS

Probably in the late thirteenth or early fourteenth century, the magnetic compass was incorporated into simple horizontal dials. The earliest known such dial is the *sandûk al-Yawâqît* completed in 1365/6,[51] but compass dials are also likely to have been produced in fourteenth-century Europe, where portable equinoctial dials were developed during the fifteenth century.[52] **Magnetic declination** was also understood, and a dial dated 1481 carries a line for its variation.[53] However, the **secular variation** was not detected until the early seventeenth century and compensation for it was only slowly introduced.[54] It was perhaps also during the fifteenth century that using a compass to measure the orientation of a wall was suggested, but the errors that arose from so using it, for example, tracing a meridian, were detected in the following century, as was the need to be far away from any mass of ferrous metal.[55]

---

43 Apian 1532, 1533a; 1533b. The latter reuses some of the contents of the *quadrant,* with the addition of a figure of the universal altitude dial showing Italian and Babylonian hours.

44 The method is as follows: knowing the date, the zodiac sign opposite the position of the Sun is determined. Then, a star (that closest possible to the anti-solar point) is sighted to determine its altitude and the bead set on the plumb line gives the nocturnal hour. Choosing a star at 180° from the Sun in solar longitude, its hour angle also differs by 180°. Thus, for $\phi = 48°$ if $\lambda = 30°$, $\delta_{Sun} = 11°$, 47; if $H = -45°$, the Sun is at 18°, 42 below the horizon (north-east). If the star has a longitude opposite the Sun, e.g. $\lambda = 210°$, then $\delta_{star} = -11°$, 47: if $h_{star} = +18°$, 42, then $H = +45°$. It is therefore 3 a.m. Since Apian's stars cannot absolutely oppose the Sun at any given date, there is a non-negligible error.

45 Schöner 1562, 75–77.

46 Chapter 1 this volume; see also Gounaris 1980.

47 On these see Van Cleempoel 2002, 14–22; Gessner 2010.

48 Oughtred 1652 included a 'Description of the General Horological Ring'. It was also described by Wynne 1682, who, on his title page, specifically names Oughtred as its inventor. Van Dyck depicted the ring dial in a double portrait dated 1639 of the Earl and Countess of Arundel, Oughtred's patrons, which shows that Oughtred's innovation was made prior to this date; see Higton 1995, 25. Also, a description of the dial is found in the numerous editions of Bobynet 1663.

49 Seller 1669 and several subsequent editions. The ring dial is discussed at 133–4 in that of 1730.

50 Bedos de Celles 1774, 345–55, describing the exceptional model made for the Cardinal de Luynes. In his 1784 contribution to Charles Bossut's *Encyclopédie méthodique: mathématiques* (i, 63–4), Jérôme Lalande noted that good watches had now made dials redundant, adding that ' ... the astronomical ring is excellent to take into the country. where there are no meridians or [public] Sundials'.

51 This volume, Chapter 5, Section 1.

52 For a selection of these see Maddison 1969, 14–15, and figs. 8–11 and on the compass 15–17.

53 The dial, attributed to Hans Dorn of Vienna, is illustrated in Maddison 1969, fig. 22, who, following Körber and other commentators, notes (15), 'it seems clear that variation was marked on the compasses of portable Sundials before the first recorded observation of changing variation at sea (Columbus, 13–17 September 1492).'.

54 Mercier 2015, 61–76.

55 Zinner 1990, 16–17. For errors in meridian layouts, see Benedetti 1574, 7. On the error in time produced by a misaligned dial, see Savoie 2001, ch. 24. If, *c.*1600, the declination is 10°, a dial not corrected for

Compass dials may be very small (finger-rings), or considerably larger (table dials). By contrast with altitude dials, they are generally, horizontal, vertical, or equatorial and use a polar gnomon. They are thus dials intrinsically linked with one of the major innovations in gnomonics. Although no precise date can be ascribed for the introduction of the polar gnomon, it is clearly a late-fourteenth- or early-fifteenth-century development and examples exist in the fixed dials on Jacobi Church, Utrecht (1463), the abbey of Alpirsbach, Germany (1477), or that of Strasburg Cathedral (1493).[56,57] All are direct south-facing dials that indicate whole hours without subdivisions.

In dials with a polar gnomon, it is the alignment of the gnomon shadow with an hour line that indicates time. Only equal hours are indicated, and these converge with the gnomon to a single point on the dial table—the centre—where the gnomon would pierce the table. Such dials are not dependent on the date; the hour angle of the Sun is read irrespective of its declination; their calculation is thus largely simplified, although exactly orienting the gnomon, especially if the plane of its use declines away from the south, can be complex.

While compass dials were theoretically known to early Arab writers on gnomonics, no discussion exists in any Arabic treatise from the ninth to the fifteenth centuries, and no dials with a polar gnomon are known before that of Ibn-al-Shâtir.[58] Although this pre-dates most European examples, for want of evidence nothing can be affirmed concerning a possible transmission. That the polar gnomon was independently developed in Christian Europe during the later Middle Ages is equally probable. Its use was favoured by the appearance of the solid-weight driven clock indicating equal hours, and the twelve-hour count from noon soon dominated. An exception, however, occurred in Italy, where the equal hour count continued to be reckoned from, or near, Sunset, and for which the position of the extremity of the shadow of the gnomon among the hour lines remained the time indicator. The calculating of dials became simpler when a polar gnomon was employed as it made the earlier method of calculating the hour lines by means of three shadow points redundant, while the fact that they emerged as straight lines facilitated drawing them.

Combined with the compass, then, the polar gnomon gave rise to easily used dials of a straightforward construction that could be realized both in very simple forms or as exceptional, highly decorative prestige pieces made of precious materials.[59] In their early treatises, Münster (1531) and Fine (1532) had already discussed dials using a magnetic compass, which would continue to be popular until at least the end of the eighteenth century, and there were many kinds. Notable among them is the diptych dial composed of two rectangular leaves, which, being opened, stretch a thread between them that acts as a polar gnomon. This allows time to be read either on the lower horizontal leaf (fitted with a compass), or on the southern or equinoctial upper leaf. Such dials are often adjustable over a range of latitudes. The best-known of this class are those produced in Nuremberg, Paris, and Dieppe.[60] Other commercial pocket dials would be conceived and made during the later seventeenth and eighteenth centuries. Among the best known is that today carrying the name of Michael Butterfield (1635–1724) and which was known to contemporaries as a 'Sundial with compass'; it features a gnomon adjustable for use with several (usually four) hour scales drawn for different latitudes.

Dials like these cover only a limited range of latitudes, whereas equinoctial dials fitted with a compass can be truly universal. The hour scale, with equal divisions of 15°, is engraved on a ring that can be adjusted for latitude against a co-latitude scale engraved on an arc on one side of the instrument. The gnomon is mounted on a cross-bar and can be turned from north to south depending on whether the dial is used in winter or summer. Handy lists of the latitudes of major places—which could be established according to the needs of the client—are usually found engraved on the reverse of such dials. Although Augsburg was particularly noted for them, these dials were made throughout Europe in styles and materials ranging from the modest to the luxurious; however presented, they were robust, effective instruments, and were simple to use throughout the world.[61]

For the construction of dials with polar gnomons, tables giving the angular distance between the hour lines and the meridian were essential.[62] An early printed example is that by Johannes Stöffler in his *Calendarium Romanum* from 1518. Earlier, Jean Fusoris[63] used spherical geometry to create tables of angles for his dials. The

---

this would have an error at 45° in winter of some forty minutes at solar noon.

56, On which, see Ungerer & Gloria 1933, 73–107.

57 Zinner 1964, 13, implies that it is present in Germany from 1401.

58 See Chapter 5.2. Janin 1972a. For lack of interest in the polar gnomon among Arab writers, see King 1999 297–300.

59 Numerous museums and private collections contain exceptional dials. See the catalogues by Delalande 2014 (private); Higton et al. 2002 (Greenwich); Turner 2007 (Florence); Schechner 2019 (Chicago).

60 A general survey is offered in Lloyd et al. 1992. The fundamental work on Nuremberg dials is Gouk 1988; for Dieppe dials, see Mercier 2014, 45–65.

61 Although Bedos de Celles 1760, 279–80, 300, makes energetic criticisms of Butterfield dials, as did Julien Le Roy in Sully 1711, 315–18, and of certain ring dials, he gives an excellent description of the making of the universal equinoctial dial. For all these dials, see Higton 2001 and the many descriptions of them in the catalogues cited in n. 58.

62 A southern vertical Sundial is exactly the same dial symmetrically drawn with respect to the forty-fifth parallel. For example, a southern vertical dial plotted for latitude 49° has the same layout as a horizontal dial for 41° (but with the numbers rotating in the opposite direction). At 45° latitude, the layout of a horizontal Sundial is the same as a southern vertical Sundial.

63 On whom see Mirot 1900, Poulle 1963. BnF MS lat. 15104, includes explanations of a cylinder dial established by Furoris at Cluny, and the results of calculations for a vertical dial. Fusoris' dials are subdivided to halves or thirds of hours. Equivalent to the calculation of a polar gnomon

simple modern formula for a horizontal dial, $tg\ H' = \sin \phi\ tg\ H$, or that for a south-facing vertical dial, $tg\ H'' = \cos \phi\ tg\ H$ (where ϕ is the latitude and H the hour angle of the Sun, H' and H'' being the tabular angles) was a problem for early diallists because of the lack of tables of tangents, which were not available until the second half of the sixteenth century, after the work of Joachim Rheticus (1514–74),[64] although the trigonometric relation using the tangent of the angle did not immediately impose itself.

Tables of sines, and later tables of tangents, are found in many gnomonic treatises, and their use was facilitated by the invention of logarithms by John Napier (1550–1617) in 1614–19. They were available in France from 1626, thanks to Didier Henrion. New tables were produced in England by Henry Briggs (1561–1631).[65] Once the fundamental work on spherical trigonometry by François Viète[66] (1540–1603) was added, all the tools needed for gnomonical calculations were available.

Although the sources for fifteenth-century gnomonics remain obscure, an idea of its capabilities c.1500 is available in the *Collectanea mathematica praeprimis gnomonicam spectania*, the *Practika*[67] of Georg Hartmann (1489–1564). Hartmann drew his dials using the Vitruvian analemma and the trigon of signs. Following Dürer, he produced multiple dials showing the different kinds of hours, a vertical declining dial (rare at his period), a cylinder dial with Italian hours, crucifix dials, diptych dials, and his remarkable Ahaz dial.[68] All kinds of dials could now be imagined and realized. The solid basis of knowledge that had been established by the circulation of manuscript texts, such as those of Fusoris and Hartmann, was now reinforced in print, while Sundials emerged from the universities to become more widely used in the teaching of applied astronomy—a function they would retain in the following centuries.[69]

Most challenging of the dials favoured in the sixteenth century—the 'apotheosis' of gnomonic difficulty—polyhedral dials would be widely developed (Figure 94, back) The earliest depiction of one occurs in an influential work by Dürer[70] (1471–1528), where a ten-face polyhedral is shown, with nine of the faces occupied by dials with polar gnomons. Dürer's image was quickly copied by Münster in Germany, Fine in France, and, in developed form, by Andreas Schöner (1528–90). Thereafter,

examples of these complex dials were produced both as large, semi-monumental structures in stone placed externally, such as those designed by Nicholas Kratzer (1487–1550) in Oxford, the numerous examples that characterize dialling in Scotland, or various examples in France and Germany,[71] or as smaller portable table models. Examples of the latter, all luxuriously made, include the dial made by Kratzer for Cardinal Wolsey c.1525, a dial by Hans Koch dated 1578 (which incorporated twenty-five dials),[72] or the intricate painted-on-wood dials of Stefano Buonsignori (d. 1589) in Florence, which mainly date from the 1580s.[73] All posed the thorny problem of calculating the declining and inclining dials that occupied the majority of their surfaces.[74] They commanded admiration, as noted in the *Mercure galant*[75] for 1678:

> In earlier times the art of arranging various faces on a single stone block on which different kinds of dial could be drawn, one to each face, was admired. They embellished the gardens of palaces and mansions.

Two technical problems had to be overcome—the calculation of a declining or inclining dial, and also the determination of the exact orientation, the latter a problem more pressing for dials on public buildings. In what may be the fullest, but also the most incomprehensible, of Renaissance treatises, Andreas Schöner discusses all the most complex dials. However, his diagrams are perhaps the most complicated and unreadable in all the literature of dialling.[76]

In the same year as Schöner's work, Federico Commandini's (1500–75) publication of Ptolemy's *Analemma* provided an important advance in the field. This has no relation with the analemma construction given by Vitruvius for Sundials, and in itself offers no applications for dial making, although Commandini accompanied his edition with examples of this application, which were then taken up by Clavius.[77] Meanwhile, the architect Danielo Barbaro (1514–70) had published an edition of Vitruvius with long commentaries.[78] Discussing Vitruvius' dialling passages he gives his own, condensed version of Ptolemy's *Analemma*, which

---

dial, is a trigonometric table from 0° to 180° giving the value of a chord for every 15'.
64 The earliest table of tangents (though not so called) in Western Europe is that of Regiomontanus 1490. A table with the tangents given every 10° was the work of Rheticus 1551, on which, see Roegel 2010a. They were subsequently corrected by Pitiscus (see Roegel 2011).
65 See Roegel 2010b.
66 Grisard 1974.
67 Lamprey 2002.
68 Turner 1999; Severino, 2009; Severino & Colombo 2009.
69 Mosley 2019. An explicit description of the pedagogic value of Sundials is given by Semphill 1635, ch. 11.
70 Dürer 1525, 109–13.

71 For Kratzer and his dials, see Pattenden 1979a; for his life, see North 1978; Verdet 1985. For Scotland, see Somerville 1990.
72 Zinner 1957, 415; Eichholz-Bochum 2010.
73 For Wolsey's dial, see Evans 1901, Pattenden 1979a, 14; for Koch, see Zinner 1957, 415; for Buonsignori, see Turner 2007, 118–26.
74 In 1564, Menher discussed the problem and included an engraving of a polyhedral dial in his work that perhaps inspired his pupil Michel Coignet, who, in 1590, made a twenty-five-face dial for the latitude of Anvers. Turner 1990, N° 234a; Delalande 2014, 70–81.
75 *Extraordinaire du Mercure galant*, quartier d'octobre, IV, 1678, 152–71.
76 Delambre 1814–27 (1819, 601–11) attempted a detailed analysis of the work, but even he failed fully to understand it.
77 Clavius 1581, Bk. VI, 528–74. The most attractive figure in Commandini is exactly reproduced by Clavius 1581, 530.
78 *M. Vitruvii Pollionis De architectura libri decem*, F. De Franceshi & J. Criegher, Venise, 1567. There are several issues of Barbaro's edition of Vitruvius, some with illustrations by Palladio.

he associates through their shared name with that of Vitruvius.[79] Giovan Baptista Benedetti (1530–90) also employed an original analemma method alongside spherical trigonometry and was the first to evoke reflex dials and the role of refraction. His book and its figures are difficult to understand and are addressed to an advanced reader;[80] it ends with the description of an instrument for drawing a conic arc on a plane.[81] In effect, the conic described by the end of the shadow of a gnomon on a plane surface constitutes another way of approaching gnomonics—one that was used, but strictly mathematically, by Francesco Maurolico (1494–1575). He studied conic sections on differently inclined planes in his *De lineis horariis tres libri*, deriving the hyperbolic form of the diurnal arcs in European latitudes, which can develop into parabolas, ellipses, and even, in equinoctial dials, into circles. Starting from Apollonius, Maurolico thus produced a compendium concerning conic sections, which would be reflected in Clavius' *Gnomonices*.[82] Austere and highly specialized, Maurolico's work contains the seed of a new approach to dials by considering them as conic sections. These would later be expressed in purely algebraic terms[83]

Sixteenth-century dialling is summed up in the work of Christopher Clavius (1538–1612), who published four works on the subject. The first is a voluminous, theoretical synthesis; the second a modest, but practical tract that includes a novel description of dialling scales. In the third, apparently, he is the first author to use the tangent of an angle to transfer it. The fourth features an extensive set of tables that allow the position of the height and altitude of the Sun to be known for a range of latitudes at different dates. This is a sort of ephemerides completed by tables giving the angle between the hour lines and the meridian— an essential complement of the trigonometrical construction of dials.[84] These tables had been preceded by a set prepared by his pupil Teodosio Rossi (= Rubeus, 1565–1637) for the latitude of Rome in 1593 (Figure 95).[85]

In 1628 Rossi would calculate the well-known quadri-concave dial in in the Quirinal Garden for Urban VIII's architect Francesco Borromini.[86] Two complex dials had earlier been conceived by Clavius for this garden, and despite the difficulty of his books, they were obligatory references for Jesuit authors such as Kircher, Riccioli, or Bettini, or other Italian authors such as Valentino Pini (?–1607)[87] or Muzio Oddi (1569–1639). Clavius' works were also important for their descriptions of the construction of astrological lines on dials. Oddi's treatise clearly describes a considerable variety of dials (although only with Italian hours). In the first edition of his work, he discusses the finding of a meridian by the position of three shadow-points.[88] His second edition[89] includes a discussion of the rarely treated subject of reflex dials in order to investigate the dial of Ahaz that so fascinated Early Modern diallists and biblical commentators.[90]

Alongside the learned treatises of academic mathematicians, a more useful vernacular literature developed from the mid-sixteenth century onwards that underpinned and reinforced the growing popularity of the subject. The following list shows only the first works known in the main European languages:[91]

> 1537 German translation of Münster 1531, Basle
>
> 1556 Claude de Boissières, *La propriete et vsage des Quadrans nouuellement exposé*, Paris
>
> 1565 Giovanni Battista Vimercato, *Dialogo della descrittione teorica et pratica de gli horologi solari*, Ferrara
>
> 1575 Pedro Roiz, *Libro de reloges solares*, Valence
>
> 1593 Thomas Fale, *Horologiographia, The Art of Dialling . . .*, London
>
> 1666 Philip Lansbergen, *Beschrijvinge der vlakke Sonne-Wisjers*, uit het Latijn vertgaald door Jacob Mogge

The rise in popularity of dialling led to an increase in its practice among both gentleman-amateurs and professional mathematical practitioners. For both, a range of dialling tools—mechanical aids—were developed. They depend upon reproducing the apparent trajectory of the Sun at the scale of the dial. To do so, they had to be oriented towards the south and set in the plane of the celestial equator. This, being divided into equal sectors of 15°, the latter can be projected onto any desired plane surface. Variation in the declination of the Sun either side of the equator can also be simulated using a variant of the trigon. The intersection of the declination circles with the receiving surface gives rise to hyperbolic day arcs. The earliest description of such a mechanical instrument was given by Clavius,[92] who ascribed invention of the device to the Spanish Joannes Ferrerius. The instrument is an equatorial table fitted with a mobile sector that simulates solar declination throughout the year. The 15° angles situated in the equatorial plane can then be projected either on a horizontal surface or on a wall, even one that declines. The whole device is oriented by two compasses.

---

79 Losito 1989.
80 See Turner 1987b, 311–20. Delambre 1814–27 (1819, 612–25). Gunella 2019, 56–7 and annexes.
81 Field 1997, 187–90.
82 Clavius 1581, 58.
83 Chasles 1837, 345.
84 Clavius 1581; 1586; 1599; 1605.
85 Rossi 1593. Essentially this is composed of tables giving the height of the Sun as a function of the hour and the season and the different values linked with the different kinds of hour
86 On this dial, see Camerota 2000.
87 Pini 1598 is a very traditional work, its constructions based on the analemma, but ends with descriptions of dials drawn on rings, breviaries, a cross, even a knife.

88 Oddi 1638, 18–21.
89 Oddi 1638.
90 On reflex dials, see Schönberger 1622, 23–110; Ozanam 1697, vol. 5, 65–71; Kircher 1646, 606–23; Maignan 1648; and Leybourn 1682/1700, 158–70. For modern studies of the questions, see Sachse 1895; Sadler 1995; Dupré 2003; Mills 1995; and Mersmann 2015.
91 Cf. the remarks on dialling literature in Chapter 25, this volume.
92 Clavius 1586.

Figure 95 Double face dial by Theodore Rubeus (1565–1637, 1588). Each face carries four gnomons indicating the hour in standard and Italian hours. That shown is drawn for latitudes 36°, 39°, and 48°, one with separate morning and afternoon gnomons. Collection and photo: Observatoire de Paris.

Throughout the seventeenth and eighteenth centuries, devices of this kind would enjoy popularity since they freed the making of Sundials from calculations, although it was still necessary to know the latitude and be able to find South. An instrument known as the sciathere (from the Greek σκιοθηρικοσ, shadow catcher) would be envisioned and made throughout the seventeenth and early eighteenth centuries. Notable examples are those devised by Galluci (1538–1621), de Floutrières, Sarazin, and Maignan,[93] and their instruments would be described in the popularizing works of Bion, Sainte Marie Magdeleine, and Penther.[94] A slotted, polar cylindrical instrument was original to Gaston Pardies (1636–74). This was internally illuminated to project hour lines and the zodiacal curves of the Sun onto any surface.[95] A further solution was to project the hour lines from a horizontal dial, which could be inclined to an angle corresponding to the latitude required, with its gnomon correctly pointing towards the pole. The intersections of the hour lines of the dial with the plane on which a further dial was drawn are located and linked to the foot of the gnomon of this plane. However, none of these mechanical methods could rival in accuracy those using trigonometry.

DIALLING IN THE SEVENTEENTH CENTURY

The invention of azimuth dials, the construction of large meridians, the development of reflex dials, and the publication of some exceptional works on gnomonics characterize the seventeenth century. It was the most prolific period for publications on the subject, and methods of dial making also changed. Dials projected geometrically as made popular by Commandini and Clavius gradually ceded to more accurate trigonometric methods.[96] Dials became more austere, the complex grids of Italo–Babylonian and unequal hours giving way to dials where the quest for precision was primordial.

---

93 Galluci 1590; de Floutrières 1619; Sarazin 1630; Maignan 1648, 256–7.
94 Sainte Marie Magdelaine 1641, 190–1; Bion 1716/1725, 353 & pl. 30; Penther 1768, 36, and pl. 15.
95 Pardies 1673 and several other editions. Bion 1716/1725, 356, strongly recommended the device for non-planar surfaces

96 A good example of a work transitional between the two is Hume 1639, 1640 where analemmatic geometrical methods derived, but simplified, from Clavius and Schöner (Sawyer, Schilke, & Severino 2009, 106–7) cohabit with the use of conic sections and spherical trigonometry.

Azimuth dials use the angular direction of the Sun in the plane of the horizon counted from geographical south to show the hour with knowledge of the latitude and the date being given. There are two kinds: those with a fixed gnomon and those with a mobile gnomon (analemmatic dials). Of the first group, particularly interesting are dials that employ stereographic projection onto the plane of the user's horizon. Known by at least the thirteenth century to Islamic mathematicians, its gnomonic possibilities were investigated in sixteenth-century Germany, and were known in England[97] where, towards the end of the century, William Oughtred combined two dials on a single plate—a classic gnomonically projected dial with polar gnomon (the shadow aligns itself along an hour line) with a dial with a vertical gnomon drawn in stereographic projection (the shadow cuts two circles).[98] The interest of combining the two dials is that it thus becomes, in principle, self-orienting—when the two dials show the same hour, the instrument is correctly aligned in the meridian.[99] A dial designed for an informed clientele, it is also in this form typically English, although there exist a few French examples.[100] Largely forgotten from the early eighteenth century, this style was revived in the mid-twentieth century as a navigational aid, allowing the azimuth of the Sun to be found and true north maintained.[101]

Shortly after Oughtred's double dial had become established, a new device appeared: the elliptical or double azimuth, analemmatic, dial. The earliest trace of this is in two short tracts by J. L. sieur de Vaulezard (d. c.1648),[102] mathematician and tobacco-lover. He set himself the challenge to devise a portable direction dial that did not need to be oriented with a compass. To achieve this, he combined an hour-angle dial with an azimuthal dial employing **orthographic projection** to draw them. The advantage of this is that the day and hour ellipses can be drawn as a single ellipse on condition that the position of the gnomon can be changed. Dials on this principle were developed by Samuel Foster (?–1652), perhaps independently, but along different lines. Foster showed how an ellipse can be equivalent to all others. He gives a formula that expresses the position of the hour points on the ellipse in relations with its centre ($tg\ H' = tg\ H/\sin \phi$), and one that expresses the displacement of the gnomon as a function of the date ($R \cos \phi\ tg\ \delta$). Although a number of early eighteenth-century portable analemmatic dials by English makers, notably Edmund Culpeper, are known, by 1775 it had been forgotten and was 'rediscovered' by Jean-Henri Lambert (1728–77), a consequence perhaps of geographical distance and a failure of knowledge transmission in the Republic of Letters.

The classic analemmatic dial conceived as a portable instrument, would have a new life as an amusing fixed dial in which the gnomon was provided by a person. Although complex in theory, an analemmatic dial can be laid out fairly simply on the ground on a large scale but without a gnomon, this being supplied by a person standing on the appropriate position of a date scale (Figure 96). Such dials probably originated in the late eighteenth century, that of the church of Brou, Bourg-en-Bresse being particularly well-known and probably the inspiration of several others,[103] like those at Dijon (1854),[104] Montpellier (1927), and Avignon (1931).

Tables, the result of lengthy calculations, are a fundamental numerical tool in both astronomy and dialling. Associated with trigonometrical tables and with logarithms, they reduce the amount of laborious calculation involved. Astronomical tables, for example, those giving the height of the Sun at a given latitude as a function of date, are essential in geometrical constructions, bringing together the numerical results of four formulas of spherical trigonometry.[105]

Dialling tables have been essential since at least the fourteenth century, for they abridge the calculation of trigonometrical quantities. Garnier noted of his voluminous tables that with their help alone a dial could be drawn in half an hour.[106] Early sixteenth-century tables are modest, contained within the texts of Münster or Fine, while separate volumes of tables first appear from Clavius in 1605. They enable the angles of the hour lines with the noon line to be found directly using a protractor or the tangent of the angle. Throughout the seventeenth and eighteenth centuries, tables became more specialized: tables for refraction, for the hour limits on vertical declining dials, for the equation of time, for solar declination, as well as those for the latitudes and longitudes of towns appear. Some treatises, such as those of Salodius (1617) or

---

97 Davis & Lowne 2009, 76–81; Sawyer, Schilke, & Severino 2009, 15–18.
98 Oughtred 1636, 1652. Janin 1979; d'Hollander 1999, 290–1.
99 But see the critique of this function, which cannot be applied close to noon, in Lowne 2001.
100 For a detailed census see Davis & Lowne 2009. An example in slate is located in the Musée de Vannes see Cornec & Segalen, 2010, 169–70. The dial is described in Deschales 1674, v, 176–186 and Ozanam 1697, v, pl. 8.
101 Spencer 1973; Turner 1985, 195, 200–1; Sasch 2010, 36; Barnfield 2011; Collin 2013, 21–38.
102 Vaulezard 1640, Vaulezard 1644, much expanded. Janin 1974 provides a survey of the history of this form of dial, and all texts concerning it are collected and translated in Sawyer 2003.

103 Archinard 2005.
104 An earlier dial (1827) was destroyed during the revolutionary upsets of 1848. Moreau 1914.
105 If ϕ is the latitude of place, δ the declination of the Sun, H the hour angle (solar hour in degrees), the height is calculated by: $\sin h = \sin \phi \sin \delta + \cos \phi \cos \delta \cos H$. The azimuth is obtained from $tg\ A = \sin H/(\sin \phi \cos H - \cos \phi\ tg\ \delta)$. The angle of an hour line with the meridian is calculated from $tg\ H' = \sin \phi\ tg\ H$ for a horizontal dial, and by $tg\ H'' = \cos \phi/(\cos D \cot H + \sin \phi \sin D)$ for a vertical dial declining from D.
106 Garnier 1774, whose tables run from 43° 18' (Marseille) to 51°, by single degrees. For a given latitude Garnier offers the tabulated angles of the hour lines with the substyle line for a vertical dial with declination varying from 0° to 90° by 1° steps.

Figure 96 A human gnomon analemmatic dial at Usseaux (Piémont). Photo: Serge Gregori.

Carafa (1689), are almost entirely composed of tables, this being particularly the case in Italy, where drawing the hour lines for dials showing Italian hours, which do not converge (as do those showing astronomical hours and depend on the length of the gnomon), required at least two points as for the unequal hour dials of Antiquity. Such sets of tables may contain two elements labelled 'latitude' and 'longitude' of the shadow point as a function of date, these being the rectangular coordinates of the tip of the shadow of the gnomon.

Outside Italy some eighty tables were brought together by Johann Gaupp (1708), who copied Flamsteed's equation table of 1672, while numerous tables are incorporated in Leybourn (1682). This, which enjoyed several editions, is a complete study in which Leybourn (1626–1716) describes both geometric and trigonometric methods of construction and explains the making of reflex dials, dials housed inside a building, and refractive dials. Numerical examples are provided, and tables of sines and tangents supplied. As publisher, Leybourn produced a small work by George Serle describing how to make a dial without calculation using a dialling scale.[107] This rule was graduated along one edge with a scale for drawing the hours lines giving their distances apart by a law of tangents, and along the other with the distances of the same hour lines from the centre of the dial as a function of latitude. Although one of the earliest mentions of such dialling scales is found in Clavius,[108] it was most widely used in England, where it was fully explained by Samuel Foster in 1638. Several models exist, although they do not entirely abolish calculations and it is difficult to establish if they were really used or were only an ingenious mathematical embellishment. In Paris, Thomas Haye produced a detailed description of them, and their history was recounted in the later eighteenth century.[109] Such rules came back to attention in the twentieth century.[110]

While calculating dials with polar gnomons by spherical trigonometry was practised from the fourteenth century onwards, determining the relations of the angles of the hour lines and the quantities implicated in the placing of the gnomon remained complicated until the early seventeenth century, when more modern

---

107 Serle 1657; Sawyer 1995.

108 Clavius 1586, 75–86, where he attributes invention of the rule to Jacob Kurz von Senftenau (Curtius, 1554–94), the Imperial Pro-Chancellor to Rudolf II.

109 Haye 1716, 1726, 1731. Von Castillon 1784.

110 Cousins 1969, 204–12, a section probably written by his co-author J. G. Porter; Sawyer 1997; Ziegeltrum 2018. Sagot 1988.

trigonometric methods began to be deployed. Bartholomew Pitiscus (1561–1613), who first used the word 'trigonometry', devoted a short chapter of his book to the calculation of particular cases of inclining and declining dials using spherical trigonometry.[111] Trigonometry also figured largely in the work of two near contemporary mathematicians, Pierre Hérigone (1580–1643) and Pierre Bobynet (1593–1668). If the work of Hérigone is addressed to the mathematically competent, and that of Bobynet addresses an audience at a more elementary level, both employ a general method known as latitude-equivalent. Bobynet, who avows that his work depends on that of a predecessor, produced three clear[112] and very well-illustrated successful books. He makes full use of logarithms for calculating trigonometric quantities and explains for a horizontal dial where the angle H' between the hour line and the noon line is a function of latitude ϕ and the hour angle H of the Sun that 'as the sinus total is to the sinus of the latitude, so the tangent of the horary distance or of the degrees of the same hour proposed on an equinoxial dial, is to the tangent of the angle made with the meridian of the required hour line at the centre of the dial'—that is to say, $tg\ H' = \sin \phi\ tg\ H$.

For vertical declining dials, Bobynet uses the **sub-style line** to construct the other hour lines (Figure 97). He affirms, as had Thabit ibn Qurra eight hundred years before him, that all inclining and declining dials can be reduced to a horizontal dial for a different place, or, in modern terms, are 'latitude equivalent'. Any inclining–declining planar dial moved parallel to itself will continue to indicate the same time as if it had remained at its original site. Otherwise stated, a planar dial of any orientation or inclination set up at a certain latitude φ functions as a horizontal dial at another 'equivalent' latitude or 'equivalent' longitude. The method was insisted on by Jean Picard in a short, posthumous text entirely based in spherical trigonometry, and described more fully by Dionis de Séjour (1734–94).[113]

MERIDIANS

Generally, a meridian in gnomonics is a strip of brass a few centimetres wide, usually graduated, and oriented along a true, geographical, north–south line, constructed inside a closed building. It functions with an **oculus** or **aperture-gnomon** set in the roof or an opposing wall, which allows a more or less elliptical spot of sunlight to fall on the line close to noon when the Sun is on the meridian of the place of construction. The higher the oculus is placed, the larger must be the amplitude of the meridian. This led to churches, cathedrals, and palaces becoming the habitual sites of such instruments. This was particularly the case in Italy, where the first meridian was laid out by Paolo Toscanelli (1397–1482) in Florence in 1475. Meridians, however, did not become common until the seventeenth and eighteenth centuries.[114] If the primary purpose of most early meridians was to determine precisely the dates of the solstices and the equinoxes, and so the length of the **tropical year** (a parameter essential to the accurate establishment of the calendar), and if throughout the eighteenth century it would be the essential instrument for astronomers studying the variation of the obliquity of the ecliptic, its role in gnomonics was to identify the exact moment of noon. Use of meridians in astronomy was an application of gnomonics, just as it was when they were employed by clockmakers for checking their products.

Establishing true noon for astronomers, clockmakers, and the general public was the primary purpose of the many public, and some private, meridians that were erected in the later seventeenth and eighteenth centuries. They were needed because even well into the eighteenth century, many people still failed to understand why the increasingly accurate clocks of the post-Huygens era did not agree with the Sun, but had to be adjusted throughout the year.[115] Meridians, therefore, moved from the realm of learning into that of public utility and were erected at public expense for public use. To take a few examples from the many available throughout Europe, the meridian constructed by the clockmaker Henry Sully (1680–1728) in the church of St Sulpice in Paris in 1727–28 was intended for public time regulation: 'It is as incommodious as bizarre', he explained, 'that in a city like Paris there is such a great discordance and uncertainty in the true time of day as has always been the case up to now'. A 'City meridian' was erected in the Place des Changes, Paris in 1738, and one was created in Blois by order of the city governors c.1755. Some private meridian dials were also created.[116] A meridian is shown emblematically in the frontispiece to Bedos de Celles (1760).

Since their primary purpose was time control particular care was taken in the construction of meridians although these were generally quite simple. Meridians were also fashionable. One of the most striking was erected in the Cour des Cerfs, Versailles in 1737.[117] Like that drawn in the clock room, its function was to aid adjusting the numerous clocks in the palace. It was so elsewhere, as Jacques Cassini noted, 'in all the royal houses that the King uses, there are meridians erected by his order and under his supervision . . . but all these meridians . . . were drawn only to mark the exact moment of Noon, and to correct clocks'.[118] In the

---

111 Pitiscus 1600, 247–67. Some of Pitiscus' examples would later be used in Morgan 1652, 75–84 and in Hume 1640, 168.
112 Compare, for example, the calculation of quantities concerning the polar gnomon of a vertical declining dial in Hérigone, 734–5 and Bobynet, 29–31.
113 Picard 1693; Dionis de Séjour 1761.

114 Heilbron, 1999. Meridians in general are inventoried in Gotteland 2008.
115 Turner 2015b.
116 Gotteland 1988, 100–6; d'Espagne 1757 (the authorship of this anonymous tract is given in a minute of the Blois Town Council, identifying d'Espagne as author of both the meridian and its explanation); Savoie & Turner 2014.
117 Janin 1972b.
118 Cassini 1732, 453.

Figure 97 Vertical declining dial, Sorbonne Paris. Drawn in 1676 by Jean Picard (1620–82), its ruinous state necessitated a complete refurbishment in 1899 when a mean time curve was added. The legend may be translated as 'Like shadows, so our days'. Photo: Serge Gregori.

early nineteenth century, the master clockmaker Antide Janvier (1751–1835) described the construction of meridians in his horological guides.[119] Some large-scale meridians constructed in Italy in the late eighteenth and early nineteenth centuries, such as those at Milan (Cesaris 1786), Palermo (Piazzi 1801), Catane (1839–1841 and 1843), by insisting on the primacy of noon, would speed the obsolescence of Italian hours. The most far-reaching official use of meridians, however, was perhaps that decreed in Belgium in 1836, when the astronomer Adolphe Quetelet (1796–1874) was charged to realize forty-one meridians in the major cities of the country for use by the newly implanted railways.[120] Only ten of them, however, were finally constructed.

Solar time was legal time in the eighteenth century. Every town had its own time, which differed from that of others by the difference of longitude between them. When it was midday in Strasbourg, it was only 11.29 in Paris and 11.11 in Brest. Clocks had to be adjusted to remain in concordance with solar time, an adjustment generally effected with the help of tables of the equation of time. From the second half of the eighteenth century onwards, clockmakers in France militated for the adoption of a standard mean time,[121] for finding it was not difficult. The idea of reading the mean noon directly on a meridian incorporated the equation that originated with Grandjean de Fouchy (1707–78). He imagined a curve in the form of an elongated eight (improperly called an 'analemma' in English),[122] but since the values of the increasing declination are not symmetrical with of the decreasing declination, the 'eight' is slightly unbalanced in relation with the noon line that it embraces, and which it cuts four times a year when the equation is zero. Nothing is known of the circumstances of this innovation by Grandjean which probably occurred around 1730, only some circumstantial remarks by near contemporary diallists such as François Rivard,[123] Antoine Deparcieux,[124] and Bedos de Celles.[125] The 'figure of eight' equation diagram, however, was not immediately successful primarily because of the difficulty of its execution evidenced by the mean-time equation drawn by Gaspard Monge c.1780 at Charleville-Mézière, in which right and left are reversed. Additionally, the mean-time equation varies slowly across time and so after a few decades becomes obsolete. Mean-time, moreover, did not generally impose itself until well into the nineteenth century.[126]

In eastern France 'industrial' meridians appeared during the second half of the nineteenth century produced by the clockmaking company Ungerer. These were large, directly south-facing metal plaques fitted with an aperture-gnomon showing true solar time, while others produced in Alsace by the clockmaker Urbain Adam (1815–81), showed mean-time. Over 183 examples are known to survive, and their success attests to a continuing social need for setting a clock or watch to an arbitrary standard. More amusing, but equally efficacious, were the 'noon-guns' that became popular from the later eighteenth century onwards. In these a horizontal dial was combined with a converging lens, adjustable for latitude, set above a small canon charged with powder set on the noon line. At midday, therefore, the focused Sun rays would ignite the powder, the explosion of which gave an aural indication of noon as the Sun crossed the meridian. The best known of such devices is that installed in the gardens of the Palais Royal, Paris (and which gave rise to a plethora of miniature imitations for tourists), which operated throughout the nineteenth century, its creator, the clockmaker Rousseau enjoying a flourishing trade in domestic versions for private houses.[127]

DIALLING IN THE EIGHTEENTH CENTURY

More dials were probably constructed in the eighteenth century than ever before. It is the apotheosis of construction methods and innovations. Among the many books published during this period, two dominate the field for clarity and completeness—Bedos de Celles and Johannes Gaupp.

François Bedos de Celles (1709–79) was a Maurist monk as well known for his construction of organs as for Sundials.[128] Nevertheless, his *Gnomonique pratique* is one of the most successful works on the subject.[129] It is concrete and highly instructive, supplying examples that combine geometry and numerical calculation without theory—the work of an author experienced in the subject who had made dials and does not write just as a theoretician.[130] He does not hesitate to describe the tools indispensable for Sundial making such as the square, the rule, the beam compass, the level, and a false gnomon, tools often ignored, even disdained, by other writers.

---

119 Janvier 1810, 187–97; Janvier 1811, 36–43.
120 Van Boxmeer 1995–98.
121 Berthoud 1802, 180.
122 Daniel 2005; Savoie 2008, 41–62.
123 Rivard 1742, 296.
124 Deparcieux 1741, 94.
125 Bedos de Celles 1760, ix.
126 Genève 1780; London 1792; Berlin 1810; Paris 1816, Dohrn-van Rossum 1996, 346. However, the date for Paris may be 1826: see Hartmann 1827, 5.

127 Gotteland 1988. *BBT*, 'Rousseau'.
128 Steinhaus & Beugnon 2008.
129 Even in the mid-nineteenth century Bedos de Celles inspired a pocket manual on the subject (Boutereau 1845), although the latter's second source (Sternheim 1842) led him into complex, and rebarbative, geometrical constructions.
130 Two dials can be attributed with certainty to Bedos de Celles at the abbey of Saint-Denis; several dials on the château of Dennainvilliers (where he was the guest of Duhamel du Monceau), and less certainly, the dial of the abbatial of the Couture, Mans. See Savoie 2003.

The plates accompanying the text are all remarkably clear and well drawn, the whole book being accompanied by the indispensable tables. Bedos devotes twenty-five pages to a problem often scamped in gnomonic texts, that of determining the orientation of a wall. In large part the precision of a dial depends on this being correctly found. Firstly, he proposes the meridian method, that is, measuring the angle between the wall and a meridian previously laid out; secondly, a method in which the solar azimuth must be calculated at the moment of measure, a method related to that known today as the board or table method.[131] Here, once again, Bedos reveals himself a fine teacher, offering advice and numerical examples to guide the reader through this essential preliminary. He castigates use of the magnetic compass and concludes with a remark that is still applicable:

> If only a very small number of [well-made dials], are seen among a prodigious quantity of dials, it is almost always because the declination of the plane has not been established with all the care needed, and that uncertain methods have been used.

Another notable aspect of the book is the long chapter on the recently invented mean-time meridian increasingly required for setting clocks and watches. In the preface to his second edition, Bedos courteously refers his readers to earlier important works on the subject such as de la Hire, Ozanam, Rivard,[132] and notably Antoine Deparcieux (1703–68). In this work the illustrations are once more remarkable, but Deparcieux differs from Bedos by his very theoretical treatment. The many tables that complete the work, however, could still be cited (like those of Garnier) 150 years later by Guillaume Bigourdan.[133]

Johannes Gaupp (1667–1738) wrote a work that is notable among German texts on dialling. It is remarkably well illustrated,[134] and tables occupy about a quarter of its length. Not only does Gaupp deal with almost all kinds of dials, but at the end he also includes separate sheets to be cut out and assembled as elaborately decorated instruments. Largely inspired by the works of Kircher, Ozanam, and de la Hire, whom he willingly acknowledges, he also refers to Georg Philip Harsdörfer (1607–58),[135] to whom he attributes the well-known 'roof' or 'invisible' dial—a vertical dial with no other indication than an hour point. Above it is a polar roof scored with hour lines through which light rays project the 'digital' numbers of the hours, which progress along the wall throughout the day.

Among the original dials erected during the eighteenth century, one of the more remarkable was that realized in 1764 by the astronomer and keeper of the library of Sainte Geneviève, Alexandre-Guy Pingré (1711–96) on the Médicis column in the Paris food market. It was designed at the request of the provost of the Paris merchants, Jean-Baptiste de Pontcarré de Viarme (1702–1775), so that stall holders could know the time throughout the day. Pingré[136] placed fifteen horizontal gnomons on the summit of the column, which is 1m 54cm in radius, each of 1m 44cm. The column, smoothed down for nearly 3m of its height, carried a diagram of hour lines and date curves, while at the end of the gnomons a brass numeral plate was fixed. The full hour was read when the image of the numeral on the end of the gnomon fell on the appropriate line on the column. This highly original dial was not easy to read. It was destroyed in the nineteenth century.

With sundials ever more frequently made and visible during the eighteenth century, interest developed in their history and development. If Jérôme Lalande (1732–1807) never wrote a full treatise on sundials or meridians they nonetheless commanded his attention.[137] He observed regularly with the meridian of St Sulpice in company with his mentor, Le Monnier; he did the same with the meridian of Santa Maria del Fiore in Florence—'the largest astronomical instrument in the world';[138] he collaborated in the reconstruction of the analemmatic dial at Bourg-en-Bresse, and wrote clear syntheses on Sundials and meridians for the *Encyclopédie Méthodique*.

Lalande's pupil Jean-Baptiste Delambre (1749–1822) also treated of Sundials, though rather as a theoretician and historian than as a practitioner. By origin a classicist, he combined knowledge of the ancient languages with astronomy and mathematics to write a wide-ranging history of the subject.[139] He devotes several, not

---

131 Savoie 2001, 65–9.

132 Bedos de Celles was in direct competition with this highly successful work (five editions between 1742 and 1767).

133 Bigourdan 1922, 93. Deparcieux at the beginning of his career constructed several dials including meridians at the Louvre and in the Rue Neuve de Luxembourg. See Grandjean de Fouchy 1770, 157. The innovative iconography of his treatise influenced other authors such as Lory 1781 who treats of the classic dials uniquely by spherical trigonometry.

134 For Gaupp, see Hofbauer & Solombrino 2009 and Sonderegger 2013, 26–34. The frontispiece to Gaupp's book, as he states in his Preface, is drawn from a work by Henric Bierum of 1676. It shows two workmen cutting vertically a graduated polar axis thus showing that a dial results from the intersection of a plane with the polar axis. Pedagogically advantageous, this way of considering dials perhaps underlay the work of James Ferguson (1710–76), who described a 'universal Dialing Cylinder' (Ferguson 1760 and at least sixteen other editions and issues). The principle of the 'cut' cylinder is also to be found in Deparcieux 1741, planche VII, fig. 96.

135 Harsdörffer 1651, ii, 324. Harsdörffer affirms having seen such a dial in Ingolstadt some twenty-six years earlier. On the principle of the dial, see Zach iii, 56–66. On a similar dial constructed at Besançon by J. L. Bizot or Bisot (1702–81), see the *Mercure de France*, January 1758; Lalande 1803, 358, 393, 466; Savoie et Goutaudier 2017.

136 Pingré 1764. Savoie 1998.

137 His *Bibliographie astronomique*, (Lalande 1803, 943–44), contains a useful table of gnomonical authors.

138 Lalande 1769, ii, 184–8.

139 Delambre 1814–1827.

always objective, chapters to ancient work on dials, underlines the writings of Ptolemy on the analemma, and was one of the first to bring out the importance of Arabo–Persian dialling. He was, in fact, the first historian of dials not simply to survey the various works on it, but to examine them minutely, rework their examples, and present them in modern form; to study the construction methods and to show what they added to the development of the subject. In this he differs from Montucla, who rather compiled than analysed.[140]

In the second half of the eighteenth century, some totally forgotten dials reappeared following excavations at Pompei, Herculaneum, and elsewhere. These soon attracted specialist studies notably by Zuzzeri (1746), Martini (1777), Antonini (1781–90), Calkoen (1797), Delambre (1814), and Davies (1818). Surveys of ancient dials were thereafter included in the more general works of Montucla (1798–1802) and Delambre (1814–27), surveyed by Drecker (1925), and inventoried by Gibbs (1976) and Bonnin (2012).[141]

NINETEENTH-CENTURY DIALLING

From the early 1800s, dials begin a period of decline and wireless telegraphs in the second half of the century would destroy their everyday usefulness, even if their pedagogic functions remained intact. Few new treatises were published about dials, the study occupying only a secondary place in more general works—encyclopaedias and textbooks of physics, astronomy, or mathematics. Some forms of dials, such as portable altitude dials, ceased completely to be studied, and depictions of dials declined in quality. The calculation of dials enthroned the analytical solutions found in the later eighteenth century where the date curves and the straight hour lines were summed up in the formula $y = f(x)$ or as rectangular coordinates. Dials were considered as any plane oriented or declining on which the equation of the conic described by the shadow of a vertical gnomon could be determined.[142]

Dialling, however, would be revitalized by the new field of descriptive geometry associated with Gaspard Monge (1746–1818). Since the Renaissance, the study of perspective gradually had begun to emancipate itself from painting, and its uses in dialling examined. Girard Desargues had already reduced perspective, dialling, and **stereometry** to a single common principle, which meant that the problem of the shadows cast was not entirely new.[143] Monge's new process was a development of these projective methods, which it used to represent with exactitude a body, existing in space and volume, in two dimensions.[144] It was primarily in the Ecole Polytechnique with the work and teaching of Jean Nicolas Pierre Hachette[145] (1769–1834) and Théodore Olivier[146] (1793–1853), Monge's successors, that this geometry was developed. The dialling constructions that can be derived from it remain, however, difficult to exploit,[147] because the hour lines and the day arcs are considered as resulting from the intersection of planes in space (for example, of a cone with a plane), which makes the construction particularly arduous and somewhat imprecise at the edges of the diagram.[148] Such methods remained marginal, deployed only in the most advanced engineering schools in France. From a practical point of view, trigonometric methods were better adapted and preferred, especially because the texts of descriptive geometry treat Sundials summarily in a single chapter organized around the principal horizontal, vertical, and inclining examples without pausing on the particular cases of orientation and inclination, as had always been done since the sixteenth century. From now on, dials are treated in general terms, and special cases such as azimuth dials and altitude dials are ignored. That said, dials had been constructed by the rebatement of planes well before Monge. The geometers of this period simply codified an existing usage.

If many towns in the early nineteenth century adopted mean solar time by which to set their public clocks, this remained a local time based on the local meridian. In rural areas, however, time was often still the true solar time of Sundials. Since the difference between the two was at most sixteen minutes, by many it was considered indifferent for everyday life, while the concept of the equation of time remained a mystery for most people.

---

140 Montucla 1798–1802.
141 For the historiography of dials see also Chapter 26 this volume; Turner 1989a, 308–9.
142 See, for example the chapter written by Berroyer in Biot 1811, 51–113, or Littrow 1831.
143 For Desargues, see Taton 1981; Field & Gray 1987, esp. ch. 9. For later work see s'Gravesande 1711, who, after his chapter on shadows succinctly discusses the use of the rules of perspective in dialling, 193–200. The principle is to imagine that the eye is placed at the top of the gnomon of a horizontal dial and that the eye's visual rays, assimilated to those of the Sun allow the horizontal dial to be projected onto other planes.
144 Chasles 1837, 355–7; Taton 2000, 305–23.
145 Hachette 1828, 260 and the splendid drawing (plate 17 O) of the equation dial drawn by Deparcieux on the College of Navarre in 1747.
146 Olivier 1847, 209–61. Olivier closes his chapter on dials with a commentary on Picard's work on the application of spherical trigonometry for the resolution of dialling problems. For a survey see Severino 2007.
147 The work that best displays the rupture between the algebraic and the descriptive methods is that of Mollet 1820 and many other editions. Pillet 1921, 229–35, offers a good example of horizontal and vertical declining dials drawn by the descriptive method. Unfortunately, his mean-time curve is erroneous.
148 The complexity and the limits of the method are illustrated by the diagram of the vertical declining dial drawn at Juvisy. See Roguet 1912.

At the International Meridian Conference of Washington in 1884, the meridian of Greenwich was adopted as the point of origin of longitude throughout the world and of universal time. An additional correction had therefore to be made to pass from true, or mean, solar time to universal time—the longitude of place counted from Greenwich.[149] The mean-time of Paris became legal time for the country in 1891,[150] which implied that to convert solar time into legal time it was necessary to add the longitude of the place counted from the meridian of Paris. This system was used for twenty years until 1911, when the Greenwich meridian as base was finally accepted.[151] Thereafter it was only necessary to add the longitude value from Greenwich to local solar time. The subsequent introduction of an hour advance[152] (UT + 1 h) in 1916 and more general modifications of legal time have only added to the complexity of converting from true solar to legal time, three corrections being now needed: the equation of time, longitude, and the advance on universal time.

The development of railways and postal services accelerated the unifying of time in industrial countries. It was in 1911 that the aberrant system whereby the internal and external clocks in French railway stations were set five minutes apart was abolished. Railway clocks were set by wireless telegraphy diffusing Paris time to the provinces from 1880 onwards—a system replaced by radio-telegraphy from 1911. These new methods of time distribution, combined with the availability of public and private clocks accurate to the second, relegated Sundials definitively to the category of 'objects from the past'. To construct them became merely a pedagogical or leisure activity.

It was, nonetheless, exactly during this period that a high-precision equinoctial dial—sometimes totally inappropriately referred to as an 'helio-chronometer'—was developed. It consists of an equatorial table fitted with an optical **alidade** that focuses sunlight onto an equation curve. Once correctly aligned in the meridian and adjusted for latitude, the Sun is sighted so that a spot of sunlight falls onto the equation curve for the appropriate date. Local mean-time, or true solar time, can then be read on the equatorial arm. Equation dials of this kind were developed by Jean Marie Victor Guyoux (1793–1869), curé of Montmerie-sur-Saône from 1834, and there are something over thirty known examples dated between 1831 and 1867.[153] Guyoux's dial and its adaptations, however, derive from mechanized tabletop equatorial dials that date back to the late seventeenth century, when Michel Bergauer (pre-1671–early eighteenth century), Claude Dunod (pre-1672–1716), and Franz Anton Knittl (1671–1744), perhaps following the example of Habermal and Clavius, fitted them with an alidade and gears for reading to minutes. Philipp Hahn (1739–90) further developed forms of such dials.[154]

Several variants were developed from the Hahn-Guyoux model. The best known is the 'solar chronometer' of Victor Fléchet patented in 1861 with, nineteen years later, a portable model.[155] The English astronomer George James Gibbs (1866–1947) devised[156] and patented in 1906 a model incorporating a mechanical correction system for the equation of time.[157] Gibbs' dial was widely diffused as just under a thousand examples were sold worldwide, especially in regions where access to legal time was difficult, such as in Russia, South America, and Australia. With these admirable devices to show mean-time easily, the limit of precision, close to thirty seconds, that a dial can show was reached. Some fixed versions, that is, not requiring the intervention of their user, would also be developed—equatorial dials integrating the correction for the equation of time directly into a profiled gnomon. The first of these was conceived by John Ryder Oliver (1834–1909) in 1892 and would give rise in the twentieth century to some original artistic creations.[158]

At the same time as sophisticated equation dials began to appear in public and private parks, a series of more humble dials were painted on village houses, farm buildings, and some churches in the Queyras, the region of Briançon. They were produced by a Piedmontese house painter, Giovanni Francesco Zerbola

---

149 At mean midday, the mean Sun culminates on the meridian when its hour angle is zero. This signifies that it is 0h mean time. So, when '12h Greenwich Mean time' is said to indicate noon, this signifies that twelve hours have elapsed since the mean noon and that it is therefore mean midnight, that is 0h in universal time.
150 *Bulletin des lois de la République Française*, 42, 12th series, 1891, 313: 'Legal time in France and Algeria is the mean time of Paris'.
151 *Journal Officiel de la République Française*, Friday 10 March 1911, 1882. 'Legal time in France and Algeria is the mean time of Paris' retarded by nine minutes twenty-one seconds, this being the exact difference between Paris and Greenwich, but the formulation avoided hurting national hubris. See, in general, Gapaillard 2011.
152 *Bulletin des lois de la République Française*, viii, 1916, 905.
153 Pommier 1978, 283–6. Rieu 2014; Mayette 1889, 213–74, Gagnaire 2004. Predecessors of Guyoux's device were the drawing room mechanical equinoctial dials made by Philipp Hahn. See Hamel & Müsch 2018, 242–5.
154 On Hahn's dials, 'Öhrsonnenuhr', see Hamel and Müsch 2018, 242–5; Zinner 1976, 88–91; Väterlein 1989, 368–73. For Bergauer, Dunod, and Knittl, see Abeler 1977, 67, 141, 344; Zinner 1976, 246, 299, 413–14.
155 *Chronomètre solaire universel et portatif*, Paris, 1879. Some slightly naïve accounts of these dials can be found in *La Nature*, 1875, 128, and 1893, 69–70. 'Heliochronometers' were the first Sundials to be patented in both France and the United States of America.
156 *La Nature* 1908, 15–16.
157 Parsons 2016, 47–8. A variant way of mechanically correcting for the equation of time was employed in the fifty or so equatorial dials produced by the Bollée company, Le Mans in the second half of the nineteenth century. See Ferreira 2004, 2005.
158 UK Patent n° 1660, 'Improvement in Sundials', 1892. Brix 1981.

(or Zarbula, c.1810/12–post 1870),[159] who produced some 110 known dials in a period of about forty years. His dials, executed in *fresco*, are highly decorative and distinctive, usually incorporating a device and surmounted by exotic birds and, frequently, a square and a compass, all highly coloured and traversed by a straight line for the equinoxes. The examination of Zarbula's dials shows they are all drawn using a geometric method without their author engaging in any calculation, not even to measure the orientation of the wall on which they were drawn. That the 45° parallel crosses the region certainly simplified their layout, which was based on determining the substyle line by the shortest shadow of a temporary gnomon and folding over the equatorial plane.[160]

To the north of the Queyras, in the region of Isère, a further local group of about 100 dials are known, produced by Hyacinthe Pascalis and Liobard towards the end of the eighteenth century, and particularly during the Revolutionary period, mainly in rural areas on farmhouse walls, so continuing a tradition of dialling that can be traced back several centuries.[161] Evidently there were many such 'open air diallists', as they can be called to distinguish them from professional instrument-makers in the towns, but we have little information about them. Some perhaps were travelling artisans, while others were masons or painters— all possessed of a modicum of geometry and skill that allowed them to work in the remote regions of Europe far from the main axes of communication. Concerning the origins of their skills, nothing is known and the works of the learned, such as La Hire, Deparcieux, or Bedos de Celles, were probably neither available nor of use to them. Perhaps for this reason simplified brochures and manuals for constructing vertical dials (the commonest type) appeared, even if these were not always accurate.[162] Many dials indeed are imprecise, drawn, 'by approximation'. Accuracy was not a determining element in their making. The indication of noon was, in many cases, all that was needed for the rhythms of civil life and the ringing of the bells. Only with the arrival of the railways would the population at large begin to be confronted with need for more exact time.[163]

If the constructions of Zarbula and Liotard astounded the inhabitants of remote, rural areas in their time, other diallists travelled the world. Enrico d'Albertis (1846–1932) produced 100 magnificent marble dials,[164] one of the most celebrated being that placed at the entrance to the Arsenal in Venice. Europe, for evident historical reasons contains more dials than elsewhere in the world, a *corpus* that has greatly developed since the late Middle Ages. It continues to be enriched with new creations, but also suffers from the disappearance of dials through negligence when house facades are cleaned or through the lack of interest displayed by proprietors. Studies made in France since 1972 have inventoried some 32,000 dials, the great majority (seventy-five per cent) being vertical dials (eighty-six per cent if ecclesiastical canonical dials are included). Horizontal dials provide seven per cent of the total followed by other, disparate, dials (analemmatic, equinoctial, armillary, polar, spherical, and the like). Meridians, whether horizontal (54), vertical (366), or industrially produced (183), make up three per cent of the total. Polyhedral dials account for one per cent of the total, with some 229 known. Most of these dials are installed on houses (fifty-three per cent) with another thirty per cent on religious buildings. Dials on public buildings (1581) are slightly more than those on country houses (1348).[165]

DIALLING IN THE TWENTIETH CENTURY

The first general history of Sundials, of which the theme was the mottoes that adorn them, was produced by Margaret Gatty (1809–73) and was published in four successively enlarged editions.[166] Mottoes or devices for Sundials stretch back to the sixteenth century. They are usually in Latin, the dominant subjects being citations from the Scriptures, philosophical or moral exhortations, or celebrated phrases such as *carpe diem* (seize the day) from Horace.[167] Many of them, as the latter, evoke the passage of time, some the coming of death, while others refer to the Sundial itself as symbol. Humorous mottoes are also to be found, some employing puns, as are professional, often mercantile, thoughts; there are mottoes that refer to the profession of the owner of the dial, and familial, patriotic, historical, even political, mottoes. Alongside Latin, the most common language of expression, vernacular mottoes are to be found, and even some in a local patois. They are an indispensable aspect of a dial with their reflections on life in general, and thus give it a function other than that of marking time. New mottoes have been continually created, and collections of them have been published since at least the

---

159 Zerbola seems to have been called Zarbula for the first time in Blanchard 1895, 29–32.
160 Gagnaire 1999; Gagnaire 2000.
161 See Mazard 2011, 31–5; Avenier 1999, 7–22.
162 These pamphlets of twenty or so pages are often anonymous with attractive, universalist titles, for example, *L'horoliographie universelle, Méthode générale, très juste, très courte & facile pour faire toutes sortes de montres solaires dans tout l'Univers, Pour l'usage & la facilité des compagnons tailleurs de pierre & maçons qui sont sur le tour de France*, 1768; or E. Monot, *Précis de gnomonique à l'usage des curés et des desservants*, Paris, 1863, a small brochure in which the explanations are neither clear nor acceptable.
163 Souchier 2018, 220–8.

164 Mesturini 2015.
165 For an analysis of dials found on the country houses of France, see Savoie 2017, 60–6.
166 Only the first edition (1872) is by Margaret Gatty herself. Subsequent editions in 1889, 1890, and the last by Eden & Lloyd 1900.
167 *Odes* I, xi, 7.

seventeenth century,[168] their origins being discernible in collections of emblems, such as those of André Alciat (1531) or Gabriel Rollenhagen (1611), so popular in the later Renaissance. Systematic collecting of such devices began with Margaret Gatty from 1835 and the Baron Edmond de Rivières (1835–1908),[169] to culminate in the publication by Charles Boursier of some hundreds of the several thousand now known.[170]

The use of dials in society was hardly discussed before the unconventional work of the American writer Alice Morse Earle (1851–1911), a historical evocation of Sundials, their mottoes and symbolism, and their placing in gardens.[171] The need to preserve the European gnomonic patrimony, however, was discerned from the early 1960s by Ernst Zinner, who discussed not only the value of collections in museums, but also, and more particularly, the treasury of dials left outdoors in the many countries of Europe. Portable pocket dials now became the object of private collections, such as that of Lewis Evans (1853–1930), who wrote the chapter on portable dials in the revised 1900 edition of Margaret Gatty's book, or that of the Belgian poet Max Elskamp (1862–1931).[172] The creation from the early 1970s onwards of associations of Sundial enthusiasts[173] was a strong motive force for both the safeguarding and the creation of an inventory of dials throughout the world. If producing dials has always been a pastime for the leisured, their role in teaching basic mathematics and astronomy has also always been well established, and their use continued into the mid-twentieth century.[174]

The use of individual computers from the 1980s onwards has not only had repercussions on the speed of calculations, allowing thousands of coordinates to be obtained in a few seconds that earlier would have taken considerable amounts of time, but has also influenced the theory of dialling—particularly the development of mathematical formulae. Previously, preference was given to equations that allowed ease of calculation with logarithm tables. Using them, a skilled operator could calculate any equation, albeit at the cost of long and tiresome manipulations. The power of numerical calculation by modern computers allows not only the use of heavy trigonometric equations, but also the resolution of problems intractable without its aid, for example, transcendent equations or systems of resolution by approximation. The inevitable result has been the appearance of calculating programmes (with images) for Sundials with which even a neophyte can calculate and visualize his dial in a few seconds. The development in the 1970s of *raytracing* has moreover allowed the shadow thrown by any volume on any surface to be simulated in three dimensions with total realism. It is certainly because of such facilities that contemporary artists have taken over Sundials as a subject allowing them to express themselves in both sculpture and design[175] while playing mathematically with light and shadow.

The eruption on the market of dials as banal consumer objects produced 'in series', has had equivocal effects. Certainly, it has contributed to bringing the use of dials back into fashion, but the lack of full explanations and often the incorrect tracing of such dials have degraded their image. Indeed, in the first half of the twentieth century the literature of gnomonics was much reduced, as there was little need seen for the object after the introduction of the speaking clock in Paris in 1933, with others elsewhere. It was nonetheless in this context that Hermann Michnik (1864–?), a German diallist, published an almost unnoticed article describing a potent novelty—the bi-filar dial.[176] an instrument in which the shadow point that indicates the hour is obtained from the intersection of the projection of two orthogonal threads placed at different heights. The law governing the distance between the hour lines of a dial depends solely on the relative height of the two wires, such that it is possible to draw, for example, a vertical or horizontal dial with the hour lines at 15° distance (homogenous dial). In this case there is a north–south wire placed at distance a from the dial table. The east–west thread is placed at height b so that $b = a \sin \phi$. Equally, a dial for latitude ϕ can be left horizontal and be moved to latitude ϕ' by changing it into a bi-filar model. Ignored for a considerable time, the bi-filar dial has given rise to a whole family of original dials in the second half of the twentieth century,[177] using, for example, hyperbolic threads, chains, a single wire, or gnomons of diverse shapes. Some of these variants have

---

168 For example, by Le Moyne 1666, 38, who defines a device as 'a metaphorical expression in the manner of a tacit similitude composed of words and figures, to express some great design, some beautious passion, or some noble sentiment'. *Cf.* the earlier collection by Parmenter 1625.
169 De Rivières 1877–1885.
170 Boursier 1936 includes some calculations for vertical dials at the end of his book (162–8), that are totally erroneous. The most recent inventory of devices in France made in 2005 runs to 3,500 examples. See also Escuder 2005.
171 Earle 1902.
172 Both would leave their collections to public museums—Evans to the Museum of the History of Science, Oxford, and Elskamp to the Musée de la Vie Wallonne, Liège. For both, see Turner 2021.
173 The Commission des cadrans solaires de la Société Astronomique de France was founded 13 December 1972, presided by the astronomer Jean Kovalevsky (1929–2018), and then by the diallist Robert Sagot (1910–2006). The British Sundial Society was founded in 1989, that of the Netherlands in 1994, and the North American Sundial Society in 1994/1995.
174 For example, Faye 1854, 196–203; Danjon 1980, 74–6.

175 For some fine examples see Lennox-Boyd 2005 and the studies in Bouchard 2015.
176 Michnik 1923. Only Drecker, 1925, 105, immediately understood the potential of the idea. Two bi-filar dials by Michnik himself (1925) are held in the Deutsches Museum, Munich, one with Italo–Babylonian hour lines (n° inv. 58435) the other drawn for unequal hours (n° inv. 58436). Michnik had long been interested in dials, on which he published in 1914, examining, in particular, the nature of the unequal hour lines on ancient dials.
177 Sawyer 1978, 334–51. Collin, 2003, 12–31. Over 100 articles have since been devoted to bi-filar dials.

given rise to true dials.[178] It should be underlined that the bi-filar dial is (with the analemmatic dial and the magnetic azimuth dial) the first veritable innovation in gnomonics since the seventeenth century.

Another true invention, however, is the latitude-independent dial invented by J. G. Freeman in 1978. Mathematically interesting, it has had little success because of the complexity of its construction and use. The last major twentieth-century innovation dates from 1987—a digital Sundial using the theory of fractals. This uses direct light traversing a three-dimensional structure that projects numbers. The idea derives from the English mathematician Kenneth John Falconer.[179] A digital dial using this principle, which can be compared with a Venetian blind superimposed in several directions to allow sunlight to pass at certain hours, was constructed and patented in 1996 by Hans Scharstein, Werner Krotz-Vogel, and Daniel Scharstein.[180] At no other period have so many dials (all types taken together) been constructed as in the twentieth and twenty-first centuries, and it is this amount that bears witness to their continuing interest. Ancient mathematical instruments that can claim such long survival are rare.

---

178 Notably the bi-filar dials of Rafael Soler on Majorca at Camp de Mar and at the University of the Balearic Isles, on which, see Soler 2009.

179 Falconer 1987, 24–7.
180 United States Patent n° 5590093, 31 December 1996.

CHAPTER NINE

CLOCKS AS ASTRONOMICAL MODELS

SECTION ONE: 'THE HEAVENS DAILY IN VIEW': PLANETARY CLOCKS IN EUROPE, FOURTEENTH TO SIXTEENTH CENTURIES

Karsten Gaulke, Michael Korey, and Samuel Gessner

MORE THAN JUST CLOCKS

Planetary clocks, or planetary automata, were rare masterpieces of technical ingenuity and astronomical learning. Though numerous tower and table clocks with astronomical indications survive from the early modern era,[1] we single out as 'planetary clocks' those few designed to emulate the motion of the stars and planets according to received astronomical theory. Such clocks were the preserve of high clergy, princes, and emperors, and notice of about a dozen of them has reached us, along with tantalizing hints of several more. A planetary clock would ideally show the position of all the heavenly bodies visible to the naked eye and trace their motion through the firmament in real time, provided that a keeper synchronized the hour hand with the Sun by day or with a star at night. In a word, a planetary clock offered a compact, real-time image of the heavens: a microcosm. No wonder, then, that when signifier and signified are reversed the moving universe has often been compared with a perfect clock.[2]

Such exceptionally complex technical devices present a particular historical challenge. In order to understand their creation and how they were judged by contemporary scholars, patrons, and artisans, one must first understand some of the technical aspects of the sophisticated 'science of the stars' which then held sway.

To this end, recall that the constellations appear to move in synchrony in the night sky, as if fixed to the inside of a vast crystal sphere spinning daily around us. Careful observation reveals that only a few heavenly bodies, namely the five naked-eye planets (Mercury, Venus, Mars, Jupiter, and Saturn) and the two luminaries (the Sun and the Moon), appear to have their own additional movement against this background field of fixed stars. Each of these seven classical 'planets', as they were often collectively called, seems to wander eastwards relative to the stars, at times speeding up or slowing down, or even (for all but the luminaries) occasionally going backwards. In the course of a year the Sun traverses a closed circular path along the celestial sphere, the **ecliptic**; the other planets all wander within a narrow band around this, the zodiac, which is conventionally divided up into twelve equal segments ('signs'). In the *Almagest* (*c.*150 CE), Claudius Ptolemy offered an impressively successful array of geometric models to account for the vicissitudes of 'true' planetary motion. His models combined various uniform circular motions—only these were accepted by Aristotelian philosophy—with appropriate offsets and well-chosen parameters in order to predict the positions of the planets with remarkable accuracy.

INSTRUMENTS OF PLANETARY MOVEMENT

Ptolemy's theory was received and refined over many centuries by successive generations of mathematicians and astronomers writing

---

1 See Chapter 4 and Chapter 5.
2 Oresme 1377, II.2; see also Pomel 2019, 27, and Popplow 2007.

in Greek, Arabic, and Latin, for whom the theoretical models largely served for the computation of tables to predict the positions of the planets. At least since the eleventh century, a class of specialized, analogue mathematical instruments known as *equatoria* emerged alongside these tables. They consisted of rotatable graduated disks and radially turned arms or threads, with which planetary positions could readily be found. Such instruments offered both a visual representation of Ptolemy's geometric models and a means for the approximate calculation of the planets' positions; they could also serve as a convenient tool to double-check the laborious computations otherwise required.[3] Certain of these instruments used metal gears, for which the earliest technical description is dated around 1300.[4] Others were even made self-moving through the incorporation of a clockwork mechanism, becoming planetary clocks in the sense described above.[5] Next to nothing of these early mechanisms survives, and for those now known, the extant records are often not extensive enough to determine whether they were ever built or indeed succeeded in showing the true, as opposed to merely the mean, motion of the planets. Table IX gives a selection of examples and their key sources.

Planetary clocks thus lie at the nexus of timekeeping and mathematics.[6] This makes their study particularly challenging—and particularly rewarding. Two outstanding examples are the mid-fourteenth century planetary clocks conceived by Richard of Wallingford and Giovanni de' Dondi (both lost), which were introduced in Chapter 3, Section 3.[7] The latter is known through de' Dondi's own description, *Tractatus astrarii*, preserved in twelve manuscripts which include arresting diagrams (Figure 98).[8] The instrument had a heptagonal frame, with one planetary dial in its own zodiacal ring occupying each lateral face and all the dials powered by a common going **train**. A central, horizontal twenty-four-hour wheel engaged all seven sides and distributed its movement to the dials of the seven planets. Adjusting the hand for the Sun repositioned the central wheel and therefore kept all the planets synchronized. This disposition does not reflect the spatial relationship of the planets in the cosmos, in which they were imagined to be carried round in nested spherical shells ('orbs'), all moving through one and the same zodiac on the celestial sphere, itself spinning daily. Astronomers, at any rate, were used to thinking about the planets separately. Indeed, de' Dondi indicates in the preface to his treatise that he was inspired by Campanus de Novara (*c.*1220–96), whose *Theorica planetarum* contains descriptions of separate equatoria, one for each planet.

In the second half of the fifteenth century, visual representations such as these became fashionable in European princely circles. Indeed, a number of fifteenth- and early sixteenth-century rulers interested in the astral sciences, such as Philip the Good of Burgundy (1396–1467), Lorenzo de' Medici (1440–92), and the Habsburg Emperor Charles V (1500–58), commissioned or acquired refined *astraria* of their own (all lost), some of which were even referred to by this term, possibly as an explicit homage to de' Dondi. Only four planetary clocks survive from the sixteenth century, preserved today in Paris, Vienna, Kassel, and Dresden. They lie at the focus of the present account.

The first is of unknown date and origin, but by the mid-century was in the possession of Cardinal Charles de Lorraine (1524–74), for whom it was modified by the mathematician Oronce Fine (1494–1555) in 1553 (Figure 99). Elector Ottheinrich of the Palatinate (1502–59) commissioned the second clock from Philipp Imser (1500–70), professor of mathematics and astronomy in Tübingen (Figure 100). The third was made for Landgrave Wilhelm IV of Hesse-Kassel (1532–92), and the fourth for Elector August of Saxony (1526–86), both under the direction of Eberhard Baldewein (*c.*1525–93), a tailor and autodidact *mechanicus* of remarkable talent for instrument and clockmaking.[9] For brevity, we will refer to these four, respectively, as *Fine*, *Imser*, *Baldewein I* (Figure 101), and *Baldewein II* (Figure 102), and we will similarly use italics to refer to other (no longer extant) clocks by the names of their makers.

The four clocks have a broadly similar outer form, featuring dials on each of their vertical faces and, above these, a crowning, clockwork-driven celestial globe. Common to all, beyond a dial or dials for the position of the seven planets in the zodiac, is a calendar dial showing fixed (and often also moveable) feasts, as well as a dial displaying various hour systems. *Imser* may present the earliest use of an analogue day-and-night indicator, in which two

---

3 See Poulle 1980a and Falk 2016; for introductions, see also Evans 1998, 403–10, De Solla Price 1955, 119–33, and Kennedy 1960.

4 'Fiat columpna ...', preserved in the early Ambrosiana manuscript H.75 sup and four other mss. See North 1966, and Poulle 1980a, 641–53; it is uncertain whether what is described was ever built. Medieval, geared equatoria have a remarkable Hellenistic precursor in the Antikythera Mechanism, recovered from a shipwreck in 1901 and still the subject of investigation, see Jones 2017b.

5 Planetary clocks are treated from a comparative perspective in Zinner 1957, 31–40; Maurice 1976, vol. 1, 53–69, vol. 2, figs. 207–24; King & Millburn 1978, 62–89; Leopold 1986, 60–7; and Lloyd 1958, 9–24, 39–60. The relation of the clocks to astronomical theory is analysed by Poulle 1980a, 483–732.

6 Poulle 1980a.

7 North 1976, 20. de' Dondi considered the clockwork drive to be a topic already so well known that he deliberately omitted it from his treatise—Poulle 2003, 20.

8 For a comparison, see most recently Dresti/Mosello 2020.

9 Nothing is known about Baldewein's training. His primary occupation was as a builder (*Baumeister*) and palace chamberlain (*Lichtkämmerer*) for the Hessian court in the castle at Marburg. For his life and works, see Vogt 1942.

Table IX Significant early planetary clocks known only through written sources (selection)

Date	Author, owner, or text incipit	Place of Origin	Primary source	Principal secondary literature	Comment or special feature
Early 14th cent.	'Fiat columpna . . .'.		Oxford, Bodleian Library, Canonici misc. Lat MS 61; Milan, Ambrosiana, MS H.75 sup., and MS 35 sup.	North 1966; Poulle 1980a, 641–53.	Epicycles of the outer planets possibly jointly aligned by weights.
1st half 15th cent.	Abbot Engelhard II	Reichenbach	Reinhard of Tegernheim, 'Est autem astronomia', Munich BSB, Clm 83, fol. 124–36.	Poulle 1980a, 509.	Monastic setting in the Benedictine abbey of Reichenbach.
15th cent.	Anon. (Milan sketches).	Germany	Milan, Ambrosiana, C 139 inf., fol. 71–7.	Poulle 1980a, 706.	Fixed Sun pointer on a moving zodiac.
1447–55	Henri Arnaut de Zwolle, Clock for the Duke of Burgundy.	Burgundy	Paris, BnF, Lat. 7295, fol. 48–52, 61–70.	Le Cerf & Labande 1932; Poulle 1963, 209–13.	Also showed the motion of the eighth and ninth spheres. Identity between the built clock and the treatise doubtful.
1499	Giampaolo and Gian Carlo Ranieri.	Venice		Simoni 1971.	Reference claims that epicycles of the planets were visible.
15th cent.	'Ad laudem et gloriam'.		Brussels, Bib. Royale, MS 10117–26, fol. 147r/v.	North 1976, vol. 3, 235–6; Poulle 1980a, 730–2.	Unfinished description. All planets on one zodiac dial.
1476	Regiomontanus, 'Astrarium'.	Nuremberg	Regiomontanus, *Haec opera fient in oppido Nuremberga Germaniae ductu Ioannis de Monteregio*, BSB Rar. 320.	Zinner 1957, 19; Maurice 1976, 56; Poulle 1980b, 335.	Uncertain whether finished.
c.1480	Pulmann & Henlein, 'Theorica planetarum'.	Nuremberg	Johann Neudörfer, *Nachrichten von Künstlern und Werkleuten [. . .]*, 1547 (first published in Lochner 1875, 66, 71).	Zinner 1957, 19; Maurice 1976, 58; Poulle 1980b, 335.	
1488	Guillaume Gilliszoon, Clock for the Duke of Milan.	Milan	*Liber desideratus*, 1494, sig. a7v.	Poulle 1980a, 507–8.	All planets on one zodiac dial.
c.1500	Anon., Clock for Emperor Maximilian I.		Paolo Giovio, *Historiarum sui temporis libri XLV*, 1554, book 40, fol. 258v.	Maurice 1976, 67.	Depicted motions of the Sun, Moon, and planets, according to Giovio. Sent to Sultan Sulaiman in 1541 (by Emperor Ferdinand, Maximilian's grandson).
Early 16th cent.	Anon., Clock kept at Cuntz Drohtzieher's house.	Nuremberg	Munich, Bayerisches Nationalmuseum, drawing, inv. NN 1263 (earlier Z 1083).	Bassermann-Jordan 1905, 8, 45; Maurice 1976, 56–8.	Hexagonal or pentagonal case, sold to Cardinal Albrecht IV of Brandenburg in 1529.
c.1540	Petrus Apianus, Clock for Emperor Charles V.			Maurice 1976, 67; Zinner 1956, 39	Perhaps only a hand-cranked metal version of the paper volvelles in Apianus' *Astronomicum Caesareum*.

Continued

Table IX Continued

Date	Author, owner, or text incipit	Place of Origin	Primary source	Principal secondary literature	Comment or special feature
1547–50	Janello Torriani, 'Microcosm' for Emperor Charles V.		Marco Girolamo Vida, *Cremonensium Orationes III*, 1550, 53–7; Gasparo Bugati, *Historia universale*, 1570, 1025–6; Ambrosio de Morales, *Antiguedades de la ciudades de Espana*, 1575, 91–4.	Poulle 1980a, 505–6; Zanetti 2017, 270–96.	Showed calendrical dial, Sun, Moon, and planets. Surmounted by crystal terrestrial globe. Likely true motion and possibly with trepidation mechanism. Commissioned by Charles in Ulm.
1554–62	Janello Torriani, 'Crystalline' for Emperor Charles V.			Poulle 1980a, 505–6; Zanetti 2017, 168, 290.	Mechanism visible through rock crystal.
1540s	Johannes Homelius, Clock for Emperor Charles V.			Schelhorn 1731.	Charles paid 1,000 gilders.
1553–6	Steffen Brenner	Copenhagen		Maurice 1976, 67.	Destroyed 1728.
1559	Anon., Clock for King Christian III of Denmark.			Maurice 1976, 67.	Given to Czar Ivan IV (the Terrible) in Moscow, who declined the present.
before 1572	Bartholomeus Scultetus, Clock for Georg Mehl of Strolitz.	Görlitz	*Gnomonice de solariis, sive doctrina practica tertiae partis astronomiae*, 1572, fol. '(:)jj'.	Zinner 1956, 39.	Instrument called 'Plansiphaerium ex automato des himmels Lauf und Unterscheid der Zeit', could be just an astrolabe dial.
1574	Christian Heiden, 'Planetenwerk' (small and large).	Nuremberg		Maurice 1976, 67.	Given to Emperor Maximilian II, who commissioned a larger one for 2,000 gilders.
1574	Andreas Schellhorn	Schneeberg		Maurice 1976, 68.	Probably only mean, not true motion.
1579	Jacob Cuno	Frankfurt (Oder)	*Brevis descriptio artificiosi, noui, & astronomici automati horologici, cuius simile antehac non exstitit […]*, 1580.	King & Milburn 1978, 67.	Offered Duke Albert V of Bavaria to build it, received payment from Emperor Rudolf II in 1583, but did not finish the work before he died.
1588	Georg Kostenbader	Strasbourg	Extant at Gaesbeek castle, near Brussels, but original inner mechanism not preserved.	Maurice 1976, 68; Poulle 1980a, 678–9.	All planets, Sun, Moon, and Dragon on one dial. Surmounted by a celestial globe. Probably only mean, not true motion, but Venus mechanism provided with an epicycle.
1600	Georg Roll	Augsburg	Daniel l'Hermite, *Iter Germanicum*, 1637.	von Stetten 1779, 185.	Attribution to Roll by von Stetten uncertain.

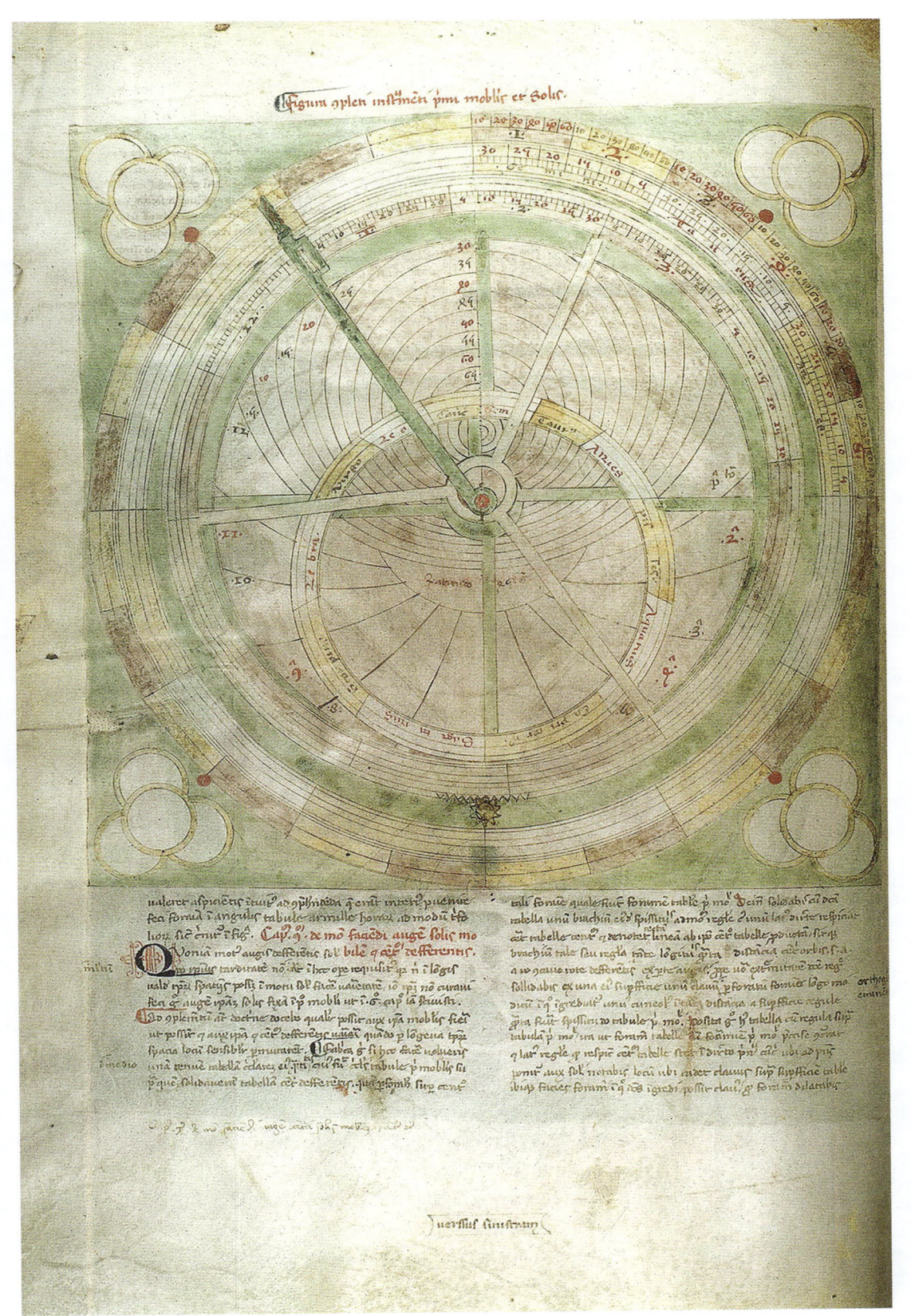

Figure 98 This dial, one of seven on Giovanni de' Dondi's *astrarium*, combines the hour of the day, the *primum mobile* (daily motion of the celestial sphere), and the Sun's non-uniform annual course through the zodiac. Reproduced from the earliest of twelve manuscripts of the *Tractatus astrarii* (© Biblioteca Capitolare di Padova, Cod. D.39, fol. 10v.).

graduated leaves, black and silver, fold into one another automatically to indicate the changing length of day and night according to the seasons (for a given geographic latitude), a device also shown on the two *Baldewein* clocks. The position of the Sun is often displayed on an **astrolabe** dial, i.e. a dial displaying selected fixed stars and the ecliptic, with which the daily westward motion of the Sun and its yearly eastward path could be shown. The Sun appears also in *Baldewein I/II* on the celestial globe, where its inclusion marks a significant advance in the history of mechanical clockmaking.[10] In connection with the Sun and the Moon, the indication of full and new Moons as well as possible eclipses was of paramount importance. All but *Fine* have an indication of the age of the Moon (up to 29½ days), as well as a scale for its latitude, that is, the small amount by which the Moon is above or below the ecliptic; this was usually tabulated as a function of the Moon's momentary distance from the lunar nodes, traditionally called the Head and Tail of the Dragon (and indicated as a rule by two extra, opposing pointers which move at the slow pace of one revolution in *c.* eighteen years).

REALIZING WONDROUS MOTION

Making a mechanism that shows movements of such complexity is no mean feat, given that each of the seven planets has its own peculiarities. All need to be represented by pointers or cursors with individual speeds and yet drawing mechanically on a common impulse, usually a daily revolution from noon to noon. At a basic level, planetary clocks must represent the correct relative speeds of the planets in the zodiac; the Sun's annual cycle must correspond both to a little more than twelve circuits of the Moon and to a fraction of a mean revolution of Saturn (nearly thirty solar years).

The most common reference for the values of such periods was the Alfonsine Tables, which enabled the computation of planetary positions for any given moment in time—with parameters in some cases slightly different from those used by Ptolemy. These tables were originally developed in the thirteenth century at the court of Alfonso X of Castille, from where their influence spread widely across Europe over the next three centuries. The tables, for instance, give the tropical period of Mars to be 686d, 22h, 24m, and 47s.[11] *Fine* approximates this tabular value via a gear train of 3/14✶70/50✶150/54✶69/10✶120, which results in a period of 690d, an error of 0.44 per cent.[12] *Baldewein II* uses a train of 143/287✶157/41✶360, yielding a period of 686d, 20h, 47m, and 56s, or less than 0.01 per cent off the Alfonsine target value. The corresponding train on *Imser* yields 12/3✶8/12✶23/15✶168 = 686d, 22h, 24m, and 0s, which is astonishingly only 0.0001 per cent away from the desired value.

As impressive as these fractional approximations are, they represent only a minor part of the ingenuity deployed. For, unlike on common clocks, the hands of true planetary clocks are generally not designed to move as uniformly as possible, but to change their angular velocity according to a pattern given by the Ptolemaic models. The simplest of these describes the Sun's progress through the heavens as seen from the Earth[13]. In the course of a year, the

Figure 99 Planetary clock, unknown maker, probably southern Germany, early sixteenth century (?), modified in 1553 by Oronce Fine. The oldest of the surviving planetary clocks, and the only one originally weight driven. Photo: © Bibliothèque Sainte-Geneviève, Paris.

---

10 Bertele 1961, 14.
11 In the common Parisian variant of the tables.
12 The first gear has fourteen teeth and turns thrice per (solar) day.
13 Indeed, the interval between the spring and autumn equinox (marking the Sun's course from Aries to Libra) is several days longer than that between the latter and the next spring equinox (from Libra back to Aries).

Figure 100 Planetary Clock by Philipp Imser with Georg Emmoser, c.1554–61. Of the four intact planetary clocks surviving from the sixteenth century, only this one has a common dial for all seven classical planets. Photo: © Technisches Museum Wien.

Sun traverses the ecliptic, but does so at a non-uniform pace. Figure 103 shows two approaches used to account for this uneven annual motion. The first, an 'eccentric' model, invokes uniform circular motion, but around a point displaced appropriately off centre. In the second, the Sun's position is predicted by uniform motion along a smaller circle (the **epicycle**), the centre of which

Figure 101 Planetary clock by Eberhard Baldewein, Hans Bucher, Hermann Diepel, and collaborators, c. 1556–62. Landgrave Wilhelm IV of Hesse-Kassel had this clock built for himself and was actively involved in its conception and construction. © Astronomisch-Physikalisches Kabinett, Museumslandschaft Hessen Kassel.

itself is carried uniformly along another circle (the **deferent**). Provided that certain geometric conditions are met, the two models yield identical predictions for the position of the Sun in the zodiac.

Close examination of the mechanisms of the extant planetary clocks has recently revealed three different ways for producing the Sun's non-uniform motion. The oldest of the surviving machines (*Fine*) uses a direct transposition of the geometric components of the eccentric model. A thin, flat arm just in front of the separate solar dial plate rotates uniformly around the offset point, and a peg fixed perpendicularly to the end of this arm goes through the slot of the main hand of the dial, which is only loosely suspended at the dial's centre as it turns, the main hand is dragged by the peg about the centre and indicates the position of the Sun in the zodiac. The *Imser* clock adapts what would be the geometrically equivalent epicycle model, but with an innovative twist, with the epicycle mounted on the Sun-opposed side of the deferent.

Imser is unique among the surviving four clocks in having all its planetary pointers displayed on a common zodiacal scale, and the mechanical innovation in the construction of the Sun's motion within this clock probably reflects Imser's response to the special challenges concomitant with building the requisite compact mechanism, which features a battery of nested, concentric tubular axes turning hands for all the planets and the lunar nodes.[14] Finally, the two other extant planetary clocks (*Baldewein I/II*) make use of a zodiac-centred gear with unevenly spaced teeth; a **worm** is used to drive this specially divided gear, whose tooth gaps subtly vary in width from one tooth to the next, on the order of less than one per cent.[15]

The intricate behaviour of the other planets, including their occasional retrogradation, is captured by more elaborate models

---

14 Gessner, Korey, & Gaulke 2020.
15 Gessner, Korey, & Gaulke 2020.

Figure 102 Planetary clock by Eberhard Baldewein, Hans Bucher, Hermann Diepel, and collaborators, 1563–68. The clock was commissioned from Landgrave Wilhelm IV by Elector August of Saxony. © Mathematisch-Physikalischer Salon, Staatliche Kunstsammlungen Dresden.

in the *Almagest*.[16] The motion of Mars, for instance, is generated by a combination of deferent and epicycle, but the former is itself now eccentric, and the motion of the epicycle's centre is uniform only if gauged from a special point: the **equant** (Figure 104).[17]

---

16 We restrict ourselves here solely to considering their key motion within the plane of the ecliptic.

17 Similar models apply to the other planets, with additional subtleties introduced for the Moon and Mercury; the centres of their deferent circles are not fixed, but wander slowly, each along its own small circle.

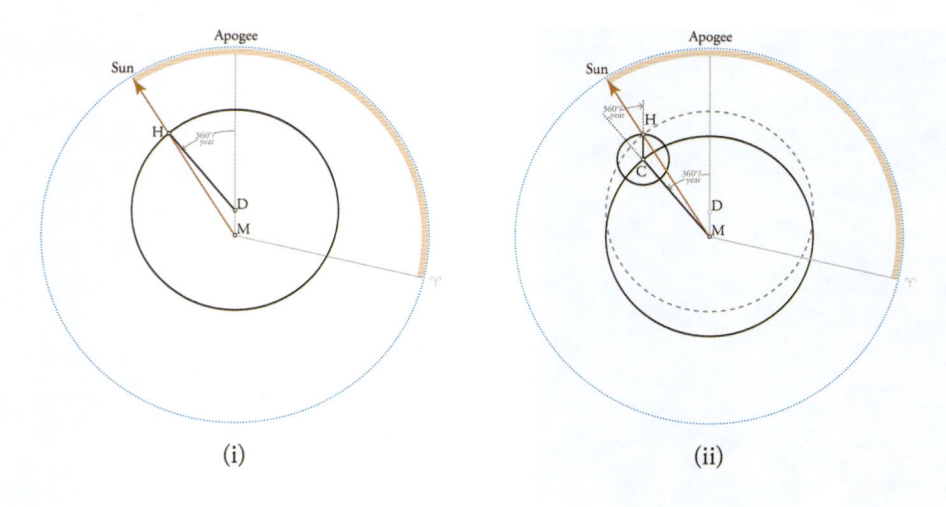

Figure 103 Equivalent models to predict the position of the Sun in the ecliptic (blue circle), as seen from the Earth at M. In (i) this position is generated by a point H moving with uniform angular velocity around a circle eccentric to M; in (ii), the point C moves with uniform angular velocity around a circle centred at M, while H circles C at the same pace in the other direction. © The authors with Claudia Bergmann.

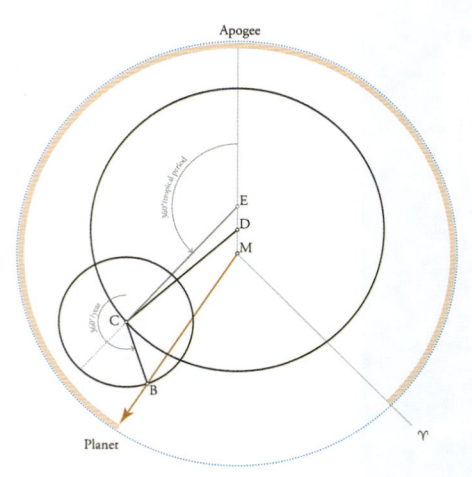

Figure 104 Ptolemy's model to predict the motion of an outer planet (e.g. Mars) through the zodiac, as seen from the Earth at M. The point B revolves uniformly on the 'epicycle' centered in C; in turn, C circuits the 'deferent' circle with centre D, moving with uniform angular velocity with respect to the E, the 'equant' point. The combined motion generates the planet's position, specifically its ecliptic longitude. © The authors with Claudia Bergmann.

Constructing a geared representation is therefore a tremendous challenge. An ostensibly uniform input impulse must be transformed into a specific, non-uniform output motion—one that simultaneously reflects an eccentric deferent, the epicycle, and the equant! Three main approaches have been used to achieve this, often combined in a single work. The first, a peg-and-slot mechanism, mentioned earlier for the solar movement on *Fine*, is used in all known planetary clocks, in particular, to transmit the movement of a peg on an epicycle to a centrally mounted hand and thus to generate retrograde motion. Such a mechanism is also used to implement materially the idea of the equant point, i.e. to dissociate the point from which a rotation is perceived as uniform from the point (the deferent centre) around which the radius of revolution is constant. Indeed, *Dondi* and *Imser* both use a slot in a spoke of what might be termed a (uniformly turning) 'equant wheel' to generate the proper pace at which the centre of the epicycle turns around the centre of the deferent. That this 'literal' approach to building a mechanism with three closely proximate centres of motion (at M, D, and E in Figure 104) is not without pitfalls can be seen in the case of *Fine*; as Poulle observed, its unknown maker inverted the proper role of peg and slot in the connection between the equant and deferent wheels. He also only managed to have the peg on the rim of the epicycle rotate at a pace proportional to the pace around the deferent centre, not uniformly with reference to the clock frame, as required.[18]

The second type of mechanism, found in both *de'Dondi* and *Wallingford* uses toothed wheels of deliberately non-circular shape in order to modulate angular velocity by effectively altering the radius of revolution. Wallingford's mechanism for the Sun's annual course, for instance, incorporates a crown wheel with orthogonal teeth of equal separation engaging an elongated **pinion**; the distance between the teeth is kept constant while the radial distance from the gear arbor varies. Finally, the third approach turns this on its head and arises on both the Kassel and Dresden clocks in the collaboration of Baldewein and Wilhelm IV: the use of a special, non-uniform tooth division.[19] For example, in the planetary gear trains, to realize the *effect* of the equant point without materializing an additional centre of motion there, a specially divided circular gear turns about the centre of the deferent, but does so as if it were turning uniformly as seen from the equant.[20]

Table X summarizes the features of those few planetary clocks for which reliable information on their technical make-up is known, namely the four extant clocks alongside the well-documented ones of Wallingford, de' Dondi, and Lorenzo della Volpaia (1456–1512). The differing approaches adopted not only reflect varying degrees of collaboration among the actors involved in the construction of these technical masterpieces—princely patrons, learned astronomers, and artful craftsmen (with these categories sometimes overlapping)—but also offer a further, mechanical contribution to the centuries-old reception and refinement of

---

18 Poulle 1974, 76–7.
19 An unevenly toothed wheel is used only once in *de' Dondi* (as the fixed central wheel to drive the Venus epicycle). Poulle 1974, 77.
20 So the claim in Poulle, Sändig, Schardin, & Hasselmeleyer 2008, 138–49.

Ptolemaic planetary theory. Seeking to express knowledge about planetary motion in the form of a geared mechanism or a planetary clock not only permitted the positions of celestial bodies to be determined with greater ease, but also allowed for testing of innovative ideas, visualizing geometrical concepts, and communicating the complexity of astronomy—all features of what Michael Shank has nicely characterized as 'mechanical thinking'.[21]

THE QUESTION OF TRADITION

Table IX and Table X facilitate comparison and provoke questions. For example, does the recurrence of such features as the peg-and-slot mechanism mean that this way of realizing non-uniform motion was, in each instance, discovered anew, or were later makers aware of earlier clocks? A tradition of reception has often been posited in the secondary literature but has received a satisfactory answer in only a few cases. It is clear from visits and reports on it that de' Dondi's *astrarium* had great renown for over 200 years and that it physically existed until the mid-sixteenth century, but not whether it was then still in working order. The noted astronomer Johannes Regiomontanus (1436–76) was so impressed upon seeing it in 1463 that he mentioned it in his introductory lecture in Padua and affirmed that such a device was under construction in the workshop in Nuremberg in his 1474 trade list.[22] No planetary clock was completed by the time of his death, but archival sources from the first half of the sixteenth century report several attempts by Nuremberg artisans to produce such a clock, and these efforts possibly go back to Regiomontanus' undertaking.[23]

Janello Torriani (*c.*1500–85), later court engineer to Charles V, also probably had personal knowledge of de' Dondi's *astrarium*. It is likely that he was commissioned by Francesco II Sforza in the 1530s to restore the clock, as it had been defective for several decades, and that this offered him the opportunity to study its technical configuration. It is not known whether de' Dondi's *astrarium* or the lost planetary clock by Gilliszoon de Wissekerke (4th quarter 15th century), which Torriani likewise repaired, served as the model for the *Microcosm* that he completed for Charles in 1556.[24]

Table X also highlights several parallels between the clock by de' Dondi and those of Baldewein's workshop. Correlation is not causality, however, and it remains uncertain whether Wilhelm and Baldewein even knew of the existence of the *astrarium*, let alone its technical details. At any rate, by 1565 they *did* know of the *Imser* clock, as Baldewein reports in a letter to Wilhelm after a visit to the Marburg workshop by a nobleman who had seen *Imser* and who discussed the advantages and disadvantages of having the planets' longitudes be shown separately or on a common zodiacal dial.[25] In spite of these and other intriguing single instances, present knowledge does not permit painting a fuller picture of the degree of transmission from one maker to another.

THE THORNY PATH OF CONSTRUCTION

Developing a mechanical device according to such requirements would challenge even the best artisans, astronomers, and mathematicians—and indeed, it drove some of them literally to the end of their wits and to blows. The conflict-laden construction of *Imser* bears witness to this. Unable to deliver the planetary clock within the period agreed, and instead requesting further money, Imser was forced by Ottheinrich to accept help from the clockmaker Gerhard Emmoser. This soon led to a complete stalemate when Imser accused Emmoser of sabotage and the latter locked the disassembled clock into a chest—just one of the outlandish episodes that occurred during construction.[26]

As planetary clocks had to meet both horological and astronomical demands, their construction required skills from several different fields. Further, as each such clock was individually designed to fulfil the specific goals and ideas of its maker(s), only in the rarest of cases was there a blueprint at hand that could be used to design the gears and displays. In each instance, the challenge was to translate the rules of planetary movement established in the models of Ptolemy and his successors into a functioning mechanism. A profound understanding of astronomy and mathematics had therefore to be combined with prodigious mechanical ability. As these skills drew upon divergent cultures of knowledge, their concomitant presence in a single person or in a team working together without friction was the exception rather than the rule.[27] This may help explain why relatively few planetary clocks were ever built.

A survey of the sparse extant sources describing the manufacture of planetary clocks shows that most were made by a division of labour, though a few were constructed by single individuals—either astronomers who had acquired fundamental mechanical abilities, or at least the capability for 'mechanical thinking', or craftsmen who had taught themselves the necessary elements of

---

21 Shank 2007; see also de la Solla Price 1980 and Keller 1985.
22 'Hec opera fient in oppido . . . ' [1474], BSB, Ink R-58.
23 The design of Fine bears clear affinity to the sketch of a planetary clock sent by a Nuremberg artisan to Cardinal Albrecht IV of Brandenburg in the 1520s, discussed in Maurice 1976 i, 56–8, ii, fig. 23.15. Furthermore, a passage in Ramus 1569, 31, suggests that Fine had access to a planetary clock that had been seized as booty during the 'German wars'. Zanetti 2017, 288 suspects it may have been taken from the abandoned camp of Emperor Charles V at Metz in 1553.
24 Zanetti 2017, 278.

---

25 STA Marburg 4a 31–3, Baldewein to Wilhelm, 28 September 1565.
26 Rau 1962, 29. Chandler & Vincent 1980a.
27 The need for close collaboration between clockmakers and mathematicians is stressed in the foreword to description of the famous Strasbourg cathedral clock: Dasypodius 1580a, b[i]–bii.

Table X Well-documented planetary clocks. Clocks showing true motion of the planets on which there is sufficiently detailed technical information to permit sustained comparison. The four preserved clocks that retain most of the original mechanism appear above four lost clocks that were described in detail by their makers. Features common to all clocks such as size (25–35cm per dial, except for the larger (*Wallingford*), a celestial globe (all but *de'Dondi* and *Wallingford*), and an astrolabe dial (all but *Volpaia* and *Imser*) are not included in this overview. For tooth counts and planetary parameters, refer to the secondary literature. The abbreviations used in the columns labelled (*) and (**) are explained in the running text.

Biography				Visible outside			Mechanism inside			Primary sources	Principle secondary literature
Date of completion	Maker(s)	Patron/commissioner or first owner	Place(s) of manufacture	Separate dial for each planet	Ptolemaic model shown (*)	Special dials, scales, and indications	Energy source	How uneven motion is realized (**)	Comments	Inventory no. or shelf mark	Reference (R) or critical edition (ce)

EXTANT

c. 1500; 1553	anonymous; Oronce Fine	unknown; Charles de Lorraine, Cardinal of Giese	Southern Germany; Paris	✓	✓	Separate solar dial, only rising lunar node, no latitude information, special seasonal hour system	spring	P, E		Paris, Bibl. Ste. Geneviève, Boinet nr. 106	R: Hillard & Poulle 1980; Poulle 1980a, 576–92
c.1561	Philipp Imser & Gerhard Emmoser	Otheinrich, Count (later Elector) of the Palatinate	Tübingen	X	X	Joint planetary dial with solar pointer and lunar nodes, analogue planetary latitude dials, calendar, seasonal duration of day/night, mechanical realization of seasonal hours shown by automatic planetary divinity figurines.	weight	P, E	Novel device to have the epicycle turn uniformly: 'wiggle' mechanism.	Vienna, Technisches Museum, 11.939/22	R: Rau 1962; Chandler & Vincent 1980
1562	Eberhard Baldewein & Wilhelm IV	Wilhelm IV, Landgrave of Hesse-Kassel	Marburg, Kassel, Giessen	✓	✓	Solar cursor in ecliptic of celestial globe, Sun also depicted on the lunar dial along with lunar nodes, values for latitude computation displayed, calendar, seasonal duration of day/night, hand for minutes of time.	spring	P, E, U	Novel device to have the epicycle turn uniformly: compensation gear with uneven tooth gaps.	Kassel, Astron.-Physik. Kabinett, U 63	R: Gaulke 2007, 132–41; Lloyd 1958, 46–57
1568	Eberhard Baldewein & Wilhelm IV	August, Elector of Saxony	Marburg, Kassel, Giessen	✓	✓	Solar cursor in ecliptic of celestial globe, Sun also depicted on the lunar dial along with lunar nodes, values for latitude computation displayed, calendar, seasonal duration of day/night, hand for minutes of time.	spring	P, E, U	Novel device to have the epicycle turn uniformly: compensation gear with uneven tooth gaps.	Dresden, Mathem.-Physik. Salon, D IV d 4	R: Poulle, Sändig, Schardin, & Hasselmeyer 2008

Table X Continued

LOST BUT WELL DOCUMENTED

Date	Name	Location	(*)	(**)	Drive	Features	Description	MS	References
1330s	Walling-ford	Richard of Walling-ford, Abbot of St Albans	St Albans	?	?	weight	P, U (ovoid)	An astrolabe dial with a moving rete and the position of Sun and Moon.	Poulle 1980a is convinced that the clock was only solilunar, without the other planets.	Oxford, Bodleian, MS Ashmole 1796	ce: North 1976
c. 1381	de' Dondi	Owned by Gian Galeazzo Visconti from 1381	Padua	✓	✓	weight	P, E, U (ovoid)	Solar cursor on an astrolabe dial, with a rete void of stars.	Novel device to have the epicycle turn uniformly: ovoid gears to drive the lunar epicycle and eccentrically mounted gears for epicycles of superior planets.	Padua, Bibl. Capitolare, MS D 39 (and 11 other MSs.)	R: Poulle 1988; Bedini & Maddison 1966 ce: Poulle 2003, Bullo 2003
(i) 1484 (ii) 1510	della Volpaia	(i) Matthias Corvinus, King of Hungary (ii) Lorenzo de' Medici	Florence	X (multiple zodiac rings within one dial)	✓	weight	P, E (ovoid for Mercury and Moon).	Rotating terrestrial globe, with indication of the hemisphere illuminated by the Sun.	Centre of the deferent and the equant were likely fused, thereby the movement of the epicycle was even.	Venice, Marciana, MS It 5363; Florence, Magliabechi, XIX 90; Florence, Laurenziana, MS Antinori 17	R: Simoni 1971; Poulle 1980a, 653–61; Brusa 1994; Strano 2005

(*) ✓ = deferent and epicycle are depicted; (**) P = peg-and-slot mechanism; E = eccentrically mounted gear; U = unevenly divided gear.

astronomy and mathematics. Recent historiography has invoked the term *Vitruvian artisan* for these two types of persons in whom mechanical and mathematical skills are present in equal measure.[28] The first category is typified by Giovanni de' Dondi, professor of medicine at several Italian universities; it is likely that he acquired the necessary practical skills to build the *astrarium* from his father Jacopo de' Dondi, who along with his profession as city *medicus* was also devoted to astronomy and the construction of (turret) clocks.[29] Alas, too little is known about the fabrication of the *astrarium* to be able to say that de' Dondi built it alone. Henri Arnault von Zwolle, Regiomontanus, and Gilliszoon were similar personalities.[30]

For the second category of Vitruvian artisan there are three representatives: Lorenzo Della Volpaia, Janello Torriani, and Jost Bürgi (1552–1632). Della Volpaia was the scion of an important family of Tuscan instrument makers[31] who sold his first planetary clock to the king of Hungary in 1488. Torriani, who constructed at least one planetary clock for Charles V (the *Microcosm*) and possibly another (the *Crystalline*), never attended university, but received a profound mathematical and astronomical education from his father, Giorgio Fondulo, a professor of medicine.[32] Bürgi possessed even more remarkable mathematical and astronomical knowledge and, from 1579, was court clockmaker to Wilhelm in Kassel and successor to Baldewein. Bürgi, about whose training nothing is known, made innovative contributions to mathematics, including a system of logarithms; through close contact with astronomers such as Christoph Rothmann (late sixteenth century), and Nicolaus Reimarus Ursus (1550–99), he acquired significant knowledge of the theoretical and practical portions of astronomy. Between 1587 and 1605 he built at least two planetary clocks, as well as a complex table clock for an innovative display of Copernican lunar theory (the *Mondanomalienuhr*) and several clockwork-driven celestial globes with true solar movement.[33]

Both the mechanically cogitating *astronomus* and the astronomically and mathematically learned *mechanicus* reflect a way of thinking that extended well beyond the manufacture of complex planetary clocks. Indeed, both contributed to a fundamental development toward the end of the sixteenth century, as astronomers grew ever less satisfied with building geometric constructions whose sole purpose was to 'save the phenomena' and increasingly came to ponder the effective causes of planetary motion. In this context the metaphor of the cosmos as a clockwork-driven machine acquired ever greater value. Natural philosophers such as Johannes Kepler (1571–1630) and René Descartes (1596–1650) would draw richly on this ferment, with great epistemic gain, at the beginning of the following century.

USE, USERS, AND SETTING

Though the planetary clocks discussed in this chapter are exceptionally important and mark a pinnacle of contemporary knowledge and mechanical ingenuity, we know remarkably little about their actual use. The preface to de' Dondi's treatise on the *astrarium*, for example, underscores his aim to create what might be termed a functioning, philosophical instrument. So many contemporaries are said to be ignorant of, or to doubt, the complex Ptolemaic theory underlying the motion of the planets, the apologetic argument goes, that if he were able to build a working version out of such base materials as metal gears and weights, how much easier must it have been for the Creator to realize these movements in the heavens! It is interesting to note how God and Aristotle are invoked, but in the entire treatise neither a patron nor an intended location is mentioned. Apparently, no later source among the many who recorded seeing the *astrarium* in the ducal library at Pavia clearly describes how or whether it was commonly used (other than perhaps to show visitors that it *was* in use—or later, when it no longer functioned, that it *had been* in use as long as a master such as de' Dondi was at the helm).

Analogously, there is no shortage of rhetorical claims for the utility of the four surviving sixteenth-century planetary clocks: the Latin couplets engraved atop *Baldewein I/II* serve not only to celebrate each clock's owner (and for the former its co-creator), but also to offer a selection of the celestial indications which can be read off the various dials and which this 'noble work reveals through its own motion'. A surviving user's manual penned by Imser offers further claims about how his clock offered the pleasure of keeping the full heavens daily in view, 'presented to the eyes so that not only much diversion but also great usefulness' can be gained.[34]

One thing is certain, however, at least for the clocks produced in Kassel. There was a clear expectation that they *would* be used regularly. Indeed, in a letter to Wilhelm during the construction of the clock for Elector August, Baldewein notes that he expects it will be used often, not 'only a few times in a year and then carefully locked away again right away, as with a cup or other rarity ... but will stand in a room on a dusty floor ... and will suffer from rough handling'.[35] So these clocks were clearly capable of being used. Were they, however, and if so, what for?

---

28 Zanetti 2017, 11–22. Vitruvius is also evoked (with other Ancient mechanicians) in Dasypodius 1580b, 50–1, and elsewhere.

29 Medieval and Renaissance physicians were expected to be learned in the astral sciences, as illness was often linked to the influence of the heavens; see Siraisi 1990; Schechner 2008, 204, but *cf.* the references in Brockliss & Jones 1997. Some such as de' Dondi or Arnaut von Zwolle also made planetary clocks: White 1975, 209, 305; Zanetti 2017, 126. For short accounts of the two de' Dondi see Turicchia 2018, 98–100.

30 For Arnault and Gillizoon see Wickersheimer 1979, iii, 115; i, 233, & iii, 103

31 Turicchia 2018, 293–6.

32 Zanetti 2017.

33 Maurice 1980a; Leopold 1986.

34 ÖNB (Vienna), Cod. 10783: Philipp Imser, *Außlegung und Gebrauch des newen Astronomischen Uhrwerks*. 1560, fol. 4v.

35 STA Marburg 4a 31–3, Baldewein to Wilhelm, 22 April 1565.

One source confirms August's interest in consulting planetary positions with such a device. Soon after finally receiving his planetary clock (*Baldewein II*) following five years of waiting, he considered commissioning the noted Augsburg instrument maker Christoph Schissler (1531–1608) to produce a further instrument to show the true movement of all the planets. August's original enquiry does not survive, but Schissler's reply notes that the Saxon elector already had a 'substantial work' to this end, though one that could not easily be moved overland—this must refer to *Baldewein II*, which weighs a quarter-ton and is nearly 1.5m high.[36] What was now desired was a portable instrument 'that could be taken anywhere without being'. On a single zodiacal dial the true longitude and the rising and setting of all seven planets would be shown; whether a planet was stationary or undergoing retrograde motion, as well as the momentary distribution of the planets across the twelve **houses** with favourable or unfavourable **aspect** relations, such as conjunction or opposition, was all to be immediately readable. It would even be possible and 'pleasurable', as Schissler noted, to determine the occurrence of an eclipse.

In fact, these are all features of *Baldewein II*—though in some cases one must combine the information on its individual dials, for instance, to determine the aspect relations among the planets. Evidently August sought not only a compact device, but one easier to use, wanting simply to turn the lunar pointer to its proper position in the zodiac in order to have all the other hands moved to their respective positions. Both patron and maker may have been out of their league here, as for his part Schissler boldly claimed that he could fit the entire mechanism behind the dial into a housing only six inches thick, if it were to be hand-cranked, or only two to three inches more, if the mechanism were to be driven by Clockwork, 'which would make it even more useful and enjoyable'.[37]

Reference to the houses, aspects, and the eclipses make clear the astrological importance of such a device for August, which he wished to have with him at all times. Indeed, each of the four surviving planetary clocks offers a self-contained heavenly tableau, displaying the positions of the planets in the zodiac at any instant and allowing their momentary place among the stars to be envisioned on the globe on top or on the astrolabe dial. In which house each planet lies is also shown,[38] as well as whether its position is above or below the horizon, near the ascendant or midheaven, or elsewhere. In short, each of these clocks provides all the data required for the casting of a horoscope, without the need for tedious calculation.[39] A subtle mechanism driving three stories of moving figures atop *Imser* also shows automatically which planetary divinity governs the current **unequal hour**.

Already in Ancient Mesopotamia the positions of the planets were thought to augur specific events. Scholars in the Latin Middle Ages reactivated astrological categories from Antiquity and the Islamic world, and interest in astrological judgements in no way faded in sixteenth-century Europe. Explicit support for promoting mathematics and astronomy to read the signs of the heavens was given in the Protestant territories by none other than the *praeceptor Germaniae* Philipp Melanchthon (1497–1560). August's Saxony, Wilhelm's Hesse, and Ottheinrich's Palatinate were all archly Lutheran. To interpret these signs one must first know the planetary positions in the sky at the desired moment of inquiry. Conventionally, such information would be offered in the form of pre-calculated positions in an ephemeris (Figure 105, left). How might planetary clocks be preferable to such?

Several answers are possible. First, planetary clocks make *visual* the information in the tables, giving an immediate and precise display of places and aspects without need of calculation. Second, they *accelerate* time's arrow, making it possible for the (princely) user to view the state of the heavens at a past or future moment; for this he needed only to disengage the clockwork drive, while keeping the planetary trains engaged, and then turn the going train to the desired instant. This possibility of retrospective or prospective inquiry is made explicit in the surviving manuscript tables by the Wittenberg mathematician Melchior Jöstel (1559–1611) for *Baldewein II*; unlike standard ephemerides, these special tables include not just the true positions of the planets in the zodiac at regular intervals, but also the two components of each planet's motion needed to set it properly in place on the corresponding dial (Figure 105, right).[40] These tables were computed at the end of 1589, but include monthly values for years beginning

---

36 Sächs. Staatsarchiv, HStA Dresden 10035 Loc. 4418/1, 194 ff., Schissler to August, 17 June 1570. This letter has been cited and partially transcribed by Bobinger 1954, 62–3, but without the key passage relating it to an existing work in the elector's possession. The plan was abandoned, as a marginal note from August explains, so that it is not clear that Schissler ever made planetary automata, as Bobinger 1954, 61, claims.

37 Imser had tremendous difficulty constructing a single-dial mechanism—see Rau 1962. As the epicycles are not depicted on the face of the common zodiac dial, retrograde movement is not visible at a glance from outside *Imser*. Nor would it be on Schissler's proposed device, in spite of his claim otherwise, if we can judge by the drawing he appended to his letter; on this drawing (reproduced in Poulle 1980a, vol. 2, 20) the house division is rigid, and no ecliptic is shown.

38 On *Imser* via the adjustable wire cage around the celestial globe celestial globe; on *Baldewein I/II* upon the turning astrolabe dial. Fine added a (non-mechanical) mathematical dial in 1553 to the existing clock in order to effect a house division, also including a peculiar type of unequal hours.

39 An astrolabe could be used for a quick approximation to the house division; more precise results are accessible, e.g. via oblique ascension tables, which reduce the trigonometric operations required to addition/subtraction and linear interpolation. On the noble use of instruments to avoid calculation, see Turner 1973b.

40 That is, to turn the deferent to the calculated 'true centre' and the epicycle arm to the 'true argument'. For more on Jöstel's tables, see Korey 2007b, 110–13.

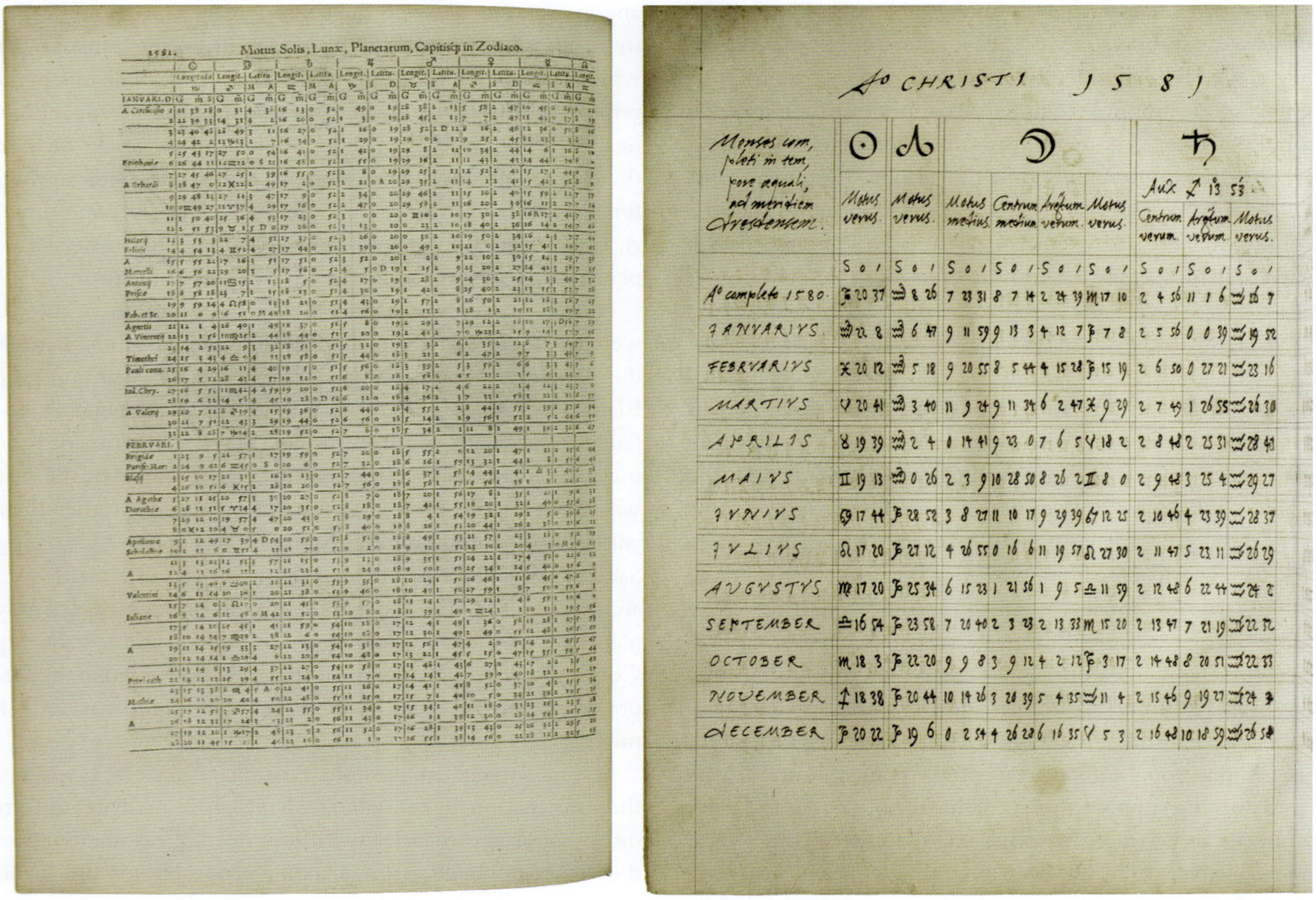

Figure 105 Two types of ephemerides: (left) a standard printed example, with daily positions of the planets (Cyprian Leowitz, *Ephemeridum novum atque insigne opus*, 1557); (right) a manuscript, with monthly values specifically calculated for setting the 'deferent' and 'epicycle' positions on each dial of the clock *Baldewein II*, yielding planetary positions equivalent to those in the printed table (Melchior Jöstel, *Tabulae centri et argumenti*, 1589). Photo: © SLUB Dresden, Astron.4 and Mscr.Dresd.C.51: https://digital.slub-dresden.de/werkansicht/dlf/6873/1/, and Mscr.Dresd.C.51: https://digital.slub-dresden.de/werkansicht/dlf/104548/1/.

in 1581, when the previous such tables left off, and for times in the future (until 1612). Third, in their hand-cranked mode, the planetary clocks offer the chance to *interpolate* between the discrete points in an ephemeris, as would be needed in an astrological 'election', for instance to find the moment within a forthcoming interval in which the heavens would be most propitiously aligned for an action. Finally, and perhaps most importantly in a courtly context, these clocks are in principle *self-contained*, offering the sovereign an overview of the entire cosmos at a single glance, at any time, independent of conditions of daylight or weather, and without having to consult others. Thus, while tables offer pre-calculated positions, a planetary clock provides them in a form truly befitting a prince, with its marvellous, self-moving ability nearly providing it with a soul. In the words of Rothmann: 'All that which is customarily taught in astronomy one can marvel at in-person on what is, in a matter of speaking, a living'.[41] In short, although we do not know much about the specific use of these Renaissance planetary clocks, what we do know points to how they were especially appropriate embellishments of a princely court, offering their sovereign users a model of knowledge and a ready means to access astral data in a precious form appropriate to their station.

Finally, the early modern princely court was not only a key node in the power relations of the era, but also a central stage for technical innovation, where the invocation of Archimedes or Vitruvius flourished both symbolically and in practice. The patronage of such efforts could redound to the benefit of both artisan and prince. In particular, engagement with planetary automata furthered comprehension of the heavens, perceived as complex but orderly; visualized and predicted the interrelations of the planets, knowledge of which was often attributed astrological significance; and promoted skilled mechanical craftsmanship as a sign of the grandeur of the prince and the level of technological sophistication of his territory.[42]

---

41 Rothmann 1589, 71.

42 Moran 1977, 213–24.

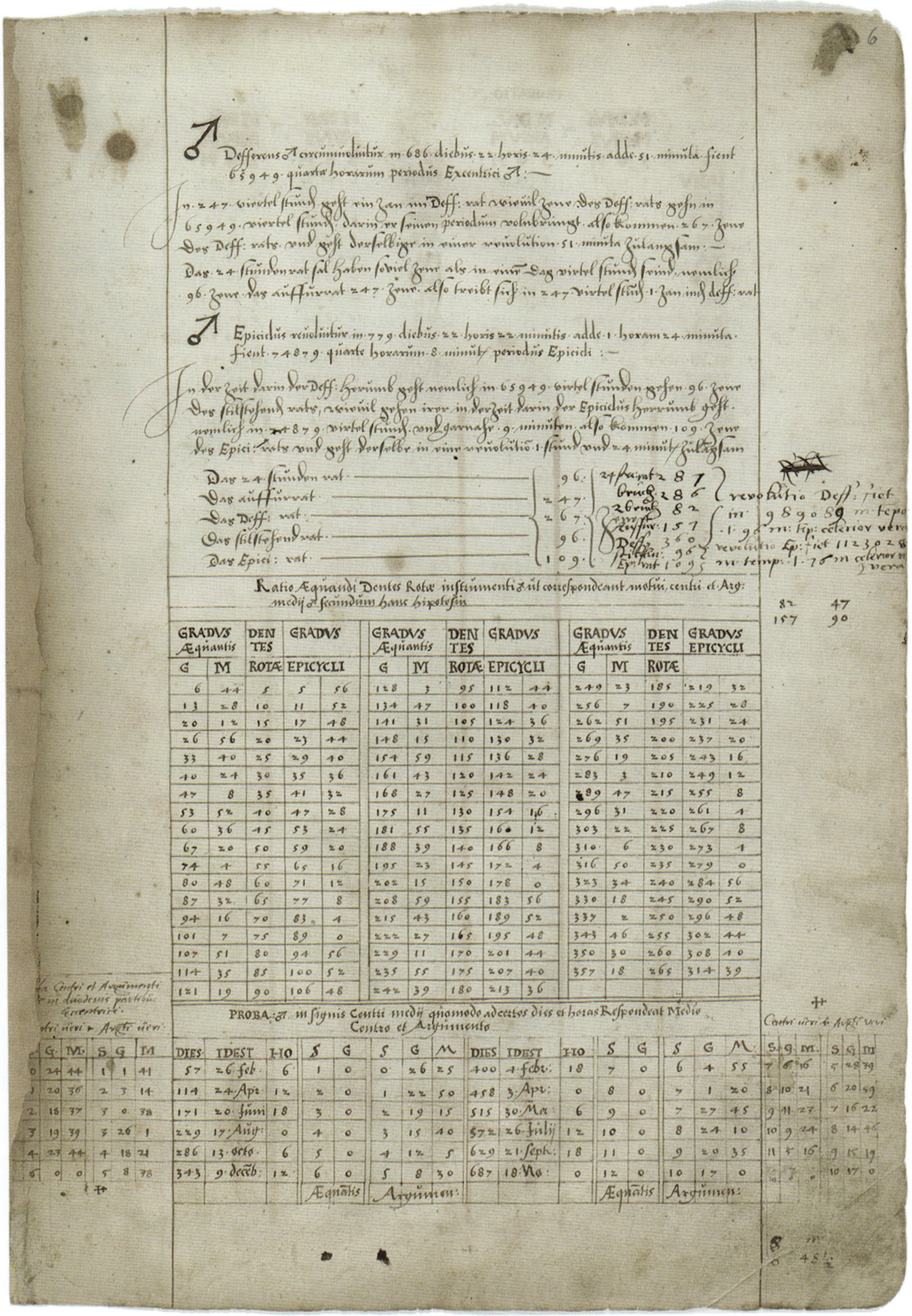

Figure 106 Calculation of various gear combinations to represent Saturn's motion on a planetary clock, c.1560. Target periods from the Alfonsine Tables are given at the top of the page; Wilhelm's own calculation of the effect of introducing a 360-tooth gear for the deferent wheel (as later adopted in *Baldewein II*) is at the right margin. Photo: © Murhardsche Bibliothek, Kassel. 2° Ms. Astron. 16, fol. 4r: https://orka.bibliothek.uni-kassel.de/viewer/image/1378193947325/8/.

'BRINGING ORDER TO THE HEAVENS': PLANETARY CLOCKS AND THE SELF-FASHIONING OF A PRINCE

As sixteenth-century European rulers sought to epitomize their role as territorial sovereigns with magnificent palaces and collections of rare and remarkable objects, planetary clocks were particularly apposite. On account of the scientific and technical complexity combined in them with high decoration, they were exceptional art objects supremely qualified for entry into the emerging *Kunstkammern* (chambers of the arts) or other court collections. As mathematical models of the cosmos, they made manifest the order established by the Creator and were thus apt symbols for the sovereigns' claim to maintain orderly rule over their dominions. It is therefore not surprising that Charles V possessed at least two such planetary clocks at the end of his life. What explicit meaning he ascribed to these is not recorded—curiously, one was held to be suitable as a diplomatic gift to the Ottoman Sultan Sulaiman I (1494–1566)[43] in a precarious situation as a —and that holds for nearly all the owners of such objects.

An exception is Landgrave Wilhelm IV of Hesse-Kassel. Unlike all other contemporary princes, he was feverishly active as an observational astronomer. *Baldewein I* was not simply a planetary clock in his possession. Wilhelm was in large part the *spiritus rector* behind its construction, as he was for *Baldewein II*. Surviving correspondence for the latter reveals that he calculated the gearing himself and discussed each important step in the construction with Baldewein (see Figure 106).[44] As discussed later, both his exceptional astronomical knowledge as well as the planetary clocks and clockwork-driven celestial globes produced under his direction played a central role in his self-fashioning, as he strove to give importance to his geopolitically rather insignificant territory within the Holy Roman Empire.

Writing about Wilhelm's astronomical ascent in the manuscript *Observationes Stellarum Fixarum* of 1589, the court astronomer Christoph Rothmann made the symbolic meaning of the landgrave's celestial automata explicit: 'After a modest start, in which he took measure of the foundation of ast-ronomy, he advanced so far that he could convey the movements of all the heavenly bodies using automata. Although he [Wilhelm] did not yet direct the ship of state, by means of his acumen he appeared to emulate the Prime Mover, as if he himself were guiding the heavenly lights—and thus to show what could be expected from him in a later phase of life'.[45] Rothmann stylizes *Baldewein I* as a material witness to Wilhelm's exceptional abilities to guide celestial bodies rightly on their path through the heavens, in emulation of the Prime Mover. This analogy finds its counterpart in the iconography of the clock, for above the celestial globe is a figure that is referred to in contemporary sources as the 'Saviour': God the Father, garbed as an earthly king, holds the orb and cross in his left hand and raises three fingers of his right hand in a blessing.

The few sources extant from the time of the construction of *Baldewein I* show how important Wilhelm's first planetary clock was to him, and why. In February 1558 he requested that the goldsmith Hermann Diepel hurry to complete the celestial globe and other of the clock's components, explaining that the electors of Saxony and Brandenburg would soon be visiting Kassel upon returning from a 'Fürstentag' in Frankfurt and must see this clock.[46] Although the visit was not made, surviving correspondence shows that Wilhelm and his father, Landgrave Philipp the Magnanimous, particularly wished to use the occasion to show the elector of Saxony, Hesse's most important ally and the richest Protestant ruler in Central Europe, the progress that Hesse had made in the 1550s in military, technical, and cultural spheres.[47] The passage of the two electors through Hessian lands would offer the opportunity to demonstrate that Hesse had recovered from defeat and was once again a political force to be reckoned with. Philipp had numbered among the leading Protestant rulers until he was routed in battle against the Catholic emperor, Charles V, in 1547, forced to agree to have nearly all Hessian fortifications dismantled, and imprisoned for many years under humiliating conditions. *Baldewein I* was to symbolize the Hessian re-emergence as much as the newly erected fortifications, the enlarged Renaissance palace in Kassel, and the waterworks in the princely pleasure garden, all intended to be shown during the visit.

The significance of the court seat in Kassel as a centre of astronomy was proclaimed to the queen of France and the learned world in 1567 by the French philosopher Petrus Ramus: Wilhelm, wrote Ramus, had transplanted ancient Alexandria to Kassel, for like Ptolemy before him, the landgrave had acquired outstanding astronomical instruments and was taking pleasure in using them daily to observe the heavens. In particular, Ramus praised *Baldewein II*, then under construction, as an *automaton astrarium*, clearly referencing de' Dondi.[48]

Much previous historical writing has seen the celestial measurements at the Kassel observatory and the construction at the Kassel court of mechanical models of the heavens as two different means by which Wilhelm pursued his interest in astronomy.[49] In fact, they were inextricably linked and served Wilhelm's aspiration to be the acknowledged expert among the princes of the Empire for all matters of astronomy and astrology, one who could not only measure the course of the stars, but also could interpret every kind of celestial appearance.[50] His clockwork-driven celestial clocks and globes served as the most important vehicle for Wilhelm's propaganda. For, unlike on all other such Renaissance machines, the positions of the most prominent fixed stars were not depicted on the basis of received measurements but rather drew

---

43 Giovio 1552, 374.
44 The description of this image in Lloyd 1958, 46, 53, & pl. 45, is misleading because the tooth counts given in the manuscript do not coincide exactly with those used in *Baldewein I*. It is also uncertain that Andreas Schoener was the author of the underlying manuscript. The correction at the right can be securely ascribed to Wilhelm.
45 Rothmann 1589, 71.

46 Hallo 1930, 665. Previous literature has given 1559–62 as the period of manufacture. This letter shows, however, that parts of *Baldewein I* (explicitly the **meridian** ring, the rete, and a celestial globe) were already in the final stages of production by 1558.
47 STA Marburg, III (Philipps Söhne), 2802, 66r. Brief Philipp an August, 1 March 1558.
48 Ramus 1567, ii, 287–8. STA Marburg, 4a 31–3, Wilhelm to Baldewein 18 March 1567 (attachment).
49 E.g. Hamel 1998, 42–3.
50 Gaulke 2010, 109–22. Wilhelm largely undertook the astronomical measurements himself until the 1570s, and then directed them until the end of his life.

on positions that had been determined with the landgrave's own instruments—and in part by himself.[51]

This already appears as a leitmotif during the manufacture of the two planetary clocks. For instance, in 1561, Baldewein wrote that only now that the 'new observations' had been sent to him could the goldsmith Hermann Diepel finally begin to engrave the celestial globe.[52] Five years later, as the globe for August's clock was to be finished, Wilhelm sent Baldewein a table with newly measured positions of the fixed stars and insisted that the stars be engraved in his (Wilhelm's) very presence, writing that even Baldewein could not fathom how 'ferociously important' this was to him.[53] A programmatic inscription on the 1582 clockwork-driven celestial globe by Jost Bürgi reveals why the visibility of the new stellar positions was so central to Wilhelm: 'After Landgrave Wilhelm, ruler of the Hessians, observed that the times of the stars did not match their [recorded] positions, he set the stars again in their proper place, down to the degree and minute. The true measure of the hour, the motion of the heavens, was thereby'.[54]

The message is clear: *this* wise ruler was not only able to put terrestrial affairs in order, as other sovereigns did, but to right the heavens, thanks to measurements which were precise down to minutes of arc. Wilhelm's astronomical observations should therefore be seen less in a modern sense as a research programme aiming to reform a discipline than as the gesture of a ruler embodied in such cunning mechanical devices. Observation and instrument-making thus merge into what might properly be called 'princely astronomy'. That Wilhelm's strategy of self-fashioning was ultimately successful can be seen by the fact that other princes of the Empire, even those far above him in rank, asked *him* to interpret the astrological meaning of the impressive nova of 1572 and the bright comets of 1577 and 1585, as well as to assess the proposed calendar reform of 1582.

In the two decades following the completion of *Baldewein II*, Wilhelm had his clockmakers primarily make freestanding clockwork-driven celestial globes, which, like the globes topping *Baldewein I/II*, showed both the uniform daily rotation of the stars and the true annual course of the Sun in the ecliptic. From the mid-1580s, however, there are hints that Wilhelm was increasingly interested in the construction of complex cosmological models. Ursus, later to become Imperial mathematician, visited the landgrave in 1586 and noted that Wilhelm was occupied with building a celestial globe, inside of which a system of planets was to be mounted as a true representation of the cosmos. At this point, Ursus noted, Wilhelm could not decide whether the system was to be Copernican or Ptolemaic. In the former, Wilhelm was troubled by the requisite double movement of the Earth (about its axis and the Sun) as well as the vast empty space between Saturn and the fixed stars (inferred from the lack of any observable stellar parallax); in the latter, Ptolemy's multitude and intricacy of eccentrics and epicycles, their lack of harmony and proportion, displeased him.[55] Although it is not known whether Wilhelm had this globe completed, the idea alone marks a milestone, as it points away from the mechanized mathematical models of planetary positions (as in *Baldewein I/II*) towards a category of instruments developed in the late seventeenth and eighteenth centuries known as 'orreries'.

Of equal importance is that the construction of such models evidently offered Wilhelm the possibility to test whether a theoretically formulated cosmological model could at all function mechanically—a further illustration of how productive an engagement with 'constructability' could be, to use Shank's phrase.[56] After Ursus handed him a rough sketch of a hybrid heliocentric system working model. The landgrave ordered Bürgi to build a working model, about which we learn that 'after this was completed and properly functioned, on account of [Bürgi's] skill, His Highness saw it and was moved to astonishment'. Having such contrivances built allowed Wilhelm to test the reality of various cosmological hypotheses; he proceeded in exactly the same manner with the geo-heliocentric system of Tycho Brahe and with the lesser-known heliocentric system of Rothmann, in which both the Earth and Sun were immovable. A model of the latter is preserved and ascribed to Jost Bürgi; it is a table clock with heliocentric indications for the planets and a separate lunar dial.[57] Not surprisingly, the landgrave never specified which worldview he favoured, as a false choice would have marred his halo of unsurpassed wisdom and severely damaged his reputation.

Wilhelm's idea of creating a globe as a celestial sphere containing a system of planets was unknowingly also picked up by the young Kepler in 1598. Two years before, in his first publication, the *Mysterium Cosmographicum*, Kepler had proposed a heliocentric system in which the number of planets and their relative distances from the Sun could be determined by a clever nested sequence of the five Platonic solids. The discovery of how well this scheme matched the observed orbital periods of the planets intensified Kepler's conviction, already emerging during his university studies, that the heliocentric world view was not just a convenient hypothesis without claim to reality, but an *imitatio naturae*.[58] Convinced of the importance of his discovery, Kepler sought to construct a clockwork-driven, three-dimensional model of this

---

51 Gaulke & Beck 2019, 114.
52 STA Marburg, 4a 31–3, Baldewein to Wilhelm 25 January 1562.
53 STA Marburg, 4a 31–3, Wilhelm to Baldewein 17 December 1566.
54 CNAM inv. 7490; Latin text in Leopold 1986, 127. Mesnage 1949.
55 Leopold 1986, 188 (esp. n. 20).
56 Shank 2007, 5. Indeed, this is a unifying feature of various attempts at 'mechanical thinking' listed in Table IX and Table X.
57 Kunsthistorisches Museum, Vienna, inv. KK 846. Oestmann 2019, 42.
58 Kepler Gesammelte Werke, Vol. 13, Letter 99 (Kepler to Mästlin, Graz 1/11 May 1598), 228.

cosmological system for his sovereign, Duke Friedrich I of Wurttemberg. Inside a crystal sphere for the fixed stars the planets with their true movements around the Sun were to be shown; the mean radii of their orbits were to correspond with those Kepler had found earlier using nested Platonic solids. In a remarkable letter to his teacher, Michael Maestlin, of mathematics at Tübingen, Kepler criticizes all the Renaissance astronomical clocks of which he is aware, noting that they are overburdened with ornament and that their hands show on various dials the movement of the planets in the zodiac, but that these have nothing to do with the true cosmos, for 'it is obvious that the heavens are not run by hands, as on a clock'.[59] In other words, Kepler distances himself from the conception of planetary clocks at the heart of this chapter in favour of cosmological models serving a decidedly pedagogical purpose. This viewpoint was picked up by Wilhelm Schickard (1592–1635), Christiaan Huygens (1620–95), and later the makers of orreries, who sought to represent the order restored to a world now enriched, yet made a bit unruly, by the telescopic revelation of the moons of Jupiter and Saturn. With the discovery of such new worlds, the question as to the true nature of the cosmos became ever more pressing and the era of *astraria* came to an end.

---

59 Ibid, 223.

SECTION TWO: NINETEENTH AND TWENTIETH CENTURIES

Denis Roegel

INTRODUCTION

After the great astronomical clocks of the fourteenth and fifteenth centuries, and the remarkable court masterpieces of the Renaissance, activity slackened until the mid-eighteenth century when Bernardo Facini (1665–1731), Johannes Klein (1684–1762), Johann Georg Neßtfell (1694–1762), Claude-Siméon Passemant (1702–69), Francesco Borghesi (1723–1802), David Ruetschmann (David a Sancto Cajetano, 1726–96), Michael Fras (Aurelius a San Daniele, 1728–82), Philipp Matthäus Hahn (1739–90), and many lesser figures produced new and original works.[60] Some of the clocks they built had intricate features and often planetary displays, sometimes even including the satellites of Jupiter and Saturn.[61] In the nineteenth and twentieth centuries, new astronomical clocks were produced, and there were technical advances.

HOW ASTRONOMICAL CLOCKS EVOLVED

The study of astronomical clocks reveals a continuous evolution as the known universe expanded. Almost always information was shown on circular scales. As clockmakers learned more about the world, they wanted to depict it in genuine models of the cosmos. Up to 1780, however, the solar system, if displayed, went no further than Saturn. The more recent planetary clocks sometimes added the satellites of Jupiter or Saturn, and sometimes tried to provide a more accurate display of the irregular motion of the planets using different planes and with different speeds.

But in 1781 the limits of the solar system were ruptured. On 13 March, William Herschel discovered Uranus, the seventh planet from the Sun. Smaller 'planets' followed during the first years of the nineteenth century with the discovery of Ceres by Piazzi in 1801, of Pallas by Olbers in 1802, of Juno by Harding in 1804, of Vesta, again by Olbers in 1807, and many others in the second half of the nineteenth century. The eighth planet, Neptune, was discovered in 1846 by Galle using Le Verrier's predictions. Little by little, the astronomical universe became richer and provided more work for astronomical clock makers. Meanwhile, independently of the contents of the universe, there was progress in optical measurement, in celestial mechanics, with the great works of Laplace and others, which eventually provided a more accurate picture of the heavens, and in particular of its periods. At the same time, clockmakers were striving to find better gear ratios for the complex motions they wished to represent, and they also tried to find simpler and more accurate solutions to some difficult problems, such as the uneven motion of the planets. All this went in tandem with progress in mechanics, particularly in gear theory.[62]

The calendar also proved to be a challenge. Clocks had sometimes been made to take account of the varying lengths of months, but leap years often required manual corrections. Many astronomical clocks displayed elements of the *computus*, useful for determining Easter, but these clocks could seldom account for **secular years**. Some clockmakers tried to implement the Gregorian calendar more fully, and this culminated with nineteenth-century breakthroughs concerning the ecclesiastical calendar. But perhaps of greater importance was a changing society, in which people and ideas could travel more and more easily. In the mid-nineteenth century, people could cross Europe by train, and journals, sometimes illustrated, could bring marvels from around the world to all corners of a single country. The incentive for innovation was thus sometimes much more due to ideas than to the technology itself. And these ideas were often made accessible in world fairs, especially after 1850. It was not uncommon to display a clock first at a fair, and only afterwards send it to its final destination.

THE GREAT MILESTONES OF PUBLIC ASTRONOMICAL CLOCKS

The history of astronomical clocks is first a history of clockmakers. But the clockmakers who made astronomical clocks usually only made a very small number of them, and one or two of their clocks usually stand out. Moreover, the same ideas and techniques were often used by a clockmaker in several of his clocks, so that we are absolved from listing all the astronomical clocks by each and every clockmaker.

The eighteenth-century clocks mentioned above had little influence on other clockmakers, because they were only locally known, and their authors, generally, did not publish detailed descriptions of the gearing. The nineteenth-century revival was one caused by independent minds, but minds whose work reached

---

60 For fuller surveys of astronomical clocks, see Lloyd 1958, King 1978, Lehr 1981, and Glaser 1990.
61 Oechslin 1996.
62 A good survey of gearing intricacies for astronomical purposes is White 2017.

farther than those of their predecessors. The first major figure is Antide Janvier (1751–1835), who built an astronomical sphere at a young age and went on to specialize in the construction of complex clocks, as well as astronomical clocks, some of them displaying the times of tides and other celestial features. His most famous, and complex, clock, the 'sphère mouvante', is partly described by Berthoud.[63] Janvier's 'horloge à sphère mouvante et à planisphères' was started in 1789 and completed in 1801.[64] This was not Janvier's first clock with a moving sphere, as he had actually constructed a dozen of them before. But the 1801 clock was far more complex, and in 1802 Janvier obtained a gold medal at the third exhibition of French industry in Paris. Janvier wanted to build the most complex such clock ever (Figure 107).

The clock has four levels: the pedestal, a square-sectioned column with four dials, a glass cylinder containing the movement of the clock, and topping everything, the celestial sphere. The movement of the clock is a marine clock based on Berthoud's principles. It contains a **balance wheel** with a half-period of one second, and the seconds hand advances every two seconds. The clock is spring driven and the varying force exerted is equalized by a **fusee**. The motion of the movement is transferred both upwards to the celestial sphere and down to the four main dials. The sphere is an armillary sphere with a horizontal zodiac strip and it is enclosed in a glass globe. The planets from Mercury to Uranus rotate inside the sphere and around the Sun, which is set in its centre. The upper part of the movement cylinder contains the main gears for the celestial sphere and drives the planets using concentric tubes. Uranus is mounted on a ring fixed on the outer tube. Then come Saturn, Jupiter, Mars, and down to Mercury. The Earth is fixed on a rectangular frame which performs one turn in one year. Inside this frame, the Earth is given its obliquity and keeps the same orientation throughout its revolution. That is not to say that the precession of the equinoxes is not considered, but the revolution periods are the tropical ones, and the signs of the zodiac are those of the moving zodiac, not those of the 'fixed' stars. Above the Earth frame, there is a fixed wheel of 235 teeth, around which turns a pinion of nineteen leaves. This pinion carries the Moon which then rotates around the Earth in about one synodic month. Two other wheels are used to show the position of the Moon in its orbit and with respect to its nodes. Finally, the Moon is moved up or down, depending on its latitude on the ecliptic.

Beneath the movement there are four faces, each with two dials. The first side shows mean time, solar time, the day, and the month (the latter two on a large ring of which only a fragment is visible at any one time). According to Janvier's description, leap years are taken into account. The solar time is shown with a special hand, read on an oscillating dial. Still on this first side a dial shows the motions of Mercury and Venus as seen from the Earth. On this quite unusual dial, the Earth and the Sun are stationary. Mercury rotates around the Sun in 116 days and Venus in 584 days, to compare with the orbital periods of 88 and 225 days. The mechanism behind this face is spring driven and it drives the other sides.

The second side shows the motions of the satellites of Jupiter, the longitude of Jupiter, its parallax, the day of the week, and the hour of passage of Jupiter on the meridian. The lower dial also shows the day of the week and even subdivides the days with the planetary symbols, M (for Morning), XII (for Noon), and S (for evening).

Although these two sides are new designs, the technology behind them is fairly simple, with a zodiac rotating with Jupiter's orbital period, and with an oscillating thread moved by an eccentric **cam** for its position as seen from the Earth. No correction is given for Jupiter's satellites.

The third face is certainly the most remarkable, not because of its dials, but for how the indications are obtained. The main dial of this face shows a view of the plane of the equator with the Earth seen from the north pole. Three hands show the positions of the Sun, the Moon, and the shadow of the Earth. The outer ring depicts the signs of the zodiac. The Earth rotates in one sidereal day. Moreover, an intermediate section shows the positions of the lunar nodes and their precession over 18.6 years. What makes this dial so particular is that the Sun and the Moon are shown with their true, and not their mean, motions. These features are obtained by the use of a number of tilted wheels that help project a motion set in the ecliptic into the equator. This principle, described by Berthoud, certainly influenced Schwilgué in his reconstruction of the Strasbourg clock, as we discuss later. Moreover, Janvier made use of a crank mechanism in order to have the Moon's distance vary. Schwilgué would later use a special cam for that purpose. A second dial on the third face shows solar time, the days of the month, and the months. The lengths of the months are taken into account, including leap years, but not the special case of secular years. Finally, the fourth side is mainly devoted to the motion of the Moon and its effect on the tides. The times for the high tide are given for sixty ports and the displays are based on cams approximating the influences of the Sun and the Moon.

Jean-Baptiste Schwilgué (1776–1856) was another major and influential figure. Born in Strasbourg, at an early age he became fascinated by the then-motionless astronomical clock in Strasbourg cathedral. A first clock had been built in the fourteenth century; it was replaced in the sixteenth century, but, badly maintained, it had ceased functioning by the late 1780s. Family folklore tells us that when Schwilgué was about ten years old, he told a guard of the cathedral that he would one day make it work. After learning the watchmaker's trade and opening his own workshop in 1796 in Sélestat, he went on to become controller of weights and measures, and professor of mathematics. In 1813, he built his first tower clock and little by little he moved his way closer to the Strasbourg clock. In 1816, he found a way to obtain the date of Easter mechanically, and in 1821 he built a small model exhibiting his solution. He was able to show this model (now lost)

---

63 Berthoud 1802, i, 207–41.
64 Augarde & Ronfort 1998; Hayard 2011; Janvier 2019.

Figure 107 Janvier's free-standing astronomical clock built between 1789 and 1801 (private collection). Photo: Courtesy of Alidade Ltd, London.

to Louis XVIII. Schwilgué's idea was to replace calendar tables, which periodically had to be recomputed and replaced, by a more perpetual contrivance that could be fitted into the Strasbourg clock (Figure 108). In 1827, he published a first project for renewing the great astronomical clock (already motionless for forty years), and in the same year he moved back to Strasbourg in order to organize a factory manufacturing balances and tower clocks. It was only in 1837 that Schwilgué obtained a contract for the reconstruction of the clock. It was built between 1838 and 1843, having been inaugurated at the end of 1842, slightly unfinished. There are two detailed published descriptions of the clock.[65]

Although Schwilgué had wished to reconstruct the astronomical clock in Strasbourg cathedral at an early age, he did not work entirely by himself. In fact, he particularly wanted to update the clock and make it a window into the scientific and technological knowledge of his time. His purpose was therefore not so much to embody a fanciful idea, but rather to showcase the best available possibilities. He read widely and was influenced by such classic writers as Lepaute, Berthoud, and Janvier; the latter's influence can be felt in the design of the 'apparent time' dial, which displays the irregular motions of the Sun and the Moon. In it, Schwilgué made use of a tilted wheel to project the motion of the Moon from its plane into the equator. Such a construction had been expounded by Berthoud in 1802, when he described Janvier's masterpiece.

Schwilgué's tilted wheel is not visible to the public (and neither are Janvier's tilted wheels), but he made two other parts apparent through glass windows, and these two components are at the same time the major innovations that should be credited to Schwilgué in this clock. The contract Schwilgué obtained in 1837, when he was sixty-one years old, was to do necessary repair work, not for constructing a new clock. But Schwilgué went his own way and decided to replace the entire works. The old mechanisms from the sixteenth century, as well as the rooster from the fourteenth century, were moved to storage, and Schwilgué put new works inside the sixteenth-century wood-covered stone structure. The clock is now famous, like those of Prague, Lyon, and others, for its complex indications and, above all, for its automata. But Schwilgué was more interested in the astronomical and calendrical elements and they are the only ones described here.

Seen from the front, the clock has three major parts—a central one of several stories, 18m tall, a staircase on the right, and a column topped by a rooster on the left. The column is actually the shaft for the five main weights which are rewound manually once a week. The main works of the clock are located on the first level, behind the orrery. There, one finds the going train, with a one-second pendulum, the quarter-striking train, the hour-striking train, and a train for the motion of the figures of the four ages of life. Above these is another train for the motion of the apostles and the cock crow.

The motion of the going train is transferred to the orrery, slightly above it, and from there to the lunar sphere. The orrery depicts the mean motion of the planets from Mercury to Saturn and does not include Uranus. It is likely that Schwilgué meant to focus on the planets visible with the naked eye. This clock was the first large public clock with a Copernican system. The orrery also shows the motion of the Moon around the Earth. In common with the separate lunar sphere located above the orrery, it only shows the mean motion of the Moon, as there would be little point in showing minute variations at such a distance from the public.

The motion of the going work is also transferred to the ground level. From there, it reaches a complex structure—the 'mechanism of apparent time'. It also leads to the sidereal train, from which the motion is transferred to the celestial sphere at the front of the clock. This sphere shows the sky as seen from Strasbourg and it has intricate gear work that integrates the precession of the equinoxes. The celestial sphere, which represents the sky as seen from the outside, moves inside a cage that turns in one sidereal day. But inside the cage, the celestial sphere performs one rotation in 25806 years.

The 'mechanism of apparent time' is used to produce the motions of the Sun, of the Moon, using a special cam that lengthens or shortens the lunar hand. The purpose of this unusual feature is to ensure that the Moon goes over the Sun only when a solar eclipse occurs, and behind the shadow of the Earth (opposite the solar hand) only when a lunar eclipse occurs. Eclipses can only occur at conjunctions or oppositions, and only when the Moon is near the plane of the ecliptic. In order to display the eclipses with some accuracy, Schwilgué needed to have accurate positions of the Sun and the Moon, all in the plane of the equator. The manner in which he displayed this information was certainly influenced by Janvier's earlier solutions, although in some cases he differs from him. For instance, instead of using several tilted wheels or eccentric cams, as Janvier did, to obtain certain corrections, Schwilgué decided to provide a more general implementation that could be used to produce any correction which can be expressed as a sum of periodic functions. In order to obtain the accurate position of the Sun at the equator, Schwilgué needed to obtain true solar time, and therefore he needed to correct mean-time for the equation of time. And the equation of time could be expressed as a sum of several periodic functions. The details are rather more complex, but that is the general idea. Next, Schwilgué needed to have corrections for the Sun, for the Moon, and also for the hidden cam which adjusts the lunar hand. These corrections are obtained by a set of cams located in the window at the right of the clock. These cams tabulate various functions. They are driven from the sidereal time work and their sums are transferred to the 'mechanism of apparent time' where they are added to the mean motions. Eventually, the positions of the Sun and the

---

65 Ungerer 1922; Bach *et al.* 1992.

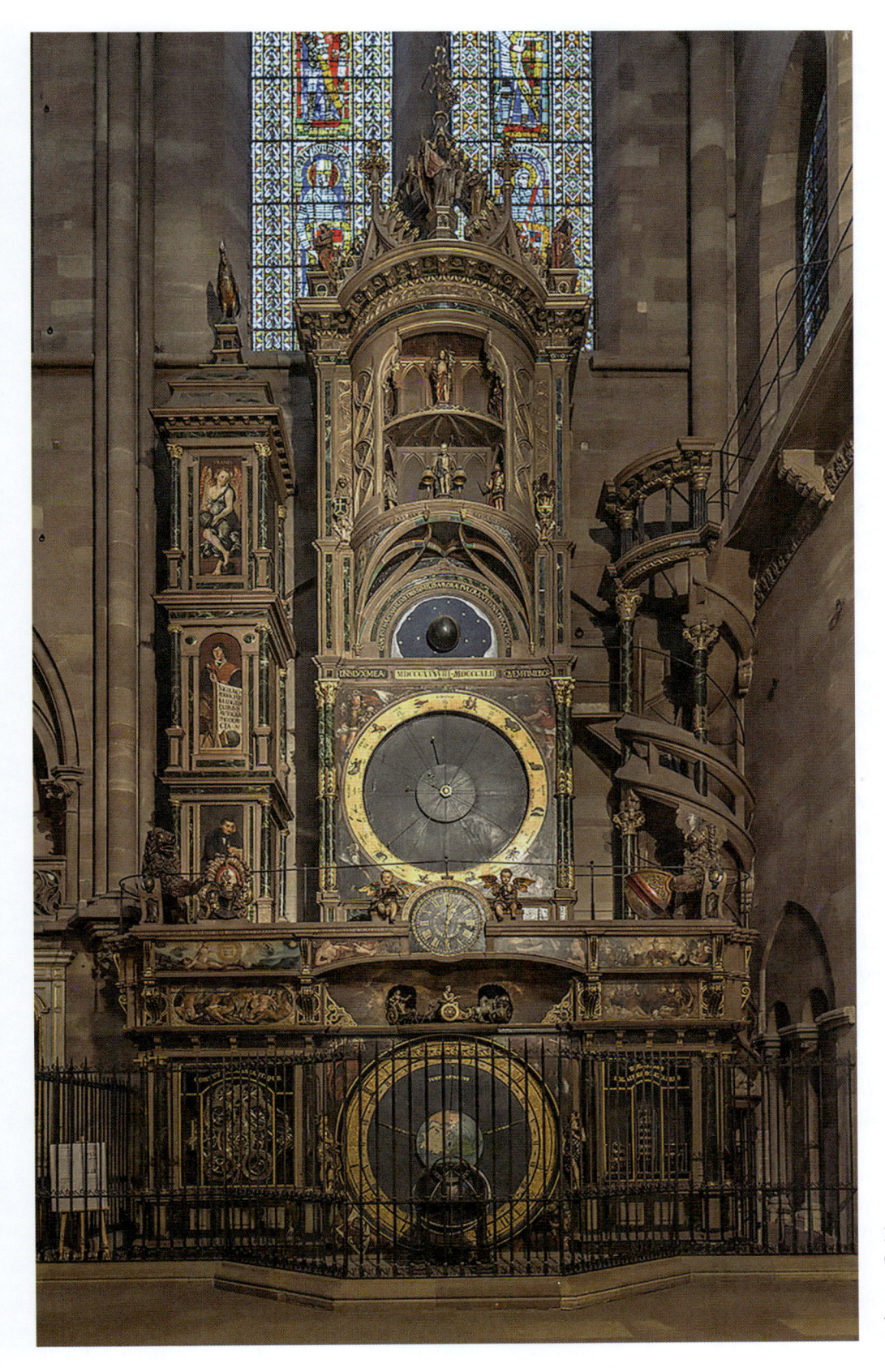

Figure 108 The Strasbourg astronomical clock. © David Iliff, Wikipedia. https://fr.wikipedia.org/wiki/Horloge_astronomique_de_Strasbourg#/media/Fichier:Strasbourg_Cathedral_Astronomical_Clock_-_Diliff.jpg.

Moon are obtained with a good accuracy. We have to stress, however, that the Strasbourg clock only displays eclipses in general and cannot take their precise location into account. When a solar eclipse is shown, it is likely that there is an eclipse somewhere on Earth, or a near eclipse, but there may be no eclipse visible from Strasbourg.

Schwilgué's clock also shows some calendrical data. The central dial has a ring with a slot for each day and this ring moves by one position each day around midnight. It also takes into account leap years, while another mobile ring marks the moveable feasts such as Easter. These are obtained from the mechanism in the left window and is summarily described later.

Once the cathedral clock was completed, Schwilgué continued constructing tower clocks (about 500, mostly in the East of France or in Germany), as well as making various mechanical instruments. After his death and his son's withdrawal, the factory was taken over by the Ungerer family.

Unlike several other clockmakers, Schwilgué only made one astronomical clock. In the 1840s, he had been contacted by Césaire Matthieu (1796–1875) Cardinal-bishop of Besançon who wished to have his own astronomical clock. Schwilgué had constructed a tower clock for one of Besançon's churches in 1846, but he declined to be involved with a new astronomical clock, probably because of his age. Matthieu then recruited Constant Flavien Bernardin (1819–1902), who had exhibited a complex astronomical clock in Paris in 1849. Bernardin built a new clock between 1850 and 1855, and it was exhibited at the 1855 universal exhibition in Paris.[66] The clock, however, did not please the bishop and in 1857 he asked Auguste-Lucien Vérité (1806–1887) to design another. Vérité was a well-known clockmaker from Beauvais, and he had invented new **escapements** and worked on synchronizing clocks with electricity. Matthieu may have met him at the 1855 exhibition. His first small astronomical clock was made for the château of Frocourt[67] in 1855 but is no longer extant.

The clock built by Vérité between 1858 and 1863 at Besançon probably gives a good idea of what Bernardin's clock looked like. Indeed, we know that the two clocks were about the same size, and that they had more or less the same dials, including one showing tides. Unfortunately, there is only a textual description of Bernardin's clock. Vérité may have borrowed the structure of Bernardin's clock and probably introduced his own mechanisms or improved those of Bernardin.

The Besançon clock is 5.80m tall, 2.5m wide, and 90cm deep and is located on the first level of the bell tower.[68] It has a **remontoire** escapement and a grid-iron pendulum. It has about seventy dials and is said to have 30,000 parts. There are in fact sixty-four dials on the case itself, to which one should add the orrery dial, the four dials of the bell tower, and four dials that used to be electrically connected to the clock. This is one of the rare astronomical clocks that also rings the bells of a church. The Strasbourg clock, for instance, is not linked to the cathedral bells, and neither is the Beauvais clock described later.

Seventeen dials give local time as well as the time in sixteen places around the world, such as Calcutta and Pékin. On the front, two dials display whether the year is a leap year or not. The first dial shows ordinary leap years, that is, every four years. The second dial shows which centuries end with a leap year and which do not, for eleven centuries. A larger dial at the centre of the clock shows a calendar for the entire year, with a fixed 29 February. The hand presumably jumps from 28 February to 1 March when the year is not a leap year. Five dials give the elements of the ecclesiastical **computus** and are advanced by one unit each new year. The date of Easter is not computed, but merely displayed in advance for nineteen years within the dial of the epact. The small medallions have to be replaced every nineteen years.

There are also dials showing the length of the day, of the night, and the equation of time. They are all obtained by cams and not by the complex mechanisms used by Janvier or Schwilgué. Two dials give the number of solar and lunar eclipses for each year, based on the Saros approximate period of eighteen years, eleven days, and eight hours. They offer a very bad approximation since the hands move in exactly eighteen years. The representation of eclipses was somewhat improved in the Beauvais clock, but it is still far from good. The Besançon clock also has an orrery that shows the motions of the planets up to Saturn, as well as the Moon. They all move with their mean motion. The most unusual dials are those showing the tides for eight French ports throughout the world. The times of high tides are apparently obtained each day by a mere offset, but the waves of the dials move by a combination of two oscillations whose regular motion is based on conical pendulums. Another dial located inside the cathedral shows the phase of the Moon, the day of the week, and the day planet.

After Besançon, Vérité was asked to construct an astronomical clock for his birthplace—Beauvais. The Beauvais clock (Figure 109), even larger than that of Besançon, was constructed between 1865 and 1868.[69] The case is about 12m tall and has many similarities with Besançon. In particular, there are many biblical automata figures, a rooster automaton, and various chimes. The clock is said to contain 90,000 parts, although no one seems to have checked this figure. Like the Besançon clock, it has a remontoire escapement and a grid-iron pendulum.

The clock has forty dials on the front and three dials on the left and right sides. The main dial on the front gives the mean or civil time (depending how it is set) for twenty-four hours. Two sets of nine smaller dials give the local time in nine cities east of Paris and of nine cities west of Paris. The clock in Besançon also shows the time in many places, a feature which does not add much complexity. Other clocks have sometime added similar profusions of almost identical dials, each one merely offset from the others by a constant value.

A set of dials surrounding a special dial giving the elements of the computus is placed under the large central dial. The values of the solar cycle, the dominical letter, the golden number, the epact and the indiction are given on concentric circles, and not on separate dials as at Besançon. There are five hands, and these advance by one position each year. The epact and the dominical letter are not perpetual: they had to be adjusted in 1900 and will have to be again when the sequences of epacts or dominical letters are interrupted at certain secular years. The hands are advanced on 1 January and allow for the secular leap years.

Around the computus dial, there are eleven other dials showing sidereal time, the equation of time, solar declination, length of

---

66 Bernardin 1855.
67 Six kilometres south of Beauvais, built in 1850, destroyed in 1980.
68 Vérité 1860.

69 Miclet 1977; de Mercey 2016.

Figure 109 The astronomical clock in Beauvais cathedral.© Chatsam, Wikipedia. https://fr.wikipedia.org/wiki/Fichier:Horloge_astronomique_de_Beauvais_09.JPG.

the day and night, the seasons, the signs of the zodiac, time of sunrise and sunset, the ruling planet for the day, and the day of the week. All of these indications are in fact very simple and some of them are obtained simply by using cams. The annual calendar is shown in two places. First, it is shown in the left panel around three smaller lunar dials displaying the phase of the Moon, the age of the Moon, the mean time of passage of the Moon at the meridian of Beauvais. The calendar is a large ring showing 366 days, including 29 February. The year is read with a long hand and the saints are shown. The hand skips 29 of February if the year is not a leap year, but years such as 1900 or 2100, which are not leap years, require manual correction.

The calendar is shown in a similar way in the right panel, this time around dials giving the age of the world, the current century, and the year. These dials also show whether the year is a leap year or if the current century ends with a leap year. This calendar has 366 days with a permanent 29 February, and also displays the saints but is used solely to indicate the date of Easter and other mobile feasts. The mobile feasts shown are all after February, so that the permanent 29 February is not a problem. Easter is determined using a special wheel which tabulates the dates of Easter over a period of 300 years. Such a wheel was not used in the Besançon clock, which only gives a list of nineteen Easter dates. It is astonishing that Vérité did not implement some more perpetual scheme, given that he must have known about that of the Strasbourg clock.

Finally, there are two dials, one showing the times of sunrise and sunset, the solar and civil times, and the other showing the phases of the Moon and its passage at the meridian of Beauvais. Some dials replicate indications given on other dials, albeit differently. On the left side of the clock, a dial shows the eclipses. The Earth sphere is shown at the centre of the dial with its inclination varying throughout the year. The Sun is in a fixed position above the Earth. A dial below the Earth shows if the Sun is above the equator or below, although for some reason the winter solstice is marked in boreal declinations. The Moon revolves around the Earth and can be put in the path of the Sun if, at the same time, a tabulated drum indicates that an eclipse should happen. This drum seems to give the eclipses for a period of 370 years. It is unclear if Vérité chose this value intentionally, as solar eclipses recur on average in the same place approximately in this period, but the period varies with latitude. It is also unfortunate that the list of eclipses was not published, as this may have been of interest to find Vérité's sources. In any case, although the representation of eclipses is better than in the Besançon clock, it is still not perpetual.

The left side also shows the tides at Mont-Saint-Michel, using moving waves as well as the time of high tide. The motion of the waves is obtained by a combination of oscillations and two conical pendulums, as at Besançon. Under the Mont-Saint-Michel dial, another dial shows the sky as it is visible at the antipodes of Beauvais, using a rotating planisphere.

On the right side of the clock, an orrery shows the heliocentric mean motions of the planets from Mercury to Saturn. Below, a second tide dial shows the tides in Mont-Orgueil. And finally, another planisphere shows the sky visible from Beauvais. Although the clocks of Besançon and Beauvais gained some fame, they also show that Vérité was no astronomer and that he apparently thought eclipses would recur after a certain time. Although the design of his clock improved between Besançon and Beauvais, it seems that he did not want to build a mechanism as complex as that created by Schwilgué in Strasbourg.

The brothers Albert (1813–79) and Auguste Théodore Ungerer (1822–85) were first employees, then the successors, of Schwilgué. They expanded the company's activities and installed many more tower clocks. It is estimated that between 1858 and 1989, when it closed, the Ungerer company installed about 6,000 tower clocks. The company also maintained the Strasbourg clock. Between the 1940s and the 1970s, Ungerer constructed several public astronomical mechanisms on the initiative of Henri Bach (1909–1991), its chief engineer. At the beginning of the 1930s, Ungerer also constructed a new large astronomical clock for the cathedral of Messina in Sicily. The cathedral had been destroyed during the 1908 earthquake and was then reconstructed.

The clock is heavily inspired by the Strasbourg clock and many mechanisms were almost directly copied.[70] Some of them were adapted or simplified, but there are also new mechanisms for the various automata. The technical side was chiefly the work of Ungerer's engineer, Frédéric Klinghammer (1908–2006), whereas the design is that of Théodore Ungerer (1894–1935). There are three astronomical components to the clock. First, the calendar dial which is divided in 368 parts, as in Strasbourg. The January/February sector is moved automatically to make 29 February appear when necessary, and the dial moves by one slot each day, moving correctly from 31 December to 1 January. However, the ring carrying the mobile feasts has to be set manually and is not set by a special mechanism, as in Strasbourg. It is interesting to observe that Klinghammer later constructed a model of the Strasbourg's computus,[71] thereby showing that he could have made one for Messina. The Messina calendar dial also shows the year under its centre. Above the calendar is the orrery, which replicates the structure in Strasbourg, but adds Uranus, Neptune, and Pluto. The Messina clock may well be the first astronomical clock to display the motion of Pluto. Finally, above the orrery, we find the lunar globe, also based on that of Strasbourg.

Some other smaller or lesser-known public, astronomical clocks include:

- In Ploërmel (France), Gabriel Morin (1812–76), known as Father Bernardin, built an astronomical clock between 1850 and 1855 as a teaching instrument. This clock shows the motions of the planets up to Uranus, as well as the four moons of Jupiter. It was extensively restored in 1920, so that most, if not all, of the gears have been replaced. The features of the original clock seem, however, to have been preserved, so that Morin's spirit is still alive in the current clock.
- In Lierre (Belgium), the clockmaker and astronomer Louis Zimmer (1888–1970) completed an astronomical clock in 1930 that was set in an old city tower, now named the Zimmer tower. The clock is made up of a central dial showing the time, surrounded by twelve dials showing the phase of the Moon, the golden number and the epact, the equation of time, the sign of the zodiac, the solar cycle and the dominical letter, the day of the week, a terrestrial globe, the month, the day of the month, the seasons, the tides, and the age of the Moon. There is also another clock, shown at the 1935 universal exhibition in Brussels, and now exhibited in an annex of the tower. It weighs two tons, and has ninety-three dials and fourteen automata, one of them showing the precession of the equinoxes.
- In Haguenau (France), Ulm (Germany), and Schramberg (Germany), the tower clockmaking company, Philipp Hörz, built three similar astrolabe clocks at the beginning of the twentieth century.
- In the Orly-West airport (Paris), the Ungerer company installed a mechanism in 1970, with an astrolabe and showing the motion of the Sun and the Moon around the Earth.
- In York Minster, the astronomer Robert d'Escourt Atkinson (1898–1982) installed an astronomical clock in 1955, as a memorial to the airmen killed in the Second World War. This clock contains a map of the sky, and a representation of the ecliptic partially hidden by the horizon. It also shows sidereal time (on three dials) and GMT (also on three dials).

RECONSTRUCTIONS OF EARLIER CLOCKS

Only very few public astronomical clocks have traversed time without alterations, damage, or repair. The following is a list of the main old public astronomical clocks that have undergone major or complete changes in recent times.

---

70 Redslob 1933; Ungerer 1934.
71 Flores *et al.* 2007.

The reconstruction of the Strasbourg clock is not included, because it differs significantly from other clocks in the list. The clocks reconstructed or modified in Prague, Lund, Olomouc, and elsewhere are all early clocks that have been restored using more or less modern mechanisms to produce approximations of the ancient dials. For instance, the new clock in Lübeck has a zodiac dial and a calendar dial and shows the motion of the Sun and the Moon. Some of these clocks also have automata, but there are no real 'modern' displays. In Strasbourg, on the other hand, the clock was modernized to such a point that, even though it is housed in an old case, it shows very complex and new features. This is not the case for any of the clocks listed here:

- Prague (Staroměstský orloj) (Czech Republic): the clock originally built in 1410 was subject to important transformations in 1866, and many parts were replaced, in particular the escapement.[72]
- Lund (Sweden): fifteenth-century clock, repaired in 1923.[73]
- Hampton Court (UK): the sixteenth-century mechanism was replaced in 1831 by a late-eighteenth-century mechanism and in the late nineteenth century by an even more recent movement.
- Olomouc (Olmütz): destroyed during the Second World War and replaced by a reconstruction, 1947–55.
- Münster (Germany): originally from the sixteenth century but given a new mechanism by Korfhage in 1929–32.
- Lübeck (Germany): new clock completed in 1967 by the clockmaker Paul Behrens, after the loss of the previous clock.
- Venice (Italy), Piazza San Marco: the eighteenth-century clock was modified in the nineteenth century and again in the 1990s by the clockmaker Alberto Gorla (born 1940) in order to restore (to a certain extent) the original eighteenth-century configuration.
- Macerata (Italy): the lost sixteenth-century clock was reimagined by Alberto Gorla and a new clock was installed in 2015; this clock displays the motion of planets in a geocentric manner certainly inspired by the reconstruction of Lorenzo della Volpaia's planetary clock, also made by Gorla, under the guidance of Giuseppe Brusa (1921–2011); the Volpaia clock has six dials for the various planets, but these dials could easily have been constructed on a single axis.[74]
- Mantova (Italy): the remains of the eighteenth-century clock were replaced in 1989 by Alberto Gorla, with a reconstruction of the fifteenth-century clock.
- Bourges (France): the astronomical clock in the cathedral of Bourges was designed in the early fifteenth century by the instrument maker Jean Fusoris; in the 1990s, a replica of the clock was built, and it is the one currently working, while the old mechanism is on public display.

---

72 Böhm 1866; Horský 1964.
73 Wåhlin 1923.
74 Brusa 1994.

SMALLER CLOCKS

Public astronomical clocks are outnumbered by smaller ones. Many clockmakers, and some amateur mechanics, spent many years of their lives constructing unique clocks. Some of these clocks were made of wood and others even use Meccano parts. Most of these clocks are undocumented and remain unanalysed in museum collections, in storage, in private collections, and some of them are of course lost. It would be pointless to try to list them all, as new clocks are regularly discovered, and as little is usually known of their details or location. A few emblematic and noteworthy cases, milestones among the smaller clocks, are discussed here.

Usually, astronomical clocks are unique pieces, and series production of several identical copies is uncommon. An exception, however, are the several tellurians made by Zacharie-Nicholas-Amé-Joseph Raingo (1775–1847) at the beginning of the nineteenth century. These are in the form of Doric rotunda, with a dial and pendulum to the front, and a tellurian on the top. This tellurian has the Sun at its centre and the Earth and Moon rotating around it. About thirty Raingo orrery clocks exist, with a number of variants. It seems that these tellurians were based on designs by Antide Janvier, and some were perhaps even made by Janvier.[75] Table astronomical clocks can be seen as teaching instruments and there have been other examples, for example, the 'Pendule cosmographique Mouret' (patented 1865), far simpler than Raingo's clocks in that it only has a large Earth globe that rotates in one solar day and also oscillates during the year.[76] More recently, in the 1960s, Athelstan Spilhaus (1911–98) designed an astronomical clock marketed as the 'Spilhaus Space Clock' (US patent 3248866). The clock was motor-driven and had plastic gears. And in the 1970s, the Ungerer company designed an astrolabe clock that it sold in many copies, having standardized its construction.

As mentioned earlier, some clocks were first exhibition clocks, before finding a final home. However, in certain cases such clocks were made only for exhibition. This was the case with Detouche's astronomical clock shown at the 1855 Paris universal exhibition (now in F. P. Journe's collection), Collin's clock at the same exhibition (now in the Grand Palais, Paris), Victor Fleury's astronomical regulator, probably exhibited at the 1867 exhibition in Paris, and a number of other more or less complex clocks. These exhibitions were, of course, opportunities for clockmakers to show their skill. After the 1840s, the new Strasbourg clock, as well as other public astronomical clocks, stimulated renewed interest. Considered marvels of the world, they were things for the modern traveller to see. Some of the clocks made after that time were meant to be models of the Strasbourg clock, and there are indeed several such models that are more or less faithful. These models usually only reproduced the general external features of

---

75 Hayard 2011, 349–51; Quill 1960; Chamberlain 1941, 274–8; Archives Nationales F12.2462/I/8 dossier Raingo.
76 *Revue chronométrique*, ix, 1876–7, 84–8; Turner 2007, 140–1.

the Strasbourg clock, or were inspired by it, without actually reproducing all the details. These were in fact unknown to most clockmakers before the publication of Ungerer's book in 1922. A model, such as that of Richard Bartholomew Smith (1862–1942) in the Powerhouse museum, Sydney, was constructed for the Australia Centenary in 1888 and Smith could only have seen engravings or read some general descriptions of the Strasbourg clock. It is remarkable that some clockmakers embarked on such copies without knowing the details of the works. But it also shows that the mechanical details were not that important to the public.

The tradition of Strasbourg clock replicas goes back as far as the sixteenth century. Isaac Habrecht, maker of the second Strasbourg clock, also constructed two models, one now in the British Museum, the other in Copenhagen. However, there is no exact copy of the Strasbourg clock. The original clock was more a source of inspiration and even when details of the movement were known, they were seldom replicated. For instance, around 1930 the Ungerer company made two smaller models of the clock, one which was in King Farouk's collection and the other in Philadelphia. In each the gears were simplified and sometimes even replaced by mock-ups. The makers of these imitations may be divided into two groups: those who, without knowledge of the details of the Strasbourg clock, tried more or less to imitate it, and those few who knew the details, but only borrowed some ideas or tried to do better. Several 'apostolic clocks' were made, for instance, one by John Fiester (1846–1921) between 1867 and 1878. Following in a long tradition he toured the clock in the United States as the 'ninth wonder of the world'.[77]

The Dane Jens Olsen (1872–1945) designed the complex astronomical clock located in Copenhagen city hall.[78] The clock (Figure 110) was completed in 1955 and is partly based on Schwilgué's clock in Strasbourg, which Olsen had seen at the end of the nineteenth century. Olsen's clock is one of a very few clocks that implement the computation of Easter following all the rules of the Gregorian calendar.

Olsen's clock is made of eleven trains laid out in three panels inside a glass case. All the dials are on one side, and all the mechanisms can be seen on the other. In a certain way, this layout was what Schwilgué had wished, since his initial project was to build a clock where all the works would be visible. Olsen's clock is weight driven and rewound manually. The central panel contains the main train with a **gravity escapement**. A large dial shows mean time and a smaller dial beneath shows sidereal time. Each of these dials has a separate smaller seconds dial. The elements of the computus are shown below and their work is heavily based on that of Schwilgué. The five dials of the computus give the dominical letter (in fourteen positions, as at Strasbourg), the epact, the solar cycle, the indiction. and the golden number. Below these five dials is a perpetual calendar, not circular, but with twelve columns, one for each month. At the end of February, a twenty-ninth day may appear if the year is a leap year. Next to the day numbers are strips with the names of the week, and these strips are also adjusted in accordance with the dominical letter. Finally, another set of strips shows the phases of the Moon, in accordance with the value of the epact. These moon phases are not the exact astronomical ones, but cyclic phases based on the mean motion of the Moon and the value of the epact. There may be a difference of one or two days between these moons and the actual Moon. The mobile feasts are not superimposed on the days but are actually part of the day-of-the-week strips. This means that there are two days for Easter (one for March and one for April), only one of which is visible at any one time.

The panel on the left shows the year, the day of the week, the day of the month and the month in seven openings. Three smaller dials placed inside a larger dial show true solar time (on a twenty-four-hour dial), mean local time (also on a twenty-four-hour dial), and the equation of time. A dial below this set shows a south-pole centred planisphere around which a ring with the hours turns. This makes it possible to read the mean time for any place on Earth. This planisphere extends up to the latitude of Scotland, and Denmark is near the edge of the map. Another dial shows the times of sunrise and sunset, in both solar time and mean-time. The panel on the right is symmetrical to the left one and also gives calendrical and astronomical information. At the bottom, a dial gives the Julian day and the number of years in the so-called Julian period, which starts on the fictitious 1 January 4713 BCE. Above the Julian period dial, three larger circular dials give astronomical data. The dial at the top shows a planispheric view of the sky, as can be seen on rotary maps for viewing the sky. Another dial shows the mean motions of the planets from Mercury to Neptune around the Sun against the zodiac. A final dial is similar to the 'apparent time dial' in the Strasbourg clock, and shows the motions of the Sun, the Moon, the lunar **apsides** (perigee and apogee), and the lunar nodes in the plane of the ecliptic, but not in the plane of the equator as in Strasbourg.

Olsen did not reuse all the features of Schwilgué's clock, perhaps because he did not know all the details. For instance, he did not make use of any tilted wheels. Although legend records Olsen hiding inside the Strasbourg clock to study it, nothing supports this claim. The first detailed description of Schwilgué's clock only appeared in 1922. However, Olsen could see most of the workings of the computus and the corrections to the motions of the Sun and the Moon, and this may have been enough for him to design his own clock. He did not, however, replicate the cams of Schwilgué's equations. Instead, he replaced each equation by a crank mechanism and was able to add the equations in a manner

---

77 Acquired by George Danner, Mannheim, PA, it was purchased with much else from this collection by Milton Hershey for his private museum in 1935, where it remains.

78 Mortensen 1957.

Figure 110 Jens Olsen's clock, Copenhagen. © Alphalphi, Wikipedia clock.
https://fr.wikipedia.org/wiki/Horloge_astronomique_de_Jens_Olsen#/media/Fichier:Jens_Olsens_front.jpg.

similar to that used in Lord Kelvin's tide-predicting machine of the 1870s.[79]

Another clock inspired by the Strasbourg clock is that of Daniel Vachey (1904–91). This French clockmaker built a small astronomical clock between 1938 and 1967,[80] which is now exhibited in the Musée Internationale d'Horlogerie in La Chaux-de-Fonds in Switzerland. It basically consists of three parts, with a central panel whose upper part has openings for various animated figures, including one with a boat for the tides. At the top is a rooster that crows at noon. The lower part of the central panel shows the time, the phase of the Moon (in a manner similar to that used in Strasbourg), and a dial for the mobile feasts linked with the date of Easter, but not Easter itself. The panel on the left contains the indications of the calendar, in particular, those of the computus. Except for the indiction, the layout follows that of Schwilgué's clock, and the mechanism is very similar. The date of Easter is given between the elements of the computus, on a dial numbered from 1 to 31, but serving for both March and April. A large dial in the middle of this panel shows the date of the month. Above it are dials with the day of the week, the planet of the day, the month, the type of the year (common or leap year), and the year itself, with its figures obtained from four concealed rings, as at Strasbourg. The calendar and the computus are perpetual and take into account the exceptions of the Gregorian calendar.

The panel on the right shows an orrery with the planets from Mercury to Saturn, as well as the Sun. Above it are two smaller dials, one for the declination of the Sun and the other for the equation of time. A larger dial shows a rotating map of the sky. Finally, four smaller dials give the times of sunrise and sunset, the time in the various time zones on Earth, the position of the Sun in stereographic projection, and the apparent motion of the Moon around the Earth. The clock works on springs that are electrically rewound.

The clock designed in the 1980s by Hans Lang (1924–2013) is a modern astronomical clock, now exhibited in the German Clock Museum in Furtwangen,[81] and it has some similarity with Olsen's clock. In fact, Lang was familiar with the clocks of both Schwilgué and Olsen. Lang's clock contains five panels (Figure 111). The central panel shows the official time and several other

---

79 For Kelvin (then William Thomson), see Tait 1983.
80 Verhoeven 2010–2012.

81 Lang 2004

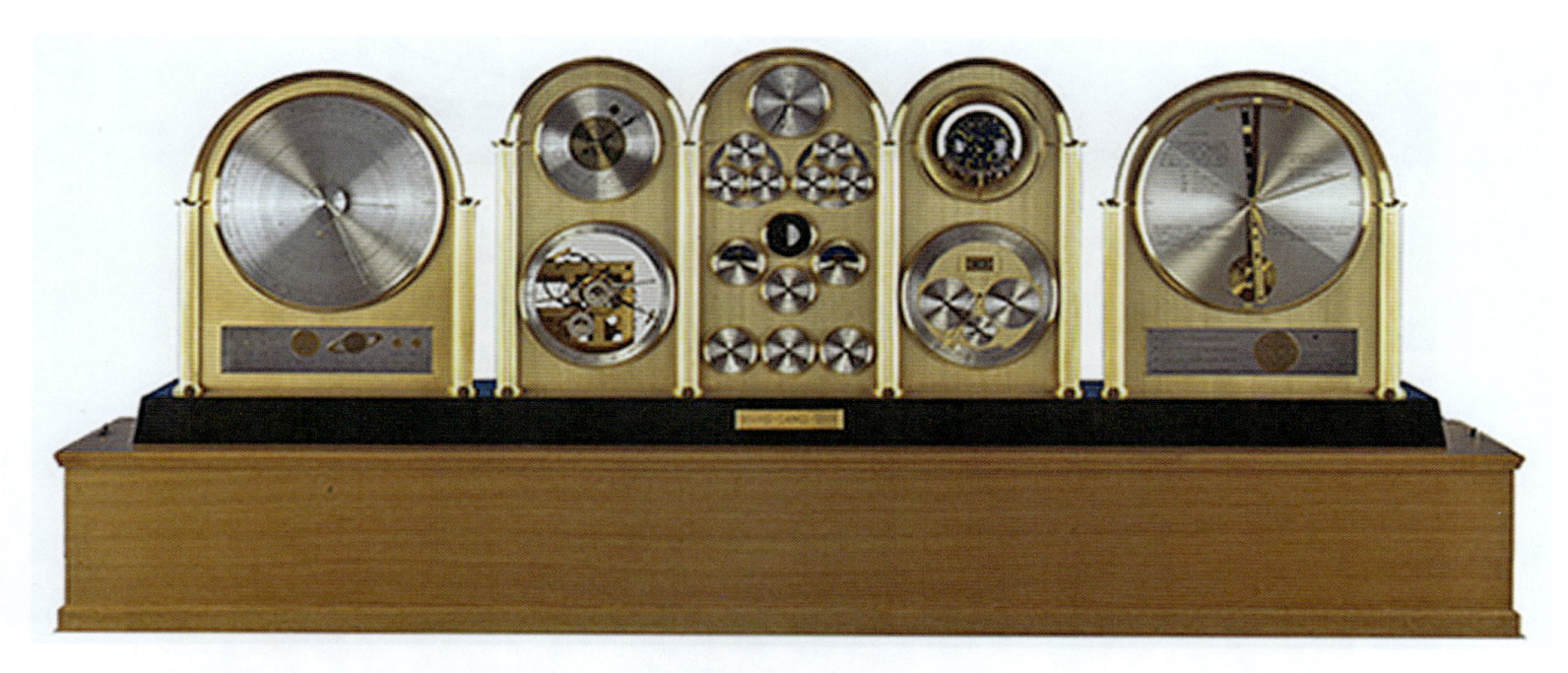

Figure 111 Hans Lang's clock in Furtwangen.© Deutsches Uhrenmuseum, Wikipedia: https://commons.wikimedia.org/wiki/File:Hans-Lang-Uhr.jpg.

times, including the sidereal time, the equation of time, the times of sunrise and sunset, as well as the age of the Moon. Lang went so far as to include a correction for the secular acceleration of the Moon, with a cam pre-computed for 10,000 years. However, it is possible that this cam exaggerates the effect of the acceleration, as Lang writes of a difference of 2.5 days in the year 12,000. It is also very interesting to note the use of mutilated wheels, apparently in order to reach certain specific ratios, and not for calendrical purposes.

The panel on the right of the central panel shows a celestial sphere that takes the precession of the equinoxes into account, and below it is a calendar dial with the year, the day of the month, the month, and the day of the week, all accommodating leap years and secular leap years. It also displays, somewhat surprisingly, the date in Elisabeth Achelis' 'world calendar' created in 1930.[82] Lang's purpose seems to have been to ensure that his clock will still be of use long into the future. The panel on the left of the central panel gives geocentric views of the motions of the Sun, the Moon, the line of apsides and the line of the **nodes**, as for Olsen's clock. The motions are shown in the ecliptic and the Sun and the Moon are always at the same distance from the centre of the dial, but there is a central disc showing the latitude of the Moon, useful for ascertaining a solar or lunar eclipse. Below this dial is another dial displaying the motions of Venus and Mars in a heliocentric perspective. This mechanism allows for the large (0.0934) eccentricity of the orbit of Mars, whereas that of Venus is considered circular. A panel located to the left shows the motion of the planets from Mercury to Neptune around the Sun. A fifth panel located to the right shows the geocentric motion of Jupiter's satellites and is reminiscent of a similar dial in Janvier's masterpiece.

The so-called Türler clock is a complex astronomical clock imagined by Franz Türler, designed by Ludwig Oechslin, and constructed by the clockmaker Jörg Spöring.[83] The clock was completed in 1995 and for twenty years it was exhibited in the Türler store in central Zürich. When the shop closed in 2017, the clock was removed to the Musée International d'Horlogerie. The clock aimed at being a model of the cosmos and is composed of five separate movements and dials mounted on a stone pedestal, in which the weights run, and is surmounted by a sphere. A first dial shows civil time and contains four smaller dials, one giving the day of the month, another the month, a third the day of the week, and a fourth the year. The year is given on four rings set in the same plane that advance slowly and whose reading needs some practice. These four rings themselves enclose a smaller seconds dial. There is a perpetual calendar, which allows for the lengths of the months and the leap years. A second side is an orrery showing the motion of the planets from Mercury to Pluto. Each planet turns around the Sun (absent) in a dial whose external border shows the signs of the zodiac. The centre of each planet moves circularly but carries a small disk with an arrow showing (approximately) the position of the planet, and above all introducing an epicyclic motion. For Pluto, this arrow is actually a long pin and Pluto may be closer to the Sun than Neptune for about twenty years of its orbital period, as was the case between 1979 and 1999.

---

82 For which, see Achelis 1937.

83 Türler *et al.* 2013.

A third side is the tellurian dial showing the motions of the Earth and the Moon around the Sun. The Sun's axis is tilted, and the Sun rotates. This dial can also be used to read the sign of the zodiac and the current day and month, although it may be off by one. A fourth side is the 'horizon' and shows a local south perspective with the motion of the Sun and the Moon at variable declinations. Above these four dials/sides is a globe made of six layers. The innermost part is the Earth sphere, which is followed by three glass shells, one carrying the Moon, another one carrying the Sun, and another one representing the sky, followed by the zodiac as a wire frame, and finally by a frame with the horizon. The Earth turns together with the outer horizon in one sidereal day. The zodiac is motionless. The Moon is slightly tilted on the ecliptic. And the celestial sphere makes one turn in 25,794 years.

Another noteworthy clock is that developed around 2000 by Hans Scheurenbrand (1938–2015) for the Festo company in Esslingen, Germany.[84] This clock has two main panels, one for the calendar and the other for the motion of the planets and the Moon. The calendar panel shows civil time, the day of the month, the day of the week, the month, the year, as well as the time in sixteen cities of the world. The calendar takes secular leap years into account. The astronomical panel of the Festo clock is a geocentric representation of the sky, with the Earth at the centre and the planets from Mercury to Saturn revolving around it. The sky is depicted in stereographic projection and the ecliptic appears as an eccentric circle, as on **astrolabes**. There are also hands for the Moon and the line of nodes. What makes this mechanism quite special is that the planets are shown with their irregularities, and Mercury and Venus for instance oscillate around the Sun. This clock was also in part inspired by Schwilgué's clock.

Mention should also be made of the several reconstructions of de' Dondi's astrarium[85] and of the reconstruction of Lorenzo della Volpaia's clock by Giuseppe Brusa and Alberto Gorla[86] noted above. Outside of Europe, there is also the 'myriad clock' made by Hisashige Tanaka in 1851, perhaps the most complex Japanese astronomical clock.[87] Finally, we mention the 'Clock of the Long Now';[88] this clock-in-progress will use a 300-pound titanium pendulum and the Sun will be used both to power the clock and to regulate it. 'The Long Now' clock features an orrery showing the planets from Mercury to Saturn, but the motions are produced by digital mechanical adders, not by gear ratios.

---

84 Scheurenbrand et al. 2005.
85 Starting in the 1960s, several reconstructions were made based on different versions of de' Dondi's manuscript by Luigi Pippa, Milan, Thwaites & Reed, London, and that which has the only claim to authenticity, because it is based on a critical edition of all the manuscripts by Emmanuel Poulle who supervised the reconstruction, now in the Paris observatory.
86 Brusa 1994.
87 This volume, Chapter 2, Section 2, and illustrated on the outer back wrapper.
88 Brand 1999.

TECHNOLOGY

It is interesting to examine and organize the clocks from the perspective of their functions, and to distinguish in particular uniform and non-uniform astronomical, and calendrical functions.

The most elementary astronomical clocks usually only display the motion of the Moon. Most advanced clocks show the motion of the Earth and the Moon around the Sun, sometimes with a tilted plane for the orbit of the Moon. Even more advanced clocks show the motions of the planets up to Saturn, and sometimes beyond. Although very diverse in the way the celestial objects are depicted, these displays all share some common features—for instance, that the objects almost always follow circular paths and have uniform motions. A clock displaying only the phase of the Moon on a dial may do so by using a disc with two Moons, of which only one is visible at any one time, and this disk may do one rotation in around fifty-nine days. A tellurian showing the motion of the Earth and the Moon around the Sun, such as those made by Raingo, actually has the Earth rotate at constant speed around the Sun, and the Moon rotate at constant speed around the Earth. Astronomical clocks showing the motion of the planets usually have these planets follow circular paths at regular intervals.

In all these cases, the basic astronomical function is to display some circular motion having a determined period. There is no fundamental mechanical difference between displaying the motion of a planet, that of the Moon, or that of an astrolabe dial. Generally, only the periods change. In order to achieve the desired results, a base motion is used, for instance, an arbor performing one turn in one hour, and this motion is geared down. The task of the maker is to find the best trains to display the motions adequately. We know since Antiquity that the celestial motions harbour non-uniform motions. For instance, the planets appear to have an erratic motion, and although they mostly move with regularity, there are times when they go backwards. Such non-uniform motions have been implemented as far back as the fourteenth century by Dondi, but the makers of astronomical clocks do not usually go to the trouble of displaying non-uniform motions for the planets. Non-uniform motion, however, insinuates itself in other forms. For instance, the times of sunrise, sunset, the lengths of days, and the equation of time are typical examples of non-uniform features. Most of the time, these features are obtained using appropriate cams. These motions usually have a period of one year and by using a cam of the appropriate shape revolving in one year, one can easily obtain the times of the given feature. These cams do not even need to be computed but can easily be obtained from tables in astronomical almanacs. The cams themselves move uniformly.

A few rare astronomical clocks try to provide a more faithful representation of the motions of the Sun and the Moon, for instance, in order to display the eclipses accurately. This is the case of the Strasbourg clock, which uses a number of cams to obtain accurate positions. The use of uneven gears in order to produce non-uniform motion, such as in Dondi's astrarium, does not seem to have been prevalent after him, cams being much simpler. The cams used in Strasbourg and elsewhere almost all move uniformly or are based on uniform motions. The general idea at work here is that some motion (not necessarily periodic) can be decomposed as a sum of periodic motions of various periods, following either the ideas of Joseph Fourier (1768–1830), or similar but more intuitive ideas about the composition of motion. The superimposition of several periodic motions goes back as far as Ptolemy's models for the motions of the planets, although, strictly speaking, these decompositions are not in a Fourier series.

Some astronomical clocks, such as Janvier's masterpiece and Vérité's great astronomical clocks, show the tides, or rather approximations of the times of high and low tides, for various places, and sometimes approximations of their intensity. These displays are also based on uniform input motions. However, Janvier's clock uses cams, whereas Vérité's clocks use compound circular motions to produce the equivalent of cams. As a good example of this duality, we can consider the equation of time. The usual way to obtain the equation of time is to use a kidney-shaped cam, and a roller on the cam moves an arm that eventually moves a hand. This cam had, of course, to be computed, or at least drawn, using tables of the equation of time. Around 1800, Janvier introduced a novel way for obtaining the equation of time, merely by implementing mechanically its underlying meaning, namely as a projection of the motion of the Sun in the plane of the equator.

Another example of this duality is the way Schwilgué and Olsen implement the irregular motion of the Moon. Schwilgué used cams tabulating the various equations necessary (the anomaly of the Moon, the evection, the variation, etc.), but Olsen dispensed with cams and used circular motions approximating the lunar irregularities, again much as in ancient astronomy. In fact, the originality of some displays is not always how the motion is produced, but how the information is displayed. Of course, the two are not totally distinct and different displays require different mechanisms. For instance, when Vérité displays the tides, he has waves going up or down depending on the intensity of tides.

Besides the time and the motion of celestial bodies, astronomical clocks often display calendrical information. Some of this information may be in form of tables or other non-moving data. For instance, the second Strasbourg clock gave the elements of the computus over a century. In fact, many eighteenth-century astronomical clocks have dials showing separately the elements of the computus. The basic calendrical information comprises the days, the months, and the years. But even simpler than that are the days of the week because they repeat without interruption. Such a display is readily implemented by a wheel making a seventh of a turn each day, perhaps triggered by another wheel making one turn in a day. The displays themselves can vary and one could have a hand on a dial, or an aperture showing the day of the week. In certain cases, the motion of the hand is even made uniform, and the position of the hand traverses an entire day in twenty-four hours.

Showing the day of the month—the date—is much more complex since the months are not of equal length. The most basic calendar clocks need to be adjusted at the end of each month other than when the month lasts thirty-one days. More elaborate clocks have a perpetual calendar and compensate for months of thirty days, for February's most common length of twenty-eight days, and for leap years. Calendrical clocks with compensation for the dropped leap years of the Gregorian calendar are excessively rare. Months, however, are no more difficult than days, as one merely has to link the passage to the first day with the passage to a new month. And similarly, changing years can be associated by moving to a new month of January. In certain clocks, the days, months, and years are simply rolled out. Taking again the Strasbourg clock, we have a large calendar dial at the centre that has a small division for every day of the year. In fact, it has 368 divisions. The month of January ends with 31 January, the month of March with 31 March, the month of April with 30 April, and so on. But February may end either with 28 February or with 29 February. This is made possible by using a mobile sector encompassing the months of January and February. Before each leap year and at the end of each leap year, this sector is automatically moved so as to make 29 February appear when otherwise it would be hidden. In the case of a leap year, there are two additional slots between 31 December and 1 January, hence the 368 divisions mentioned. The names of the months are written around the dial. Such a display dispenses with the need to have elaborate mechanisms for the end of the months, but it does require a special mechanism for advancing appropriately the dial on 1 January, as well as a mechanism for moving the mobile sector. These mechanisms allow for leap years, including the fact that 1900 was not a leap year.

Years could be displayed in the same fashion, with a hand moving on an appropriate dial. But the year can also be shown in an opening, using either a counting mechanism such as an odometer with coaxial drums, or, in the case of Strasbourg, coplanar concentric rings. The start of a new year both causes the day and month calendar to advance to 1 January, and the year to advance by one unit. The most sophisticated calendrical data that we considered here is that of the computus. Before 1842, astronomical clocks only gave the elements of the computus (golden number, epact, solar cycle, dominical letter, and indiction) in a recurrent manner. For instance, an eighteenth-century clock might have a ring or dial with the values of the epact for the eighteenth century: 0, 11, 22, 3, . . . , 7, 18, and then the same values would be repeated. Often, such dials would show at the same time the golden number and the epact, for instance, golden number 1 and epact 0, golden number 2 and epact 11, and so on, and there would be no way to shift one series from the other. Such a scheme could work for up to three centuries, but would then become invalid. For instance, as soon as 1900 was reached, the epact associated with the golden number was 29 and no longer 0, and the entire eighteenth-century dial would become useless.

The elements of the computus were useful to find the date of Easter, as the dominical letter determined the positions of Sunday and the epact determined the first full moon of spring. Easter was then set as the Sunday immediately after the first full moon of spring, the beginning of spring being set on 21 March in this definition. Hence, the date of Easter can take place between 22 March and 25 April. The Strasbourg clock was the first to automate the determination of Easter. Schwilgué built a prototype (now lost) of the mechanism in 1821. His idea was, in fact, rather simple. The mechanism is based on a large wheel making one turn at the end of each year. During this rotation, five other dials are updated. The dials of the golden number, of the solar cycle and of the indiction are of course straightforward, since they are merely advanced by one position. Things become slightly more complicated with the dominical letter and the epact. For instance, the dominical letter dial has a hand with fourteen different positions, two for each possible first day of January, one for common years and another for leap years. Usually this hand moves by two positions, but it will move by three positions when going to, or when leaving, a leap year. Epacts are even more complex, as some exceptions in secular years need to be considered, the epact being increased by ten, eleven, twelve, or thirteen units. Eventually, after these data have been obtained, two special arms are released and their positions depend on the values of the dominical letter and the epact, and hence on the place of Sunday and the place of the full moon. At the same time, there is a special ring on the central calendar dial carrying a small strip for Easter. The clock moves this strip up to 3 May. Then, the computus mechanism restores the initial position, and at the same time moves the date of Easter one day backwards. Eventually, this process puts the day of Easter on the right day, assuming the computus was initially correctly set. Although it is often claimed that Schwilgué's computus is an ancestor for computers, it is actually a machine that stores information and that updates it every year. The date of Easter is not computed *ab novo*.

CONCLUSION

It is unfortunate that for the vast majority of astronomical clocks detailed technical descriptions are lacking. Even such a clock as that at Besançon, the mechanism of which is almost entirely visible, and which went through several restorations, has not yet been described in detail. The Beauvais clock was restored in 2011, but the description subsequently published about it also lacks information that the researcher might want to have, such as details of the eclipse mechanism. Clocks are also sometimes included in museum catalogues without having been properly studied. Sometimes restorers write restoration reports, but without considering the needs of researchers, and oftentimes these reports are not made available to the public and sometimes not even to researchers.

In fact, it is likely that this chapter does not do justice to all the ingenious devices which have been developed, and some of the clocks not described may well be more interesting than some of those included here. Clockmakers and other amateurs have been full of imagination. In discussing astronomical clocks there has been far too much focus on the accuracy of gear ratios, when this is in fact not as important as one might think, particularly because almost every astronomical clock ends up being neglected and not correctly set.

Clocks should be described in their entirety, not merely by giving gear ratios and comparing them, anachronistically, with the modern celestial periods. Levers, cams, teeth shapes, dimensions, and in fact all the minute details that are useful in the construction of a clock should be recorded for a proper understanding. Perhaps this summary of what is known today will be an incentive for others to produce thorough technical descriptions of the many interesting astronomical clocks of which, at present, we know next to nothing.

CHAPTER TEN

MUSICAL AND AUTOMATON CLOCKS AND WATCHES

SOUND AND MOTION IN TIME-TELLING DEVICES

Sharon Kerman

From ancient times the act of telling time has been closely linked to motion, accompanied by sound. Sundials visually demonstrated the Sun's movement, providing a sensorial means of apprehending time's passage. Ancient water clocks were provided with doors that opened on the hour, animated figures, and spheres that produced sounds as they dropped into metal receptacles. The earliest tower clocks in Europe had no dials and thus relied exclusively on the acoustic indication of the hour; the first striking clocks seem to have appeared in Europe by the early fourteenth century. Sound and movement have thus always been intrinsically linked with timekeeping instruments, at first publicly announcing the hour, and in later centuries, in the private enjoyment of pleasant moments of entertainment.

On a practical level, mechanical timekeepers share the same basic requirements with automata and mechanical musical devices: a source of energy (such as air, liquid, weights, or a spring); a device to ensure the regularity of that power (such as an **air brake**, **fly-regulator**, or **speed governor**); a means of transmitting the energy to the moving parts (in mechanical musical and **automaton** pieces this takes the form of various rods, levers, valves, cams); and a defined series of actions to be carried out (such as a succession of gestures or a musical programme of notes to be played).[1]

The 'programme' generally took the form of a cylinder or barrel, fashioned first from wood and later from metal, to the surface of which a number of pins (also called pegs or studs) were securely fastened at right angles to its axis. As the motor made the cylinder rotate, the pins activated levers fastened to a key-frame, so causing organ pipes to open, to receive air produced and sent from a bellows mechanism to produce a note, or to close, thus ending the note. Similarly, they could activate the striking of bells in a carillon, or other musical mechanism. Other types of 'programme' include cams (brass discs whose circumference is cut out to reflect the desired movements of birds), the piston, and the bellows valve in singing birds, or the motions of the musician's hands in musical automata, such as those by Roentgen & Kintzing or Jaquet-Droz. Later still, the musical programme of mechanical musical instruments was contained in a simple perforated sheet or zigzag cardboard books.

In order for a melody to be pleasing the pins had to activate the right notes and guarantee they sounded for the exact length of time required. The correct placement of the pins in a barrel, cylinder, or disc—that is, the transcription of a piece of music to the language of mathematics and mechanics—was a delicate task. As Alexander Buchner noted:

> ... it is necessary to determine the exact point on its surface where the pegs of different shapes and sizes must be hammered in or pressed in with pliers, so as to engage with the levers controlling the pipes or strings of the instrument and thus sound notes of the required tone, sequence, and duration. The cylinder was made of many sections of absolutely seasoned wood, usually oak, which were firmly glued together to maintain perfect shape. The pins, which were originally of hard wood and later always of metal, varied in shape at different periods.

Engramelle, in his *La Tonotechnie ou l'Art de noter les cylindres* (1775), considered the crucial part of the cylinder pinner's art to be the

---

1 Haspels 1987, 27.

correct determination of the positions of the pegs on the cylinder surface so that they opened the valves at the right time and kept them open as long as the note was intended to sound.[2]

A handful of sixteenth-, seventeenth- and, eighteenth-century authors described the construction of mechanical organs and their cylinders, among them Salomon de Caus (1575?–1626) in his *Les raisons des forces mouvantes* of 1624; Athanasius Kircher (1602–1680) in his *Musurgia Universalis* of 1646; and François Bedos de Celles (1709–79) in his *L'art du facteur d'orgues* of 1766–78.

SURPRISE, ADMIRATION, AND AWE

Animated figures and their musical accompaniments were not just representations of passing time; they offered a fascinating and easily grasped spectacle that additionally conveyed cultural information and religious values. To a largely illiterate medieval populace, for example, the figures of the Three Wise Men bowing to the Virgin and Child were tangible embodiments of religious teachings. When a familiar and beloved tune rang out from a tower clock, audible for miles around, people were alerted to stop what they were doing and pay attention to the imminent striking of the hour; the knowledge of time was collective, shared by all who heard the bell.

To spectators, music and automaton action driven by a clockwork mechanism appeared at first glance to be independent, autonomous phenomena. Brought about by no human agent, they might easily be thought to be magical manifestations. In earlier times, the feelings of surprise and awe engendered by musical and automaton clocks and watches were among the most powerful reasons for the appeal of these 'enhanced' timepieces.[3] The English physician and mystical philosopher Robert Fludd (1574–1637) emphasized the importance of that effect of surprise:

> '... thus it becomes possible to have music without a musician or action of any living being: it will be splendid, graceful and an impressive marvel for those partaking, or in the presence of a festive meal, to hear unexpected music without the presence of any moving being, from some corner of the dining hall ... '.[4]

The feeling of astonishment described by Fludd continues to be operative even today, despite the fact that modern audiences are no strangers to machines, electronic devices, or even robots. The powerful, almost primeval, sense of awe that exists seemingly independently of the rational brain remains an essential component of the attraction of musical and automaton timepieces.

AUTOMATA AND SOUND IN EARLY WATER CLOCKS

Water-powered clocks were the earliest timepieces to be enhanced by sound and moving figures. Several important scholars exerted a lasting influence on the design and construction of such clocks, powered by water, air, and occasionally other materials, such as sand. Ctesibios (mid-third century BCE), a Hellenistic mechanician at Alexandria, was the author of the earliest treatises on the use of compressed air, or pneumatics. He invented a hydraulically powered pipe organ and made important improvements to clepsydras. Philo of Byzantium (end of third century BCE) wrote treatises on automata and water clocks. Hero of Alexandria (first century CE) enlarged on Philo's work, also studying pneumatics and automata. In the preface to his *Pneumatics*, Hero clearly states his desire to provoke astonishment and wonder, in addition to more utilitarian goals: '... By combining air, water, fire, and earth, and through the union of three, or even all four, of these elements, various compositions may be brought into play, some of which satisfy the most basic needs of human life, and others that provoke admiration and surprise'.[5] Both Philo and Hero described automaton 'theatres' in which the illusion of life was created by fluid powered mechanical scenes with animated figures; these were intended to induce a sense of awe bordering on the supernatural.[6]

Several impressive clepsydras are recorded in which sound and motion played an important part. The Christian rhetorician Procopius of Gaza (c.465–528 CE) described a monumental water clock that stood in the centre of Gaza, striking the hours loudly enough to be heard from a great distance. Prominently featured on its façade was a Medusa's head with eyes that rolled terrifyingly every hour, as well as a figure of the sun god Helios. There were twelve doors under the Medusa's head; every hour one of them opened and the figure of Hercules appeared, each time carrying the attribute of one of his twelve labours.[7]

Another very famous water clock, presented to Charlemagne around 807 by the Abbasid Caliph Hārūn al-Rashīd (763 or 766–809), incorporated opening doors and acoustic signals that marked the passing of the hours. Described in the contemporary *Frankish Annals* as being 'wondrously constructed of brass by mechanical art',[8] it remained legendary—so much so that centuries later it was described at length by John Gifford (1758–1818) in his *History of France*:

> 'The dial was composed of twelve small doors, which represented the division of the hours; each door opened at the hour it was intended to represent, and out of it came the same number of little balls, which fell one by one, at equal

---

2 Buchner 1992, 16–17.
3 Haspels 1987, 29.
4 Fludd 1617, 245; Haspels 1987, 29.

5 Argoud & Guillaumin 1997, 24; for a survey of the work of all three, see Drachmann 1948; Gille 1980.
6 Lebrère 2015, 31–53. This volume Chapter 3, Section 4.
7 Hill 1984, 232; Dohrn-Van Rossum 1996, 28.
8 Davies and Fouracre 2010, 133.

distances of time, on a brass drum. It might be told by the eye what hour it was by the number of doors that were open; and by the ear, by the number of balls that fell. When it was twelve o'clock twelve horsemen in miniature issued at the same time and, marching round the dial, shut all of the doors.'[9]

In the Far East, an impressive hydro-mechanical astronomical tower clock over thirty feet high was designed and constructed by the polymath scientist and poet Su Song (1020–1101 CE), who also wrote one of the most important Chinese horological treatises of the time.[10] Completed in 1094 in Kaifeng, the clock incorporated a rotating armillary sphere and a celestial globe. Each of its five storeys had a door from which automaton figures emerged ringing bells and displaying tablets indicating the time. The clock was powered by a water wheel eleven feet in diameter, with thirty-six scoops along its perimeter; a complex gear system linked the figures to the water wheel. Several years after Su Song's death, the city of Kaifeng was invaded by the Jurchens, who dismantled the clock and took its components back to their capital (now Beijing) in CE 1127. Less than a century later, however, the celebrated clock had disappeared.

The tradition of water clocks with animated figures and metal balls that dropped onto cymbals or drums was perpetuated and amplified by Persian and Arabic scholars, whose translations of the important scientific works of antiquity preserved them for future ages. The tradition was summed up in the work of the polymath engineer Ibn al-Razzāz al-Jazarī (1136–1206), chief engineer at the Artuklu Palace, the residence of the Artuqid rulers of eastern Anatolia during the eleventh and twelfth centuries. In 1206 al-Jazarī wrote his *Book of Knowledge of Ingenious Mechanical Devices*, which contained descriptions of many of his inventions and provided details on their construction. While many of the devices were practical items, such as clocks or waterwheels, a number included fanciful embellishments in the form of moving figures. Al-Jazarī's Castle Clock was an eleven-foot-tall hydraulic clock with animations that included the orbit of the Moon, the movements of the Sun, and the signs of the zodiac. On the hour, the corresponding number of doors opened automatically, and bronze balls released from the beaks of gold falcons fell into copper vases. Three times a day water flowed from a reservoir into a wheel with an attached camshaft, causing five mechanical musicians to begin playing their instruments.[11]

Al-Jazarī was influenced by the work of the brothers Muhammad, Ahmad, and al-Hasan Banū Mūsà ibn Shākir, ninth-century scholars in Baghdad. While the Banū Mūsà did not describe any musical clocks, the inventions in their *Book on the Description of the Instrument which Sounds by Itself*, for example, that of a mechanical organ powered by weights and water, served as inspiration to later inventors. Among their designs was a revolving pinned cylinder that provided the music for an automaton flute player.[12]

EUROPEAN CLOCK TOWER CARILLONS

In medieval Europe clocks were installed in high towers, with bells that could be rung not just to indicate the passing hours, but also to call people to work, to inform them of important gatherings, or to sound the alarm in case of danger.[13] Since the earliest public clocks had no dials, people initially learned the time acoustically, the hours being sounded by human time-announcers either manually striking a bell, or by giving a trumpet blast after having climbed to the top of the tall towers to do so. These towers became beloved landmarks and, in the case of civic belfries rather than those belonging to the Church, symbols of the independence and self-determination of the towns that possessed them. The addition of a carillon, considered one of the essential attributes of a progressive city, was a further source of pride, furnishing proof of the town's prestige and cultural importance. Cities rivalled with each other to possess the finest carillon. The prestige of carillons was so great they were often treated as treasures that could be used for bargaining during negotiations, or as spoils of war. The large bell and striking automaton of the cathedral were seized after the looting of Courtrai by Philip II of Burgundy in 1382 and removed to his capital, Dijon, although the bell broke on the way and had to be recast. In 1711, the magistrates of Brussels emphasized both the status and the commercial importance of their carillon when they ordered that a new one be purchased to replace the instrument destroyed in a 1695 bombardment:

> 'It is for the honour of a court town like Brussels to have as one of its ornaments a perfect carillon which can serve not only for the satisfaction of the townspeople but also to give diversion to strangers who are often attracted to a town by the harmony of a carillon, which both adds to the town's renown and also increases its business.'[14]

Just as towns vied with each other to possess the finest carillons, they also hurried to mechanize them in order to keep up with the latest technical innovations. Automatically functioning carillons, a natural corollary to the mechanical clock movement itself,

---

9 Gifford 1793, 171.
10 Su Song 1090, Needham 1959, Needham et al. 1986. This volume Chapter 2, Section 2.
11 Hill 1974, 17–51.

12 Wiedemann 1909, 164–85; Hill 1998; this volume Chapter 3, Section 1. Other flute players, and earlier makers of them, are described by Jazarī, see Hill 1974, 170–6.
13 Dohrn-van Rossum 1996, 129–33.
14 Rice 1914, 119.

should have made human time-announcers redundant, but did not entirely do so. The Kraków announcer is active to this day.[15]

Since the sounding of the hour could easily take people unawares—or even be lost in the multiplicity of bell signals diffused in late medieval towns—by the late sixteenth century it had become a common practice to play a short tune before the hour as a warning that the bell was about to sound.[16] According to the twentieth century carillon specialist William Gorham Rice, '… this mechanism, striking the small bells just before the hour, announced that the heavy hour bell was about to sound'. It was not long before more than four bells were used, and as the number increased the mechanism was arranged to play a little tune. Thus we reach the eight or ten bells of the Flemish 'voorslag', or 'forestroke', obviously so-called from its play before the hour. To possess a 'voorslag' was an indication of municipal progress and the principal Flemish towns were soon thus equipped. Owing to this periodic playing, which before long preceded the strokes of the half-hour as well, bell music came to be a distinctive feature of the Low Countries. As prosperity increased and as taste developed, still more satisfactory musical effects were sought, and 'Bells were added to the *voorslag*; all the intervals of the chromatic scale were supplied; and the barrel of the playing device was enlarged until each quarter hour had its share of notes, and the hour tunes lasted a minute or more.'[17] Some examples of early clocks with tunes played automatically on bells are the Abbey of Sainte Catherine near Rouen (1321), which played the hymn *conditor alme siderum*;[18] Beauvais cathedral (1302–4); Strasbourg Cathedral (1352–54), and Toledo Cathedral, which had both bells and automata.[19] By the end of the fourteenth century, bell towers had become very popular in the Low Countries, with examples in Middelburg (1371) and Saint Rombold's tower in Mechlin (1372),[20] and indeed, throughout Early Modern Europe.

Like all mechanical musical instruments, mechanical carillons required a power source, a means of regulating that power, and a programme. In European tower clocks, power was furnished by weights, and the 'programme' was contained on a very large-diameter barrel that was provided with pegs that fitted snugly in its surface. As the cylinder turned, these pins activated levers that lifted the hammers, or clappers, that struck the bells. These large cylinders often had a grid traced on their surface to facilitate the correct placement of the pins. In some cylinders the iron pegs that fit into the holes could be moved, thus allowing new tunes to be programmed as desired, although the art of reprogramming—or repinning—mechanical musical cylinders could be accomplished only by specialists. By the seventeenth century the use of pinned cylinders had become so widespread that Gaspar Schott (1608–66) proposed a system of classification in his 1664 *Technica Curiosa*, a compendium of contemporary technology.[21] Schott proposed dividing the programmes of mechanical musical instruments into 'monomusi-immutabiles' (with a single melody that could not be altered); 'polymusi-immutabiles' (with more than one melody that could not be altered); and 'pantomusi-mutabiles' (reprogrammable cylinders with more than one melody).

MECHANICAL FIGURES IN CATHEDRAL CLOCKS

Animated figures were a vital component of monumental clocks. While they helped to focus attention on the passing of the hour, their more important function was as a form of sensorial storytelling, transmitting narratives both spiritual and secular. It is not difficult to imagine how animated displays of religious doctrine, astronomical indications, and music would have blended to create a unique and memorable experience for contemporary spectators. One illustration of this is a clock (destroyed during the French Revolution) that the Abbot Pierre de Chastellux had installed around 1340 in the Abbey of Cluny. It indicated:

> … the year, the month, the week, the day, the hour, and the minutes; an ecclesiastical calendar designated the holidays and times of services each day. The positions, oppositions, and conjunctions of the stars, the phases of the Moon, the movements of the Sun, were shown in this clock. The complicated mechanism allowed one to see, appearing in a niche one after the other, for each day of the week, the mystery of the Resurrection, the figure of Death, Saint Hugh, Saint Odilon, the Feast of the Holy Sacrament, the Passion, the Virgin. Each one, at midnight, ceded its place to the next. The hours were announced by a rooster that flapped its wings and crowed twice. At the same time an angel opened a door and welcomed the Virgin; the Holy Spirit, in the form of a dove, descended upon her head, and the Holy Father blessed her. All this took place amidst a harmonious carillon of little bells, and the bizarre movements of fantastic animals that moved their tongues and their eyes. The hour sounded, and all these figures retreated inside the clock'.[22]

Probably the most famous early automaton cathedral clock is that at Strasbourg. The current clock dates from 1843 and is the third rebuilding. The original clock, completed in 1354 by

---

15 So was the announcer at Dôle until the early twentieth century. For the fourteenth century time-trumpet of Strasbourg Cathedral, and that of Dôle, see Turner 1990, 100 N°s 184, 185.
16 Starmer *c.*1908, *passim*; Starmer 1904–5, 43–51; Starmer 1907–8, *passim*; Ord-Hume 1995, 61.
17 Rice 1914, 107.
18 Chéruel 1844, 148; Grisel & Bouquet 1643, 470; Dohrn-van Rossum 1996, 101.
19 Turner 2014, 929; Pérez Álvarez 2018, 65–103.
20 Rice 1914, 109.

21 Schott 1664, 685–9.
22 Lorain 1845, 203.

an unknown maker, incorporated several automaton figures. A rooster (which is the only extant component of the original clock) was re-employed in the second clock (finished in 1574) and is today the oldest known surviving automaton figure. Apart from being a classic symbol from the natural world of awakening and paying attention, the rooster was also a visual reference to the words Jesus spoke to Peter: ' . . . this night, before the cock crow thou shalt deny me thrice',[23] which perhaps explains its frequent appearance in medieval clocks. At noon in the first Strasbourg clock, the bird flapped its wings, opened its beak, stuck out its tongue, and crowed, the sound being activated by a bellows mechanism linked to a reed. Spectators of the day surely perceived this as an exhortation to lead a righteous life so as not to betray Christ. The original Strasbourg clock featured additional animated drama in the form of the Three Wise Men who paraded before the figures of the Virgin Mary and the Christ child, and bowed down to them, to the accompaniment of carillon music.

The first Strasbourg clock, in poor repair by the sixteenth century, was replaced by the second, designed by the mathematicians Christian Herlin (d. 1562) and Conrad Dasypodius (1531–1601), and built by the clockmakers Isaac (1544–1620) and Josias Habrecht (1552–75). Their work, which incorporated the original automaton rooster (it continued to function until struck by lightning in 1640) was completed in 1574. The clock included numerous astronomical indications, automaton representations of the days of the week (riding by in chariots), and the striking of the quarter hours by figures representing the Four Ages of Man—Infancy, Adolescence, Manhood, and Old Age. Dasypodius, who placed himself in the tradition of Hero of Alexandria, explained his use of these animated figures in a text examining both the work of his illustrious predecessor and his own:

> ' . . . since the magnificence of mechanical works is very much increased by extra adornments . . . we have . . . joined to our machine these four statues of the ages, using automata . . . it seemed to us to be necessary to add these for the sake of delight and wonderment, not however without a certain particular significance'.[24]

The animated scenes and music incorporated into church and cathedral clocks thus filled a dual function: they furnished edification while eliciting powerful emotional reactions. Figures of the Apostles, the Three Magi, the Virgin, and the Christ Child encouraged devotion, while reminders of mortality in the form of Father Time, a skeleton wielding a scythe, or the Four Ages of Man incited onlookers to lead a pious life before it was too late.

Musical automaton clocks were evidently such a familiar spectacle for his contemporaries that the fifteenth-century moralist and religious author Alexander Carpenter used the metaphor of a weight-driven musical clock with automata as a grim reminder of the transience of human life:

> The clock-keeper places one pin in a certain wheel, and when it reaches a certain point in the clock immediately the mechanism is released; and then all the bells strike and the figures in the semblance of clerks and priests pass by in procession chanting . . . After the weight has grounded everything immediately stops . . . So, spiritually, God who is guardian of man puts a pin in him, he ordains a limit to his life which he cannot pass . . . When his corpse is cast into the earth immediately the tumult ceases and the dead man passes into oblivion.[25]

JACKS AND JAQUEMARTS

Around the fifteenth century, as the humanist currents of the Renaissance shifted the focus from God to man, a new kind of automaton made its appearance in tower clocks. Perhaps a reminiscence of the human bell-ringers that had sounded the hours in the first turret clocks, these monumental personages depicted real people: everyday figures such as men-at-arms, shepherds, or simply outstanding citizens and their families. While their names varied: 'Jack' in England, 'Jaquemart' in France, 'Jan' in Flanders, and 'Hans' in Germanic countries, their task was always the same. At the appointed moment, these mechanical giants pivoted, hoisted their massive hammers, and struck (at least in appearance) the huge bells that announced time. Among the most famous surviving examples are the Jacks of the Saint Mark's clock tower in Venice. Installed in 1497 along with the bell, the huge Jacks were meant to symbolize the passage of time, for one is young and the other an old man with a hoary beard. They are clothed in sheepskins and, although originally gilded, over time they have acquired a deep patina that appears dark from a distance, earning them the nickname of 'the Moors'. Two minutes before the hour—a sign that he lingers in the past—the older man strikes the bell. Two minutes after the hour his young companion does the same, announcing the start of the new hour. The Jacks are accompanied by other automaton figures: the Virgin and Child, who are seated beneath them on a semi-circular gallery just below the lion of St Mark, with the Roman numeral hours and minutes appearing in apertures to their left and right. Twice a year, for Epiphany and the Feast of the Ascension, a procession of the Three Magi led by a trumpeting angel emerges and parades before the Virgin and Child.

Dijon's Notre-Dame cathedral is inhabited by another of the oldest known Jaquemarts (restored in the seventeenth century). Originally from the clock of Courtrai, whence it was abducted by the Duke of Burgundy in 1382, the figure began to be called 'Jaquemart' around 1458.[26] In 1651 the debonair pipe-smoking Jack was given a wife named Jacqueline, or Jacquette. The city

---

23 Chaucer, 'The Nun's Priest's Tale' (line 34) discussed in this volume, 'Introduction', n. 9; Matthew 26.34; Mark 14.30, who has Peter denying before the cock crows twice.
24 Dasypodius 1580b/2008, 151. For general accounts of the three clocks, see Ungerer 1922; Bach *et al.* 1992.
25 Carpenter, *Destructorum viciorum*, 1497, cited from Beeson 1971, 109.
26 Picard 1921, 77–82.

council deliberations reveal that this was actually a marriage of reason, embarked upon for the purpose of sparing the bell: '... If one wanted to add another figure to the Jaquemart ... so as to relieve the bell, which, always being struck in the same place, is becoming very worn'.[27] Around 1714 a son called Jacquelinet arrived, and in 1884, a daughter named Jacquelinette. The children strike the quarter hours on two smaller bells (once for the quarter, twice for the half hour, and so on), while their parents share the task of sounding the hours.

MUSICAL CLOCKS IN THE HOME

The use of the **mainspring** as a power source made it possible to produce smaller timepieces suitable for private use, and thus in the fifteenth and sixteenth centuries wall-mounted and table clocks began to adorn the homes of the wealthy. Symbols of status and privilege, these were all the more impressive and desirable when they contained built-in entertainment in the form of music and/or automata. Early examples of these smaller clocks often retained elements reminiscent of monumental tower clocks. Two carillon-playing examples from the late sixteenth century, both now in the British Museum, are examples of the lingering influence of turret clocks in timepieces made for domestic use.[28]

The first of these was made in 1589 by Isaac Habrecht, one of the creators of the second Strasbourg cathedral clock, and it clearly shows the relation between monumental church clocks and the earliest smaller-sized clocks. Standing over five feet tall, more suited to a princely residence than the home of a wealthy citizen, it was closely inspired by the Strasbourg clock, with four balcony carousels upon which several types of automata perform their actions. On the uppermost level a Christ figure appears in a doorway and a skeleton symbolizing Death strikes the hours on a bell. Below, the quarter hours are sounded by figures representing the Four Ages of Man. Further down there is a parade of angels (replacing the original procession of the Three Magi before the Virgin and Child) that follows the striking of the hour. Above the dials the days of the week—symbolized by deities in chariots drawn by fantastical animals—ride by on a carousel.[29] The second, a lantern-form carillon clock made by Nicholas Vallin (c.1565–1603), a Flemish refugee in London, is smaller and dates from 1598. There is a separate **train** for the carillon music. The hours are struck on a large bell at the top of the clock, while at the quarters a tune is played on thirteen bells that are visible under a pediment, with a different melody to signal each quarter hour. The brass-covered, pinned wooden barrel is reprogrammable, allowing the tunes to be changed as desired.[30] Another example of a musical timepiece diminutive enough for domestic use is a quarter-striking musical table clock signed 'I*H' (probably Isaac Habrecht); it dates from around 1590 (Figure 112). Its unusual disc musical movement and eight bells are concealed in the base; the pinned disc (not commonly used until the nineteenth century) is reprogrammable, making it one of the earliest such disc movements known.[31]

Clocks such as these gave their owners more than just knowledge of the time. They were, above all, a source of private enjoyment, no doubt greatly enhanced by the knowledge that few others could possess the like—an unambiguous sign of influence and social status.

MUSICAL AUTOMATON CLOCKS FOR RENAISSANCE PRINCES

In the sixteenth and early seventeenth centuries many of the most remarkable clocks were produced in Augsburg. The city had become the leading centre for the production of a specific type of musical automaton clock intended for the aristocracy, and in particular for the courts of the Habsburg Holy Roman Emperors Charles V (1500–58) and his grandson and great-nephew Rudolf II (1552–1612). They included elaborate miniature silver gilt warships known as 'nefs', manned by tiny sailors, soldiers, and musicians. Nefs had been a common accessory on aristocratic tables since the Middle Ages, where they were used to contain table utensils, napkins, and spices. Traditionally placed before important guests at banquets, nefs were considered to bring good luck.[32] The mechanical nefs created by the Augsburg-based artisan Hans Schlottheim (after 1544–c.1625) performed elaborate sequences of sound and motion while travelling down banquet tables. Not simply sources of entertainment, they were symbols of power that provided tangible proof of the immense wealth of the prince who had commissioned them. Such astonishing and exorbitantly expensive items were far beyond the reach of even the members of the elite upper classes. This was the period during which the *Kunstkammer*, or cabinet of curiosities, flourished in Europe, and these collections of items, both natural and manmade, were highly prized by their aristocratic owners as irrefutable demonstrations of their wide-ranging knowledge, power, vast wealth, and influence. Ornate and complex musical and automaton clocks such as mechanical nefs filled a similar function.[33]

One spectacular Schlottheim nef, made around 1580–90, incorporates a small clock. The hours and quarters were sounded by sailors who pivoted in the crow's nests of the ship, striking them with hammers (for they are, in fact, inverted bells). Trumpeters played, accompanied by a kettledrum. As the ship proceeded along

---

27 Thomas 1904, 74.
28 Thompson 2004, 48–51 and 56–7.
29 British Museum, no. 1888,1201.100; a similar example is in Rosenborg Castle, Copenhagen.
30 British Museum, inv. 1958,1006.2139; a replica is in the Speelklok tot Pierement Museum in Utrecht, Holland. For Vallin, see Lloyd & Drover 1955.

31 See Turner 2014b, 930–1. The clock was offered at the Sotheby's London sale 'Treasures, Princely Taste' held on 3 July 2013, lot 5.
32 Levenson & Massing 1991, 238.
33 Kepler/Frisch 1858–1871; Morsman, 2006; Korey 2007b.

Figure 112 Four-train, spring-driven quarter-striking and musical clock carrying the initials 'I. H.' with sub-stage reprogrammable pinned disc for the music. Late sixteenth century. Photo by Colin Crisford, courtesy of Alidade Ltd London.

the table, a tune was played on a reed organ (*regal*) with drum accompaniment, which was furnished by a drum skin stretched across the base of the hull. As the masts twirled, the heralds ceremoniously advanced to announce the seven electors (including Augustus, Elector of Saxony) who paraded before the seated Emperor Rudolf II. Then the main cannon fired, triggering as it did so the other smaller cannons, in what must have been an astounding grand finale.[34] The function of clockwork automata used at banquets evolved over time, and eventually they played a less ceremonial role, becoming accessories in what must have been rather rowdy princely drinking games—the person in front of whom the automaton stopped was obliged to empty his goblet.

At more or less the same time, other clocks, perhaps less awe-inspiring but certainly just as amusing, were being produced. Still so expensive as to be accessible only to a few, these animated table clocks were a specialty of southern German clockmakers.[35] Made in the form of lions, bears, dogs, birds, and elephants, or human figures like bear tamers, archers, or horsemen, these precious clocks were treasured by those who could afford to own them. Generally some part of the figure went into motion to mark the hour or the quarter hour; sometimes it was the eyes that moved from side to side with the oscillations of the **balance**. An eagle made in Augsburg *c*.1630 waved its sceptre when the clock struck the hour and its eyes and beak moved on the quarter hours.[36] Precious animated clocks continued to be made in southern Germany throughout the seventeenth century. One quarter-striking clock made in Augsburg, meant to roll down a table or another flat surface, represents the goddess Diana riding in a leopard-drawn chariot. On one side a clock dial indicates the hours from I to XII, while the dial on the other side has a twenty-four-hour cycle of twice twelve hours. As the chariot advances, the leopards leap forward while moving their heads, a monkey raises its paw, and a bird guarding the rear of the chariot flaps its wings. When the chariot stops, the huntress's eyes, which had been flitting from side to side in rhythm with the balance, cease their motion and appear to focus on her prey as she shoots an arrow, using her articulated middle finger.[37]

---

34 British Museum, no. 1866, 1030.1; similar nefs are in the Vienna Kunsthistorisches Museum and the Musée National de la Renaissance, Ecouen, France. Kepler/Frisch 1858–1871; Fritsch 1999. See Haspels 2006, 196–201 for more on the Vienna nef mechanism.
35 Vincent, Leopold, & Sullivan 2015, 54–7.
36 Metropolitan Museum, New York, accession no. 29.52.14.
37 Vincent, Leopold, & Sullivan 2015, 75–87.

Another remarkable early domestic musical and automaton clock was made in Augsburg around 1625 by Veit Langenbucher (1587–1631) and Samuel Bidermann (1540–1622) together with the latter's son, also named Samuel. It incorporates a sixteen-note pipe organ, a sixteen-string spinet, and *commedia dell'arte* figures that dance when the music sounds the hours. One of the most musically unusual and elaborate automatic musical clocks of its period to have survived to the present day (the cylinder features Bidermann's characteristic 'L' shaped pins), it offers a manual choice of orchestration: strings only, strings and organ, or organ only. The clock plays three tunes, which were probably written by Hans Leo Hassler (1564–1612), who had taught the elder Bidermann.[38] However, by the middle of the seventeenth century, production of musical and automaton clocks in southern Germany had all but ceased because of the devastation of the Thirty Years' War (1618–48) and the plague epidemic of 1627–28.

MUSICAL MATURITY

By the mid-seventeenth century many remarkable musical and automaton clocks were being produced for a sophisticated and well-off clientele in cities such as London, Paris, Berlin, Prague, and Vienna. The quality of the musical performances became primordial and during a roughly hundred-year span beginning in the early eighteenth century, many clocks playing excellent music were made, some of them further embellished with animated figures.

The majority of the clocks of this period had organ movements in which the pinned cylinder, in turning, triggered levers that activated valves admitting a flow of air into the pipes. A bellows system including air reservoir and feeder insured a stable and adequate air supply. Such musical mechanisms consumed a great deal of energy and therefore usually required a separate train. One reason why the organ mechanism came to be preferred over the carillon was that it tended to be more satisfying as music.[39] It permitted polyphony and harmony, in which two or more notes are sounded at once, as well as notes of longer and varying durations, unlike carillon mechanisms, in which there is no means of prolonging a note. Once struck, bells continue to reverberate, particularly in the lower register, resulting in a muddy sound. As a result, some makers, such as Charles Clay, fitted carillon clocks with dampers, but the problem, inherent to bell mechanisms, remained. An additional constraint was presented by repeated notes, since a certain lapse of time was required to allow the levers and hammers to return to their original position before they could be activated again. This led to the practice of doubling the hammers, common by the mid-eighteenth century. Not all bells had to be provided with more than one hammer, however; that was determined by the music to be played.

In addition to carillon and organ clocks, compound clocks featuring various combinations of organ and carillon or string instrument mechanisms were made, though less frequently. Clocks playing stringed instruments such as the dulcimer, the violin, or the spinet in addition to organ pipes or a carillon are indeed quite rare. This is probably due to the greater amount of maintenance they required, for strings often broke and had to be changed, and stringed instruments required frequent tuning and were in general much more sensitive to changes in temperature and humidity. On the other hand, such movements were often very pleasing musically and were not subject to the excessive reverberation that marred musical clarity. Nevertheless, few clocks with stringed mechanisms appear to have been made, and even fewer have survived to the present day. Some musical clocks of the time were an ambitious combination of fine cabinetmaking, exceptional goldsmithing, fine painting, and music. Major composers such as Handel (1685–1750), Haydn (1732–1809), Mozart (1756–91), Beethoven (1770–1827), and C. P. E. Bach (1714–88) wrote music for clocks with mechanical musical mechanisms. In fact, Joseph Haydn worked directly with his colleague at the Esterhazy court, the librarian Joseph Niemecz (1750–1806, in religion Primitivus), writing pieces especially for the small organ clocks that Niemecz built as a hobby.[40]

English horologists, such as Huguenot Claude Duchesne (c.1670-1733), were among the foremost makers of musical clocks during this period. Charles Clay (1695–1740) was also a maker of high-quality musical clocks housed in imposing sculpted cases. For these cases he often adopted innovative forms, abandoning the traditional rather sober longcase style for a more voluminous, architecturally inspired case that was meant to be placed in a central position so as to be admired from all sides, draw the eye, and capture attention. Clay, who catered to an aristocratic clientele—he presented a musical clock to Queen Caroline in 1736[41]—had a preference for precious materials like mahogany, silver, and gold, and worked with the most fashionable artists of the day, including Jacopo Amigoni (1675–1752), who painted the fine mythological scenes that decorated the clocks, the goldsmith Edward Amory, and musicians such as Handel, Francesco Geminiani (1687–1762), and Arcangelo Corelli (1673–1713).[42] John Pyke (c.1696–1762), clockmaker to the Prince of Wales, and his son George (c.1725–77), organ-builder to His Majesty, made fine musical clocks in the same vein as Clay. John Pyke finished the 'Temple of the Four Grand Monarchies of the World', a clock that Clay had been unable to complete before his death. Several Pyke

---

38 Metropolitan Museum, accession number 2002.323a–f. For the dispute over barrel-pinning that set the following generation of the two families at loggerheads, see Groiss 1980b, *passim*; for a description of the clock, see Haspels, 2006, 165–7.

39 For the sound of such organ clocks, listen to those from 1750–1810 recorded from the Museum van Speelkloek, Utrecht, on Stemra CD STP 002.

40 Ord-Hume 1982 provides a detailed survey.
41 Ord-Hume 1995, 115.
42 For recent surveys of Clay's work, see Murdoch 2013; Turner 2014b.

organ clocks were sent to the Far East; in 1933 a musical clock with automata made by George Pyke was described by Harcourt-Smith in the Palace Museum in Peking.[43] A large clock made c.1765 by George Pyke—now in Temple Newsam House, Leeds, England—plays eight tunes on a pipe organ movement; its dial plate is painted with a village scene, and it incorporates several animated figures.[44]

Many other fine makers of musical clocks during this period deserve mention, among them Cornelius Engeringh, a maker of organ clocks in Dordrecht, Holland, who was active during the late eighteenth century. One popular type of Dutch musical clock was the longcase that often featured prominent maritime scenes with animated ships rocking on the waves and windmills with turning blades, which spoke to Netherlanders' pride in their trading prowess and maritime skills. In Sweden, Per (also Pehr or Petter) Strand was an excellent organ builder active in the late eighteenth and early nineteenth centuries. He furnished musical movements for many clocks, particularly longcase ones in the Gustavian style (Figure 113).

In the mid-eighteenth century in Germany, Frederick the Great of Prussia (1712–86) was a great lover of music and a promoter of the art of horology in his kingdom. Thus it is not surprising that during his reign and those of his successors Frederick William II and Frederick William III a number of excellent-quality musical clocks were made. Noteworthy makers of the period include the king's watchmaker Louis George (fl.1769–c.1800) (a descendant of French Huguenot refugees) who often purchased both finished and unfinished musical clocks from Pierre (1721–90) and Henri-Louis (1752–91) Jaquet-Droz and their associate Jean-Frédéric Leschot (1746–1824) in La Chaux-de-Fonds. The court clockmaker Christian Ernst Kleemeyer (fl.1766–1812) and Kleemeyer's student Christian Möllinger (1754–1836) were also well-known makers of organ clocks.

Elegant and complex pieces of furniture made by the father and son team of cabinetmakers Abraham (1711–93) and David Roentgen (1743–1807) sometimes incorporated clocks by Peter Kintzing (1745–1816), and on occasion musical movements provided by the organ-builder Johann Wilhelm Weyl (c.1768–1813). Musical clocks by Roentgen and Kintzing were much sought after by aristocratic clients during the late eighteenth century. The Austrian diplomat Count Mercy-Argenteau, who represented the Empress Maria Theresa at the court of Louis XVI, owned a longcase clock by Kintzing and Roentgen. It had organ and dulcimer movements and played compositions by Christoph Willibald Gluck (1714–87), who had been Marie Antoinette's music teacher. Toward the end of the eighteenth century and well into the following one, musical clocks were a specialty of the Black Forest region of Germany. These clocks, fitted with organ, bell, or string movements (*Flötenuhr*, *Glockenspiel*, and *Harfenuhr*, respectively) generally made use of the wood that was readily available in the region. They were often further embellished with automaton figures in the form of dancers, acrobats, or musicians.

In France one of the most renowned clock and mechanical musical instrument makers of the time was the German Michel Stollenwerck (c.1700–68). Having settled in Paris around 1730, he made technically complicated clocks and musical movements, particularly carillons, for himself and other makers. Held in high regard by his contemporaries, he was considered to have 'a particular gift for arranging carillons of singular accuracy and exact tone'. The praise awarded him by Engramelle echoes the practices of the sixteenth- and seventeenth-century Jesuits who sought to impress the Chinese with animated musical clocks:

> As for the instruments with chimes, or carillons ... those of Stolkverck [sic] have enjoyed the highest reputation ... products of this kind, which can be transported into the remotest lands, will complete the taming of barbarian Nations, who have already been filled with admiration upon seeing some instruments by Marchal & Stolkverck ... Some carillons by Stolkverck transported to China, Mongolia, Turkey, and amongst the Hurons, have left the sovereigns of those vast countries enchanted with admiration.[45]

Other beautiful musical clocks were made in Paris during this period. One example, from the mid-eighteenth century, has a finely sculpted case in the form of a patinated bronze rhinoceros created by the renowned bronze caster Jean Joseph de Saint-Germain (1719–91), and a clock movement made by François Viger (1704–84). Its base contains a carillon.[46] This kind of clock, with a finely sculpted gilt bronze animal atop a musical movement lodged in its base, was greatly appreciated. Bronze casters were often inspired to produce magnificent sculptures of animals such as elephants, bulls, lions, and rhinoceroses.

The city of La Chaux-de-Fonds in the Swiss Jura mountains was one of the major horological centres of the time, able to boast of many highly skilled artisans who specialized in music and automata and who produced a great number of noteworthy musical clocks. Perhaps the most famous of all was Pierre Jaquet-Droz, who first won fame when, in 1758, he set out to present an extraordinary clock to King Ferdinand VI of Spain. The 'Shepherd clock' struck the hours and quarters, and executed a complex programme of musical performances, including a nine-bell carillon that played six tunes after the hour sounded (with manual or automatic tune change), and a nine-pipe serinette with two bellows, also playing six melodies. The complex sequence of automaton actions, coordinated with the music, included the flute-playing shepherd (with dog and bleating sheep), a singing bird perched on a Cupid's hand, a lady following a musical score at her balcony while taking pinches of snuff, and a pair of see-sawing Cupids. In the central portion of the dial the equation of

---

43 Harcourt Smith 1933.
44 Dawe 1974, 68–70.
45 Wilson *et al.* 1996, 196.
46 Louvre Museum, no. OA10540.

Figure 113 Long case organ clock by Per Strand, 1794, Sweden, with twelve cylinders that include works by Haydn and Mozart. Museum Speelklok, Utrecht.

time, the length of each month, the date, the age of the Moon, the zodiac signs, and the four seasons were shown. The sale of this clock to the Spanish king ensured both Jaquet-Droz's reputation and his financial stability for some time.[47] Pierre Jaquet-Droz's career path was indeed emblematic of the period: in his youth, when no clockmaker could succeed without aristocratic patronage, he was able to win recognition, fame, and financial security through his marvellous automaton clocks with music. The Spanish king's patronage provided him with the capital to construct the android automata he created along with his son Henry-Louis, and during the latter part of his career, to orient his firm's production toward smaller-volume pieces that were easier to make and to transport, and which, more importantly, met the demands of a new clientele.

Several other skilled makers of musical clocks were active in the region at the same time, among them the Robert family, highly regarded as makers of musical clocks. The company was founded around 1770 by Josué Robert (1691–1771), watchmaker to the King of Prussia since 1725, and his sons David (1717–69), who married Pierre Jaquet-Droz's sister, and Louis-Benjamin (1732–81). According to Alfred Chapuis, Josué Robert et fils was the largest firm working in the La Chaux-de-Fonds region during the last quarter of the eighteenth century; Chapuis further noted that 'from 1781 to 1809, the inventories of J. Robert & fils mentioned over 50 musical clocks, either finished or not yet completed … with organs, dulcimers, carillons, harpsichords, serinettes and "canary cages"'.[48] Later iterations of the company included Robert et fils, Josué Robert et fils et Cie, and Robert & Courvoisier. Other noted makers of musical clocks in the region were Samuel Roy and his sons, also of La Chaux-de-Fonds, the Courvoisier family, and David-Guillaume Engel.[49]

Any discussion of eighteenth-century mechanical music and moving figures must include the work of Jacques Vaucanson and Jaquet-Droz. Despite the fact that their automaton creations (often called '**androids**', although Vaucanson's famous 'digesting duck' is to be excepted from this term) did not tell the time, they were enormously influential for future makers of horological, musical, and automaton pieces. The son of a Grenoble glove-maker, Jacques Vaucanson (1709–82) was fascinated by machines from childhood and devoted his life to inventing and improving them. In the 1730s he began work on an automaton flute player that he presented to the French Royal Academy of Sciences in 1738. Perhaps influenced by the materialist thinking of such philosophers as René Descartes (1596–1650) and later Julien de La Mettrie (1709–51), Vaucanson set out to explore the mechanisms of human movement. Although his automata were displayed to the public (in Paris in February 1738, and in London in 1742),[50] Vaucanson was more interested in gaining the approbation of the learned members of the Academy of Sciences than in being a showman. His 1738 memoire *Le mécanisme du flûteur automate* provided a detailed explanation of the construction of the automaton, which truly played its instrument (unlike most musical automata, in which the music is produced by a separate musical movement). The flute player executed twelve melodies. The cylinder pins and bridges engaged with a **keyframe** that had fifteen levers:

> These fifteen Levers answer to the fifteen Divisions of the Barrel, by their Ends which have the Steel elbows or lifting Pieces; at an Inch and a half Distance from each other: When the barrel turns the Bars of Brass fix'd upon its divided Lines meet with the Lifting pieces, and keep them raised a longer or shorter Time, according as these Bars are longer or shorter: And as the Ends of all those lifting Pieces, make one right Line, parallel to the Axis of the Barrel, cutting all the Lines of Division at right Angles; every Time that a Bar is fix'd at each Line, and that all the Ends of those Bars make amongst them also a right Line, and parallel to that which is form'd by the lifting Pieces of the Levers, each End of a Bar (as the Barrel turns) will touch and raise at the same Time the End of a Lever; and the other Ends of the Bars likewise forming a right Line parallel to the first, will, by the Equality of the Length of the Bars, each let fall its Lever at the same Time. One may easily see by this, how all the Levers may act, and at the same Time concur to the same Operation, if it be necessary.[51]

In 1739 Vaucanson showed a second mechanical musician, a Provencal shepherd who played approximately twenty different tunes on a three-holed recorder, a *galoubet*, that he held in one hand, while he beat a drum with the other. This android, like the flute player, actually played his two instruments.

A generation later, in the early 1770s, Pierre and Henri-Louis Jaquet-Droz and their associate Jean-Frédéric Leschot made three android automata, including a lady playing a keyboard instrument that is usually referred to as a harpsichord, although it is in fact composed of organ pipes. Like Vaucanson's flute player, the nearly life-size figure actually plays her instrument, performing five melodies on a twenty-four-key instrument. Her jointed fingers depress the keys of a keyboard that forms two segments of a circle, so that the strings are always struck at the same angle no matter the location of the key. To shorten the transmission distance, the commands run from the lady's elbows to her fingertips. Four separate movements are lodged underneath the lady, with a powerful **bellows** mechanism providing the air for the organ

---

47 The 'Shepherd clock' is now in the Royal Palace Museum in Madrid, 10,003,042. See Colon de Carvajal 1987, 29–30, N° 10. For Jaquet-Droz's journey to Spain with six clocks including the 'Shepherd', see Tissot 1982; Girardier 2012; Girardier 2020.
48 Chapuis 1917, 350.
49 Chapuis 1917, 351; Girardier 2017, 157–73.

50 Vaucanson 1985 (first edition 1738), XI; Altick 1978, 65; Doyen & Liaigre 1966, 49–174.
51 Vaucanson 1985, 14 of the Desaguliers translation.

pipes. The brass cylinder, placed in the musician's stool, controls the action of her fingers. The cylinder has rows of pins on either side that correspond to the keys on the key frame: twice five, to drive the movements of the fingers of each hand, while a central **cam** section on the cylinder controls the lateral motions of the figure's forearms A separate train controls what is sometimes called the 'life mechanism', including the up and down movement of the chest that suggests breathing, as well as movements of the lady's eyes and head. This mechanism can be started even before she begins to play, heightening the illusion of life. When the musician has finished playing, she inclines her head and bust in a discreet bow.[52] The three Jaquet-Droz automata made a tour of Europe; in 1775 they were in the rue de Cléry, Paris, and the following year in London.[53]

A third android musician that also truly played its instrument was made in 1784 by Roentgen and Kintzing—a departure from their more usual fine furniture, which sometimes included musical clock movements. The dulcimer player automaton was made for Marie-Antoinette; she gave it to the Academy of Sciences, whence it passed to the Conservatoire National des Arts et Metiers, where it may be seen today. The mechanism is lodged underneath the musician's seat; its brass cylinder is non-changeable and plays eight melodies on a forty-six-cord dulcimer (the strings are doubled, making 2 × 23 notes). The figure, which also has motions of the eyes and head, is driven by a brass cylinder with a row of eight pins on either side (controlling the movement of the hammers) and a central section of sixteen cams that control the sideways movements of the automaton's forearms, which are jointed at the elbows and thus serve as pivot points.[54]

The *tableau animé*, or moving picture, was another type of musical clockwork piece that was appreciated by eighteenth century elites and became accessible to a wider audience during the following century. These were framed pictures with animated scenes that are reminiscent of those that often adorned seventeenth- and eighteenth-century musical clocks. Desmares (*fl.* pre-1739–post-1759), a 'méchanicien-machiniste' at Versailles and Paris, was a notable maker of such pieces in the eighteenth century. One of his moving pictures, which dates from 1759 and belonged to Mme de Pompadour, depicts the Château of Saint-Ouen, with a fisherman, washerwomen near a river on which boats are sailing, a cart and horse, and other figures. The cries of the animals are imitated by bellows on the reverse.[55] Another animated picture made by Desmares in 1739 has three dials and several apertures set in its imposing carved and gilt wood frame; they indicate the time, the date, and the days of the week, and the month. Painted in oil on metal, it includes approximately sixty animated figures: men sawing, masons hewing stone, a windmill, children playing on a seesaw, and the like. This tableau animé was once in the collection of Joseph Bonnier de la Mosson (1702–44), General Treasurer of the Etats de Languedoc.[56]

Tableaux animés remained popular throughout the nineteenth century, though by that time they had been much simplified and were generally made of cheaper and less durable materials. Some offered simply a flat, painted surface including a tower with a clock. Such pieces usually played a tune on the hour, but the music could also be manually activated by a pull cord. Other moving pictures consisted of three-dimensional pasteboard towns with clock towers, windmills, bridges over which trains and passengers passed, and three-dimensional ships that sailed on crushed paper or chamois-leather seas.

THE EIGHTEENTH CENTURY: ENLIGHTENMENT AND SHOWMANSHIP

With the Age of Enlightenment came a powerful current of liberty and rationalism that transformed society across Europe and in the New World; musical and automaton clocks were not exempt from its influence. The elaborate and precious pieces of former centuries, enjoyed by the elite and made to order by craftsmen who depended on noble patronage, progressively became more accessible, particularly to the rising merchant class. Those who were not able to buy them could admire them at public exhibitions open to anyone who would pay the admission fee. Such exhibitions became quite common during the eighteenth and early nineteenth centuries. They were much appreciated by the public, as attested by numerous advertisements and engravings that appeared in newspapers of the time. One impressive piece shown to the public in the early eighteenth century was the 'Theatre of the Muses' made by the London clockmaker and creator of musical automata Christopher Pinchbeck (*c.*1670–1732). A contemporary newspaper advertisement described it as being 'most beautifully composed of Musick, Architecture, Painting, and Sculpture, with such diverting Variety of moving Figares, that renders it the most entertaining Piece of Art that has ever yet appeared in Europe'. The music—composed by 'Mr Handel, Corelli and other celebrated Masters'—was played on the 'Flute, Flageolet, German Flute and Organ', and its tunes were accompanied by 'the sweet Harmony of an Aviary of Birds' and diverse animated scenes.[57]

In 1772 James Cox (*c.*1723–1800), a London purveyor of elaborate musical and animated pieces who dealt extensively with China, opened a museum. It included twenty-three pieces, some of which combined clocks with mechanical music and

---

52 The Jaquet-Droz 'Musicienne' is today in the Neuchâtel Musée d'art et d'histoire. See Haspels 2006, 239–43, for more information.
53 Altick 1978, 66; Girardier 2012, 24–5.
54 For further details on the dulcimer player, see Haspels 2006, 182–5.
55 Paris, Musée des arts et métiers (inv. No 01407-2).
56 Gersaint 1744, N° 618. Paris, Musée des Arts Décoratifs (inv. No 3131).
57 Ord-Hume 1995, 102; Altick 1978, 60.

automata.[58] Public curiosity and love of novelty also drew spectators to Henri-Louis Jaquet-Droz's 1776 'Spectacle Mécanique' in Covent Garden. In addition to the three famous Jaquet-Droz androids (the harpsichord player, the writer, and the draughtsman), his mechanical spectacle included the 'Grotto', a legendary creation that is now lost. This was an elaborate mechanical scene measuring about four square feet, with a palace and gardens and a rocky Swiss landscape, all peopled by numerous animated figures performing various tasks. Among them were a shepherd playing the flute (complete with an echo effect), a shepherdess playing a duet with him on the guitar, a young girl playing minuets on a hammered dulcimer while two others danced, a barking dog, bleating sheep, birds that sang and flew through the air, flowing fountains, and a mill with a running stream. While we have no precise indication of the type of musical mechanisms employed, one assumes there would have been serinette and dulcimer movements; we do know that the Grotto prominently featured a clock set in its bas relief-adorned façade. Jaquet-Droz claimed the piece embodied a 'contrast of art and nature'—presumably because it depicted the natural world through mechanical means—thus echoing the *artificialia* and *naturalia* displayed in Renaissance curiosity cabinets. This was a spectacle meant to provoke wonder, incite curiosity, and even increase knowledge, but rather than being reserved for the court, it could now be enjoyed by ordinary citizens.[59]

Public enthusiasm and appreciation for displays of 'pieces of mechanism' incited many others to offer similar shows. They included the London associate of H. L. Jaquet-Droz, Henry Maillardet (1745–1830), who opened an 'Automatical Museum' in London at the turn of the century;[60] his brother Jean David Maillardet (1748–1830), who travelled with mechanical pieces during the early years of the nineteenth century; and Thomas Weeks (or Weekes c.1740–1834). Founded around 1790 near Piccadilly in London, Weeks' museum is said to have included several pieces formerly in James Cox's museum. Jacob Frisard (1753–1810) and Johann Nepomuk Maelzel (1772–1838)[61] were among the many others who became showmen of mechanical pieces. These exhibitions frequently included musical and automaton clocks among their delights.

DIPLOMATIC MESSENGERS

Throughout the centuries musical and automaton clocks and watches have served as prestigious gifts of state for several reasons: first, the high cost of their production and the enormous skill of their makers, which heightened their rarity and therefore their prestige; and next, their unfailing ability to elicit surprise and awe in onlookers, which made them the most fascinating and delightful sources of entertainment then available. One of the most famous of these diplomatic gifts was a late sixteenth-century organ clock made by Thomas Dallam (c.1575–after 1620). Queen Elizabeth I sent it as a present to the Ottoman Sultan Mehmet III. The queen hoped to impress and ingratiate the sultan and thus to obtain trading rights. In 1599 Dallam journeyed to Constantinople to deliver the 'Great and Curious Present', which would have to be assembled, adjusted, and perhaps repaired, upon arrival. The sixteen-foot high and six-foot-wide clock was truly spectacular. It included a talking head that told the time; birds that sang and fluttered among silver bushes; the figure of Queen Elizabeth I raising her royal sceptre as the planets revolved around her; a twenty-four-hour clock that also indicated the position of the Sun and the phases of the Moon; and an organ with a six-hour bellows reserve that executed five tunes automatically and could also be played manually.[62] Dallam described the presentation of the clock as follows:

'Firste the clocke stroucke 22; than The chime of 16 bels went of, and played a songe of 4 partes. That being done, tow personagis which stood upon to corners of the seconde storie, houldinge two silver trumpetes in there handes, did lift them to their heads, and sounded a tantarra. Than the muzicke went of, and the orgon played a song of 5 partes twyse over. In the tope of the orgon, being 16 foute hie, did stand a holly bushe full of blacke birds and thrushis, which at the end of the musick did singe and shake their wynges. Divers other motions thare was which the Grand Sinyor wondered at.'[63]

This astonishing clock was apparently much appreciated, for the sultan asked Dallam to stay in Constantinople, offering him the choice of three concubines or virgins from his harem.

From the mid-fifteenth century onwards, the Ottoman Empire was an important destination for lavish diplomatic gifts, which were often musical and automaton horological pieces. After the 1453 fall of Constantinople to Ottoman conquerors, an uneasy truce reigned between West and East, involving an annual 'tribute' or honorarium paid to the Sublime Porte by the Hapsburg Emperors. This practice of appeasing the Ottomans by means of spectacular gifts was continued in the sixteenth century by the Emperor Ferdinand I (1503–64), and by the Emperors Maximilian II (1527–76) and Rudolf II (1552–1612). The impressive pieces for the tribute were often made by Augsburg clockmakers. In 1578, the Turkish honorarium included a gilt ostrich whose eyes, beak, and wings were animated. In 1584, Hans Schlottheim received a commission from the emperor for an elaborate mechanical galleon—probably a sort of nef—that was to be part of the

---

58 Cox 1772, 2; Pointon 1999; Smith 2000; this volume Chapter 16.
59 Perregaux & Perrot 1916, 102–5; Altick 1978, 62.
60 For the Maillardet, see Fima-Leonardi 2019.
61 Many of the latter's mechanical music machines are described in Steblin. See also Bingham & Turner 2017, 25–30.

62 Bicknell 1996, 72–3.
63 Bent 1893, 67–8.

honorarium. However, because Schlottheim was slow to complete the piece, and additionally because his price was considered exorbitant, it was not included in that year's tribute.[64] While the balance of power between the Ottoman and the Roman Empires shifted with the 1606 peace of Zsitvatorok, musical and automaton pieces continued to be sent even as late as the early nineteenth century to the Ottoman Empire as a means of securing the favour and cooperation of Turkish officials.[65] Parallel to the use of horological items as gifts, there was a flourishing commercial trade on the part of English makers, for example, which began modestly in the seventeenth century and had substantially grown by the end of the following century. From the end of the eighteenth to the first quarter of the nineteenth centuries, a major supplier of these clocks was the firm of Thwaites (later Thwaites and Reed).[66]

THE 'INCHANTMENTS OF MECHANISM' IN THE FAR EAST

Beginning in the late sixteenth century, complicated European clocks—and particularly those with music and automata—served as a particular kind of diplomatic gift for Jesuit, and other, missionaries in China. Matteo Ricci (1552–1610) arrived in Peking in 1601 bearing gifts for the Wanli Emperor (1563–1620), including a harpsichord, a map of the world, and two striking clocks. The Jesuits hoped that the fascination with musical automaton clocks (and the vast trove of scientific and technical knowledge that had permitted their construction) would convince the Chinese to accept Western religion—propagation of faith through science. Ricci explained that the Chinese were more impressed by the animations of European clocks than by their timekeeping qualities: 'They marvel more to see a machine that moves by itself and sounds the hours than as an artifice to tell time.'[67] Another important, complex automaton clock was brought to China from Lisbon in April 1618 by Nicolas Trigault (1577–1628). He described it as being 'a clock the like of which we have never seen for its ingenuity, splendour and value ... each time the twelfth hour sounded in the upper part of the clock, the history of Christ's birth could be seen below, marvellously enacted by little figures in gilded copper'.[68] That clock was probably modelled after the 'Nativity Clock' made around 1584 by Hans Schlottheim, which was destroyed during the 1945 bombing of Dresden.

The Chinese Emperors Kangxi (1654–1722) and Qianlong (1711–99) were enthusiastic collectors of musical and automaton clocks. Kangxi established a workshop devoted to Western-style clocks in the palace.[69] The Jesuit clockmaker Valentin Chalier (1697–1747), who was sent to China during this period, was impressed by the sheer number of these pieces:

'... As for clocks, the imperial palace is stuffed with them. Watches, carillons, repeaters, automatic organs, mechanized globes of every conceivable system-there must be more than four thousand from the best masters of Paris and London, very many of which I have had through my hands for repairs or cleaning.'[70]

Musical and animated clocks and watches (Figure 114) were much in demand in the Far East and India during the eighteenth and nineteenth centuries. This commerce originated for the most part in England, thanks to the importance of London as a horological centre, and the British East India Company's dominant role in marine transport.[71] This passion for European musical and automaton clocks and watches gradually expanded beyond the Chinese emperor and his immediate circle to well-off merchants and high-level civil servants. Among the noted makers of such pieces in England during the late eighteenth and early nineteenth centuries were Henry Borrell (1795–1851), William Carpenter (1770–1817), and Timothy Williamson (fl.1768–88), and best known of them all, James Cox.[72]

By the late 1760s Cox had gained recognition for his animated musical pieces, probably subcontracting their actual construction to workmen in London, including the Belgian expatriate John Joseph Merlin (1735–1803), who was Cox's principal mechanic until 1773, and in the 1780s, with 'Merlin's Mechanical Museum', became himself a showman.[73] Cox thought to avail himself of the

'taste of the Orientalists for brilliancy, by adding to its design, a higher finishing of the workmanship, the inchantments of a mechanism the more likely to captivate those people for their having to them the charm of novelty; the whole combined with the sweet harmony of sounds; and even with utility itself.'[74]

THE OTTOMAN TRADE

The Turkish market remained active throughout the late eighteenth and early nineteenth centuries, with several well-known British watch and clockmakers as principal suppliers. Among

---

64 Mraz 1980, 37–48; Kugel 2016, 25.
65 Talbot 2016, 66–8.
66 White 2012, 277.
67 Rienstra 1986, 37.
68 Pagani 2001, 32–3.
69 Pagani 2001, 36–9; this volume Chapter 2, Section 2, and Chapter 16.
70 Pagani 2001, 61.
71 Smith 2017, 175–92, and this volume Chapter 16.
72 White 2012, 211–60; this volume Chapter 16.
73 French & Wright 1985, *passim*.
74 Cox 1772, 1.

Figure 114 Venus Binding Cupid's Wings. Pair of repeating pocket watches with music and automata for the Chinese market. Enamel painting attributed to Jean-Abraham Lissignol after Louise-Elisabeth Vigée-Lebrun (1755–1842); movement by Piguet & Meylan, Geneva, c. 1815. Patek Philippe, Museum, Geneva.

them were George Prior (1735–1814) and his son Edward (fl.1815–c.70), Henry Borrell (fl.1790–1840) who also worked for the Chinese market, Markwick Markham (fl.c.1729–41),[75] and Justin (1712–97) and Benjamin (1747–1811) Vulliamy. A large mahogany organ clock by the latter, which has a Turkish numeral dial and plays seven automatically changing tunes on the hour, is in the British Museum.[76] The similarity of the musical mechanisms observed in many English clocks may be explained by the fact that Thwaites & Reed furnished high-quality movements to the several makers who made pieces for the Ottoman market. An entry in the Thwaites & Reed Day-Books from 9 March 1797 details a musical clock with several animal automata that was prepared for George Prior:

a small organ with 2 stops of metal pipes & Double Bellowes to play a piece of Turkish Music in 5 parts ... the cupalo of the Case to carry Round a Ring of 6 Figures all of which move as they pass Round. Viz. A Lion to open his Mouth & move his Tail. A Tyger to open His Mouth, A Bear to move His Head. A Hyena to move His Tail. A Horse to move His Head. A Stag to move His Head. All done by Leavers inside the Body of each Animal & all properly painted ....[77]

The Jaquet-Droz firm also made musical and automaton clocks for the Near East. In 1782, Councillor François de Diesbach admired a clock 'destined for Cairo' in the Jaquet-Droz workshop. It played several tunes, imitated the sound of the violin by means of 'small elastic wheels made of horn' that rubbed against cat-gut strings and featured animated scenes including flowing fountains.[78] The 1789 *Almanach de Gotha* reported that the Jaquet-Droz workshop contained ' ... artificial canaries hopping about in cages and singing different songs, moving their beaks, throats,

---

75 The Markwick Markham business probably ended with Robert Markham's death in 1741. The name, however, was retained in use by other London makers for the sake of continuity in the Ottoman market. See White 2012, 345.

76 British Museum, no. 1958, 1006.2172.

---

77 Ord-Hume 1995, 118; White 2012, 72.

78 Chapuis, 1917, 113.

and bodies naturally. He sends them to Constantinople and has already sold a large number to the Seraglio of the Grand Turk'.[79] By the nineteenth century, while musical and automaton pieces continued to be in demand in the Orient, the nature of the objects offered on those markets had changed; they were smaller in size and less opulent than those of the previous century.

SINGING BIRDS

Although the mechanical singing bird has been part of the tradition of mechanical musical clocks since Antiquity, during the eighteenth century a new kind of singing bird was developed that looked, sounded, and moved very much like a real bird. These mechanical singing birds arose from a hobby that began in the seventeenth century: that of raising canaries and teaching them to sing. This was a mind-numbing and time-consuming task, for bird owners had to play tunes over and over on a flageolet until at last the canary learned to repeat them. Proof of the popularity of this pastime, the word *seriner* (*serin* meaning canary in French) entered the French language, designating any tiresome and mindlessly repeated action. Eventually a small mechanical organ was invented, which could mechanically produce music at the turn of a crank: the *serinette*. The *serinette* was equipped with a pinned cylinder, a set of organ pipes with valves that were activated by levers, and a bellows system to generate the air flow.

Small mechanical organ movements, often called 'serinette' movements, were used in mechanical singing bird pieces toward the middle of the eighteenth century. Such items, prized by the upper classes, were sold by merchants of luxury goods. One of the lots offered at the 1749 sale of the belongings of the *savante* Emilie du Châtelet (1706–49) was 'A cage in the form of a chandelier, in which there is a Canary that whistles six different tunes'.[80] Mechanical singing bird cages were often associated with clocks. The most prestigious makers of mechanical singing birds in the second half of the eighteenth century were Jaquet-Droz and Leschot. Beginning in the third quarter of the century, they produced a number of large bird cages that were meant to be suspended from the ceiling so that the large clock face on the bottom could be read when one looked up. The inhabitants of the cage, one or sometimes two birds, warbled on the hour or at will. The bird 'sang' popular tunes played by a *serinette* movement; it generally opened its beak, flapped its wings and tail, and spun around. The movements of the bird were linked to the *serinette* mechanism so that music and motion appeared to be coordinated, reinforcing the appearance of life. Jaquet-Droz also made 'Neuchâtel'-type clocks; one example is now in the Musée des Monts in Le Locle, Switzerland.[81] Made during the third quarter of the century, it has a nine-bell carillon with sixteen hammers. Placed in the lower portion of the clock, it plays seven tunes, while the song of the bird (six tunes) perched on top of the clock is furnished by a nine-flute *serinette* that is lodged in the clock's upper portion. After the carillon has finished playing the bird performs one of the six *serinette* melodies, turning and fluttering as it sings.

As time went on and fashions changed, certain disadvantages became apparent: a set of pipes or bells, along with their pinned cylinder, took up considerable space. Another drawback was that, though they could give a pretty rendition of popular tunes, these movements did not offer enough agility convincingly to imitate birdsong. In carillon movements, hammers could be doubled to permit repeated notes, but even so there was a time lag, and organ pipes could not come close to approximating the trills and glissandos of the song of real birds. These were no doubt among the reasons that *serinettes* gradually fell out of favour, although they continued to be used from time to time, even as late as the early nineteenth century. They were superseded by a new mechanism that appeared around the third quarter of the eighteenth century.

This mechanism, much smaller and simpler, was a whistle with a sliding piston. It comprised a small hollow cylinder with a highly polished interior. A piston, with a disc, fitting snugly inside the cylinder moved up and down inside it, thus changing the effective length of the tube, much as notes are produced by a trombone slide. Low notes require greater length, and high notes a shorter length of tube. Movements of the piston were commanded by a series of cams, metal discs whose outer profile contained the programme controlling its up and down trajectory. Air was provided by a bellows system, with its own set of cams governing the air supply, and thus the beginnings and ends of the notes. By *c.*1800 the whistle with sliding piston had become the most common form of mechanical bird song. A 1794 letter from Jean-Frédéric Leschot refers to singing bird pieces with 'only one flute', a clear allusion to the whistle with sliding piston; Leschot stated that he no longer made any other type.[82]

The whistle with sliding piston was indeed widely used by the Jaquet-Droz and Leschot in their prestigious creations. It had probably been designed by Jacob Frisard, Jaquet-Droz's principal subcontractor. A specialist in mechanical singing bird mechanisms, Frisard often prepared his cams in a unique way: he made them in graduated sizes so that when placed together they formed a continuous spiral, resulting in an unbroken song from start to finish.[83] This new mechanism had two important effects: it allowed singing birds to warble what Jaquet Droz and Leschot called their 'natural' song, which was generally sandwiched between two popular melodies of the time: tune, natural song, tune. Secondly, it allowed the creation of much smaller musical pieces, for the mechanism could easily be fitted into a great number of miniature items and *objets de vertu* such as watches, rings, bracelets, perfume flasks, miniature cages, snuffboxes, and

---

79 Kurz 1975, 91.
80 *Affiches de Paris*, 11 December 1749, section 'Ventes ou inventaires'.
81 Musée d'horlogerie du Locle, Château des Monts, MHL inv. 1.

82 Bibliothèque de Genève, MS suppl 961–4, letter from JF Leschot to Henri Maillardet, 18 July 1794.
83 Mayson 2000, 15.

even tiny pistols that shot out warbling birds in the place of bullets; often tiny watch movements were fitted in these pieces as well.[84] From a musical point of view the result was pleasing, with a clear and audible sound; visually it was delightful.

Perhaps the most popular type of decorative piece was the snuffbox, or *tabatière*. This very fashionable item of the late eighteenth and early nineteenth centuries was a luxurious and stylized version of the boxes that snuff users carried in their pockets. Made of gold or silver, often adorned with finely painted enamel scenes and precious stones such as pearls or rubies, these idealized snuffboxes were often fitted with watches. *Tabatières* were among the Jaquet-Droz's most popular specialties in the final decades of the eighteenth century. One can imagine the pleasure of surprising friends by pulling a box from one's pocket, from which a tiny bird popped out as if by magic, warbled a tune, and then vanished, as suddenly as it had appeared.[85] It was to this sort of pleasure that John Roger Arnold referred in 1791 when he wrote to his colleague Abraham-Louis Breguet, then visiting London, with a request: 'my father hopes you will be so kind as to come see us in Cornhill this coming Friday morning and bring in your pocket the little bird, since he very much desires to show it to some ladies'.[86]

WATCHES WITH MUSIC, AUTOMATA, AND SINGING BIRDS

Watches with miniature singing birds, just as charming as singing bird snuffboxes and even easier to wear or carry, were also produced for a privileged clientele in the final years of the eighteenth century. Although by this time watches had become a common personal item, they remained relatively costly, only to be owned by the well to do. How much more desirable and exclusive, then, were watches that contained their own built-in entertainment in the form of music and animated scenes! In an era of still unjaded senses, when the nerves were not frazzled by an overload of noise and images, how astonishing these small timepieces—easily portable and just as easily set in motion—must have been as sources of instant amusement. Similarly to clocks, in which bells initially served to sound the hour but from which it was only a short step to make them play brief melodies, striking watches were provided with tunes played on bells fitted inside the case (Figure 115). The principle was familiar: a small, pinned cylinder, in turning, activated levers that raised hammers, which struck the bells.

Watches playing chime tunes were being made by the late eighteenth century and became fairly common over the course of the following decades. The earliest musical watches produced their melodies through small sets of miniature bells, generally very thin and fairly wide in diameter. Space was optimized by stacking the bells in a row, from the smallest to the largest (Figure 116). In this manner, five or six bells could be placed within the case, each nested inside another bell that was slightly larger and therefore lower in tone, and still the case did not generally exceed half of an inch.

Although the limited number of bells allowed only simple tunes to be played, better-made watches had two hammers per bell, allowing quick repetitions of the same note. Watches with bells generally played a tune on the hour, though the music could often be played at will by means of a slide on the band.

One example of such is a watch now in the Louvre Museum, which was made around 1768 by John Archambo (1699–1777), a London watchmaker of Huguenot descent. It plays two tunes on six bells with twelve hammers; the names of the melodies—'Mirliton' and a 'Trumpet Tune'—are engraved on the tune-changing rosette.[87] Other examples of carillon watches of the period are signed by Joseph Martineau, Timothy Williamson, and William Carpenter, all of whom worked in mid- to late-eighteenth-century London.[88] The account books of the Jaquet-Droz contain mention of this type of watch: in March 1784, for example, they sold 'two large carillon watches' to James Cox and son.[89] One such piece was recently acquired by the Geneva Musée d'Art et d'histoire; signed 'Pierre Jaquet-Droz'. It played five melodies on the hour or on demand, initially on five bells with five hammers (however, due to a later intervention, the piece now has only four bells and four hammers).[90] Watches with singing birds represented the height of contemporary luxury; the small number of surviving examples indicates their scarcity even at the time of their creation. The earliest known watches of this type, presenting an enamelled or feathered bird perched on a branch above the small dial, and containing a whistle with sliding piston to provide the bird song, have generally been attributed to Jaquet-Droz.[91] A later watch with whistle and sliding piston by Piguet & Meylan, in which a bird is perched on a multi-colour gold bouquet, is in the Basel Historical Museum.[92]

In the early years of the nineteenth century, Jacob Frisard sought success as an independent constructor and exhibitor of mechanical pieces. On 3 March 1809 one of his singing bird watches was examined by the prestigious 'Comité de Mechanique' of the Geneva Société des Arts. They reported:

> Mr Frizard presented to the committee a gold watch garnished with pearls ... which has the particularity that when one pushes a slide on the band, one sees a small bird appear,

---

84 Very small *serinettes*, however, did exist; one was sold at Sotheby's New York on 11 June 2015, lot 104.
85 Mayson 2000; Bailly 2001.
86 Chapuis & Gélis 1928, ii, 86.

87 Louvre Museum, inv. No. OA8390. Cardinal 1984, 145 N° 156.
88 Ord-Hume 1995, 196–7.
89 Jaquet-Droz account books, personal archives.
90 Musée d'Art et d'histoire, Geneva, inv. H-2019-15.
91 Antiquorum sale, 31 March–1 April 2001, lot 423.
92 Historisches Museum, Basel, 1982.1001; Saluz 1996, 124.

Figure 115 Musical watch with châtelaine, Patek Philippe, Museum, Geneva.

Figure 116 Movement of the watch in Figure 115 showing the nested set of bells. Patek Philippe, Museum, Geneva.

which warbles a tune and its natural song, and when it has finished, returns to its place inside the watch case.[93]

Mechanical singing birds were often used as visual and musical enhancements in small decorative objects. These precious *objets de vertu* are of a seemingly endless variety, including miniature bird cages, *nécéssaires*, and perfume flasks. Two examples, today in the Musée d'horlogerie in Le Locle, Switzerland, are noteworthy: a delicate miniature gold and enamel cage with a feathered bird that flutters, turns and whistles a tune, and a lyre-form pendulette signed Jaquet-Droz et Leschot. When the clock strikes the hours, the singing bird perched atop it begins to sing and turn. As it does so a panel opens in the base, revealing animated silver swans that swim across a mirror lake; in the background a waterfall is charmingly simulated by rotating spiral-cut glass rods.[94]

At the turn of the century Geneva was the principal centre of production of musical and automaton watches and *objets de vertu*. With its large population of highly skilled, independent-minded artisans, the city possessed a long and fiercely defended tradition of innovation, sophistication, and excellence. These conditions encouraged the creation of a myriad of highly decorative watches, embellished with engraving, beautiful enamel scenes, half-pearls, and four-colour gold work. These watches, which continued to be in great demand throughout the nineteenth century, often took the form of flowers, animals, seashells, musical instruments, and many other amusing shapes; many were further enhanced with music and animated scenes. One especially popular type was the 'Jaquemart' watch (known in France as 'Martin/Martine') in which automaton figures on the dial appeared to strike bells to sound the hours and sometimes the quarters, with the sound actually being produced by a bell or gong inside the watch case. These were, in essence, tiny versions of the mechanical Jacks that struck turret clock bells for the general public, now miniaturized so as to be slipped inside the pocket or attached to a waistcoat. The figures were produced in a variety of styles, such as Europeans, Africans, Native Americans, oriental figures, and others. However, the subjects of automaton watches are so numerous and varied that it is difficult to offer an exhaustive list; a few examples follow here. One very unusual automaton watch, a demonstration of the strong Protestant faith shared by most of Geneva's horological community, depicts a religious theme. Inspired by the biblical episode of Moses bringing forth water for the people of Israel, it depicts Moses as he strikes the rock with his staff; the rock opens to reveal

---

93 Archives de la Société des Arts, Comité de Mécanique 1787–1821, 93.

94 Musée d'horlogerie du Locle, Château des Monts, MHL inv. 24 and MHL inv. 6, respectively.

a flowing waterfall from which the Israelites thirstily drink.[95] A second automaton watch, made in Geneva c.1800–10 by Meuron & Cie, has a skeletonized dial that reveals its movement. The vari-coloured gold scene on the dial features a man, a fox, a monkey, and a parrot. The man and the monkey strike the hours when the repeat is operated.[96] Another example, made during the same period and known as 'The Dutch Kitchen', is generally attributed to Pierre Simon Gounouilhou (1779–1847) of Geneva. It shows a lady spinning in her kitchen as chickens roast on the hearth (a dog running on a wheel turns the spit), and water (a twisted glass rod) flows from a fountain.

Other watches depict automaton dancers, tightrope walkers, harpists, violinists, and many other musicians, often made of engraved, varicoloured gold, with articulated limbs and heads. Alongside these musical and choreographic activities, and the more mundane pursuits of the various washerwomen, cooks, and blacksmiths, there existed another type of automaton watch. These were 'erotic' watches, shown only to the initiated. In them, at the owner's demand, animated figures engaged, tirelessly and repeatedly, in sexual intercourse. This sort of watch was produced from the late eighteenth century onwards. The Jaquet-Droz company made them; in the 1780s their account books enumerate a number of watches featuring what they referred to as 'lewd figures' ('*figures lubriques*'), which they sold to James Cox, who presumably exported them to the Orient. The specific nature of these early erotic watches, though almost always concealed by the cuvette or hidden behind a sliding metal curtain, was sometimes hinted at in their case decoration: the billy goat, for example, was a symbol of unbridled libido that was recognized by contemporary connoisseurs.[97]

COMB AND CYLINDER MOVEMENTS

In the late eighteenth century a new type of mechanical musical movement was invented that completely revolutionized mechanical music and the production of watches and other items with music and automata by the early years of the following century. Presented to the Geneva Société des Arts in 1796 by Antoine Favre (1734–1820), the physicist Marc-Auguste Pictet (1752–1825), then vice-president of the Society, described this invention as:

... a carillon without bells that plays two tunes and imitates the sound of the mandolin, housed in the lower part of a tabatière of ordinary dimensions; [the commission] considers this invention a new and ingenious idea. The sonorous pieces that replace the bells are very easy to install and to tune; they are not liable to crack as are the bells of carillons. The nature of the sound they produce is less likely to cause prolonged vibrations, thus eliminating the disadvantage of bells, which create a confusing effect when notes are played in quick succession. The easy execution of this new instrument is such that ... it would take more time simply to adjust adequately a number of bells than to construct the same number of these new musical pieces. This invention will be especially precious in animated horological pieces because it offers the double advantage of being easily executed and of saving space.[98]

The system described by Favre consisted of a series of tuned steel blades that vibrated when plucked by pins set in a turning disc or cylinder, so producing musical tones. This type of movement could offer a wider range of notes within a much smaller space than carillon or organ movements. Thus it became possible to produce thinner watches, as well as a wide variety of other items, both large and small. It appears that Favre was not the only one to have invented what came to be known as the 'comb and cylinder' movement due to the arrangement of the steel blades, which were generally laid out in a row like a comb. A few prior pieces with tuned steel blades have been recorded, including several musical watches constructed in Nancy around 1770 by Michael-Joseph Ransonet (1705–78).[99] By the following generation, such movements were frequently used, particularly in Geneva, which was then the primary centre of the 'novelty' horological industry. Antoine Favre did not have the means to invest in the exploitation of his invention. That task fell to his colleagues, among them Jean-Frédéric Leschot, Henry Daniel Capt (1773–1841), and Capt's early business partner Isaac Daniel Piguet (1775–1841). Although the latter was credited by some with the invention of the comb and cylinder movement (in Piguet's 1815 petition to become a Geneva burgher, his peers stated: 'he is the first to have produced here musical pieces, of which he is the inventor'),[100] it is more likely that he and Philippe Samuel Meylan (1772–1845) were simply among the first to grasp its technical, artistic, and commercial potential.

Several kinds of movements with tuned steel blades were employed: the '*barillet*' movement, in use before 1810, in which pins set in the barrel lifted and released the stacked steel teeth; in

---

95 Fondation Edouard et Maurice Sandoz, on view at the Musée d'horlogerie du Locle, Château des Monts; another example is in the Patek Philippe Museum in Geneva.
96 British Museum, no. 1958, 1201.287.
97 For a general account of such watches, see Carrera 1980, who illustrates on 50–1, a watch that unites the themes of suggestiveness and singing birds. He also discusses (136ff.) the application of such automata to snuff boxes.

---

98 Saluz 1996, 17: Report of the Commission of the Geneva Société des Arts, 7 March 1796.
99 Blyelle 2011.
100 Requête de Bourgeoisie, 4 August 1815, Archives d'Etat de Genève; Saluz 1996, 20.

the earliest pieces a limited number of blades were used. A second type, the '*sur plateau*' movement, which came into use around 1810, had a flat disc with protruding pins that in revolving plucked the tips of the steel teeth that were fanned out around the periphery of the case. This arrangement, which optimized the space available inside the case, resulted in a very slim movement yet could offer a wide range of notes. An ingenious improvement that enhanced musicality by increasing the number of notes that could be played consisted in placing the pins on both sides of the revolving pinned disc, with a double set of teeth arranged both above and below the disc. Alternately, small groups of teeth could be set on either side of the disc. A third type, which gradually became the predominant one, was the 'comb and cylinder' movement, in which a revolving pinned cylinder acted on a straight row of tuned steel teeth.[101]

Initially the teeth had to be individually made and were attached separately to the base plate by means of tiny screws. Progress in metalworking soon allowed groups of two or three teeth to be manufactured as a unit. These units were attached side by side to the plate, forming what was called a 'sectional comb'. Eventually it became possible to manufacture an entire comb by cutting slits into a single piece of steel; the teeth were arranged from highest to lowest, with the lower teeth being longer than the higher ones—that is, their slits penetrated more deeply into the steel plate. If rapid repetitions of a particular note were required, that note could be included two or more times in the comb. Further changes improved musical quality as weights were used to give depth to bass notes, and dampers were added to the tuned teeth to eliminate undesirable noise.[102] As of around 1820, the multi-tooth comb laid out in a straight line gradually became the most common. This is the type of movement one sees in music boxes, both large and small, in many musical watches and objects, and even in some larger musical clocks. The advent of the comb and cylinder musical movement made it possible to accelerate production and lower production costs. It was, therefore, particularly well adapted to the growing industrialization of the following century.

SERIAL PRODUCTION IN THE NINETEENTH CENTURY

The social and political upheaval of the late eighteenth century had a profound influence on the horological industry. By the early nineteenth century, with the emergence of a more democratic society, musical and automaton clocks and watches could no longer be considered playthings for the elite—one-of-a-kind pieces made on commission for wealthy patrons. Nor could they remain merely symbols of status and privilege. Methods of production were changing as well, with the increasing presence of machines in the workshop and the gradual generalization of serial production. The pieces of the period, while still not cheap, became accessible to a much wider audience. By the second half of the century, the majority of musical and automaton pieces—clocks, musical boxes, snuffboxes, watches—were being fitted with comb and cylinder movements, which were easily and cheaply produced.

A new generation of makers of musical and automaton clocks and watches appeared. Among the most important and prolific were the Piguet & Meylan firm. Founded by Isaac Daniel Piguet (1775–1841) and Philippe Samuel Meylan (1772–1845), both from the Swiss village of Le Chenit in the Vallée de Joux, the company was active from 1811 to 1828. Piguet & Meylan specialized in luxury watches with music and automata and exported much of their production to the Orient. While not the creators of the genre, Piguet & Meylan nevertheless perfected the serial production of the musical watch with automata, though the series remained relatively small and still entailed a great deal of individual work by many skilled artisans. The firm was one of the most influential in early nineteenth-century Geneva and played a leading role in the expansion and development of the market for musical and automaton pieces.

One example of the superb work of Piguet & Meylan is a pair of heart-shaped watches made for the Chinese market (for which pieces were generally offered in mirror pairs).[103] Decorated with pearls, turquoises, and fine enamel scenes, each watch opens to reveal an automaton scene with a boy playing the pan pipes, a girl playing the mandolin, and a windmill turning in the background. The music is played by a nineteen-tooth comb on a 'plateau' movement.[104] One of Piguet & Meylan's most popular pieces was the 'barking dog' watch. On its dial there was a realistically chased gold dog that moved its head, apparently barking furiously at a gold swan slightly above it on the dial (other variations include a cat, or doves). The dog's 'bark' sounded on the hour and the quarter hours. The sound was produced by a bellows connected to a small whistle, with an opening in the band to allow the sound to escape.

A new generation of singing bird makers, who adapted their creations to new styles, appeared in the nineteenth century. Among them were the Rochat dynasty, founded by David Rochat (1746–1812), a master clockmaker from Le Brassus in the Vallée de Joux, Switzerland. Three of his six children were makers of singing bird pieces. Towards 1800 the firm of 'D. Rochat and sons' was created. David Rochat's sons moved to Geneva in 1813 after their father's death. They were: François Elisée Rochat (Jacques François Elisée, 1771–1836), who worked mostly at Terreaux de Chantepoulet 39 in Geneva, assisted by his son Ami François Napoleon (1807–75); Frédéric Rochat (David Henri Frédéric, 1774–1848), who became a member of the influential

---

101 Saluz 1996, 143.
102 Ord-Hume 1995, 194–203.

103 For many examples of these see, Mirror 2010.
104 Patek Philippe Museum, inv. S-133 a + b.

Société des Arts around 1832, was aided by his brother Samuel and later by his sons Louis (b. 1795) and Antoine (b. 1799), working at 76, rue de Coutance in Geneva; and Samuel Henri (1777–1854), who never married and seems to have worked with his brothers and nephew all his life. The Rochats made singing bird pieces of exceptional quality. They were able to adapt successfully to the post-Revolutionary landscape, while continuing to remain faithful to the tradition of excellence they had learned as subcontractors to the Jaquet Droz.

Charles-Abraham Bruguier senior (1788–1862), who also made music box movements for a time, was the founder of the Bruguier dynasty of singing bird makers. The Bruguiers produced singing bird pieces at 87 rue de Coutance, Geneva, beginning around 1823 and continuing until the late nineteenth century. Blaise Bontems (1814–93), who originally came from eastern France, settled in Paris in the mid-nineteenth century. His son Charles (b. 1848), Charles' son Lucien (1881–1956), and Blaise's nephew Alfred Bontems (1856–1936) all made singing bird pieces until after the death of Lucien, when the business was sold to the Reuge Company. The Bontems produced many singing-bird cages and bushes featuring many birds perched on a small tree, placed under a glass globe; several also feature clocks.

While the heyday of singing bird *tabatières* and other pieces was the eighteenth and nineteenth centuries, a handful of makers remained active in the twentieth century. These include Karl Griesbaum (1872–1941) and his sons Mathias (1902–74) and Karl Joseph Griesbaum (b. 1916), and the Eschle company (founded c.1925), both in Triberg, Germany; Charles-Armand Marguerat (1887–1931), the brothers L. and F. Cattelin (active from the late nineteenth century to the early years of the twentieth century), and Antoine Salmon (c.1876–1951) of Geneva, the Reuge firm in Sainte Croix, Switzerland (founded 1865); and E. Flajoulot (active 1st half 20th century) in Paris. The pieces made by these firms and individuals, some of which feature clocks, reflect the materials and production practices of the time. Unlike the luxurious singing bird pieces of the eighteenth century, the objects produced during the twentieth century were more affordable and tended to be less technically complex, with later pieces often featuring cheaper materials like plastic and inexpensive metals.

THE TWENTIETH CENTURY

The main developments of the twentieth century included the production of very large automatically playing instruments, and a variety of new forms of 'programming'. The music played by many of these large pieces was usually very loud, reflecting their frequent use in train stations, restaurants, and other public places. As such, these instruments often featured clocks prominently to inform their audiences of the time, and sometimes with coin-operated musical entertainment. Electricity was increasingly used as a source of energy for machines that had to function for long periods of time. Among the different types of 'programmes' that came into use were metal discs with protruding shells that, in revolving, activated the teeth of tuned metal combs. These could be produced in various sizes for different kinds of machines. New discs could easily be purchased, and changing the discs was simple, thus making it easy to add new tunes to the machine. Musical programmes could also be recorded on perforated cardboard books that folded, zigzag style, or on perforated paper rolls.

CONCLUSION

Music and automata have accompanied clocks since nearly the beginning of horological history. In the clepsydras of Antiquity they marked the passing of the hours both visually and audibly, sometimes embodying allusions to well-known cultural references such as the legend of Medusa or the labours of Hercules. In medieval European cathedral clocks, mechanical music and animated figures enhanced the presentation of religious doctrine to a largely illiterate audience. During the Renaissance animated musical clocks served as a tangible sign of luxury and ingenuity. When, during the Renaissance, clocks became smaller, they provided entertainment and status to the fortunate citizens wealthy enough to possess them. As they became increasingly affordable items for personal use, watches evolved into sources of personal entertainment, first in the form of carillon watches, and later with automaton scenes. Singing birds and other musically accompanied animations offered endless entertainment when placed in smaller items such as watches, snuffboxes, and other *objets de vertu*. While their forms and audiences have varied widely over the centuries, these pieces have always been prized, not for their horological precision, but for their ability to generate surprise and delight. These qualities have always overshadowed their timekeeping capacities. Throughout their history, musical and automaton clocks have had in common one important function—that of generating astonishment and awe in the minds of their beholders.

CHAPTER ELEVEN

THE QUEST FOR PRECISION
ASTRONOMY AND NAVIGATION

Jonathan Betts

Precision timekeeping was born from the needs of astronomy and navigation. However, timekeeping in one form or another was, and is, essential to almost all modern scientific endeavour: without accurate timekeeping, practical research and the consequent advances would have been impossible.

It was only slowly during the fifteenth and sixteenth centuries that clocks became viable tools for astronomical measurements and observations. The idea to use clocks for this purpose came up quite early; a note in the register of the cathedral of León (northern Spain) about the lunar eclipse of 17 February 1421 states: 'This day, six hours after midday, the moon began to darken, and the eclipse continued until half past seven or later, so the eclipse lasted two clock hours'.[1] It is unclear however whether the cathedral clock was actually used for this observation. The first definite instance of the use of geared clocks for astronomical purposes is an observation that was carried out in 1484 in Nuremberg by Bernhard Walther (1430–1504), who kept records of his frequent astronomical measurements. On 16 January 1484, he notes:

> I observed Mercury with a well regulated clock, which gave the time precisely from noon the day before to noon. I saw Mercury in the morning in contact with the horizon, and simultaneously I put a weight on the clock, which had 56 teeth in the hour wheel. It made one complete turn and 35 teeth more, by the time the centre of the Sun appeared on the horizon, whence it follows that Mercury on that day rose one hour 37 minutes before the Sun, which almost agrees with calculation.[2]

Walther also noted the use of the clock during a lunar eclipse on 8 February 1487. His use of the clock is interesting since clocks at that time did not typically have minute hands. He overcame that problem by counting the teeth of the hour wheel in the open **going train**. The hour wheel of his clock had fifty-six teeth and since he seems to have counted full teeth only, the accuracy of his observation of celestial longitudes was limited to about +/− a quarter of a degree.[3]

Although Walther had to count the teeth on the hour wheel to derive a minute's reading of his clock, in principle, minute hands and separate minute dials were already known at his time. The Almanus Manuscript from around 1480 shows a clock with a minute dial.[4] It was spring driven and showed the minutes in intervals from 5 to 60 as the dial 'for the sunrise'. While this specific example might mostly have served mundane purposes, clocks were also used for observational purposes in Italy, as witnessed by the note 'when the clouds prevent the direct observation of the stars, one can, with a well-set clock, find their positions on the astrolabe'.[5] Minute indications, however, were not widespread, and certainly not widely used for astronomical observations at the time. Moreover, with the **verge** and **foliot** controlling such a clock, the minutes indication would have been doubtful.

After the middle of the sixteenth century we have more detailed information on the requirement for precise mechanical clocks for astronomical observations, and also the oldest preserved pieces. In

---

1 Pérez Álvarez 2015a.
2 DeB Beaver 1970, Walther's Mercury observations are famed: they were used by Copernicus (1473–1543) for his heliocentric model in *De revolutionibus orbium coelestium* (1543), since he had never succeeded in observing Mercury—which he attributed to the hazy atmospheric conditions surrounding Frauenburg in western Prussia.
3 Matthes & Sánchez-Barrios 2017.
4 Leopold 1971, clock 4.
5 Paolo Tritii, MS I.20.Sup, Veneranda Biblioteca Ambrosiana, Milan.

1557, Paulus Fabricius wrote in his *Encomion Sanitatis* that Andreas Wolf, 'Rentmeister' (Quaestor) of Regensburg, had a timepiece that showed minutes, seconds, and 'even shorter time intervals'.[6] Another archival note on seconds hands—and the first one that proves its use in astronomy—is from star observations in Kassel[7] and informs us that a seconds clock or watch was in use on the observatory that Wilhelm IV of Hesse-Kassel had constructed for determining the positions of the stars in 1586. Furthermore, in the same year there is a note of 'a clock with weights which shows the hours, minutes and seconds'.

A watch from the 1550s, now preserved in the Kunsthistorisches Museum, Vienna,[8] is the earliest known with a minute hand and may have been used in astronomical observations. It has a flat cylindrical case of 60mm diameter. It is marked 'TW' (or 'WT') with an engraving on the inside of the upper lid, which may refer to the German clockmaker Wenzel Tobias (d. 1616), who was active in Prague.[9] The twenty-four-hour dial in the centre has a small hand and is marked twice 1 to 12 with half-hour divisions. The central disc has the hour hand attached directly to it, reading the hours on the innermost numbered ring of the dial. This is surrounded by four concentric rings, each equally divided into fifteen minutes. The outermost ring has 0 to 14, in Arabic numerals, the second ring from the outside has 15 to 29 and following. The hand is constructed like the rule of an astrolabe to allow precise reading at each radius; it does not point to the correct minute number, but rather has a fiducial edge that moves across the dial four times each hour, and hence the moment when it reaches a new minute (marked by a line) it can be read off very accurately. The specific division to allow each minute to have an angle of 24° on the dial is unique for the time and allows reading this watch to a precision of around ten seconds. This makes it the first known example of a watch from which seconds could be read.

Currently, the oldest preserved clock with a seconds hand is only a few years older, c.1560.[10] It is a weight-driven clock with three separate dials showing (from the bottom) hours, minutes, and seconds. Although the authenticity of some parts of the movement has been questioned,[11] the dial is certainly authentic to the period. Given the peculiar arrangement of the dial, this may have been a timekeeper for use in astronomical observations. Being weight-driven, it stood a better chance of approaching an accuracy to match the precision of its minute indication, though the veracity of seconds indication could only have been trustworthy in the very short term.

The astronomical observations in Kassel in 1586 were carried out with clocks that according to one contemporary measurement deviated from a sundial by less than one minute from noon to noon,[12] an accuracy that Tycho Brahe found hard to believe given his own experience with the inaccuracy of his clocks, 14.5 minutes during the same time period,[13] a figure much more in line with modern experience of the longer-term stability of these early clocks. However, for short periods, with careful adjustments, such clocks were probably able to maintain better timekeeping than this, and from the 1550s such clocks began to serve as useful instruments for astronomical use. The next step—beyond the precision of movement execution in Bürgi's clocks—was only possible when the pendulum was introduced.

Astronomical observations aided by clocks became a near-global phenomenon by the last quarter of the sixteenth century. Following observations in Germany, France, and Italy, observers in the Spanish empire undertook 'the first coordinated worldwide attempt at systematic astronomical observations.'[14] Following the Treaty of Tordesillas in 1494, knowing with certainty the location of newly discovered territories had significant geo-political implications.[15] Hence, improved accuracy in terrestrial longitude measurements became a political necessity. The court of Philip II of Spain (1527–98) initiated a lunar eclipse project (1577–88) to augment geographical longitude accuracy by measuring lunar eclipses. Since the eclipses would take place near dusk in some parts of the Indies and hence the shadow might not be visible clearly for the measurement, the instructions clarified that the observers should have a 'geared clock' on hand.

An astronomer working in the astronomical tradition of the Islamic world started to make use of mechanical clocks for observational purposes at a similar time. From 1573 to 1580, Taqī al-Dīn al-Rasid (1526–85) was chief astronomer of the Istanbul observatory built by him for Sultan Murad III (1546–95). The aim of the observations made from this observatory was the improvement of Ulugh Beg's (1394–1449) astronomical tables. Taqī al-Dīn was very aware of the sources of error in observations, quoting Ptolemy: 'I would have been able to establish a great regularity in method if I was able to measure the time precisely.'[16] This is why one of his key initiatives was building a clock. In his *Sidrat al-muntāh* he says 'We built a mechanical clock with a dial showing the hours, minutes and seconds and we divided every minute into five seconds'.[17]

---

6 Winter 1980.
7 Gaulke & Korey 2007.
8 Inv. nr. KK 1572; see Matthes 2018, ill. 6.7–6.9.
9 Abeler 2010, 616.
10 Nuörnberg, Germanisches Nationalmuseum, WI 318. Eser 2014, cat. 44 and references therein.
11 Coole & Neumann 1972.

---

12 '... with our minutes and seconds clock, which is very accurate in giving time/and from noon to noon often it differs by less than one minute ... which we use to determine longitudinal angles'; Dreyer 1913–29, vi, 51.
13 'The clock has moved too fast by 14.5 minutes by the next noon compared to the previous noon and I don't know where the error comes from', Dreyer 1913–29, xi, 183.
14 Portuondo 2009.
15 The treaty of Tordesillas settled geographical disputes between Portugal and Spain by establishing that any newly discovered territories west of a line from the north pole to the south pole at 46°37′ were Spanish and those east of that line Portuguese—leading to the colonization of Brazil by Portugal and most of the other territories by Spain.
16 Tekeli 1963, 71–122.
17 Tekeli 1966.

From this we know he is the first recorded astronomical user of a clock that is more precise than showing minutes.

LONGITUDE

The need for accurate star charts was not just of academic interest. In the early sixteenth century the practice of astronomy had a serious practical mission. Late in the previous century the beginnings of ocean exploration had revealed an urgent need for more certain ways of navigating accurately when out of sight of land. Navigators looked to astronomers to solve what would, in the coming two centuries, become a notoriously vexed issue—the determination of the *longitude*, one's east–west position, at sea.

Establishing a ship's position north–south of the equator, the *latitude*, was relatively straightforward by observations of the Sun at noon (or the pole star at night, in the northern hemisphere). Theoretically, finding longitude when out of sight of land was also relatively simple. It was well understood in sixteenth-century Europe that a comparison of one's local time (established by an observation of the Sun, or the stars around the pole star at night) with the time at one's home port, on the reference meridian, *at that same instant*, would provide the difference in local time between the ship and the home port. Local time difference is the same thing as the difference in longitude: each hour of local time difference around the globe is equivalent to fifteen degrees of longitude, so time difference between the ship's current position and that at home can be used to calculate how far around the Earth one is from home. Combined with one's latitude this provides the ship's position at sea. This is assuming, of course, that an accurate chart of the oceans and coastlines was available, which itself was only possible once longitudes could be determined accurately. Knowing the time at home at the instant of the local time observation was, of course, the difficult issue.

Two solutions for this were proposed in the early sixteenth century. In a 1514 treatise on geography, Johann Werner of Nuremburg published the basis of what became known as 'the lunar distance method', and in 1530, the Louvain professor Gemma Frisius (1508–55) discussed the theory of an alternative solution, using a marine clock. Both solutions enabled the longitude to be found by determining home time remotely on-board ship. The theoretical marine clock of Gemma Frisius would be set before leaving port and would be kept carefully to indicate home time throughout the voyage, but no portable clock nearly accurate enough was available at the time.[18] Werner's lunar distance method employed the night sky as a celestial 'home time clock', the stars providing a kind of dial, and the Moon moving against that background in a very regular and predictable way throughout the month, providing the 'hand'. For this theory to be possible, of course, accurate tables of the motions of the Moon were necessary—a task which proved far more difficult than anyone supposed.

By the end of the sixteenth century the maritime nations of Europe were increasingly aware of the serious nature of 'the longitude problem'. As greater numbers of merchants and adventurers turned from coastal navigation to the open seas, losses of both life and precious cargo began to multiply. Ships were wrecked on unexpected and ill-charted shores and shoals. They also sometimes failed to find accurate passage into the correct currents and winds and sat becalmed or were forced to adopt longer routes. Either might result in supplies of water and food running out, often with terrible consequent disease and death. In known waters, many captains, therefore, chose the safer option of sailing due north or south to the required latitude, and then 'running their latitude down' until reaching their destination. This, however, could double the length of a voyage and, again, place the crew at risk of running out of supplies. Furthermore, sailing such a well-established course placed the ship at risk of ambush and piracy.

In addition to these dangers, competitive colonial and commercial interests were also pressing and advancement of marine navigation in their service became imperative. The nation whose ships proved fastest to reach far-flung trading posts and could first safely explore and colonize new lands enjoyed a tremendous advantage in the scramble to increase economic prosperity and expand its boundaries of empire. Similarly, fleets that could quickly, safely, and strategically position themselves enjoyed a much greater chance of success in warfare. The governments of the seafaring nations of Europe were acutely aware of the implications of being first to determine longitude at sea accurately and each made considerable efforts to encourage resolution of the problem.

In 1567 Philip II of Spain offered a substantial financial reward for a solution, followed in 1598 by the offer of 6,000 ducats and a pension of 2,000 ducats from Philip III, his successor. In the early seventeenth century, both the Portuguese and the Venetian governments offered rewards for a means of finding longitude and in 1606 another large one—100,000 florins—was put up by the Dutch government.[19] In the mid-1630s the Italian astronomer Galileo Galilei (1564–1642) was in negotiations with the Dutch for a reward for his method of finding longitude by using Jupiter's moons to determine 'home time',[20] while an official commission was established in France in 1634 to investigate a possible solution, concentrating on the lunar-distance method proposed by the astronomer-mathematician Jean-Baptiste Morin (1583–1656).[21] In 1667, Louis XIV created the Paris Observatory, appointing Jean Dominique Cassini (1625–74) as its first astronomer in order to promote astronomical research and improve the cartography of

---

18 Gould 2013, 11

19 Ditisheim *et al.* 1940, 16.
20 Van Helden 1996, 92; Bedini 1991, 9ff.
21 Ruellet 2016 (with full references to earlier work), 173–213. For other methods proposed, see Marguet 1931, 97–103; Turner 1996, 117–19.

France. England followed the French path but aimed the astronomical remit squarely at improving marine navigation, and in 1675 Charles II founded the Royal Observatory at Greenwich, just outside London. At the time, few dared even to speculate that a sea-going clock might seriously be the answer: the sole function of the Greenwich observatory at its foundation was to advance astronomy as a means of improving navigation and, specifically, to resolve ways of determining longitude at sea. To be useful this meant knowing one's position to an accuracy of about half a degree of longitude (at the equator this represented a distance of about 55km). For a marine clock to provide this accuracy, it needs to be correct to within +/- two minutes, and this perhaps after a voyage of many weeks. In the sixteenth century it was well known that no clock was capable of such accurate timekeeping: Frisius' theory would remain only that for over two centuries. Thus, Europe's observatories were founded primarily to develop an *astronomical solution* to the problem. At the same time there occurred one of the greatest improvements in timekeeping technology— the introduction of the pendulum as a timekeeping element in clockwork.

Galileo Galilei's use of a swinging pendulum for timing observations of heavenly bodies, and his famous design for a mechanism to drive it while counting the swings, must represent the first of all pendulum mechanisms. In fact, Galileo himself never saw it in physical form and it was his son Vincenzo (1606–49), in collaboration with a young locksmith Domenico Balestri (*fl.* mid-seventeenth century), who created much of the design in metal in 1649. This oscillation-counter, however, remained unfinished until about 1656 when it was completed as a clock by Johann Philipp Treffler (1625–98) in Florence.[22] Although not surviving, the clock mechanism was drawn by Galileo's pupil Vincenzo Viviani (1622–1703), and it is on that drawing that modern reconstructions of Galileo's mechanism have been based. The drawing definitely shows a clock mechanism controlled by a pendulum. It could not have been a 'pendulum swing counter' with the pendulum pushed by hand; the **escapement** is specifically designed to impulse the pendulum.

Viviani, who lived with Galileo and witnessed these events first hand, recorded them in a letter written in 1659, though some scholars doubt Viviani's claim that the clock was fully complete. It is suggested that his description of a working clock was actually of an unfinished movement which, without weight or spring, had to have the **great wheel** driven manually.[23] In either case, the concept was clearly one of a mechanical mechanism, intended eventually to be driven by weight or spring and to have a dial, either indicating numbers of swings of a pendulum or the passing of time.

In fact, Galileo was motivated by other scientific projects, including a determination to solve the longitude problem astronomically. Like Galileo, the great astronomer Johannes Hevelius of Danzig is also known to have used a cord and weight as a manually swung pendulum from about 1650 for timing his observations, and like Galileo, Hevelius was at that time attempting to design clockwork to keep a pendulum swinging but, before he succeeded, a fully developed form of pendulum clock was created by 1656 by the extraordinarily talented Dutch scientist Christiaan Huygens.

Opinion is divided as to precisely when Huygens conceived the idea and where his inspiration came from. It was suggested at the time that Huygens had heard of Galileo's design, something hotly denied by Huygens, but historians have argued, following Galileo's correspondence with the States General of Holland in 1636 and with Admiral Lorenzo Reael in 1637 in which he proposes a pendulum controlled *numeratore del tempo*, that Huygens would have known of Galileo's design for a pendulum clock some years before completing his own.[24] However, it is generally accepted that it was Christian Huygens who designed the first practical, working pendulum clock. The great horological historian Granville Hugh Baillie sums it up well by saying: 'It is, then, fairly certain that Galileo conceived the idea of a pendulum clock before Huygens, and it is possible that others did also, but there is no question that Huygens was the first to make a pendulum clock and publish it to the world'.[25]

As for when Huygens' design first saw physical form, opinion is also divided. Until recent years, received wisdom has generally been that Huygens first conceived a complete design for a domestic clock with pendulum control in December 1656, the date based on a letter sent by Huygens to Ismael Boulliau in 1657, in which he states the precise day when he finalized the design for a pendulum clock. In June 1657, a patent was then granted by the States General of Holland to Salomon Coster and both he and the London clockmaker John Fromanteel, who travelled to the Hague to work in Coster's workshops in the same year, have generally been regarded as the makers of the first pendulum clocks of Huygens' type. However, Sebastian Whitestone has argued that Huygens, in fact, had at least one experimental long pendulum clock made earlier in 1656, probably by Isaac Thuret in Paris, which enabled him to conclude by December 1656 that the design was successful, and that the Coster clocks were the first production run of *domestic* clocks but were not the first *pendulum* clocks.[26]

Described by Huygens in his seminal treatise *Horologium*, published in The Hague in 1658, his design for a pendulum clock effected an extraordinary advance in timekeeping. Because the swinging pendulum (Figure 117) employed gravity as its *restoring force* (the force that tends to cause the oscillator to want to return to its central position), the time of its swings was much more constant. From an accuracy of about a quarter of an hour a day, clocks could now keep time to within a few *seconds* a day. Indeed, the pendulum invention is considered by many to

---

22 Bedini 1991, ch. 3.
23 Edwardes 1977, 19–20; Whitestone 2017.

24 Robertson 1931, 88; Edwardes 1977, 24; Morpurgo 1958.
25 Baillie 1951, 58.
26 Whitestone 2008.

be the most important advance in the whole history of horology. Unlike Galileo, who had seen his pendulum as useful in astronomy and had concentrated on the astronomical alternative for solving the longitude problem, Huygens always believed his invention of a pendulum-regulated clock might also be a way to find longitude. Pendulum timekeeping was of the right *kind* of accuracy but initially he believed that only longer pendulums (of over three feet in length and beating seconds) would be sufficiently accurate, and the inconveniences of such a design delayed further development by him. It was only after Scottish nobleman Alexander Bruce showed interest in his ideas that progress was made in the early 1660s. Bruce commissioned an experimental marine timekeeper with a short pendulum and its performance at sea was sufficiently acceptable to convince Huygens to join in his investigations. By December 1662 Bruce had commissioned two experimental marine clocks controlled by pendulum from the Hague clockmaker Severyn Oosterwyck and tested them on a voyage across the North Sea. The crossing was very rough. One of the clocks fell off its mounting and was damaged: the other stopped, although they were able to start it again when the weather calmed and its going thereafter was felt to be encouraging. The damaged clock was replaced by another made by the London clockmaker John Hilderson, and the experiments continued. In 1663 the two clocks were taken on a voyage in the forty-gun *Reserve* by Captain Robert Holmes (c.1622–92) to Lisbon, and again the results were quite promising. The following year Holmes took the clocks on a longer voyage to the west coast of Africa, returning with tales of their spectacular performance, the good news being published by the Royal Society.[27] It is believed that the remains of at least two of these important early timekeepers have survived. One, signed by Oosterwyck, resides in the collections of the Scottish National Museum, and the other is now in the Royal Museums Greenwich collection.

In spite of these promising trials, it seems that interest in furthering the experiments in England was limited and in 1666 Huygens moved to France. Here he joined the Académie Royale des Sciences, and the development of designs for marine timekeepers controlled by short (nine-inch long) pendulums continued, the Paris clockmaker Isaac Thuret being a close collaborator. Over the next four years three sea trials took place, the second, undertaken in the Mediterranean, being the most successful, while the others were either disappointing or inconclusive in their results. In 1673 Huygens published *Horologium Oscillatorium*, his magnum opus, in which he described his longitude experiments up to that time and included designs for marine timekeepers as yet untried. These included a clock with a 'triangular' pendulum suspended from two points, widely separated, ensuring that it would only swing in one plane. However, it was only some years later that

---

27 Recent research, however, has revealed that the clocks had not gone nearly as well as Holmes claimed, and he was falsifying the evidence to ingratiate himself with the members of the Royal Society. Jardine 2008, 283–9.

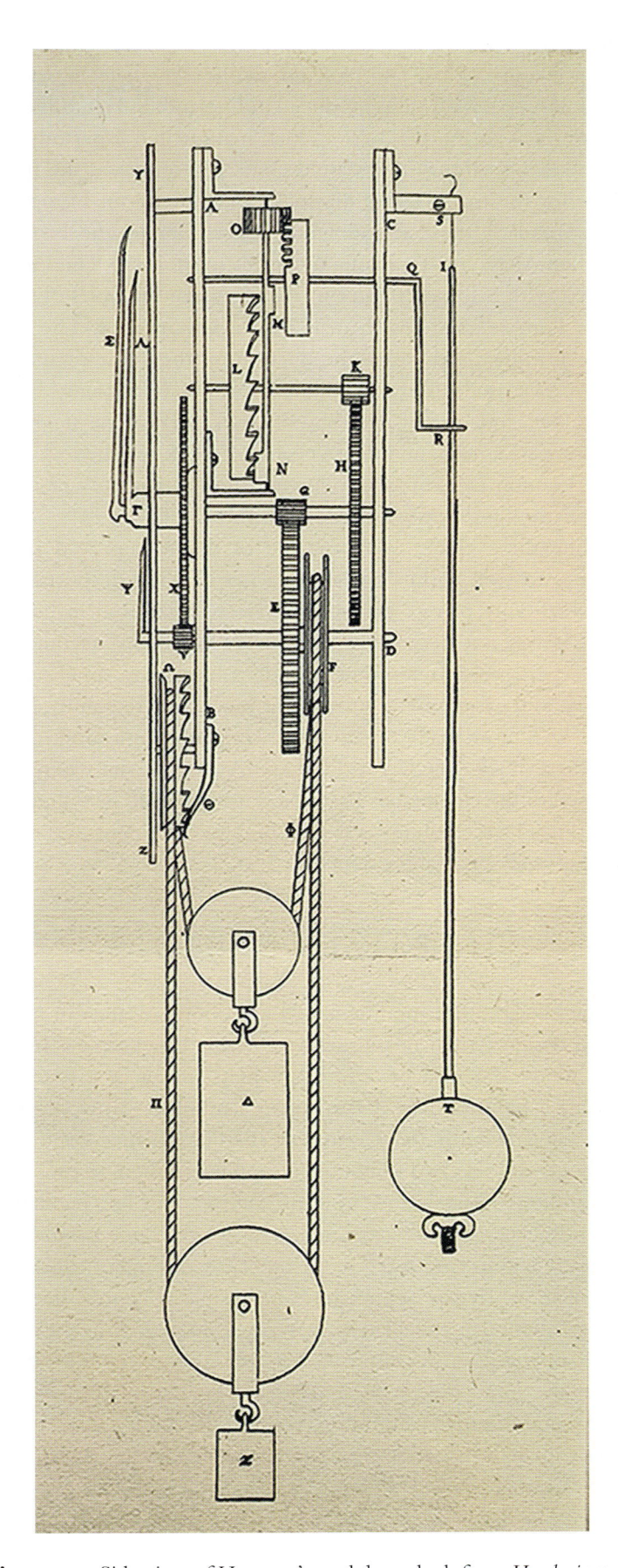

Figure 117 Side view of Huygens' pendulum clock from *Horologium*, 1658. Photo: Jonathan Betts.

Huygens's experiments on marine timekeepers began again, by which time he had returned to The Hague.

As well as in France and Holland, the new 'high-tech' pendulum timekeepers were also produced in England. In fact, this early development of the pendulum clock inaugurated a period over the following century that would establish London as the finest centre for the manufacture of clocks and watches in Europe. It was particularly the celebrated Fromanteel family who began it all, when Ahasuerus Fromanteel sent his son John to The Hague to work with Salomon Coster in 1657 to make examples of these clocks alongside the Dutch maker for a few months. The Fromanteels immediately began production for the domestic market of their own model of the pendulum clock, both in spring-driven table clock and weight-driven wall clock form. The next few decades were a new and intense period of clock production, with many London clockmakers following the Fromanteel's practice, particularly the group associated with Edward East.

While these clocks were capable of much improved timekeeping, to well within one minute per day, it is important to emphasize why private customers were so interested in buying them. It was not because of their new 'high-tech' accuracy as timekeepers *per se*. It was owing to their long-term timekeeping stability, as they required setting right much less frequently. Such clocks were now made to run for a week or even a month, and with these clocks, owners had the luxury of knowing when they returned to the clock after that period it would still be telling the right time. That capability was new, highly convenient, and made the new clocks extremely desirable.

THE BALANCE SPRING

Meanwhile, another monumentally important advance in horological design would revolutionize portable timekeepers. In the early 1660s, the idea of applying a light spring as a restoring force to a **balance** was occurring to a number of investigators in England and on the Continent. Earlier, balance-controlled watches and clocks had a light spring applied to them to cause the 'rebounding' of the balance in its oscillations, enabling regulation of the timekeeping,[28] but this idea was not the same as applying a consistent restoring force to an oscillator throughout its cycle of oscillation. By the application of a spring to control the balance throughout its oscillation, it was reasoned that, in the same way as a pendulum was controlled by gravity—giving it a natural frequency—one could replicate the same feature in an oscillating balance and improve its timekeeping, too. In about 1660, Hooke drafted a patent in London for applying a slender, light spring to the balance of a watch to give the oscillating balance a restoring force and hence its own natural frequency. Appropriately, Hooke described his spring as providing an *artificial gravity* to the timekeeper. However, he did not find time to develop this idea much beyond theory, and no practical workable design resulted from his plans at that time. There were also European contemporaries of Hooke's who had similar ideas. Huygens noted that in November 1660 the watchmaker Martinot, in Paris, had told him of a design of his for 'a spring in lieu of a pendulum'. In 1665, in answer to a letter from the soldier and natural philosopher Robert Moray (1608/9–73), Huygens stated that—also in 1660—the Duc de Roannez and his friend, the physicist Blaise Pascal (1623–62), had described a similar invention to him but that the means by which the spring was applied rendered the design defective.[29]

In early 1675, however, it was Huygens himself who came up with a practical design for a balance spring. He had the brilliant idea of winding a slender steel ribbon into a volute—a flat spiral—with its inner end attached to the axis of the balance and its outer end attached to the frame of the movement. This was successfully fitted to the balance of a watch in 1675 and the stage was set for a whole new breed of accurate portable timekeepers. When Huygens sought to patent the invention, both Robert Hooke and the French physicist, the Abbé Hautefeuille (1647–1724), strongly objected: both claimed priority, but neither could reasonably claim to have conceived of the spiral balance-spring applied to the axis of the balance. And it was this design that was quickly adopted by watchmakers in England and on the Continent, elevating the performance of pocket-watches from perhaps +/- fifteen minutes a day to just a minute a day or so.

Putting a balance spring in watches was similar to applying the pendulum to clocks. In fact, the early watches with a balance spring were often known as 'pendulum watches' for this reason, but the improvement in accuracy was not as dramatic as in clocks and was not nearly stable enough to make them a contender for navigational use. Huygens realized this too and, after experimenting with marine-timekeeper designs employing a balance (of wheel form) with balance spring (made for him by The Hague clockmaker Johannes van Ceulen), he concluded their timekeeping would still not be stable enough. He also realized, significantly, that the steel balance spring that provided the 'artificial gravity' was affected by temperature change, getting weaker as it got warmer, and so he went back to experimenting with pendulums for a marine timekeeper. It has been suggested that Huygens's design for one with balance and balance spring was nevertheless pursued by Isaac Thuret: a description of a timekeeper, said to be based on Huygens's work and intended for marine use, was published in 1999.[30] However, there is neither convincing context nor provenance for this timepiece, and its authenticity is heavily in doubt.

After a year experimenting unsuccessfully with an oscillator in the form of a metal ring suspended horizontally on three threads and swinging in a rotating mode (that is, rising and falling as the threads twisted and untwisted), Huygens returned to his idea for the triangular pendulum, which he had not yet tested. Two clocks

---

28 Described in Chapter 5, Section 1.

29 Defossez 1946, 196–7.
30 Plomp 1999.

were fitted with this new oscillator and were mounted in **gimbals** to ensure they remained horizontal. The movements, which were spring driven with a **fusee**, were also fitted with a **remontoire**, a device to ensure that the impulse to the oscillator was as uniform as possible. These clocks were tested in 1686 on a voyage to the Cape of Good Hope and, although the results were not as good as expected, Huygens concluded that they showed real promise. The clocks were tested on another voyage to the Cape in 1690–1, returning in 1692, but this time the results were not as good, and Huygens had to conclude that the design had obviously not made progress. Having now lost faith in his triangular pendulum design as well, he next pinned his hopes on yet another gravity-controlled design of oscillator, his 'perfect marine balance'. In this new model, Huygens employed an oscillating balance but with gravity applied to it as a restoring force. After much thought and experimentation, he settled on a design incorporating a balance mounted in a vertical plane, with a mass suspended on a line acting near the centre of the balance. As the balance swings, the mass is lifted up and down, the weight of it on the thread providing the restoring force, although the design was more subtle than this description makes it sound. Christiaan Huygens was not just a good designer of mechanisms: first and foremost, he was a mathematician and physicist with a profound understanding of the theory and geometry of the pendulum and the forces acting on it. In fact, just having a constant force performing the function of gravity on the oscillator is not precisely what is needed for the best possible timekeeping, and Huygens tried to design his marine balance to be theoretically perfect in its concept.

Huygens understood that, for an oscillator to keep perfect time, its swings had to be *isochronous*—literally, from the Greek, *of the same time*. In other words, all the swings of the oscillator, whether large or small, had to take the same time. In order for it to behave in this way, its restoring force has to have a linear relationship with the motion of the oscillator: that is, the restoring force must increase proportionally as the oscillator moves away from the position of rest. The way Huygens achieved this was by suspending the thread between 'cheeks' mounted on the centre of the balance. As the balance swings increased, so the thread acted at a greater and greater radius from the centre of the balance. The new perfect marine balance was constructed for Huygens, according to his direction, by another clockmaker, Barent van der Cloese. By April 1694, this clock was complete, and he was carrying out experiments with it that continued until the middle of the following year. There is also a reference in Huygens's correspondence to the conversion of an existing clock to his new balance and escapement.[31]

However, that was where the experiments ended, as Huygens died in July 1695. Whether he would ever have succeeded with this new design is doubtful. It may have performed well when stationary, but could not have done so in the longer term in any sea trials, since we know that relying on gravity as a restoring force when the oscillator is subjected to vertical disturbances will always introduce instabilities in timekeeping. Even so, Huygens's contribution to the world of horology was immense: his invention and publication of the practical pendulum-clock design, his publication of the world's first statement on oscillator theory, and his creation of the first workable balance spring for portable timekeepers are all matters of monumental importance in the history of clock- and watchmaking, and the last of these, the invention of the spiral balance spring, would be an important element in the eventual successful design of the marine timekeeper in the following century. These achievements alone mark Christiaan Huygens out as one of the greatest contributors to the science of horology.

Ironically, however, in terms of the story of the portable precision timekeeper, there was also a negative effect consequent on his generally great achievement. In spite of several decades of careful experimentation and intelligent design, this widely admired scientist did not manage to produce a specification for a practical marine timekeeper. This, in turn, undoubtedly contributed to the growing opinion, including that of Isaac Newton, that clocks were unlikely to provide an answer to the problem of finding longitude at sea. A particular conviction of Huygens, confirmed after his experiments with short-pendulum marine timekeepers, was that to ensure stable timekeeping, such a device absolutely must have a large and low-frequency oscillator. Indeed, it was the generally held view among horological longitude projectors and seemed logical enough: watches were, simply, the worst timekeepers. Even the highest-quality watches, fitted with a balance spring and made to the best possible standards, could not be guaranteed to end a day's running within a minute of the correct time: in bigger clocks daily timekeeping was always much better and it seemed that the bigger they were, the better it was. Clocks with long, seconds-beating pendulums could keep time (given a reasonably constant temperature) to within a few seconds a week. Of course, this was only as long as they were fixed and not moving, but the absolute necessity for this with a pendulum timekeeper was simply not fully understood at the time.

Now that accurate clocks were available another fundamental question of astronomy arose—the Equation of Time, the difference between apparent solar time, as shown by a sundial, and mean time, or clock time. This had been known to astronomers from Ptolemy onwards, but needed to be explained and recalculated in mid-seventeenth-century Europe. Huygens, and shortly afterwards Flamsteed, set themselves to calculate and publish new accurate tables of the equation, to allow users of the new pendulum clocks to set them accurately using the Sun—usually via a sundial.[32] The availability of this data led to the creation of

---

31 OCCH, x letter 2891, 709–710.

32 For Ptolemy, see Neugebauer 1975, vol. 1. For the equation in the late seventeenth/early eighteenth centuries, see Daniel 2005; Davis 2003; Kitto 1999; Smith 1686; Turner 2015b.

clocks designed to indicate the Equation, first by London clockmakers such as the Fromanteels, and later Thomas Tompion and Joseph Williamson, with such clocks then becoming a speciality of a number of French makers during the eighteenth century.

One concept, the importance of which both Huygens and Hooke had recognized from the beginning of pendulum development, was that longer pendulums, swinging through much smaller arcs, not only required minimal energy to maintain their swings, but tended to be more stable timekeepers than shorter pendulums swinging larger arcs. Huygens fully understood this concept, which was also demonstrated in the experiment by Robert Hooke at the Royal Society in 1669, where a thirteen-foot pendulum was maintained by a watch movement. The problem with the use of long pendulums in clockwork in the 1660s was that the standard escapement used at the time, the verge and crown wheel, required a very large arc (over sixty degrees) to operate reliably—admissible in a small clock, but impractical with a long pendulum. Thus, apart from a few clocks designed by Huygens which employed gearing between the verge and the pendulum to reduce the pendulum arc, such clocks were not made during the first decade of the pendulum's existence.

Although the precise dates and creators remain uncertain, new designs for escapements, which allowed for smaller arcs and hence long pendulums, appeared in London-made clocks towards the end of the 1660s. The first to appear was probably that known as the cross-beat, with long-pendulum examples by Ahasuerus Fromanteel and Joseph Knibb known to have been made at this period. The design of the cross-beat escapement may well have been influenced by Jost Burgi's original realizations from the 1590s, and employed separate **pallets** linked together, and acting in unison. However, the escapement was complex to make and was soon superseded by what is known as the anchor escapement, much simpler to make and adjust. This was developed in the late 1660s in England, although who should be considered its originator remains debateable.[33] The anchor escapement allows a much smaller swing of the long pendulum, and hence a much-improved performance: ordinary clocks with seconds pendulum are capable of keeping time to within a few seconds a week, given a reasonably constant temperature, and the long pendulums was indeed a great improvement. But the anchor escapement itself is simply a convenient and practical way to keep a long pendulum swinging—the critical improvement was the long pendulum itself, not the anchor escapement. When considering the merit of a timekeeping system the first and most important factor to consider is the specification of the oscillator, not what impulses it. The means of impulsing the oscillator, the escapement, is important of course, but only of secondary importance. A well-chosen oscillator impulsed by a poor escapement will invariably keep better time than a poor oscillator, however well designed the escapement.

Although the verge escapement remained the norm in spring-driven clocks made throughout Europe for the following century, from 1670 the anchor escapement quickly became universal in English longcase clocks, superseding the verge.

On the foundation of the Royal Observatory at Greenwich in 1675, it was thus long pendulum clocks that were commissioned from Thomas Tompion. One of the first successful tasks of these clocks was to determine that the Earth is, in fact, spinning at a regular rate. This had always been assumed, but unless it was certain, the positional astronomy—the mapping of the stars—which was Flamsteed's principal task, would not be possible. We know today that the Earth's rotational rate is not actually wholly stable, but it was within Flamsteed's parameters, and allowed his work to proceed. The Greenwich clocks were year-going and were each controlled by a long pendulum of two-seconds period, that is, of thirteen feet in length, suspended above the movements, and with a very small amplitude.

Another special feature of the Greenwich clocks was that the escapements were of the **dead-beat** type. Almost certainly first invented in about 1675 by Lancastrian Richard Towneley (1629–1707), Flamsteed's friend and correspondent, the dead-beat escapement was an improvement over the anchor escapement as it does not cause the escape wheel (and the rest of the going train) to slightly reverse at every swing of the pendulum. The dead-beat escapement was thus in place in observatory clocks from the late 1670s, and the accounts that survive suggest the clocks were usually capable of timekeeping to within +/- ten seconds a day. However, this was no better than a well-made longcase clock with anchor escapement, and the dead-beat was not generally taken up for precision clocks until the 1720s, when Tompion's partner George Graham further developed the design.

Meanwhile, the longitude problem persisted. In a pamphlet of 1714 publicizing the need for a solution, the influential mathematicians William Whiston (1667–1752) and Humphry Ditton (1675–1714) observed that:

> Watches are so influenc'd by heat and cold, moisture and drought; and their small Springs, Wheels, and Pevets are so incapable of that degree of exactness, which is here requir'd, that we believe all wise Men give up their Hopes from them in this Matter. Clocks, govern'd by long Pendulum's, go much truer . . . .[34]

Of the small number of horologists who actually believed a timekeeper solution was possible at all, each appears to have held this view; so it is understandable that Huygens and those horological pioneers who followed him all believed the same. But, as Huygens found out, the large-clock solution was also flawed, Whiston and Ditton also noting:

---

33 Chapter 5, Section 2 at n. 14.

34 Whiston and Ditton 1714, 16.

But then the difference of Gravity in different Latitudes, the lengthening of the Pendulum-rod by heat, and shortening it by cold; together with the different moisture of the Air, and the tossings of the Ship, all put together, are circumstances so unpromising, that we believe Wise Men are almost out of hope of Success from this Method also.

Whiston and Ditton may not have been professional clockmakers, but their views were very influential in their day. High-profile supporters of Newton's ideas, Whiston had been Lucasian Professor of Mathematics at Cambridge University, while Ditton was Master of Mathematics at Christ's Hospital, and their opinions were therefore bound to be taken seriously. So, in the early years of the eighteenth century, it is hardly surprising that most of the scientific world continued to look to astronomical observatories for the solution to the longitude problem.

Astronomy, however, was not looking likely to provide a ready-made solution either. By the time of Whiston and Ditton's pamphlet, the Royal Observatory at Greenwich had been in business for almost forty years, with little prospect of the completion of the accurate star charts and lunar tables that it had been founded to provide, in spite of Flamsteed's intense work in observations and calculations over the preceding decades. Even in the late 1720s, by which time Edmond Halley (1656–1742) was the second Astronomer Royal at Greenwich, the lunar-distance method for determining longitude was nowhere near developed and it was said that Halley despaired of ever seeing it achieved.[35] Indeed, it was to be half a century before the lunar-distance method became practicable and the enabling data, in the form of the *Nautical Almanac*, generally available for mariners to employ it at sea.

At the beginning of the eighteenth century, British commercial and political ambitions were growing and interest in solving this increasingly acute navigational problem intensified in London. The famous maritime tragedy off the Isles of Scilly in 1707 is sometimes cited as a prime example resulting from not knowing one's longitude when at sea. In fact, this was not the cause, but the tragedy undoubtedly contributed to British public perception that something had to be done to improve navigation at sea. Thus, when a few years later, in 1713, the House of Commons received a request by Whiston and Ditton for funding to support a scheme to solve the longitude problem, the result was an offer of substantial reward. The British government's inducement was enshrined in the last Act of Parliament given royal assent by Queen Anne before her death in 1714. The Longitude Act involved three possible rewards. The greatest was £20,000 (equivalent to several million pounds today) and the successful method had to provide longitude to within half a degree on an official six-week trial voyage to the West Indies. The clearly stated requirements of the Act provide sufficient information to calculate how accurate a timekeeper would have to be to achieve that goal. At the time, an average voyage to the Caribbean took about six weeks (approximately forty-two days) and half a degree of longitude is equivalent to two minutes (120 seconds) of time. Therefore, it can be seen quite easily that a longitude timekeeper would qualify for the full reward if it were able to keep time to better than about +/- 2.8 seconds a day. At once the conundrum is evident: on the one hand, a good pendulum clock was easily capable of this given a reasonably constant temperature, but Huygens and others had positively proved that pendulums would not work at sea. On the other hand, any smaller, balance-controlled timekeepers, including pocket-watches, were incapable of better than a variation of +/- one minute a day, so that alternative was also universally discounted.

While the offer of such large financial reward attracted a great deal of interest, much of it misguided,[36] there were neither immediate solutions nor early results of any kind in terms of horological hardware. Nevertheless, in the years following the Act, one important attempt to create marine timekeepers was that made by the London-trained clockmaker Henry Sully (1680–1728). Inspired by the work of Christiaan Huygens and convinced that greater support was likely on the Continent than in England, Sully spent most of his horological career in France. There, his research resulted in four prototypes for a marine timekeeper, and a version of one of these made in 1724 is now in the collections of the Royal Museums, Greenwich. Study of this and two other surviving examples reveals Henry Sully to be a fine craftsman and a very resourceful and intelligent designer. However, the clocks he made all followed Huygens's last design and had a balance without balance-spring, but instead were controlled by gravity. Also following Huygens's strictures on the design of the oscillator, the balance was relatively large and slow moving, all of which resulted in the design failing to perform. Had Sully not died in 1728, aged only 48, he might have gone on to create something of real practicality; however, he would have needed the vision and courage to make a complete change in his philosophy.

Over half a century was to pass before significant longitude money was paid, and both the lunar distances and the timekeeper method were successfully developed.

The first viable chronometer also represented the first of all precision watches and was one of the major technological achievements of the eighteenth century. This is, of course, the story of John Harrison (1693–1776).

Harrison (Figure 118) tells us he heard of the longitude prize in 1726. It is clear that from the moment he did so, he was determined to try and solve the problem and the rest of his life was principally directed to this end. By this date, he and his brother James had set to work on a series of remarkable precision longcase-clocks with movements made almost entirely of wood. It was timely and experiments with the pendulum clocks were useful to see how far they could push the capabilities of a fixed-clock design

---

35 Harrison 1775, 19–20.

36 Gingerich 1996.

Figure 118 Portrait of John Harrison engraved by P. J. Tassaert (1732–1803) 1768, showing a regulator with his gridiron pendulum and H3 behind him, and H4 on the table beside him. Photo: Jonathan Betts.

before attempting a marine timekeeper. A remarkable feature of the design of these clocks was that Harrison eliminated the need for lubrication, ensuring a much-improved long-term stability for their performance. This was principally achieved by his choice of materials, the bearings mostly being made of the naturally dense, waxy, tropical hardwood *lignum vitae* in combination with brass, while his special 'grasshopper' escapement was designed to deliver impulses to the pendulum with a straight, pushing action, and no sliding, thus also avoiding the need for oil on those parts.

Harrison quickly identified the most significant bar to accurate timekeeping in a mechanical clock: the effects of temperature change. Almost all clocks tend to run slow when they get hot, chiefly because metals expand and because metal springs get weaker. With pendulum clocks it is mainly because the pendulum gets longer. The rod, which constitutes the main length of the pendulum connecting the suspension at the top with the 'bob' (the swinging mass) at the bottom was usually made of brass or steel, both of which expand with heat. Harrison solved this problem in 1726, by inventing a pendulum which, instead of having a single rod of metal, was made up of alternating rods of brass and steel. This ensured that the effective length of the pendulum (the distance between its point of suspension and its centre of gravity) remained the same, whatever the temperature. Owing to this compound rod having the appearance of a grid, Harrison's new type became known as a *gridiron pendulum*. Once he had fitted these gridiron pendulums to his longcase clocks, he tells us they achieved the astonishing accuracy of a variation no greater than one second in a month, a performance far exceeding the best London clocks of the day.

By the mid-1720s London clockmakers had made advances themselves, and George Graham, Tompion's successor, created what would become the standard precision pendulum clock of the eighteenth century—the type which became known as a *regulator* (Figure 119). This design was fitted with an improved version of Tompion and Towneley's dead-beat escapement, and initially with Graham's own version of a temperature-compensated pendulum. First conceived in 1721 and using a bob in the form of a jar containing a column of mercury, it predates Harrison's pendulum by four years and was the first design for a temperature-compensated pendulum. However, when Graham heard of Harrison's type, he preferred it as being more consistent in action and adopted the gridiron for his regulators in future years; his principal maker of regulators, John Shelton, continued to make what became the 'industry standard' regulator after Graham's death in 1751.

By the mid-1730s, a further design for a compensated pendulum had been invented by the noted London clockmaker John Ellicott (1706–72). Like the gridiron, this employed rods of brass and steel to provide an element of differential expansion, but in his case Ellicott used levers in conjunction with a rod of brass and a rod of steel to magnify the effect and control the position of the pendulum bob accordingly.

From the mid-1750s, following the advice of John Smeaton (1724–92), gridiron pendulums were occasionally constructed with a combination of zinc and steel,[37] requiring only five rods as zinc has a greater coefficient of expansion. This design eventually saw considerable use in regulators at the end of the century. Also, from the 1740s, following an idea of the Earl of Ilay (later 3rd Duke of Argyll, 1682–1761), pendulum rods in some regulators were made of wood, usually seasoned, straight-grained pine, removing the need for compensation.[38] However, it was either Harrison's or Ellicott's pendulum that was used in the majority of regulators constructed in England, and on the Continent, for most of the eighteenth century.

As for practical scientific applications of these early regulators, there was much fundamental research needing their use. In the early 1730s, clocks by Graham were used in continuing experiments on the period of a seconds pendulum at different latitudes, the different rate of the clock enabling the experimenters to estimate the relative difference in gravity and thus to determine the exact shape of the Earth. It was following this work that it was determined that the Earth is in fact an oblate spheroid, with the globe flattened at the poles. The papers given to the Royal Society were published by the Society in its *Philosophical Transactions* and these, with various equivalent publications of the Academie Royale des Sciences in Paris, are a rich source of scientific horological information. For example, an important and seminal work published in 1733 in the Memoires of the Academie Royale was the paper by François Joseph de Camus (1672–1732) on the correct forms of wheel teeth to be applied in horology; Camus' work has been the foundation stone of horological gearing theory ever since.

Empowered with the very encouraging performance of his 1720s pendulum clocks, John Harrison began thinking about marine timekeeper design. Only one logical course of action seemed clear: he would have to consider ways of making his precision pendulum clocks 'portable'. Following all his horological predecessors, Harrison believed a practical marine timekeeper would have to have a large and relatively low-frequency oscillator. But this progressive and apparently logical approach did not provide the successful design, although it was perhaps a necessary step towards discovering it. By about 1728 Harrison had formulated a plan for his first marine timekeeper and took his ideas to London, in the first instance approaching Edmond Halley, the Astronomer Royal at Greenwich. Halley felt unable to judge the soundness of Harrison's plans and suggested that he see George Graham, who, after a frosty start, was highly impressed and encouraged Harrison to carry on.

On returning to Barrow, Harrison spent the next five or six years, partly aided by his brother, constructing his first marine timekeeper (H1). This timekeeper is largely made of brass and it is interesting that Harrison, and perhaps his brother, were by this stage very able to work in this material, requiring distinctly different skill sets from that of the woodworker. It strongly suggests that

---

37 Wright 2019, 343–59 for an extant example by Henry Hindley from *c.*1756.
38 Desaguliers 1744, 301; Cumming 1766, 89; Ludlam 1769.

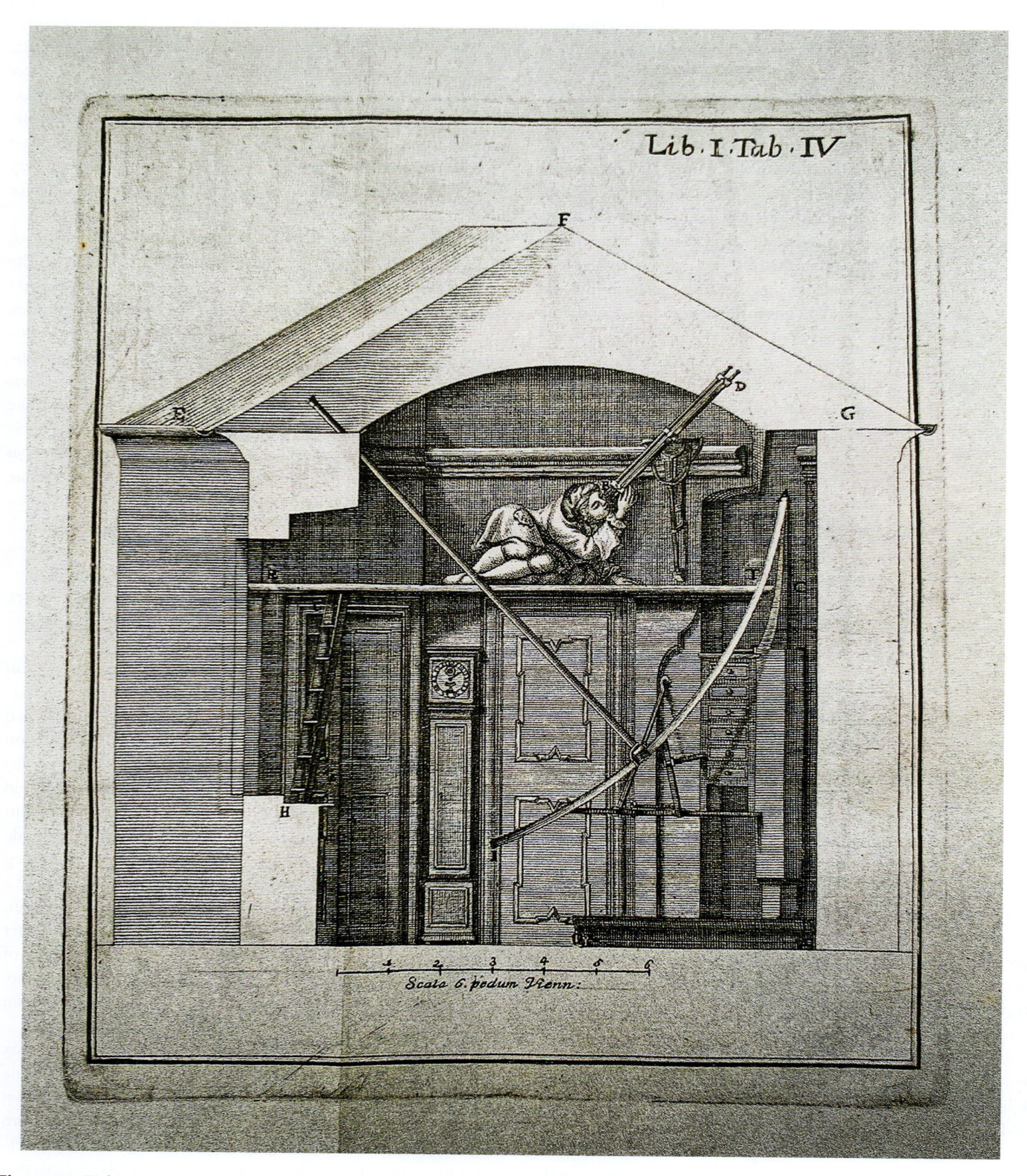

Figure 119 Eighteenth-century astronomy in action. A Graham regulator shown in the observatory of G. I. Marinoni (1676–1755), in Vienna in 1745. Image: Jonathan Betts.

Harrison had indeed had some experience working with a clockmaker in the past, probably in Hull, or that he took advice from such a craftsman. An important invention incorporated in H1 was Harrison's version of the device known as **maintaining power**. This is, of course, essential in a marine timekeeper. Harrison's version was completely automatic, and would be employed in almost every fusee chronometer and precision watch made thereafter.

THE FIRST COMPENSATION BALANCE: 1728

One of the most significant of all the inventions in H1 is the temperature compensation. With balance-controlled timekeepers, the effects of temperature change are doubly problematic. First, an oscillating balance will expand with rising temperature, giving it a greater moment of inertia, which causes the timekeeper to lose time. More important, the steel balance spring will get weaker as the temperature rises, reducing the restoring force. H1 represents the first moveable (as opposed to pendulum-controlled) timekeeper to have the refinement of a temperature-compensation mechanism. What is even more extraordinary is that, as originally designed, the compensation mechanism was incorporated *within* the balances themselves, and the compensation balance would become one of the fundamental tenets of good chronometer design.

Many clockmakers and men of science saw H1 in London in the 1730s and it is not unreasonable to suppose that, although the actual design had not proved reliable, the concept of the compensation balance was discussed on or after such occasions. According to Gould, the great French pioneer Pierre Le Roy (1717–85) was among those who then visited Harrison to inspect H1.[39] As discussed later, it was Le Roy who should be credited with having built the first successful compensation balance, made for his own timekeeper of 1766. Gould quotes Le Roy as describing H1 as 'of a very ingenious construction'.[40] Harrison always believed in this principle and in 1775 published the statement that the compensation should be in the balance,[41] an assertion which the younger pioneer John Arnold says was the inspiration for his own patent taken out just a few months afterwards. The compensation balance was fundamentally important—it would ultimately be adopted in modern chronometer design—and it is John Harrison who should be credited with the fundamental concept.

Following preliminary but rigorous tests on a barge on the River Humber, Harrison felt ready to have the machine more formally tested and H1 was taken to London in 1735.

The timekeeper was publicly displayed by Harrison to the scientific community at George Graham's premises and later in Harrison's own house, and it was widely regarded as a wonder of its age. Perhaps surprisingly, far from discouraging visitors and keeping this extraordinary device a secret, Harrison seems to have been happy to show it to anyone with an interest. It was a clever decision, as it brought the whole scientific community on-side, ready to give him support, which would certainly help his case when dealing with the Longitude Commissioners. No doubt, details of many of the inventions in the large timekeepers became known in this way and were thereby adopted beyond Harrison's workshop. Equally, having well-informed horologists and scientific gentlemen commenting on his designs would have presented him with a range of opinions on possible improvements.

Armed with the support of many of London's scientific elite in the form of a certificate vouchsafing the great potential in H1, a semi-official trial of it was considered worthwhile. Thus, Sir Charles Wager (1666–1743), the First Lord of the Admiralty, himself arranged that Harrison and his timekeeper should be sent to Lisbon on board the warship *Centurion* in May 1736. After H1's official baptism by a week-long voyage of tempestuous weather, Proctor wrote from Lisbon that Harrison 'was sea sick withall, but … seems satisfied that the motion of the ship was not in the least detrimentall to its keeping true time.' However, the ship's log reveals that the timekeeper did not perform as well as Harrison hoped and it appears to have been very much affected by the rough seas experienced during the passage.[42] At Lisbon, H1 was transferred to the *Orford* and on the long return voyage, in much calmer conditions, Harrison used it to correct a very serious misreading of the ship's longitude. On 30 June 1737, the Commissioners finally met in formal session, for the first time, to hear how H1's trial had gone and to inspect this model of 'high technology'. The news the Board received was evidence that, after all, a marine timekeeper might just prove to be a practical solution to the longitude problem.

In spite of its good performance on the *Orford*, Harrison was evidently not fully satisfied with H1's timekeeping. Realizing that it was not likely to perform better in a further trial, he did not ask for one but instead requested financial assistance from the Board to make an improved timekeeper. Recognizing that there appeared to be real potential in the design, the Commissioners awarded him an immediate £250, with the promise of another £250 on completion of the next version.

Harrison's second version was together and running quite quickly—probably within a year. However, not long after it was completed and under test, Harrison made the devastating discovery that the design of the dumb-bell balances was flawed and that it would be necessary to start all over again with one incorporating circular balances instead. On the second visit to the Longitude Commissioners in January 1741, when George Graham spoke on

---

39 Gould 2013, 45, 83, and 91

40 He does not, however, give his source, and every attempt to trace it has so far failed.

41 Harrison 1775, 103.

42 Ereira 2001, and Journal of the *Centurion*, 1 July 1735–2 October 1736, TNA ADM/L/C/82.

Harrison's behalf, he very cleverly concentrated on the exciting new design for the third timekeeper, and it seems that no reference was even made to H2. Having Graham present a highly positive update on Harrison's work was probably all that was necessary for the Commissioners to retain confidence in Harrison and grant the next instalment of support money. The *wording* of the decision of 16 January 1741, as recorded in the official minutes, is particularly interesting, as it suggests how the Commissioners themselves interpreted the Act at that time. In agreeing to award Harrison a further £500 to develop the third timekeeper, they wished to make it clear that the money being granted to him must be considered as part of the full amount of the award, should the timekeeper qualify after trial. They specify the trial (the 'Experiment') as being the voyage to the West Indies, and then state that (emphases added):

> ... if upon Experiment it shall be found that the said Machine will contribute to the finding out of the Longitude & *that thereby* the said Mr Harrison *shall be entitled* to receive any of the Rewards mentioned in the Act of the 12th of Queen Anne, that then, & in such case, the sum of £500 hereby granted to him & such other sum or sums of money as he hath heretofore received on account of other Engines invented by him for finding out the Longitude shall be deducted out of such Reward.

The meaning here seems clear. First, the single voyage they proposed to the West Indies was to be 'the trial' that is to determine entitlement. Second, if Harrison was found to be entitled, then it was those rewards specified in the Act of Parliament to which he was entitled. This strongly suggests that if Harrison's timekeeper had been sent out at that time and had performed on that trial to within the requirement of half a degree, the Commissioners were committed, on the official record, to awarding him the full amount of £20,000. In later years, however, the Commissioners would adopt a very different view on how Harrison might qualify for any reward.

In fact, it would be many years before the third timekeeper would be ready for any kind of trial, and the Commissioners were not called upon to grant any award money based on a successful trial to the West Indies. Nineteen years of painstaking labour were to follow and the whole period would prove to be a terrible mid-life crisis for Harrison, while also providing him with some extremely useful learning opportunities. There is every reason to believe that, soon after his second meeting with the Longitude Commissioners in 1741, he had his third timekeeper running. A complete redesign from his second, it incorporated large brass wheels instead of dumb-bell balances and placed one above the other, as opposed to side by side. Right from the beginning the timekeeper presented problems in its performance and through virtually the whole of the 1740s and 1750s Harrison struggled with the design. It was undoubtedly the most problematic and troublesome of all the projects he attempted. While some of his most important technical discoveries were made as a result of his development of H3, for almost twenty years it stubbornly refused to perform to the necessary accuracy, in spite of continual adjustments, redesigns, and rebuilding. In the end, although Harrison learned a great deal from this Herculean endeavour and incorporated a number of brilliant inventions into H3, its ultimate role was solely to convince him that the solution lay in another design altogether.

H3 may not have been a successful marine timekeeper, but it did incorporate two highly significant innovations that went on to serve in many applications in later technologies. In order to compensate for temperature changes, Harrison invented a new and revolutionary device, the temperature-sensitive **bi-metallic strip**. This was, and still is, a tremendously versatile invention: most households today contain appliances utilizing it as a thermostatic control or temperature-sensitive switch. Another vitally important invention that Harrison created for H3 was the *caged roller-bearing*—the ultimate version of his anti-friction devices. The design was the predecessor of the caged ball bearing, a device used in virtually every machine made today with rotating components. H3 had another great claim to fame as far as Harrison's development was concerned. It may itself not have performed well but certain features proved incredibly useful to him as they enabled him to understand better the principles with which he was dealing. In working on the diameter of the balances of H3, Harrison tells us how he discovered that the result of increasing the radius of the mass in the rim of the balance in a simple linear series (i.e. in equal steps) increased the energy stored in the balance much more rapidly; in fact, quadratically (by a square law), which he had not expected. This understanding would prove extremely useful a few years later when designing the balance for his longitude watch. Working on the spiral balance spring in H3 was also very revealing for Harrison, as he learned that flat-spiral springs of a small number of turns do not produce linear torque, i.e. they do not obey Hooke's Law.

The education that resulted from Harrison's continued experimentation on H3 not only provided him with much new knowledge and a better understanding of dynamics, but also a gradual realization that the fundamental scale of H3's oscillators was part of the problem. At least as early as 1750 this was becoming clear to him and, while continuing his struggles with that timekeeper, he began to consider a completely different approach. Just as his situation was becoming critical, in the early 1750s Harrison made the breakthrough he was desperately seeking. It was not, after all, a large timekeeper like H3 that would solve the longitude problem, but something on a much smaller scale. The successful design, developed into his fourth timekeeper and known today as H4, appeared to be simply a large watch. The view generally held during the first half of the eighteenth century was that watches were simply not capable of sufficiently good timekeeping. As will be shown, around this period there do appear to have been one or two experiments on watches, in attempts to improve their performance, but these also seem to have confirmed the old view that they were unpromising. So what was the route by which Harrison succeeded, and what was it about his design for a precision watch that induced more stable timekeeping?

Following Huygens' work, others finally recognized that a pendulum would never be sufficiently stable as an oscillator on a moving ship. However, the concept of a heavy, low-frequency, low-amplitude oscillator was still firmly held as the most promising for development by all in the next generation of horological projectors. The appearance of seconds-beating balances, as seen in the work of Sully and Harrison (and later by Thomas Mudge in London, and Pierre Le Roy and Ferdinand Berthoud in France) was thus an absolutely logical progression from that thinking. The problem with that simple theory—one easy to see with the perfect vision of hindsight—is that the typical disturbances to which such a timekeeper would be subject when on a moving ship were precisely those which upset it most. By contrast, a timekeeper with a different specification of oscillator in which the balance beats five times in a second and has a total arc of swing of, say, 240°, will be much less easily disturbed by external motion. This was, incidentally, observed as far back as 1722, when Jean de Hautefeuille (1647–1724), in a work on portable timekeepers, noted that a watch with a large balance went very well lying static, on the table, but very badly when hunting, whereas a watch with a light balance went relatively well while hunting, but no better when at rest.[43]

Thus, to achieve the high frequency and high amplitude needed for a portable timekeeper, the choice came back to watches, which already employed reasonably high-frequency balances. Of all the horological products being made in the eighteenth century, watches were also easily the most common kind of timepiece and represented by far the most important part of the contemporary international industry. The common pocket-watch of the period, with an ordinary verge escapement, had a balance of relatively high frequency (commonly 17280 beats in an hour, or a little under five beats a second) and an arc of swing of about 100°: but, as already noted, these watches were relatively poor timekeepers, even the best being incapable of variations under about one minute a day. What, then, was the problem that seemed to render them incapable of stable timekeeping? Evidently it was not primarily one of disturbance caused by external motions, as the relatively high-amplitude and high-frequency specification of the oscillator defended against this. The answer is that there was another very significant disturbance to which all mechanical oscillators are subject. This is one introduced from *within*, by the escapement itself, and is caused by the variation of forces which act directly on the oscillator. There are, however, ways of minimizing these effects. Principal among these is the concept of using an oscillator that has relatively high stored energy within it. If the oscillator has a great deal of energy stored in proportion to the energy delivered at each impulse, then the effects of the escapement, and any variations in the impulses delivered by it, are proportionately reduced. In the example of pendulum clocks, where the pendulum was given high stored energy by, for example, making the bob heavier, the arc of swing larger and the overall length greater, the pendulum's motion was much less affected by this internal disturbance. In his pendulum clocks Harrison described this factor as the 'Dominion' of the pendulum over the wheels, and this quality was a major contributor to the good performance of pendulum clocks.

The same protection can be applied to the balance of a watch. Indeed, there is a greater need for such protection in a watch as it is spring driven and, even with a fusee, the force delivered to the balance will vary more than in weight-driven clocks. So, how does one increase the stored energy in a watch balance? There are four ways, but all have limitations, and an optimum level has to be found:

1. The *mass* of the balance can be increased.
2. The *diameter* of the balance can be increased, which like the mass, also increases its moment of inertia.
3. A greater *amplitude* in the balance will also store more energy in the oscillator.
4. Increasing the *frequency* of the oscillations of the balance, by increasing the restoring force (i.e. fitting a stronger balance spring), will equally store more energy in the oscillator and will equally render it less affected by external motion.

Increasing the amplitude and the frequency of the oscillations increases the velocity of the balance rim, which, like increasing the diameter, has a particular benefit. The energy E stored in any moving body of mass m moving at velocity v is defined as $E = 1/2\, mv^2$, from which it can be seen that increasing the velocity of the moving body also increases the stored energy as a quadratic function, and thus much more effectively than simply increasing the mass of the balance.

However, there was a problem. Ever since the introduction of the balance spring in the mid-1670s, there had existed within the watchmaking profession a cardinal rule in the design of the balance and spring. This required that the oscillator should always be designed so that the watch was *self-starting*: if it stopped, perhaps because allowed to run down, it must automatically start itself once the drive was restored to the wheelwork. Watches made in the period before the balance spring were, by their nature, always self-starting, and the idea that the new 'high-tech' watches with balance springs would need to be shaken or twisted to start them would have been anathema to both maker and customer. It is certain that by the mid-1750s John Harrison knew he was onto a very exciting new lead in the improvement of precision watch-work, but where the first hints for his development came from is still uncertain.

George Graham had introduced the only significant improvement in watch-work in the eighteenth century up to that time, that of the development in about 1726 of the **cylinder** (or 'horizontal') escapement from earlier designs by Thomas Tompion. An improvement over the verge, the standard watch escapement of the day, the cylinder had the advantage that it was a form of

---

43 Hautefeuille 1718.

'dead-beat' escapement and therefore did not cause recoil. This avoided some of the extra friction that recoil necessarily introduced but, more importantly, without the recoil the cylinder escapement was not subject to the large, gaining escapement error which comes with the verge. The cylinder escapement also allows a slightly larger amplitude for the balance but did still have considerable friction in its action. The cylinder-escapement watches of the period employed a balance of similar specification to those with a verge escapement, and the effect of friction was still able to take its toll on the timekeeping. Cylinder-escapement watches, though marginally better than the equivalent verge watch, did not, in fact, manage significantly better timekeeping.

Another escapement that was briefly popular in English watches in the 1730s and 1740s was known as the 'tumbling-pallet' escapement. Invented in France and announced in July 1729[44] by Painel de Flamanville, who was secretary to the Comte de Colovin, the envoy for Russia to the Swedish court, it was also a frictional-rest, dead-beat escapement employing an escape wheel similar to that in a conventional verge escapement, but with drum-like pallets with parallel impulse faces. Thiout remarks, hugely optimistically, that he thought watches with this escapement might keep time to within some seconds in a month, but it is unlikely they would have gone better than the equivalent cylinder-escapement watch, if the balance was of the same type.

There are only a few indisputable facts about what happened concerning watch improvements in this period, but these can be used as chronological landmarks in discussing what might have occurred, leading up to the creation of H4. What is undoubtedly highly significant information in this story comes from a remark made in a paper read to the Royal Society on 4 June 1752 by John Ellicott. The paper described and discussed the types of temperature-compensated pendulum he had invented. By that time, the concept of temperature compensation in clock pendulums had been known for over twenty years and, by the middle of the century, the real benefits of employing such a sophisticated pendulum were becoming clear. Ellicott evidently decided at this point to stake his claim to his own, more recent version of this 'state-of-the-art' device, which was a rather less reliable system than Harrison's or Graham's and employed levers within the bob of the pendulum. At the end of his address, Ellicott remarked:

> Before I conclude this paper, I beg leave to acquaint this honorable Society, that in the year 1748, I made the model of a contrivance to be added to a pocket watch, founded upon the same principles, and intended to answer the like purpose, as the pendulum above described; and at a meeting of a council of this Society, on the fifteenth of February last, I produced a watch (which I had made for a gentleman) with this contrivance added to it, and likewise a model, by which was shown to the gentlemen then present what effect a small degree of heat would have upon it. But as I have not yet had sufficient trial of this watch, I shall defer giving a particular description of this contrivance, till I am fully satisfied to what degree of exactness it can be made, to answer the end proposed.

From this comment, we know that in February 1752 Ellicott was in possession of a pocket-watch fitted with temperature compensation, almost certainly the first of the kind ever made. Ellicott was a very successful and ambitious horological retailer and one with known scientific interests. The success of clocks with temperature compensation was well established and Harrison's prototype marine timekeepers were also by then well-known both for incorporating temperature compensation and apparently being significantly improved by it. So it was perhaps inevitable that, sooner or later, someone would try out the same device in a watch, just to see if its performance might also be at all improved. It seems that person was Ellicott. Having so publicly staked his claim to this production, he now had to put his new piece of technology to the test. There were, no doubt, others in the Society and beyond who followed the progress of the experiment: if it succeeded, the commercial and scientific implications were great.

By this date Ellicott employed many top craftsmen to make his watches and clocks for him and we know that some of his best watches were, in fact, made by Thomas Mudge. It therefore seems highly likely that it was Thomas Mudge who made the watch shown by Ellicott to members of the Royal Society in February 1752. If so, it is interesting to speculate what kind of temperature compensation it would have had. The most up-to-date concept in such a device was the bi-metallic curb Harrison had created for H3, described publicly in 1749 at the Royal Society. A few years later, in 1755, one of the watches Mudge made for the King of Spain, Mudge's No. 260, was given temperature compensation consisting of a brass and steel bi-metallic curb, but coiled into a spiral and acting on a separate curb mounted on pivots, and having curb pins embracing the balance spring (in fact, one of two on the balance in that particular watch). So, what happened to this important and, after his presentation to the Royal Society, relatively well-known experiment of Ellicott's? As he stated, he was careful not to make predictions of final performance or give 'a particular description' too early. In the event, no more was heard of the watch and it seems highly likely that it did not come up to expectations since, if it had, Ellicott would certainly have made public claims to a successful design, as he did with his pendulum compensation.

Some later references to this experiment support the view that it was not a success. Over ten years later, in 1763, when Harrison was complaining that people did not believe his marine watch H4 was going well, he stated: 'But still they say a watch is but, or can but be a watch, and that Mr Ellicott has tried what a watch will do and that the performance of mine, though nearly to truth itself,

---

44 In the *Mémoires pour l'histoire des sciences et des beaux arts . . .*, 1330, and the *Mercure de France*, 544–5.

must be altogether a deception.' Undoubtedly a direct allusion to the experiment Ellicott made and announced at the Royal Society in 1752, this suggests that it was well known and that the results were not encouraging.

During this same period, John Harrison also designed an experimental watch and had it made for him by the trade watchmaker John Jefferys (d. 1754). The watch was significantly different in terms of technical specification from anything that had been made before. Harrison provided it with a new kind of oscillator having more stored energy while running and coupled with a form of temperature compensation. In doing so, he seems to have been the first maker to succeed in significantly improving the performance of a pocket-watch. The good going of the Jefferys watch surprised Harrison himself, and he tells us that it was this timepiece which gave him the clue he needed on how to proceed with designing a new longitude timekeeper.

The watch survives (despite having been roasted in a safe during the Blitz in 1940). It has temperature compensation in the form of a miniature bi-metallic curb acting on the balance spring much in the same way as in H3, but on a smaller scale. The fusee also has Harrison's automatic maintaining power fitted to keep the watch going while being wound. For quality of construction, though this is perfectly sound, it does not compare with the beauty and fineness of one of Mudge's watches of the period. The strongest feature is in the specification of the oscillator, which is less affected by both external and internal disturbance by having greater stored energy in it when running.

Four features are probably largely responsible for the improved performance of this watch. First, the escapement that Harrison instructed Jefferys to fit was an adapted form of the Flamenville escapement. Being of this kind, it allowed the balance to have a larger amplitude, and by altering the form of the pallets somewhat, the design was intended to induce isochronism in the system. Second, the system was designed so the balance frequency was higher than usual and beat five times a second (a watchmaker's *18,000 train*, as it makes that number of beats in an hour). Third, the balance was somewhat larger in diameter than the average verge watch. Fourth, the balance was made of gold. As already explained, by making the balance larger and heavier, it had a greater moment of inertia and carried more energy in proportion to the impulse. It was thus less affected by variations in impulse delivered. However, because of the higher energy specification of the balance, the watch was not self-starting—something which must have seemed to many a backward step in design, even though not actually a problem in practice.

Once tests were under way with the Jefferys watch, Harrison knew he had made the break-though he needed: the watch performed better than he had expected and gave him the clues he required to embark on a new direction in timekeeper design. H3 would not be forgotten: the evidence in that timekeeper and in Harrison's writings reveals that he was now 'backing both horses' and design alterations continued unabated on H3 through the mid-1750s. Nevertheless, it seems clear that, at least in his own mind, Harrison was sure the successful path lay in developing the smaller watch designs, rather than continuing with large clocks. In June 1755, when he was before the Board of Longitude again asking for support, ostensibly for further development of H3, he also told the Commissioners that he needed money

> ... to make two watches, one of such a size as may be worn in the pocket & the other bigger ... having good reason to think from the performance of one already executed according to his direction (tho not brought to perfection) that such small machines may be render'd capable of being of great service with respect to the Longitude at Sea ....

With the £500 Harrison received from the Board, and another grant following a further meeting in November 1757, he was able—in 1759—to complete his fourth marine timekeeper, the longitude watch known today as H4. In most respects, this large silver-cased watch (13cm in diameter and weighing 1.45kg) appears completely different from the earlier large machines. However, H4 shows many elements of Harrison's technological thinking that followed through from them, and it is only in its adoption of the new scale and form of oscillator that it differs fundamentally from them. One of the most significant keys to H4's success was in having a balance of 56mm diameter, beating five times a second (i.e. having an 18,000 train), and running at an arc of about 240°.

At a Longitude Board meeting on 18 July 1760, Harrison asked for a trial of H3, hoping that it would be possible to send 'the watch' (H4) as well, as its timekeeping had far exceeded his expectations. In March 1761, the Board agreed and told John's son William (1726–1814), now in partnership with his father, to prepare for a trial voyage to Jamaica in charge of the timekeepers. This first trial was not well planned or organized by the Commissioners, in spite of the Royal Society providing some structure for how it should be conducted. For their part, the Harrisons incorrectly assumed that the Board fully understood what would be required for a fair and open trial of a marine timekeeper. It would become clear, however, that there were fundamental misunderstandings about how the results should be interpreted.

While in Portsmouth, awaiting instructions to board a Royal Navy ship bound for the West Indies, William was able to check the performance of H4 at the Naval Academy there and set it to correct mean time. After innumerable delays, on 18 November 1761 the *Deptford* sailed for the West Indies with William and H4 only: H3 did not accompany them. William Harrison and H4 arrived at Jamaica on 19 January 1762, the official end of the trial, and it was clear to all that the timekeeper had performed exceedingly well, although the actual result would only be determined by careful calculation after returning home. The transatlantic return voyage was an appalling experience for William, who was obliged to make it in the small sloop *Merlin*, which endured a very rough passage. He was determined to keep H4 running and dry since he was very conscious that, in order to qualify for reward, the Act of Parliament required that the timekeeper should prove to be 'practicable and useful at sea'. As William perceived it, the performance of the watch had already qualified it for the full award, but it was important not to compromise this by any suggestion that it was unduly affected by rough weather. On return to Portsmouth the error of the watch, according to William's reckoning, was just 1 minute 54.5 seconds after a period of 147 days. Calculations then carried out on the going of the watch during the official trial showed it to have been in error by just 5.1 seconds during its whole timespan. It was a remarkable achievement that was spoilt by one crucial oversight.

The Harrisons had failed to discuss and agree with the Commissioners the **rate** of the watch. Even very accurate, reliable timekeepers do not usually keep *exact* time. It is extremely difficult to adjust a clock or watch so it does not gain or lose at all. As long as the amount is regular and *predictable,* it will have no effect on the machine's ability to keep time. However, for testing timekeepers this concept was then entirely new and, without a clear agreement to accept the rate, the trial was next to useless. It seems inconceivable that the Harrisons did not make it abundantly clear to the Commissioners, and to everyone concerned in the trial, that the timekeeper had a rate and that this would have to be allowed for. But it seems they did not and, at their meeting in June 1762, the Board announced that they were dissatisfied with the trial on a number of counts. Not surprisingly, top of the agenda was the question of applying a rate and there is no doubt there was suspicion that Harrison had retrospectively created one to suit his purposes—an imputation which would have infuriated him.

From this point on, the attitude of the Board toward the Harrisons began to harden. Harrison had now switched his focus entirely onto a watch and there was, no doubt, deep prejudice about the capability of any watch to perform well—hence Harrison's bitter remark about the perceived view of watches and the reference to Ellicott's having 'tried what a watch will do and that the performance of mine, though nearly to truth itself, must be altogether a deception.'

In March 1763, the original Act of Parliament was qualified by a new one concerning the longitude. The new Act did not supersede the original of 1714 but made clearer provisions for how Harrison would qualify for reward. One of the clarifications was to specify that the method and its technology be fully explained and then published. Harrison was thus aware that, to qualify for the full reward, he would in due course be required to provide an explanation of the mechanism within the timekeeper to a group of fellow watchmakers and provide a written description. To this end, in 1763 he drafted some notes entitled *An Explanation of my Watch . . .,* in which he attempted to describe the physics and technology behind his design and explain why it performed so well. This manuscript[45] makes fascinating reading. While Harrison's writing in it exemplifies his decidedly verbose and endlessly qualifying style, it is not as wholly incomprehensible as some commentators have suggested.

At the August meeting of the Board, Harrison reluctantly agreed to a second trial of H4 to the West Indies, this time to Barbados. When the Board met on 4 August 1763 to make the necessary arrangements, the first item on the agenda was the dreaded question of deciding H4's rate. It was finally agreed that Harrison should be allowed to provide his own statement of what the rate should be. William Harrison made his declaration of H4's rate (that it gained one second a day) to the Admiralty on 24 March 1764 and he and a companion, Thomas Wyatt (mid-eighteenth century), departed with H4 in the *Tartar* from Spithead on the 28 March. On arrival in the West Indies William found, to his dismay, that the official astronomer sent out separately by the Board was Nevil Maskelyne (1732–1811). He was to carry out the observations that would contribute to a determination of the performance of the watch: yet, as Harrison knew all too well, Maskelyne was an ardent supporter of the lunar-distance method.

In the event, this second trial was another astonishing success for H4, with results nearly as good as those of the first. The average computation put the watch's error at just 39.2 seconds after the voyage of forty-seven days. This was *three times* better than the performance needed to qualify for the full £20,000 longitude award. Whatever may have been said and done before, the Board should now have recognized that the requirements of the 1714 Act had been met. However, on the return to England, and in spite of the good results, which the Board accepted, they were still not yet ready to pay anything. First, they stated that they would pay half the total, £10,000, but only after John Harrison had given a full explanation of H4's mechanism, under oath, to a specially appointed panel of experts. The details would then be published for the benefit of the world at large. Second, the Board observed that the watch was a single example and, even if it could continue to perform as well as it had on the trials, others of the same kind should be made and tested before the full award was made.

While the Board's concerns about accepting a single watch as sufficient evidence to qualify for the full award were understandable, that is all that the 1714 Act literally required. Inconvenient as its wording had been, it was the definitive statement on how the methods should be judged and the monies awarded. It was, naturally enough, the wording of the Act which the Harrisons were using as the basis for their claim to reward and, as they consistently pointed out, it was the wording on which the Longitude Commissioners themselves had encouraged Harrison to persist in his endeavours. Understandably, by that perfectly

---

45 Clockmakers' Company, London MS 3972/1. Bromley 1977, 97 N° 972.

logical thinking, they believed they were now entitled to the full £20,000.

Nevertheless, the Commissioners, some of them just small children when the 1714 Act was passed, saw its terms as a historical inconvenience, and considered it their duty to try and 're-interpret' them as best they could to ensure that such a large sum of money was not given away for a method which might turn out to be unworthy. The Commissioners were, inevitably, either politicians or high achievers in, generally, military or academic fields. Most of them would have been conscious of how posterity might judge them if they neglected such a considerable responsibility. At this stage, the Harrisons' relations with the Board were at an all-time low but, realizing that they would get nowhere if they did not compromise, both father and son finally agreed to sign the oath and disclose the inner workings of the watch.

The panel of six experts consisted of three well-respected practical watchmakers, Thomas Mudge, William Matthews (c.1723–c.76), and Larcum Kendall (1719–90), the scientist/mathematicians William Ludlam (1717–88) and John Michell (1724–93), and the London instrument-maker John Bird (1709–76). The meeting could scarcely have been a relaxed affair under these circumstances but, to add to the tension, the Board instructed Nevil Maskelyne, recently appointed fifth Astronomer Royal and thereby now one of its *ex-officio* members, to oversee the presentation. It seems that, after six days of study, cross-examination, explanation, and the answering of questions, the disclosure of the mechanism was completed to the satisfaction of all (though several of the experts would express their doubts later) and a certificate was signed to that effect on 22 August 1765.

The Board of Longitude met again on 28 October and granted Harrison enough money to make up the first half of the full reward. In return, they required possession of H4 in order to commission the making of a copy, as further evidence that the timekeeper could be reproduced. They also asked Harrison to recommend someone suitable to do this and he suggested Larcum Kendall. Kendall had been apprenticed to John Jefferys and was, no doubt, known to Harrison through that connection.

THE PRINCIPLES OF MR HARRISON'S TIMEKEEPER

The description of H4, entitled *The Principles of Mr Harrison's Timekeeper*, was finally published in 1767 and it did have a very significant impact on those interested in marine timekeepers. Some criticized Harrison and the publication for being obscure: others even suggested that he had intended to withhold the secrets within the timekeeper. But the important features of H4 are all described in *Principles* and Harrison was undoubtedly trying his best to describe why his oscillator, and the design of the escapement, were important. The use of an 18,000 train was clearly stated, as was his observation that a small balance would be better beating even faster, with Harrison recommending a 21600 train for pocket-watches—something re-introduced in the mid-twentieth century, in wrist watches, as an entirely new improvement.

The message in *The Principles of Mr Harrison's Timekeeper* was better recognized by some. With H4 on the scale of a watch, watchmakers, as opposed to clockmakers, were now taking an interest. Within just a few weeks of publication, a French edition was released and distributed widely among watchmakers in France. Ferdinand Berthoud himself remarked: ' ... every village watchmaker now has plans for a marine timekeeper and we shall have to see what this fervent activity produces ... '.[46] There was indeed fervent activity, some of it to no avail, but one or two followers took note of the essential elements in Harrison's work. Combining these with ideas probably originating in France, would see the marine timekeeper evolve into something truly practicable, an instrument that would become known as the *marine chronometer*.

H4 was then put on trial at Greenwich, but owing to the fact that it had not previously been overhauled and adjusted, and because it had grossly unfair treatment while under Maskelyne's care at the Observatory, it did not perform well. From 1772, Kendall's copy of H4, K1, was given the most exacting imaginable trial when it was issued to Captain James Cook (1728–79) for his second voyage to the Pacific. It performed magnificently. Through the pages of his own log of the second voyage we can read of Cook's steady conversion towards belief in the timekeeper. He used K1 on both his second and third voyages (the latter of 1776–80) to help chart many Pacific islands and north-eastern Pacific coastlines, amply demonstrating the reliability of both the concept and the hardware.

By 1772, further finishing and adjusting of the second watch, now known as H5, provided one further Harrison timekeeper. As a last resort, John decided to appeal to the highest authority in the land, King George III, who had been following the Harrisons' progress with interest. William Harrison requested an opportunity for H5 to be put on trial by the King himself at his private observatory. It went on trial at Richmond from May to July 1772. Its daily rate over the whole ten weeks averaged out at less than a third of a second per day. Harrison now formally approached the Prime Minister, Lord North, with the full story. The Harrisons then petitioned Parliament, with 1772 (13 Geo. 3) Supply, etc. Act 1772 c. 77 then duly receiving Royal Assent. This awarded John Harrison £8750 which, in addition to all the sums he had already received from the Board, including expenses, in fact came to a total of £23,065. Within three years, on 24 March 1776—his eighty-third birthday—Harrison died at Red Lion Square.

---

46 Gould 2013, 9

HARRISON'S LEGACY

In terms of his contribution to marine timekeeper design, Harrison not only showed that such a device was practicable, but also he introduced the most important element which led, not just to the successful marine chronometer, but to the very concept of the precision watch itself. Without the correct specification of balance, introduced by Harrison, the precision watch could never have succeeded. The importance of what he demonstrated is shown clearly when one considers that there was no design for any successful portable timekeeper on the scale of H4, before the 1760s, but, by contrast, in the years following the publication of *Principles*, one sees a growing number of designs for marine timekeepers and there are scarcely any that are *not* being produced on the scale of H4, or close to it. Apart from some of the timekeepers made by Ferdinand Berthoud in France, and two known to have been made by Willem Snellen (1727–91) in the Netherlands, no further examples with large, slow-beating balances, were made after *Principles* was published. One of the major stumbling blocks to reducing the cost of replicating Harrison's model was the use of his escapement and remontoire. For the origins of how the escapement in the later marine chronometer was developed, it is necessary to cross the Channel.

Until the publication of *Principles* in 1767, Harrison was, as far as is known, the only horologist actively working on designs for a marine timekeeper in England. However, there had for some years been others working on the matter in France. Ever since Henry Sully's valiant attempts to build on the designs of Christiaan Huygens, there had been interest in France in finding solutions by marine timekeeper to the longitude problem, and parallel research strands had been running in Paris since that time. Sully's close confidant there had been Julien Le Roy (1686–1759), the celebrated Parisian clockmaker and *Horloger du Roi*. In 1737, after Sully's death, Le Roy had published the important third edition of Sully's book, *Règle artificielle du Temps*, much enhanced by his own additions. An early general work on horology, the *Traité General des Horloges* ... was then published by Jacques Allexandre,[47] referring to Sully's research and, following the third edition of Sully's book, Antoine Thiout published his highly important *Traité d'Horlogerie* ... .[48]

Since the middle of the previous century, clockmakers in England and on the Continent had gradually become aware of the need to reduce friction in the movements of clocks and watches, and specifically to reduce friction in the escapement. The introduction of jewelling into watches in England at this time was very much part of a growing awareness of the need to avoid friction disturbing the oscillator, thereby improving timekeeping. Similarly, in about 1710, the London clockmaker Daniel Delander introduced a new form of escapement, known as the *duplex*, in which the functions of locking and impulsing were separated, and impulse was delivered close to the line of centres between pallet and wheel in order to reduce the friction. This was in fact a very fine escapement and deserved to be taken up in much greater numbers, but evidently it was not fully understood and did not see large-scale production. Nevertheless, well-informed clock- and watchmakers continued to search for improved ways to reduce friction and diminish disturbance to the oscillator as the century progressed.

Something of a landmark in this respect are some notes on escapements by Antoine Thiout in his treatise of 1741. Here he shows three escapements (one of which first appears in Sully's book, *Règle Artificielle* ...) in which he says the oscillator has half of each vibration independent of the escapement,[49] the implication being that such an escapement would be better.

It was Le Roy's son, the Paris watch and clockmaker Pierre Le Roy (1717–85)—also appointed *Horloger du Roi* on the death of his father—who appears to have been the first to look again at this concern about frictional contact and consider the issue in its much broader sense. In fact, the problem is not simply one of frictional contact: it is a more general issue of dynamic disturbances as a whole. For this reason, reducing friction is only part of the problem: there are distinct advantages in having an oscillator which avoids contact with the escapement as completely as possible. This new concept became known as *detachment*. It was to prove one of the central features of the modern chronometer and the evidence suggests that Pierre Le Roy, in Paris, was the one of the earliest to understand and to name it.

Strictly speaking, the *detached* escapement, which would soon also be fitted to precision pendulum clocks in England and later in France, is a form where the escapement impulses the balance or pendulum during a very short interval, in the middle of its cycle, leaving it to swing freely for the remainder of its oscillation on either side. The escape wheel is prevented from turning for most of the escapement cycle by a pivoted piece known as a **detent**. As the balance passes the central position of its swing, a piece attached to it moves the detent aside and the escape wheel is released. At this point, a tooth of the escape wheel strikes a pallet on the staff of the balance, delivering a short sharp push (an impulse) to the balance as it passes, and the wheel is then locked on the detent again. The theory states that as long as this delivery of energy occurs at the centre of the swing, it will minimize the disturbance of the timekeeping of the balance. Another critically important factor in the design of the detached escapement for a balance-controlled timekeeper (but one which Le Roy does not appear to have fully appreciated) was that it provided the potential for a much larger amplitude in the oscillator—potentially more than double that of frictional-rest escapements, which would further isolate it from external disturbance and allowing for higher stored energy.

---

47 Allexandre 1734.
48 Thiout 1741.

49 Thiout 1741, 106, 110, pl. 43.

Virtually all developed marine chronometers made in subsequent years would have a kind of detached escapement. In 1748 Le Roy deposited at the Académie des Sciences a sealed packet containing a drawing of an escapement that was clearly intended to provide an element of detachment. Like the Thiout escapements, it does not allow for freedom of the oscillator at both extents of its swing and is, at best, only a slight improvement over the escapements shown in Thiout's work of seven years before. But the description accompanying the drawing makes it clear that Le Roy understood the need for detachment and was heading in the right direction.

The concepts of reduced friction and of detachment also appear in London at this time. In about 1755, Thomas Mudge created an extraordinary table clock with an elaborate lunar dial and with what has been described as the first detached lever escapement. However, the escapement in this large clock of Mudge's does not appear to have been intended as a lever escapement *per se*, when he constructed it. Mudge certainly recognized the value of not having the oscillator in constant contact with the escapement as a way of reducing friction, but detachment was not the *raison d'etre* of the whole concept. When stating his philosophy on marine-timekeeper design in later years (in 1763, when he was due to inspect H4 and was wishing to record his own ideas so as to avoid accusations of plagiarism), Mudge made it clear that he did *not* approve of the concept of detachment. As he wrote, 'the less space the balance moves beyond the arc that is necessary to its scaping, the greater quantity of motion can be procured by the same force from the **mainspring** ... [and] ... These are the reasons why it appears to me necessary that the arc it scapes should be large ... '.[50] However, Mudge clearly came to realize the benefits of the lever escapement and stated unequivocally in later years that it was the best escapement for use in a pocket-watch. The lever escapement would be further developed by Mudge from the late 1760s and become the most popular and significant escapement in history, being the one fitted in virtually all twentieth-century mechanical pocket- and wristwatches up to the present day.

There appears to be a clear, but isolated, reference to the concept of a detached escapement in England at this period. In the archives of the Royal Society of Arts, London, there is some correspondence concerning new clock designs by the engineer and inventor, Humphrey Gainsborough (1718–76). Humphrey, a man of clever and wide-ranging intellect, was one of two horologically and scientifically minded brothers of the celebrated artist Thomas Gainsborough. In a letter of 5 April 1759 to the newly founded Society of Arts, Gainsborough revealed to the members his plans for a new pendulum clock, evidently having it in mind as a potential longitude timekeeper. Describing the escapement, he states:

... the pallets ... I have contrived in such a singular manner that they keep the wheel dead ... and at the same time permit the pendulum to have undisturbed vibration, ... neither being retarded by any attrition from the teeth upon the said pallets ... before it has completed its entire oscillation, but left in its motion altogether as free as it would be were it to swing naturally without either wheel or pallets.

He goes on to ask for financial support for his plan 'for helping the world to the knowledge of true time, even at sea'.[51] There cannot be a clearer description of a detached escapement, revealing that in 1759 there was some English appreciation of its potential value. A brief reply from George Box, the Secretary of the Society, encouraged him to proceed to construction and to undertake tests on of performance, but nothing more is currently known about the clock and it was perhaps never made.

THE *MONTRE MARINE* OF PIERRE LE ROY

Meanwhile, Pierre Le Roy was continuing his research and in 1754 had deposited with the Académie another sealed package containing the description of a more advanced marine timekeeper. This had a balance beating half seconds, a remontoire, and temperature compensation. In 1763 he presented to the Académie a second large marine timekeeper, no less than three feet high, a smaller version being presented in 1764. Although the latter was tested for over a year, no results of the performance have survived and it was only in August 1766 that Le Roy, in presenting an entirely new design for a marine timekeeper to Louis XV, really made his mark. According to Gould, this timekeeper, which is today in the collections of the Musée des Arts et Métiers, 'stamps [Le Roy] for all time as one of the very greatest masters of horology who ever lived'. Principally based on the elements contained in this timekeeper, Gould goes on to claim that 'He stands alone, the father of the chronometer as we know it'.[52] Le Roy is undoubtedly one of the great pioneers of the modern chronometer, but closer study of his work shows Gould's lionizing of his contribution to be much too simplistic.

Le Roy's design was described in a *mémoire* sent to the Académie in 1767 as a contender for a prize offered by them for a marine timekeeper. With some alterations, this article, *Mémoire sur la meilleure Manière de mesurer le Temps en Mer*, was then published in 1770. In this submission, which did indeed win Le Roy the Académie's double prize in 1769, he proposed three important elements of the modern chronometer: a detached escapement, a balance spring which has isochronous properties, and a balance incorporating temperature compensation. Whereas

---

50 Thomas Mudge, 'Thoughts on the Means of Improving Watches ...', II, in Mudge 1799, 4–16.

51 Royal Society of Arts, PR.GE/110/7/54 (Alt Ref: GB/4/54).

52 Gould 1923, 94.

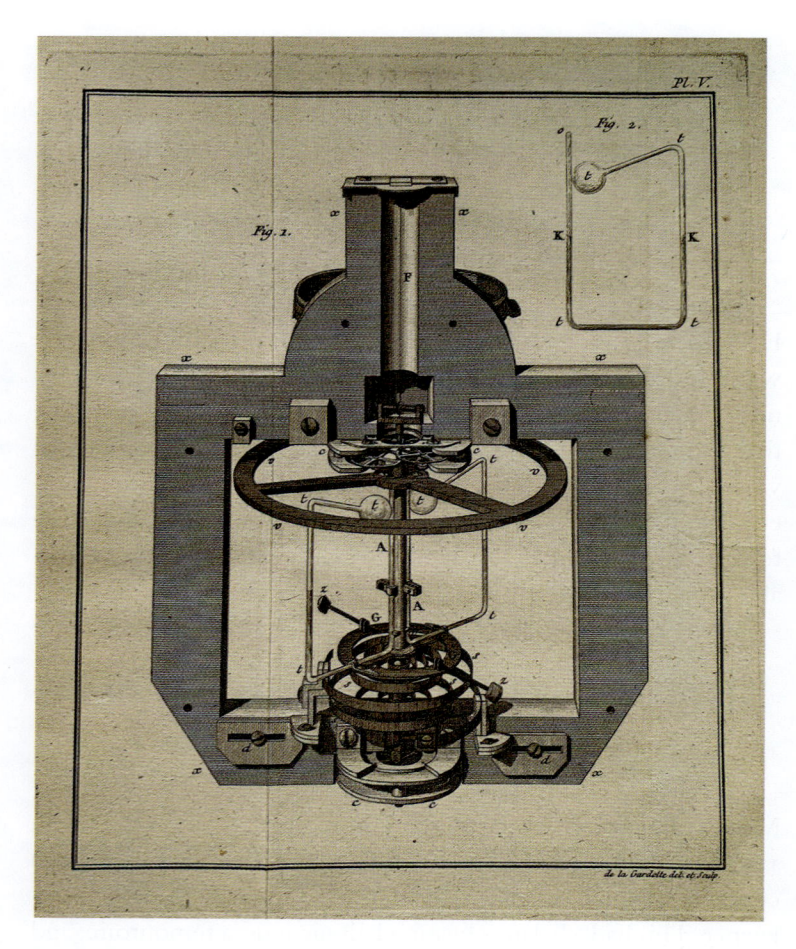

Figure 120 *Montre marine* by Pierre Le Roy from his *Mémoire...*, 1766. Photo: Jonathan Betts.

Harrison's description in *Principles* left much to the imagination, Le Roy's exposé, by contrast, is a model of clarity and completeness. The escapement, which impulses the balance across the central position of the balance's cycle of swing, is more detached than anything seen before, and, as Gould observes, must represent the foundation of all the detached escapements that followed (Figure 120).

AN ISOCHRONOUS BALANCE SPRING

A second ground-breaking observation in Le Roy's description is his statement about the two balance springs he used in the timekeeper. By experimenting with spiral springs, he found that the isochronous characteristics of a spiral balance spring differed depending on where its outer fixing point was placed. What he had in fact witnessed was the changing characteristics of the spiral spring when its *relative points of attachment* are changed. However, Le Roy explained what he had witnessed as being due, not to the points of attachment of the spiral, but simply to the *length* of the spring material. Therefore, his published statement in the *Mémoire* that in every spring there is a *length* that will produce isochronous behaviour is not strictly correct. Neither would any of the relative points of attachment in a plain spiral spring produce a wholly isochronous result. However, the more general observation that the spring itself can be manipulated to improve isochronism was perfectly correct. This was the first time that anyone had suggested that a solution to isochronizing a balance and spring might be managed by changes to the spring alone, a hugely important realization and one which undoubtedly inspired later pioneers to focus on the spring itself for that purpose.

Having the temperature-compensating device within the balance itself is undoubtedly an extremely important element of the developed chronometer. By this means, the balance spring is not altered by any compensation device to modify its force when the temperature changes. As described earlier, there is no doubt that John Harrison was the first person to design balances with the compensation incorporated within them, and he was clear from the earliest creation of H1 (in the late 1720s) that this was the correct principle. But Harrison failed to make the concept work in practice, whereas Le Roy's design of compensation balance in his *montre marine*, using alcohol and mercury thermometer tubes to change the acting radius of the mass in the balance, was a highly ingenious version and, unlike Harrison's, was the basis of a workable design.

Perhaps even more important, Le Roy can be credited with having first created the concept of a compensation balance with *bi-metal rims*, truly the basis of the form used in the modern chronometer. The concept, he tells us in his *Mémoire*, came to him after seeing Harrison's bi-metallic compensator in the description of H4 published in the latter's *Principles*. In Le Roy's design, the rims of the balance are formed out of bi-metals, which then automatically change the moment of inertia of the balance when the temperature changes. The design is clearly shown in the *Mémoire*, published in 1770, and represents the first example of this important concept. Le Roy does not, however, show the balance rims bearing any weights: after experimenting with the design he tells us he discarded it, having found that the rims suffered permanent changes after heating and became inconsistent in their action.

There was, however, one element of the modern chronometer which Le Roy's *montre marine* did not incorporate and without which its performance would probably always have been limited. In common with all the designs which had preceded it, the *montre marine* used a relatively large, slow-moving balance, swinging through a relatively small arc of under 90° and beating half-seconds. As Harrison had discovered, in order to avoid instability when subjected to external motion, the balance should beat faster and at a considerably larger amplitude. Another potential problem for this portable timekeeper was that the balance was five inches (125mm) in diameter and weighed over a quarter of a pound (153g). Such size would present problems, as the bearings needed to support the balance would create excessive interference and consequent variations in timekeeping. Added to this, the balance in the timekeeper was suspended on a slender harpsichord wire and proved too vulnerable to breakage to be reliable. Le Roy was certainly heading in the right direction and, had he reduced the scale of his balance and increased its frequency and amplitude, he would undoubtedly have created a much more practical design. But he was adamant that a smaller version of his timekeeper with detached escapement could not perform as well and that only one on this scale would do.

Le Roy made two examples of his *montre marine*, which were tested at sea in 1767, 1768, and 1771. In the early trials they appear to have gone well, especially the second timekeeper, performing nearly as well as H4 in its trials, but with occasional strange glitches in rate, probably caused by external disturbance owing to their low frequency and amplitude. In the later trials, their going was not so consistent as they suffered from a considerable acceleration in their rate, and the wire suspension proved too delicate to be reliable. Le Roy's *montre marine*, with its detached escapement, balance springs which have some isochronizing characteristics, and a pioneering compensation balance, is undoubtedly of very great significance in the history of precision horology. However, it simply cannot be described as the whole basis of the developed, 'modern' chronometer as Gould and many other twentieth-century writers on the subject suggest.

Pierre Le Roy's horological career might have developed more fully, and been more productive, had he not had a contemporary professional rival in the horological community in Paris. This was Ferdinand Berthoud (1727–1807), a Swiss who came to Paris in 1745 to make his horological fortune, and who began designing marine timekeepers in 1754 when he was just twenty-seven years old. In contrast to Le Roy, Berthoud's nature was both ambitious and confident and his practical work—and his presentation of it—impressed his contemporaries.[53] While both craftsmen had in common a great interest in experiment and horological development, Berthoud's research tended to be less focused, more wide-ranging, yet perhaps less penetrating than that of Le Roy. Nevertheless, it was Berthoud, as the more persistent and determined writer, who worked to ensure his achievements were better known. In this he continued the established French tradition of publishing horological research for the benefit of all and thereby became one of the most celebrated of all French authors on the subject.

As has been shown, up to 1767—the year in which details of H4 were published for the world to see—the few men brave enough to experiment with and build marine timekeepers had concentrated on improving large-clock designs. Although H4 had shown itself, in both its official trials, to be wholly 'practicable and useful at sea', it is certainly true to say that the solution as it appeared in England was still not particularly practicable in a more general sense, because H4, and copies of it, were such complicated, and therefore expensive, things to make. Larcum Kendall's copy, now known as K1, had cost the Board of Longitude £500. It was clear, at least to the Board and some aspiring longitude pioneers, that a version of H4 was needed which performed as well but was simpler and easier to make, and therefore cheaper to produce. In fact, Kendall had himself been specifically commissioned by the Board to try and make a simpler version. The result was K2, which cost £200 in 1771, and K3, which cost £100 in 1774, although both failed to perform as well as Harrison's original, and K1, had done.

However, some watchmakers in England did take notice of Harrison's designs for H4 and, with his work and that of Berthoud and Le Roy becoming better known and understood, this period proved to be the beginning of a whole new chapter in the story of portable timekeepers. A quite different epoch began at this exclusive end of the horological trade, when intuitive and entrepreneurial craftsmen grasped the initiative and made their fortunes in this new high technology. It was thus a commercial imperative, with stiff competition among the 'second generation' of chronometrical pioneers, which drove this next stage. In the following thirty years the marine chronometer was to evolve into its developed form, prove its efficacy, and become an article of commerce. In parallel with this change in the kind of practitioners interested in future involvement, there was a shift (in England) from publicly supported science and research into the hands of the practitioners themselves. This radically new approach to technological development saw the makers largely responsible for funding their own innovation, but the profits to be made were considerable and few complained of the hard-nosed new commercial climate in this growing industry.

JOHN ARNOLD (1736–99)

In April 1767, within days of the publication of *The Principles of Mr Harrison's Timekeeper . . .*, the Astronomer Royal Nevil Maskelyne presented a copy to a resourceful and ambitious young watchmaker named John Arnold, a thirty-one-year-old Cornishman then working in London. Arnold had set his sights on improving Harrison's timekeeper and, as will be seen, in the course of the next few years contributed more, in detail, to the developed chronometer than any other maker.

There is some evidence that Arnold began his involvement in designing marine timekeepers straight after he was presented with the copy of *Principles*. A timekeeper evidently by Arnold and dating c.1768, current whereabouts unknown, has technical specifications suggesting that it followed, and built upon, features described in *Principles*—evidence which is supported by similar features in other marine timekeepers made by Arnold in the early 1770s. It seems Arnold was thus closely following Harrison in the scale of his timekeepers, but another significant feature of all these early examples is that they were fitted with a detached escapement. It is possible that Arnold had conceived the need for detachment from *Principles*, where Harrison, referring to his escapement in

---

53 On Le Roy's comparative modesty, Alfred Beillard wrote that Le Roy's qualities 'made him withdraw into a modesty that deprived him of a reputation that his genius would otherwise have made universal' (Beillard 1895, 145); see also Ditisheim et al. 1940: 'Le Roy, shy, a little obsequious in his manners . . . nevertheless fills us with the sympathy and admiration for [one] who does not want to humble himself'.

which the escape wheel has a very low mechanical advantage over the balance, remarks that 'it must be allowed, the less the wheels have to do with the Balance the better'. It is also possible that, by the late 1760s, word of similar designs created by Pierre Le Roy in France had reached parts of the London scientific and horological community. There was considerable exchange of technological and scientific information between France and England during the eighteenth century. Corresponding members of both the Royal Society and the Académie Royale des Sciences reported regularly on meetings and papers, and it is not unreasonable to imagine that the description of Le Roy's detached escapement in his timekeeper, presented to the Académie in 1767, was known to Arnold soon afterwards.

Two early Arnold machines embarked with Admiral Robert Harland (1715–84) in 1771. He was most impressed with their performance. The Royal Society then commissioned three timekeepers from Arnold to join K1 accompanying James Cook (1728–79) on his second voyage to the South Seas in the following year, returning in 1775. Two of these early Arnold timekeepers from Cook's second voyage are still at the Royal Society (the whereabouts of the third is uncertain) and their construction shows how much they were influenced by Harrison's published concept for H4. The size and scale of the movement as a whole, and of the balance and spring in particular, are exceedingly close to that in H4 (within a few millimetres), and the compensation employs a very similar bimetal construction. The movement is also housed in precisely the way Harrison recommends. However, Arnold's oscillator was impulsed by an escapement, which, although crude, was of the detached type, enabling him to dispense with the complex remontoire used in Harrison's design.

Unfortunately, however, owing to teething problems with the design, Cook's timekeepers by Arnold did not perform well, especially when compared with the superb performance of K1. But during the 1770s Arnold developed a form of pivoted detent escapement of a much more sophisticated design, and the timekeepers that he made in the later years of that decade properly established his reputation. In 1775 he took out his first patent, which was for improvements in the design of marine timekeepers.[54] In the years that followed many other patents were taken out for improvements in chronometry: Arnold's was the first and, perhaps, one of the most notable, because it encompassed both a compensation balance and the cylindrical (helical) balance spring, two of the most significant features of the developed marine-chronometer movement. With his astute, exclusive patent protection, Arnold was able to exploit the advantage of these refinements for the next fourteen years. Indeed, his name would dominate the world of chronometry for the rest of the century.

The helical spring (sometimes popularly called a cylindrical or coil spring) patented by John Arnold was designed to work in torsion (rotation around its longitudinal axis) rather than in a straight, axial pulling direction (i.e. in extension, as in H1 and H2). This form of balance spring, being in the form of a cylindrical helix, has a much-reduced 'point-of-attachment effect' compared with that of a flat-spiral spring. It is consequently much more isochronous in its action. The patent of 1775 also included a design for a compensation balance. Again, it is possible that this was inspired by Harrison's writings. Some years later, Arnold himself pointed out that 'the correction piece of my timekeeper is in the balance itself, which Mr Harrison says in his book would be the greatest possible perfection.' Arnold (speaking to the 1793 Select Committee on Mudge's timekeeper) was referring to Harrison's *A Description Concerning Such Mechanism . . .*, published[55] just a few months before Arnold took out his patent for the compensation balance. It is possible however, that he had also heard of Pierre Le Roy's compensation balance as described in Le Roy's memoire.

THE EAST INDIA COMPANY

In 1775, the East India Company ship *Grenville*, carrying an Arnold chronometer, sailed to Madras (now Chennai), India. On board was Alexander Dalrymple (1737–1808), later the company's Hydrographer. Dalrymple realized the great economic benefits of having on board such an instrument for accurately finding the longitude, quite apart from those of safe navigation for ship, cargo, and crew. Ships would also be enabled to sail faster for longer, on the shortest practical courses and without uncertainty of their precise location, which over several weeks or months of sailing would significantly shorten voyages, lowering the cost of transportation. Vessels equipped with chronometers would therefore have a distinct commercial advantage over those without and strengthen the overall capability of the Company. Dalrymple soon made himself known to Arnold, and the resulting relationship between the two men ensured both Arnold's continuing commercial success and the establishment of the marine timekeeper in East India Company service as a practically proven method for finding longitude at sea. Dalrymple's and the East India Company's enlightened attitude to the use of chronometers contrasts with the official government view, held into the early nineteenth century. Chronometers were, of course, sent on all British expeditions of discovery at this period but the whole concept of a practical marine timekeeper was still regarded as decidedly unproven, and influential men like Nevil Maskelyne continued to state publicly that chronometers were, at best, a complement to lunars for safe navigation.

This was a period which saw the growth across Europe of new institutions founded to further knowledge in the exact and natural sciences. With this rapid growth came a commensurate increase in the need for up-to-date instrumentation, including timekeepers and regulators. In parallel with this was the desire for European

---

54 British patent N° 1113, John Arnold, 30 December 1775.

55 Harrison 1775.

navies to begin supplying their ships, particularly those undertaking voyages of exploration, with marine chronometers to supplement the lunar-distance method of determining longitudes. The need was satisfied in various ways, and generally instrumentation was obtained from makers in either London or Paris. However, clock and watchmakers in other countries also began to design and produce their own regulators and chronometers. In Dresden, for example, Johann Heinrich Seyffert (1751–1817) produced several astronomical regulators for use in the Mathematical-Physical Salon; the contemporary clockmaker Peter Kinzing (1745–1816) of Neuwied created precision regulators following French patterns, and in later years Dresden clock- and watchmaker Friedrich Gutkaes (1784–1845) produced exceptionally fine regulators for observatory use in Germany. Spanish attempts to establish a school of watch and clockmaking in the eighteenth and nineteenth centuries were less successful, with early trainees sent to Paris and London successively dying of the plague, attacks of which beset the Iberian Peninsula at this period.[56]

In some cases, the acquisition of instruments formed part of a more general quest by those seeking to buy and learn about latest developments, who would personally visit workshops and other institutions to exchange information and learn of current best practice. The Danish mathematician and astronomer Thomas Bugge (1740–1815) toured Holland and England in 1777 to meet fellow scientists and to buy instruments for his work at the University of Copenhagen. He kept a record of his travels.[57] Among a number of most interesting descriptions of horological businesses, Bugge visited John Arnold in London and described his work on marine timekeepers and precision clocks. There were also one or two interesting figures in this world of eighteenth-century science who grasped the commercial initiative and operated as international agents—middlemen between customer and supplier. The most notable of these during the second half of the eighteenth century was the Portuguese polyglot Jean Hyacinth Magellan (1722–90).[58] London-based Magellan was not only a go-between in sales of hardware but also published on the new technologies, including descriptions of the latest designs for scientific instruments. He was one of the first to publish widely on the horological concept of detachment and took the opportunity to describe several of the pioneering timekeepers.[59]

Then there were those in official positions, such as Nevil Maskelyne at Greenwich, who saw it as very much part of their wider role in the international scientific community to encourage practitioners in the instrument-making and horological trades. As already described, on appointment as fifth Astronomer Royal in 1765 he became an *ex-officio* member of the Board of Longitude and responsible for the trials of new designs of clocks and chronometers by those seeking rewards. But Maskelyne voluntarily extended this role to include unofficial trials of prospective timekeepers and regulators for other institutions, acting as intermediary between his fellow scientists in European observatories and London instrument and clockmakers.

Among those joining the full-time experimenters in the 1770s was the celebrated Thomas Mudge. In about 1767, Mudge designed and built a small portable clock that contained a miniaturized version of the balance-controlled dead-beat escapement in the lunar clock. The clock contained a much smaller, H4-type, higher-frequency (18,000 train) balance, running at a considerably larger amplitude than in his earlier clock. Running at that higher frequency and amplitude, the escapement takes on a whole new significance, and certainly qualifies as the first proper detached-lever escapement. No doubt encouraged by the performance of this little clock, in 1768 Mudge made a watch for King George III, which the King then presented to Queen Charlotte and which is consequently known today as the 'The Queen's Watch'. This watch, which also incorporated a high-frequency, high-amplitude balance, also had a small version of the same lever escapement and is also reported to have performed very well. Mudge was certainly aware of the importance of the detached escapement by this date but he still resolutely refused to accept that detachment was the better solution for a marine timekeeper and spent the rest of his life working on an alternative plan, based on an even more complex interpretation of Harrison's design. Nevertheless, by this period, news of the importance of detached escapements was beginning to filter out into the wider international community, thanks to the encouragement and championship of patrons such as Maskelyne, Magellan, Dalrymple, Sir Joseph Banks (1743–1820), and Hans Moritz von Bruhl (1736–1809), and through the growing number of related publications serving the needs of the similarly expanding scientific community.

With the detent escapement gradually proving its worth in marine chronometers, the lever escapement was almost never used in them. Nevertheless, the lever escapement was, in its own way, an equally important invention in other areas of precision timekeeping. Mudge was absolutely correct in stating that it was very robust and tolerant of hard, physical use. It was thus ideal for use in precision pocket and deck watches, and for a new breed of hybrid timekeeper known at the end of the nineteenth century as a *chronometer watch*. Indeed, the lever became the standard form of escapement for watches of all kinds. Mudge never exploited its true potential himself, although he was aware of it, and it was his supporter, Count von Bruhl, who eventually persuaded another very talented watchmaker, Josiah Emery, to produce a series of watches with the device, the performance of which impressed all who owned them. By the time of his death in 1794, Emery had sold about thirty-three watches with Mudge's lever escapement. Indeed, it was a close inspection of one of these watches by a number of talented French watchmakers which is said to

---

56 Gould 2013, 101–4, following Duro 1879, 83–199.
57 Pedersen and de Clercq 2010.
58 Betts 2003, 2004, 2007a, 2007b.
59 *Cf.* Rozier 1793, 376–81.

have prompted that fraternity to join with Ferdinand Berthoud in designing chronometers in the 1780s.

In 1771 Mudge had retired from his London business to his native county of Devon, partly owing to poor health but ostensibly to concentrate on timekeeper design. The design he produced was, in essence, similar to Harrison's but of even greater complexity. Not only was the timekeeper complex by design, but it also employed a type of constant-force escapement, following a design by the Swiss astronomer Johann Jacob Huber (1733–98), that was both complex and difficult to make, requiring the very finest workmanship to ensure good performance. Mudge's first timekeeper was tested at the Royal Observatory late in 1774, and again in 1776–77. Between April and October 1777 it went exceedingly well and greatly impressed Maskelyne, the Astronomer Royal. Its performance during the middle of that trial would not be beaten in chronometer trials at Greenwich for almost 100 years.

Following the repeal in 1774 of the 1714 Act of Parliament relating to longitude, new, much tougher conditions were applied, including one that stipulated that if the proposal related to timekeepers, there had to be *two* instruments that should be tested together. In 1777 Mudge therefore produced a pair of one-day going chronometers, known as 'Blue' and 'Green' (owing to the colour of their shagreen-covered cases), but in formal trials, neither performed as well as Mudge's first. Nevertheless, in 1793 and after much petitioning and canvassing for support, his son Thomas, a lawyer, managed to get Parliament to grant his father £3,000 as a reward for his work. Then, to prove the efficacy of the design, Mudge junior established a manufactory in Kennington, London, to make copies of his father's timekeepers. Removed from Mudge senior's craft skill in making them, these chronometers never performed sufficiently well, and also proved too complex for repair and adjustment by other makers.

Meanwhile, the pioneers working towards rationalizing and reducing the complexity of chronometer design were having greater success. John Arnold's pioneering business making longitude timekeepers grew rapidly during the 1770s. Perhaps the most famous of all the timekeepers Arnold ever made was his watch No. 1/36, made in 1778. This large eighteen-carat-gold pocket-watch,[60] which is almost certainly the first pocket-watch to incorporate a compensation balance, was sent to Greenwich for trial in 1779 and performed incredibly well, astonishing Nevil Maskelyne and his assistants. As a result of its spectacular performance, Arnold published the trial results.[61] By name, watch No. 1/36 is the very first chronometer, and must rank as another of the great milestones in horological history.

Keeping total production in-house was a valuable element in increased flexibility, efficiency, and secrecy. This appears to have been the case at Well Hall, Arnold's main manufactory in Kent. Thus, an Arnold chronometer movement of the 1780s would have been produced almost entirely under one roof, including the frame, barrel and fusee, train, and all ancillaries. One extremely important innovation by Arnold, made and patented during the early 1780s, was his conception of 'terminal curves' for the helical balance spring. By putting incurved ends on the balance spring during manufacture, Arnold discovered it was possible to make one that was almost completely isochronous in its effect. It was a brilliant concept and would be used in virtually all successful chronometers in the nineteenth century. Arnold patented the design in 1782, along with several other of his new ideas, and stated of these curves that they are 'attended with the property of rendering all the vibrations of equal duration'.[62] By the early 1780s he was thus concentrating on improving the design of the helical balance spring and also on new, improved versions of his compensation balances.

It was at this very period, in 1781, that a new escapement, the *spring detent*, made its appearance. In practice this is a much simpler and more satisfactory design than the pivoted detent, and the spring-detent was the final form that went on to serve in tens of thousands of chronometers, being used worldwide for well over 150 years. The detent itself, as its name suggests, is mounted on a spring blade fixed to the frame of the chronometer. It has no pivots and therefore almost completely dispenses with any friction when being moved aside. The spring will still require energy to deflect it and thus needs to be as light as possible, commensurate with safely locking the escape wheel. However, as it has no pivots, and thus avoids the varying friction associated with them, the long-term stability of the timekeeping is improved. The authorship of the design was, then, naturally of some significance, but was contested from the beginning. Indeed, the question of who first invented it constitutes one of the most famous controversies in horology—the dispute between the great John Arnold and the thirty-one-year-old trade watchmaker Thomas Earnshaw (1749–1829), who always claimed Arnold stole the idea from him.

Earnshaw's and Arnold's designs are slightly different but, in both, the escapement has the advantages over the old, pivoted detent in that there are no pivots to oil and the spring detent can be made slenderer, reducing its inertia. In Earnshaw's design, the spring detent receives the tooth of the escape wheel and locks it with the spring in compression. Arnold's design acts in the opposite direction, with the force of the locked tooth taken by the spring in tension. Arnold's escapement was rather more complex to construct, and it was thus Earnshaw's that was eventually adopted as the norm by almost all chronometer makers after the 1830s.

After more than 230 years it is unlikely that the controversy can be satisfactorily resolved. Careful reading of the case, however, inclines one to believe there is some basis in Earnshaw's claim. In addition, it should be said that Arnold neither verbally claimed the original invention, nor complained that his patent had

---

60 Now in the Royal Museums Greenwich, Inv ZBA 1227.
61 Arnold 1780.

62 British Patent No. 1328, John Arnold, 2 May 1782.

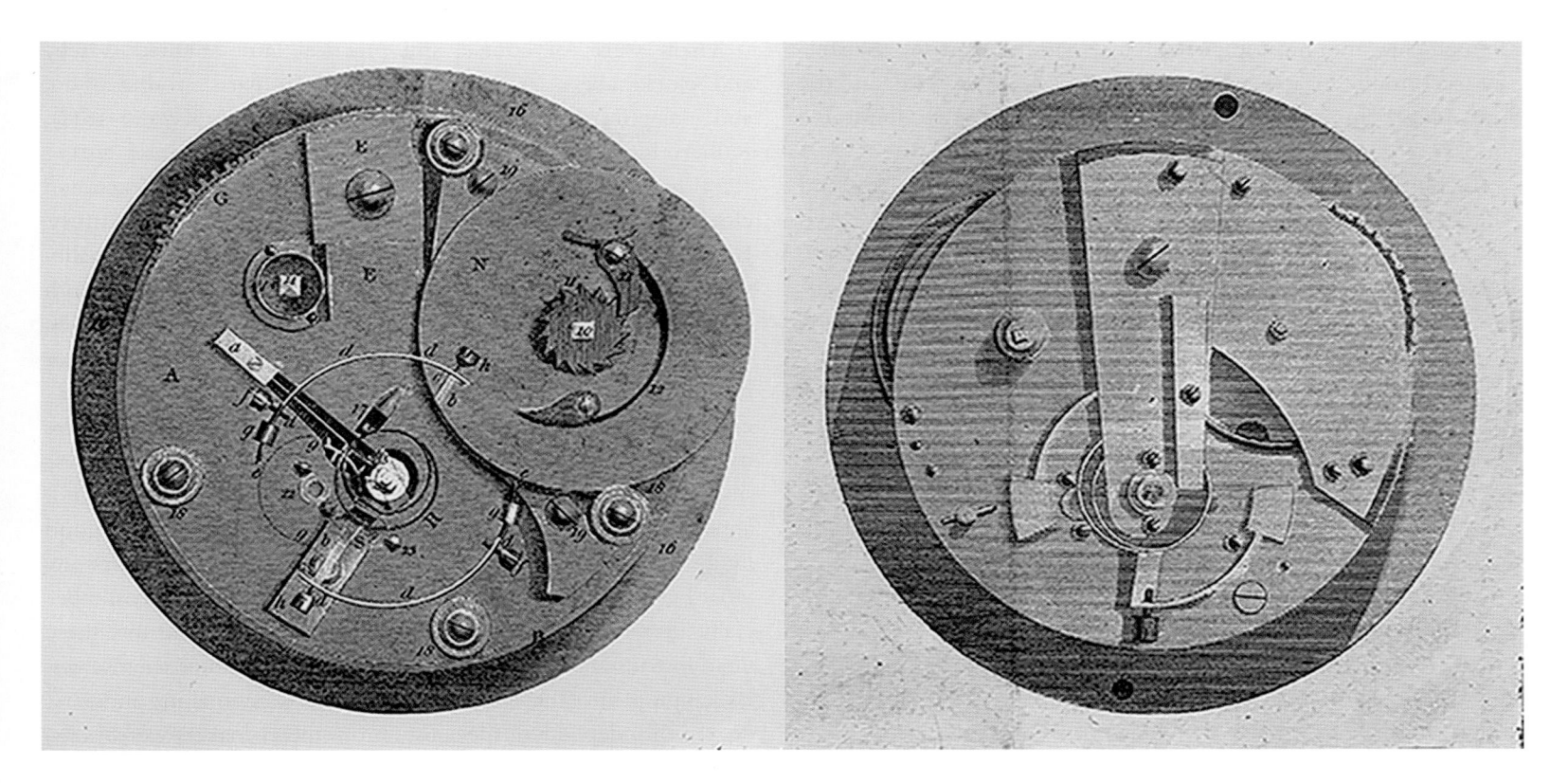

Figure 121 The marine chronometers of John Arnold and Thomas Earnshaw as published in 1806. Photo: Jonathan Betts.

been contravened when Thomas Wright's (taken out for Earnshaw) appeared just one year later. And on the several occasions when confronted with the allegation of copying Earnshaw's idea, he always hedged and never expressly denied it.

Wright's patent for Earnshaw's spring-detent escapement included one or two other inventions, most notable being the compensation balance with turned rims and brass fused to the steel. Arnold's bi-metallic rims were made as a straight piece, with the two elements soldered together (for which they were temporarily held by one or two rivets) and then bent afterwards into the circular shape. Earnshaw's design, by contrast, had the brass element fused onto the edge of a steel disc that was then turned into perfectly circular bi-metal rims. This was altogether a better, more uniform construction and Earnshaw's balance was to be the accepted form in the later, developed chronometer.

During the last two decades of the century, and especially after Arnold and Earnshaw's patents had expired in 1796–97, a number of other watchmakers of note, including George Margetts (1748–1804), John Brockbanks (1747–1805/6), Josiah Emery, and others began making marine chronometers. Like Mudge junior's productions, none ever worked well enough to win Longitude reward money, but many were sold and helped determine longitude in this pioneering period. The message from the experience of all these post-Harrison pioneers was becoming clear: what was needed was a standard design and one that did not require highly specialized craftsmen to make it. The evidence suggests that, as a practical watchmaker, it was Thomas Earnshaw who realized that the perfect method was already in place:

it was that used for making pocket-watches. Since the seventeenth century English watches had been made in two stages, first an embryonic 'rough movement' (almost exclusively made in Lancashire), then transported to the destination (mostly London, later also Liverpool) for finishing and retailing. There was no reason in principle why marine chronometers could not be made the same way and it seems Earnshaw was the first to exploit the idea. Thus it was John Arnold and Thomas Earnshaw, in that order of precedence, who oversaw the pioneering 'post-Harrison' development of the English chronometer, and in 1805 the contribution of both were recognized by award of £3,000 from the Board of Longitude (Arnold's posthumously to his son) (Figure 121).

MARINE CHRONOMETRY IN FRANCE

In France, the introduction of marine chronometers had taken a very different path. Until the 1790s Ferdinand Berthoud was virtually the only craftsman supplying the French navy with timekeepers. Once Ferdinand had details of Le Roy's detached escapement in the early 1770s, he began to employ the principle in his own timekeepers and, with somewhat smaller balances swinging at larger amplitudes, they began to perform well, proving of real use in longitude-finding on a number of voyages of exploration. With government funding, Berthoud was enabled to introduce a steady supply of marine timekeepers for use by the

French navy and, most importantly, he had the time and the ability to continue his experimentation, publishing his work and the reasoning behind it in a regular series of descriptive, illustrated books.

In 1808, Louis Berthoud (1754–1813), Ferdinand's nephew and successor, wrote to the Director of the Conservatoire des Arts et Métiers in Paris on the subject of French chronometry and, looking back, described the 'state of the art' in France in the mid-1780s. In this interesting document, Louis Berthoud stated of the French chronometer in 1785:

> All navigators know that at this period there existed in France no other machines for finding longitude at sea than those of M. Ferdinand Berthoud [and] no pocket watches except three English watches by Emery. The advantageous reports that were made, with reason, of these works, seemed to have deprived the artists ... of all courage ....

Louis goes on to remark that he, too, should have been more discouraged, given the small scale of Emery's design and the very fine construction necessary, but was urged to continue with developing timekeepers on the Emery-type model by 'leading savants and navigators'. He tells us that having tried to make examples of this watch with the lever escapement, but using only brass and steel, he realized it would really only work properly with jewelled parts, which he was unable to fit. Therefore, he adapted the design by fitting a pivoted-detent escapement instead of a lever, and it was with this small-scale but complex and beautifully made model that Louis began to form the basis of his marine chronometer designs. These then continued into the early years of the nineteenth century and served the needs of the French navy, and a smaller number of private customers well. However, Louis Berthoud's chronometers, and those of the next generation of French makers, including Breguet and Henri Motel, remained essentially national products and were never made in anything like the same numbers as the instruments being produced by the London chronometer makers during the nineteenth century.

The demand from the French navy and French merchant fleets in France was considerably more modest than in Britain, and the needs of the Navy were already satisfactorily provided for by the appointment of the *Horloger de la Marine*. In addition, the French economy was far less reliant on maritime trade, so it seems there was not the same commercial imperative motivating watchmakers in France to switch to the new product. In addition, in Paris there was a well-established and highly remunerative trade selling clocks and watches to private customers. France still entirely led the way globally in style and fashion in the decorative arts, and the horological industry in Paris was principally serving the needs of a much larger international market for clocks and watches in the latest taste.

There were, of course, still practical, and talented watch- and clockmakers, as well as retailers, in Paris around 1800, and exceptional figures like Jean-Antoine Lépine (1720–1814), Robert Robin (1742–99), and Abraham-Louis Breguet (1747–1823) presided over a positive revolution in precision watch and clock design during the second half of the eighteenth century. Breguet was perhaps foremost in creating pieces of great quality and technical novelty during the last quarter of the eighteenth century and first of the nineteenth, and that period might be considered as a French horological 'Golden Age'. However, for most of the retail watchmakers, looking at the esoteric and literary example of Ferdinand Berthoud and his followers, perhaps making marine chronometers was also seen as a different profession altogether and one which, without a government stipend, appeared risky when conventional watchmaking provided a good living. In England, by contrast, once London watchmakers saw what Earnshaw, in particular, was making in the mid-1790s, especially after 1797 when his patent expired, they realized that these were instruments they could relatively easily produce themselves. By tapping into the established London watchmaking industry, dozens of independent specialist chronometer makers were established in the city by 1800 and, with London quickly reputed as the place to buy good and reasonably priced chronometers, the success of the English chronometer trade was guaranteed for decades to come.

In parallel with the development of detached escapements during the second half of the eighteenth century, London clock- and watchmakers continued to find ways to provide a more constant delivery of impulse in both clocks and watches. Undoubtedly inspired by Harrison's use of a train remontoire to achieve this, a number of designs took that idea one stage further and incorporated the pure delivery of impulse from the remontoire into the escapement itself. The Swiss astronomer Johann Jakob Huber already mentioned was probably the first in the mid-1750s, introducing Thomas Mudge to the design for a constant force escapement which he would later employ in his marine timekeepers in the 1770s. However, before this, Mudge applied the concept of the constant force escapement to the pendulum clock during the 1750s, creating an escapement which used gravity to impulse the pendulum. This design would spawn a number of similar designs, one in 1766 by Alexander Cumming (1733–1814), and versions using spring pallets instead of gravity arms, one first made in 1806 by William Hardy (d. 1832), for which he was awarded a prize in 1820 by the Society of Arts and which was purchased by the Royal Observatory for use as the Sidereal Standard clock. Regulators with a similar type of spring pallet escapement were also made by Thomas Reid (1746–1831) from 1811. While these types of precision clock escapement certainly qualify as 'constant force', they are not 'detached', in spite of Hardy's description of his as being so. It must also be stated that while they were reckoned to represent the 'state of the art' in precision pendulum clocks when made, they were difficult and expensive to make and did not always provide the enhanced performance expected of them.

In truth, the vast majority of regulators used in the world's observatories and laboratories continued to employ the dead-beat escapement, being simpler to construct and, technically, a 'known quantity'. It was only in the second half of the nineteenth century that truly superior designs for precision pendulum clocks were developed.

POSTSCRIPT

In the 1740s, in parallel with his development of marine timekeepers, John Harrison created a highly unorthodox new design for a precision pendulum clock. Wholly different from what had been considered optimum design, it employed a relatively light pendulum swinging through a very large arc (over ten degrees) impulsed by a recoil escapement—all features traditionally considered bad for precision clocks. Harrison claimed in 1775 that the design was far superior to the standard Graham-type regulator and stated it should be possible to maintain an accuracy within one second in 100 days. As far as is known, the only contemporaries of Harrison to experiment along these lines were the celebrated Vulliamy family who, in the second half of the century, made at least three clocks similar to Harrison's. However, the specification and proportions of the escapement and pendulum in these clocks were very different from Harrison's and the clocks appear not to have performed particularly well. Until recently, therefore, Harrison's claim was dismissed as fantasy, but recent experiments on modern clocks employing Harrison's system correctly suggest he was right. It now seems likely that, had others made pendulum clocks according to that plan, eighteenth- and nineteenth-century time standards could have been considerably more accurate than they were.[63]

ACKNOWLEDGEMENT

The author is grateful to Dietrich Matthes who provided some text for the early part of this chapter.

---

63 McEvoy & Betts 2020.

CHAPTER TWELVE

DECIMAL TIME

Anthony Turner

Decimalizing time was born from the political and ideological desire to emancipate France from its religious and monarchical past. Leap-frogging a millennium, the regimes of the Revolution found warrant in Antiquity—particularly in the decadal calendar of Ancient Egypt—for such an adventure.[1] At the same time, the explanations and advocacy in Europe of decimal systems of calculation, from Simon Stevin (1548–1620) onwards, by theoreticians of practical mathematics offered a technical basis.[2] Decimalizing time, however, is generally absent from early proposals for decimal use, although the advantages of decimal calculation were obvious to practitioners of the cognate discipline of astronomy. William Emerson wrote:

> It would be much for the ease of calculations if the sexagenary account was laid aside, and the decimal one substituted in its room. For there are many reductions in the one, that make it exceeding tedious, which are entirely avoided in the other. But that tyrant Custom has already got possession of the former and is likely to keep it.[3]

In the slightly earlier *Encyclopédie*, the author of the lucid article on decimals thought that 'it is much to be desired that all divisions; viz the pound, the penny (*sou*), the *toise*, the day the hour &c should be ten by ten; this division rendering calculation much easier and commodious'.[4]

It was this desire of rationality in measure that would lead Gilbert Romme and his associated framers of revolutionary time to insist that not only the year, but also the day should be decimally counted.

> The division of the hour into sixty minutes and of the minute into sixty seconds, is incommodious for calculation and no longer corresponds with the new division of astronomical instruments [. . .] In consequence, to render the system of decimal numeration complete, the Convention decrees that the day shall be divided in ten parts, each part in ten others and so following to the smallest commensurable portion.[5]

The desire to have a completely rational system, one that was primarily organized in the interest of astronomers, geodesists, and navigators, is here clearly expressed. A glimmer of practical sense, however, is shown by the fact that the decimal day would only become obligatory a year later on 1 Vendémiaire an III (22 September 1794), whereas the decimal calendar was immediately applicable from the date of the decree 4 Frimaire An II (24 November 1793). Indeed, realization that decimalizing

---

1 See *Décret* (N° 1838), 'The sacred traditions of Egypt, which became those of all the East, brought the Earth out of chaos under the same sign as our Republic, and established there the origin of things and of time', 8. *Cf.* 11–13, 18.
2 A survey of the decimal idea pre- and post-Stevin is offered by Sarton 1935. For Stevin in general see Elkhadem et al. 2004, *passim*. For *De Thiende* (1585) editions and translations, see Smeur 1965. For advocacy of decimals in seventeenth-century France, see Goldstein 2014; for suggested practical applications in mid-seventeenth-century England, see Webster 1975, 412–20; for the following decades, see Willmoth 1993, 74–6. A number of surviving instruments attest to some use of decimal scales at this period: two instruments by the itinerant craftsman Johann Eggerich Frers (Sotheby's 1994 lot 472; von Mackensen 1991, 106); simple theodolites by Jacobus de Steur (Leiden 2nd half 17th century), N. Huguet at Orleans and an unsigned and undated example (all Whipple Museum Cambridge, see Brown 1982, N°s 56, 58, and 61); a graphometer by N. Huguet, Paris (Frémontier-Murphy 2002, 286–7).

3 Emerson 1769, ix–x.
4 *Encyclopédie* 1751–72, iv, 669–70. Decimalizing time had earlier been discussed in Maslot 1718, ch. 3.
5 *Décret*, 16.

the day, despite the optimistic suggestions contained in the original decree,[6] was rather less simple than so dividing the year, rapidly led to a second decree establishing a public competition for finding a way to arrange clocks and watches to show decimal divisions, 21 Pluviose An II (9 February 1794).[7] Participation in the competition closed on 1 Messidor (9 June); a jury for the entries consisting of Ferdinand Berthoud (1727–1807), Joseph Louis Lagrange (1736–1813), Jean Baptiste Lepaute (1727–1802), the physicist Jacques Alexandre Charles (1746–1823), Antide Janvier (1751–1835), and the younger Lépine and Claude Mathieu the elder (fl. 1754–94) was tardily constituted on 4 Fructidor (21 August), only to report 14 Frimaire (4 December) that none of the pieces submitted was worthy of a prize. Five months later the decree establishing the decimal day was indefinitely suspended.[8]

Despite the fact that some notable clockmakers such as Robert Robin (1742–99) had participated in the competition for designing clocks and watches, the problem was less to produce a decimal timekeeper *de novo* (Figure 122) than to convert existing instruments. Robin estimated that there were some 15,000,000 watches circulating in France. Gradually the impracticality of a decimal day was realized by the members of the Convention. Existing clocks and watches were not to be converted as easily as was at first thought—particularly if they chimed. Decimal products would be unsalable outside France, and most French citizens were themselves indifferent, even hostile, to the system. For civil life, decimal time was not needed. Even so, many clock- and watchmakers attempted to provide it. Clocks incorporating the thirty-day decimal calendar are not uncommon, those that offer decimal hours and minutes rather more so. Watches allowing a simultaneous reading of duo-decimal and decimal hours were produced in some numbers, but those showing decimal hours only are less frequent.[9] Sundials using decimal hours were hardly made at all—only four are known to survive, three of them dated 1794.[10]

For the research of *savants* the decimal system offered clear advantages. An immediate stimulus to decimal production was given by the Convention requiring a decimal clock for its debating chamber, a clock supplied in the form of a long-case **regulator** on 8 October 1793 by Robin.[11] At the same time, Louis Berthoud was at work on a decimal regulator for the Paris observatory, made decimal watches for Jean Charles, Chevalier Borda (1733–99), and a decimal marine chronometer for Lavoisier.[12] Slightly later, c.1818, Abraham-Louis (1747–1823) and Antoine-Louis (1776–1858) Breguet developed, together with the astronomer François Arago (1786–1853), a clockwork counter for attachment to the eyepiece of a telescope that could measure a transit to tenths of a second and to hundredths of a second by approximation.[13] The astronomers and geodesists of the Revolution and their immediate successors continued to think and calculate in decimals, their expectations and deceptions being clearly expressed by Laplace.

> Finally, the uniformity of the entire system of weights and measures demanded that the day be divided into ten hours, the hour into a hundred minutes, and the minute into a hundred seconds. This division, which will become essential to astronomers, is less advantageous for civil life where there are few occasions to use time as a multiplier or a divider. The difficulty of applying [decimal division] to clocks and watches, and our foreign commercial relations, have led to its use being indefinitely suspended. Even so, it is believable that in time decimal division of the day will replace that actually used, which contrasts too much with the division of the other measures not to be abandoned.[14]

That the metric system for weights and measures finally became obligatory in France on 1 January 1840 can only have encouraged decimalists, and decimal calculation and measurement maintained itself through the middle decades of the century, particularly in geodetic, cartographic, and astronomical work, to gather strength in the last quarter of the century.[15] In 1881 the physicists succeeded in establishing the 'C. S. G. system' (centimetre, second, gram). It remained to decimalize time and the circle. The question of how the day and the circumference and their subparts were to be divided was hotly debated. Should 400, 200, 100 divisions, 40, 20, or 10 hours be the basis? Should the 360° circle and the twenty-four-hour day be retained but subdivided decimally?[16] The pressure for some such system, however, was

---

6 'Above all, clock and watchmakers should seek a way to make the old movements of clocks and watches serve the new decimal division with the least change possible'. *Décret*, 21–2.
7 Reprinted in Droz & Flores 1989, 117–18.
8 For a detailed account of this episode, see Cardinal 1989, 65–80.
9 Cardinal 1989 offers a wide-ranging pictorial survey of such clocks and watches. Further examples are shown in Droz & Flores 1989. A recent contextual survey is Shaw 2011.
10 Mayall 1982 describes a remarkable presentation horizontal dial in Sèvres porcelain; Bosard 1989 a horizontal dial of polished granite. A very brief account of the four known eighteenth-century examples, and two modern 'centenary' products is given by Marquet 1989.

11 *Description de la pendule astronomique décimale ... et à sonnerie décimale présenté à la Convention Nationale, par Robert Robin ...*, Paris l'An II [1793–1794], reprinted in Droz & Flores 1989, 207–17.
12 For Borda's watch N° 26-2440, see Sabrier 1993, 108–17, which gives a clear account, from Berthoud's notebooks, of the problems involved in creating such a watch, 476–84; for the gimballed marine watch made for Lavoisier, 133–7, 509–12; for the regulator, 153–5, and for a second, 175. Several other decimal instruments are recorded by Sabrier. For the decimal watch 2433 of 1792 also made for Borda, see Chayette & Cheval 2010, lot 217.
13 Breguet 1997, 260–1.
14 Laplace 1796, xiv.
15 For a summary of this development see Turner 2002b 58–60; for a more detailed account, see Pasquier 1900, 68–75.
16 The seven major systems proposed are conveniently summarized in Pasquier 1900, 75–6.

Figure 122 Decimal watch with virgule escapement and centre seconds, signed on the dial 'A de Elyor, Palais d'Egalité N° 88' (anagram of Leroy). Private collection. Photo: Jean-Baptiste Buffetaud, courtesy of Chayette & Cheval, auctioneers, Paris.

considerable, and the period of the centenary of the Revolution offered a suitable occasion. In October 1896, the Minister of Public Instruction in France established a special commission to study the question. The conclusions were that a twenty-four-hour day should be divided decimally; that the circumference should be divided into 400 grades subdivided decimally; that an international conference should be organized to validate these conclusions; and that some preliminary practical trials should be made using these units. Action was immediate. The organization of an international conference began immediately, and the naval Captain Emile Guyou (1843–1915) set about arranging a nine-month campaign in which five warships were despatched to navigate purely by decimal means.

It was in this positive context that decimal horological instruments were once more produced. As early as 1875, Antoine d'Abbadie (1810–97)—a notable advocate for full decimalization—ordered two decimal regulators for his private, and entirely decimal, observatory at Hendaye, one from Dent in London, the other from Collin in Paris.[17] By the turn of the century, the most fervent advocate of decimalizing time and the circle was Joseph-Charles-François de Rey-Pailhade (1850–1934), a mining engineer holding a doctorate in medicine and with interests in the history of scientific instrumentation—above all, he was President of the Committee for the propagation of decimal methods. Rey-Pailhade was a ceaseless proselytizer for full decimalization, and a practical designer of decimal instruments. Partisan of a ten-hour division for the nychthemeron such as had been used by Louis Berthoud in his watch 2433 for Borda, from 1893 onwards, in the wake of Laplace, Rey-Pailhade campaigned for the extension of decimals to the hour and the second, insisting on their greater convenience for the measurement of very short intervals in astronomy, physics, practical engineering, navigation, and sports. To this end, he commissioned the making of decimal watches (Cémètres as he called them) from well-known makers such as Modeste II Anquetin (1817–1909), L. Leroy & Cie, and F. Olivier & fils,[18] assisted at the rediscovery of two decimal regulators by Louis Berthoud (which were remounted in the meteorological Observatory directed by Emile Guyou in the Park Montsouris in Paris), and in the Ecole d'Horlogerie de Paris, instigated a competition for the design and making of decimal watches through the pages of the journal *L'Horloger*.[19] Furthermore, he persuaded both the Nautical Club of Nice (of which he was president) and that of Cannes to offer decimal watches as prizes in their regattas and to employ decimal methods for timing the courses of the competitors. All this seemed of sufficient commercial interest for Leroy & Cie to develop a low-cost decimal watch: 'Private industry has given a terrible blow to the hoary sexagesimal citadel; the house of L. Leroy, one of our most esteemed, is currently making ... an excellent decimal watch with three hands for a price, accessible to all budgets, of only 50 francs'.[20] The watch was delivered in late 1909 but cost sixty francs. The following year, Rey-Pailhade could list the names of thirty-eight owners of *cémètres* known to him and the watch was on sale in Paris, Nice, Toulouse, Angoulême, and other main towns in France.[21]

The advances made in propagating decimal measures in the first decade of the twentieth century were made possible in part by the success reported by Emile Guyou of the trials he made of them at sea. The horological instrument he employed, equivalent to the marine chronometer in standard sailing, was the *tropomètre*, so named 'to recall that it is not an elapsed interval that it supplies to the navigator, but the real angle through which the Earth has turned in relation to the mean Sun'.[22] The tropomètre could also be built as a sidereal timepiece. Both had three hands showing decagrades (0–40); decigrades (0–100); and milligrades (0–1,000). The tropomètre thus divided the nychthemeron to 400,000 parts and had a beat of two milligrades, the equivalent of 0.432 sexagesimal seconds. For Guyou's trials the watchmakers by appointment to the marine produced eighteen instruments, of which seventeen when tested met the required standard and six were purchased.[23]

Guyou's officers had no difficulty in navigating decimally, and indeed found some advantages in so doing, but Guyou did not think that decimalization of the day was practical for everyday activity. Nor indeed was it, despite the presentation at the 1900 International Exhibition of a decimal mantel regulator following the ideas of d'Abbadie.[24] Decimal subdivisions of the minute and the second, however, would increasingly find applications in the precision timing of scientific, military, sporting, and other everyday activities. Hipp chronoscopes (Figure 202), widely employed in psychology laboratories at the turn of the century, measured reaction times to tenths and hundreds of seconds.[25] Electric chronographs could read to thousandths of a second,[26] while mechanical counters with or without chronograph and stop functions reading from one-fifth to one-hundredth of a second were, and are, a speciality of the Heuer company in Bienne.[27] Time and motion study in industry, largely dependent on the stopwatch,

---

17 Mercer 1977, 481; Turner 2002b, 29–30.
18 A fuller list of makers is given in Rey-Pailhade 1899a, 8.
19 An early presentation is Rey-Pailhade 1894. A wider-ranging lecture on the subject given 23 June 1898 is augmented in its published version by a substantial annotated bibliography; see Rey-Pailhade 1899b. From its foundation in 1905 *L'Horloger* was to provide Rey-Pailhade with a reliable platform for spreading news about decimal watches and the organization of competitions for their design.

20 Rey-Pailhade 1909a, 5; for a fuller description of the watch and its uses, see Rey-Pailhade 1909b, *passim*.
21 Rey-Pailhade 1910, 4.
22 Guyou 1903, C6.
23 Guyou 1903, C6. See also Guyou 1902, *passim*.
24 Planchon 1900, but this is not mentioned in the survey of decimal timepieces in the exhibition given by Borrel 1901, 22–5.
25 See Schraven 2003.
26 Chaponnière 1924, 35.
27 Ibid xv. For the company see Pritchard 1997, i, H30–4.

employed instruments showing tenths of a minute, the dials being graduated to 100 (i.e. 0.01 minute), and the hand revolving once per minute.[28]

The dreams of Rey-Pailhade and like thinkers of completing the metric system by a total decimalization of the day and the circle were disputed in their time. The progress they seemed to be making in the first decade of the twentieth century was destroyed by the First World War. Time would remain sexagesimal, but since its smallest unit, the second, was, as Henri Poincaré had emphasized in his report on the deliberations of the 1896 commission, the basic unit for physicists, mechanicians, and electricians in need of ever finer measurements as well as ease of calculation, it would come to be treated decimally in practice. Slow empiricism[29] eventually found a compromise solution for the decimalization of time that rationalist thinkers and government commissions had failed to achieve.

---

28 Barnes 1940, 255–6 (1st edn., 1937).

29 Cf. Guye 1955.

CHAPTER THIRTEEN

CLOCK- AND WATCHMAKING FROM THE NINETEENTH TO TWENTY-FIRST CENTURIES

INDUSTRIAL MANUFACTURE AND WORLDWIDE TRADE

SECTION ONE: THE MIXED FORTUNES OF BRITAIN

James Nye

At the end of the eighteenth century, the horological industry in Britain was at its zenith, measured both in terms of its share of world trade—Britain was producing about half the world's output of watches[1]—and in its overall reputation. It would continue to produce some of the finest horology for another century, but the trade would decline and suffer in various ways from the early nineteenth century onwards.

William Pitt's tax on clocks and watches introduced in 1797 dealt a terrible blow to the industry, and strenuous efforts were made, successfully, to repeal it soon after. The Napoleonic Wars in the early nineteenth century shielded Britain from imports, giving clock- and watchmakers an advantage, but this rapidly disappeared with the end of war in 1815, demobilization, famine, and harvest failure.[2] This was also a period in which copies of London watches—in reality usually Swiss watches (including 'Dutch fakes')—continued to arrive in British (and many other) markets, often bearing the names of London makers, but of inferior quality in cases and movements compared with true London-made watches—a practice that dated back to the second half of the eighteenth century. It is here that we see a long-term conflict emerging within an industry in which not everyone's interests were aligned. There was money to be made from importing and distributing foreign-made watches, and evidence can be found of strong complicity from parts of the British trade. For example, giving evidence to a parliamentary committee in 1817, Henry Clarke (1780–1865) explained he knew of a British watchmaker that had made unsigned watches:

> ... which were then ready to have any name placed on them, and also many other watches upon which he put the names, & c. of his foreign correspondent; by this aid of the reputation conferred by well-made English watches, in a few years the foreign manufacture was fostered, and sufficiently established to enable the foreigner to carry it on for the future by his own resident workmen, collected in different countries. The British watchmaker showed to me one of the watches made in that foreign establishment [...] He kept it, as he said, for a memento of his own folly, in making in the manner aforesaid, about 20,000 watches for that correspondent, whereby he had ended his own career by bankruptcy, arising from his having sold his watches to that person for little more than the cost prices paid to his workmen.[3]

---

1 Davies 2012.
2 Landes 1983/2000, 274–7.
3 *Report from the Committee on the Petitions of Watchmakers of Coventry*, Parliamentary Papers, 1817, 67. Landes 1983/2000, 276 discusses the phenomenon of the watches 'by' Wilter in relation to this evidence.

The committee heard a great deal of evidence of members of the British trade facilitating the flouting of regulations, and generally supporting foreign competitors, whose success in turn tended to undermine the domestic market for the British trade. It would not be the last time that there would be a clear division between those interested in manufacturing in Britain, and those interested in importing to satisfy British demand but with foreign products.

For decades after the first arrival of Swiss and other imported watches, the British manufacturing industry appears to have kept its head in the sand, essentially believing there would always be a market for the highest quality, largely hand-finished English watches and clocks that it produced. The trade had long argued for protection through import tariffs and duties, but the economic philosophy of the time was that of laissez-faire trading, not protectionism. It would remain the same story for nearly the entire period through to the 1950s and the beginnings of the death throes for any kind of large-scale British horological industry.

Over the course of half a century, from the 1790s to the 1840s, English watch output fell by roughly one-half, from around 200,000 per annum to fewer than 100,000 in 1841.[4] Against this backdrop in the decline of the traditional trade, the 1840s saw the attempt by Pierre Frédéric Ingold to introduce into Britain systems of volume production for watches that would come near the creation of interchangeable parts. The British Watch and Clock Making Company was established at 75 Dean Street in Soho as part of this project but it was vigorously opposed by the trade, failing to secure Parliamentary approval in 1843 as a result of highly effective lobbying.[5]

In the following decade, the trade once again amply demonstrated its unwillingness to countenance change. John Bennett (1814–97) ran a successful jewellery business from retail premises in Cheapside, from which he sold large quantities of imported clocks and watches. He very publicly criticized the English trade for its deficiencies by comparison with the Swiss, particularly in working practices. He favoured the training of women in the families of watch and clockmakers, and generally compared Swiss practices favourably with what he considered to be inferior practices in England. All of this excited the ire of the British trade, which once again expressed a strongly protectionist sentiment in favour of the continued employment of large numbers of watchmakers (chiefly men) in a system of intense division of labour.[6]

As Alun Davies has noted, the Great Exhibition of 1851 offered a useful opportunity for international makers and their agents and importers to develop the British market. He says:

> Swiss exports to Britain alone rose from 42,000 in 1852, to 70,000 in 1854, to 90,000 in 1855 and to 160,000 in 1861; the aggregate annual values of Swiss watch sales in Britain climbed from £174,000 in 1850 to £565,000 in 1860. A decade after exhibiting at the Crystal Palace, Switzerland sold more watches in Britain than the entire British watchmaking industry made.[7]

It was not only Swiss watches that began to take hold of the market. American clocks aimed at the mass market, well-designed and manufactured efficiently down to a price, were imported in volume and avoided direct competition with the finest handcrafted products of traditional British makers.[8] The flows were not wholly inbound. Discussing 'Contemporary Horology' in 1863, the *Horological Journal* commented, 'The Geneva houses have numerous representatives in London, whence they export to all the English colonies, and to the utmost limits of the East.'[9]

For many years, the *Horological Journal* published annual summaries of trade statistics with an accompanying commentary. The summary from 1879 revealed exporting houses 'regularly for many years in receipt of large orders from India for watches, not sending out a single watch in 1878'.[10] Exports to Spain and Latin America 'usually of considerable extent' were described as 'nil'. The 1880 summary is highly characteristic of a long-term trend in both the underlying data (imports rose inexorably), and the tone:

> Short of absolute extinction, the English watch manufacture certainly reached as low a point as possible during the year just passed. All kinds and grades of work were alike affected. Even the marine-chronometer trade, in which we have no foreign competitors, shared the general slackness.[11]

The British makers lost more than their domestic markets. By the 1880s, important traditional colonial export markets (Canada, Australia, South Africa, New Zealand, etc.) were largely captured, particularly by the Swiss and Americans, the latter in particular.[12] Traditional British export markets were not only lost to firms such as Waltham from America but were still being lost to other foreign (probably Swiss) watches still being passed off as British. The *Horological Journal* reported in 1887 that '... these watches in most instances had never seen England at all, but were sent direct to their destination from the place where they were fabricated. As often as not the English hall-mark itself was fictitious, so that no precautions operating here would stop such villainous practices.'[13]

The British Horological Institute moved to a new building in London's Northampton Square in 1879, where William Parker (1815–88), Master of the Worshipful Company of Clockmakers, spoke at the opening ceremony:

---

4 Davies 2009, 638.
5 Davies 2009, *passim*; this volume Chapter 13, Section 2; for Ingold's career in general, see Berner 1932.
6 Midleton 2007, 312.
7 Davies 2012, 592.
8 This volume, Chapter 13, Section 2.
9 BHI 1863.
10 BHI 1879, 58.
11 BHI 1880, 63.
12 Davies 2012, in particular 410, n. 12, which quotes US Consular Reports for the statistics.
13 BHI 1887, 154.

... he had been forty years a manufacturer in Clerkenwell, and he remembered the clock trade, which was now lost to that parish. French clocks came in and English clocks went out, because prices came down, and our artisans did not take pains to compete with the French market. He feared the watchmaking trade was now similarly going away from us.[14]

Regulations for the hallmarking of watch cases, imports, and duties contained numerous structural deficiencies and inconsistencies. The manufacturing trade kept up a persistent wail over more than a century for enhanced protections for the home industry, but the tide was in favour of the importers, and for the smugglers who found it easy to evade regulations and for whom occasional penalties offered limited sanction.[15] In the final years of the nineteenth century, the figures for watch cases hallmarked at Goldsmiths Hall for the London area showed a steady decline, from levels of around 25,000 gold cases and 90,000 silver cases in 1880 to 6,000–7,000 gold and just 5,000 silver cases at the turn of the century. Hallmarking regulations in part served to encourage a gradual downmarket shift in the trade, in both price and quality.

Over the latter part of the nineteenth century, one area of positive news for the British market was the continued excellence of its chronometers, for which demand remained sufficiently strong despite working practices remaining unchanged over the long term. This continued broadly to be the case until after the First World War, when oversupply finally eclipsed demand.

It was not only the Swiss whose products entered the British market with mass appeal. Robert Ingersoll of the United States secured an order for one million watches on his first sales visit to Britain in 1901, and in 1905 established a British business selling a range of watches for just five shillings, with sales surpassing fifteen million watches by 1906.[16] In that year, Thomas P. Hewitt (1848–1933) of the Lancashire Watch Company could claim:

... Switzerland and America have to a great extent monopolised the whole of our Colonial markets for watches, and driven us out of the foreign markets. America is selling watches as cheaply in Canada, Australia, and South Africa as in the United Kingdom itself, even crediting our Colonial buyers with the rebate that Colonial Governments give to British manufacturers for British-made goods.[17]

The Edwardian period saw the continued rise of large-scale imports of German clocks. According to Alun Davies, in 1913:

... the collective annual value of clock exports from Germany to world markets amounted to £753,450—some twenty-five times more than the value of equivalent British clock exports, at £30,610. The mutual trade in each other's clocks simply underlined Germany's overwhelming superiority. In 1912 Germany sold clocks worth £294,750 to Britain, whilst in the following year Britain exported a mere handful of clocks, worth £870, to Germany.[18]

The First World War created massive disruption in the British market, and while import flows were understandably interrupted during hostilities, the trade figures for the 1920s were once again dominated by German clock imports. This was despite the imposition of the 'McKenna duties' ($33^{1}/_{3}$ per cent duty on imports) from 1915, which were maintained until 1924, repealed, but then later reintroduced. Over time, British manufacturers believed their German competitors were gradually granted increased export subsidies that negated British import duties, and this may prove to be correct when fully researched, but there were many other factors involved, including differentials in productivity and efficiency, standards of living, wages, and the effects of changing rates of currency exchange.

BRITAIN IN THE TWENTIETH CENTURY

After the First World War, there followed a half-century in which various reinventions occurred in British horology. There were important isolated cases of significant technical advance, but for the mass market it was a story of several major attempts at building volume production businesses by firms such as Smiths, Ingersoll, and Timex. While there were occasional successes, ultimately the overall result was failure. Through a combination of factors, including further war, unfavourable exchange rates, strong foreign competition, a lack of competitiveness, and a failure to innovate, the industry had faded away by the 1980s. The firm that had the most significant impact throughout the period was S. Smith & Sons (Figure 123), and it is perhaps most instructive to treat it as a case study for the wider industry.[19]

The twentieth-century history of horology in Britain is frequently intertwined with the history of the motor car industry. In the wake of the Great War, the motor industry enjoyed significant expansion. In Britain it tended to be populated by assemblers, as opposed to vertically integrated manufacturers like Ford. It therefore became a target for the supply of instruments by firms such as Rotherham & Sons and H. Williamson of Coventry, and, notably, S. Smith & Sons. Williamson and Rotherham were survivors of a British manufacturing capacity in watches, while Smiths was a retailer of watches sourced from others, with their best pieces

---

14 BHI 1879, 34.
15 Davies 2016.
16 BHI 1969, 27.
17 BHI 1906, 192.

18 Davies 2011.
19 Nye 2014.

Figure 123 S. Smith & Sons shop at 68 Piccadilly, London c.1920.

made by firms such as Nicole Nielsen. The need for speedometers, as well as car clocks and other instruments, offered a new market for firms that could not compete with the wave of inexpensive imported clocks and watches arriving from Germany, the USA, and Switzerland. While Germany's economy recovered rapidly in the 1920s, thanks to measures such as the 1924 Dawes Plan, the British economy struggled. Smiths had expanded vastly to support the war effort and found itself struggling to keep afloat, and to keep expanded or newly built factories fully loaded as the immediate post-war boom rapidly switched to bust. It took the rest of the 1920s to recover, during which time car clocks dominated its horological production, gradually rising to the order of 2,000 per week.

Stung by a 1927 lawsuit in relation to its use of imported Swiss **escapements** in its car clocks, Smiths moved to create a local capacity, which took several years, but resulted in the creation of the All British Escapement Company (ABEC), which made platform escapements. That this was jointly owned by Smiths and the Swiss firm of Jaeger LeCoultre is strong evidence for the deep ties between individuals in Switzerland and Britain that were to characterize future developments for the industry in Britain.

On the technical side, Smiths had Robert Lenoir (1898–1979), a French watchmaker trained at the Technicum in Le Locle, as their chief horological engineer. In developing ABEC and the Chronos factory in North London that would manufacture many Smiths clocks, Lenoir knew he could not spend years developing fine hand-skills in a workforce of raw recruits. Instead, he chose to de-skill operations. A good example of this was a device he co-patented in 1932 that enabled the adjustment of watch **balances** to time using a photoelectric cell, in place of a skilled observer and operative.[20] Operations were split down into small routines, and highly specialized jigs were used for accuracy. Tasks were automated, reducing labour costs and making ABEC viable. It established volume production of platform escapements by the end of 1931. By 1932, 30,000 escapements had been made, and production expanded rapidly.

Smiths were at the forefront of growing the scale of the horological industry in 1930s Britain. When a major press feature was published on the industry in 1933, it claimed £3.5 million of capital had been raised over four years, involving 15,800 employees.[21] Starting from a very low base, clock production had risen to over 800,000 units in 1932, with an estimate that it could rise to 2.5 million per annum. Note that the industry believed this

---

20 Nye 2014, 83.
21 British Clock Manufacturers Association, 1933, 4.

was despite desperate dumping of foreign clocks by firms, particularly in Germany, vastly below cost. And Germany was not the only threat—the newly formed British Clockmakers Association (later the BWCMA) was also aware of Seikosha in Japan, already producing a million clocks per year with just 2,000 staff.

Another important feature of the late 1920s had been the gradual standardization of the electricity grid following the adoption of the Electricity Supply Act of 1926, which provided an accurate time-base for an entirely new type of clock, employing a synchronous motor. Just half a million houses had electrical supplies in 1919, but from 1926 numbers began to increase rapidly, reaching eight million in 1938 (two-thirds of the housing stock), though the majority of houses only had a lighting circuit, and perhaps one socket.[22] From the early 1930s, there was widespread advertising for synchronous clocks. Despite the small numbers of houses with a full complement of wall sockets, this was an expanding market for clocks. The synchronous clock had no delicate escapement. It simply plugged in, the hands were set, and away it went, potentially for many years and with no need for the regular servicing required by conventional clocks. One particularly noteworthy engineer in this period, Ernest Ansley Watson (1867–1975), who was named on more than 100 patents between 1911 and 1948, was by late 1930 engaged on the design of a synchronous clock motor for Smiths. It was his fundamental design that was to serve the firm for many years. The production of synchronous clocks rose quickly to 4,000 per week by January 1932.[23] Smiths had recreated a clockmaking industry in a new and modern guise.

Looking at the larger market for conventional mechanical clocks, the growth was equally spectacular. From just 24,000 clocks produced in Britain in 1930, the numbers reached 636,000 per annum inside two years, and a formidable proportion of this was down to Smiths and its new factories. This growth occurred despite the onset of the Depression and the simultaneous dip in housebuilding between 1930 and 1932. From 1933, however, housebuilding in England and Wales recovered, and the market for clocks followed suit.

Despite success in building domestic production, the question of imports remained of critical importance. In 1930, Board of Trade figures revealed that out of a total of 3.7 million clocks imported, nearly ninety per cent came from Germany. Soon after Hitler's appointment as Chancellor in 1933, and despite active lobbying against it, Britain and Germany concluded a new Anglo–German trade agreement designed to increase British coal exports to Germany. But the quid pro quo exacted by Germany included reductions in tariffs for certain German imports to Britain—and these included clocks. For most categories, the duty rate fell from 33 1/3 to 25 per cent, which did not help British makers.

It was not only a question of reduced import duties proving problematic. When Britain left the gold standard in September 1931, sterling suffered a short-lived devaluation. Despite this, the unit price for German clocks actually fell in 1932.[24] By the mid-decade the price of German clocks in Britain fell far below the price of the same items in Germany, reportedly as much as forty-five per cent below cost in some cases, part of an export drive that was clearly sponsored by the German state.

How could British manufacturers make any headway against such determined competition? In part, the survival of the industry relied upon the synchronous clock, as this was not subject to any serious foreign competition. Mass advertising and an appeal to the public to 'Buy British' will also have had some effect, but patriotism was not enough.

In 1936, after five years of operation, ABEC reported making its millionth platform escapement. It was also the year in which Smiths introduced a new and updated synchronous motor for its clocks, named the 'Bijou', which was to become an industry standard. Sixty and more years later, clocks are still found in operation which house a Bijou movement, probably never having been serviced.

The marketplace was substantially contested. From the 1930s, the British Watch and Clockmakers' Association tended to speak for manufacturers and exporters, but it did not have a monopoly on representing the industry. The Horological Section of the London Chamber of Commerce was also important, and both before and after the Second World War it had influential members who were exclusive agents for various foreign factories which exported large volumes to Britain. It was in their personal interests to see British import quotas removed, licensing abandoned and import duties lifted. In watches, Switzerland managed to export more than five million units to Britain each year from the mid-1930s, and these were the cheapest possible. By contrast, the average price the Swiss achieved for watches exported to Germany in 1936 was nearly four times higher than in Britain. The *Chamber of Commerce Journal* noted in mid-1937 that 'Britain is becoming practically the dumping ground for the lowest-priced watches'. British acceptance of this was a problem in later years.

Just as in 1914–18, the Second World War created massive disruption, and import flows of vital components, such as watch jewels needed for aircraft instruments, largely dried up. Much watch- and clockmaking capacity was repurposed into instrument and fuse-making, but from early on there was significant forward planning for the post-war period and the revitalization of the industry. Close links were forged between Smiths and government agencies and officials, not least when the managing director of Smiths, Allan Gordon-Smith (1881–1951), was given a senior role in the Ministry of Aircraft Production from 1940.

Turning to the post-war period, with strong assistance from the new Labour government Smiths invested in new factory capacity to produce both clocks and watches in geographical regions that required support. The firm started producing alarm clocks in a former aircraft instrument factory in Carfin, Lanarkshire, the heart

---

22 Hannah 1979, 188–9.
23 Anon. 1931.
24 Gordon-Smith 1938, ix.

of an important development area. They were producing 1,000 alarms per day by the end of 1945.[25] By March 1947, Carfin was exporting a third of its production, and produced its millionth alarm clock in August 1947. It was to this same factory complex that Smiths moved its wartime synthetic watch-jewel operation in 1947, forming Synthetic Jewels Ltd. Within five years, it was producing three million jewels per year.

Ideas that had first been hatched during the war came to fruition for Smiths in its aftermath. The production of watches was divided between two projects: high-grade watch manufacture at Cheltenham (with significant help from friends at Jaeger LeCoultre), and volume production of pin-pallet watches at Ystradgynlais in Wales, in partnership with Ingersoll, with whom Smiths formed the Anglo-Celtic Watch Company in May 1945. Smiths opened a second factory in Wales, to which it moved its Enfield clock operation in 1947. Yet everything was against a backdrop of serious shortages of both material and labour.

In Wales, under heavily improvised conditions, Smiths achieved 10,000 units per week by late 1948. But the target was double this and growing, while the economics were poor. Since inception, Anglo-Celtic had lost money, as it failed to achieve volumes and efficiencies. Likewise, at Cheltenham, stubbornly high production costs continued to necessitate an internal subsidy in order to achieve a workable retail price. In essence, British horology generally was being supported not by the logic of a sensible return on capital, but by the political will of an interventionist government keen to rebuild a post-war economy.

By 1948, sentiment towards Germany had changed. It was no longer a prospective enemy, unlike Russia. For a time, however, the government continued to subsidize watch, balance spring, and jewel manufacture, but this commitment started to wane by 1950. The Conservative election victory of October 1951 began the dismantling of the protectionism that had characterized the post-war Labour government. Nevertheless, the Cold War backdrop influenced the minds of some civil servants. In 1950, the Ministry of Supply reported that 'the manufacture of alarm clocks on a substantial scale would give us, in the event of an emergency, a large manpower pool which would immediately be used for the manufacture of fine precision mechanisms, such as fuses, etc.'[26] Knowing that Britain imported 2.5 million alarm clocks each year before the war, sufficient confidence emerged that plans to enlarge domestic production were drawn up. Not only could thirty-hour alarms be sold domestically, but also they could be exported in vast numbers—but to do so economically, they needed to be made on a colossal scale. The result was claimed to be one of the most up-to-date flow production factories in the world, at Wishaw, in Scotland. In reality, important elements of the design were copied from the German firm of Junghans as a result of the surveying of immediate post-war Germany by teams that included Smiths' specialists. Smiths created a plant (Figure 124) that could produce 90,000 clocks per week, or 4.5 million clocks per year. It opened in September 1951.

The major feature copied from Junghans was an innovative solution to the problem of adjusting the timing of the large volume of clocks involved. This involved a specially constructed overhead conveyor carrying on average 50,000 clocks, stretching round the entire site. Its speed was governed by a controlling clock, and the conveyor made a circuit once in twenty-four hours to an accuracy of fractions of a second. A finished clock was placed on the conveyor and would return to the same spot exactly twenty-four hours later. If it showed a different time from the 'standard time' of the conveyor, a quick adjustment could be made—a simple yet remarkably elegant solution.

The clock industry was still working within government restrictions on the availability of simple raw materials, such as steel. With the rearmament programme that coincided with the Korean War, less material was available for non-essential supplies, and there were also Board of Trade price control orders. From July 1952 the Ministry of Supply agreed to subsidize directly the manufacture of high-grade watches, but the clock and watch division of Smiths was in trouble, as, overall, it made no money. New restrictions on imports imposed in South Africa, New Zealand, and Australia damaged export sales.

In May 1953, Smiths watches received an extraordinary fillip with the successful climbing of Mount Everest by Edmund Hillary and Tenzing Norgay. While Norgay wore a Rolex watch to the summit, Hillary was wearing a Smiths 'De Luxe', made in Cheltenham, for which Smiths had developed a special low-temperature lubricant. But for the industry, this good news was rare.

With the post-war development of the idea of an 'Ideal Home', automatic labour-saving controls came into vogue. Seeking a way of keeping factories fully loaded, by the late 1950s Smiths began focusing less on clocks and watches and more on oven timers and other household appliance timers. There were still some further developments in the watch departments, such as the automatic wristwatch developed by Richard Good (1926–2003) and Andrew Fell (1919–88), which appeared in 1961, but overall the tide had turned.

It was the production of alarm clocks that funded the rest of the project at Smiths, allowing the creation and subsidy of its watch business. An industry target set after the Second World War was the production of five million alarms per year, twice the pre-war British consumption.[27] That target was met by 1950, but sales immediately began to decline, although throughout the 1950s, fifty per cent of clocks produced in Britain were thirty-hour alarms. In 1953, the Germans could produce them at fifteen shillings each, landed in Britain, after all taxes and duties, against Smiths' price of nineteen shillings. Some sales of British-made clocks were maintained to the mid-1960s, but then came under renewed attack.

---

25 Nye 2014, 132.
26 Hooper 1950.

27 Anon 1967, 8.

Figure 124 Production line at Smith's alarm clock factory, Wishaw, Scotland, c.1951. The continuous conveyor, capacity 50,000 clocks, is visible in the background.

The Swiss exported a record forty-seven million watches during 1964, while the Russians sent more than £250,000 worth to Britain in the second half of the year. On a 1966 trip to Hungary, John Boyd of the Amalgamated Engineering Union executive had found that alarm clocks were being exported at the 'quite fantastically low figure of one third of their actual prime cost'.[28] This was crippling for the Wishaw facility, which reduced its output, thus losing economies of scale. Smiths could not keep its factories fully loaded. It was no longer sensible to produce platform escapements at the Chronos Works, and in 1966 these were abandoned at the end of a thirty-five-year run. Anglo-Celtic and Enfield were faring little better in Wales. To promote trade with Russia, the government nearly doubled the permitted value of imported Russian watches to £400,000 for 1966. In early 1968 it was increased further to £500,000, with direct damage to Smiths and Ingersoll.

By the end of the 1960s, British domestic mass-production of synthetic jewels, balance springs, and complete clocks and watches was largely uneconomical, and Smiths and other makers began winding down or even closing their operations.

From mid-1968, Smiths' withdrawal from the horological industry commenced with the closure of Synthetic Jewels at Carfin, and then the winding down of the Enfield clock operation in Wales. Mechanical spring-driven clocks were giving way to battery-driven versions using the new 'Sectronic' movement, which could be volume-produced in Cheltenham. Smiths closed its balance-spring factory there after a thirty-year run.

By March 1971, Smiths had ceased high-grade watch production at Cheltenham, although a small team continued to work on the development of quartz clock and watches. It was to develop an innovative quartz watch—the Quasar—launched at the Basle Fair in 1973, but abandoned soon after, probably in acknowledgement that Seiko had developed a much superior watch in the calibre 38 series, while Omega had already successfully launched the Megaquartz.

WHERE DID IT ALL GO WRONG?

Why was it all not enough in the end? The answer lies in understanding the relationship between wider political objectives and the commercial imperatives that drove Smiths and others.

In the 1930s, private sector attempts to grow the industry were hampered principally by German dumping. Awareness of this influenced wartime planning for a post-war industry. There was

---

28 Anon 1967, 8.

a strong coincidence of interest between the industrialists and certain government departments—particularly the Ministry of Aircraft Production and its effective successor, the Ministry of Supply—which accepted the strategic argument of the need for a light-engineering capacity that could be turned to war-ends in times of emergency. But the development of weapons gradually left behind the world of clockwork fuses.

Under the post-war Labour government, policymakers saw value in creating a new horological industry with an emphasis on the needy development areas of Wales and Scotland. The establishment of a Timex factory in Dundee in 1946 was another example of government support for the industry, although Timex ceased operations there in 1983 after a long struggle to keep the factory in operation, including diversification into the manufacture of camera and computer parts. Government plans also included education, with the establishment of the relatively short-lived National College of Horology in 1947.

But political priorities can shift quickly. All Smiths' high-grade watches were sold at a loss, ultimately subsidized by the government. Smiths mistakenly saw the immediate post-war government backing of the industry as establishing a permanent commitment in which tariffs and quantitative restrictions on foreign imports would remain permanently in place, and at effective levels. This unjustified belief of Smiths was emblematic of the failure of the whole industry to adjust to a changing climate. Some in the trade believed Germany should be made to pay for its past sins. This left them out of touch with mainstream political thinking by 1948 at the latest. There was much characterization of the clock and watch sector as an 'infant industry' in British government documents throughout the late 1940s and 1950s. It would never grow to be an adult, since incoming officials had to deal with the realpolitik of rebuilding Germany.

The situation in Germany was wholly different. There, a large pool of trained talent, lower wages, and a flexible and extended working week provided enormous productive capacity, and a British–German horological trade war was decisively won by Germany in the 1950s. This outcome was repeated two decades later with the success of the German quartz clock industry, which anticipated the success and eventual dominance of the Far Eastern producers, and which ultimately caused firms such as Smiths to concede defeat and to cease production.

The common thread, observable over much of the nineteenth and twentieth centuries, is that British domestic manufacturers stubbornly continued to demand interventionist protection for British watch- and clockmaking, whether in tariffs, quotas, or any means that would reduce the effect of foreign competition and imports, rather than restructuring their operations to meet changing market situations, adopting new technologies, or increasing efficiency. The manufacturing trade therefore remained in structural opposition to a strong long-term trend towards free trade and market-based governmental policies, which led to its inexorable decline.

BRITAIN IN THE LATE TWENTIETH CENTURY AND BEYOND

By way of a coda, there is an irony in the outcome for British horology in the half century since Smiths and Timex wound down their operations. The British trade long believed that there would always be a market for the very finest, largely hand-crafted objects (mainly watches) with exquisite finishes, employing fine escapements and including complications. With the benefit of hindsight, that can be seen to be true.

George Daniels (1926–2011) was the best-known pioneer of a resurgence in British watchmaking from the 1970s onwards, based on a reversal of the age-old principle of the division of labour. Instead, Daniels's aim became to master many skills and to be able to claim that much of the work in his finished watches had been completed 'in-house'—a phrase that has become the touchstone of twenty-first-century high-end watchmaking. Supported by other highly talented craftspeople alongside him, such as Derek Pratt (1938–2009), Daniels recreated the mystique of the highly desirable, largely hand-finished British watch. In his wake, makers such as Roger Smith (b. 1970) and the long-standing chronometer-making firm of Charles Frodsham & Co have continued and developed this market further, and there is now a strong worldwide appreciation and desire for the very finest British wristwatches. However, it is important to note that an industry based on this principle cannot support a large workforce. Its products are highly prized (and valued) partly in recognition of the very limited production runs that are implied by high levels of hand-finishing.

But while there is a strong market for items produced in very limited numbers there remains no prospect of volume production of either clocks or watches with mechanical escapements in Britain—it is the luxury trade only that can be expected to succeed in the appreciable future. One intriguing note in conclusion is the continued and deep appreciation of the heritage value of the historic British horological industry, which can be seen in the way that some modern firms, keen to historicize their entirely new products, seek the borrowed authority of famous British names from the seventeenth to the nineteenth centuries and create new companies and brands that revive those historic names. It is not just East, Graham, Arnold, and others that have enjoyed such attention—it is an international phenomenon, with names such as Jaquet-Droz, Oudin, and Leroy (among many others) similarly rising again.[29]

---

29 Donzé 2017a, for the phenomenon among Swiss names.

SECTION TWO: AMERICAN HOROLOGY AND ITS GLOBAL REACH

Michael Edidin

A horological distribution network of agents, wholesalers, and retailers centred on Europe was in place at the beginning of the nineteenth century.[30] The network grew and expanded, but structurally it did not change over the period considered. In contrast, manufacture of clocks and watches outside of Europe changed over the nineteenth century, evolving from individual workshops, often using imported materials, to automated vertically integrated factories producing for both internal and overseas markets. Clocks and watches are discussed in separate sections, reflecting differences in the modes of production, sources of materials, and markets for each. Despite these differences, an openness to global trade and ideas was common to all horological production in the United States and in westernizing Asian countries.

AMERICAN CLOCKMAKING

By 1800 clockmaking was well established in the United States. It was typically local—clocks were made for wealthy or 'comfortable' patrons in the vicinity of the clockmaker's establishment. The making was divided, as with its English and German forbears, into movement construction and case construction. The movement makers went down two different paths. Along one, with some investment in imported tools, it was possible to produce a complete movement from domestic materials, though some of these, particularly brass for clock plates, were scarce and expensive. One way to lower costs was to use the minimum amount of metal for plates. Another, was to fabricate the movement, including the plates, from wood. Locally made clocks with brass **trains** were typically eight-day going and were expensive in their time.[31] Clocks with wooden plates and trains are usually thirty-hour going and cost less than their brass counterparts. Both English and German immigrants made clocks locally and analysis of the mechanisms of surviving clocks points to differences in approach to details, for example, **striking** mechanisms reflecting different European practices.

The second path of American clock-making used imported kits of castings, wheel trains, and other parts which were either supplied finished (for example, painted dials) or were finished by the local clockmaker. The paths described are averages for town and country makers. Clockmakers in big cities, such as Boston, New York, and Philadelphia, departed from these averages in both the quantity and sophistication of their work. A (relatively) large output was maintained by shops comprising significant numbers of journeymen and apprentices. When they left the master's shop, these craftsmen seeded practices and designs into the rest of the country. Simon Willard of Boston (1753–1848) is the outstanding example of a sophisticated city clockmaker. His own shop became the core of a complex of specialist suppliers, for example, case makers. Besides public clocks of exceptional design and finish and novel shelf clocks, Willard devised and patented (1802) a uniquely American clock, the banjo clock, a wall clock of novel design. This design was widely used and elaborated both by Willard's trainees and by independent clockmakers in many New England towns.

Using the methods summarized, a clockmaker might produce four to five clocks a year, at most. For example, the account book of Daniel Burnap (1759–1838), an important Connecticut clockmaker of the late eighteenth century, records fifty-one (brass movement) clocks sold between 1787 and 1805.[32] The typical price for these movements (eight-day going and hour striking) was ten British pounds, or about $33. For comparison, a skilled worker's wage was $0.50–1.00 a day, a loaf of bread was $0.12–0.20, and a barrel of flour was about $6.00.

Workshop-based production of limited numbers of expensive clocks gave way to industrialized production first of wooden and then of brass clocks. Eli Terry (1772–1852)[33] was a key innovator here. He began as an apprentice in Burnap's shop, training to be a traditional clockmaker. While in that shop he was also exposed to wooden clock production by another maker in the town. Setting up on his own, Terry fabricated both brass and wooden clock movements using some hand-powered tools. The tools increased production of wooden clocks enough to raise problems of distribution; according to his son and biographer, Terry himself would periodically go out peddling wooden clock movements at $15 to $25 a movement.

In 1802 Terry built a new shop with a water-powered milling machine and lathe and began supplying clocks to wholesalers. This shop produced 200–300 clock movements a year. In 1807 Terry contracted to supply 4,000 clocks to wholesalers. He opened another water-powered factory with a multi-blade water-powered saw that speeded the production of wheels, built jigs and fixtures for clock assembly, and devised templates and go/no-go gauges

---

30 See this volume Chapter 13, Section 1, Chapter 13, Sections 3–6; 17. *Cf.* Landes 1983/2000, ch. 14 *et seq.*
31 A survey of such clocks is given in Hohmann 2009.
32 Roberts 1970, 4–5.
33 Roberts 1970, 13–14.

for measuring tolerances. It took a year to build the machinery. The second year the factory produced 1,000 clocks and the third year it produced 3,000 clocks, fulfilling the contract. The 4,000 wooden clock movements were cased by contractors as wall or tall clocks and distributed by peddlers making rounds on horseback or by wagon. The clocks were within the means of prosperous farmers and were sold on credit, as well as being bartered for cash crops.

Terry's production methods and products were taken up by his former apprentices and other Connecticut clockmakers. Expanding production drove prices down. A letter of 1836 to a wholesaler offers cased thirty-hour shelf clocks for between $4.50–$5.50 and eight-day wooden work clocks for $6.50 to $8.00, depending upon the decoration of the case.[34] The relatively small size (about 2ft high) of the clocks and their low prices opened markets beyond New England. Connecticut clocks were peddled as far west as Ohio and throughout the south as far as Georgia and South Carolina. Clocks for sale in these regions were usually shipped from the factories or middlemen and labelled as produced in Connecticut. However, when some states enacted 'buy local' laws, manufacturers resorted to shipping clock parts and cases, and assembling and casing the movements in local 'factories' in, among other states, Virginia, Ohio, and Tennessee.

The market for clocks used credit and financing at many levels, from purchase of raw materials through purchase of movements and cased clocks for consignment to peddlers, to retail purchases from these peddlers. This financing collapsed under the weight of the depression of 1837. Debtors failed, bringing down peddlers, middlemen, and manufacturers, particularly manufacturers of wooden clock movements. Though wooden movement production continued into the 1840s, it was supplanted in domestic and international markets by simplified brass movements, both thirty-hour and eight-day. The simplified thirty-hour clocks, patented by Chauncey (1793–1868) and Noble Jerome (1800–61), soon dominated the market.[35]

Both wooden-movement and brass-movement clocks were produced by a constellation of specialist factories and suppliers concentrated in a few towns. Records for Bristol, Connecticut for 1810–50 list over 100 different manufacturers of movements, cases, or fittings. Most of these lasted only a few years and then closed, among other reasons from insolvency, quarrels between partners, or fires. Those that persisted increased output to the point that in 1850, eleven Bristol, Connecticut clock factories, employing about 300 hands, produced 200,000 clocks for both the domestic and the export markets.[36] As production increased and clocks became more reliable, distribution moved from specialized agents to general merchants, further opening the domestic market. Competition for this market drove manufacture of an ever-increasing variety and elaboration of cases. At peak when production concentrated in large integrated factories a single firm, E. N. Welch offered 200 different models (over a few decades) and another, Waterbury clock company, offered 500 different models of case.

Distribution of wooden clocks was limited by the sensitivity of movements to moisture. Damp air swelled the train wheels, leading to jams or to bad timekeeping. Overseas export of these clocks was not then possible. Brass clocks were not sensitive to damp and in 1841 or 1842, Yankee clockmakers began exporting to England. There is some dispute about who gets the credit for this, but no question that the export business boomed. American manufacturers established agencies for the distribution of complete clocks, and later for the assembly of complete clocks from imported parts. These were often labelled to suggest that they were of British origin, rather than retailed by British agents. Liverpool directory entries show an agency for Jerome clocks as early as 1853 and for the successor, New Haven Clock Company, as late as 1930. The export trade expanded well beyond England. In 1860 the pioneer maker of thirty-hour brass clocks, Chauncey Jerome, wrote 'Metal clocks can be sent anywhere without injury. Millions have been sent to Europe, Asia, South America, Australia, Palestine, and in fact to every part of the world'.[37] Later in this section we discuss the impact of these exports on indigenous Asian clockmaking.

The Connecticut clock industry evolved from a complex of movement makers, case makers, and suppliers of specialized parts and fittings to a group of larger, vertically integrated manufacturers developing and producing a great variety of movements and cases. Paths to the consolidation, and strengths, varied. Chauncey Jerome's company entered manufacturing through movement-making and through a series of mergers and joint ventures became part of the New Haven Clock Company, also founded (c. 1850) as a movement manufacturer. By 1860, the company, now making its own cases, employed about 300 workers and was producing 170,000 clocks a year. On the other hand, the E. Ingraham Company was developed by Elias Ingraham, a master case maker, who in less than twenty years' time received seventeen design or manufacturing patents for clock cases. Both New Haven and Ingraham, as well as Waterbury, Sessions, and other Connecticut manufacturers continued to improve movements and to develop new movement types. Case designs drove drafts of some of these new movements. For example, a change from weight-powered movements to spring-powered movements allowed production of smaller clocks, some standing less than half the twenty-seven-inch height of the early shelf clocks. The earliest known have Swiss **mainsprings**. Labels on some later shelf clocks boast of their springs having originated in New York, where spring making was developed by French Huguenot immigrants.

American clock producers thrived into the 1920s and even the 1930s, developing highly portable **balance wheel** clocks, including alarm clocks and even large cheap watches. Spring-driven

---

34 Roberts 1970, 31.
35 Jerome 1860.
36 Roberts 1970, 34.

37 Jerome 1860, 61.

clocks were supplemented by electric motor-driven clocks; both types of movement were fitted to a wide variety of cases. In 1938 Ingraham was producing 100,000 spring-driven and electric timepieces (including watch-sized mechanisms) a day! Most of these were cased at the factory, but movements were also sold to small manufacturers for specialized time use or for casing in new and different styles of case. Factory case shops also produced wooden cases for radios, drawing on the experience of decades of woodworking and finishing.

The Second World War saw the clock factories turn to war production, especially fuses and then transit back to clockmaking. This post-war manufacturing gradually decayed, with sales falling due to pressure from less-expensive imported movements and cased clocks. In the 1960s factories either closed or were merged into conglomerates that retained company names as brands. Only one American clock producer survives today: the Howard Miller Clock Company was started in the 1920s to build cases for imported, European, movements. This business model proved sustainable, combining finished movements with cases that drew on a long tradition of furniture making, combined with innovative new designs.

AMERICAN CLOCKS IN ASIA AND ASIAN CLOCKMAKING

In 1851 a report by an American resident in China estimated that about 100 clockmaking shops produced about 1,500 clocks annually in all of China. The report also commented on Chinese tastes for elaborate and gilt cases even for common clocks.[38] American clock manufacturers indeed exported considerably to China, offering options of elaborate gilding on cases for this market. This report was written just when the old trading system of a few 'factories' of Western merchants at a single port was giving way to the opening of multiple ports and the penetration of traders to the interior of the country. By the 1880s, importing and marketing goods was shared by western mercantile houses established in China and local factors and merchants. These were tied together and to the manufacturers by a web of credit and financing and by technical improvements in communications—the telegraph and advances in steamship design and construction.

Edward Ingraham (1887–1972), head of E. Ingraham in the 1950s, commented that the firm had done a considerable business with the Orient, particularly with China. The Chinese taste for elaborately gilt cases was supported by a group of thirty gilders, whose sole work was decorating cases for the Chinese market. Ingraham cited a letter of the 1880s estimating an annual market for 300,000 clocks.[39] When, if ever, this total was reached is not clear. In 1875 the value of all US exports to China was $1,000,000, a tiny fraction of a total $605,000,000 exports. By 1900 it was $15,000,000, still just one per cent of total American exports. In turn, clocks were a very small part of total exports.

The indigenous Chinese industry responded to imported clocks by producing inexpensive clocks whose locally made cases were fitted with movements copied from American and European clocks. These were in direct competition with imported clocks. The label on one such, a shelf clock of around 1900, reads:

> This factory was established ... to make new-style sitting clocks and beautifully crafted hanging clocks capable of working for 8 days. We advise the accuracy of the clock, richness and high quality of the materials used, solidity of the machine and durability ... With all these qualities the clocks are equal to the best foreign-made instruments.[40]

The label goes on to offer clocks in a wide variety of cases, again following the American emphasis on case design and novelty. Despite the description of the producer as a 'factory' the scale of production is not clear. Were hundreds or tens of thousands of clocks locally produced each year?

The westernization of Japan in the second half of the nineteenth century opened the country to imported clocks and drove the establishment of clock factories producing western-style clocks for the plethora of government offices and for the westernizing middle classes. Imports were financed by credit for foreign trading houses, who then wholesaled them to Japanese merchants. By 1877, about 250,000 clocks, mainly American, had been imported, and 700,000 by 1887. Japanese manufacturers, aided and encouraged by the government, responded by setting up factories producing clocks at a price that competed with the imports. One source notes that, from 1887, domestic clockmaking capacity doubled every ten years. By 1907, more clocks had been made domestically than had been imported and there was a significant Japanese export of clocks to other Asian countries.

Factories ranged from those like Seikosha, an integrated plant using the latest machinery and constantly developing new movements and styles, to assemblers who bought in components made by specialist factories and even by individual outworkers. By 1912 there were twenty major companies, some designing their own movements and exporting half of their production. Seikosha was the largest of these. By the 1930s they employed 2,000 workers producing 300,000 watches and 980,000 clocks a year.[41]

In Japan as in the rest of the world, the Second World War brought militarization of horological industries and production of military timepieces as well fuses and other fine mechanical devices. After the war, clock development continued along lines seen elsewhere—development of electric, electromechanical, and quartz oscillator movements for internal use and for export. Korea,

---

38 Bedini 1956b, 215.
39 Bedini 1956b, 219.

40 Edidin 2000, 804.
41 History and data on Japanese clock imports and domestic production are from Uchida 2002.

occupied and colonized by Japan in the years 1910–45, was late to clock manufacture, but an industry developed there by the early 1960s.[42] The first products were electromechanical clocks for the domestic market, but a plethora of manufacturers soon expanded to traditional movement making. Movements were made in a variety of styles, mainly copies of Japanese, American, and French designs and cased for both local and western tastes. Export-driven production ramped up to the level of hundreds of thousands per annum by the mid-1970s, followed by a race to the bottom as importers drove prices down, manufacturers cut costs and lowered quality to compete, and customers turned away from the poor-quality clocks that resulted. The last efflorescence of mechanical-clock production ended in the mid-1980s loss of export markets and bankruptcies.

Currently, the horological industry is fragmented and reminiscent of the state of the trade in 1800. A variety of manufacturers, large and small, in Europe and Asia, produce quartz, electric, and mechanical movements. These movements are cased in literally thousands of different cases, ranging from individual kits to factory produced longcase clocks to modern wall clocks. An American clockmaker of the early nineteenth century would recognize the global pattern of trade in movements and fittings supplying local shops with materials to assemble and case complete clocks.

AMERICAN WATCHMAKING

English and European watchmaking of 1800 depended on local networks of specialists, each producing one or a few parts of the hundreds of parts of a common verge watch of the day. These parts were assembled and integrated by a specialist finisher, then cased by other specialists to produce a complete watch. This system effectively prevented the export of kits, homologous to the rough collections of plates and parts available to American clockmakers. American watches of 1800 were either imported complete or were finished in this country from imported rough movements. A few watchmakers modified the imports to use unusual **escapements**, but without the outworker network, the labour cost of producing a complete watch was prohibitive.[43]

The nineteenth century saw the development of an American watchmaking industry, based on invention and improvement of machines to make parts for watches which, in turn, were designed to reduce the number of parts needed and the number of operations for their production and assembly. In the first stage of development there was some two-way traffic across the Atlantic; some watches and watch parts were still imported, and some factory-produced watches exported. In the second stage, the stream of exports overwhelmed imports, displacing much of the trade in complete English and Swiss watches. In a third stage, beginning in the 1880s, the sale of conventional watches was overtaken by sales of dollar watches, made with the techniques of, and by, the great American clock manufacturers. In the twentieth century, imported watches again gained dominance as American manufacturers lost their innovative momentum and steadily fell behind Swiss competition.

The bane of all watch production was the cost in time and money of the system of specialist outworkers and the lack of standards and interchangeability that prevailed in watch production. The promise of an industrial revolution loomed large for watchmaking in Europe and America. Realization of the promise was the key to American achievements and to the later downfall of American watch manufacturing.

The revolution developed in stages, constrained by the available technology and the need for accuracy and reproducibility (precision) orders of magnitude better than needed for wooden or brass clocks. It would be decades before machinery was developed to work at the smaller scales and tighter tolerances needed for machine production of wheel trains, escapements and the multitude of small screws needed for watch assembly.

The idea of watchmaking by machinery was in the air in Europe and America. In late eighteenth-century France, Frédéric Japy (1749–1812) pioneered machinery to produce rough movement plates, the largest scale of a watch. The Swiss Pierre Frédéric Ingold (1787–1878), who was trained in traditional watchmaking, conceived and created machines to drill watch plates reliably (and so accurately pitch wheel trains), and a press to both punch out wheels (instead of cutting them from a solid disc) and to finish them. Japy was successful. Ingold was not. He had the imagination and skill to conceive of the machines and the watches they could produce but lacked the capital to sustain his project.[44] Bankrupt, he sold his machinery and migrated to the United States. Though he spent a decade in the United States, there are almost no traces of his work in that time and no evidence that he met with American inventors who were on their way to successful development of industrial watchmaking.

American watchmaking entrepreneurs built on the experience of New England metalworking industries, particularly Connecticut clockmaking, and on the experience with automated machinery at the US government Armory in Springfield, Massachusetts. Capital for development was supplied largely by firms in the watch and jewellery trades. Henry Pitkin (1811–46) and his brothers, the first people to produce an American watch using machinery, were apprenticed to a Hartford Connecticut silversmith and watchmaker. Henry concentrated on watchmaking and repair. In

---

42 The history of modern Korean clockmaking is drawn from Hovey 1986.

43 Watches and watchmaking in pre-1800 America are illustrated and discussed in Newman 2020.

44 Ingold organized two companies for production of machine-made watches, the first in France and the second, The British Watch Company, in England. Both failed, but the British Watch Company produced some hundreds of watches. Ingold's life and work are described in Carrington 1978. The politics and economics of the failure are discussed in Davies 1996. See also Berner 1932.

the course of his work he developed the idea for a watch and the machines to make its parts. With his brothers' support, Henry developed the tools and machines to produce a ¾-plate watch with plates punched out of rolled brass, American-made **lantern pinions** rather than solid pinions, and conical pivots running against hardened steel screws instead of hole jewels. About 350 of these 'American Lever' watches were produced in East Hartford, starting in 1838 and approximately another 500 after a move to New York City in 1841. These are truly the first American watches made using machinery to ensure uniformity and reproducibility. The watches were sound, but they were not well-received by the trade and the Pitkin brothers never attempted to raise money for larger-scale manufacturing.[45] Despite this practical failure, the Pitkin watch venture is important both as a pioneer and as the origin of a key inventor of machines for watch production, Nelson P. Stratton (1820–88). Stratton, apprenticed to Pitkin, built the machines of the Hartford workshop to Henry Pitkin's designs. Stratton likely had further exposure to machines for accurate and reproducible production of parts through his brother-in-law, who worked at the Springfield Armory.

Experience, imagination, skill, capital, and luck finally led to the flowering of mass production of American watches.[46] Aaron Dennison (1812–95), trained in clock and watch repair, recruited the skills and capital that built the first effective machine-made watch factory adjacent to the Roxbury, Massachusetts shop of the clock and instrument maker Edward Howard (1813–1904). Howard and his father-in-law Samuel Curtis (1785–1879) supplied the capital, Dennison the vision and designs for watches and machinery to make them. Initially, neither the design of the watch nor the machinery to make it was a success. However, less-ambitious but better-designed machinery and a watch design that was essentially a simplified English full-plate watch of the period led to success. Some parts—hairsprings, jewels, and dials—were imported from England, but the rest of the watch was made in Roxbury.

Production rates increased as machine design improved. In the 1880s Edward Howard reminisced about the problems of building reliable machinery: 'The company kept on building tools, trying them, laying them aside and making others better adapted to the end sought.'[47] By 1854, 100 workers at Roxbury were producing six to ten watches a day, 1,800 to 3,000 annually. This is approximately ten times the annual output of an English watch producer and two to three times the annual output of prolific Liverpool makers.[48] Howard and Dennison had to work hard to sell this number of watches to wholesalers, but the watches did find acceptance in the trade to the point that a large New York City firm, Fellows and Schell, invested $20,000 in the company in return for an exclusive agency for the Boston Watch Company.[49]

In 1854 the Boston Watch Company moved to a new factory in Waltham, Massachusetts and continued to expand production. In 1856 a general economic depression reduced sales to the point that the company was bankrupt and was sold at auction. A consortium led by Royal E. Robbins (1824–1902), a New York wholesaler, bid successfully for the company assets. Robbins worked assiduously to sell the ever-increasing production of the firm, recapitalized as the American Watch Company, and at the same time, to reduce production costs and to match output to demand in slack times. Robbins' ruthless wage-cutting, financial juggling and focus on just two watch models of mediocre quality got the company through the stormy times before the outbreak of the Civil War in 1861, at which point demand for an inexpensive watch drove up sales and profits. William Keith (1803–after 1883), a former treasurer, and later president, summed up the period '… the enterprise of American watch-making was begun, carried forward, and ultimately became a great success, during the period between 1850 and 1865, a period of political commotion and business fluctuation and discouragements, especially for an untried and generally considered doubtful enterprise … '.[50]

Aaron Dennison, now superintendent of the new company, wanted to continue to innovate new models and to experiment generally. but this was not compatible with Robbins' focus and in 1862, Dennison was fired. He went on to found another watch manufacturer, Tremont Watch Company. When this failed, Dennison moved to England where he ultimately had success, founding the Dennison Watch Case Company, which thrived throughout the nineteenth and twentieth centuries.[51]

A few years before Dennison was fired, a number of master mechanics, notably Nelson P. Stratton, Charles van der Woerd (1821–88), and Charles Moseley (1828–1918), left voluntarily to join the Nashua Watch Company, a firm established to make modern, ¾-plate, precision watches. Financed by a local jeweller who worked briefly at Waltham, the company built the machinery and some of watches, then failed for lack of enough financing to cover the costs of machinery and of watches in process. The watch design, and the human skills and machinery to make it, were bought by American Watch Company and integrated into the factory as a separate department, producing their highest-grade watches.

Nashua's brief history epitomizes the factors in founding watch companies. Capital was often supplied by local jewellers or wholesalers and expertise brought in by a few dozen skilled mechanics and their disciples, all with experience at Waltham. For example, the Newark Watch Company was financed by Fellows and Schell, the New York wholesalers who had invested in the Boston Watch Company. Their machines were designed by Napoleon

---

45 Rosenberg 1963.
46 This section, particularly price and production data, is largely based on Harrold 1984 and Harrold 2007.
47 Keith 1883, 21–22.
48 Edidin 1992, 670.
49 Moore 1945, 19.
50 Keith 1883, 52.
51 Priestley 2009; Priestley 2018, 313–16.

P. Sherwood (1823–72) who had worked at Boston Watch Company and at E. Howard and Company. Another New York jewellery company and importer, Giles Wales & Co., flush with wartime profits, started the United States Watch Company, using the skills of a Waltham alumnus, James H. Gerry (fl. 1860–90), and nearly a dozen other experts who left Waltham to join him.

Neither Newark nor U.S. Watch ever made sales equal to their costs in machinery and labour. A third Waltham clone, the National Watch Company of Elgin, Illinois, succeeded through a combination of expertise, strong financing, careful cost control, and production of quality watches. Their machine makers and designers were lured from Waltham and included Charles Mosely, the Boston Watch Co. and Nashua pioneer and innovator. Financing came from a midwest lumber baron and cost controls halved the prices of Elgin watches by the mid-1870s, with the rest of the watch industry forced to follow suit.[52]

THE DOMESTIC MARKET FOR AMERICAN INDUSTRIAL WATCHES

The boom in watch manufacturing coincided with a boom in the American economy after the Civil War. An expanding middle class bought much of the increasing output of the major watch factories after this output trickled down through layers of factory-appointed distributers, wholesalers, and retailers. Each layer added cost to the production cost of a watch. On the other hand, production costs decreased steadily as improved machinery—first semi-automatic, then fully automatic—was introduced. The seven-jewel American Watch Co. product of 1860 cost $40 at retail. A seven-jewel Elgin of 1910 cost $4–5. Over the same period, prices dropped tenfold and production increased by about the same factor. The 1910 price, when output was millions of watches a year, was still several days' wages, so even the least expensive jewelled watch remained out of the reach of working people.

Market saturation and periodic depressions were literally the death of a dozen or so watch producers. The survivors maintained sales by reducing costs and by introducing new models and grades. Cases, sold separately, were also elaborated and styled. Merchandizing of both watch movements and cases was controlled by a 'watch trust', which made sure that all steps of distribution made their profits, on both watch movements and cases. Some second-line makers, notably the Hampden Watch Company of Canton, Ohio, controlled by case maker John Dueber, sold directly to retailers and incurred the wrath, and the lawsuits, of the watch trust.

The trust and conventional watch manufacturers did not succeed in maintaining production and profits. Some wholesalers began selling to general merchants who were not bound to maintain prices. Dueber-Hampden, supplying cased watches, next began selling (through a nascent Sears Roebuck) direct to consumers, as did other firms, often general mail-order suppliers and not specialists. Profits dropped continually under all these pressures.

THE EXPORT MARKET FOR AMERICAN WATCHES AND MACHINERY WATCHES

The American Watch Company dominated the export market. Well capitalized, with good connections in England through both Robbins and Dennison, the company made a strong foray into the English and colonial market, selling dozens of models, including a few specifically designed for the English market. The watches were well received. At the low end they were cheaper than English watches and more reliable timekeepers. At the high end, deluxe models were more affordable, and often more stylish, than competing English hand work. Emphasizing the American penetration of the English market, a picture of advertising at a London railway terminal (Figure 125) in the mid-1870s shows a prominent wall poster for American Watch Company.

The American Watch Company was driven to export by the collapse of its domestic market in the panic of 1873. The factory closed for four months during which watches were designed specifically for the English market. An agency, Robbins and Appleton, was opened in London to coordinate dialling, casing, and shipment of Waltham watches to Britain and its colonies.[53] The earliest advertisements for American Watch Company watches appear in English newspapers in May 1874. Advertising was not confined to Britain. There are no Waltham advertisements in Australian newspapers before 1 June 1875. In the following months literally hundreds of Waltham advertisements appear, all in the name of the London agency. In the 1870s, twenty-five per cent of production was exported. The major distribution network was in England, where several companies and franchises, including case makers, were maintained well into the twentieth century. By then Waltham exports had fallen to about eight per cent of total sales but covered much of Asia and South America. There was even a sales office in Tokyo. Elgin National Watch Company also targeted England in the 1870s, like Waltham, producing models specific to English tastes. However, this market never became as important for Elgin as for Waltham, perhaps because of the costs and complications of exporting from the midwest.

---

52 Hoke 1991 gives an overview of the variety of American watch manufacturers.

53 Moore 1945, 54–5.

Figure 125 This picture of advertising in a London railway station was drawn in the year that the American Watch Company began its export drive to Britain. A large poster, third from the left on the station wall, boasts the company's output and offers both traditional key-wind and innovative stem-wind watches through its office in Hatton Garden. Frontispiece to H. Sampson, *History of Advertising*, London 1874.

MACHINERY

Failure of a watch manufacturer released machinery for sale. Numbers of these machines, as well as watchmaking machinery built by free-standing machine makers, found homes overseas. Dennison brought machinery and rough movements to England after Tremont Watch Company failed. There he constituted the Anglo-American Watch Company—which failed. Newark machinery migrated west to Chicago, then to California, and ultimately, about 1895, to Osaka Japan, to a company that failed after about ten years. American machinery was also used for manufacturing Swiss watches, notably by the International Watch Company of Schaffhausen, starting in the 1860s. American watchmaking machinery was exported as late as 1930s when the complete Dueber-Hampden watch factory was sold to the Soviet Union.

DECLINE AND FALL

We noted earlier that acute competition for markets and increased efficiency of production drove down the prices of conventional watches and reduced profits. The next blow to profits was delivered by the clock factories of Connecticut. In the late 1870s, at the peak of conventional watch production, these pioneers of machine-made horology began producing dollar watches, using stamping, not machining, techniques and omitting all jewels in the watch design. The watches were distributed by direct mail-order and through a variety of retailers, all outside the traditional distribution and sales network. The watches were bulky, mediocre time keepers and not repairable, but they were cheap, and they sold by the millions, bringing in a new market, but also taking volume from jewelled watch producers. By about 1910 production of dollar watches was about double that of jewelled watches—four million a year. They ever after dominated the American market for mechanical watches, reaching their apotheosis in the Timex wristwatch produced from 1945 by a direct descendent of the Waterbury Clock Company.

A second pressure on American watch sales came from the Swiss, who gradually increased the use of machinery in their watch production. They further began innovating machinery and tools capable of working to ever-finer tolerances which became the norm as fashion, and production, shifted from pocket-watches to wristwatches. First, Swiss pocket-watches encroached on the higher-end market for jewelled pocket-watches; imports surpassed American production by 1905. Then, the Swiss swamped American producers, who were slow to shift to wristwatches. By 1930, seven million Swiss watches, almost all wristwatches, entered the country while domestic production of both pocket- and wristwatches was about one million.

The decline and death of Waltham was accelerated by a shift in control from sales and technical managers to financiers. These drained capital from the company and reduced investment in machinery. Wristwatches were made on fifty-year old machines, instead of on modern tooling and these never competed well with Swiss watches made with modern machine tools. The successor firm to the American Watch Company ended as a brand name on Swiss-made watches. Elgin, less heroic but better managed than Waltham, was badly hurt by the depression of the 1930s, staggered on into the 1950s, making serious investments in electromechanical watches, but finally ceased manufacturing and persisted as a brand name on Swiss-made mechanical, electromechanical, and quartz movements.

A LATE TWENTIETH-CENTURY FLOWERING AND ANOTHER DECLINE

Development of miniature coils and small batteries during the Second World War opened the way to the production of battery-driven watches. These used conventional balance oscillators impulsed by magnetic fields and were at least as accurate as the best detached lever watches. Development of these watches in the United States was driven by Elgin in partnership with Lip of France, and Hamilton, with some relationship to Epperlein in Germany. Elgin's watch was a technical failure and never reached the market. Hamilton's watches sold well from their introduction in 1957 into the 1970s. This success was based on movements with adequate timekeeping in striking designer cases.

The American boom in semiconductor design and fabrication made available timing modules for watches using quartz crystal oscillators at orders of magnitude higher frequencies than balance oscillators. Some of these modules were exported and used by the Swiss in their development of their quartz oscillator watches. In the US, Hamilton produced a luxury quartz watch in 1970, the Pulsar. Less-expensive quartz watches soon appeared, all limited to display on demand because of the high power consumption of their light-emitting diode (LED) displays. A few years later a liquid crystal display (LCD) watch was developed by Optel, a spinoff of an RCA project to build flat screen televisions. LCD displays had much lower power consumption than LED, so could show the time continuously. With improved displays and the technology for quartz watches widely available from a number of US companies, quartz oscillator watches rapidly evolved from luxury to consumer items. As in the nineteenth century, so in the twentieth century, improvements in production led to a price race to the bottom, driving American producers out of the business or reducing them to branding movements and watches produced in Asia.

Today only the small are flourishing. A number of twenty-first-century American watch manufacturers design, produce, and case movements in the US, notably RGM and Weiss. Others, such as Shinola, assemble mechanical watches from Swiss components. The American watch industry has returned to its state in 1820, local, at times eccentric, and selling to a limited market.

SECTION THREE: THE HOROLOGICAL ENDEAVOUR IN FRANCE

Joëlle Mauerhan

Recent French horology can be examined from three perspectives:

1. the traditional distinction between 'great clock' and 'small clock' making;
2. the location of production: regional centres collaborating or competing between themselves and often indifferent to frontiers; and
3. the mechanical, electrical, and electronic innovations, accepted with greater or lesser difficulty within them.

THE PARIS–ST NICOLAS D'ALIERMONT AXIS

Paris, the nineteenth-century centre of the luxury trades, was active in most branches of horology. At the 1889 exhibition commemorating the centenary of the revolution and with Gustave Eiffel's tower as its star, it was said that 'Paris is the principal centre for the making of tower clocks, marine chronometers and precision timepieces, domestic clocks, carriage clocks, eight-day clocks, alarm clocks, and various counters'.[54] Parisian production ran to 5,000 pocket-watches a year, high-quality pieces mainly for export,[55] and a non-negligible number of repairers looked after the watches of a clientele from the whole of the country.

But rivalled by Besançon, Paris abandoned industrial watch production. Nonetheless, as the arbiter of the conditions of exportation and external relations, it remained the nerve centre of the horological industry. Major manufacturers had to have a branch there, even indeed the personal residence of their owner, as was the case of Jules Japy (1846–1914) from Montbéliard or Ernest Lipmann (1860–1943) from Besançon. Moreover, Paris was the obligatory route towards Great Britain. The leading makers, such as Leroy and Breguet, opened offices in London and during the 1860s several Besançon makers opened a shared retail outlet there for well-made, but standard, watches.

Clocks were also a part of the capital's fame. In the previous century, founders and cabinetmakers were renowned for the quality of the cases in which utilitarian domestic clocks were presented. This reputation was exploited with the production of copies and models in the same style, while following new trends. Clocks were decorated with moralizing and symbolic sculptures,[56] and driven by new small round movements—first an artisanal production in Paris, later manufactured industrially in series at Monbéliard in the Franche-Comté and St Nicolas d'Aliermont in Normandy.[57] The capital itself made a speciality of precision, aided there by an important production of machine-tools and precision instruments particularly for geodesy and navigation, both rapidly expanding strategic activities stimulated by colonialism. Precision horology and mechanics had a natural environment there.[58] In 1881, in conformity with this orientation, a school of horology—a school for the precision arts—was opened, which even attracted Swiss watchmakers eager to learn the latest improvements. In this environment of high competence, electrical horology emerged alongside the major field of telegraphy in the mid-century, while the vast field of metrology, in need of precision timers, would rapidly become a semi-independent production.[59]

PARISIAN PARTNERSHIP WITH ST NICOLAS D'ALIERMONT

St Nicolas d'Aliermont was noted in the eighteenth century for the production of a longcase clock, the 'St Nicolas', an equivalent of the **Comtoise**, but the regional market for it became too limited, leading to difficulties for a local industry. This led to government intervention and the arrival in the village in 1807 of the Parisian clockmaker Honoré Pons (1773–1851) to revitalize the industry. Grouping the individual manufactories of the village and introducing small machines he had himself devised, after the manner of Japy, he effected a remarkable revival, which led the local industry first to develop the making of movements for Parisian mantel clocks, and then to develop the construction of carriage clocks.[60] St Nicolas became a serious rival to Montbéliard. The Normandy–Paris partnership was sufficiently strong for Paris makers to order the making of **blancs** for marine chronometers. The makers of St Nicolas obtained good results and became competitors with English makers. London, like Paris,

---

54 Trincano 1940, 25–6.
55 Parisian watchmakers in themselves are not further studied in this chapter, but the *Revue Chronométrique* is a major source for their activity from 1860 onwards.
56 Dupuy-Baylet 2006.
57 See this volume Chapter 13, Section 6.
58 Schiavon 2013.
59 For some examples see Chapter 23, Section 2, 3, 4, and 5.
60 Allix & Bonnert 1974, 90ff; Cournarie 2011; Chavigny 1997.

Figure 126 Workers 'clocking-in' using a punch clock of the type devised by Arthur Lambert established at Saint-Nicolas d'Aliermont. The business closed in 1987, although AGT-Systèmes continued making time control systems there until 2011 © Collection du Musée d'horlogerie de Saint-Nicolas d'Aliermont.

was specialized in precision as well as marketing. Nonetheless, this development from longcase clocks, to movements, to marine chronometers left the industry weakened. Further restructuring was needed and effected in the twentieth century, first with the switch to armaments manufacture during the First World War, then with a move into the making of alarm clocks with the Bayard Company, and of various recording instruments, notably the Lambert clocking-in clocks (Figure 126). By the 1960s, however, the local industry was on its knees, its trajectory today memorialized in a local museum.

THE ROLE OF THE EXHIBITIONS

Participation in the numerous commercial exhibitions that punctuated the nineteenth century was popular with Parisian and other manufacturers. Obtaining a medal, or even a mention, offered commercial advantage.[61] The Universal Exhibitions, which owed something to the 'Exhibitions of the Products of French Industry' organized from 1798 onwards, were intended to display a panorama of national industries, vaunting the progress made in the sciences (Figure 127) and in technical manufacture, affirming contemporary faith in 'progress'. Basically nationalist, especially in France, where the civilizing role of the country in its colonies was put to the fore, the presence of invited foreign countries favoured commercial exchanges, and marked a step towards world trade.

The first Universal Exhibition was that of London in 1851, for which a glass and iron structure—the Crystal Palace—was specially built. Paris followed in 1855 and other cities of Europe did likewise. In 1876, at Philadelphia, for the anniversary of the Declaration of Independence, America astounded Europe with some 1548 acres of exhibition space, whereas Paris in 1867 used only 837 acres. In Philadelphia ten million visitors idled their way through an enormous display of novelties and merchandise. It was the beginnings of a mass culture based on advertising, publicity, and discounted goods. For Frederick le Play (1806–82) and Michel Chevalier (1816–79), the leading administrators of the early Universal Exhibitions in France, paternalism, free trade, and the condition of the working classes were preoccupations. Workers' delegations were invited to the exhibitions. In their reports they expressed their amazement at the new machines and displayed their anxiety when confronted by the rational organization of German and American factories, their fear of foreign competition given their own out-of-date tools, and their lack of machine tools, the degrading effects of producing cheap goods, and the supposed indifference of their employers.[62]

---

61 Hilaire-Perez et al. 2012.

62 Such reports could be seen as polemics against conditions in France

Figure 127 Laying a transatlantic cable in 1899, from Georges Dary, *A travers l'électricité*, Paris, 1900, 81.

PUBLIC CLOCKS

Despite growing German competition, the making of turret clocks was still a major speciality in nineteenth-century France, their making having been abandoned in Switzerland. Although it remained geographically a scattered production, four main areas of activity can be designated: Paris, Eastern France, Normandy, and the Franche-Comté.

J. Bernard Henry Wagner (fl. 1790–1851) emigrated from Germany to Paris, where he established a turret clockmaking workshop in 1790. This business was continued by his cousin Bernard Henri Wagner, who sold it to Armand François Collin (1822–95) in June 1852, who in turn sold it to the father and son team of Château in 1884. It remained active until 1937. A report on the Wagner business at the time of the 1844 exhibition was a eulogy— 'The name of Wagner is to large clockmaking what the names of Berthoud and Breguet are to precision horology'.[63] To the East, in Alsace and Lorraine, the notable business of Schwilgué–Ungerer was rivalled in Nancy by that of the Gugumus brothers, both of whom had worked in the Schwilgué concern, and by that of Urbain Adam (1815–81). To the North, Augustin Lucien Vérité (1806–87) furnished his region with public timepieces, invented a constant-force **escapement**, and created the astronomical clocks of Beauvais and Besançon,[64] while in the Franche-Comté Louis Delphin Odobey (1827–1906) created his turret clock factory in 1858. Continued by his three sons, and grandson, it remained in activity as a leading manufacturer until 1964, producing some four to five thousand clocks. In the High Jura, industrial methods were applied. Developing from the earlier industries of wiredrawing and nail-making, horology developed in the area around Morbier, Morez, and Foncine-le-Haut with a national and international distribution. Electrification and new methods of time distribution however severely affected it and from the mid-century onwards the making of eyeglasses largely replaced family clock production.[65] Maintenance of public clocks, still necessary, enabled the survival of some enterprises, such as that founded by Théophile Prêtre in 1780, which would eventually in 2014 be certified as a 'Living Heritage Business'.

DIVERSIFICATION AND INDUSTRIAL HOROLOGY

In the mid-century, clockmakers became interested by the new technology of electricity. In Paris, the Henri Lepaute company produced large electric clocks, and diversified into lighthouse equipment and railway regulators. Vérité was responsible for installing the electrical time-system at the Gare du Nord in Paris. Paul Garnier was a particularly audacious maker. He developed effective synchronized electric clocks, thus participating in the beginnings of time distribution services, and sought simple, horologically based solutions for the production of a variety of inexpensive instruments, which brought him industrial giants as clients. For the Schneider concern in Le Creusot he developed a counter for the revolutions of machines;[66] for the railways he produced a counter able to measure both the fatigue and the speed

---

as much as objective assessments of the competition. See, for example, the scream of alarm concerning the industry of the Black Forest emitted in the Rapport 1876, 322–8, by one delegation. See also Alexandre 1867.
63 Rapport 1844. There were, however, two Wagner clock businesses: that of Bernard Henry, and that of his nephew Jean. See Robert 1878, 77.
64 Miclet 1976, 47–8; Gros 1913, 218–21. For the astronomical clocks, see this volume, Chapter 9, Section 2.

---

65 Olivier 2002.
66 Hurrion 1993, *passim*.

of locomotives. In parallel, for the astronomer Urbain Le Verrier (1811–77), he built an electromagnetic regulator used during the establishment of the exact longitude difference between Paris and Le Havre.

Industrial horology was born from such kinds of research. Today it is dominated by Bodet, founded in 1868 by Paul Bodet (1844–post-1910) at Trémentines (Maine et Loire) to build turret clocks, a business that in the twentieth century has evolved to meet the need for time distribution in railway stations, airports, banks, and schools. Preserving its horological activity, in 1989 it purchased the firm of Ungerer and developed a specialty in the restoration of public clocks. Always diversifying, it has developed a branch for time control and synchronization using ultra-precise measuring instruments and today employs 780 people in France and three foreign branches.

THE COMTOISE CLOCK

In parallel with the development of turret clockmaking in the High Jura, a distinctive longcase clock appeared—the Comtoise. Developed from the late seventeenth century onwards, only the movements were made in Morez and Morbier. These, sold by pedlars throughout France, were locally cased in local styles. With the development of a rational, industrial production in the mid-nineteenth century, however, standardized pine cases were also supplied with the movement, their violin shape becoming almost identical but with a host of minor variations.[67] The clocks were exported to Spain, Portugal, North Africa, and South America—manufacture in the Jura was at its height.[68] From the end of the First World War, however, it was in decline. The longcase clock was obsolete, its makers increasingly converted to eyeglass making. Nostalgia, however, led to the return of the model in the 1970s, with copies being produced by Romanet and Odobey, the latter becoming (together with Vedette, a Savoy maker of alarms, small clocks, and wall carillons) the leading French maker of domestic clocks, outstripping Bayard in St Nicolas d'Aliermont. The heritage movement in the decades around 2,000 has not left the Comtoise clock untouched. In Besançon, Vuillemin continues to produce traditional movements—housing them in modernist cases aimed at an artisanal exclusive market, made by Philippe Lebru with his Utinam company. The Comtoise, like the mechanical watch, is now being rethought.

TOOLS AND MACHINE-TOOLS

From its beginnings at Beaucourt at the end of the eighteenth century, the Japy business has been noted for its machine-tools applicable to horology. A bronze medal was awarded them at the 1801 exhibition of the products of national industry. Clockmaking was then the principal activity before diversification into general ironmongery products. Thanks to its aggressive marketing techniques, a paternalist policy, and, from its beginnings, a culture of innovation, by the 1830s the company ranked among the leading exporting industries. With the blancs produced in Beaucourt, **pinions** in Berne and Seloncourt, and with its proper production of clocks, small clocks, and watches, it swept the board with prizes in the international exhibitions, and its prestige extended to the companies that derived from it—Roux, Marti, Beurnier, l'Epée, Maillard-Salins, Vincenti, and Dodillet, to name but a few. During the Second Empire, Japy counted among the leading French employers.[69]

Attentive to current developments, the principals at Japy interested themselves in the latest scientific discoveries and grew closer to Paris. They manufactured the horological parts needed for new devices, such as in telegraphic apparatus. Mechanician-horologists at the frontiers of research, often far ahead of simple clockmakers,[70] theirs was the model on which a mechanized production of watches was developed in America. In the twentieth century, however, they failed to adapt and disappeared, although, until recently, a nostalgic specialized production of Japy-type traditional carriage clocks was continued by L'Epée at Ste Suzanne, a satellite town of Montbéliard.[71]

With a marketing strategy that was almost as audacious as that of the makers of Comtoise clocks, horological tools were exported from Montécheroux throughout Europe and the Americas. It resulted from the adaptation of an existing industry. In about 1776, the Chafaudian Jonas Brand persuaded the cutlers of the region to abandon making knives in favour of horological tools, which required the same techniques of forging, filing, and polishing. Success was rapid. Forty craftsmen engaged in the early years of the nineteenth century had become 400 by 1876. It remained, however, an artisan activity, not developing on an industrial scale until the 1880s, with the first machines appearing in the 1890s. But the reputation of Montécheroux tools with their elegant wood handles was high. They were considered the finest until well into the twentieth century.

The Haute Savoie, for long a part of the kingdom of Piedmont-Sardinia, had also long been exploited by the Swiss. The makers of Geneva drew on the skills of craftsmen in Cluses and some twenty other small towns and villages in the Valley of the Arve. Pluri-ambivalent in mechanics and agriculture, the Savoyards subcontracted to supply wheels and pinions for watches to the

---

67 Caudine 1992.
68 The industry has been studied in detail in recent decades. See Buffard 2019.

69 Mayaud 1991.
70 Blanchard 2011.
71 On Japy see Lamard 1984/88; on L'Epée, see Bonnin 2015/16.

finishers in Geneva. Difficulties appeared in the course of the centuries. The Savoyards continued in the production of **verge** escapements at the moment when the **cylinder** imposed itself. The problems were aggravated by the great fire that devastated Cluses in 1844. To revive activity, a technical school was opened in 1848 with the double aim of freeing the region from its dependence on Geneva, and to introduce the making of entire watches. The aim was not attained. The manufacturers continued producing wheels, pinions, and winding-stems for Geneva, while instruction focused on general mechanics. Under the Second Empire, after Savoy had been ceded to France, for political reasons the school was accorded national status, thus going one-up on Besançon, where the school of horology would remain municipal for some years more.

The Arve valley developed its orientation towards mechanics in the twentieth century, specializing in profile turning with primary markets in the automobile and armaments industries. In 1900 a technical centre for mechanics and profile turning was set up,[72] and in 1990, 9,700 salaried workers were numbered in the area. A museum maintains the collective memory marked by the workers' strike of 1904, the subject of a novel by Louis Aragon, *Les Cloches de Bâle* (1934).

HOROLOGY IN THE JURA

Ever since the eighteenth century, a vast horological region covered the two flanks—French and Swiss—of the Jura Massive along the 230km of their frontier. In France, horology was practised from Saint Hippolyte, at the foot of the Black Forest, as far as Cluses in the Haute Savoie. It is easy to be a horologist in the High Doubs, at Maîche or at Morteaux, and in the High Jura, at Morez and Morbier. In Switzerland, horology extended through the cantons of the Vaud and Neuchâtel. In this web, Besançon played its role.

The gradual extension of horological work from Swiss centres into the entire Jura was aided by traditional local working in iron, and by the possibility of dividing up the work, and to work, customarily, in the home. The movement took place at a moment of considerable population expansion—forty per cent growth in the first half of the nineteenth century—which led to some families seeking work in the towns or leaving the region. The opening of new roads towards Switzerland in the 1830s confirmed old habits of mobility. The Maîche plateau, on the left bank of the Doubs in the heart of the Franche Montagne, was sucked dry by the Geneva manufacturers in search of low-cost manual workers (Figure 128). In 1840, some three hundred families in Charquemont worked for La Chaux-de-Fonds and the region specialized in the making of the **assortiment**, the cylinder, and its **balance**. At the beginning of the twentieth century, the work was still subdivided

Figure 128 Germaine Jacot, Frambouhans, sur le Plateau du Hauts Doubs. Cliché Henri-Louis Belmont, 31 October 1968. For a modest salary Germaine polished cylinders, mounted on a bow-shaped holder, with emery paper all day whether at home or in the street.

into fifty-eight successive operations. In such a context an artisan had but one aim—master a series of operations so as to become independent and be accepted as a maker in the network. But once established, such workshops, often limited to the family circle, might have only a brief life. The best rooted would convert either to watch assembly, or to case making. Four new centres thus appeared: Damprichard, Charquemont, Bonnétage, and Maîche.

The Morteau valley developed a more industrial profile, which was confirmed by the opening of the Frainier factory in 1864. While maintaining strong relations with the Swiss, it moved closer to Besançon while borrowing the production methods of the Monbéliard region. Low-cost watches with metal cases were produced for a wide-ranging clientele in colonial countries, the near east, and Russia. Plateau and Val weathered the interwar crisis better than Besançon, enjoyed a lull after the Second World War, but failed to resist the arrival of quartz watches. However, the teaching of horology being abandoned in Besançon in the 1990s left the field to Morteau. Successful pupils found work easily in Switzerland like a great many other frontier workers crossing the border every day, although employment could vary with commercial conditions. In the twenty-first century, although still close to Switzerland, the High Doubs can assure the continuation of establishments such as Pequignet, and Herbelin more easily. The making of cheap watches having departed to Asia, medium-to good-quality watches have become the path, whether mechanical or quartz. At Morteau and Villers-le-Lac the local horological inheritance is defended with passion[73] around a traditional figure of the peasant/watchmaker perhaps more mythical than real.[74] Heritage, however, has become the sustaining core of the 'Horological Country', an administrative entity which groups the villages into a coherent and closely linked network. But the culture of the High Doubs, almost more Swiss than French, is markedly different from that of Besançon, where the watchmakers were rivals to the Swiss and so a competitor with the High Doubs. A seesaw effect may be the sign of these tensions. When all is well in the High Doubs, Besançon has problems, and vice versa.

---

72 Absorbed into the CETIM, technical centre for the industries of mechanics in 2019.

73 Droz 2018.
74 Ternant 2004.

BESANÇON

The horological centre of Besançon[75] was created by order of the Revolutionary government. It was intended to compete with Swiss manufacturing centres, the importation of whose products was too costly. Led by Laurent Megevand, an entire colony of emigrants from the Neuchâtel region left their native mountains for Besançon. The factory established in 1793 rapidly gave place to assembly work. The making of blancs was also quickly abandoned. *Ipso facto* Besançon came to specialize in finishing and watchcases. The Besançon factory was intended to produce standard watches for everyday use, luxury watches being reserved to the factory established at Versailles at the same time. By the middle of the nineteenth century, however, the Besançon makers had turned to the production of shoddier goods—the **patraque**—that booksellers gave away to their customers. Exhibiters from Besançon were few in the first Paris Universal Exhibition of 1855.

The fatter years of the Second Empire saved the situation. Better-quality products were once more made. In 1860 the Chamber of Commerce, the Doubs Department, and several bankers united to cock a magnificent snook at Paris. Besançon, its head in the clouds, inaugurated its own Universal Exhibition. The city proclaimed itself the capital of French horology, affirmed itself capable of supplying the nation with good, solid watches, and to take its place on the international market. In the decade 1880–1890 Besançon supplied two-thirds of the watches sold in France, although the 300,000 units produced per year weighed little against the 3,000,000 then annually produced in the world. The city was at an apogee and even Swiss manufacturers were anxious. The creation of a school for horology, a chronometric observatory, and a course in chronometry at the university sustained this ambition.

It was not to last. The efforts in Montbéliard and Cluses to manufacture complete watches, and the success of the Morteau makers in making low-priced watches, shook the complacency of those in Besançon, poisoning relations between the watchmakers of the Franche-Comté. Besançon had also to cope with disagreements among its industrial leaders. The majority of them, descendants of the first Swiss colonists, were partisans of traditional methods and ignored new production methods, refusing the American model discovered at the 1876 Philadelphia exhibition. The forward-looking group, favourable to the American style of mechanization, was largely composed of a Jewish community recently established and centred on Geismar and Lipmann, future leaders of the industry in the twentieth century. The situation was stirred up by Auguste Rodanet (1837–1907), a leading Paris maker, the agent for Patek Philippe in France, president of the prestigious Paris school of horology (founded 1880), and mayor of the second arrondissement. 'In Besançon', he announced, 'and I repeat it, the finishers are not watchmakers'.[76] Was it an act of revenge? Ernest Lipman had issued a poster incorporating Rodanet's head proclaiming that 'I do all appropriately and on time thanks to my Lip chronometer', which was displayed throughout the capital. Rodanet won the lawsuit that he instantly instituted. In reply, Lipmann produced a new poster in which Rodanet's head was replaced by a dial figuring a man-in-the-moon face. Journalists had a field day with the dispute.

In Besançon the economic crisis of the 1880s continued until the outbreak of the First World War and business casualties were high. Deprived of blancs by the Swiss reacting to the Besançon attempt to compete with them, the city could not establish an autonomous industry. It became a city of cases, but of standard type, Geneva monopolizing the luxurious models in gold and silver and supplying in return only those of base metal. For watch components, Besançon was locked into the tightly structured organization of the Jura industry. For an ordinary watch, the blank was purchased in Montbéliard, for one of standard quality in Villers-le-Lac, and of good quality in Cluses. Top quality, however, could only be obtained in Switzerland from the Val de Travers or the Val de Joux. All these exchanges were accompanied by intense activity in smuggling.

The aftermath of the First World War was aggravated by the crisis of the 1930s. While the Swiss industry, supported by the Federal government and the banks, reorganized itself rapidly, favouring the regrouping of manufacturers and the standardization of products, the dispersed French makers, with obsolete structures, tried only slowly to compensate for their backwardness in modernizing. They were moreover obliged to consider the rise of the wristwatch and to confront German, American, and Japanese competition, as well as that of the Swiss. In the 1930s, France was entirely reliant on its domestic market. Internationally, it was in last place with only 0.9 per cent of production exported, compared with 93.2 per cent for the Swiss, 3.2 per cent for Germany, and even 2.6 per cent for the United States.[77]

But the availability in Besançon of skilled artisans permitted a reorientation towards small mechanical productions that were less affected by economic crises. The objectives changed. In 1924 the Compteurs de Montrouge, a Parisian enterprise founded in 1872 for the making of gas meters, moved to Besançon. The relative position of horology from then on continued to diminish, despite a final break in the clouds during the 1950s when the Compteurs with a thousand employees, and the Lipmann enterprise with its successful 'Lip' mark (Figure 129), brought a last moment of success to Besançon.

RECONVERSION

When the idea of using a quartz crystal resonator in watches developed during the 1950s, Lipmann was the only manufacturer in France ready for the change. In 1947, Fred Lip, the young director of the company, opened a research laboratory to develop an electric watch. It appeared in 1952 and was followed by a

---

75 For a fuller general account see Mauerhan 2018.
76 Cited in the Besançon journal *Le Petit Comtois*, 3 November 1889.

77 Daclin 1968, 20.

Figure 129 Advertisement for Lip, 1907. The business was established in Besançon in 1867 and moved towards production by Ernest Lippmann in 1903 using the abbreviation 'Lip' adopted by his son Frederick as the company name. Reproduced from *Horlogerie Ancienne*, xxxxiv, 1993, 31.

series of original calibres, with the quartz watch arriving in a final flourish. But despite the help of the Technical Centre for Horology (CETEHOR), industrialization of the Besançon quartz watch failed. The company passed into the hands of the Swiss firm Ebauches SA. The closure in 1973 provoked a strong union reaction—'the Lip affair' was reported worldwide. The development of microelectronics and the failure of the quartz watch mark the end of watchmaking in Besançon. Despite its laboratories, partly concerned with time-measurement, the university failed to develop a fruitful exchange with industry; notwithstanding, the Institute of Chronometry and the School of Mechanics together gave rise to an engineering school, the present National Higher School of Mechanics and Micro Technology.

Thirty per cent of Lip's production was devoted to small precision mechanics, armament manufacturers being the principal market. This was a hotbed from which emerged the engineers and technicians who made Besançon into a major centre of microtechnology in France in the closing decades of the twentieth century. Diversification, begun in the 1920s, bore fruit. A similar development occurred in Switzerland and by 2020 the traditional horological Jura had been transformed into a region of microtechnology bringing together French and Swiss in a similar activity. Now watchmakers called on the microtechnicians, especially in matters of surface treatments. University institutions were also solicited for innovative materials, as much for luxury mechanical watches as for quartz watches and related items. The research microtechnicians in Besançon thus found themselves furnishing new technology for the major Swiss brands, even in such marginal areas as wristwatches in leather or metal—a horological speciality that led to an important development of leatherworking in the High Doubs. Today, the respectable number of some 15,000 people work in the horological sector of the Doubs, this figure covering frontier workers, parts makers, including strap and hand makers, as well as such microtechnicians as serve the horological industry.[78]

At the turn of the century, with the challenge of microelectronic techniques, the major companies brought design centre stage. They aligned themselves with the luxury industries: jewellery, gold- and silverwork, table arts, all preoccupied with appearance.[79] Paris endeavoured to rival Geneva, the home of prestige horology. The arms, in this struggle, long since sharpened in Switzerland, are heritage and the longevity of the mechanical watch, an object of desire that can signal social status. But in this revaluation, the services of the microtechnicians in making watches ever more complicated to respond to clients' dreams, were ashamedly hidden away. In 2019, the connected watch is triumphant; the luxury mechanical watch as a star offers the only path for traditional horology: the survival of the quartz watch seems in doubt.

Some young French watchmakers have taken up the challenge, desirous of reinventing horology and creating their own brands—some five hundred brands are born every year.[80] For these innovators a relation with the Swiss is inevitable, as is a passage via the skills of Besançon. Distribution methods are also under reconsideration following the example of the larger groups who offer their products in shops that are equally art galleries. An example of such a one has been present in Besançon since 2018 run by the Utinam-Philippe Lebru brand, witness to this necessary mutation.

CONCLUSION

In the course of the last two centuries, French horologists have attempted the adjustments necessary to avoid extinction, though not always avoiding it. Most often, however, successful diversification without loss of skills has given birth to new, younger branches that have rapidly become autonomous, and which have led to the birth of new areas and new technologies. With this capacity to generate new paths of development, horology retains an important place in the technical systems of the nineteenth and twentieth centuries. Despite a troubled history, France, and with it Besançon, has maintained a position therein.

---

78 Doubs Chamber of Commerce and Industry, June 2019.
79 These final paragraphs rely on press reports and on unpublished interviews conducted by the author.
80 Doubs Chamber of Commerce and Industry, June 2019.

SECTION FOUR: THE CHALLENGE OF THE SWISS AND THEIR COMPETITORS

Johann Boillat

INTRODUCTION

The performance of industrial districts is a problem general to economic history.[81] In theory, these 'local systems of production', or 'flexible production systems', are characterized by the concentration in a particular region of a large number of, usually, small- or medium-sized businesses, the specialization of these firms around a product, and voluntary mechanisms of coordination, more or less complex, between them.[82] Their cooperation is modified by changes in the conditions that underlie its structure according to region and period. Among the institutional and political factors identified by recent research are: the level of centralization of banking and finance, the nature of the retail distribution network, the effectiveness of state policies concerning rationalization and mergers-acquisitions, the shape and intensity of antitrust regulation, the degree of political tolerance for associative direction, or again the balance between centralization and the autonomy of local administrations.[83]

For its part, the history of Swiss horological competitiveness is generally explained in two ways. The first is concerned with the structure of the production system, while the second focuses on product innovation. Analysis of the first offers several parameters explicative of the continuity of horological districts. Among them are the development of employers' organizations,[84] the institutionalization of negotiations between employers and employees,[85] the nature of state intervention,[86] the structure of the banking system,[87] and the transmission of knowledge at a family[88] or professional[89] level. The second addresses the question of research and development around the product, its components, and the system of its production. This analysis seeks to explain the development of Swiss horology through successive technical breakthroughs in time measurement.[90] So many studies spend time on the integration, more or less rapidly, of innovative products or production methods, thus allowing them to discern the success or failure of a district. At an international level several authors have displayed the chronology of horological competition from the industrial era onwards[91]; at a national level competition has been examined through research processes, whether individual,[92] company based,[93] or communally undertaken.[94] At the company level, several studies underline the comparative Swiss advantage, whether in luxury watches,[95] precision watches,[96] or more everyday watches.[97] This last development is particularly evident in the post-Second World War period when world horology, like most branches of the international economy, needed to respond to the development of the microprocessor.[98] On this basis the main lines of development of Swiss horological district since the beginnings of the industrial era in terms of structure and in terms of performance can be discerned.

STRUCTURE

Etablissage, the production system of Swiss horological regions, is highly complex.[99] In theory it is characterized by a systematic delocalization of the production of individual parts between a multitude of dispersed workshops, then by the assembly of these, in centres of highly variable size, into semi- or entirely finished products.[100] This model, which depends on an important network of suppliers, employs a large proportion of specialized workers maintained by massive recourse to domestic work and helps in the

---

81 With thanks to Philippe Pegoraro, Head of the Economics and Statistical Service, Federation of the Swiss watch industry, Biel/Bienne, Switzerland; the Bibliothek des Deutschen Museums, Munich.
82 Lescure 2006a, 1–7.
83 Zeitlin 2008, 231.
84 Henry Bédat 2006; Donzé 2006; Boillat 2013a.
85 Boillat & Garufo 2013, 55–67; Garufo 2015; Lachat 2017.
86 Perrenoud 1993, 209–40; Froidevaux 2000b, 65–78; Boillat 2010, 89–136; Boillat & Garufo 2012, 209–26.
87 Froidevaux 2000a, 251–70; Bolhalter 2016; Boillat 2019a, 177–99.
88 Jequier 1972; Landes 1979, 1–39; Jequier 1983; Picard 1993, 85–105; Gagnebin-Diacon 2006. Donzé 2007; Mahrer 2012; Hanloser 2015.
89 Cardinal 1999a; Fallet & Cortat 2001. Donzé 2008, 5–28; Fallet & Simonin 2010; Munz 2017.

90 Attinger 1973, 563.
91 Landes 1983; Glasmeier 1991, 469–85; Aguillaume 2006, 190–217; Donzé 2014. Raffaelli 2019, 1–43; Donzé 2020.
92 Veyrassat 2000, 69–76; Veyrassat 2001, 367–83.
93 Pasquier 2005, 313–32, Pasquier 2008a; Pasquier 2008b, 151–69; Pasquier 2008c, 76–84.
94 Perret, Beyner, Debély, & Tissot 2000; Forrer, Le Coultre, Beyner, & Oguey 2002.
95 Sougy 2007, 71–84; Sougy 2018, 7–28; Yagou 2019, 78–107.
96 Fallet 1995. Tonnerre 2019, 129–44.
97 Landes 1990, 227–36; Schulz 1999; Donzé 2015b; Fracheboud 2016, 381–400.
98 Trueb 2005. Trueb 2006. Bielefeld 2007. Trueb 2008. Trueb, Ramm & Wenzig 2011.
99 Blanchard 2011, 114.
100 Fallet 1912, 295–9.

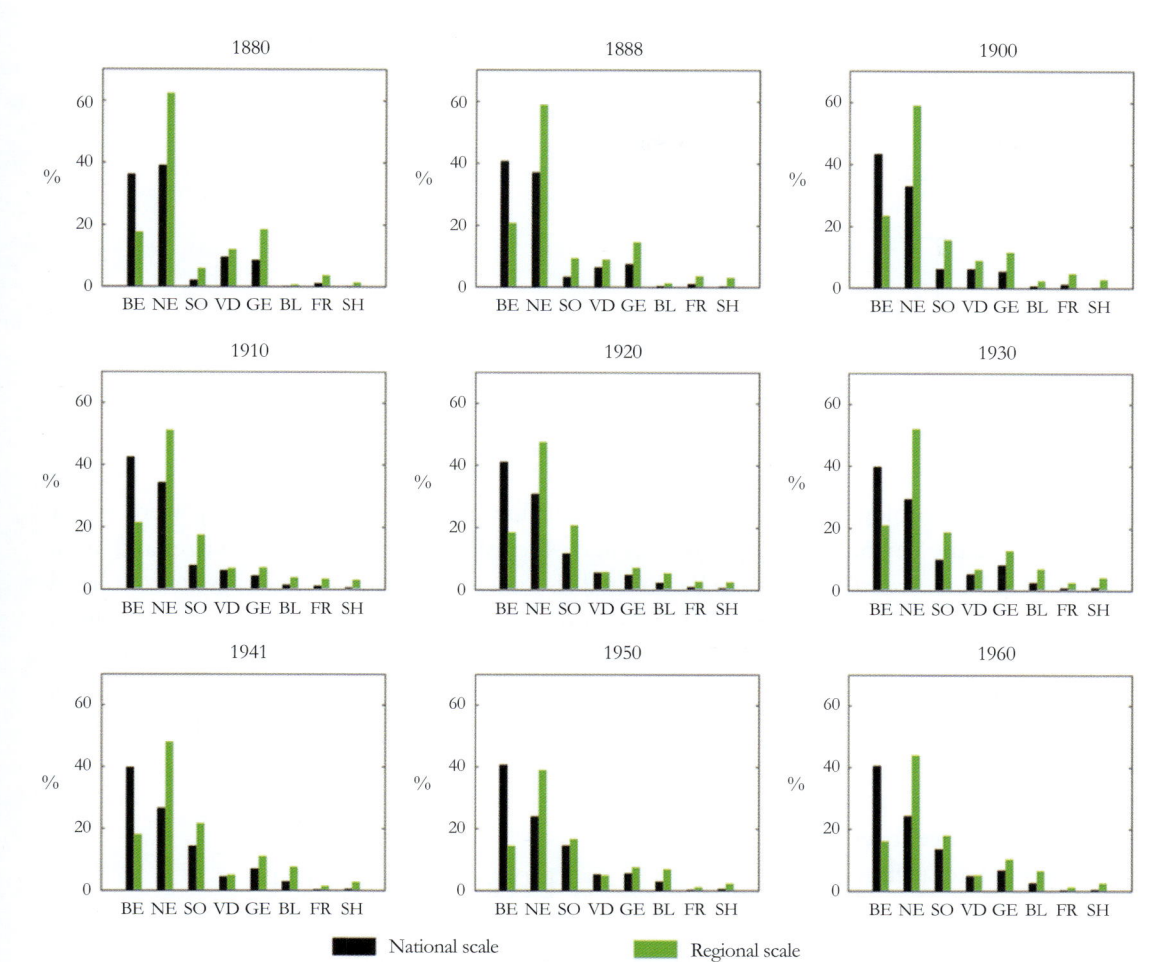

Figure 130 National and regional distribution of Swiss watchmakers, 1880–1960. Source: Siegenthaler 1996, tables F10a,b.

affirmation of a strong worker identity.[101] Two elements, however, derange this concept: one geographical, the other technical.

Most investigations of the flexible horological production system ignore the previous localization of factors concerning production. Historical statistical analysis, however, allows the system to be located geographically through the spatial distribution of horology-jeweller workers on a national scale, according to their proportion in the local, secondary individual cantons—on a regional scale (see Figure 130). A first investigation shows that that most of the workplaces for horology and jewellery are located in eight cantons: Bern (BE), Basel-Country (BL), Geneva (GE), Neuchâtel (NE), Solothurn (SO), Vaud (VD), Schaffhausen (SH), and Fribourg (FR). Three of them alone—Bern, Neuchâtel, and Solothurn—account for four-fifths of Swiss watchmaker-jewellers. A second investigation shows that the latter three cantons have the highest employment numbers for the period 1900–60, with almost half of employment in the Neuchâtel canton, and between fourteen and twenty-one per cent of those of Bern and Solothurn. This is far beyond those of other regions such as the Vaud (four and six per cent) or Geneva (six to twelve per cent).[102]

This allows us to clarify two essential points. Firstly the divisible nature of *établissage* explains the establishment outside the region of subsystems articulated around one or two manufacturers and their exclusive suppliers. This is the case notably in the Broye of Fribourg, in the French-speaking Valais or in the Jura of the Vaud, Basel, and Schaffhausen, as in the Italian-speaking regions of the country.[103] Secondly, official statistics make no distinction between watchmakers and jewellers. It is this initial amalgamation that explains, for the greater part, the peculiar position of the 'Geneva fabric', evidently active in horology but of which an important proportion of its employees is absorbed by the non-horological luxury trades.[104] But it is precisely the homogenous nature of economic activity that constitutes the essence of an industrial district.[105] In our case therefore, the Swiss horological industrial district can be defined as the areas delimited by the cantons of Bern, Neuchâtel, and Solothurn, in which *établissage* represents the major part of employment in the secondary sector both for absolute and relative value, for a period of nearly a century.

---

101 Loertscher-Rouge 1977, 143–99.
102 Siegenthaler 1996, tables F10a, b.
103 Fallet 1991, 117–22; Berlinger-Konqui 1991a, 173–8; Berlinger-Konqui 1991b, 179–86; Ackermann 1991, 99–104; Heller 1991, 155–60; von Felten 1991, 167–72.
104 Sougy 2007, 71–84; Sougy 2018, 7–28.
105 Zeitlin 2006, 450; Zeitlin 2008, 224.

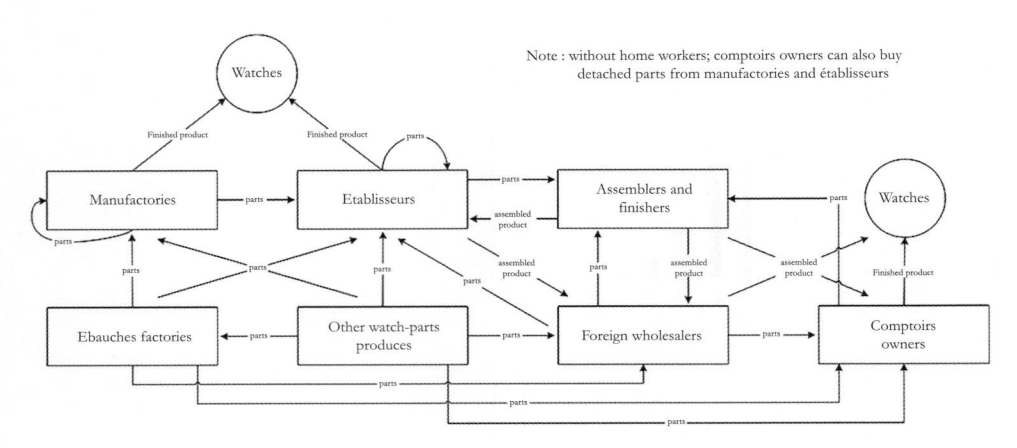

Figure 131 The system of *établissage* Source: Boillat 2013a, 45.

From a technical point of view, the ability of the Swiss horological industrial district to adapt has been questioned for the period 1870–1970. If this has been demonstrated for the mid-nineteenth century,[106] it is more difficult to admit it, historiographically, for the later period. The essence of the confusion is that, for some authors, the application of mass production to the Swiss industry following the Universal Exhibition of 1876 in Philadelphia[107] led to the disappearance of the traditional structures to the advantage of factories, refusing in a certain sense the mechanization of *établissage*.[108] On the contrary, it has since been shown that this system is particularly well adapted to technical progress in the sense that the fabrication is only one of the results of technical and commercial changes in the original structure.[109] Swiss horological production depends on six different categories of manufacturers (see Figure 131). The first is composed of watch factories, centres of production that bring together most of the elements needed to make a watch and its case. They are about seventy in number. The second category covers 'makers of horology', called *établisseurs*, who can be numbered in the hundreds. It is made up of brands, the owners of which purchase **blancs** at various stages of completion, together with all the parts, allowing the finished product to be produced *intra muros*. The third group is that of the makers of the blancs. They could supply directly to the factories, but their main customers were the *établisseurs*. The manufacturers of blancs were the essential centre of the manufacturing district, even if a great variety of practices can be observed since some manufacturers made blancs for watches of varying quality. The fourth group of producers worked in the market for spare parts, itself exploded into dozens of sections to cover all the high variety of pieces. Finally, two other groups have to be added to these manufacturers. These were retailers who could very well sell either finished or half-made products in which they could find a good source of revenue by added value.[110] They were either foreign wholesalers, furnishing themselves in Switzerland with parts to be finished elsewhere, or retailers whose agents contented themselves with purchasing finished products for their own local resale. This industrial structure allowed hundreds of small, specialized businesses concentrated in a small region to produce a very large variety of objects, of different size, and of different quality.[111]

PERFORMANCE

Crises are part of the motive force of the Swiss horological industry. Generally speaking, they are provoked by three kinds of shock with different, though not mutually exclusive, origins. They may be firstly of a technical nature, generated by the incursion of process or product innovations; secondly, provoked by geopolitical instability leading to a (sometimes brutal) closing off of external markets; thirdly, economic disturbances that can lead to a loss of purchasing power among foreign consumers. An analysis of the changing volume of Swiss watch exports between 1885 and 2000 (Figure 132) allows the various crises that have occurred during the industrial era to be situated. Two particularly painful periods may be underlined: the interwar period, and the decade 1975–85.

These statistics also show that after each major crisis the industrial horological district was able to bounce back. This phoenix-like ability to arise from the ashes can be explained by the fact that political and economic leaders in the Jurassian Arc institutionalized micro- and macroeconomic mechanisms which in the long term allowed improvement in the overall quality of national production, rather than having recourse to competition through prices. Three distinct periods can be distinguished.

---

106 Veyrassat 1997, 217–18.
107 David 1877. Bodenman 2011.
108 Barrelet 1987, 400–2; Liengme Bessire & Barrelet 1996, 52–3; Glasmeier 2000, 101–2; Koller 2003, 167–8; Donzé 2004, 81–2.
109 Pfleghart 1908, 50–200; Fallet 1912, 299.

110 Bairoch 1996, 214.
111 Boillat 2013a, 45.

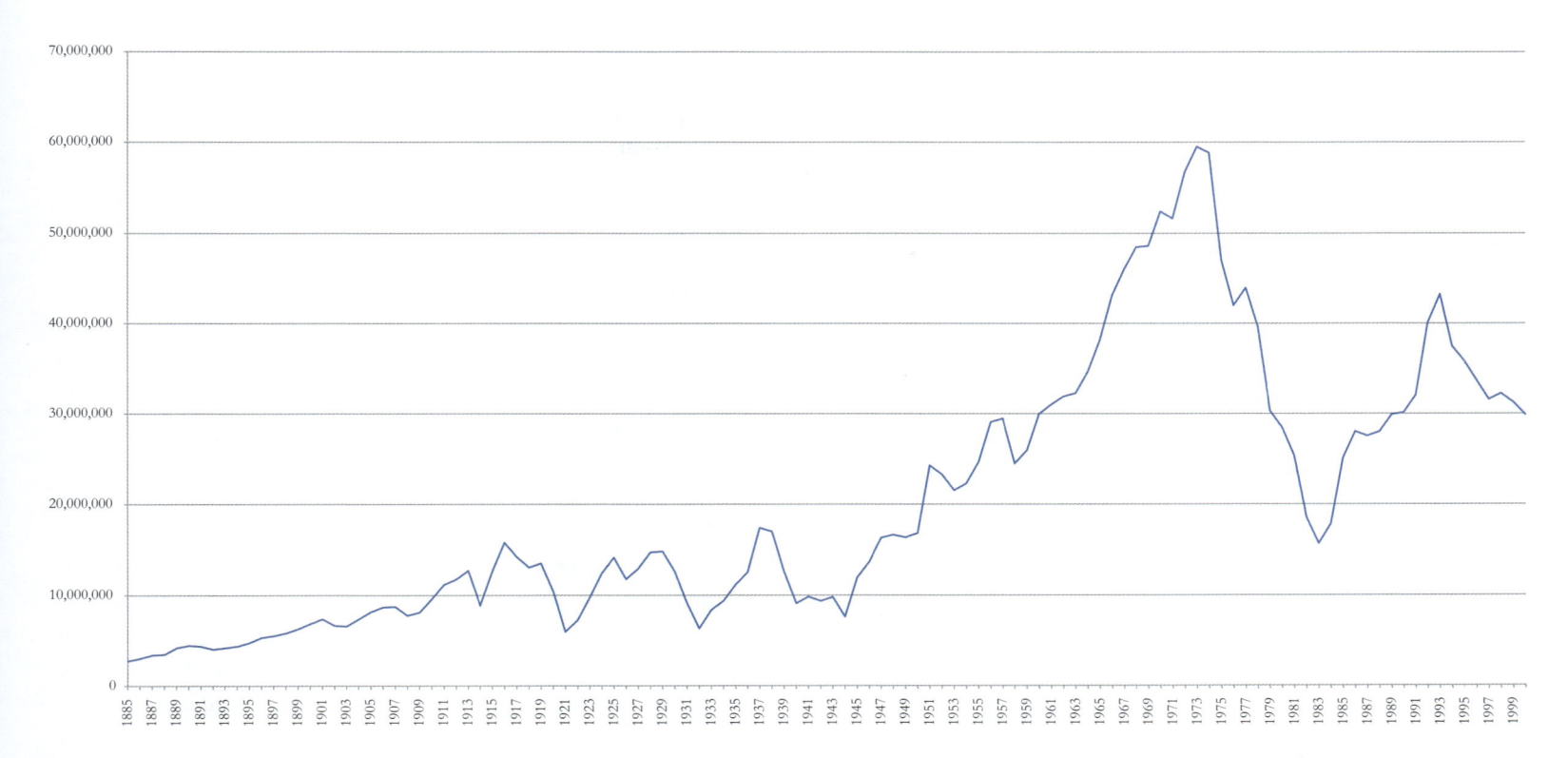

Figure 132 Exports of Swiss watches, 1885–2000 by volume Source: Swiss Watch Industry Federation, FH, Biel/Bienne, Switzerland.

PRECISION IMPORTED AND QUALITY INCREASED, 1850–1900

In terms of its general volume of production, Switzerland dominated world markets in the mid-nineteenth century.[112] In precision chronometry, however, the situation was slightly different as, for the needs of its Empire, the making of marine chronometers had continued to advance in Britain. The situation began to change following the 1855 Universal Exhibition in Paris. In their report to the canton of Neuchâtel, the Swiss delegates underlined the importance of taking administrative measures that would allow the making of chronometers to be implanted.[113] Among the measures envisaged was the founding of a chronometric observatory at Neuchâtel[114]; it would supply the exact time determined astronomically to manufacturers for over a century,[115] and was the cornerstone for the development of Swiss precision horology.[116]

From the decade 1870–1880 onwards, American competition began to make itself felt. This, as the Philadelphia Exhibition of 1876 showed, was capable of producing standardized technical objects at prices that defied all competition. To overcome this 'Philadelphia shock', Swiss makers followed the path of mechanizing the *établissage* and developing the interchangeability of parts.

These innovations in production, however, led Swiss watchmakers into protectionist measures. The firm aim for Swiss makers was to distinguish themselves from the new horological nations by the series production of higher qualities pieces, thanks to the establishment of collective structures. The first employers' association, charged to defend the interests of the makers, was created in 1876—the 'Intercommunal Society of the Jura Industries', renamed in the early twentieth century as the 'Swiss Chamber of Horology' and then in 1982, as the 'Federation of the Swiss Watch Industry'. Over time, other sections of the manufacturing sector, according to their industrial speciality or geographical position, adhered to and enlarged the association.[117] Thanks to its lobbying, federal legislation was enacted, controlling the quality of the precious metals (gold and silver) used in horology. A first law of 1880 created assay offices throughout the country and obliged makers to punch their watchcases. In 1886, a second law regulated the commerce of scrap gold and silver, of which the horological industry was simultaneously a producer and consumer of importance.[118] The whole system was reinforced in the watchmaking cantons by a network of horological schools.[119] By 1900, therefore, both the volume and the quality of Swiss production increased.

---

112 Donzé 2020, 30–1.
113 Richard 1856.
114 Département de l'Instruction publique, 1912, 10–11.
115 Boillat 2019b, 41–50.
116 Fallet 1995.
117 Boillat 2013a, 91–6.
118 Boillat 2016, 23–45.
119 Fallet & Simonin 2010.

1900–50: RETRIEVING INNOVATION AND PROTECTING PRODUCTION

The first half of the twentieth century was characterized by a great instability, the result of an accumulation of disturbing factors. If Switzerland remained the leading horological nation in terms of its volume of production, American and European (French and German) competition affirmed itself, while from the 1930s onwards Japanese rivalry developed.[120] On the technical level the period is marked by the appearance of three new materials from foreign sources.[121] The first came from the Bureau International des Poids et Mesures in Paris, where Charles-Edouard Guillaume (1861–1938),[122] in collaboration with the Société de Commentry-Fourchambault & Decazeville,[123] created two new nickel based alloys, Invar[124] and Elinvar (1918).[125] These, thanks to their low coefficient of thermal expansion, revolutionized mechanical watchmaking and led to the award of a Nobel prize to their inventor in 1919.[126] Elinvar thereafter linked into research on Beryllium undertaken almost simultaneously in Germany and the United States from 1919 onwards.[127] The remarkable properties of the metals drew the attention of Reinhard Straumann (1892–1967),[128] who, in collaboration with the German company Heraeus-Vacuumschmelze in Hanau-am-Main, researched further. His efforts were conclusive and led to the patenting of a new anti-magnetic, non-oxidizing, and very hard balance spring—the Nivarox in 1931.[129]

Commercially, the period is marked by a profound change in consumer taste, the pocket-watch being definitively abandoned from the mid-1930s onwards in favour of the wristwatch (Figure 133), which first found favour with the military during the First World War before slowly becoming adopted in civil society.[130]

Two major economic crises also distinguished the period. The first, generally under-estimated in historical accounts, ran 1921–3 as the speculative production of watches in response to demand from the belligerents of the recent war had to be adapted to a smaller civil demand. These difficulties of adjustment were, moreover, aggravated by geopolitical factors linked to the closing of imperial markets and monetary stabilization in the new European states. The crisis of 1929 in its turn led to a long-term contraction of the volume and value of international commerce, which affected Swiss watchmaking. Specifically, commercial relations with the leading client, the United States, deteriorated following the imposition of protectionist taxes.[131] To alleviate this situation the Swiss Chamber of Horology adopted two complementary strategies in the later 1930s.

The first operated at the microeconomic level. Banks and employers together encouraged concentration in the key manufacturing industries following a well-established schema: the development of employers' associations leading to **cartelization** that, in its turn, favoured mergers and acquisitions.[132] *Etablissage* thus came under the domination of a cartel financed by the banking sector and organized around two holding groups. The first of these, from 1930 onwards, combined the two most important makers of the period (Omega and Tissot) within the Swiss Society of Horological Industry SA. The second, the General Society of the Swiss Horological Industry SA controlled the providers of the constituent parts of watches (blancs, balances, balance-springs, and **assortiments**). The latter group, thanks to its financial power, was able to purchase the rights to such French and German innovations as Invar, Elinvar, and Nivarox, integrating them into the Swiss system through new Swiss patents.[133] On this new industrial base, the Swiss Chamber of Horology launched a national campaign to standardize the most important constituent parts.[134]

The second strategy adopted operated at a macroeconomic level with the aim of preserving the Swiss competitive advantage. The Swiss Chamber of Horology acted as intermediary in bringing the industrial district under direct state protection from the mid-1930s onwards.[135] The main provisions were a prohibition on exporting horological machines, overseeing of the firms (creation, diversification, and expansion), and control of a specific commercial practice, the sale of complete but unassembled movements (*chablonnage*), which had developed with the interchangeableness of parts, was encouraged by the interwar crises, and enabled foreign companies to assemble Swiss watches locally themselves, so avoiding the high customs dues levied on complete watches.[136] But the development of the practice from 1929

---

120 Donzé 2020, 68–70.
121 Boillat 2020.
122 Gorgé 2013. On Guillaume see Guillaume 1922; Cardinal 1998.
123 Déré, Duffaut & de Liège 1996, 3–23; Lambret & Saindrenan 1996, 39–47.
124 Deutsches Patent- und Markenamt (hereafter DPMA), Munich. Patent CH17746, 'Dr Charles Edouard Guillaume à Sèvres (Seine-et-Oise, France), nouvel échappement'. Deposed 20 October 1898. See also British Patent GB4,699 for 1904 concerning non-magnetic balances.
125 DPMA/Patent CH82,081, 'Société des Fabriques de Spiraux Réunies, Petit-Saconnex (Genève, Suisse), et Charles Edouard Guillaume, Sèvres (Seine-et-Oise, France). Spiral compensateur pour chronomètres de montres'. Deposed 4 June 1918. Published 1 September 1919.
126 Guillaume 1922 is his prize oration.
127 Boillat 2018, 47.
128 Müller 2012.
129 DPMA/Patent DE578,39, 'Reinhard Straumann in Waldenburg, Schweiz. Feder aus Nickeleisenlegierung, insbesondere für thermokompensierte schwingsystem'. Deposed 19 April 1931. Published 24 May 1933.
130 Béguelin 1994, 33–43; this volume Chapter 18.

---

131 Bolli 1957.
132 Bouwens & Dankers 2013, 1121.
133 Boillat 2020; Galvez-Behar 2007, 35–47; Galvez-Behar 2019, 327–36.
134 Carrel 1936.
135 Boillat 2010, 89–136.
136 Boillat 2012.

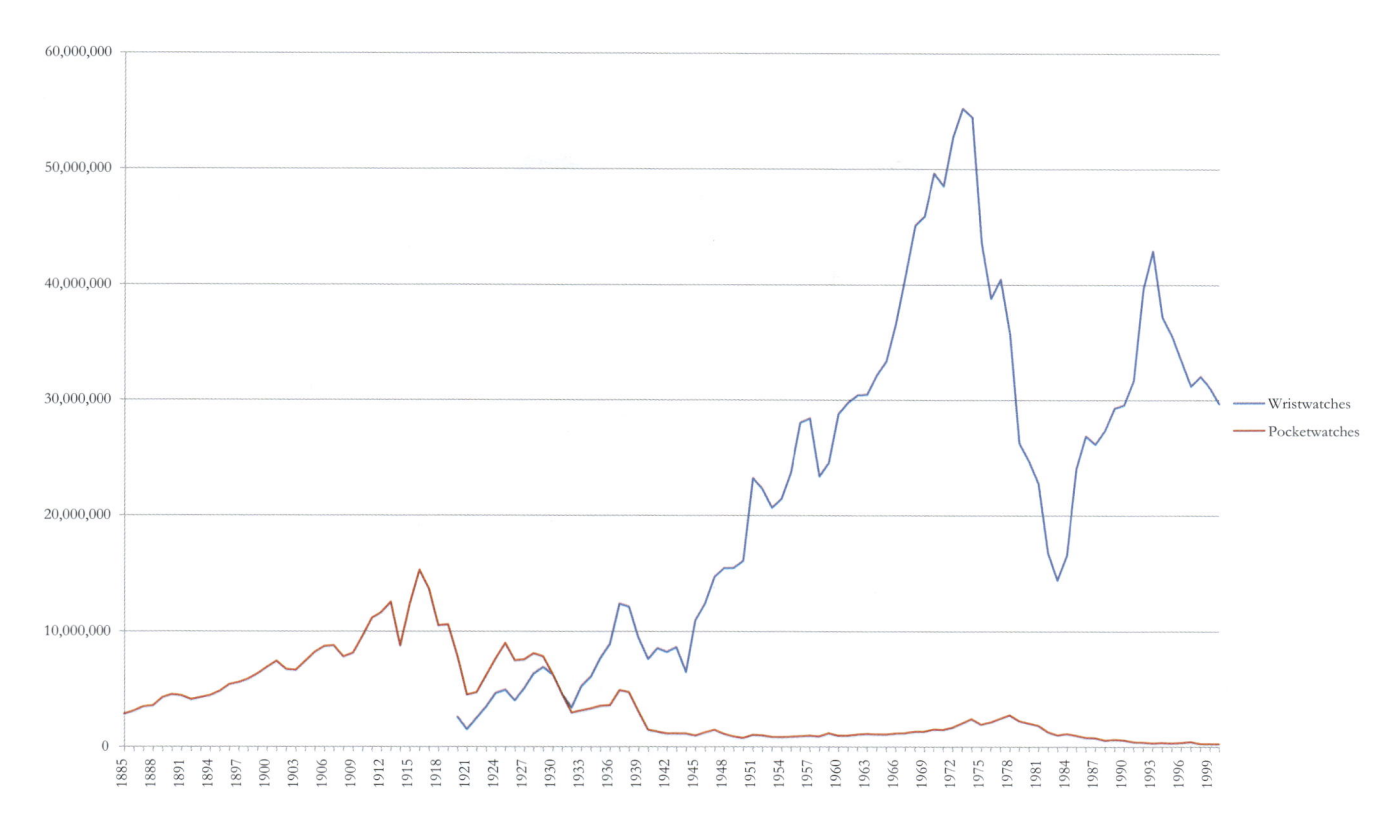

Figure 133 Exports of Swiss watches, wristwatches and pocketwatches compared, 1885–2000 by volume Source: Swiss Watch Industry Federation, FH, Biel/Bienne, Switzerland.

onwards contributed to a technology transfer towards competing nations (primarily Germany, France, Poland, Japan, and the United States). In consequence, between the 1930s and the 1970s, that is, between the two most serious crises of its existence, the horological district was protected by instruments that, for political-economic reasons, maintained groups dependent upon a particular technique or product, thus preserving a dense network of non-competing small manufactures, and hindering an international division of the work.

SINCE 1970: OVERCOME THE OBSTACLES AND EXPORT VALUE

The global monopoly of Swiss horology was progressively called into question from the end of the 1960s.[137] Switzerland had to face up to international competitors in the form of Time-Timex, Bulova, and Hamilton in the United States, Junghans in Germany, and Hattori-Seiko and Citizen in Japan.[138] The effects of this competition were felt from the mid-1960s onwards when Swiss producers encountered competition in the American market from Japanese watches.[139] The high value of the Swiss franc caused a slowdown in sales to America, production for which employed tens of thousands of workers and represented the major part of the country's exports in the 1970s.[140] Alongside this, the protective system inherited from the interwar period between 1951 and 1961 became a system of promotion in 1971 ('Swiss made').[141] The stakes in the redefinition of the role of the state in the Swiss horological economy were not understood by the players in an industrial district still largely populated by small businesses.[142] Increasingly fragile, the Swiss industry was incapable of resisting a series of external shocks: the rise in the cost of energy (1973), the floating of the exchange rate (1974), and above all, the series production of quartz, Japanese watches. The latter was the final blow, leading the sector to the brink of a system rupture. In a single decade, this structural reversal led to the disappearance of two-thirds of jobs in the sector, and the closing of nearly half Swiss horological enterprises.[143]

The first riposte was industrial in nature, in the late 1970s and early 1980s. For employers it meant using the results of the Electronic Horology Centre in Neuchâtel, which had developed a first quartz watch (Figure 135).[144] This communal research organism, founded in 1962 and financed by the employers' associations, would become the starting point for a collective reorganization

---

137 Uttinger & Papera 1965, 200–16; Jequier & Landes 1982, 60–75.
138 Donzé 2020, 94–5.
139 Donzé 2012, 275–89.
140 Donzé 2020, 136–8.
141 Boillat 2013b, 52.
142 Boillat 2013a, 46.
143 Glasmeier 2000, 200.
144 Forrer, Le Coultre, Beyner & Oguey 2002; Bielefeld 2007, 317–21.

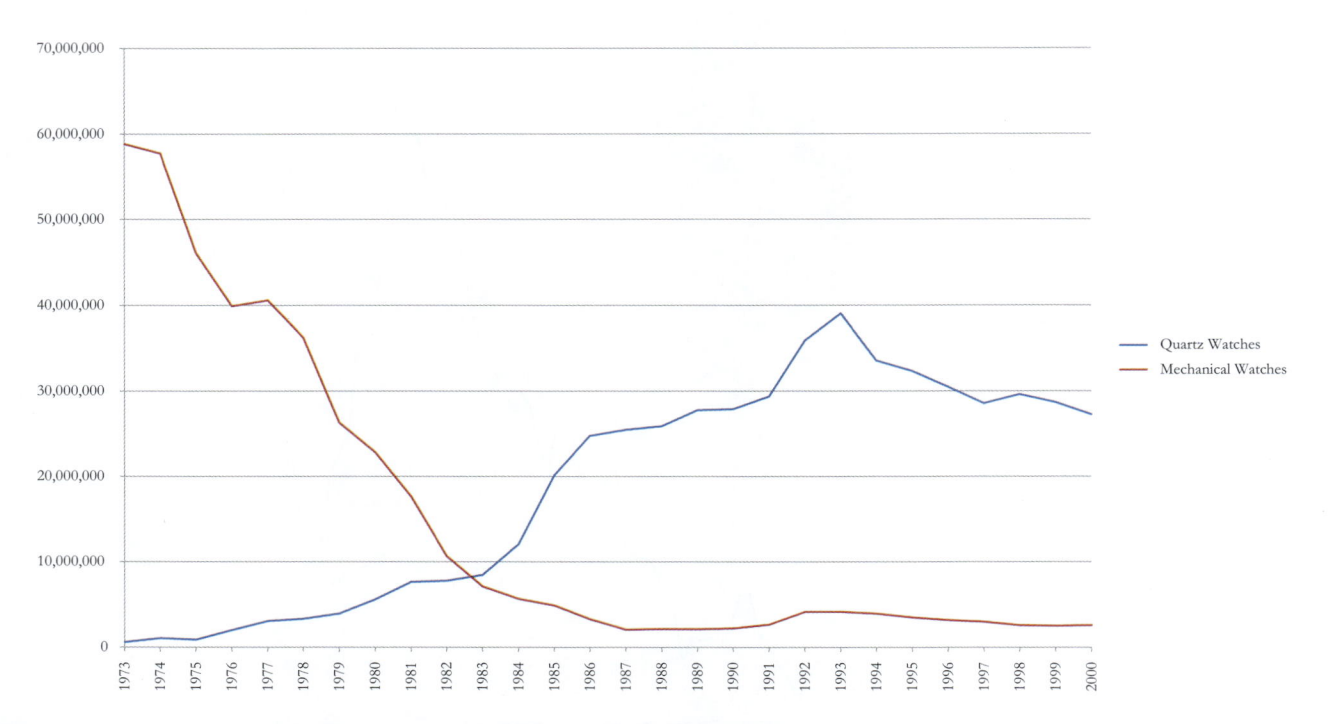

Figure 134 Exports of Swiss mechanical and electronic watches; 1973–2000 by volume Source: Swiss Watch Industry Federation, FH, Biel/Bienne, Switzerland.

capable of surmounting the intellectual barriers posed by the new technologies.[145]

If the use of artificial jewel-stones is not new in horology,[146] synthetic quartz represented a major technological rupture since it allowed the production of more accurate watches at a lower price, whereas earlier the precision of horological products was proportional to its cost. The Swiss problem was to proceed, following the example of their Japanese competitors, to produce the new product in quantity and to sell it in international markets.[147]

The way to innovation was made possible by a vast restructuring of the industry. Firstly, a double wave of mergers and acquisitions took place, imposed by the banks. The first concerned the makers of strategic components and led to the appearance of new, large-scale units of production, such as ETA, SA (blancs),[148] and Nivarox SA (balances, *assortiments*, and balance-springs).[149] The second operated on the older holdings such as ASUAG (Rado, Mido, Certina, Longines) and the SSIH (Omega, Tissot, Lemania), which were reassembled into a single entity in 1983, the Swiss Company of Microelectronics and Horology, which became the Swatch Group in 1998.[150] This change favoured first the reduction, and then the standardization, of calibres, provoked technical control of the finishers, and led to the externalization of part-making by the older manufacturers to Grenchen, in the German-speaking part of the district.

At the same time, changes in the industrial make-up of Swiss horology continued at a technological level in the mid-1980s. The Swiss Company of Microelectronics and Horology launched a new quartz watch, the movement produced in quantity was standard, but the flexibility of the case design allowed the watch to be seen, and used, as a fashion accessory.[151] This was the 'Swiss Watch', better known as the *Swatch*.[152] A transformation occurred in 1983 when, for the first time in its history, Switzerland exported more electronic than mechanical watches (Figure 134).

The reinforcement of the industry continued during the 1990s and succeeded in transforming the *établissage* into a productive system integrated in the global economy. However, the period is marked by a twofold change. On the one hand, for the parts of little added value, which had no influence on the label 'Swiss Made',[153] an international division of labour was operated, notably towards Asia.[154] On the other hand, attracted by the macroeconomic promotion of the Swiss territory, foreign multinational concerns, active in the luxury trades, established themselves in the district, resuscitating brands that they sought to render autonomous by grafting onto them embryos of *établissage*. The presence of these new players favoured the integration of the Swiss horological district into the process of globalization.[155]

---

145 Raffaelli 2019, 35.
146 Vaupel 2015, 278.
147 Donzé 2012, 133–6.
148 Pasquier 2018.
149 Boillat 2020.
150 Pasquier 2018.
151 Garel 2015, 34–40.
152 DPMA/Patent CH643,704A3, 'Montre électronique à affichage analogique. Eta S.A. Fabrique d'Ebauches Grenchen, Jacques Muller, Reconvilier, Elmar Mock, Biel/Bienne'. Deposed 6 March 1981. Published 29 June 1984.
153 Boillat 2010, 133.
154 Donzé 2015a, 295–310.
155 Markusen 1996, 297.

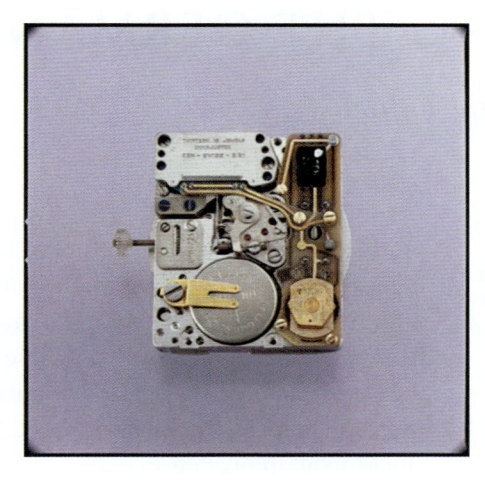

Figure 135 The first Swiss quartz watch. DPMA/Patent CH18,799/66, 'Centre Electronique Horloger S.A., Neuchâtel. Montre à quartz. Max Forrer, Neuchâtel, Armin Frei, Hauterive, Rolf Lochinger, Zug et Henri Oguey, Peseux, sont mentionnés comme étant les inventeurs'. Deposed 30 December 1966. Published 31 July 1972.

So the establishment (1971), and then the reinforcement (2017), of 'Swiss Made' had two results. Firstly, thanks to an advertising policy based on brands and their past, the annual value of Swiss horological exports passed from six to over twenty billion dollars for the period 2000–15.[156] Secondly, officially labelling the territory also aided in the international promotion of Swiss products whether horological or not.

CONCLUSION

Retracing the history of Swiss horology in the perspective of competition brings out its extraordinary longevity. From the mid-nineteenth century to the beginning of the twenty-first, Switzerland managed to retain its leadership faced by such competing nations as France, Germany, England, the United States, Japan, and China. The investigation shows that, in periods of crisis, the Swiss horological industrial district has had recourse to an elaboration of mutually agreed cooperation mechanisms, allowing an improvement in national production while banishing selfish behaviour. Product or production innovations, often of foreign origins (United States, Japan, France, Germany), have been integrated at the same time as industrial flexibility has developed. The creation of professional schools and the establishment of a standard measure were the principal results of the 'Philadelphia shock'. Cartelization by the industry leaders, protection of the instruments of production by the state, and the oversight of *établissage* by the banks are some of the several responses found by employers for the perils of the interwar period. During the 1970s, state promotion of the horological region, fusions, and mergers, together with shared financing of research, were some of the elements that allow an explanation of the renaissance of an industry integrated into the world economy through an international division of labour and by the arrival in Switzerland of multinational businesses in the luxury trades.

This inborn plasticity of Swiss horological production should not obscure the stakes for an industry historically made fragile by economic, geo-politic, and technical shocks. Maintenance of this flexibility seems to be linked with three factors. One, economic, concerns the role of the state as much for regulating non-competitive positions in the lucrative market of movements as for defining the norms controlling the label 'Swiss made'. The second is technical and concerns understanding of the disruption caused by the fourth industrial revolution in the domain of connected objects. The third, and last, is geo-political, for it poses the question of the capacity for resistance of the Swiss horological industry in general to the international instability caused by the environmental and energy crisis of the twenty-first century.

---

156 Marti 2016, 237.

SECTION FIVE: DEVELOPING THE GERMAN INDUSTRY
Sibylle Gluch

NINETEENTH-CENTURY CLOCKMAKING: CENTRES AND OVERALL DEVELOPMENT

The nineteenth century brought major changes to the clock- and watchmaking crafts. Traditional centres such as Augsburg and Friedberg had lost their relevance. The Black Forest became the preeminent German clockmaking area, and new sites such as Freiburg in Silesia and Glashütte in Saxony emerged. The old guild system dissolved and gave way to economic freedom;[157] traditional clock- and watchmakers abandoned actual production of timepieces and became repair- and service-craftsmen.[158] Factories replaced smaller workshops and the cottage industry as industrialization brought new production methods and different work structures, while at the same time creating a mass market for timepieces. Over the course of the century, the German clock industry situated in Freiburg in Silesia and the Black Forest region developed into a world market leader. However, the evolution, particularly of the latter area, was triggered by a profound crisis.

In the eighteenth century the Black Forest region saw the establishment of a cottage industry that produced cheap wooden clocks.[159] These clocks, characterized around 1800 by **anchor** escapement, pendulum, and **lantern pinions**, reached acceptable precision at unrivalled prices. In combination with an adroit distribution system that relied on the separation of production and sale, the industry proved extremely successful: Black Forest clocks were exported all over Europe as well as to Russia, Turkey, and the United States.[160] The system remained stable until the middle of the nineteenth century with no noticeable changes in either products or production methods. In the 1840s it was increasingly challenged by new clocks coming from the United States with movements made of parts stamped from rolled brass. These brass movements were of reasonable quality and could undercut the so-far unbeatably cheap wooden Black Forest movements. In consequence, clocks from the USA superseded Black Forest clocks in all important markets, and the traditional cottage industry with its numerous small clockmakers plunged into a deep crisis. Remedy was sought by prolonging work hours in order to produce a greater number of clocks. In addition, the severe hardship intensified attempts by government to stimulate industry and the economy. Germany's political fragmentation caused these endeavours to take different forms with varying degrees of success. For example, in 1850 the Grand Duchy of Baden founded a school for clock- and watchmaking in Furtwangen in order to standardize and modernize the craft. The enterprise met with only limited success; the school closed down in 1863. A more comprehensive programme was established by the kingdom of Württemberg, which, in 1848, founded the Central Department for Industry and Trade in Württemberg.[161] This state department purposefully supported the introduction of polytechnic schools, aimed at increasing the theoretical and technical understanding of craft apprentices. In addition, it created collections of samples—machines, devices, and instruments—that could be studied by professionals, and promoted the foundation of training workshops for clockmakers. Most importantly, the department created and regulated ways of obtaining credit for the inception of businesses. While many German states adopted one part or other of this programme, only a few followed such a comprehensive scheme. As a result, the Württemberg clock industry supplanted the Baden part of the Black Forest region, dominant in the eighteenth and early nineteenth centuries. This success, however, would have been impossible without an alteration of production methods.

Until the middle of the nineteenth century, the Black Forest clock industry consisted of an increasing number of individual makers working at home. Division of labour existed, for instance, regarding the production of cases, chains, and dial plates, but generally reached only a modest degree. Productivity was accordingly low. Around 1800, one maker assembled about four clocks in six days.[162] This number hardly increased over the next decades. Competitiveness could only be reached by abandoning individual workshops, deploying machines, and introducing factory pro-

---

157 In the different German states the guild system was differently organized, and clock- and watchmaking was sometimes treated as a free trade. Economic freedom occurred at different times in different places. In Württemberg, for instance, King Frederick I abolished guild constraints for clock and watchmakers in 1809. Zubal 1989, 391.

158 Cf. the petition submitted by the Württemberg watchmakers to the Elector in 1803. The watchmakers complained that they were 'reduced to the repair of watches. Because it is notorious that the factories in Geneva, Locle, La Chaux de Fonds etc. deliver so many pocket watches and pendulum clocks every year, and at such low prices, that even the most reputable watchmakers can hardly complete [verschließen] a few new watches every year, and that he who is forced to make new watches must inevitably perish.' Quoted from Zubal 1989, 396.

159 For a comprehensive account of the Black Forest industry, see Bender 1975 and Kahlert 2007.

160 Kahlert 2007, 102.

161 Also for the following, see Schmid 2006a, 71–2.

162 Bönig 1993, 336.

duction. One of the first attempts at central organization and rationalized work was the Aktiengesellschaft für Uhrenfabrikation, Lenzkirch, founded in 1851 by Eduard Hauser (1825–1900) and Ignaz Schöpperle (1810–82).[163] One year earlier, in 1850, Gustav Becker (1819–85) had established his enterprise in Freiburg in Silesia, which functioned according to the same modern principles. While Lenzkirch did not receive noticeable support from the government in Baden, Prussia (of which Silesia was a part) assisted Becker in several ways, among other things by granting interest-free loans and placing government orders.[164] Both companies produced high-quality massive movements for pendulum wall clocks—Lenzkirch inspired by French, Becker by Viennese models. The late nineteenth-century rise of the general Black Forest clock industry, however, relied on an entirely different concept.

In the second half of the nineteenth century the Württemberg part of the Black Forest became the dominant centre of clockmaking. The success of its industry mainly resulted from the introduction of American-style movements. The material basis of these movements was rolled brass sheets. Parts such as plates and wheels were stamped. The production process rested on interchangeable parts, fabricated on successively arranged machines, and operated by semi-skilled workers. As the stamping technique depended on high power, steam-engines entered into the fabrication of clocks, which was now based on a specialized equipment that required large investments and thus necessitated series production in order to be economic.[165] Wilhelm Jerger (1845–1921) and Erhard Junghans senior (1823–70) were the first to embark on the production of these new kind of clocks (Figure 136): Jerger started in 1866, Junghans in 1867, the latter being in the fortuitous position of having a brother in the United States who studied the American production methods and purchased and shipped the required machines.[166]

Others, such as Friedrich Mauthe and Thomas Haller, followed. By the end of the nineteenth century most companies worked according to American technology; in 1906, ninety-five per cent of the estimated yearly production were American movements, which had replaced the traditional wooden Black Forest clocks.[167] Parallel to this development, independent clockmakers vanished. Production now clustered in workshops and increasingly in factories. Also, traditional outwork, initially used even by factories for the manufacture of specific clock parts, declined. In line with the increasing centralization and rationalization of the production process, the quality of the American-style movements improved. While the first generations distinctly differed from the high-quality massive movements as produced by Gustav Becker in Freiburg and the Lenzkirch clock factory, later generations constantly improved in quality. In consequence, the manufacturers of the more expensive solid clock-movements had to adapt their product range in order to compete and also in order to comply with changing customer preferences.

Early factory production of wall and table clocks emulated American models such as Ogee and Cottage clocks.[168] These low-priced timepieces were affordable for broad segments of the population. Bourgeois customers, however, preferred the more prestigious weight driven regulator based on the Viennese model as fabricated in Freiburg/Silesia. This predilection changed towards the end of the century; weight-driven regulators were replaced by cheaper spring-driven ones. The latter, usually with **striking** work and medium-length pendulums, were produced in great numbers and varying case forms in the Black Forest region, and also in Freiburg after 1880. They became a standard part of the central-European living room until the beginning of the Second World War.[169] In addition, the alarm clock developed into a fundamental product of the German industry. Cheap, transportable alarm clocks in round ('Baby') or rectangular ('Joker') cases of sheet metal came into fashion in the United States in the second half of the nineteenth century. Junghans, surely the most influential company in the 'Americanization' of the German industry, introduced the production of this type of clock; its alarm clock movement 10, developed around 1875, became the model for the majority of German clock companies.[170] By 1900, Junghans became the world's largest clock company with alarm clocks accounting for seventy per cent of annual production.[171] While alarm clocks with spring and **balance** became a mass product, sold to all layers of society, the clock industry also produced more specialized types of timepieces such as telltale watches with paper strips for time registration, patented in 1856 by Johannes Bürk (1819–72) and Michael Vosseler (1807–84).[172] Yet the strength of the German industry, which during the late nineteenth and early twentieth centuries rivalled the United States as market leader,[173] rested on spring-driven wall, table, and alarm clocks produced according to the American system. However, these clock types were initially extremely difficult to sell.

---

163 General information on the German clock and watch industry, companies and persons involved as well as further literature, in Schmid 2017.
164 Kahlert 2007, 199–203.
165 Schmid 2006a, 72.
166 Schmid 2006a, 73–4. The brothers Erhard and Xaver Junghans jointly founded the Gebrüder Junghans company in 1861. Preparations for the production of American movements started in 1863. Xaver Junghans left the company around 1869 and returned to the United States. Schmid 2017b, 307.
167 Kahlert 2007, 216.

168 Schmid 2011a, 199.
169 Kahlert 2007, 265.
170 Schmid 2011b, 202.
171 Schmid 2011a, 202.
172 Kahlert 2007, 228.
173 Graf, 2019, 241.

Figure 136 Gebrüder Junghans, advertisement, *Deutsche Uhrmacher-Zeitung*, no. 15, 1889 Photo: Johannes Eulitz.

Already in the eighteenth century the Black Forest clock industry was characterized by a separation of production and distribution. Forwarding agents, or 'Packers', sent the clocks to an agreed destination, where they were sold by pedlars and small trading companies operating in a particular area.[174] The system changed to retail sale during the second half of the nineteenth century. The pedlars vanished, and wholesale dealers supplied to retail traders with permanent shops. This specialized trade, as well as established clockmakers outside the industry that served as retailers, regarded the American-style movements as inferior and refused to stock them. Hence the clock industry was forced to find new distribution and sales channels as well as to launch innovative marketing campaigns. Initially a number of companies, such as Junghans and the Hamburg-Amerikanische Uhrenfabrik, played with an American image.[175] Junghans later stressed its foundation in craft and tradition combined with innovative power. In addition, companies opened their own retail outlets in Germany and abroad, which at times also served to circumvent customs barriers. As a novelty the industry started to supply department stores and the mail-order business, just emerging at this period, where customers would not expect to be served by experts in the field. In general, the German industry relied on exports, perhaps originally fuelled by the rejection of its products inland. This orientation caused problems in times of global economic crisis as would occur during the first half of the twentieth century. However, before exploring this topic further, it seems apt to have a look at the watchmaking branch of the industry.

GERMAN WATCHMAKING IN THE NINETEENTH AND EARLY TWENTIETH CENTURIES

Until the First World War, Germany was a leading exporter of clocks, but competition from Switzerland stifled the market for watches.[176] In the eighteenth century there had been some attempts to set up a watch industry—for instance, in Pforzheim, on the northern edge of the Black Forest, then belonging to the Grand Duchy of Baden,[177] but this was of little consequence. Production was limited and the whole industry relied heavily on Swiss imports of watch parts as well as **blanks**.[178] Further attempts were made in the middle of the nineteenth century. In 1850 the Baden government founded the Grand Ducal Baden Clock and Watchmaking School, which was supposed to introduce the making of watches to the region in addition to the traditional clocks. The venture failed after a short period of time, and the school was closed in 1863.[179] Also in 1850, Eduard Eppner (1812–87), originally from Halle, moved his fabrication of watch parts to Lähn in Silesia, where it developed into a full-grown watch factory. In 1869 the company relocated to Silberberg. Eppner & Co. succeeded in manufacturing lever and cylinder watches of simple and medium quality. Yet the total output hardly exceeded 70,000 watches until about 1890, when the company seems to have abandoned watch production and concentrated on turret clocks and

---

174 Kahlert 2007, 99–103.
175 Also for the following, Lixfeld 2011, 182–6, 188–9.

176 Kahlert 2007, 263.
177 For a comprehensive account of the Pforzheim industry, see Pieper 1992.
178 The parson Philipp Matthäus Hahn, whose workshop is known for astronomical clocks and watches as well as for calculating machines, had some dealings with the Pforzheim industry. Cf. Pieper 1992, 137–9.
179 Saluz 2015, 129–33.

clock systems.[180] While Eduard Eppner initially experienced considerable difficulties in securing financial help from the Prussian government, Ferdinand Adolph Lange (1815–75) in Saxony was more successful. With substantial loans from the government of Saxony, Lange founded a cottage industry in 1845, in Glashütte in the Ore Mountains, which produced a small series of high-quality lever watches.[181] The enterprise flourished with several small, specialized firms springing up; the annual output, however, was marginal in macroeconomic terms. Substantial German production of pocket-watches developed only towards the end of the century, not in Saxony, but in Thuringia.

In 1861/62 the brothers Georg (1827–81) and Christian (1832–79) Thiel founded their metalworking factory at Ruhla.[182] In 1874 they started making mock watches for children; a couple of years later the toys were equipped with a simple mechanism. The article proved highly successful and was exported in great numbers, mainly to Great Britain, but also to the United States, France, Italy, and even Switzerland.[183] In the course of time, the children's play watches progressed into simple pocket-watches; the 'Fearless', with a winding period of ten to twelve hours entered the market in 1892. Development continued into the early years of the twentieth century, culminating in the model '1904', named after the year of its release. The '1904', with a duration of thirty hours, became the pattern for some of the pocket-watches of the Black Forest.[184] Here, companies such as Thomas Haller AG, Schlenker & Kienzle, and Junghans sought to add watches to their product range.[185] Haller in particular became a credible rival for Gebrüder Thiel. Having started watch production in 1895, production rose to 4,000 watches per week by 1897, though Gebrüder Thiel were producing 4,000 per day.[186] The distribution of these extremely low-priced simple watches met with the same difficulties that had hampered inland sales of Black Forest clocks with American movements. Hence, the manufacturers adopted similar retail strategies. A large part of the production went into export mainly to the United States and Great Britain. In order to facilitate sales, Thiel recruited contractually bound representatives, and later opened branch offices in several countries such as Austria and Great Britain.[187] In addition, department stores and mail-order business were supplied. From 1907, pocket-watches were complemented by wristwatches, which initially contained small pocket-watch movements. Gebrüder Thiel were the first in Germany to start series production of wristwatches in 1908.[188] In the 1920s Kienzle developed the movement 51 with pin pallet escapement, twenty-five million of which had been sold by 1975.[189] Junghans was the only clock factory to embark on the production of high-quality watches; from the 1930s onwards the company presented ambitious self-developed calibres with Swiss lever escapements.[190] Pforzheim also reappeared on the scene. Starting with the production of watchcases and the import of watch parts and movements blanks from Switzerland, the Pforzheim watch manufacturers advanced to the production of entire movements for pocket- and wristwatches in the 1930s. In general, the German watch industry matured until the Second World War; by that time German companies furnished fifty per cent of their home market. On a global scale, however, the German watch industry was of negligible significance: around 1913 its Swiss counterpart accounted for ninety per cent of worldwide export business.[191]

THE CLOCK AND WATCH INDUSTRY DURING THE FIRST HALF OF THE TWENTIETH CENTURY

As with other industrial and economic sectors, the German clock and watch industry was affected by the general changes and dramatic events of the first half of the twentieth century—automation, industrialization, and rationalization, as well as inflation, global economic crisis, and the two world wars, hit the industry, though by contrast it did benefit from armament and warfare-related work. On the whole, the German industry thrived until 1914, despite increasing competition that put companies under pressure. At the same time, production increased as new fully automatic machines of high capacity, for filing, sawing, and super-finishing, came into use.[192] Simultaneously, operational processes were organized in highly effective ways, with machines equipped with individual drives, a stronger division of labour, and faster transfer systems within the factories, even automatic feed. As a consequence, production figures soared: the old-established firm of Schlenker & Kienzle, for instance, manufactured 2.35

---

180 The company existed into the 1940s. Cf. Saluz 2015, 118–29.

181 On Lange and the general development in Glashütte, see Meis 2011, v. 1, 94–175. Ullmann 2015, 102–14.

182 For further information, see Schreiber 2017 (personal dates of the brothers Thiel: 24), Kamp 2019, 220, and Kamp, Paust, & Mleinek 2015 (I did not have access to this latter volume.).

183 Schreiber 2017, 39.

184 Kamp 2019, 224.

185 Thomas Haller (1854–1917) took over business from his father around 1880. According to Schmid, Haller started to sell cheap pocket-watches in 1895. Schmid 2017b, 217. Schlenker & Kienzle was founded in 1883. The company altered its name in 1922 to Kienzle AG. Schmid 2017b, 344–51.

186 Latzel 2009, 33.

187 Schreiber 2017, 97–102.

188 Kamp 2019, 225–6. Latzel 2009, 27.

189 Kahlert 2007, 271.

190 Kahlert 2007, 271. After 1945, Junghans strove to produce high-quality wristwatches. To this end the company aimed at constructing its own movements, and established, among other things, divisions for the fabrication of jewels and springs. Success was achieved from the 1950s onwards. See Schmid 2017b, 317.

191 Kahlert 2007, 263.

192 Also for the following: Bönig 1993, 368.

million clocks in 1908, compared with 470,000 in 1898; Junghans, the largest factory of the period, manufactured 4.2 million clocks in 1911.[193] But growing competition, as well as the move to mass production, increasingly limited profits. Hence, from the late nineteenth century onwards, the industry witnessed a process of consolidation. In 1899, several Freiburg companies merged creating the Vereinigte Freiburger Uhrenfabriken AG, including the former Gustav Becker company (Figure 137).[194] In the Black Forest area several associations were formed, such as the Uhrenfabrik Villingen AG in 1899/1900 and the Vereinigte Uhrenfabriken Gebrüder Junghans and Thomas Haller AG in 1900.[195] Crisis, however, came in the 1920s.

The First World War substantially decreased the output of timepieces. Because of material shortages firms adopted substitutes, such as iron for clock movements, or else stopped production entirely.[196] As discussed later, a number of companies profited considerably by entering the armament business. On the whole, the German industry quickly regained its strength: in 1924 production and export figures outstripped those of 1913.[197] However, a period of inflation led to a decline in the domestic market. Simultaneously, and more importantly, countries such as Great Britain, Germany's most important export market, erected customs barriers. The German industry, which in the 1920s exported more than seventy per cent of its annual production, plunged into a severe crisis,[198] which was further exacerbated by the global economic crisis taking hold in 1929. The clock industry reacted with mergers. In 1927, Junghans, the Hamburg-Amerikanische Uhrenfabrik, and the Vereinigte Freiburger Uhrenfabriken formed an association, which in 1930 developed into the Uhrenfabriken Gebrüder Junghans AG (Figure 138).[199] Companies such as Kienzle Uhren AG and Thomas Ernst Haller AG followed suit and, one year later, created associations of their own.[200] The Baden part of the Black Forest was hit hardest. Unlike their Württemberg counterparts, Baden firms focused on the more expensive massive clock movements and used elaborate artisanal production methods. In times of economic hardship their comparatively expensive products did not sell.[201] So, during the 1930s most Baden companies, including the long-established Uhrenfabrik Lenzkirch, went out of business. Though the domestic market recovered in the late 1930s, export figures, which had declined by fifty per cent in 1932, remained constantly low until the Second World War.[202] In consequence, the German clock industry as a whole shrank, although some firms achieved significant sales during the war by converting to military production.

THE SECOND WORLD WAR

Several clock and watch companies profited during the war, either through sales of military timepieces, such as pilot's watches, as did A. Lange & Söhne in Glashütte and the Laco-Uhrenfabrik in Pforzheim,[203] or through the supply of fuses, as the industry was increasingly converted to the production of materiel. The latter line of business yielded very large profits, in particular, for Gebrüder Thiel and Junghans, but also for many other companies. As early as 1903, Thiel, in cooperation with the Friedrich Krupp AG, one of the most important German steel and armaments groups, had started the development of clockwork-operated mechanical fuses.[204] In 1915, Thiel, again in cooperation with Krupp, intensified their efforts, which ultimately led to the foundation of a specialized subsidiary enterprise, Gebrüder Thiel Seebach GmbH. The company made significant profits, which increased after the National Socialist takeover: turnover of about 12,085 Reichsmarks (RM) in 1925 rose to almost eight million RM in 1934 and increased further to about 17.5 million RM in 1935. By the outbreak of war, turnover was more than twenty million RM.[205] The most important producer in the Black Forest area was Junghans; already in about 1905 the company had engaged in the development of fuses, and from 1914 onwards dedicated growing parts of its production capacity to their fabrication.[206] By the end of the First World War, Junghans was delivering 30,000 fuses per day, which outstripped the production at Thiel of up to 5,000 fuses a day.[207] During the Second World War, by 1942/43, ninety per cent of the company's manufacturing capacity involved the fabrication of fuses, while only ten per cent went into timepieces for military and civil purposes.[208] Both companies, Thiel and Junghans, used forced labour.[209] After 1945, the equipment and facilities of Gebrüder Thiel Seebach GmbH were dismantled by Soviet troops and the company deleted from the commercial register.[210] Junghans also suffered from the dismantling of machinery

---

193 Kahlert 2007, 259.
194 Schmid 2017b, 45–50.
195 Though Junghans and Haller separated three years later. Kahlert 2007, 261. In 1899 the Uhrenfabrik Villingen Maurer, Pfaff & Maier changed into the 'Aktiengesellschaft' (public limited company) Uhrenfabrik Villingen. In 1900, a merger with the company of Wilhelm Jerger occurred. Schmid 2017b, 621–2.
196 Graf 2019, 242.
197 Kahlert 2007, 270.
198 Graf 2019, 243.
199 The group abandoned the Freiburg site in 1932. Kahlert 2007, 273.
200 Kahlert 2007, 272. Schmid 2017b, 219–20 and 347.
201 Graf 2019, 245.

202 Graf 2019, 246.
203 Lange & Söhne also presented the Führer, Adolf Hitler, with watches, Meis 2011, 90–7.
204 For Thiel, see Schreiber 2017, 76–84, 91–7 (77).
205 Schreiber 2017, 96.
206 Kahlert 2007, 276. There is still no comprehensive account of this part of the company's history. Cf. Schmid 2017b, 316.
207 Graf 2019, 243.
208 Kahlert 2007, 277.
209 And so did many other companies as, for instance, in Glashütte.
210 Schreiber, 226.

Figure 137 Vereinigte Freiburger Uhrenfabriken AG, incl. vormals Gustav Becker, advertisement, *Deutsche Uhrmacher-Zeitung*, no. 10, 1903, supplement Photo: Johannes Eulitz.

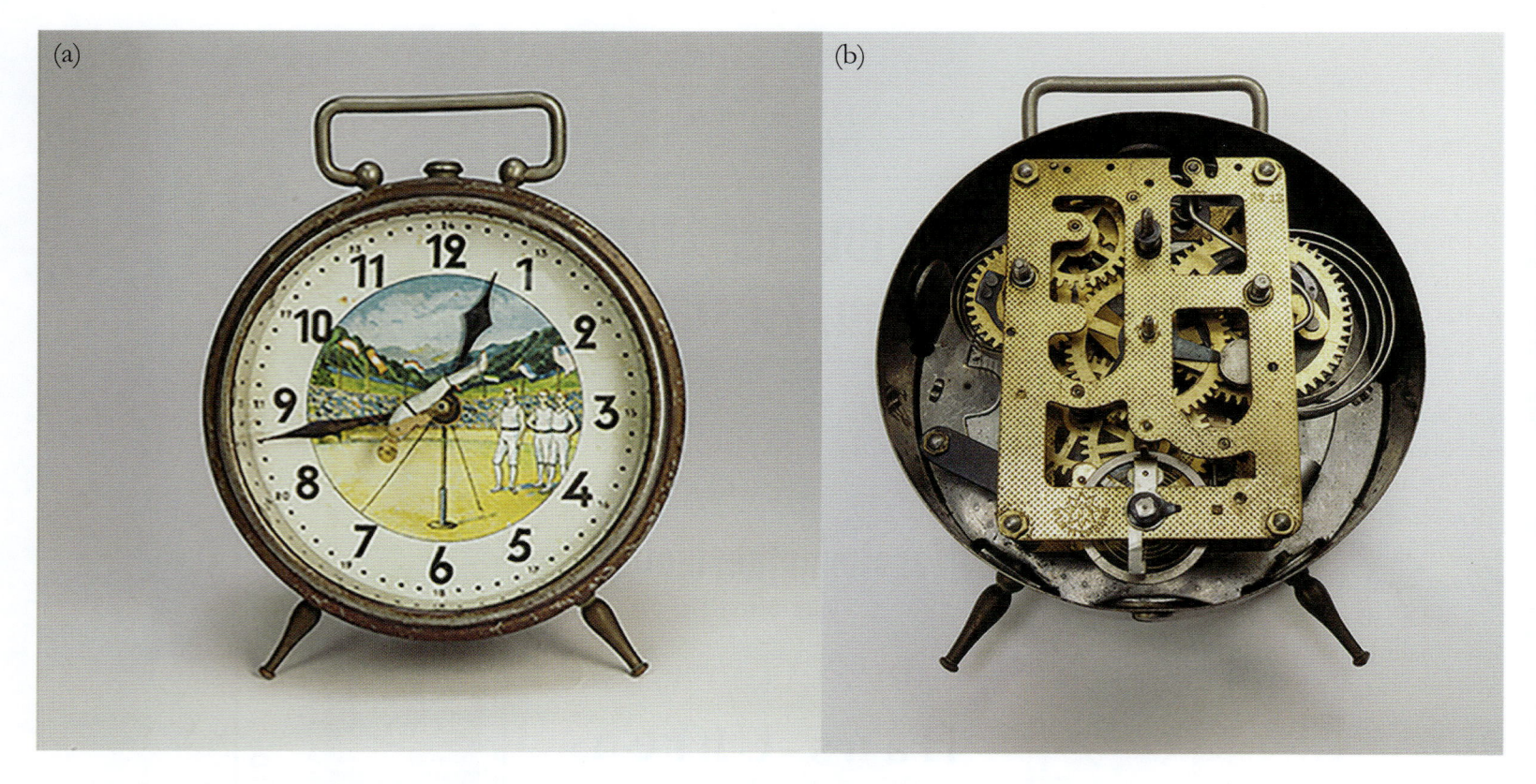

Figure 138 Junghans, alarm clock 'gymnast', 1932, private collection Photo: Johannes Eulitz (a) front; (b) back.

and plant by the French occupying forces, but ultimately continued the production of fuses.[211] After the takeover of the company by the Diehl group in 1957, fuse activity was increased. In 1984 the company separated the production of timepieces and fuses, and in 1999 the fuse business was transformed into a specially founded enterprise, Junghans Feinwerktechnik GmbH, which was renamed Junghans Microtech GmbH in 2007.[212] By that time, Germany had been reunited for almost two decades. From 1949 to 1989, the clock and watch industry had been shaped by the politics of two independent countries: the Federal Republic of Germany (FRG) and the German Democratic Republic (GDR).

THE GDR CLOCK AND WATCH INDUSTRY

The development of East German industry was defined by central regulation in the socialist planned economy. On the one hand, production had to cover domestic demand; on the other, it served to obtain much required foreign currency through the export of timepieces to West Germany and other capitalist countries.[213] This export business, in turn, ensured the continuous technical development of the politically cut-off East German industry, the beginnings of which, incidentally, were marked by extensive restructuring.

After the initial dismantling of machines and equipment by the occupying Soviet forces, which hit the Glashütte firms in particular, individual companies were nationalized.[214] By order of the Soviet Military Administration (SMAD), the first state watch company, called Produktionsgemeinschaft Precis, was founded in 1946 in Glashütte. Together with other old-established and now nationalized companies such as A. Lange & Söhne (renamed Mechanik Lange & Söhne VEB), Precis was merged into a nationally owned company, VEB Mechanik Glashütte Uhrenbetriebe, in 1951.[215] The company comprised branches for timepieces and mechanical engineering, thus mirroring the traditional structure of the Glashütte industry. Gebrüder Thiel GmbH in Ruhla

---

211 Graf 2019, 248 and company website: https://junghans-defence.com/de/unternehmen/historie/ (accessed 2 March 2020).
212 Diehl sold the timepiece business in 2000 and concentrated on the so-called defence business. See company website: https://junghans-defence.com/de/unternehmen/historie/ (accessed 21 February 2020). A short account on Diehl in: Schmid 2017b, 108–9.

213 Ruhla, for instance, exported wristwatches to the Federal Republic of Germany (FRG), Belgium, and the Far East, while alarm clocks mainly sold domestically. *Cf.* Kamp 2019, 228. Glashütte and Ruhla watches could be found in the catalogues of the West German mail-order company Quelle, where they were traded as Quelle's own brand 'Meister-Anker'. *Cf.* DIK 1997, 41.
214 Meis 2011, 98–9.
215 For Glashütte see the comprehensive webpages of Hans-Georg Donner: https://www.glashueteuhren.de/historische-entwicklungen/ (accessed 22 February 2020). The abbreviation VEB stands for Volkseigener Betrieb, i.e. nationally owned company.

became VEB Uhren- und Maschinenfabrik. After the war, the company was owned by the Soviet Union, but was transferred back to the GDR in 1952.[216] As in Glashütte, its divisions for timepieces and apparatus engineering had been formed in the nineteenth century. With the foundation of VEB Feingerätewerk Weimar in 1950, the GDR established a second location for the watch industry in Thuringia. The initially minuscule factory with twenty-four workers produced alarm clock glasses and office supplies. Over time, the factory grew substantially; the product range diversified and included alarm and other clocks, as well as optical instruments such as photometers. The making of gold-plated watchcases, as well as high-quality dials and hands, started in the late 1950s and early 1960s in order to replace imports from Western Germany (FRG) and France, and was of special importance. After an uncertain period, the three establishments in Glashütte, Ruhla, and Weimar were formed into the state combine VEB Uhren- und Maschinenkombinat Ruhla. In 1978 yet another renaming followed to create the VEB Uhrenwerke Ruhla, which was reallocated to the newly created state combine VEB Kombinat Mikroelektronik Erfurt. In this form, the company existed until 1990, when it was dissolved and privatized by the Treuhandanstalt (Trust Agency). Thus, in the Eastern part of Germany the clock and watch industry underwent radical changes in its organizational structures. Its products, however, displayed a remarkable consistency.

Before the war, timepieces from Glashütte and Ruhla differed in quality and price. The GDR industry largely retained these features. While Glashütte fabricated high-quality, multi-jewelled movements, Ruhla produced simpler, cheaper ones. VEB Mechanik Glashütter Uhrenbetriebe continued the former product range of A. Lange & Söhne, and made men's wristwatches, marine chronometers, and deck watches.[217] Towards the end of the 1950s, the company developed a heated airplane clock with chronograph function that went into series production until 1967. By that time, the portfolio also contained lady's wristwatches and diver's watches as well as automatic winding watches. In 1964 Glashütte released one of the flattest wristwatches of the time, the so-called 'Spezimatic' (Figure 139). At the same time, Ruhla struggled with economic difficulties.[218] After the war, the company had started mainly with wristwatches and alarm clocks, some of the models being revived from the Gebrüder Thiel collection. However, these movements did not meet contemporary taste that required smaller dimensions and date displays. Facing liquidation, the Ruhla factory succeeded in the development of a new wristwatch model (Figure 140), the successful calibre 24 with a pin pallet escapement, also used for small travelling clocks: over 115 million pieces of this calibre were produced between 1963 and 1990.[219] Generally from the 1960s onwards, production methods were characterized by increasing automation and rationalization. The variety of movement models diminished; production focused on standard calibres offered in different cases and with different features.[220] In the late 1960s, automatic assembly machines entered production. One of the first companies worldwide to use them, Ruhla accomplished fully automated assembly of an entire mechanical watch movement in 1978.[221] By that time, its product range also included chess clocks, and from 1984 onwards timers, while the Glashütte engineering branch produced a variety of elements, such as power regulators and time switches for domestic appliances. Both sites turned to crystal-controlled clocks in the 1970s.

As early as 1964, Glashütte had presented an electric wall clock, the 'elektrochron', a contact-controlled movement with battery drive and balance oscillator.[222] The first crystal-controlled clocks followed, rather late internationally, in the second half of the 1970s, with a marine chronometer and living room clocks of the 'Piezochron' type.[223] Ruhla started making analogue quartz wristwatches in 1976, after a first unsuccessful attempt in 1971.[224] Towards the end of the 1980s the company also presented a radio-controlled clock, though at this point they were overtaken by history. On 3 October 1990 the GDR became part of the Federal Republic of Germany (FRG). The VEBs in Glashütte, Ruhla, and Weimar were broken up and subsequently privatized. At all locations independent companies emerged, which had to develop strategies in order to survive in a global and highly competitive market. While the socialist planned economy with its notorious shortages of commodities and resources had hampered technological advancement, it had simultaneously protected and subsidized the East German industry, which already in the 1980s could not obtain export prices that covered production costs.[225] West German industry, on the other hand, had to face free-market competition with its increasingly fiercer conditions.

---

216 Kamp 2019, 226. Schreiber 2017, 223–34. Generally on Ruhla: Kamp, Paust, Mleinek 2015.
217 For Glashütte and its products see Donner: https://www.glashuetteuhren.de/die-uhrenfabriken/glashuetter-uhrenbetriebe/ (accessed 22 February 2020). Herkner 1988; Herkner 1994/95.
218 For Ruhla see Kamp 2019, 226–8.

---

219 Kamp 2019, 228.
220 Donner: https://www.glashuetteuhren.de/die-uhrenfabriken/glashuetter-uhrenbetriebe/ (access 22 February 2020).
221 Kamp 2019, 234.
222 Donner: https://www.glashuetteuhren.de/kalibueruebersichten-modelle/glashuetter-uhrenbetriebe/wanduhr-kaliber-49-11-tischuhr-kaliber-49-12/ accessed 22 February 2020).
223 Donner: https://www.glashuetteuhren.de/kalibueruebersichten-modelle/glashuetter-uhrenbetriebe/kaliber-71-quarz-marinechronometer/ and https://www.glashuetteuhren.de/kalibuerueber sichten-modelle/glashuetter-uhrenbetriebe/piezochron-kaliber-1-48/ (accessed 22 February 2020).
224 Kamp 2019, 236.
225 The Glashütter Uhrenbetriebe, for instance, produced watches at 10.70 marks (cost price), but had to sell for 7 marks. DIK 1997, 42.

Figure 139 VEB Glashütter Uhrenbetriebe (GUB), men's wristwatch, caliber 75 'Spezimatic', 1968, private collection Photo: Johannes Eulitz [front and back].

Figure 140 VEB Uhren- und Maschinenfabrik Ruhla, men's wristwatch, caliber 24, c.1975, private collection Photo: Johannes Eulitz [front and back].

THE FRG CLOCK AND WATCH INDUSTRY

As might be expected, most of the old established Black Forest companies picked up business after the war. While the industry had recuperated by the middle 1950s, it did not reach the same level of economic importance as in pre-war times.[226] The advent of cheap quartz timepieces made in the Far East ultimately led to the decline of the whole industry—many of the old companies filing for bankruptcy. The first decade after the end of the war, however, was dedicated to rebuilding production sites and regaining production figures.

Initial situations differed greatly. While the watch industry in Pforzheim had been completely destroyed by the Allied bombing raid on the town on 23 February 1945, the production sites in Württemberg and Baden remained largely intact. The latter were impaired, however, by the dismantling of machinery and plants by the French occupying forces. Also substantial parts of the first items newly produced went to France as compensation.[227] The overall situation stabilized with currency reform in 1948, the effects of the so-called Marshall Plan (European Recovery Program), initialized the same year, and finally with the constitution

---

226 For an outline of the West German clock industry, see Graf, 2011, 241–60, and Graf, 2019, 251–8.

227 Graf 2011, 242.

of the Federal Republic of Germany in 1949. A period of rapid economic growth followed, the 'Wirtschaftswunder'. During this time, the clock industry in the Black Forest region and the watch industry in Pforzheim and Schwäbisch Gmünd, both in Baden-Württemberg, recovered. In 1949, output reached eighty-five per cent of that of 1936.[228] Figures fully regained pre-war levels in the early 1950s, and then rose constantly until 1960. As before the war, the German industry was dominated by clocks. Compared with Switzerland, which held the largest market share, but also with Japan, the Soviet Union, and the USA, German watch production remained insignificant, and annual output stagnated from 1960 onwards.[229] The clock industry, on the other hand, recaptured its foremost position on the world market: in 1970 the FRG fabricated thirty-eight million clocks, followed by Japan with twenty-three million, and the United States with twenty-one million. Alarm clocks constituted sixty-one per cent of German production—two-thirds of them for export.[230] With a share of only 1.1 per cent of the total national economy in 1963, however, the industry was of minor macroeconomic significance.[231] Owing to high labour intensity and high wage costs it experienced comparatively low economic growth.[232] Already in the 1960s a crisis became apparent that intensified from the mid-1970s.

By that time, the German industry was hit by the worldwide recession that coincided with the technological shift to crystal-controlled timepieces. The first symptoms had, however, shown much earlier, particularly in the market for watches. The period up to the end of the 1950s was characterized by an excess demand that paved the way for the German companies. When demand slackened, the industry simultaneously faced new competitors, such as Japan, China, and Hong Kong, resolutely entering the market.[233] At the same time, European countries such as France and Great Britain sought to build up their industries behind tariff walls and subsidies.[234] Not surprisingly, given the numbers of competitors, prices fell constantly, and the trade complained about dumped goods.[235] As the production of small series became economically unviable, the Pforzheim industry steadily abandoned the making of blanks, returning to imports from Switzerland; some companies ordered watch parts from the low-wage countries of the Far East or even set up branches there.[236] Attempts at rationalizing and standardizing production as well as at introducing a general quality control for German wristwatches suffered from the particular structure of the German watch industry, mainly composed of small- and medium-sized companies. Thus, despite notable technological advances, the already minor German watch industry underwent a continuous decline: production figures dropped by more than half between 1977 and 1991,[237] while the international market share became negligible, at only one per cent in 1982.[238] As export markets vanished, import figures exploded; in 1980 15.1 million watches were shipped to Germany.[239] This crisis, but some two decades later, also struck the clock industry.

Unlike the watch industry, the clock industry could compensate for falling prices by raising production for a time, and by intense technical development. However, the drop in demand in both internal and external markets seriously affected the clock companies, which were, moreover, challenged by the introduction of new synthetic materials and crystal technology.[240] Polymer materials appeared in the clock industry towards the end of the 1960s; by the mid-1970s they prevailed. While they facilitated a reduction in production costs, they also necessitated new machines and technological know-how, which several of the longstanding companies could not master—for example Uhrenfabrik Villingen, Josef Kaiser GmbH, and Mauthe Uhren in Schwenningen had to close in 1974 and 1975, respectively.[241] Nevertheless, German companies played a leading role in the implementation of electrical and electronic technology in clocks. Making seven million electronic clocks per annum by 1976, Germany was the world's largest producer in this area, followed far behind by Japan with 1.9 million clocks. By the end of the 1970s,

---

228 Kahlert 2007, 278.
229 Kahlert 2007, 279. Pieper 1992, 290. Cf. world production of watches and clocks in 1971, in: UJS 1973, 7. In watches, Germany was in sixth place with 7.9 million, just ahead of China (6.5 million). The largest producers were Switzerland (72.4 million watches), Japan (25.5), the Soviet Union (22.5), the USA (20.1), and France (12.5).
230 All figures from Kahlert 2007, 279.
231 Graf 2019, 252.
232 Graf 2019, 252.
233 Pieper 1992, 289–91.
234 Great Britain had been the most important German export market before the Second World War. After the war, it absorbed a mere two per cent. Pieper 1992, 284.
235 UJS 1973, 7. From January to May 1982, import figures of watches increased by 31.2 per cent, yet the monetary value fell by 5.5 per cent. See, also for further literature, Graf 2008b, 72.
236 Graf 2011, 256. Of the thirteen factories for blank movements in 1960, only five existed in 1973. See Pieper 1992, 295–300 (300). One exception was the Deutsche Uhrenrohwerke (DUROWE), which existed until 1983.
237 7.1 million watches in 1977, 3.1 million watches in 1991. Pieper 1992, 352. Cf. also Graf 2019, 255–7.
238 In 1976 the share of the world market was 3.4 per cent. See: Graf 2008a, 70–1.
239 These figures increased annually. In 1981, 11.15 million watches were imported from Hong Kong alone. Graf 2008a, 72. The industry's dependency on exports and its consequences was addressed early on. Cf., for instance, UJS 1973, 7. For a later analysis, see Ew 1983.
240 Graf 2011, 259.
241 Graf 2011, 251. Graf, 2019, 253. The Uhrenfabrik Villingen, Josef Kaiser GmbH was founded in 1914; Mauthe Uhren was one of the oldest firms of the Black Forest. Its founder, Friedrich Mauthe (1822–84) started industrial production of massive clock movements in 1868; production of American movements followed in 1886. See Schmid 2017b, 421–6.

forty-five per cent of Germany's annual production of clocks contained crystal-controlled movements.[242] However, the rapidity of technological development as well as the ensuing costs proved beyond the means of medium-sized companies. The government therefore funded inter-company research projects, while the industry intensified cooperation.[243] Joint ventures facilitated the evolution of fully automated, large-scale production that was successfully implemented by the mid-1980s. Yet the possibilities of automation and rationalization were exhausted after a short period. In view of rapidly falling prices and ever-increasing competition from the low-wage countries of the Far East, the German industry faded to economic insignificance: in 2009 the remaining Black Forest firms employed a mere 1,369 people compared with 32,000 in 1970.[244] Most of the old established firms had closed, despite the fact that, from the 1970s to the 1980s, companies such as Kundo, Gebrüder Staiger, and Andreas Haller had proved highly innovative.[245]

Before the rise of electronics, the German industry followed well-trodden paths and produced timepieces that were in line with those of pre-war times. Customers had already developed a taste for less expensive, smaller clocks.[246] Thus, after 1945, longcase clocks and regulators went out of fashion, while a variety of smaller forms appeared, adapted to use in any room of the house. Alarm clocks continued to be the main product. In addition, the German industry was recognized for its production of timepieces for public spaces and technical appliances as well as specialized timepieces such as recording clocks, short-duration timers, and counters.[247] Fundamental changes became first noticeable in the fabrication of turret clocks: traditional manufacturers were replaced by large companies in the electrical sector such as Siemens and the Allgemeine Electricitäts-Gesellschaft (AEG) that constructed electric clock systems and increasingly also electric clocks for domestic use.[248] In general, from the late 1950s onwards, electric timepieces appeared in higher numbers, while mechanical timepieces slowly fell behind.[249] During this period the Pforzheim watch industry presented a number of innovative timepieces. For example, Helmut Epperlein, owner of the Uhrenwerk-Ersingen (situated near Pforzheim) developed in cooperation with the American Hamilton Watch Company the first electrically driven wristwatch with balance. The Epperlein-Elektric 100 and its American counterpart, the Hamilton Electric 500, were released on the market in 1957.[250] One year later they were followed by the Laco-Elektromat from Lacher & Co, an electric wristwatch with a balance motor (Unruhmotor) and electrodynamic drive.[251] A third company, the Pforzheimer Uhren-Rohwerke GmbH (PUW), developed a high-beat movement (schnellschwinger) series production of the Porta-elechron started in 1966.[252] By that time, crystal-controlled timepieces started to appear. During the initial phase of experimentation, the Pforzheim watch industry managed again to compete internationally alongside the United States and Japan. Thus in 1971 the Arctos Uhrenfabrik Philipp Weber KG presented the first German quartz wristwatch.[253] Together with other firms, Arctos then devised a liquid crystal wristwatch with digital display, while PUW succeeded in the further development of the Porta-elechron in an electronic controlled watch. Another PUW calibre, the number 632 with Lavet-type stepping motor, developed in cooperation with Junghans, became one of the best-selling quartz watches in Germany.[254] By then, however, the experimental stage had been superseded by a process of standardization, which resulted in mass production and price erosion and ultimately led to the decline of the fragmented watch industry. The much more adaptable clock industry, on the other hand, met the challenges posed by the large concerns in China and Japan with innovative technical solutions of its own—at least for a while.

In 1967 Junghans created the first quartz clock for everyday use: the Astro-Chron, a table clock, sold for a deliberately high price.[255] The development of lower-cost types mainly rested with smaller companies. Thus in 1971 Gebrüder Staiger Uhren in St Georgen, which had already distinguished itself with clock movements made of plastic, presented the first affordable quartz clocks (Figure 141). These first quartz timepieces underwent a continuous process of further development. Companies such as Staiger, Kundo, and Andreas Haller worked to simplify and standardize the movements in order to reduce production and assembly costs. Ultimately, by the end of the 1970s, quartz movements could be produced at relatively low costs, and crystal clocks started to supplant electric ones. The technology was fully developed by the mid-1980s.[256] Another novelty of the period was the digitally coded time transmission for radio clocks, patented in 1967 by Wolfgang Hilberg, who also developed the first prototypes.

---

242 Graf 2008b, 70.
243 Graf 2008a, 74–5. Graf 2011, 255. Graf 2019, 255.
244 Graf 2011, 259. The unit cost of 12.91 DM per clock in 1970 dropped to 4.14 DM per clock in 1985. See Graf 2019, 257.
245 All in St Georgen in the Black Forest: Kundo, formerly Kieninger & Obergfell, founded in 1899; Andreas Haller, founded in 1925; Gebrüder Staiger, founded in 1898, started production of timepieces in 1953, and merged with Kundo as Kundo-Staiger in 1992. Schmid 2017b, 211–12, 374–7, 582.
246 Also for the following: Graf 2019, 246.
247 Kahlert 2007, 271. Graf 2019, 248.
248 Kahlert 2007, 271. Graf 2019, 248.
249 Graf 2019, 252.

---

250 Pieper 1992, 326–8. On the company: Schmid 2017b, 132–3.
251 Pieper 1992, 329–31.
252 Pieper 1992, 331–3. On the company: Schmid 2017b, 477–8.
253 Also for the following: Pieper 1992, 334–45. A short account of the firm's history in: Schmid 2017b, 631–2.
254 Pieper 1992, 365.
255 The Astro-Chron cost about 800 DM—a rather high price for wage levels at the time. It was mostly accounted for by the palladium coated case. See, also for the following: Graf 2011, 252–6.
256 Graf 2008a, 74. Graf 2011, 256.

Figure 141 Gebrüder Staiger, quartz movements, c.1971–c.77 Courtesy of the 'Technik und Event Museum St. Georgen'.

The clock industry took a late interest in radio-controlled timepieces. They became widely available only in the second half of the 1980s. Kundo and Junghans were the market leaders in this area.[257] Generally, the period was marked by increasing automation as well as intensified cooperation. Thus Kundo and Gebrüder Staiger founded a subsidiary enterprise called Uhrentechnik Schwarzwald GmbH (UTS) in order to cover the investment costs for a fully automated production line.[258] They accomplished fully automated production in 1985. In consequence, production figures soared—from 3,000 pieces per day in 1980 to 60,000 in 1989.[259] From the middle of the 1990s onwards, UTS also supplied Junghans, after the latter company had abandoned its own production of clock movements. These united efforts brought market dominance to the German industry, which, however, did not last. At the turn of the millennium, the German clock companies lost their technological lead to the Far East. Most of the companies mentioned above no longer exist: Andreas Haller, Kundo Gebrüder Staiger, and others have all become part of history.

Today the German industry is of no significance on an international scale and carries but little weight in the national economy. Few companies survived the technological and structural changes of the twentieth century. The transition from mechanical to electronic timepieces, from small series to fully automated massproduction, brought fundamental alterations in the profession: With the emergence of the first factories in the nineteenth century, the traditional clock or watchmakers have been gradually replaced by specialized workers. The making of electronic timepieces involved entirely new components, produced by specialists in the electronics industry. Only in the high-price niche of mechanical timepieces is some of the trade's former distinction maintained, but even here highly complex machines are indispensable. Characteristically, the clock and watch industry developed international ties from an early point in time—be it by import and export activities, the foundation of subsidiaries abroad, the migrations of journeymen, or the educational or rather espionage travels of manufacturers in the later nineteenth century. Some occurrences of the twentieth century mirror those of the nineteenth. Thus the cheap quartz watches from the Far East went through new distribution channels and were sold outside the established specialized trade, much as the American movements of the previous century. However, today global interconnectedness reaches a level that at times complicates even the designation 'German industry', as most of the companies at the high-end market are part of international groups typically based in Switzerland or France.

---

257 Kamp 2019, 237.
258 Graf 2011, 255. Schmid 2017b, 609.
259 Graf 2011, 255.

SECTION SIX: THE PENDULE DE PARIS, FROM THE WORKSHOP TO THE FACTORY, 1800–1910

Françoise Collanges

'Hands start [the work], tools help, instruments make it more perfect, and machines shorten the time [of making].' Ferdinand Berthoud in *Encyclopédie*, 1751–80 xxi, 2.

The nineteenth century *pendule de Paris* had very specific features. Its cases demonstrated all the rich diversity of the crafts developed for luxury items in the French capital, gathered under the name of *articles de Paris* in the middle of the century.[260] It is thus mostly a decorative item (Figure 142), which moved from a high-luxury object in the eighteenth century, to the world of the *demi-luxe*, cultivated by the lower-middle classes.

This change was mostly due to the industrial capability of achieving a clock movement which could be precise enough for daily life, mass-produced, and adaptable to as many decorative cases as possible. The first move to standardization for French clocks was thus linked to the choices made by some manufacturers to focus on a similar frame: for most of the nineteenth century, the iconic French *pendule* consisted of a drum movement, with two **trains**, **striking** every hour and half-hour on a bell. The vast majority of the striking systems employed a **count wheel**.[261] The mechanism was fixed to the dial and attached between two bezels linked by two straps so that the whole assembly could be fitted in any case pierced with a round hole of the appropriate diameter.

Although square or rectangular movements inherited from eighteenth-century cartels still existed, and **rack striking** was used,[262] the bulk of the production focused on 8.2cm (3in) diameter, round calibre movements throughout the century (Figure 143). Then smaller calibres appeared, adapted to simple timepieces and clocks in which decoration entirely superseded the time-keeping functions.

Ebauche, **blanc**, and *blanc-roulant* are terms normally used for the mechanisms produced by the factories. Their meaning evolved during the nineteenth century, following changes in the trade. Initially, *blanc*, like *ébauche*, described the assembled plates, bridges, and spring barrels. *Blanc-roulant* included the wheels and **pinions**,[263] a meaning that *ébauche* took on (mostly for watches), as the century progressed.[264] During the eighteenth century, several unsuccessful attempts to create manufactories were made. Their first aim was to produce *ébauches* that could be finished in Paris.[265]

From 1800 to the First World War, the French *pendule* industry rose from artisan production to a prominent industry, before becoming a branch—not always financially sustainable—of bigger industrial companies. This occurred through a century of technical innovation, leading to the standardization of a luxury product, deployed as an affordable marker of social success.

THE BEGINNINGS OF MECHANIZATION IN CLOCKMAKING: FROM THE 1770S TO 1850, BUILDING UP AN INDUSTRY

The French Revolution, after creating a decade of economic and political turmoil, laid the foundations for new economic development in the next century. Once the guild system was abolished,[266] entrepreneurs could seize new opportunities to develop clockmaking into a flourishing industry. While mass-production of movements started to be organized outside Paris, assembling and finishing was a major activity in the capital. If the *blancs-roulants* were bought from the provinces, fitting the **escapement**, finishing the parts, finding the right case, and assembling belonged to the capital. Louis-Gabriel Brocot (1791–1872) and his sons Antoine-Gabriel (1814–1874) and Louis-Achille (1817–1878) are good examples of how most clockmakers evolved in Paris during the first half of the century (Figure 144). They maintained a firm focus on experimenting with specific technical features, like escapements, suspension, and regulator systems, obtaining several patents between 1840 and 1877.

---

260 Dequidt 2013, 96: definition given by the *Chambre de Commerce de Paris*, 1851.
261 Chavigny, 2007, 5–21, description of the parts.
262 Rapport 1850, 508
263 Travail 1856, 387, 'In horology a *blanc-roulant* is a clock movement composed of two plates forming the cage, the wheels and pinions needed for the mechanism, the count wheel . . . the bridges, all the pieces of the under-dial work, etc, all made according to the principles of the art'. However, Berthoud, in the article 'Mouvement, terme d'Horlogerie'

(*Encyclopédie* 1751–80, x, 842, republished Manuel 1827, 8), used '*mouvements en blanc*' to describe a watch or clock mechanism that was only a **blank** with wheels and pinions but unfinished, the fusee not cut, the escapement and pivots not finished, the brass neither polished nor gilded.
264 Lamard 1984/1988, 50 and 321–2.
265 Dequidt 2013, 102–3: attempts in 1718, 1787–90, and 1797. Augarde 1996, 89–91.
266 Guilds were abolished in France by three statutes between 1789 and 1793.

Figure 142 Mantel clock with figure of Napoléon, Musée de l'horlogerie, Saint-Nicolas d'Aliermont, Inv. 90.04.01.

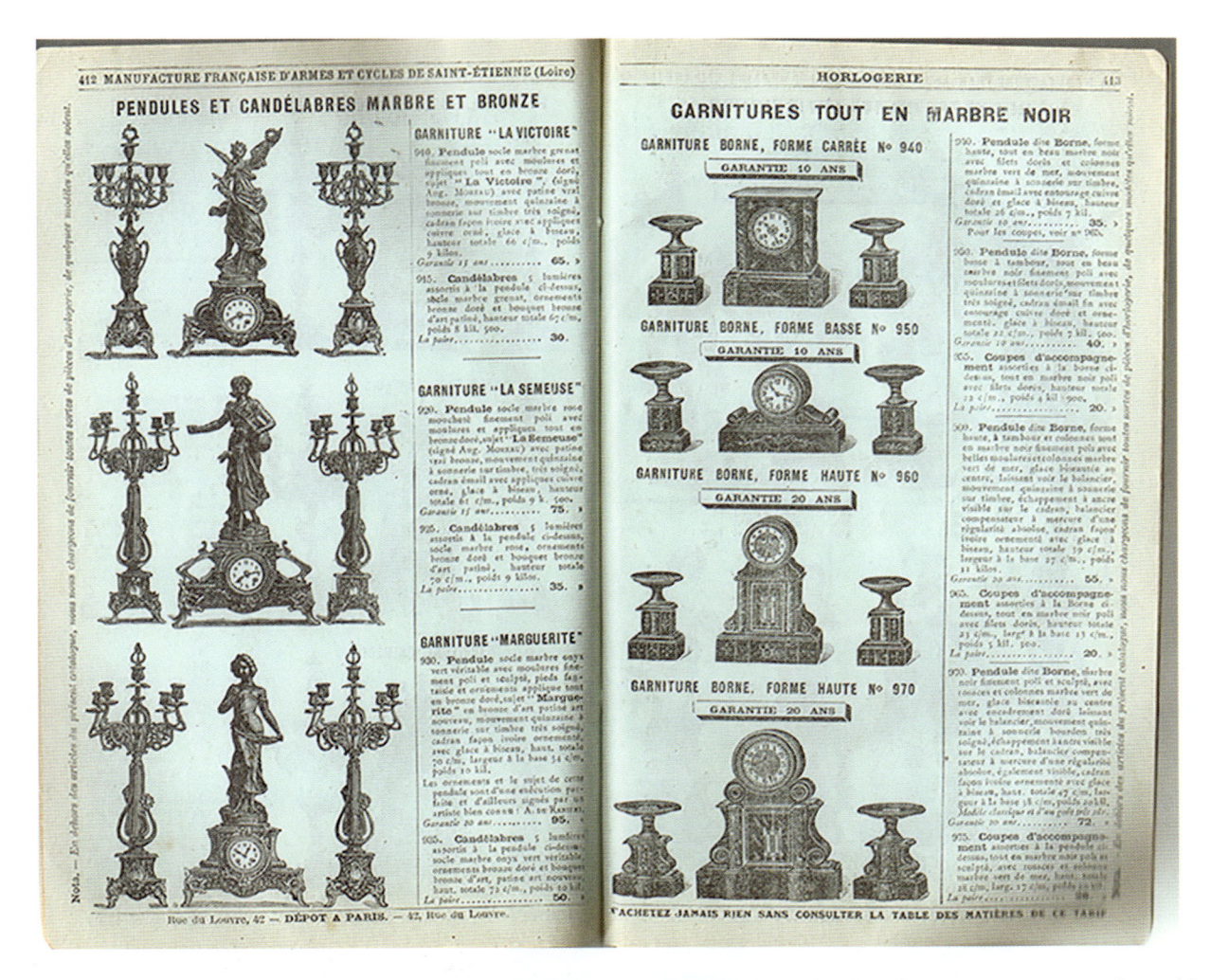

THE PENDULE DE PARIS, FROM THE WORKSHOP TO THE FACTORY, 1800-1910

Figure 143 Pages from the *Manufrance* catalogue for 1908 showing a selection of marble clocks fitted with Paris movements and matching candelabra or bowls.

For the Brocot suspension,[267] for example, the patents show major innovation between 1840 and 1852. Those which followed seem more like a way of keeping control of the invention and simplifying it to support mass production. In the same way, Louis-Gabriel and his son Louis-Achille experimented and patented several escapements in 1826 (a first version of a **dead-beat** escapement with pin pallets by Louis-Gabriel), 1839 (a double-wheel visible escapement), and 1842 (a single wheel visible escapement). From then on, innovation slowed down, and trading took precedence.[268]

This illustrates a general feature of Paris manufacture: simultaneously the city offered competition between highly qualified clockmakers making high-quality pieces in the tradition of the eighteenth century, and a rising clientele seeking decorative, fashionable pieces. Research in technical features thrived during the 1820s and 1830s, followed by a need to produce and exploit these efforts. Establishing huge factories was easy in rural Normandy and the Jura, but in Paris, manufacturing specific parts and finishing seemed a more reasonable aim for clockmakers who had their customers in the street in front of their shops.

Frédéric Japy (1749–1812) trained in watchmaking in Switzerland before returning to his home country, near Montbéliard. At that time, he bought the plans of Swiss machines from his former employer and established his first manufactory outside the town, in order to avoid guild interference. In 1793, he embraced the ideals of the Revolution when the area became French and developed his watch production.[269] He applied new industrial principles: workers were gathered in one building, part production and assembly were all done on site, and management of the workmen was paternalist. Machines enabled the use of a less skilled workforce and increased production by making each part easier and quicker to produce. Consequently, in 1806, Japy had a monopoly on the production of watch *ébauches*.[270]

---

267 Chavigny 1991, 96–102: After this first invention phase, Antoine-Gabriel took over the industrial manufacturing of the Brocot suspension and settled in a new factory near Paris, in Charenton-Le-Pont from 1853.
268 Chavigny 1991, 118–38 on escapements: Louis-Achille continued experimenting on escapements until 1861, but with limited practical effect.

269 Lamard 1984/1988, 42 and 54.
270 Lamard 1984/1988, 66.

Figure 144 Achille Brocot, black marble clock with visible escapement and compensated pendulum, private collection.

In 1806, his three elder sons, Frédéric-Guillaume (1774–1854), Louis (1777–1852), and Pierre (1785–1863) took over and started clock production on the same principles. Machines were developed on a new site in Badevel, built between 1814 and 1817, but production tests continued until 1819.[271] The aim was to produce a *blanc-roulant*: plates, pillars, wheels, and pinions, all made and assembled on-site. Finishers could then make the escapement and add the pendulum, bezels, and hands to match the case.

Machines were designed to produce several types of components. The main feature was the use of the *balancier*, so-called probably because it was an adaptation of a process used for cutting coins from metal sheets, developed by Jean-Pierre Droz (1740–1823).[272] In 1851, the report on horology at the London exhibition in the Crystal Palace showed that production processes in the mid-century still relied on the same basic principles as earlier. Brass and steel were bought raw, then worked on site. Pinion and wheel cutting was automated, batches of twelve wheels being cut in one go. All the machines were designed and made in the factory.[273] The energy source was water until 1843, when it was replaced by steam.[274]

On Frédéric's death, a conflict between the three elder sons and their two younger siblings resulted in splitting the company. Between 1817 and 1833, clock production was managed by a separate company owned by the younger sons, Jean-Charles and Frédéric.[275] After 1833, Japy Frères took over the Badevel factory, invested in new machines, and started the production of drum movements. Japy Frères, an early example of an emerging industrial company, developed in a very specific way compared with its competitors. Diversified production was considered as early as 1806, with tests for producing wood screws. This also started their interest in hardware production (nuts and bolts, hooks, corkscrews, and others).[276] Welcoming innovation, the Japy brothers supported inventors, and many future clock company owners started in their factory, like Auguste Lépée (1827–31) or Jean Vincenti (before 1817–23), who both worked for Japy before they created their own companies in Sainte Suzanne and Montbéliard.[277]

The 1830s showed the first rise in local competition. Vincenti started his own company in 1823 for finishing watches, then went into association with Roux. When he died in 1833, Roux focused the business on clock movements as they had just invested in machines and expanded during the 1830s.[278] In 1837, Louis Japy, a grandson of Frédéric, established his own brand in Berne and produced clock movements, watches *ébauches*, and finished movements.[279]

In the first half of the nineteenth century, Japy and the Montbéliard manufacturers took over the mass-production of *pendules*. The Japy Company, started as a one-man factory in the late eighteenth century, was governed by nine directors from the third generation of the family in the mid-century. In 1849, Japy Frères employed 3,000 workers on six industrial sites,[280] but clock production had already turned into a background activity, as better profits were obtained in hardware and kitchenware goods.[281] Competition had risen steadily, first with the Normandy factories initiated by Pons, then by new clock companies appearing in the Jura, mostly thanks to the activity generated by Japy.

However, in the 1840s, the company seemed to lose its innovative edge and started focusing more on sales and controlling the market than on improving the quality of its products. Machines were kept long in use,[282] exports were developed, and the focus of the business moved away from clocks. In 1843, the Paris trading house started selling all the other Japy products and in 1855 at the International Exhibition in Paris the company displays were shown in several categories. For clocks, however, Japy remained the leader for cheap reliable movements. Their *blanc-roulants* sold at 9–10 francs a piece, and could be fitted with complications on demand, whereas their Normandy competitors (Pons' successors) offered *blanc-roulants* at 13–20 francs.[283] The same source indicates that before 1814, when Japy brought the price down to fifty francs, *blanc roulants* sold for 120–140 francs. The massive price drop effected in half a century was the result of mechanization and standardization in delocalized factories. In 1851, Japy Brothers advertised at the Crystal Palace exhibition a production of 90,000 movements a year, used for clocks, lamps, and telegraph machines.[284]

---

271 Lamard 1984/1988, 110.
272 Travaux 1857, 13: short history of the machines designed to cut, trim and laminate metal sheets and round bars. Droz's machine seemed to work well, trimming circular shapes with a circular saw. Although he worked in Paris, Droz was born in La Chaux-de-Fonds and designed his cutting machine between 1783–7, the period when Frédéric Japy was training in the same area.
273 Travaux 1854–1873, 23–4.
274 Lamard 1984/1988, 140.
275 Lamard 1984/1988, 81, 105, and 112. They were not very successful, but they developed alternative productions of lamps and metronomes for a while before the factory came back to the parent company.
276 Lamard 1984/1988, 107–9.
277 Lamard 1984/1988, 93–4.

278 Rapport 1850, 508: silver medal for M. Roux, Montbeliard, successor Vincenti & Cie, making clock movements, parts, and lamps. Formerly awarded medals in 1834, 1839, and 1844. Lamard 1984/1988, 94. In 1834, the company was producing 780 movements a month.
279 Lamard 1984/1988, 127.
280 Lamard 1984/1988, 125.
281 Rapport 1850, 387: Japy Frères exhibited in the ironmongery (*quincaillerie*) section.
282 Lamard 1984/1988, 143: in 1847, some machines built in 1815 were still working and the Badevel factory did not see much investment, except the change from water power to steam engines.
283 Le Travail 1856, 387 and 392.
284 Travaux 1854–1873, 1851, 25. In 1849, they had given the consistent figure of 48,000 clock movements a year; see Rapport 1850, 387.

Figure 145 Movement signed by Pons, Musée de l'horlogerie, Saint-Nicolas d'Aliermont. Inv. 80.13.04.

PONS AND THE INDUSTRY IN NORMANDY

In 1807, the French government representative in Normandy, confronted with the collapse of economic activity in some areas, requested that a clockmaker from Paris, Honoré Pons (1773–1851), be sent to Normandy to restart the clock industry.[285] The area near Dieppe, around the city of St Nicolas d'Aliermont, had developed a cottage industry for longcase clocks during the eighteenth century.[286] Watches had never been made. Pons (Figure 145) therefore developed a production of *blancs-roulants* for *pendules* of high quality, shipped to Paris to be used by finishers and clockmakers.

Pons' manufactory was a successful attempt at mass-production of movements, based on division of labour and machines. However, it remained strongly linked to parts and finishing made in family workshops and failed to scale up to factory level as fast as had Japy Frères. When Pons' business was taken over by his successors, Boromée Delépine & Lanchy, it was already behind its competitors, mostly because it stuck to producing on-demand movements for Parisian clockmakers, which made division of labour and mass-production virtually impossible to develop.[287]

At the industrial exhibition in Paris in 1849, the Norman production was still acclaimed for its quality, but Japy had already won the battle of mass-production, cutting down prices.[288] Other small manufacturers carried on producing in this area, with easy access down the Seine river to the Paris market, but had to develop specialized products like chronometers (Delépine & Conchy), meters (Emile Martin), alarm clocks (L. Cailly aîné), or parts (Croutte & Cie).[289]

FROM A NATIONAL TO AN INTERNATIONAL MARKET

In 1849, the industrial exhibition in Paris gave an overall view of the production of French *pendules* in France: 80,000 *pendules* were produced as finished products in Paris, two-thirds of them for export. This success was explained by the quality of their bronze decoration, their low cost, and the reliability of their

---

285 Annuaire des cinq départements de la Normandie, 1841, 239–41.
286 Cournarie 2011, 20–8.
287 Bayard 1999, 4: quoting an article from the *Revue Chronométrique* in 1859.
288 Rapport 1850, 509.
289 Rapport 1856, 422.

blanc-roulants.²⁹⁰ However, general thinking was that clockmaking had changed, and that the target should now be good-quality clock movements at lower prices, and not technical innovation.²⁹¹ This illustrates a shift from a technical approach to industrial innovation to a more scientific one: applied sciences brought to industry the stern look of rational methods and fully thought through processes, against a more romantic view of invention and experimentation inherited from the previous century.²⁹²

Japy Frères was competing at the Exhibition with smaller manufacturers (Vincenti, Marti, Delépine and Canchy, Louis Japy fils, Emile Martin, A. Croutte & Cie, Ph-H Paquet, L. Cailly aîné are named in the 1855 section for *blancs-roulants*). They announced a production of 60,000 clock movements, against 36,000 for Vincenti successors, 30,000 for Marti, and 500 to 600 for Delépine Conchy. In 1856, the figures announced at the 1855 Universal Exhibition were similar. Japy also advertised a production of nine million screws a year, showing that aside from *blancs-roulants*, their position in the market for parts was important.

In 1851, the French report on horology at the London exhibition provided an insight into the production methods in Japy manufactures. It still mentioned the production of square movements, but mainly described a production of drum movements, fully integrated in the factory, from casting the plates to assembling. All the cutting, planning, drilling, and piercing operations were performed on machines invented by the company, using an unskilled workforce. Ninety thousand movements for clocks, but also lamps and telegraph machines, were produced per year.

By this point, the production of French *blancs-roulants* had reached its height: in fifty years standards had been organized, at least by brand; production was mechanized and entirely managed in the factory; Japy was dominant but there was still room for smaller brands, all supplying finishers, mostly based in Paris, who chose movements, fitted them in cases, and retailed them.

1850–1900: TOWARDS CRISIS

In 1851, for the first time, European industries came face to face in the brand-new Crystal Palace in London, in the first International Exhibition. The international market appeared very good for French clocks²⁹³ as English horology struggled to reach an integrated manufacturing system, and the German clock industry was in crisis as the market for cheap wooden clocks was hit by new products from the USA.²⁹⁴ French clocks, targeting mainly luxury and middle-class consumers, and relying on the fashion of *articles de Paris*, were not yet affected by American competition. The mid-nineteenth century was, therefore, a period of exceptional prosperity for the clock industry in France. While the new imperial regime installed by Napoleon III offered peace and support for economic expansion, decorative objects from Paris, *les articles de Paris* were sought after and the market for *demi-luxe*,²⁹⁵ including French *pendules,* expanded fast. In France, more than seventy per cent of the people with middle-class income could afford one or several clocks (Figure 146),²⁹⁶ and the French *pendule* found its main market in series-produced, middle-range production. Abroad, the regime supported free trade and an expanding American market offered great possibilities for massive exports. This lasted for clocks, longer than for watches, mostly because of the decorative value of French clocks.

The London exhibition in 1851 triggered the organization of the 1855 French exhibition. This showed plainly that French *pendules* were sought after for their aesthetic quality, for their reliability, and for their reasonable cost, thanks to the industrial production of the mechanisms and the efficiency of Paris cabinet- and bronze makers. The main issue for the factories was to produce enough while maintaining the quality and preventing a fall in prices.²⁹⁷ For this, companies carried on investing. Japy Frères targeted mostly commercial activities by expanding points of sales, but also by renewing the machines in the main *pendule* factory in Badevel, improving automation.²⁹⁸ The companies also urged improvements in transport, and railway lines were perceived as an efficient way to bring the Jura closer to Paris. In 1868, Montbéliard was linked to the main railway network, and Japy supported financially a branch line into its main factory in Beaucourt.²⁹⁹ This supported the expansion of other companies in the area. In 1855, the main companies that appeared in the 1830s were still active in the Jura. For example, Vincenti, owned by Roux, merged with Marti and become Marti-Roux & Cie during this period.³⁰⁰

Japy Frères, however, was still the leader and started organizing the trade in order to limit competition. If the 1850s were dominated by free trade and a rush to export, the 1860s saw protectionist views gaining ground, as well as a stronger focus on reaching the market and organizing sales networks. In 1865 the *Syndicat des Fabriquants d'ébauches et de pendules du Pays de Montbéliard* was launched,³⁰¹ in which Japy Frères and Marti-Roux & Cie agreed prices and combined their sales locations. In Paris,

---

290 Rapport 1850, 510.
291 Rapport 1850, 480: 'the fight is now settled on prices'.
292 Mertens 2011.
293 Travaux 1854–1873, 25.
294 Krämer 2019, 167–8.

295 This expression covers the middle range of luxury items, covering bronzes, silver- and goldsmithing, furniture, clocks, and watches, etc.
296 Daumas 2018, clocks are taken as indicators of material wealth and quoted in the chapter on farmers (3 *L'amélioration de la condition paysanne*) and factory workers (2 *L'émergence d'une nouvelle consommation ouvrière*) to demonstrate a rise in new types of objects in lower-middle classes.
297 Lamard 1984/1988, 206.
298 Lamard 1984/1988, 173–4.
299 Lamard 1984/1988, 161–2.
300 Rapport 1856, 421–2: Vincenti & Cie had become 'Veuve Roux', gathering 300 workers for a production of 36,000 movements a year, while Marti & Cie was advertising 30,000 movements a year. Japy brothers announced 60,000 a year.
301 Lamard 1984/1988, 94.

Figure 146 Eugène Atget, Interior of a worker, rue de Romainville, Paris 19th *arrondissement*, 1890, 78 CC0 Musée Carnavalet, Paris.

the movements provided by the provinces fuelled the business. Figures gathered by Saunier[302] seem to indicate that the horological world diversified greatly, passing, between 1867 and 1878, from 3,700 to 6,000 workers. This increase signalled the expansion of the finishing industry in Paris, with a rise in carriage clocks, alarms, and electric apparatus, which would have been finished by clockmakers.

In 1873, an internal report from Japy Frères pointed out that American competition was already strong and had conquered the British market, but that Germany would soon be a bigger rival.[303] Indeed, German factories like Junghans started to switch to American methods of production as early as 1870[304] and specialized in copying American clocks and French *pendules*. American production relied on sparing costs on materials and interchangeability of parts, based on specialized machines and a simple range of finished products. In 1889, Paul Garnier in his report on the International Exhibition in Paris again underlined this. Paris still produced parts and springs (10,000 people worked in clockmaking in Paris and its region), but German and American competition had become important, even overwhelming, and *blancs-roulants* producers were led to produce finished clocks at low prices. Japy Frères now advertised a production of 1,000 to 2,000 clocks of all types (alarms to carriage clocks) entirely finished and cased per day and per workshop. Between 1,200 to 1,500 workers were producing these items in Beaucourt, Japy's chief factory.

If American companies were using thinner and cheaper materials, leaving to French producers the better-quality market, German manufactures were also aiming at this same market for good-quality, semi-luxury items. Some Black Forest companies started to produce drum movement clocks, the most famous being A.G.U (*Aktiengesellschaft für Uhrenfabrikation Lenzkirch*), and American companies targeted the black marble clock market, which emerged after the death of Albert Prince Consort in 1861. The Centennial Exhibition in Philadelphia in 1876[305] epitomized the situation and was perceived as a turning point.

At this stage, French industry missed the point. In 1886, a technical report was written for Japy about American production.[306] It pointed out that interchangeable parts should be a target to optimize production. This meant changing the whole chain of

---

302 Allix & Bonnert 1974, 86.
303 Lamard 1984/1988, 163.
304 Lixfeld 2011, 176–8: After observation trips to the United States between 1870 and 1872, Junghans introduced mass production on a large scale and became the biggest factory in the world thirty years later.

---

305 Lamard 2004, 569.
306 Lamard 2004, 579–80.

production: a reduced number of calibres had to be defined in order to be assembled from parts cut on specialized machines, instead of using wide-range machines that could be set up to produce a variety of calibres. Implementing this major change would take years for Japy Frères, entrenched in methods and habits built over fifty years of production.

1885–1910: NEW ENTREPRENEURS AND AREAS OF PRODUCTION

Building on specific fashions, like marble clocks or sculptures cast in spelter to imitate bronze, new companies appeared from the end of the 1860s. In the Jura, a new man like Fritz Marti was perceived as a skilled newcomer trained in a bigger company who managed to set up a good-quality production on his own in a very short time. However, the overpowering Japy Frères & Cie controlled competition through its syndicates.[307]

In Normandy, new companies developed in a more unruly way, created by local men like Albert Villon (1867), Gustave Denis (1874), or Armand Couaillet (1892).[308] Often starting with the production of *blancs-roulants*, these companies developed in family partnerships (Couaillet Frères, Denis Frères) and started to narrow the scope of their products, like carriage clocks for Couaillet,[309] or alarm clocks. Albert Villon embodied the new type of industrial entrepreneur, and in 1893, he described his company as producing 20,000 *pendules* (movements or finished clocks) and 130,000 alarm clocks of all types, sold worldwide.[310] Taking younger partners from 1885, he built up the business with them, then retired in 1902, leaving Paul Duverdrey and Joseph Boquel to take over the company completely. This some years later became Bayard, which wholly specialized in alarm clocks after the First World War.[311]

Clock activities also developed North of Paris, seen with the end of the Brocot dynasty, whose companies were bought by the Thièble brothers in 1885. Coming from the North of France, Adolphe (1817–82) and Alexis-Joseph (1820–84) Thièble were initially mechanics and land owners.[312] They settled their factory in their home town, Ruyaulcourt (Pas-de-Calais), seemingly from 1856. The north of France had attractive mining resources: steel and coal, both useful for French clock companies, and also black marble quarries at the Belgium border. Specializing in producing parts, they patented their own type of pendulum, with a setting screw in the middle of the bob, in 1865 and the acquisition of the Brocot businesses signified their importance in the clock parts market, closer to Paris than the Jura. Thièble Frères continued to improve the Brocot suspension to make it easier and cheaper to produce,[313] but failed to patent their innovations. Technical improvements seemed not to have been considered essential to the business, perhaps because mass-production and meeting demand was more crucial for the company's survival and expansion.

Finally, a last area in the country became involved in production. In 1860, the Savoie region bordering on Switzerland and Italy passed from the King of Sardinia to France. The area around Cluses had developed a watch- and clockmaking cottage industry producing parts from the eighteenth century, mainly exported to Switzerland and Germany.[314] With Savoie joining France, the production found a new market and diversified. Still producing parts, the area engaged in the manufacturing process later than the others, but specialized in precision-turned parts, especially pinions. The new target of producing interchangeable parts implied new machines, but also workers trained to match tolerances, and thus, for example, able to take precise measurements. As home-apprenticeship still existed, further means of practical education were thus needed, so that books and manuals on practical clock and watchmaking developed. At first based on reworkings of Berthoud's book from 1763,[315] they took a more usable form in 1827, with *L'Art de l'horlogerie enseigné en trente leçons,* and in the 1830s with the Manuel Roret, *Nouveau Manuel Complet de l'horloger,* published shortly afterwards.[316]

At the end of the century, the need for more specialized knowledge increased—a need exploited by Manuels Roret[317] by the publishing of two separate books on horology: the *Nouveau Manuel Complet de l'horloger,*[318] reworking Lenormand's book, now aimed at a wider audience and introducing electric and pneumatic clockmaking, and, in 1882, *Le Nouveau Manuel Complet de l'horloger rhabilleur,* for the repairer.[319] Meanwhile, Claudius Saunier, publisher of the *Revue Chronométrique,* wrote two different volumes: *Le Traité d'horlogerie moderne théorique et pratique*

---

307 Lamard 1984/1988, 182.
308 Cournarie 2011, 51–2.
309 Cournarie 2011, 148–55: time lines for some of the main companies in St Nicolas since 1850.
310 Bayard 1999, 8 and 11.
311 Bayard 1999, 6 and 14.
312 Profession given by their civil status: Their father, Joseph (1790–1859) was a blacksmith. Alexis' son, Léon (1858–1915) was the first defined as 'manufacturer of horological parts' then 'manufacturer'. Information from the family tree researched by J. C. Hombert, for Léon Thièble: see https://gw.geneanet.org/hombert62?lang=fr&n=thieble&nz=hombert&oc=0&p=paul+leon&pz=achille+louis&type=tree

---

313 Chavigny, 1991, 110–11.
314 Barillet 1894, 173–6.
315 Berthoud 1763: in two volumes, the book presented first several types of clocks and watches, then explained the theoretical background needed for clock and watchmaking.
316 S. Lenormand's book was published at least in 1837, 1850, and 1896.
317 Mertens 2009. Roret started to publish encyclopaedic manuals by specialism in 1822.
318 Lenormand et al. 1896.
319 Persegol 1895: First edition seems to date from 1882, it was republished several times in the twentieth century.

and *Le guide-Manuel de l'horloger*.[320] This increase in the number and variety of technical books shows a diversifying interest. While the first Manuel Roret had sought a wide public, later volumes were addressed more specifically to the general amateur and to the shopkeeper-repairer (*rhabilleur*). In the meantime, Saunier was writing for the craftsman, especially in his *Guide-manuel*, which is very detailed about machinery and tools.[321]

The need for technical schools also emerged as early as the 1860s, but there was limited success. The pressure exerted by American and German producers showed that, even if less skilled, the French workforce needed to know the basics in measuring to make standardized products. The newly acquired Savoie provided the first clockmaking school, created in 1848 in Cluses, to support the redevelopment of the area after a fire that had destroyed the town four years earlier. It became an Imperial school in 1863.[322] At the end of the century, several schools had been organized in the Jura (a Japy company school, and in Besançon, in 1862, a training focused on chronometry), in Paris, and around (Anet in 1871, and Paris in 1880), with varied success. Cluses specialized in metal turning, with a special focus on clock- and watchmaking, Anet on repairs. The others tried to provide skilled workers, qualified for hand-finishing parts and in the setting up of complex machines, as they were made to be adapted for the production of varied parts.

However, after 1890, the crisis generated first by the rise in competition, then by the closing down of new markets,[323] created a new set of issues for the workmen. In a need to lower wages, companies like Japy or Rolez tried to use cheaper labour, mostly children and women, and turned toward the production of interchangeable parts by more simple machines, hence following the American manufacturing model. This seemed to put technical schools in jeopardy. The Cluses school drifted away from clockmaking to precision turning after 1892 but demonstrated that training workers to respect standards and tolerances was better performed in a technical school than in home apprenticeships. It also showed that the skills acquired by the students could be employed in a wider circle of industries than just clock- and watchmaking.

Between 1850 and the First World War, major changes occurred in the organization of the trade. Department stores had appeared as early as 1825 in Paris, but their main expansion started after 1860. From then until 1900, twenty were built or rebuilt in Paris alone.[324] This changed the structure of retailing, leaving less room for clockmakers in the sale of clocks, and encouraging them to focus more on finishing and repair than on sales. The contact between the buyer and the clockmaker rapidly faded, the mechanical characteristics of the clock being less and less a topic for discussion. The next blow happened in 1885 when a new company was created in the centre of France, the 'Manufacture française d'armes de Saint Etienne', or 'Manufrance'. Initially established to trade in hunting guns, the company quickly diversified into bicycles, and then into all kinds of items, including clocks. It started mail-order selling in France, from a catalogue published every year (Figure 143). In 1906, the catalogue for clocks and watches, bronzes, and optical items displayed more than fifty models of alarm clocks and *pendules*, with warranties and a repair service. Choosing a clock was here clearly an aesthetic choice, made from a description, without seeing the item first.

Clocks and watches also became objects for prizes in raffles or charity events[325] or a tool to drag in new customers for other goods, as the marketing technique of pyramid selling exemplified. French as well as German industries tried to put an end to this practice, which devalued the objects sold and promoted low-quality production.[326] Following these changes in the market, large companies like Japy developed their production into fully finished clocks and the following step appeared logically to sell and ship them directly. After 1900, clockmakers tried to stop big companies increasingly selling directly to private customers and professional structures appeared, like local syndicates for clock, watch-, and jewellery makers, in each French department and advocates of their needs like the journal *La France Horlogère*, which started publication in 1901, aiming at creating a union for French clock and watchmaking.[327] More directly linked to jewellery shops, clockmakers' activity turned more and more to the finishing and repair of clocks made of interchangeable parts produced by specialized companies.

CONCLUSION

From the late eighteenth century to the First World War, French clockmaking contributed to a major change in the later industrial revolution. After a troubled period following 1789, the new republican government, followed by Bonaparte's First Empire, the country moved at high speed through the several phases leading to modern industrial power. Clockmaking contributed to this movement, either by the initiatives of a new type of entrepreneur, like

---

320 Saunier 1869 and Saunier 1870. They were followed in 1874 by a third volume, *Recueil des procédés pratiques usités en horlogerie: formant la deuxième partie du Guide-manuel de l'horloger*.
321 Saunier pleaded for the development of technical schools as early as 1841, as he published then a *Mémoires sur la possibilité de créer une école d'horlogerie sur de nouvelles bases, et avec le seul concours moral des autorités; proposition faite à la ville de Mâcon*.
322 Brevet & Chapiro 1981, 29–37.
323 Early twentieth century, troubles in South America and the Boer War seemed to have been a major issue for the survival of a company like Jules Rolez Ltd, which was wound down after 1909.

324 Marey 1979, 256.
325 *Manufrance* 1906, 410: clocks in bronze imitation were described as 'particularly fitted for gifts or prizes in shooting, military or sports competitions, as well as raffle and charity events.'
326 *La France horlogère* 1901, 2 and for the German reactions, 1909, 199, 24.
327 *La France horlogère* 1901, 1.

the Japy family, or by the support of central administration, as in Normandy, where the clockmaking industry was rescued so as to drag the area out of the crisis generated by the Revolution.

This led to a very dynamic period of innovation, mass-production of *pendules* being a field of exploration for industrial process and economic growth. This movement reached a climax between 1850 and 1870, illustrated in the International Exhibitions in London and Paris and by massive exports through the world. The Prussian invasion and the Commune in 1871 were the warning signals of a growing crisis, which could be observed in clockmaking through the struggles—and even the end—of some major actors of the century, while new categories of investors and industrialists started to develop. The *pendule* industry receded, weakened by strong international competition and social agitation among its workmen. Clockmaking, which started industrial production in the early nineteenth century, then became a branch of larger sectors, and the clockmakers, at first at the cutting edge of industrial expansion, were replaced by entrepreneurs coming from other areas, balancing clock production with others and trying to master the market by controlling competition more than producing better or new types of clocks. From inventing new machines and processes to manufacture movements in the early 1800s, to mass-producing interchangeable parts for a limited number of models in the early twentieth century, the French *pendule* companies did succeed, at some cost, in adapting to and resisting strong international competition and to changed values as a luxury object turned into an essential household item, or even a marketing gift.

CHAPTER FOURTEEN

PRECISION ATTAINED
THE NINETEENTH AND TWENTIETH CENTURIES

Jonathan Betts

By the turn of the eighteenth/nineteenth centuries, the development of precision timekeeping had reached a relatively stable plateau. By this time, the importance of detachment for an oscillator was clearly understood and the vast majority of portable precision timekeepers now employed the detached **detent escapement**. While efforts had been made to apply this improvement to fixed, land-based, precision pendulum clocks, the design had generally been found to be too difficult to work reliably, and too tricky to maintain, and the traditional **dead-beat** escapement, as established by George Graham *c.*1725, was still the norm for use in clocks for astronomical observatories and other scientific applications. In spite of regular attempts to improve the performance of precision pendulum clocks by enhancing the design of escapement and compensation pendulum by makers and designers, and in spite of encouragement from astronomers and scientific bodies such as the Board of Longitude and the Society of Arts in London, little real improvement was made. The Graham-type dead-beat escapement regulator would largely remain the standard internationally for much of the nineteenth century.

In some cases, when very carefully made, the dead-beat regulator was found to perform exceptionally well. A case in point is the fine regulator made by Thomas Earnshaw for the Armagh Observatory in 1789. The movement was provided with extensive jewelling to escapement and pivots and was fitted in a sealed dust-proof case to prevent the infection of the lubricants.[1] During the 1820s the Armagh Astronomer Thomas Robinson (1792–1882) was very impressed with the regularity of the clock, but claimed to have observed small variations in its timekeeping related to changes in barometric pressure. He therefore fitted small mercury tubes to the pendulum in an attempt to compensate for this effect.

This was the first example of such a compensation, if one ignores Harrison's incorporation of air-density compensation, integral to his pendulum system.[2]

Another eighteenth century development for precision timekeepers, the 'constant force' escapement, had been applied to precision pendulum clocks, with some success, but while designs continued to be developed and tested for most of the nineteenth century, these too were generally found to be too complex and difficult to maintain and were not generally made in any numbers. One of them, the **gravity escapement,** originally designed by Thomas Mudge, was developed further in the nineteenth century by a number of talented designers and makers, including James III Harrison (1767–1834) and James IV Harrison (1792–1875),[3] William Nicholson (1753–1815), Henry Kater (1777–1835), James Gowland (1835–80), and James Bloxam (1813–57) for use in precision pendulum clocks. Again, none of these designs ever saw large-scale production, but a development of Bloxam's gravity escapement by Edmund Beckett Denison, later Lord Grimthorpe, would prove highly successful when applied to turret clock movements, where variations in the original driving force were great owing to the exposed nature of the dial and hands of the clock. Grimthorpe's double three-legged gravity escapement (1851) was later fitted to the celebrated Great Clock at Westminster, familiarly known as 'Big Ben'. The escapement proved highly reliable and its timekeeping accurate, going on to be fitted to large numbers of turret clock movements, not only in Britain, but also in the USA and occasionally in Europe, until the middle of the twentieth century.

---

1 Betts 1989.

2 See the postscript to Chapter 11.
3 McKay 2010, 132.

The early nineteenth century saw the chronometer come of age in Britain and significant numbers of watchmakers, especially in London and Liverpool, turned their attention to the production of marine chronometers. There were, of course, still practical and talented watchmakers, as well as retailers, in Paris in the early nineteenth century, and exceptional figures like Jean-Antoine Lépine (1720–1814), Robert Robin (1742–99), and Abraham-Louis Breguet (1747–1823) had presided over a positive revolution in watch design during the second half of the eighteenth century. Breguet and his contemporary Antide Janvier (1751–1835) were foremost in creating pieces of high quality and technological novelty during the last quarter of the eighteenth century and the early nineteenth, a fruitful period for French horology. However, for most retail watchmakers, looking at the esoteric and literary example of Ferdinand Berthoud and his followers, making marine chronometers was perhaps also seen as a different profession and one which, without a government stipend, appeared risky when conventional watchmaking provided a good living. In England, by contrast, once London watchmakers saw what Earnshaw (in particular) was making in the mid-1790s, and especially after 1797 when his patent expired, they realized that these were instruments they could relatively easily produce themselves. By tapping into the established London watchmaking industry, dozens of independent specialist chronometer makers became established in the city by 1800 and, with London's reputation quickly established as the place to buy good and reasonably priced chronometers, the success of the English chronometer-making trade was guaranteed for most of the nineteenth century.

Into the nineteenth century, French production of chronometers remained both in the hands of a small number of specialists and, following Berthoud's example, the subject of scientific and theoretical improvement, rather than commercial technological simplification. For example, a typical Earnshaw chronometer movement and dial of the first quarter of the nineteenth century consists of 128 separate parts. The equivalent chronometer from the 1830s by the Paris maker Jean-François Henri Motel (1786–1859) has 270, every one of which has a more highly finished surface, representing hundreds of hours of work. State-funded, and with attention to every detail, chronometers made by Motel and his master Louis Berthoud were absolute masterpieces of the designer's and finisher's craft but, in both short- and long-term comparisons, they generally performed no better than Earnshaw's and it was inevitable that commercial pressure would see the French model simplified as the century progressed. For the most part, however, the English and French chronometer makers were not directly in competition and each satisfied the different requirements of their clientele.

The creation of precision pendulum clocks by the most celebrated French craftsmen at this period is marked by a similar difference from their English counterparts. Rarely were French precision pendulum clocks produced in a very simple form. Often with hour striking and sometimes with complex calendar and equation work, the buyers for such clocks were generally from a very different, much wealthier, bracket than their English counterparts. However, even with the simplest regulators produced by the top Paris makers, the style and proportions of case and dial, and the quality of their construction, were invariably first rate. From a technological point of view however, such precision regulators were rarely ground-breaking in principle, mostly employing a form of dead-beat escapement, usually either of the Amant pin-wheel type (with all pins on one side of the wheel) or of the Lepaute pin-wheel type (with pins on alternate sides), and almost always using the gridiron pendulum. An exception to this general rule is a small number of outstanding precision regulators made by Janvier and by Breguet. In a particularly spectacular design, first produced by Janvier and then by Breguet, two pendulums are used, one in front of the other, each driven by a separate movement and swinging in anti-phase with the other. A tour de force of design, the pendulums of these extraordinary clocks have the ability, when carefully adjusted, to remain in synchronization with each other, a dynamic stabilizing effect from the motion of each transmitted to the other through the medium of the pendulum support, causing their oscillations to remain linked. The result of this is that any slight instabilities of timekeeping within the two pendulums will be averaged out between them, improving the performance of the whole, and rendering the oscillating system generally less prone to external disturbances. Fine versions of these designs have been produced in more recent times by Maurice Blake (1980s), Stephan Gagneux & Beat Haldiman (2000), Frisch & Shauerte (2003), Derek Noakes (Buchanan 2006), and David Walter (2010).

In French chronometry, it was Abraham-Louis Breguet who conceived an entirely new plan for the movement. Breguet recognized the need to rationalize the design of the marine chronometer in order to make it less expensive, more robust, and more consistent in its performance. From the beginning of the nineteenth century he had been working on radical new ideas for the instrument and at the Paris Exhibition of 1819 he published a detailed description of a new model for a twin-barrel chronometer. Breguet had been working on this new *calibre* for several years, and it was clear that a great deal of rational thought had gone into the design. Although he died in 1823, his son was able to sell nearly a hundred and thirty chronometers to the French navy over the next ten years, despite the title *Horloger de la Marine Royale* going to Henri Motel.

While French chronometer making continued with a relatively small output, by 1800, the making of marine and pocket chronometers in London had become an industry in its own right. The centre for this trade focused in Clerkenwell, just to the north of the City of London. It had long been a centre for clock- and watchmaking: since the 1600s the trade had been gradually migrating from around Blackfriars on the Thames, north through Hatton Garden (still the centre of the London jewellery trade today), and into Clerkenwell, though the retail outlets were naturally concentrated in the City and West End of London.

The countryside north of Clerkenwell began to be engulfed by urban development in the early nineteenth century and many in watchmaking, finishing, and its allied trades continued migrating northwards into built-up Islington and Pentonville as the century progressed. For most of the century, however, the centre for those specializing in watch and chronometer work, and in the making of clocks and **regulators,** remained in Clerkenwell. The multitude of small suppliers such as spring makers, **balance** makers, hand makers, dial engravers, and box makers were all well established in the area, as were larger-scale clockmakers such as A & J Thwaites, who provided a practical clockmaking and clock repair service for many of the businesses purporting to be makers, but who were, in fact, retailing clocks made by others. Thwaites were able to produce clocks of a variety of qualities, but for the best-quality precision work, firms such as John George Holmden, also in Clerkenwell, would produce top-quality finished regulator movements for the trade. Sometimes such movements were made entirely in-house, but regulator frames and **trains** were also procured from Prescot in Lancashire (much in the same way as rough movements for chronometers were supplied from there for finishing by the London trade). Thomas Leyland was perhaps the most celebrated of the regulator movement makers from Lancashire. In this way, the majority of precision clocks retailed by famous names such as Vulliamy, Barraud, Dent, etc., were actually supplied from specialist outworkers, and were only cased up and finished (sometimes not even that) by the high-end retailer.

By the early 1820s the marine chronometer had effectively come of age: even the Board of Longitude and the Royal Navy finally began to accept that it was the wholly practicable solution they had been looking for. The instrument had become standard issue for all Royal Navy vessels sent on voyages of exploration, but there was also now a more regular demand for good chronometers, not just for scientific expeditions. As the quality of cartography improved (which chronometers, of course, assisted), so did the value for the navigator of being able to identify his position on the charts and to return to that position accurately and efficiently, should he need to. The East India Company, especially its officers, was keen to use chronometers for longitude determination on routine long-haul commercial voyages, and the Admiralty was by this time also issuing chronometers to most of the Navy's larger ships. By the middle of the century the use of chronometers on board ocean-going vessels was almost universal. This was despite the fact that, with the gradual introduction of steam propulsion, ocean passages were taking less time and the periods over which a navigator needed to establish his position out of sight of known lands was significantly diminishing.

One challenge faced by chronometer and precision watchmakers was the use of steel in the instrument, especially in such sensitive parts as the balance and balance spring. The deadly duo of rust and magnetism— either will spell disaster for performance at sea if they affect the steel parts of the balance and spring—were always on the minds of watch- and chronometer makers. Rust would obviously affect the mass of the balance itself, while rusting of the spring would have a huge effect on its elasticity and ruin the rate of a chronometer almost as soon as it had begun to appear. Magnetism was becoming a particular problem as ships made with iron or steel hulls became the norm, and the problem was exacerbated late in the century as ships began to incorporate heavy electrical machinery with powerful electromagnets. The presence of magnetic forces in close proximity to the balance and spring will modify and destabilize the frequency of the system, and if the steel in the balance and spring itself becomes magnetized, it will cause permanent instabilities in the rate of the chronometer. The problem had been understood for many years: it was specifically why both Thomas Mudge and John Arnold made their early marine-timekeeper balances of brass rather than steel, and Arnold was describing non-ferrous compensation balances as early as 1791.[4] In these balances he combined the brass in the rim segments with platina.[5] Platina was a platinum alloy sourced mostly in South America, as a side product of the gold-mining industry, which had few practical uses at the time. It is indicative of Arnold's broad and enquiring mind that he should have searched out the material and applied it in balances for this purpose. In the early nineteenth century, platina would be separated into its constituent parts and one of the products, palladium, would eventually find a real use in the chronometer industry.

From the late 1770s, Arnold, and later his son John Roger, used balance springs made of hard-drawn gold, the resistance to oxidation being a critical element in that choice. However, even when drawn, gold springs are relatively soft and, having found that their elasticity was non-permanent, they were not used much after Arnold senior's death. The talented and eccentric chronometer maker J. G. Ulrich (1795–1875), who, incidentally, claimed to have observed middle-temperature error (MTE) years before anyone else but was unable to prove it, was one of a number of makers to produce completely non-ferrous balances in the nineteenth century to avoid these problems. In France, Abraham-Louis Breguet experimented with making balance springs of glass in the early nineteenth century, and in the 1820s they were made by the Glasgow maker James Scrymgeour (fl. 1816–37). In the 1830s, balances of the kind were also tried successfully by the firm of Arnold and Dent, and the results published in the *Nautical Magazine* in December 1836.[6] Then, in 1853, a paper was read to the Society of Arts in London by F. H. Wenham (1824–1908) on the making of chronometer balance springs in glass.[7] The astronomer the Reverend George Fisher (1794–1873) was one of the first to attempt to explain and quantify the effects of magnetism on chronometers, publishing his findings in 1820.[8]

---

4 Arnold 1791, vii.
5 For earlier work by William Lewis (1708–81) on the gold–platinum alloy communicated to the Royal Society and which Arnold could have drawn, see McDonald & Hunt 1982, 37–40.
6 Arnold & Dent 1833, 1836.
7 Wenham 1853. F. H, 'On a method of constructing Glass Balance Springs and their application to Timekeepers', *Journal of the Society of Arts*, 3 June 1853, 325–7.
8 Fisher 1820. Fisher, as headmaster from 1834 and later overall Principal of the Greenwich Hospital Schools, was a significant figure in British navigational training.

This work led to criticism and alternative theories from other commentators, including the highly experienced navigator and scientist William Scoresby (1789–1857) and, a few years later, from the American chronometer maker William Cranch Bond.[9] The noted chronometer maker E. J. Dent (1790–1853), while still in partnership with John Roger Arnold, patented the idea of coating balance springs with varnish to protect them, and in 1840 took out another patent that included the idea of gilding a steel spring to protect it.[10] Unfortunately for their inventors, neither idea proved viable as the coatings introduced instabilities and inconsistencies in the longer-term characteristics of the springs.

With the chronometer now established as the practical means of finding longitude at sea, increasing numbers of these instruments were being purchased and issued to ships by the British Admiralty. Up until 1820 the general issue of the Royal Navy's chronometers had been overseen by the Hydrographer of the Navy but in 1821 it was decided a more official arrangement for their management was needed. In that year, John Pond, the sixth Astronomer Royal, was put in charge of Navy chronometers and from then onwards all purchasing, issuing, and managing of government chronometers was undertaken at the Royal Observatory, Greenwich. From then, too, the issue and receipt of chronometers at the Observatory was recorded in ledgers, a system which continued for over a hundred years until the mid-1930s when a card-index system superseded them. If a chronometer were purchased by the Admiralty and became the responsibility of the Observatory's chronometer department, it would be described as 'taken on charge'. The ledgers and cards survive at the Observatory and are a rich source of information about the service life of most of the marine chronometers used by the Royal Navy. At the suggestion of the chronometer maker, W. J. Frodsham, the Premium Trials were introduced at Greenwich in 1822, the first of their kind anywhere in the world. Seen as a way of encouraging chronometer makers to improve the performance of their instruments, the trials offered premiums, i.e. enhanced prices paid by the Admiralty to purchase the chronometer, of £300 and £200 to the best two submitted. A third premium (£100) was introduced in 1829. With the marine chronometer now so well settled into the equipment of the Royal Navy, and the *Nautical Almanac* published annually as a means of finding longitude by lunars as a supplement to it (and for use in determining local time with the chronometer method too), 1828 finally saw the Board of Longitude disbanded.

In order for a chronometer to be useful it must be possible to determine the 'home-time' (in the case of Britain, Greenwich Mean Time) from it. No chronometer will keep absolutely perfect time, and for accurate navigation it is vital to know its *error* (by how much the time it shows differs from GMT at that particular moment) and its *rate* (by how much that error will change each day). To this end, as a further support for the process of using chronometers, in 1833 John Pond arranged for the Greenwich *time ball* to be mounted on a mast on the roof of the Observatory. One of the first public time signals in the world, the time ball was a means of providing accurate Greenwich Mean Time (GMT) on a daily basis to all shipping in the London docks and those passing on the river below. One comparison with the ball will determine the chronometer's error, and daily comparisons will then provide the chronometer's rate.

The time ball still operates to this day at the Observatory, rising half way up the mast at 12:55 p.m. to signal that the one o'clock signal is soon to occur. At 12:58 p.m., it rises to the top of the mast, and at precisely 1 p.m. it drops back to the bottom of the mast, the instant of its beginning to move marking the instant of one o'clock. The more logical time of 12.00 noon was avoided because the astronomers were too busy observing the noon Sun crossing the meridian—effectively 'finding the time' at that point. In subsequent years Britain and many other countries adopted the idea of using time balls and time signals for setting chronometers at many other ports across the globe.

Liverpool, as it developed into a major British port at the beginning of the nineteenth century, became a rival centre to London for the finishing of watches and marine and pocket chronometers. The Prescot movement makers were close to the city and, with the new and very vibrant sales of chronometers to the growing merchant fleets in the Liverpool docks, it was very natural that a thriving horological industry would soon appear. The first half of the century was a heyday for the Liverpool makers, mostly those firms that were also in the much-larger business of producing watches for export. In 1843 a new observatory was built at Liverpool to service the growing need for chronometer rating and testing, and to provide a source of correct Greenwich time for setting the instruments. The first Director was the talented astronomer John Hartnup senior (1806–85), who had served under George Airy, Astronomer Royal from 1835, and who would continue to liaise with him in the following years. It was Hartnup at Liverpool who first designed and introduced an oven specially for testing chronometers in different temperatures (a design then copied by Airy for Greenwich in 1850- see Figure 148). Companies like Robert Roskell, and Litherland, Davies & Company, had their roots in Liverpool watchmaking from well before 1800 and, once Earnshaw's spring-detent design was introduced, were able to adapt to producing the new Earnshaw-type chronometer movements. In fact, there is evidence that Roskell in particular already had strong links with Earnshaw's business in London and may well have acted as intermediary for him with the Prescot movement-making trade. There was, no doubt, a considerable amount of trading and exchanges of material and services between London and Liverpool during the eighteenth and nineteenth centuries.

The Board of Longitude being disbanded, encouragement for further improvements to the chronometer became the business of the Admiralty and the Royal Observatory. By 1835, however, the

---

9 Barlow 1821; Scoresby 1823; Bond 1833.
10 British patent N° 7,067; British patent N° 8,625 discussed in Mercer 1977, 46–7; 112–14.

Premium Trials were failing in their purpose. Because the premiums were simply competitive and awarded to the chronometer that performed best without tight specifications for performance, there was little incentive for real improvement. The trials were thus ended for the time being, though notices that were sent round to chronometer makers encouraging them to keep trying to improve their products stated that government reward money might be offered for improved designs.

By this time, a new and altogether more dynamic (seventh) Astronomer Royal was in place: George Biddell Airy (1801–92). Airy was a veritable force to be reckoned with—a first-rate mathematician, astronomer, and administrator. His rule of the Royal Observatory between 1835 and 1881 was absolute and more was achieved in that time than in almost any other period in its history. Airy had a particular interest in precision timekeeping and applied his mathematical brain to the theory of the pendulum. In 1826, while still Plumian Professor at Trinity College, Cambridge, he wrote his now famous treatise *On the Disturbances of Pendulums and Balances, and on the Theory of Escapements*,[11] in which, by mathematical analysis, he determines the effects on the rate of a given oscillator by the presence of the most common escapements. In simple terms, he proves mathematically that any impulse delivered to a swinging pendulum or balance before its position of rest will tend to increase the rate of the oscillator, and any impulse delivered after the position of rest will decrease its rate, to the same extent. Equally, any subtraction of energy (for example, friction or physical obstruction) experienced by the oscillator before its position of rest will tend to decrease the rate, and any subtraction of energy or physical obstruction after the position of rest will increase its rate. Known today as 'escapement error', these effects are not really errors, but merely physical properties of the various escapements and, to a greater or lesser extent, are a constant, being corrected for in a one-off adjustment when the clock is regulated. Airy, however, concluded that those escapements with larger escapement errors, such as recoil escapements like the anchor and verge, are likely to be more unstable as timekeepers, and those with smaller escapement errors, such as the dead-beat and detached escapements like the detent chronometer and lever escapements, will be better. Practical experience demonstrates that this is partly true, although it is not by any means as clear-cut as Airy's theory suggests. For example, experiments have shown that the real difference in practice between the dead-beat escapement and an anchor escapement with reasonably small recoil is almost nothing—far from the situation predicted by Airy's treatise. Nevertheless, following his calculations, Airy proposed that the ideal precision pendulum clock ought to be one with a detached detent escapement, and he published a description and drawing of such a design in the treatise (Figure 147).

A small number of these escapements were then made by E. J. Dent in small half-second pendulum regulators during the 1840s, and Dent's would go on to create Airy's ultimate version of this form in the regulator No. 1906, made for the Royal Observatory in 1870. Such was the good performance of this clock that Airy was able to observe variations caused by barometric pressure changes, and the clock was soon enhanced with a mercury tube barometric compensator, echoing the practice of Thomas Robinson at Armagh half a century before.

New annual chronometer trials were introduced by Airy at the Royal Observatory in 1840 without premium prices being offered, but by this time sales to the Navy were virtually guaranteed for successful chronometers and with them came the much-prized right to describe the company as 'Maker to the Admiralty'. Indeed, there is evidence that, over the years, a number of retailers commissioned first-rate chronometers from top manufacturers, but with the retailer's name on, for submitting to Greenwich specifically with the aim of making a sale in order to use that accolade in subsequent advertising.

In 1835 the chronometer maker John Sweetman Eiffe informed Airy that he had detected an insufficiency in the compensation provided by an ordinary compensation balance. He found that, when fine-tuning the compensation, it could be made correct for temperatures at the extremes but that the chronometer was always then too fast at the middle (ambient) temperature. What Eiffe had observed was what became known as middle temperature error (MTE), a problem that needed a solution if makers were further to improve the performance of marine chronometers. In fact, this phenomenon had been remarked on, though not understood, by the chronometer makers Arnold & Dent a couple of years before in a letter published in the *Nautical Magazine*, when they attributed the problem to a 'want of affinity existing between the metallic particles which compose the balance-spring'. They go on to say, significantly, that 'This is a subject which is well worth the attention of the scientific chronometer maker, since it involves the existence of difficulties in the most important part of the adjustments of chronometers'.[12] Indeed it was, and it did receive a great deal of attention over the next half a century, some most ingenious and elegant compensation balances being created in attempts to overcome the problem.

The fundamental reasons for MTE have been much misunderstood in the past and many published explanations have been misleading. With an uncompensated balance-controlled timekeeper, when the temperature rises the balance oscillates more slowly owing to the balance arms having expanded (greater radius of gyration and hence a greater moment of inertia in the balance), and chiefly because the balance spring becomes weaker (lower modulus of elasticity). The compensation balance, with bi-metallic rims, is intended to compensate for this in the following way. If the temperature rises, the rim segments move the weights on the balance rims in closer to the centre. This counteracts the effect of the expanding arms and reduces the moment of

---

11 Airy 1826.

12 Arnold and Dent 1833–1836, 224.

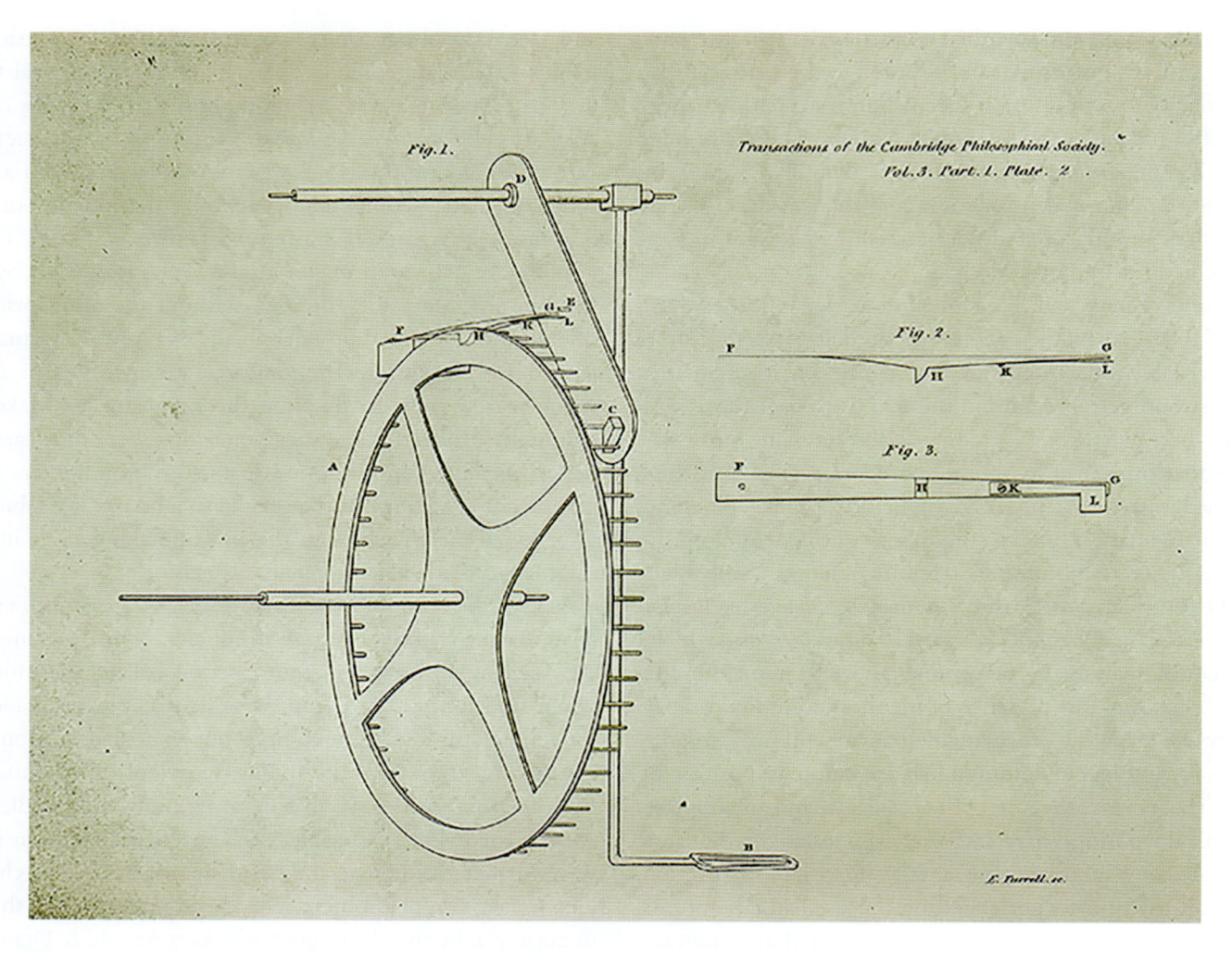

Figure 147 George Airy's 1826 design for a detached detent escapement for a precision regulator. Image: Jonathan Betts

inertia of the balance still further to compensate for the effect of the weakening spring. Looking at the action of the balance, as the temperature steadily rises, the rim segments produce an approximately linear progression of the compensation weights in towards the centre of the balance. This means that for a specific change in temperature, the radius of gyration of the balance will also change a specific amount: a linear change in temperature produces a linear change in radius of gyration. A linear change in the radius of gyration of the balance does not produce a linear change in the moment of inertia, as the moment of inertia of the balance changes as the square of the radius of gyration (it is a quadratic function). However, *the effect on the timekeeping* is inversely proportional to that change, and thus the effect of the compensation in the balance itself is a linear one.

It is generally believed that the problem of MTE occurs as a result of the other factor in the equation—the changing elasticity of the balance spring. The theory says that as elasticity itself changes in a linear fashion as the temperature changes, then the *effect* of that linear change on the period of the oscillator is as a quadratic function—the period of the oscillator changes as the square of the restoring force—the changing period is non-linear and MTE results. In recent years, however, it has been recognized that there are other factors at work in addition to the non-linearity of restoring force. The most recent analysis[13] suggests the explanation above is only partly correct and that by far the largest element in MTE comes from a non-linearity in the changing elasticity of the balance spring itself. Woodward cites the Swiss physicist and metallurgist Charles-Edouard Guillaume (1861–1938), who stated that 'Amongst several things which have been shown to contribute to the overall effect, one is certainly predominant; it is the very form of the variation of elasticity of steel with temperature, which has an important quadratic term represented graphically by a marked curved line'.[14]

Whatever the cause, all are agreed that the variation exists in ordinary compensation balances associated with carbon-steel balance springs. It was also realized that one way to compensate for this change would be to have a balance that moved its weights inwards with increasing rapidity as the temperature rises, and the opposite as it falls. Eiffe's note to the Astronomer Royal was to be just the beginning of a whole new direction in chronometer development during the next fifty years. Two of Eiffe's chronometers, with a newly designed balance intended to compensate for

---

13 Woodward 2011.
14 Woodward 2011.

Figure 148 The chronometer ovens at the Royal Observatory for testing the rates of chronometers at tropical temperatures. Photo: B.T. Archives

MTE, were tested by Airy at the Observatory in 1836 and performed well. In late 1839 or early 1840, Airy was approached by Robert Molyneux, who informed him he had invented a balance for MTE compensation, which Airy realized was virtually the same as Eiffe's. Molyneux took out a patent on his design, while Eiffe received Admiralty reward money of £300.

Now that chronometers were being produced that aimed at more subtle corrections for temperature, Airy decided to include tests over a range of temperatures. In January 1841, as part of this new aspect to the trials, he commissioned from Robert Molyneux a new and very special kind of chronometer, the *chronometrical thermometer*.[15] The chronometrical thermometer was designed to do exactly the opposite of a standard chronometer when subjected to temperature change. The compensation balance had the two elements of its bi-metal rim segments reversed, so that instead of compensating for temperature changes, the bi-metal rims caused the balance to respond to temperature change in a highly exaggerated manner. Airy felt that the resulting rate from this instrument would provide a much more complete, holistic measure of the potential effects of temperature change on a chronometer under those conditions. The intention was to maintain the chronometrical thermometer running alongside all chronometers under test and keep a record of its performance as an adjunct to the record of the temperature during the trial. The instrument was considered a great success and was employed in just about every trial at Greenwich thereafter.

Keeping the chronometer makers informed of the new requirement for temperature testing, Airy circulated a notice, dated 26 November 1841, to the effect that chronometers would, at his discretion, be tested 'through a variety of temperatures, from the lowest that can be obtained without artificial means, up to that of 100° Fahrenheit.' News of Eiffe and Molyneux's successes appears to have inspired considerable interest, and from the 1840s a succession of new designs for compensation balances appear. Success was varied, but after about 1860 the performance of chronometers with these balances in the Greenwich trials steadily improved.

Another factor that can greatly affect the rate of a chronometer with a conventional compensation balance is centrifugal force acting on the balance rims when it is oscillating. Under its influence, when the amplitude of the swings of a balance increase, there will be an increased tendency for the free ends of the rim segments of the balance to be thrown outwards. This means that systems subject to the effect would be much less isochronous, tending to

---

15 Royal Museums Greenwich, ref: ZAA0126. Betts 2017, 352–4.

run considerably slower in the larger arcs. The phenomenon had little exposure in the horological literature until Victor Kullberg published a clear explanation showing that it was a very significant factor in balances with thin bimetal rim segments.[16]

Although not a subject with a high profile in early writing on chronometry, it was understood by a few makers from that period and efforts were made to reduce the effect. John Arnold was probably the first to create such a balance to compensate for it by using a white-metal alloy as the outer element, with steel as the inner one, in his bi-metallic balances. These balances have sometimes been confused with Arnold's non-ferrous type referred to earlier, but those are quite different in their function. The white-metal alloy was harder and consequently stiffer than brass and had a greater resistance to flexure when the balance was oscillating. In all probability Arnold experimented with several such alloys for both non-ferrous and for 'anti-flexure' rims. One alloy that John Roger Arnold described to Breguet in 1796 had two parts silver to one part zinc. It was probably intended for one of the non-ferrous types, in conjunction with brass as the other element in the bi-metal.[17] The alloy used for one of Arnold's 'anti-flexure' rims, in the balance on Arnold No. 356 of about 1810, was approximately fifty-five per cent copper, twenty-five per cent silver, fifteen per cent zinc, three per cent tin, and two per cent lead.

While English chronometers were still the favourite of the world's navigators, and their makers justly proud of their position in the horological pecking order, things were changing rapidly elsewhere. Since the eighteenth century the Swiss had been pre-eminent in the large-scale production of watches to supply an increasing demand for medium-quality models from Europe's 'blue-collar' classes. The Swiss had evolved a finely subdivided and highly organized system of labour, which enabled watches to be produced inexpensively and in large quantities. It was not exactly mass-production, but during the 1830s, the Swiss innovator Pierre-Frédéric Ingold (1787–1878) had tried to introduce machine production into the watchmaking industry in France. Having failed in that enterprise, in the early 1840s Ingold attempted the same in London. However, in spite of the support of several well-known watch- and chronometer makers (including Thomas Earnshaw junior), there was great suspicion and opposition from the trade and the enterprise also failed in England. Determined not to give up, Ingold then took his revolutionary ideas to the USA. While also rejected there, American mass-production nevertheless started soon afterwards, undoubtedly inspired by Ingold's vision, with successful and inexpensive machine-made watches starting to be produced in the 1850s. Alarm bells should, by then, have been ringing in the corridors of the English horological industry.

One far-sighted Englishman tried to alert the traditional Clerkenwell makers to the commercial threat these inexpensive products posed. John Bennett, a highly successful watch and clock retailer and importer, warned the trade at several public meetings in Clerkenwell in the late 1850s that retailers like him would have to continue buying foreign produce if the British would not tool-up and make cheaper watches and clocks, but he received only criticism for his remarks. In fact, by the middle of the nineteenth century most of the high-end retail trade in London were buying clocks and watches from France and Germany, and it should have been clear to all what the market wanted.

In some ways, perhaps, the chronometer-making industry played a negative role in this blind refusal of English horology to face the reality of the future. By comparison with the commercial production and sales of pocket-watches, the business of making marine chronometers was tiny. Nevertheless, it was very prestigious and high profile, and its end-product—the 'aristocrat' of the nation's entire horological output—was the one most preferred by other nations' shipping. It is possible that the enduring success story of this small part of the English horological industry gave watchmakers an unwarranted confidence and inspired them to emulate and maintain a standard which was simply unsustainable in the wider watchmaking world. A similar phenomenon was occurring in clock sales in the second half of the nineteenth century, with American and German manufacturers mass-producing inexpensive and successful domestic clocks, which were imported into Britain and sold in great quantity in the growing market catering for small households.

For these, and other reasons, the English clock- and watchmaking trades mostly refused to adapt to the new mass-production techniques practised in America, Switzerland, and Germany. The high-quality, labour-intensive hand-made watches were what the English trade was best at and it would not 'lower its standards' to make cheaper products. The imports were, of course, seen as a threat and many attempts were made over the years to introduce embargoes and tariffs on mass-produced watches from Switzerland. This pattern of resistance to change—in methods and attitudes—would become a familiar one in Britain's decline from industrial power.

Clerkenwell clock- and watchmakers many times petitioned London's Clockmakers' Company, demanding that it put pressure on the government to ban low-value foreign imports, but to no avail. As a result, in 1858, they formed the British Horological Institute (BHI) as a 'union' body, one of whose aims was to defend their trade against foreign imports. They might as well have tried to stop the Sun rising: for in turning their backs on necessary progress they effectively signed the death knell of the English horological industry, which went into inexorable decline thereafter. While high-quality, expensive products continued to be made for a steadily diminishing luxury sector, the market for less-expensive watches was overwhelmed by the competition from America and Switzerland.

The marine chronometer, and the high-precision astronomical regulator, however, were an exception to this tendency. There was no mass market for them, and it was only by mass-producing that makers would have been able to reduce significantly the price of a good chronometer or regulator. Being essential scientific

---

16 Kullberg 1887.
17 Mercer 1977, 303.

instruments, pricing was also less competitive than in the general trade of pocket-watches and clocks. So the BHI, based in Northampton Square close to the heart of Clerkenwell, was soon also the meeting place for Britain's chronometer and precision clockmakers, some of the greatest names not only being members but serving as officers on its Council over the years. As the Institute from its inception published its monthly magazine *The Horological Journal*, it has also provided us with the most wonderful record of the activities of those craftsmen, as well as their opinions and (occasionally) their practices.

From the mid-nineteenth century, in parallel with a steady development in temperature compensation, marine chronometers entered a period of consolidation and proliferation. By the end of the 1850s one can reasonably say that the design of the English chronometer, in both movement and box, was settled, and that a 'standard form' established, which was replicated in the following eighty years. Pocket chronometers, which since the 1780s had been used by some navigators, were by now mostly used as 'deck watches' for carrying the time up on deck, or as 'hack watches' for carrying the time from ship to shore. Nevertheless, well-made pocket chronometers of the late eighteenth and early nineteenth centuries were often found to have plenty of life left in them and were often converted to small box chronometers.

Industrial exhibitions, regular events in France during the first half of the century and then internationalized following the Great Exhibition in London in 1851, proved to be a valuable forum for chronometer and precision clockmakers. A number of new inventions made their first appearance at these 'trade shows' during the 1800s, and those lucky enough to be awarded prize medals invariably proclaimed the distinction proudly on the dials of their products. The French had taken the earliest initiative in organizing such exhibitions and, once other countries began to host similar shows, they invariably contributed, playing an important and valuable role in revealing to the world the results of their practice and industry. Similarly, in scientific and technological publication (and in stark contrast to the long tradition in English horology for keeping trade secrets), the French view was wholly outward-looking. Following the great, liberal tradition begun by Henry Sully, and continued through the eighteenth century by Thiout, Lepaute, and Berthoud, French nineteenth-century practitioners engaged in an ongoing scientific study of the principles of horological design, codifying and publishing their results for the benefit of all. Today, we are fortunate to have inherited a very considerable horological literature published in France in the nineteenth century, both from official sources and independent ones. The publication, for example, of *Recherches sur les Chronomètres* by the French Hydrographic Service in a series of volumes between 1859 and 1895 saw a considerable amount of important research made available to the international horological and scientific community.[18] Independent journals such as *La Tribune Chronométrique*, launched in 1852 and edited by Pierre Dubois, saw much valuable information circulated, as did the *Revue Chronométrique*, which began with Claudius Saunier as editor in 1855. This, which became the official organ of the Société des Horlogers the following year, is comparable to the BHI's *Horological Journal* (published from 1858).

The difference between the approach to chronometer making on either side of the English Channel was marked. While many French horologists wished to discover and explain why some designs worked better than others, and took a holistic view of their products, the equivalent English craftsman, usually more focused on a particular specialization and happy to put out more of the work to other specialists, was generally content that those designs did work. He was perhaps more concerned with producing a reliable movement for profit and would be just as likely to keep the discovery of improvements to himself to ensure commercial gain as he might be to share it for the greater good. An example of the difference between the two philosophies is shown in the choice of tooth forms for wheel trains of chronometers and regulators. On the one hand, correct 'cycloidal' tooth forms for perfectly smooth horological gearing were calculated and published in France as far back as 1733 by the distinguished mathematician Charles Camus (1699–1768), and many later French publications in the nineteenth century reiterated and expanded on the theory.[19] On the other hand, in England as late as 1903, one finds the chronometer makers Johannsen's specifying a tooth form 'more like an ash leaf' as sufficient for their needs.[20] Similarly, while in France the theoretically correct forms of terminal curve on balance springs were calculated and illustrated by Edouard Phillips in 1864, and the effects of different points of attachment of a balance spring are carefully codified and published by Louis Lossier in 1890, most English chronometer makers had found the optimum point of attachment and terminal curve empirically and applied their own rule, without feeling the need to understand why it worked, nor to share their knowledge with anyone else.[21]

Nineteenth-century chronometer and precision clockmaking in Europe was of course not confined to Britain and France: other European countries had begun to develop their own traditions by the late eighteenth century. Since the middle of the previous century the Jürgensen family had made their name as fine watchmakers in Copenhagen, and the 1800s saw a great dynasty of Jürgensen chronometer makers in the city. The most celebrated perhaps was Urban Jürgensen (1776–1830), who had studied in Switzerland but who was also inspired by the work of Breguet and Arnold, both of whom he had visited at various times. There were one or two isolated examples of pioneering eighteenth-century German chronometer designers, too, such as Johann Georg Thiele of Bremen. Following directly after Harrison's fourth marine timekeeper, Thiele's design was complex, highly original, and well-made, but did not perform well when sent to Maskelyne to be tried at Greenwich. The most notable of the pioneering

---

18 Service Hydrographique de la Marine, *Recherches sur les Chronomètres*, Paris, 1859–95.

19 Camus 1735 but the paper was read to the Académie in 1733.
20 Betts 2017, 81.
21 Phillips 1864; Lossier, 1890.

chronometer makers in Germany was Heinrich Johann Kessels (1781–1849), who studied with a 'celebrated' chronometer maker (identity uncertain) in London and with Breguet in Paris. From the early 1820s Kessels worked in the Hamburg area, and while not producing the same quantity of some of his London contemporaries, turned out very fine pocket and marine chronometers, and precision regulators. He was an early contributor to an important German tradition of chronometer making, which developed during the nineteenth century, and trained a number of students who went on to be fine makers in their own right, including Andreas Hohwu (1803–85) and Joseph Thaddeus Winnerl (1799–1886). On Kessels' death, his business was taken on by Friedrich Moritz Krille (1817–63), said to have been working with him for nine years previously, and who continued his practice of fine chronometer and clockmaking. Krille's successor, in turn, was another fine craftsman, Theodor Knoblich (1827–92).

However, by the second half of the century, at the time Knoblich and his contemporaries in Germany became established, it was increasingly the norm to obtain whole chronometers, or at least partly finished movements and boxes, from London. Only Adolph August Kittel of Altona still produced his instruments himself.[22] At this period the German Hydrographical Institute was established in Hamburg, very much along the lines of the Royal Observatory, Greenwich, and began testing chronometers for the German navy. At the end of the century this aspect of the German chronometer trade was to change, and partly inspired by the growing arms race at that time, a new and exclusively German instrument would be developed. The celebrated firm of Lange, in Glashütte, would play a significant role in that story.

In the second half of the nineteenth century, while chronometer production in Europe moved steadily towards the English model, regulator production by many of the celebrated London makers began to incorporate productions from France. Thomas Leyland, the finest of the regulator movement makers in Prescot, died in 1861, and from that point it seems likely that London firms like Dent increasingly bought movements from companies like Jacob in Paris for finishing and retailing under their own name in London. Mercer cites a letter written in August 1839 from E. J. Dent to a Paris correspondent asking him to speak to 'Monsieur Jacob' about the prices for the supply of jewelled movements and plain movements for regulators.[23] This was almost certainly Jean-Aimé Jacob (1793–1871), who founded a firm in St Nicholas d'Aliermont making chronometers and regulators.[24] Following the closure of Dent's Pall Mall premises in the 1970s a number of unfinished imported regulator movements, said to have come from the firm, appeared in London. Still packed in straw, these regulator movements bore mock dials made of card printed with a conventional clock dial and hands on the front (perhaps to evade import tax of some kind on scientific instruments?)

and printed with the name Jacob. French regulators signed by Jacob himself with movements identical to the type produced by Dents during the 1870s and 1880s suggest that the majority of Dent's regulators (Figure 149) made at this time were actually of French origin. Top retailer/manufacturers such as Charles Frodsham were also selling precision clocks with entirely French-made movements in the second half of the nineteenth century.

A European development that enjoyed widespread adoption in the chronometer industry was the creation of new alloys of palladium for balance-spring material. Palladium is an element found in alloy form with platinum. In the very early 1800s William Hyde Wollaston (1766–1828) and Smithson Tennant (1761–1815), two English chemists, analysed the alloy. Between them, by 1804, they had isolated five separate metals from it: palladium, rhodium, iridium, osmium, and platinum. Wollaston's interest was principally in the soft, silver-white metal platinum, not only as a focus for his research but also for its commercial applications. Throughout the first quarter of the century Wollaston sold considerable quantities of the metal to Edward Troughton (c.1753–1836), the top London instrument maker, for the scales of his instruments.[25] Palladium is similar in appearance to platinum and is highly resistant to oxidation, but it is less dense and has a lower melting point. Pure palladium is also quite soft, suffering from the same problem as gold when used as a balance spring material, and experiments by E. J. Dent using palladium in this application led him to abandon the idea.

Nevertheless, when alloyed with other metals and hard drawn to provide an enhanced elasticity, palladium offers a much more permanent option for a non-oxidizing, non-magnetic spring material. However, it was not until the 1870s that the Swiss watchmaker Charles-Auguste Paillard (1840–95) developed the idea of using the metal for making balance springs. Paillard had experienced problems with the rusting of steel balance springs while working as a watchmaker with his uncle in Brazil, and he began to look for alternative materials for these and for non-magnetic balances. On his return to Switzerland in the 1860s he continued his research, and by 1877 had developed an alloy of palladium as an alternative. The new spring material was described simply as palladium but Paillard produced, and patented, several alloys of that name and the palladium itself was in fact alloyed with a variety of other metals. The alloys for use in non-magnetic balances combine palladium with copper and iron, while alloys for use in spring material also include silver, gold, platinum, and nickel, with steel substituting for iron.[26] The creation of these new materials was announced in the *Horological Journal* in July 1879 and, in spite of a few negative opinions expressed in its correspondence pages, the use of Paillard's new palladium alloy for making balance springs was soon adopted by some of the top London chronometer makers. Kullberg's, for example, adopted it almost at once and hardly ever used steel again for their chronometer balance springs. Not

---

22 Oestman 2012.
23 Mercer 1977, 86–7.
24 Roberts 2003a, 81; Roberts 2003b, 178–84.

25 Chaldecott, 1987, 91–100.
26 Britten 1929, 110–15.

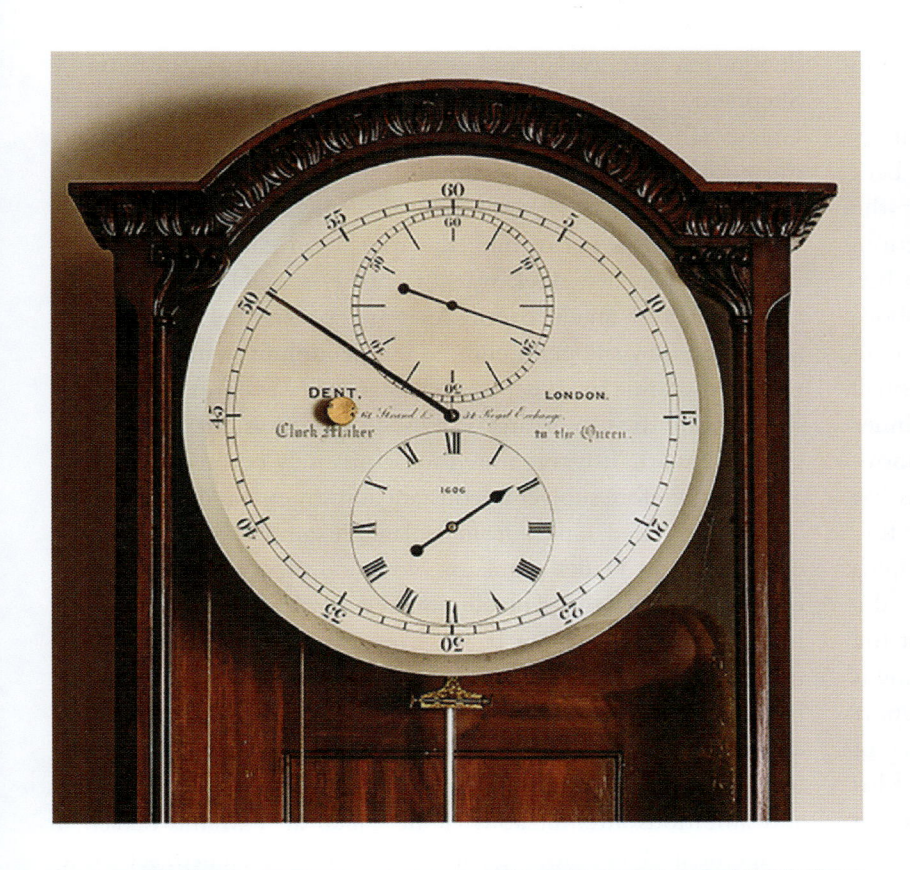

Figure 149 A typical regulator by Dent, made during the second half of the nineteenth century. These used movements supplied from Jacob in France. Photos: Tobias Birch and Jonathan Flower.

only was the spring material almost completely non-magnetic and resistant to oxidation, but it was also less prone to acceleration in new instruments and less subject to MTE.

A feature that appeared on the balances of British Admiralty chronometers in the 1870s was a device known as 'Airy's bar'. Invented by Airy, and intended as a fine adjustment for the ordinary compensation, Airy's bar was not an MTE correction, as is sometimes believed. It comprised a brass or steel cross-bar, friction-mounted on the balance staff, within the balance, above the arms but below the balance spring. At the ends of the bar are fixed curved steel blades, having attached at their other ends small brass subsidiary compensation weights that bear against the inner sides of the bi-metal rims, gently pressing against them and moving with them as the temperature changes. Turning the cross-bar on its friction mounting will swing these subsidiary weights further round towards, or away from, the main compensation weights and the end of the bi-metal rims. This increases or decreases the amount of compensation but should not significantly affect the radius of gyration, and the balance should retain its adjustment for mean-time. Airy describes the device in his Annual Report of 1875, mentioning that he had introduced it to a number of chronometers, and it was described by William Ellis, one of his assistants, in the July issue of the *Horological Journal* that year. In his Annual Report in 1876 Airy stated that the device had proved successful and that 'in future all chronometers sent in to the annual trial are to be so fitted'. In fact, not all (but about three-quarters) of the chronometers in the 1877 trial were fitted with it. However, many makers felt it was unnecessary and resented the stipulation, as it was more work to do; besides which, many of the more successful designs of balance were incompatible with it. In his 1877 Annual Report Airy remarks that 'The supplementary compensation, to which allusion was made in the last two Reports, has now been repeatedly tried and I think I may say that in the makers' hands it is successful', but adds: 'For the convenience of makers, it was judged prudent in the competitive trial now advancing, not to insist on the application of this apparatus to every trial-chronometer, but to intimate that the Government would be unwilling to purchase chronometers not fitted with it.' In fact, the appearance of the device gradually diminished in the trials over the next four years and, as long as the performance of the chronometer was good enough, the absence of it never seems to have been a serious consideration when the Royal Observatory made purchases for the Navy thereafter.

In spite of the need for improved accuracy in astronomical regulators, and a variety of different new designs of escapement and pendulum proposed during the first half of the nineteenth century, these either failed to perform or were simply too difficult to maintain, and the dead-beat regulator remained the standard through most of the century. However, an important creation for the Royal Observatory, and first seen at the Great Exhibition in London in 1851, was Charles Shepherd's new patent electric clock system, the first electric controlling and secondary clock system. (Shepherd was rightly challenged by the Scottish electric clock pioneer Alexander Bain, who had patented his electric clock designs ten years earlier and certainly had priority of invention.) Shepherd's system was purchased by the Astronomer Royal at Greenwich and would serve for the rest of the century as Britain's mean solar time standard, sending electric time signals throughout Britain for over sixty years, using the railway system's telegraph lines. As well as performing the role of an electric controlling clock, the primary clock was, of course, a regulator in its own right, and, with an electrically reset gravity arm, held by a detent, and released by the pendulum, as the pendulum swung by, it was potentially a fine timekeeper. However, as it was controlling many dials around the Observatory and sending signals out on the telegraph network, it was required to be precisely correct at all times and was therefore subject to constant adjustments and corrections.

As mentioned, apart from isolated examples such as Airy's escapement in Dent's regulator No. 1906 of 1870, the detached escapement was not generally applied to clockwork during the nineteenth century—not because more accurate regulators were not needed or desired by astronomers or scientists, but no doubt because of difficulties with the reliability and fragility of such escapements. Early in the century, at the Royal Military Academy in Woolwich, William Congreve had made a hugely complex and ambitious attempt at what he called an extreme detached escapement clock, although it was not truly detached, let alone extremely so. Equally, the wonderful constant force regulators made by Bond of Boston in the United States in the mid-1860s, beautiful as they are, cannot claim to be detached either, as the gravity arm that impulses the master pendulum is carried by the pendulum during the extent of its swing on one side. It is, however, a very fine and ingenious design and performed very well at Harvard Observatory and at Bidston Observatory, Liverpool during the second half of the nineteenth century.[27]

While in practice these designs were proving difficult to make and adjust, research continued on theoretical means for improving the quality of precision pendulums. Taking the concept of detachment to its logical extreme, the ideal appeared to be what became known as a *free pendulum*, i.e. a pendulum that had no work to do except simply receive its maintaining impulse. The first pendulum clock of this kind, though by no means perfect in its design, was that built in the late 1860s by William Thomson, later Lord Kelvin (1824–1907). Developed in 1865 and perfected in about 1869, the clock does appear to be the first that qualifies as a free pendulum design, as the master pendulum swings entirely freely, merely receiving impulse via an escapement controlled by a constantly running conical pendulum, the rate of which is itself controlled by a half-seconds pendulum. However, the design was also very complex and difficult to adjust, and it seems only three examples were ever made by James White of Glasgow, Lord Kelvin's instrument maker.[28]

---

27 Saff 2019.
28 Aked 1973b.

Until a published description of the clock by Charles Aked in 1973, it had generally been accepted that the first clock to qualify as a free pendulum timekeeper was the 1898 design of R. J. Rudd,[29] in absence of knowledge of Kelvin's of thirty years before. Rudd's design, however, is more satisfactory, being the first to employ a secondary pendulum properly synchronized to the main pendulum. However, like Kelvin's, it still relies on a geared mechanism to impulse the main pendulum, and there was clearly a further improvement to be made by providing a more constant force directly to impulse a free pendulum. Within twenty years Shortt's free pendulum clock system appeared (Figure 150).

There was, however, an alternative route to impulsing a precision pendulum—one conceived by the celebrated German physicist and clockmaker Sigmund Riefler (1847–1912) of Munich, patented by him in 1889. Riefler's design, employing his own special escapement, impulses the pendulum *through* its suspension spring, entirely avoiding the need for physical contact with the pendulum itself. This is described in detail in Chapter 19.

A similar escapement to Riefler's was invented in 1899 by Ludwig Strasser, ten years after Riefler's, and produced by the firm Strasser & Rohde. The difference in Strasser's design is that the main weight of the pendulum is taken by a conventional suspension spring mounted firmly on the top support of the clock, and impulse is delivered via a second double-bladed spring mounted above the support, but connected to the top of the pendulum below the main suspension spring. The second spring above the support is connected to an escapement and is flexed to and fro, in the same manner as in the Riefler, as the pendulum swings.[30] Riefler's success in suspending a precision pendulum in a reduced pressure tank was taken up a few years later in the early twentieth century by the great French chronometer, clock-, and watchmaking firm of Louis Leroy et Cie. The precision pendulum clocks made by the firm, using a form of spring pallet escapement, are described in Chapter 19. The performance of the clocks was said to be of a very high order, but in 1920 there appeared what might be described as the ultimate form of free pendulum clock, the performance of which had not until then been attained. It was the creation of the designer William Hamilton Shortt, working in conjunction with the manufacturer Frank Hope-Jones: the Shortt free pendulum clock, which is also described in detail in Chapter 19.

In common with the Leroy precision pendulum clock, the Shortt Clock incorporated another highly important development in horological science: the pendulum with a rod of *Invar*. This was a new nickel–steel alloy created in France by the Swiss physicist Edouard Charles Guillaume (1861–1938) in 1896 and which had a virtually zero coefficient of thermal expansion. This alloy revolutionized pendulum design, as pendulums with rods of invar only needed the smallest residual compensation. Following this invention, Guillaume created another nickel-steel alloy—Elinvar—that proved to have an almost zero thermoelastic coefficient and, when used in the production of watch and chronometer balance springs, would almost entirely negate the need for temperature compensation in the balance.

So important were these inventions that Guillaume was awarded the Nobel Prize for them—the only one ever awarded for work directly related to horology. A similar alloy to Invar was designed to substitute for the steel element in the bi-metallic rims of compensation balances. Because this alloy's expansion is virtually nil, the bi-metallic rim segments using it, in conjunction with brass, have a greater motion for a given change in temperature than the older steel/brass design. This enabled shorter rim segments to do the same work for a given change in temperature, and thereby reduced centrifugal effect in the bi-metal balance.

More important, however, such balances had a further benefit when used in conjunction with a steel balance spring. The rate at which the new alloy expanded and contracted with temperature was not only small, but it was also non-linear and enabled the compensation to follow the non-linear effects of the changing elasticity of the steel balance spring. Thus, in combination with a steel balance spring, the design of the Guillaume balance had almost completely removed MTE. Because the materials of the balance itself incorporated the means by which MTE was removed, balances of this kind were referred to as *integral* balances.

The introduction of Guillaume's nickel–steel alloys ushered in an entirely new era in the horological world: the new material was seized upon by a number of makers of marine chronometers and immediately gave them improved performance in their instruments. Naturally, the Swiss had long been interested in the making of precision watches but, perhaps because not a maritime nation and because the larger marine-type chronometer was unfamiliar to them, they did not branch out into wholesale manufacturing of them until relatively late in the nineteenth century. Nevertheless, many fine Swiss pocket chronometers were produced, especially during its second half, most notably in the Jura region, in the great watchmaking centres of La Chaux-de-Fonds, Neuchâtel, and Le Locle. In fact, the Swiss were well advanced in the concept of testing the rate of chronometers at astronomical observatories, the first official trials beginning at Geneva in 1818 with a prize of 800 florins offered for the successful contestant. The Neuchâtel observatory was founded in 1858 and adopted a system of chronometer trials based on proposals and trials run over the preceding thirty years by the splendidly titled 'Patriotic Emulation Society'. From 1865 this observatory offered prizes for the best timekeepers in various categories, the first two prizes in 1868 going to the firm of Ulysse Nardin of Le Locle. This company was the first in Switzerland to realize the great business opportunity for sales of marine chronometers in Europe, organizing a system for their manufacture, which came close to mass-production. In 1902 the company introduced a new model, similar to the British type, with an Earnshaw spring-detent escapement and a wheel

---

29 Hope-Jones 1949, 1951, 152–8.
30 Roberts 2003a, 136–8.

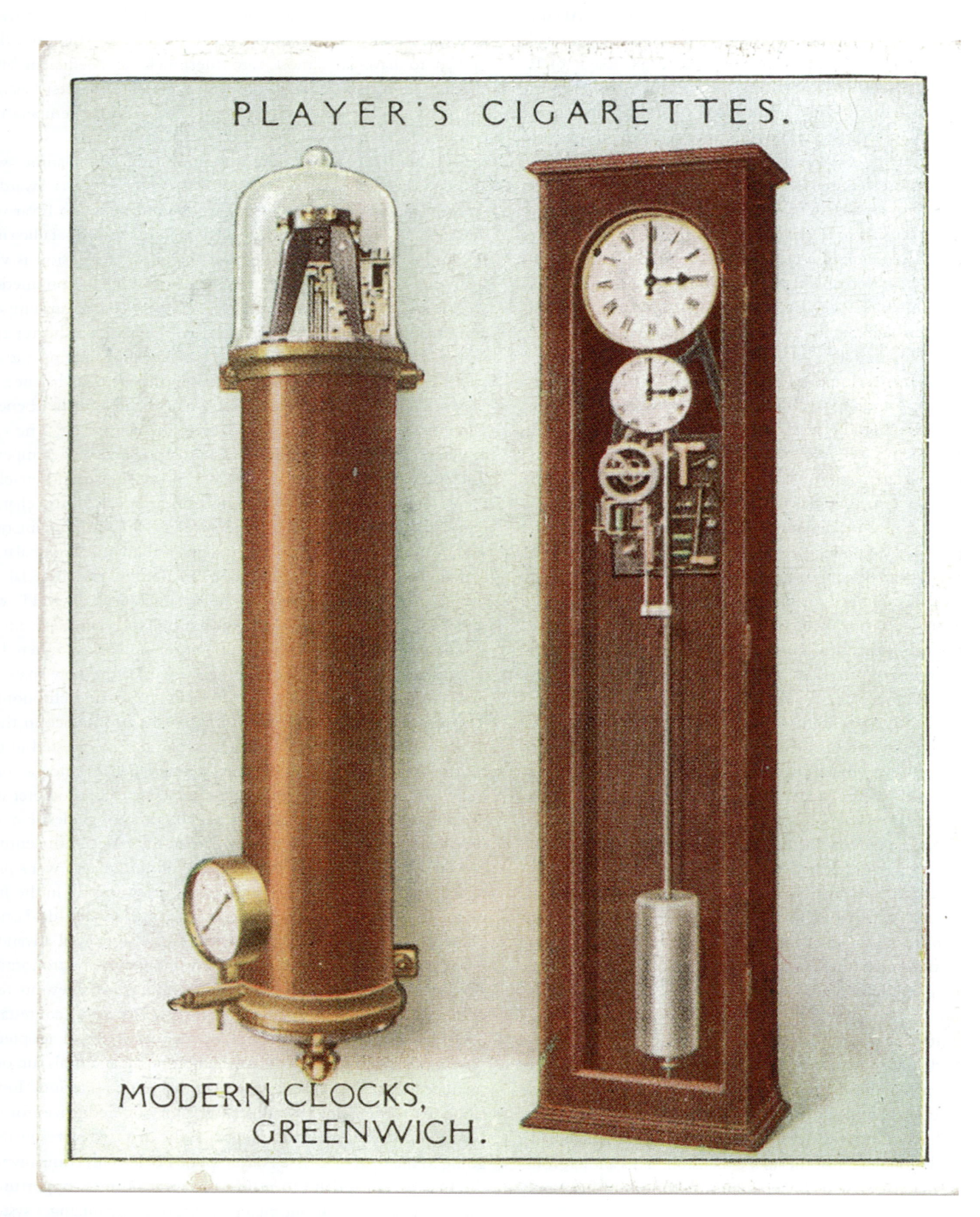

Figure 150 'State of the Art'. In the 1920s the super-accurate Shortt free pendulum clock was significant enough to feature in a sequence of horological cigarette cards. Image: James Nye

train identical to what was now the standard model. However, the production method was made much more efficient and standardized by having the movement parts formed using press tools and jigs, ensuring that the parts needed the minimum of hand work to finish and fit them. The *Marine Grande Format* design of Ulysse Nardin was so successful that it was first bought, and then copied, by the Japanese when establishing a better-equipped and increasingly powerful navy during the first three decades of the twentieth century. It would later also form the basis of the Hamilton Watch Company's highly successful Model 21, used by the US Navy during the Second World War.

The late 1890s saw the first small signs of the problems that the marine chronometer and its manufacturers would face in the coming years. Reliable and relatively inexpensive chronometers were now being made in considerable numbers by European and USA manufacturers, and the world's navies and merchant fleets were approaching a stage where their needs were satisfied.

In 1898 Guglielmo Marconi (1874–1937) made his first radio communication across the English Channel between England and France, followed by the famous transatlantic signals between England and Newfoundland in 1901. The implications for international time distribution were obvious, the US Navy being the first to broadcast low-powered signals from Navesink, New Jersey in 1904, which led—from January 1905—to the first regular daily time-signal transmissions from Washington DC (at noon, Eastern Standard Time). Germany followed in 1907 with France broadcasting Paris Mean Time daily, from the Eiffel Tower, after May 1910. These radio time-signals were not, however, quite the answer to a navigator's prayer that they might seem, and wireless telegraphy never effectively replaced the use of chronometers on board ship. Nevertheless, both wireless telegraphy and, from the 1920s, radio communication provided easier means for rating and correcting errors in ships' chronometers during voyages. This, combined with significant nineteenth-century improvements in marine propulsion (leading to voyage times becoming shorter), meant that the need for marine chronometers with the very best long-term stability began to diminish.

To make matters worse (from a manufacturing viewpoint), the instruments already in use had generally been so well constructed that they had remarkably long life-spans: with a little updating of box and perhaps balance and spring, many early nineteenth-century instruments were still doing service a century later. The First World War saw a new demand from the combatant navies, but to a large extent this had been anticipated and Royal Navy ships were at least reasonably well supplied. As a result, the Greenwich trials, current in one form or another since the 1820s, ceased in 1916. Thereafter, those wishing to have a chronometer rated in England were obliged to send it to the National Physical Laboratory which had, since 1884 (as the Kew Observatory), been rating timekeepers for commercial watchmakers supplying the civilian market.

Another factor in the demise of the marine chronometer resulted from improvements in the lever escapement and the balance and spring, as fitted to pocket watches, coming from all European manufacturers during the late nineteenth century. Pocket-watches with this escapement, especially those fitted with the Guillaume (integral) balance, were now performing nearly as well as marine chronometers, and a number of navies began purchasing this cheaper alternative for navigational purposes. Some were used as deck watches, but others, known as 'Chronometer Watches' in Admiralty ownership, were intended to substitute for the use of a chronometer on lesser ships. The watch factories in the US also now started producing models fitting this description, but mounted in gimbals and boxed like a marine chronometer, and at a very reasonable price. During the latter years of the First World War, the Admiralty purchased a large number of these chronometer watches from the firm of Waltham, in Massachusetts, but were then obliged to dispose of the majority of them, mostly unused, soon after the war ended.

After 1918, demand again fell and many of the British chronometer makers, facing competition from companies like Ulysse Nardin in Switzerland and the hardships of economic depression in the 1920s and early 1930s, followed their watch-making counterparts and went out of business. The Astronomer Royal's correspondence includes a number of very sad letters from independent chronometer specialists desperately offering their services at this time, all politely refused by the Observatory staff.

In 1923, Commander Rupert T. Gould of the Admiralty Hydrographic Office published his comprehensive and accessible book, *The Marine Chronometer, its History and Development*, a work which, under the circumstances, must have seemed a kind of valediction for the chronometer. In fact, the instrument was not obsolete in the 1920s, but was still essential equipment on board ship. As well as discussing the 'modern' (i.e. standard) chronometer of the period, which Gould characterized as being a typical two-day instrument by Kullberg, he looked to the future and discussed Ditisheim's new design, stating that 'It supplies convincing proof, if any were needed, that the development of the chronometer is by no means finished'.[31] As it turned out, he was correct and it would be more than sixty years before ships were navigating safely without having a chronometer or two on board, in spite of the development of other navigational aids.

The early development of radar equipment by the US Navy in the 1920s, and by Robert Watson-Watt in England in the mid-1930s, began by aiding local navigation and detection, and led to the creation of low-frequency, modern position-fixing systems like Decca, Loran, and Omega. These greatly aided off-shore position-fixing but did not supersede the use of a timekeeper of some kind in ocean navigation. Another more general issue that maintained the need for a ship to carry a chronometer were the continually improving safety standards at sea as the twentieth century progressed. The need for shipping companies to demonstrate to passengers and to insurers an increasing awareness of 'due

---

31 Gould 2013.

diligence' in navigational practice saw navies and merchant fleets worldwide continuing to issue them for safe navigation, in spite of the back-up of radio and other systems.

However, there was virtually no large-scale industry anywhere in the world still producing chronometers and, in 1939, as war broke out in Europe, and new instrumentation was needed for warships, it was too late for most British manufacturers to be revived. Nardin in Switzerland were largely unable to get chronometers to England and only a handful were now able to supply the war effort, including Thomas Mercer, and Johannsen & Co. Technically, Victor Kullberg's firm was still in business, but Sanfrid Lunquist, his successor, was hardly making any chronometers and would mostly prove useful to the Admiralty in repair work alone. On the other hand, the German watch and chronometer industry was still producing, encouraged by the Third Reich during the 1930s. In 1905 the Hamburg Chronometer Works had been founded in cooperation with local shipping companies, an enterprise that became the very productive Wempe company of chronometer manufacturers. Since 1907, Lange & Söhne had been using Guillaume balances especially made by Richard Griessbach in Glashütte, and a highly successful 'German chronometer' was established. These companies now concentrated on military and marine instruments, and the Kriegsmarine was well supplied. In 1942, the Wehrmacht and the German air ministry decreed that the movement design for marine chronometers should be standardized and that the standard (*Einheits*) chronometer-movement frames should be produced by Wempe in Hamburg.[32] As for Britain, the Royal Navy fell back on its reserves of old instruments, borrowed from private owners, and bought in from Switzerland when it could.

On the question of how Admiralty chronometers were used at this period, it is interesting to note from some of the service records gleaned from the Observatory ledgers and ledger cards relating to NMM examples that Royal Navy chronometers were by no means solely issued to ships for marine navigation. One regular location where they were also used were the various points round the world entered as 'C.S.O', for 'Composite Signals Organisation'. These were the British government's highly secret wireless-intercept listening stations, gathering radio and telegraph traffic and feeding it back to the UK, where the encoded messages mostly found their way to the Government Code and Cypher School (G.C. & C.S.) at Bletchley Park for deciphering. The C.S.O. stations evidently used the chronometers to record the precise time (GMT) at which a given message was intercepted.

When the United States entered the war in 1941, Britain was in no position to supply them with some of its chronometers, as it had done in years gone by, and the only large-scale-production instruments that US industry had been manufacturing were of the gimballed chronometer-watch design. These small instruments were all very well *in extremis* but were not a very reliable or professional alternative to proper marine chronometers, and it was recognized that renewed manufacturing of these was the only answer. The American firms of Elgin and Hamilton were involved in taking this further (though not always without discord between them) and nearly 10,000 marine chronometers were produced by Hamilton during the war years. The Hamilton Model 21 was not only the solution to that urgent wartime need: it also proved to be the ultimate development of the mechanical marine chronometer and, in terms of performance, stands as probably the finest-ever design of the mechanical form of the instrument.

By about 1950, quartz technology—which had been available in larger clock form since the 1930s—was in turn nearing a scale suitable for marine use and experiments were under way on both sides of the Atlantic to produce a quartz marine timekeeper. It was over fifteen years before reliable quartz marine chronometers began to supersede mechanical versions in the major western navies, but during that period (and without global conflict) the remaining 'mechanicals', including very large numbers of Hamilton Model 21s, supplied most needs.

Immediately following the Second World War, several European chronometer-making firms began to consider moving over to the design and production of electrically powered chronometers. In 1923, Louis Dubois, a graduate of the Technicum at Le Locle, adapted a standard Mercer-type marine chronometer to have an electrically rewound remontoire. The train of the movement was adapted to receive power from a one-minute spring remontoire, based on the design patented by David Perret in Switzerland in 1900. The new design did not immediately lead to manufacture, but a very similar one was taken up by the German firm of Wempe and put into production by them in 1962. Known as the Type 907, it was the same in principle as the Dubois instrument, having a standard marine-chronometer balance and escapement with the upper part of the train, and with the power supply adapted to electrical remontoire. The year 1953 saw the patenting of a new concept for the marine chronometer, in which the high-grade compensation balance was retained but the mechanical escapement and train were dispensed with. Instead, the balance was impulsed by electrical coils, the switching for the impulse determined by the balance itself, with a magnet on the staff of the balance inducing current in a sensing coil. Marius Lavet of Etablissement Leon Hatot was the creator of this fine design, which was marketed as the Chronostat by Leroy of Paris during the 1950s. In the post-war period from 1945, the Royal Navy also began to consider moving over to wholly quartz-controlled marine chronometry, and in the late 1950s placed a contract with the Sperry company to develop such an instrument, with half-seconds indication and employing germanium transistors in the dividing circuits. Unfortunately, no sooner had this work started than the germanium transistor was superseded by the silicon type and the project was abandoned, probably in about 1960–61. Only a handful of prototypes from it survive.

---

32 Meis 2011.

Following these early attempts at creating a design for a working electronic/quartz chronometer, successful development of other quartz models took place from the 1960s, more or less simultaneously, in Europe and the Far East, until satellite-based global positioning systems (GPS) superseded them for general navigational uses in the 1990s. From the 1960s into the early 1980s, the later stages of the Cold War inspired development of the GPS in the USA and GLONASS in the Soviet Union, for improving satellite positioning and missile guidance. Both systems were in continuous parallel development and 'full operational capability' for both was only finally declared in 1995 and 1996, respectively. Now with the accuracy of GPS pinpointing to within a centimetre, the marine chronometer is no longer an essential instrument for accurate navigation.

It was the demand for improved frequency standards for the development of radio broadcasts during the early 1920s that had led to the introduction of quartz clock technology by Horton & Marrison in the USA in 1928, further improvements being introduced by Louis Essen at the NPL in the UK in the 1930s. By the 1940s, quartz clocks were capable of timekeeping stability of about a second in two years. Thus, the development of quartz timekeeping technology not only rendered the portable mechanical timekeeper (the chronometer) redundant: in their earlier, somewhat larger, land-based form, quartz clocks soon outperformed even the Shortt free pendulum and would be adopted as principle time standards in most world observatories after the Second World War.

However, it would be untrue to say that the precision electromechanical clock was wholly superseded when quartz technology arrived. In the Soviet Union following the Second World War, for example, a whole new step was taken in advancing the precision of the pendulum clock, with a design by the Russian physicist Feodosii Mikhailovich Fedchenko (1911–89), described in detail in Chapter 19.

Working on further improvements to the stability of the quartz clock, Harold Lyons (1935–91) and associates at the National Bureau of Standards in Washington DC, developed the earliest form of atomic clock in the late 1940s, turning their attentions to developing a caesium atomic clock in 1949. At the same time, parallel research in the UK with Louis Essen (1908–97) and Jack Parry (1923–95) led to the first caesium atomic clocks coming into regular use at the NPL in 1955. Developments in this field are covered in more detail in Chapter 19. In conjunction with the use of Hydrogen Masers as a higher-stability, short-term time standard, Hewlett Packard caesium clocks became the standard source for UTC in most contributing countries from the 1970s. These were in use until the late 1990s, when research on caesium fountain technology was begun, leading to great increases in accuracy and stability of precision clocks. For example, in 2011, improvements to the NPL-CsF2 caesium fountain clock operated by the UK's National Physical Laboratory resulted in an evaluated frequency uncertainty reduction from $u_B = 4.1 \times 10^{-16}$ to $u_B = 2.3 \times 10^{-16}$, the most stable performance for any primary national standard at the time. At this frequency uncertainty, the NPL-CsF2 was expected to neither gain nor lose a second in about 138 million (138×10^6) years.

CHAPTER FIFTEEN

RESPONDING TO CUSTOMER DEMAND
THE DECORATION OF CLOCKS AND WATCHES FROM THE RENAISSANCE TO RECENT TIMES

Catherine Cardinal

Once clocks and watches began to be widespread from the sixteenth century onwards, they were completely integrated with the decorative arts. Doubtless, their users thought them practical objects for which improvement was indispensable, but that did not exempt them from also being agreeable objects to behold and even contemplate. Gold- and silversmiths' work, furniture, jewels, clocks, and watches were intended to be integrated harmoniously into a setting. For watches, this was dress, accompanied by other jewels and accessories; for clocks, this constituted the entire domestic space. Apart from their horary function, clocks and watches deployed contemporary aesthetic language. Clock- and watchmakers were able to satisfy their clients with the aid of *ornamenistes*, goldsmiths, jewellers, engravers, enamellers, joiners, and bronze founders. But the decoration of clocks and watches was not simply a function of artistic fashion: technical innovations, different production methods, customer taste, and changes in lifestyle also influenced the development of the form and decoration of personal and domestic timepieces.

RENAISSANCE AND MANNERIST PERIODS: GOLDSMITHS' AND JEWELLERS' WORK

Private clocks spread through Europe from the later fifteenth century onwards among wealthy and cultivated customers, an interest manifested in portraits with moral and symbolical connotations. Two from the 1530s show Henri II of France (1519–59), the other Georg Gisze (1497–1562), a successful young merchant, both depicted beside small brass cylindrical table clocks typical of the period.[1] Clocks and watches in this period, which put decoration to the fore, display their close links with goldsmiths' and jewellers' work. Many designers, often goldsmiths, helped to diffuse models through their publications. Here, the success may be noted of such German and Italian *ornamenistes* as Peter Flötner (1485–1546), Virgil Solis (1514–62), and Francisque Pellegrin (14?–c.1562). Noteworthy also is the influence of the French architect Androuet de Cerceau (1515–85) and his compatriot, the goldsmith-engraver Etienne Delaune (1518–83).[2]

French table clocks are most commonly made in the form of aedicules—square, hexagonal, or cylindrical—surmounted by a pierced dome when they are fitted with a bell—and supported by pilasters or short columns (Figure 74). The cases are frequently engraved with allegories of the seasons, the days of the week, the virtues, mythological subjects, or with strap-work. Among the earliest is a clock by Noel Dauville dated 1544, which presents the days of the week (without Sunday), on its six faces.[3] Two clocks signed by founding members of the Paris Corporation (1544) are also worth noting. One by Antoine Beauvais (*fl.* 1536–51) is hexagonal presenting the days of the week after models by

---

[1] Henry II, Musée Condé, Chantilly; Gisze (Holbein, 1532), Gemalde Galerie, Berlin.
[2] Rohou 2019. A wide-ranging collection of ornament designs from the sixteenth to the eighteenth centuries from throughout Europe is offered by Guilmard.
[3] Deutsches Uhrenmuseum, Furtwangen inv. K-1296, Cardinal & Vingtain 1998, N° 42.

Figure 151 Table clock by Florant Valeran with symmetrical interlace decoration, c.1550. Musée International d'Horlogerie, La Chaux-de-Fonds, inv. I-1. Photo. MIH.

the master 'I. B.'[4] The other, by Florent Valeran (f. 1542–62), has engraved strap-work decoration.[5] Both pieces carry mottos: that by Beauvais showing *in via virtutis nulla est vi* (no way is impassable for virtue)[6] and on that by Valeran (Figure 151) *je patiente en ma loyauté ferme en atandant l'heure de mon heure PDV* (I am patient in my firm loyalty while waiting the moment of my time).

A later clock by Nicolas Plantart (1st quarter 17th century) also illustrates the fashion for mottos characteristic of the period and the use of moresco-type 'arabesques'.[7] The motto of the Montmorency family, *Terror et error Montmorency*, is repeated in the superimposed cases for the clock and the alarm. The death's head above the alarm illustrates the prevalence of *memento mori* in the period. Designs by Etienne Delaune (1519–83), frequently used by goldsmiths and instrument-makers, appear on several clocks sometimes combined with designs from other artists. Thus a cylindrical clock by Simon I Gribelin (pre-1588–1633) of c.1600 shows four allegories after Delaune: Peace, Abundance, War, Famine—the pierced dome is engraved with hunting scenes derived from models by Virgil Solis.[8] A square tower clock that has preserved its leather case gilt with the arms of Gaston d'Orléans is remarkable for its decoration after engravings by Delaune.[9] The pierced dome shows Apollo, Jupiter, and Urania; the side panels are inspired by Delaune's sequence *Grotesques à fond noir* depicting biblical scenes. Also to be mentioned is a cylindrical table clock decorated after the well-known *Suite des mois* (1568), four medallions showing March, May, July, and September, representing the four seasons.[10]

Archival sources allow the luxurious nature of clocks ordered by the élite to be appreciated. Charles the Bold (1433–77), the Duke of Burgundy, owned 'a gold clock set on six lions with several images around, set with thirty-nine pearls and twenty-seven rubies, and in the summit there is a salt on which there is a gun'. This calls to mind the royal clock salt decorated with pearls and rubies belonging to the Goldsmiths' Company of London thought to be made by the Paris goldsmith Pierre Mangot between 1530 and 1535 and doubtless presented to Henry VIII by François I.[11] In the list of clocks by Jean Naze (1539–81), sold by lottery after his death, the finest was in silver and rock crystal, decorated with an image of Mirrha (mother of Adonis), and furnished with an astrolabe.[12] It may be compared with a surviving clock by Naze contained in a rock crystal

---

4 Musée national de la Renaissance, Ecouen, E. Cl. 1468, Chapiro et al. 1989, N° 11; Cardinal & Vingtain 1998, N° 86.
5 Musée international d'horlogerie, inv. I-1; Cardinal & Piguet 1999, N° 76.
6 Ovid, *Metamorphoses*, xiv. 113.
7 Musée international d'horlogerie, inv. IV-162; Cardinal & Piguet 1999, N° 77.
8 Musée Paul Dupuy, Toulouse, inv. 18 072. Hayard 2004, 92–6; Rohou 2019, cat 67.

9 Musée du Petit Palais, Paris, inv 1406. Cardinal & Vingtain 1998; Rohou 2019, cat 75.
10 Victoria & Albert Museum, inv. 379–1906. Rohou 2019, cat 43.
11 Laborde 1851, ii, 128. The clock is N° 3140 in the inventory of Charles' goods. Pending publication of the proceedings of a colloquium devoted to the Mangot salt in 2018 see Jagger 1983, 16–18.
12 Vial & Côte 1927, 76–7, 213–16.

case inserted in a temple-like gilt silver frame.[13] The cylindrical case, held within three Corinthian columns of a clock signed by the Blois maker Isaac Poitevin (c.1600), is also made of rock crystal.[14]

The great age of German clockmaking extends from the early sixteenth century to the first quarter of the seventeenth century. The diversity of the cases produced is astounding. The most frequent take the form of a drum with a horizontal dial or of a tower. Others are to be found in the shape of a monstrance, a crucifix, or a book (Figures 73 and 74). Their decoration, engraved on brass or sculpted in gilt bronze, is made up of Moresque designs, *grotesca*, hunting scenes, or allegorical and mythological figures derived from contemporary pattern books such as those of Virgil Solis.[15] Among the more spectacular pieces are the astrolabe clocks of Metzger (1564), Bohemus (1568), and others.[16] Bookclocks, easily transported, were popular, decorated with engraved trailing fronds, and garnished with a sundial, a lunar calendar, and an alarm. Fine examples are those signed by Hans Schnier at Speir (1583), and Hans Koch at Munich (c.1580).[17]

Automaton clocks offered a quite different field for decorative elements.[18] Particularly appreciated in the courts of northern Europe, they also made part of the annual 'tribute' paid from the Holy Roman Empire to the Turks,[19] and were deployed to favour the spread of Christianity in China. One example is a spectacular clock incorporating an animated scene of the Nativity surmounted by a gold globe in which God the Father appears. This was supplied in 1618 at the expense of Ferdinand of Bavaria (1577–1650), Prince-Elector of Cologne.[20]

The variety and fantasy of automaton clocks is considerable. Reflecting contemporary fascination with the exotic, strange animals were particularly popular. Pride of place, appropriately, went to the lion shown standing, lying, or walking. His eyes would roll at every beat of the movement, his jaws open and his tongue emerge when the hours were struck. Several examples survive such as one by the Augsburg maker Hans II Buschmann (post-1591–1662).[21] This is mounted on an ebony base that incorporates dials for the hour, the age and phase of the Moon, and the day of the week. Every quarter of an hour the eyes and jaws move and (a highly unusual element) a small sand glass, placed on a shield on which the lion rests a paw, is turned. Indian elephants, sometimes carrying a sort of palanquin with soldiers, and the dromedary, also inspired makers. Other incarnations are more astonishing, such as a crowned eagle that raises a sceptre as the hours strike and rolls its eyes and opens its beak on the quarter hours, or a parrot that beats its wings and rolls its eyes.[22] Mythological personages may also be depicted such as Minerva in a horse drawn chariot, or Diana in one drawn by panthers.[23]

The forms of watches were no less diverse. Apart from small drum watches, spherical models inspired by pomanders, oval, and hexagonal models, **'form' watches** offered endless variety to an international clientele. Most surviving drum watches are German, although French examples are known, with engraved or cast and chiselled cases.[24] One is dated 1542 and has the makers initials 'C W', perhaps for Caspar Werner (*fl.* 1527–post-1548). The oldest known dated French watch, by Jacques de la Garde (Figure 152), is spherical, but two others, one dated 1530, are German.[25]

Oval and octagonal watches remained in use until the mid-seventeenth century. Silver or gilt brass were used, either singly or in combination, for their cases, which were adorned, like the dials, with scenes or designs derived from Delaune, Theodore de Bry (1528–98), or their followers such as Antoine Jaquard (1st half 17th century), and Michel le Blon (1587–1656). Allegorical figures, mythological and religious scenes, and **grotesque** decoration were all used, with Delaune an especially popular source. A Blois watch by either Paul I Cuper (1520s–1612) or Paul II Cuper (1557–1625) reproduces Delaune's 'Narcissus admiring himself at the fountain' and his 'Orpheus charming the animals', while a watch carrying the name of Jean Barberet, Paris represents Mars after a design in the *Divinités sur grotesques à fond noir*.[26]

But watches, whether worn hanging from a chain around the neck or at the waist, could have more extravagant forms, presented as jewels in precious materials. Hard stones were much appreciated, especially rock crystal, which allowed the otherwise hidden movement to be seen, as well as jasper, agate, amber, and cornelian. Pearls and precious stones were also used. An inventory of the jewels of Elizabeth I of England drawn up in 1587 reveals watchcases decorated with diamonds, coloured stones, hard stones, and pearls. Similarly, Marguerite de Valois had a small watch garnished with diamonds and rubies used in **ronde bosse,** that was repaired for her

---

13 Cardinal & Vingtain 1998, N° 43.

14 Deutsches Uhrenmuseum, Fürtwangen, inv K1300. Cardinal & Vingtain N° 47.

15 For examples see Brusa 1981; Maurice & Mayr 1980.

16 Kunsthistorisches Museum, inv. n°852, 'Jeremias Metzker Urmacher in Augspur 1564'; Metropolitan Museum of Art, inv. 17.190.634 a–d, 'Me fecit Chasparus Bohemus in Vienna . . . 1568'. Vincent, Leopold, & Sullivan 2015, N° 1.

17 Ashmolean Museum WA 1947.191.7 in Thompson 2007, N° 4. Musée du Louvre, OA 10 144 in Cardinal 2000, N° 9.

18 Chapuis & Droz 1949, 77–96; Maurice & Mayr 1980, 27–48; Kugel 2016.

19 Kurz 1975, 30–53; Mraz 1980, 37–48.

20 Chapuis & Droz 1949, 82–3.

21 Maurice & Mayr 1980, 228, 256–7, N° 86.

22 Eagle: Metropolitan Museum of Art, inv. 29.54.14, Vincent, Leopold, & Sullivan 2015, N° 9; parrot: Bayerisches Nationalmuseum, Maurice & Mayr 1980, N° 79.

23 Minerva: Kunsthistorisches Museum, Vienna, Maurice & Mayr 1980, N° 108; Diana: Museo poldi Pezzoli, Milan, inv 1149, Brusa 1981, 137, N° 17.

24 They are fully surveyed in Matthes 2018.

25 Musée du Louvre OA 7019, Cardinal 2000, N° 19; Ashmolean Museum, Oxford, WA 1947.191.1, Thompson 2007, N° 1; Walters Art Gallery, Baltimore, inv. 58.17.

26 Musée du Louvre, OA 7022 and OA 7028. Cardinal 2000, N° 81, N° 88.

Figure 152 Spherical watch by Jacques de La Garde, Blois, 1551, Openwork case decorated with masks, crescents, and interlace work. Musée du Louvre, Paris, OA 7019—Documentation du Département des Objets d'Art.

in 1579 by Maurice Bernard Ferry.[27] Nor were such luxury watches confined to royal households; in 1569, Yves de Bernon, seigneur of Sardes and Guyencourt sold a time-piece 'enclosed in gold set crystal, ... with rubies and diamonds around and a pearl in the head'.[28] An early seventeenth-century star-shaped case with a movement by the Genevan maker Duboule offers another example of these luxury pieces: the two openwork enamelled covers of the case being set with diamonds and emeralds.[29]

Clear or opaque **champlevé** enamels could be associated with precious stones. A good example is a watch by Caspar Cameel (*fl.* 1623–46) in Strasburg, which is decorated with fifteen diamonds heightened by champlevé motifs worked around an agate cameo.[30] Enamel could also be used in ronde bosse and a watch by Nicholas Vallin (*fl.* 1565–1603), London, which forms a 'lesser George',[31] provides a remarkable example.[32] A square watch by Nicolas Lemaindre shows how engraved surfaces could be heightened by a transparent enamel applied over engraved fronds and figures below. The same watch illustrates the delight in mottos discussed above.[33]

Judging by the number that have survived, crucifix watches in rock crystal were especially popular and examples signed by makers from various centres such as Josias Jolly (Paris), Jean Vallier (Lyon), Zacharias Fonnereaux (La Rochelle),[34] Jean Rousseau, Jean Cusin, or Henri Ester (Geneva) are well known.[35] Watches with tulip-shaped cases, testimony to the wild fashion for this flower in the mid-seventeenth century, are sometimes executed in

---

27 Archives nationales, Paris, KK, 165 f. 454^(r-v).

28 Cardinal 1985, 230.

29 Musée national de la Renaissance, Ecouen, E. Cl. 20,706; Chapiro et al. 1989, 76 N° 61; Cardinal & Vingtain 1998, N° 104, Cardinal 1999, N° 4.

30 Musée du Louvre, oA 8292. Cardinal 1984, N° 33; Cardinal & Vingtain 1998, N° 105.

31 A pendant that knights of the Garter were expected to wear on their daily dress.

32 The Metropolitan Museum of Art, New York, inv. 17.190.1475. Vincent, Leopold, & Sullivan 2015, N° 6.

33 Musée du Louvre, OA 7026. Cardinal 2000, N° 52.

34 Musée du Louvre, oA 679 et OA 7040; Cardinal 2000, N° 40 & N° 41. Musée du Petit-Palais, Paris, inv. 1412; Cardinal & Vingtain 1998, N° 99.

35 British Museum, inv. 74, 7–18, 28; Metropolitan Museum of Art, inv. 17.190. 1577. Musée du Louvre, OA 7054, Cardinal 2000, N° 45. Musée national de la Renaissance, Ecouen, E. Cl. 18,515.

Figure 153 Case of a form watch in the shape of a turkey. Photo Jean-Baptiste Buffetaud, courtesy of Chayette & Cheval, auctioneers, Paris.

rock crystal,[36] more frequently in plain metal,[37] or enamelled.[38] Cases in the form of shellfish were also favoured,[39] but animal form (Figure 153) cases appear to have been specific to Geneva.[40]

THE BAROQUE: NEW FORMS AND DECORATIVE OPULENCE

The style designated 'Baroque' prevailed in Europe and its colonies from the early seventeenth to the mid-eighteenth centuries. It is identified by a rich polychromic decoration and high fantasy, its products characterized by opulent, flowing forms of monumental dimensions produced in striking materials such as gold, silver, pewter, marble, stucco, ebony, hard stone, ivory, and tortoiseshell. Strongly informed by both Ancient and contemporary Italian models, the decorative arts put mythological and allegorical figures to the fore alongside architectural motifs. The potent commerce of art in later seventeenth-century France, particularly the collections of ornamental designs published at Paris, propagated French styles in the Low Countries and the German principalities. The decoration of watches further illustrates two basic tendencies of the period: inspiration from nature linked with a growing passion for flowers and gardens, and a fashion for depicting historical scenes. Engraving, chasing, repoussé work, and enamelling were the ideal mediums for this iconography, with the latter, already popular on contemporary jewellery, becoming particularly favoured for the decoration of watchcases.

In the first third of the seventeenth century, abstract styles of vegetable or floral arabesques were in favour, especially that known as the 'pea pod' design. The engraved designs proposed between 1619 and 1625 by Jean I Toutin, Pierre Nolin, Stéphane Carteron, Jacques Hurtu, Alexandre Vivot, and Balthazar Lemercier are representative of this decoration.[41] The technique of enamel in ***resille sur verre*** is found on a number of cases;[42] champlevé enamelling was used for naturalistic floral decoration

---

36 E.g. Jacques Sermand (Musée du Louvre OA 7063 Cardinal 2000, N° 26); Metropolitan Museum of Art, inv. 17.190.1015.
37 E.g. that by Jean Baptiste Duboule, Musée national de la Renaissance, Ecouen, E. Cl. 18,515.
   Chapiro *et al.* 1989, 77, N 65.
38 Josias Jolly, Musée du Louvre, OA 7032; Cardinal 2000, N° 25.
39 Cardinal 2000, N° 31–7.
40 A dog by Jacques Joly, a lion by Duboule, a pelican by Henry Ester. British Museum, inv. 88, 12–1, 206; Cardinal 1985, 190. Ashmolean Museum, WA 1947. 191.54; Thompson 2007, N° 10. Victoria & Albert Museum, inv. 788–1864.

---

41 Bimbenet-Privat & Fuhring 2002.
42 E.g. Victoria & Albert Museum, London inv. 2553–1855. Cardinal 1985, 136, pl. 99, right.

after c.1630.⁴³ Polychrome enamel in light relief covered the metal of a case pierced with flower designs to produce a delicate effect.⁴⁴ Shortly before 1630 a new process was developed, that of painting on enamel. To produce a varied palette of colours, mineral oxides were reduced to a fine powder and mixed with a flux. Slightly diluted in a medium such as oil of aspic, the powders were applied with a fine brush on an opaque white enamelled ground. At the application of each colour the work was placed in a furnace at about 800°C. The result was a painting comparable, except in miniature, to one in oils, offering similarly delicate shades of colour and a similarly subtle model.

Floral bouquets painted on enamel were as realistic as those to be found in the plates of botanical works. The enamel painters took their models from the collections of Jacques Caillart (1ˢᵗ half 17ᵗʰ century) and especially from François Lefebvre (*fl. c.*1635), *Livre de fleurs et de feuilles pour … l'art d'orfèvrerie* (1635). Minutely executed on a white ground, the flowers were separated one from the other in order to bring out their individual characteristics. The Victoria & Albert Museum in London holds two such watch-cases,⁴⁵ both of which can be compared with a watch in the British Museum in the form of a perfume flask set with garnets and decorated with flowers on a white ground.⁴⁶ Towards the middle of the century, naturalistic designs, after Jacques Vauquer (1621–86) and Gilles Légaré (1617–63), comparable with contemporary still life paintings, became fashionable. Covering the entire surface of the case, flowers in full bloom press one against the other in an apparent disorder. A remarkable example is offered by a watch by John Adamson (*fl.* 1686–98) decorated with a variety of flowers (Figure 154).⁴⁷

The craftsmen of Paris and Blois also used painting on enamel for historical scenes.⁴⁸ Their customers liked discovering miniature versions of works by fashionable painters: episodes from the life of Christ and the Virgin, subjects drawn from the love affairs of the gods (Figure 155) and the heroes of Antiquity, and battle scenes.

Representing various episodes of a story on the several faces offered by a watch was characteristic of production between 1635 and 1660. Following engraved versions, the enamellers reproduced paintings by Simon Vouet (1590–1649), Sebastian Bourdon (1616–71), Charles Poerson (1652–1725), Jacques Stella (1596–1657), and Laurent de La Hyre (1606–56). Sometimes they worked directly from original works such as the *tondi* of Bourdon and Poerson.⁴⁹ These small works in circular form were perfect models for enamellers who copied them onto the cases of round watches having a diameter of five to six centimetres. Many watches are known with a version of Bourdon's *Abduction of Helen* painted on the back of the case, and on the cover with *The Judgement of Paris*, a *tondo* by the same artist known today only by a drawing. Watches by different makers employ these same scenes after Bourdon, for example Elias Weckherlin (pre-1646–89) in Augsburg (now in the Louvre), Mathias Riepold (late seventeenth century) in Regensburg (now in the Rijksmuseum), Jacques Goullons (pre-1626–71) in Paris (at the Patek Philippe Museum), Grégoire Gamot (pre-1628–73) in Paris (now at the Metropolitan Museum), and a box with a later movement (ex-Time Museum, Rockford).⁵⁰ The various surfaces of some dozen watches offer almost identical versions of scenes from Heliodorus' story of *Theagenes and Charicleia* as painted by Charles Poerson (1609–67).⁵¹

Cases signed by enamellers are extremely uncommon. Only Robert Vauquer (1625–70), Jean II Toutin (1619–post-1660), and Henri Toutin (1614–84) appear during this period. The signature of Henri Toutin appears twice on the paintings that adorn a spectacular watch by Antoine Mazurier (mid-seventeenth century) commemorating the marriage of William II of Orange with Mary Stuart in May 1641.⁵² Robert Vauquer signed a watch by Jacques Goullons of Paris showing the *Battle of the Milvio Bridge* by Jules Romain (?1499–1546), and military scenes after Antonio Tempesta (1556–1630).⁵³ Jean II Toutin left his name attached to three cases carrying mythological scenes after Cornelius Poelenburgh (1594–1667).⁵⁴

The advantages that painting on enamel offered—precision of line, a variety of subtle colours, resistance to light and humidity— were immediately capitalized upon for portraits. Isaac Gribelin

---

43 E.g. Victoria & Albert Museum, London inv. M81–1913. Cardinal 1985, plate 96, right.
44 E.g. Victoria & Albert Museum, London inv. 362–1855. Cardinal 1985, plate 96, left.
45 Musée international d'horlogerie, inv. I-1219 et I-1331; Cardinal & Piguet, 1999, N° 127 et 126.
46 British Museum, inv. MLA 1874, 12–14. 1. Cardinal & Piguet 1999, N° 95.
47 Musée international d'horlogerie, inv.I-1116; Cardinal & Piguet 1999, N° 135.
48 For an introduction to enamel decoration at this period see Cardinal 1985, 130–155 and Cardinal 1999. A more detailed discussion is in preparation by Bull, Cardinal, & Turner.

49 See Cardinal 2018a.
50 Musée du Louvre, OA 8320, Cardinal 1984, N° 39; Rijksmuseum, BK-NM-636; Patek Philippe Museum, S 446; The Metropolitan Museum of Art, inv. 17.190.1625, Williamson 1912, N° 38. *Masterpieces from the Time Museum*, Sotheby's, New York, 2 December 1999, N° 36.
51 Examples: Musée du Louvre, OA 8318, Cardinal 1984, N° 21; Patek Philippe Museum, Genève, inv. S 200; Musée Paul-Dupuy, Toulouse, inv. 18 229, Hayard 2004, 163–5.
52 Rijksmuseum, inv. NM 638; Cardinal 1985, 138, repr. 104.
53 Museo Poldi Pezzoli, Milan, inv. 5909; Cardinal & Galli 2017, N° 8, 140–143.
54 Musée du Louvre, OA 7075; Cardinal 2000, N° 116. Patek Philippe Museum, Genève, inv. S 178. Walters Art Gallery, Baltimore, inv. 58.136; Cardinal 2002, repr. p. 44. A fourth case in the Octavius Morgan Bequest in the British Museum was destroyed during the Second World War.

Figure 154 Watch by John Adamson, London, c.1680, the case painted on enamel with flowers in the style of Jacques Vauquer. Musée International d'Horlogerie, La Chaux-de-Fonds, inv. I-1116. Photo MIH.

(*fl.* 1616–51), Jean Petitot (1607–91), Jacques Bordier (1616–84), and Louis du Guernier (1614–59) all excelled in this genre. Boxes and watches with portraits were frequently used as presents. Two pieces signed by Jacques Goullons, noted for the luxurious watches he sold from his shop on the Ile de la Cité, Paris, are distinguished by the high-ranking personages depicted. One of them, dated 1642, shows Louis XIII and Richelieu,[55] the other, probably made between about 1645 and 1648, carries an equestrian

---

[55] Victoria & Albert Museum, inv. 7543-1-1861; Cardinal 2018b, 193 & 198.

Figure 155 Watch by Nicolas Morel with the case painted on enamel by J. Loyseau, (a) *Venus and Adonis;* (b) *Susanna and the elders.* Musée Patek Philippe, Genève, Inv. S-1064. Photo: Colin Crisford.

portrait of Louis XIV at about the age of ten and doubtless belonged to him.[56]

Pierre Huaud (1612–80), and his sons Pierre II (1647–98), Jean Pierre (1653–1723) and Ami (1657–1724) also painted on enamel and continued their predecessors' art by developing a characteristic style marked by pointillism and strongly contrasting colours (lapis blue, vermillion red, and saffron yellow). They supplied cases to watchmakers throughout Europe. Some are signed by the eldest of the brothers who became miniaturist to the Elector of Brandenburg, and dozens are signed by Jean Pierre, by Ami, or both, since they formed a partnership in 1682. Their subjects are varied, copied primarily from engravings of works by Simon Vouet and Laurent de la Hyre. Examples include a case signed 'Huaud l'aisné pinxit à Genève' showing a combat on horseback after Philippe Wouvermann (1619–68); another by Jean Pierre depicts *Juno and Iris,* after Vouet. A box signed by the brothers together shows *Aurora and Cephalus,* after Vouet, and two *Judgements of Paris,* after La Hyre, one of which can be attributed to Pierre II, the other to the brothers together.[57]

The 'grand style' that developed in France during the last third of the seventeenth century, a baroque style tempered by the rules of classical art, monumental, rich but rigorous, offered the 'Versailles' model to all Europe, its propagation helped by the quantities of Huguenot artists and craftsmen forced into exile, especially to London and Amsterdam. *Ornemanistes* of French origin had a particularly strong influence on the decorative arts, shown by the large number of engravings they produced and the number of their followers in other countries. Their designs can be found on tapestries, carpets, parquet floors, marquetry work on furniture, ceilings, staircases, iron gates, small arms, clocks, and watches. Jean Lepautre (1618–82), the creator of some thousand designs, developed a style dominated by the use of classicizing acanthus leaves. Influenced by him, the designer-engravers of

---

56 The Metropolitan Museum of Art, inv. 1975.1.1244; Vincent, Leopold, & Sullivan 2015, 78–83; Cardinal 2018b, 193 & 198. Inventories made in 1639 & 1652 of Goullons' stock and the corpus of preserved watches are representative of the variety of fashionable subjects found on painted enamel watches. Cardinal 2018b, 196–201.

57 Musée du Petit-Palais, Paris, inv. O. Tuck 240 et inv. O. Tuck 233. Cardinal 1999, N° 51 et 52.

Musée du Louvre, OA 8435, OA 8326, OA 8440. Cardinal 1984, N° 97, N° 20, N° 98. For a representative collection of thirteen Huaud enamelled watches see Bull 1978.

ornamental motifs incorporated this foliage into their designs, which were often specifically intended for application to clocks and watches. Pierre Bourdon (fl. c.1700) in Paris, Daniel Marot (1661–1752) in the Hague and London, and Simon Gribelin (1662–1733) in London designed clock cases, watch cocks, and hands where acanthus leaves spread sinuously and luxuriously and, accompanied by other flowers, especially the tulip, over the back plates of (mainly) English table clocks.[58]

With the application of the long pendulum to domestic clocks, a typical form of floor-standing clock emerged in England.[59] It would remain in use, with modifications, until the early twentieth century, appreciated for its elegance of line and the way in which it could be made to harmonize with other furniture. Usually between 1.80m and 2.50m in height, it was invariably made up of three parts: a cube-shaped base, a long trunk, and a hood surmounted by a fronton (Figure 78). The case might be veneered in ebony, walnut, olive wood, pear, or mahogany. Following current fashion it could be ornamented with life-like marquetry flowers or with lacquered panels with orientalist decoration.[60] Mouldings, pilasters, barley-twist columns, and gilt bronze mounts completed this decoration. A clock by Ahasuerus Fromanteel (1660–65) with the arms of the Duke of Norfolk, one by Thomas Tompion in a case adorned with rosaces and marquetry work, and a large clock with complications by Daniel Quare, perhaps made for Hampton Court Palace, in a walnut case with gilt bronze mounts are worth mention.[61] A fine series of longcase clocks in the Clockmakers' Company collection includes one by Tompion (c.1672) in ebony veneer with a carved and sculpted fronton and a silver hour circle with gilt cherub; also included is an equation clock by George Graham (N° 779, c.1750), distinguished by a plain case in mahogany, its form underlined by simple mouldings and with a silvered dial with gilt corner pieces.[62] A **regulator** by John Ellicott from the same period is in walnut veneer with inlaid light wood stringing.[63]

The longcase clock was also favoured in the Low Countries and Germany. Particularly decorative examples were produced by Dutch clockmakers such as one by Jacob Hasius (c.1682–post-1747) of Amsterdam, with a marquetry decoration of naturalistic flowers, barley-twist columns, and an open sculpted fronton emphasizing the dial.[64] Another example, c.1750, is an impressive carillon clock (3.05m high) in burr walnut veneer with marquetry decoration surmounted by Atlas supporting the world and two *fama*.[65]

Bracket or table clocks, 30–55cm high, could be veneered in ebony or walnut (Figure 79). The basic form was that of a simple rectangular structure surmounted by a flat dome to which a carrying handle was attached. It could be decorated with silver, gilt brass or bronze affixes. A series in the Clockmakers' collection displays its development.[66] A clock by Samuel Knibb, c.1665, has a soberly moulded ebony case and a triangular fronton that contrasts with the dial plate richly engraved with flowers. One by Thomas Tompion, c.1675, in ebony, has gilt bronze mounts and dome with a bronze handle.[67] Attention is drawn to the dial with a silvered hour ring set off by gilt, relief-chiselled, and engraved spandrels. Towards 1720–30, an arch was added to the dial plate to carry additional indications.[68] It accentuated the architectural aspect of the clocks, as did the balusters and flame carriers set on the corners of the dome piece and the columns placed on the angles of the case.[69] A clock by Joshua Gibbs (c.1675–post-1720) is a good example of the type.[70] The 'English clock' model spread throughout the continent, especially to Italy, the Low Countries, and Germanic regions, as is shown by many clocks signed by makers from Milan, Munich, Augsburg, Vienna, Prague, Basle, and Winterthur.[71]

In France, from the 1660s, table clocks that could also be mounted on wall brackets, the *religieuses*, appeared (Figure 76). They were made in ebony, walnut, or pear, carried on hoop feet, and reflected contemporary architectural forms. They were framed by pilasters or colonettes either fluted or in barley twist, with the dial face surmounted by a triangular or incurved fronton decorated with fire pots, all imitating the façade of a baroque building. The dial, set on a black or red velvet ground, was placed in an arcade simulating a porch. A group of clocks in the Musée

---

58 For back plate engraving see Dzik 2019, especially 94–95 (acanthus leaf), 156–157, 210 (Lepautre). For the major influence of Marot, see Baarsen et al. and for his life Murdoch 2008 in *ODNB*. Gribelin who engraved watchcases in his old-fashioned Louis XIII style, became a member of the Clockmakers' Company in 1685. Loomes 2014, 230.
59 Pendulum clocks by Ahasuerus Fromanteel were on sale in London from October 1658. The name *grandfather clock* often applied to them derives from a massively popular song 'Grandfather's Clock' written in 1876 by the American songwriter Henry Clay Work (1832–84).
60 Two remarkable early eighteenth-century longcase clocks that illustrate this kind of decoration are preserved in the collections of the Musée d'horlogerie, Le Locle, Cardinal, & Mercier 1993, 94–5.
61 British Museum, Tait 1983, 57–8, repr. inside back cover and fig. 62.
62 Clutton & Daniels 1975, N° 543, N° 554.
63 Musée international d'horlogerie, IV-628; Cardinal & Piguet 1999, N° 8; Cardinal & Mercier 1993, 45.

64 British Museum; Tait 1983, fig. 62.
65 Musée international d'horlogerie, IV-354; Cardinal & Piguet 1999, N° 9. The dial is marked for the Dutch East Indies Company, *VOC* (Vereenigde Oostindische Compagnie).
66 Clutton & Daniels 1975, 81–3.
67 Clutton & Daniels 1975, N° 559, N° 562.
68 Bassermann-Jordan 1905, 248–9, 286–300.
69 If the answer to Fabian's question 'could it have been Wren?' who influenced the design of mid-seventeenth-century English architectural clock cases must remain in the negative, his presentation of the topic usefully underlines the close relationship between the two. See Fabian 1977.
70 Musée international d'horlogerie, IV-498; Cardinal & Piguet 1999, N° 10; Cardinal & Mercier 1993, 44.
71 See Plomp 1979.

des Arts décoratifs in Lyon, notably by Jacques Thuret (1669–1738), Mathieu Marguerite (d. post-1702), Balthazar II Martinot (1636–1714), perfectly illustrates the type.[72] In the last third of the century the use of brass, pewter, ebony, and red tortoiseshell, the addition of a flat dome surrounded by a balustrade, and the multiplication of gilt bronze motifs, reveal a taste for a more ostentatious presentation exemplified in two clocks, one by Nicolas Gribelin (1637–post-1700) of Paris, the other by Jacques Huguet (d. 1733) of Paris.[73] Another example, by Joseph Baronneau (d. c.1711), is distinguished by its case in tortoiseshell and brass inlay, its brass cartouche engraved with the maker's name, and its skirt dropping from the bottom of the case—an addition typical of the 'Grand Style'.[74]

The introduction of the balance-spring[75] for watches led also to an increase in the size of their cases and their cocks (Figure 86). The thick, round-bellied cases of the French 'oignon' (Figure 88), made of gilt brass or silver offered space for a rigorously symmetrical decoration of trailing scrolls of acanthus leaves, lambrequins wherein birds appear perching, grotesque masks, or mythological or allegorical figures. A striking and repeating watch specially made in 1709 for the academician and collector Léon Louis Pajot, Comte d'Onzembray (1678–1754) by François Joseph de Camus (1672–1732) has a pierced, gold repoussé case engraved in the style of Bourdon, the band carrying four insets representing the seasons, the back harking back to the earlier fashion for mottos with 'Rien de bas ne marest' (nothing below arrests me), surrounding an eagle looking towards the Sun.[76]

While the single-cased thick verge watches were characteristic of French and some Swiss production, pair-cased watches were typical of English work from c.1680 onwards. The inner cases of striking watches were adorned with richly engraved pierced friezes following fashionable designs, notably those of Simon Gribelin (1666–1733) in whose *Book of Ornaments usefull to Jewellers, Watchmakers and all other Artists* the acanthus leaf is abundantly deployed. Two watches by Daniel Quare and George Graham have silver cases that illustrate this style.[77] Another watch by Quare is distinguished by two silver cases pierced and engraved with matching friezes of grotesque masks, birds, and acanthus leaves. They were no doubt originally protected by a third outer case.[78]

Cases for non-striking watches were generally plain but could be finely engraved. A watch of c.1680 by the Huguenot maker Daniel Le Count (pre-1676–1738), has an interesting gold case featuring a portrait of Charles II in a medallion with the royal supporters and framed with acanthus leaves.[79] A watch by Thomas Tompion has a gold case that is signed and dated 1683 by the expatriate Geneva engraver Abraham Martin (fl. 1682–1713), and decorated with trailing acanthus leaves inhabited by animals in the style of Gribelin.[80] Outer cases covered in leather, tortoiseshell, shark skin, or ray skin were often further decorated with small gold or silver nails arranged in scrolling patterns.

From the 1680s onwards, the designs of Jean Bérain (1640–1711)—the 'bérinades'[81]—dominate contemporary ornament. In the midst of porticos formed of symmetrical trellises and arabesques on several levels, Bérain placed apish figures, Chinese style scenes, or perching birds. His connections with the clock- and watchmaking world through his son-in-law Jacques Thuret should be underlined, for his influence is noticeable on the decoration of watches and the marquetry of clock cases.

With their pediments, coiling columns, fluted pilasters, vases, and statuettes, clock cases align with contemporary architecture. In France, the king's cabinet-maker, André Charles Boulle (1642–1732), created several different models.[82] His multiple talents as cabinet-maker, marquetry-maker, chaser, founder (he had his own foundry), gilder and sculptor allowed him to create pieces that combined gilt bronze statues harmoniously with joiner's work.[83] Bronzes were sometime supplied to him by his neighbouring sculptors in the Louvre Galleries such as Nicolas Coustou (1658–1739), or François Girardon (1628–1715). From the early 1700s, several clocks show his collaboration with the architect-designer Gilles-Marie Oppenordt (1672–1742).[84] Boulle was one of the first craftsmen in France, if not the first anywhere, to produce marquetry-decorated cabinets for longcase clocks. One such, containing a movement by Antoine Gaudron, dated to c.1675, may be among the earliest.[85] Towards 1680, Boulle completed another for Pierre Duchesne (d. post-1702), which was delivered to Louis XIV, and of which the marquetry is in copper and pewter on a tortoiseshell ground. It is very close to another housing a movement by Gaudron.[86] One of Boulle's most successful models was 'Love Overcoming Time', based on a design by François Girardon, which was widely used until the 1730s when

---

72 Musée des Arts décoratifs, Lyon, Mazur 2008, N° 1 to 6 (inv. MAD 884, 1053, 803, 2083, 860, 680).

73 Musée des Arts décoratifs, Paris, inv. 2014 12 1 and inv. 5353.

74 Musée d'horlogerie du château des Monts, Le Locle; Cardinal & Mercier 1993, 92.

75 Discussed in this volume, Chapter 6.

76 Musée du Louvre, OA 8310. Cardinal 1985, 157, repr. 120. Cardinal 1984, N° 69.

77 Musée international d'horlogerie, inv. I-568 and I-398; Cardinal & Piguet 1999, N° 161, N° 166. The outer case of the watch by Quare is in leather decorated with small silver nails (piqué-work); both cases of the Graham watch are pierced and engraved.

78 Victoria & Albert Museum, inv. 1362-1904. Cardinal 1985, 165, repr. 127.

79 Ashmolean Museum, WA 1947.191.84. Thompson 2007, N° 18.

80 Ashmolean Museum, WA 1947.191.86. Thompson 2007, N° 19.

81 See *Ornemens inventez par J. Berain...*, particularly the clock designs, and the interiors with clocks placed on a chimney-piece.

82 Ronfort 2009, 459–95.

83 Boulle c. 1724.

84 Ronfort 2009, 79–82. Some clock designs by Oppenordt are reproduced in Maurice 1967.

85 Edey 1982, N° 27.

86 Duchêne: Ecole nationale supérieure des beaux-arts, Paris; Augarde 1996, 195 and Ronfort 2009, 71, repr. 11. Gaudron: Getty Museum, Malibu. Wilson et al. 1996, N° 1, 3–9.

it appears in paintings by Jean François de Troy. An example containing a movement by Jacques III Thuret can be dated to the 1720s, but a drawing by Boulle shows another version that can be dated to c.1690.[87] The dial for this model was derived by Boulle from François Duquesnoy (1597–1643). Collaboration between Boulle and Oppenordt appears in two floor-standing clocks (Figure 156), one made c.1719 for the Count of Toulouse, for which a preparatory drawing has survived, and the other is a 'clock of the four continents'.[88]

The taste for clocks as furniture, fine pieces of joiners' work decorated with bronzes, was widespread in the early eighteenth century. This being so it is not astonishing that the idea emerged of incorporating clocks into a piece of furniture—a desk, tables, bookcases, dressers. Such pieces were usually specially commissioned, specifically some from Boulle's workshop, for example, a large wardrobe in ebony and marquetry with applied bronzes, and the centre of which is built around a floor-standing clock signed by Gaudron. The dial is contained in a cartouche formed by three cherubs carrying the emblems of love and learning. The shaped door is partly glazed and incorporates a winged mask of time surmounted by a sand-glass.[89]

ROCOCO: FORMS FREED AND INSPIRED BY NATURE

Rococo decoration developed in France in the early eighteenth century and spread through Europe from the mid-1730s onwards. Its success largely resulted from the works of the painters François Boucher (1713–70), Jacques de la Joue (1687–1761) and the goldsmiths Juste-Aurèle Meissonnier (1695–1750) and François Thomas Mondon (1709–55). The aesthetic they espoused was based on asymmetry, on oblique, sinuously curving lines, and jagged outlines. Shells, waterfalls, flowers with cartouches composed of chicory leaves, palm leaves, and putti were preferred motifs. The style manifested itself particularly in the working of metal: vases, trays, terrine dishes, tobacco boxes, watches, and clocks offer many examples. It is noteworthy that sculptors, founders, and goldsmiths are among the leading exponents of Rococo art. Meissonier, creator of several clock models,[90] has already been mentioned, as has Mondon, a watch- and snuffbox chaser, Thomas Germain (1673–1748), Jacques Caffieri (1678–1755), and Charles Cressent (1685–1768), sculptors, founders, and chasers.

In the course of the period clocks became indispensable items of furniture in every comfortable home, placed on mantelpieces or hung on the walls of drawing rooms, in studies and bedrooms[91] Earlier forms were continued while subject to the new style, asymmetrical and full of movement. Sumptuous clocks were created by the Paris masters adorned with marquetry in precious woods, brass, and tortoiseshell, with gilt bronzes, or with oriental lacquers. The Regent's well-known cabinet-maker Charles Cressent supplied two versions of a seconds beating clock of curved profile decorated with gilt bronzes. The more elaborate was adorned with a figure of time and an allegory of the winds; the simpler version offered only heads of the Zephyrs beneath the dial.[92] For a clock with astronomical indications by Jacques Jérôme Gudin (1732–post-1789) of Paris, Pierre Duhamel (1723–1801) produced a highly luxurious case of flowing asymmetrical lines accentuated by applied gilt bronzes, and richly adorned with floral marquetry.[93] Jean Pierre Latz (1691–1754), was a cabinet-maker particularly attuned to the Rococo style as is shown by a majestic case for a seconds-beating clock by Michel Stollenwerck (d. 1768).[94]

Derived from an ornament characteristic of the rococo, the tablet or cartouche wall clocks worked in a single piece of bronze (cartel) were widespread (Figure 82).[95] Like Boulle, Cressent produced a popular model, *Love and Time*, of which many examples survive.[96] Standing clocks whether placed on a mantelpiece or a mural console, shared the same sinuous forms: volutes, floral garlands, shells, putti, animals, and mythological figures which make up a rich bronze decoration to which marquetry panels were sometimes added. Two carillon clocks with bronzes probably designed by Jean Claude Duplessis offer a good illustration.[97] A Swiss musical clock by Pierre Jaquet-Droz (1721–90) at La Chaux-de-Fonds with a case by the Paris case maker André Foullet (fl. 1735–75) incorporates a scene from La Fontaine, the fox and the stork, and showcases the international nature of the style with its volutes and leaf decoration and its floral incrustations in tortoiseshell, horn, mother-of-pearl, and bronze.[98] Fascination with the exotic, and the discovery of wild beasts, underlies the return of a vogue for clocks in which a rhinoceros, a dromedary, or an elephant carries the movement. The founder Jean Joseph de St

---

87 Wallace Collection, F 43 and F 55. Hughes 1994, 30–1.

Inv. 723 D 6. Hughes 1994, 30–1; Ronfort 2009, 332–3.

88 Musée du Louvre, OA 6746. Ronfort 2009, cat. 76. Bibliothèque de l'Arsenal, Paris, ARS Inv. 1943-196. Turner 2017, N° 128.

89 Wallace Collection, F 429. Hughes 1994, 24–5.

90 *Œuvre de Juste Aurele Meissonnier peintre sculpteur architecte & dessinateur de la chambre et cabinet du roy*, c.1736 with reproductions in Maurice 1967; for an astronomical floor clock after a Meissonier design, Augarde 1996, 63.

91 A multitude of paintings attest to the importance of the clock in interior decoration (cf. this volume Chapter 25), while the numbers listed in inventories are a further proof of their presence.

92 Elaborate model, Pradère 2003, 176–99; 294–305; simplified model Musée des Beaux Arts Lyon, inv. MAD 1161, Mazur 2008, N° 13.

93 Musée des arts et métiers, Paris, inv. 4148; catalogue JB, 1949, 142.

94 Art Museum, Cleveland, inv. 1949.200; Augarde 1996, 18, repr. 6.

95 A good illustration is provided by the painting by Boucher *Le Déjeuner*, 1739 (Musée du Louvre), where in a rococo drawing room a *cartel* is hanging on the wall.

96 For example, The Metropolitan Museum of Art, inv. 1971.2016.27, Vincent, Leopold, 2015, N° 36. Musée du Louvre, OA 9586, Alcouffe 2004, N° 30. Rijksmuseum, inv. BK-18,018. Wallace Collection, F 92, Hughes 1994, 42–3. See also Pradère 2003.

97 Wallace Collection, F 96 and 97. Hughes 1994, 48 and 50.

98 Musée international d'horlogerie, inv. IV-20, Cardinal & Piguet 1999, N° 88.

Figure 156 Floor standing pedestal clock, 'the four continents', signed Julien Le Roy (dial) and Gilles Martinot à Paris (movement), c. 1720. Ebony, brass inlay on tortoiseshell, gilt and chased bronze. H. 2m 90. Bibliothèque de l'Arsenal, Paris, Inv. ARS. 1943-196.

Germain (1719–1791),⁹⁹ made a speciality of them. A good example is provided by a carillon clock for which the rhinoceros in patinated bronze could have been modelled on that exhibited at Paris in 1749.¹⁰⁰

Chinese scenes in bronze or in porcelain were fashionable in the mid-century. Saint-Germain produced a model in bronze showing two Chinamen supporting the case of the movement, which is surmounted by a native American with a feathered headdress.¹⁰¹ Pastoral scenes in the style of Boucher, Chinese scenes in Saxony porcelain, or in bronze varnished to imitate lacquer, and porcelain flowers from Vincennes or Mennecy, offer many variations.¹⁰² **Cartonniers**, chests of drawers, bookcases, and desks could all, in their most opulent versions, incorporate a clock movement, so increasing their use and sophistication. A bookcase of this type is shown in the portrait of Mme de Pompadour in her study by Boucher, and a movement by Etienne Le Noir adorned a **cartonnier** with lacquered Chinese decoration (1746), by Bernard II van Risenburgh (1700–60).¹⁰³

Whether chased or enamelled, French watches were accommodated to the new styles and as in the previous century offer a panorama of contemporary painting, while their close relationship with snuffboxes is clear throughout the century. Volutes, shells, and flowers frame scenes painted on enamel after engravings of the work of Antoine Watteau (1684–1721), Boucher, Lancret, Pater, Charles van Loo (1705–66), or Jean-Honoré Fragonard (1732–1806). The Louvre collection alone features watches portraying *L'Education de l'Amour*, *Le Printemps*, *Le Panier mystérieux* by Boucher, *Le Conteur* by Watteau, *La Musique* by Van Loo, and *Le Colin-Maillard* by Fragonard.¹⁰⁴ Matching **chatelaines** allowed the watch to be carried suspended from a belt or girdle.¹⁰⁵ The success of watches thus decorated in French fashion is attested by a report in an English newspaper:

> I HAVE known an Englishman at Paris give a watch in a plain case made by Mr MUDGE, the scholar and rival of Mr GRAHAM, for one of Mr Le Roy's which he himself confess'd was inferior in real value, because it was enclosed in a more elegant outside, and more than one of Mr. GRAHAM'S has been exchanged with money to the bargain; for others of more taste in the case, and less truth in the workmanship within, at the same shop in Paris.¹⁰⁶

The well-to-do in the reign of George II appreciated cases set with hard stones such as jasper and agate mounted on engraved or repoussé cases in gold. Precious stones and pearls accentuated the colourfulness of the designs. The Olivier collection in the Louvre features a series of watches, some with matching chatelaines, that illustrate the refinement and the variety of these watches (Figure 115).¹⁰⁷ Several models for this kind of watch, witnesses to the fantasy of English Rococo, were published during the 1740s by William de la Cour (1700–67).

Similarly attractive to the mid-eighteenth customer were watches with outer repoussé cases carrying religious or historical scenes.¹⁰⁸ Looking like miniature sculptures, these cases in gilt metal, silver, or gold present figures either in low-relief, or almost in the round when they are detached from the background. Many of the subjects derive from paintings known through engravings and skilfully reinterpreted in three dimensions. Several goldsmiths specializing in this work include George Michael Moser (1706–83) and Henry Manley (1698–1733), both of whom were particularly outstanding.¹⁰⁹

NEOCLASSICISM: PURE FORMS AND CLASSICAL MODELS

A new design tendency, fundamentally opposed to the Rococo, appeared in Europe from the mid-eighteenth century onwards. An increasing influence from the work of the classicizing Andrea Palladio (1518–80), combined with the classical vestiges visible in Rome (and elsewhere in Italy) and the discovery of Herculaneum and Pompei influenced the taste of artists, cultivated amateurs, and art theorists. Notably in Britain and France, a neoclassical style imposed itself, and stretched across Europe and its colonies, its partisans inspired by ancient models favouring the use of austere forms and noble ornament. Closely linked with the engraver Piranesi (1720–78), William Chambers (1723–96) and Robert Adam (1728–92) introduced a neoclassical approach to all forms of English interior decoration. In France, Charles Nicolas Cochin (1715–90), Jacques Germain Soufflot (1713–80), Charles Louis Clérisseau (1721–1820), and Louis Joseph le Lorrain (1715–59) were in the vanguard of a considerable group of artists following the new style. A subtle equilibrium was established between sober lines and graceful ornament that would dominate the decorative arts, particularly those of cabinet-making, porcelain, and goldsmithing. Luxurious clock cases were created to house movements constructed by preeminent makers (Figure 157). Mantel clocks, much appreciated exports, were particularly well served by the bronze makers, certain of whom, thanks to their training in sculpture and drawing, were able to create their own models without disdaining patterns supplied

---

99 Augarde 1986, 131, 138.
100 Alcouffe 2004, N° 34, 78–81.
101 Musée des arts décoratifs, Lyon, inv. MAD 1041; Mazur 2008, N° 15. The Metropolitan Museum of Art, inv. 2019.283.70.
102 Verlet 1987, 122–32; 180–3. Much of Verlet's work is devoted to the analysis of clock decoration.
103 Getty Museum, inv. 83 DA.280. Wilson et al. 1996, 78–85, N° IX. Another example, Wallace Collection London, Hughes 1994, 52–3.
104 Cardinal 1984, pp. 119–28.
105 For numerous examples see Cummins 2010, chs. 2–4.
106 *The Spectator*, xiii, 1753. We are grateful to Jeremy Evans for this reference.

---

107 Cardinal 1984, 143–151.
108 For a general, detailed discussion of them see Edgcumbe 2000.
109 Edgcumbe 2000, 70–83 (on Manley); 85–132 (on Moser).

Figure 157 Hour and half-hour striking clock, 'Leclerc à Bruxelles', bronze attributed to Jean André Reich (1752–1817). in the form of a dromedary in gilt and patinated brass and bronze, in which the Rococco taste for exotic animals is married to a neo-classical base. Photo Jean-Baptiste Buffetaud, courtesy of Chayette & Cheval, auctioneers, Paris.

by designers, architects, and sculptors.[110] Such was the case of Robert Osmond (1711–89), Jean Louis Prieur (1732–95), Etienne Martincourt (1735–post-1791), François Vion (*c.*1737–post-1790), Pierre-Philippe Thomire (1751–1843), and Pierre Gouthière (1732–1813).[111]

Symmetrical arrangements dominated by straight lines featured intermingled allegorical and Antique subjects with the decorative motifs of the neoclassical vocabulary: rose medallions, Greek key patterns, palm leaf mouldings, laurel garlands, gadroons, sphinxes, ram's heads, and lion snouts. Etienne Martincourt was the author of a model carrying allegories of geography and astronomy that was realized by Charles Le Roy.[112] Signing clearly on the base, Jean Louis Prieur claimed paternity in a clock case representing 'awakening' with a cherub aroused by the crowing of a cock.[113] A model by the bronze caster François Vion, 'Sorrow', also found popularity.[114]

The taste for the Antique favoured the appearance of clocks in the form of vases, lyres, porticos, truncated columns, and pyramids. A representative display is offered in the Wallace Collection in London.[115] Clocks inspired by the balloon flights of the Montgolfier brothers in 1783 were fashionable for a short period,[116] while wall clocks in particular adopted classical motifs intended to enhance their symmetrical form. If the making of clock cases shows the role played by bronze casters, that of sculptors and designers is also made clear, for example, by a clock ordered by the city of Avignon in honour of the Marquis of Rochechouart. Realized by Pierre Gouthière (1732–1813) it featured an allegory of the Rhone and the Durance rivers, but was made after a model by the sculptor Louis Simon Boizot (1743–1809).[117] A clock by Robert Robin (1744–99) delivered to Marie-Antoinette in 1788, 'Two vestals carrying the sacred flame', is the work of Pierre Philippe Thomire (1751–1843), the design having been provided by Jean Démosthène Dugoure (1749–1825), and the clay pattern by Louis Simon Boizot (1743–1809).[118]

Some celebrated sculptors, such as Augustin Pajou (1730–1809), Etienne Maurice Falconet (1716–91), Clodion (1738–1814), or Jean Antoine Houdon (1741–1828), provided designs that were adapted for clocks by the founders, gilders, and chasers. Pajou conceived the model of the clock cast for Frederick V of Denmark in 1765. Crowned by the figure of Fame carrying the royal arms, it is decorated with allegories of Commerce, Navigation, Agriculture, the Arts, and the Sciences.[119] The design of two Venus clocks is attributed to Falconet.[120] A clock presenting Uranus after a design by Houdon was realized in the Lepaute workshops around 1765,[121] while that conceived by the sculptor Laurent Guiard (1723–88) for Madame Geoffrin, 'Study', which depicts a seated woman reading, was particularly successful. Many examples exist, such as one cast by Edme Roy (d. 1764), furnished with a movement by Ferdinand Berthoud.[122]

Ornamental designers, in particular François Joseph Bélanger (1744–1818), Richard de Lalonde (*fl.* 1770–90), Jean Démosthène Dugourc (1749–1825), Henri Salambier (1753–1820), Jean François Forty (*fl.* mid-eighteenth century), and Jean Charles Delafosse (1734–89), offered fashionable models, collections of which were sometimes published, for clocks.[123] The architect François Joseph Bellanger (1744–1818) who constructed the 'Bagatelle' for the Comte d'Artois, also designed the interior decoration and so a clock with two winged female sphinxes that was delivered by Lepaute in 1781. It may be compared with two other sphinx clocks by Lepaute.[124]

A number of well-known cabinetmakers included the making of clock cases among their specialities. As in contemporary furniture in general, marquetry was allied with gilt bronze when smooth surfaces, veneered in ebony or mahogany, were not preferred. Balthazar Lieutaud (1720–80), Nicolas Petit (3rd quarter 18th century), and Martin Carlin (*fl.* 1746–80) produced floor standing clocks in which mechanical precision and the neoclassical style were in perfect harmony. Lieutaud perfected the making of longcase clocks in which the decorative richness of the gilt bronzes was balanced by the rigour of the geometrical form. A remarkable example is the clock made in collaboration with Berthoud for the Duc de Praslin.[125] Also with Berthoud, Lieutaud made two cases similarly ornamented in collaboration with Caffieri, one veneered

---

110 In the decoration of drawing room clocks, the displacement of marquetry work by bronze figures, already noticeable during the Rococo period, was accentuated in the second half of the century, accompanying a fashion for theme clocks.
111 See in Ottomeyer & Pröschel 1986, vol. II, the articles by Roland de l'Epée, on the Osmond family, 539–48; Christian Baulez, on Gouthière, 561–642; David Harris Cohen on Thomire, 657–68.
112 Getty Museum, inv. 73.DB.78; Wilson et al. 1996, 115–23.
113 Alcouffe 2004, 137–8, the movement signed by Joseph Léonard Roque, free in 1770.
114 Ottomeyer & Pröschel 1986, i, 247. Musée Carnavalet; Augarde 1996, 242. Musée des arts décoratifs, Cardinal 1983, 25, but there described as 'Venus and Cupid'.
115 Others are to be found in the catalogue of Dupuy-Baylet 2006.
116 For an opulent example in the Musée des Arts Décoratifs Paris, see Cardinal 1983, 29.
117 Wallace Collection F. 258, Hughes 1994, 68–9.
118 Musée des arts décoratifs, inv. MIN. INT.ss (N° 2); Verlet 1987, 326–7. A variant is conserved in the Getty Museum, attributed to

---

the sculptor Jean-Guillaume Moitte, Wilson 1996, N° XVII, 124–31; a similar version is in the Musée international d'horlogerie, La Chaux-de-Fonds. Cardinal & Piguet 1999, N° 91.
119 Amalienborg palace, Copenhagen. Verlet 1987, 34.
120 Wallace Collection, F 260 and F 261. Hughes 1994, 72–6.
121 The Metropolitan Museum of Art, inv. 19.180.2; Vincent, Leopold, & Sullivan 2015, N° 39.
122 Wallace Collection, F 267. Hughes 1994, 54–5.
123 Maurice 1967 reproduces designs by Delafosse (2768), Salmabier (1775), and Forty. See also Guilmard 1881.
124 Wallace Collection, F 269. Hughes 1994, 86–7. Metropolitan Museum, Edey 1982, 82–5.
125 Royal Palace, Madrid. Augarde 1966, repr. 19.

with tulipwood and kingwood, the other with ebony.[126] Both are crowned by an imposing sculpture of Apollo in his chariot. In a simpler version, this is replaced by a classical urn.[127]

In his furniture *à système*, David Roentgen (1743–1807) did not fail to incorporate clocks into his more elaborate creations. Such was the case with the desk fitted with a musical clock delivered in December 1779 to the Prussian court. Variant models were sold to Charles of Lorraine, Louis XVI, and Frederick William II of Prussia. The marquetry, after designs by Janvier Zlick (1730–97), offered allegories of the arts that were allied to ingenious mechanisms. Doors and drawers opened automatically at the touch of a button and secret compartments allowed confidential papers and precious objects to be concealed. Supported by Father Time on his knees, the eight-day clock was fitted with a calendar and played music every two hours. Roentgen was probably assisted in the mechanical part by Jean Christian Krause (1748–92).[128]

Neoclassical styles maintained supremacy over the decorative until at least the 1830s. Charles Percier (1764–1838) and Pierre François Fontaine (1766–1853) in Paris and Thomas Hope (1769–1831) in London display the continuing influence of Ancient Egypt, Greece, and Rome. Winged lions, eagles, griffons, sphinxes, caryatids, altars to love, vestal virgins, and laurel crowns are omnipresent in the decorative arts. Sold to the Princess Mary in 1807/1808 an Egyptian style clock in black marble and gilt bronze, decorated with sphinxes by Vulliamy and son, is witness to a prevalent Egyptomania.[129]

Mantel clocks played a major role in fashionable interior furnishing. 'In houses where guests are received, three luxury items are now *de rigueur* if the hosts wish to appear as respectable people: firstly bronzes, clocks, statuettes, candelabras'.[130] The sculptural element of clocks took precedence over their time-telling function.[131] In 1814, the journalist John Scott (1784–1821) was in ecstasy before the shops of the Palais Royal and the decorative clocks there displayed.[132] In low or high relief, in symmetrical compositions, classical figures depicted the gods and heroes of mythology, the personalities of Ancient history and literature (the Horatians, Julius Caesar, Homer, Virgil, Socrates), the Bacchantes, allegories of time, of peace, of abundance, or of the muses (in particular Clio and Urania).[133] The design by Thomire, *The Genius of the Arts*, was particularly attractive to consumers.[134] Distributed in domestic dwellings, such clocks offered lessons in history, morality, and learning. At the same time, subjects derived from everyday scenes of the arts and crafts were also a source of inspiration: water-carriers, knife-grinders, barrel-organ players, gardeners, hunters, and readers all had an assured popularity.[135] Apart from the prolific Thomire, the most active bronze-casters were Pierre François Feuchère (1737–1823), Antoine Jean Ravrio (1759–1814), and Gérard Jean Galle (1788–1846).[136]

In contrast with domestic clocks, in which the decoration tended to mask its function, more precise mantel and floor standing clocks (regulators) normally had restrained cases, often in mahogany (Figures 119, 149). Pilasters or Doric columns carry a straight entablature, emphasized by mouldings, sometimes surmounted by a triangular pediment. Gilt bronzes discretely enhance the architectural forms of such clocks signed by makers like Robert Robin, Abraham-Louis Breguet, Antide Janvier, the Lepaute dynasty, or Aristide Pons.[137]

An elegiac sense of Antiquity seen through the prism of the ruins of Pompei, a sense of nature, and a sentimental reaction to them mark the decoration of watchcases. Allegories of love (Figure 193), flowered urns, and the emblems of gardening and of music dominate the decorative repertoire. Such subjects appear on the back covers in a round or oval medallion framed by drapes, garlands of flowers, or laurel leaves held together by ribbons. Chasing in gold of two or three colours was widely used for such decoration and many examples, mainly signed by French or Swiss makers, have survived. Equally appreciated was the restrained decoration obtained from applying translucent enamel to an engine-turned, *guilloché*, metal surface, responding to a taste for geometric compositions. Turned, to obtain patterns of concentric circles, radii, or barley pearls, the surface was covered with enamel, the colour of which was heightened by the turned pattern below, recalling with refinement the sparkling of polished stones. Occasionally scenes painted on enamel, opaque enamels, spangles, pearls, or diamonds enriched such work. For example, a watch produced by the partnership of Louis Esquivillon and De Choudens (1776–1800) combines violet translucent enamel on the engine-turned ground with a garland of leaves interlaced with roses in cloisonné enamel, and a medallion painted

---

126 The Frick Collection; Edey 1982, 69–75. Musée national du château de Versailles. Augarde 1996, 266.

127 Wallace Collection, F 271; Hughes 1994, 56–7. The Metropolitan Museum of Art, inv. 1982.60.50; Vincent, Leopold, & Sullivan 2015, N° 42.

128 Staatliche Museen, Berlin, O-1962.24; Koeppe 2012, 134–9. Roentgen also produced several pieces of furniture *à système* with the clockmaker/mechanician, Peter Kinzing. See Fabian 1983.

129 Victoria & Albert Museum, inv. M119-1966.

130 *Journal des Dames et des modes*, 31 January 1825.

131 Dupuy-Baylet 2006, 20–6. The collection of the Mobilier National, Paris offers an exceptional selection of contemporary models. That of the Patrimonio nacional, Madrid, is also representative of the variety of the bronze casters' products. See Colon de Carvajal 1987.

132 Scott 1815 121–2. 'The shops of the Palais Royal are brilliant… Nothing can be imagined more elegant and striking than their numerous collections of ornamental clock-cases: they are formed of the whitest alabaster, and many of them present very ingenious fanciful devices… Others were modelled after the most favourite pictures and sculptures: David's Horatii and Curatii, had been very frequently copied'.

---

133 In a gallery of 1365 examples, Niehüser 1999 displays all the variety of the French figural clocks.

134 Château de Malmaison, Chevallier 1991, N° 23.

135 Examples in Niehüser 1999.

136 Ottomeyer & Prôschel 1986, 667–709.

137 Several examples in the Musée des Arts et Métiers, Paris, see *Catalogue JB* section 3-3, N° 4, 13, 21, 25, 28, 30, 31, 32, 51.

on enamel showing Cupid armed with his bow beside an altar.[138] A self-winding watch by Jaquet-Droz offers a harmonious combination of an urn of flowers set against an Alpine background, the whole scene painted on translucent enamel, framed in champlevé enamel.[139] Indeed, painting on enamel was one of the most widely used techniques between about 1780 and 1830. It allowed scenes loosely inspired by Antiquity to be represented in the same delicate colours as were appreciated in the decoration of interiors. An enamel by George Michael Moser (1706–83) in brown cameo representing the Vestals decorates a watch by Edward Ellicott.[140] Another offers a copy in *grisaille* of the work by David Allan (1744–96), *The Origin of Painting*.[141]

Painting on enamel in this period was able to display all its artistic possibilities with the appearance of relatively flat cases of between 5–6cm diameter. Scenes, simply framed by a ring of half pearls, were copied from engravings reproducing, among others, the work of Angelika Kaufmann (1741–1807), Benjamin West (1738–1820), or Gavin Hamilton (1723–98). A watch by Piguet and Meylan decorated after Kaufmann's *Jupiter and Callisto* is exemplary.[142] Fitted with automata and music to accompany the painted enamel case, it is typical of Geneva production in the first third of the nineteenth century when some two hundred enamellers were active. Most worked anonymously in workshops supplying cases to English and Swiss makers specialized in supplying watches to China.[143] Capt & Piguet (1802–11) and Piguet & Meylan (1811–28) were the best known of the workshops. A few painters on enamel of exceptional ability emerge from the general anonymity. Jean Louis Richter (1766–1841), who specialized in landscapes; Jean François Victor Dupont (1785–1863), a portraitist; and Jean Abraham Lissignol (1749–1819), who devoted himself to historical scenes—all acquired personal reputations. Watches realized in pairs (Figure 114), with the scenes adorning them in mirror image, carry reproductions after engravings of contemporary or earlier paintings, revealing a Chinese taste for European art. The repertoire was made up of historical subjects, Alpine landscapes, and 'Chinese bouquets'—flowers arrangements often enlivened by doves. Music and automata were often added, and the watches could be incorporated into snuffboxes or perfume bottles. A rare surviving example is evidence of such pieces in Western diplomacy: a pair of watches offered to the Emperor Jianqing in the name of George III by Lord Amherst in the course of a commercial mission. Still preserved in their original case with the royal arms, they are adorned by facing reproductions attributed to Lissignol of *Affection* and *Innocence* after engravings of the works by Francesco Bartolozzi (1727–1815).[144]

Genevan workshops at the end of the eighteenth century were also distinguished in the making of watch as jewels of almost infinite variety. Worn as brooches or as pendants, small watches in the form of hearts, flowers, fruit, butterflies, beetles, harps, mandolins, or violins seduced purchasers with their opaque and translucent enamels in contrasting colours offset by chased gold mounts. The most luxurious of such cases were enriched with diamonds, coloured stones, and pearls.[145]

The work of Abraham-Louis Breguet (Figure 158) is in strict consonance with the styles and taste of his period. The cases of his watches have a discrete elegance that derives from the contrast between gold and silver subtly underlining form. They often carry 'barley-grain' engine-turning that combines an aesthetic effect with the practical advantage of not tarnishing the case with finger imprints. The spherical pendant with its suspension ring subtly recalls the pure circular form also deployed in the hands

Figure 158 Gold and silver Breguet 'tact' watch N° 1668, derived from the 'souscription' model, 1808. Photo: Jean Baptiste Buffetaud, courtesy of Chayette & Cheval, auctioneers, Paris.

---

138 Musée du Louvre, OA 6233; Cardinal 2000, N° 167, repr. 175.
139 Musée international d'horlogerie, inv. I-494, Cardinal & Piguet 1999, N° 197.
140 Musée du Louvre, inv. OA 8582, Cardinal 1984, N° 267.
141 Victoria & Albert Museum, inv. 1924–1898, Cardinal 1985, 195.
142 Musée international d'horlogerie, inv. I-1579, Cardinal & Piguet 1999, N° 194.
143 Chapuis 1919; White 2019; Patek 2010.
144 Antiquorum 2008, 14–19.
145 Patrizzi & Sturm 1979.

made with an open disc at their extremities. The dials, in harmony with the cases, are in silver or gold, and engine-turned in the centre.

Even the watches with complications displayed in small eccentric dials, pierced openings, or sectors, follow the principal of formal unity. In thus uniting beauty and function in a functional aesthetic, Breguet had an important influence on his peers in France such as Lépine, Robin, or Bazile-Charles Le Roy (1785–1839), and on his pupils Mugnier (*fl.* 1800–23) and Charles Oudin (1768–1840).[146]

HISTORICISM: REFERRING TO THE PAST

From the 1820s onwards, the neoclassical vocabulary of decoration increasingly ceded to one that was historicist (Figure 159), based on past styles from the Middle Ages to the later eighteenth century. Encouraged by trends in the theatre, historical novels, and works of popularizing history, Gothic and Renaissance 'revival' invaded architecture and the decorative arts. Eloquent witness is provided by the International exhibitions from 1851 onwards where gold and silver plate, horology, jewellery, armour, windows, and furniture were displayed in Gothic, Renaissance, Baroque, Rococo and Louis XVI styles. Over-abundant ornamentation, the complexity of form, the use of imitation materials, and industrial manufacture gave them an air of pastiche that was already underlined by artists and art critics of the time.

In France, as in Britain, faith in artistic progress based on the imitation of old models by the use of new industrial processes, was propagated by the bourgeois. Flaunted on chimney breasts, the ostentation of clocks affirmed the artistic taste of their owners. It was a production concentrated on Paris. In his report on horology at the 1889 exhibition, Paul Garnier remarked that Paris horology ' ... owed to its unity with other art industries, marble-working, joinery, and especially bronze-casting, the maintenance of supplying clocks the world over for a long period'.[147]

The second quarter of the nineteenth century was marked by the consecration of neo-gothic. Already nascent in Britain in the later eighteenth century, it developed in France during the early nineteenth century notably because of the works of Chateaubriand (1768–1848). Clocks appeared in the shape of bell towers decorated with rose medallions, trilobes, and fantasy beasts. 'Cathedral decoration' promoted to the limit by the Duchess de Berry (1798–1870), led to the copying in both England and France of celebrated monuments. But it could also be playful. A clock by Baullier & Fils (mid-nineteenth century), depicts a love scene beneath a gothic arch.[148]

From the mid-century onwards, Renaissance and eighteenth-century styles came more to the fore. A gilt bronze and enamelled clock at the 1855 exhibition by Levy frères, Paris, in a clear neo-Rococo style was even purchased by what would become the Victoria & Albert Museum in London.[149] At the 1867 exhibition, Gustave Baugrand (1826–1870) the Emperor's jeweller, showed a Renaissance-style clock which in its combination of different materials and techniques was intended to evoke the refined luxury of earlier models. Ivory, rock crystal, gilt silver, lapis lazuli, engraved glass and metals, chasing, and painting on enamel were all used.[150] In 1878, J. Berlioz confirmed the success of models from the past:

> For several years, fashion has been for furnishing in the style of the Renaissance, Louis XIV, Louis XV, Louis XVI. Assuredly nothing better can be done to rest the eye from the inflexibility of the First Empire. Horology, especially for what concerns clocks, has followed prevailing taste. At the exhibition fine Boulle-type examples were to be seen either with brass and tortoiseshell marquetry, or with bronzes elaborately worked and delicately chased.[151]

The decoration of watches, like that of jewellery, reflected these tendencies of romantic taste. Dandies, male or female, enjoyed carrying on a chain extra-slim watches decorated with champlevé enamel depicting polychrome flower bouquets or garlands on a black ground. Ladies tried carrying watches in heavy bracelets enamelled with flowers or arabesques.[152] Such a watch carries the inscription 'Nicette, to her dear sister, Paris 5 July 1857' and is decorated with champlevé enamels, pearls, and hard stones.[153] Japanese scenes appear in watch decoration from the 1860s onwards executed in delicate cloisonné enamels. With the help of the enameller Antoine Tard (*fl.* 1867–89), the jeweller Alexis Falize (1811–1898) produced watches with matching chatelaines much remarked in the international exhibitions.

But the fashion for Japanese styles was not characterized by the nostalgia for Renaissance and Rococo styles evident from the mid-century. A production specific to Vienna not only illustrates the passion for styles from the past, but also the eclecticism of nineteenth century design. Already reputed for their work in the eighteenth century, the Viennese goldsmiths now covered pieces based on earlier models (cups, horns, salt cellars, perfume bottles, table clocks), in relief or champlevé enamel. The vogue for 'Viennese enamel', which was highly decorative with its polychrome, fantastical, designs, culminated in the 1870s and 1880s.[154] The best-known manufacturers, employing numerous

---

146 For these innovations and the diffusion of Breguet's style see Daniels 1975; Cardinal 1997; Breguet 1997; and Rigot 2017, who discusses imitations and forgeries of Breguet's work.
147 Garnier in Picard 1891, 694.
148 Musée des arts décoratifs, Paris, inv. 17 741.
149 Victoria & Albert Museum, inv. 2650–1856.
150 Cardinal 1983, 108.
151 Berlioz, 'L'horlogerie' in Lacroix 1878, vi, 345–6.
152 For an elegant example see Fallet 2012, 137.
153 Musée international d'horlogerie, inv. I-1276, Cardinal & Piguet 1999, N° 273.
154 Speel 2008, 141–9.

Figure 159 Historicist watch made in imitation of a seventeeth century crucifix watch. As shown by its attendant label, it comes from the collection of the noted watchmaker Léon Leroy who believed it to be genuine seventeenth-century work. Photo Jean-Baptiste Buffetaud courtesy of Chayette & Cheval, auctioneers, Paris.

anonymous workmen, were Hermann Ratzerdorfer (*fl.* 1842–94), Herman Böhm (*fl.* 1866–1922), Ludwig Politzer (1841–1907), and Simon Grünwald (last quarter 19th century). In particular, their workshops produced clocks and watches with enamels copied from seventeenth- and eighteenth-century mythological, allegorical, pastoral, and romantic scenes.

In Paris, jewellers also met the taste of their clients for the Renaissance. At the International Exhibition of 1878, Alphonse Fouquet (1828–1911) and Lucien Falize (1839–1897) stood out for their jewels influenced by sixteenth-century ornamentalist designs. Fouquet received a gold medal. He presented one of his pieces to the Musée des Arts décoratifs—a watchcase with a matching chatelaine showing a sphinx, cast in gold, chased, and engraved.[155] Like Falize, he received a 'Grand Prix' for his creations inspired by Delaune and Collaërt. Among them were several clocks and watches.[156] Watch- and jewellery makers produced watches with enamel or chased decoration in gold based on Renaissance and eighteenth-century French styles. Frederick Boucheron (1830–1909), for example, liberally interpreted sixteenth-century models to produce watches with matching chatelaines decorated with mythological figures, mascerons, and sphinxes. Boucheron also specialized in watches and chatelaines in the Louis XV, or Louis XVI style, decorated with painting on enamel, or trophies in multicoloured gold. At the 1867 exhibition he was awarded a gold medal for these products.[157] Patek Philippe also responded to this taste. From the 1851 exhibition onwards, the company offered watches with cases painted in enamel or made in engraved and repoussé gold evoking the styles of the eighteenth century.[158]

If the market for decorative clocks was concentrated in Paris, the decoration of watches took place in numbers of workshops in Geneva and the Canton of Neuchâtel. From the 1830s, and for the rest of the century, engraving came back into favour with customers. Surviving watches, the designs for them used in the workshops, and collections of models, reveal the diversity of the decoration proposed and realized. These were generally based on Rococo and neoclassical models. Entanglements of flowers, milfoil, exotic vegetation, or shells cover cases or frame landscapes, portraits, monograms, or reproductions of paintings.[159]

ART NOUVEAU: NATURALISTIC DECORATION

'Art nouveau' in France and elsewhere, 'Liberty' or 'Modern Style' in Britain, 'Jugendstil' in Germany and Austria, 'style nouille' in Belgium developed in the turn-of-century decades as a reaction against academic eclecticism and the tyranny of aged styles. It also opposed the standardization of machine-produced work. Art nouveau was a movement conducted in a deliberate and thought-out manner so as to create a new style and to save artisanal qualities. New journals, the international exhibitions, salons, and art criticism all witness a revival in the decorative arts from the 1880s onwards. Its language everywhere is similar: long, serpentine forms derived from nature and women were allied with asymmetric compositions. Audacious mixtures of materials were employed. Nevertheless, if inspiration from the natural world is clear in the French and Belgian versions, in those of England and Austria, line is simplified, and the forms are more geometrical. A manifest taste for enamel, translucent, cloisonné, and champlevé also prevailed at the end of the century and found a perfect medium of expression in watches.

Enhanced by engraved or chased gold, by hard stones and pearls, enamel decorated 'form' watches in the shapes of birds, insects, or flowers were popular. On hanging watches, or those attached to a brooch, they depicted poetic subjects derived from nature. Noteworthy, for example, was that in the form of a flower that gained a silver medal for Clemence frères, La Chaux-de-Fonds at the 1889 exhibition, or the *pine cone* shown by René Lalique (1860–1945) at the 1900 exhibition.[160] The leading Swiss houses, Omega, Vacheron & Constantin, and Patek Philippe all followed the dictates of the new style which also informed the production of floor and table clocks. They are instantly recognizable from their curving lines, naturalistic, often symbolic, ornament and the variety of, sometimes uncommon, materials employed. 'Day and Night', a clock in wood, brass, and pewter by Japy Frères is representative with its allegorical decoration. But the 1906 exhibition of decorative arts in Milan heralded the end of this style.

ART DECO: COLOUR AND GEOMETRICAL ABSTRACTION

An international exhibition at Paris in 1925, the 'Exhibition of Modern and Industrial Decorative Arts', displayed the omnipresence of a new decorative language in harmony with the modern movement in painting. The works shown united refined traditional artisanal techniques with modernity expressed in geometric forms heightened by strongly contrasting colours. The style combined a long-seated taste for luxury with a resolutely modern aesthetic. It is closely linked with the contemporary movements of cubism, futurism, and that of the supremacists. Very varied materials, such as ebony, silver, hard stones, enamel, lacquer, ivory, sharkskin, and eggshell were employed.

Wristwatches, pendant watches, pocket watches, and handbag watches (Ermeto) offered contrasting colours in geometric

---

155 Musée des Arts décoratifs, Paris, inv. I 4851F. Vever 1906–1908, 43–7.
156 Vever 1906–1908, 158–62.
157 Néret 1992, 13–21. Cardinal 1983, repr. 110.
158 Huber & Banbery 1993, 126–41.
159 See, for examples, the extensive collection in the Musée international d'horlogerie, many illustrated in Cardinal & Piguet 1999. See Chapuis 1944, 102–15, pl. 82–7, 130–1, 140–1.

160 Musée international d'horlogerie, inv. I-433, Cardinal & Piguet 1999, N° 278. Musée des arts décoratifs, Paris, inv. 9370.

patterns or stylized motifs.[161] Brightly coloured cloisonné or champlevé enamels, hard and soft stones such as coral, turquoise, onyx, or jade all supplied the decorator's palette. The Geneva workshops of Henry Blanc and Albert Weber were strongly represented in the Paris exhibitions, alongside such other Swiss manufacturers as Omega, Longines, Ebel, Juvenia, Zenith, and Optima. With similar success, the Art Deco style was applied to small clocks for bedside tables and studies. A good example combines a stylized floral decoration in enamel with straight lines and dial in 'eggshell' enamel.[162] The style also adapted perfectly to the mystery clocks produced by Cartier, of which the architectural cases sumptuously combined translucent materials such as rose-coloured quartz or rock crystal with hard materials such as coral, lapis lazuli, onyx, or jade.[163]

THE INTERNATIONAL MODERN MOVEMENT: THE DEVELOPMENT OF A RATIONAL ÆSTHETIC

The protagonists of the modern movement, mainly architects, favoured functionalism and rationality above all. In Great Britain and Austria in the later nineteenth century, even while Art Nouveau was at its height, some artists, such as Christophe Dresser (18734–1904), in *The Art of Design* (1862), advocated this approach. From 1910 onwards, architects contributed to the search for functional objects that could be produced in quantity, thanks to a simplification of their geometrical form and the rational use of new materials. The best known of them are Walter Gropius (1883–1969), Mies van der Rohe (1886–1969, Le Corbusier (1887–1965), and Gerrit Thomas Rietteveld (1888–1964). They all expressed their ideas through teaching, periodical publications, and the founding of groups to establish with force the language of industrial design. Notable were the *Bauhaus* school directed by Gropius at Weimar, the group *De Stijl* led by Theo van Doesburg (1883–1931) and Piet Mondrian (1872–1944), and Le Corbusier's periodical *L'Esprit nouveau*.

Rationalism coincided with the development of watches intended for production on an industrial scale. The extensive research made concerning wristwatches reflects the strict functionalism required by the new doctrine (Figure 178). Cases should have clean, geometric lines and espouse the form of the arm. Among the best-known results are the Polyplan by Movado (1912), the Santos (1904), the Tank (1917), and the Octagonal (1910), all by Cartier, and the Oyster (1926) from Rolex. An exhibition organized in 1932 by the Museum of Modern Art in New York gave its name *The International Style* to that which from the 1930s to the 1950s would characterize architecture and the decorative arts in both Europe and the United States. The international nature of the style was in part the result of the emigration to the United States in the years preceding the Second World war of such partisans of rationalism as Gropius and Mies van der Rohe. There they encouraged creativity, delineated already by the celebrated architect-designer Frank Lloyd Wright (1869–1959).

Intended to serve industrial progress, the rationalist style was based on the simplification of form, the reduction of volumes, the visibility of structure, and the use of primary colours. Modern materials, economical and easy to work, responded to this conception by facilitating series production. Specialist salons, dedicated sections in international exhibitions, and professional revues, all display the influence of the 'international style' on wristwatch design. Designers produced a high variety of forms, but all based on pure, simplified geometrical shapes. Exemplary of these are the models 'Prince Brancard' by Rolex (1928 onwards), the 'Curviplan' by Movado (1931), the 'Reverso' of Jaeger-Lecoultre (1931), and the 'Parallelogram' and the 'Baignoire' by Cartier (1936 and 1957). The triumph of the wristwatch was consecrated at the 1937 Universal Exhibition, where highly varied forms with uncluttered lines and volumes showed in some instances their paternity in contemporary sculpture. In 1947, Nathan George Horwitt (1898–1990), designed a watch perfectly in accord with the precepts of the rationalist movement. With a black dial punctured at 12 o'clock by a gilt disc, it has been produced by Movado ever since.

POST-MODERNISM: VARIETY OF PRODUCTS AND INDIVIDUALIZATION

From the 1950s onwards, rationalism, now reduced to an uncreative professional code, was increasingly challenged. Alongside the functionalist current inherited from the modern movement, a new tendency came to dominance which considered the variety of life styles, the different classes of consumers, and the diversity of social habits. Design became a reflection of an ever-changing world. In the 1960s the ease of shaping plastic provoked a proliferation of shapes and colours. 'Pop' culture infiltrated all areas of the consumer society. Objects produced were practical, inexpensive, and disposable. From the 1970s the methods of industrial production allowed for an easier and more rapid response to changes in taste. The products offered by designers became increasingly varied tending to satisfy individual taste. Form was not conceived solely in terms of function but was also to express a social image and be a personal symbol. Fundamental technical changes apart, the wristwatch displays many aspects of this post-modern aesthetic. The simple case-lines betray an influence from contemporary architecture and abstract art. Watches shown in the Swiss pavilion at the Brussels International Exhibition in 1958 showed particularly daring design work. In the 1970s audacious materials and

---

161 Chapuis 1944, planches 142–52. See the several colour plates in the fiftieth anniversary number of the *Journal Suisse d'horlogerie*, 1926.
162 Musée international d'horlogerie, inv. IV-644, Cardinal & Piguet 1999, n° 100.
163 Nadelhoffer 1984, 245–54, pl. 65–73.

Figure 160 Tissot Le Locle 'Rockwatch', 1986 in which a single piece of granite supplies both dial and case. Musée International d'Horlogerie, La Chaux-de-Fonds, inv. VII-10. Photo MIH.

shapes were deployed linking back to decorative styles and questioning the principle of functionalism. From the 1960s on, Piaget distinguished itself by producing watches in audacious forms set with stones such as opals, turquoises, onyx, coral, lapis-lazuli, and others. Corum stood out with its 'Buckingham' watch, the dial covered by a peacock feather and, even more so, by the 'Golden Bridge' in which the sapphire case allowed the movement to be seen, carried between two bridges, and engraved with leaf designs.[164]

With the exception of a few houses, design became somewhat muted in the 1970s. Since then, however, wristwatches have held an important place in contemporary design. Painters, architects, and sculptors take them as a base for imaginative creation, deploy varied art movements, and exploit modern materials to find a diversity of forms and uses.

The plastic watch *Swatch*, commercialized since 1983, displays spectacular design success made possible by a new conception of the movement and an efficacious marketing strategy. Dozens of models were offered to enthusiastic customers who eagerly awaited the renewal of the collection, and who wanted watches, often designed by well-known artists, produced in a limited number. Seduced by their variety and originality, purchasers collected these pieces sold at affordable prices. The company also showed that 'design' could be served by technical innovation when it launched, in April 1997, the 'Swatch Skin' with innumerable variations. Inspired by the 'Delirium' produced by Ebauches SA in 1979, it provoked astonishment by the thinness of the case (3.9mm), made possible by having the back serve as the backplate.

The end of the twentieth century was marked by the launching of watches in which originality came from the use of unexpected new or natural materials. In 1985, Tissot launched a watch in stone, the 'Rock Watch' (Figure 160), which would have an enormous success with its case carved from different Alpine granites (green, rose, grey, black).

In 1990 Corum made a series of watches with the dial made from a fragment of the meteorite Peary discovered in the early nineteenth century. The same year Rado stood out for its DiaStar Ceramica in which the strap was made of high-tech, unscratchable ceramic and the dial was fitted with a glass in sapphire.[165] In parallel, dials were commissioned from contemporary artists such as Jorg Hysek, who imagined a series of colourful abstract designs before developing his own watch brand in 1997.[166]

---

164 Musée international d'horlogerie, I-1750 and I-1930. Cardinal & Piguet 1999, N° 375, N° 376.

165 See 'Les montres du XXe siècle' special issue of *Montres Passion*, February 2000. Some sixty watches are described, providing a panorama of the significant models of the century.

166 Antiquorum 2004. For other examples of contemporary design, and the frequent lack thereof, see Friis-Hansen 1988; Griffiths 1985.

CHAPTER SIXTEEN

EIGHTEENTH-CENTURY CLOCK EXPORTS FROM BRITAIN TO THE EAST INDIES

Roger Smith

INTRODUCTION

At the start of the eighteenth century, clock and watch producers in Britain, France, and Geneva (for watches) were developing a substantial international trade, not just within Europe, but with the Ottoman empire, and European settlements overseas.[1] These markets would attain considerable importance over the course of the century, especially for Britain which established a leading position as its general commerce expanded, but it is difficult to gain a general understanding of their size and value. They involved many individual merchants with little official oversight, and surviving records are too fragmentary to show how a particular market developed. There are, for example, numerous contemporary statements about the importance of the Ottoman market, but little quantification until the end of the century, when a French observer noted that Britain dominated this trade in the period 1787–97, annually exporting an average of 1,160 dozen (13,920) watches valued at £266,400 sterling.[2]

In south and east Asia—the 'East Indies'—the demand for mechanical clocks was still very limited at the beginning of the century. Within this vast area, European trade with Japan and the Indonesian archipelago was monopolized by the Dutch East India Company, which occasionally sent presents of clocks to some of the local rulers, including the Chinese emperor, but was uninterested in developing a major trade in such articles.[3] The Indian subcontinent offered a larger market for clocks, which seems to have grown substantially during the century, but the fragmentation of political power meant that imports were dispersed among many Indian states and came from several sources, making detailed analysis of the trade as a whole difficult. Fortunately for historians, the clock trade to China was conducted under unusual conditions, which produced abundant documentation for much of the century. Furthermore, the basic statistics can be illuminated through studying the clocks themselves. Surviving examples of these distinctive clocks display the remarkable inventiveness and flexibility of contemporary European clockmaking, while the Chinese-made clocks they inspired are evidence for the global spread of European technology in this period.

Being ship-borne and conducted over extreme distances, all European trade with the East Indies involved high risks and long delays, leading states to grant national monopolies of the trade to 'chartered' companies, like the British East India Company (hereafter EIC, or the Company). These were not only larger and more securely funded than private firms, but were more formally organized and usually longer-lived, producing detailed accounts of their activities over an extended period. Furthermore, the important position some companies attained within their national economies brought close scrutiny by governments and private investors, while commentaries and debates in the periodical Press shed further light on the bare statistics and policy statements in official archives.

There is also much information about how the trade was conducted when it reached China. Most European trade

---

1 Clocks dominated these export markets for most of the century. Unless stated otherwise, references to clocks and clockmaking include watches and related mechanical 'curiosities'.

2 For contemporary statements, see Kurz 1975, 72–3. Figures for 1787–97 are from Beaujour 1800, ii, 18.

3 For presents from European embassies to the emperor, see Jamieson 1883.

channelled through Canton (now Guangzhou), which became the sole official point of entry after 1757.[4] Once there, the trade formally became the responsibility of a dozen or so officially appointed Chinese 'security' merchants, known to Westerners as the 'Hong' merchants, who acted as securities for the payment of port duties and other levies on foreign ships, purchased the bulk goods brought by the Europeans, and supplied the tea, silk, porcelain, and other goods to be taken back to Europe. As a minor but troublesome part of their duties, they were also required to buy such Western clocks and other 'curiosities' as senior officials in Canton wished to include in the city's annual tribute to the emperor.[5] Since the Hong merchants were central to the main trade of the EIC, their activities and financial health were closely monitored by the EIC's Canton supercargoes ('supracargoes' in Company parlance) and reported to the Court of Directors in London. Taken together, the records created by, or relating to, the EIC in London and Canton allow the Anglo–Chinese clock trade to be studied more closely than any other major export market for European clocks in the eighteenth century.[6]

ORIGINS OF THE CLOCK TRADE WITH CHINA

This study is not concerned with the early history of clocks in China.[7,8] It is sufficient to note that the first European clocks to reach China in the late sixteenth century were welcomed as exotic foreign novelties, and this remained their main appeal for another two centuries. Any surviving knowledge about earlier Chinese achievements had no effect on the demand for imported clocks, and it was under the influence of these foreign products and foreign clockmakers that a horological industry eventually developed in China, at first in the Imperial Palace Workshops, and later in Canton, where imported clocks were prepared for their long overland journey to Beijing.

The first European clocks reached East Asia as diplomatic gifts from Catholic (mainly Jesuit), missionaries seeking permission to proselytize from Asian rulers. St Francis Xavier took a clock to Japan as early as 1549, and Matteo Ricci played a major role in introducing Western clocks to China.[9] His appointment to maintain the emperor's clocks and advise him on astronomy and Western science was inherited by his successors in the China mission, and they and the European rulers who supported them presented further clocks to later emperors, while others were brought by the few European embassies to reach Beijing in the seventeenth and eighteenth centuries.[10] It is also likely that cheaper clocks were given to Chinese merchants and officials by individual European merchants throughout this period, but these are rarely recorded.

The missionaries presumably maintained workshop facilities in Beijing from soon after their arrival, but it was not until the accession of the Manchu Qing dynasty that mechanical clocks gained the special position at Court that they were to retain until the early nineteenth century. The Kangxi emperor (r. 1661–1722) was particularly interested in clocks, and around 1690 formally established an Imperial Clockmaking Workshop in the Forbidden City.[11] According to a European visitor to Beijing in 1721, '[t]he making of clocks and watches was lately introduced under the protection of the present Emperor [Kangxi] who, at his leisure hours, amuses himself with whatever is curious either in art or nature'.[12] The Imperial Workshop employed Chinese craftsmen (around 100 in 1736) supervised by European missionaries, including some who were trained clockmakers. It operated until the early nineteenth century, providing a route by which clockmaking and other aspects of European science and technology were introduced at the heart of the Chinese empire.[13] However, while there was some limited use of clocks for timekeeping in Court circles under Kangxi and his successors, most clocks, even within the Imperial Palace, were valued chiefly as curiosities and status symbols.[14] This attitude was made clear in the official Index to the Emperor's 'Grand Library' (1782), which stated that:

> In regard to the learning of the West, the art of surveying the land is most important, followed by the art of making strange machines. Among these strange machines, those pertaining to irrigation are most useful to the common people. *All the other machines are simply intricate oddities, designed for the pleasure of the senses. They fulfil no basic needs.*[15]

---

4 But Russian trade with China was conducted overland via the border town of Kiakhta.
5 The security merchant system was not fully established until the 1730s. Cheong 1997, 104; also Van Dyke 2005, 12.
6 Especially British Library: India Office Records (hereafter IOR), Canton Factory records (series R/10 and G/12); and Minutes of Court of Directors (series B/-).
7 Smith 2008 for a brief survey. Pagani 2001 is valuable for clock design and Chinese clockmaking. White 2012 provides a well-illustrated survey of British clocks exported to Turkey and China. There is valuable information from Chinese records in Guo 2011. I am grateful to Simon Bull for alerting me to this important article.
8 See Chapter 2, Section 2.

9 For Jesuit missionaries taking mechanical clocks to Japan see Hiraoka 2020. For China, see Chapter 2, Section 1; Zhang 2012, 567. Clocks and watches as diplomatic gifts were also widely used by European states both between themselves and in their relations with the Ottoman Empire. Mraz 1980; Eisler 2014/15, 28–30.
10 Pagani 2001, 48–57 gives details of the clockmaker-missionaries.
11 Pagani 2001, 36–7.
12 Bell 1763, 103. Also Zhang 2012, 568–9 for early references to the workshop.
13 Bonnant 1960, 31. For the missionaries, see also Pagani 2001, 48–57.
14 For the attitude of the emperors to clocks, see Pagani 2001, 58–67; also Landes 1983/2000, ch. 2. For Japan, see Hiraoka 2020, 216.
15 Cipolla 1967, 89, (author's emphasis).

In this situation, there was little incentive for the knowledge of mechanical clockmaking held in the Imperial Workshop to spread more widely, and even within the Workshop the main focus seems to have been on maintaining and altering the growing collection of imported clocks. Some new pieces were made, but by 1769 automata were taking priority, with the emperor's clockmaker de Ventavon noting that the Qianlong emperor (r. 1735–96) expected him to produce 'not really clocks but curious machines and automata'.[16] Working within these constraints, the few missionary-clockmakers in Beijing had little opportunity to keep abreast of contemporary Western developments and could not satisfy the Court's demands for ever more novel and complicated clocks.

This provided an opportunity for European exporters, and clocks sent to China in trade soon replaced the Imperial Workshop and diplomatic gifts as the main source for the Imperial Palace. In 1736, it was already said to contain more than 4,000 clocks, watches, and similar articles 'from the best masters of Paris and London'[17]; by the end of the century, the first British embassy decided not to bring ornamental clocks for the emperor, because '[i]t would have been vain to think of surpassing, in public presents of this kind, either as to workmanship or cost, what had already been conveyed to China through private channels'.[18]

These Western exports could look quite different from clocks sold to European customers. Though they all contained clockwork, the simple term 'clock' can be misleading, and contemporaries used vaguer terms like 'curiosities' or 'toys' or resorted to circumlocutions like 'curious and expensive pieces of mechanism ornamented with jewels'. By the later eighteenth century, the pidgin-English term 'sing-song' was often used, derived from Cantonese words meaning 'self-sounding bells'.[19] This usefully emphasized their difference from conventional clocks, and stressed the importance of mechanical chimes and music, but since it may suggest a degree of triviality to modern readers, 'clock' will continue to be used in this chapter, although it should be broadly interpreted.

The export of clocks from Europe to the East Indies in the eighteenth century, and especially from Britain to China, was closely linked to the growth in general trade between these regions. For much of the century, trade was largely one sided. While Asian products like tea, silk, and porcelain were in great demand in Europe, there was little demand in south and east Asia for traditional European staple exports like woollen cloth, which might have to be sold in barter arrangements or at a loss. Europeans, therefore, relied heavily on silver bullion to buy the goods they wanted, supplemented by a miscellaneous range of products and raw materials, including goods from other parts of Asia. The problem was particularly acute with trade to China, where silver constituted some ninety per cent of British exports in 1721, and while the actual value of the EIC's silver imports to Canton rose and fell, it was around £700,000 in 1786, when the demand for tea had increased.[20] Apart from its availability depending on the political and economic situation in Europe, the export of silver was disliked by traditional political economists as a weakening of state power, while merchants were concerned that its use meant that profits were only made on the return cargo from China. There was consequently great interest in identifying alternative exports, and the desire of Chinese elites to acquire European clocks provided an opportunity for anyone prepared to take the risk of producing and exporting the specialized articles needed for this market.

As both a growing mercantile power and a major clockmaking country, Britain was closely concerned in this new branch of exports, and by the 1750s, most clocks were taken to China in the EIC's ships. The strengthening of Britain's position in India after the battle of Plassey (1757), and the decline of French competition with the end of the Seven Years' War in 1763, led to further growth in Anglo–Chinese trade, and with it, the incentive to export clocks. This was matched, if sometimes imperfectly, by increased demand for them at Canton, where growing European trade produced larger customs receipts and associated revenues. The traditional way for the emperor to share in these benefits was through the annual tribute of 'curiosities', which Canton sent to Beijing, and this included many of the finest imported clocks. Consequently, they have sometimes been seen as presents from the foreign merchants to the emperor, but although the foreigners financed the tribute through levies on their ships, and the profits made by Chinese merchants, their contribution was indirect and involuntary.

As a result of this flow of tribute, the imperial collection of musical clocks and similar articles was, by the end of the century, valued at over £2 million sterling by the British ambassador Lord Macartney, and 'all made in London'.[21] In fact, his estimate may have been too low, and he was mistaken in thinking that the articles were all made in London, but it now dominated the export trade, as both a centre of clock production and the main port of departure for general trade between Europe and the East Indies. However, these clocks also had a less benign significance for wider Anglo–Chinese trade, which merits further examination.

---

16 Cipolla 1967, 89.

17 Bonnant 1960, 32, quoting Father Chalier, superintendent of the Imperial Workshop. This huge figure included mechanical music and perhaps scientific instruments. The Palace Museum now contains few clocks earlier than mid-eighteenth century, or from France before the nineteenth century.

18 Staunton 1797, i, 43.

19 'Tse ming tsung' in Braga 1961, 69 and n. 23. 'Zimingzhong' in modern Mandarin.

20 For 1721, Morse 1926–1929, vol. 1, 165. For EIC silver exports to China 1760–1800, see Pritchard 1936, 399, app. 9.

21 Cranmer-Byng 1962, 261.

TRADING ARRANGEMENTS AT CANTON

Britain was later than some of its rivals in sending ships regularly to China, but its trade grew rapidly during the eighteenth century, driven particularly by the increasing demand of British consumers for tea. This led to growing clock exports, but although they were carried on EIC ships, they were not exported by the Company itself but were part of the private or 'privilege' cargo that its ships' officers were allowed to carry as encouragement to act efficiently and responsibly.[22] The Company also hoped that the proceeds from the sale of private cargo could be used for its own purchases, in return for bills of exchange payable in London.

In fact, it took a while for clocks in China to feature strongly in private British trade with little evidence of significant British clock exports in the early years of the century. However, it was not long before clocks, even in small quantities, revealed an alarming ability to threaten the Company's general trade at Canton, because of the leading role they played in the annual tribute of foreign 'curiosities' sent to the emperor. The usual procedure for acquiring these articles involved the chief Customs administrator or 'Hoppo' and other officials examining any clocks brought by a newly arrived European ship, before instructing the Hong merchant acting as security for that ship to buy those selected for the tribute.[23] The merchants were supposed to be reimbursed with allocations from official funds—customs revenues for much of the century and the 'Consoo' Fund extracted directly from foreign shipping from the 1780s. The allocation was set at 50,000 taels (around £17,000) per year in the early years, changing to 30,000 taels (£10,000) after 1726, where it seems to have remained for many years.[24] With the establishment of the Consoo Fund and the further increase in foreign trade towards the end of the century, the allocation for clocks and watches rose to 100,000 taels in 1793/4 and even 200,000 taels in 1806/7, though it is unclear whether these last amounts were actually spent.[25]

However, much of the allocation (estimated at half in 1754) was needed for transporting the tribute to Beijing, and the remainder was insufficient to pay for everything selected by the officials for the emperor or (illicitly) for presents for their own sponsors at court.[26] The Hong merchants, therefore, had to supply clocks at a heavy discount, and by 1754 it was believed that they received back only twenty-five per cent of the price they paid the European vendors, and this may have fallen to only five per cent by 1784.[27] The obligation to buy these curiosities, therefore, involved the Hong merchants in heavy losses, which they tried to recover through charging the Europeans higher prices, getting port charges increased, and periodic moves towards establishing a cartel. Given their importance for the EIC's own trade, the threat to the financial soundness of the Hong merchants and their attempts to recoup their losses were continuing problems for the Company.

Clocks could also create more immediate difficulties for the Company, as shown in 1723, when the trade was still in its infancy and before the Hong merchants had become directly responsible for buying tribute clocks. On that occasion, the Hoppo refused to pay the price set by a British vendor for a clock he wanted for the emperor and stopped EIC ships from loading cargo until the clock was delivered to him. Although the dispute concerned private trade, the Company was forced to intervene and accept the Hoppo's offer of £320 while paying the vendor his full price of £467.[28]

That happened during the short reign of the Yongzheng emperor (r. 1722–35), who was less interested in acquiring foreign curiosities than his father or son and is not evidence of an extensive trade. However, the situation had changed thirty years later, when the importation of curiosities at Canton, mostly in British ships, had become of 'great value', and since the Hong merchants were now responsible for buying them, their potential impact on general trade was even greater. This was demonstrated in 1754, when the Hong merchants refused to stand security for EIC ships because of the potential losses of having to deal in such articles.[29] The Company's supercargoes asked the Chinese authorities to end the merchants' boycott, stressing that they wished 'only to be upon the footing of other Europeans, whose ships did not bring any of these curiosities with them'.[30] Although that was done, the Hong merchants were also instructed that in future they should buy clocks collectively, rather than individually. That effectively established a Hong cartel to the detriment of the EIC's own trade, so when a supercargo returned to England in 1757, he persuaded the Company's Directors to place an immediate ban on the export of 'clocks and valuable toys' to China in private trade.[31] A rough draft of the explanation given to the Company's officers is valuable for its clarity:

> ... our trade at Canton has suffered greatly by the Importation there in Private trade of Clocks and certain valuable Toys which represent Temples or Gardens in which there generally are some Clockwork or Musical Contrivance. When the Hoppo's Officers see them the Merchant who is Security for the Ship that brings them, must buy them and either give them away to the Mandarins or sell them

---

22 For private trade by EIC commanders, see Bowen 2007.
23 For the Chinese view of these arrangements, see Guo 2011.
24 Cheong 1997, 225.
25 IOR, G/12/162, 65–7. But note that in 1814, the Hong merchants claimed to spend £25,000 annually. G/12/189, 121–2.
26 Morse 1926–1929, vol. 5, 13; Cheong 1997, 226.
27 For twenty-five per cent, see Morse 1926–1929, vol. 5, 10. For five per cent, see Quincy 1847, 176–7.
28 IOR, G/12/24, (10.1.1723 O.S.), 87–9, (transcribed in Kessler 2016, 218–19).
29 Morse 1926–1929, vol. 5, 12.
30 IOR, R/10/3, 210 (29 July 1754); 350–1 (8 August 1755).
31 IOR, B/74, 421.

for less than a Quarter of their value. Wee have therefore to prevent the Inconveniencys attending this Practice prohibited the Export of such Articles for the present on any of our China Ships . . . .[32]

Watches were not originally included, on the purely pragmatic grounds that 'being carried in the Pocket & if not seen by the Hoppo's Officers [they] cannot be hurtful,' but this proved to be over-optimistic, and in 1760 the ban was extended to include watches and small toys.[33]

Curiously, the ban was revoked just four years later.[34] No explanation has been found, but British trade to the East Indies was expanding rapidly, increasing the demand for 'alternative' exports to replace silver. Indeed, entrepreneurs like James Cox (c.1723–1800) were already organizing the production of clocks for this market, perhaps ostensibly for India. The clock trade was therefore resumed, leading to a boom, which, within a few years, had glutted the restricted market. By 1770, the position was so serious that the Hong merchants took the extraordinary step of writing directly to the EIC's Directors in London to ask that further exports of 'curiosities' in the Company's ships should be banned.[35] This time the Directors were slower to respond, but in 1771 the Company's supercargoes wrote to support the request. With unconscious irony, they pointed out that their recent success (at considerable expense) in securing the dissolution of the Hong merchants' cartel or Cohong had weakened the ability of individual Hong merchants to sustain losses caused by having to buy clocks, thereby threatening the EIC's own trade.[36]

The Company finally took action in late 1772, when it announced a ban on the carriage in private cargoes of 'clocks, toys or other articles ornamented with jewels'—later defined as clocks valued at over £100.[37] The ban only took effect from 1773–4, but then apparently remained in force for the rest of the century. However, it proved easy to evade through undervaluation or simple concealment. When a leading Hong merchant complained in early 1776 that clocks were still imported 'to a considerable amount', he was reminded that they had been banned, but by continuing to buy them, the Hong merchants had themselves 'promoted the clandestine methods by which they were introduced.'[38] This smuggling could be on a massive scale: ten 'tons' (or 400 cubic feet) of clockwork in fourteen cases were reported in 1781 as having been illicitly loaded on a Company ship before it left England, which then disappeared before official unloading at Canton.[39] In these circumstances, strict enforcement was impossible.

It may be wondered why the EIC did not control this troublesome trade by bringing it within its own monopoly. However, the production and sale of clocks required a commercial agility and flexibility that was difficult for a large corporation, which preferred to arrange long-term contracts with large-scale producers. As Lord Macartney was to note at the end of the century in his observations on trade with China:

> With regard to toys, jewellery, etc., commonly called sing-songs . . ., it is better to leave them at large as objects of speculations [sic] for private traders, whose habits of industry and individual activity are better calculated than the magnificent system of a great commercial body for a traffic in such articles.[40]

MANUFACTURE

This characterization of clocks as objects of speculation for private traders raises questions about how the various stages of this trade were conducted and by whom. The unusual demands of the market meant that those prepared to risk becoming an exporter were rarely trained clockmakers. Like all overseas traders, they faced significant problems in obtaining timely market intelligence, suffering possible damage or loss on the voyage, arranging sales in distant markets, and remitting the proceeds. For the clock trade to the East Indies, these problems were compounded by the specialized nature of the articles, the extreme length of the voyage, and the unpredictability of the market. Dealing with such problems required strong financial backing, and it helped to have experience of trading with this region or connections with those who had. It is therefore no coincidence that the leading exporters during the peak years of the 1760s–1790s were wealthy or well-connected goldsmiths, jewellers, and toy men, like James Cox,[41] Timothy Williamson (1729–99), John Duval & Sons (fl. 1740s–90s), and Francis Magniac (1751–1823). Such men were able to organize the design and manufacture of the elaborately gilded and jewelled cases typical of these export clocks, and they often had prior links with the East Indies trade through the officers of EIC ships, who sold them precious stones and curiosities and bought 'toys' and other articles for export. These officers were essential since they could arrange to export the clocks in their privilege cargo and knew how best to sell them and remit the proceeds to London.

---

32 IOR, D/106, (not paginated), Instructions to China Council.
33 IOR, G/12/195, Miscellaneous Documents (n.d. but January 1760), 22–3; B/76, 62 (10 September 1760).
34 IOR, B/80, 230 (17 October 1764).
35 IOR, B/86, 173 (19 September 1770), consideration of letter from Chinese merchants 14 February 1770. The letter also complained about excessive imports of woollen cloth (E/1/53, letter 50).
36 IOR, R/10/7, 79, China Council to Court of Directors (8 August 1771).
37 IOR, B/88, 285 (27 November 1772).
38 IOR, G/12/59, 13 (13 February 1776).

39 IOR, G/12/77, 52–3 (17 July 1783). The commander was Patrick Lawson, notorious for flouting regulations. See Bowen 2007 for smuggling by ships' commanders.
40 Cranmer-Byng 1962, 260.
41 For Cox, see Smith 2000, 353–61.

Exporters could obtain their clocks in various ways. Earlier clocks exported to the Far East were similar to those made for the domestic market, and standard types continued to be exported throughout the eighteenth century. However, the growing importance of this market led, by mid-century, to the emergence of clocks specially designed to appeal to Asian customers. Specialist clockmakers gradually emerged, signing their products for the exporter if desired, though unlike the Ottoman market, signatures were apparently not seen as a guarantee of quality in China.[42] It was also possible to obtain a case from one supplier and fit it with a dial and movement from another. However, until these specialist suppliers emerged, and even later for the most distinctive and expensive clocks, exporters had to be more directly involved in their design and manufacture, and some tried to control much of the production themselves. James Cox had a large workshop in Shoe Lane where, in the 1760s–1770s, he employed jewellers, gold workers, mechanicians, and watchmakers to produce clocks and smaller articles in a distinctive 'house style'[43] (Figure 161). However, it is clear from surviving pieces that Cox also bought movements, cases and even complete clocks from independent suppliers, and even articles made in-house might have some ancillary tasks put out to independent craftsmen.

Other major exporters with the necessary facilities and knowledge might also carry out tasks in their own workshops, but most relied on the system of 'outworking' traditional within London's clockmaking and other luxury metal trades. This allowed a producer to finance and direct the manufacture of complex articles by a network of independent craftsmen operating from their own small workshops. The system relied on trust to ensure the quality of workmanship and proper use of materials and would come to seem outdated when large-scale production within integrated workshops and factories became the norm in many industries. However, for trades where little power-driven machinery was needed and the craft networks were already in place, the traditional system offered important benefits for the producer, permitting an elaborate subdivision of labour and rapid adjustment to changing market demands. It also reduced financial pressure if the situation became unfavourable, since payment to independent outworkers and suppliers could be delayed longer than for those directly employed. Since the problem of supervision was effectively subcontracted, the system allowed a producer to assemble a very large workforce, and Cox claimed to have employed from 800–1,000 workers in the seven years up to 1773—more than were employed at Matthew Boulton's famous Soho factory.[44]

Although most of the manufacturing was done in London, the workers included migrants from all over the British Isles and Continental Europe, and as the trade became well-established, some foreign craftsmen set up London workshops to supply it. Some of these had formerly worked for Cox, either directly, like the mechanicians John Joseph Merlin (1735–1803) from near Liège, and Samuel Rehe (c.1735–99) from Germany, or as independent suppliers, like the Prussian clockmaker Frederick Jury (fl. 1768–91), who made a pair of peacock automatons for Cox, one being now in the State Hermitage Museum, St Petersburg.[45] Above all, there was Henry-Louis Jaquet-Droz (1752–91) from La Chaux-de-Fonds, Switzerland, who first came to London in late 1775 to exhibit his own automata.[46] From the early 1780s, his main customer was Cox, who was then re-establishing his export business following his 1778 bankruptcy and could no longer maintain his own manufacturing facilities. In 1783, Jaquet-Droz left his London workshop under the control of his fellow countryman Henry Maillardet (1745–1830) and moved to Geneva to organize the production of the watches and smaller articles, which were increasingly in demand for this market. After his death in 1791, his Genevan partner Jean-Frédéric Leschot (1746–1824) continued to supply London exporters, along with other Swiss-based makers. In London, Jaquet-Droz's workshop had closed in 1790, but Maillardet continued as a supplier on his own account, in collaboration with his brothers in London and Fontaines. Around the same time, another immigrant Swiss clockmaker, Jean-Henry Borrell (originally Borel) (1757–1840), set up in London. Although Borrell worked mainly for the Ottoman trade, some of his clocks went to China, and they display features in the movements and case-decoration which suggest the use of Swiss suppliers, perhaps including his brother in Couvet[47] (Figure 162).

The easy movement of skilled labour to London from other parts of Britain and Europe gave the system of outworking much of its flexibility, but a financial crisis could rapidly reverse this movement. During the credit crisis of 1772, the London Press announced that 'goldsmiths, jewellers, and lapidaries are now emigrating to America, their branches of business having of late

---

42 It has been suggested that the Chinese objected to makers' signatures, but while many later clocks were anonymous, most earlier clocks were signed.

43 See Smith 2018, 73–99, for Cox's workers. Many of his smaller articles are illustrated in White 2012, 165–79.

44 *HCJ*, 34, 1773, 210. Cox's claim is consistent with his wage-bill and sales. Boulton claimed to have 700–800 employees at Soho in 1770. Goodison 2002, 26.

45 For Jury and the Peacock, see Zek and Smith 2005, 699–715.

46 For the Jaquet-Droz enterprise, Perregaux & Perrot 1916 is still interesting, but see Girardier 2020 for a thorough analysis.

47 For Swiss makers and the London export trade, see Smith 2017, 184–91. Charles de Constant noted that the English ordered clocks and watches for this trade from Geneva and Neuchâtel. Dermigny 1964a, iii, 1238, n. 4.

Figure 161 Cabinet musical clock, signed James Cox, London c.1766. Photo: © Bonhams, London, reproduced by permission.

Figure 162 Musical clock, signed Henry Borrell clock, London, c.1795. Photo: © Christie's Images, reproduced by permission.

so greatly decayed.'[48] Clock exporters to the Far East were especially vulnerable, given their reliance on long-term credit and the difficulties they already faced at Canton, so it is not surprising to find a London jeweller noting in September 1772 that 'Mr. Cox ... dismissed such a vast number of hands when he stopt payment that very many in the various branches & their families are destitute & some gone & going to our settlements abroad.'[49] It took several years for the trade to recover fully.

DESIGN

Clocks made for the Eastern market are difficult to categorize, varying greatly in size, complexity, and decoration.[50] Some were simply conventional musical clocks with added ornamentation in an 'Oriental' taste, while others were more imaginative in form, but all reflected the belief that they were valued in the East for the purpose of conspicuous display, rather than timekeeping. The designs have been described as 'chinoiserie' and often use elements of that style,[51] but they are very different from clocks designed for European decorative schemes in the chinoiserie taste, being meant to appeal to what Cox called the 'warmer conceptions' of Asian customers, and 'the taste of the Orientalists for brilliancy'.[52] Towards the end of the century, a European merchant who had lived in Canton recommended gilt-metal cases, flower bouquets set with imitation coloured jewels, and rotating glass rods imitating waterfalls and fountains—all features that are commonly found on clocks for this market made in both Europe and Canton.[53] Makers also responded to the Chinese preference for pairs of clocks and watches, and for features like centre-seconds hands, multiple dials, visible **balances**, and moving ornaments, which revealed the running of the mechanisms.

However, the originality of many of these clocks did not lie in their surface decoration or added ornamentation, however exotic or brilliant, but in their general form, and especially the way in which the separate elements of clock, automata, and mechanical music were combined.

As early as the 1750s, a musical automata clock in the shape of a 'Chinese Temple' was sent to India [sic], but it was the growth in the trade from the 1760s that led exporters to commission the fanciful clocks now considered typical.[54] The most 'advanced' designs showed a whimsicality of form to rival chinoiserie and rococo engravings, combining a wide variety of incongruous elements in lofty compositions, with no regard for relative scale, for example, the Goat clock by Cox (Figure 163). For even greater impact, a large clock might be flanked by smaller clocks and vases of jewelled flowers with their own mechanisms or placed within a pavilion or on an elaborate stand. These assemblages rarely survive intact; however, see the 1772 engraving of Cox's Chinese chronoscope (Figure 164) for a prime example of such work.

Some 'pieces of mechanism' were not primarily clocks at all. Qianlong's interest in automata has already been mentioned, and automated ornaments were often added to clocks, while moving beasts featured prominently on clocks like the Chronoscope,[55] which has an elephant moving around a gallery. A small number of life-sized automata were also made for this market, like Cox's Silver Swan, now in the Bowes Museum,[56] and the pair of Peacocks made for Cox by Jury, of which one was sold at Canton in 1792, and the other is now in the Hermitage Museum.[57] Towards the end of the century, copies of the Jaquet-Droz androids were sent to the East Indies, including a Musician sold to Cox in 1784, and another Musician, which Henry Maillardet sent to India in 1796.[58] A small Writer automaton in the Palace Museum now attributed to Timothy Williamson (d. 1799) is probably by Jaquet Droz or Maillardet, and may be the *petit anglois* that wrote praises to the emperor in Chinese characters and was modified by de Ventavon in 1785–6.[59]

The producers of these clocks knew that their bizarre designs would puzzle European observers, but as Cox explained rather defensively:

> A curious spectator may find here wherewith to satisfy himself, in considering the difference between the European and Asiatic tastes ... He must be little acquainted with the nature of things, that would judge of these pieces, which were calculated for the Indian and Chinese markets, by the austere rules of our European Arts.[60]

---

48 *Public Advertiser*, 15 August 1772, 3.
49 TNA, Chancery Masters Exhibits, C108/284. Letter from Arthur Webb, 21 September 1772. I am grateful to Helen Clifford for this reference.
50 The variety of designs can be seen in the Palace Museum handbooks; also White 2012, chs. 7–9.
51 Pagani 2001, especially ch. 4, which provides a good discussion of the designs, though the present author cannot agree that they would have looked equally at home in Europe.
52 Anon. 1772, 28–32, B1–B2, 'Advertisement'.
53 Charles de Constant, quoted in Chapuis 1919, 57–8.
54 The EIC gave a Chinese Temple clock to the Nawab of Bengal in 1757. BL, IOR, D/106, (unpaginated), 'Sundry Presents shipp'd on the Warren'.
55 From an article on Cox's Spring Gardens Museum. *British Magazine*, I, 1772, 40–5. See also a drawing of a Bull clock on stand in the Victoria & Albert Museum, London. Print-room, album E.3985–1906.
56 Smith 2016, 361–5.
57 Zek and Smith 2005, 699–715.
58 Chapuis 1951, 36; Wilkinson 1987, 128. For the Maillardet automata, Fima-Leonardi 2019.
59 The dial is signed by Williamson; Cordier 1922, 15, letter Raux to Bertin (17 November 1786); also Chapuis 1957, 54–7.
60 Anon. 1772, B1–B2, 'Advertisement'.

Figure 163 Goat musical clock, signed James Cox, London, c.1770. Photo: © Partridge, reproduced by permission.

Figure 164 *The Chinese Chronoscope*, engraving, 1772. *British Magazine*, vol. 1, 1772, 40–5. Photo: © Private collection, reproduced by permission.

The reference to the austerity of contemporary European taste is a reminder that chinoiserie, Rococo, and Baroque elements were being used for these clocks at a time when advanced European taste was abandoning these styles for the simpler lines of neo-classicism. It was not until the 1780s that neo-classical motifs were generally adopted, and clock cases became more restrained in form, while retaining the lavish gilding and jewelling thought appropriate for the East Indies, like the clock in Figure 162.

In practice, Western clocks produced for the East Indies show little awareness of authentic 'Asiatic tastes', apart from the superficial aspects already mentioned. Designers had few opportunities to acquire a deeper knowledge, but there was one occasion when direct Chinese input might have been obtained. In 1769, a Canton modeller of clay portraits named Chitqua came to London on an EIC ship. He met various leading artists during his stay, and in early 1771 met Cox's foreman-mechanician Merlin and was shown some of the clocks for the Spring Gardens Museum.[61] It seems likely that he was asked for his opinion and offered advice, but such opportunities were rare. Otherwise, it appears that exporters were not even concerned to distinguish between clocks suitable for China and those for Mughal India, sending clocks featuring turbaned 'Moors' to China, where they entered the Imperial collection, while as already noted, a clock described as a 'Chinese Temple' could be given to an Indian ruler.

The explanation for this apparent neglect may be found by looking at the clocks produced in China. While the earlier clocks made in the Imperial Workshop broadly followed European designs, they subsequently became more Chinese in form and decoration, apparently under instructions from Qianlong, who was personally involved in the designs. That might suggest that he would also have liked the tribute clocks from Canton to follow Chinese designs, and indeed, there is some evidence in the Chinese archives that he may even have wished to commission clocks to specific designs from Europe.[62] However, the emperor seems to have been inconsistent in his preferences. While many European clocks in the Palace Museum display Chinese alterations and additions, an edict of 1749/50 ordered Cantonese clockmakers to copy European examples, and clocks made (or recased) in Canton seem in general to have stayed closer to European forms, while using traditional Chinese motifs in their decoration (Figure 19). If European producers ever learned of the emperor's preferences, they had little effect in standardizing the multiplicity of designs being produced. Nevertheless, Qianlong continued to collect European clocks, and on at least one occasion instructed the Canton Customs Office that 'Clocks bought for the imperial court must be made in the West'.[63] Overall, the evidence suggests that imported clocks were not expected to conform to traditional Chinese aesthetics, but were valued in China as exotic curiosities, appealing to a taste for *occidenterie*.[64]

The finest clocks for this market were very expensive, though never reaching the fabulous sums of £20–30,000 and even £100,000 mentioned by contemporary observers. These seem to have believed that the cases were made of gold and set with real gemstones,[65] but they were usually of gilt-metal with jewelling of coloured glass pastes, though smaller articles might use rolled gold strips, gemstone chips, and panels of semi-precious stone.[66] The larger export clocks might have cost around £300–£500 earlier in the century, similar to comparable clocks made for the domestic market, but with the emergence of more specialized pieces in the 1760s, 'values' or prime costs (the distinction is not always clear) rose to about £2,000 for the finest examples, like the Peacock automaton. If two or more pieces were combined in a single assemblage, the total cost could even reach £5,000, like the 'Horse and Tent' that was a main prize in Cox's Museum Lottery: it incorporated two small Rhinoceros clocks and was further flanked by vases of jewelled flowers, all with music and automata.[67] Sale prices in Canton could be even higher if the market were buoyant: Cox claimed that his first 'Chronoscope', sent to China in 1769, was sold to the Emperor for £15,000.[68] As late as 1787, a clock seen in Canton was believed to have cost £10,000 in London,[69] but it was evidently unsold, and by that date, Chinese taste was probably turning away from the larger pieces towards smaller mechanical articles. While often ingeniously made and finely ornamented, these smaller articles cost much less than the larger pieces: a pair of scent flasks containing mechanical singing birds and watches supplied to Cox in 1787 by Jaquet-Droz and Leschot in Geneva cost only £237, compared with £1,000 for the uncased mechanism of a Musician android with which Jaquet-Droz's London branch had supplied him in 1784.[70]

---

61 *Gentleman's Magazine*, xli, 93 (28 February 1771). In Zoffany's portrait of Royal Academicians 1771–2, Chitqua's head is third from the left, behind the standing figures. Royal Collection: RCIN 400,747.

62 Zhang 2012, 569. Guo 2011, (n/p) Section II/3, 'Commissioned Orders to the West'.

63 Yang 1987, 138, no. 83; Liao 2002, Foreword, [5].

64 For *occidenterie*, see Kleutghen 2014, 117–35.

65 Farrer 1903, 233–4, Wedgwood to Bentley, 21 November 1768; Archenholtz 1785, 107.

66 Seeing Cox's museum, the jeweller Arthur Webb noted that 'the machinery is curious but the stones all different coloured paste.' TNA, Chancery Masters Exhibits, C108/284. Letter Book 7 July 1772).

67 Cox 1775, 2–4. A clock of this description by Cox is in the Palace Museum.

68 Cox 1775, vii.

69 Blancard 1806, 416. He 'presumed' it to be a present from the EIC to the emperor.

70 AVN, Fa. Jaquet-Droz, Account books: entries December 1784 and 13 October 1787. On these smaller articles see also Chapter 10, this volume.

EXPORT AND SALE

The crucial role played by officers of the EIC's ships in exporting these articles in their 'privilege' cargo has already been noted. Most private trade was conducted by the commanders, who were entitled to much more cargo space than junior officers (56.5 'tons' by 1770, equivalent to 64m^3) and had better access to credit. The goods were usually obtained on credit often using respondentia bonds,[71] or were carried on commission for the real owners. In the peak years 1768/9–1772/3, commanders carried nearly ninety per cent of all consignments of clocks and jewellery valued over £500: indeed, clocks and related articles were essential for a commander wishing to maximize his private investment.[72]

For much of the period, the same officers arranged sales and remitted the proceeds, with the detailed arrangements varying over time and according to destination. In India, the cheaper clocks could be sold directly to European or Indian merchants or consigned to a European warehouse for sale on commission.[73] Selling the more valuable clocks might take longer, involving resident agents who, for a major exporter like Cox, might be leading private merchants and Company officials. They would arrange to show them to wealthy customers across India, including princes like the rulers of Bengal, Arcot, and Oudh, who are named in Cox's publications.[74] Following sale, the proceeds could be remitted to London by Company bills of exchange (usually payable one year after sight), by loans to ships' commanders, or even by directly investing in private trade.

Sales in China were very different. The involvement of the Hong merchants has already been discussed, and even after lesser Chinese merchants ('shopkeepers' to Europeans) were allowed to deal in other private trade goods, imported clocks remained a Hong responsibility. This ensured that Chinese officials could examine the clocks as soon as a European ship arrived and choose those required for the annual tribute to Beijing. Other clocks were presumably bought for resale to other collectors or for stock, but unfortunately the European records contain little about this retail market.

If trade were buoyant, the agents would receive cash, which could be remitted to London in the ways already outlined, but if the clocks did not sell quickly, they had a major problem. For much of the century, Europeans could not stay in Canton outside the trading season, and when their ships left, the officers were reluctant to leave behind fragile goods unattended in a humid climate. There was also no guarantee that their clocks would find buyers in the following season, when other ships (and other agents) would bring a new selection. Therefore, they were keen to find a buyer at almost any price, rather than have to take the clocks back to London, which would incur further costs and could be disastrous for the consignor. During the serious downturn in the trade in the early 1770s, sales were made on extended credit or barter, but neither was ideal. The precarious financial position of many Hong merchants made lengthy credit risky, even at very high interest rates, while goods received in barter might be unsuitable for the European market and have to be sold elsewhere.

These problems, or just the wish to obtain larger returns than were possible with EIC bills of exchange, led to some of the larger exporters becoming involved in general trade from Canton, both to Europe and within the region, but this required expert knowledge. Since many of James Cox's consignments in the 1760s–early 1770s predated the regular availability of Company bills at Canton, his agents used them (from Bengal) only in late 1770 and early 1771. The bills, totalling £7,828, were repaid in London in three instalments between July 1772 and August 1774, indicating how slow this method was. Although they were obtained in India, the bills may have represented the proceeds of sales at Canton which had been reinvested in the Country trade.[75] Certainly, Cox had large investments in both India and China before his bankruptcy in 1778, and it was to recover these that his younger son John Henry Cox (c.1757–91) first went to the East Indies, sending his father bills from Canton in 1777–8. He was allowed to go back in 1781 to collect large debts still owed to his father, and once there, he persuaded the supercargoes to let him stay for several years, becoming the agent for his father's renewed trade and for other clock exporters. He then made some use of Company bills to remit the proceeds to his father, but this was evidently not his main method, and it is likely that he also invested in return cargoes to Europe and in the Country trade, in which he was active on his own account and as agent for private British merchants in India.[76]

A similar expansion from clock exports into general trade can be seen with Francis Magniac, whose sons joined J. H. Cox's former firm in Canton in the early nineteenth century, and the temptation was not confined to exporters in the East Indies trade. Isaac Rogers (1734–post 1817), a leading maker and exporter for the Ottoman market, described his activities to a parliamentary committee in 1817 as '... the making of clocks and watches and other mechanical works and manufactures, and exporting the same, and other merchandize to various parts of the world, receiving in return or as payment the products of those places.'[77] There

---

71 High-interest loans secured on the cargo.
72 Author's conclusions from analysis of the PTLs, to be discussed later.
73 For disposal of private cargoes in India, see Bowen 2007, 69–74.
74 Anon. 1772, 28–32. A musical clock by Cox was owned by the Nawab of Bengal in 1788. *Calcutta Chronicle*, 15 May 1788. I am grateful to Professor Peter Marshall for this reference.
75 Perhaps including the proceeds from three pairs of clocks valued at £3,000 that Cox wanted to send to China in 1769. IOR, B/84, 461 (8 February 1769).
76 IOR, E/1/66, 421 (3 May 1780). Bills for larger sums were sent to Cox's assignees. For J. H. Cox's career, see Williamson 1975, 1–35. His agency house, which was independent of James Cox & Son in London, eventually became the firm of Jardine Matheson. For its early years, see Greenberg 1951.
77 *BPP* 1817, vol. 6/504, 39.

is no evidence that the East Indies were among those places, and producer-exporters seem usually to have focused on just one of these distant markets, presumably because of the specialist knowledge needed to engage in the general trade of the region. If Cox and Magniac showed only limited interest in the Ottoman empire, their counterparts, like Isaac Rogers and George Prior, saw themselves as Levant merchants.[78]

CLOCKMAKING IN CANTON

J. H. Cox was not technically trained, and as main agent for imported clocks at Canton, he would have recognized the need for skilled assistance in preparing clocks for sale after their voyage, including tasks beyond the competence of local Chinese craftsmen, who seem to have worked mainly on the cases. For this reason, clocks sent to the East Indies had sometimes been accompanied by their own clockmaker. Indeed, the Italian polymath Count Algarotti (1712–64) had claimed in 1739 that, since the Chinese could not make mechanical clocks, English ships at Canton 'have always some journeyman watchmaker on board', who could repair broken clocks that the Chinese brought to be exchanged, reselling them as newly arrived.[79] That was an exaggeration, but such arrangements were certainly made for particularly valuable clocks, even for those going to India, where European clockmakers had already settled. In 1768, Cox announced that his first Chronoscope would shortly leave for India, 'provided a proper person can be procured to attend the same to the place of its destination, and there to continue to keep it in repair.'[80] James Upjohn's son, who accompanied his father's clocks to Canton in 1776–7 to help with their sale, probably had a similar role.[81] These ad hoc arrangements became unsatisfactory once clocks could be left in Canton to be sold later, and J. H. Cox arranged for a European clockmaker to join him. The first was Félix Laurent (c. 1755–1802), formerly with Jaquet-Droz, who was in Canton 1785–1794, and he was followed at J. H. Cox's former firm by several other Swiss and British clockmakers until the 1820s, when it gave up trading in clocks.[82]

Whether transient or resident, solitary European clockmakers in Canton presumably had local assistance. That would have helped to introduce a knowledge of European clockmaking techniques in southern China, but how quickly and to what extent is unclear. European observers disagreed about the levels of skill available in China. Algarotti's assertion in 1739 that Chinese craftsmen could not make, or repair, European clocks was repeated in 1772 by the Russian scientist Peter Simon Pallas (1741–1811), at least for watches. He advised against exporting European watches via the Russo–Chinese border mart of Kiakhta: 'The Chinese receive watches quite cheaply from Canton. They even bring to Kiakhta watches that are out-of-order; they sell them at miserable prices to the Russians because they have no one who can repair them.'[83] However, by the 1790s, Sir George Staunton, who accompanied Macartney's embassy, could refer to the ability of Canton craftsmen to 'mend and even make watches', while another member of the embassy claimed that they could make a watch or clock 'as well as in London, and at one third of the expense', though **mainsprings** were imported.[84]

The emperor seemed to be of the same opinion. He rejected the suggestion that the scientific instruments brought by the embassy as presents were too delicate to be moved after being set up by the embassy's own craftsmen, with assistance from the Imperial Workshop, and noted that the official who conveyed the embassy's message had

> ... not held the office of Superintendent of the Canton Customs, [so] his acquaintance with Western clocks ... is not extensive and he is inexperienced in such matters. Now the Tribute Envoy [Macartney] has seen that there are people in the Celestial Empire who are versed in ... clock repairing and are now helping alongside those who are setting up the articles, he can no longer boast that he alone has got the secret.[85]

This, however, was clearly a political response rather than a considered assessment, and it is curious that, while the emperor mentioned the help provided by the Imperial Workshop's craftsmen, he referred to unfamiliarity with Western clocks in Canton when emphasizing the official's inexperience. That may suggest some doubts about the Imperial Workshop's current capabilities (de Ventavon had died in 1787). Any doubts would have been strengthened two years later, when a pair of complicated automata that a Dutch embassy brought as presents were damaged on the way to Beijing. One was repaired by the embassy's own clockmaker, but the missionaries declined to repair the other 'since there was no-one in Beijing who could do it', and it was suggested that it would have to be returned to Canton, presumably for repair at Beale's shop where it had been bought.[86] Similarly, in

---

78 Rogers and Prior attended a meeting of Levant Merchants, 31 March 1790. Guildhall Library, London, broadside 7.43. Some Ottoman-market clocks by Cox are known, including an elephant automaton in the Topkapi Museum, Istanbul.
79 Algarotti 1769, 89: letter IV, Petersburg, 30 June 1739.
80 *Gazetteer & New Daily Advertiser*, 21 December 1768.
81 Leopold & Smith 2016, 144, 156. He travelled as a midshipman.
82 Laurent's successors included Charles Henry Petitpierre (probably), Frédéric (de) Goumois, James Lindley, James Ilbery, and Edouard Bovet.

83 Gauthier de la Peyronie 1793, 212.
84 Staunton 1797, iii, 385; Barrow 1804, 306–7. Watch-glasses were also imported in large numbers.
85 Singer 1992, 46, quoting a Court Letter 28.8.1793. The official, Zhengrui, escorted the embassy and presumably supported its message.
86 Guignes 1808, vol. 2, 1. The clockmaker was Charles-Henri Petitpierre, who had also accompanied the British embassy. Beale was J. H. Cox's former partner and successor.

1807 the fall in the price of 'curious mechanical objects' at Canton was attributed to there being 'no-one in China who can repair them when anything happens to put them out of order',[87] while valuable smaller articles like watches were sent back to Europe for repair.[88] On this evidence, there seems to have been a relative decline in the technical skills of the Imperial Workshop by the end of the century, even before the last missionary supervisor was expelled in 1811.[89] By that time, Chinese craftsmen in Canton had certainly learned to make complete clocks, but in spite of what some foreign observers believed, it is unlikely that they yet possessed the advanced skills needed to make or repair the more complicated clocks and smaller items like watches, imported from Europe.

THE PEAK YEARS

British clock exports to China evidently grew strongly in the first half of the eighteenth century to achieve 'great value' by 1754, but the growth is unquantified. Thereafter, the trade is better documented. Chinese records show that the number of clocks sent to the imperial court increased quickly after 1759, with the Canton Customs sending between forty and fifty clocks to the emperor each year, and that over 2700 clocks were purchased by Customs officials for the emperor during the whole of Qianlong's long reign, most of which were thought to have been made in Britain. These figures relate only to the imperial collection, but by the end of the century the market must have spread well beyond Court circles, since the Canton records show that 1025 chime clocks, watches, and snuffboxes inset with watches were imported in 1791 alone.[90]

This growth can be observed in the EIC archives, but the data pose problems. As private cargo, clock exports should have been recorded in the Company's Private Trade Ledgers (PTLs) showing the nature and value of such consignments, the volume of space they occupied, and the names of suppliers or 'tradesmen'.[91] The ledgers survive in full for the sailing seasons 1768/9–1774/5 and 1779/80–1781/2, and show that the annual values of all clock (or 'clock and jewellery') exports on EIC ships (almost all on China-bound ships) in this period ranged widely, from almost nothing to £21,000. However, the figures must be treated with great caution since low-volume, high-value articles like clocks were particularly prone to the under-recording and undervaluation that affected all private exports. The chronic problem of concealment or smuggling has already been mentioned, but even when consignments were declared, the Company's surveyors evidently had problems with assessing their true value. It is, for example, difficult to accept that twelve cases of clocks filling 128 cubic feet of cargo space declared by a commander in 1768/9, were worth only £70.[92] For these reasons, actual values of clock and jewellery exports must have been much greater than those recorded in the PTLs.

The records of imported goods kept by the Company's supercargoes at Canton for several years might have provided a check on these figures, but unfortunately, they are not comparable. Unlike the PTLs, they give numbers of clocks and watches imported but not their value or volume; and in the few seasons for which both London and Canton records survive, there is disagreement over which ships were carrying clocks. However, if the PTLs seriously understate the value of such exports, they can still provide useful information about fluctuations in the trade during the years of its greatest importance for Anglo–Chinese commerce. Table XI shows the declared values for all private exports on EIC ships sailing to China in the seasons 1768/9–1771/2 (column C), with corresponding values for clock and jewellery exports to China (column D), and to all destinations including India (column F). Jewellery is included because many entries combine the two categories, and it is clear that 'jewellery' often included clocks and other mechanical objects ornamented with jewels.

The figures confirm that the clock trade was dominated by the Chinese market. From a level that was already high by the late 1760s, the declared value of clock and jewellery exports to China peaked in the early 1770s, at over £20,000 in 1770/1 and £14,000 in 1771/2, when there were slightly fewer ships. In those years, such articles represented thirty-five per cent of all private cargo declared on China ships in 1770/1 and thirty-one per cent in 1771/2, even though not all China ships carried them, at least openly, and they were particularly likely to be under-recorded and undervalued. Furthermore, they represented an even higher proportion of *European* goods reaching China in private trade, since most other private exports would have been unloaded in India and replaced with Asian goods for the rest of the voyage to China.[93]

The table also shows that such exports peaked in the years immediately *after* the Hong merchants' complaint in 1770 that they were already overstocked.[94] This reflects a general problem for exporters to the East Indies in having to organize the following season's exports before knowing how successful the previous season had been—a problem particularly acute for specialized clocks, which could take many months to commission and produce. It is likely that the boom years of the mid-to-late 1760s had encouraged exporters to invest heavily in production, and they no doubt

---

87 Renouard 1810, 134.
88 Bonnant 1964, 43, n. 5.
89 Pagani 2001, 56.
90 Liao 2002, Foreword, 4, 6; Yun 2008, 103.
91 IOR, H/18–22. The 'tradesmen', who were probably the real owners/exporters, were not always named.

92 IOR, H/18, 37, *Britannia*. Two tons on *Cruttenden* were valued at £5,000. *Britannia's* clocks were supplied by 'Cocks', probably James Cox who had previously supplied the same commander. For undervaluation by commanders, see Bowen 2007, 58–62.
93 Morse 1926–29, v, 167. Goods worth £69,000 were added in India, plus an unspecified part of the silver.
94 I owe this important observation to Professor Huw Bowen.

Table XI EIC Private Trade Ledgers 1768/9–1781/2

A[a]	B[b]	C[c]	D	E	F	G
Sailing season	China ships	All Private Exports—China ships (£)	Clocks & jewellery—China ships (£)	D/C (%)	Clocks & jewellery—All EIC ships (£)	D/F (%)
1768/9	16	69,657	11,670	17	13,470	87
1769/70	15	54,498	9,282	17	10,139	92
1770/1	15	59,714	20,647	35	20,961	99
1771/2	13	45,995	14,280	31	14,330	100
1772/3	10	32,136	4,365	14	5,515	79
1773/4	3	12,505	90	0	120	75
1774/5	4	14,721	43	0	67	64
1775/6	[7]	–	–	–	–	–
1776/7	[10]	–	–	–	–	–
1777/8	[9]	–	–	–	–	–
1778/9	[7]	–	–	–	–	–
1779/80	14	36,092	4,257	13	4,357	98
1780/1	12	37,654	3,063	8	3,178	96
1781/2	13	45,369	1,250	3	1,250	100

[a]The table shows data for the ten complete sailing seasons from London covered by the surviving PTLs. Data for seasons 1775/6–78/9 is missing. Season 1781/2 is not quite complete but includes all China ships. [b]Column B shows the number of ships leaving London for China in each season. For 1775/6–1778/9 the figures in brackets are taken from other EIC records. Most China ships also called at Indian ports where clocks may have been unloaded or even loaded, and ships were sometimes rerouted, but column G shows the extremely high proportion of clock exports from London that were consigned on China-bound ships. [c]Except for 1768/9, figures in column C for All Private Exports on China ships are from Pritchard 1936, App. 2A, 236–7. For comparison, in 1771/2 the Company's own exports included British goods valued at £152,000 and £192,000 in silver.

wished to forestall any response by the EIC to the Hong merchants' petition by maintaining a high level of exports for as long as possible. These did not fall significantly until 1772/3, but in fact the Company was slow to react, and when the ban on jewelled clocks was eventually implemented in 1773/4, its scope was vague, though later clarified to cover clocks valued at over £100 each.

The values shown in the official statistics can be supplemented with other contemporary records. The Hong merchants' 1770 petition said they held unsold stock worth over a million taels (about £330,000). Given that the trade started to take off in the mid-1760s, this suggests that they had accumulated unsold goods averaging £60,000 each year, in addition to the clocks bought for the emperor's tribute and any other sales.[95] A similar picture emerges from the exports of James Cox. It was claimed in early 1772 that 'in the course of a few years' he had 'executed to the amount of near half a million [pounds sterling] for the different markets of the East';[96] while in 1773, a Parliamentary committee considering Cox's petition to sell his Museum by lottery was told that he had exported clocks with a value of some £740,000 to the East Indies over the previous seven years, of which about seventy-five per cent had been sold. The same committee heard from a ship's commander, who had exported clocks for Cox in the difficult year of 1771, that he had sold his goods in China for £36,000 and might have sold others worth £40,000 but for the severe shortage of money there.[97] Taken together, the evidence from both buyers and sellers suggests that the real value of clocks carried to China on EIC ships may have exceeded £100,000 per year by the late 1760s–early 1770s. This was at a time when the Company's own exports of British goods to Canton averaged £150,000 per year, or about £218,000 if Indian goods are included (seasons 1767/8–1772/3).[98] Although clock exports may have increased in absolute terms in later years, other exports, particularly from British India, increased even more, so this was probably the peak of their relative importance for Anglo–Chinese trade.

The virtual disappearance of declared clock exports in the mid-1770s was not entirely due to the EIC's ban. Credit crises in both London and Canton in the early 1770s affected several major clock exporters, and the Company's own financial difficulties forced it to cut back its trade with China in the years 1773–5, recovering slowly thereafter. With clock sales dependent on the revenue that general European trade generated for Chinese Customs and merchants, any reduction in that trade was damaging. Even before the EIC reduced its sailings, the falling demand for clocks at Canton was being blamed on 'the diminution of the trade of

---

95 See n. 35. So much unsold stock suggests that not all purchases had been for the emperor's tribute.
96 Anon. 1772, 28–32.
97 HCJ, 34, 1773, 210. Evidence of Alexander Jameson, who blamed the scarcity of money on the absence for two years of the ship that brought silver from Manila. Cox's agent in Bengal also blamed poor sales on 'the present scarcity of money in that part of the world,' presumably due to the 1770 Bengal famine.
98 Pritchard 1936, 394, app. 4. Silver exports are excluded.

the East India Company with the Chinese.'[99] Even so, smuggling evidently continued, since an undeclared consignment of valuable clocks and jewels caused problems with the Hoppo for the Company's supercargoes in 1775.[100]

Overt clock exports—ostensibly limited to 'clocks not ornamented'—revived with the recovery in general trade, and reached a modest level by 1779/80 when the PTLs briefly resume, before falling again during the final years of the American War of Independence. In the two seasons immediately before and after the return of peace in 1783, very few Company ships visited Canton, and much private trade, including clocks and watches, was diverted into neutral carriers like the (Austrian) Imperial Company. Its ship *Count of Cobenzel* sailed from Britain in May 1782 for China via Trieste, carrying fourteen chests of watches and watchwork from London; and in Madeira it loaded 'about 60 cases of rich clocks and boxes of watches . . . executed by some of the best workmen at Vienna, Paris, Geneva and in England'.[101] With the recovery in the EIC's own trade from 1784, there was a boom in British clock exports, and it was claimed in 1785 that these had been worth £1,000,000 sterling in the previous year.[102] That seems unlikely, but the market was soon glutted and collapsed, exacerbated by the decision of Qianlong in 1785 to stop the flow of clocks and other tribute from Canton, ostensibly because of the burden this imposed on the Hong merchants.[103]

DECLINE

Fortunately for the trade, the emperor's order was short lived, but demand for the more elaborate clocks seems to have remained depressed. This was evident in 1792, when numerous musical clocks, automata, and jewels formerly belonging to James Cox were sold in London. The fifty-five lots, dating from before his 1778 bankruptcy, included grand display pieces such as a peacock automaton like the one now in the Hermitage Museum. Their combined invoice prices totalled £52,000, and although auctioned in London, the various items were already at Canton, saving buyers the trouble and cost of exporting them. Even if they needed cleaning and repair, the fact that such exceptional pieces sold for only £12,000 suggests they were no longer in demand in China.[104] The lack of interest may also have reflected the emergence of alternative exports. Finding these had become urgent following the slashing of British import duty on tea from 119 per cent to 12.5 per cent by Pitt's Commutation Act of 1784. By reducing prices and eliminating smuggling, this greatly increased the demand for the Company's own tea. Financing these purchases at Canton required more than the export of silver, woollen cloth, and miscellaneous European products could provide. The solution was found in greater exports from British India, mainly raw cotton at this time, which were carried to Canton by private merchants who loaned the proceeds to the EIC in return for bills of exchange on London.[105]

On the demand side, the declining appeal of expensive clocks in China led in 1802 to the main recipients of imported clocks at Canton, Beale & Co., refusing to sell them for their London correspondents.[106] Their reluctance is explained by Renouard de Sainte-Croix, whose comments about the fall in the price of 'curious mechanical objects' have already been noted. In 1807, he recorded that Beale & Co. held a large number of *pièces mécaniques* which had been unsold for many years because the Chinese had lost interest, owing to the difficulty of having them repaired when put out of order by the damp conditions.[107] That may well have been a factor, but the death in 1799 of Qianlong, the greatest collector of such pieces, may also have played a part. However, while his successor Jiaqing (r. 1796–1820) was less interested in Western clocks and curiosities, large sums continued to be spent on buying them for his annual tribute for several more years. As late as 1810–11, the Hong merchants' resistance to official demands that they buy expensive clocks 'at any price & to an enormous amount' once again delayed the loading of EIC ships.[108] That produced another request to London for such imports to be banned. The Directors agreed, but the renewed ban could have had little effect by 1814, when the Hong merchants complained of having to spend $100,000 (£25,000) every year on clocks and watches.

The real turning point came in 1815, when Jiaqing published a sumptuary edict against clocks, condemning them as useless things on which 'the valuable property of our country is gradually almost totally expended'.[109] Thereafter, they were replaced by ginseng in the emperor's annual tribute, and European clockmakers had lost their most important 'customer' in China.[110] They also now faced competition from locally made clocks at half the price, and in 1824 the main trade agents at Canton, now named Magniac & Co., gave up that part of their business, symbolizing the end of a century when highly ornamented clocks and automata had played an important, but also disruptive, role in British trade to China.[111]

---

99 Evidence from Captain Jameson, see n. 97.
100 Morse 1926–29, ii, 14.
101 The Imperial Companies of Ostend and Trieste were heavily dependent on private British merchants. For the *Count of Cobenzel*, see *Parker's General Advertiser*, 18 October 1782, and *Morning Post*, 2 June 1784.
102 *General Advertiser*, 14 June 1785, 'East-India Trade. Ninth Irish Resolution'.
103 Extract from a Swedish Supra Cargo '. . . 25 February 1785, per True Briton', in *Morning Chronicle*, 26 November 1785; also Cheong 1997, 226–7.
104 Christie's, 16 February 1792. The vendors were the assignees under Cox's 1778 bankruptcy.

---

105 Opium overtook raw cotton from 1824. Cheong 1979, 6–8.
106 Greenberg 1951, 151. They received clocks worth £25000 in 1802 (78).
107 Renouard 1810, iii, 201–2.
108 IOR, G/12/175, (31 January 1811), 87–8; G/12/189, (13 February 1814), 121–2.
109 Greenberg 1951, 86.
110 Cheong 1997, 227.
111 Greenberg 1951, 87.

WATCHES

Towards the end of the eighteenth century, as imported clocks lost their position as status symbols for the Chinese elite, there was a growing demand for smaller mechanical objects like watches, snuffboxes, and scent flasks containing musical movements and miniature automata. Although some watches were used as timekeepers in Court circles, they were primarily articles of personal adornment, often worn or carried in pairs[112] (Figure 114). By the end of the century, this fashion had spread more widely in Chinese society, encouraged by the increasing availability of cheaper watches. In spite of the growing market, this led to a chronic problem of oversupply, perhaps exacerbated by producers trying to replace traditional markets now hit by war in Europe. In 1794, J. F. Leschot of Geneva, former partner and successor of H. L. Jaquet-Droz, noted that watches were returning unsold from China, and four years later, he blamed continuing problems in Canton on excessive exports of low-quality watches.[113] By 1804, it was being claimed that 'the gaudy watches of indifferent workmanship, fabricated purposely for the China market and once in universal demand, are now scarcely asked for', but that was too sweeping, and it is clear that significant numbers of watches continued to be exported to China.[114]

With watches so easy to smuggle, records of their importation at Canton are probably even more unreliable than for clocks, but EIC ships clearly had no monopoly. While they carried 1,304 watches in 1802, 377 watches arrived in American and Swedish ships; and in 1803, when 1183 watches were imported, nearly half (522) came in Danish ships.[115] According to Renouard de Sainte-Croix, EIC, Danish, and Swedish ships brought 2,203 watches in 1804, 502 in 1805, and 730 in 1806, and he commented in 1807 that Swedish and Danish ships had brought watches in such 'prodigious' numbers, virtually all made in Switzerland, that they were sold in Chinese shops for five piastres (about £1 5s) per pair.[116] These would have been mainly the products of the Neuchâtel Jura and neighbouring regions, though many no doubt claimed to be made in London or Paris, leading the London Clockmakers' Company to complain in 1809 that such fraudulent articles had considerably reduced the 'formerly very lucrative and extensive' trade in British clocks and watches to China, Russia, and Turkey.[117]

The success of Swiss makers in capturing these distant markets, particularly China (which they retained for the rest of the century), deserves consideration. It has been claimed that they were more willing to adapt their clocks and watches to Chinese taste than the British, and in support, it has been stated that the leading British producer, James Cox, 'did not make any modifications for the Chinese consumers', since his aim in publicizing sales to China was to increase the appeal of his 'chinoiserie' clocks to domestic customers.[118] Unfortunately, this interesting theory is not supported by Cox's published statements,[119] and in fact, there is no evidence that British designs were less successful than those of their competitors in the eighteenth century. The Swiss had long copied British and French clocks and watches to compete in their traditional markets, and even in the early nineteenth century, many of the 'English' watches sold in the Ottoman empire were Swiss copies. In 1806, Achard *père et fils* of Geneva sold 600 watches in Smyrna copied from one by the English maker George Prior, at a price of 116 francs each (about £5), compared with 168 francs (£7) for the London-made watch.[120] It is true that Swiss manufacturers later developed special kinds of watches for the Chinese market, like the 'Chinese duplex' associated with Bovet of Fleurier, but even these may have been derived from watches first made by the London firm of Ilbery (Figure 165).[121]

Swiss success therefore cannot be attributed to the greater cultural sensitivity of their designers, and superior marketing is unlikely since most sales in Canton were conducted by British merchants until they switched to more profitable goods in the early nineteenth century. The real problem for British makers was, as the Achard example suggests, their inability to compete with the Swiss on price, especially once cheaper watches and other small articles became fashionable in China. In this period before extensive mechanization, lower prices mainly reflected the lower cost of skilled labour in Geneva and the Jura. This was not due simply to their regular use of cheaper female labour, or to British watchmakers clinging to outdated practices, but to Britain's higher cost of living. That was the constant complaint of the London Clockmakers' Company, but even cheaper British manufacturing centres were uncompetitive. It was a Coventry watchmaker who told a parliamentary committee in 1817 that 'We cannot now compete in price with the foreign manufacturers, on account of the high price of the necessaries of life.'[122]

---

112 See Blancard 1806, 416, for the Chinese habit of wearing a pair of identical watches.
113 BGE, MS suppl. 962, 70, Leschot to Frisard, 14 November 1794; and MS suppl. 964, 29–30, Leschot to Duval & fils, 27 April 1798. I owe these references to Sandrine Girardier.
114 Barrow 1804, 181.
115 IOR, G/12/140, 63, 67–8, 147 for 1802. Figures for 1803 are from Chinese Customs registers quoted by Charles de Constant. Dermigny 1964b, 308–9.
116 Renouard 1810, iii, 153, 160, 169, 201–2.
117 Atkins & Overall 1881, 275, 'Memorial to government', 5 May 1809.

---

118 Pagani 2001, 122–4. The suggestion (114) that Charles de Constant contributed significantly to Swiss success is unconvincing.
119 Anon. 1772.
120 Babel 1916, 519.
121 Chapuis 1919, 180–4. The relationship between Ilbery and Bovet remains to be clarified. See also, this volume, Chapter 2, Section 2.
122 Atkins & Overall 1881, 275, 278, 290, 296. Also *BPP* 1817, vol. 6/504, 81.

Figure 165 Gold enamelled duplex watch for the Chinese market by William Ilbery, London N° 6042, *c*.1815. Photo: Jean-Baptiste Buffetaud, courtesy of Chayette & Cheval, auctioneers, Paris.

CONCLUSION

Throughout the eighteenth century, the European trade in clocks to south and east Asia was closely linked to the development of general trade, particularly between Britain and China. In the search for exports to offset the cost of products like tea and silk, ornamental clocks proved to be a valuable supplement to silver, but they would not have been so important if Britain, which came to dominate the general trade, had not been a major clock producer. Britain's European competitors continued to rely heavily on silver for their declining share of trade with China, exporting few clocks or watches until some became carriers for Swiss products at the end of the century.[123]

Exporting clocks could be very profitable for the private trade of the EIC's ship officers, but the effect on the Company's own trade was more problematic. The sale of clocks in Canton generated cash that the Company could sometimes borrow for its own purposes, but their purchase for the emperor's annual tribute created political complications, particularly after the financial burden was imposed on the Hong merchants. Since the same merchants monopolized the EIC's own trade, the strain on their finances and their attempts to protect themselves created problems for the Company throughout the eighteenth century.

The private exporters also faced unusual challenges. Apart from the cost of making these specialized objects, the fluctuating demands of a restricted and distant market made sales unpredictable. If increased demand in one season led to investment in higher production, conditions could change radically before the increased output reached Canton at least two years later, producing a glut that could be disastrous for heavily indebted exporters. They also faced periodic attempts by the EIC to support the Hong merchants by banning or restricting clock exports. However, while bans could be damaging in the short term, they could usually be circumvented to allow trade to continue, if in modified form. Towards the end of the century, the expansion of general trade and discovery of alternative exports produced a relative decline in the importance of the clock trade, but its absolute value remained high, bolstered by the growing market for smaller and cheaper articles like watches. Real change only came when the Imperial Court lost interest, culminating in Jiaqing's sumptuary edict of 1815 and the removal of clocks from his annual tribute. These steps deprived imported clocks of the special position they had held in China for almost a century, and, when combined with

---

123 Cheong 1997, 283. The Dutch company was an exception since most of its exports to China came from its Indonesian possessions.

growing competition from cheaper clocks made in Canton, led to a collapse in British production for this market. Although some European clocks continued to be sent to China, the trade would, in future, be dominated by Swiss firms selling cheap watches to a much wider range of customers.

The exotic and expensive clocks produced for this market have always had their detractors. A contemporary critic dismissed them as 'gaudy trifles' and 'of no sort of use',[124] and similar views persist. However, such criticism is irrelevant in assessing the value of a trade that, in Cox's words, gave 'employment to thousands,' and brought 'a large revenue to the kingdom, in return for the labour of our elegant artificers.'[125] He might have added that some of that revenue would be used to expand general British commerce in the East Indies. The success was more remarkable when it is remembered that the prior investment required for production and the long-term credit needed for export were not specially provided by a 'great corporation' or even by the state (as might have happened elsewhere in Europe), but rather by private individuals.

Similarly, the many skills required for design and manufacture were provided by London's existing systems of outworking and subcontracting, but the impact of the industry extended well beyond the city. There were suppliers in other English production centres, like Lancashire and Coventry, and by the end of the century, London was receiving musical clock movements and unfinished repeating watches from Geneva and the Swiss Jura, as well as finished clocks and watches for this trade.[126] The flexibility of this system allowed the London clock industry to respond effectively, if not always painlessly, to changing demand from distant markets like the East Indies for nearly a century.

As well as relying on the movement of components, the trade encouraged the migration of skilled workers, sometimes over considerable distances. Clockmakers, jewellers, and merchants from across Britain and other parts of Europe were attracted to London, where the former learned the latest techniques and styles, and the merchants made essential contacts. This knowledge was spread more widely when they eventually returned home or moved elsewhere, sometimes setting up new industries to collaborate or compete with London, as in Switzerland. Some clockmakers even travelled to India or China with the clocks they had helped to make, passing knowledge of European clockmaking techniques to local assistants and collaborators. In China, the presence of foreign clocks and clockmakers in Canton helped to create a local industry capable of producing complete ornamental clocks by the end of the century. If these could not rival the finest European products, they were much cheaper, and they satisfied the wider market for clocks that emerged in China, as the traditional market centred on the emperor declined.

---

124 Staunton 1797, i, 47–8.
125 Anon. 1772, A1–2, 'Preface'.

126 For use of Swiss clock movements by Cox and Borrell, see Smith 2017, 187, 191; and for unfinished watches, see Atkins & Overall 1881, 275.

CHAPTER SEVENTEEN

PUBLIC CLOCKS IN THE NINETEENTH AND TWENTIETH CENTURIES

Marisa Addomine

SURVIVAL OF ANCIENT SCHEMES, CREATION OF NEW SOLUTIONS

Turret clocks in their earliest 'birdcage' frame form made in wrought iron, with iron wheels and **lantern pinions,** survived in minor centres and in isolated communities until the beginning of the twentieth century. Survival in this context does not merely mean that old clocks were kept and preserved, but that they continued to be made in the old way, following an uninterrupted tradition. This was the result of lack of contact with more advanced technical communities, of the poor tools available, and, in general, of extremely limited resources: some villages in the mountains and in remote European rural areas were reached by electricity only in the 1960s, and illiteracy was still common in the southern, south-western, and eastern peripheries of Europe before the Second World War. In many of these cases, clockmaking began locally with a single, self-taught artisan—perhaps a blacksmith with a minimum of arithmetical competence, who simply copied existing clocks in his region.

However, for the more advanced urban centres, from the end of the seventeenth century the use of the pendulum spread through Europe and in the course of the eighteenth century practically all existing turret clocks were updated. They gained precision, most of them being modified with the addition of a minute hand, but lost forever their original **escapement** configuration. The cost of the modification was not a major one and the advantages were noticeable.

The great innovations introduced in the eighteenth century in turret clock design were the flatbed layout and the pinwheel escapement. The flatbed was a totally new concept, so its adoption had an impact on new clocks, which, from this time, were made according to this advantageous design. The pinwheel escapement was chosen for new clocks and was also applied for the improvement of existing clocks. A typical pattern of transformation for a good, ancient, wrought-iron clock was: verge-and-foliot → verge-and-pendulum → anchor-and-pendulum → pinwheel-and-pendulum. The pinwheel escapement, in its many variants, was relatively easy to make and to maintain; it proved to be robust, required little care, and gave good results in terms of precision. Increasingly, turret clock wheels were made of cast bronze, which made their construction quicker and simpler and took less time and effort to achieve a higher overall quality at a reduced cost. Lantern pinions were gradually abandoned and were replaced by machine-cut leaf-pinions.

The industrial revolution had an impact on horology, offering new technology, new materials, and a new approach to design and manufacture. The response varied from region to region, creating differences in the local design of frames and in the general layout of movements. A later, but fundamental, innovation stemming from advancing industrialization was the concept of the interchangeability of parts. Once ironworkers could cast parts from moulds intrinsically identical, at least to the degree of precision that frame parts of a turret clock would require, series production could take place. Public clockmakers understood immediately the immense potential of cast iron, and it gave birth to new clock frame shapes and gear work configurations. In the meantime, quality bronze casting for wheels of large size became a standard technique, limiting handwork to finishing and adjusting operations.

The production of identical cast parts in batches and in advance, with fixed distances between **arbors** and hence standard wheel sets, gave birth to the idea of having stocks available off the shelf. As a consequence, the old process starting with a written proposal for a new turret clock and ending maybe one year later, when clocks were wrought by hand, wheel after wheel, was completely transformed into an assembly process of standard pieces—a process that still required skill—but one much faster and less prone to uncertainties. The customer's needs could normally be satisfied by a model in a catalogue of one of the major makers.

NEW FRAME STYLES

As a general remark, from the mid- to late-eighteenth century, screws, nuts, and bolts appeared increasingly often as fixing elements in turret clock frames, and traditional iron wedges disappeared. With the introduction of flatbed structures, thanks to the standardization of screws, elements were joined in what we consider a 'modern' way.

France was the cradle of flatbed clocks: the earliest frames were often made from industrially made iron bars that were shaped, cut to proper length, and mounted in a rectangular shape. Similar structures can be found in those areas where French horological influence in the period was strong, such as the southern Netherlands (which would become Belgium), and the French-speaking areas of Switzerland and Northern Italy. As an example, in Piedmont, close to the French border, the Jemina family,[1] still active in clockmaking today, made good-quality flatbed clocks following the purest French style.

In England, apart from experiments and isolated trials, several different styles were adopted. These can be identified as the posted frame, which could be simply interpreted as the immediate transposition of a bird-cage structure obtained using cast iron elements, which was common between 1790 and 1850; the so-called double-framed, sometimes mis-described as 'chair frame', in which the wheels were housed in a higher frame and the barrels were in a lower, external frame, obtaining the typical 'chair' structure popular in England between 1730 and 1860; and the plate-and-spacer type, with two twin frames, one for the front and one for the back, kept parallel at proper distance by pillars. This design was introduced in 1800 but remained in use for the rest of the century. The flatbed model was also extensively used, the celebrated Westminster great clock—Big Ben—being perhaps the most famous example of its kind, and the largest in the world (Figure 166).[2]

In France, clockmakers advanced beyond the simple, early, plain flatbed created in the second quarter of the eighteenth century to a more advanced structure in which the drums were placed side by side. This layout is well documented in contemporary horological literature.[3] Starting from this configuration, the going train was placed in an upper position, having its wheels mounted in a pyramidal, or triangular, structure, with the escapement in the higher part of it, while drums and other wheels were deployed on the flat base.

The passage to flatbed structure and the introduction of industrially made metal bars lowered the costs of turret clocks. In France, the Morez-Morbier region in the Jura Alps was one of the most important manufacturing areas in terms of the number of items produced per year and specialized in low-cost timekeepers. In that area some interesting experiments helped to optimize production and simplify construction, with a consequent reduction of costs and time to market. The Lepaute-type flatbed was the standard choice until the mid-nineteenth century. However, no documents reveal which company really introduced the cast iron triangular-shaped structure, which, with its variants, dominated the turret clock market until the mid-twentieth century. Several companies were active in the area during the period and more or less contemporaneously started offering the new type of timekeeper. The leaders, among many, were Odobey (1858), Prost (1840), Cretin-L'Ange (1830) (Figure 167), and Bailly-Comte (1804).[4] When trying to establish a priority, in such a context, it is important to remember that many of these suppliers were also producing for other retailers or anonymously. Morez-Morbier clocks could be quite large, often had three or four trains and the best of them have a high-quality finish. The **striking** system was frequently by **rack**-and-snail, but sometimes the traditional **count wheel** was still employed.[5]

The triangular structure proved highly successful, being economical, robust, and compact, and became the most common form employed in France during the later-nineteenth century and first half of the twentieth century, with imitators all over Europe and beyond. This was not always recognized. Not more than twenty years ago, while helping to classify a collection of nineteenth- and twentieth-century turret clocks from the production of a family of north Italian makers who intensively used the French triangular style for their timekeepers, the heirs of the family learned some surprising news. They were told that the shape of the clocks made by their grandfathers was not invented in their family, as their fathers had been claiming for a lifetime, but had simply been borrowed—not to say *copied*—from classical French models!

Two main design tendencies can be identified in German public clockmaking in the nineteenth century once the use of cast iron had become widespread. Some makers adopted the plate-and-spacer structure, mentioned earlier. This design, for example, was typical of the production of such companies as Weule of Bockenem (Lower Saxony) (Figure 168). It did not pay homage to the then-fashionable flat layout but was compact. Plates and spacers were kept in place by nuts and bolts and assembly was a simple process. This structure was typically used for two-train movements.

French triangular structures inspired what developed into another typical German solution: the triangular, but modular,

---

1 For whom see Turicchia 2018, 160.
2 For details, see McKay 2013.
3 Lepaute, 1755 and 1767, ch. XII; *Encyclopédie* 1751–80, plate II, a–e and related text.

4 For an introduction to them see Buffard, Dulain, & Renaud 2013.
5 Turret clocks were also made in Paris and Normandy, see Chapter 13, Section 3.

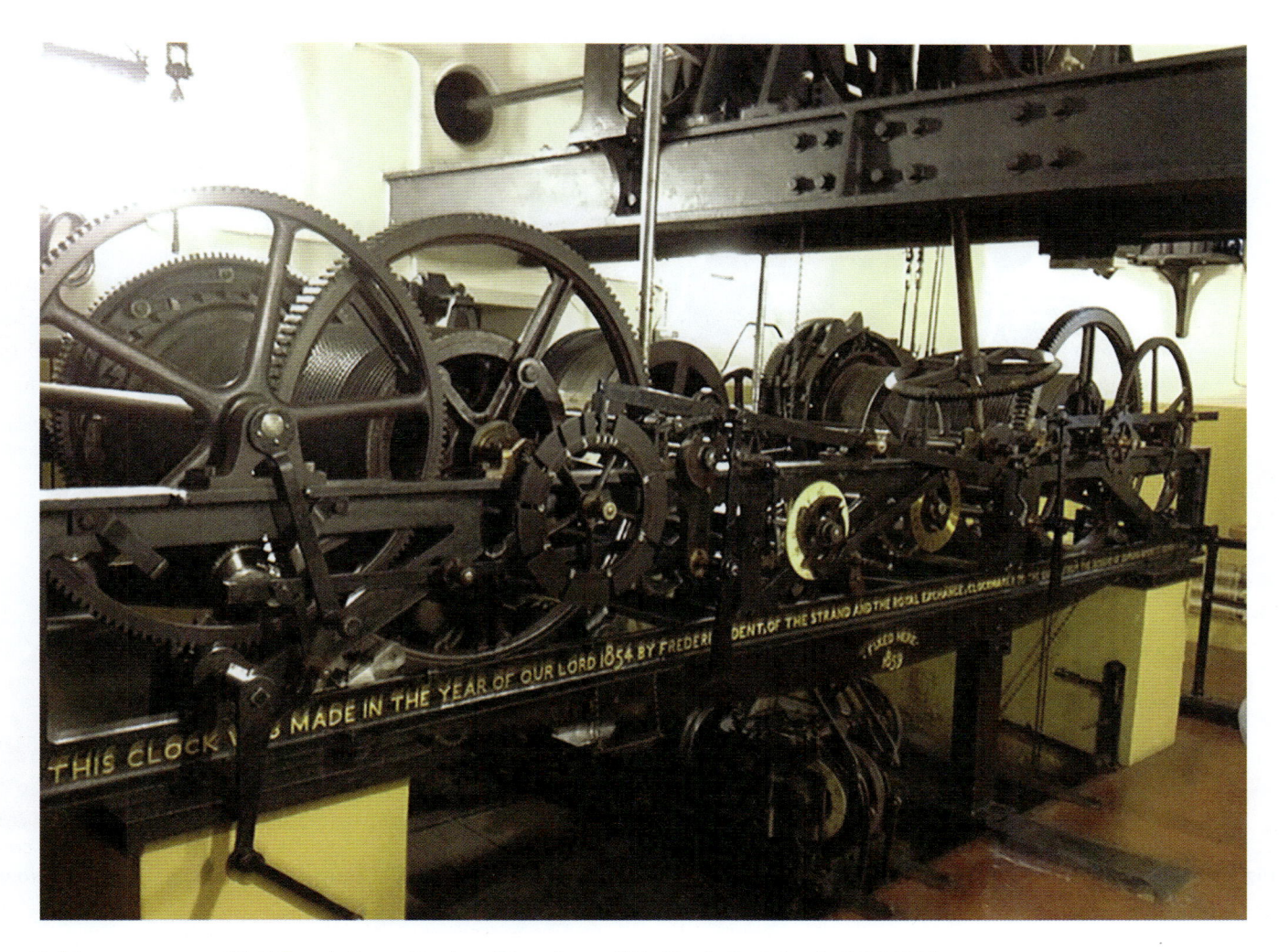

Figure 166 The movement of E. J. Dent's Great Clock at Westminster ('Big Ben'). Paulobrad, CC BY-SA 4.0 <https://creativecommons.org/licenses/by-sa/4.0>, via Wikimedia Commons.

flatbed turret clock. In this configuration, each train is kept in a separate module designed to be placed in a row with others, with mechanical junctions at fixed positions. Such an arrangement was not only easily accessible for maintenance, but also allowed the construction of a clock adapted to the individual needs of the customer, with a minimum number of different parts held in stock. The German tradition of striking quarters required, in most cases, a third train; sometimes, when buyers could afford it, a fourth when chimes were added, the so-called *Angelus*. In the early and middle periods of the nineteenth century there was considerable interaction in design between Eastern France, in particular from Strasbourg where Schwilgué was active, and the western regions of Germany, which would remain within the commercial catchment area of the Ungerer company until at least the end of the century.[6]

German makers distinguished themselves in turret clockmaking thanks to several high-quality brands. Weule, already mentioned, was one of the most distinguished companies in the later nineteenth and early twentieth centuries. Johann Friedrich Weule (1811–97) was apprenticed as a turner and as a watchmaker in Braunschweig and Kassel. He started his own business in Bockenem in 1836 but did not begin making public clocks until 1847, when he built his first for the market church at Goslar. His clocks generally used the plate-and-spacer style. In the following generation, under the direction of his sons Friedrich (b. 1855) and Wilhelm (b. 1856), the business expanded considerably and in 1879, when four timekeepers were shown at the Sidney International Exhibition, it had twenty employees and an average yearly production of about forty clocks. In 1886, the company annual reports, besides announcing a new bell foundry, described a newly introduced steam system, a workforce of forty, and the inauguration of more modern buildings. In 1898, larger plants and another foundry were announced.

In 1900, the first Weule clocks equipped with electric winding systems appeared. The First World War compelled the company to adapt some of their manufacturing plants to war needs. After 1918, the company was managed by Friedrich Weule, a representative of the family's third generation, but a merger with Ulrich, a company from Thuringia, gave rise to a new company, Ulrich & Weule. In the following years there were other mergers until the business was finally taken over by Bornumer Wilhelmshütte, its activity

---

[6] For the different constructions employed, illustrated by both French and German examples, see Ungerer 1926, ch. 2.

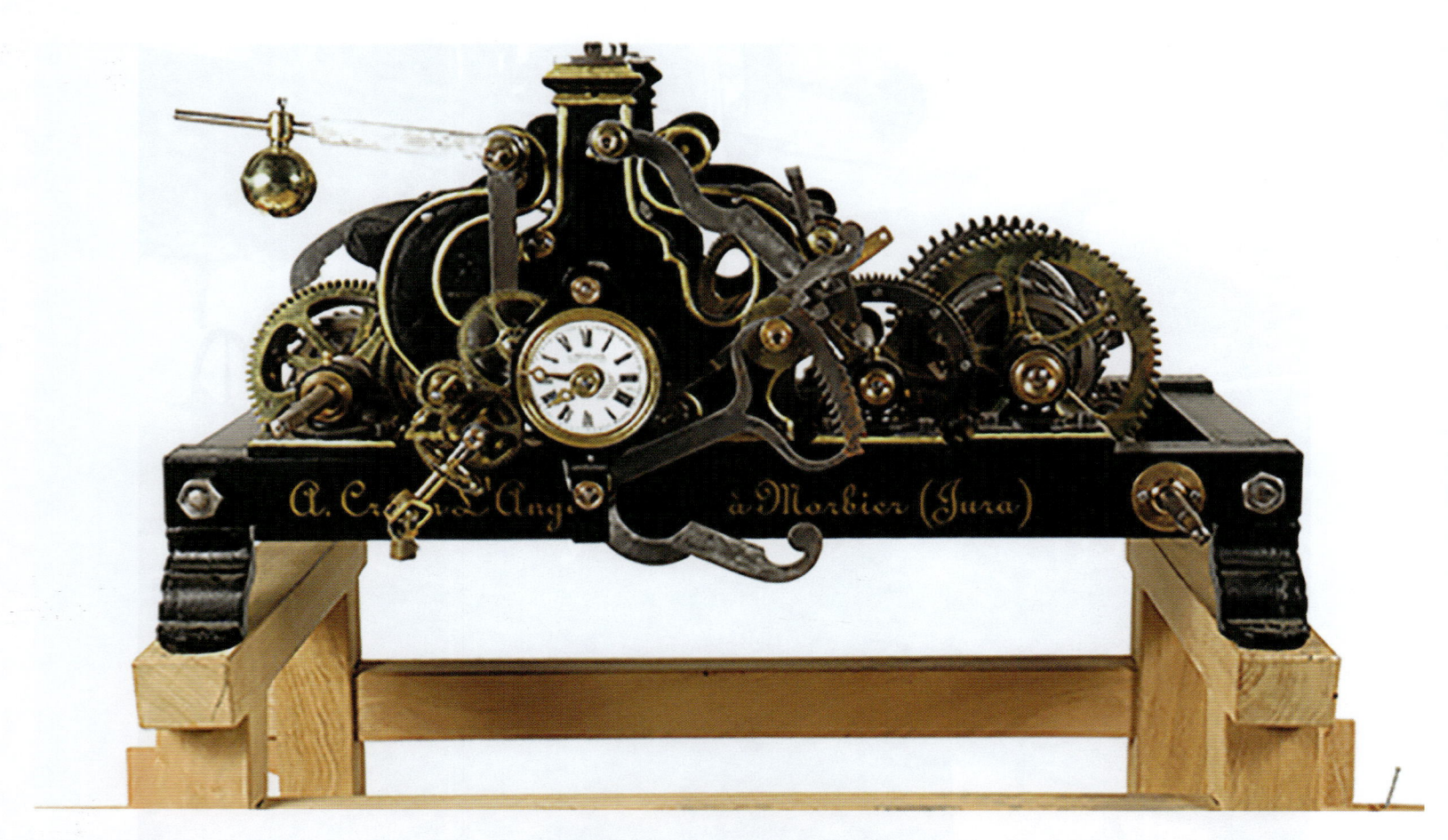

Figure 167 Flat bed tower clock by A Cretin l'Ange, Morbier, c.1900. Photo: Jean-Baptiste Buffetaud, courtesy of Chayette & Cheval, auctioneers, Paris.

coming to an end in 1965. In 1979, the Weule factory buildings in Bockenem were demolished.

Italy was only created as a nation-state in 1861. Before then it was split into different territories, with different governments, histories, and completely separate cultural influences. These ranged from areas under German control, now near the Austrian border, to areas where Spanish influence was dominant, like Sicily. It is, therefore, very difficult to discern a typical profile of Italian turret clockmaking. Essentially, most production was local, where clocks were made by craftsmen who were often far from contact with the innovating centres in horology. Frequently, when an Italian public clock bears a date—which is not all that often—it surprises the researcher, as the clock looks older than it is. This archaic look is a consequence of the gap between provincial makers and any horological literature, and because 'modern' texts were mainly in English, French, or German—languages that an Italian clockmaker could seldom read. Such makers continued making wrought iron bird-cage frame clocks, where the pendulum and the pinwheel escapement were the only modern touches. In style such clocks could have been conceived two or three centuries before.

Only a few companies were organized as good workshops and produced clocks following up-to-date design and technology. Practically all of them followed French layouts, with the exception of the Solari family at Pesariis in the region of Friuli, now close to the Austrian border but in the eighteenth century, part of the Austro-Hungarian Empire. In the eighteenth century, members of the Solari family worked in German workshops, and so adopted German styles in the manufacture they established in 1725. Surviving for over two centuries, in the twentieth century the company developed a revolutionary public clock (discussed later). Their turret clocks were inspired by German prototypes of the plate-and-spacer style.[7]

Most important public clock factories were in Northern Italy, with a concentration in the area of Recco, not far from Genoa, some good makers in Lombardy, and a very few makers in southern Italy. An excellent workshop, which started in the nineteenth century and is still active in the Southern region of Basilicata, is that of the Canonico family (founded 1879). Other good Italian makers in the nineteenth and twentieth centuries were Cesare Fontana (*fl.* 1860-c.1911), Antonio Jemina (*fl.* 1832–75), Miroglio (2nd half 19th century), Terrile Brothers (late nineteenth century), the Trebino family (founded 1824), Frassoni (1825–1980), Granaglia & Co (*fl.* 1830–c.1900), Augusto Bernard (*fl.* nineteenth century)—a Swiss maker summoned to Naples by the king to develop and improve local clockmaking—Alfonso Curci (2nd half 19th century), and Domenico de Vita

---

7 On members of the Solari family see Turicchia 2018, 264–5.

Figure 168 A typical Weule clock, made in 1888. https://commons.wikimedia.org/wiki/File:WeuleUhr1888_P1260783.JPG.

(late nineteenth century), to name some of the better known.[8] Others, like the makers in the region of Marche, Pietro Mei (*fl.* 1825–65) and Antonio and Luigi Galli (mid- to late-nineteenth century) reached good levels in terms of craftsmanship and mechanical ingenuity, but never developed beyond being a local workshop.[9]

In the nineteenth century, the most important Italian producer in terms of quality and quantity was by far the Frassoni family in Rovato Bresciano in Lombardy. Their activity began in the early nineteenth century, but some sixty years later this family-run company employed 150 workers, and had its own internal foundry, a privately run indoor school for mechanics, and was active as supplier of good-quality mechanical components for other companies. The analysis of what remains of their archives also shows that in the 1930s they had modern catalogues, an advanced marketing technique, good methods for preparing technical proposals, and their organization was undoubtedly an advanced industrial one for the time. The Frassoni company experimented with the application of electricity for auto-rewinding, for synchronized networks of public clocks with a master clock and slave units, and tried several methods for the auto-lubrication of wheels. They developed innovative patents and experimented to find new solutions.[10] Unfortunately, due to the unexpected death of the last male heir, the company ceased activity in the second half of the twentieth century and most of their archives and of technical tools were dispersed.[11]

In the rising American clock industry of nineteenth-century America, the Howard Company in Boston, Massachusetts, should be mentioned. A company with wide mechanical competence, it was founded in 1858 by Edward Howard (1813–1904) and Charles Rice starting from the technical assets of a former company, the Boston Watch Company. Their initial activity focused on clocks, watches, regulators, and marine chronometers. Edward Howard had former business experience in producing quality clocks in a company of his named Howard & Davies in 1842, joined in 1843 by a third partner named Luther Stephenson, after whose arrival they entered the turret clock market.

The history of the Howard Clock Company, under its several appellations, is complex. It ended in 1980, the last turret

---

8 For details concerning them see Turicchia 2018.
9 I thank Mr Pietro Sebastianelli of Montecarotto for sharing information about makers Pietro Mei and Antonio Galli.
10 Addomine and Pons 2009.
11 I thank Mrs Sonia Genocchio Frassoni and her daughters for sharing their family memories.

clock having been produced in 1964. The company produced all types of timekeepers and many other mechanical items, but for about a century it was one of the finest producers of turret clocks in the United States. From a mechanical point of view they are not comparable, especially in terms of finish, with the best European production; nevertheless, they were extremely well engineered. In a very typical American approach, their design was oriented toward ease of maintenance, simplicity, and the use of good-quality materials. Clock frames and their stands, especially those of the late-nineteenth-century production, were often richly decorated with painted friezes in the Belle Epoque style, giving a delicate artistic touch to the ensemble.

NINETEENTH-CENTURY ESCAPEMENTS

As domestic clocks and pocket watches became increasingly precise, the public clock was no longer the sole time reference for a whole community, except in some special cases. Nonetheless, they retained civic and symbolic importance, and some evolution occurred in the mechanics of turret clocks, especially in their escapements.[12] Without entering into details, it is to be noted that the quest for a free escapement had a strong impact on high-quality turret clocks and was the chief element that established the superiority of the Westminster clock by Edmund Beckett Denison (1816–1905) in terms of precision.

In Denison's own words:

> A gravity or remontoire escapement is one in which the impulse is not given to the pendulum directly by the clock-train and weight, but by some other small weight lifted up, or a small spring bent up, always through the same distance, by the clock-train at every beat of the pendulum. And the great advantage of them is that the impulse is therefore constant; for the only consequence of a variation in the force of the clock is that the remontoire weights are lifted either faster or slower, which does not signify to the pendulum, as the lifting is always done when the pendulum is out of the way. If this can be managed with certainty, and without exposing the pendulum to some material variation of friction in the work of unlocking the escapement, which it must perform, its motion and therefore its time must be absolutely constant, since there is nothing to disturb it.[13]

Several variants of the **gravity escapement** were applied to turret clocks, with three, double-three, four, five, and even six legs. An interesting and different solution found on some of the best late-nineteenth-century German public clocks is the so-called Freischwinger escapement, used in some of the turret clocks made by Mannhardt in Munich. It mediates the flow of energy to the pendulum with high efficiency, in terms of low friction and absence of recoil.

Apart from these quality escapements, produced by excellent makers and requiring constant care and adequate maintenance, nineteenth- and twentieth-century turret clocks display every variant of the pinwheel escapement; pins on the same surface of the escape wheel, on opposite surfaces, or incorporated inside the pendulum rod, as well as different positions and geometries for the **anchor**, from the simplest to the most refined—all were tried, with differing results.

BIG BEN: THE MOST FAMOUS CLOCK IN THE WORLD

Big Ben, properly speaking, is not the name of the Great Clock of the Palace of Westminster, but of its great bell in the Elizabeth Tower. Nevertheless, for millions, even billions, of people worldwide it is synonymous with this most famous clock. Much has been written about it and its long and controversial history.[14] These few lines simply summarize its main horological features.

The Palace of Westminster has housed a clock since *c.*1290. When Charles Barry's design was chosen in the second quarter of the nineteenth century for the renewal of Westminster Palace after the fire, which largely destroyed it in 1834, his plans included an imposing tower and a magnificent clock on it. Benjamin Lewis Vulliamy, then Clockmaker to the Crown, was asked to design this clock. Having one's name associated with the supply of what was going to be the king of clocks generated a real war among potential suppliers. A tender was finally put out, asking the bidder to offer a clock different from the original Vulliamy design. One was submitted by Denison, the young lawyer and amateur horologist. However, the winner was E. J. Dent, an established company well known for its chronometers and watches. The specifications were highly exacting: at a time when a turret clock was considered good when its precision was to within thirty seconds per week, the new clock was, however, to be accurate to better than a second per day. Its size was also a challenge. The new Westminster clock would have four dials, each of which, over twenty-two feet in diameter, was illuminated from behind. The size of a turret clock is a direct function of the size of the bells it has to ring. This clock had to ring both the quarter bell and the hour bell, the latter weighing fourteen tons. The accuracy requested for the moment of the first blow on this bell was within a second. The giant size and incredible performance required (in

---

12 See Ungerer 1926, 53–62.
13 Denison 1854.

14 Most of the information in this paragraph comes from McKay 2010 and from conversations with Chris McKay, whom I thank.

terms of precision) outlined in the remit were a real challenge even for the most knowledgeable maker. The clock was finally made between 1852 and 1854, although after its completion a number of problems arose owing to troubles in the casting and mounting of the bells, and the final installation was delayed by several years.

From a technical point of view, the most impressive mechanical feature of Big Ben is its precision, obtained mostly thanks to its double three-legged gravity escapement, which was not installed immediately. This escapement and its extraordinary performances were the final result of a very long quest—a mechanical challenge which began well before the clock itself and that had involved such masters of horology as Thomas Mudge, Alexander Cumming, Henry Kater (1777–1835), James Gowland (mid-nineteenth century), James Bloxham, and James III Harrison. In the various editions of his book, Denison described different versions of it. A four-legged prototype was devised, but the final version, chosen for the Westminster clock, used the double three-legged structure.[15] It was installed on the Great Clock in 1860.

Another important issue for precision was pendulum compensation. Even if this aspect was normally taken into consideration for **regulators** and not for turret clocks, in such a big clock the effects of temperature on the length of the pendulum rod could not be neglected. Big Ben has a compensated pendulum of a peculiar type, using concentric tubes:

> A central rod of iron descended from the suspension spring and a tube of zinc sat on a collar at the bottom. An iron tube descended from the top of the zinc tube, and the bob was suspended at the bottom of that tube. Holes in the tube allowed access for the air to enable the compensation pendulum to react to rapid changes in temperature.[16]

The giant cast iron movement is 4.7m wide and weighs five tons. The pendulum length is 4.4m, corresponding to a two second period; its weight is 310kg. Chiming occurs each quarter hour. The worldwide famous tune, known as the 'Westminster Chimes', should more properly be called the 'Cambridge Chimes', being written in 1793 for a clock in the University Church of St Mary the Great, Cambridge.[17] The tune was chosen by Denison, who was familiar with it from his youth, as a graduate of Trinity College, Cambridge, and a keen bell ringer.

THE TRADITION IN STRASBOURG

The extraordinary astronomical clock in Strasbourg is a potent rival to the Westminster clock as the best known in the world. Succeeding clocks there had created strong local interest in horology for centuries. In the nineteenth century this was reaffirmed by Jean-Baptiste Sosimé Schwilgué (1776–1856).[18] An outstanding mechanician, he was fascinated by the second Strasbourg clock, which had been derelict since the 1780s. He vowed to make it work again but achieved his dream by entirely replacing it.[19] Incorporating the making of more standard public clocks into his production, Schwilgué's company made some five hundred of them, some still working today. Two of his employees, the brothers Auguste-Théodore and Albert Ungerer, became partners with Schwilgué, and carried on his company, which took their name after 1903. They continued in the tradition of making high-quality turret clocks, among which is the Messina astronomical clock in Sicily, inaugurated in 1933, considered to be the biggest astronomical clock in the world, with a planetary dial, several automata, musical effects, and a perpetual calendar. The company ceased operation in 1989, having produced over 6,000 public clocks.

FLORAL DISPLAYS

Public time-telling has been associated with flowers in two different ways. The first one dates from the eighteenth century and is linked with the name of the renowned Swedish naturalist Carl von Linné (1707–78), better known in Latinized form as Linnæus, the author of the classification of fauna and flora that introduced the concept of binomial taxonomy. He described his 'floral clock' for the first time in his *Philosophia Botanica* (1751), under the name of *Horologium Floræ*, i.e. the clock of Flora, the ancient Roman goddess of vegetation. Thanks to his studies of plants, he could select twelve species that opened or closed their flowers at given hours of the day, using them as time markers in a dial. This was only a theoretical proposal, and several alternatives were offered for the same hour of the day, considering different latitudes, since differences in climate affected the growth of plants in different locations. Some passionate gardeners tried to put Linnæus' proposal into practice, especially in the nineteenth century, but no one seems to have achieved very good results.

More technology is involved in the other combination of flower and clocks: a clock with a large single hand driven by a concealed movement, with a beautiful dial composed by flowers at ground level. It appeared for the first time in Scotland, created by John McHattie (d. 1923), Edinburgh Parks Superintendent, with

---

15 McKay 2010, ch. 10.
16 McKay 2010.
17 McKay 2010, 89, 119; Ord-Hume 1995, 37.

18 Kintz 1982.
19 Schwilgué 1857. For details of his clock see Chapter 9, Section 2; Ungerer 1922; Bach *et al.* 1992.

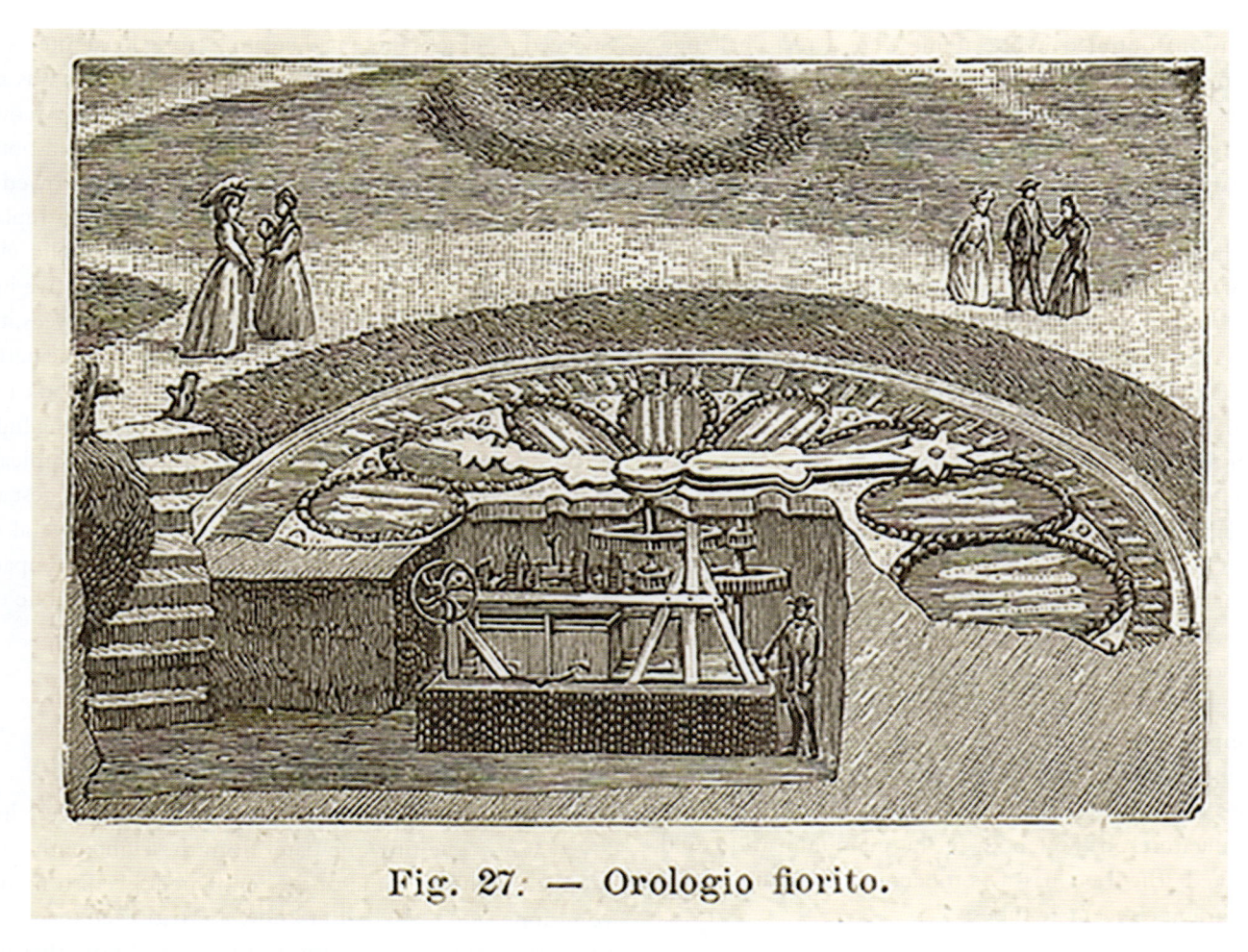

Figure 169 A floral clock run by a turret clock movement concealed in an underground chamber. Image from *Enciclopedia Moderna Illustrata*, 1929.

the advice of the jeweller James Mossman (late nineteenth–early twentieth century). The mechanical parts were executed by James Ritchie & Son (*fl.* 1805–1920s), and the clock was inaugurated on 10 June 1903 in the gardens of West Princes Street. In the following year, a minute hand was added and in 1905, thanks to a little pipe organ, the clock also played a cuckoo sound each hour. The movement driving the hands was an old turret clock from the parish church of Fife. The dial was periodically modified, not only to change the perishable floral decoration, but also to adapt the overall design of the dial to current trends, or to reflect current events, such as the Jubilee of Queen Elizabeth II, the centenary of Robert Louis Stevenson (the author of *Treasure Island*), or the end of the Second World War.

These floral clocks were initially operated by standard turret clocks, hidden from view, either in an underground chamber beneath the dial (Figure 169), or placed at some distance from the dial driving the hands by long mechanical links to the **motion work**, concealed in underground pipes. From Scotland they became popular throughout the world, and they can be found in many parks and gardens, even if they are now more easily powered by electrical systems.

DIFFUSION OF TIMEKEEPERS, THE DEATH OF PUBLIC CLOCKS

At the beginning of the history of public clocks, we encountered a need-to-know time and to share time in societies where most of the population did not have a personal timekeeper. Furthermore, most clocks and watches would not have been precise enough to offer a synchronized service. From the eighteenth century onwards, and especially in the nineteenth, clocks and watches were made in large quantities, leading in the later nineteenth century to true mass production. Even turret clocks, thanks to the new technologies, could be provided more economically. As a result, clocks spread everywhere. Schools, colleges, even minor town halls, all had their own clock.

Many shops, especially clockmakers and jewellers, had a dial on the road, typically at a crossroads, to attract the attention of those passing by: some of them also had little jacks, adding fun by blowing trumpets or drumming on the hour. Repairing old clocks, some well over a century old, was no longer viable if new ones could be installed at an equivalent, or nearly equivalent, cost.

Public clocks, apart from those with exceptional features, like the extraordinary astronomical clocks, lost their glamour and their appeal: they no longer represented the power or pride of towns and cities.

In many cases, important clocks, even astronomical ones, were dismantled: there were two reasons for this. The loss of technical and scientific knowledge about these complicated contrivances made their restoration a challenge that very few clockmakers would undertake. In most cases, local makers were unable to repair the old movements and would not agree that an outsider should do the job. So, for the sake of modernity, they concealed their incapacity by declaring that the clocks were unrepairable, quickly dismantling the masterpieces of the past and guaranteeing new work for themselves. In other cases, unfortunately, it was simply a matter of business. If the renewal of church clocks were financed at a national level, some companies could obtain substantial profits, and indiscriminate campaigns for the replacement of old timekeepers were carried out. Despite the developing conservation movement throughout Europe in the later nineteenth century, no concept of horological heritage had yet developed: hundreds, perhaps thousands, of movements (and sometimes their dials) were scrapped, to be replaced by anonymous, modern clocks. The example in England of Benjamin Lewis Vulliamy is notorious. Although a knowledgeable maker, he had a radical attitude as a restorer towards existing clocks, intervening often in a totally disruptive manner. Even if little documentation of his interventions has survived, he destroyed many, now forever lost, important clocks.[20] In the nineteenth century, turret clocks definitely lost their importance. Low-priced pocket-watches were in the hands of the layman, most houses had at least one domestic clock, and clock dials were almost everywhere. The decline of ancient turret clocks had definitely begun.

By the end of the century another social phenomenon occurred—the depopulation of villages and the ageing of those who remained. These facts, and the growing pressure for social insurance, the need to pay a wage for services which previously were performed as a volunteer contribution by local members (daily clock winding), or the shrinking of the budgets (especially of minor religious communities) all led to reconsidering the clock in terms of cost. As a church clock was no longer indispensable for the community, most of them were left inactive. In consequence, they were forgotten and exposed to burglars, rust, damp, damage, and birds. For others, especially when the community liked to keep alive the tradition of bell ringing, different solutions were found. A few heroes, however, still go on today—as in Clusone, Lombardy—where volunteers have established a weekly shift table for winding every day their 450-year-old astronomical clock, the pride of the town.

A REVOLUTIONARY WAY OF DISPLAYING TIME

The Italian company of Solari should be recorded for an important invention in public clocks. They may have been inspired by a system already used in some American desk clocks patented in 1902, the 'Chronos' clocks or 'Ever-Ready Chronos Clocks'. These clocks had a lantern-shaped case, in which the current time was shown by figures on tickets revolving around a vertical axis.[21]

Solari developed this concept using it in an electric clock and applying it to tickets revolving around a horizontal axis, improving the reliability of the turning system, which was the weak point of the 'Chronos' clock. Starting from there, they created the flip displays used in the second half of the twentieth century all over the world in railway stations and airports for departures and arrivals until the introduction of digital monitors. Fratelli Solari's flip clocks, mostly used for indoor timekeepers, were also made in large size and used in public installations: an example of such an early digital time reading system is still visible in the Florence Santa Maria Novella Railway Station, dating to 1932.

AUTOMATING CLOCK-WINDING OPERATIONS

Having an automatic winder eliminates the need for the daily activity of raising the weights by the clock keeper. Several attempts at automation have been made since the introduction of motors: at present, only fully reversible systems, i.e. systems whose installation can be removed without causing any permanent alteration to the clock, are allowed on ancient turret clocks, in the spirit of the full respect of clocks as heritage items. Specialist companies offer different systems, from the so-called Huygens Endless Chain, to the very common Monkey up the Rope, with the motor winding itself up the chain when it has descended to a certain level, to the Differential (or Epicyclic), the Remontoire, the Power Assisted Winder, and others as well. These systems are different in principle and their suitability must be evaluated by an expert as a function of the clock and of the overall installation.[22]

ELECTROMECHANICAL AND ELECTRONIC SYSTEMS

Another almost daily task of the clock keeper was time adjusting. This was a major challenge that was more complicated to resolve because it needed a solution that was not a simple electrification of a mechanical problem. In this case, the solution had

---

20 Cf. the remarks on this subject at the end of Chapter 21.

21 For these see Shenton 1985, 266–7.
22 For earlier systems see Ungerer 1926, ch. 7; McKay 2013.

to wait for a complete change of paradigm: the passage from the mechanical clock to the electrical clock and to the availability of a synchronization signal, coming, in Europe, from the Frankfurt DCF77 Radio Signal. This has been available since 1959 and has allowed an easy way to reset electronic devices to correct time, also avoiding human intervention when summer time variation is required. The evolution from mechanical to electromechanical and then to electronic systems was gradual and could well be exemplified following the evolution of models in the Italian factory of Trebino, makers of mechanical turret clocks in Recco, near Genoa, since 1824, and still active today. As a first intervention, old existing mechanical clocks were simply modified to avoid human intervention for weight lifting, introducing motors and Galle chains in place of the ropes and drums of the time and striking trains. The second step in this evolution saw clocks change, becoming smaller in size, simplified, and essentially reduced to their crucial role of timekeepers, i.e. suppliers of the time base. Where pendulum and escapement were still present, they were placed in the upper part of a small closet, which also contained electromechanical programmable switches, and some circuitry. It was a hybrid solution, widely used in Italy during the 1950s and 1960s. After that, mechanical clocks practically disappeared and electronic cards in racks at first, and then PLCs and small programmable systems, later replaced them all.

REPLACING, KEEPING, RESTORING

In several situations a compromise is the best possible solution. In most cases, clock owners cannot afford the cost of a clock keeper, but still need to offer time indications to the public. Sometimes it is the high antiquity of the clock—and the need to preserve it from mechanical stress and from any risk—that prevents operation. At the same time, for the sake of tradition, as a tourist attraction, or for whatever other reason, hands must show the correct time on the dial and bells must be rung. In these situations, the correct decision is to remove the original clock—possibly giving it a professional, but conservative, restoration—and place it in a protected environment—maybe in the same structure or building—in a position where it can be admired by visitors and studied by scholars. All operations can be performed by modern systems, providing that the external appearance and displays are preserved for passers-by. In the past century there was a different approach towards restoration: a tendency to bring the clock movement back to its—presumed—pristine condition, i.e. for pre-pendulum clocks converted to pendulum to be returned to a, more or less invented, verge and foliot escapement system. Nowadays, a completely different attitude prevails. Conservative restoration, after necessary cleaning and consolidation, tends to preserve the movement, together with the modifications it has undergone through the centuries, and the availability of digital techniques (for example, three-dimensional modelling) offers scholars the tools of virtual reconstruction, hence exploring possible alternative structures without any modification to the physical item.

CHAPTER EIGHTEEN

WRISTWATCHES FROM THEIR ORIGINS TO THE TWENTY-FIRST CENTURY

David Boettcher

The wristwatch was not an invention, per se, as the idea of strapping a timekeeper to one's wrist stretches far back in time. But the evolution of the wristwatch in its own right, from an initially expensive curiosity to an everyday necessity, took many years and changes in ways of daily living, technology, and fashion.

Portable spring-driven clocks were invented in the fifteenth century. As soon as they were made small enough to be carried about and worn as an ornament, it was inevitable that someone would do just that. In turn, someone had one made that was smaller, finer, and better decorated; and they then would themselves be outdone, and so the trend continued. Finally, someone carried a small portable clock (now called a watch) by hanging it from their belt or on a chain around their neck—or even on their wrist, which was a visible and noticeable place where the watch could be seen by others.

Following this logic, the wristwatch itself was an inevitable outcome. But its wide adoption, by both sexes, in all classes of society, required not just the technical capability to make watches small enough that they could be comfortably worn on the wrist; it also required fundamental changes in economic circumstances and fashions. For the wristwatch, dramatic technical breakthroughs played a lesser role. The trend for wristwatches began with the creation of expensive novelty items for the extremely rich and ended in their mass production for daily wear by large numbers of people. This story begins in the sixteenth century, when ownership of small portable clocks began to spread through the wealthier classes of society.

THE FIRST WRISTWATCHES

One of the earliest references to something that can be called a wristwatch concerns a New Year's gift to Queen Elizabeth (1533–1603) in 1572 from Robert Dudley (1532–88), Earl of Leicester. It was a richly jewelled gold armlet, having ' ... in the closing thearof a clocke'.[1] As far as we know, the item no longer exists but it clearly contained a small spring-driven clock or watch, and was worn on the arm, forearm, or wrist—presumably somewhere where the watch would be easily visible.

In addition to this watch, it is known that Queen Elizabeth had a watch set in a finger ring. This was not only a timekeeper— it also served as an alarm; a small prong gently scratched Her Majesty's finger at the set time. It would not have been a precision timekeeper, but it was certainly a tour de force of miniaturization for the sixteenth century, and it shows that it was possible even at that time to make watches small enough to be worn on any part of the body.

Some of the first wristwatches for which details survive were created as expensive and impressive jewellery novelties for wealthy ladies, rather than for practical purposes.[2] An account book of Jaquet-Droz and Leschot of Geneva mentions in 1790 'a watch to be fixed to a bracelet'. When Eugène de Beauharnais (1781–1824) Duke of Leuchtenberg married Princess Auguste-Amélie (1788–1851) in 1809, the Empress Josephine (1763–1814) presented her daughter-in-law with two bracelets, one containing a watch, the other a calendar. These were made by the Parisian jeweller Nitot, now Maison Chaumet. In 1810 the Parisian watchmaker Abraham-Louis Breguet (1747–1823) was commissioned by Caroline Bonaparte (1782–1839), the Queen of Naples, to make a wristwatch, which was completed in 1811. Patek Philippe made a key-winding lady's bracelet watch in 1868 for the Countess Koscowicz of Hungary.

---

1 Hopper 1864, 349–50.
2 Jaquet & Chapuis 1970, 105.

At the Great Exhibition of 1851 in London, A. Bacher of Stuttgart exhibited 'a small watch set in a bracelet, after the Swiss fashion, which is much admired by the fair visitors.'[3] At the International Exhibition of 1862 a Swiss bracelet watch exhibited by Courvoisier was stolen.[4] The police report records that it was worth £45, and that it was subsequently pawned by the thief for £10, both considerable sums at the time.

Generally, these early wristwatches were worn by aristocratic ladies. There was no 'wristwatch industry' supplying wristwatches to the general public. It is said that men did not wear wristwatches because they considered them effeminate, akin to bracelets, which were only worn by women. But there is little evidence to support this. Wealthy men often wore finger ring watches, such as the half-quarter repeating watch in a finger ring presented by John Arnold (1736–99) to George III (1738–1820) in 1764, and finger ring watches made by, for example, John Gregson (?–1807) from the 1770s onwards. In 1876 moreover it was reported that the Duc de Penthièvre, who had a passion for watches, wore them in his waistcoat buttons, and had ordered a set for shirt and wrist studs.[5] While men would wear a finger ring watch or watch studs, they apparently would not wear a watch on the wrist, even if they could afford to do so.

The cost of watches of any sort remained relatively high and beyond the financial reach of most people until the nineteenth century, when the coming of 'railway time' meant that carrying a watch became more of a necessity than simply a display of wealth. The need for the average person to know the exact time dramatically expanded the market for watches. The mass production of watches, first in France and Switzerland, and subsequently in America, both reduced their price and increased their availability. The middle classes began regularly to carry watches, and every gentleman carried a pocket-watch. In the early nineteenth century this was held in a pocket inside the breeches, with a ribbon or fob attached to the watch and hanging on the outside of the waistband. By the mid-nineteenth century, the pocket-watch was usually tucked into a pocket of the waistcoat, attached to the end of a chain known in England as an 'Albert', after Prince Albert (1819–61), the consort of Queen Victoria (1819–1901), was portrayed wearing one. Ladies might carry a small pocket-watch, or a fob watch attached to a chatelaine pinned to clothing.

In the final quarter of the nineteenth century two events occurred that led to use of the wristwatch more widely in society: military men began to use watches as part of their field equipment, and it became fashionable for ladies to wear wristwatches.

MILITARY DEVELOPMENTS

In the 1880s the increasing availability and decreasing cost of watches reached a point where military men realized that their practical use in the field enabled strategic manoeuvres to be coordinated by time instead of by visual or aural signals. It became possible to arrange for attacks on defended positions to occur simultaneously from all sides without any signal that could alert the enemy, thereby increasing the element of surprise.

The watches that were in everyday use were pocket-watches. However, it soon became obvious that a watch on a wrist could be checked regularly, quickly, and easily while on horseback, rather than having to fumble about in a pocket. Watches were adapted for wrist wear by placing them in a leather cup attached to a wrist strap. These leather watch holders were called 'wristlets', although this term persisted and was sometimes used for the true wristwatches that followed.

The British Army was involved in various overseas campaigns during the second half of the nineteenth century. Figure 170 shows an officer of the 2nd Battalion Seaforth Highlanders wearing a wristlet in 1888. Richard Edwards searched through thousands of photographs in the National Army Museum and Royal Artillery Firepower Museum and observed an increasing number of wristlet watches in photographs taken in India and Burma between 1885 and 1887. From this he concluded that wristlet watches began to be used by British Empire forces in India or on the Northwest Frontier around 1885, and that between 1885 and 1887 their use became widespread.[6]

WRISTLET MANUFACTURE

The first leather wristlets were probably made in the field by the regimental harness maker or saddler, but as demand increased, wristlets began to be produced commercially. A design for one of these was registered with the British Board of Trade on 18 March 1887 in the name of John Barwise, watchmaker, at 40 St. James' Street, Piccadilly, London.[7] The well-known London watchmaking firm John Barwise was founded in 1780 by John Barwise (1756–1820) and carried on by his son, also John (1795–1869), until he was ruined in 1846, having invested in Pierre Frédéric Ingold's (1787–1878) failed attempt to introduce machine manufacturing of watches into England.[8] The name survived after the Barwise family left the business, but its ownership in 1887 is unknown. The business was later taken over by Birch & Gaydon, which itself was acquired by Asprey in 1959.

---

3 *ILN*, 21 June 1851.
4 *Clerkenwell News*, 14 May 1862, 2.
5 *The Watchmaker, Jeweller & Silversmith's Trade Journal*, 5 October 1876.

6 *AH*, xxxvii, 2016, 397–8.
7 Board of Trade records, Registered Design 70,068, 18 March 1887.
8 *AH*, xxxv, 2014, 621–34.

Figure 170 An officer of the 2nd Battalion Seaforth Highlanders wearing a wristlet in 1888. Photo: National Army Museum.

An article in *Invention Magazine* of September 1888 describes the 'blocked watch wristlet' introduced by Messrs. A Garstin & Co., as 'one of those original creations of fashion which are certain to recommend themselves to the fair sex'.[9] The possibility that a wristlet watch might be worn by a man is not even considered by the writer of the article, who was evidently unaware of military practice at the time. A. Garstin & Co. of Queen Square in the City of London was a long-established leather goods manufacturer. It seems logical that Barwise would have approached a prominent leather goods manufacturer such as Garstin to make up wristlets to his registered design, who then made an improved design of his own. To increase their attraction to male customers, Garstin advertised wristlets under the name 'The Indiana—As worn by Her Majesty's Officers in India',[10] which had a compass mounted on the wrist strap.

The design of a later model of wristlet, still often seen today with the number 217622 stamped inside the cup that holds the watch (Figure 171), was registered by Garstin in 1893.[11] It is very similar to the earlier models and shows more clearly what the soldiers wore on their wrists.

GIRARD-PERREGAUX GERMAN NAVY WRISTWATCH

There is a story that in 1879 the German Emperor Wilhelm (1797–1888) visited the Berlin Trade Fair and saw some experimental wristwatches made by Girard-Perregaux, and that subsequently 2,000 wristwatches in fourteen-carat gold cases were supplied to the German Imperial Navy.[12] Thus, Girard-Perregaux were named as the first manufacturers of men's wristwatches in significant volume. However, there is no evidence that any of this happened: no records exist of such wristwatches being exhibited at the Berlin Trade Fair. Furthermore, there are no records of such watches being used by the German Navy, nor pictures of any being worn. The archives of Girard-Perregaux for the relevant period no longer exist, and finally, naval service watches would not have gold cases. There have been many attempts to locate one of these wristwatches or evidence for their existence, but

---

9 *Invention Magazine*, 8 September 1888, 863.
10 *WJS*, 1 October 1891, xxii.
11 Board of Trade records, Registered Design 217,622, 2 September 1893.

---

12 *HJ*, August 2003, 291.

Figure 171 Wristlet watch. Photo: David Boettcher.

none has been forthcoming. A total absence of evidence would be surprising if thousands really were produced, and the story is most likely apocryphal.

WRISTWATCHES BECOME FASHIONABLE

Reference to ladies wearing watches in leather wristlets when horse riding or hunting first appeared in the *Horological Journal* in December 1887, the same year that Barwise registered his design of wristlet.[13] Since British soldiers in northern India wore watches in leather wristlets from 1885 onwards, this suggests that soldiers adopted the wristlet watch first, and that the ladies followed. However, the awkward and bulky leather wristlet watch is not something that would have spontaneously been offered to ladies at the time—a jeweller or watchmaker wanting to offer a wristwatch that would appeal to female customers would never have created such a design.

It seems most likely that one or more ladies had seen wristlet watches worn by military officers returning from active service and copied the idea. The anonymous author of the *Horological Journal* article in 1887 would not have encountered men wearing wristlets on active service overseas. Upon their return home, these men would have reverted to pocketwatches. Seeing only ladies wearing wristlet watches, the anonymous writer must have assumed that it was a female fashion.

Jewellers soon responded to the new fashion of ladies' wristwatches with designs that were more feminine in appearance, and advertisements for metal bracelets by the French company Le Roy et fils first appeared in the British press in 1887.[14] In later adverts (Figure 172), Le Roy et fils claimed that they were the 'originators of the bracelet watch'. This 1890 advertisement is particularly interesting because it mentions patent watch bracelets that 'Never Want Winding'. The patent was granted to Le Roy et fils in May 1890 for a bracelet watch that was wound automatically by the action of opening its rigid metal bracelet when putting it on or taking it off.[15]

---

13 *HJ*, December 1887, 50.

14 *ILN*, 3 December 1887, 657. Note that Le Roy & fils, then in the Galerie Montpensier, Palais Royal are to be distinguished from their imitators, Leroy & fils in the Galerie Valois, Palais Royal.

15 British patent GB 189,004,313, 'Improvements in Watch Bracelets', priority date 23 April 1890.

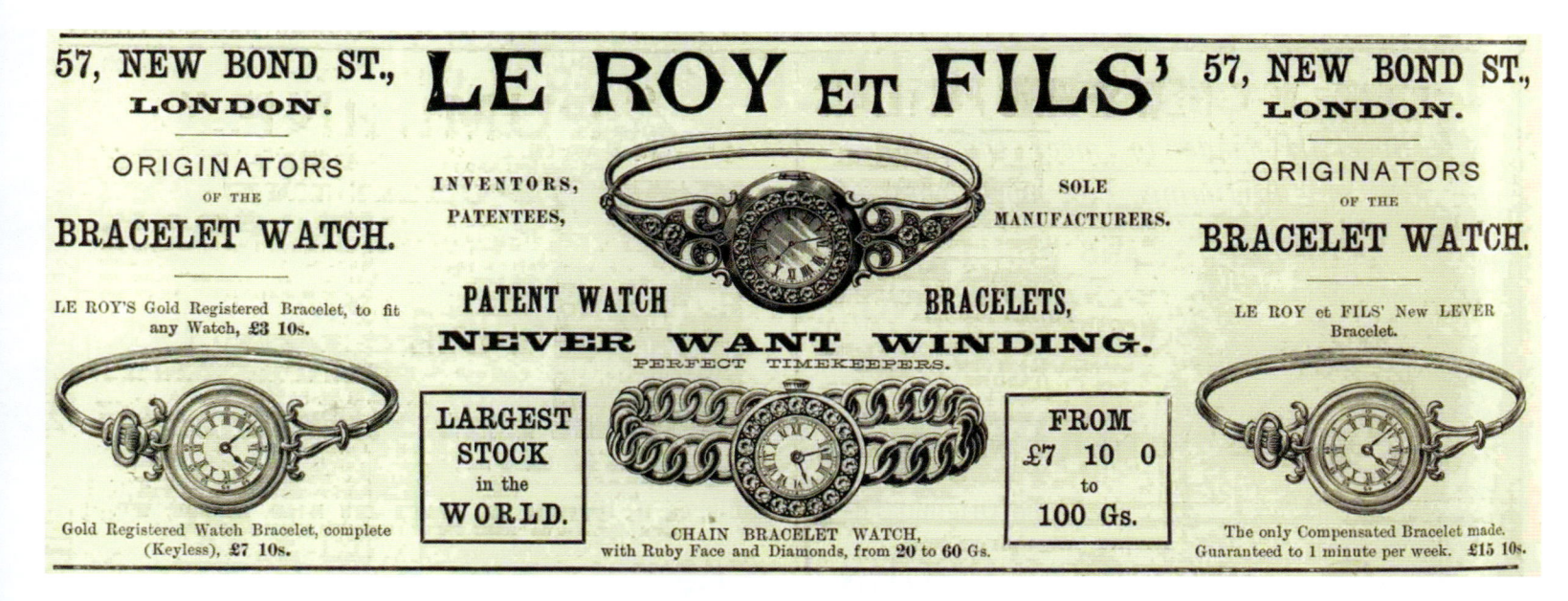

Figure 172 Advertisement for Le Roy & fils, London. Photo: The Mary Evans Picture Library.

The London shops of Le Roy et fils were retail outlets, and the wristwatches were made in France. However, it took longer for the fashion of wearing wristwatches to cross the channel. In August 1888, the Paris correspondent of the *Daily Telegraph* reported that 'some months ago' it had become fashionable for ladies to carry small discreet watches, at first on their card-cases or in the handles of parasols, and more recently ' ... embedded in a bracelet of morocco leather which is worn around the wrist.'[16] This is clearly a description of the wristlet watch, which had evidently become fashionable in Paris in the summer of 1888.

THE BICYCLE CRAZE

In 1885 John Kemp Starley (1854–1901) invented the first successful modern bicycle, which replaced the penny-farthing. Starley's model had equally sized front and rear wheels, a steerable front wheel (self-aligning owing to a caster), and chain drive to the rear wheel. This bicycle was much easier to ride than the penny-farthing and was taken up by both men and women in great numbers, resulting in a cycling craze in the 1890s. The wristlet watch became popular among cyclists as a practical way to carry a watch so that it was easily visible without removing hands from the handlebars.

However, despite attempts to persuade civilian men of the utility of wristwatches, the prejudice against them persisted. In 1889 Mr Louis Platnauer, the Consul in Birmingham for Portugal and a horological inventor, advocated their use as being very practical in the winter, when a watch in an inside pocket was difficult to reach under an overcoat.[17] The reporter thought that the idea might catch on with men, and remarked that among ladies, watch wristlets, whether in the form of leather, silver, or gold bracelets, were selling well, and that military officers found them useful. Shortly after this another article commented scathingly on the 'new compass-watch bracelet' (most likely the Garstin *Indiana*), which it considered an attempt to extend the 'recent fashion among women of wearing watches on the wrist' to men.[18] The article opined that this would not persist once the novelty had passed but did concede that it might be useful when riding a bicycle.

LATER BRITISH MILITARY CAMPAIGNS

Mappin Brothers, a company formed by four brothers out of a long-established family firm in Sheffield in 1852, was one of the first companies to advertise a man's wristwatch. In June 1893 an advertisement for their 'Skirmisher' wristwatch (Figure 173) said that it had been specially made to meet the requirements of officers in timing the movement of troops, as well as for other pursuits such as hunting, yachting, and cycling.[19] This is the earliest appearance in a newspaper of an advert for a wristwatch aimed specifically at men, but it is notable that it promotes the wristwatch as being suitable for specific activities, rather than as an item of everyday wear, which men were evidently not prepared to accept. The benefits of wearing a wristwatch to a mounted soldier or someone riding a horse or a bicycle are self-evident, but there was no such need for a civilian man in everyday life to wear a one.

---

16 *WJS*, 1 August 1888.
17 *WJS*, 1 January 1889, 114–15.
18 *WJS*, 1 October 1889, 79.
19 *ISDN*, 10 June 1893, 523.

Figure 173 The Mappin skirmisher, 1893. Photo: The Mary Evans Picture Library.

One of the Mappin brothers, John Newton Mappin (1835–1913), left the original business in 1860 and started his own business, Mappin & Company at 77–78 Oxford Street, London. He was joined two years later by his brother-in-law George Webb (1834–81), creating the company Mappin & Webb. The Mappin & Webb 'Campaign' wristwatch was said to have been first used in great numbers at the battle of Omdurman. This was fought on 2 September 1898, when a British army commanded by General Sir Herbert Kitchener defeated the army of Abdullah al-Taashi as part of British efforts to re-conquer the Sudan. The Mappin & Webb 'Campaign' watch comprised an Omega pendant watch mounted in a leather wristlet.[20]

The Second Boer War (1899–1902) was fought in Southern Africa between the British and the descendants of Dutch settlers called Boers (farmers). The Boers operated as self-organizing commando units and were used to life in the saddle and to hunting with a rifle; they knew the terrain and were highly motivated. Against such a mobile adversary, British officers found it advantageous to employ tactics of precision timing to coordinate troop movements and synchronize attacks. In 1901 the Goldsmiths and Silversmiths described their 'Service Watch' as 'The most reliable timekeeper in the World for Gentlemen going on Active Service or for rough wear.'[21] A testimonial in the advert dated 7 June 1900 reads:

> Please put enclosed Watch in a plain Silver Case. The metal has, as you can see, rusted considerably, but I am not surprised, as I wore it continually in South Africa on my wrist for $3\frac{1}{2}$ months. It kept most excellent time, and never failed me. Faithfully yours, Capt. North Staffs. Regt.'

The watch is a pocket-watch, but the captain's statement shows that he had worn it in a leather wristlet strapped to his wrist.

In 1912 a catalogue of Wm. Potter listed the huge variety of leather watch wristlets that were available from the company, in seven different sizes and ten different leathers, ranging from cow hide to crocodile, and each leather available in a range of colours.[22] This was long after many watchmaking companies were making true wristwatches for both men and women. The leather wristlet remained available until the Second World War despite, or perhaps because of, its bulky and macho appearance.

---

20 *HJ*, August 1998, 275.

21 Goldsmiths 1901, 14.
22 Harris 1988, 359.

LEATHER WRIST STRAPS

A significant development was the introduction of the leather wrist strap, attached to the watch by fixed wire lugs soldered to the case. These made the wristwatch more acceptable than the previous bulky wristlets worn by military men. Although watches with fixed wire lugs are often referred to as 'trench watches', they were being used long before the First World War. The wire lugs of early men's wristwatches and trench watches were made to take a leather strap about 12mm wide. The reason why the lugs were so narrow is unknown. It may be simply that, at the time, watch manufacturers were better at designing mechanical, rather than aesthetic, aspects and simply scaled up in size the leather strap wristwatches made for women.

A wristwatch made by Rotherham & Sons of Coventry in a gold case hallmarked in 1898 or 1899 is among the earliest known with wire lugs for a leather strap.[23] The watch is 34.5mm diameter, a man's size, and the lugs appear to be original rather than a later addition, making it a very early example of a watch having a leather wrist strap directly attached to the case.

Another claimant to the invention of the leather wrist strap was H. Williamson Ltd, also of Coventry. In an interview in 1933, an erstwhile employee, Mr W. E. Tucker (1862–1935), told an interesting story of what was, so far as he knew, the origin of the wristwatch:

> It was during the Boer war that we received a watch belonging to an officer in South Africa, which he said he wanted to wear on his wrist. I suggested putting loops on the case and sewing straps on to them. This was done, and we were struck with the idea and had it registered. It was some time before the idea took on, but eventually it became extremely popular. Prior to that there had been a strap which went round the wrist, and contained a pocket in which an ordinary watch was kept. The introduction of the wristwatch, of course, has been largely responsible for the decrease in watch sizes, and, with the fair sex, it has eliminated all other types of watch.[24]

Wristwatches with this design of wrist strap were advertised by Williamsons as 'the most serviceable wristwatch produced'. There is no mention of whether they were intended to be worn by men or ladies. However, while the plain style of the watches could be worn by either sex, their small size, a statement of their elegance, and their availability in all colours suggests that a female audience remained the principal target.

---

23 Penney 2012, 391.
24 *WCM*, 10 July 1933, 153.

DIMIER BROTHERS

The Anglo–Swiss company Dimier Brothers & Co./Dimier Frères & Cie had offices in la Chaux-de-Fonds and in London, where, from 1868, they had been a large and important watch-importing company.[25] In February 1903 the company registered a design of a wristwatch (with fixed wire lugs and a leather strap) at the Board of Trade in London as Registered Design (RD) number 405,488.[26] The same design was registered in Switzerland as Modèle Déposé, the Swiss/French equivalent to 'Registered Design', No. 9,846 on 29 July 1903.

The British registration is purely pictorial and contains no descriptive matter apart from the index in the ledger, which says 'watch wristlet'. The Swiss registration document shows a picture of the design and contains a brief description—'montre à bracelet-courroie'—which translates literally as 'wristwatch with bracelet-belt'. The addition of word courroie, or belt, distinguishes this design from a metal bracelet wristwatch, which ladies had worn for many years. Thus the specific design features being registered were the use of a leather wrist strap like a belt, and by implication the wire lugs that were used to attach the watchcase to the leather strap.

An interesting feature of this design is that the centre section under the strap flares out into a circle that covers approximately the same area on the wrist as the watchcase. Since there is no description, we can only guess its purpose. It could have been to prevent the watchcase from touching the wrist—perhaps concerns about allergies— or perhaps it was simply a unique feature conceived to facilitate registration of the design. Once a registered design number had been secured, it could be used in advertising to intimidate potential rivals. Whatever the reason, the flared section and fixed wire lugs meant that the buckle had to be stitched into the strap after it was fitted to the watch. The author has in his collection a man's Omega wristwatch from 1905/6 with just such a strap and permanently attached buckle.

CARTIER

In 1904 the Parisian jeweller Joseph Louis Cartier (1875–1942) was the first to attach projecting horns or fixed lugs on a wristwatch he created for the pioneer aviator Alfredo Santos Dumont (1873–1932). Cartier followed this in 1906 with the Tonneau wristwatch, and in 1912 with the Tortue. The straps were attached to bars fixed between the horns. This design of wrist strap attachment remained unique to Cartier until spring bars were adopted.

---

25 Culme 1987, i, 120–1.
26 Board of Trade records, Registered Design 405,488, 20 February 1903.

E. J. PEARSON AND SONS

The London-based company of E. J. Pearson and Sons, harness makers and saddlers at 275 and 277 St. John Street, traced its roots back to 1804. The company was taken over by Edward John Pearson in the 1880s,[27] who made wristwatch straps to the Dimier Brothers' design.

In 1907 either Dimier Brothers or Pearson and Sons registered the design of a buckle that fits onto a wristwatch strap without being stitched into it, which overcame the problem of the flared centre section.[28] The gap of four years between the registration of the strap design in 1903 and the registration of the detachable buckle in 1907 is notable, as it suggests that, before 1907, there were very few leather strap wristwatches being produced. Thus, while the inconvenience of having to stitch each buckle into a strap after it had been threaded through the lugs of the watch was not intolerable, by 1907 the market for wristwatches with wire lugs was starting to accelerate.

At around the same time, Dimier Brothers attempted to gain some control over the nascent wristwatch market by reinterpreting what they had registered in 1903. An announcement published in *La Fédération Horlogère Suisse* in October 1907 translates as:

> To avoid trouble and misunderstandings, we inform Gentlemen makers of watchcases of gold, silver and metal, and Gentlemen watch manufacturers of Switzerland, the curved handles for wristwatches are our registered design No. 9846 dated 29 July 1903. We will pursue anyone who manufactures watches with these handles, without having previously made arrangements for a royalty to be paid to us, and that does not send his watchcases to our factory in La Chaux-de-Fonds to have our registered mark stamped in the case back.[29]

That early Swiss wristwatches frequently have 'Déposé 9846' stamped inside the case back shows that this threat was taken seriously. On 27 August 1908, E. J. Pearson and Sons registered two designs for leather wristwatch straps with the British Board of Trade.

One of these designs, called Simplex, was given the Registered Design number 529,337.[30] In plain view this is identical to Dimier Brothers' design with the centre section beneath the watchcase flaring out into a circle. In the Pearson design the flared section is a separate piece of leather underneath a parallel-sided strap, which passes through the lugs of the watch. The flared section is stitched to the strap at the buckle end, and at its opposite end it has a loop through which the strap passes to hold the two pieces together. The benefit of this design is that the buckle can be permanently stitched in, and the strap can be easily attached to, or removed from, a wristwatch.

The second design was named 'Victor' and given the Registered Design number 529,336.[31] It is identical to the Simplex except that the under part is not flared; it has straight sides and is the same width as the top strap. This design is known today as the British military G10, or the NATO strap. Pearsons were the largest manufacturer of watch straps in Britain and would have been the obvious company to supply straps to the British military when wristwatches were officially issued. No doubt they would have been delighted to see the Victor adopted as the standard military design. From 1912 Dimier Brothers advertised to the Swiss watch trade that they could supply leather bracelets of English manufacture, undoubtedly Pearson, for wristwatches.[32]

MAKING WRISTWATCHES

Sometimes it is said that early wristwatches were converted pocket-watches. Although this might seem logical, there is no evidence that supports this assumption, despite it being widely-held. Furthermore, conversions of pocket or fob watches into wristwatches by simply soldering on wire lugs are not nearly as easy or as common as is often thought.

Attaching fittings or wire lugs to a case to take a bracelet or strap can be simple enough, it is more difficult and less satisfactory to convert a pocket or fob watch into a wristwatch in this way than simply to make a wristwatch in the first place. Undoubtedly, there *were* conversions of pocket-watches into wristwatches, because specimens exist. But these conversions did not precede true wristwatches; the time and reason for them is discussed in a later section.

The principal benefit of a wristwatch is that it is easily accessible, not tucked away in a pocket, and it can be read without using both hands, meaning that wristwatches were usually 'open face', without a hunter lid. A wristwatch with a hunter lid may be easily accessible, but it occupies both hands when reading the time; one to bring it into view, the other to open the lid.

If a fob watch were to be converted into a wristwatch, it would be a Lépine or open face watch. Lépine watches (Figure 174, left) have the pendant and crown at 12 with the small seconds at 6 o'clock.

Turning an open face Lépine watch through ninety degrees so that the crown is in a convenient position on the wrist and leaving

---

27 Culme 1987, i, 361.
28 Board of Trade records, Registered Design No. 499,803, 11 April 1907.
29 *LFHS*, 3 October 1907, 625.
30 Board of Trade records, Registered Design No. 529,337, 27 August 1908.
31 Board of Trade records, Registered Design No. 529,336, 27 August 1908.
32 *LFHS*, 2 March 1912, 116.

Figure 174 Lepine and savonnette watches. Photo: David Boettcher.

space for the lugs to be soldered on to the sides—which are now top and bottom—immediately reveals several problems:

- The 12 has rotated with the watch and is at the side next to the crown.
- The sub seconds dial is on the opposite side of the dial to the crown.
- The long pendant and bow are not required in a wristwatch.
- There is engraving on the back of the case.

Pocket-watches converted to wristwatches usually show signs of all these. The 12 and small seconds are in the wrong place, or if there are no small seconds, the dial feet might have been cut off so that the dial can be rotated. The bow has been removed and the pendant and winding **stem** cut down, usually quite obviously. The position of a small seconds display cannot be moved and would appear at 9 o'clock, and thus watches often were chosen that have no seconds display. Nothing is done about the engraving on the back of the case, confirming that it was originally a pocket-watch. Wristwatches, even early ones, do not have engraving on their case backs. This would have been an unsatisfactory way of making wristwatches.

The key to the successful wristwatch was the paradoxical use of savonnette (hunter) movements in Lépine (open face) cases. A hunter movement has the correct layout for a wristwatch (Figure 174, right), but the hunter lid is an unwanted item in a wristwatch.

Making wristwatches by converting fob watches is both a bad idea, and unnecessary. A manufacturer wishing to make wristwatches would simply put small hunter movements into purpose-made Lépine open face cases that had wire lugs to attach a wrist strap.

EARLY WRISTWATCH MOVEMENTS

The European manufacture of bracelet watches became dominated by the Swiss and the French, who gained a monopoly on the production of the very small movements. Purpose-made wristwatches for men, as opposed to pocket-watches strapped to the wrist, were created by utilizing existing designs of movements that had originally been created for fob watches; again, the Swiss and French took the lead because few English manufacturers produced the small movements needed.[33]

Early ladies' bracelet wristwatches had movements with cylinder or Roskopf pin-pallet escapements. These were adequate, but not excellent, timekeepers and were not compensated for the effects of temperature. If a higher-quality wristwatch—with a jewelled **lever escapement** and temperature compensation—were required, then at first a small pocket-watch in a leather wristlet was the only choice. Lever escapement movements were made progressively smaller and, in the early twentieth century, replaced cylinder escapements entirely in higher-quality wristwatches.

Wristwatches are almost always stem wound and set; sometimes the stem is pulled out by the crown to put the keyless mechanism into the hand-setting mode, and sometimes a push-pin in an olivette on the side of the case is depressed with a finger nail for hand setting. Key wound wristwatches are converted pocket-watches.

In the nineteenth century, watches compensated for the effects of temperature on timekeeping by cut bi-metallic **balances**, which automatically varied their moment of inertia as the temperature changed; this countered a reduction in the elasticity of

---

33 David Penney in private correspondence with the author, May 2014.

the balance spring as the temperature increases. These were difficult and expensive to make. The discovery in 1897 by Paul Perret (1854–1903) of the thermoelastic anomaly of Invar, i.e. that its elastic modulus increases with temperature, began a collaboration with Charles Edouard Guillaume (1861–1938) that resulted in temperature-compensated balance springs.[34] These were used to provide temperature compensation in wristwatches fitted with plain brass balances.

RED AND BLUE FOR 12

When the wristwatch was introduced, the unusual combination of a hunter movement in a Lépine case meant that its face was unfamiliar, which could cause confusion when reading the time. Every watch owner was familiar with the usual location of the 12 on the dial of a watch; an open face watch had the 12 below the pendant; on a hunter watch the pendant was next to the 3 o'clock and the 12 was at the top of the dial. However, a wristwatch was neither of these; it was a hunter movement in an open face case. To draw attention to the fact that the 12 was not where it was expected to be, the number 12 was coloured red, or sometimes blue.

THE STRUGGLE FOR MEN'S ACCEPTANCE

The reluctance of civilian men to wear wristwatches persisted for many years. In 1913 a Swiss trade paper wrote that wristwatches accounted for one-third of the total production of watches, and that the fashion was 'no longer exclusively feminine'. Many 'young people have adopted it, particularly in the military and sports world'.[35] However, the inference here is that, apart from some young men engaged in military or sporting activities, the wristwatch did indeed remain an exclusively feminine item.

The vast majority of the wristwatches that were actually sold, as opposed to simply being in a manufacturer's catalogue, were ladies' wristwatches. The idea of a man wearing a watch on his wrist may have gained acceptance with military men, sportsmen, and motorists, but not yet with the wider public.

From the Boer War onwards, wristwatches for men, with leather straps attached to their cases by wire lugs, were made by a number of European manufacturers (Longines, Omega, IWC). These were sold mainly to military men. Civilian men remained resistant to the idea of wearing a wristwatch, and men's wristwatches failed to gain general public acceptance.

---

34 *HJ*, November 2018, 493–501.
35 *LFHS*, 4 October 1913, 571.

WRISTWATCHES IN EUROPE AND AMERICA

The use of wristwatches in everyday life, rather than as exotic and expensive items of jewellery by the nobility, began with military men wearing watches in leather wristlets. When ladies copied this, watch manufacturers soon produced more feminine bracelet watches. Swiss, French, and American watch manufacturers could easily have produced wristwatches for men, but civilian men in Europe and America, like those in Britain, did not wear wristwatches unless it was for some specific purpose. The advertisement (Figure 175) from 1911 by Kirby Beard, the French distributor of Omega watches, illustrates this perfectly.

Two of the wristwatches illustrated are described simply as ladies' wristwatches (montre-bracelet pour dames), whereas the wristwatch for men (pour homme) is specially adapted for motoring, flying, tennis, yachting, touring, 'and is without equal for horse riding or the army'. In America, the situation was the same. American watchmaking companies such as Waltham and Elgin made suitably sized movements and could manufacture wristwatches for men. Again, while these were used by the military, the demand from civilian men did not exist.

THE FIRST WORLD WAR

The event that changed fashion and persuaded civilian men that it was acceptable to wear a wristwatch was the First World War. British Army officers had been wearing watches on their wrists since 1885 and an officer wearing a wristwatch was familiar to regular soldiers. But before the war, the British Army was relatively small—in 1914 it numbered around 250,000 men distributed throughout British colonies across the globe. Even if every one of its officers of wore a wristwatch, many years would have passed before British society as a whole became accustomed to the practice. However, during the First World War, huge numbers of civilian men entered armies all across Europe; Britain mobilized around eight million men, most of them to the army. These men saw their officers routinely wearing wristwatches under the most trying of circumstances. The volunteers who joined Kitchener's New Army took a keen observational interest in military practice and soon adopted the term 'a proper wristwatch' for a smartly turned-out officer (Figure 176).

The idea that a man could wear a wristwatch entered the consciousness of the new recruits, and that of the conscript armies that followed. The many officers who remained in uniform while on leave back in Britain ensured that men wearing wristwatches were seen by the wider general public.

Newly commissioned officers were given a kit allowance from which to purchase their uniform, sword, revolver, mess kit, great coat, etc., and they looked for guidance about what they would need in the field. An experienced officer, Captain B. C. Lake

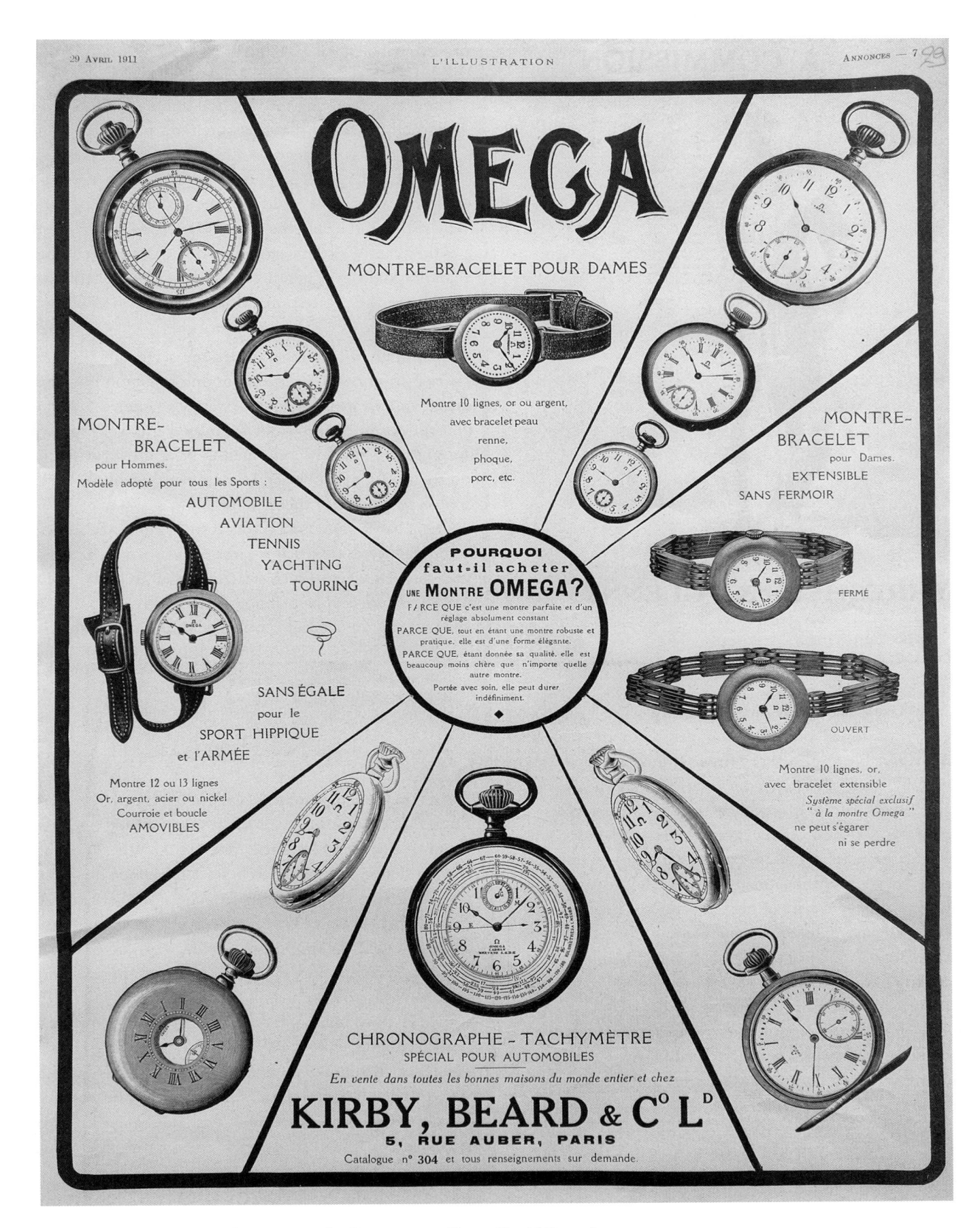

Figure 175 Kirby Beard advertisement for Omega, 1911. Photo: David Boettcher.

Figure 176 A proper wristwatch. Photo: David Boettcher.

of the King's Own Scottish Borderers, produced a handbook for newly commissioned officers. In it was a list of suggested items, and it is notable that the very first is 'luminous wristwatch with unbreakable glass'; this item appears ahead of such otherwise indispensable officer's items as 'revolver' and 'field glasses'.[36]

During the war, for the first time, popular newspapers and magazines carried large numbers of adverts showing wristwatches for men aimed at newly commissioned officers. A wristwatch was purchased by an officer as part of his kit using his 'kit allowance' and was as much a part of his equipment as his uniform, his sword, and his pistol. Because such watches do not carry the pheon symbol (broad arrow) associated with government-issued items, they are often incorrectly described as 'private purchases'.

Seeing officers regularly consult their wristwatches (and which, thanks to radium luminescent paint, glowed brightly at all times of day and night) fascinated the new recruits, and many of them determined to get a wristwatch for themselves. In his memoir of the First World War, George Alfred Coppard MM (1898–1984) remarked 'At a post parade I was happy to receive a wristwatch from a dear aunt and uncle of mine. It was the first watch I had ever owned.'[37]

Corporal Coppard was not the only 'Tommy' (slang for a common soldier in the British Army) who wanted a wristwatch. The 1916 Annual General Meeting of H. Williamson Ltd was told that:

> The public is buying the practical things of life. Nobody can truthfully contend that the watch is a luxury. In these days the watch is as necessary as a hat—more so, in fact. One can catch trains and keep appointments without a hat, but not without a watch. It is said that one soldier in every four wears a wristlet watch, and the other three mean to get one as soon as they can. Wristlet watches are not luxuries; wedding-rings are not luxuries. These are the two items jewellers have been selling in the greatest quantities for many months past.

The public in Britain soon became accustomed to seeing officers and Tommies in uniform coming home on leave from active service and wearing their wristwatches. After the war was over, thousands of veterans were demobilized and went back to civilian life. These ex-soldiers continued to wear the wristwatch that had cost them a lot of money and served them faithfully, surviving the terrible conflict with them.

Seeing battle-hardened veterans wearing wristwatches changed the public perception that wearing a wristwatch was for women. After the war, sales of wristwatches began to grow. In December 1917, the *Horological Journal* noted that the ' . . . wristlet watch was little used by the sterner sex before the war, but now is seen on the wrist of nearly every man in uniform and of many men in civilian attire.'

TRENCH WATCHES

The term 'trench watches' was soon coined for wristwatches purchased by officers to wear in the cramped and often damp conditions of the trenches of the First World War. The features that came to define a trench watch are a luminous dial, unbreakable crystal, and the narrow fixed-wire lugs of the earliest leather strap wristwatches. Dennis Harris (1920–2014) speculated that the narrow strap could accommodate a wider separate back pad, which he calls a 'wrist support'.[38] This is certainly more comfortable to wear than a narrow strap alone.

Then, as now, military men did not economize on kit on which their life might depend. Wristwatches were made with a number of case designs. The cheapest wristwatches had cases with jointed (hinged) opening backs, which were simply small versions of pocket watchcases, sometimes with double backs. A more robust design was the Borgel screw case, a patented design where the case was made in a single piece and the movement screwed in from the front, rendering the watch less prone to the ingress of dust and damp. Francois Borgel (1856–1912) of Geneva was

---

36 Lake 1916, 177.
37 Coppard 1969, 44.

38 *AH*, Summer 1988, 363.

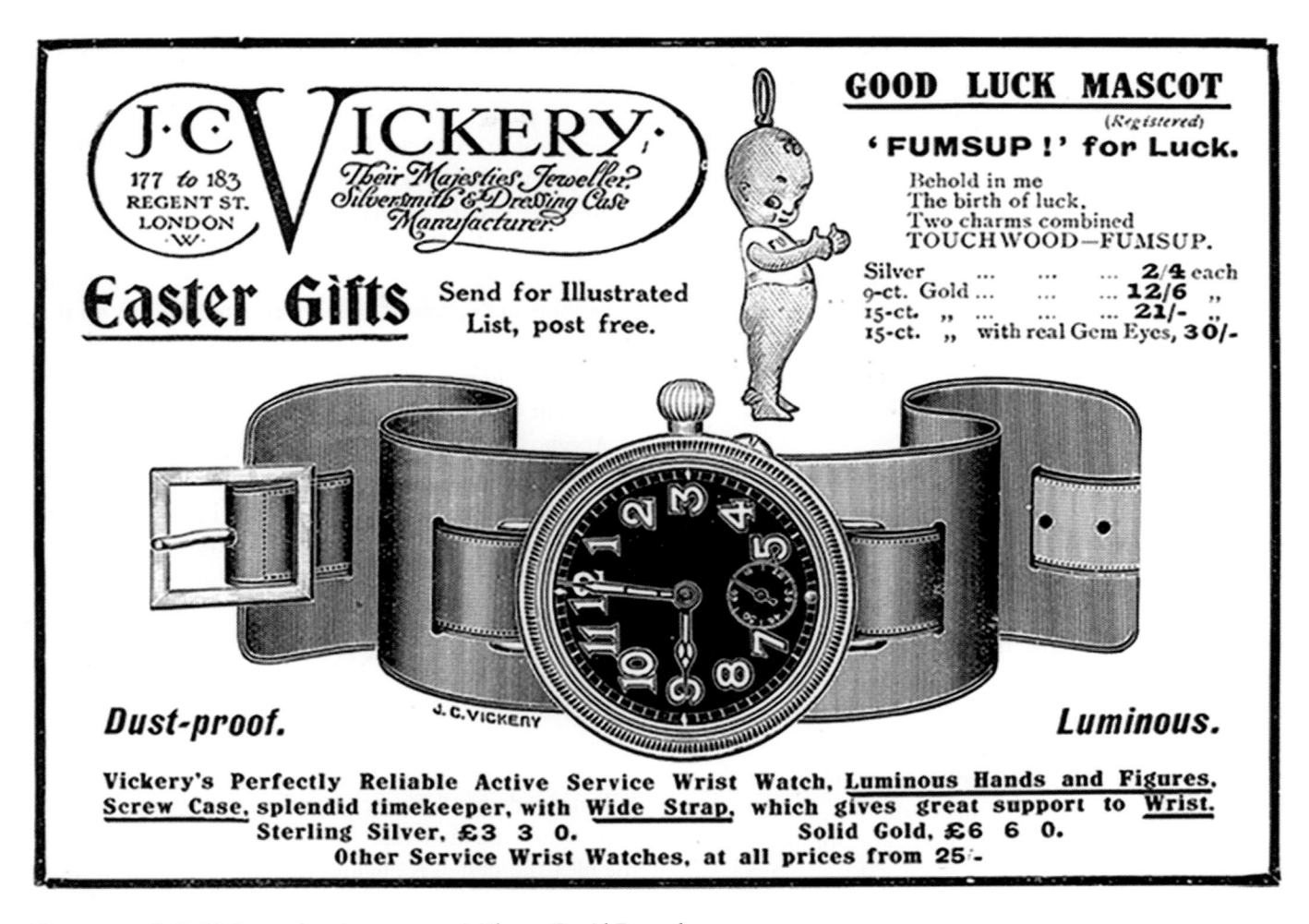

Figure 177 J. C. Vickery advertisement, 1916. Photo: David Boettcher.

granted a patent for this case design in 1891.³⁹ A wristwatch with a Borgel case was significantly more expensive than one in an ordinary case, but it was the best wristwatch available at the time for life in the trenches, and many of the watches advertised for service use had Borgel cases. Wristwatches from the time of the First World War with Borgel cases survive in much better condition than those in ordinary jointed cases, showing the wisdom of that choice. The watch in the Vickery advert (Figure 177) has a Borgel screw case.

In addition to the wristwatches purchased by officers as part of their kit, the War Department issued wristwatches to other ranks who needed one to perform their duties. These were principally signallers operating in forward trenches who were required to note the time on dispatches but could not be expected to use a pocket-watch owing to the exposed and cramped conditions. Several types of case were tried, and a screw back and bezel case manufactured by the Dennison Watch Case Company of Birmingham was chosen as the most suitable. Many of these watches were collected and disposed of after the war, making them scarce today. They usually had nickel cases, which were corrosion resistant but cheaper than silver, and carried the pheon and an alphanumeric reference indicating from which store they were issued.

LUMINOUS DIALS

Radium was discovered by Pierre (1859–1906) and Marie (1867–1934) Curie in 1898. It was found that it could be mixed with a phosphor—zinc sulphide doped with an activator (usually copper)—and a binder (such as a clear varnish) to make a luminescent paint. The zinc sulphide phosphor glowed brightly when hit by radiation from the radium and continued to glow day and night without needing exposure to sunlight, even when stored away in a drawer.

This radioluminescent paint was first used on clocks, especially alarm clocks, where the ability to read the time at night was an obvious advantage. Luminous dials were applied to some watches before the First World War, but during the war, when many attacks took place at first light, a luminous dial was a necessity in a trench watch. The numerals laid out on the enamel dial were hollow Arabic style, and the hands were skeletonized, to take the luminous paint.

The dangers of radioactive paint were not initially appreciated, but became apparent when radiation was found to cause illness,

---
39 Swiss patent CH 4001, 'Nouvelle Boîte de Montre', 28 October 1891.

such as affected the dial painters known as the radium girls. Original radium luminous paint no longer glows in the dark because the fluorescent material has long since been worn out by the constant bombardment of radiation, but the radium, which has a half-life of about 1600 years, is still very nearly as radioactive as when it was new. In addition to the alpha and gamma radiation emitted by radioactive decay of the radium in the paint, radon gas is given off, which is itself an alpha emitter. It is important to take suitable precautions if there is a danger of exposure to such paint.

MESH GUARDS

An obvious concern for a wristwatch is that the wrist, being prone to be knocked against things, is a vulnerable location for a delicate instrument. The glass crystal was therefore in constant danger of being broken. The first response to this, well before the war, was to make a perforated metal guard that could be slipped over the watch to protect the glass. Today these are often called 'shrapnel guards', but at the time they were referred to as a 'mesh guard' or 'watch protector'. A number of rival designs were patented before, and in the early years of, the First World War, indicating that the vulnerability of the glass was a concern, real or imagined, from the early stages of the wristwatch. During the war, mesh guards were produced in large quantities, but were made redundant when unbreakable materials for watch crystals were introduced.

UNBREAKABLE CRYSTALS

The need for an unbreakable material for watch crystals became urgent in the early years of the First World War. A notice appeared repeatedly in the Swiss watch industry newspaper *La Féderation Horlogère Suisse* from January 1915 onwards asking which factory could supply celluloid watch glasses. The company Kaeser, Moilliet & Co., Schweizerische Celluloidwaren Fabrik, who before the war made novelties in celluloid, responded and a Swiss patent was granted to them in 1915 for watch glasses, especially for wristwatches, made from 'water-clear' celluloid, Zellon, or any other celluloid-type product.[40]

Celluloid is a generic name for cellulose-based plastics that were invented in the nineteenth century. Nitro-celluloid, which was used for photographic films, is tough, flexible, and transparent; its principal deficiency is that it is highly flammable. Zellon was a trade name for cellulose acetate, a less flammable type of celluloid, registered by Kaeser, Moilliet & Co., described as flame retardant and glass-hard.

Celluloid watch crystals, called 'UB' for UnBreakable, came into use in trench watches in late 1915 or early 1916. Celluloids are not as stable as glass and, used flat as a watch crystal, could shrink over time or in cold weather, and fall out. To overcome this a technique was developed of fitting the crystal to the bezel under tension. A circle slightly larger than the bezel was cut out and then pressed into a dome shape, which was released into a groove in the bezel so that some residual tension was retained. In cold weather this held the crystal firmly in place.

Watch crystals made from celluloid during the First World War have long ago yellowed and deteriorated, mostly to the point at which they have been replaced. Occasionally a deeply yellowed specimen is found, with radiation burns from the radioluminescent paint on the hands and numerals.

Acrylic 'glass' (polymethyl methacrylate) was invented in 1928 and first brought to market in 1933 by the Rohm and Haas Company as Plexiglas. In Britain it was made by ICI and called Perspex. This is used to make the acrylic UB crystals that replaced their celluloid forerunners and are still used today.

CONVERSIONS

Wristwatches made by converting pocket- or fob watches did not, as is sometimes supposed, precede purpose-made wristwatches. When the demand for wristwatches arose in the nineteenth century, watch manufacturers responded by making wristwatches; the idea and design is obvious and there were no technical barriers. The manufacture of new wristwatches could not match the surge in demand brought on by the First World War. It was at this time that the vast majority of conversions of pocket-watches into wristwatches were performed, and many such conversions were done by jewellers; most conversions are very obvious.

However, manufacturers naturally wanted to fulfil the demand and sell as many watches as possible, of whatever style, and stores of fob watches of suitable size were converted into wristwatches. These professional conversions are less easy to identify. The author has a Longines trench watch with fixed wire lugs and skeletonized numeral hands carrying radium luminous paint, which is still strongly radioactive. It has a silver case with hallmarks for 1908 to 1909, and the factory archives show that it was made in 1909. Longines records show that it was sent to Baume & Co. in London until 12 March 1915. But Longines did not make wristwatches with luminous dials like this in 1909. Even if they did, such a wristwatch would not have been kept in stock until 1915. The watch was made in 1909 as a small open face fob watch. It did not sell and remained in stock until the pressing demand for wristwatches in 1915 caused it to be converted into a wristwatch in the Longines factory.

HUNTER WRISTWATCHES

Sometimes wristwatches are seen with hunter lids; a metal lid that completely or partially covers the front of the watch. These are

---

40 Swiss patent No. CH 70,925, 21 June 1915, Kaeser, Moilliet & Co., Schweizerische Celluloidwaren Fabrik, of Bern-Zollikofen, Switzerland, 'Uhrenglas, insbesondere für Braceletuhren'.

not common, and for a very good reason. Although having a lid to protect the glass seems rational, especially for a trench watch, it defeats the purpose of a wristwatch, which was created so that time could be easily read by glancing at the wrist. Both hands are needed to read a hunter wristwatch, which can be a nuisance. Half-hunter or demi-savonnette wristwatches, with a small opening in the centre of the lid, are slightly better, but the small amount of visible dial means that the time is more difficult to read than with an open face wristwatch, and to read the time accurately it is necessary to open the lid. The 'Army wristwatch protector' was a sprung clamshell device that completely enclosed the watch and gave it, in effect, a hunter lid. A patent for this invention was granted in October 1914 to Charles Adolf Schierwater of Liverpool. Once unbreakable crystals became available in 1915, the extra protection offered by a hunter lid was unnecessary, and the drawbacks of covered wristwatches meant that such devices found few customers.

WATERPROOF WRISTWATCHES

Since the beginning of the portable timekeeper or watch, it was obvious that the movement needed protection, and the cases in which they were enclosed were steadily improved to keep out dust and damp. Today a well-sealed watchcase is termed 'water resistant', but in earlier times the term 'waterproof' was often used. The first watches with cases that were sealed against the ingress of dust and damp were made in the nineteenth century. A watch was exhibited at the Great Exhibition of 1851 suspended in a glass globe filled with water and gold and silver fish.[41]

Although the Borgel case provided good resistance to dust and damp, the first wristwatch that was advertised as truly waterproof was the 'Submarine' made by Tavannes in 1915 (Figure 178). In 1917, the *Horological Journal* gave some interesting details about how the watch came into being.[42] Two submarine commanders requested a wristwatch suitable for their work. It had to fulfil certain conditions: be watertight, non-magnetic, and luminous. The Submarine wristwatch fulfilled all of these conditions. It was made watertight with a screw-on back and bezel with gaskets, and the hole where the stem enters the case was sealed by a gland around the stem compressed by a hollow nut.

The cases of Tavannes Submarine wristwatches are stamped with the Swiss federal cross and 'Brevet', indicating a Swiss patent. The patent was granted on 16 September 1915 with the number CH 70,307 to Kramer-Guerber & Cie., watchcase makers in Bienne.[43]

The Submarine wristwatch was advertised in *The Scotsman* in March 1916: '"SUBMARINE" is the name of the first WATERPROOF WRIST WATCH made; Non-Magnetic, Intensely Luminous Hands and Dial. Sole Agents, BROOK & Son, 87 George Street, Edinburgh'.[44] On 6 April 1916 Brook & Son published a notice in *The Scotsman* recording comments in a letter received from a naval officer; 'The Submarine Wrist Watch purchased six months ago has been a great success.' This implies that the first Submarine wristwatches were supplied in October 1915, two years before the *Horological Journal* report. It seems likely that the naval officer was one of the two submarine commanders.

Waterproof wristwatches from other makers followed the Submarine. During the First World War, Fortis manufactured a watch called the 'Aquatic' to a design created by inventors Paul Ernest Jacot and Auguste Tissot, who were granted a patent on the design on 1 February 1916.[45] In America, the Depollier 'Waterproof Dustproof' wristwatch was made by Jacques Depollier & Son to

Figure 178 Tavannes SUBMARINE, the first waterproof wristwatch. Photo: David Boettcher.

---

41 *ILN*, supplement 5 July 1851, 34.
42 *HJ*, December 1917, 41.
43 Swiss patent CH 70,307, 'Boîte de montre à fermeture hermétique', priority date 5 March 1915.

44 *The Scotsman*, Saturday, 4 March 1916.
45 Swiss patent CH 71,715, 'Montre avec emboîtage hermétique du mouvement', priority date 27 August 1915.

a design that originated with Auguste Jaques of La Chaux-de-Fonds, Switzerland. A patent for this design was granted to Jaques and Depollier on 22 February 1916.[46]

An alternative to the stem sealing gland used in the Submarine are crowns that screw down onto the case to make a watertight seal. The screw-down crown was invented in the nineteenth century by the American Ezra C. Fitch (1846–1929), who was granted US patent No. 237,377 on 8 February 1881. This design was used for waterproof pocket-watches by the American Watch Company of Waltham, of which Fitch became president in 1883. The first successful application of a screw-down crown to a wristwatch was made by the Rolex Watch Company in 1927. The managing director, Hans Wilsdorf (1881–1960), named the watch 'Oyster' for its ability to survive under water.[47]

BARS REPLACE WIRE LUGS

During the 1920s the fixed wire lugs of most wristwatches were gradually superseded by horns with bars between them, as found on Cartier's earliest wristwatches. Bars that were solidly fixed to the horns of a watchcase were very secure, but they had to be used with either a one-piece pull through strap, or a two-piece strap that was stitched onto the bars. Although the one-piece pull through strap was favoured by the military for extra security, it was seen by watch designers as increasingly old fashioned. The alternative of a two-piece stitched-on strap was equally undesirable.

A significant advance in convenience came with the introduction of 'spring bars', which comprise a hollow tube containing two pins separated by central spring. The pins enter holes made in the case horns, and the spring holds them in place. To change the strap, one of the pins is pressed into the tube against the push of the spring and disengaged from the horn. This allows a two-piece strap to be fitted very quickly and easily.

The inventor of the spring bar is unknown. Although several names have been put forward as the first to use them with a wristwatch, none claimed at the time to be the first user. Spring bars were in use during the nineteenth century as jewellery fittings, so any claim for a patent on them was impossible by the time they came into widespread use on wristwatches.

With the introduction of the watchcase with moulded horns, and wrist straps attached to these by spring bars, the modern external form of the wristwatch case was established, although it was still open to a wide variety of shapes and decorations.

SHOCK RESISTANCE

A watch on the wrist is in a very vulnerable position and is often knocked against hard surfaces. The pivots of the balance staff of a watch have a very small diameter to minimize friction and are therefore delicate and prone to breaking if the watch receives such a shock. To overcome this problem, shock protection systems were devised. Abraham-Louis Breguet (1747–1823) invented what was probably the first system, which he called the 'pare-chute' or 'suspension élastique'.[48] When wristwatches became popular for military and sporting activities, it might be thought that shock protection would be essential, but this was not the case.

Shock protection systems for wristwatches began to be fitted in the 1920s. One of the first designs, widely used in wristwatches made by the General Watch Co. under the brand Helvetia, was granted a Swiss patent on 31 October 1930.[49] It was not until after the Second World War that they began increasingly to be fitted to wristwatches. Today, shock protection is standard on every mechanical wristwatch. At the time, however, there was little demand from the public for shock protection. Watch owners knew that watches were delicate and liable to break, and broken balance staff pivots were so common that watch repairers kept stocks of spare staffs and could replace one very quickly.

Even in a military context, shock protection was not regarded as essential. The military committee that drew up the 'Watch, Wrist, Waterproof' specification in the 1940s debated whether the benefit from reducing broken balance staffs by shock protection was not offset by the increased time required for regular service of watches with shock-proof settings.[50] The introduction of a clip-in spring in the Incabloc shock-resistant setting actually made it quicker and easier to service than an old-style setting. Still, it was not until 1947 that an amendment to the specification included a requirement for shock protection.

Before the Second World War, vigorous amateur pursuits like diving and mountain climbing were the realm of a few and not followed as widely as they are today. Even if people outside the watchmaking profession realized that shock protection was technically possible, they probably did not think about it. It was advertising, rather than demand, that brought about the widespread fitting of shock protection to wristwatches. Initially more in America than in Europe, consumers were persuaded that they needed a watch with shock protection, and to pay more for one with it. Once shock-proof watches entered the public consciousness, early owners were able to show them off, and any watch that lacked shock protection was deemed old fashioned.

All shock-protection systems work in the same way. The balance staff bearings, instead of being mounted rigidly in the plates, are held in place by light springs. When a shock occurs, the springs

---

46 US patent No. 1,172,601, Watch Waterproof Case, priority date 9 November 1915.
47 Rolex 1946, i, 17–18.

48 Daniels 1975, 331–2.
49 Swiss patent No. CH 143,073, 'Dispositif amortisseur de chocs pour axes de balanciers de montres', priority date 24 August 1929.
50 *HJ*, October 2012, 461.

allow the bearings and staff to move slightly, and a stronger part of the balance staff contacts a fixed part of the housing to take the shock. Once the shock is over, the springs return the bearings to their usual positions. The best-known system is Incabloc, developed from an invention of Fritz Marti (1892–1983) while working at Fabrique Election in La Chaux-de-Fonds, who was granted a patent in 1930.[51] Incabloc was a clever design that was assembled with easily fitted lyre-shaped springs that, together with good marketing, gained prominence for the system. After the patents expired many other shock-protection systems were created on the same principles.

Shock protection is a real benefit as owners can treat their watches less delicately. The only downside is for collectors of older watches without shock protection. The days of the local watch repairer who could quickly replace a balance staff have long gone, and replacing a broken staff is now an expensive operation!

AUTOMATIC (SELF-WINDING) WRISTWATCHES

Pocket-watches that wind themselves from the movement of their wearers date back to the eighteenth century. A bracelet wristwatch that was wound by the action of opening the bracelet when putting it on or taking it off was patented by Le Roy et fils in 1890, and a number of self-winding wristwatches were made by Léon Leroy in the 1920s.[52]

The first widely successful application of automatic winding to a wristwatch was made by John Harwood (1893–1964), a watch repairer from the Isle of Man. Harwood conceived the idea of automatic winding in the trenches during the First World War. He aimed to remove the need to wind the watch in muddy and cold conditions, and to eliminate the stem, which required a hole in the case through which damp and dust could enter. Harwood and his business partner Harry Cutts were granted a British patent for the invention on 10 July 1924.[53] The corresponding Swiss patent (No 106,583) had been deposited on 16 October 1923 but was not published until 1 September 1924. The Harwood automatic winding system uses a segmental weight that is pivoted at the centre of the movement and swings through a 260-degree arc as the wearer moves their wrist or arm, winding the **mainspring** through a **train** of gears. This is called a 'bumper' design because the weight hits a spring buffer at the end of its travel, which the wearer can feel as a bump. The watch was made without a crown and stem and could only be wound by moving so that it wound itself. The hands were set by rotating the bezel.

Harwood formed the Harwood Self-Winding Watch Company to finance advertising, promotion, and sales support for the watches, and to hold the patents. Fortis assembled most Harwood automatic wristwatches, with smaller numbers being assembled by Blancpain and Selza. An A. Schild movement was used as the base, with the automatic winding gear added as an assembly. The Harwood automatic wristwatch was launched at the Basle fair in 1926.

The Harwood company collapsed in 1931 during the Great Depression and, although the patents still existed, the company could no longer exercise them. Thus, other companies were free to copy and develop their own versions.

Emile Borer (1898–1967), nephew and ultimately successor to Hermann Aegler (1874–1944) and head of research and development at the Aegler Bienne factory, improved on Harwood's design by converting the weight into a rotor. This improved durability and feel for the wearer, although it only wound in one direction.[54] Felsa introduced bi-directional winding in 1942.[55]

Automatic winding means that, as long as the wearer is reasonably active, it is no longer necessary to wind a wristwatch manually every day, and the crown is used principally to set the time. Automatically wound watches are more consistent in accuracy than manually wound ones because, while the watch is worn, the tension of the mainspring is more constant. Automatic winding is also a benefit to watches with screw-down crowns because there is less wear to the screw threads, and fewer opportunities to forget to screw down the crown after winding.

COMPLICATIONS

Almost since they were first made watches have been fitted with different 'complications', such as indications of the date, the age and phase of the Moon, the state of wind of the mainspring, the current time in a second time zone, and others. These add to the cost, so most wristwatches are simple pieces that show only the time, but the date has wide practical appeal and features on many wristwatches.

Date indications on wristwatches initially took the form of an extra hand that pointed to the current date on a circular track. In 1930 the Swiss wristwatch manufacturer Marlys was granted Swiss and French patents on a date indication where a number is shown through a small window on the dial.[56] This patent was refused in Britain, most likely because such indications had been used on clocks for centuries.

Another popular complication is a chronograph function, used for recording the duration of events such as races. The modern wristwatch form usually has a sweep centre seconds hand

---

51 Swiss patent CH 141,098, 'Dispositif d'absorption de chocs et de protection de pivots d'axes dans les mouvements d'horlogerie, compteurs etc.', priority date 27 July 1929.
52 *HJ*, July 2002, 234.
53 British patent No. GB 218,487, priority date 7 July 1923.

54 *HJ*, August 2002, 287.
55 *HJ*, September 2002, 312.
56 Swiss patent no. CH 130,200, priority date 6 January 1928, published 30 November 1928.

indicating fractions of a second, with subsidiary dials to record longer periods. Chronograph watches had been invented in the nineteenth century—the first wristwatch with a chronograph was made by Longines in 1913 using the 13.33Z calibre movement.[57] All wristwatch chronographs were manually wound until 1969, when automatic movements were introduced almost simultaneously by Zenith, Seiko, and Heuer.

DIGITAL READOUT WRISTWATCHES

In the 1930s during the Art-Deco period, streamlined cars and trains and everything modern was in vogue. Dispensing with the old-fashioned hands, wristwatches with digital displays became fashionable. These were not truly digital devices; the minute and seconds hands were simply replaced by discs carrying numerals—a nineteenth-century invention. The minute and second discs usually rotated continuously and showed the current reading through a small window. The hour disc was often made to advance in a jump on the hour, and such watches are called 'jumping hour'.

LATER MILITARY WRISTWATCHES

In the period before the Second World War, the British Army recognized a need for wristwatches and hurriedly acquired them from a number of Swiss suppliers under the designation ATP, believed to stand for 'Army Trade Pattern', which was engraved onto the case backs. These were wristwatches that were available 'off the shelf' and were accepted without modification, provided that they met certain requirements, principally those established during the First World War of having a luminous dial and unbreakable glass.

This situation was clearly unsatisfactory, and the Army began to develop an official specification for wristwatches, which resulted in the issuing of British War Office Specification No. R.S./Prov/4373A 'Watches, Wristlet, Waterproof' (W.W.W.).[58] Service wristwatches manufactured to this specification started delivery in May 1945 as the war was drawing to a close. The Royal Air Force (RAF) developed its own specification for pilot's watches, the 6B/159 (known informally as the Mk VII) and 6B/234. In 1948, work began on developing a specification for a General Service (GS) wristwatch suitable for adoption by the three Services; Army, Air Force, and Navy.[59] This culminated in the 'provisional' Defence Specification No 3 'General Specification for Wrist Watch, General Service' (DEF-3), published on 1 September 1951. This specification underwent several revisions but remained the basis upon which all wristwatches were purchased by the British military for land, sea, and air use.

PILOT'S NAVIGATION WATCHES

In the 1920s, as long-distance air flight developed, wristwatches with special features were developed to enable pilots to perform air navigation. In 1927, the US Navy officer Phillip Van Horn Weems developed the Weems System of Navigation. Weems and Longines designed a wristwatch with a rotating disc at the dial centre, enabling the seconds hand to be synchronized with a time signal. Charles Lindbergh suggested an improvement that allowed the hour angle to be determined more easily, a necessary part of finding longitude based on Greenwich Mean Time. In the late 1930s Weems, then Chief of Navigation and Aviation at Longines-Wittnauer, developed another 'second setting' wristwatch with an external rotating bezel.[60]

DIVERS' WRISTWATCHES

Although 'waterproof' watches had been made in the nineteenth century, and waterproof wristwatches during the First World War, the water pressure, and hence depth of submersion, that they could withstand was not tested or quantified. The first wristwatch that was specifically tested and qualified for diving was introduced in 1932 by Omega, called the Marine, with the Omega reference number 679.[61]

The design was based on a Swiss patent granted to Louis Alix of Geneva on 1 July 1931.[62] The patent was also taken out in France, Britain, the USA, and Germany. The design overcame the problem of making the winding stem waterproof by the simple expedient of sliding the whole watch inside a second outer casing, against which a widened part of the watchcase sealed with a gasket, the two parts being held together by a substantial spring clip. It was also the first watch to use artificial sapphire for the crystal, chosen because it is ten times stronger than glass and therefore better able to resist water pressure at diving depths. In 1936 an Omega Marine was sunk to a depth of 73 metres in Lake Geneva for thirty minutes. In May 1937, the Swiss Laboratory for Horology in Neuchâtel certified the Omega Marine as being able to withstand a pressure of 13.5 atmospheres, equivalent to a depth of water of 135m. These were the first tests to establish the depth capabilities of a watch and qualify its suitability for underwater activities.

---

57 Linder 2007, 566–7.
58 *HJ*, October 2012, 460 *et seq*.
59 *HJ* October 2012, 460.

60 Lachat 2017a, 109–21.
61 *WCB*, January February 2012, 34–45.
62 Swiss patent CH 146,310, 'Boîtier étanche de montre-bracelet', priority date 10 March 1930.

SCUBA diving remained a specialized area open only to a few enthusiasts and military personnel until after the Second World War, when the first modern dive watches were created within a few months of each other—the Blancpain Fifty Fathoms, followed by the Rolex Submariner.[63] Both of these wristwatches had similar features: a waterproof case with screw back and crown, luminous dials that could be read in gloomy conditions underwater, and a new feature—a rotating bezel that showed the diver how long they had been underwater, which they could compare with the anticipated duration of the air in their tanks. To avoid accidental movements of the bezel misleading the diver into thinking that they had been submerged shorter than they really had, movement of the bezel was made unidirectional so that it could only turn in the direction that was safe.

VACUUM WATCHES

In 1959 Hans Ulrich Klingenberg was granted a patent for a wristwatch with a vacuum inside the case.[64] The crystal was held in place solely by the pressure difference between the external atmosphere and the internal vacuum. The vacuum had the benefit of removing oxygen and moisture from inside the case, so that the crystal would not fog in cold conditions, and the metal parts and lubricating oils and greases would not oxidize, which is a major cause of oils breaking down. Klingenberg licensed his patented design to Glycine and several wristwatch manufacturers and used the income to set up the Vacuum Chronometer Corporation in Nidau. Vacuum wristwatches were very successful from the 1960s until the 1980s. A rival design called the Airvac was created by Jaquet Girard, with cases made by Ervin Piquerez S.A.

Another benefit of a vacuum inside the case of a mechanical wristwatch is that the reduced drag on the balance decreases energy losses and increases the 'Q' factor of the oscillator. Development of mechanical wristwatch movements was almost completely halted after the 1970s by the dominance of the quartz movement, but in recent years has resumed. The Cartier ID Two concept wristwatch has a 99.8 per cent vacuum inside its case, which reduces the energy consumption of its balance oscillator by thirty-seven per cent and significantly increases its Q factor.[65]

ASTRONAUTS' WRISTWATCHES

Watches to be used inside a space capsule in the same atmosphere used by the astronauts have no special requirements. But if they are to be used during 'extra vehicular activity', space walks, or on the surface of the Moon, they must be worn on the outside of the space suit to be visible, and so must be able to withstand both the vacuum of space and extreme temperatures of hot and cold.

In 1962 NASA began to search for a wristwatch that could be worn by the Gemini and Apollo astronauts. They purchased 'over the counter', from a Houston Jewellery Shop, wristwatches made by several different manufacturers. The successful wristwatch had to be waterproof, shockproof, and antimagnetic, and the specimens were subjected to a number of tests, including high and low temperature and vibration. The wristwatch selected as a result of this testing was the Omega Speedmaster Professional. As a consequence, the case backs of Speedmaster Professionals are engraved 'Flight Qualified by NASA for all Manned Space Missions', the only watch ever to be so qualified.[66]

Omega Speedmaster Professional watches were used on all manned space flights during projects Mercury, Gemini, Apollo, Skylab, Apollo-Soyuz, and the Space Shuttle, as well as on the surface of the Moon and during spacewalks. On the Apollo 13 mission, Command Module Pilot Jack Swigert's 'Speedy' was used to time a rocket burn that put the damaged spacecraft into the centre of an atmospheric re-entry corridor that returned it safely to Earth.[67] The Fortis 'Spacematic' wristwatch was worn by seven astronauts of the US space programme. In 1994 the Fortis 'Cosmonauts Chronograph' was chosen as Russian Cosmonauts' equipment. It was worn on several space walks outside the International Space Station, and during the Mars 500 test mission.[68]

ELECTRICAL AND ELECTRONIC WRISTWATCHES

Electrical devices have been used in clocks since the nineteenth century to rewind them or to provide impulses to pendulums by magnetic field. For wristwatches, the principal problem in using electricity was its source, which at the time had to be a battery small enough to wear on the wrist.

The first application of electricity in a wristwatch was made by Hamilton in the USA, who worked with a battery manufacturer to make the small batteries used in the Hamilton Electric 500 wristwatch. In Europe this was followed by the French company Lip, with a design patented in 1952. In both of these designs the battery simply acted as the source of energy instead of a mainspring, the balance carrying an electrical coil that is magnetically impulsed.[69]

---

63 https://www.hautehorlogerie.org/en/brands/history/h/blancpain/.
64 Swiss patent No. CH 355,742, priority date 23 June 1959.
65 Cartier ID One and ID Two 2012, Press pack produced by GS Presse, Issy-les-Moulineaux.

66 Richon 2007, 596 *et seq.*
67 Lovell and Kluger, 1995, 282–5.
68 *HJ*, October 2012, 469–70.
69 *AH*, September 2016, 408–9.

The next development was replacement of the balance by a tuning fork in the Accutron wristwatch developed by Max Hetzel and his team at Bulova. This had an electrically stimulated tuning fork that vibrated at 360 Hz, which, via jewelled ratchet pawls, drove a tiny wheel with 300 teeth, and which, in turn, drove the rest of the train. Hetzel had created a prototype in 1953 and applied for a patent on the tuning fork resonator.[70] The watch was announced to the public on 25 October 1960.[71]

In its turn, the tuning fork was displaced by a piezo-electric quartz crystal. The quartz crystal in a wristwatch is a mechanical vibrator like a tuning fork, but much smaller and oscillating at a much higher frequency. The first quartz wristwatch crystals operated at 8,192Hz and then 16,384Hz. Today they usually have a resonant frequency of 32,768Hz. The electrical oscillator signal which maintains the vibration is divided electronically by successive factors of two until it reaches 1Hz, which is used to create an impulse that moves the train forward. The Seiko Astron was the first quartz wristwatch to be sold to the public in the autumn of 1969, although it was not a commercial success.[72] This was soon followed by a Swiss wristwatch developed by an industry collaboration in the Centre Electronique Horloger (CEH).[73]

The quartz crystal design also lent itself to fully electronic watches with no moving parts. At first the displays were power hungry Light Emitting Diodes (LEDs), usually red, that were soon replaced by the Liquid Crystal Display (LCD), which requires only a tiny fraction of the power.

The development of electrical, particularly electronic, wristwatches enabled countries without a long history of making mechanical watches to wrest the leading position away from Switzerland. The early entry of Seiko into the quartz wristwatch market was a sign of things to come. Seiko, together with Citizen and Casio, precipitated the 'quartz crisis' of the 1960s and 1970s, when a large proportion of wristwatch manufacture shifted from traditional Swiss mechanical watches to quartz watches that were cheaper and more accurate, and also embodied the zeitgeist of the era of the silicon chip. Japan, China, and far eastern countries now dominate the manufacture of electronic watches of all price ranges, as well as making high-quality mechanical watches.[74]

The source of electrical power has also undergone development, with conventional batteries being supplanted with accumulators charged by solar cells built into the watch dial, or by electrical generators powered by a moving weight in the same way as the winding mechanism of an automatic mechanical wristwatch. Further developments of electronic wristwatches enabled them to receive synchronizing time signals broadcast by radio transmitters, and then from satellites in the Global Positioning System (GPS). The latest technology is the smart watch, effectively a miniature computer worn on the wrist.

TECHNICAL DEVELOPMENTS IN MECHANICAL WRISTWATCHES

The jewelled lever escapement and its cheaper alternative, the Roskopf or pin-pallet escapement, quickly became the almost exclusive movements used in wristwatches.

The pin-pallet escapement movement was regarded as poor quality and used only in cheap wristwatches, but was extremely successful, nonetheless. The 'dollar watches' produced by the Waterbury Clock Company for Ingersoll eventually gave birth to Timex, which became the largest manufacturer of wristwatches, thanks to their low-cost pin-pallet movements.

The jewelled lever escapement dominated the market for high-quality wristwatches throughout the twentieth century. The movement looked much the same, with an oscillating spring balance controlling the rate at which a mainspring drove the wheel train and hands. One of the most significant innovations was the addition of automatic winding, discussed earlier. However, there was also much development of the traditional parts of the movement, a lot of it invisible.

After Paul Perret's untimely death in 1904, Guillaume continued work on temperature-compensation balance springs alone, developing harder alloys that eventually resulted in the Elinvar temperature-compensation balance spring.[75] In 1933, Dr Reinhard Straumann (1892–1967), technical director of Thommen S.A., working in conjunction with Heraeus Vacuumschmelze GmbH. invented an improved auto-compensating balance spring material called Nivarox.[76] When joined with a beryllium bronze Glucydur balance, Nivarox provides excellent temperature compensation and is still the most widely used temperature-compensation balance spring.

In the early 1940s the Elgin National Watch Company of America noted that about thirty per cent of watches returned under warranty had a problem with their mainspring, so they commissioned Dr Oscar E. Harder (1883–1956) and Simon A. Roberts to develop a better material.[77] The research resulted in a cobalt chrome alloy that was named Elgiloy. In 1946 mainsprings made from Elgiloy were installed in 100,000 wristwatches, not a single one of which was returned with mainspring trouble. So successful was the new material that mainsprings made from it, and similar alloys developed since, are called 'unbreakable'. Straumann and Heraeus Vacuumschmelze developed a similar alloy called Nivaflex. This can be treated to achieve a very high-yield strength,

---

70 US patent 2,888,582, 'Tuning fork oscillator', priority date 19 June 1953.
71 *HJ*, November 1960, 714.
72 *HJ*, October 1994, 556–7.
73 Vermot *et al.* 2014, 900.
74 Further discussed in this volume Chapter 2, Section 3 and Chapter 2, Section 4.

75 *HJ*, November 2018, 497–8.
76 *HJ*, December 2018, 538 *et seq*.
77 *HJ*, October 2017, 20.

which enables a thinner mainspring to produce the same torque as previous materials, enabling a longer spring to be fitted in a given barrel. All this results in a considerable gain in the time a watch can run between being wound.

Research into advanced materials by the Centre Suisse d'Électronique et de Microtechnique (CSEM) funded by a number of Swiss watchmaking companies resulted in the production of silicon balance springs and other watch parts that are much lighter and harder than steel or brass, non-magnetic, resistant to corrosion, and often require no lubrication. Silicon parts are machined using an ion beam, which can produce complex shapes and very precise dimensions.[78]

WRISTWATCHES IN THE TWENTY-FIRST CENTURY

Wristwatches have ceased to be a necessity of everyday life. The time is shown accurately on devices everywhere, the mobile or smart phone has become the new 'pocket-watch', and smart watches have brought a new level of functionality to the wrist. But the manufacture of traditional wristwatches continues. Wristwatches are fashionable accessories, even if they are not used to tell the time, and they make good presents. Many people appreciate the intricate mechanism of a mechanical watch, and the engineering and craftsmanship that goes into creating it. Some appreciate the impression of wealth and status that an expensive wristwatch conveys.

Wristwatch companies allocate large budgets for advertising and brand ambassadors, and wristwatches are promoted as part of a luxury lifestyle image. To serve their diverse markets, wider and wider ranges of wristwatches are made, from the simplest and cheapest battery-powered quartz watches to mechanical wristwatches with incredibly complicated and intricate mechanisms, created in 3D computer models and carved out by robotic micro-machines. Some of the products of these modern manufacturing methods are finished by hand in the finest of traditions. There are even a small number of individual makers who alone, or with a small team, create wristwatches after their own visions.

The wristwatch was born from the need of its wearer to carry the time with them—a need that has now vanished—and the wristwatch as a timekeeper is therefore essentially obsolete. But instead of declining and disappearing, and now free from the requirement simply to exist, the wristwatch has taken on a new life in many diverse forms that reflect the personality, lifestyle, ideas, nostalgia, or interests of their wearers. The future for an obsolete piece of technology has never looked brighter!

---

78 Vermot et al 2014, 712–13.

CHAPTER NINETEEN

ELECTRICITY, HOROLOGY, AND NETWORKED TIME

James Nye and David Rooney

INTRODUCTION

This chapter introduces a set of inter-related and overlapping themes in the history of horology that broadly relate to the use of electricity in enabling standardized precise time to be measured and distributed in modern industrial societies.[1]

It starts by examining the electrical communication networks that were used from the nineteenth century onwards to disseminate precision standard time signals over large geographical areas, as well as changes in the primary time standards of astronomical observatories and physics laboratories brought about by the use of electricity. Then it looks at ways in which the infrastructures of electricity and compressed air were used in the distribution of time across clock networks, at the city-wide scale, and in individual premises. It offers some comparisons between the situations in different geographical territories. Finally, the chapter examines the widespread use of precision timekeepers based on electrical and related technologies in everyday life.

Four significant ideas tend to recur throughout this chapter. The first is the multiple motivations for distributed precision time networks, specifically about who needed standardized time, when, and why. The second idea is about automation, or the reduced requirement for human effort or intervention in timekeeping, and the changes this brought in society. The third recurring idea is the role electricity played in fostering and facilitating a desire for increased accuracy and precision in timekeeping. The fourth idea is the disappearance from view of standardized precision time owing to the cheapness of its acquisition and the universality of its presence.

STANDARDIZING TIME WITH ELECTRICITY

Demands for standardized time

Over much of the extensive period covered by this book, there was little concerted demand for precise, standardized time, or for the distribution of time over wide geographical areas. But this started to change by the early years of the nineteenth century, influenced by the diverse demands of industrialization, imperial expansion, and urbanization, and catalysed by the development of network technologies like railways and electric telegraphs. Interconnectedness, in all sorts of ways, increased in this period, with people in distant communities becoming more closely linked. Common 'operating systems' for these growing networks relied increasingly on uniform standards and one of the most important was time. People needed access to accurate and precise time that was standardized over wide areas, and electricity provided solutions that transformed both horology and society. For Europe and North America this story has been extensively explored.[2]

An early demand for accurate and standardized time came from the maritime sector. From the late eighteenth century, developments in navigational practices increasingly led to a requirement for ships' captains to know the time precisely and accurately in order to set marine chronometers correctly before setting sail. In parallel with this, makers of navigational timekeepers also needed access to accurate time in order to adjust and set their products correctly. At first, their only option was to send somebody to the

---

[1] This chapter has benefited greatly from close review by Jonathan Betts, Derek Bird, and Johannes Graf, from whom valuable suggestions for improvements were gratefully received.

[2] Bartky 2007.

nearest astronomical observatory (such as Greenwich) to ask for the time.[3]

In the 1830s, the first modern railways started construction. The Liverpool & Manchester Railway, which opened in September 1830, was the world's first modern, high-speed inter-city railway using steam locomotives for haulage of both passengers and freight. Huge numbers of railways followed in the ensuing two decades. While interconnected coaching networks had already highlighted the need for coordinated timetabling, it became clear that a common time standard for any railway was also critical for its safe operation. Numerous local times, inaccurately kept, would hamper the growth of this ambitious industry.

The popular account of time standardization has commonly claimed that the near-universal nineteenth-century civil adoption of standard time was chiefly the result of the emergence of railways. This has been revised in more recent scholarship, which is not to say the railways played no role (quite the reverse), but that there were many other factors also at work.[4] The railways by their very nature operated down narrowly defined corridors that crisscrossed countries across the globe. These corridors were natural pathways for telegraph lines, and it is clear that information, including time, travelled in the same narrow bands down which trains processed.

But, if railways boomed in various major economies by the mid-nineteenth century, the fact that national statutes for the adoption of universal time standards only emerged towards the end of the century is suggestive that other factors were at play—the building of railways did not immediately impose universal time standards across wider geographies. Indeed, in a mountainous country such as Switzerland, where the establishment of a rail network was both difficult and expensive, it was the independent creation of the telegraph network that drove the standardization of time and its distribution across the country.[5]

Instead, industrialization in a wider sense was a major driver of time standardization. As large-scale workforces grew, the slow move towards legislation to safeguard rights for those workforces, including a right to be paid for hours worked, relied on time standardization and its widespread distribution. So, too, did the liquor licensing legislation that emerged from Victorian concerns over alcohol consumption in the working classes.

Beyond these concerns over imperial expansion, railway safety, factory conditions, and temperance, finance (as always) was another important factor in the move towards time standardization. Early clients for time distribution systems included bankers' clearing houses, stock exchanges, and institutions where the emergence of 'time-stamping' was often critical. When was a bargain struck? When did markets open or close? Research has demonstrated that the effect of the undersea cable between the USA and Britain being laid in 1866 (Figure 127) was observable in the increased speed with which price-sensitive information crossed the Atlantic—a reduction from 10–13 days to just 0–2 days.[6] Against such a backdrop, market players were now striving to exploit arbitrages that lasted hours and minutes, not days or weeks. Time, very pressingly, was money.

By the mid-nineteenth century, then, we can plot a growing population of people and industries for whom a uniform and accessible time standard was desirable—and the underlying reasons usually related as much to notions of safety, morality, or certainty, as mechanization or industrialization, and the finance that underpinned it became more sophisticated.

Non-electric time signals

As industrialization and empire building had increased during the eighteenth century, so had the demand for shared knowledge of time; time signals of many types started to become embedded in daily life and, throughout the nineteenth century, were supplemented by novel technologies. Whether for the control of hours worked, or the hours of operation of plant, or the safe running of railways, or the closure on time of licenced establishments (and much more besides), a broad framework emerged from the mid-nineteenth century in which hierarchies of standards were defined and understood.

While many clocks were set by traditional methods (such as checking a Sundial) right up to the late nineteenth century (or even beyond), it was natural that populations would become used to taking the time from supposedly reliable sources further up the technological hierarchy. Church clocks had long been one such source, but cities offered compelling evidence from the earliest times that different public clocks often disagreed even within a limited area. Consequently, certain single public clocks in towns and cities became established with reputations for particular accuracy or trustworthiness.

Big Ben in London, installed in 1859, is only the best known of such nineteenth-century clocks, renowned for being kept accurate within tight tolerances. For clockmakers needing an accurate source of time for the process of rating and regulating clocks in their workshops, a convenient map was produced, showing concentric rings radiating out from Big Ben at the centre. The rings each represented a given number of seconds of delay by which the strike of Big Ben would be delayed as it radiated out across London.[7]

As maritime empires grew and navigational technology based on timekeeping was developed, networks of time grew up around the world's coasts and ports, providing ship captains and their navigation staff with accurate and precise time to rate and set their chronometers before they set sail. At first these were local, ad hoc arrangements. From the 1830s, a growing network of

---

3 Betts 2017.
4 Rooney and Nye 2009.
5 Messerli 1995, 70–2.

6 Hoag 2006.
7 De Carle 1947.

time-finding institutions along with coastal time signals—balls, flags, guns, whistles—was being built.[8] Alongside these maritime networks, urban time distribution for the marine chronometer industry was required. Hand-carried networks, such as that operated by the Greenwich Observatory astronomer John Belville (1795?–1856) and his family, thrived for several generations, as trust was hard won and long lasting.[9]

If the nascent railway industry demanded standardized time to be distributed across its networks, at the start it had to follow the lead of urban industrial practice. The hand-carrying of watches to distribute time and to regulate the delivery of the Royal Mail was an eighteenth-century Post Office practice that was adopted in the 1830s by the railways and, while it was low-tech, it allowed time to spread from the headquarters to the ends of branch lines as the time on the watches was used to correct clocks throughout the network. Railway clocks in general moved towards a reputation for reliability, as the railway authorities in many countries invested in systems to ensure accuracy, precision, and synchronization. Soon, electricity, in the form of telegraphy, started to offer a new technological solution.

Electric telegraph time signals

The public electric telegraph was first introduced in 1837 by William Cooke (1806–79) and Charles Wheatstone (1802–75) in a trial on a section of the London and Birmingham Railway between Euston and Camden. The new technology developed rapidly in the 1840s with the expansion of the railway network, along whose lines the new electric cables commonly ran.[10] The two technologies were symbiotic: railways demanded instantaneous communication for signalling and operational purposes, and telegraphs needed lines of route—wayleaves—along which to be run.

Railways were, therefore, early adopters of time signals transmitted along telegraph lines and, in 1852, this time network was given a major boost by the installation at the Royal Observatory, Greenwich, of Charles Shepherd's (1830–1905) new clock that transmitted electrical impulses every second or less frequently according to need. In 1849, the Astronomer Royal George Airy (1801–92) wrote:

> Another change will depend upon the use of galvanism; and, as a probable instance of the application of this agent, I may mention that, although no positive step has hitherto been taken, I fully expect in no long time to make the going of all the clocks in the Observatory depend on one original regulator.[11]

> The same means will probably be employed to increase the general utility of the Observatory, by the extensive dissemination throughout the Kingdom of accurate time-signals, moved by an original clock at the Royal Observatory; and I have already entered into correspondence with the authorities of the South Eastern Railway (whose line of galvanic communication will shortly pass within nine furlongs of the Observatory) in reference to this subject.

Airy's electric (or 'galvanic') clock system acted both as an internal time-distribution network for the Observatory itself, a type of system which is considered later in this chapter, and as the originating time standard for an electrically distributed network of time signals, by which countless local clocks or clock networks could be corrected or synchronized. This included clocks on the railway networks as well as a growing number of other recipients of its time signals.

The Greenwich time ball began to be dropped automatically by the Shepherd clock, as, later, did the time ball at Deal, and then many others (Figure 179). Big Ben used the electric telegraph to report back on its accuracy. Telegraph offices across the country received time signals from Greenwich, and in the 1870s the Post Office began selling time signals derived from the Greenwich signal to a wide range of commercial and private customers. Thus, after only a few years, the time from a single electromechanical clock at Greenwich had wormed its way across Britain. We return presently to the development of urban time networks, and electric observatory and impulse transmitter clocks.

Telephone speaking clocks

Electric telegraph networks in numerous countries were joined in the later nineteenth century by new cable networks for telephones. For many years after their introduction, telephones were used as an ad hoc time distribution network, as subscribers called telephone exchanges in great numbers not to make a connection, but to find the time. By the 1930s, telephone service providers in partnerships with national observatories were looking for more systematic ways to provide this service while relieving pressure on exchange operators and introducing potentially lucrative revenue streams.[12]

In 1933, the Paris Observatory installed a telephone speaking clock that could give observatory time to up to 12,000 callers per day, spoken by the Paris radio celebrity Marcel Laporte (1891–1971). In 1934, one was set up in The Hague, spoken by the Netherlands schoolteacher Cor Hoogendam (1910–2009). In 1936, a speaking clock for Poland used the voice of the stage and screen actor Lidia Wysocka (1916–2006). The same year, the British General Post Office switched on a speaking clock using the voice of the telephone exchange operator Ethel Jane Cain

---

8 Howse 1997.
9 Rooney 2008.
10 Blyth 2014.
11 Howse 1997, 95.

12 Rooney 2008, 148–61.

Figure 179 Time ball and signal station, Hong Kong Observatory, *c.*1910 (private collection).

(1909–96), who had been selected the previous year in a nationwide 'Golden Voice' competition designed to maximize publicity for the new service. Numerous other national telephone providers set up similar systems in the following years and changing the 'voice' was often an event of great public interest.

Telephone speaking clocks offered a service that other electric time-signal networks could not: they provided accurate and precise standard time, in many cases to a fraction of a second, continuously and on demand, rather than infrequently at set times of the day. This, combined with the personification of their voices, enabled them to become part of everyday life to a much deeper, more personal, and long-lasting extent than other time services. This has meant that many speaking clocks are still being used regularly today, despite a profusion of other ways to get time.

Time signals by radio

By 1910, there were many systems in place worldwide, both at institutional levels but also cross-country and overseas, for the transmission of time signals using cabled electrical systems. However, less than a century after telegraphy revolutionized information transfer, radio did the same again.

From 1886 onwards Heinrich Hertz (1857–94) was instrumental in developing wireless transmission, and in 1899 Guglielmo Marconi (1874–1937) demonstrated transmissions across the English Channel, and then in late 1901 across the Atlantic. Before long, this new technology was enabling precise time signals to be transmitted over long distances using less fixed infrastructure than was needed for cable networks, as well as being accessible to mobile bodies like ships at sea; this aided greatly in navigation by allowing more frequent chronometer calibrations. As was the case with time signals finding their way down the recently established corridors of rail and telegraph lines from the 1840s onwards, so in the early years of the twentieth century time rapidly found its way in among the other data being transmitted across the newly exploited airwaves.

The USA experimented with time transmissions from New Jersey in September 1903, followed by a regular time service from Washington, DC in January 1905. Germany followed suit, in 1907, with transmissions from Norddeich, which became regular from 1910. As early as 1903, Gustave-Auguste Ferrié (1868–1932) had proposed that the Eiffel Tower be used as a mast for transmitter aerials, and he succeeded in transmitting time signals from May 1910, which were available all over Europe (Figure 180). France had the initiative and maintained it, particularly through securing agreement in 1912 for the foundation of the Bureau International de L'Heure (BIH) in Paris, coordinating time transmissions worldwide.[13]

The BIH rapidly established a global network of time transmitters: at San Fernando (Brazil), Arlington (near Washington, DC, USA), Mogadishu (Somalia), Manila (Philippines), Timbuktu (Mali), Massawa (Eritrea), and San Francisco (USA), which provided fairly comprehensive cover for international shipping requiring time signals.[14]

In succeeding decades, the transmission of time signals by radio widened significantly. Navigators and other specialists used coded signals in the tradition of wireless telegraphy, which, as the name suggests, was an extension of the practices of the cable telegraphs that preceded it. The wider public had to wait until the introduction of broadcast radio for time signals meaningful to them.

One of the best-known radio time signals, the 'six pips' broadcast by the BBC and used or adapted in many countries worldwide, was devised by the British electric clock pioneer Frank Hope-Jones (1867–1950) and first transmitted on 5 February 1924 (Figure 181).[15] At the Greenwich Observatory, an astronomical regulator by Dent, numbered 2016, originally made for observing the Transit of Venus in 1874, was adapted with three sets of electrical contacts so arranged as to be closed by the escape wheel to provide six impulses of electricity every fifteen minutes. These impulses were transmitted by direct wire connection to the BBC headquarters then at Savoy Hill, on the Thames Embankment, where they were converted into audio tones, or 'pips', and made available for broadcast when selected by the engineers at the 2LO transmission station in Marconi House on London's Strand.

Many systems for the production of six-pip time signals were exported from Britain, often to Commonwealth countries, but a number of installations went further afield. For example, in 1938, the Synchronome company supplied a Shortt-type clock numbered 63 (see later) to the Sternberg Astronomical Institute in Moscow, together with a complete apparatus for transmitting six-pip time signals. In the twenty-first century, nearly thirty countries have regular radio time signals which are similar or analogous to the six pips.

The ultimate development of time by radio came with the introduction and maintenance of powerful radio transmissions of **timecode** worldwide that would allow for the automatic wireless correction of large numbers of watches and clocks. Germany led the way from 1973 with timecode transmissions from Mainflingen, near Frankfurt, and semi-professional receivers (for example, by Hopf Lüdenscheid) were available soon after. Raymond Phillips wrote to the *Standard* in London on 19 September 1912 with a prescient vision for the future of radio time synchronization: 'the systems of tuned oscillations and directive aerials now in vogue might be superseded by a time factor principal whereby the receiver and transmitter are brought into sympathetic action only at appropriate moments'.[16]

Since their introduction some 73 years later in 1985, it has been a feature of radio-controlled watches and clocks that they are free-running, only corrected daily or hourly by the radio time signal. It should therefore be clear that they do not offer the accuracy suggested in advertising that often describes them as 'atomic' clocks or watches—it is the underlying signal by which they are corrected that is in fact accurate, and between corrections these clocks and watches can drift by substantial amounts depending on the quality of their quartz mechanism.

Electricity and time standards in observatories

As electricity offered new ways to disseminate standard precise time over wide areas, so it also transformed the primary timekeepers in astronomical observatories themselves, at the top of the hierarchy of accuracy. We have already mentioned the Charles Shepherd electrical (or 'galvanic') clock system installed at the Greenwich Observatory in 1852 that set running a chain of developments in electrical horology. At the same time, across the Atlantic, a series of innovations were being made by the Bond family of Boston, Massachusetts—William Cranch Bond (1789–1859) and his sons Joseph (1823–61), George (1825–65), and Richard (1827–66)—who were closely associated with the Harvard College Observatory.[17]

The same year that the Shepherd clock started telegraphing time signals from Greenwich, the Bonds introduced a similar service from the Harvard observatory. The previous year, and of great significance to the world of astronomy as it introduced the 'American method' to Europe, the Bonds had used the Great Exhibition, London, to reveal their drum chronograph: an electromechanical device that aimed to reduce human error in the recording of star transit times. This was an important example of how electricity could be applied to a scientific timing process in order to mitigate a human element, increasing accuracy.

In the ensuing decades, numerous innovators sought to harness the potential of electricity in the ever-more precise and accurate measurement of time. Here, we can discern an unusual situation when considering the development of electrical horology around the world. Whereas in most cases, the design and uptake of electric clocks and time distribution systems varied, depending very much on local cultures of time, astronomers were a more homogenous

---

13 Miles 2011, 196–207.
14 Hope-Jones 1913.
15 Rooney 2008, 126–47; Miles 2011, 204–5.

16 Quoted in BHI 1912, 26.
17 Saff 2019.

Figure 180 Samuel Begg (1854–1936), 'Greenwich Time by Wireless from Paris'. Original caption read, 'The greatest scientific marvel of the present age for family use: wireless signals from the Eiffel Tower give correct Greenwich Time in a London house'. Image from *Illustrated London News* (December 1913).

Figure 181 Clock for sounding 'six-dot' time signal across Clerkenwell, with illuminated dots. Image from a Synchronome Company catalogue, 1929. Courtesy Antiquarian Horological Society.

group: scientists who communicated in a shared language and who placed similar value upon the finest scientific instruments for their work. We therefore observe that broadly the same family of astronomical regulators, drawn from a small palette, equipped large numbers of observatories in the twentieth century, irrespective of location. We next survey the key developers of observatory regulators that made use of electricity, following the lead set by Charles Shepherd, the Bond family, and other pioneers of the 1850s.

Riefler and Leroy pendulum regulators

The firm of Clemens Riefler (1820–76) of Munich, Germany was highly prolific, making pendulum regulators over a long period once it passed into the control of his son Sigmund Riefler (1847–1912), who was a highly gifted engineer.

The finest Riefler clocks employed an **escapement** in which the impulse was delivered through the suspension spring, avoiding the need for physical contact with the pendulum itself. The pendulum's full weight is supported by a double-bladed suspension spring, but that assembly is not rigidly attached to the upper support. Instead, it uses knife edges sitting upon hard stone bearings. The knife edges are mounted on pallet frames with jewelled **pallets** which, in turn, engage an escape wheel. Pressure from the escape wheel normally holds the pallet frame across to one side but, as the pendulum swings and approaches the central position, the pallet releases and the pallet frame swings until the other pallet locks, impulsing the pendulum through the suspension spring as it does so. The cycle continues on the opposite swing. This escapement requires considerable care and knowledge in setting up and adjusting, as it is prone to 'tripping', but once correct it is reliable.

Whether or not Riefler's design can be considered a form of free pendulum is an interesting question. It is true that the pendulum rod and bob are not interfered with at any stage in its operation. However, from a dynamic point of view, the shifting of the pendulum suspension to and fro could be considered a significant interference with the pendulum's motion, and strictly speaking disqualifies it from being considered as free. In addition to this innovative and successful impulsing system, Riefler used a mercury-filled compensation pendulum of his own unique design, later employing **Invar** instead. He also fitted the clock into a sealed metal tank of reduced and constant pressure, largely removing variations caused by temperature and atmospheric pressure (the escapement was driven by an electrically reset weighted arm on the wheel below the escape).

It was a highly successful design and the regulators made by Riefler at the end of the nineteenth century out-performed anything which had gone before, with claimed accuracies in the order of a few seconds a year. Isolating the oscillator atmospherically in this way was, of course, not a new idea. Several theoretical designs for clocks running in a vacuum had been proposed by others, including Christopher Wren, John Ward of Chester (1648–c.1727), and a writer signing himself Jeremy Thacker of Beverly, Yorkshire.[18] But, by using electricity as a power source, Riefler's design was the first successful application. Key to Riefler's success was the rigour with which he approached the design and manufacture of his regulators. For example, he tempered and then carefully tested each of his **Invar** pendulum rods in order to establish its performance, matching the compensation for each pendulum according to the performance of its rod.

From France, the firm of Leroy & Cie produced a run of sixty-four 'constant pressure' pendulum clocks housed in vacuum tanks similarly utilized in observatories and manufactured between 1912 and 1957. Of very high-quality construction, the movements of these regulators employed a form of spring pallet escapement similar to those of William Hardy and Thomas Reid, but with the spring pallets attached directly to the pendulum and projecting above the suspension, engaging with the escape wheel in the movement situated above the pendulum support. As with the Riefler, the escape wheel of the Leroy was driven by an electrically reset weighted arm on the wheel below, and the pressure in the tank reduced to about 600mm of mercury.

Shortt free-pendulum regulators

A highly important time standard used in large numbers of observatories and laboratories worldwide between 1921 and the Second World War was the Shortt 'Free Pendulum' clock.[19] This was the brainchild of William Hamilton Shortt (1881–1971), a railway engineer and (originally) amateur horologist who joined Frank Hope-Jones in 1912 to assist in the design and further development of electric clocks. After nearly a decade of development, he arrived at a revolutionary design that embodied two important features. One of these was common to a range of existing clock designs: infrequent impulsing, as first demonstrated in the electrically impulsed pendulum timepiece of Matthäus Hipp (discussed later) and much championed by Hope-Jones. Warren Marrison (1896–1980), pioneer of quartz crystal clocks, commented:

> The advantage is that no friction effects of driving mechanism are coupled to the pendulum except during that minimum time required to impart energy to it. Actually, in theory, the phase error introduced by one large impulse after n free swings is exactly the same as the sum of the phase errors for n small impulses ... But experience has shown that a pendulum is actually more stable when the sustaining mechanism is detached from it the greater part of the time.[20]

The second feature was of using two pendulums, which was introduced to Shortt by Hope-Jones, first conceived by William Thomson (1824–1927) in 1865 and later developed by Robert James Rudd (1844–1932) in 1898. In Rudd's clock, one pendulum regulated the timekeeping of a subsidiary clock, the job of which was to release a mechanism that would impart an impulse at periodic intervals to a primary pendulum, say each half-minute. The primary pendulum was as near 'free' as possible. It had no mechanical connection to any part of the clock save for its suspension and the occasional impulse from a mechanism released by the secondary pendulum.

However, this impulsing mechanism had a subsidiary function. The end of its interaction with the free pendulum was defined by that pendulum's position, and the instant the impulse was complete, it operated a synchronizing device, correcting the phase of the secondary clock if necessary and thereby ensuring the secondary clock's timekeeping was locked to that of the free pendulum. Thus, the secondary clock commenced the process of impulse, but the timekeeping of the entire system was derived from the free pendulum. Put another way, the primary clock was free to dictate the time and the secondary clock did all the work.

Shortt refined this idea significantly by separating the two pendulums into two distinct clocks and placing the free pendulum in an evacuated tank. Considerable effort went into ensuring a good support for the free pendulum, including isolation from temperature and barometric effects, as well the standard techniques for compensation, such as using Invar for the pendulum rod, certified by the National Physical Laboratory.

The secondary clock was a largely standard Synchronome controlling clock (which is described later), and its standard resetting impulse each thirty seconds was placed in series with the electromagnet that unlocked the latch of the gravity arm of the impulsing mechanism in the free pendulum clock tank, thus initiating the impulse to its pendulum. At the end of the free pendulum impulsing phase, the electrical contacts for the resetting electromagnets would close, the timing of this depending entirely on the geometry and position of the free pendulum—thereby defining a precise time that was dictated solely by the free pendulum. The circuit for the resetting action of the gravity arm in the free pendulum clock was placed in series with a synchronizer in the secondary clock. This synchronizer had the effect of advancing the phase of the secondary clock pendulum, which was rated to run naturally slow.

In theoretically perfect conditions the secondary clock phase would shift either side of that of the free pendulum clock, running

---

18 Betts 2017, 15–16. For Ward, see Turner 1996, 124–6.
19 Miles 2011, 168–95.

20 Marrison 1948a, 343; Marrison's assertion in the last sentence has been challenged in the work of Martin Burgess and the Harrison Research Group—see McEvoy and Betts 2020.

slightly ahead immediately after a correction and falling back steadily thereafter. After thirty seconds it should still be just ahead and the synchronizer would miss the secondary pendulum, but by the following thirty seconds its phase would have fallen back sufficiently for a hit, and a correction to take place, the secondary pendulum having its phase shunted forward again. The synchronizer was thus described as a 'hit-and-miss' device. With the secondary clock designed to produce seconds pulses, and with the errors by reference to the time of the free pendulum being vanishingly small over thirty seconds (and always corrected rather than accumulated), the system could provide very precise time signals for observations in observatories or laboratories.[21]

The first Shortt clock was used from 1921 by Ralph Sampson (1866–1939), the Astronomer Royal for Scotland, at the Royal Observatory in Edinburgh, and thereafter Shortt clocks were adopted by numerous observatories worldwide, with 100 clocks being produced between 1921 and 1956. Some clients therefore continued to order the clocks well after the Second World War, although superior time standards were by then readily available.

It was with the introduction of the Shortt clock that the decisive break emerged between timekeeping and the rotation of the Earth. The Moon revolves around the Earth in a plane at about an angle of five degrees different from the plane of the Earth's revolution around the Sun. Gravitational effects as a result cause a slight irregularity in the motion of the Earth, with a period of 18.6 years in an otherwise 26,000-year cycle known as the precession of the equinoxes, reflecting essentially a cyclic wobble in the Earth's axis. Observations of the stars are influenced as a result, as the viewing platform—the Earth is not entirely stable. Errors arising from the difference between using a stable timekeeper and the Earth's motion in observations were first resolved in data derived using Shortt clocks in the 1920s.[22] In essence, the Shortt clock was more stable than the Earth on which it was mounted.

Clocks by Riefler, Leroy and Shortt were used extensively worldwide, and their use tended to reflect the diplomatic and political reach of (in particular) Germany, France, and Britain across the globe. In the case of Leroy, around twenty-five clocks remained in France, and a further twenty were delivered in Europe. Reparations following the First World War also resulted in high-precision German clocks being installed at some international locations. In one example, Belgrade Observatory was founded in 1909 but lost its instruments in the War. In the 1920s, a fine Strasser & Rohde precision regulator was installed, but the Director, Vojislav Mišković (1892–1976), organized a time service from 1935 onwards which used a combination of six Riefler and Leroy clocks.[23] National time has always been embroiled in national politics.

Soviet pendulum regulators

Synchronome in Britain shipped sixteen Shortt clocks to Russia and the Soviet Union between 1930 and 1951, but as early as the mid-1930s, Ivan Ivanovich Kvarnberg (1869–1937), assisted by Nikolay Christophorovich Preipich (1896–1946) of the St Petersburg/Leningrad Étalon factory, began work on their own version of the Shortt clock. The Étalon factory was the research laboratory for the Russian Chief Bureau of Weights and Measures (VNIIM). It produced all the finest instruments for the Soviet Union after the October revolution.[24] Though work was interrupted by the Second World War, the resulting clock, produced in small numbers from 1952, achieved a diurnal variation of just one millisecond.

The Soviet Union was also the home of the last and most precise pendulum clocks ever to be manufactured in a production run, namely the Fedchenko ACH-3 astronomical regulator. From 1948, Feodosii Mikhailovich Fedchenko (1911–89), of the Kharkiv State Institute of Measures and Measuring Instruments, began attempting to develop an **isochronous** pendulum through experimentation with a variety of designs and dimensions of suspension spring. A serendipitous error in the set-up of one suspension spring assembly led to the design of what is effectively a tuneable suspension, where its effective length changes through the arc of swing. If correctly tuned, the pendulum beats isochronously. This led to the ACH-1 and ACH-2 clocks in 1956–8, and finally the ACH-3 in 1958 (Figure 182), which contained the most refined impulsing mechanism, and which is reported to have achieved an accuracy equivalent to that of precision quartz clocks—keeping time to within one second per year.[25]

Impulsing was achieved in broadly the same way as the method devised by Alexander Bain (1810–77), which is considered later in this chapter, using electromagnetic coils and permanent magnets, but with considerable sophistication. The pendulum carries two prismatic magnets at its lower end, with a fine gap between them. The magnets pass close to a pair of very fine coils, wound concentrically, one of which is a sensing coil and one a driving coil. The sensing coil is energized by the approaching magnetic field of the moving pendulum and, through transistorized circuitry, causes the switching-on of the driving coil with an amplified signal, thus imparting an electromagnetic impulse to the pendulum. The design allows for the sensing and driving action to take place at the ideal position for impulse, and the absence of any mechanical parts or friction allows for the minimal use of energy. The mercury cell used to drive the clock would last for more than three years and could be replaced without stopping the clock.

The design, which was often used as a controlling clock driving numbers of secondary clocks, ran in parallel with the development of quartz clocks in the Soviet Union during the 1960s, especially

---

21 A simplified schematic diagram of the Shortt system appears in Hope-Jones 1949, 165.
22 See for instance Jackson 1929.
23 Todorović and Milić Žitnik 2019, 168.

24 Arutyunov, Brodskii, and Fomichev 1968.
25 Feinstein 1995.

Figure 182 Fedchenko ACH-3 astronomical regulator clock No. 17, 1966. Photo: Dr Crott Auctioneers.

Figure 183 Staff adjusting National Bureau of Standards ammonia-molecule clock, c.1949. Photo: National Institute of Standards and Technology Digital Collections, Gaithersburg, MD, 20899.

for use in applications where simpler, more robust timekeepers were required, such as television studios and the more remote outstations of the Soviet Union.

Tuning-fork timekeepers

In the period before quartz crystal timekeeping, fine chronometers using mechanical balances and escapements understandably remained crucial at sea, while on land the precision pendulum clock dominated the field of time standards. Quartz pioneer Warren Marrison made an important generalization in his explanation of the development of the quartz crystal clock: 'the constancy of rate obtainable depends on two kinds of properties: those which concern the inherent stability of the governing device itself, and those concerned with the means for sustaining it in oscillation'.[26] The emergence of time standards that could improve on the performance of the finest precision pendulum clocks of the 1920s required the use of an oscillator quite different from the pendulum, and a very different mechanism for maintaining its oscillation. The quartz clock was the next major step, but the use of tuning forks formed an important part of the developmental arc.

In all clocks there is an oscillator, and a vibrating tuning fork can be used in such a capacity. Indeed, tuning forks formed an important part of laboratory equipment where a short-term frequency standard was required. Its invention dates back to the early eighteenth century, but in 1857 Jules Lissajous (1822–80) developed an electrically maintained tuning fork, which therefore contained elements comparable to those in a clock (in other words, an oscillator that could be maintained in oscillation by the provision of energy to match the losses occurring through oscillation). The tuning fork was, however, generally used to calibrate other instruments, or as a time base in metrology other than timekeeping. It could not compete for stability with precision pendulum clocks in the late nineteenth century, owing to fundamental problems associated with high-speed switching across mechanical contacts.

However, with the emergence of the electronic valve, or vacuum tube, which could be used to replace the mechanical electrical contacts used in maintaining oscillation, the tuning fork clock became a practicality, and examples were in use from 1918 in various locations, including the National Physical Laboratory in England. In the 1920s, considerable improvements were made to the design and construction of tuning fork oscillators, such as in using the alloy **Elinvar** to make the forks themselves in order to reduce the effect of temperature on frequency. Improvements in fork shape, mountings, and materials all contributed to the production of tuning fork clocks that out-performed precision pendulum clocks.[27] However, such tuning fork-controlled timekeepers generally remained in the world of the laboratory. It was to be developments in other fields, including the development of the electronic transistor and miniaturization techniques, that made the practical application of the tuning fork to horology possible, as is discussed later. In the world of national time standards, it was the quartz crystal, not the tuning fork, that finally supplanted the pendulum after nearly three centuries at the top of the hierarchical tree.

Quartz crystal timekeepers

As the foregoing account shows, and as is further discussed in the section on electric clock systems, the use of electricity in clocks from the first half of the nineteenth century focused on the rewinding of otherwise traditional mechanical clocks, or to power electromagnetic impulsing of once again traditional oscillators, principally precision pendulums. The development of quartz crystal timekeeping involved a fundamental change of technology, in that the swinging pendulum was abandoned as the underlying oscillator.[28]

Arising from the field of pyroelectricity, a term coined by David Brewster (1781–1868) in 1824, experimentation proceeded for several decades to relate the electric or 'magnetic' charge of crystals to other factors, such as changes in heat. Charles-Augustin de Coulomb (1736–1806) conjectured that an electric charge might result from pressure on a crystal. Experiments to prove this were unsuccessful until 1880, when the brothers Pierre Curie (1859–1906) and Jacques (1856–1941) proved that the compression of certain crystals could give rise to electrical poles of opposite sign at the extremities of their **hemihedral** axes. The two crystals that exhibited the strongest effect were quartz and Rochelle salt.

Gabriel Lippman (1845–1921) took this work further by predicting the corollary of the Curies' discovery: that applying an electric charge to a crystal would cause a dimensional change. By the end of 1881, the Curies had verified this. Crystals that exhibit this phenomenon were described as piezoelectric.

The field of piezoelectricity remained largely confined to laboratory study for several decades thereafter. As is frequently the case, it was against the backdrop of war that new developments and applications emerged. In the First World War, the need for communication with submarines, and their detection, were spurs to new work. Paul Langevin (1872–1946), a successor to Pierre Curie at the Ecole Municipale de Physique et Chimie Industrielles, began experimenting with radio transmissions using a **transducer** formed of a steel–quartz–steel sandwich and a newly available eight-valve R6 Radiotélégraphie Militaire amplifier, engaging in sea tests in February 1918. Signalling distances of eight miles underwater were achieved.

---

26 Marrison 1948b, 406.
27 Marrison 1948a.

28 Much of this account is drawn from Katzir 2003 and Voore 2005.

While Langevin had been working on submarine detection in Europe, Walter Guyton Cady (1874–1974), a professor of physics at the Wesleyan University in Connecticut, was experimenting in the same field, and embarked on a career focused on piezoelectricity—coming to be known as the pioneer of the subject. In 1921, Cady was granted what is now regarded as the first patent for a quartz-crystal-based oscillator, in which the crystal was used as a mechanically tuned feedback path. His priority, ultimately settled in court, was disputed by Western Electric, whose employee Alexander McLean Nicolson (1880–1950) had been experimenting with Rochelle salt, producing microphones, gramophone pick-ups, and loudspeakers based on the piezoelectric effect, described in US patent 2,218,845, filed on 10 April 1918 but only issued in 1940.

This activity paved the way for significant work in the 1920s on the development of quartz oscillators as frequency standards. These were used to support the field of radio transmission and the measurement of frequencies that could not easily be achieved with typical pendulum clocks, which usually beat seconds. Astronomers were important clients who normally drove the development of precision pendulum clocks, but it was the scientists and engineers working in radio communication in the 1920s who drove the development of quartz crystal oscillators, through their need both to measure and control radio frequencies. The first quartz clock was constructed at the Bell Telephone Laboratories in New York in 1927.[29]

In 1932, Adolf Scheibe (1895–1958) and Udo Adelsberger (1904–92) published an account of their first quartz clock, and then in 1936 described the application of quartz timekeeping to astronomy, at the Physikalisch-Technische Reichsanstalt (PTR), where it became possible to observe variations in the Earth's rate over periods as short as a few weeks.[30] It became common for international institutes from the early 1930s onwards to maintain quartz-based timekeeping systems as the basis for the maintenance and transmission of a wide range of frequency standards. An example would be the US Bureau of Standards in Washington, DC, broadcasting a number of precisely controlled carrier frequencies and seconds pulses. Warren Marrison commented in 1948 that these signals 'are so precise, and so convenient to use, that they may be employed for the high precision intercomparison of quartz clocks across the Atlantic and for studies in astronomical time'.[31]

With the development of portable quartz clocks from the late 1930s onwards, this new high-precision timekeeping technology could be used in geophysical experimentation, and thus assisted in measurements of gravity on both land and sea, generally significantly improving the accuracy of geophysical experimentation and metrology.

Experimentation continued in the development of quartz clock technology as scientists sought the best configuration of the numerous technical elements involved. One area of significant development was the form of the quartz crystal used. At the National Physical Laboratory (NPL) in Britain, David W. Dye (1887–1932) experimented with 10cm-diameter annular rings. These delicately mounted rings, housed in a temperature-controlled oven, led to clocks that were more accurate than could be assessed by reference to a pendulum. At the PTR in Germany, experimentation centred on a different cut of crystal, in a bar cut in the direction of an electric axis, and which could be supported very differently, producing fine results.

After Dye's death, his colleague Louis Essen (1908–97) took over his experimental programme, incorporating insights from the PTR's approach to the form of the electrodes. Essen constructed smaller quartz rings of six centimetres diameter, with the electrodes of cylindrical form inside and outside the cylindrical faces of the quartz. A notable outcome was that he increased the frequency from the 20kHz of the earlier form of ring to 100kHz, with advantages for use in measurements in the higher part of the radio frequency range then coming into use.

From 1928, General Radio in the United States offered a quartz frequency standard, the C-21-H model, though rather than promoting any timekeeping function the emphasis was on its use as a laboratory standard for frequency measurement.[32] The first production-run high-precision quartz clock built primarily as a timekeeper and not as a frequency standard was the CFQ model produced from 1938 by Rohde & Schwarz in Germany. From October 1939 onwards, two of these clocks were used by the Deutsche Seewarte in Hamburg, responsible for establishing German standard time. The two quartz clocks were used in conjunction with four precision pendulum clocks, and in particular the quartz clocks were utilized in a new system for distributing time signals. Around 1942 Rohde & Schwarz produced the first portable quartz clock, though battery size remained an issue. Essen later recalled that, early on, issues over the size of power supplies made the commercialization of early quartz clocks problematic.[33]

With the increase in accuracy of timekeepers, ushered in by the Shortt free pendulum clock and made orders of magnitude better by the development of quartz crystal timekeepers, irregularities in the rate of rotation of the Earth meant that our planet eventually lost its place at the top of the hierarchy of time standards, at least for the scientists. After the Second World War, the field of microwave spectroscopy, which had advanced at great speed owing to the increased pace of wartime development, offered a path for both increased precision and accuracy in timekeeping, leading to the replacement of observatories by physics laboratories as the world's primary timekeeping institutions. For the first time since Antiquity, physicists took over the control of timekeeping from astronomers.

---

29 Horton and Marrison 1928.
30 Scheibe and Adelsberger 1932; Scheibe and Adelsberger 1936.
31 Marrison 1948c, 466.

32 This section relies heavily on Graf 2015.
33 Essen 1977, 3.

Electricity and time standards in physics laboratories

The fundamental shift in timekeeping that took place during the twentieth century was the move from considering the rotating Earth as our time standard, to using the properties of matter itself to define time, by measuring the frequency of atomic transitions.

The basic requirement for an atomic frequency standard (AFS) is, in the words of a leading figure in the industry, the 'utilization of an atomic resonance that is, in principle, an inherent and invariant natural property. In practice, to implement an AFS, one must devise an apparatus that allows one to observe an atomic resonance in a detectable manner and with minimal external disturbance'.[34] Since the 1940s, scientists and experimenters have worked to develop this basic model of a clock that exploits fundamental atomic properties. The subtleties of their solutions are remarkable, and the assessed error of each new frequency standard has reduced by at least an order of magnitude with each development.

The first atomic clock, built in 1949 by Harold Lyons (1913–98) at the US National Bureau of Standards (NBS), was no more accurate than the best contemporary quartz clocks, but was ground-breaking for its use of a new concept in how time might be derived (Figure 183). As an NBS publication later explained, it relied on the discovery that 'ammonia molecules strongly absorb microwave radiation when this radiation precisely matches the molecules' natural resonance frequency of 23,870 million hertz'.[35] The microwaves were generated using a quartz clock and a frequency multiplier, with corrective tuning, to ensure maximum absorption. This meant that, in the short-term, accuracy could be no better than the underlying quartz clock, but long-term accuracy was determined by the ammonia absorption line.[36]

Louis Essen at the NPL travelled to see the NBS clock and was greatly influenced by it, immediately urging his employers to prioritize research and development in the 'atomic clock' field, in which it had been understood for some time that the caesium atom might offer significantly better prospects than ammonia, principally derived from the higher frequency involved. Support at the NPL was several years in coming, but in 1955 Essen and his colleagues succeeded in producing a clock in which the underlying time base was the transition between two hyperfine states of a particular caesium isotope.[37]

Atoms of vaporized caesium that are circulating in an oven emerge if their trajectory takes them through a narrow slit. The electrons of an atom spin in a magnetic field. Using an external magnetic field, it is possible to reverse the direction of spin of the electron. The process is not gradual, but binary: an electron spins either in one direction or another.

When the caesium atoms emerge from the oven, they enter a tubular vacuum chamber and pass through a magnetic field, then through a cavity, and then through another magnetic field. The first magnetic field has the effect of deflecting the atoms in two opposite directions, dependent on the direction of rotation of their electrons when they enter the field. One of the two sets of atoms will continue through a further fine slit in the cavity in the centre of the vacuum chamber, and from there pass through the second (identical) magnetic field, where once again, those electrons that have changed polarity will be deflected away. Atoms that successfully pass through the centre line then reach a detector, formed of a hot platinum or tungsten wire, which gives up an electric charge before emitting the atoms again, which are now drawn to an electrode, allowing for the measurement of a resulting electric current.

The cavity itself plays a vital role since it forms part of a resonator device, excited by radio frequencies that are generated through the multiplication of the frequency of an underlying quartz clock. The fine-tuning of the frequency in the cavity resonator allows for the finding of a frequency at which the optimum number of caesium atoms hit the detector at the end of the tube. At this stage, the quartz oscillator acts as a time standard, its accuracy derived from the stability of the caesium atoms.

Considering the improvements in accuracy of various techniques of time measurement in a 1977 lecture, Louis Essen usefully provided a summary:

> The precision of an astronomical measurement of a group of star transits used for measuring the day, is 0.015 seconds. Early forms of the pendulum kept uniform time to about 0.15 seconds per day but this was improved in later forms to 0.015 seconds. Quartz clocks enabled astronomical measurements to be averaged over long intervals of up to year and improve time measurement to 0.001 seconds per day or expressed as a fraction 1 part in 10^8.[38]

While Essen's assessment of pendulum accuracies only applied in the short term, with the caesium clock developed at the NPL in 1955, Essen and his colleagues now had an accuracy of 1 part in 10^{12}.[39]

This and successive clocks, including those in the USA, were large, complicated, and expensive. From early on there was realization of a potential demand and rationale for a practical and cheaper alternative—a clock that would offer a performance well in excess of that of a standard quartz timekeeper, adequate for a wide range of metrological applications, but not offering the same degree of precision afforded by the caesium clock.

---

34 Riley 2019, 1.
35 National Bureau of Standards 1978, 2.
36 National Bureau of Standards 1950, 105.
37 Jones 2000.

38 Essen 1977, 1.
39 In other words, its daily accuracy could be expressed by dividing the 86,400 seconds in a day by 1,000,000,000,000, meaning the clock was accurate to within 0.000000086 seconds (or 86 nanoseconds) per day.

It appears to have been understood as early as the late 1950s that the newly emerging caesium frequency standards were impractical for anything other than the most rigorous scientific experimentation. A number of scientists working in various laboratories in the USA contributed to the development of an alternative, the rubidium standard, which was accurate to a sufficiently high level for a wide range of applications, but which could be produced more easily and at a significantly reduced cost compared with that of a caesium clock.[40]

Rubidium clocks operate in a broadly similar fashion to caesium clocks. A beam of light is emitted by a rubidium discharge lamp and passes through a cavity. Microwaves are used to interact with the beam of light, which eventually reaches a detector at the end of its path. The frequency of the microwaves is determined by a quartz clock and is tuned by feedback. Like caesium, rubidium exhibits two hyperfine states, and will move between the two states when the microwave power is at the transition frequency. At that frequency, the detector will see a detectable change in the light intensity. As long as the detector sees the trace of the transition between hyperfine states of the atoms, the underlying time base is being held to a stable and highly accurate frequency.

Rubidium clocks can be manufactured for a fraction of the cost of caesium clocks and have therefore become widely used as time standards in a variety of infrastructures, from navigation satellites to telecommunications networks. But research has not slowed in the development of atomic clocks at the top end of the field. In the twenty-first century, the long-time model of the atomic frequency standard began a move from microwaves to visible light, an increase in frequency that has the potential for greatly increased precision and accuracy of time measurement. Absorption spectroscopy continues to be the technique by which the frequency of atomic transitions is measured, but new elements such as **strontium** and **ytterbium** are replacing caesium and rubidium in the most state-of-the-art atomic clocks. Most recently, the optical lattice clock has moved to centre stage as a focus for developmental work, though no clocks of this design are yet contributing to the setting of International Atomic Time. Beyond optical clocks, work is already taking place on the potential use of nuclear transitions as the underlying constant to be measured.

The emergence of clocks that offer accuracy to the order of 1 part in 10^{18} offers new potential for scientists in that such clocks are susceptible to changes in the gravitational potential of their specific location. Identical optical lattice clocks in two different locations cannot theoretically present the same passage of time, owing to the different gravitational fields in which they are placed. Loosely analogous to gravimetric pendulums used for long periods in surveying (for example, for oilfields), such clocks can now be used to measure aspects of the very environment in which they sit.

NETWORKS OF PRECISION TIME

The growth of network infrastructures

The rapid increases in industrialization and urbanization seen in the nineteenth and twentieth centuries brought with them the growth of network infrastructures—the underlying systems and services that support an economy such as water supplies, sewerage, gas, electricity, transport systems, and communications. Also on the increase over this period were the physical and operational networks of working life—factories, offices, educational establishments, government, bureaucracy, legislation, regulation, and so on. The extractive industries reshaped environments on an ever-larger scale. The growth of engineering—civil, mechanical, structural, electrical, chemical, etc.—made the world more connected and ever more complex. In societies around the world the built environment changed radically.

This networked landscape of infrastructures, once exceptional but increasingly commonplace and overlooked, is the context in which time standardization and distribution emerged. It provided both the motivation and the means to spread and use accurate time across the modern world. Standard time provided (and still provides) the operating system to coordinate and synchronize the infrastructures of modern life, and therefore the study of networks of precision time in the nineteenth and twentieth centuries informs a wider study of modern societies. This part of the chapter looks at the technologies of networked time, primarily that of electricity, but starting with an analogous infrastructure—compressed air.

Pneumatic clock networks

Despite the main focus of this chapter being on electrical horology, it is worth first covering an alternative technology that shared important characteristics with some of the electric time distribution systems: the pneumatic technologies that advanced during the eighteenth century.[41] The French inventor Jacques Vaucanson (1709–82) was a noted pioneer who was responsible for a wide range of remarkable automata and developer of the metal-cutting slide-rest lathe. He also made a major contribution to the development of the Jacquard loom, which used punched-card technology to enable one weaver to produce intricate designs without the assistance of 'draw boys'. For his pneumatically driven automata he devised the process for making India-rubber tubing.

Following such work, by the nineteenth century, compressed air had become a vital infrastructure underpinning industrial development. It powered industrial machines. It operated communications networks, sending messages across cities, or cash and receipts around department stores. It provided motive power for

---

40 See Riley 2019 and its extensive bibliography.

41 This account draws heavily from Read 2005; see also Dikeç and Lopez Galviz 2016.

Figure 184 Gilbert, 'The Pneumatic Clock on the Place de la Madeleine'. Image from *Nature* (8 July 1880).

hoists and lifts, pipe organs, air brakes, self-closing doors, and so on. The pneumatic drill emerged in 1861. Compressed air was sufficiently commonly used in many major cities that 'mains' pipelines were laid under streets, alongside gas pipes, to provide the means to drive industrial machinery. Pneumatic telegraph systems, providing trunk communications between major postal and telegraph offices, were developed in various European cities from the 1850s.

Against this backdrop, Carl Albert Mayrhofer (1835–1901) introduced a pneumatic time-distribution system in Vienna in 1877. At least one report suggests he moved to pneumatics when he met resistance to his previous design of an electrical time distribution system.[42] He assigned his pneumatic patent for the new system to his fellow Austrian, Viktor Antoine Popp (1846–1912), and his partner, one of the owners of Gebrüder Resch, from Ebensee (the largest Austrian clock manufacturer). Looking to expand into France, these two formed the Compagnie des Horloges Pneumatique and successfully demonstrated their system to the Parisian authorities in 1879, leading to its widespread introduction across Paris from 1880 onwards (Figure 184).

As in comparable electrical systems (which are considered presently), a central control station regulated the transmission of the pneumatic time signal out to a large network of subsidiary clocks. A compressor system charged a series of large air vessels. A clock mechanism, synchronized by the Paris Observatory, allowed the flow of air to the circuit of clocks through the opening of a valve for twenty seconds each minute. The pressure wave flowed out through a system of main trunk lines, off which smaller-bore subsidiary pipelines branched. The system was vented to air for the balancing forty seconds of each minute.

Subsidiary clocks generally comprised a simple bellows mechanism, expanded upwards by the incoming pressure wave, thereby lifting a gravity arm, on which an indexing pawl was pivoted, engaging with a sixty-tooth wheel. With each pulse, a tooth would be gathered, and as the system vented to air, the gravity arm returned to rest, advancing the minute hand one division.

Contemporaneous with the work of Mayrhofer and Popp, Hermann Wenzel (1830–84), a German immigrant to San Francisco, obtained a US patent (number 140,666) in July 1872 for a pneumatic clock system.[43] His colleague, another German immigrant, August(us) Hahl (1843–1923) later acquired the business, developing an improved system that became relatively widespread in the USA, and which seems to have been favoured

---

42 Bush 1880, 62.

43 Wenzel 1883; Stephens 1948.

in a variety of municipal buildings, notably in courthouses.⁴⁴ The Wenzel/Hahl system comprised a conventional mechanical clock that released a separate **train** each minute, powered by massive weights, using the release of their stored energy to operate air pumps that used a network of pipes to advance secondary clocks (analogous to the Magneta system of Martin Fischer of Zurich (1867–1947), which generated its own electricity using energy stored through the lifting of heavy weights, described later). The self-contained nature of the design, which did not rely on a separate compressor station, offered advantages, but the low-pressure design was compromised by the slightest leak. In common, therefore, with electrical clock systems, an important potential weakness was the fragility of the network.

Electric clock networks

The 1870s and 1880s saw the first attempts in large cities to provide networks of standardized and accurate time. In London and Berlin, for example, electric systems were installed that employed the increasingly common model of a central controlling clock and remote secondary clocks, which received their time from the central clock.

The rhetoric of the terms used to describe the component parts in such systems varied greatly over time and from place to place. While 'master' and 'slave' are in common modern currency, this was not always so. In German, for example, there was the use of 'mutteruhr' and 'tochteruhr' (mother clock and daughter clock). One English manufacturer, Gent & Co, insisted over the long-term in using 'transmitter' and 'impulse clock'. Charles Shepherd, whose system was installed at the Greenwich Observatory in 1852, used the terminology 'motor' clock and 'sympathetic' clocks. The terms 'master' and 'slave' are obviously problematic.⁴⁵

It is also worth reflecting on what is meant by 'electric clock' in this context. While the history of electrical horology offers numerous examples where mechanical escapements were abandoned altogether, it is important to note that many of the most successful designs retained important elements that were mechanical. Most English controlling clocks throughout the twentieth century were **'gravity escapement'** clocks relying on stored (mechanical) energy being given up, but with the resetting action and transmission of time electrical.

Finally, given the diversity of designs of electric clocks worldwide in this period, it is worth considering the fundamental advantages in deploying electricity as a means both to keep the mechanism in motion, but also to distribute information, sometimes at a distance. Warren Marrison explained:

> An astronomical clock, in addition to having as nearly constant a rate as can be attained, should also be able to operate over long periods of time without change or interruption. The reason for this is that many of the phenomena that are of interest in time measurement occur in continuous succession and the greatest amount of information can be obtained only by the use of clocks with which measurements can be made in unbroken sequence.⁴⁶

Horologists aiming for precision and improved performance typically aim at reducing sources of disturbance or interference, and therefore aim at solid mounts (for example, attaching clocks to bedrock in a deep cellar) and stability of temperature and pressure—generally seeking to remove potential sources of error. Periodic winding of a clock is an obvious source of disturbance—indeed, opening the case door of a precision pendulum clock has an immediate and detectable effect on rate. In the cases of electrically rewound (or electrically impulsed) clocks, it is a significant potential enhancement for performance that the clock case, or even an air-tight tank containing the clock, can be kept closed or sealed.

Furthermore, the time delivered from the clock, in sharply defined electrical pulses, each minute or half-minute, or perhaps even each second, can be derived from electrical contacts, and can be used in circuits that travel appreciable distances. In an observatory or laboratory, a precise time base can be used for experimentation or observation with human or personal error eliminated. Indeed, taking the example of observatory clocks in the late nineteenth or early twentieth century that were in every other way purely mechanical, the addition of electrical contacts that could provide precisely defined seconds impulses, at a distance and yet instantly, for a variety of purposes, significantly enhanced the utility of the clocks.

There are wider advantages, too. Mechanical clocks impose a regular and frequent maintenance burden. They need to be rewound and, in most cases, also set to time. This might be daily or weekly, or less frequently with some high-quality clocks, and of course each individual clock requires the same regular and direct personal attention. An installation of ten clocks requires ten times the maintenance of a single clock.

Domestic clocks are one subset of this problem, where the clock owner will most commonly also have undertaken such duties, but of more significance to the economy was the maintenance burden applying to the case of public clocks. These clocks were usually difficult and sometimes dangerous of access. In the case of early church clocks, the archival records of payments to a clock winder have, on occasion, been important items of evidence in the history of a church clock, but they bear witness also to the daily or weekly ritual of the climbing of tall buildings, possibly with very poor lighting, and without the benefit of the safety structures and measures that we now take for granted.

---

44 Gensheimer 2005, 57–9.
45 On the problematic conceptual terminology of masters and slaves in horology and engineering, see Eglash 2007.

46 Marrison 1948d, 588.

Of still more significance was the regular maintenance burden applying to the winding and setting to time of mechanical workplace clocks in factories and offices, which could number in the hundreds or thousands for large companies or institutions. Workplace clocks were often positioned relatively high on walls, necessitating the use of ladders or complicated rear access requirements. This work was unlikely to have been carried out by employees of the firms but would instead have been contracted out to specialist clockmakers, unless the organization concerned were large enough to warrant a maintenance department. Either way, whether with specialized in-house staff or those contracted to operate winding rounds, there were cost and administrative burdens that companies would have been keen to reduce.

And it was not just fixed workplaces. In an age of the sea, large ocean-going vessels, especially those travelling east to west and requiring regular time changes, imposed their own set of burdens where tens, if not hundreds, of clocks were involved. Ship officers would not have wished to see ratings engaged in attending to clocks on board but would have had little choice, so the ship's complement would need an additional hand.

In this technical, economic, and organizational context, inventors and promoters of electric time systems offered a compelling alternative solution. The advantage they offered largely lay in system design, employing a single central source of time (the controlling clock) and any number of subsidiary clocks that simply showed the time dictated to them. The single controlling clock could be set up in a location convenient for maintenance and was under the control of a single person, reducing the risk of damage or tampering.

As electric time systems and their power sources (such as batteries) developed, supervision and maintenance were gradually reduced to the point that, from the 1920s onwards, engineers' visits became rarer—simply to perform a spring or autumn hour change, or to service the controlling clock and set it to time. Surviving service records frequently confirm that systems could maintain a rate sufficiently good that long service intervals were acceptable. Generally, the cost of such visits formed part of an annual charge made for the provision of an entire system.

Electricity provided particular advantages in respect of attendance time-recorders. In smaller establishments, where a single time recorder could be positioned conveniently near the entrance for the use of all employees, a key-wound mechanical 'time clock' sufficed. As factories increased in size and covered a number of floors and perhaps more than one building, it was in the interests of both management and employees to have a number of time recorders spread around the site, close to people's workplaces. The only way such machines could be made to 'click over' on each minute simultaneously (thereby avoiding disputes) was by central electrical control. Fully electrical time recorders were developed, often with automatic stamping and, in one system (of the International Time Recording Co) hourly synchronization with the controlling clock for both time recorders and wall clocks was provided. A particularly useful adjunct to an electric clock system was a device for controlling sound signals (such as bells and hooters) to define start and finish work times, lesson changes in schools, and so on. Such devices, known by various names usually including the word 'program(me)' or 'signal', ensured that sound signals were synchronized with clocks and time recorders.

Overall, then, the introduction of large distributed-time systems greatly reduced the need for personal involvement in the provision of time across large numbers of individual locations, with a variety of attendant benefits. Staff costs, however apportioned or derived, were decreased and overall safety increased, even if this were less of a concern than it would later become. This economic advantage also brought new financing models. For the centuries over which clocks and watches offered unique and solitary presentations of time, unconnected to outside inputs other than their owners who wound and set them to time, it was understood that such objects were purchased by their owners outright, or perhaps on rare occasions purchased on an instalment basis. Ownership was clear. But the emergence of electrical timekeeping systems heralded a new phenomenon. The commodification of time as a utility to be provided to commercial premises or institutions led to the rise of the rental of complete clock systems.

Right from the beginning, the accounts of Synchronome in London from the late 1890s reveal a reliance on rental over sales income.[47] Much later, firms such as Dictograph or Telephone Rentals in Britain supplied large numbers of extensive time-distribution systems for factories, schools, hospitals, and other institutions on ten- or fifteen-year contracts (which were on occasion rolled over for a second term). In place of the old, regular, weekly winding round, firms such as these made only occasional visits to their clients' sites, typically for daylight-saving clock changeovers, for dealing with systems reported 'fast' or 'slow', for battery servicing, or for routine maintenance of time recorders, such as ribbon changes.[48]

A more crucial change, though, was the fact that with electric clock systems running from a single controlling clock, the presentation of time across an entire site could be both uniform and correct. Even if errors crept in, the uniformity provided by a single time base for an entire site would preserve the rights of workers whose hours were monitored (by clocking in and out) against a uniform measure.

One extraordinary example from the textile industry in Britain may be representative of significant numbers of factory systems across many countries, the records of which have been lost. At the John Smedley textile mill in Matlock, Derbyshire, a Gent electrical time system was installed in the early twentieth century, possibly before the First World War. This operated a variety of subsidiary dials around the mill premises, both inside buildings and as waterproof external dials. The system could operate

---

47 Synchronome 1899.
48 Bird 2012, 19.

start and cease signals, but more than that, the cable network was extended to run through all the adjoining workers' cottages, with subsidiary dials above the fireplace on each parlour mantelpiece.[49] Thus, the entire workplace community shared a common time, leading to greater efficiency, but also a sense of security—wages dependent on a time system were protected by a shared standard.

An interesting consequence of the widespread adoption of electric clock networks throughout industry, commerce, transport, broadcasting, and other fields was that the manufacture, installation, and servicing of such networks increasingly became the province of electrical engineers, telephone engineers, and electricians—these networks were not the world of the clockmaker, but of a very different sort of technician.

The early history of electric clock networks

The first clocks to employ electricity in some way were developed from about 1840 onwards. From the start, the field developed down a number of different developmental paths. Among the many goals addressed by inventors on several continents, two stand out. The first was the maintenance of an oscillator in motion—essentially keeping a clock going. In what way could electricity reduce or replace the reliance on springs or weights? The second goal was the transmission of time. How could electricity be used to distribute the time across a series of linked clocks, especially over considerable distances?

An influential author in this field was Frank Hope-Jones, who set up the Synchronome company and who wrote prolifically on everything relating to electrical horology. Like other electric clock pioneers, Hope-Jones felt the need to engage in classification at an early stage of his surveys,[50] and divided the field into four main categories:

1. Synchronizing systems—any sort of system in which a clock, driven by any means, might be corrected regularly by reference to a central source of accurate time.
2. Independent self-contained clocks—clocks in which springs, weights, or any other mechanical store of energy is replaced by electricity.
3. Circuits of many subsidiary dials—in which a central master clock transmits regular minute or half-minute signals to secondary dials, which are themselves inoperative without those signals.
4. Synchronous motor clocks—clocks that are essentially simply motors plugged into the fixed frequency mains supply and which rely for their timekeeping on that frequency.

Such classifications are, of course, prone to flaws. For example, category 2 is insufficiently broad, since there are large numbers of designs in which weight- or spring-driven clock mechanisms are simply rewound using electricity, rather than electricity replacing the source of motive power. But putting such flaws aside, Hope-Jones's taxonomy divides early electrical horology into sensible categories.

In the case of domestic clocks, only the goal of making them independent of traditional winding and sources of energy was achieved in the first century or so of electrical horology. The unified presentation of time was not achieved until the emergence of synchronized electricity transmission grids, operating on a fixed frequency, which made domestic clocks accurate through the technology of synchronous motors, rotating at a speed dictated by the grid frequency. Around the globe, such grids emerged piecemeal over the twentieth century, and therefore the take up of synchronous motor clocks proceeded at varying rates. The USA led the field, largely as a result of the work of Henry Warren (1872–1957), who is widely regarded as the pioneer of synchronous electrical timekeeping.[51] We return to the history of synchronous motor clocks later.

In the case of non-domestic clocks, meaning clocks for any sort of institution, inventors were looking both to remove springs and weights and to synchronize time between many clocks. Similar ideas emerged in several locations over the next decade, and broadly speaking three principal geographical groups of electric clock systems emerged over time: those developed in Britain, Continental Europe, and North America. Electric clocks were doubtless to be found elsewhere around the globe, even in the nineteenth century, but research to date on other geographic regions is lacking. In several cases, the systems in use were imported from the most likely sources (for example, from Britain in the case of Australia). The Indian sub-continent, Africa, and Latin America are major locations where electric clock systems emerged, but so far it appears that the technology was imported from obvious colonial origins.

A further generalization is that, while balance clocks certainly emerged, pendulum clocks were by far the most common form for electric clocks, prior to synchronous motor clocks emerging in volume in the first quarter of the twentieth century. In view of their general stability, particularly where they could be securely fixed to masonry, pendulums were an obvious choice for clock systems that had both accuracy and stability as a goal. Balances tended notably to occur in two scenarios. The first was for aesthetic reasons, usually where a large and decorative balance could be employed. The second scenario was the specific case of marine clock systems for use on ships at sea, where practical considerations meant balances provided the obvious solution.

While the first practical electric clocks were not developed until the 1840s, it is clear that there was experimentation in electrical horology from early in the nineteenth century. As

---

49 Reported in Nye and Bird 2001.
50 Described, for instance, in Hope-Jones 1949, 6.

51 See Warren's February 1937 lecture, reproduced in Aked 1997.

in all fields, scholars have striven to identify where priority lies in the invention of early electrical clocks, and there were important contributions involving inventors such as Carl August von Steinheil (1801–70) and Giuseppe Zamboni (1776–1846).[52] Steinheil's 1839 patent is very likely the first for an electric clock.[53] However, these actors had limited lasting influence on the development of the field. In the 1840s and later, the field advanced significantly through the efforts of pioneers such as Alexander Bain and Charles Shepherd in Britain, and, in France, Jean-Paul Garnier (1801–69), Paul-Gustave Froment (1815–65), Louis Clément François Breguet (1804–83, grandson of Abraham-Louis Breguet), and Constantin-Louis Detouche (1810–89) working with Jean-Eugène Robert-Houdin (1805–71). The latter two exhibited together with success at the 1855 Exposition Universelle.

A stumbling block for early inventors was the unreliability and cost of rather unwieldy early cells or batteries, but in 1866 George Leclanché (1839–82) invented the Leclanché cell, which was rapidly adopted in telegraphy and in electric clocks. It was a significant improvement on previous cells, in weight, lower cost of raw materials, and ease of use.

Alexander Bain (1810–77)

Alexander Bain, a prolific Scottish inventor, is often regarded as the principal originator of electrical horology, following his 1841 British patent numbered 8783. This covered wide ground, but importantly described the means of maintaining a pendulum in motion using electricity as well as the transmission of time from a central ('parent') clock to subsidiary ('companion') clocks. Bain organized the production of electric clocks that drew their power from Earth batteries (zinc and carbon combinations, buried in the ground, delivering approximately 1.1 volts), and these were ground-breaking devices, in more senses than one.

Bain effectively abandoned most traditional horology with these clocks, beyond the use of a pendulum as the oscillator. There was no conventional train of wheels through which energy escaped from a coiled spring or falling weight via an escapement. Instead, the motion of the pendulum was maintained by electromagnetism, and employed the familiar principle of the attraction of opposite magnetic poles. An electromagnet, if energized through the closing of a switch, has a north and a south pole, comparable to the poles of a permanent magnet. With the electromagnet forming part of the pendulum (usually the bob) and arranged to swing close to a fixed permanent magnet, a simple switching arrangement allowed for the pendulum to be impulsed, with the bob being attracted by the fixed permanent magnet. The pendulum effectively became part of a motor, and some of the energy from the swinging pendulum was robbed to advance the motion-work of the dial using a simple advancing pawl arrangement.

Such a design is an example of an entire ancestral line in electrical horology, but importantly, many other designers chose a quite different path, retaining much of the train of a traditional clock, and simply replacing the spring or weight with another source of motive power. The reasoning for the emergence of these two broad pathways is explored from various angles in this section.

Charles Shepherd (1830–1905)

Charles Shepherd was responsible for key innovations in electrical horology, including laying the foundations for standard British practice in some of the fundamentals of controlling clock design.[54] Probably based on his roots in chronometer making (shared with his father) he understood a pitfall of Bain's approach, in which varying battery voltage would lead to changing pendulum amplitude, and thereby inferior timekeeping. Shepherd's early controlling-clock design employed a spring, latched by a detent under the action of electromagnets, switched as the pendulum swung in one direction. At the extremity of its return swing, the pendulum unlatched the detent, and the spring delivered an impulse to the pendulum—this had the advantage of delivering an unvarying impulse, even though the impulse was delivered over a large part of the pendulum's travel, and in one direction only.

While Edmund Beckett (later Lord Grimthorpe, 1816–1905) criticized Shepherd's system, others had very different (and positive) opinions of Shepherd's work.[55] The installation that George Airy, the Astronomer Royal, ordered for the Royal Observatory at Greenwich comprised a controlling clock driving one large dial (the famous 'gate clock') and three smaller dials. Importantly, for the controlling clock, Shepherd moved to using a pivoted gravity arm, again latched electromagnetically. With this development, he anticipated an important element in many later British designs. Becoming the Observatory's mean solar time standard between 1852 and 1893, Shepherd's clock operated as the heart of distributed Greenwich Mean Time for Britain. Given Beckett's involvement in the design of Big Ben, it is therefore ironic that its performance should be measured for many years by reference to time distributed using Shepherd's clock as its source.

Jean-Paul Garnier (1801–1869)

An early example of an alternative route emerged in France in 1847 with the work of Jean-Paul Garnier, who introduced a system for railway station clocks. Like Bain, with a strong interest in telegraphy, Garnier distributed time around a cable network from

---

52 See, for instance, Aked and Rizzardi 1975.
53 Aked 1976.

54 This section relies heavily on Shenton 1994 and Vaughan 1994.
55 Denison 1854.

a central source to remote dials located in station buildings and on platforms. However, his system employed traditional key-wound clocks in which a standard wheel train was arranged to close sets of contacts in order to transmit the time, usually sending a pulse of current each minute.

It was common to almost all distributed time systems that the pulses to advance the network were sent each minute (for example, the standard for much of Continental Europe), or half-minute (the standard for Britain, from the twentieth century onward). Initially, there was no perceived requirement for the presentation of time to be resolved to any greater precision for such public clocks (though, of course, the moment of advance of the minute hand was in itself a precise indicator).

Wilhelm Förster (1832–1921)

In Germany, an important actor was the Berlin astronomer Wilhelm Förster, who believed strongly that a punctual society was an ordered society.[56] He was in part influenced by trips to Britain during which he witnessed the system of sympathetic pendulums of Robert Lewis Jones (discussed later). Recognizing increased demand for accurate time, whether from clockmakers or commerce generally, he saw it as an obvious civic duty that the Berlin Observatory should provide a free public time service. But beyond this, he was unusual in having an ambition that public clocks should present time not to the nearest half or whole minute, but to the nearest second.

On 20 July 1869, a subsidiary clock made by Bernhard Theodor Friedrich Tiede (1833–1907) was erected in front of Berlin's Chamber of Justice building in Lindenstrasse, joined in succeeding years by a further five public clocks. These were sympathetically driven by electric impulses derived from the precision regulator in Förster's nearby observatory.

It proved to be too ambitious to present time to the second, as the physical and operational obstacles were overwhelming, but Förster's broad goal to distribute accurate civic time around Berlin continued to be carried forwards, though over time the infrastructure shifted from being organized by the public sector to a private-sector approach with the creation of the Central-Uhren-Gesellschaft in 1889, using technology developed by Mayrhofer. This operation gave way in 1892 to a new company, the Urania-Uhren- und Säulen-Kommandit-Gesellschaft, backed publicly by Förster. Around Berlin they erected a series of columns ('Urania towers'), each containing an independent clock but also connected via telegraph lines to the observatory, to allow for time correction. The Urania towers presented local Berlin time, GMT, the weather, the phase of the Moon, maps, and railway timetables, and were lit by gas lamps controlled by astronomical time-switches (discussed later). Eventually, by the early years of the twentieth century, nearly fifty towers had been built.

But commercial success eluded the Urania-Uhren company, despite the sale of advertising space on the lower panels of the towers, and from 1895 the distribution of Berlin time was taken up by a new firm, Normal-Zeit GmbH (Figure 185). In a bravura display of technological innovation, many Normal-Zeit subsidiary clocks were actually independent clocks but corrected regularly by an incoming signal—in addition to which they also reported their status back to the central station, and this could be read from a ticker-tape. Thus, the control centre was supposedly alerted early to a fault either on a circuit or with an individual clock. The growth of the system was remarkable—from some 500 clocks installed in 1895 to 2,000 two years later. By 1905 some 5,000–6,000 people were connected to the system, and the business had expanded as far afield as Bremen, Antwerp, Copenhagen, Helsingfors, and Stockholm—each city with several thousand clocks installed. While highly innovative and brilliant in conception, the system was also noted for its clocks failing regularly as the physical network demands were too great. This was a problem that became common across ambitious networked time systems in cities.

John Alexander Lund (1835–1901)

In London, a similar concept had been employed since the early 1870s following the designs of John Alexander Lund.[57] Here, however, there was a realization that a vast number of conventional key-wound mechanical clocks already existed, populating countless rooms in the offices of commercial establishments, and sitting high on the walls of countless factories across the City of London and beyond.

Lund's business model involved exploiting this significant existing investment in a widespread timekeeping infrastructure. As described, Lund's potential clients were used to a working environment in which their premises were visited weekly by representatives of the clockmakers on weekly winding and setting rounds. Lund's proposition was that his firm, for a regular fee, would take over the responsibility for the winding of such clocks, but crucially, that the setting to time would be accomplished remotely by means of a regular hourly pulse sent over a telegraph line.

The pulse (probably of one-second duration) was used to energize electromagnets which in turn activated a pair of pivoted pins, projecting through a slot cut into the dial at the twelve o'clock position. The pins would come together either side of the minute hand, having the effect of zeroing its position at the top of each hour. Exploiting the respectable short-term accuracy of typical clocks, this had the effect of removing the possibility for accumulated errors—thus, the clocks now kept accurate time. In the absence of seconds hands (as was typical for the clocks in question), the clocks were always right.

---

56 This section relies heavily on Graf 2013.

57 Nye and Rooney 2007.

Figure 185 Control room at Normal Zeit, Berlin, c.1900–10, from which time was distributed around the city. Individual street clocks reported back their status (private collection).

William Gardner (1842–98) and Chester Pond (1844–1912)

The Lund system was patented in the USA but did not come into common use. Instead, in the late 1860s, a system devised by William F. Gardner (US patent 307287) became popular, in which forcible correction of standard clocks using a heart-shaped cam occurred electromagnetically at twelve noon each day, controlled by the signal for the drop of time balls at the Naval Observatory, Washington DC, and at the Western Union building in New York.[58] However, this was insignificant by comparison with the system originated by the US patent of Chester Henry Pond (numbered 308,521) in 1884 for an electrically-rewound clock. This led to the formation of the Self-Winding Clock Company (SWCC) in 1886, which bought in large numbers of movements from Seth Thomas before starting its own manufacture in 1892.[59]

The clocks employed straightforward designs of trains and escapements that could be volume-produced, and which were highly dependable, with the advantage that they were automatically rewound by rotary motor. With the addition of a synchronizer using the Western Union network for the networked transmission of daily time signals, this system was widely adopted across the USA through much of the twentieth century. By 1958, more than 50,000 SWCC clocks were installed across the country.[60] The entire system maximized the benefit from low cost and simple manufacture. It was relatively reliable, it removed the need for monitoring, and it derived its time from a trustworthy external source.

The challenges of electric clock systems

While the underlying logic of electrically distributed time was clear and the technological solutions often highly ingenious, the practical realization of electric clock networks frequently left much to be desired. Cable networks were prone to failure, whether by accident or occasionally sabotage. Weather damage, road works, water ingress, the effects of the wind—these and more

---

58 Saff 2019, 96, 98.
59 Bartky 2000, 167–80.

60 Saff 2019, 99.

were all problems encountered by early city-wide schemes that relied on long runs of cables though the urban landscape.

If practical problems abounded, so did financial ones. The designers and promoters of early urban time-distribution systems were relying on the commodification of accurate time. From the 1840s, a desire for accurate and synchronized time had emerged as a safety issue for the railways. The world of finance came to place a value on standard time. Establishments selling alcohol subject to licensing laws and strict operating hours were also potential clients for accurate time. But for most of the population in the mid-nineteenth century (and for many decades beyond), there was no great desire or need for accurate and visible urban time. The techniques used over hundreds of years for knowing the time continued to suffice for much of the population, and therefore the range of potential paying customers willing to subscribe to any time service was limited.

Therefore, electric clock companies had to encourage a culture of public time-mindedness through publicity and public relations. The fact that public clocks generally had a poor reputation for showing the right time, or for working at all (at least in London) was exemplified in the correspondence in *The Times* in 1908, when the Standard Time Company, an offshoot of John Lund's firm set up to exploit his synchronizing system, sought to promote its own products as a counter to the presence of 'lying clocks'.[61]

Wired, city-wide schemes for time distribution would eventually become stable, reliable, and widely used, at least in some countries, and many remain in use in the early twenty-first century. But this stability tended to emerge only once the development of reliable smaller-scale systems had been advanced, and the systems ultimately employed were those of well-established firms, as opposed to standalone companies formed for the sole purpose of supplying a given city with time.

Other factors came into play, too. The standardization of national electricity grids allowed for the creation of synchronous motor clocks, each taking their timekeeping from the controlled frequency (and therefore time base) of the relevant grid. While this field is usually considered a story of domestic clocks, and will be discussed in that context presently, many large public clocks around the world from the 1930s onwards adopted mains synchronous motors as their source for accurate time, and it is still often the case that the old movements of large public clocks are replaced with synchronous motors by owners who do not wish to invest in the proper care and maintenance of their historic assets.

The further development of electric clock systems

While city-wide systems of electric time distribution were hampered by a range of recurring problems, systems designed for use in a single building or institution enjoyed significantly greater success, as the entire environment was under localized control. In particular, the network elements—the physical cable looms, and so on—were not subject to the same risks as in a complex urban environment, and it was for systems intended for indoor use that the most research and development took place.

Different experimenters approached the field with different ideas and goals. In Britain, Robert Lewis Jones (1806–92), a railway station manager from Chester, pioneered the use of sympathetic pendulums from the late 1850s. Originally using accurate key-wound clocks, he kept their pendulums in phase by using signals transmitted from one clock (switched in time with the beat of a pendulum) to energize a coil in the second clock that would electromagnetically hold its pendulum in sympathy. This idea was taken up by Frederick James Ritchie (1828–1906) in Edinburgh in the 1870s and formed the basis of installations in several observatories, and for a city-wide system in Edinburgh.[62]

Frank Hope-Jones argued that systems of this sort led others to adopt the principle of 'forcible correction', and he was dismissive of a trend in which electromagnets were enthusiastically adopted since they could 'pull this clock back and forward just as we want', ignoring, as he complained, horological laws.[63] That said, holding the pendulums of many clocks in lockstep with a single time source from which synchronizing pulses were generated was widespread and successful in many contexts. Indeed, Hope-Jones lauded the French systems on these lines devised by Charles Féry (1865–1935), Leroy & Cie, Marius Lavet (1894–1980), and Brillié Frères, the last of which he considered was 'performing duties of the highest responsibility in the Paris Observatory' and 'widely adopted on a commercial scale'.[64]

It became standard practice in important clock installations across Europe in the mid-twentieth century, using French, Swiss, and German systems, to use two clocks kept running in parallel and synchronized together, one being a backup in case of failure of the other.

Systems of electrical time distribution were more in evidence in Continental Europe in the latter part of the nineteenth century than in the United Kingdom, and once relatively standardized overall designs were established, generally by the early twentieth century, they tended to remain unchanged for several decades. Several noted Continental European designs abandoned standard mechanical trains, as described shortly, but it was common in Germany and Switzerland (though rare in France) to retain much of the **going train** of a conventional clock while introducing an electromechanical method for rewinding it, just as the Self-Winding Clock Company did in the United States—with great success. This might be as simple as the addition of a motor to lift a driving weight at periodic intervals, or an electromagnetic

---

61 Rooney 2006.

62 Hope-Jones 1949, 20–1.
63 Hope-Jones 1949, 21.
64 Hope-Jones 1949, 21.

rewind of some sort—maintaining almost entirely the form of a conventional key-wound clock.

Long-lived designs of this sort were produced in Germany by firms such as Siemens & Halske, Bohmeyer, AEG, and Wagner, and in Switzerland by Moser Baer and Landis & Gyr. Typically, such clocks employed a second train of wheels. This can be seen as an adaptation of the **striking** train of a conventional clock, where normally the going train would release the second train at periodic intervals to sound the hours. In these Continental European clock systems, the second train is usually released once each minute, often still with a 'warn' element a few seconds before time, but in this case the job of the second train is to close the contacts that will transmit an impulse to the secondary dials in the circuit.

At different times, different makers either adopted the approach of distributing impulses of a single polarity, or their designs reversed the polarity at each impulse. The choice of a design in which the polarity reverses stemmed in part from the difficulty in deriving the firm, reliable closing of contacts in the controlling clock in order to energize the impulse circuit. If the contacts closed slowly or imperfectly, they might 'chatter', resulting in multiple impulses being delivered. If the subsidiary clocks employed a reciprocating armature (or, later, a rotary action) in which the polarity needed reversing at each impulse for them to advance, this problem was overcome.

Over time, it became typical in Continental European clocks (though certainly not universal) for successive impulses to change polarity, such that for any given connection cable leaving the controlling clock and travelling around a circuit of connected subsidiary clocks, it will first carry a (say) positive signal, and then a negative signal at the next minute, and will then continue to alternate at successive minutes. While early Continental European subsidiary clocks typically used a reciprocating armature, a different design emerged, first patented by Heinrich Grau (1857–1922) of Kassel in 1880[65] and commercialized by the firm of Carl Theodor Wagner (1826–1907) of Wiesbaden.[66] This resembles a motor in having a magnetic rotor with a series of north and south poles, arranged to pivot inside a circumference formed in part by the curved pole faces of a pair of electromagnetic coils. As the polarity of incoming signals alternates, so the north and south poles of the coils change. The permanent magnetic poles of the pivoted rotor are repelled by their corresponding electromagnetic poles, since like poles repel, and this motion is converted through the motion-work to the advancing of the hands.[67] Such subsidiary clocks are generally referred to as 'rotary dials', or 'polarized dial movements'.

Rotary dials were not unique to Continental Europe. For example, George Bennett Bowell (1875–1942) adopted a similar system in the United Kingdom from around 1906, though he succeeded in designing rotary dials that could function with signals of the same polarity at each impulse.[68] Over time, rotary dials became more commonplace in Britain, most likely to achieve quietness of operation—they are generally substantially quieter than the alternative 'impulse dial' movements that characterized most British systems.

Other benefits, besides quietness, were claimed for reverse polarity systems, for example, that the continuous changing of polarity would tend to prevent certain mechanical problems that might arise. Put simply, over long-term use, a rotor would be less inclined to stick in the wrong position (perhaps as a result of dirt or aged lubrication) if the impulse changed each time. A reverse polarity system also prevented the magnetizing of armatures, though if good-quality soft iron was used, this could not occur. There were potential pitfalls with alternate polarity systems, which might require heavier duty cabling for parallel (rather than series) operation (at greater expense) but also if an impulse were missed by a secondary clock, owing to voltage drop over long distances on a large site, the succeeding impulse would be ignored, thus doubling any error.

Other ideas emerged that were designed to avoid such issues. For example, Siemens & Halske arranged each minute to send more than one pulse of the same polarity, within a few seconds, but to a reverse polarity subsidiary clock, working on the principle that if by chance the first impulse only partially advanced the rotor of the subsidiary dial, the next would ensure its positive action.

While different designs of device were used for reversing the polarity of impulses transmitted, one Continental design was relatively widely adopted by several manufacturers and is typically described as a 'polwender'. It is a switch in which a contact revolves, usually as part of the release of the second train, through 180 degrees, each minute. As the contact travels from its start position to its next position, it closes a circuit by brushing along a spring contact arranged to one side, and an impulse is generated, say of positive/negative polarity. However, when the train is next released, and the contact revolves through a second arc of 180 degrees back to its original position, it brushes against a second fixed spring contact. The wiring to the polwender is such that the impulse generated is now of reversed polarity—negative/positive.

While electrically rewound clocks retaining much traditional wheelwork were common in Continental Europe, an alternative design, by Matthäus Hipp, was widely used and highly successful. The principal clock design for which Hipp is best known has some common ground with that of Bain, in that both dispense with a conventional train. The pendulum is not impulsed through a conventional escapement giving up energy stored in a weight or spring via a train of wheels. Bain impulsed his pendulum using electromagnetism and used the resulting motion of the pendulum to advance the motion work. While Hipp also used the pendulum to drive forwards the hands, a key difference is that

---

65 Anon 1922.
66 Hope-Jones 1900.
67 Anon 1883.

68 See Remelec 1900.

the impulsing of the pendulum was infrequent, with many swings of the pendulum between each impulse.

Hipp's original design employed a weight-driven train and derived the infrequent impulses from this, but he moved to the delivery of the impulse by an electromagnetic coil, typically fixed below the pendulum, acting on a soft-iron armature attached to the bottom of the pendulum rod. It was the elegant method of switching power to the electromagnet that is perhaps Hipp's longest-lived legacy and is derived from the ingenious method used in his mechanical clock for unlocking the train and delivering an impulse.

In one design, a freely pivoted vane or trailer is attached to one end of a light spring, the other end being fixed rigidly to a support, to one side of the pendulum. The pivoted vane naturally hangs downward under gravity. The light spring holding the pivoted vane carries one of a pair of electrical contacts, the second positioned close by, with only a minimal air gap. A typically pyramidic-shaped piece of agate or steel, with a notched top and mounted on a suitable block, is fixed to the pendulum. The notched agate piece is arranged such that, as the pendulum swings, the pivoted vane lies exactly in its path. Between impulses, the vane freely passes over the notch each time the pendulum swings by.

As the arc of the pendulum reduces, there arrives a moment where the vane fails to clear the notch completely, and its tip remains held in the notch as the pendulum comes to the end of one excursion. As the pendulum reverses direction, the vane causes the flexible spring, on which it is mounted, to flex, and this small motion closes the air gap between the contacts, which therefore meet, and an electrical circuit is made. This energizes the electromagnets below the pendulum, imparting an impulse, as the soft-iron armature on the pendulum is attracted by the electromagnet below. This impulse restores the arc of the pendulum and the vane again passes over the notched block each swing. A common alternative design (for example, employed on the most widely used Hipp-style clocks in Britain) reversed the positions of the pivoted vane and the agate or steel block, with the vane on the pendulum and the notched steel or agate piece on a fixed block.

The significance of Hipp's arrangement is that an impulse is only taken when needed: it is a feedback arrangement. The usefulness of this can be appreciated if one considers its use in a turret clock, where different weather conditions (strong winds or accumulated snow, both likely to cause additional losses of energy through the train) may occur. But the design also ensures a reasonably constant arc of the pendulum, somewhat enhancing the timekeeping qualities, and the system proved remarkably effective in the realm of precision clocks.

Hipp collaborated with Adolphe Hirsch (1830–1901) at the Neuchâtel observatory from the 1860s onward in the measurement and distribution of standard time. A tank regulator clock using Hipp's design was evaluated over long periods by Hirsch, who recorded its performance. In 1891, Hirsch reported on a run of more than seven years of observational data. He concluded that the average diurnal variation of the clock was of the order of plus or minus three-hundredths of a second. However, he further concluded that, at this level of observation, errors in the methodology of measurement were probably contributing to the result, indicating the clock's capacity for accuracy was even better.[69] Probably the only clocks achieving a finer rate at this period (and indeed until the arrival of the Shortt clock in 1921) were those of Sigmund Riefler of Munich.

The development of electrical horology in the United Kingdom involved some important individuals, but on balance was relatively slow and only accelerated from the beginning of the twentieth century. Alexander Bain's important patent of 1841 described a wide variety of methods of applying electricity to clocks and will have served to discourage other entrants to the field. Various pioneers had operated earlier but had not made deep impressions. As well as Bain (and Charles Wheatstone to some degree alongside him), Charles Shepherd was the other important actor, and his machines were well-known through their use in the grand glass buildings of the Great Exhibition in 1851 and subsequently at the Greenwich Observatory. But their apparent unreliability was the spur to comments such as a now-infamous remark by lawyer and architect Edmund Beckett, who claimed that:

> These clocks never answered in any practical sense; nor would anything but the strongest evidence, independent of the inventor, convince me that any independent pendulum directly maintained by electricity can succeed in keeping good time for any considerable period.[70]

Given the slow development of electrical horology in Britain, Continental European systems were commonly imported into the country in the late nineteenth century. One example was the Van der Plancke system installed at Mostyn House, a school located on The Wirral, in northwest England. Frank Hope-Jones and George Bennett Bowell examined the system in about 1894 and convinced themselves they could devise something better. The result was the 'Synchronome switch' (the credit for which Hope-Jones admitted lay with Bowell), which, either itself or in variants by others, lay at the heart of a range of important British systems for the next three-quarters of a century.

An Achilles' heel for many early electric clock designs lay in their electrical contacts and in the quality and performance of the cells used to drive them. It was undesirable for timekeeping performance to vary with battery voltage, but more serious were issues caused by poor contact pressures, and especially by electrical arcing. In an inductive circuit that includes an electromagnetic coil, a magnetic field is generated when current flows. As the circuit is broken, and the magnetic field collapses, voltage

---

69 Hirsch 1883.
70 Beckett 1850, 157.

rises rapidly and can cause a spark across the gap between the contacts as they part company, potentially causing serious damage to the contact surfaces, which gradually vaporize and are deformed, leading to failure.

Adopting best practice imported from the telegraphy world, where such problems were well understood and combatted, electric clocks rapidly adopted spark-quench solutions to dissipate the rising voltage harmlessly and to preserve contact life. Problems with contact pressure derived from the use of the relatively slow-moving parts of a clock mechanism to close the contacts.

A solution to this problem was devised by George Bennett Bowell, who worked with Frank Hope-Jones in the Synchronome business in the late 1890s. Rather than Bowell, it was Hope-Jones who very publicly championed both the importance and the priority of the 'Synchronome switch', and whose name is therefore commonly associated with it. The relevant British patent of 1895 (numbered 1587) details 'a new device whereby all the energy required to maintain the swinging of the pendulum or **balance wheel** is mechanically transmitted through the electrical contact surfaces, thus ensuring absolute reliability with regard to the electrical circuit'. They employed a pivoted gravity arm raised by the action of electromagnets, thus storing energy. It gradually gave up this energy as it engaged with the clock train, moving downwards from its initial raised position. At the limit of its downward travel, the gravity arm would cause the contacts to close, thus energizing the electromagnets again, attracting an armature that raised the gravity arm causing it to re-engage with the train again, repeating the cycle.

In fact, Bowell's and Hope-Jones's first clocks embodying the Synchronome switch were not wholly reliable, in part because the electrical contacts still came together slowly.[71] It was only once Bowell had left the business to pursue other designs that Hope-Jones brought together a series of key design elements that led to a successful combination.

Hope-Jones was critical of most systems (other than his own) at one time or another. For example, one target for his ire was the Swiss Campiche-type system installed at St Columba's Church on Pont Street in London, which he criticized for what he felt to be an unnecessarily large accumulation of wet cells. As an inveterate self-promoter, and later author of one of the standard works in English in the field of electrical horology,[72] it is vital to tread carefully in using Hope-Jones as a source. Hope-Jones's style landed him in trouble, for example, in proposed patent infringement legal action from Gent & Co, whose Isaac Hardy Parsons (1868–1940) and Alfred Ball (c.1867–1932) had patented a design in 1904 that included the main features of a controlling clock which characterized what became standard British practice: the electrically reset gravity-escapement clock.[73]

Owing to Hope-Jones's prolific lecturing and publishing output, we are left with a clear record of his claims for the priority of Bowell's design of the Synchronome switch, and a modified form of this appeared in the Parsons and Ball patent (no. 24,620).[74] Crucially, their patent included four elements that were adopted not only by Gent & Co, but by Synchronome, and then later by Gillett & Johnston, English Clock Systems, and others in succeeding decades. These were: (i) a gravity lever, normally supported on a latch; (ii) its release at half-minute intervals by the pendulum, through the medium of a countwheel; (iii) an impulse delivered by the gravity lever via a roller running down an inclined plane; and (iv) electromagnetic resetting of the gravity lever through the contact surfaces. The circuit is arranged such that the electromagnets are in series with any subsidiary dials connected to the controlling clock. As the circuit is closed, all the clocks are indexed.

There are various historic forerunners for elements of this 'standard' design, though it remains uncertain how much the later inventors were directly influenced. Hope-Jones's correspondence with Silvanus Thompson (1851–1916), an authority on electricity, reveals that Hope-Jones borrowed a variety of German books on horology from Thompson.[75] From these, he expressly acknowledged the work of Sebastian Geist (1817–1908) of Wurzburg, and in particular the notion of a gravity arm, held latched, but giving its impulse when released via a roller to a pallet mounted on the pendulum.[76] While a roller and pallet were involved, the design, adapted and used briefly by Hope-Jones, was poor, and did not embody the important elements in Parsons and Ball's 1904 patent, namely, the roller running down an inclined plane.

The adoption of a countwheel used to count out the vibrations of a pendulum between impulses has interesting precedents. In his writings, Hope-Jones identifies Campiche of Geneva, and Lowne and Palmer of London, as the pioneers in the use of such a device. Argued by some to be entirely coincidental, there are also intriguing parallels between the broad design of controlling clock embodied in later British practice and a design of Johann Mannhardt (1798–1878), the gifted turret clockmaker based in Munich. In Mannhardt's 'freischwinger', a countwheel is indexed one tooth at a time by an advancing pawl, on one excursion of the pendulum. Once each revolution, the countwheel unlocks a latch that holds a gravity arm in place between cycles. Upon being unlocked, the arm falls through gravity and a pivoted roller attached to the gravity arm is allowed to run down an inclined plane attached to the pendulum. The energy thus released provides an impulse to the pendulum, either each minute or half-minute. The unlatching of the gravity arm also has the action of releasing the going train, which now performs a resetting action, lifting the gravity arm back into a latched position, with the short run of

---

71 Miles 2011, 53.
72 Hope-Jones 1949.
73 Miles 2011, 64–5.

74 Ball 1930.
75 Documents 41, 42, and 45 in the Silvanus P. Thompson Pamphlet Collection 'Electric Clocks III', *IET* Archives: UK0108 SPT/P/I/065
76 Hope-Jones 1949, 123.

the train communicated to the motion works through leading-off rods.

One might generalize that the electrically reset gravity-escapement clock was the most common form of controlling clock in systems manufactured in the United Kingdom over the first seven decades of the twentieth century. But what of electric clock systems in territories besides Britain, Continental Europe, and the USA? Unfortunately, knowledge of the development and use of electrical horology and time distribution systems in substantial parts of the world remains limited at present, with little published coverage, certainly in English. The vast regions of Africa, Asia, Australasia, the Middle East, South America, and the former Soviet bloc remain largely uncovered in commentary to date. Vanessa Ogle has discussed different calendrical systems and the historic conflicts between 'Western' time and time set by traditional Islamic methods, particularly in cities such as Istanbul and Beirut, but the actual use of time distribution systems in such locations is uncertain. Late-nineteenth-century newspaper articles in Beirut 'reported with awe on newly installed systems for controlling the distribution of time via networks of clocks in large European cities', suggesting these were not yet locally implemented.[77]

The promotional literature of many of the major Western manufacturers offers evidence that large numbers of installations were manufactured for export markets, and this tends to suggest that time distribution systems manufactured in the traditional centres (Britain, Germany, France, Italy, and so on) were exported widely across Africa, South America, Asia, the Middle East, the Caribbean—and even to remote locations such as Iceland. British firms appear to have been particularly successful in this activity, exploiting links from a former Empire and then later in the Commonwealth. By the 1920s and 1930s, British installations were found worldwide in large numbers, and by the late 1960s and early 1970s, Gent & Co were represented by agents in more than 40 countries.

By contrast, the spread of German-made time distribution systems of a standard sort (in other words, not astronomical tank regulators) appears to have been relatively limited and focused on north-western Europe. In the case of C. T. Wagner of Wiesbaden, for example, it boasted in 1929, after seventy-five years of operation, a long and impressive list of installations in a huge variety of institutions across Germany and near neighbours, but only a handful internationally, all concentrated on Moscow and Leningrad with a small number of outliers such as Buenos Aires, Caracas, Monterey, and Sydney.[78] In the same year, another important German manufacturer, Siemens & Halske, could only boast a small number of distant international installations in their catalogue, including Moscow, Yokohama, Kobe, and Tokyo.

One remarkable system, patented in 1899 in Zurich, Switzerland, did achieve widespread adoption in many countries worldwide. This was developed by Martin Fischer, and was marketed under the name Magneta, and later Inducta. A standard key-wound train released a second train each minute. Energy stored, usually in a substantial weight, was used in the second train to revolve the armature within a magneto generator from a rest position against one banking spring until stopped by a second banking spring, moving through a relatively limited arc. This rapid movement of the armature, pivoted within the two poles of an arrangement of large horseshoe magnets, generated a brief high-voltage impulse, used to distribute to secondary clocks. At the succeeding minute, the armature of the magneto moved in the opposite direction, generating an impulse of opposite polarity. The system claimed two major advantages over other designs. Firstly, it involved no contacts in the closing of a circuit to deliver impulses—the current was induced by the action of the magneto, and it therefore avoided any of the classic pitfalls associated with physical contacts. Secondly, it involved no batteries, also a source of problems for other system designs.

Magneta established operations in Britain, France, Germany, and the USA, and succeeded in exporting installations further afield such as Canada and Australia. Many installations were located in prominent locations, such as Washington DC's Union Station and the Plaza Hotel in New York. Marine versions appeared on many well-known ships, such as the *Titanic*, *Olympic*, *Lusitania*, and *Mauretania*. Magneta systems were also widely used by the British Post Office, particularly before the First World War.

Eastern Europe

In the case of the Eastern bloc, it is clear from the flow of intriguing electric clocks that have gradually emerged and been acquired by collectors in the West since the 1990s that there were indigenous and innovative designs for controlling clocks. These have not yet, however, been studied and written up in the Western horological literature.

But it has also been apparent for a long time that many designs familiar in the West had corresponding models produced in Eastern-bloc factories. For example, Pragotron clocks originally from Czechoslovakia are near identical to those of the German Telefonbau und Normalzeit. When the German Democratic Republic formed, the nationalization of enterprises saw the creation of the umbrella enterprise Rundfunk-und Fernsehtechnik (RFT), which included parts of Siemens, so controlling clocks produced by VEB–RFT Fernmeldewerk Leipzig were essentially Western German Siemens clocks, if engineered down to a lower cost through the use of less expensive materials and finishes.

---

77 Ogle 2015, 120–48, quotation on 127–8. See also Georgeon 2012.
78 Wagner 1929.

Africa and South America

Time distribution systems found across the African and South American continents in the twentieth century appear to have been supplied largely by European manufacturers, but evidence is limited and based largely on occasional references in promotional literature to installations supplied by any given company, and also the flow of disused clocks into modern secondary markets. In particular, British, German, and French systems were installed in countries with natural language and trading links—for example, Leroy & Cie supplied three precision clocks in North Africa, and nine in Latin America.

Synchronome shipped four Shortt clocks to Helwan Observatory in Cairo between 1924 and 1950, while another went to Nairobi in East Africa in 1936. Two went to the Cape of Good Hope Observatory in the late 1920s. Of the British firms, Gent & Co were probably the most prolific in Africa, concentrated in significant deliveries of installations in South Africa, in major banks, post offices, railway stations, town halls, libraries, and other public buildings. Riefler sent precision clocks to the Cape of Good Hope and Johannesburg observatories.

India

There is a growing scholarly discussion of timekeeping in India, or more distinctly of 'Indian time politics'. As part of a wider analysis of colonial rule by the English, the building of clock towers and the introduction of time systems are argued to be political statements, intended to be reminders of the superiority of the Raj, the value of punctuality, and the imposition of order where there is chaos.[79] There is a significant literature dealing with the broader cultural issues at play: of empire, subjugation, religion, philosophy, the rhythms of life, and so forth.

Like other large territories, India had to face issues of time standardization in the late nineteenth and early twentieth centuries.[80] In material terms, the clocks and systems that were introduced, typically from the 1857 Mutiny onwards, were imported from Britain. Large numbers of clock towers were erected with turret clocks imported from well-known English makers. The three obvious arenas in which electrical horology emerged, from early in the twentieth century, were in the post office network, on the railways, and in observatories. It would appear that it was also English firms that dominated this field.

British electric clock companies rapidly established agencies in India, before the First World War, and their catalogues offer some evidence of significant installations. For example, Gent & Co of Leicester supplied clocks for major stations on the Bombay, Baroda, and Central India Railway, the East India Railway, and the Great Indian Peninsular Railway, as well as clocks for town halls, hospitals, and legislative buildings.[81] It also supplied the new observatory commissioned in 1906 by the Nizam of Hyderabad (1866–1911) with a regulator clock in a vacuum tank, to its 'Thornbridge' pattern.[82] On a larger scale, in 1951, it was reported that a new system comprising 900 secondary clocks had been installed across the Central Secretariat government buildings in Delhi, including 100 in Parliament, 100 in the All India Radio offices, forty-five in Government House, and 400 in the north and south blocks of the Secretariat itself, an installation which was almost certainly by Gent.[83] James Murray, of Old Court House, Calcutta, was agent for Synchronome in the 1930s, supplying clock systems to the local jute mills.[84] Between 1930 and 1954, Synchronome supplied Shortt clocks to Dehradun Observatory, Nizamiah Observatory, and Alipore Observatory (Calcutta). A synchronous clock showing hours and Ghati was installed at the Benares College. Involved in importing anything railway-related, Heatly & Gresham of Britain acted as agents for the Silent Electric Clock Company.[85]

Asia and Australasia

As with Africa, historical research into electrical horology in Asia and Australasia is to date very limited, though Graeme Davison highlights the slow adoption of distributed time systems in Australia from the mid-nineteenth century.[86] As might be predicted, at the precision end, astronomical clocks were delivered to observatories. Riefler appears to have delivered a single clock in China but delivered seventeen clocks to Japan between 1905 and 1935, to Kyoto, Yokohama, and Tokyo. Leroy delivered one of its early clocks to the Zikawei observatory in Shanghai, then a second in 1933, and delivered three clocks to Indochina in the late 1920s. Synchronome delivered three Shortt clocks to Japan between 1924 and 1929, and four to Australia between 1926 and 1957. As the Shortt neared the end of its production run in the mid- to late 1950s, Beijing in China ordered a remarkable eight clocks.

For public time distribution, the Western manufacturers seem to have had less success in exporting their systems to Asia, though Gent & Co were proud to advertise the four eight-foot (2.4m) electrically driven turret clock dials they had supplied for the tower above Kowloon Canton main railway station in the 1920s.[87]

As for locally developed and manufactured electrical horology products, inexpensive volume-produced domestic quartz-controlled clocks and watches were a feature of production in Asia

---

79 See a summary of current scholarship in Salamé 2016.
80 Prasad 2015, 134–64.
81 Gent & Co 1932; Gent & Co 1938; Gent & Co 1939.
82 Gent & Co 1911.
83 BHI 1951.
84 Miles 2011, 17.
85 Silent Electric Clock Co 1911.
86 Davison 1993.
87 Gent & Co 1929.

in the latter part of the twentieth century through to the present. More sophisticated products existed in the marine world, both in the form of marine chronometers and marine time-distribution systems. Marine systems from firms such as Citizen and Seiko appear to have been widely fitted to the international merchant fleet in the latter part of the twentieth century, offering the accuracy of quartz control but replicating the functions of traditional controlling clocks in being able to add additional impulses to ship-wide subsidiary clocks as ships travel east, and to retard the system on westward journeys. As a result of ship-breaking, many of these have now found their way to collectors in the West, though there has been no systematic publishing of any details to date.

In Japan, Seiko developed sports timers and a portable quartz crystal-controlled chronometer, which was used at the Tokyo Olympic Games in 1964. In the late 1960s, an improved chronometer was introduced, the 951-11, and this was used by several navies worldwide, but suffered from a synchronous-motor-driven seconds display that was hard to set correctly. This was superseded in the late 1970s by the QM-10, and then by the QM-20 in 1986. These utilized integrated circuits and half-second stepping motors for the presentation of seconds, and were regarded as high-quality products, for example, being selected by the Dutch Royal Navy.[88]

In Yantai in China, formerly known as Chefoo, a large industrial manufacturing base grew up during the twentieth century, centred on the port. A spring-detent marine chronometer was produced from 1957, and in 1967, research and development led to the production of the SY1 quartz observatory clock, followed by the SY2 marine chronometer. It is a relatively close copy of a Seiko model, but offered an ingenious method, using sensing coils, for deriving precise seconds from the synchronous motor display.

PRECISION TIMEKEEPERS IN EVERYDAY LIFE

So far, this chapter has looked at the role of electricity in measuring and distributing precise standardized time across technological networks on a variety of scales, focusing particularly on time-distribution systems. An important related story is that of the emergence of standalone electric timekeepers in everyday life, whether as end-user clocks and watches for domestic or personal use, or as embedded timing devices in other infrastructures such as electric lighting.

Domestic electric clocks

In the second half of the nineteenth century there were worldwide applications for patents for a wide variety of electric clocks that were not intended for commercial or industrial use, but in a domestic setting. These were frequently for standalone devices, not intended to transmit pulses to subsidiary movements. There are many survivors of such clocks, but only a small subset appears to have been produced in meaningful quantities. An exception would be clocks by Hipp of Neuchâtel, since appreciable numbers can be found in museums and collections worldwide.

However, from the turn of the century, numbers of electric clocks for the domestic market increased, though these tended to be expensive and exclusive items. For example, their desirability is perhaps indicated by the appearance of electric clocks in the lists of wedding gifts published in British newspapers in the period before the First World War, now readily searchable online.

The Eureka electric balance clock from 1909 is a good example of a high-value item from the pre-First World War period. It employs an electromagnet, at the centre of a large balance, which is switched by the mechanical closing of a pair of contacts in each oscillation. Using a by-now familiar principle, one of the poles of the electromagnet on the balance is arranged to swing very close to a soft iron bar, and as the electromagnet is energized on each swing, a powerful attractive force is created between the electromagnet and the bar, and an impulse is provided. In the 1890s, Cauderay (1844–1932) of Lausanne had also used a large balance wheel, electromagnetically impulsed, but with the switching controlled by a **Hipp toggle**.

One of the most successful domestic electric pendulum clocks emerged in France in the 1920s—the Bulle clock.[89] More than 300,000 of these clocks were produced between the two World Wars, in an enormous variety of case styles for wall and mantel. This employs the same principle as Alexander Bain's clocks from the 1840s. The pendulum bob is formed largely of an electromagnetic coil, arranged so that its hollow centre passes around a fixed, curved, soft-iron magnetic bar. This has the unusual arrangement of a single south pole to the centre, and north poles at either end. Current from a battery is routed to a pin attached to the pendulum, which engages in a pivoted fork mounted on the movement frame, offering a simple switching arrangement. One tine of the fork is insulated, the other conductive. When a circuit is made, the coil of the bob is energized, and an impulse is delivered. As in a conventional mechanical clock, the arc of the pendulum increases until a steady state is maintained, in which the losses to gravity as the bob reaches the limit of its excursion match the impulse derived electromagnetically.

An important figure at Bulle was Marius Lavet, who left in 1923 to join the French company Établissement Leon Hatot, which, like Bulle, produced hundreds of thousands of domestic clocks

---

88 This section draws substantially from Read 2015, 357–60.

89 Belmont 1975.

over succeeding decades under the name ATO. ATO electric clocks used broadly the same principle as those by Bulle, except that the pendulum carried the permanent magnet, and the coil was fixed to one side.

Later, Lavet was responsible for an important patent in 1953, for the application of a transistor in both balance and pendulum clocks, which enabled electronic switching without the use of mechanical contacts. This offered tremendous advantages in horology, removing the longstanding problems associated with mechanical switching as well as offering switching frequencies in the range of millions per second if required. In the words of electric clock historian David Read, this allowed for the creation of 'portable clocks and watches using high frequency electronically maintained tuning forks and quartz oscillators. It marked a turning point in the world of timekeeping and countless examples of clocks and watches by other makers round the world are marked ATO licence or ATO patent'.[90]

While large numbers of the resulting clocks were inexpensive and volume produced, capable of being used as 'battery insertion' movements in a vast array of potential clock cases, one result was the creation of a new generation of precision clocks. These took the form of marine chronometers, notably the Chronostat models I, II, and III, which were supplied by Leroy & Cie to the French Navy and other demanding clients, and also marketed by Mercer in Britain to Commonwealth countries from 1959.[91] Using transistors, tuning-fork controlled timepieces (discussed later) emerged as an inexpensive option, with Bulova developing tuning-fork watches, and Schatz and Jeco developing clock movements.

Prior to the Second World War, many clocks produced in a wide variety of countries employed essentially the same principle as the Bulle or ATO or used a Hipp-toggle arrangement to deliver impulses either to pendulums or balances. With the advent of synchronized electricity grids, another radical new alternative presented itself. A key figure in the development of this field was Henry E. Warren of the United States, who founded the Warren Telechron Company in 1916, later acquired by General Electric in 1943.

Warren's approach was influenced by the observation that independent clocks typically varied in time, and that the creation of building-wide systems of a single controlling clock with a large number of subsidiary clocks had significant advantages. He considered that there were two main networks that connected very large numbers of buildings and households: the telephone system, and the electrical power network, largely used then for the provision of electric light. Some ninety per cent of electrical power was distributed as alternating current. As Warren acknowledged in a lecture given to the Clock Club in Boston in February 1937, several others before him had considered the idea of using alternating power as a means of distributing time, but crucially, no one had taken any steps to regulate the frequency of mains power to a stability that would make time distribution a practical possibility.[92]

Warren succeeded in several notable achievements. Alternating-current motors did not necessarily spin at precisely the speed suggested by the numbers of alternating cycles generated by the power-station turbines. The designs of early motors allowed for a degree of 'slip'. Warren, on the other hand, devised alternating-current motors whose speed would exactly match the speed of the turbines and were thus 'synchronous'. This was achieved by using a rotor at the centre of his new motors made of hardened steel and permanently magnetized, arranged to revolve in the rotating electromagnetic field determined by the alternating frequency of mains power. Such motors were not useful for applications demanding significant torque but were well-suited to carrying forward clock hands.

Warren's second notable achievement was to devise an instrument that could be used in the control room of a power station to keep the frequency of the turbines at a fixed target number of cycles per second. Since one-second pendulum clocks can be produced to keep time to within a small number of seconds per day, or better, their accuracy lies in the order of 1 part in 10^6. As a degree of accuracy against which to match the speed of turbines in a power station, this is a fine target.

In essence, Warren offered an instrument that illustrated time as derived from an accurate clock as well as time derived from the frequency of the electricity being generated in the power station. This allowed the power station operators to observe errors in the average frequency far smaller than the electrical instruments then available could discern. If the operators managed the speed at which their turbines were spinning with an eye on Warren's clock, they could maintain very close to a constant frequency for the power grid, with the result that any synchronous motor clocks connected to that grid would show accurate time (assuming they were set right in the first place). The power station turbines connected to Warren's clock acted as the controlling clock in this system, and the synchronous clocks in people's homes were effectively subsidiary clocks, but rather than receiving occasional impulses, their progress was controlled constantly over the electricity grid.

Warren's final important achievement was to persuade electric supply operators of the utility of managing frequency to this degree of stability. It imposed a new and additional burden (and therefore cost) on their operation. He claims the success came down to a combination of the public spiritedness of operators and an eventual realization of the commercial upside. Continuously connected clocks, by definition always switched on, consumed power constantly, and therefore represented revenue. Each clock consumed a very small amount of power but taken in aggregate

---

90 Read 2015, 346.
91 Betts 2017, 669.

92 Reproduced in Aked 1997.

Figure 186 Unknown, 'Time from the Mains', The illustrator imagines the synchronous link from the power station turbines to the domestic mantel clock. Image from *Wireless World* (27 May 1931).

it became significant. The clocks offered an entirely stable and predictable load profile, and the overall additional return to the power companies vastly outweighed the cost of maintaining a stable frequency.

In the late 1930s, Warren estimated that more than ten million synchronous clocks were operating in the United States. Such clocks started to be introduced in Europe in the 1920s, in each country depending upon the introduction of widespread synchronized grid systems. In the United Kingdom, it took until the early 1930s for the synchronous clock to be adopted, but take-up was rapid (Figure 186). This was not only domestically, but also institutionally, so that in some cases mains frequency was used to control the rate of pendulum clocks, and commercial premises might also choose simply to use large numbers of individual synchronous clocks, rather than time distribution systems.

From the First World War onwards, S. Smith & Son (much later Smiths Industries) rapidly became Britain's main horological manufacturer.[93] In the two years to 1932, Smiths went from starting manufacture to producing more than 180,000 units per annum. Interestingly the British experience, probably repeated elsewhere, was that the horological trade was little involved. Dennis Barrett (1901–89) of Smiths reported that 'electricians are selling 75% of all synchronous clocks sold in this country'.[94] It was also unnecessary to produce high-quality wheels and **pinions**, to the same standard as in traditional horology, as the train of a synchronous clock has a very low load—a further cost saving to the manufacturers.

Clock historian Charles Aked commented that the reputation of synchronous clocks suffered badly in the middle years of the twentieth century, dominated by two features of the Second World War: grid damage and the resulting outages. Further, over the much longer term the significant effects of load shedding—controlled shutdowns of power to protect the system from over-demand—resulted in wide variations in frequency.[95] System operators had more important goals than maintaining frequency control. While load shedding was a feature for several European countries in the post-war period, owing to a wide range of economic factors, it became a feature of grid operation in a wide range of countries in the 1970s (Britain and the USA included), with the result that synchronous clocks became less useful than hitherto. Pendulum-based time distribution systems, therefore, enjoyed something of a revival, evidence of which can be seen in the example of Gent & Co's records, where the most prolific period was 1947–70, during which the firm installed more than 8,000 systems, by comparison with the 3,000 systems installed between 1920 and 1939.[96]

This coincided with the emergence of quartz-controlled domestic clocks, whether as alarm clocks (an important type of synchronous clock) or as general household or kitchen timepieces. From the 1960s, Germany and Japan were to lead the way in domestic quartz clocks, with the Astro-Chron released by Junghans in 1967, though this was an expensive item and not produced in large quantities, followed later the same year by the transistorized Crystron from Citizen. In 1971, Staiger introduced the Chrometron CQ 2000, the first inexpensive and truly volume-produced quartz crystal clock movement. This used a 16kHz crystal in a large can, controlling an electronic circuit with nine transistors, two integrated circuits, and fifteen resistors, all individually mounted. A moving coil transducer converted electronic pulses into mechanical movement to advance the hands in one-second steps. It was a forerunner of large numbers of quartz-controlled battery insertion movements produced worldwide over succeeding decades.

At the 1985 Basle Fair, Junghans and Kundo introduced quartz clocks combined with a radio receiver. These clocks

---

93 Nye 2014, 78–105.
94 Anon 1936.

95 Aked 1997.
96 Information derived from long-term survey data collected by the AHS Electrical Horology Group.

were designed to run with a very slight losing rate and were corrected when the error between a received radio time signal and a clock's own time moved outside a tolerance of 60ms. Adjustments for spring and autumn time changes could be automated. This stemmed from the fact that the Mainflingen radio signals (commonly referred to as DCF77) do not consist of a simple pulse, but are coded signals containing not only the correct time but also the date. The receiving clocks are therefore significantly more sophisticated than earlier forcibly corrected examples—they have to decode an incoming signal and act upon it.

Electric wristwatches

There had been experimentation in the electrification of watches from relatively early in the twentieth century, but such experiments were beset by classic problems, particularly associated with the unreliability of switched contacts required to make millions of operations, and insufficiently advanced battery or cell technology. After the Second World War, though, considerable advances were made in various locations worldwide.

Electric watches drew attention from the 1950s onwards, with models such as the Hamilton Electric 500 in the USA, the related Epperlein 100 watch in Germany, and the Lip watch from Besançon in France. Early electric watches such as these used the same principles as some early electric clocks—the use of the action between an electromagnet and a permanent magnet. For example, in the Hamilton watch, an electromagnetic coil was added to the oscillating balance, while permanent magnets were fixed to the plates. With mechanical switching, operated by the balance, closing the circuit to deliver current to the coil, the magnetic field of the moving balance interacted with the field of the permanent magnets, providing an impulse. Other designs (for example, the Lip) placed the permanent magnet(s) on the balance and the electromagnetic components on the plates, but the broad principle remained the same.[97]

Once again, the concept addressed the same desideratum as most electrical horology: removing the need for external winding or human agency. Unlike an automatic mechanical wristwatch, which will stop within days as its mechanical power reserve is exhausted, an electric watch will run for the life of its battery. While this might not be for years, as with much later watches, the power reserve was significantly longer than in their mechanical equivalents. Accuracy was also respectable and probably better than many comparably priced mechanical watches, though they could suffer from contact problems.

While Hamilton and Lip both sold electric wristwatches in large numbers, Bulova developed a tuning-fork controlled watch to the design of Max Hetzel (b. 1921). The origins of the use of a tuning fork as an oscillator in the nineteenth and early twentieth centuries have been discussed in an earlier section. Experimenting much later, from 1952 onwards, Hetzel chose the tuning fork as his oscillator, operating in by now well-charted waters, but he employed the latest technology for the switching of current to the electromagnetic coils that keep the tines of the fork oscillating.

A key component was the new Raytheon CK722 transistor he could use in his oscillator circuit. Initially operating at 200 cycles per second, rising to 360Hz, Hetzel succeeded in devising an indexing system sufficiently miniaturized for watch operation. This involved microengineering, which has continued to surprise succeeding generations. The motion works are advanced by a pecking pawl system that vaguely resembles that employed in many subsidiary clocks. An advancing pawl pushes a wheel forward through engagement with a stepped tooth, while a locking pawl prevents the wheel from counter-rotating. In the case of a typical Bulova Accutron advance wheel there are 320 teeth, in a wheel just 2.4mm in diameter, and each tooth is only 10μm deep.

The Accutron was accurate, dependable, and, thanks to good advertising and associations with the 'space race' through use by NASA, successful. It has continued into the twenty-first century as a highly sought-after collectible and yet still relatively practical wristwatch. Other manufacturers produced tuning-fork watches, notably Omega, which used higher frequencies to achieve chronometer status, but the name Bulova remains inescapably linked to the type through its Accutron models.[98]

It was clear from the experience of firms such as Hamilton, Lip, and Bulova that there was a future for wristwatches driven by battery power. They combined the practicality of long-term operation, without intervention, with a sense of modernity promoted by advertising, and were therefore considered highly fashionable. But it was quartz crystals that were to bring electrical horology truly to the masses.

There were two main centres for the development of quartz-controlled wristwatches in the 1960s: the Japanese firm of Seiko, and in Neuchâtel, Switzerland, where a consortium called the Centre Electronique Horloger brought together the resources of a large number of watch companies including Rolex, Omega, Patek Philippe, Jaeger le Coultre, and Longines. Watches were launched by both camps within months of each other but using different technologies to 'step down' from the 8,192 Hz oscillation of the quartz crystal to the required low frequency. Seiko employed a stepper motor, which became an industry standard approach, while the Beta 21 designed by CEH used a vibrating oscillator (in some senses echoing Bulova's Accutron).

There is some scholarly debate over priority. Seiko's Astron was launched in December 1969 before any of its competitors launched their products using the Beta 21, but it is also well-documented that the Astron was launched with just 100 units, which were essentially laboratory test pieces, prone to significant problems, and therefore not a 'production' series, unlike the Swiss

---

97 Weaver 1982, 50–62. Doensen 1994, 89–111.

98 Doensen 1994, 112–22.

watches. The Beta 21 in its various guises has therefore become highly collectible.

The CEH consortium did not pursue the quartz path further, and most of the Swiss owners developed their own proprietary quartz calibres. Seiko continued with significant product development, and as early as 1971 launched its 38 Series, definitely a mass-produced and highly successful calibre.

The further development of the quartz watch played out against a complicated backdrop of economic factors. The worldwide financial upheaval of the 1970s caused the Swiss Franc to appreciate, thus reducing further the competitiveness of the Swiss, who already suffered from structural deficiencies in their watch industry. Buyers of watch movements were keen to break through the cartelized monopoly of the established Swiss industry (principally ASUAG), and the new low-cost quartz movements that were developed, principally in Asia, offered an ideal opportunity. As Pierre-Yves Donzé has described, while mechanical watchmaking clearly made a long-term recovery, nevertheless the emergence of inexpensive and mass-produced quartz movements precipitated a much-needed rethink, particularly in the Swiss industry. An important result there was the creation in 1983 of what later became Swatch Group, under the direction of Nicolas Hayek (1928–2010), charged by the main Swiss banks with reorganizing and merging the major watch-producing companies to which they were overexposed in credit terms.[99]

Time-switches

In commercial life, there are many ways in which the phrase 'time is money' may be relevant. Labour cost is usually an important component part of the financial performance of any enterprise. The proper control and monitoring of employee hours worked is vital to the employer, but it is equally important to the workforce that hours worked are accurately recorded. It is therefore apparent that a common time base across a factory installation is desirable. We have already seen that electric time-recording systems became commonplace throughout industry and commerce.

Beyond the measuring and control of workers' hours, other elements may require control based on accurate timekeeping. In the twenty-first century context of climate change, there is an understandable focus today on energy efficiency and the elimination of wastage, but long before this entered the public consciousness there has been a longstanding drive to control the use of energy sources in daily use, whether in a commercial or domestic setting.

Lamplighters were traditionally employed to tend to gas lighting, but their function was replaced, when possible, with clockwork gas controllers, eliminating costly labour. Electricity, like gas, has always been an expensive commodity, and therefore the automatic timing of its use is also desirable. Early time-switches, designed to control the provision of electricity, and incidentally of gas, were generally mechanical and spring-wound. Just as with parking meters, such time-switches required periodic winding, but by the early twentieth century electrically rewound clocks were being provided with programme wheels, allowing for the setting of switching on and off times for a wide variety of purposes.

As to when time-switches emerged first, there are references to lighthouses and buoys being equipped with 'electric clockwork' devices in the early 1880s, designed to switch off their incandescent lamps during daylight, automatically, and unsupervised over periods of several months.[100] An 1898 account in the *Electrical Review* (27 July) describes an 'automatic timeswitch' manufactured by Hubbard & Mortson in Connecticut which was 'claimed to be the first appliance that has successfully accomplished the two-fold movement of turning on and off an electric switch at any time of the day or night at which the clock may be set'. It was used 'chiefly for controlling lights in show windows, burglar alarms and light circuits'. In Europe, priority is attributed to Ernst Kuhlo (1843–1923), director of the power station in Szczecin/Stettin, who is thought to have designed electric time-switches in the early 1890s, brought into volume production later that decade in Berlin by the firm Paul Firchow Nachfolger.[101]

Early twentieth-century catalogues from a wide range of manufacturers worldwide illustrate that there was a demand for the automatic switching of lighting. Some switches were relatively crude and had to deal with high voltages. They were designed to lower a two-part, or pronged, conductor into two small cups of baths of mercury, thus making a circuit. Over time, designs evolved that used a sprung element in order to effect rapid switching.

Until the 1930s, the time base for such controllers would be a pendulum clock. With the widespread emergence of unified electrical grids, and the adoption of synchronous motors in clocks, time-switches followed suit. By now they were incorporated in a wide variety of locations, such as street lighting posts. Over the course of the twentieth century, as domestic central heating became commonplace, it was standard to fit a synchronous motor time-switch to control hot water and central heating times. Many public spaces, such as stairwells, would be lit during the hours of darkness by lamps controlled by a time-switch. In Germany, for example, from early on there were regulations that compelled landlords to light stairwells adequately throughout the year, accompanied by sanctions in the case of accidents arising from inadequate lighting (Figure 187).[102]

In the context of lighting, whether of shop windows or stairwells, a feature of time-switch development from early on was the

---

99 Donzé 2015a; 2015b, 122–30. This volume Chapter 13, Section 4. For quartz watches themselves, Doensen 1994, 133–77.
100 Douglass & Gedye 1911, 642.
101 Bohnenstengel 1911, 2.
102 Bohnenstengel 1911, 1.

Figure 187 Aronwerke AG (Berlin) catalogue for staircase-lighting timers, 1930 (private collection).

introduction of an 'astronomical' function. For any given latitude, the length of daylight hours can be mapped out in the form of an appropriately shaped annual cam, on which a follower in the time-switch will rest. The varying motion of the follower on the cam across the seasons will advance or retard the physical position of the actuating pins or other setting elements that define switching times, thereby keeping lighting on during varying periods of seasonal darkness.

Over the latter course of the twentieth century, synchronous domestic clocks would gradually be unplugged, moved, and eventually very largely abandoned. But synchronous motor central heating and stairwell time-switches are generally hidden from view, less accessible, and less likely to be changed, with the result that some survive in use two decades into the twenty-first century, in some cases after more than half a century of operation, essentially uninterrupted.

With the emergence of quartz technology and solid-state electronics, particularly towards the end of the twentieth century, time-switches have tended to become smaller, simpler, and purely electronic, carrying out a programme that can be set and held in the memory of the device. The use of semiconductors eliminates the presence of mechanical and electromechanical elements, and indeed the problems associated with the long-term multiple operation of physical contacts.

Network timing

Having considered electrical horology in a wide range of contexts, we now conclude with a comment about its disappearance, not having been eliminated but having become invisible owing to ubiquity. Remarkably accurate time standards have now become deeply embedded in everyday life.

The complex and fragile telegraph lines and early hard-wired networks used for time distribution from the mid-nineteenth century gradually came to be replaced or significantly upgraded. The emergence of the radio transmission of time was significant, potentially increasing precision and accuracy and reducing costs. The widespread use of wired systems for distributed time within institutions led to economies and efficiencies. Subsidiary clocks became very simple—they did not have to cope with the high energies transferred through conventional clock trains. Indeed, they did not need to be precision instruments at all, in the same sense that standard mechanical clocks must be if they are to keep good time.

Subsidiary clocks could be made inexpensively and be engineered to work consistently over countless millions of operations, with service intervals at large multiples of those of their mechanical cousins. The distributed time systems of the twentieth century, using a single central source of time and potentially many hundreds of subsidiary dials, offered a potentially profitable product for their makers, sellers, distributors, agents, and such like. The marginal profit from each additional subsidiary clock was significant.

The move to volume production of quartz-controlled watches and clocks from the 1970s onwards significantly cheapened access to accurate time for widespread populations. Inexpensive and often very accurate quartz modules came to be embedded in a wide range of household devices, including the emerging personal computer.

From the 1990s onwards, these computers increasingly became networked via the Internet, and therefore (potentially) frequently had their onboard clocks synchronized to centralized timeservers—computers specifically designed simply to provide accurate time, located in centres such as the US National Institute of Standards and Technology. Distant consumers came to have a direct connection into a cascading hierarchy of time standards, at the top of which were located clocks of very high accuracy—yet such a connection came without any incremental cost to the user.

Rubidium frequency standards are now routinely deployed, for example, in GPS satellites, and generally in communications networks, to facilitate calculations and processes that demand high accuracy and precision. For large numbers of people, time sufficiently accurate for all their daily purposes is available via their telephone or computer, and large numbers of devices around them (such as televisions, radios, cars) will have access to accurate time embedded into their design. It has become a distinctive choice to select a timekeeper (a stand-alone watch or clock) that offers the single function of telling the time. Electrical horology has evolved to become 'network time' and we rarely pay it any attention.

CHAPTER TWENTY

WOMEN IN HOROLOGY

Joëlle Mauerhan

For centuries, in Europe the education of women was considered to be useless and was generally neglected, although this could vary at different dates and in different places. Women were, like children, placed under the authority of men: husband, brother, or son. The accounts throughout this chapter run from the affirmation of the Irish essayist Richard Steele (1672–1729) that 'a women is a daughter, a sister, a wife and a mother, a mere appendix to the human race'[1] to that of Michelle Perrot, who states that 'the notion of feminine work is linked to the idea that is held of the place of women',[2] with one recurring question as to how 'feminine' work can be distinguished from 'masculine' work.

The new horological corporations of the sixteenth and seventeenth centuries[3] cemented the working community. They controlled apprenticeship and product quality, oversaw the making of masterpieces, regulated the market by constantly opposing competition, and governed women's work—especially that of widows. If there were women employed in some medieval corporations, this was increasingly reduced during the sixteenth and seventeenth centuries when they were relegated to effecting the least interesting and most unskilled work that required no special training. In Geneva, the corporation statutes laid down that 'All women and girls are forbidden to do horological work on pain of a fifty écus fine and the confiscation of their work and tools: they are permitted only to do piercing, make chains and other things that do not concern the wheelwork or parts of a watch'.[4] Train-setting, finishing, and repairing were reserved to men, with only the most ordinary and least-gratifying work left to women.

Nonetheless, some apprenticeship contracts are known in which the making of pull-strings for striking, keys, and small wheels are to be taught alongside lace-making.[5]

The rigid rules of Geneva were slightly more supple when a woman worked within a family workshop[6] because the family unit was fundamental to the community. A wife, elder daughter, or sister could assist the master in the organization and direction of his workshop. She may have looked after the shop—a major responsibility, as, according to the Paris and Blois statutes, 'no master may work unless he maintains a shop open onto the street'.[7] Surveillance was therefore essential. As most workshops often were established in very ordinary houses, the family and professional life mingled with the premises serving indifferently as habitation, workshop, and point of sale.[8] Wives and daughters lived there together with parents, servants, and work-ladies. In Paris, Françoise Dérissart, the widow of a chain-maker, took in a fourteen-year-old to teach him the trade.[9]

The hearth was thus an important area of economic production. The formula was efficacious—it employed women without remunerating them, without much recognition, rather in the manner of ordinary physical labour. It had the benefit of being mixed, each sex having a defined and complementary role, although limits were set: 'it is forbidden to all masters and others to teach their wives or daughters, or to have them taught, the craft of horology under pain, for those who are masters of being deprived of their mastery, and for those who are not, a fine of 50 écus'.[10]

---

1 Quoted from Hufton 1991, 27.
2 Perrot 1978, 8.
3 See Chapter 1, Section 1.
4 Statutes of the Geneva watchmakers' corporation 1690, arts. 40 & 41, Eyraud 2004; statuts de 1785, art. 24, Babel 1916, 563. Gélis 1950, 82, records the name of one such 'mistress maker of chains', Lucrèce Dunant, the wife of Jacques Caillate (1658–1709).

5 Cf. mauerhan 2018, 80-85
6 But compare the comments by Ogilvie 2019, 275–6.
7 Statutes of the Paris and Blois corporations of clock- and watchmakers, Franklin 1888b, 182; Fourrier 2001, ii, 83. This volume Chapter 7, Section 1.
8 For examples from Lyon, see Bayard 2004.
9 *Cf.* Dequidt 2014, 125.
10 Statutes of the Geneva Corporation 1690, art. 24, in Eyraud 2004.

Marriage was a necessity for women—without it, survival was difficult. It was a partnership to which the wife brought a dowry or knowledge acquired from a previous husband. Marriage—and remarriage—assured the stability of businesses or built strategic alliances between them. Thus, marriage alliances existed between the two leading dial-painting families in Halifax in the early 1800s, and prominent roles were played in the businesses by the daughters Leah and Sarah Whitaker.[11]

The only exit from dependence on the male was widowhood. Only then did women, usually for the first time in their lives, obtain juridical autonomy. Male mortality being high, widows were numerous. Generally, they were authorized by the corporations to continue trading, although conditions varied; sometimes they could do so only for a limited period, as they were expected to remarry.[12] They could continue the term of an already-bound apprentice—a matter of importance as apprentices were the least-expensive source of labour—that is, unless the apprentice deserted, hardly believing 'that a widow could do as well as her husband'.[13] A male presence in the workshop was often made obligatory. The female owner should have 'in her house expert, faithful, men',[14] and she was encouraged to marry another master clockmaker or an apprentice capable of becoming one. The purpose was to keep the business within the body of the corporation and protect its secrets. In her marriage contract, therefore, the wife was required to specify, because she was a widow, that she traded on her own behalf, that she was proprietor of the workshop, and that the tools in her workshop were recorded. Closely surveyed, a widow could be tricked by the community. Outside family workshops women could work as day-labourers, saleswomen, or pedlars. Guild control continued to be exerted, as in Lyon, where the officials inspected goods and verified that they were not stolen.[15] It should be noted, however, that this clause applied equally to male retailers, and that, in general, the statutes of horological corporations concerning women differed little from those enforced by the controlling guilds of most other trades.

UNRULY LADIES

During the eighteenth and early nineteenth centuries, during the Revolution, the status of women became less rigid. In Europe, wherever horology was practised, women protested against the restraints imposed on them by the guilds. A Parisienne, the widow des Bourgets, who had succeeded in establishing a watch factory in Charenton on the condition that her nephew signed her requisitions, wrote, somewhat bitterly to the Minister of the Interior, Calonne: 'I will not hide from your dignity that my nephew is only a name borrowed so as to obtain the decrees of the Council'.[16] Some took account of such aspirations. For example, the English clockmaker John Gregson (1750 or 1760–post-18?), who had purchased a protected post in France as a 'clockmaker to the King following the Court', accorded full powers to his wife Elizabeth Jefferson to administer his affairs when he was travelling. She was charged 'to undertake and have made all works of horology, agree the prices of the said works, and to accept in payment all notes, recognizances, and other instruments'.[17]

One women, however, was more audacious. In 1770, Marguerite Elisabeth Aumerle wrote a letter to her creditors from the Petit Châtelet where she was imprisoned:

> As soon as I was married I worked successfully for the space of eleven years. I created a manufacture of eleven workers who remained constantly with me during this period. You can verify this activity in the trade of horology, as I was the only woman in Europe who as I dare to claim was thus skilled. This condition was turned into another by the will of my husband, he was chosen to be at the direction, which has made me lose the first and the second. I thought on doing this third and last alone that I should be happy, but destined no doubt to be the plaything of fortune, losses, thefts, illnesses, and a husband who dogs my footsteps, have led me to make an arrangement with you.

Starting from account, archival research has enabled Camille Dejardin to reconstruct Marguerite Aumerle's career. She was born in Paris in 1727 and married William Blakey (*c*.1714–post-1788), son of an English clockmaker celebrated as the inventor of elastic bandages. At the age of forty-six, when she directed her workshop of eleven workers, she petitioned for the right to live separated from her husband. She had, however, to abandon her business and to accompany him when he accepted the direction of a file factory in the Essonne, although she then retailed bandages and horological materials in Paris. Finally freed, she opened an 'English Shop' and traded with the provinces and abroad. In financial difficulty, she was imprisoned for bankruptcy and released to mount another business. In doing so, she had to make an agreement with a male partner, although she remained at the head of the enterprise. All trace of her is lost after 1777. Subject to even heavier constraints than those of a widow, she did not capitulate, but instead exploited such possibilities as were available to her.[18]

Was Marguerite Aumerle an exception, or were there other women like her? Detailed genealogical investigations have unearthed information about Marie-Françoise Wetzel (1817–1886).[19]

---

11 Robey 2019.
12 Ogilvie 2019, 251–60. In Lyon, the period in which a widow could continue working was limited to a year. Statutes of 1660, art. XVI, in Vial & Côte 1927, 25.
13 Hufton 1991, 55.
14 Statutes of the Lyon corporation of clockmakers, 1660, art. XVII in Vial & Côte 1927, 26.
15 Statutes of the Lyon corporation of clockmakers, 1660, art. XIX in Vial & Côte 1927, 26

---

16 Dequidt 2014, 123.
17 Archives Nationales, Paris. MC/ET/XXVII/409.
18 Dejardin 2019.
19 Prenot-Guinard 2019, 40

Figure 188 Marie-France Wetzel (1817–86) is shown dressed as befits a member of the rural bourgeoisie of the Franche Comté, posed against a similarly appropriate decor. Photo: © Private collection.

Born into a horological family at Morteau (Doubs), Wetzel (Figure 188) married the director of the horological school in that town, himself a wholesaler of horological tools at Les Gras. Widowed, she moved to Pontarlier, close to the Swiss frontier, region where horology was already prosperous. Remarried to a tax-collector, she participated in the 1855 Universal Exhibition in Paris where, the only woman candidate for a recompense in the class of horological tools, she was awarded a silver medal. We know also of Victorine Lucie Marie Bailly-Comte from Morez, where, in 1869, she deposed a patent application for 'a pendulum system giving fixed seconds, adaptable to clocks of all kinds'.[20] Further archival research will eventually reveal the place women have occupied in the technical and industrial history of horology.

MINOR TRADES FOR GIRLS

From the end of the eighteenth century, women appear in all horological activities, and although difficult to measure, women made up between fifteen and thirty per cent of the work force.

In Geneva during this time (c.1785), they obtained access to **regulation** (**timing** and **rating**), thus creating a substantial break in a masculine monopoly[21] and to several other, less prestigious, branches of manufacture. In the context of a division of labour ever more specific, new activities became feminine specialities alongside those of the cutting and piercing of bridges and cocks. For the most part, these concerned the preparation of the **ébauches**, with the activity of finishing remaining in male hands. But for every new branch that opened to women, the high number of candidates led to wages being reduced and the task thereafter identified as 'feminine'.[22] This devaluation led to girls being apprenticed to those parts of horology disdained by men. If some apprenticeship contracts concern girls, at the popular levels of society, in order to survive, women were condemned to work without training in disagreeable conditions. Female salaries, far lower than male salaries, were considered as belonging to the father, husband, brother, or son who supplied a roof. And if the qualities proper to women required only a modest remuneration, it was *natural*; no apprenticeship was needed, so legitimately, *little hands* were devoted to *little tasks*, for example, polishing, making hands or hinges, and the most dangerous tasks, such as gilding.

Mercury gilding was employed for both clocks and watches in the eighteenth century. Organized in groups, and often in pairs, women made up by far the majority of workers in this speciality. Mortality was high and the social cost heavy. In Geneva the Société des Arts, and in Paris the Académie des Sciences, organized competitions to find measures of protection against poisoning, but it was not before the perfecting of electrochemical gilding that the danger could be fully countered. However, odours arising from silvering and enamelling also provoked mephitic troubles. Philanthropists and the avatars of public health launched many cries of alarm, but the corporations were evasive.

The old corporate world, however, was breaking up; the tocsin sounded. In France, a first reform by Turgot, Louis XVI's Minister of Finance, authorized women to be received as masters. In its wake, the law, Le Chapelier (1791), established freedom of work. The Revolutionary government oversaw the future of girls and widows, the organization of female labour, and of their training. Decisions made concerning the establishment of a national horological manufacture in Besançon, 1793–5, applied equality of the sexes.

The watchmakers of the new manufacturer in Besançon came from the Neuchâtel region under the guidance of Laurent Mégevand (1754–1814). They were accompanied, besides their wives, by single women, such as the case polisher Félicité Brandt (Figure 189). Quickly numbering some four hundred women for eight hundred men, apprenticeship in the traditional 'feminine' branches was available, and hitherto masculine preserves, such as pinion making, were opened to them. Every girl, at the end of her apprenticeship, received a bonus 'a gratification of five hundred

---

20 Buffard 2019, 138.

21 Babel 1916, 119 and cited in Lachat 2014, 160.
22 Hufton 1991, 38.

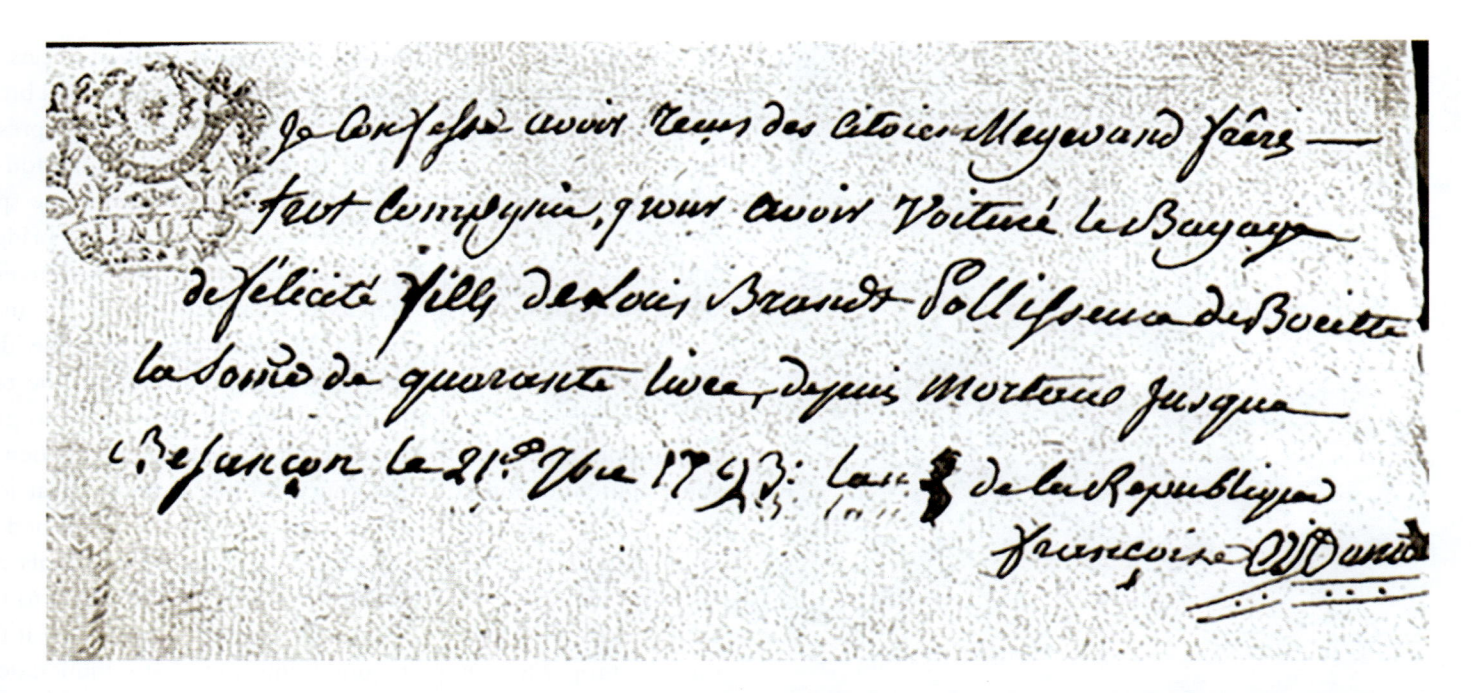

Figure 189 Transport ticket for Félicité Brandt who, the daughter of Louis Brandt, case-polisher, is apparently travelling alone. She has left Switzerland in response to the call of the French Revolutionary government for horological workers in Besançon. Photo: © Archives Départementales du Doubs.

livres to help them set up'.[23] Male apprentices were excluded from this aid. With this money women could establish their own business, as did the dial painter Adelaïde Villeneuve. This effort, motivated as much by the Revolutionary ethos as by a concern to retain the Swiss in Besançon and to draw the Bisontins into marriages with them, was effective. In the year IV [1795], Besançon counted forty per cent of its population as watchmakers. Laurent Mégevand vaunted horology that offered resources 'precious both for the physique and the morale. The orphan, the foundling, and above all the sensitive and delicate sex will find there numerous occupations appropriate to their physique, which putting them out of need, preserves the honest soul from prostitution'.[24]

FOR MEN A CRAFT, FOR WOMEN TASKS

The presentation of female labour in the nineteenth century was subject to the precedent set in the eighteenth. Women were defined, above all, as housewives and as mothers in a period when a low birth rate was perturbing, and infant mortality remained high. For Jules Simon (1814–96), a leader of the moderate republicans during French Third Republic, and a ferocious opponent of women working, horology was perfectly adapted to the female sex: a special occupation that corresponded to weak physical force and innate capacities. His presentation of so-called feminine qualities assuaged consciences: these 'little hands' with their skilful fingers would be so useful for polishing, engine-turning, redrilling—so many activities thought to be simple, requiring neither dexterity nor study. In the 1875 directory of horology at Besançon, the 1564 female workers of the 5150 total listed are at the bottom of the hierarchy, the only exception being the springers, who occupied half of the available positions, although this was still only fifteen per cent of qualified posts. Women were particularly numerous in polishing, for this took place during all the twenty-odd finishing operations and the completing of the raw blanks.

Methodically, employers noted these skills directly developed from domestic skills, for all girls learned how to sew, and used them to give carefully delimited, manual, repetitive, simple tasks to women—the tasks that required little knowledge but much speed.[25] The vocabulary employed in manufacturing companies such as Longines even today tends to demean female work—'men have a skill, women a task, even when they do the same thing'.[26]

In the society of their own times, women horological workers held a privileged position as they could gain a daily salary twice that obtained by a cotton spinner. They affirmed this gratifying status with pride, repudiating labouring status by adopting bourgeois dress habits. Even so, Jacques David (1845–1912), reporting on American horology on the occasion of the 1876 World Fair in Philadelphia, and probably used to less attention among his countrywomen, would report that:

Women workers employed in horological manufactures are of an irreproachable morality. They are generally characterised by their seriousness and their assiduity. They take

---

23 Arrêté de Bassal, 21 Brumaire an II [11 November 1793].
24 Arrêté du Comité de Salut Public, an II [1793/94].
25 Guibert 1966.
26 Lachat 2014, 161–2.

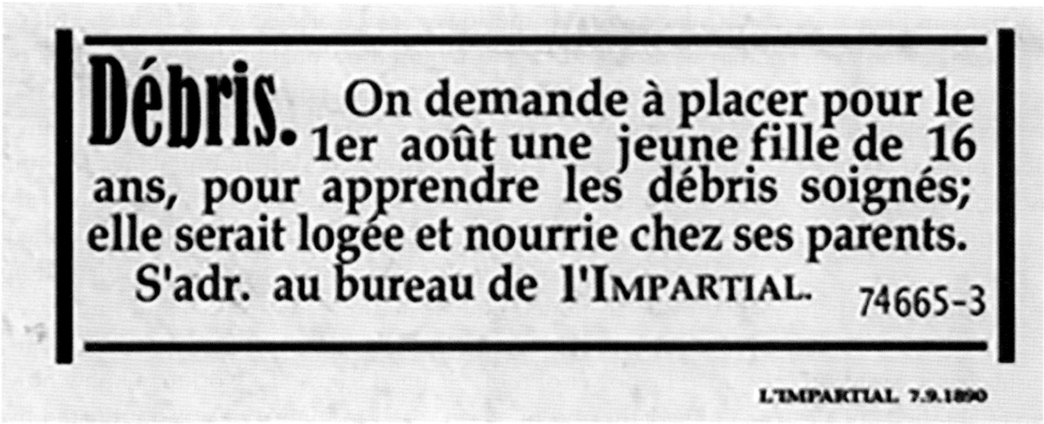

Figure 190 Advertisment to place a young girl for the polishing of very small watch parts such as winding stems and clicks. Journal *L'Impartial*, 7 September 1890. Photo: Reproduced from Muller & Simonin 2006.

care of their appearance, and their dress, although simple, is often elegant. Fashionable hair styles, flowers necklaces, broaches, and rings are worn, and there is a certain refinement in their factory dress which generally consists of a large apron, not in its own nature very elegant, but which they know how to embellish.[27]

Precision mechanics visiting New England were similarly impressed: there they found that women were courteously treated and not the victims of hurtful sallies. The question of morality was preoccupant in a century that separated males and females in the workshops. Fear of a descent into prostitution was everywhere present. In the towns it seemed less menacing among horological workers than among the immoral workers in factory 'prisons'. In the little towns and villages of the Doubs mountains, women in horology, celibate mothers at their head, seem to have enjoyed relatively freer conditions.[28]

Worries about morality were extended to children. In England, industrial enquiries show that, alongside women, children were a cause of concern: 'The race must be protected against degeneracy, the fruit of industrialisation; the child against delinquency and immorality, the consequence of working mothers'.[29] Even so, the use of child labour was not prohibited. A good example of the methods employed to set them to work is found in the town of Christchurch (Hants). Here, c.1790, the watchmaker Robert Harvey Cox (1754–1815) established a fusée-chain manufactory. To man it he approached the overseers of parish workhouses offering a small sum of money in exchange for the services of boys and girls over nine years to whom he undertook to give breakfast and teach reading and writing as well as chain-making, the latter being remunerated (Figure 190). Such initiatives were not uncommon at the time. From 1798 onwards the Christchurch establishments were discussed in the English press, and a few years later fifty children were said to be executing tasks requiring great dexterity, but which were dangerous for their eyesight. But generally, the journalists approved. Boys and girls supplemented the household income, however modestly, and learned the value of work and obedience. On the death of Robert Cox, his widow continued the business, moving it to Wimborne, Dorset, and there grown women worked alongside the children.[30] In France and Switzerland, as in England, legislation was gradually introduced during the second half of the nineteenth century controlling the working hours for women and children. Such legislation, however, did not apply to work in the home or in very small workshops where large numbers of women were to be found, nor did it treat of the inequality of salaries.

EMPLOYERS' STRATEGIES AND WORKERS' ABNEGATION

At the end of the nineteenth century the economist Paul Leroy-Beaulieu (1843–1916) noted that salaries for women were only half, if not a third, of those for men. He became indignant when some of his contemporaries applauded such inequalities and dared to affirm that women ate less and had fewer needs than men. Vaunting 'feminine' qualities offered a sop to the conscience, but many argued for maintaining low remuneration—a woman's salary was considered as only a complement to the familial income. In Besançon, an apprentice in spring-making was paid 2.50fr for a woman, 3fr for a man; a female worker was paid on average 5fr a week, a male worker 15.5fr per week. The salaries for both fluctuated from one workshop to another. In 1890 pay for equivalent work varied from 2.50fr to 4.50fr. Children were generally paid half as much as women.

Despite this, according to the militant Aline Valette (1850–99), clock- and watchmaking enterprises were those into which women threw themselves without restraint.[31] In the last quarter

---

27 David 1877, 31.
28 Author interview with Martine Prenot-Guinard, Pontarlier, 2019.
29 Dieboldt & Zylberberg-Hocquard 1984, 15.
30 Newman 2010, ch. 6.
31 Dieboldt & Zylberberg-Hocquard 1984, 43.

of the nineteenth century, the level of female activity in France—hitherto lower than elsewhere in Europe—surpassed earlier levels to reach thirty-seven per cent of the work force in 1906. Directory and census analysis shows these figures to be applicable at least to industrial horology. However, women listed under their husband's occupation do not appear in the census under their own activity until the very end of the nineteenth century, and unpaid and unofficial work in the home was not recorded.

Work in the home continued to dominate. ***Etablissage***, which was outside corporate control, continued in France, Switzerland, and Germany, although the factory model displaced it earlier in England. The family unit prolonged the system established during the previous century: wives and daughters continued to be the assistants of the master clockmaker, whether it was for the making of Comtoise clocks at Morez in the high French Jura,[32] or watches in the Valley of the Joux, in Switzerland, where women played a role in maintaining the continuity of such prestigious manufactures as Audemars Piguet.[33]

The Franco-Swiss horological region abounds in examples of women working at home or in small workshops. In 1901 a French survey found an exceptionally high percentage of women engaged in the piercing of hard stones for the pivots in watches—work for which Septmoncel was reputed—in the making of keys and hands (slightly less than fifty per cent), and small parts (slightly over thirty-three per cent).[34] The same enquiry revealed a higher percentage of women working in small localities than in the large horological centres. This can be explained by the importance of ancillary work in rural areas where, as Laurence Marti illustrates from the history of her own family, women engaged in agricultural tasks became horological artisans in evenings and nights to augment their income.[35]

But mechanization was advancing in the Jura. Factories and manufacturers were modelled on the rational models discovered by Europeans in America at the time of the Philadelphia exhibition in 1876. The division and simplification of work and the multiplication of steam engines are among the elements that down-graded skilled work and so facilitated the employment of women and children in the industry. David again opines that 'Women are much appreciated in these establishments and all is arranged to make as much as possible of it accessible to them. As elsewhere they are content with a modest salary, and for much of the work they are as good as men'.[36]

To put women to work on machines was far from using their 'special innate qualities', but that was of no concern. What was considered an intrinsically feminine task could be changed according to need. As always, the most important thing was to restrict women to the lowest-paid, piece-meal activities—those that were repetitive and required no qualifications.[37] A study of the reports on labour made at the time of the 1867 Universal Exhibition displays the strategies of employers who used machines to reinforce their control and remove skill and intelligence from labour. At the dawn of the second industrial revolution, manufacturers chose to replace skilled male handwork by the unqualified work of the more pliant women. This caused a transformation of the social model. The structure based on a family unit was replaced by a class structure, and was not effected without tension. The power of tools changed hands: if a workman knew how to use a lathe and master it, the machine-tool was dominated by the employer and the divide between the sexes could only widen.

Already in 1862, this problem preoccupied workers who felt its effects:

> We are not . . . in a gender conflict. It is a means of lowering wages, a worker at a lower price . . . if women are eternally inferior, will not the workman be also? To accept unequal wages for the same work is to accept a process that leads to a general devaluation of wages.[38]

In the 1870s workers' associations, only recently established in France as in Switzerland, put the question of women's work on the agenda of their first meetings. Influenced by the writings of the philosopher Pierre Joseph Proudhon (1809–65), for whom the male role was to support women whose place was in the home, the discussions ended in discord. Then in 1909 the question was again discussed in the Congress of watchmaker-jewellers. It was against their will that the members of the congress, frozen in traditional masculine positions, decided to defend women's rights. Aline Valette wrote in *La Fronde* that 'industrialised women are everywhere exploited without having the arms of male workers to defend themselves'.[39]

Besançon, however was a somewhat reformist city. There, women workers aspired to a greater liberty and responsibility. In 1906, ninety-five women were numbered among the owners of the three hundred small workshops in the city (that is, about two per cent of the minor horological entities). These businesses continued the traditional industrial structure with a complement of one to five persons but without sufficient capital

---

32 Buffard 2019.
33 Aubert 1993, 50.
34 'Enquête sur l'emploi des femmes en horlogerie', *La France Horlogère*, 124, 1912.
35 Marti 2001, *passim*.
36 David 1877, 30

37 Lachat 2014, 176.
38 Typographers' report on the London exhibition of 1862, cited by Rancière 1975, 5–22.
39 This was perhaps one of the reasons why, in 1873, ten women workers from Longines emigrated to found a community in Argentina. Influenced by the anarchist ideas of Mikhaïl Bakounine and Errico Malatesta that they wished to put into practice, they insisted that they were neither young women in search of husbands, nor prostitutes, but citizens faithful to their trade who individually could not make a watch, but together could achieve something. A recent novel by Daniel de Roulet, *Dix petites anarchistes* (2018), following Eitel 2018, is based on this episode.

Figure 191 Domestic assembling and rating in Besançon, 1908. Photo: © Jacques Boyer/Roger-Viollet, n 5565–5.

to last for more than a few years, although adaptable to other forms. When male-directed businesses regrouped case makers, women's businesses specialized without delay in springing and gilding. Among the ninety-five women employers, two stand out: Augustine Guerrin spring-maker, declared bankrupt in 1897, and Adelaide Bourthioux. The latter developed her business remarkably and a 1901 inventory of her workshop revealed a plethora of machines: piercing and cutting lathes, tools for cutting bosses, hollowing winding buttons, polishing arbors or screws, and more.

FEMINIZATION IS NOT EQUALITY

Throughout the twentieth century there was an increase in female employment. In Switzerland between 1895 and 1944, the number of women employed in the industry rose from 5600 to 17800, an increase of 320 per cent. This was intense competition for men, and, at the demand of the unions, the federal authorities reserved priority of employment to them. On the eve of the recession of the 1980s caused by the quartz crisis (and when dismissals affected women first), female employment had attained fifty-five per cent of the workforce, a rather exceptional long-term progression in the Swiss economic landscape.[40] With a return to prosperity, forty-eight per cent of the horological workforce was female in the early twenty-first century.

In French factories during the First World War, women replaced the men who had gone to fight, and their return home was not without tension. Many women remained in their jobs. Industrialists, following the example of Jaeger in his Paris factory,[41] were obliged to modernize their machinery and to rationalize the work process, subdividing it ever more. Watchmakers protested, as a Swiss account of a visit to the Lipmann enterprise in Besançon reveals: 'We had difficulty in accepting the sight of women doing work reserved for men ... but there, everything evolves'.[42] In 1930, fifty-six per cent of Lipmann's manufacturing workforce was female. The Vichy government would legislate (without success) to send women back to the home. The aftermath of the Second World War was calm, but the innovation of quartz plunged the French horological industry into a long and deep depression.

---

40 Lachat 2014. For the situation in Besançon before 1914 see Zunino 1989.
41 Omnès 1997.
42 *L'Impartiel*, 1922, cited from Muller & Simonin 2006.

Nonetheless, in the Jura, now the nerve centre of European watchmaking, the position of women changed little. In 1979, 'the extraordinary resistance of old production methods which even today display an exceptional faculty of adaptation', was noted.[43] In the *Etablissage* system, work in the home always mixed labour with family, thus removing workers from union demands. Although a powerful force restricting wages, until the last years of the twentieth century this mode of production offered an efficient response to fluctuations in demand and supply.[44] In the factories and larger workshops, the least well-paid tasks remained feminine. Without fail, employers fashioned feminine labour for their own needs. Compelled to expand so as not to lose their share of the market, they recruited immigrant Italian women unlikely to rival male watchmakers. Swiss women workers keeping work in the home for themselves, it was the Italians who filled the least-qualified posts, and so were the worst remunerated in the factories during the 1950s and found on the production lines then introduced.[45] In 1982, thirty-five per cent of the Swiss horological population was immigrant.

Dangerous work had always largely been effected by women. Throughout the eighteenth century, gilding had destroyed their health. In the twentieth century, the painting of dials and hands using radium caused scandal. Marie Curie's discovery of radium in 1898 had, at the time, been considered a blessing. The radioactive metal was seen as a source of heat for clothing, a way of recovering youth through beauty products, and more. Watchmakers could not bypass such a tempting field. In France, the Lipmann enterprise confidently affirmed priority in its use, and in the United States, the US Radium Corporation employed 'radium girls' (otherwise known as 'ghost girls' because they glowed in the night).[46] In 1918, in Switzerland, the enterprise Monnier transformed into Monnier-Radium proposed 'watches that cannot be extinguished' to its customers. Radium painters needed no qualifications and were paid by the piece: they could work at home or in a factory.[47] Advertisements burgeoned: 'skilful and conscientious radium painter required'.

The radium girls would put a point on their brushes with their lips, a practice that caused an outcry from the medical profession: 'some women have received in diluted doses throughout their lives more [radium] than was needed to kill a man in Hiroshima'.[48] In England the Brandhurst Company with a radium laboratory carried out severe controls. In the United States, scandal erupted and the radium girls, led by the dial painter Catherine Wolfe, took legal action that led to new legislation and the creation of the Federal Agency for the Protection of American Workers. In Switzerland, the first legislation concerning protection against radioactivity appeared in 1963, when thirty-three enterprises declared employing radium painters.

Access to professional training became one of the great goals of the twentieth century. The fallacious hope insisted that, by acquiring male knowledge and skills, women would have their own qualifications recognized and that these equal skill levels would automatically give them an equal position and salary. In Besançon, to maintain the war economy, the horological school was opened to women. The first, Marie Doyet, obtained her certificate the following year. The school was then, however, far behind most French or Swiss schools, which had been mixed from their foundation. In the interwar period, Besançon girls were directed into the channels traditionally reserved to them. Excluded from a complete training in watchmaking, some of them turned to pivoting or jewel-setting; the majority, however, opted for springing. They were brilliantly successful, as is demonstrated by Hélène Jaccard, a timer working for Petolat-Anguenot, who during the 1930s, obtained five gold medals, five silver medals, and seven bronze medals in rating trials.[49] However, that was the limit. Precision timing was reserved to men. If national education tended to integrate women into its framework, industry kept them in the least-qualified jobs.

Apart from the timers, few women attended horological schools.[50] At the end of the twentieth century, eighty per cent of them had no training, although eighty-three per cent of their male counterparts had qualifications.[51] The unionist M. T. Sautebin[52] denounced this situation in 1982, saying that if 'some men are to be found in the tasks of direction and development, most are to be found in the machine shops or the research departments where no women appear'. He evoked female workers grouped together in large workshops, 'dumb, riveted to their mini-chain turning imperturbably on the bench, or staring through their binoculars to solder electronic modules'. Working with binoculars, like radium, was denounced by the medical profession. This was the moment, however, when horology migrated towards the Far East. In Hong Kong, most of the numerous jobs created were for young women without any particular qualifications: no professional training systems having been seriously considered.

---

43 Levy-Leboyer 1975, 46.
44 Unpublished investigation effected by Jean Buathier for the Musée du Temps, Besançon, 1986.
45 Interview with Daniele Mariani by Francesco Garufo, *Swissinfo*, July 2011.
46 Boquen 2019.
47 'Les radiomineuses, ...', *Le Temps* 8 June 2018, citing Monnier and a study carried out by the University of Berne.
48 Maurice Cosandey, formerly Head of the Radium Control Service, Geneva, cited in *Le Temps* 2018 (n. 46).

49 Briselance 2015, 540.
50 Lachat 2014, 155.
51 Figures derived from the Fédération des travailleurs de la metallurgie et de l'horlogerie.
52 Sautebin 1982, 12.

'WOMEN CROSS THEIR ARMS, THE COUNTRY LOSES BALANCE'

Female demands came out into the open between 1970 and 1990, partly because of the shock caused by electronics and the quartz watch. Despite the disinclination of unionists to help women, two social movements affected them: the 'Lip affair' in France and the 'strike of women' in Switzerland.

In the Lip factory in Besançon in 1970, women made up the majority of the productive work force. Trained very briefly on the job, they were massively employed at the bottom of the hierarchy. In the venerable horological tradition they were considered as interchangeable executants, identified not by their task, but by the capacities considered specifically feminine—agile hands, good eyesight, and the rest. The family firm was presided by Fred Lip, a direct descendant of the founder. He positioned himself as the father of his workers, with a paternalism that could only reinforce masculine power. During the uprising of 1973 which attained international notoriety and destroyed the business, the French Democratic Confederation of Work (CFDT), a labour movement that guided the strike, put a group of male unionists forward in the media. Women, not being considered as worth including in the protests, were relegated to subaltern tasks, maintaining the canteens, the unofficial sale of watches, and the distribution of tracts. However, as the majority of those affected by the movement, in the course of the conflict they progressively asserted themselves: 'We, the women in this struggle, we have something to say'. They wrote publications denouncing the inequality of work conditions and salaries. The first-hand narrative of Monique Piton was followed by a collective work by an active committee of women, *Lip au féminin*. Nonetheless, class status predominated over gender,[53] however disturbing this was, and became a source of fragility for the solidarity of female workers. In 1976, when militants attempted to start up the enterprise once more, women were evacuated from all influential posts.

The Swiss 'women's strike' in 1991 was born in the Valley of the Joux by a group of women workers protesting against the non-application of a law adopted ten years earlier decreeing male/female equality. Salaries having failed by far to adhere to this, a strike was declared under the slogan of 'Women cross their arms, the country loses balance'. The numerous manifestants, joined by some men, demanded not only the full application of the decree, but also the same opportunities for training, a division of domestic tasks, the end of marital violence, and much else. Demonstrations were many. In 1995 a new law was voted without greater success, while in 2018 and 2019 celebrations of the 1991 strike recalled the same unchanged demands. For those writing in *Le Temps*,[54] 'the women's strike of 14 June 1991 was the most important political manifestation since the general strike of 1918'. As such, it overturned a view of Swiss history that suffered from the myth of consensus. A comparative study of the 'Affaire Lip' and the Swiss women's strike is overdue and worth undertaking, given that relations between the Swiss and Besançon watch workers were close, and that the union in Besançon was founded by Adrien Graizely (1845–1925), native of Neuchâtel on the other side of the frontier.

In the twenty-first century, Swiss companies and their workshops and after-sales service departments in Besançon remained little concerned with valorizing female labour even though this increased. The Swiss union *Unia* in 2019 claimed fifty-two per cent of horological workers to be feminine but with salaries inferior by 24.8 per cent to men, although Hervé Munz counts only 45.02 per cent. In 2016, Munz found that 'the widespread employment of women workers goes hand-in-hand even today with reducing the cost of salaries'.[55] Women are always little qualified, trained on the job, and accepted with difficulty. For *Unia*, 'women are too often but the "little hands" of horology'. Munz goes further—he notes that this description derived from the nineteenth century has given place to the more disdainful 'these good women'. Clichés remain: 'Women are accorded a better sense of organisation, to be more diplomatic, and more skilful in repetitive and minute activities'.[56] Even so, some efforts are being made to integrate women into the direction of the larger Swiss horological concerns. But the question posed at the opening of this chapter remains pertinent: what still today distinguishes a task, a craft, or an activity as masculine or feminine?

---

53 Cros 2018.
54 5 December 2019.
55 Munz 2016, 73.
56 *L'Heure by Fleurier*, 2018.

CHAPTER TWENTY ONE

KEEPING CLOCKS AND WATCHES
MAINTENANCE, REPAIR, AND RESTORATION

Jonathan Betts

REPAIR—THE 'POOR RELATION' IN HOROLOGY

What were the after-sale lives of clocks and watches? Who cared for them, day by day, and when they needed professional attention? The mechanical nature of the majority of clocks and watches over the centuries has required regular winding, setting to time, and occasionally adjusting for correct timekeeping. All these actions were performed by the role of the 'clock *keeper*'—who often, though not always, was the owner. Almost all mechanical clocks and watches suffered from friction, required lubrication on some of the working surfaces of the movement, and were sometimes subject to mechanical failure. Therefore, they needed regular cleaning, repairing, and relubrication—the work of the 'clock *repairer*'. With careful maintenance, many timepieces continued in use far longer than their makers might have expected and were often subject to technical and aesthetic 'updating' to make them suitable for continued use in line with their original purpose.

It is certain, therefore, that for the period the average clock or watch lasted, many more hours of professional work went into its repair and maintenance than ever were needed for its construction. So, although the designers and makers of clocks and watches have generally commanded attention in the history of horology, there has, in fact, always been a considerably larger body of practical horologists busy keeping the objects functioning; perhaps it is they who should be regarded as the more important, in terms of horological employment and business. Less glamorous and usually less remunerative than the sale of new watches and clocks, the vital role of the repairer has been little recognized over the centuries, and almost ignored in modern historiography.

Even so, this 'poor relation' has been defended. In 1788 the French watchmaker Pierre Vigniaux wrote 'One normally regards repairing as a quick and easy job; Specialists and real artists have a very different view …'.[1] Another French watchmaker, J. P. Etienne in Nancy, stated in 1810:

> It is a prevalent prejudice, even among some watchmakers, that repairing is the least estimable part of the art. Knowledge and talent are certainly needed to construct the parts of a clock and make them able to work together … but far more is needed for an able repairer: he should master the theory and the practice of the art so as to be able to know the faults of a clock, and to remedy them; he should even have more knowledge of physics … the repairer, without knowing the theory of levers and friction cannot work safely on a machine that is out of order … in a word he must master the art at a high level and himself be able to make a perfect clock.[2]

Across the channel, the notoriously grumpy chronometer maker Thomas Earnshaw complained to the Board of Longitude in 1800 that, through lack of financial aid from them, he had to resort to: '… mending common clocks and watches for my support …'. Earnshaw notes that, on awarding him £500, the Chairman of the Board, Sir Joseph Banks 'in a sort of taunt, further said to me: "and that, Mr Earnshaw, will enable you to leave off the mending of old watches"'.

---

1 Vigniaux 1788, 219.
2 Etienne 1810, 1–2.

EARLY PERIOD

From early days, timepieces—even sundials—have needed keepers and repairers, but this chapter is restricted to Europe from the mid-thirteenth century onwards. In at least some early examples, the craftsmen who had made the clock remained in the locality and acted as both keeper and repairer. This was, for example, probably the case with the clock positioned over the rood screen and made by the Austin Canons at Dunstable Priory in 1283.[3] Such a mechanism would have needed regular attention and those who constructed it were on hand both to wind and repair it. Ely Abbey archives record that in 1302 one of the monks had responsibility for ensuring the clock was kept in order and running accurately to mark the time of assembly for prayer by night and by day. In Rouen, a beneficed priest, Raoul de Carville, was charged with care of the cathedral clocks c.1414/15.[4] Similarly, the well-known astronomical clock conceived and partly made by Richard of Wallingford would have been wound, set, maintained, and repaired 'in house'. If it were the sacristan's job to wind and set the clock (invariably the setting was to local time), repairs and maintaining the movement might have required more practical ability. There are many references to regular payments made to clock keepers who were sometimes also able to act as clock repairers. From the later fourteenth century, by which time more clocks were being provided for cathedrals and larger churches, travelling craftsmen sometimes provided the new clocks, but once they had installed the movement it was down to other practical people, either local, if available, or from further afield, to maintain and repair it. For example, in 1373 the clock at Queenborough Castle in Kent was attended by the clockmaker John Lincoln, who travelled from London for the five-day job. In 1567 the Royal clockmaker Nicholas Oursiau travelled to repair various clocks including those at Hampton Court, Westminster Palace, and Oatlands.[5] The Westminster Palace clock, built c.1368, had already needed substantial repairs in the mid-1420s, which were carried out by Geoffrey Dalavan, whose widow was obliged in 1428 to submit an itemized account in order to receive payment for her late husband's work. Less than 100 years old, the clock was evidently in a fairly poor state, needing repairs to the striking fly, barrel ratchet spring, foliot suspension, a pinion, and the renewal of all the weight lines. Dalavan's widow was also due 'his rewarde for the sayd yere', suggesting he had been employed as clock keeper, and probably winder. Interestingly, there were also costs for making a protective cover for the clock—an early example of a clock cabinet to protect the clock from dust, dirt, and draughts.[6]

But repairs were not necessarily undertaken by dedicated clockmakers. At Exeter Cathedral, in 1284, the bell-founder Roger de Ropford and his heirs were charged with providing the bells and repairing the organ and the clock.[7] In other examples during this period, apart from generic 'repairs', payments were made for 'painting' the clock, almost certainly repainting the dial—one sees similar regular payments for the painting of sundials. Typically, payments were also made for new cords for the weights and for upgrading of the clock with a clock case or extra striking work. Such clock movements were sometimes updated or altered early in their working lives—no doubt, these early designs were still evolving, with the makers and repairers learning as they worked. The archives of St Paul's Cathedral record a clockmaker, Bartholomew, working there in 1286, presumably either installing or repairing a clock movement and, fifty-eight years later, 'Wauter Lorgoner de Suthwerk' making a dial for the clock with an automaton angel, and taking 'les veux ustimentz'—old parts no longer serviceable—as part of his payment. These parts were surely of iron, a valuable commodity, and would have been for recycling.[8] Similarly, if a clock movement were replaced, the clockmaker would certainly have recycled the ironwork of the old clock and doubtless many early clocks disappeared as a result of this practice.

Early spring-driven clocks, such as those described by Paulus Almanus c.1475–85, were often designed to run for just twelve hours, to be wound at the beginning and end of the day. By the time Paul inspected this group of clocks, however, one at least had already been converted to run for twenty-four hours.[9] Complex public clocks, such as the great astronomical clock at Strasbourg, received constant repair and updates, the celebrated Habrecht dynasty of clockmakers being the first of a series of craftsmen to work on, and even replace, parts of that monumental structure over the years.[10] At the time, such alterations were wholly uncontroversial and generally simply seen as improvements; the concept of historical integrity would only enter public consciousness in the late nineteenth century.

As for the lubrication and cleaning of clockwork in this early period, naturally occurring oils and greases were the obvious choice, animal fats rendered from the carcasses of sheep and cattle being the most successful at lubrication, at least in the short term. But these lubricants broke down readily, going solid, rancid, and corrosive after just a few months, and most iron clocks would have needed regular cleaning and relubrication. Most, doubtless, were simply given a wiping down before reoiling, the residual oils forming a useful semi-protective layer on the iron parts of the movement. Some movements, however, had their old lubricant removed by heating, either by boiling or burning off. The wardens' accounts in 1689/90 for the church at Braunton, Devon, include a payment of one shilling to 'John May for Wood for the

---

3 Beeson 1971, 13, the earliest of many examples given of the construction, repair, and replacement of English church clocks before 1500.
4 De Beaurepaire 1892, 311.
5 Bodleian MS: Rawlinson A. 195 C.263–270. On Oursiau see the entry in BBT.
6 Post & Turner 1973.

7 Beeson 1971, 14.
8 Madden 1855, 3. *Cf.* Chapter 4, 'Early survivors' and Chapter 7 at n. 64.
9 Clock 18 in Leopold 1971, 170–2.
10 Ungerer 1922, 19–22.

Clocke to burn off the oyle and dirt thereof' and a similar amount is recorded in 1725/26 for the church clock at Gittisham, East Devon, for 'Boiling ye Clock and Brasses'.[11] From 1400 onwards many clocks were installed in the churches of towns and villages throughout Europe and every one of them will have needed maintenance and repair. The fraternity of clock keepers and professional repairers would continue to grow for the following half millennium.

THE SEVENTEENTH CENTURY

The introduction of the pendulum as a controller in clockwork in 1656, and that of the balance spring in watches in 1675, was certainly responsible for an exceptionally high number of alterations and changes to clocks and watches. Such was the improvement in timekeeping they brought that the new devices were, where feasible, 'retrofitted' to the majority of existing watch and clock movements and must have represented a very significant part of the 'after-sales service' offered by many clock- and watchmakers in the major European centres. The phenomenon has been clearly illustrated from English lantern clocks.[12] A typical example for public clocks is offered by Melchbourne church, Bedfordshire, where, in 1733, Joseph Eayre of St Neots repaired the clock and converted it to pendulum control, providing a case to protect the pendulum at the same time. This kind of upgrading was doubtless both remunerative for the clockmaker and a most satisfactory improvement for the owner in the usefulness of the watch or clock. Even the type of complex astronomical table clocks made in Renaissance South Germany, principally intended as items of status and indicators of intellect and temperance, were so improved, the updating often including the conversion of twenty-four-hour dials to twelve hours, which by the mid-seventeenth century had become the norm in domestic clocks. Astronomical clocks such as these seem to have continued in use, and their calendrical and astronomical indications referred to, long after their creation. English clocks with astronomical indications, created before 1752 and thus with the calendar based on the Julian reckoning, have almost always had their calendars brought up to date following the adoption of the Gregorian calendar that brought Britain into line with the rest of Europe.

Such conversions did not only occur during the period when the inventions were new but continued over many decades. For example, seventy years after the introduction of the pendulum in 1656, the London clockmaker Henry Elliott described a method for converting a balance clock '. . . to alter it to a Pendulum either long or short'.[13] Similarly, William Derham in his famous book *The Artificial Clockmaker* gives clear instructions on altering and updating 'old Balance Clocks'.[14] Such was the ubiquity of this practice that today only a small handful of pre-pendulum, balance-controlled clocks survive in original condition, as do unaltered pre-balance spring watches.

EARLY HOROLOGICAL PUBLICATIONS FOR THE CLOCK KEEPER

With pendulum-controlled movements capable of keeping excellent time over long periods, and usually constructed to run for a week (sometimes even a month or more), there was a great surge in the demand from private owners for domestic clocks. The improved performance of balance spring watches brought about a similar increase in demand from those wealthy enough to afford such items. One result was the first book printed in English concerning clocks and watches and how to appreciate and care for them. The 1675 *Horological Dialogues* by the London clockmaker John Smith was intended for the intelligent lay reader '*shewing the nature, use, and right managing of clocks and watches . . .*'.[15] Written as a dialogue between a clockmaker and an enquirer, all the essentials are explained in simple, accessible terms in this remarkable pioneering text. Following advice on choosing a good watch or clock from the many options available in London at the time, Smith provides chapter and verse on safe packing and transporting one's purchase. With the movement covered with paper or cloth and protected with straw stuffed all around it, it is interesting to note, for example, that Smith advises transporting a longcase clock with the movement fixed in the trunk with the pendulum attached and tied to the backboard. On setting up at home, the case should then be fixed in two places against the wall—very sound advice, largely ignored over the following centuries by owners and causing countless unnecessary stoppages!

Advice is given on regular cleaning of a working clock, proposing a twice-yearly oiling of the movement and an annual clean of the movement, although without details as to how that should be carried out. While this would be far too often today, it may well have been necessary in 1675, given the poor quality of the oils available. Further good advice is to keep an eye on rope-driven clock movements to ensure the rope does not begin to fray 'for when once [they] begin to grow any whit worn they foul a Clock exceedingly'. Another very good point, as true today as then, is that longcase clocks are so well made that they will continue to run and give little sign of a problem, even though they have poor lubrication: 'This I fear will be the fate of those long swing eight day Pendulums, so much at this day celebrated, for so long as they keep going well other things are not considered, nor no regard is had to those things that should continue them in that placide estate of motion'. On the subject of having the clock overhauled,

---

11 Ponsford 1985, 48.
12 White 1989, 402–87, who also offers advice on how to recognize recent alterations and on best conservation practice.
13 Elliott 1726, 11.

---

14 Derham 1696/1734, cited from the 3rd edn, 1714, 61–8.
15 Smith 1675, subtitle.

he insists the owner should take the clock to the clockmaker and not expect him to visit them to do the work *in situ*: ' ... those that desire this are certain to have their business never well done', and that they should allow him a decent period of testing in his shop before return. Sound advice is also given on winding clocks, and Smith's practicality as one experienced in setting clocks is further evidenced by his instruction to take the minute hand past the point of correct time and move it back onto that point, to take up any backlash 'the not understanding of this hath bread great mistakes between many a gentleman and his watch'. Smith recommends the use of a single sundial for checking a clock: '... 'tis seldome known that two sun-dials go true together', and ideally at the same hour of day, in case the dial itself has inaccuracies in division.

Eleven years after publishing this excellent book, John Smith published a second, *Drawn up Chiefly for the use of the Gentry, on order to their more true Adjusting, and right Managing of Pendulum Clocks, and Watches*, in which he explains the concept of the equation of time and how best to set one's clock or watch correctly, using a sundial.[16] Eight years later he produced his third and final textbook, *Horological Disquisitions* ... in which he gives further explanations for the equation of time and provides a table by which it will be possible to set a clock at regular intervals so that it accords with what the sundial says. This relates to the prevailing belief that it is the Sun which naturally provides 'correct' time, that the sheer regularity of pendulum clocks is not what is ideal, and that it needs correcting to 'follow' the Sun.[17] At the time, this view on the correct setting of a clock to time was common throughout Europe but during the eighteenth century was turned on its head, with mean solar time becoming the norm, and solar time seen as an inconvenience. Printed equation tables were developed to show the difference between the two allowing the owner to set his clock to mean time using the sundial.

The book goes on to provide more advice on setting up one's clock correctly, especially for those without access to a professional clockmaker, providing great detail in setting the clock in beat, ensuring the pendulum is free of the case, and that the pulleys are hanging correctly. For bringing the clock to mean time, Smith even goes as far as to recommend dividing the rating nut on the pendulum into seconds per day. Then, in an extraordinary further refinement, he proposes providing a movable register against which the arc of the pendulum can be read, from which any diminution of the arc may reveal a problem with lubrication and the need for the movement to be cleaned. As for setting clocks correctly to time, here Smith provides further helpful information. Doubting (as in *Dialogues*) the truth of sundials, he recommends, if one is used, that the time always be read at noon, where inaccuracies in the engraving are irrelevant. He then describes a better 'noon indicator' of his own invention, which employs two plates having small holes aligned which, as the Sun crosses the meridian, will give a brief flash of light on a black board behind.

An even more accurate method was sent to him by the astronomical clockmaker Samuel Watson (post-1640–?post-1726): one aligns a notch fitted against the side of a window frame with a neighbour's wall or chimney, and one is able to observe the passing of specific stars at night to very great precision. This was perhaps the first publication of this method of time determination, using buildings as a giant transit instrument, although Robert Smith states that it was first used by Christiaan Huygens.[18] It was described by several other authors,[19] including Nicholas Saunderson (1682–1739), from whose lectures John Harrison (1693–1776) probably learned of it.[20] Perhaps a more readily available, and reliable, system of time determination was to observe the passage of stars or the Sun in the same way, but using a transit telescope. This was also described by Robert Smith in his *Opticks*, as earlier by Derham.[21] However, for most owners of clocks and watches, if local 'public' clocks such as those on the church or in the tavern were not considered sufficiently reliable, the use of a good sundial would always provide local time to an accuracy close to one minute and was the routine time standard up to the mid-nineteenth century.

REPAIRS AND OVERHAULS

Something concerning seventeenth- and early eighteenth-century repairing practice can be derived from activity in Tompion's workshop.[22] Even a manufactory like his was prepared to take on repairs and overhauls for clients, even on objects not of his own production. On his various visits to Tompion, Robert Hooke brought a number of clocks and watches in for repair on behalf of other people. In 1695 Constantyn Huygens records in his diary taking a clock by Severyn Oosterwijk and a pocket watch to Tompion for repair.[23] However, he was not happy with Tompion's work, took the clock to Jonathan Lowndes for attention, but soon returned it to Tompion as 'it had not been repaired well'. Six years earlier Huygens had, however, been happy with Lowndes' work, as in January 1689 he had taken a watch to Lowndes for repair, which he did '... very conscientiously and beautifully, for the price of 15 pounds sterling', a major repair job at such a price; most of Tompion's repairs were usually charged at shillings rather

---

16 Smith 1686.
17 Discussions of this problem in White 2009; Turner 2015b.

18 Smith 1738, II, 325, but the method is far older; see Poole 1915.
19 Smith 1694; Sully 1737, 87; Thacker 1714, 16–17, the latter likely being a spoof by the polymath satirist John Arbuthnot (1667–1735).
20 Described by him in 1730, Clockmakers' Company Library, London, ms6026.
21 And would be again in the third quarter of the nineteenth century. See Clark 1882; 1888.
22 Various repairs are recorded in Evans, Carter, & Wright 2013, for example (240), a detailed bill to 'Mr King' 11 March 1706/7.
23 Evans, Carter, & Wright 2013, 87–8.

than pounds. Given the large number of repairs going through Tompion's business, there may have been two craftsmen, perhaps journeymen, employed full time on such work.[24]

There are few exact details about repair work in Tompion's workshop apart from references to standard jobs such as 'stopping holes' in the movement. Stopping of holes involved the removal of wear in the pivot holes, where necessary. Today the process used is known as 'bushing' and involves the broaching out of the worn hole (re-centring at the same time) and inserting and riveting-in a hardened-brass sleeve or 'bush'. It is unclear when the practice of bushing was introduced, the earlier practice being to plug the hole completely using a depthing tool to mark and redrill the correct position for the pivot hole.[25] The term 'stopping holes' suggests that this is what was being done. An interesting reference in the 1699 correspondence of William Winde reveals a process Tompion used for patinating a chased gold watch case by boiling it, presumably in some form of acid. Winde decided against having this: 'I found they pute corroding lickeours, wch might doe some damage to ye graving and for that reason I whould not have it done, besides thier is so much moneyes saved [it cost 8 shillings] and the color it gettes by boylinge is but for a small thyme'.[26]

Evans supposes that Tompion loaned watches to customers while theirs was being repaired, noting that Samuel Whichcote of Fleet Street did this, and on one occasion was obliged to place an advertisement for the return of the one loaned.[27] As will be seen, this 'loan service' certainly occurred later in the century and must surely have been a widespread and very sensible practice by makers undertaking repairs.

In the early eighteenth century we learn more about care and repair practice from Henry Sully's *Regle Artificielle du Temps* . . . , especially from the third edition, enlarged after Sully's death by Julien Le Roy. Sully states that watches should never be opened up '. . . unless there is an absolute need to adjust the balance-spring *(resort reglant)*; that being necessary, care must be taken that hair- and wig-powder or other dirt does not fall into the work; indeed, they should be looked after like the most precious jewels'. From the same period, a notebook written by the French watchmaker Jean Helot (mid-1650s–1728) provides a mass of interesting and useful information relating to the watchmaking and repairing trade at the time, one of the main observations being the reluctance of owners of watches to pay for high-quality repairs and maintenance.[28]

CONTRACT MAINTENANCE

By the early eighteenth century in England, larger houses and city institutions that owned a number of mechanical clocks were beginning to adopt a practice for the maintenance of their clocks. This practice would continue, in one form or another, to the present day becoming what we know as the concept of contract maintenance. Recognizing the specialist nature of the maintenance of these complex and sometimes wayward machines, this sensible idea employed the clockmaker to look after all that pertained to them. For an agreed sum, a clockmaker would ensure the timekeepers in a given house or organization were cared for, serviced, and performed reliably. In many cases, especially in later years, this included weekly visits to wind the clocks, meaning the owners could relax in the knowledge that the clocks were looked after and would continue to work and tell the correct time. In 1700, Tompion was paid £13 for thirteen years' care and repair of the clocks at the Inner Temple, an early example of this practice, although this appears to have been a nominal sum and it is unlikely to have included the weekly winding.[29] Maintenance was also a large part of the duties of 'royal' clockmakers throughout Europe.[30]

ADVICE FOR OWNERS

For those who perhaps owned just one or two clocks and watches and needed advice on caring for them themselves, one of the first 'self-help' guides, *How to Manage and regulate Clocks and Watches*, appeared in France in 1759. Written by the young and ambitious Swiss-born Ferdinand Berthoud (1727–1807), and similar in intention to Smith's *Horological Dialogues*, this book was more detailed and clearer in its descriptions, but was also as verbose as his voluminous later works. Berthoud begins this laboured, but useful, work by explaining why owners need to follow guidelines on the subject and adds that 'It would be no less useful to watch and clockmakers, because the pains they take to make good watches are completely wasted if those to whom they sell them do not know how to operate them'.

After explaining the equation of time necessary for the owner to set his clock or watch correctly by a sundial, Berthoud describes the movements of both and explains how they work. He then looks at the causes of variations in the timekeeping of both clocks and watches and states what he believes one can expect from a good specimen. A good spring-driven pendulum clock, he reckons, can keep time to within a minute over a fifteen-day period, and, he supposes, a weight-driven regulator with temperature compensation and a long pendulum will stay within a minute over a year. 'As for the accuracy we can expect in the common

---

24 Evans, Carter, & Wright 2013, 94.
25 Information imparted to the author during training (Hackney College 1972–5) by ex-Clerkenwell clockmaker Charles Allen (c.1915–c.2003). John Robey (private communication) concurs with this, commenting that the depthing-tool was exclusively a clock repairer's tool, as the wheels in English clocks were originally 'topped-in' during construction and had no need for a depthing tool.
26 Evans, Carter, & Wright 2013, 94.
27 *General Advertiser*, 30 May 1751.
28 Turner 2011.

29 Evans, Carter, & Wright 2013, 97.
30 See Augarde 1996, 28–9 for France; Jagger 1983 for Britain.

or ordinary watch, we must not complain when it has but a one minute daily error'. These figures conform approximately with modern day experience of testing and rating clocks of these types, although the likelihood of a standard regulator (i.e. with **deadbeat escapement**) staying within a minute over the course of a year are slender. An important point Berthoud makes for owners is to recognize the difference between a watch that is simply gaining or losing consistently (in which case, it can be adjusted by the owner using the regulating disc) and one that keeps time erratically, in which case, it needs repairing or, if it is just a poor watch, replacing. Advice is given on acquiring a good watch or clock (in short, go to a good maker and pay more), followed by notes on choosing who to repair and maintain one's watch (go to a good repairer and pay more). On this point Berthoud states '... at the same time he should also have it cleaned at least every three years', further noting that:

> There are people whose pocket is so warm, that in a very little time the oils of the watch dry up; which causes the watch to vary and later to stop ... Those who are in this situation should have their watches cleaned more often or protect their watch from this excess heat. One does this by lining the watch pocket.[31]

A final chapter repeats all the advice in another (six-page) summary. If the reader has not by then taken on board the simple messages contained in the book, it is difficult to know what more Berthoud could have done! In his next work, the monumental *Essai sur L'horlogerie* ... of over 1,000 pages, Berthoud includes some twenty pages on the causes of stoppages or variable timekeeping in watches, including a separate chapter on clocks, and the appropriate 'trouble-shooting' for each. The correct cleaning of pivot holes is remarked upon, but unfortunately no clue is given as to his methodology for cleaning clocks and watches.

LUBRICATION

A later section in Berthoud's comprehensive work discusses the effects of friction and the variable viscosity of the lubrication but does not include any details of the types of oil used, nor any comparative considerations. In fact, references to lubricants in eighteenth-century reference works appear to be non-existent. Frustratingly, even Henry Sully, describing in 1716 his invention of oil sinks and discussing the importance of protecting the oil and preventing it from spreading, does not actually tell us what oil it is, nor where it comes from.[32] Similarly, J. A. Lepaute, summarizing the work of Sully and Julien Le Roy and stressing the 'absolute necessity' of oil, says nothing of what oil should be used, and although Alexander Cumming also considers the subject at length, he, too, is silent on the response to the question.[33]

REPAIR SERVICE

Again, although archival references to clock and watch repairs are plentiful, very few give clues as to the specific practice for clock and watch cleaning and repairs. It is certain that most retail clock- and watchmakers had some kind of workshop on their premises, and at least simple, everyday repairs and overhauls were often undertaken there, especially in the provinces. In a surviving early eighteenth-century manuscript notebook by the Lancashire watchmaker Richard Wright of Cronton, there is an entry for the costs of repair work to a number of clocks and watches. For example, repairs to the wheels and pinions of a spring clock movement are charged 5/6 (five shillings and sixpence), with an additional charge of 2/6 for 'twice cleaning', perhaps before and after repairs, making 8/- (eight shillings) in total.[34] A 1795 regulation booklet for The Society of Watch and Clockmakers of Leicester contains a detailed price list for new clocks and watches and includes seven pages for all kinds of repair work. For example, cleaning a turret clock cost 10/6 (ten shillings and sixpence), cleaning a thirty-hour clock was 1/6 (one shilling and sixpence), a three-part (chiming) spring clock was priced at 7/6 (seven shillings and sixpence), and a new spring for a three-part spring clock (including cleaning) cost £1/4/- (one pound four shillings).[35]

It is equally clear, however, that many businesses, especially those in larger cities such as London or Paris, did not carry out much repairing themselves, perhaps either lacking the space or the staff. Just as with many high street jewellers and watchmakers today, they preferred to put the work out. No doubt there were a number of businesses, large and small, that undertook what became known as 'jobbing' work (repairs and cleaning, as opposed to making new clocks and watches), but the records of very few companies have survived. An advert published in 1788 in the London newspaper *The World* tells us a little of this practice and, under the title of 'Intelligence of Public Utility', offered a cut-price service:

> The long delay, vast expence, and other inconveniences occasioned by putting Jobbs in the Jewellery and Watch branches into capital sale-shops, which to the keepers thereof, are more troublesome than profitable, having to send them to work-men at great distance; and the frequent loss of articles in transferring them from one place to another, has induced W. RHIND to open a work-shop in Cranbourn Passage, Leicester-square, where ... watches are ... repaired, with taste, expedition, and care, 30 per cent

---

31 Berthoud 1759.
32 Sully 1711; Sully 1737, 275.
33 Lepaute 1755, 37–4 & 79; Cumming 1766.
34 Smith 1985, 620, 624
35 Hewitt 1992, 47–9.

lower than the usual prices ... Watches well cleaned at 1s 6d, little above half the usual charge. N.B. Servants having the misfortune to break any articles in these branches, will find every indulgence by applying as above.[36]

The records survive of what must have been one of the largest firms in London doing trade repairs—Thwaites and Reed of Clerkenwell. Founded by Aynsworth Thwaites (1719–1794), the firm was a major clock and turret clock manufacturer both on their own account, but on a much larger scale as trade makers and repairers. A large proportion of the work recorded in the company's day books[37] appears to be repair work for the London trade, and provides a few clues as to the practice at the time. For example, the entry for 4 January 1785 states:

Mr Ellicott

For Mr Walpole

From the House in Spring Gardens Cleand a Eight day Clock Name Eardley Norton, stopt up the holes in the back Cock & Repaird the Scapement New Silverd & Varnishd the Dial Plate & blewd the Hands
    T O

This entry shows Thwaites overhauling a clock originally made by the Royal clockmaker Eardley Norton (1728–92). The owner of the clock was one 'Mr Walpole'. However, rather than taking the clock to its maker Eardley Norton for repair, Walpole took it to the celebrated London firm of Ellicott at the Royal Exchange, then owned by Edward Ellicott, the son of the founder John Ellicott (1706–72). No doubt Mr Walpole would have been unaware that Ellicott's then passed the work on to Thwaites, whose clockmaker was probably instructed to collect the clock from Walpole's house (as the address is given in the entry). Thwaites' man would surely have understood not to mention his employer, and that on his visit to collect and to deliver the clock back again, he was representing Ellicott. On return of the clock, Ellicott would have paid Thwaites and then sent their own bill to Walpole, doubtless with a significant addition. The final part of the entry, 'T O' is Thwaites' charge to Ellicott written in code.[38] The job cost 8/6 (eight shillings and sixpence). The work carried out suggests a clock with several years of wear to it, requiring the removal of wear in the rear pallet pivot in the back cock and repairs to the escapement. Resilvering and lacquering the dial and reblueing the steel hands was another piece of restoration frequently carried out by the firm, and was done to present a fresh and improved appearance for the owner.

THE CLEANING OF CLOCK MOVEMENTS

The inclusion of the general remark 'Cleand' evidently suggests the movement of the clock was completely dismantled, and that the parts were probably the subject of some kind of active cleaning regime. Unfortunately, this seems not to be specified anywhere in the records, but comparable techniques used at the time for cleaning other functional metalwork involved the use of light abrasive powders on a brush to remove tarnish, old oil, and dirt. It is said that traditionally domestic brasswork and pewter were cleaned, even 'scoured', using the plant 'horsetail' (*Equisetum arvense*). Containing ten per cent silicon and a number of organic acids, it combined both physical-abrasive and chemical qualities, although whether it was ever used by clockmakers is uncertain. In the late 1970s this author asked the distinguished London clock restorer Dan Parkes (1919–2002)[39] whether he knew what cleaning methods seventeenth-century clockmakers employed. He replied that he believed pretty much the same as he used, which was the fine abrasive powder rottenstone suspended in paraffin.

This was certainly the practice in France later in the eighteenth century, as described by Pierre Vigniaux,[40] who describes his cleaning and repairing methods and reveals that worn pivots holes were stopped up and drilled (as in earlier practice), but with a clever twist: he recommends making an off-centre hole in the plug, and not riveting the plug in tightly at first, so that if the drilled hole proves to be at a slightly wrong centre distance, the plug can be turned to change the depth of meshing. Under cleaning, following dry brushing and the use of rottenstone to clean the teeth of the wheels and pinions, he brushes with chalk ('*Blanc Espagnole*') and then pegs out with wood, the point soaked in oil. Vigniaux's work is important, as it appears to be the first textbook describing actual repairing and cleaning processes for practitioners. Just one year later, a treatise appeared in Spain—which is also pioneering in this respect—including many examples of typical repairs for clocks and watches and how to carry them out.[41] In 1789, Jean-Bruno Savarin in Bourg (France) offered watch and clock overhaul payable by subscription, an ordinary watch costing six livres (about five shillings sterling—very roughly £25 in today's money), a repeating watch twelve livres, and twenty-four livres for a clock. A guarantee of two years was offered—probably quite an early example of such a concept.[42]

By contrast with active cleaning work, another job in the Thwaites records, charged on 22 April 1785 to Ellicott, also for Mr Walpole, records simply: '*Wiped out a Spring Striking Diall & a*

---

36 *The World*, 7 March 1788.
37 Now in the library of the Clockmakers' Company, Guildhall London.
38 The letters of AYNSWORTH represented 1 to 9, 0 being represented by a 'b'.

39 His firm, Rowley Parkes & Co., coincidentally, descended from that of Eardley Norton. Elton 2002, 28–9.
40 Vigniaux 1788, part 4.
41 Zerella e Ycoaga 1789.
42 Tardy 1972, 588. Earlier, Jean Helot had offered guarantees for a yearly subscription of an ecu and would even issue a written guarantee. Turner 2011, 796, 798.

New Line ... S O' for which the cost was 4/6 (four shillings and sixpence)—just over half the cost of the previously mentioned job. The reference to 'wiped out', where one would otherwise have expected to have seen the expression 'cleand', suggests a somewhat more perfunctory operation, where old oil and dirt are removed, but the movement as a whole is not subjected to a full cleaning regime (and perhaps without closer study and light repair). It is tempting to suggest that simply 'wiping out' might not involve dismantling, but 'a New Line' for one (or both?) of the **fusees** reveals that the movement must have been taken apart—it is not possible to replace the fusee line otherwise—so perhaps 'wiping out' simply meant wiping all the parts to remove old oil and loose dirt, but not further attentions.

While the Thwaites records are a wonderfully rich resource of historical information, the information relating to repair and restoration is fairly repetitive. A cursory sampling of the archive has not revealed useful information about *how* clocks were cleaned and repaired. What is clear is that clocks were constantly being worked on, repaired, and updated, with almost every conceivable amelioration and alteration being carried out at one time or another by Thwaites. No doubt this simply reflects what was routinely going on in Britain (and the wider world of clock repair) in the eighteenth and nineteenth centuries, and there is plenty of supportive evidence found in the many books on provincial clock- and watchmaking published in recent years. Similar regular maintenance and technical updating occurred with marine chronometers where the safety of a ship and crew depended on the reliability of the timekeeper. For example, the British Admiralty's chronometer records show that chronometers were sometimes overhauled thirty times in the course of eighty years' service, often being altered and updated as part of that work. All this reminds us that the idea that an eighteenth-century clock (for example) today being in 'totally original condition from the day it was made', is almost impossible.

WATCH REPAIRS

This same constant repair and cleaning, but of watches rather than clocks, is echoed in another historical archive, 'accidentally' preserved in the Royal household. A number of pages cut from manuscript watch-repair day-books belonging to the royal clockmaker, Justin Vulliamy, used during the 1760s and later, were recently discovered 'recycled' as packing in porcelain and in the structure of royal carriages. The pages are incomplete and have only briefly been examined to get a sense of work undertaken by Vulliamy, but they provide some interesting indicators to the business of watch repair in the second half of the eighteenth century. The entries record the name and address of the customer, the make and serial number of the watch, and the repairs required. As with Thwaites service for clockwork, just about every conceivable repair could be undertaken. Entries range from the general, e.g. 'Stopt', 'Had a Fall', 'Goes Badly', 'Put in Order', 'Clean and regulate', to the more specific, e.g. 'Spring broke', 'Balance broke', 'Stops in ye night'. An overall view of these English records suggests that the principal weaknesses in both clocks and watches under repair were little different from today's experience in clock repair and conservation: breakdown in lubrication (sometimes, but not always with resulting wear to moving parts); **mainspring** breakages, which often caused breakage of the fusee line or chain (fusee chains almost never break of their own accord); and breakages and damage caused by mishandling.

Quite major work was sometimes required, including making a whole new gold case, taking the bruises out of a case, repairing the enamel, or even re-enamelling the whole case. We know that Vulliamy employed men on his premises in Pall Mall and there was certainly some kind of workshop in the building. Although the German noblewoman Sophie von La Roche (1730–1807) visited Vulliamy in 1786 and, in describing his shop in her diary mentions dust covers for protecting the clocks on show, she does not mention a workshop, and she is unlikely to have been shown it.[43] As one might expect from such a high-end business as Vulliamy's, the clientele was both domestic and international and often aristocratic or royal. As with Tompion's business years before, the common practice was to lend the customer a watch, if needed, while a repair took place. The watches themselves ranged from what must have been fairly new timepieces, often French as well as English, but also some examples more than fifty years old, including, for example, watches by Tompion, Delander, Wise, and Goode, all still serviceable after many decades of use, and worth maintaining. Indeed, over the years, watchmakers would sometimes recondition second-hand watches and sell or part exchange. Several examples are found in the records of John Bull of Bedford,[44] and fine-quality seventeenth- and eighteenth-century watches in later silver cases are frequently seen today, their gold cases having probably been 'cashed-in' to provide funds, while also paying for a new silver case in a more modern style. An alternative recycling of still-serviceable yet orphaned watch movements was the early nineteenth-century practice of using them as the movement for inexpensive 'cottage clocks' for the mantel piece, and small sedan clocks for wall mounting.[45]

The business of watch repair and cleaning both in Britain and abroad continually increased as larger numbers of inexpensive watches became available, particularly from Switzerland, and the new century saw greater numbers of the middle and working classes being able to afford them. In the particular case of the Vulliamy firm, however, the surviving records suggest that the business of repairing and cleaning watches declined during the first decade of the nineteenth century as that company concentrated more on luxury products for their royal and noble patrons, leaving other London watchmakers to pick up the repair trade.

---

43 Hutchinson, 1992.
44 Pickford 1991, 19.
45 Allix & Bonnert 1974, 27–9.

In the second generation of the Vulliamys, Benjamin Vulliamy (1747–1811) was, from the 1770s, almost certainly the first in London to introduce small spring clocks in sculptural cases following French fashions. For these clocks, a new type of small fusee movement was created that, also following latest French practice, employed **anchor**, or half-dead-beat, escapements instead of the traditional verge with bob pendulum. This reflected a significant change in the use of domestic clocks—no longer intended to be carried from room to room (where the verge escapement was important, being much less easily upset), these clocks were for a fixed situation on a side table or a mantelpiece. Thus, for the first time the more accurate anchor, or half-dead-beat escapement, with heavier pendulum on spring suspension, was feasible for small domestic clocks, and they performed significantly better as a result.

ESCAPEMENT CONVERSIONS

This evidently set clockmakers—both in Britain and on the Continent—to thinking about how existing spring clocks might be converted to the heavier pendulum and anchor escapement. With the old verge escapements often badly worn and difficult to repair efficiently, a conversion of this kind would improve the timekeeping of their customers' clocks and would provide plenty of remunerative work for jobbing clockmakers everywhere. As with the earlier conversions from **balance wheel** to pendulum, the owners were keen to see improvements in timekeeping and were happy to pay for such conversions. Thus, from the beginning of the nineteenth century, a very high proportion of English table clocks were converted from verge escapement to anchor, with larger heavy pendulums. In France, especially following the 1840s, after Achille Brocot (1817–78) had introduced the escapement bearing his name and his clever adjustable spring suspension (which was almost universally fitted to new French clocks from that time onwards),[46] many clockmakers in France would routinely offer to 'improve' customers' clocks by converting them to Brocot's escapement and suspension. Similarly, throughout the previous two centuries, clocks had been subject to aesthetic 'improving' to bring the appearance and style of a still-useful clock up to date. Thus, cases were reveneered (sometimes even changed altogether), and mounts or feet and finials were added to echo the latest fashions in furnishing.

While sometimes this technical and aesthetic 'updating' was not carried out to a particularly high standard, in the majority of examples the work was done with skill and integrity and certainly should not be seen as examples of 'botchery', or vandalism. In Britain and on the Continent, there was no concept of the need to preserve the original integrity of a clock case or movement—changes were simply welcomed as functional and aesthetic improvements and are today simply part of the technical and practical history of those functional objects. It was only later in the nineteenth century that the idea occurred that an object should have intrinsic 'originality', and that this perceived 'purity' was something to be admired and sought.

It is often stated among antiquarian horologists today that Benjamin Lewis Vulliamy (1780–1854), the third generation of that clockmaking family, was responsible for much horological 'vandalism' in that his firm routinely removed original movements from fine seventeenth- and eighteenth-century clocks and fitted new ones. A particular *cause celebre* is the example of the fine table clock known as the Castlemaine Tompion, from which Vulliamy removed the movement in about 1845 and replaced it with a new one of his own. The clock belonged to the Duke of Grafton, who was evidently content with the proposal, requiring a fully reliable and trouble-free clock, and was happy to allow Vulliamy to take the movement as part of the payment. While Vulliamy believed the original Tompion movement was unserviceable, he evidently regarded it as a fine 'curiosity', presenting it to the Institution of Civil Engineers of which he was an Associate Member. While such an action would be considered reprehensible today, it reflects an interesting stage in the development of a connoisseur's view of 'Antiquarian Horology'. Tompion's movement was a curiosity, but as a geared machine, was not yet understood to form an intrinsic part of the art work. Respect for the 'insides', as well as appreciation for the 'applied art' that was the case, would only gradually establish itself as the nineteenth century progressed.[47]

NINETEENTH-CENTURY CLOCK AND WATCH REPAIR

With greater numbers of standard watches and clocks produced as the century progressed, many more repairs and regular overhauls occurred. This was certainly true in early nineteenth-century America; the account books of the multi-tasking and trading Dominy family of clockmakers and furniture makers of East Hampton, New York, provide typical evidence. Taking just the year 1800 as an example, the family undertook cabinetmaking and carpentry, the sale of domestic equipment, and repairs to guns, tools, and carts, but the largest single trade in the records that year is for horology, with repairs making up the greater part. The entries for the twelve months show the making and sale of one non-striking clock and over 250 clock and watch repairs.[48]

The ordinary European trade of the jobbing watch and clockmaker was also increasingly busy, with a commensurate steady increase in printed textbooks on repairing for the trade. In 1824 the French watchmaker J-J-M Ayasse, of Angers, published his *Manuel de L'apprenti horloger . . .*, giving instruction to apprentices

---

46 Chavigny 1991.

47 *Cf.* Chapter 26 near n. 74.
48 Hummel 1968, 384–90.

on watch repairing. On the cleaning of movements, he stipulates the reblueing of screws and the brushing of parts with a soft brush and burned sheep's bone. Here it is presumed he meant gilt watch movement parts; even as late as the 1970s (in this author's training) this was the process taught as an alternative to putting gilt watch parts through cleaning solutions in the modern way, and it was still the method used by some old-school watchmakers at the time. When the finish was dull (presumably for ungilt clock movements, as gilt movements would never normally have abrasives used on them), Ayasse recommends polishing with rouge and spirit of wine followed by dry brushing, and then thorough pegging out of all holes with wood. Dry brushing of watch parts is also recommended by the 'Ancien Elève de Breguet',[49] who also recommends that the pinions should all have their leaves cleaned with wood ('bois blanc'—boxwood or dogwood—'pegwood'?), 'the pivots with cork and the wheels with a very dry clean brush'. In the following decades several more textbooks appeared, principally in France, giving advice and instruction on clock and watch repairing, including those by Lenormand (from 1830), Foucher (1850), and Liman (1854). In 1860, New Yorker Mary Louise Booth, in her updated translation of Lenormand's *Nouveau manuel ...* (1850), dedicates just two pages to cleaning and repairing, echoing the view that repairing is not as easy as supposed: 'They often rub the pieces with a brush and Spanish White and remove the gilding in a short time ... so that the watch is often dirtier when they have finished than when it was brought to them'.

HOROLOGICAL OIL

As mentioned, very little is known of horological lubrication until the nineteenth century. There is no doubt that a breakdown in lubrication is one of the greatest causes—if not *the* greatest cause—for clock and watches failing to work. The importance of finding the best lubricant was understood throughout horological history, and the subject is nicely emphasized by the apocryphal remark attributed to Abraham-Louis Breguet: 'Give me the perfect oil and I shall give you the perfect watch'.[50] There must surely have been work undertaken on refining horological oils in the seventeenth and eighteenth centuries, but the earliest note that has so far been found about the processing of a special horological lubricant is by Ezekiel Walker in 1810, who describes purifying olive oil for chronometers and cites a letter from P. P. Barraud confirming the quality of the lubricant.[51] Further essays appeared in the following decades,[52] and one of the advantages claimed by Achille Benoît (1804–1895) for the platinum–silver alloys he was investigating in the 1830s was that they preserved oil in the pivots since there was less oxidation from the metal to accelerate acidification and thus the breakdown of the oil.[53]

A particularly important study was 'Considerations Pratiques sur L'huile Employée en Horlogerie' by Henri Robert.[54] In it, Robert gives the first systematic, reasoned analysis of the problem of lubrication, discussing the various types of animal and vegetable oils and describing careful trials he had carried out to determine the relative values of various lubricants, considering the effects of atmosphere, light, and the metals with which they were in contact. Robert refers to one of the few earlier works on the subject—a report to La Societé d'encouragement in 1820 by Charles-Louis Cadet-Gassicourt (1769–1821) on the subject of refining olive oil, in which he states that by boiling many types of fine oil in alcohol it is possible to extract the colourless refined oil 'elaine' (Oleine). This, he adds:

> ... you will no doubt find useful to communicate to all watchmakers ... it seems to us that the pure elaine fulfulls all the conditions they desire ... [It is] colourless, little odorous, tasteless, ... having the consistency of white olive oil and difficult to freeze. Watchmakers use so little oil that the preparation of elaine would increase their expense little. They would have, moreover, the certainty of always using the same substance.[55]

Robert appears to conclude that refined olive oil is still the best, saying 'Oil has been extracted from a great number of substances, corn, melon seeds, cucumber, grapeseed, nuts and almonds of various kinds, cocoa, hazelnut, etc., ... None of these oils can be used for watchmaking.' One of the largest horological oil manufacturers today is the Moebius Company in Switzerland, founded in 1855. No trace has yet been found of any earlier companies. By 1877, however, advertisements in the *Horological Journal* included five different companies, three of them American, offering specialist horological oils, in all the American cases, refined from whale and porpoise products.

HOROLOGICAL JOURNALS

It was only in the second half of the nineteenth century that textbooks for the repairer multiplied, and that trade journals began to appear.[56] In England, following the creation in 1858 of a trade organization, the British Horological Institute, which produced its own *Horological Journal*, information about the activities of the 'jobbing' side of the trade, hitherto almost invisible, slowly began to appear. Similarly in France, Claudius Saunier's *Revue Chronométrique*, published from 1855 until 1914, contains

---

49 Ancien Elève 1827, 554.
50 No source is known for this quip in either the Breguet papers or contemporary literature. It is likely to be of recent fabrication. Private information from Emmanuel Breguet.
51 Walker 1810.
52 Beaufoy 1822; Laresche 1828; Long 1860 describing experiments on whale blubber and olive oil in 1814–15.

---

53 Letter to the Baron Seguier, 29 June 1838, in a private collection.
54 In Robert 1852, 81–158.
55 Robert 1852
56 On which see the account in Chapter 25.

increasing information on day-to-day workshop practice including repair work, with trade journals in other countries, including the *Allgemeine Journal der Uhrmacherkunst* in Germany, appearing in the following decades.

In Germany, Erasmus Georgi published a *Clockmakers Handbook* in 1867 with 'full instructions for the making and repairing of all kinds of timepieces',[57] while in 1869 in the USA the popular and long-lasting *Kemlo's Watch Repairer's Guide* 'being a complete guide to the young beginner in taking apart, putting together, and thoroughly cleaning the English **lever** and other foreign watches . . .', first appeared. By far the most comprehensive and detailed repairer's manual at the time, the author Francis Kemlo (mid-nineteenth century) includes detail on cleaning both watches and clocks, the watches to be dry brushed with a little very fine chalk (still likely to be abrasive) followed by pegging out of pivots and using pith to clean pinion leaves. He states that some watch repairers clean movements in oxalic acid followed by a neutralizing fluid, but he feels this alternative process is unnecessary. For clocks (only American are specified), he proposes wiping parts with cloth and running cotton strips through pivot holes. As for taking out wear, unfortunately he proposes punching the brass on the worn side of the pivot hole, a process which over the years has caused untold damage to brass clock movements.

POOR WORKMANSHIP

If breakdown in lubrication can be said to be the greatest cause, throughout history, of stoppages in clocks and watches, then one might equally say that the greatest cause of substantial damage to clocks and watches is mishandling and bad practice by owners and repairers—in short: human intervention. Noted collector and antiquarian horologist Reverend H. L. Nelthropp (1822–1901) was definite: 'More Watches are ruined by the complete incapacity of the workmen, than by fair wear and tear'.[58] Sadly, to some extent, it remains true today.

Another controversial example of bad practice is the tendency in the past for watch- and clockmakers to scratch their name, a date, and sometimes more onto parts of the movement or case. The reason for this practice was sometimes to record the fact that an overhaul had taken place at a certain date, in case the timepiece was returned under guarantee, but as often it seems it was simply to make their mark and memorialize their work for the information of future generations of watchmakers. While such records are today of historical interest and can be very valuable in determining where a clock or watch was at a given date in the past, it is not something which can be condoned today and is strictly prohibited in best conservation practice such as that defined by the UK's National Trust.[59] One of the most celebrated of practical textbooks produced at this time was that by the British Horological Institute secretary, F. J. Britten (1843–1913). Appearing first in 1878, Britten's *Handbook* went to no fewer than sixteen editions and has been an important source of advice for the clock and watch repairer ever since.

CLEANING SOLUTIONS

On cleaning practice, there appears to be no evidence that mid-nineteenth-century clockmakers were routinely using what later became known as 'Clockmaker's Soup' (a solution of soft soap and ammonia in water), though ammonia solutions were used for cleaning in the jewellery and electroplating trades by that time. In his thirty pages on 'Repairing and Examining Watches', Saunier specifies dry brushing, or wet-soap brushing, followed by water rinsing and then rinsing in alcohol, stating 'The employment of essences in cleaning watches is becoming more general every day'.[60] In an 1887 volume of the *Horological Journal* the editor suggests cleaning a nickel–brass clock case in an ammonia solution, so the ammoniated cleaning method probably began to be used occasionally in the later nineteenth century. Abbott, however, did not refer to it, preferring potassium cyanide and petroleum spirits for brightening watch parts where necessary,[61] and even in 1938, when the fourteenth edition of F. J. Britten's *Handbook* was published, the advice for cleaning clock movements was washing in paraffin and to use a brush and rottenstone if brightening were required. Having said that, in 1929, Britten's son, F. W. Britten (1869–1954)[62] proposed the use of strong (880) ammonia and soft soap in water and recommended leaving all the clock parts in the solution overnight, a regime which will surely have caused significant etching to the surface of the brass parts. There is also the attendant risk in this process, with any work-hardened brass parts, that stress corrosion cracking (SCC) might occur. This is something frequently seen in thin, old, work-hardened brass parts that, at some time, have been exposed to corrosive reagents like ammonia. Traditional ammoniated 'Clockmaker's Soup' cleaning is today a controversial process, and certainly one never used in conservation practice.

SETTING CLOCKS AND WATCHES TO TIME IN THE NINETEENTH CENTURY

As the nineteenth century progressed, owners and keepers of watches and clocks benefitted from improved methods for setting and keeping their timekeepers. Various improvements in sundial design, such as the dipleidoscope by Dent (1841),[63] enabled more

---

57 Georgi 1867.
58 Nelthropp 1873, 256–60.
59 Betts 1982.

60 Saunier 1881, 340, para. 525.
61 Abbott 1893.
62 Britten 1929, 239.
63 Mercer 1977, 196–203.

accurate setting using the Sun. Additionally, as national time networks began to appear using the telegraph systems, controlled clocks and time signals in post offices, railway stations, and shop window displays gradually provided greater access to correct standard time.

NINETEENTH-CENTURY COLLECTING—AND RESTORING

The mid-nineteenth century saw many European collections formed of precious jewellery and *objets d'art*, including watches and small clocks.[64] A 'collector's market' developed, and the value of good specimens increased. Such was the value and desirability of fine pieces among collectors, that inevitably, when supply of perfect examples dried up, fakes began to appear, and altered and damaged examples were restored and sometimes sold as 'perfect and original'. Thus, the often-unholy alliance of dealer and restorer developed to feed the market, and talented jewellers, enamellers, and watchmakers began to serve this doubtful industry with a new service, that of *antiquarian restoration*. Ironically, the connoisseurship that seemingly motivated this occasional fakery was often responsible for justifying restorations, as a culture emerged that regarded altered objects as 'incorrect', 'impure', or simply 'wrong'. It became accepted that putting an object such as a fine clock or a watch back into 'original condition' was a good—and 'right'—thing to do. Whether that object then went on to be sold as 'all original' was another matter, which was better not discussed, and there is no doubt that collectors either preferred not to know or, if they did, then often developed convenient amnesia.

Feeding into this culture towards the end of the century was the otherwise excellent and wholly creative Arts and Crafts movement, which sought a return to simple honest values in the making of functional objects. The traditional craft of clockmaking by hand (as opposed to factory, machine-made examples), which was already in terminal decline across Europe, was much admired and its survival in the form of a new breed of clock restorers was welcomed and encouraged.

ANTIQUE CLOCK COLLECTORS

This movement coincided during the second half of the nineteenth century with the expansion of interest by the connoisseur into the larger world of antique clocks, with a number of fine collections formed by the end of the century. The general view that 'original condition' was highly important was reinforced, and as the better-informed collector began to admire the movements as much as the cases, the practice began in the early twentieth century of restoring and 'reconverting' clocks that only a century before (even less) had been updated with anchor or Brocot escapements. The clock-collecting phenomenon continued to grow and a considerable market for good horology developed, especially in the USA. This naturally resulted in an expanding number of dealers in antique clocks and watches throughout Europe. Values for the better English and French clocks rose dramatically, resulting in many fine examples being restored and exported across the Atlantic during the interwar years. Thus, with more focus on the beauty and technical importance of the *movements* of clocks and watches, regular cleaning and restoration to 'new condition' was becoming the norm, and the first half of the twentieth century saw the appearance of a small number of clockmaking firms specializing in antiquarian restoration of historically important clocks and watches.

At the same time, the main horological industries were now fully employing mass production techniques, and fewer businesses were taking on apprentices and training them to undertake repairs of ordinary clocks and watches. As a result, those wishing to learn in this expanding trade sought reference works, and the demand led to a number of new books on clock and watch repair. The target audience was both professional horological students and amateurs—a sector which has always played an important role in the world of horology. *The Watch Jobber's Handybook* by Paul Hasluck (1887) was followed by his *Clock Jobber's Handybook* (1889), F. J. Garrard's works *Clock Repairing and Making* and *Watch Repairing* (both *c*.1903) are publications from this period, and *Clock Cleaning and Repairing* (Bernard Jones, ed.) appeared in 1917. Further editions of manuals such as Saunier and Britten also continued to appear, adding to the available literature for those wishing to join the ranks of the professional watch and clock repairer, especially during the mass unemployment after the First World War.

A major figure in the twentieth-century world of clock and watch repairers was Donald de Carle (1893–1989). In 1933 he published *With the Watchmaker at the Bench*, the first of his many excellent books on the subject of clock and watch repair; countless watch- and clockmakers today will attest to the influence of his books on their careers. *Practical Watch Repairing* appeared in 1946, followed by *Practical Clock Repairing* in 1952, both compiled from articles published in the *Horological Journal* during the 1940s, and both going on to be reprinted and produced in many new editions. Another very distinguished, if shy and retiring, figure in the horological world was William J. Gazeley (1901–67), a first-rate watch and chronometer restorer who published *Watch and Clock Making and Repairing* in 1965. In the United States, a distinguished post-war author was Henry B. Fried, whose 1949 book *The Watch Repairer's Manual* has run to four editions since its first publication. By this date, cleaning of clocks and watches by immersion in strong ammoniated solutions was pretty much the standard method proposed, and De Carle and Gazeley recommend the process. While this was not ideal in terms of preserving the surface finishes of the brasswork and put any stressed brass parts at risk of

---

64 See discussion in Chapter 26.

permanent damage, the general repair procedures were aimed at minimizing damage to the movements and certainly encouraged best practice.

ANTIQUARIAN HOROLOGY

Following the Second World War, something of revolution took place in the world of antiquarian horology. Appreciation for antique clocks and watches, and the stories behind the objects and their makers, grew rapidly among the technical and professional classes looking for new interests in post-war Europe and America. The National Association of Watch and Clock Collectors was founded in the USA in 1943, and several equivalent European Societies followed, including the *Deutsches Gesellschafte für Chronometrie* in Germany (1949), the *Antiquarian Horological Society* in Britain (1953), and French, Dutch, and Italian societies following them. In the second half of the twentieth century there was a huge increase in the number of publications on horological subjects, including some on clock restoration and repair. Antique clock and watch dealers across Europe and the United States prospered and grew, as did an increasing number of businesses specializing in antique clock restoration, and, to a lesser extent, antique watch restoration. By the late 1970s, when interest in antiquarian horology was perhaps at its peak, the horological approach to restoration and repair of old clocks and watches was naturally informed by the deeply entrenched culture from which it had grown: 'original condition', both in structure and appearance, was still an ideal to be sought and attained in the process of restoration. With examples sometimes three centuries old or more, this often meant the removal of a significant amount of those objects' technical and aesthetic history in the process of considerable conjectural reconstruction. But this was generally seen as the right course by restorers, dealers, and collectors, whose pride in craftsmanship, pride in quality of stock (and consequent value), and pride in ownership drove the process hard.

THE CONSERVATION MOVEMENT

From the 1970s, however, dissenting voices were heard. To understand their origin, it is necessary to go back to immediate post-war Britain. As a result of widespread destruction of works of art and cultural property during the Second World War, the International Council of Museums (ICOM) was founded in 1946, *inter alia*, to support and develop a professional approach to conservation and restoration,[65] principally in the fine arts and in a museum context. Then, in 1956, the United Nations Educational, Scientific and Cultural Organization (UNESCO) founded the International Centre for the Study of the Preservation and Restoration of Cultural Property (ICCROM) as an internationally recognized body to study and improve restoration methods of works of fine and applied art, principally in the context of buildings and archaeology. Based in Rome, its first Director was Dr Harold Plenderleith (1898–1997), ex-British Museum Research Laboratory, and the centre soon created an international network of specialized experts in restoring and conserving different types of cultural objects, who then developed principles for conserving all kinds of historic objects in museum collections and for preventive conservation.[66]

In parallel with this development, in 1950, the International Institute for Conservation (IIC) was founded in the UK, 'to improve the state of knowledge and standards of practice and to provide a common meeting ground for all who are interested in … the conservation of museum objects'.[67] Partly in recognition of the need for a more conservative approach to preservation, in the following decades a number of professional conservation bodies were founded to represent specific areas and to encourage the adoption of conservation practice in the private sector. In Britain, the United Kingdom Institute for Conservation (UKIC) was founded *c*.1980 to support all conservation professionals. At the same time, those in the museum world fought successfully to achieve formal recognition and their own professional departments within the museum structure, achieving equal status with the curatorial colleagues under whom they had previously worked. In 2005 the UKIC amalgamated with a number of other UK conservation bodies to form the Institute of Conservation (ICON) representing a multi-disciplinary approach to the profession.

HOROLOGY CONSERVATION AND RESTORATION

It was thus that during the 1970s, when all interested in cultural heritage were becoming increasingly aware of a move towards preserving rather than restoring artefacts, some in the horological world started asking why horological restorers were not also beginning to change the aims and objectives of their work. Long after other disciplines had moved to a more conservative approach, preferring to aim for minimal change while putting objects into presentable and stable condition, clock and watch restorers continued to undertake wholesale conjectural restorations, while refinishing and cleaning the surface of metalwork, to ensure a 'like new' appearance.

The pressure for change slowly bore fruit, and in the four decades since 1980, horological practice has changed in some areas and many more clock- and watchmakers recognize that 'less is more' if the historical integrity of the horological heritage is to be preserved. Especially where movements are unseen inside the

---

65 See http://www.icom.museum/en/about-us/history-of-icom/.

66 http://www.iccrom.org/about/overview/history.
67 Brooks 2000.

case, while ensuring the movement is clean, the need for heavy polishing regimes is beginning to be considered both unnecessary and undesirable. Equally, where clocks and watches have been subject to conversions or updating in the past, owners and conservators are increasingly recognizing the sense of simply accepting this as part of the object's history and leaving well alone. One of the first articles discussing this change in approach, especially in a museum context, was published in 1985,[68] and in 1986 a pioneering conference, *Horological Conservation and Restoration*, was held at the National Maritime Museum Greenwich, at which the many differing views on the subject were aired and which explored the way forward. Since the 1980s, horological training in the UK, and to some extent elsewhere, has gradually encouraged this understanding of the benefits of simple maintenance and putting antique horological objects into sound working order without further restoration. For example, the clock course at West Dean College in the UK now concentrates on training with the emphasis on conservation, a significant move away from its wholesale restoration approach of former years. In recent years students from several different countries have attended its courses and many younger horologists, especially those now employed in museums across the world, adopt a more conservation-style practice. In 1995, Peter Wills, on behalf of the BHI, organized the publication of *Conservation of Clocks and Watches*, a compilation of chapters written by various horologists from different backgrounds, and which attempted to summarize the current 'state of the art' while representing a range of different approaches to conservation and restoration. In 2001, the first Professional Accreditation of Conservator-Restorers (PACR) in Britain was achieved, and a number of horologists were accredited, the PACR process now being managed by ICON. From the 1980s, correspondence in the horological press on the subject of conservation and restoration has developed and been keenly debated. The publication of general books on repairing clocks and watches has continued, and a considerable number have been published in the last thirty years. These largely repeat traditional restoration practice and there remain a significant number of dealers and restorers across the world who continue to restore to a conjectured 'brand new, original condition' and who regard evidence of a subsequent interventions to the clock as something to be erased. From the restorer's perspective, this derives from a justifiable pride and interest in maintaining the craft skills required in clock- and watch*making*, as opposed to *maintaining*, and it can be argued there are some positive aspects to restoration practice. For example, where original construction processes can be rediscovered, and the running of newly made sections of a movement might provide a more realistic idea of the performance of a timekeeper when it was new. Equally, the context of ownership, and the type of object being maintained, will probably always inform the extent to which conservation is possible, or is perceived as desirable. But it is to be hoped that more clocks and watches of significant historical importance will in future years be maintained to professional conservation standards, and that the horological sector involved in conservation and restoration globally will continue a move towards more conservative practice, allowing our horological heritage to speak for itself.

---

68 Betts 1985.

CHAPTER TWENTY TWO

ACCESSORIES IN HOROLOGY

Estelle Fallet

Horological accessories, as a collective, represent a number of useful, everyday items that have been rendered mostly obsolete by technical advances, and most of these items have become more appropriate to the fields of academic study and to both the professional and amateur collector. It is an only partially explored historical field— of used objects that constitute the material culture of the modern world.[1] Work in this area has clarified the way in which the status of such objects changes as they become not only historical evidence, but also art objects that display the uses, practices, and gestures of the past—its needs, fashions, and its methods of production with the techniques proper to them. Looking at such items 'patrimonially'[2] revives their double beauty: practical and aesthetic. This the more so as the collection (public or private) enhances all their diversity of form and decoration, materials, and techniques, thanks to the quantity of objects shown. To preserve a watch with its original case and with its key and chain is a bonus for the collector: it shows the watch complete and in the best manner. Eighteenth-century morocco or tortoiseshell-covered cases still containing a spare dial glass are exceptionally rare. Watchcases separated from their movements are present in many collections.

While the mechanics of timepieces and their development are now well-known, the way in which the watch was worn remains somewhat neglected,[3] even if works devoted to jewellery and goldsmiths' work have described how they are attached to clothing, or to the stands on which they are placed at night. In horological literature, the watch is often seen as a fashion accessory. What, then, can be said about all these auxiliary items—the ribbons, chains, **chatelaines**, hooks, keys, stands, and cases—concerning their purpose, about the way they were used, and their development?

From their very beginnings, small- and medium-sized autonomous timepieces have been 'accessorized' for their protection. Portable clocks and watches were given protective travelling cases; keys, chatelaines, and bracelets were stored for the night on watch stands or beside nightlights. The accessories that have accompanied timepieces through their evolution themselves became objects of adornment, costume accessories, and embellishments of daily life.

Pictorial sources are indispensable for knowledge of these items and even if a prevailing fashion sometimes impedes the historian's curiosity,[4] images compensate for the lack of physical evidence caused by pair cases, chains, and cases of precious metals being melted down or otherwise transformed. The technical characteristics of the watch, its various displays, varied functions, and its potential qualities—being waterproof, robust, and accurate—linked with its aesthetic qualities reinforce its status as a technological marvel of miniaturization.

THE WATCH IN ITSELF

The portable watch derives directly from the small drum-form table clocks and pomander-balls of the Renaissance. Evolution in its size and form influenced its appearance, how comfortable it was to wear, and the requisite accessories. The imposing early watches were heavy and of an inconvenient shape. They were carried, very visibly, on heavy chains around the neck. With the adoption of flatter cylindrical cases, the diameter and thickness

---

1 Pomian 1987.
2 Baudrillard 1968, 121, ' ... the pure object, devoid of function, or abstracted from its use, takes on a strictly subjective status: it becomes an object of collection ... '.
3 With the notable exception of Cummins 2010.

4 The long coats worn in the early eighteenth century, for example, completely hid gentlemen's timepieces.

both diminished. The chain was replaced by a knotted ribbon or linked to a ring or snap-hook attached to the watch. The watch then migrated from the neck to the belt—where it was associated with a hook or with the chatelaine—and then from the waist to the pocket. Flat, small-sized nineteenth-century ladies' watches fitted with a long watch-guard were fixed to a brooch on the bust, at the waist, or elsewhere. Finally, the watch arrived on the wrist where it would take many forms (round, square, octagonal, tank-shaped) and inaugurate a practical way of carrying a watch, adapted to new activities.[5] Wristwatch cases became waterproof and were protected against magnets; the materials of which they were made became more varied (for example, platinum, steel, nickel and its variants, titanium).

The watch carried on the person was, in itself, as much a dress accessory as a measuring instrument. It displayed fine craftsmanship in different materials, and the engraving, chasing, and enamelling styles were all constantly changing. Whatever changes there were in dress habits, watches always remained visible, detectable by their accessories, as markers of social status. While the watch was the only piece of jewellery generally accepted for men during a long period, diversity in watches was favoured by feminine taste and usage. How the watch was carried was strictly dependent on its use in different periods, and it was always absorbed into general technological development: the contemporary 'connected watch' can be taken as emblematic of this process.

WATCHES AND CASES

Watchcase making was always a subject of dispute between the watchmakers themselves and the goldsmiths. In Paris and Geneva, at least, the movement makers were authorized 'to make, to sell and debit all kinds of cases, gold and silver, enamelled, engraved, with all sorts of ornament, for their watches and clocks'.[6] In the seventeenth century, case makers working gold and silver were obliged to respect a defined quality of the metal and to have it hallmarked. The development of the manufacture led to the emergence of a specific trade in 'case making', best exemplified by Geneva, where the central figure was the coordinating merchant. Genevan contracts show that it was he who provided the capital and the basic materials to bring movement and case together.[7] Indeed, the merchant's name often replaced that of the watchmaker. Specialization occurred throughout the industry: in 1685, a hard stone worker devoted himself to cutting watchcases in rock crystal, another in the making of watch glasses, while in 1692 a watchmaker was identified as making only watch keys.[8] Development of the manufacture in Geneva led to the emergence of two new corporations—that of the watchcase makers (1698) and that of the engravers (1716).[9] In 1699, reconciliation was sought between the goldsmiths and the watchcase makers:

> The makers of cases claiming that the gold- and silversmiths should have no liberty but to make boxes and cases and other kinds of work that are solely gold-smiths' work, and not to meddle with those mixed with shagreen or tortoise-shell and other similar [materials]; the gold- and silversmiths maintaining the contrary that as they have always made without distinction all such sorts of work, the mastery recently accorded to the said makers of boxes and cases, cannot deprive them of it.[10]

A single watchmaker with an important order might be constrained to supply himself from several case makers, which also helped diversify supply.[11] Moreover, a case passed through many hands between its basic design and it being polished. Some trades were rarer than others. In the early eighteenth century it was affirmed 'there are not in Geneva more than three of four master chain-makers, who make chains in gold or silver for watches as good as those made in England. Those of Germany are far inferior'.[12]

In the mid-eighteenth century, Ferdinand Berthoud catalogued the various specialists who worked on watches: the case maker working in gold or silver, the outer case maker, the engraver, the chaser, and the enameller to decorate them, the chain maker, and the founders who made, turned, and polished the bells.[13] He sets out in detail how the cases passed successively from hand to hand, from workshop to workshop, from the piece in the rough to the turner, the founder, and the finisher; from the pendant maker, the ring maker, the crown maker, the finisher, the polisher to the hinge maker. In his chapter on clocks he distinguishes the joiners who made the cases 'following the indications of the master for the design', as he does also for the gilders of the bronze fittings. In all this, in eighteenth-century Paris, some craftsmen obtained their mastery with a triple title: watchmaker, case maker, and engraver.

The accoutrements of the watches, which were not necessary to the functioning of the watch, need to be distinguished from the movement itself, even if they are indispensable when it comes

---

5 *Cf.* this volume, Chapter 18.
6 *Cf.* the clause specifically authorizing this in the Paris statutes of 1646, Franklin 1888b, 193.
7 Jaquet & Chapuis 1970, 38–9 supply several examples.
8 Babel 1916.

9 For their masterpiece, the engravers were to make 'an envelope or case for a [striking] watch engraved all over in *taille douce* or [chased] in relief, or another work'.
10 RC. 198. P. 265; RC. 199. P. 195. ' ... Reglemens et ordonnances ...des Maitres monteurs de boëtes et étuis de montre ...'.
11 One silver watch, *c.* 1750 carries, apart from the name Reguillon & Bergier engraved on the backplate, two other marks: a punchmark 'IC' incuse on the inner case, and a signature engraved on the pair case. Grasset. Inv. N 1136. Musée de l'Ecole d'Horlogerie de Genève, today in the Musée d'Art et d'Histoire, Geneva.
12 Savary des Bruslons 1750, i col. 793.
13 *Encyclopédie* 1751–80.

to protecting and winding the watch. Easily damaged, the dial rapidly became fitted with a full or pierced cover set above the hand. If the watch was fitted with an alarm, or struck the passing hours, the case was pierced to let the sound escape. Normally the case was made of silver or gilt brass. Gold examples are rare, having been sold or melted down in time of need. Particularly precious watches could be fitted into faceted rock crystal cases, the block having been hollowed out, the edges bevelled, with the watch itself held in a delicate frame by slender screws and the cover held in place by claws. The dial could be decorated with **champlevé** enamel, and the forms of the watchcase could vary—for example, a cross, star, heart, shell, octagon, square, animal or insect were all possible. In Germany, as in Blois, a spherical form was also sometimes used, but whereas other forms endured throughout the early seventeenth century, the spherical form largely disappeared and was replaced by a large convex **'bassine'** form that was particularly convenient for painted enamel decoration.

The *bassine* case was characterized by being fixed in the case, together with the bezel, by a hinge. Cases for the larger, rounder watches that appeared after the introduction of the balance spring and a fourth wheel had been added to the train (the French *oignon*), which displaced the *bassine* for a time, were made in gilt brass or silver (though for *oignons* apparently never in gold). The surface of the inner case was plain or engraved. If wound from the back, a hole was worked in the case above the arbor and sometimes fitted with a rotatable cover. The case fittings also developed during the eighteenth century. The hinge connecting the bezel to the glass was placed at 11 o'clock, that of the movement beneath 12 o'clock in a collar on the movement. This could be pivoted out by moving a small locking-piece situated at the rim of the dial usually, but not invariably, at 6 o'clock. If pierced French cases were usually worked with a saw and finished with a graver, German cases were more likely to be cast and then finished with a graver or burin. In Geneva, the work of 'cutting out' in cases, cocks, clicks, and pillars was consigned to women, authorized by an order of 1690.

PAIR CASES

A further accessory, also not essential to the going of the watch, was a second outer case (Figure 192) offering additional protection.[14] These cases had to be not only resistant to shocks but also be finished in a manner appropriate to the style of the watch that they completed. For the round watches of the eighteenth century, a second case was habitual. While the inner cases are made of silver or gilt metal and are generally without ornamentation, the outer cases are decorative. For striking watches, both cases were engraved and pierced. Textile or paper watch pads may be found in the outer cases (particularly of English watches) to ensure a snug fit between the two cases. These could carry instructions for use, advertisements for retailers, or decorative designs.[15]

LEATHER OUTER CASES

Protective leather cases have been deployed from the Middle Ages onwards to protect the highly worked small objects that were often to be transported. Early gilt brass clocks and watches were protected from dust and accidents by fitted gold- or blind-tooled cases or, less frequently, in wood or metal. With the development of the pair case, a separate leather case was abandoned, but leather, (particularly in the form of **shagreen**), was used to cover pair cases in the later seventeenth century and throughout the eighteenth. Leather-working thus became a part of case making. In 1675 Noël Godet (2nd half 17th century) in Geneva was apprenticed to Jean Melland (2nd half 17th century) as a 'maker of cases for horology and shagreen cases'.[16]

Outer cases could be decorated by *cloutage*, a process also used on tortoiseshell cases. A pre-determined design was executed by piercing the leather-covered case with a sequence of small holes into which minuscule gold- or silver-headed nails were inserted. When a monogram[17] was to be realized, the letters were sometimes formed using nails with different sized heads. An alternate decoration was produced by inlaying the leather cases with gold or silver thread to form arabesques, fleurons, or lozenges. Such minute work was habitually effected by women who might also, like Catherine Caillatte (*fl.* 1663–91), 'maker of silver nails for the covers of watch cases', make the nails themselves.[18]

In the following century, the large inner cases of 55–60mm, engraved or enamelled, were protected by the pair case in metal repoussé in gold, silver, or gilt brass, or covered in dark leather, tortoiseshell or tinted horn, fish skin, **galuchat**, **shagreen**, goatskin, died horn or **vernis Martin**, and sometimes ornamented with *cloutage*. Such uncommon materials distinguished the watches within from those contained in enamelled or stone set cases.

In the probate inventory of Philippe d'Orléans (1674–1723), a watch 'with two gold cases, its gold chain and a steel hook' is listed. Four other watches were housed in shagreen cases.[19] Lacquered cases appear in the last quarter of the seventeenth century.

---

14 Late seventeenth-century English terminology named the inner case the 'box', the outer case the 'case'; see Priestly 2018, 2, whose work offers a definitive account of English watchcase making and its control. Late nineteenth-century Swiss case making is described in Hof et al. 1900.

15 Turner 2010, *passim*.
16 Jaquet & Chapuis 1970, 38.
17 Designs for these were drawn from such collections as that created by Charles Mavelot, the *Livre des chiffres* (1680), or the *Livres cucrieux et utile . . . de trois alphabets de chiffres simples, doubles & triples . . .* Paris 1724 by Nicolas Verrien.
18 Gélis 1950, 80, 220.
19 https://www.siv.archives-nationales.culture.gouv.fr/siv/media/FRAN_IR_043346/c1p74bbkcflv–1ef38a17ojm18/FRAN_0159_22104_L

Figure 192 Watch pair case, in brass, and pinned leather, Geneva (?), late seventeenth century. Ville de Genève, Musée d'art et d'histoire, Inv. N° 0599.

After several layers of varnish have been applied, the design was painted on the topmost in gold or in several colours. A final coat of clear varnish was then applied. Although such decoration was also applied to clock cases, its fragility gave it only a short period of popularity.

Other cases were executed in **filigree**, in porcelain, and far more rarely, a few were made of amber, in different rare woods, in bloodstone, or in ivory. A merchant such as James Cox in London could mount his luxurious watches destined for the Ottoman and Chinese markets in cases with bands and bezels composed of hard stones (grey or brown agate with or without veins, red jasper, and the like), all set within engraved gold borders.

Repoussé pair cases in gold and silver represent a peak in the goldsmiths' art and the same technique was used on other small items such as snuff boxes and chatelaines. The leather covered pinned outer case gave way to one in metal, the decoration of which could be harmonized with that of the chatelaine.[20] Eighteenth-century cases similarly renewed the decorative qualities of metal worked directly by engraving, piercing, and the addition of **strass**, glass, and precious stones. Miniature paintings on enamel (often portraits) were thus framed, while figurative translucent enamels harmonized with the reflected play from engine-turned surfaces.

But enamelling also developed. Because of the wear to which it was subject, enamelled watches were most likely supplied with an outer case. These were fitted with a glazed back to protect the delicate decoration while also allowing it to be seen. Around 1760, however, a new process was developed simultaneously in Geneva and Germany in which a layer of colourless enamel was placed on the decorated surface after this had been fired, and which was then highly polished (Figure 193). Enamel decoration on an engine-turned ground was very fashionable in early nineteenth-century Europe as watchcases became flatter sufficiently so that both the back and the band could be engine-turned or more simply polished. Pair cases at this period became obsolete. They remained, however, *de rigueur* for watches destined for the Ottoman market, where a third case, usually in galuchat or tortoiseshell, was joined to the two inner cases in silver or gilt brass, and where a fourth conical-shaped case in engraved silver was often locally added. For all regions of the world, watches mounted in leather carrying cases that could be attached to a saddle while riding, or even incorporated in the saddle itself, were useful, but expensive. One particularly luxurious model patented in England by Ralph Gout (1770–1836) in 1799 incorporated a pedometer.[21]

BELLS, SMALL BELLS, AND GONGS

Bells and gongs are further auxiliary items that complement the functioning of a watch or clock, and although they are not essential to its running, are an integral part of it. They are intrinsic to musical timepieces where they may be used in batteries of three to twelve—even fifteen—pieces. Breguet liked to make bells of gold for his gold watches so that case and bell had a similar sound resistance. From 1783 onwards he built repeating watches using sonorous strips of steel rather than bells. By the end of the century coiled gongs in plain or blued steel had been introduced and would become standard in the nineteenth and early twentieth centuries. They were fixed by one end only to the case and struck by two hammers. A low tone was sounded for the hours, a higher and a lower tone for the quarters. Tuning them required a fine ear, even one capable of determining absolute pitch.

DUST CAPS

A high number of eighteenth-century watches are fitted with a dust cap, which covers the movement to prevent the intrusion of dust and insects while winding the watch. It also offered a space for engraving the maker's or retailer's name, although this was not invariable. Occasionally, different names are to be found on dust cap and movement plate. English dust caps are generally independent items and may be removed by sliding back a catch, while some Swiss and French dust caps are hinged, or they may be replaced by a fixed collar running round the movement. Dust

---

20 For English repoussé work, see Edgcumbe 2000.

21 See *AH* 2007, 538–45.

Figure 193 Triple cased watch by the Frères Bordier, Geneva, c.1770, the first case in gold; the second in gold painted in enamel with 'Mars and Venus', engraved and set with strass; the third of brass with vernis Martin and silver point decoration. Ville de Genève, Musée d'art et d'histoire, ancienne Collection Reverdin, 1946. Inv. N° AD 0282.

caps were largely replaced by the hinged inner cover (*cuvette*) in nineteenth-century watches. This was almost invariably fitted to **hunting cased** watches.

CHATELAINES AND NECESSARIES

In the sixteenth century, keys, knives, rosaries, perfume censers, small mirrors, and books of hours were all attached to a metal girdle or chatelaine, while watches, as befitted their higher status, were worn apart suspended from a chain around the waist or attached at the waist.[22] Among the gifts offered by Henry VIII of England to Catherine Howard were 'three tablets of gold wherein is a clocke' and a gold pomander which contained another.[23] Such watches could be carried in the hand but were linked by a short chain to a finger ring. In the seventeenth century watches were carried at the waist suspended from a chain attached to a belt. In the early eighteenth century, however, decorative chatelaines came into use. They were carried at the waist, sometimes even two being attached.

A chatelaine is composed of a wide hook that passes over and beneath the belt, a series of decorative plaques matching the decoration of the watch itself, and from three to nine short chains to which (apart from the indispensable watch key and seal), various trinkets such as medallions, pen-knives, mirrors, perfume or smelling salts bottles, sewing or drawing implements, tape measures, and needles could be hooked. Adornments of predilection,[24] the chatelaines stimulated the skills of chasers, engravers, enamellers, and goldsmiths who produced them in multicoloured gold, with inset pearls, with pierced or gilt plaques, in agate, silver set with hematite, faceted and polished or open work steel, in tortoiseshell, ivory, enamel, and mother of pearl. By contrast, other models were realized in fabrics, embroidered silk, or sable. Chatelaines were always made to harmonize with the other items worn around them. Just as carrying two watches was not uncommon, neither was it for one watch to be carried on a chain, the other on a chatelaine. Female garments being without pockets, the use of such accessories was unavoidable, but many surviving chatelaines have lost their accessories and are disassociated from the watch, the essential element for which they were made.

The typical chatelaine composed of linked plaques continued to be made until the early nineteenth century with the attachment hook being extended to give greater security, and the chains between the plaques shortened. Between 1820 and 1835, a variant form appeared in the shape of a waist clasp with a single plaque, often in the shape of a hand, carrying the watch. From the 1830s onwards the chain linking watch and carrying plaque became thicker and shorter. If, in the mid-century, women continued to favour chatelaines, men turned to fob watches carried in a waistcoat pocket with only the chain and attachment visible. Once the stem-winding watch made keys redundant, the carrying chain became thinner. Special belts for female dress appeared in the third quarter of the century. These leather 'Norwegian belts' were fitted with rings, eyes, and chains to carry a variety of objects like pencils, pen-knives, scissors, purses, notebooks, perfume/vinaigrettes, matchboxes, and handkerchief boxes. At the same period, Asprey, the London jewellers, promoted their 'watch waist bag', in which a watch in a leather holder, similar to the wristlets soon to be developed for carrying a watch on the wrist, was attached to a belt together with a handbag.[25]

---

22 For an anonymous portrait of an unknown lady wearing both c.1560, see Cummins 2010,12.

23 BL Stowe MS 559, cited from Cummins 2010, 13.

24 Cummins 1994 provides a history of chatelaines with special attention paid to the nineteenth century, when the Victorians used them for carrying keys, sewing implements, dance cards, perfume bottles, spectacles, and purses.

25 For wristlets, see this volume, Chapter 18. The Asprey 'Waist Watch Bag' is illustrated in Cummins 2010, 92.

IMITATION WATCHES

From about 1780 onwards, it once more became fashionable, as it had been in the sixteenth century, to carry two watches. Dandies carried a watch in each pocket of their embroidered waistcoats—one attached by a chain, the other by a ribbon. The same fashion appealed to women, but as it was expensive to carry two true watches, many contented themselves with an imitation for the second, which might in fact be a mirror or a flattened flask. These accessories were made in gold or silver and enriched with stones or with miniature portraits. Some of them had a dial, purely decorative, and others contained a pin cushion. More economical, gilt or painted cases were also used. Such imitation watches on a waistcoat chain or chatelaine remained current for a good part of the nineteenth century. They provided accessory items—mirrors, powder holders, compasses, tape measures, patch boxes, and the like—that resembled the round form of a watch and, like a watch, could be carried on the person—even on the wrist. In 1878, a 'justification' for the practice was found in the habits of the Chinese: 'do we not see, in our time, the Chinese also carry two watches, one on each side of their belt in an embroidered pocket?'.[26]

SAUTOIRS OR NECKLACE CHAINS

In the early sixteenth century both men and women carried watches that were worn on long chains around the neck and reaching to the chest, or at the waist near a pouch or chain purse. Long necklace chains for watches were fashionable until the early seventeenth century but then fell victim to the sumptuary laws enacted throughout Europe until the third quarter of the eighteenth century.[27] Thereafter, the *sautoir* reappeared in feminine fashion. During the First Empire in France dresses favoured a high waistline, which a watch showed off well. Watches were engraved or enamelled and worn suspended from a long necklace chain, from a ribbon fixed by a clasp to a belt, or slipped into a pouch made of fabric or occasionally of leather. There it would join eyeglasses or a pair of scissors. *Sautoir* chains can take several forms with long twisted links (*macaroni*), round or flat buckles, and plain or enamelled. The names of queens and princesses became attached to different kinds of chain.[28] The Leontine chain was one that 'holds a middle place between a necklace and the belt hook'.[29] The watch was held by a slide on a large chain attached to the neckline by a small brooch. In effect it could serve simultaneously as necklace, watch chain, and belt hook. From the mid-century, men also adopted the neck chain, possibly the forerunner of the pocket watch carried on a pendant chain. Until the end of the century, chains remained the favoured way of securely wearing a watch.

In the latter half of the century, necklace chains, clasps, belt buckles, earrings, and brooches were brought together in matching sets, and could also be harmonized with large bracelets, some of which were developed to contain very small and flat watches. Watches were also incorporated into other objects, including face screens, fans, bouquet-mounts, and perfume pistols. Portraits show how the watch pinned visibly to the bodice moved from the waist towards the shoulder. The watch was offset by a knotted brooch pin, imposed on the long chain, whose fluid curves link them together (Figure 194). The brooch had a clasp to hold the watch in place anywhere on the chain, thus allowing adjustment. This system was renewed during the Art Deco period by changes in the way that the watch was fixed to the clip. The nurse's watch would also adopt the clip system.

Pendent watches held on a short neck chain were also worn in the mid-nineteenth century, although the watches were frequently replaced by portrait cases contain small daguerreotypes. The long chains also served to hold back ladies' sleeves. For men, the buttonhole became a convenient place to attach the chain.[30] Gold and enamel brooch watches appeared in the mid-century for women, and the button-hole watches, usually in gun metal, for men soon followed.

Figure 194 Brooch watch with key in its case. Ville de Genève, Musée d'art et d'histoire, Inv. H 2007–23.

CHAINS

The waistcoat chain appeared from about 1820 onwards, and there were many styles available, from low-cost versions in steel or other base metals, to high-cost designs in gold. The chain ran from the watch in the pocket to the owner's waistcoat and was secured

---

26 *Journal Suisse d'Horlogerie*, ii, 1878, 25–6, following the *Watchmaker, Jeweller and Silversmith*.

27 Small cases in gold and silver were, however, sometimes exempted from the laws on the grounds of their utility.

28 So the chains Eugenia, Victoria, Clotilde (from Clotilde of Savoy), or Matilda.

29 Vever 1906–8.

30 These were reported as a novelty when worn by the Duc de Penthièvre in 1876. See *The Watchmaker, Jeweller & Silversmith's Trade Journal*, 5 October 1876.

by a T-shaped attachment bar placed across a buttonhole, and a sprung or screwed hook at the other end of the chain by which the watch was attached. For heavy watches, the chain was relatively short and was more comfortable to wear. Later, the attachment bar was modified to allow for longer chains or a simple attachment in leather or a watered fabric.

In England, Prince Albert and Charles Dickens were associated with either single or double vest chains, the latter allowing two watches to be worn, or one and an accessory, in separate pockets. For men, small pendent watches, suspended from a short chain, became highly fashionable in the twentieth century. These watches were extra thin, decorated with semi-precious or precious stones, and translucent enamel, and the movement fitted to them also became more solid and reliable. Their chains were constructed of various types of links, including jet, pearls, plaits, and tresses, and they were made in gold, platinum, patinated steel,[31] or silver niello. As fashions changed, metal wristwatch straps were increasingly manufactured by jewellery chain makers as the century proceeded.

Finally, sentimental or mourning jewellery in the forms of brooches, medallions, and rings were fitted to contain tresses of hair from lovers or the deceased. Jewels using or incorporating hair were worn since at least the later seventeenth century,[32] but they enjoyed a resurgence in popularity in the 1800s. Human hair could be plaited into solid watch chains. Supplied from a flourishing market in women's hair, they were produced as flat plaits closed by a jewelled buckle, or coils closed by a clasp in precious metal. For both forms, a wide variety of patterns was employed and they incorporated slides and hasps that often featured amatory motifs.

OTHER ACCESSORIES

Placed in a trouser, skirt, or waistcoat pocket, mounted on a belt, laid on a dressing-table or desk, watch styles continued to change. From the sixteenth century onwards they ornamented the pommels of daggers and, in the eighteenth century they were incorporated into flasks, message boxes, belt buckles, and rings. In the twentieth century they enhanced elegant cigarette lighters.

The watch has also appeared on walking sticks, umbrellas, or evening bag clasps. There were buttonhole watches carried on the reverse of the lapel and sometimes supplied with a small leather pouch so that it could also be carried in the pocket. Art Deco fashion transformed the watch into a clip, while in the 1930s imitation watches came back into fashion and hid items like thermometers, spy-glasses, mirrors, compasses, calculators, tapes, metronomes, and lipstick holders. Such 'fantasy' watches responded to a desire for novelty, which resulted in cufflink and keyring watches. Among such 'accessory watches' and popular in the 1920s, the handbag, or 'hermetic', watch[33] was portable in an age of greater mobility. However, this taste for novelty weighed heavily on the industry, and one commentator somewhat sourly remarked: 'much is paid for this triviality of a moment which concentrates on one article, transforms the shop into a bazaar, freezes and becomes a museum object'.[34]

Nevertheless, hermetic watches were developed in the course of research carried out to find a better way to protect them, and which resulted in the development of the impermeable watch. Juvenia, Movado, La Champagne, Tavannes, Tissot, Longines, Omega, Zenith, and others popularized the model, which was about the same size as a box of matches. The case was composed of two elements sliding without a hinge or pivoting on a hinge. The latter system allowed the watch to be opened with only one hand. A thumb-knob on the cover allowed the time to be set. The pivoting and sliding arrangements transformed the watch into a tiny clock. The 'Captive' model by the Tavannes Watch Co. had a case in two parts that opened when the sides were pressed. The first model produced by Tissot was a sort of elegant pair of tongs, which opened to reveal the dial.

These handbag watches were covered in snake- or lizard skin, crocodile, or in Chinese-style lacquer and were convertible into mini-clocks. Occasionally, they were discretely fitted with a mirror on the back of the case. The practicality of this element increased the appeal of the accessory, which was often fitted with an eye loop for attaching it to a chain.

WATCH STANDS

As watches were delicate and precious, many owners used watch stands to ensure the safety of their watches when not being worn. Iconographic evidence shows watches hanging on walls, suspended above beds, or placed in an alcove, and some watch chains were fitted with retractable hooks for this purpose. The watch stand is a sort of extension of the watch hook. A few chatelaines with a watch stand element were used to protect the watch from coming into contact with the various trinkets attached to the chatelaine.

Known from the late seventeenth century onwards, the watch stand developed across a wide variety of forms and materials throughout the eighteenth and nineteenth centuries. In itself a decorative object, the watch stand (Figure 195) protected the watch from loss and damage. A watch could be placed on or in it in a favourable position, hanging downwards, or on a bedside table, desk, or dressing-table thus replacing small table or mural clocks. Early watch stands took the form of miniaturized clocks in carved wood, gilt or painted, or with Boulle-type marquetry decoration. Later they were produced in polychrome porcelain,

---

31 Steel-cased watches with accessories in the same material gave a new look and greater resistance to wear, partly replacing cases in precious metals. For a survey see Sabrier & Rigot, 2005.

32 Rings, bracelets, and purses of hair are noted in the *Dictionnaire de l'Académie françoise*, 1694 and later editions.

33 'La montre hermétique', *Journal Suisse d'Horlogerie*, October 1939.

34 *Journal Suisse d'Horlogerie*, 1930, 219, 245–8, 269–75.

Figure 195 Six watch stands. Ville de Genève, Musée d'art et d'histoire, Inv. AD 3832; AD 4435; AD 4383; AD 4324; AD 4412; H 2006-55.

glazed earthenware, bronze, cast iron, pressed tin-plate, zinc, brass, and were sometimes decorated with champlevé enamel, steel in silver, papier-mâché, bone, or glass. Absorbed into domestic interiors, watch stands mirrored the prevailing decorative styles. Models in porcelain, marble, or carved jade reflected the taste for Chinese themes. There were Dutch, German, and English styles, Baroque and Rococo styles, and Art Nouveau and Art Deco styles. They could take the form of a dish, an ostensory, a miniature clock, or be made in the shape of animals, allegorical figures, porches, suspensions, boxes, or be entirely abstract. With an average height of 10–45cm, watch stands had one or two openings in which the watch could be placed, or a hook on which to hang it. In the nineteenth century stands often incorporated a small dish for cufflinks, keys, or small change, a ring-holder, or a small writing box. Between about 1830 and 1860 modest watch stands in the form of small carved wood boxes or coffers were produced in the Black Forest, Brienz, and Grindelwald. Generally, watch stands were common to all classes of society, hence the wide variety of models from the simplest to the highly elaborate.[35] However, the ubiquitousness of domestic clocks in the latter part of the century and the spread of the wristwatch thereafter resulted in the watch stand falling out of use.

CASES

Generically, cases are made to protect one or several objects, including pencils, spectacles, and other items.[36] The fitted case is one in which the interior, and sometimes the exterior, is moulded

---

[35] Pertinent references for watch stands are given in Havard 1887–90, 552–553; a pictorial survey is Puraye 1982.
[36] For an introduction to boxes in general see Cummins 2006.

Figure 196 Watch case in tortoiseshell on gilt wood with cream silk and velvet lining, marked 'Eug Bornand & Cie à Ste Croix 2376'. Ville de Genève, Musée d'art et d'histoire, gift of Madeleine de Weiss, 190. Inv. AD 7646.

to the object it contains (for example, a cello case). Many cases in public and private collections today are missing their contents, but the cases themselves remain as testaments to the case maker's art and what they used to hold. Small- and medium-sized clocks, watches, sand-glasses, and sundials all had their protective cases.

That Gothic iron clocks were supplied with wooden cases to protect them from dust and moisture is known from archival and iconographic sources. Protection was the primary purpose of the case. From the Middle Ages onwards leather cases for horological devices were made by specialists. These could be plain or decorated with blind- or gold tooling, could have an opening that left the dial visible, and could be fitted with a lock and key. These, like the hinges and other fittings, could be in brass or in silver. Cases intended for transport were fitted with buckles through which a leather carrying handle was passed, or with a sewn-in rigid handle. Early cases generally mimicked the form of the object they contained, and were made of wood or pasteboard covered in leather; later cases were covered in morocco, silk, or velvet. Later, the fitted case itself was rectangular but the interior was moulded to the shape of the object or objects it was to contain, and lined with silk, velvet (Figure 196), or (particularly in Germany) with marbled paper. Printing the name of the retailer of the object on the tissue inside the case was standard from the later eighteenth century onwards, although earlier cases might contain an applied printed-paper label. The exterior of the case was reserved for the initials or stamp of the proprietor. The name of the case-maker was rarely given.

In the eighteenth and nineteenth centuries, many cases had an outer carrying case made of canvas or oilskin. Generally however, with the exception of coach watches, outer carrying cases for clocks, as for marine chronometers, were made of wood: deal, pine, or beech. From the mid-eighteenth century onwards, domestic clocks, which were not intended to be moved, were frequently protected by made-to-measure glass domes. In mid-nineteenth Paris, the Rigault family made them for both clocks and for scientific instruments.[37]

The outer cases of coach watches were usually of leather within a brass or silver mount and were sometimes studded with silver nails and set with holes allowing the sound of the repetition to escape. In the early nineteenth century, top-of-the-market watches were presented in cases that featured spaces for a key and chain, and even a spare glass for the dial. The latter became characteristic of Glashütte production. Awards and medals obtained by

---

37 BBT, 'Rigault'.

563

Figure 197 Watch pouch in crocheted cotton set with multicoloured translucent beads. Ville de Genève, Musée d'art et d'histoire, gift of Anna Gaspar. Inv. H. 2002–10.

manufacturers in the various national, international, and chronometric exhibitions of the nineteenth and early twentieth centuries were also recorded in the inscriptions printed on the lining of boxes of subsequent products.

POUCHES

Embroidered or crocheted pouches for watches became a part of feminine attire in the later nineteenth century, although rather few of them have survived.[38] Patterns for them, however, can be found in commercial knitting and crocheting catalogues. Similarly uncommon are beadwork watch pouches (Figure 197). These, composed of hundreds of miniscule beads in mosaic, first appear in the later seventeenth century and were fashionable up until c.1850. The beads were threaded on silk and carefully knotted. The process was applied to many kinds of small cases, not only pouches for watches. Those described as 'Turkish' are embroidered pouches that are partially beaded. These continued to be produced into the twentieth century.

The harmonization of dress and watch pouch is particularly clear in China, where 'trousers are held by a belt which carries a pouch intended to contain a watch'.[39] In China, where watches were frequently sold in pairs, sales were often concluded with the offer of a fitted case in morocco or tortoiseshell for the pair with additional reserves for their keys. Chinese form double cases with Chinese markings became a speciality of the leather workers of the Val de Travers. Such cases were dispatched by the dozen in soldered metal-lined wooden packing cases to protect them from the humidity of the Chinese port towns.

MODERN CASES

Modern approaches to case making developed during the later nineteenth century. Specialized businesses existed that combined the different techniques of the manufacture—from joinery to leather cutting to sewing, assembly, and application—in a single establishment, and cases were produced in all shapes and sizes; most were fitted cases. Alongside them, more ephemeral small pouches adapted to the different sizes of watches were produced in tissue or in chamois leather. One of the oldest enterprises in this area was M. Dardel in Paris, active from c.1895 onwards, where the cases encouraged the buyers to a purchase and offered the

---

38 For examples see Cummins 2010, 220–3.

39 The Comte de Escara de Lature, *Mémoires sur la Chine* (1865) cited from Chapuis (1919), 201.

company a means of publicity. The product was enhanced by its packaging. Based on Dardel's success, other leading manufacturers renewed their presentation, and boxes, associated with guarantees and records of performance, were the object of careful consideration. Case makers, like the watchmakers, became scrupulous about details, insisting on delicate cutting and jointing.

In Paris at the turn of the twentieth century, Siegel & Augustin ran a 'manufacture for cases and watchmaker-jewellers'. Their products ranged from showcases and shelving to presentation cases, trays, watch stands, carrying cases, and labels. In Geneva, the Vaudaux family founded a workshop in 1900 and specialized in the design and production of presentation watches cases, storage cases, and other high-quality leather work. Their catalogues illustrate boxes for pocket- and wristwatches; boxes were made in imitation glazed morocco of green, dark red, grey, or white, with convex tops, rounded corners, gold vignettes, and cream velvet- or satin-lined interiors. Other cases were specifically designed to receive the miscellaneous contents of pockets.

Alongside these, the traveller's case is worth mentioning. Composed of two or more parts nesting each above the other and used by commercial travellers, these cases were at first made of solid wood, with strong hinges and locks. Later, they were replaced by leather-covered models, and towards 1920 the exterior was covered in **Pegamoid**, a material more resistant than morocco goatskin. The internal trays lined with a felt-like tissue were adapted to all the standard watch sizes. Standard cases for pocket- and wristwatches were shown in the catalogue alongside shallow presentation cases used by enamellers and dial makers to present their wares. In the mid-twentieth century, fashionable brands began to personalize their cases and establish an intrinsic relation between the watch and its packaging. The case could contain the guarantee, a small catalogue, and the user's manual. On occasion, a sale might depend more on these ancillaries than only upon the watch itself. Otherwise, with the revival of the automatic watch, some makers offered cased sets with both hand-wound and automatically wound possibilities integrated. The materials used evolved from fabrics and paper to calf hide and crocodile skin. Covered resin was used for the basic form. Precious woods were obligatory for top-class horology, with compound cases that could include special tools—the indispensable additions included cleaning cloths and reserve straps. The sumptuous case made in 2008 for the new 'Marie-Antoinette' watch by Breguet was cut in oak from Versailles under the shade of which the Queen, allegedly, had rested.

Some makers propose innovative materials to their clients such as vegan 'leather' composed of recycled vegetable substances such as cork, linen, and hemp. Contemporary design favours sobriety with little variation, wood being the dominant material. Horological cases for special series are supplied by makers who also create similar cases for spirits, perfumes, or telephones. As many of these objects are now on their second lives and are no longer used for their original purpose, some have been offered in packages intended to be original and the cases themselves have been transformed into decorative objects.

Finally, many brands offer sumptuous storage cases. Jaquet-Droz and Piguet have created highly unusual leather cases in collaboration with the French workshop Pinel & Pinel. These are personalized and can hold several dozen watches together with such tools as are needed to maintain them. Jaeger LeCoultre have produced the largest of all cases, which doubles as a strong-box, for the impassioned collector. To develop these trends, schools of design in Switzerland now associate knowledge of horological products with packaging design and environmental product courses.

UNEXPECTED ACCESSORIES

The wristwatch, in its position and its being in constant movement, is always in danger of damage. Manufacturers proposed protective cases in celluloid or metal to guard it against unforeseen shocks while retaining the visibility of the dial. Hinged lids in nickelled metal were invented and used a hermetic closure. Other devices attached to the bracelet arms and included a glass protector or a grill in polished or nickelled white metal placed above the dial glass.

Other innovations were linked with the transport revolution. The development of regularly timed coaching services in England in the late eighteenth century gave rise to an accessory that rapidly became generally used throughout the century—the 'foot board clock', where the coachman could easily read the hour from his seat. Such watches, particularly those exposed to bumps and the vagaries of the weather, were mounted in wood or leather holders. The later dashboard clocks for cars and airplanes[40] were similarly mounted in leather holders that not only protected them but also assured that they were rigidly fixed.

WATCH KEYS

From the sixteenth to the early twentieth century, the indispensable accessory for a watch or clock was its key to wind it and set the hands. Although keys were essential, they had no permanent storage places in clock cases, and only occasionally in the presentation cases of watches. However, a special spot for the key was incorporated into the boxes of marine chronometers and carriage clock cases throughout the nineteenth century. Earlier than this, the key was linked to its instrument by a chain or suspended beside it on a chatelaine. Key-making was a large

---

40 For these see this volume Chapter 23, Section 4.

industry from the seventeenth to the twentieth centuries. Manufacture was concentrated in metal-working regions where brass casting for the key itself and iron or steel working for the shank were both available. The making of keys was often associated with that of spits, tie- and hatpins, or metal buttons. Some masterpieces of miniature mechanics were produced among the many keys, ranging from the cranked forms of the seventeenth century to George Sanderson's (c.1726–64) calendar keys (patented in Exeter in 1762)[41] and Etienne Tavernier (1756–1839) in Paris, c.1780, including musical keys. Keys could use the same decorative elements (Figure 198) as those on the watches they served (enamel, cameos, carnelian, precious and semi-precious stones, pressed, engraved or engine-turned silver, Berlin steel, hair, micro-mosaic, Limoges porcelain, Wedgewood earthenware, painted ivory, and others). Keys were also subject to technical improvement, the 'tipsy' key, and the Breguet key (introduced c.1789), both prevented winding in the wrong direction.

The very small keys required for ladies' watches were seldom displayed, but the larger one used for gentlemen's pocket watches, often adorned with a cut stone (citrine or amethyst) were intended to be seen. From about the 1860s a straight cylindrical key could itself form the 'T'-bar cross piece that attached watch chain to waistcoat, and the set of watch, chain, and key was thus formed. Multifunctional keys incorporating seals or associated with Stanhope lenses, and spy cameras were a further development. Keys served also as an advertising platform, while others incorporated compasses or thermometers. A major key production company was founded by David Darier (1770–1829) to make watch hands using machines of his own invention, and which, trading as Darier & Vagnon, made Breguet-type keys but with an internal ratchet. However, following the research of Adrien Philippe (1815–94) in the mid-century, the introduction of keyless winding brought an end to the great age of key-making.

WRISTWATCH CASES AND THEIR STRAPS

Development of the wristwatch coincided with the development of industrial methods of gilding, silvering, and nickelling watchcases and covers, and an increase in popularity of the composite case invented in Geneva by Hughes Darier in 1857. In this, thin layers of gold or silver were applied to each side of a thin plaque of brass by pressure and laminage using special machines devised by Darier.[42] In variant forms, the system would become widely adopted, the Denison Watch Case Company being a major manufacturer towards the end of the century. Although plating was possible in a variety of metals and on a variety of bases, gold on brass or steel was the most widespread. Maillechort was also suggested for low-cost watches (Figure 199). Competition led to the development of more varied shapes for wristwatch cases—oval, round, square, hexagonal—while research was devoted to making such cases both waterproof and robust. Gold and gold-plated cases were favoured over those made of silver before chromed cases became fashionable during the 1930s. Attractive and solid cases of stainless steel, at first reserved for high-cost watches, spread to middle-of-the-range products after about 1945, while stainless steel began to be used as the centre for composite gold cases. New materials like carbide, tungsten, or titanium (quickly abandoned but to reappear in the twenty-first century) were also tried, followed by synthetic materials.

From its origins the wristwatch had been carried on leather 'bracelets',[43] often produced by saddle makers. The preferred material was pigskin sometimes with perforated decoration, and the bracelet passed beneath the back of the case so that only the leather was in contact with the arm. The strap was glued or sewn, sometimes riveted, and closed by a pressure stud that was later replaced by a buckle. Dial glasses, as noted, could be protected by grills, and metal capsules were also used mounted on plated straps. Even before 1900 straps existed in gold, silver, niello-silver, gold plate, and steel. Later, stainless steel and chromed steel would be introduced. Straps made as chains were much in vogue in the 1940s with many variants, rigid or articulated, the 'marquise' open ring, or spring-loaded forms. The extending strap was created c.1890. In the pre-war decades, textiles like watered silk, repp, and brocade were favoured for the narrower ladies' straps, as was a simple cord. Plaited straps of nylon or perlon appeared in the early 1950s, but all such textile or synthetic material straps enjoyed only an ephemeral popularity. In about 1910, Karl I Scheufele

Figure 198 Watch key in chased pink gold and steel, Swiss ?, c.1825. Ville de Genève, Musée d'art et d'histoire, Inv. AD 2838.

---

41 Ponsford 1985, 85–86. General surveys of keys are Kaltenböck/Schwank; Droz.

42 Wartmann 1873.
43 See this volume Chapter 18.

Figure 199 Lever watch in maillechort, transformable from pocket (with chain) to wristwatch. L.U.C. Louis-Ulysse The Tribute. Chopard et Ecole d'horlogerie de Genève, 2010. Ville de Genève, Musée d'art et d'histoire, gift of Chopard, 2014. Inv. H 2014-4.

(1877–1941) invented a clip-joint for watch straps—the precursor of the modern models (gold, plated, or silver) that held all kinds of wristwatches. The patent was renewed for the Chopard company by his descendent-successor. He developed a stretcher-type system that gripped any watchcase firmly between adjustable horns by a quarter rotation of the back. It could thus convert a pocket-watch into a wristwatch.

A similar system was the fruit of a long development for Bovet & Dimier by the case makers Amadeo in 2010. This allowed a reversible wristwatch, a pocket-watch, or a pendant watch, all with their chains, to be carried at will, with a final option of using the watch as a small table clock. Interchangeability, transformation, and secret systems are constants of the history of watches.

The decades of the 1970s and 1980s were marked by the use of new materials for dials and cases. Tissot in Le Locle pioneered cases made of synthetic materials, the 'Astrolon Idea 2001' (1971), and a designer watch, the 'Carrousel' (1968), in gold plate with five bezels in epoxy with interchangeable colours. The iconic 'Gucci 11/12' watch has a rigid strap with a spring clip accompanied by thirteen interchangeable bezels. In 1986, the electronic movement made possible the development of the 'Rock' watch by Tissot (Figure 160), where the case was carved from Swiss Alpine granite, with later models in other stones. It was followed by the 'Wood watch' (1990).

CONCLUSION

Horological accessories, especially those for watches, are functionally linked with the adornment of timepieces, simultaneously protecting them and facilitating their integration into clothing. Once the watch had become synonymous with a new social use of time in which precision and reliability dominated,[44] accessories made possible various new ways of carrying the watch. Today, fashion and marketing have given rise to a new status: the watch, among handbags, eyeglasses, and the rest, has itself become a fashion accessory.

From before the Second World War, the first watches commercialized as being in themselves accessories were promoted in the jewellery and fashion sectors. Bulgari, Cartier, Dunhill, Hermès, and Tiffany all produced watches in small numbers for their range of jewellery. Swiss watchmakers like IWC, LeCoultre & Co, Patek Philippe & Co, and the Tavannes Watch Co

---

44 Landes 1983, 2000, part II, especially ch. 5.

joined them and fed the growth of creativity in fashion. These watches were not developed as part of a strategy of diversification,[45] but rather were accessories placed in showcases or in advertisements. These items complemented the main products of the label, whether clothing or leather goods. Some watches produced by jewellers were made to order for well-off clients.

Today, the nexus between fashion, luxury goods, and their derivatives allows the watch to take an increasingly important place as a fascinating accessory. Integrated into the strategies of fashion houses, watches contribute to aesthetic and technical creativity. Watches, after six centuries of existence, continue to carve out an essential place in the world.

---

45 Donzé 2017, 73.

CHAPTER TWENTY THREE

APPLICATIONS OF CLOCKWORK

Outside time-telling, geared mechanisms have many uses that claim a place in a history of horology either because they supply a time input needed to effect a task, or because the making of them was an integral part of a clockmaker's production and so contributed to his livelihood. The sections in this chapter offer studies of a selection of such activities.

SECTION ONE: PLANETARY MODELS

Jim Bennett and Anthony Turner

Distinguished from clocks incorporating planetary indications,[1] planetary models were frequently mechanized through a gear train using either a manual or a powered drive. Usually they represented the Ptolemaic or the Copernican system; popularization of the latter from the late seventeenth century onwards was a powerful stimulus to their production. Some armillary instruments, such as the Jagellonian globe,[2] in which movement is imparted to Sun and Moon rings around the stationary central Earth, may be considered planetary instruments, and this model would have considerable influence. A parallel line of development was of planar instruments—vertical or horizontal—perhaps deriving from the dials of astronomical clocks. In a design for a planar Copernican planetary instrument, 1598–9,[3] Johannes Kepler (1571–1630) proposed using coaxial gearing but this was not practically employed until Roemer designed his planetary planisphere in 1680.[4] Planar models would develop in parallel with the armillary model from the late seventeenth century.

In the same way that a globe, terrestrial or celestial, might be moved by a clockwork mechanism, so might the armillary sphere to create a *sphère mouvante* (moving sphere, see Figure 200).[5] In 1701 a model with sphere, Sun, and Moon driven by clockwork was presented to Louis XIV by Jérôme Martinot (1671–1724) and Thomas Haye (? pre-1680–post-1740) and immediately provoked an order for a Copernican model from Louis' physician. Nothing is known of this, but in 1706 a Copernican sphere conceived by Jean Pigeon (1654–1739), who had been developing it since 1690, was shown to the Académie des Sciences and subsequently purchased by the crown. It surpassed its predecessors by simulating the orbits of all the known planets and was immediately published by the Abbé de Vallemont.[6] For him its immediate precursor was the Copernican tellurian produced by Willem Janszoon Blaeu (1571–1638).[7] This was a manual instrument in which the Earth, enclosed within an armillary sphere, was pushed round a central Sun set in the plane of the zodiac band.[8]

Vallemont stresses the difficulty and originality of producing working models of the Copernican system, and since this was still only tentatively establishing itself in popular understanding, the tangible support that models offered for it made them attractive. Pigeon made five such machines, and more followed in the century as the 'moving sphere' became the quintessential French celestial model. The tradition would culminate in the remarkable creations of Antide Janvier (1751–1835),[9] a craftsman well aware of the work of his predecessors, much of which he had examined (see Figure 107).

The armillary model, however, was widespread and several English examples from the early eighteenth century are known. John Rowley (c.1668–1728) tried to adapt it to a central Sun with surrounding planets, and two such attempts were in the Earl of Orrery's instrument collection.[10] Other makers, notably Richard Glynne (fl. 1707–30) of London, took this further. A large armillary sphere by Glynne surrounds a mechanical model for Mercury, Venus, the Earth, and the Moon, worked by a hand-crank, with the outer planets, Mars, Jupiter, Saturn, and satellites present, though moved by hand.[11] Apart from the expense of the armillary construction, it was also inconsistent to combine a rotating Earth with a celestial sphere on a polar axis. The model had lost favour in England by the mid-century.

Glynne was credited by a near-contemporary writer as being first to apply planets to an armillary sphere. If so, this must have occurred at the same time as the planar instrument that would become known as an orrery was developed in the context of the manufacture in London, by Joseph Moxon (1627–1700), of Blaeu-type tellurians and Copernican spheres,[12] and of designs c.1705 by Stephen Hales (1677–1761) in Cambridge for an 'horary'.[13] The new instrument was a Sun-centred astronomical model, comprising at least part of the solar system, minimally the Sun, Earth, and Moon, illustrating their relative, radial positions in order from the centre, but without representing their distances or relative sizes. Although inexpensive examples were later designed with planets mounted on arms adjusted by hand, the first specimens were moved mechanically by wheel trains calculated to deliver the planetary periods, the whole moved either by a hand crank

---

1 Discussed in Chapter 9.
2 King & Millburn 1978, 76; Von Bertele 1961, 26–7 and 56–7 for an armillary sphere c.1650 with manual planet bands.
3 King & Millburn 1978, 92.
4 BnF inv. N° CVPL. Inv. gén. 23; cote Ge A 280. Turner 2018, 116–22; Millburn & King 1988, 109; Darnell & Nielsen 2013.
5 For an early example, c.1540–50, see Kugel et al. 2002, 140–3.
6 De Vallemont 1707.
7 Prefigured by a simpler device associated with Wilhelm Schickard (1592–1635), King & Millburn 1978, 93.
8 King & Millburn 1978, 94–5.

9 King & Millburn 1978, 295–303; Augarde & Ronfort, 1998; Rees 1819, xxv sig. 4 H 4^r– 4 I 1^v.
10 Turner 1987c, 236, 239.
11 MHS, Oxford, Inv 69–190; King & Millburn 1978, 157–9, who note its resemblance to the Blaeu-Moxon model and mention other examples.
12 King & Millburn 1978, 96.
13 King & Millburn 1978, 151.

Figure 200 An early 'sphère mouvante' by Jérôme Martinot, between 1709 and 1718. Bibliothèque nationale de France, département des Cartes et plans, Ge A-355 (Res).

or an integral clock.[14] Early examples were made by the London clockmakers Thomas Tompion (1639–1713) and George Graham (c.1673–1751). These display the annual motion and daily rotation of the Earth, the motion of the Moon around the Earth, and the rotation of the Sun. The horizontal platform represents the zodiacal plane, to which the Earth's axis is inclined, while an outer ring has scales for the date and degrees of the zodiacal signs. A ring moved by the vertical stem for the Moon gives its age, while in one of the machines, with thirty wheels in the planetary mechanism, a spring-driven pendulum clock shows solar and sidereal time.

The Tompion/Graham instrument was copied and developed by Rowley. The outer zodiac and calendar ring is carried on pillars above the surface of the drum containing the mechanism. The Sun rotates on an inclined axis and the axis of the Earth maintains a constant direction. The Moon rotates, maintaining its bright side towards the Sun, and its orbit is inclined to the zodiac ring, while the lunar assembly has a counter-rotation to compensate for the annual motion of the base plate and to reproduce the recession of the lunar nodes. This represented a considerable development, but in 1715, Rowley fulfilled an order from the East India Company for a yet more ambitious model.[15] The motions of Mercury and Venus were added, and their heights were varied to represent the inclinations of their orbits to the plane of the ecliptic. Near-contemporary accounts emphasize its complexity.[16]

Thomas Wright (fl. 1718–47), a clockmaker's son, was apprenticed to and then employed by Rowley, before eventually taking over his business.[17] He issued a trade card with an instrument close to Rowley's design of a Sun, Earth, and Moon configuration but including Mercury and Venus, as with Rowley's East India Company model. This Wright seems to have taken as 'the orrery' from which he developed both simplified and more ambitious, complex ones. In the former category was his 'Orrery Reduc'd', omitting Mercury and Venus and the diurnal rotation of the Earth. By 1732 he had extended the range in both directions. An advertisement in the second edition of Joseph Harris, *The Description and Use of the Globes, and the Orrery*, cites 'several very large Orreries' exhibiting all the known planets in the solar system but, since 'that makes the Price very great', for schools and academies one showing the motions only of Sun, Earth, and Moon may be had 'at a very moderate price'.[18] The instrument was being adapted to different purposes and pockets.

The first edition of Harris's textbook had included a print of Wright's 'Great Orrery', four feet diameter, still with a zodiacal ring raised on pillars in the manner of Rowley, but now surmounted by a brass armillary hemisphere. In the third edition (1734), an addition to the print announced that such a hand-cranked orrery had been made for the Royal Naval Academy, Portsmouth.[19] Wright now noted that his orreries could demonstrate the motions of all the planets, of their satellites, and of Saturn's ring, but did not omit to mention that, while he had supplied instruments to gentlemen, noblemen, and the king,[20] he also had 'small ones for Schools'.[21]

The 'Great Orrery' received another addition when Benjamin Cole (fl. 1720–50), who had been employed by Wright, took over the shop in Fleet Street. One instrument survives with their joint signatures.[22] In editions of Harris from the 1750s the advertisement notes that Cole had been 'Servant to Mr Wright at the Time of the above [great orreries] being made'.[23] Wright's reputation in this area had an enduring commercial value. It may have been Cole who introduced the name 'Grand Orrery' for the large, ornate instruments with all the known planets and their Moons. Inevitably, he applied it to his shop sign. For a time he worked in partnership with his son, also Benjamin (fl. 1766–82), who succeeded to the business and the same range of goods. By then, however, there were competitors in the field, who were not in the trade succession from Rowley.

Prominent among them were Benjamin Martin (1705–82), James Ferguson (1710–76), and George Adams (1709–72), each making their own distinctive contribution. It is not a coincidence that all three were involved in writing books on experimental natural philosophy for a broad public, while Martin and Ferguson were strongly committed also to public lectures. The lecturer and writer John Theophilus Desaguliers (1683–1744) also became interested in the development of the orrery, designed one for all the planets, and had it made for his own use. It was later added to the collection of George III by Stephen Demainbray (1710–82).[24] Martin became interested in orreries through his ambitions as a lecturer, a trade where demonstration and spectacle were appreciated by subscribing audiences. He used an orrery set within an armillary sphere, probably something like the instrument by Glynne described above, so it is not surprising that when he moved into instrument making and retailing, he was exercised by the cost of the large orreries, too much for his auditors, whom he wanted to convert into readers of his books and customers for his instruments.[25]

---

14 Orreries are described in great technical detail in Rees 1819, xxv, sig. 4 E 3^r-4 I 3^v.

15 Ryder 1939, 139–40; King & Millburn 1978, 155–7; Rees 1819, xxv, sig. 4 E 4^r.

16 Stone 1723, 189–91, 'nearly a hundred wheels; Ryder 1939, 140, 'above eighty different wheels'.

17 Clifton 1995, 239, 306.

18 King & Millburn 1978, 160–1; Harris 1732–1734, 'Advertisement'.

19 Harris 1732–1734, 'The Great Orrery'.

20 Morton & Wess 1993, 402–3.

21 Harris 1732–1734, 'Advertisement'.

22 King & Millburn 1978, 162.

23 Harris 1732–1734, 'The Great Orrery', 'Advertisement'; King & Millburn 1978, 163.

24 King & Millburn 1978, 170–3.

25 King & Millburn 1978, 195–201; Millburn 1976, *passim*.

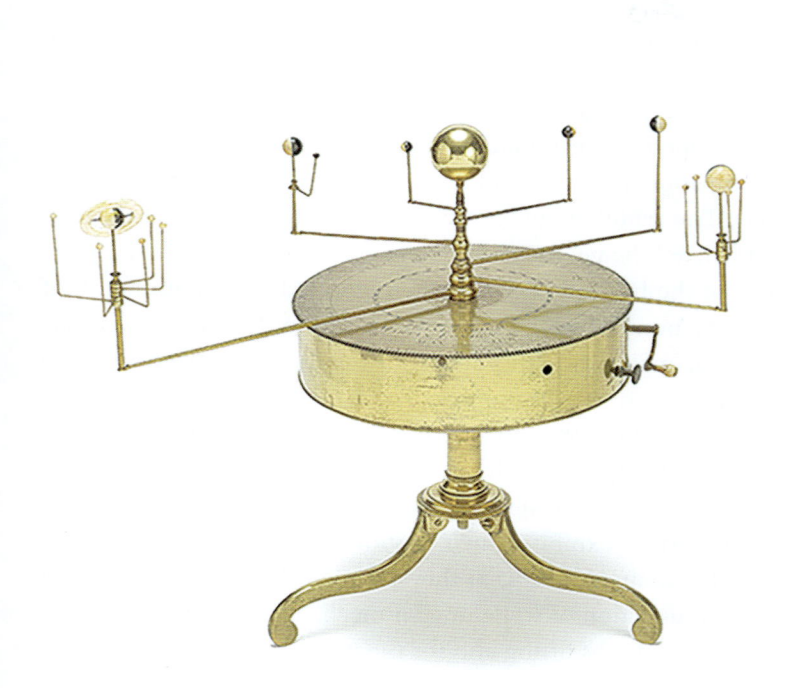

Figure 201 A manual table orrery by Benjamin Martin 1765. Left: established as a planetarium; Right: as a tellurian. Photos: courtesy of Bonhams, London.

A typical Martin instrument was a table orrery (Figure 201) on a tripod foot with a pillar stand rising to a circular drum, from the centre of which protruded a set of co-axial tubes of different heights, the central one supporting and rotating the Sun.[26] Each tube carried a horizontal arm extending to a planet, while within the drum each rose from a separate gear wheel in an inverted conical stack, with the longest planetary period, Saturn, at the top. These wheels engaged with a corresponding set of driver gears, fixed to a single arbor, this set forming an upright cone, with the gear ratios and pitches chosen to deliver the planetary periods, when any one of the driver gears was moved by a worm worked by a crank handle. The arrangement was simple, versatile, and could readily be fitted with an internal clockwork drive. In the form described, Martin called the instrument a 'planetarium', but he could use the same base to drive either a 'tellurian' (demonstrating day and night and seasonal change) or a 'lunarium' (for the relationship between the Earth and Moon). These required a toothed rim to the upper plate of the drum to drive them, while the annual motion was effected by the double-cone arrangement for the planetarium. Everything worked from the single hand-crank.

From humble beginnings in rural Scotland and little formal education, James Ferguson became a successful lecturer, writer of popular astronomical books, and designer of original and ingenious machines for demonstrating astronomical phenomena.[27] He delighted in finding original mechanical ways to illustrate more recondite astronomical phenomena than circular planetary motions accompanied by satellites. The Moon's motions had a singular fascination for him and offered particular challenges. Venus also drew his attention, as did comets, tides, and eclipses. For example, he applied what became known as 'Ferguson's mechanical paradox' in different ways. A gear train comprised three wheels of the same size; one was fixed at the centre, while the others were free to move on axes set in a frame that could be turned around the fixed wheel. Slight alterations in the numbers of teeth would produce different outcomes in the outer wheel from a slow advance to a slow regression. An orrery, operated by pulleys and cords, delivered the advance of the Moon's apogee and the regression of the nodes.[28] While only a handful of Ferguson's instruments survive, there was an enthusiastic public for his books and lectures, and he was awarded a small pension by George III.

George Adams was well placed to appropriate all these developments. While he introduced some new design features, his main strength was as a consummate businessman, workshop manager, and retailer.[29] He cultivated and served wealthy clients, understanding the role of the grand orrery in a large house or palace.

---

26 Millburn 1973, 381–91; Millburn 1976, 95–9.

27 King & Millburn 1978, 178–94; Millburn & King 1988, 39–47, 52–7.
28 Science Museum 2020; King & Millburn 1978, 191; Millburn & King 1988, 57, 104–6.
29 Millburn 2,000, 25–30, 218–24.

He was also more successful than his predecessors in exporting orreries to mainland Europe. Yet he did not neglect the aspirations of the middling sort who ventured into his shop on Fleet Street. They were encouraged through his books, describing enticing instruments for sale, and it was the Adams firm rather than Martin who made the most of the latter's design of a complete set of planetarium, tellurian, and lunarium. Adams was succeeded by his son, also George (1750–95), and this range of devices was taken on in turn by the Jones company, who acquired aspects of the Adams business, notably their books. One concession to cheaper models was the younger Adams's use of a 'hybrid' orrery, constructed mostly in wood, with the planets attached to coaxial tubes (as noted in Martin's work) and their wheelwork hidden beneath a wooden base plate, except for the Earth, whose gearing, together with the Moon, was carried above in full view. This not only saved on cost; it also made a feature of the more interesting aspects of the wheelwork.

William Jones (1762–1831) catered for interest in cheaper models by using a wooden base plate with printed paper scales, supplied with a range of planetary arrangements in combinations of push adjustment, Martin-type cranked movement, and hybrid forms[30] As W. & S. [William and Samuel] Jones, the firm pursued this line into the mid-nineteenth century and was joined by Newton, another firm that ran through a sequence of names. As William Newton & Son they displayed a different design of hybrid at the Great Exhibition of 1851, where push adjustment for all the planets was combined with a lunarium driven by pulleys and cords.

While London was probably the most active centre making planetary machines for commercial exploitation, such devices were made throughout Europe, though often as one-off pieces for show and the demonstration of skill. Examples from the Netherlands include the 'Leiden Sphere', a collaborative work of Adriaen Vroezen (1641–1706), Steven Tracy (d. 1703), and Nicolaas Stampioen (1639–1721), which included the movements of Jupiter's satellites along with the planets,[31] and the domestic planetarium of Eise Eisinga (1774–1828) built into the ceiling of the living-room of his home in Franeker in 1778.[32] German examples are provided by the several *Weltmaschine*, c.1780, made by Philipp Matthäus Hahn (1739–90).[33] These were the culmination of his mechanical predilections. Hahn, although not driven by commerce, attracted commissions and sold clocks and planetary machines to order. Notable orreries, which returned the machine to the vertical plane employed by Huygens, were made by David Rittenhouse (1732–96) in Pennsylvania.[34]

Two settings—astronomical and social—in the eighteenth century helped to naturalize the orrery—to make it a fitting accompaniment to the culture of contemporary natural philosophy. The armillary sphere was a nest of graduated circles, which constructed the geometrical system according to which the positions of the heavenly bodies were registered. This registration of the 'phaenomena' was the substance of the discipline of astronomy, unconcerned with physical or causal questions of what the heavens were or why they moved; such questions belonged to natural philosophy, not to astronomy. The outcome of the successive accounts of Copernicus, Kepler, and Newton required something different. Now the planets in motion were the objects of attention and the orrery exhibited their motions, not the reference framework for their positions, as they pursued their orbits under the continuous action of some unseen mechanism or force.

The orrery was fitting also for the public cultivation of natural philosophy. It was the perfect demonstrational prop for popular lectures, where audiences were assured that no knowledge of mathematics would be needed. To engage with the Newtonian treatment of the planets was an impossible challenge for almost everyone, but the orrery offered an experience of the planetary system that bypassed the impediments. Participants could jump straight to the outcome. Further, the appeal of popular lectures relied on demonstration and experiment: the orrery brought the universe itself into this educational programme.

One further, probably minor, aspect of the orrery's appeal may be mentioned. Some devotees clearly enjoyed its mechanism, and mechanics in general fell within the compass of lecturers and writers on experimental philosophy. Textbooks offered impressive diagrams of the gearing that operated complex examples. One grand orrery by Heath and Wing survives in a fully glazed original case.[35] The intention must have been to expose the working operation, usually concealed. Seeing a grand orrery in operation evoked fascination and delight in an audience. The Savilian Professor of Astronomy at Oxford, Thomas Hornsby (1733–1810), invited his auditors: 'Let us now take a general View of the Universe'.[36] This route to astronomical knowledge combined several sets of skills with mechanical horology at their core.

---

30 King & Millburn 1978, 205–11.
31 Dekker 1986.
32 King & Millburn 1978, 217–24.
33 Väterlein 1989; King & Millburn 1978, 232–43.
34 On which see Hindle 1964, especially ch. 2.
35 History of Science Museum 2020.
36 Hornsby MS f.198.

SECTION TWO: TIMING AND DRIVING SYSTEMS

Paolo Brenni

Many special devices have been developed for measuring or recording short intervals of time such as chronoscopes and chronographs, instruments for driving telescopes, or for simulating the rotation of the Earth. This section discusses some of the main types of such instruments and bibliographical orientations are given for more detailed accounts.

CHRONOSCOPES, CHRONOGRAPHS, AND STOPWATCHES

Chronoscopes are instruments used for the precise measurement of time intervals. They are time-measuring devices used for a non-time-telling purpose. Chronoscopes were widely used in ballistics, physics, experimental physiology, and psychology for determining the—sometimes very short—periods between the beginning and the end of an event. In these instruments time intervals are read directly on the dial or calculated from the mechanism. Chronographs are arranged to give automatically a written record of the time interval. The term is also used for stopwatches.[37]

In the nineteenth century, the progress of artillery led to theoretical and practical ballistic studies to understand precisely the behaviour of the projectiles fired. Many systems were proposed for measuring velocity. In ballistic research a wire was placed in front of the mouth of a gun and a contact was placed on the target. A chronoscope was activated when the projectile broke the wire and stopped when it hit the target. In the 1830s the French physicist Claude Pouillet (1790–1868) conceived a chronoscope that was, in fact, a ballistic galvanometer. It was based on the fact that the deflection of a galvanometric needle under the effect of a pulse of current is proportional to the duration of the pulse if the current remains constant. In the 1840s the British physicist Charles Wheatstone (1802–75) proposed a chronoscope with a weight-driven clockwork mechanism and an electric circuit including a battery and an electromagnet. When the latter was activated, it engaged an indicator, which was disengaged and stopped when the electromagnet was deactivated. The time interval was read on a dial. The electric chronoscope was greatly improved in the late 1840s and 1850s by Matthäus Hipp (1813–93).

In his system, a clockwork mechanism (supported by four columns) was controlled by an **escapement** consisting of vibrating steel lamella fixed near the **escape wheel**. The lamella made 1,000 vibrations a second and the typical sound produced by it allowed for calibration against a tuning fork. The instrument had two hands and two dials: the upper one indicated tenths of a second, the lower thousandths of a second. The driving movement was completely separated from the **motion work**. When the device was used, the former was started and only when it had reached a constant working speed would a pair of electromagnets cause the wheel **train** of the hands to engage to measure an interval.[38] This system avoids the inevitable error of the Wheatstone chronoscope arising from the inertia of the mechanism. The very first applications of Hipp's chronoscope were for physics teaching and in ballistics. With it, the time of falling bodies could be measured in a much simpler and direct way than with an Atwood machine. The typical set-up included a fall apparatus that released a metal ball. This fell onto a board which acted as an interrupter. When the sphere was released, the hands of the chronoscope started: they stopped when the sphere dropped onto the board. Like Wheatstone's apparatus the Hipp chronoscope was also used for measuring bullet velocity. The bullet activated the train of hands of the chronoscope when it broke an electric wire stretched across the mouth of a gun and stopped it when it hit a target acting as an interrupter.

In the second half of the nineteenth century chronoscopes found new applications. Since the early 1860s Hipp's apparatus had been used for measuring 'physiological time' by Adolphe Hirsch (1830–1901), the first director of the Neuchâtel chronometric observatory. This observatory systematically tested watches, certified them, and provided clock manufacturers with accurate time by means of telegraphy. It was Hipp himself who provided the electric apparatus for the transmission of time and took part in several of Hirsch's experiments. The latter tested the accuracy of chronographs and used them in determining reaction times as well as the speed of sensory impression and nerve conduction. In Hirsch's experimental setting, the sound of the ball hitting the board-interrupter started the chronoscope while the observer stopped it with a telegraph key. Hirsch's pioneering publications on the subject opened the way for use of chronoscopes in physiological laboratories.[39]

In 1874, in the first edition of his *Grundzüge der physiologischen Psychologie*, the physician, physiologist, and philosopher Wilhelm Wundt (1832–1920), known today as one of the founders of

---

37 Jervis-Smith 1911. Bud & Warner 1998, 110–11, 115–16.

38 For details of the different models, see Schraven 2003.
39 Hoff & Geddes 1960.

experimental psychology, pointed to the importance of the apparatus. Hipp's chronoscope was adopted in the Leipzig Institute for Experimental psychology established by Wundt at the end of 1879. It was used for determining reaction times in various circumstances, the duration of mental processes, and in many other experiments.[40] The basic setup of the experimental system remained the same, even if the telegraph key was replaced by a voice key and the dropping apparatus was replaced by other devices. In spite of the fact that other instruments such as chronographs (see below) were generally considered more reliable than Hipp's chronoscopes, the latter contributed to opening new methodological problems in psychological research and to the improvement of the related experimental settings. Finally, Hipp's chronoscope became the standard apparatus for psychological time measurements in the 1880s.[41] It was manufactured and improved by his successors, Albert Favarger (1851–1931) and Alfred von Peyer (1839–1926), and by other precision instrument making firms well into the twentieth century (Figure 202). The Hipp chronoscope became such an emblematic instrument of accurate and precise measurement, that its presence in a collection whether of physics, experimental physiology, or psychology, increased the importance and reputation of the collection itself.

If the German and Anglo-Saxon world widely adopted Hipp's apparatus, the most common one in France was the chronoscope invented in 1886 by the physicist and physiologist Arsène d'Arsonval (1851–1940). A spring-driven clockwork mechanism, the speed of which was kept constant by a Foucault regulator, rotated on a horizontal arbor (1 rps) by a small electromagnet. Co-axial to the first, there was a second arbor with a light hand indicating the zero position on the dial of the apparatus. A spring maintained the disk separated from the electromagnet. When the latter was activated, the disk was attracted by it and the hand began to rotate. When the current was interrupted the disk (not attracted any more) and the hand immediately stopped. The d'Arsonval apparatus was simpler and quieter than Hipp's chronoscope (the vibrating lamella were noisy and could disturb experiments) and could work for about ten minutes non-stop. Finally, it was portable and could easily be used in hospitals. On the other hand, it was not so precise (1/100 or 1/200 of a second instead of 1/1000 of a second), and the constancy of rotational speed, in spite of the regulator, was not always certain. Several other chronoscopes of different types, based on varied principles (falling weights, pendulum, hammer, tuning forks), were proposed in the nineteenth century and used for physics demonstrations, for ballistic tests, or for psycho-chronometry. In the twentieth century, the introduction of synchronous motors completely transformed chronoscopes. The armature of an alternating current motor runs at a constant speed. An indicator can be engaged with or disengaged from the armature by an electromagnetic clutch. These simple and reliable chronoscopes replaced the Hipp and all other mechanically driven models.

Chronographs provide written records of time intervals and are thus registering instruments. Many type of chronographs were developed from the mid-nineteenth century onwards thanks to the progress of practical applications of electricity, which derived from the rapid development of telegraphy.[42] They were mainly used in astronomy to determine the time of star transit, in experimental psychology to measure reaction times, and in physics for recording various events. Many different devices of this type were proposed, although the most common were drum and tape chronographs. In the former, a metallic drum covered with the recording paper was turned by a clockwork mechanism equipped with a regulating device (conical pendulum, **fly**-ball, or **air brake** governor) in order to assure a constant rotation speed. A trolley with a slow movement parallel to the axle of the drum carried one or more pens driven by an electromagnet. These could register the electric signals that came, for example, from a telegraphic key. The paper could be covered with lampblack and in this case the writing stylus left a trace in it, while with normal paper a small inking syphon pen was used. The time reference could be external or internal to the chronograph. In the first case, for example, a second pen electrically connected to a separate clock could mark every second with a trace on the drum near the one left by the signal recording pen. Other chronographs with internal time calibration had a small electromagnetic tuning fork fixed to the trolley and carried a light writing stylus on one of its prongs. The vibrations of known frequency of the tuning fork left a small sinusoidal trace on the drum, which was used as the time reference.

The tape chronograph looked like a common Morse telegraphic receiver. A paper tape was dragged by a weight-driven clockwork mechanism. Two (or more) pens were operated by electromagnets: one could inscribe on the tape the incoming signals while the other, which was connected to a clock, marked a time reference trace indicating every second. Electromechanical chronographs were common in observatories and laboratories until the second half of the twentieth century. Finally, they were replaced by electronic instruments, plotters, and computers.

But, because of a certain confusion in terminology, a chronograph can also describe a type of watch that is used as a stopwatch for displaying elapsed time.[43] They generally indicate the time in hours, minutes, and seconds like a normal watch, but also have an independent sweep seconds hand. This can be started, stopped, and returned to zero by successively pushing either a button on the band or, later, the pendant. Developed by Adolphe Nicole (1812–76) in the early 1840s, he obtained a first patent for them in 1844 (N° 10,348) and a second for improvements in 1862 (N° 1,461).[44] Watch chronographs can mark discrete intervals of time.

---

40 Serge & Pins 2014; Benscop & Draaisma 2000.
41 Schmidgen 2005.

42 Du Moncel 1878, iv, 183–302.
43 Chaponnière 1924.
44 For details, see Poniz 2021.

Figure 202 Hipp's chronoscope made by Peyer, Favarger & Cie around 1900. When the instrument was properly calibrated, measurements of up to 1/1000 of a second were possible. Extremely successful, it was used in countless laboratories. Collection of Historical Scientific Instruments, Harvard University, Cambridge, MA.

Certain only have this latter function and do not the indicate the time of the day, while more sophisticated ones use additional complications and can have multiple independent hands.

In 1815/16, Louis Moinet (1768–1853) invented a chronoscope for use in astronomy and measuring to sixtieths of a second, which anticipates some chronograph functions. Another such astronomical timer (with split seconds) was produced by Abraham-Louis Breguet (1747–1823) in 1820. In collaboration with Louis Fréderic Fatton, one of his most able watchmakers, Breguet then developed a recording or inking chronograph.[45] A month later (9 March), Nicolas Mathieu Rieussec (1781–1852) obtained a French patent for his inking chronograph.[46] Both had a nib at the end of an index, which placed a black ink mark on the dial at the moment when an event being measured ended. Both had originally been conceived to time horse races but were quickly found to have other practical and scientific applications. Alexander von Humboldt (1769–1859) and François Arago (1786–1853) used Rieussec's instrument when measuring the speed of sound at Montlhéry in June 1822. If the paternity of the instrument remains doubtful, paternity in the name 'chronograph' belongs to Rieussec, who used it in his patent application.

Various improvements were made in the following decades: the resetting mechanism of the needle of the timer was introduced in 1844. In 1870 a timer with two seconds hands was presented. The mechanism allowed for the simultaneous measuring of two events of different durations and was also capable of recording two different times and intermediate times: the '*rattrapante*' (split-seconds hand) function was born. The use of chronograph watches boomed at the beginning of the twentieth century (Figure 203) with the increasing popularity of sports such as athletics, horse, bicycle and car races, and motorboat and airplane competitions, which all required precise timing. Many firms proposed sophisticated chronographs and different systems using electric circuits, photocells, and cameras in order to eliminate human error from the time-keeping operations. However, chronographs of various types also found technical applications—for example, combined with revolution counters for calculating the rotation speed of machines. Peculiar chronographs were introduced for specific measurements. Tachymeters are chronographs with a special scale inscribed around the rim of the dial to compute a speed based on travel time or measure a distance based on speed. Pulsometers have a special dial allowing pulse-beat to be determined. Furthermore, in the first decades of the twentieth century, chronographs were essential to the scientific 'time and motion' study of industry introduced by Frederick W. Taylor (1856–1916). Workers performing their tasks were timed in order to economize their movement time and thus increase labour productivity. Today digital technologies and computers are universally used for timing professional sport competitions, scientific phenomena, or industrial processes. However, complicated and sometimes very expensive mechanical chronographs are horological masterpieces and have therefore a continuing market appeal as coveted collectibles and status symbols.

TELESCOPE DRIVES, HELIOSTATS, SIDEROSTATS, AND COELOSTATS

From the seventeenth century onwards, there have been proposals for using clockwork mechanisms to move telescopes so as to follow the apparent celestial rotation. Such drives, however, only became used systematically in the early nineteenth century when equatorial mountings were adopted for most large telescopes.[47] One of the advantages of this mount lies in its ability to allow the telescope to stay fixed on any celestial object with diurnal motion by driving the polar axis at a constant speed. The motion generated by the clock drive was transmitted by a worm wheel to a toothed circle on the polar axis. Obviously, it was impossible to use clocks with a common escapement because this would have produced a jerky movement. It was necessary to find a drive with smooth and continuous movement. The first large equatorial telescope equipped with such a device was the 24.4cm aperture refractor made by Joseph von Fraunhofer (1787–1826) for the astronomer Friedrich Georg Struve (1793–1864) and installed in 1824 at the observatory of Tartu (Dorpat) in Estonia. The weight-driven clockwork had a centrifugal friction-brake governor. This was later adopted in large telescopes made by Georg Merz (1793–1867), Fraunhofer's successor. Several types of device were used in the telescope drives to maintain a constant speed of rotation and were more or less popular in different countries. One of the most common was the centrifugal fly-ball governor, which derived from Watt's governor for steam engines. It was improved and modified several times and, together with the spring governor, was often used in English and American clock drives. In France, Léon Foucault (1819–68), Antoine-Yvon Villarceau (1813–83), and others developed air-braked governors using rotating vanes (combined with springs and weights). In the second half of the nineteenth century and in the first decade of the twentieth century, the increasing size and weight of astronomical refractors and reflectors required solid and more complex driving devices and electrical servo-mechanisms for raising their weights were introduced. Finally, clockwork mechanisms were replaced by simple synchronous electric motors and today computers connected with stepped motors are fast enough to make the necessary calculations and adjustments for tracking even in alt-azimuthal-mounted telescopes. The largest, modern reflecting (sometimes multiple-mirror) telescopes use this kind of computerized system, which continuously calculates and corrects the

---

45 Fatton, Breguet's agent in London, obtained a British patent (N° 4,707) for this on 9 February 1822.
46 According to Chaponnière 1924, 84, the only difference between the two was that the Breguet/Fatton instrument had a mobile knib and a fixed dial, while that of Rieussec had a fixed knib and a mobile dial.

47 Darius & Thomas 1989; Caplan 2012.

Figure 203 A gold-cased pocket chronograph with lever escapement signed by Longines c.1910. White enamel dial with three colour tacheometer scale, subsidiary accumulator dial at 12 o'clock, seconds at 6 o'clock, centre seconds hand with 'stop' and 'flyback' functions. Model introduced in the 1890s. Photo: Jean-Baptiste Buffetaud, by courtesy of Chayette & Cheval, auctioneers Paris.

azimuth and the elevation in order to remain aligned on the target.

Heliostats, siderostats, and coelostats are all clock-driven apparatus that reflect a light beam in a fixed and desired direction.[48] Heliostats and siderostats (which generally have a single mirror) produce in an arbitrary direction an image of a fixed point, while the image of the sky rotates around it. The first portable heliostats were probably conceived by Giovanni Alfonso Borrelli (1608–79) and Daniel G. Fahrenheit (1686–1736). In 1742 Willem Jacob 's Gravesande (1688–1742) proposed, made, and coined the name of such an instrument in order to facilitate his optical demonstrations. In the nineteenth century heliostats were commonly used in physics laboratories and lecture rooms to reflect and use Sunlight as a powerful source for optical experiments, photography, and spectroscopy. The apparatus was generally installed on a shelf on the façade of a building in front of a window. The beam of light reflected by the heliostat entered the room through a hole in the window shutter. Various types of laboratory heliostats were proposed; simple ones had an orientable mirror mounted on a polar axis connected to a clockwork drive. Because of their use as light sources, they generally employed clocks with an anchor escapement, whose jerky movement did not disturb their function. Those invented by Henri-Prudence Gambey (1787–1847), Jean-Thiébault Silbermann (1806–65), Foucault, and Moritz Meyerstein (1808–82) were among the most successful, continuing in use for several years after the introduction of strong electric light sources (Figure 204).

Large heliostats are installed in permanent solar observatories for directing a beam of light into the instruments (spectrometers, cameras) used for studying the Sun. Finally, a large array of computer-driven heliostats is used for concentrating Sunlight in solar thermal power stations.

Siderostats are generally large and sturdy, and are used in astronomy for reflecting the image of a star to a fixed telescope. In 1869 Foucault designed a widely used siderostat (Figure 205), although that the field of view rotates slowly is a drawback for research employing astronomical photography.

Coelostats (with one or more mirrors) reflect a non-rotating image of a portion of the sky. These instruments were often used with a fixed telescope during explorations and astronomical missions and for solar observatories. They allow the expensive and

---

48 Mills 1985; Mills 1986. Entries in Bud & Warner 1998.

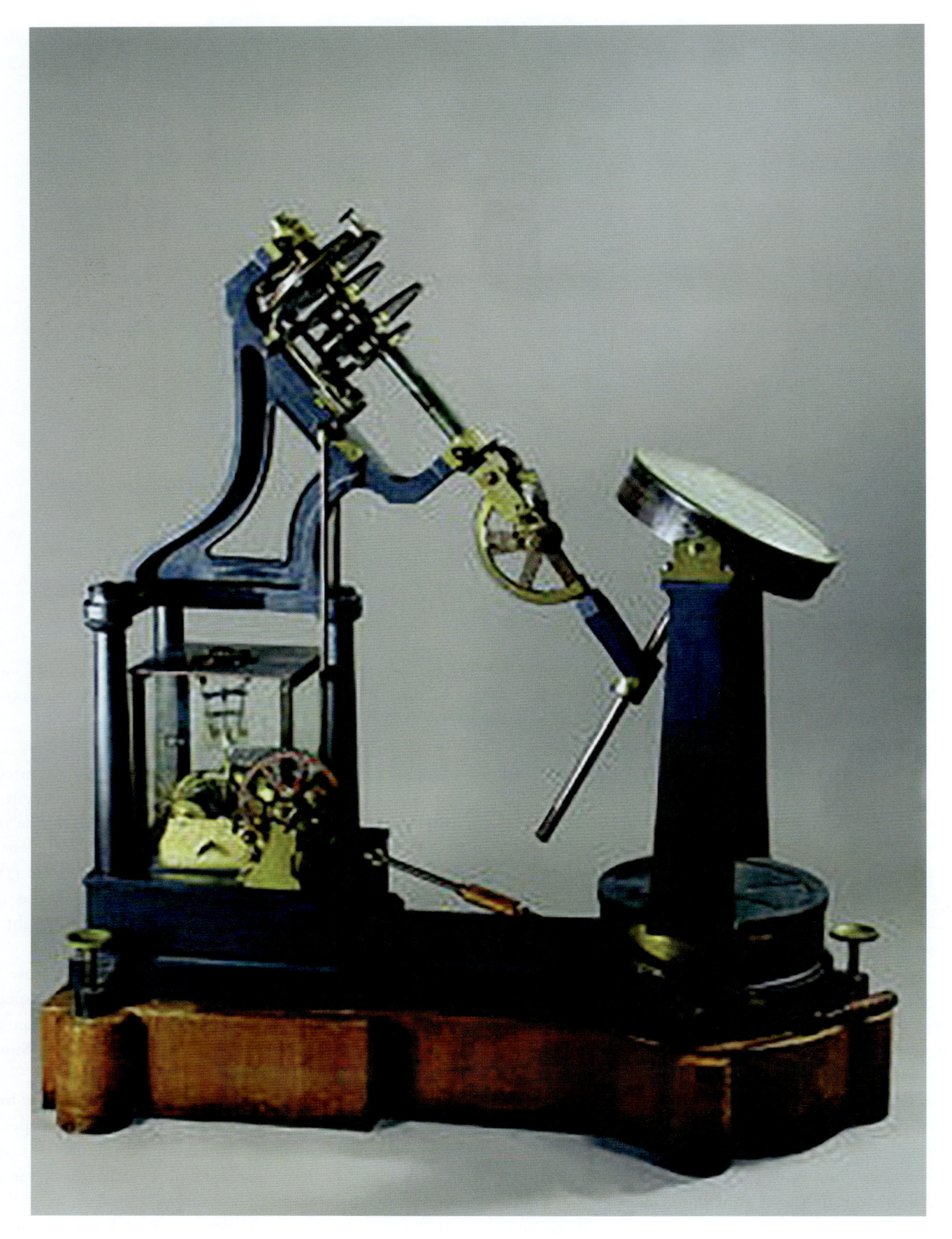

Figure 204 Foucault's siderostat made in Paris by William Eichens in 1869 with a mirror by Adolphe Martin. This type of single-mirror instrument, invented by Foucault in 1862, was very successful and proved to be particularly useful for solar observations. Observatoire de Paris.

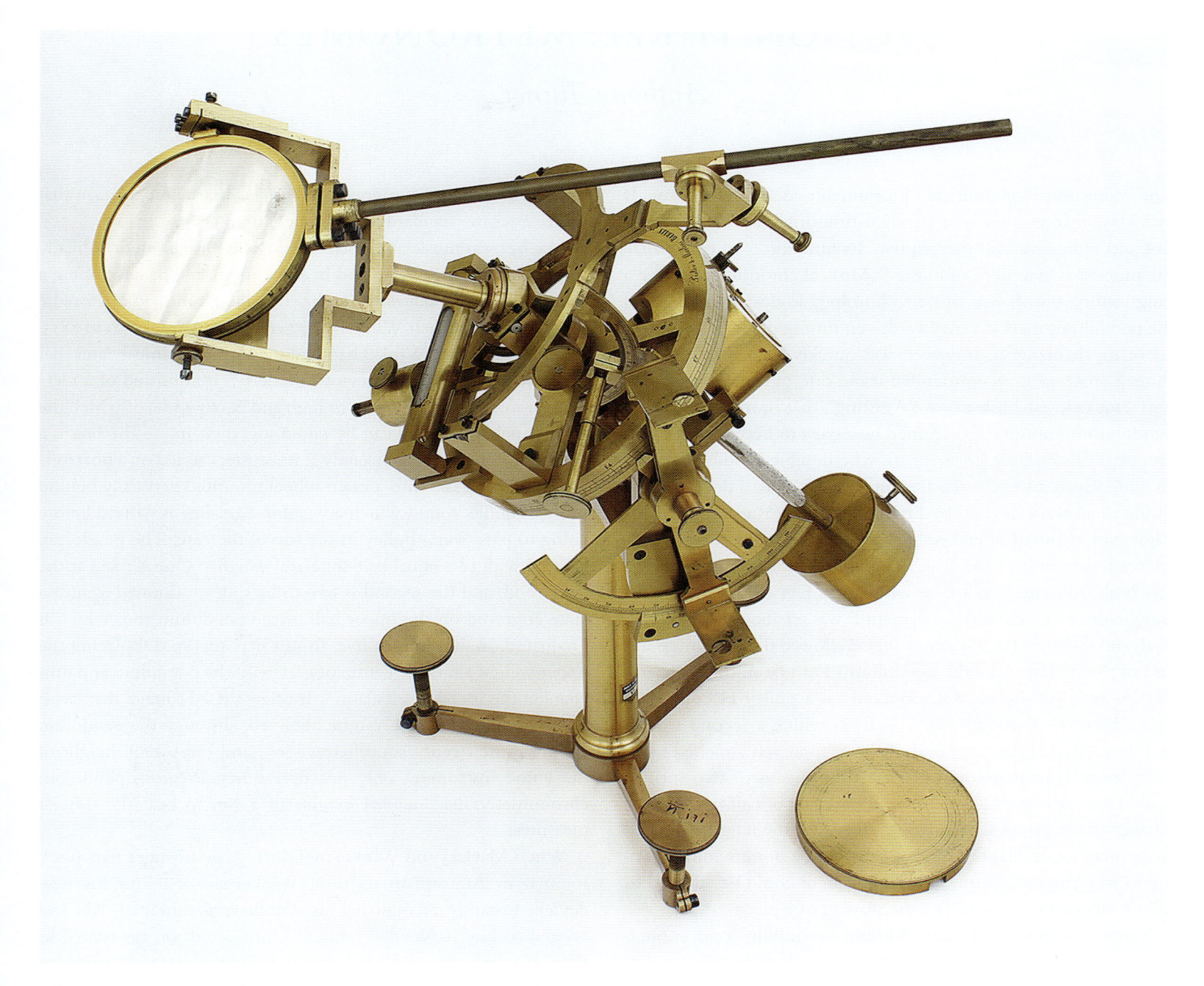

Figure 205 Gambey-type heliostat made by Pistor & Martins in Berlin around 1850. The instrument is an excellent example of mid-nineteenth century precision mechanics. However, it was expensive and so superseded by simpler, cheaper instruments. Historical Physics Collection, University of Vienna, Faculty of Physics.

complex installation of large and expensive domes, needed to protect orientable telescopes, to be avoided. As for large telescopes, before the introduction of synchronous motors, and later of computer-driven motors, all such apparatus was moved by clockwork mechanisms whose speed was kept constant by governors similar to those used for large telescopes.

SECTION THREE: METRONOMES

Anthony Turner

The metronome or 'musical chronometer' is a time-marker. It provides an indication of a standard time interval. It is descriptive and indicative, not prescriptive. Its function is to materialize the time basis of a piece of music.[49] Musical chronometers have a long history, but their relation to horology is closely linked with the pendulum, used as a gravity-driven timing device. The interest of mechanizing it, however, was quickly recognized. Henry Purcell (1658–96), recommended using one of the new 'pendulum' clocks to establish a beat. 'Assisting', that is, maintaining the pendulum in motion, also seemed necessary to Léon Louis Pajot, Comte d'Onzembray (1678–1754), who designed and had made a weight-driven *métromètre* (as he named it).[50] This, a development of the *chronomètre* devised by Etienne Loulié (1654–1702), which Pajot had acquired after Loulié's death, consisted of two parallel vertical members, A, B, C, D (Figure 206), each about five feet high, mounted on a base with screw feet for levelling it. A weight-driven clockwork movement E was set above the two verticals and the sound of its ratchet-wheel allowed the beginning and end of the vibrations of the pendulum PH to be distinctly heard. The flexible pendulum was mounted on a pulley H connected by an arbor to a second pulley G from which a second ribbon G I descended. The lower end of this ribbon was attached to an index plate L, which could be placed by means of two spring-loaded pins wherever desired against the scale of half tierces C D. Changing the position of L would thus alter the length H P and so its speed of oscillation. Since this could be done without having to stop the pendulum, speed adjustment could be effected to follow changes of beat during the playing of a piece.

Pajot's machine could run without rewinding long enough for the duration of most pieces of music and offered an aural as well as a visual indication of the beat. Pajot also explained how the mechanism could be adapted to a domestic pendulum clock. The clacking beat of the mechanical timer, however, interfered with listening to the music it regulated (unless it was used only for establishing the initial beat of the piece to be played), and therefore, non-mechanical gravity timers remained popular. In the second half of the eighteenth century, however, particularly in the decades around 1800, several mechanical timers, including one that doubled as a mantel clock by Breguet, were developed. It was in this context of active research for a reliable timer that would enable composers to indicate tempi for their works in an increasingly international musical world that Johann Nepomuk Maelzel (1772–1838) produced his *metronome*, the mechanism that was to become standard for over a century.

Maelzel was musician, mechanician, and entrepreneur. Already in the period 1808–15 he had become interested in developing a practical musical *chronometer* (as he still called it in 1810), before he met Nicolas Dietrich Winkel (1777–1826) in Amsterdam in 1815. The previous year, Winkel had designed a music time-beater that departed radically from previous models.[51] It consisted of a rectangular mahogany box, 31.9cm high and 22cm wide, of which the front panel and the lid can be raised together. Inside the box is a brass pendulum, pivoted exactly at its centre, carried on a horizontal arbor that gears into a main wheel mounted vertically behind it on a spindle, round which a weighted gut line is wound before rising to pass over a pulley in the top of the case. The pendulum is fitted with two equal half-spherical weights. One is fixed at the lower end, and the second is free to be slide-positioned against a scale engraved on the upper half of the pendulum rod, which is graduated, in divisions of five, from thirty to 180. Displacing the upper weight alters the centre of gravity of the pendulum and thus changes the speed of oscillation. It offers the advantage that, with it, very slow oscillations can be obtained, although the pendulum is only some twenty centimetres long, and the overall height of the entire instrument only just over thirty. Previous pendulum chronometers had needed heights of 2–8m to beat the slowest measures.

When Maelzel and Winkel met during the former's five-week sojourn in Amsterdam in 1815, Winkel showed him his new device. Instantly recognizing its commercial potential, Maelzel offered to buy it. Winkel refused. Undeterred, on his return to Paris Maelzel obtained a patent for a slightly adapted version of the device on 16 September 1815. On 5 December he obtained a British patent and began production of the instrument. He promoted it vigorously through advertising, deploying the (solicited) testimonials of well-known composers such as Salieri (1750–1825) and Beethoven (1770–1827), and further protected it with patents in Austria and Bavaria. It is clear that Maelzel derived the basic structure of the *metronome*, as he christened it, from Winkel, but he set the pendulum pivot lower, approximately two-thirds distant from the top. This allowed the entire mechanism to be placed deeper in the case, thus increasing stability and allowing an elegant pyramidal form to be employed for the case within which the

---

49 The account that follows is primarily based on the far more detailed description given in Bingham & Turner 2017.
50 D'Onzembray 1735.

51 Winkel's original model is conserved in the Gemeente Museum, The Hague, see Turner 1990, 227. An exact reproduction of it is shown in Bingham & Turner 2017, 74.

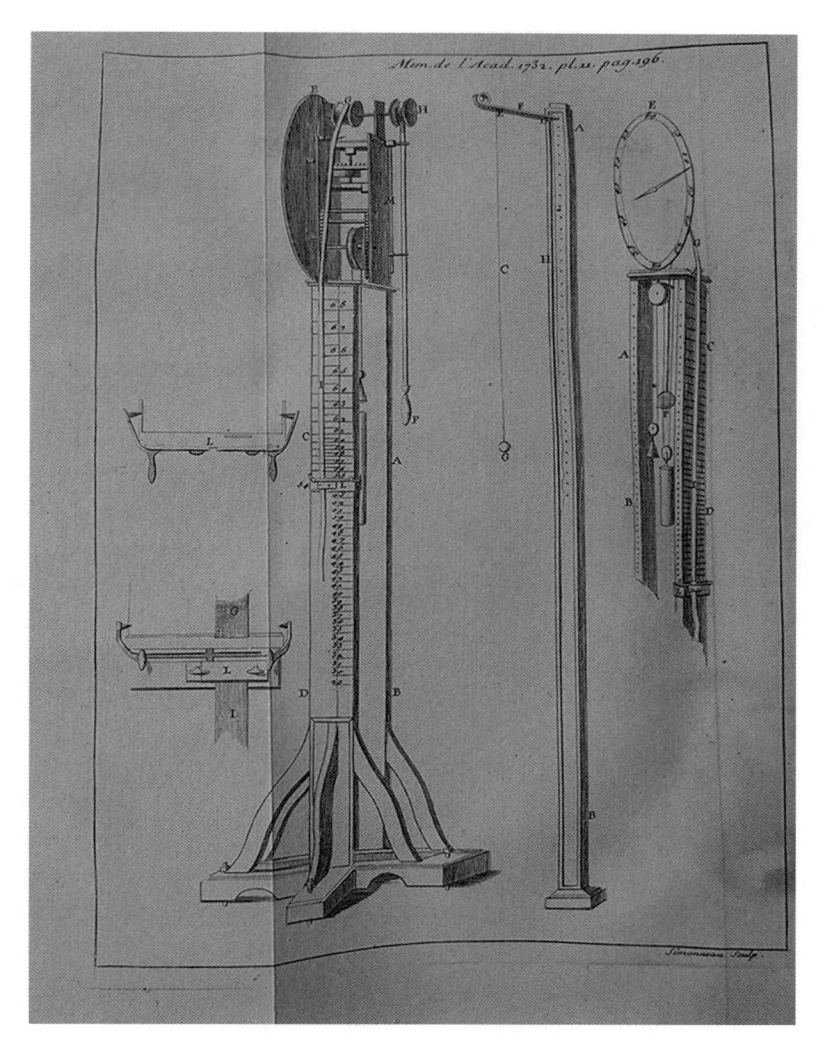

Figure 206 Loulié's *chronomètre* (centre), and Pajot d'Onsembray's *métromètre* (left and right), from *Mémoires de l'Académie Royale des Sciences pour 1732*, Paris 1735, 196. Photo: Tony Bingham.

pendulum was enclosed. Winkel's graduation for setting the sliding weight had been on the pendulum rod itself. Maelzel removed it to an independent scale behind the rod and, more important, graduated it to show vibrations per minute—a universally applicable standard.

In the following years, while energetically promoting the metronome, available as either a weight- or spring-driven model, Maelzel made some refinements to it that were embodied in extensions to his French patents in 1829 and 1834. More importantly, between 1820 and 1825, Maelzel entrusted manufacture of the device to Jean Wagner neveu (1800–75). Born in Pfalzel, he, at the age of twelve, joined the highly successful turret clockmaking workshop of his uncle Jean Henry Bernard Wagner (fl. 1790–post-1820) in Paris. Here Jean Wagner was trained by both his uncle and his cousin Bernard Henry Wagner (1790–1855). Probably it was to the Wagner business in general that Maelzel consigned manufacture of the metronome, Jean Wagner's nephew being chiefly responsible for it. The metronome joined roasting-jacks, watchman's clocks, lightening conductors, turnstiles, timers, and other applied horological devices that were manufactured by the Wagners in parallel with their highly successful turret clocks.[52] They, and their successors, would have a monopoly of the making of Maelzel metronomes (Figure 207) until the second quarter of the twentieth century.

Success inevitably breeds competition. Already in 1825, Bienaimé Fournier (fl. 1806–1828), a clock- and music-box maker in Amiens, had obtained a patent for his 'perfected metronome', the **fusee** spring movement of which was arranged to beat not only equal time, but also the different measures—two, three, four, and six, even six/eight, time—and to indicate the first beat in each measure on a bell. Primarily intended for beginners and teachers, Bienaimé's instrument was too expensive (at 100 francs) to be successful, but Wagner quickly incorporated its new features into his own production of metronomes and at half the cost. Indicating subdivisions of the beat in the bar and the different measures were questions that exercised the ingenuity of inventors and clockmakers contemporary with Bienaimé and others throughout the century. At the same time, just as a variety of non-mechanical music-timers were developed to rival

---

52 See Chapter 23, Section 6.

Figure 207 A metronome Maelzel, following Wagner's 1829 patent. Private collection. Photo: Tony Bingham.

Maelzel's popular metronome, so efforts were made to develop a spring-driven pocket instrument, usually in the form of a watch.

The stimulus for this came from military as much as musical needs. From the later seventeenth century onwards, rhythmical cadenced movement accompanied by music—the march—spread from the Ottoman Empire to German and French armies and even infiltrated court music in the operas and ballets of Jean-Baptiste Lully (1632–87). With the large, rapidly manoeuvring field armies of the Napoleonic period, pocket beat-timers could serve both musical and military purposes. At exactly the time when Maelzel was developing the metronome, two watchmakers in Besançon, Denizet[53] and Barrier, obtained a patent in 1812 for a 'marching regulator' or 'mechanical timer' in the form of a watch that could beat 40–100 strokes per minute as selected by the index on the dial. Two examples of the device, both probably made by the Japy company, are known, while in 1817 the Danish watchmaker Anders Christian Sparrevogn (fl. 1788–pre-1821), produced a watch-form music-timer employing a short pendulum pivoted at the centre of the movement not unlike the mass employed in the earliest self-winding 'automatic' watches.

Although a number of other suggestions for pocket metronomes were made in the 1820s, during the middle decades of the century interest in them waned. It revived towards the end of the century when Patek Philippe introduced a variant chronograph watch with stopwork that had an unnumbered twenty-division scale concentric with a musical score and a hand rotating once every ten seconds. The whole is controlled by a standard movement with a duplex escapement, probably subcontracted as it does not conform with Patek **ébauches**. A metronome/timer watch by Joseph Schmidt (c.1895) has an unnumbered eight-division dial, can be set on a scale of 8–100, and indicates beat on a bell. In 1906, the musician Robert David Glyn Roberts obtained a patent for 'a metronome made something like a watch', the dial of which was divided with concentric circles, each graduated for a different number of beats per minute. In 1908 Henri Coullery, director of the school of mechanics at La Chaux-de-Fonds, patented a keyless wound 'métromètre' that was probably made in his brother's watch factory at Porrentruy, although it was sold under his own name. More sophisticated was the watch-form metronome, patented in the United States 25 May 1909, by Charles A. White and Ernest R. Hunter that employed a complex system to vary the speed of a standard watch movement with **anchor escapement** to fit the music to be played. The most successful of pocket-watch metronomes, however, was the 'Cadenzia', launched in 1938 and later to be absorbed by the Heuer Watch Company, specialists in timers and chronographs.

The recurrence of complaints about Maelzel's metronome being easily deranged is sufficiently high that they cannot be dismissed as the denigrating rhetoric of rivals. Indeed, the mechanism of the standard metronome supplied by Wagner and their successors is most politely described as 'basic'. If the price of metronomes dropped steadily throughout the nineteenth century, the number of 'improved' instruments continued to grow. Half of the twenty plus innovations appeared during the first half of the century, most of the rest in the decades from 1880 to 1930. In the earlier part of the century most tended still to seek a better, or more simply, a different, way of accenting the first beat of a bar and of indicating the different speeds within it. But most of the later changes in metronome design were cosmetic—marketing ploys rather than fundamental innovations. The standard metronome maintained its position, included even in the catalogues of musical instrument-makers in the rubric 'accessories'.

It had, however, uses outside the realm of music. By 1900 these can be divided into three categories: a timing instrument, a demonstration instrument, and a piece of laboratory equipment. As the first it became particularly important, fitted with electric contacts, in physiological and experimental psychology laboratories for measuring the period of body functions and reaction times. As a demonstration instrument the metronome could illustrate the behaviour of a simple synchronous pendulum, points of equilibrium, and the effects of the displacement of the centre of an oscillating body. As such, it became a piece of classroom demonstration apparatus in colleges and high schools. In laboratories it was used for investigations into the registering of simple and reflex ideas. It could also be used for inculcating typing speeds. In the latter half of the twentieth century, however, the mechanical metronome was largely displaced by electronic devices. By the

---

53 Presumably the elder, uncle of the Jean Joseph Denizet active in Besançon in the mid-century.

late 1970s these had developed to a point where it became economically feasible to apply them to metronomes. As a result, the accuracy of beat was enhanced by a factor of ten or more. Even so, as noted in 2005, mechanical metronomes continued to be appreciated despite rivals such as the Yamaha Clickstation capable of registering eighty-six different rhythmic sequences in its electronic memory. These, however, were no longer devices using the mechanical technologies of horology.

SECTION FOUR: CAR CLOCKS

James Nye

'At 60 miles an hour the loudest noise in this new Rolls-Royce comes from the electric clock'.

ROLLS-ROYCE ADVERTISEMENT CAMPAIGN (1958–62).

The Section epigraph is taken from a successful advertising campaign for Rolls Royce in the United States, and is probably advertising pioneer David Ogilvy's most famous headline. The clock in question, by Smiths, employed an electromagnetically maintained **balance-wheel,** with a distinctly audible tick. Ironically, it was possibly one of the least reliable parts of the car—on the evidence of numerous service records showing clocks being replaced, until the eventual fitting of a reliable substitute sourced from German clockmakers Kienzle. For a long time the motor car offered a challenging environment in which to fit a reliable timepiece.

This account charts a history of little more than a century in the development of clocks, revealing much uniformity across an international canvas, given the rapid and widespread adoption of the motor car. Car clocks emerged as mechanical timepieces, largely adapting existing products—frequently pocket-watch calibres—and offered useful diversification for underused plant in various locations, both before and after the First World War. Car clocks, perhaps infrequently wound, were often not of the best quality, and many could not be relied on for accuracy, but the car assemblers nevertheless made features of them, with elements such as rim-winding, and the option of mounting them in rear-view mirrors, or even in steering wheels. Moving to electrical rewinding or electromechanical operation was a significant advance—at least clocks kept going—but by the 1970s greater accuracy and reliability were demanded, and the quartz clock offered an ideal solution.[54]

MECHANICAL CAR CLOCKS

Timepieces were fitted early in cars, and there were a number of reasons for measuring both distance and time. Taximeters, combining the function of odometer and clock in a single device, were first manufactured by the firm Carl Werner in Germany, c.1900. However, as well as it being necessary to measure time and distance in order to determine taxi fares, there was a need to calculate speed. While some drivers wanted to race, others (such as the police) wanted to restrict the speed of cars, so a means of measurement was important. The first prosecution for speeding occurred in Britain in 1896. By 1904 the British press was warning about police speed traps, and the Automobile Association was largely founded as a result. Initially, speed was calculated using a motor chronograph—essentially a stopwatch with a tachymetric dial. Monitoring the seconds hand against the passing of measured markers (mile or kilometre posts) allowed for speed to be read. But these devices had substantial disadvantages, failing in the absence of roadside markers—offering only an average—and certainly no instantaneous read-out. The speedometer emerged from the need to determine speed more easily, and often from the workshops of companies that had strong watch- or clockmaking traditions—firms such as Nicole Nielsen in Britain, which began diversifying away from the supply of watches to the jewellery trade in the 1890s. In the early twentieth century, the car was a rich man's toy and its accessories were expensive. Smart showrooms emerged where the plate glass, Turkish carpets, and mahogany of a jewellery shop offered a backdrop to the sale of lamps, horns, magnetos, and carburettors, as well as motor watches and clocks. In London, high-end firms such as Charles Frodsham and S. Smith & Son (later 'Smiths') diversified into this market (Figure 208).

The first cars were made in Germany in the mid-1880s, and the first German vehicle clocks were produced from 1893 in Schramberg by the Hamburg American Clock Company, but it is significant that their competitor, Arthur Junghans, founder of what would soon become the largest clock manufacturer in the world, was himself an early adopter of the motor car. His son Oskar produced a speedometer combined with an odometer and clock in 1905, but Junghans generally concentrated on simple car clocks. German demand not met by local production was largely met by Swiss imports, for example, from companies like Doxa.

In Britain, the burgeoning motor industry at the start of the twentieth century offered significant opportunities for other industries: cycle companies reinvented themselves as car manufacturers, and the clock and watch industry was able to diversify into instrument manufacture. Before it became standard for cars to carry an onboard battery, however, clocks needed mechanical winding. Some intriguing solutions were devised over time, but early on it was largely a question of utilizing existing designs. The most obvious route was to adapt or enlarge existing pocket-watch calibres, and to present them in new cases suitable for mount-

---

54 This account relies heavily in parts on Richon 2014, Brown 2015, and Graf 2020.

Figure 208 Interior of S. Smith & Son showroom, Great Portland St, c.1912. Dashboard clocks are reflected in the mirror at the left of this portrait of the old-fashioned retail jewellery-type showroom in which car accessories were proposed.

ing on the bulkhead or dashboard of the car. In Germany, in the absence of a mass-market watch industry, small alarm clock and other existing pin-pallet clock designs were adapted.

Keyless pocket-watches conventionally wind from the 12 o'clock position, and this meant that some changes were needed. It was often more convenient to present the winding and setting of dashboard clocks at the 6 o'clock position, and thus dials were reoriented accordingly. In 1901, Omega advertised such a clock, which could be fixed to a dashboard, or even attached to the fuel tank of a motorcycle with a leather strap. In Switzerland in 1908, George Ducommun (1868–1936) patented a new design, which became the standard Doxa eight-day movement, widely used by European car manufacturers.

While it became relatively common to extend a winding and hand-setting stem into a larger and knurled knob projecting below the clock, several makers (such as Omega, Doxa, and the German firm Gebruder Thiel) mounted the dial and movement in a heavyweight bezel, hinged out from a surface-mounted rear case, retaining winding and hand-setting from the rear of the movement, as in a conventional alarm clock. The flanged back might be at a sharp angle to the plane of the dial, reflecting the early trend of surface mounting instruments on steeply sloping car bulkheads.

Other early Swiss makers included Octo and Heuer, who developed a 'time of trip' dashboard chronograph c.1915.[55] Zenith later emerged as another prolific car clock manufacturer.

The rapid development of the motor car industry was abruptly interrupted by the First World War. For the makers of clocks and watches, output was frequently directed towards munitions manufacture—in fuses, or perhaps in instruments. In Switzerland, for makers like LeCoultre, opportunities emerged not only in increased watch manufacture for demanding military clients, but also to fuel a new demand for instruments (for aircraft, for example). It was these circumstances that brought together LeCoultre and the firm of Edmond Jaeger in Paris. By 1917, nearly half of LeCoultre's sales were represented by non-horological output. During the war, LeCoultre produced 120,000 revolution counters in the Vallee de Joux, which were sold through Jaeger in Paris.[56] Series production of these items was more profitable than

---

55 Swiss registered design 26,327 (not a patent). See *Schweizerisches Handelsamtsblatt* 1915.
56 Jequier 1998, 560–4.

watch manufacture where labour and other associated production costs were substantially higher. When the two firms formed a close association in 1917, the backers included three pilots and four aircraft engine manufacturers.

The colossal production of fuses and instruments (including aircraft and motor clocks) across Europe during the First World War was largely undertaken by newly recruited female employees operating on factory production lines, who proved themselves adept at skilled and semi-skilled work. Though many subsequently lost their jobs, the advantage of employing capable and less-expensive labour in modern industrial processes was not lost on employers after the war. In the transition from wartime production to renewed component and instrument manufacture for the motor industry, women remained significant in the production line, and were so worldwide from the 1920s onward.[57]

The 1929 crash hit firms such as Jaeger LeCoultre and Smiths, as it did every manufacturer in Europe faced with over-capacity, inflated salaries, and over-extended credit. A crisis of confidence led to cancelled orders and markets rapidly shrinking in size, while over-indebted firms sat on bloated levels of stock. Those firms that survived often needed to diversify. For Jaeger LeCoultre, the leap from mass-production of aircraft engine revolution counters to the production of simple eight-day car clocks was a small one, and the rapidly expanding car market of the 1920s offered a vital outlet. For car instruments in general, Jaeger remained the brand through which sales were made, capturing the French motor industry. Expansion into Britain followed in 1924, with the creation of Ed. Jaeger (London), in competition with Smiths, which later acquired the London subsidiary, eventually using British Jaeger as a prestige brand.[58]

In Switzerland, a variety of watch and clock firms diversified into car clocks, including the Buren factory (owned by Williamsons of Coventry in Britain), the Tavannes Watch Company, Movado (under their Ralco brand), Sandoz, Breitling, and the Cortébert Watch Company. Another prolific brand was Octo, manufactured by Fabrique d'Horloges Couleru-Meuri from La Chaux-de-Fonds, a firm that developed a range of instruments including speedometers, casing up their eight-day pocket-watch movement as a car clock. Octo and Cortébert car clocks were widely fitted over several decades.

In Italy, Borletti (which produced Roskopf-style watches) diversified post-war into instruments, acquiring the Veglia brand in 1930. Veglia Borletti car clocks were noted for their stylishness and ranged from early mechanical versions to later quartz models. They were fitted to many Italian cars through to the 1980s, and occasionally some foreign marques (such as Citroën and Mercedes Benz).

A feature of some clocks (e.g. from Ralco and Maar) was an additional and moveable bezel holding a second glass, bearing a pointer to signal elapsed time. A range of instruments were also available for the rear-seat passengers, including repeaters for the speedometer. The Omega Museum features an elegant rear compartment canteen, c.1910, with perfume bottle, cream jar, and a mirror, and a large pocket-watch-sized timepiece.

Phinney-Walker in the United States (formerly the Keyless Auto Clock Co.) were the first to produce 'rim-wound' clocks (US patent 967,428, 1910). Versions were made in Europe, for example by Smiths and Watford in Britain, and Zenith in Switzerland. Whether positioned at the 12 or 6 position, stem-winding was inconvenient. In rim-wound clocks, an additional rotating bezel (the 'rim'), with a knurled edge for easy grip, engaged with the movement to wind it. A button or slider to the side of the case (or just pulling outwards on the rim) allowed the moveable rim to disengage from the winding work and to engage with the **motion work** for hand-setting.

In the United States, the major firms were well represented, with Waltham, Elgin, and the New Haven Clock Company as the major suppliers, all essentially repurposing components or whole calibres from their existing ranges—in the case of Waltham using twin spring barrels to achieve eight-day running (as Junghans had done with their early car clocks). Over time, many well-known names in watch- and clockmaking produced car clocks, including Ansonia, Waterbury, Sessions, Westclox, and Chelsea—the last supplying the Rolls Royce plant in Springfield, Massachusetts (1921–31).

In the 1950s, a novelty item was the steering-wheel clock. Oldsmobile, among other marques, offered as an optional extra a clock from Maar of Zurich. Essentially an enormous automatic watch, the turning of the steering wheel wound the clock, but the unusual element in the Maar design was that acceleration and braking also wound the clock. Benrus also produced steering wheel clocks, offering seven- and fifteen-jewel models, available across the Chrysler range.

ELECTROMECHANICAL CLOCKS

The significant disadvantage of mechanical car clocks was their need to be wound—commentary in the clock press suggested people never bothered. Electrical car clocks emerged early in the United States.[59] Even before the First World War, Robert Hight (1877–1947) patented an electrically rewound car clock, produced in Decatur, Illinois.[60] The patent covered anti-vibration measures and even allowed for a 'dial adapted to be illuminated whenever desired', as well as a movable pointer for elapsed time calculations. In concept it resembled other electrical clocks in having a traditional mechanical **escapement,** with the **train** rewound periodically using an electromagnet.[61]

---

57 Glucksmann 1990, *passim*.
58 Nye 2014, 68–9.
59 Brown 2015, *passim*.
60 US patent 984,218 (1911).
61 See Chapter 19. For the evolution of clock illumination, see 'The Dark Side' in Brown 2015.

A handful of other US firms commenced operations before the First World War. The Thompson Electric Clock Co. used Seth Thomas mechanical movements, with the spring rewound periodically. Clocks from Stewart Warner Corporation used an electromagnetic rotary action, powered by its own 1.5V cell. Between 1913 and 1917 the Chelsea Clock Co. produced electric car clocks under the 'Boston' brand—essentially adaptations of their mechanical clocks—and these were ordered by Cadillac, Packard, Pierce Arrow, and other well-known marques, but the failure rate was unacceptable, and the firm reverted to a mechanical design.

After the First World War, other US manufacturers produced electrically rewound clocks, but it was from the 1950s that the market expanded significantly. Car numbers in the United States increased from eight million in 1920 to twenty-seven million in 1940 and rose to sixty-two million in 1960. These numbers dwarfed those in Continental Europe—for example, in the mid-1920s there were approximately 250,000 cars in Germany, and roughly twice that number in Britain, where numbers only reached just over two million by 1940.

In Europe, the first electrical car clock was probably offered by Embe-Fahrzeuguhren GmbH from Berlin, in 1924.[62] This employed a small motor, periodically rewinding the train by means of a worm-and-wheel engagement. Rather than using motor-rewinding, two other German manufacturers, Kienzle (from 1930) and Jauch & Schmid (from 1937, under the Jundes brand), used electromagnetic rewinding of their balance-wheel movements. In Switzerland from the 1920s onward, Eltric from Neuchâtel offered a clock with a high-quality movement (for example, using a compensated balance), but once again simply using electromechanical rewinding.

In the United States, General Motors originally bought in clocks from a wide range of makers, but Raymond H. Sullivan (1894–1980), an electrical engineer in GM's Delco division, filed for three patents in 1945 for clock designs, embodying some sophisticated ideas. He utilized an electromagnetically impulsed balance-wheel but did not use it to advance the train (as in the Smiths design, to be discussed), but purely for timekeeping. He included spark-quenching with a secondary winding in the electromagnet.

Also employing a balance-wheel for timekeeping, General Electric devised an asynchronous motor-driven clock in the 1950s, adapted from elements of its well-known Telechron domestic synchronous clocks. In the 1960s, Hamilton entered the field, again with an electrically rewound mechanical movement.

Ray Brown has charted the prolific career of William Greenleaf (1883–1958), linked to five separate car clock manufacturers over time (Hartford Clock Co., Connecticut Clock Co., Greenleaf Clock Co., Borg-Greenleaf Corp, and Western Clock Co./Sterling).[63] Although his designs evolved, a relatively common thread was the regular electromagnetic rewinding (each minute) of a large, jewelled watch movement.

Smiths was a major force in car instrument supply in Britain by the early 1930s. Acquiring English Clock and Watch Manufacturers in 1932, it secured the design for an electrically rewound balance-wheel clock. This design eventually gave way to an electrically maintained balance-wheel clock in which the driving mechanism had some similarities to those in early electric watches. The balance carried permanent magnets on its rim, designed to interact with a fixed coil, switched by the action of a contact pin mounted on the balance. The balance **arbor** carried a worm wheel, engaging with a pin wheel in the train. In 1958, when viewing an eighteenth-century prototype marine timekeeper by Henry Sully that employed a large balance-wheel, Alan Lloyd was heard to remark 'The clock has an unusual escapement which incidentally bears a remarkable resemblance to that in popular modern electric car clocks'.[64] Some horologists therefore now refer to the Smiths escapements (and similar escapements in battery insertion movements) as 'reverse Sully'.

QUARTZ CLOCKS

Throughout the 1960s, the German automotive supplier VDO used a standard electromechanically rewound balance-wheel movement from Kienzle, the 606 calibre, for its car clocks, taking the lion's share of Kienzle's annual production of some four million movements. But it was clear that electromechanical car clocks were inaccurate within ranges of many minutes per day, principally influenced by the possible fluctuations in temperature for a car in extreme Sun or snow. Such errors were revealed by the increasing prevalence of car radios and regular time signals.

Facing the possible loss of sales of a high-volume component for the car industry, VDO took a major strategic decision to invest significantly in the development of a quartz car clock, accurate to within seconds per day. Germany led the way in the development of quartz clocks generally in the 1960s, for example, with the Astrochron mantel clock from Junghans.[65] From 1970, VDO started volume production of quartz car clocks, with a traditional dial presentation. Initially the clocks were sold below cost in order to ensure market penetration, but VDO was rewarded with sales of approximately thirty million units per year between 1970 and 1984. And it was not only the German car industry that bought VDO clocks. Rolls Royce replaced the long-standing Smiths clock with a Kienzle unit from 1965, and then replaced this from 1977 with a VDO quartz clock.

---

62 German patent 381, 494 (1924).

63 Brown 2015, 311–17.
64 *AH* ii, 1958, 152.
65 See Chapter 19.

THE FUTURE

In common with a theme outlined in Chapter 19, cars in the twenty-first century are moving from carrying their own time base to receiving it as part of network time. 'Connected cars'—in which links to the GPS system and the Internet are required as part of providing both navigation and safety data used for insurance purposes—are becoming commonplace. The development of the humble car clock is further witness to the gradual embeddedness of timekeeping in the background of everyday life.

SECTION FIVE: THE NOCTUARY OR WATCHMAN'S CLOCK

Jonathan Betts

The purpose of the watchman's clock was to ensure that the night watchman in a large building carried out his duty patrolling the various parts of the establishment throughout the night and did not miss part of his rounds. At the time this genre of timekeeper was created, industrialization was in full swing and institutions were increasing in size and vulnerability. Thus, numbers of large buildings were increasing, and factories, warehouses, banks, prisons, and hospitals needed routine patrol for the prevention and detection of burglary, fire, and flood. In the first dedicated study of these interesting instruments,[66] Adrian Burchall quotes David Landes: 'Once the work day was defined in temporal rather than natural terms, workers as well as employers had an interest in defining and somehow signalling its boundaries. Time measurement here was a two-edged sword: it gave to the employer bounds to fill and to the worker bounds to work'.[67] Further, Burchall comments 'The watchman's clock was just such an instance of the "two-edged sword", for it was a means by which the employer sought to ensure the watchman duly performed his round and yet for the watchman was proof that he had performed it'.

The first such device appears to have been made by Whitehurst of Derby and sold by Matthew Boulton (1728–1809) of Birmingham, though neither ever overtly claimed to have invented the device. Burchall demonstrates clearly that the idea, much repeated in the literature, of its invention and production by John Whitehurst FRS (1713–88) cannot be so, but then concludes it was almost certainly first manufactured in Derby about 1790 by John Whitehurst II (1761–1834), nephew of the John who was FRS. It seems the original idea for the noctuary came from the inventor and architect William Strutt of Derby (1756–1830) in the late 1780s, although it was not put into production at the time.[68] One of the first customers for such a clock was the potter, Josiah Wedgewood (1730–95) of Stoke-on-Trent who, like Boulton and Whitehurst, was a member of the Lunar Society in Birmingham. It was at meetings of that society that many scientific and technical discoveries and inventions were formulated in the second half of the eighteenth century and it was perhaps during these meetings that the need for such a device was discussed and Strutt's idea developed.

The concept of the first watchman's clock was very simple. Instead of a simple fixed dial with moving hands, it employed a single, rotating twelve-hour dial, the hour registered against a fixed pointer on the movement. The dial was constructed in the form of a shallow drum around the periphery of which were friction-mounted radial pins, projecting at quarter or half-hour intervals. By pulling a chain or lever outside the case at exactly the appropriate times, each of these pins would be depressed into the body of the dial drum by a lever in the movement, indicating that the watchman had been present at the time, proof that he had undertaken his round correctly. Fixed inside, behind the rotating drum dial, is a ramp which, after 12 hours, forces the depressed pins back out to their projecting position, enabling the cycle to begin again. A few early examples of this type are known in which the pins are not disposed radially round the periphery, but project horizontally out of the front of the rotating dial. The clock case is locked securely to prevent any tampering with the dial pins.[69]

Wedgewood's watchman's clock exists along with two still surviving at Burghley House, and these represent some of the earliest watchman's clocks; they have weight-driven, thirty-hour movements in plain wainscot oak cases befitting their utilitarian function. In 1803 the Marquis of Exeter described his examples at Burghley House in *Nicholson's Journal* when commenting on an alternative design patented by Samuel Day (1756–1806) of Hinton, Somerset. Day's design, using counters that dropped into cells within the clock, was not as simple as the earlier model, and, although implied, was certainly not the first of its type. The Marquis stated: 'I have had two of them above four years . . . I have one in my Library . . . the other machine is placed in a building at the other end of my premises. I have always two watchmen every night and they go the round every half hour'.[70]

While Whitehurst was certainly the first to make these clocks, the London firm of Thwaites and Reed were producing them in the early nineteenth century, and twelve were made for Barraud of Cornhill, who supplied Martin's Bank with them in October 1815. The cost to Barraud was £94/10/- (ninety-four pounds and ten shillings), each clock costing £7/17/6 (seven pounds, seventeen shillings, and sixpence). If large premises were to be covered, the system required several clocks, all of which would need weekly winding (daily in the early examples), and regular overhauling. In spite of these aspects, the design was evidently successful and considerable numbers of clocks of this type were made during the nineteenth century. Whitehurst continued to produce them, as did their successors, John Smith and Sons. Thwaites in London also continued production well into the century, selling a

---

66 Burchall 1985.
67 Landes 1983, 74.
68 Craven 2015, 196–201.
69 Thomas 1979.
70 In Nicholson 1803, 58–9.

spring-driven version in 1842 for nine guineas (nine pounds, nine shillings). By that date, a version of the noctuary had been created that also incorporated a conventional dial and hands, making the clocks of greater general use. In these examples the noctuary dial was usually positioned within the dial centre, the hours on the drum just visible round the periphery within the conventional dial circle. Notable London companies, such as Vulliamy and Dutton, with government contracts for clock production and maintenance were supplying clocks of this type, sometimes incorporating twenty-four-hour dials to distinguish night and day time patrols.

In the mid-1840s the design was 'reinvented' by Thomas Fillary (1811–86), engineer at the 'House of Correction', Coldbath Fields in Clerkenwell, London. Having made several for the prison, Fillary then commissioned J. L. & W. Smith of Clerkenwell to produce them commercially, with Smith's advertising their availability 'in plain or ornamental cases' in a printed catalogue in 1849. These were **fusee** spring clocks going eight days and usually had a passing strike on the hour and half hour to remind the watchman to press the plunger on top of the clock to push the relevant pin down. The vast majority of these clocks were supplied by Smiths to the retail trade, and many appear with other clockmakers' names upon their dial. Later in the century, clocks of this type, with both weight- and spring-driven movements, and in longcase, mantel, and wall-mounted versions, were produced by several other manufacturers both in Britain and on the Continent, the German company of Winterhalder and Hoffmeyer in Schwartzwald being a notable example.

A simpler design for a watchman's clock was patented by J. F. Woods (British patent No. 13,990, September 1888) and marketed by C. Taylor & Co. of New North Road, London, in which the revolving hour drum did not have pins, but merely a narrow paper band wrapped around its periphery. The plunger, instead of depressing a pin, simply imprinted the paper band with an ink mark.[71] But a better version of this design, produced by the Howard Clock Company in Boston, Massachusetts, had already appeared in the mid-nineteenth century. In the Howard design, just one clock was employed centrally, and a series of levers, connected by cable to various stations around the building, actuated a pricker within the clock movement that marked a paper chart. Only after all stations had been actuated, and in the correct order, would the pricker mark the paper, indicating what time the round was completed. This was superseded in the 1860s by an electromagnetic version invented by J. Hamblet Jr.[72]

A further development in the mid-century was the creation of a single portable clock carried by the watchman. British patent No. 957 was taken out in 1852 by J. Rowbotham, in which a paper chart in a single portable clock carried by the watchman was marked by unique keys fixed by chain in the locations to be visited. There appears to be little evidence that this specific design saw large-scale production under Rowbotham's name.[73] However, a similar system designed by Armand Francois Collin (1822–95), the celebrated successor to B. H. Wagner and the independent company of his nephew, J. Wagner in Paris, was evidently produced and used for many years in France, as it is illustrated in a 1904 horological journal as currently in use.[74] Soon after its creation, the design was patented on Collin's behalf in Britain by Thomas Buckney (1799–1873),[75] a partner in the firm of E. Dent and Co, who went on to produce the clocks in very large numbers. With small alterations, they continued to be used in many British institutions for more than a century. As first designed, this system used a single clock movement in a sealed drum-like case mounted with a large handle on the back and with a slot on the front, behind which was a circular paper recording dial. The locations visited had receiving boxes mounted on the wall and the whole clock was pushed into the box, a projecting 'type' inside the box, entering the slot in the front of the clock and marking the paper dial with a character identifying that location.

Rowbotham's system of 1852 was developed in Germany in the 1860s by Theodore Hahn (fl. 1868–1904) of Stuttgart, who, as in Rowbotham's plan, arranged the marking of a paper chart within the clock to be actuated with a key, each individual station having a unique key for that location firmly chained in place within a protective box mounted on the wall. This protected the clock movement better from dust and dirt, which, in Collin's arrangement, may have entered when the clock was inserted into the wall receiver. Hahn's design was sufficiently flexible to allow for a variety of slightly different layouts and durations of paper chart to suit the customer's requirements, one type following the design of J. B. Schwilgue, who, in 1847, had taken out Patent No. 5,537 for his own type of watchman's clock. Hahn took out patents in the USA (seven between 1869 and 1891), Germany (four between 1875 and 1879), and two in England (1874 and 1889), as he recorded in an advertising brochure (Figure 209) in French about 1892.[76] A price list attached to the brochure reveals costs (depending on the type and number of clocks purchased) of 70–112 francs. A twenty per cent discount was offered for trade buyers. Letters of commendation in this brochure include six from US buyers, five from Britain, one from Italy, one from Belgium, one from Vienna, three in Germany, one in Denmark, one in Slovenia, one in Ireland, and one in France. The design was evidently very successful and was internationally popular.[77]

---

71 Thomas 1998.
72 Blackwell 1992.
73 Bromley 1982, 444. Although principally concerned with 'clocking-in clocks', this article also contains useful information about patents for watchman's clocks.
74 Reverchon 1904, 3–4.
75 British patent No. 1431 for 1862; Mercer 1977, 454.
76 Theodore Hahn, *Montres-Controle portative et fixes patentees Hahn*, Stuttgart, c.1892.
77 For other German guard clocks and their later development, see Schmid 2002, 2006b, 2007, 2009; Romer 2010, 2011.

Figure 209 Cover of an instruction manual/publicity publication in French translation by the Hahn Company for their watchman's guard clock. Private collection. Photo: Jonathan Betts.

The Rowbotham/Hahn system was evidently better than its Dent/Collin predecessor and finally in 1940 Dent's applied for a British patent (No. 544,119) improving on their earlier model and registering a design very similar to Hahn's of nearly fifty years before. This later model of the Dent watchman's clock would remain in use in many museums, banks, and other large institutions in the United Kingdom for much of the rest of the twentieth century. In a final evolution of the watchman's principal, the later part of the nineteenth century saw an updated version of the idea invented nearly fifty years previously by J. Hamblet Jr in the US, where electrical switches at the various stations marked a chart with an electromagnetic pen on one central clock in the building. J. T. Gent, I. H. Parsons, and A. W. Staveley took out a patent for just such a system in 1894 with Parsons and Staveley revising the patent the following year.[78] The resulting Gent's watchman's clock system employing a central clock with paper chart on a large drum below the clock dial went on to serve in many institutions in the same way as the portable Dent variety. It was only in the late twentieth century, when video recording equipment became commonplace for security in larger institutions worldwide, that the watchman's clock became redundant, and, at the end of the second decade of the twenty-first century, very few such devices remain in use.

---

78 Reynolds 2005, 28. British patents 12,716 for 1894; 15,884 for 1895.

SECTION SIX: ROASTING-JACKS

Anthony Turner

Turning meat mechanically before a fire requires only a simple gear-train but it represents one of the earliest applications of clockwork to a non-time-telling task. Whatever the motive power—human, animal, vapour, a falling weight, or a coiled spring—turning one or several arbors (spits) through pulleys or secondary gears and **pinions,** a fly—usually driven through a worm wheel and its gear—ensured a smooth turning of the spit at a controlled speed. The origins of such devices are unknown, although it is reasonable to presume that human and animal power was used originally. Children were responsible for this and other tasks in the kitchens of Charles the Bold, Duke of Burgundy (1433–67). Children were still used in the sixteenth century,[79] and animal power would continue in use probably until the mid- to late-eighteenth century.[80]

Alternatives, however, had already appeared in the fifteenth century when two forms of vapour-driven machine were in use. One of them was a smoke-jack in which a small turbine was set in a chimney and geared to turn a spit below when set in motion by the upward blast of hot air from the fire kindled to roast the meat. This was an ingenious and economical device, but the fire must have required careful control since the hotter the rising air, the faster the meat would turn, thus roasting less well. A device of this kind was sketched by Leonardo (c.1485/90), in which the rotating vanes turn a vertical arbor carrying a **lantern pinion,** which, through a right-angled gear, turns a horizontal arbor linked through a pulley to turn the parallel spit below.[81] Such a system has been associated by Needham with Tang- (CE 618–907) or even Han- (206 BCE–220 CE) period zoetrope wheels in China, and with Mongolian and Tibetan prayer-wheels. Transmission of these to Europe could have occurred in the fourteenth and fifteenth centuries when many Central Asian slaves were deployed in Italy.[82]

In the smoke-jack, the rising hot air strikes the vanes of the turbine at right angles. In a passing suggestion, Leonardo also evoked turning a spit by air striking the vanes in their own plane through use of an æolopile or ***sufflator***: 'The water, which spurts through the little opening of the vessel in which it is boiling, blows with fury and is all converted into wind; with this one may turn the roast'.[83] The *sufflator* derives from Hellenistic Antiquity and has a continuous history in Western Europe from at least the thirteenth century.[84] It must also have continued to be known and used in Arab–Islamic regions for, in *The Sublime Methods of Spiritual Machines*—a work completed in AH 959 (CE 1552)—Taqī al-Dīn described such a device in which the vane, being attached to the end of the spit, turned this directly:

> Making a spit which carries meat over a fire so that it will rotate by itself without the power of an animal. This was made by people in several ways, and one of these is to have at the end of the spit a wheel with vanes, and opposite the wheel place a hollow pitcher made of copper with a closed head and full of water. Let the nozzle of the pitcher be opposite the vanes of the wheel. Kindle a fire under the pitcher and steam will issue from its nozzle in a restricted form and it will turn the vane wheel. When the pitcher becomes empty of water bring close to it cold water in a basin and let the nozzle of the pitcher dip into the cold water. The heat will cause all the water in the basin to be attracted into the pitcher and the [steam] will start rotating the vane wheel again.[85]

Ingenious though it was, the intermittent action of the *sufflator's* blast must have led to somewhat uneven roasting. Nonetheless, the æolopile system was not forgotten and would again be suggested in the early nineteenth century.[86] It had also been mentioned by John Wilkins (1614–72), although he gives no details, but remarks that 'there is a better invention to this purpose' and hastens to describe the smoke-jack, noting that it may 'be useful for the roasting of many or great joints: for as the fire must be increased according to the quantity of meat, so the force of the instrument will be augmented proportionably to the fire'.[87]

---

79 Havard 1887–90, iv, 1491; Franklin 1888a, 175.
80 In 1723, the revd Henry Mease purchased a dog wheel for his kitchen spit at a price of 7s (Hart 1962, 107), and these were sometimes shown in kitchen scenes in eighteenth-century automaton watches. The author of the detailed account of roasting-jacks in the *Edinburgh Encyclopedia* (1830), xi, however, mentions animal power only to state that it was now disused. The use of cranked handles for turning spits is likely to have been an early development. It is shown in a woodcut in Cristoforo da Messisbugo, *Banchetti, compositioni di vivande e apparecchio generale*, Ferrara, 1549, where two, apparently argumentative, spit turners are assisted by a bellows-blower maintaining the fire.
81 Hart 1962, 249 and plate 45.
82 Needham 1965, 125.

83 White 1962, 92, n 1 citing Codex Leicester (now codex Hammer), f. 28v, translation from Christie, Manson, & Woods Ltd. 1980.
84 Needham 1965, 226–7.
85 Al-Hassan 1976, 34–5.
86 Havard 1887–90, iv, 1493; Armonville 1825, ii, 462.
87 Wilkins 1648, 171–2. Wilkins, like Thomas Powell 1661, 35, is likely to have derived his information from Cardan.

Taqī al-Dīn had been specific that self-turning spits could be made 'in several ways'; Wilkins insisted that the smoke-jack was 'much cheaper than the *other* instruments that are *commonly* [my emphases] used for this purpose'. By his time both weight- and spring-driven jacks had been in use for at least two centuries. A Venetian merchant who died in Damascus in 1455 possessed a German weight-driven jack.[88] A sketch of such is again offered by Leonardo.[89] It shows a weight-driven machine in which the barrel with capstan winding is turned by an overhead weight acting through a pulley. The barrel wheel, with right-angled teeth, meshes with a horizontal pinion, the arbor of which carries a second contrate wheel; this links through a vertical pinion mounted on the arbor of a fly-brake set above the frame. A second pinion is mounted on the arbor of the first horizontal pinion and simultaneously drives the two wheels set on the spits.

In the following years, the model would change: all the wheels were set in the same plane; worms and worm-wheels were employed; the fly could be placed within or without the frame; gears or pulleys could be used to turn the spits; the entire machine could be mounted on the chimney surround or free-stand on its own base. The variety attests to the widespread use of the device which, in the course of the sixteenth century, spread out from the notebooks of savants and the houses of the great. According to Henri Estienne, a notable agent in its popularization was the Frankfurt Book Fair:

> ... what family will not avow an immense debt to this fair, if only for having given us an instrument that accomplishes alone the main culinary operations ... for while previously it was indispensable to maintain in one's house a boy and a servant to turn the spit, the Frankfurt fair has offered us the roasting-jack which carries out this task with no less ardour and more efficaciously.[90]

Roasting-jacks and smoke-jacks existed in parallel. If the former had been installed at Ingatestone Hall, Essex, in the mid-sixteenth century,[91] the smoke-jack in John Evelyn's brother's house had already been there for a century in 1676.[92] Thomas I Geiger (*fl.* 1550–83) in Augsburg made both turret clocks and spit-jacks.[93] What was clearly an increasingly profitable market led, inevitably, to demarcation disputes between the members of different guilds wishing to control it. In 1581 in Leipzig, contention arose between the locksmiths and the clockmakers as to which of them had the right to make roasting-jacks—the dispute would drag on for eighty years.[94] A century and a half later, nothing had changed. Alexandre Beuve, a clockmaker in Rouen, was fined in 1739 for having made roasting-jacks on a plea of the locksmiths who considered the product reserved to themselves.[95]

Use of roasting-jacks then was widespread throughout the seventeenth and eighteenth centuries. In 1677/78, Joseph Moxon gave a detailed description of the construction of the 'worm-jack', concluding with a summary of the 'excellencies of a good *Jack*':

> 1. That the *Jack-frame* be forg'd and Fil'd Square, and conveniently Strong, well set together, and will screw close and tight up. 2. That the *Weels* be Perpendicularly, and strongly fix'd on the Squares of the *Spindles*. 3. That the *Teeth* be evenly cut and well smooth'd, and that the *Teeth* of the *Worm-wheel* fall evenly into the *Groove* of the *Worm*. 4. That the *Spindle Pins* shake not between the *Fore* and *Backsides*, nor are too big, or too little for their Center holes.[96]

Numerous fine examples were indeed produced. In 1736 a weight-driven roasting jack made by John I Davis for Eton College could turn up to six spits at once and remained in continual use until the 1920s.[97] Margotin in Paris *c.*1770 made a jack for the Duke of Orléans 'considered to be a masterpiece of its kind', while Roizin supplied instruments to the Royal Military School and the military retirement home and hospital of the Invalides, 'which are moved by only a small force and turn over four hundred pounds in weight'.[98] High institutional demand and the development of the bourgeois kitchen were stimulants for developing roasting-jacks, and supplied a substantial source of revenue for clockmakers, locksmiths, and ironsmiths.

This was the more so since, alongside smoke- and weight-driven jacks, spring-driven models had also been developed. Already *c.*1485, Leonardo had sketched one[99] and during his travels in Italy in 1580 and 1581, Montaigne noted that, because the Italians were excellent iron workers, 'almost all their spits turn by

---

88 Carboni 2007, 82. My thanks to Marisa Addomine for this, and the reference to Armonville.
89 Codex Atlanticus 5.v.a reprinted in Hart 1961, pl. 62.
90 Translated from *Frankfordiense emporium* (1575) after Havard 1887–90, iv, 1490.
91 Emmison 1964, 67.
92 Buchanan-Brown 1972, 453. Evelyn adds that the jack had worked night and day virtually ever since it was erected and that 'It makes very little noise, needs no winding up, and for that preferable to the more busy Inventions'.
93 Pérez Álvarez 2020, 12.
94 Ogilvie 2019, 153.
95 Tardy 1972, 56.
96 Moxon 1677/78, 39–51 (quotation from 51). See the important commentary on this text by Wright 2000.
97 Ashworth 1990, 3–4.
98 *Almanach Dauphine*, 1772 cited from Franklin 1888b, 156, 158.
99 http://codex-atlanticus.it/#/Detail?http://codex-atlanticus.it/#/Detail?detail=1051, with thanks to Dietrich Matthes.

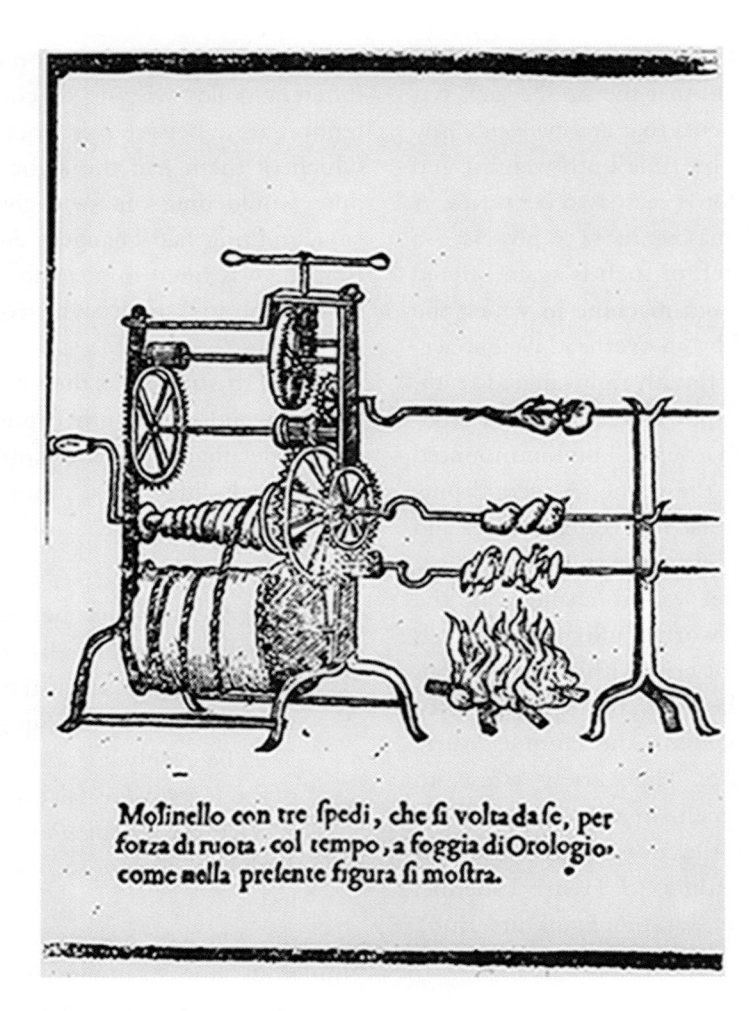

Figure 210 A spring-driven roasting-jack from Bartolomeo Scappi, *Opera dell'arte del cucinare*, Venice, 1570.

springs or by means of weights, like clocks'.[100] His observation is confirmed by what is perhaps the earliest extant printed picture of a roasting-jack published ten years earlier. It shows an instrument for turning three spits fitted with a large **spring-barrel**, a **fusee**, and a crank handle for rewinding (Figure 210).

Rather little is known about the development of the spring-driven jack before the later eighteenth century, when it seems reasonable to assume that at least some of the portable jacks produced were spring-driven.[101] This is partly confirmed by the innovations that John Joseph Merlin (1735–1803) made to roasting-jacks, and patented in 1773, which applied equally to spring- and to smoke-jacks.[102] Even so, since the power and duration of the spring would limit the load that could be turned, for heavy work in institutional kitchens smoke- and weight-driven models were probably preferred. Several innovations to these were made and a number patented in the years around 1800.[103] The most notable innovation, at least in England, in the closing decades of the eighteenth century was a development of the torsion, or 'dangle', spit. The 'bottle-jack' was a cylindrical brass canister containing a spring-driven clockwork motor. This was suspended above a fire and slowly rotated a piece or pieces of meat hung below it on an iron ring with the direction of rotation alternately reversed. Made usually in Birmingham in the early nineteenth century, it formed a major part of the general domestic metal goods manufactured by men such as John Linwood (1760–1840) of St Paul's Square, who was there succeeded by Edward Bright Bennett and Edward Holmes Bennett. The latter were specifically described as roasting-jack manufacturers in Birmingham trade directories, but they continued to use Linwood's name and his mark of an oak tree well into the second half of the century.

From the middle-class kitchens of England, the bottle-jack spread rapidly to North America,[104] and was perhaps occasionally used in Europe during the nineteenth century, although here

---

100 Montaigne *Voyage* cited from the edition by d'Ancona 1889, 34. Montaigne also noted the use of chimney smoke jacks fitted with light vanes of pinewood.
101 Havard 1887–90, iv, 1493. An English roasting-jack for bivouacs was introduced into France in 1819. Armonville 1825, ii, 462.
102 French & Wright, 1985, 66–7.

103 Ten innovatory jacks, for example, are listed in Armonville 1825, ii, 461–2.
104 More detail on roasting jacks in North America is given in Schinto 2005. See also Riegel 2006.

there were other forms. In France, the pioneer of the spring-driven jack was the immigrant clockmaker J. B. H. Wagner (or Wagener, *fl.* 1790–1820). According to Havard,[105] he brought the spring-jack into current use in France from the 1790s onwards, and it was his standing model that was still in use a century later. Wagner was the founder of a horological business that, by 1887, belonged to Chateau & Fils, successors to Collin, who had purchased the Wagner business in 1852. Havard notes some changes made to Wagner's original model, which could now (*c.*1890) run for two hours without rewinding and had been fitted with a warning bell. Even so, 'if it is good for light and medium weight roasts, it often lacks the force needed to turn large pieces'. Restaurant owners therefore remained faithful to the older weight- and smoke-driven models. This situation is clearly reflected in Chateau & Fils catalogue of 1887, where spring-driven jacks are presented alongside weight- and smoke-jacks. For the latter, however, the end was perhaps already insight. Describing the smoke-jack, Richard Jeffries (1848–87), noted:

> Upon one side of the hearth is a long vertical steel handle, brightly polished, much like the valve-handle of an engine. By this handle the smoke-jack is regulated; at a touch a small endless chain depending from the chimney causes the horizontal spit to slowly revolve. Looking up the chimney the smoke-jack fills the cavity, like a horizontal windmill perpetually revolving, driven by the heated air ascending. In how few, even of the most ancient houses, are smoke-jacks still at work. No meat is so good and richly flavoured as that cooked before a wood fire.[106]

Oven-roasting would soon replace both smoke-jacks and all their brethren.

---

[105] Havard 1887–90, iv, 1493.

[106] Jeffries 1884, 172.

CHAPTER TWENTY FOUR

HOROLOGY VERBALIZED, HOROLOGY VISUALIZED

Christina J. Faraday

Almost as soon as it was invented, the mechanical clock became a source of inspiration for writers and artists alike.[1] As the most advanced technology of its time, and for centuries afterwards, the complexity of clocks made them an attractive vehicle for the expression of a variety of ideas. Several aspects of the clock contributed to its popularity. First, the clock manifested, audibly and visually, the passage of time: marking out the moments with the movement of the hands, a ticking sound, or the striking of a bell.[2] Unlike other horological instruments such as sundials or sandglasses, the mechanical clocks measured time relentlessly, working independently of any external time marker, at least in theory. In reality, early clocks needed frequent resetting according to the more stable time registered by the sundial, and the unreliability of clocks also became a source of inspiration for anyone who wanted to call into question the trustworthiness of human reason and empirical observation.

While artists and writers were particularly drawn to the symbolic nature of the clock, particularly in the realm of *memento mori* and *vanitas* imagery, they were also interested in its operation, from its 'first mover', a weight or spring, through a variety of greater and lesser wheels, which ultimately caused the hands to move or the hammer to strike. This causal chain-of-command sequence was ripe for metaphorical application and is described in detail in a variety of texts from the fourteenth century onwards. Poets and artists alike were fascinated by the clock's ability to self-regulate, which was emblematic of human self-control and wisdom. As a self-moving, yet man-made, object, the clock was used to symbolize both the skill of God—the ultimate clockmaker—in making man, and man's ability to rival his creator's powers. Finally, related to this was the way the clock seemed to portray its interior workings through external movements, which could be symbolic of the workings of the human body and the relationship of the body to the soul.

THE MEDIEVAL PERIOD

Among the earliest references to the clock in literature are those by Dante Alighieri (1265–1321) in his *Paradiso*, begun *c.*1308 and finished in 1320.[3] In these important early examples we already find a connection being made between human and divine intellect and the clockwork mechanism. Dante uses the word *orologio*, an ambiguous term that could refer to a range of clock types, whether a traditional water clock or a solid-weight clock with an oscillating controller.[4] He mentions a clock when he describes the call to

---

[1] I thank the following people for their suggestions, guidance, and enthusiasm: Vanessa Braganza, Bob Frishman, Richard Ketcham, Oscar Nearly, and Anthony Turner.

[2] According to the *Oxford English Dictionary*, the first use of the phrase 'tick tock' appears in William Thackeray's *Vanity Fair* (1847): (*OED* 1972, ii, 3317, 9). The variant forms 'tick-tack' and 'tick-tick', however, were used much earlier, for example, in John Aubrey's account of the Elizabethan mathematician Thomas Allen, whose maids 'hearing a thing in a case [his watch], cry *Tick, Tick, Tick*, presently concluded that that was his Devill, and ... threw it out of the windowe into the Mote (to drowne the Devill)', Aubrey 1898, i, 28. Other examples are given in *OED*, 1972, ii, 3317, 9.

[3] An earlier example is found in the Jean Meun[g] continuation of the *Roman de la Rose.*, composed 1268–75. This includes striking clocks in a list of musical instruments. Although it has no metaphorical overtones it is crucial evidence for early development of domestic clocks and is discussed in Chapter 5, Section 1, this volume.

[4] Moevs 1999, *passim*.

matins: 'Then, like a clock that calls us in the hour / when the bride of God rises to sing a dawn / song to the Bridegroom, that he may love her',[5] and later compares the circular dance of the blessed in paradise to the movements of a clockwork mechanism:

> So Beatrice, and those happy souls became
> spheres spinning on fixed poles, flaming, as
> they turned, like comets.
> And as the wheels in the mechanism of a
> clock turn so that, to one who watches, the first
> seems to be motionless and the last to fly:
> so those carols, differently dancing, allowed
> me to judge their richness, being fast and slow.[6]

By the mid-fourteenth century the clock appears in a range of literary compositions and their accompanying manuscript illuminations. The most extended homage to the clock from this period is found in *L'Orloge Amoreus* (The Clock of Love), c.1368 by Jean Froissart (1337–1410). Froissart, who clearly took great delight in his understanding of this novel mechanism, compares himself, the lover-poet, to a clock: 'Je me puis bien comparer a l'orloge' (1): 'I may indeed compare myself to the clock'.[7] He goes on to anatomize the clock, comparing each part and its role in the mechanism to an aspect of the human soul in love. Beauty sets the love-clock in motion, like the weight, *le plonk*, which starts the mechanism. Beauty wakes joy, which Froissart associates with *la corde* or chain, which activates the first wheel. The first wheel, originator of all the clock movements, is compared to desire, originator of love, while the second wheel, or **foliot**, acts to regulate movements of the first wheel. According to Froissart, desire is unruly, requiring moderation to prevent the lover from breaking social norms: the foliot symbolizes the necessary restraint placed on desire. Finally, *la sonnerie*, the striking mechanism, symbolizes the poet himself, who gives voice to his inner feelings through his poetry.[8]

Froissart's identification of the foliot with moderation or restraint is representative of a wider interest in the self-regulation of clocks. This interest caused the clock to become a popular attribute of allegorical figures of Temperance, an association that continued long into the early modern period.[9] Previously shown with jugs of water and wine, associated with sobriety and the tempering of extremes, in the fourteenth century Temperance acquires a new iconography which emphasizes more clearly the virtue's restraining role.

The first instance of Temperance being associated with a timepiece occurs in Ambrogio Lorenzetti's fresco *The Allegory of Good Government* (Figure 211), painted in 1337–9 on the walls of the Sala dei Nove in Siena's Palazzo Pubblico. Here, in a part of the fresco repaired after Lorenzetti's death in 1348, Temperance is shown with a sand-glass.[10] As well as being the first instance of the association of Temperance with timepieces, this also constitutes the earliest known representation of a sand-glass. The iconography can be explained by the association of Temperance with measure, sand-glasses being seen to measure out discrete sections of time. By the end of the fourteenth century, however, it was not just measure but restraint that gave Temperance her closest association with the increasingly well-regulated mechanical clock.

Christine de Pisan's *Epistre d'Othéa*, c.1399, written in the form of a letter of advice to a fifteen-year-old nobleman, exists in several manuscript editions, many of which contain beautiful illuminations. One of the most iconic images from this manuscript tradition shows Temperance, also thought of as a goddess, adjusting and maintaining a clock. In a manuscript of 1407–9 in the Bibliothèque National de France (MS 606, f.2v) and a slightly later manuscript of c.1410–14 in the British Library (MS Harley 4431, f.96v), Temperance is shown floating down from the clouds to adjust a clock shown from behind, its gears, weights, and bell clearly visible. In another manuscript of the same text (BNF MS 848, f.2v), Temperance stands in front of a clock whose mechanism has been drawn in careful detail, with three weights, an **escapement** device, and a bell complete with hammer.[11] Images of this kind appear from the earliest manuscripts of the text onwards, so we can reasonably assume that they met with the author's approval. Indeed, Christine glosses the images as follows:

> Temperance should be called a goddess likewise. And because our human body is made up of many parts and should be regulated by reason, it may be represented as a clock in which there are several wheels and measures. And just as the clock is worth nothing unless it is regulated, so our human body does not work unless Temperance orders it.[12]

By the middle of the fifteenth century, the association of Temperance with the clock had become even more direct. A striking iconographic tradition traceable in several French illuminated manuscripts of c.1450–c.70 shows Temperance with a variety of technological attributes illustrating different aspects of her nature. These included a bit and bridle representing restraint, a pair

---

5 Dante 2011, 212–13, Canto X, 139–42.
6 Dante 2011, 478–9, Canto XXIV, 10–18.
7 Froissart's poem is edited by Dembowski 1986. It is quoted and translated by Wright 2007, 178. For an earlier presentation of the text with translation and a detailed horological commentary, see Robertson 1931, 53–64.
8 Wright 2007, 188–9.
9 A late English example is found in Richard Day's *A Book of Christian Prayers* (1578 and later editions) where an illustration of Temperance shows her with a clock, trampling on a vomiting man symbolizing intemperance. In c.1614 this horological image of Temperance was copied into plasterwork on an overmantel at Postlip Hall, Gloucestershire, this time without the vomiting man.

10 Gibbs 1999, *passim*.
11 Willard 1962, 151.
12 Tuve 1963, 289.

Figure 211 Temperance holding a sand-glass, 1338–40. Detail from Ambrogio Lorenzetti, fresco. Palazzo Pubblico, Siena, Italy.

of spectacles for perspicacity, and a windmill trampled underfoot, illustrative of Temperance's mastery of fickle, wind-blown passions. Chief among her attributes, and longest-lasting in subsequent representations, was her 'horological hat', a clock worn on her head, symbolic of self-regulation. Émile Mâle was one of the first to deal with what he described as the 'new iconography of the virtues', citing their appearance in a manuscript of *c.*1470 containing Jehan de Courtecuisse's French translation of the Pseudo-Senecan treatise *De Quatuor Virtutibus* ('On the Four Virtues').[13] Although nothing in the text explains the complex technological iconography of Temperance in this manuscript, the illustration features a verse of ten lines interpreting the image:

> He who is mindful of the clock
> Is punctual in all his acts.
> He who bridles his tongue
> Says naught that touches scandal.
> He who puts glasses to his eyes
> Sees better what's around him.
> Spurs show that fear
> Make [sic] the young man mature.
> The mill which sustains our bodies
> Never is immoderate.[14]

As Rosemond Tuve and Lynn White Jr point out, this poem is surely a later attempt to interpret the complex iconography, rather than a programme that informed the illuminator of the manuscript.[15] Several pieces of evidence suggest this, not least that Mâle's manuscript comes fairly late in the Temperance-with-clock tradition. In fact, one of the earliest known examples of this iconography is found twenty years earlier, in a manuscript of 1450 now in the Bodleian Library (MS Laud Misc. 570). This is an illustrated compilation containing both Christine de Pisan's *Epistre d'Othéa* and an anonymous translation of John of Wales's thirteenth-century text *Breviloquium de Virtutibus* ('A Short Account of the Virtues').[16] Tuve highlights the fact that the iconography appears fully formed in Laud 570, even though neither of its texts makes reference to this complex imagery.[17] In addition, the same iconography is found in mid-fifteenth-century manuscripts with other texts, including the French translation of the Pseudo-Seneca noted by Mâle, and Nicole Oresme's translation of Aristotle's *Nichomachean Ethics*.[18] This all suggests that a textual explication may once have existed elsewhere, but that the iconography had become standard enough to feature independently in a variety of texts on the virtues by 1450.

As for their connections with temperance, an earlier tradition also associated clocks with wisdom more generally. The *Horologium Sapientiae* (Clock of Wisdom) of *c.*1334 by the Dominican friar Heinrich Suso (1300–66), was a remarkably popular devotional work, translated into Dutch, French, Danish, Swedish and finally English over the next two centuries. As the title suggests, the clock is here used as a metaphorical vehicle for a devotional program aimed at helping devotees organize and regulate their lives:

> Hence the present little work tries to expound the Saviour's mercy as in a vision, using the metaphor of a fair clock decked with fine wheels, and of a dulcet chime giving forth a sweet and heavenly sound, exalting the hearts of all by its complex beauty.[19]

It was divided into twenty-four chapters, perhaps to represent the hours in the day, and was envisaged by the author as an alarm clock 'to waken the torpid from careless sleep to watchful virtue'.

Early manuscript editions of this text do not contain illustrations of clocks, but after 1406 their inclusion becomes standard. In a manuscript of the mid-fifteenth century we see this tradition reach a new pitch. Here, Wisdom is shown in conversation with Suso in a room full of horological instruments, including sundials as well as mechanical clocks (Figure 212).[20] In one hand Wisdom holds open the text she is explicating, with the other she appears to adjust the mechanism of a large twenty-four-hour clock. On the table to the right is a table clock, part of our earliest firm evidence for the existence of spring-driven, fusee-regulated mechanisms.[21]

In a similar vein is the illustration in a French edition of Suso's *Horologium*, the *Horloge de Sapience*, *c.*1461–5, which shows the manuscript's owner, Louis de Bruges, Lord of Gruuthuse (1427–92), observing King Solomon as he anachronistically repairs an elaborate clock, the ground around him littered with tools.

The presence of Solomon, the embodiment of earthly wisdom, further reinforces the relationship between clocks and wisdom. White has suggested that the illumination specifically shows Solomon 'repairing' the clock, rather than adjusting it, as the manuscript post-dates the adoption of equal hours. According to White, earlier images of the kind discussed above, in which Temperance is shown interacting with a clock mechanism, may allude to the earlier practice of adjusting the clock in line with the seasonal changes to the length of unequal hours.[22]

It was only a small step from the clock of wisdom to the vision of God as the clockmaker. In 1377 Nicholas Oresme (1320–82)

---

13 BnF MS 9186, fol. 304r; Mâle 1931, 311–16.
14 Translation from White 1969, 214.
15 White 1969, 214; Tuve 1963, 279.
16 Tuve 1963, 264.
17 Tuve 277ff.
18 Rouen Bibliothèque Nationale, MS 927, fol. 17v.

19 Suso, as translated by White 1969, 211.
20 Brussels, Bibliothèque Royale MS IV, III, fol.13v. For discussion, see Monks 1990.
21 Michel 1960, *passim*; Spenser 1963, *passim*.
22 White 1969.

Figure 212 Heinrich Suso, *Horologium Sapientiae*, just before 1450. Brussels, Bibliothèque Royale MS IV, III, fol. 13v (detail). Photo: Heritage Image Partnership Ltd Alamy stock photo.

composed his *Livre du Ciel et du Monde* ('Book of the Heavens and of the Earth'), a commentary on Aristotle's *De Caelo* in which the French philosopher expanded on the ancient trope of the *machina mundi*. By observing the regular motions of the planets, Oresme deduced that their movements were regulated by some mathematical equilibrium of motion and resistance, not unlike the regulated mechanism of the clock:

> These powers and resistances are different in nature and in substance from any sensible thing or quality here below. The powers against the resistances are moderated in such a way, so tempered, and so harmonised, that the movements are made without violence; thus, violence excepted, the situation is much like that of a man making a clock and letting it run and continue its own motion by itself. In this manner did God allow the heavens to be moved continually according to the proportions of the motive powers to the resistances and according to the established order.[23]

The three thinkers who played the most prominent roles in promoting the mechanical clock as an imaginative and inspirational object in the first century of its existence, Jean Froissart, Christine de Pisan, and Nicholas Oresme, overlapped in more than just their chronology. Christine and Oresme were both members of the court of Charles V of France and may have known each other. Froissart, although he spent most of his life at courts abroad, was born in France and attended Charles V's coronation.[24] At the centre of these connections, Charles V may be credited with providing some horological inspiration for those in his entourage. He is remembered for having installed striking clocks in his palaces at Vincennes, St Pol, and the Louvre, and deliberately sought out foreign experts, such as Thomas de

---

23 Oresme 1377, 286, cited in Mayr 1986, 38–9. Mayr notes that Oresme also made this comparison in an earlier treatise *De Commensurabilitate ve*

Incommensurabilitate Motuum Celi (c.1350s): 'For if someone should construct a material clock would he not make all the motions and wheels as nearly commensurable as possible? How much more [then] ought we to think [in this way] about that architect who, it is said, has made all things in number, weight and measure', Edward Grant (ed.), Madison, 1971, 292–5. Compare with Henry of Langenstein (1325–97) who also likened God-as-creator to a clockmaker, Mayr 1986, 40.

24 Mayr 1986, 40.

Pisan, to join his court and work on such projects.[25] Already in the first centuries after the invention of the mechanical clock and its introduction in Europe, it was inspiring a range of poets and artists with its novelty, complexity, and symbolic potential. As clocks became more widespread in the sixteenth and seventeenth centuries, their increasing familiarity encouraged an even broader range of creative interpretations in literature and the visual arts.

THE EARLY MODERN PERIOD

Like the real examples that inspired them, references to clocks proliferated during the sixteenth and seventeenth centuries. In this period, the *memento mori* aspects of horological symbolism became increasingly prominent. As a physical—visual and/or audible—manifestation of passing time, clocks, sand-glasses, and sundials were an obvious choice for anyone wishing to express the transitory nature of human life. As a finite measure of passing time, sand-glasses were particularly well suited to this symbolism, but other horological devices were also frequently deployed. This helps to explain, for example, the common conjunction of clocks or sand-glasses with skulls in portraits, particularly popular in sixteenth- and early seventeenth-century English portraits of the 'middling sort'.[26] Take, for example, two portraits of father and son Jacques and Jacob Wittewronghele, now in the collection of Rothamstead Research Ltd, Harpenden, UK. In the 1574 portrait of Jacques, a table clock engraved with the motto 'ut hora sic fugit vita' (as the hour, thus life flies) floats ambiguously on top of the sitter's shoulder, its golden colour echoing the table and yellowing skull beneath, on which Jacques rests his hand. His son, Jacob, also rests his hand on a skull, while behind him we see a wall-mounted weight-driven clock.

We also find clocks in the portraits of aristocrats and courtiers, particularly in Italy and Spain. Here the *memento mori* aspects of the clock's symbolism are usually downplayed, in favour of allusions to wealth and status. In Titian's 1538 portrait of Eleonora Gonzaga, a clock and a sleeping dog accompany the Duchess of Urbino, combining with her fine clothes and jewels to remind us of the sitter's status and importance. Similarly, in all three versions of Diego Velazquez's portrait of Mariana of Austria (1652–3), the rich setting and clothing are supplemented by a gilt table clock in the shape of a tower, alluding to the sitter's wealth, but also perhaps her prudence, drawing on the long association between clocks and temperance. In a more mysterious portrait by Annibale Carracci (c.1585) we find an African woman holding a clock, gazing straight at the viewer. The presence of another woman's arm and collar to the right suggest that the picture was once a much larger portrait, subsequently cut down: it may be that the African woman is represented as a servant, even a slave, to the woman now missing from the picture, a reminder that for the wealthiest citizens a person could be as much a possession as a gilt clock, both depicted to show off their owner's wealth.[27]

In Titian's portrait of Pope Paul III and His Grandsons (1545–6) the clock may pick up on these themes of wealth and status, but in the company of the visibly aging Pope and his vigorous (illegitimate) descendants, it also reminds the viewer that time is running out, emphasizing the theme of succession and, perhaps, alluding subtly to the dynastic manoeuvring that characterized the end of the Pope's reign. *Memento mori* aspects of horology also found expression in literature. As French Huguenot Philippe de Mornay put it in 1587:

> Again, what is death? The uttermost poynt of moving, and the uttermost bound of this life. For even in living we dye, and in dying we live, and there is not that step which we set downe in this life, which dooth not continewally step foreward unto death, after the manner of a Dyall or a Clocke, which mounting up by certeine degrees forgoeth his moving in moving from minute to minute.[28]

The piety inherent in the act of attending to the passing hour was also reflected in religious literature. Many Christian texts, both Catholic and Protestant, provided prayers to be said when the clock struck, while accounting for one's use of time by reference to a clock or watch became a common refrain. According to the lesser-known Puritan cleric Henry Greenwood, 'it is the duty … of every man, to imitate that person, that vigilant person, that carried alwaies about with him in his pocket a little clocke, and when hee heard it sound, hee would instantly examine himselfe how he had spent that houre'.[29] Perhaps the most famous example of all is William Shakespeare's twelfth sonnet, a meditation on the 'wastes of time' and the unstoppable 'Time's scythe', which begins 'When I do count the clock that tells the time'.

It is for these reasons that we find clocks so frequently displayed in *vanitas* still-life paintings, perhaps most recognizably in Netherlandish art of the sixteenth and seventeenth centuries. In Harmen Steenwyck's *Still Life: An Allegory of the Vanities of Human Life* (c.1640), now in the National Gallery in London, a delicately lit arrangement of skull, book, shell, and musical instruments on a

---

25 Dohrn-van Rossum 1996, 217–20. Dohrn-van Rossum casts doubt on the existence of Charles V's 'legendary decree' that the churches of Paris should regulate their tolling by reference to his clocks, but states that Charles's installation of striking clocks is attested by several independent sources.
26 For more on English clock portraits, see Faraday 2019b, *passim*.
27 Sold TEFAF Maastricht 2017, visible at: https://www.apollo-magazine.com/pick-fair-tomasso-brothers/.
28 De Mornay 1587, 241.
29 Greenwood, f.C4v. For prayers to be said when the clock strikes, see, for example, Bentley 1582, 365; 998. Trigge 1602, 531; the Jesuit Gaspar de Loarte 1579, ff.41v; 116v–117r and Wilson 1622, 552.

Figure 213 Antonio de Pereda, *Allegory of Vanity*, 1632/36. Oil on canvas, Kunsthistorisches Museum, Museumsverbande, Vienna.

table encircles a watch, painted with its lid open and almost touching the cheekbone of the skull. These motifs also inspired artists of other cultures, for example, the Spanish artist Antonio de Pereda, who took the Netherlandish *vanitas* theme and elaborated on it in his *Allegory of Vanity,* now in the Kunsthistorisches Museum in Vienna (1632/36) (Figure 213). A winged figure points to a globe, and beyond it a gilt table clock, while next to a pile of skulls and a snuffed-out candle sits a sand-glass, the sand completely run out.

One English painting of the early seventeenth century makes use of a visual pun or rebus to further cement the association between the clock and mortality. On the reverse of a triptych now in the Victoria & Albert Museum in London showing the family of Henry and Dorothy Holme (1628), the words 'wee must' appear above a clock dial, supplying the macabre punchline to the pun: 'die-all', accompanied by an image of the Redeemer: 'Yet by (Christ) live all'.[30]

As a symbol of fleeting time, and by extension the vanity of worldly endeavours, clocks and sand-glasses also came to be associated with the reclusive and/or contemplative life of monastics and of the Church Fathers like Jerome and Augustine. Saint Jerome is often shown in his study with a sand-glass or even, anachronism notwithstanding, a mechanical clock. A well-known Flemish series of Jerome paintings associated with the studio of the Flemish artist Pieter Coeke van Aelst, ultimately descending from Albrecht Dürer's clockless painting of the same subject, show the saint variously with a sand-glass and a gilt, wall-mounted weight-driven clock in the background. Other engravings of Jerome in his Study by Dürer also show a sand-glass and skull, an allusion to the contemplative life and the 'living death' characteristic of the cloistered life.[31]

Sandro Botticelli's fresco of St Augustine for the Church of Ognissanti in Florence (*c.*1480) is one of the most famous portraits of St Jerome's fellow and contemporary Church Father. Here a clock, resembling Italian clocks dating from *c.*1450, is shown behind the saint's head. But far from simply symbolizing the self-disciplined life of the saint, this timepiece also plays a key role in isolating the legendary moment which Botticelli has chosen to represent. The hands of the clock show that the time is

---

30 Unknown English Artist, *Triptych Portrait of Henry and Dorothy Holme*, 1628, oil on panel. Victoria & Albert Museum, London W.5-1951. 115.6cm × 95.9cm.

31 For more on the 'living death' of monastic life, see Olson 2013, 19, 61

approaching the end of the twenty-fourth hour of the day: counting from the previous sunset, this would have indicated the hour of Compline in a monastic setting. According to legend, Augustine, settling down to write to Jerome, hears the saint's voice, accompanied by a sweet odour and a bright light: marking the exact moment that Jerome died in Jerusalem.[32] In all these examples, clocks are just one aspect of the highly familiar, contemporary settings chosen for these saints by their artists: part of an attempt to bring home the Church Fathers' immediacy and continuing contemporary value.[33]

No doubt partly inspired by its *vanitas* resonances, the clock came to be compared with the human body itself: complex, fragile, breakable. The horological analogy took further inspiration from the clock's reputation for unreliability and frequent need for repair. As the anonymous translator of the Spanish Dominican Luis de Granada put it in 1599:

Why is a clock so oftentimes disordered and out of frame? The reason is, because it hath so manie wheeles and points, and is so full of artificial work, that although it be made of yron, yet every little thing is able to distemper it. Nowe, how much more tender is the artificial composition of our bodies, and how much more fraile is the matter of our flesh, than is the yron whereof a clocke is made? Wherefore, if the artificial composition of our bodies be more tender, & the matter more frayle: why shoulde wee wonder if some one poynt among so manie wheeles have some impediment, by reason of which defect, it stoppeth and endeth the course of our life?[34]

Indeed, death itself was often compared with the taking in of a clock for repairs by its maker. A striking example of this comparison is found on the Exeter Cathedral monument to Lady Dodderidge (d. 1614). In the northeast corner of the Lady Chapel she has a recessed wall-tomb. Dodderidge is shown reclining in a full-length three-dimensional effigy, propping herself up on a cushion, her hand caressing a laurel-crowned skull. On the tomb chest, the right-hand panel displays this poem:

As when a curious clock is out of frame
a workman takes in peeces small the same
and mending what amisse is to be found
the same rejoynes and makes it trewe and sound
so god this ladie into two parts tooke
too soon her soule her mortall corse forsooke
But by his might att length her bodie found
shall rise rejoyned unto her soule now cround
Till then they rest in earth and heaven sundred
att which conioyeed all such as live then wondred.[35]

The conceit may have been inspired by John Donne's elegy for a young woman (1610): 'But must wee say she's dead? May't not be said/That as a sundred clock is peecemeal laid,/Not to be lost, but by the makers hand/Repollish'd, without errour then to stand.'[36]

This view of the body as a complex clockwork mechanism, in frequent need of repair, may have inspired the inclusion of a clock in two portraits of physicians: one Scottish and one English. Dr David Kinloch (1614) and Sir William Paddy (c.1600) shared their profession, their inclination to write poetry, and ultimately their positions as royal servants, both having tended James VI and I when he was in Scotland and England, respectively.[37] The portrait of Kinloch is a near-contemporary copy of an original, painted after his service to the King had ended. Curiously, the original portrait shows an ink well on the table, a motif which has morphed into a clock in this later version. He holds a carnation and studies a book, which displays Greek and Latin texts upside-down, the Latin (*vita brevis, ars longa*) being a variation on an aphorism of Hippocrates, the ancient Greek physician.

Approaching the portrait of William Paddy, we likewise feel as if we have interrupted the physician in his study. He stands at

---

32 Lightbown 1989.

33 For more on the use of anachronism, see Burke 1969; Woolf, 2005; Margreta de Grazia, 'Anachronism', in Cummings & Simpson 2010, 13–32; Nagel & Wood 2005, 403–15 at 409; Nagel & Wood 2010; for specifically horological anachronism, see Faraday 2019a, 194–202.

34 De Granada 1599, 81. Cited in Mayr 1986, 45. The relative unreliability of the clock compared with the sundial forms the basis for an extended anti-Catholic satire by Thomas Scot published in 1622: 'Solarium' describes a disagreement between a church clock and a sundial over which of them is more accurate. The sundial represents the truth of Scripture, punningly reflecting the 'Sonne [/sun] of Righteousness', and the clock stands for the Church, which has fallen out of step with the dial over the centuries. The triple-crowned weathercock, metaphor for the Pope, tries to intervene in the clock's favour, but the sexton eventually resets the clock, and 'humbles' the weathercock by removing its crown. Scot 1622, G8v-H8r.

35 Cited in Mayr 1986, 209–10, n. 58.

36 John Donne, 'A Funeral Elegie' (1610), in Grierson 1964, 222, lines 38–40. Other poems by Donne that mention clocks include the 'Obsequies to the Lord Harrington, Brother to the Lady Lucy, Countess of Bedford', in Grierson 1964, 250–1,with a long passage of twenty-four lines (130–54) ending 'Why wouldst not thou, then, which hadst such a soul, / A clock so true, as might the Sunne controul, / … / … stay here, as a generall/ And great Sun-dyall, to have set us All?'; and 'An Anatomy of the World. The First Anniversary', in Grierson 1964, 211 at lines 129–30: 'Alas, we scarce live long enough to try / Whether a true made clocke run right, or lie'.

37 Unknown Artist, *Dr David Kinloch* (1614), oil on canvas. University of Dundee, Tayside Medical History Museum Art Collection. 89 x 76cm; Unknown English Artist, *Sir William Paddy* (1600), oil on panel. St John's College, Oxford University. 210.8cm × 129.5 cm. See Yagüi-Beltrán & Adam 2002; *passim;* Lauren Kassell, 'Paddy, Sir William (1554–1634)' in *Oxford DNB* [online] Jan 2008 (2004). Available at: http://www.oxforddnb.com/view/article/21080. Accessed 28 November 2019.

a table, on top of which rests an ink-well, a clock with a winding key, and an anatomy book. We can just make out the page at which Paddy has paused to acknowledge us: it shows various iterations of the skull, clearly inspired by the ground-breaking anatomy textbook published in 1543 by Andrea Vesalius. Paddy is presented to us as a diligent gentleman-physician, and the clock on the table contributes to the impression of status, wealth, and reliability. In both portraits, however, the clock may also allude to the developing association of the clock with the human body. The fact that Paddy's clock is shown with its winding key beside it could perhaps be read as a reference to the physician's role in maintaining and regulating the body.

Another late sixteenth-century portrait of a physician and theorist, the Oxford-based Aristotelian John Case, also makes use of a skull, a skeleton, and a horological instrument, this time a sand-glass, in combination with vanitas inscriptions to emphasize the sitter's piety and awareness of his mortality (Figure 214). Yet the painting can also be seen as referring to the sitter's profession, both as a physician and a lecturer: Case may be shown in the process of teaching an anatomy lesson from the skeleton in the foreground, timing his lecture with the sand-glass in the upper-right corner.[38]

Just as the human body was considered a microcosm of the larger universe, in this period references to clockwork as an organizing principle of both the human or animal body and creation at large fuelled the development of the 'mechanistic' philosophy, 'the study of natural phenomena as if they were actions of machinery', which came to prominence in this period.[39] Often we find a continuation and further development of the 'God-as-clockmaker' trope, now referred to as the 'argument from design'. Poets and theologians alike took up where Oresme had left off, and the model was even used by Joachim Rheticus (1540) to defend Copernicus' radical heliocentric view of the solar system:

> Since we see that this one motion of the Earth satisfies an almost infinite number of appearances, should we not attribute to God, the creator of nature, that skill which we observe in the common makers of clocks? For they carefully avoid inserting in the mechanism any superfluous wheel or any whose function could be served better by another with a slight change of position.[40]

We find the same sentiment in the work of Philippe de Mornay in 1581:

> Sure, the sky is as the great Wheele of a Clocke, which sheweth the Planets, the Signes, the Houres, and the tides, every one in their time, and which seemeth to bee his chiefe wonder, proveth it to bee subject to time, yea, and to be the very instrument of time. Now, seeing it is an instrument, there is a worker that putteth him to use, a Clock-keeper that ruleth him, a Minde that was the first producer of his moving. For every instrument, how moveable soever it bee, is but a dead thing, so farre forth as it is but an instrument, if it have not life and moving from some other thing than it selfe.[41]

De Mornay goes on to compare man, a microcosm of creation, with a clock: 'O man, the same worke-master, which hath set up the clocke of the heart for halfe a score yeeres, hath also set up this huge engine of the skies for certaine thousand of yeeres'. De Mornay's words were an intimation of what was to follow.[42]

The comparison of the universe and living beings—human and animal—to clockwork machines, which had by this point been in use almost since the clock was invented, found its apogee in the 'mechanical philosophy' that dominated the mid-seventeenth century. Although, as Otto Mayr notes, such 'mechanical' philosophers were free to choose from any number of machines then in existence for their model of the universe, systems 'based on clocks outnumber all others'.[43] René Descartes (1596–1650) was one of the most significant proponents of this version of natural philosophy. Fascinated by self-moving machines, a category which included the clock, Descartes saw the bodies of humans and animals in terms of automata. Each system was subject to an internal hierarchy of organs, culminating in humans with the uniquely free and sovereign soul, and in animals with a brain which could initiate only built-in, pre-programmed actions, or as Descartes put it, 'naturally and by such spring forces as a clock'.[44]

Other aspects of the clock appealed to writers and painters alike during the various social, political, and religious upheavals of the early modern period. English Protestant sitters found in the clock a particularly apposite symbol of their belief in the doctrine of 'justification by faith alone': the idea that only belief in God would enable one to reach heaven. As sermons and portraits from the time suggest, the clock's unidirectional chain of command, from the weight through the wheels to the hands on the clock dial,

---

38 In the collection of St John's College, Oxford, HM07. Available at: <https://artuk.org/discover/artworks/john-case-223386> I am grateful to Anthony Turner for this suggestion.
39 Mayr 1986, 56.
40 Georg Joachim Rheticus, *De Libris Revolutionum Copernici Narratio Prima* (1540) translated in Rosen 1959, 137–8.
41 De Mornay 1587, 95–6.
42 The 'argument from design' continued into the late nineteenth century in devotional writing, e.g. Macmillan 1898, vi: 'they [the essays] may all help to illustrate the spiritual revelation of God in Christ, by the revelation of God in Nature, and prove that, as the dial of a clock reveals the unseen movements within its case, so the visible world reveals the workings of the invisible'. So too did using the clock and the hours of the day as a peg on which to hang edifying homilies in the manner of Suso (Harrison 1848). I am grateful to Anthony Turner for these references.
43 Mayr 1986, 57.
44 Descartes 1974, vi, 56–9, in *Discours de la Méthode*; ix.1, 67 in *Meditation No. 6*; iv, 575, in Letter to the Marquis of Newcastle 23 November 1646.

Figure 214 Unknown artist, *Dr John Case*, late sixteenth century. Oil on panel, 89.5cm × 68.6cm. The President and Fellows, St John's College, Oxford.

the clock represented the fact that outward signs of goodness—for example, the performance of charitable deeds—were only signs of inner grace, and not (as Catholics believed) themselves able to affect one's ultimate spiritual fate. This idea is made particularly clear in the portrait of William Chester, member of the Drapers' Company of London and one-time Lord Mayor. He is shown with a wall clock, the time registered as 1—an optimistic counterpoint to the more common choice of 12. On the lower weight perches a skeleton with the motto 'Deathe at Hand'; on the upper weight the risen Christ and the phrase 'Hope to Live'. Chester, renowned for his obstinate Protestant beliefs even during the Catholic reign of Mary I, uses the clock as an explicit symbol of his soul and the nature of his faith.[45]

Horological devices were also an obvious means to express the notion of the 'times out of joint' (*Hamlet*, I v 186–90). In Jan Steen's 1663 painting 'The World Turned Upside Down', now in The Hague Museum, a wall clock watches over a chaotic scene of revelling and misbehaviour. Mounted towards the top of the picture to the right, a monkey (generally symbolic of sin and human error) plays with its weights. The clock's associations with temperance and *memento mori* themes are thus subverted by the monkey's abuse of the clock: taken as a whole this may allude to the revellers' intemperance and lack of attention to their mortality and may even suggest that human misbehaviour sets time itself awry. In this sense, the painting could not be more different from Holbein's now-lost painting of 'Thomas More and his Family' (1527), surviving in the form of preparatory drawings and later copies. Here order reigns, the family sits quietly or engages in the intellectual and pious activity of reading. Above the scene we see a weight-driven wall-clock, part of the family's well-appointed interior to be sure, but in its placement (in later versions) directly over the head of Thomas More, perhaps also symbolic of his well-regulated household, and the readiness with which his family members obeyed his command as the 'first mover' of the group.[46]

Of course, for all their (variable) utilitarian value, horological instruments were not without their sceptics. In Rabelais' *Gargantua and Pantagruel* (c.1532–c.64) we read of the fictive Abbey of Theleme, where:

> ... because in all other monasteries and nunneries all is compassed, limited, and regulated by hours, it was decreed that in this new structure there should be neither clock nor dial, but that according to the opportunities and incident occasions, all their hours should be disposed of; for, said Gargantua, the greatest loss of time that I know, is to count the hours. What good comes of it? Nor can there be any greater dotage in the world than for one to guide and direct his courses by the sound of a bell, and not by his own judgment and discretion.[47]

Other critiques were more grotesque even than Rabelais. 'The Human Sundial' (c.1540), a wood-block print by the German artist Peter Flötner (1491/2–1546), shows a prostrate peasant, one foot propped up on a diptych sundial, the other on a sand-glass.[48] Along the bottom of his thighs we see the numbers I to XII inscribed in Roman numerals, while a gnomon is formed by sticks protruding from the man's anus and mouth. Flötner was from Nuremberg which, along with Augsburg, was the prolific centre of luxury timepiece production in the sixteenth century, and this rich, if indecent, print has most often been read as a satirical commentary on the popular taste for luxury time-telling devices of that period.[49] The accompanying text recounts the peasant's daily activities, eating, drinking, playing cards, and sleeping, suggesting that despite the many elaborate methods for calculating the time available to citizens, such devices cannot prevent them from wasting what time they have.[50]

The scatological aspects of the print—the pile of excrement on the cushion in the foreground (pierced by the engraver's awl, Flötner's artistic signature), echoed in the ampoules of the sand-glass to the left, and finally oozing out of the human sundial himself—suggest that the users of these advanced horological devices can control time no more than they can control themselves, parodying human folly and pretension to celestial levels of knowledge while inhabiting messy, grotesque, and fallible bodies.[51] No shadow is cast by the gnomon on the man's thighs, a detail which may relate to the fact, noted by Suzanne Karr-Schmidt, that the print itself was envisioned as a functioning sundial, the text suggesting that it may be pasted to a wall and provided with a gnomon to this end. In this way the print becomes, not just a parody of Nurembergers' obsession with time-telling devices, but a satire on interactive sundial prints and paper instruments themselves.[52]

REVOLUTION AND EMPIRE

Like many other aspects of culture, clocks were caught up in the various political, industrial, and colonial upheavals that dominated the eighteenth and nineteenth centuries. As clocks became more accurate, particularly in the wake of the invention of the pendulum clock in 1656, their association with punctuality and accurate time regulation became more prominent in literature and in art.

---

45 For more on this, see Faraday, 2019b.
46 Clocks were often used as metaphors for the well-regulated household, in which wife and children follow the commands of the male householder; see Faraday 2019b, 260. For the suggestion that Thomas More is the equivalent of Temperance 'wearing her horological hat', see White 1969, 202, n. 86.
47 Adler 1990, 60.
48 Geisberg 1974, iii, 794, no. 829, reproducing a copy in Wolfenbüttel.
49 Chipps-Smith 2013, 182–4; Grossinger 2002, 174–5.
50 See Grossinger 2002, 174–5, for a translation of the text.
51 Karr-Schmidt 2017, 228–35; for the relationship of prints with artistic ingenuity and bodily metaphors see De La Verpilliere 2018, 114–21.
52 Karr-Schmidt 2017, 228–35.

For example, in Jonathan Swift's *Gulliver's Travels* (first published 1724) Gulliver's pocket watch is regarded by the Lilliputians as:

> ... either some unknown Animal, or the God that he worships; But we are more inclined to the latter Opinion, because he assured us (if we understood him right, for he expressed himself very imperfectly) that he seldom did anything without consulting it. He called it his Oracle and said it pointed out the Time for every Action of his life.[53]

Jonathan Turner has noted the colonial overtones of Swift's comparison of the watch 'chain' and the chains of slavery, while the changes of scale in the encounter between Gulliver and the 'less temporal minded' Lilliputians is emblematic of British colonial power and typical of Western views of 'lesser' non-Europeans.[54] Here the watch stands for 'advanced' imperial time in contrast to the 'primitive', natural timeline 'disposed to colonial domination'.[55]

As clock accuracy increased, the extent to which it was allowed to dictate the daily schedule also grew. The significant role that the clock could play in the owner's life is euphemistically alluded to in Laurence Sterne's *Tristram Shandy* (published between 1759 and 1767), where the main character's misfortunes are blamed partly on the inauspicious moment of his conception, his mother having distracted his father by asking 'have you not forgot to wind up the clock?'.[56] In addition to his misfortunes, Shandy's dilatory approach to narration is perhaps partly to be blamed on his conception under a stopped clock. Later in the novel he tries to visit 'the wonderful mechanism of the great clock of Lippius of Basil' in Lyons but discovers that 'Lippius's great clock was all out of joints, and had not gone for some years'.[57] A 1760 pamphlet called *The Clockmakers' Outcry*, purportedly written by an anonymous clockmaker, but perhaps penned by Sterne himself, complained about the euphemisms which had corrupted the profession: 'Our manners and speech at present are all *be-Tristram'd*. [...] The directions I had for making several clocks for the country are counter-manded; because no modest lady now dares to mention a word about *winding-up a clock*'.[58]

Besides the general schedule, as the eighteenth century progressed the clock played an ever more-important role in dictating the rhythms of the working day. The elevation of punctuality and diligence to a special virtue was another consequence of this development and is reflected in art and literature of the period. The image of the clock in the marketplace, surmounting (and, it is implied, superseding) the sundial below it, plays an important role in William Hogarth's image *The Four Times of Day—Morning* (1736). In this image the clock of Covent Garden market stands at 6:55am: for the boys on their way to school and the woman on her way to church the day is just beginning, but for the revellers at the infamous coffee house-cum-brothel run by Thomas and Moll King, the night is coming to an end. Turner suggests that Hogarth is using the clock to 'bring [...] together both the "polite" and "impolite" worlds of the market, showing the confrontation between time patterns of the industrious middling kind, the low prostitutes and beggars, and high aristocratic libertines'.[59]

Similarly, Hogarth's satirical engraving 'Masquerade Ticket' (1727, Figure 215), is a mock ticket on which a debauched ball is depicted, where we find masked aristocrats making a 'sacrifice to Priapus', and trying out two 'lecherometers' taking the form of barometers. Watching over all this lewd behaviour is a large clock with two candelabra, registering the time 1:30 a.m. The parts of the clock are labelled: 'Nonsense' on the pendulum, which Ronald Paulson reads as suggesting 'that in this room there is nonsense at every instant', while the minute hand is labelled 'Impertinence' and the hour hand 'Wit', 'showing their relative frequency'.[60] At an altar dedicated to Priapus in the left niche, revellers are sacrificing Father Time, who seems to be making a lunge for the offerings on the altar.

The print as a whole has been identified as a satire on the infamous masquerades of Swiss count John James Heidegger, who received patronage from George II and who became his Master of Revels. Coinciding with the year of George II's accession, the engraving shows the beasts of the royal arms lolling indecorously either side of the clock dial, perhaps an allusion to the indignity of the King's support of such excesses. Yet although they appear as an object of ridicule in these Hogarth images, aristocrats also engaged with the new emphasis on punctuality and attentiveness to the clock: Philippe Bordes has suggested that the clock painted by François Boucher in the background of his portrait of Madame de Pompadour in 1756, now in the Alte Pinakothek in Munich, alludes to the popular virtue of punctuality in the period.[61]

In other contexts the disregarding of time could be a mark of dominance and dedication. In Jacques-Louis David's 1812 portrait of Napoleon, beside a low-burning candle, the clock tells us the time is 4:13am (Figure 216). In a letter to the Scottish nationalist, Alexander, Marquis of Douglas, the patron who commissioned the portrait, David writes:

> I have shown [Napoleon] in the condition most habitual with him—that of work. He is in his study, after a night spent composing his Code Napoléon. The candles flickering out and the clock striking four remind him that day is about to

---

53 Swift 1726, part I, ch. 2, 41–2.
54 Turner 2018, 15–19.
55 Turner 2018, 19.
56 Anon 1760, 6.
57 Anon 1760, 467; 479; see also Parker 2000, 147–60.
58 Anon 1760, 40–2, cited in Fawcett 2016, 133–4.
59 Turner 2018, 99–100.
60 Paulson 1965, i, 133–4.
61 Reported in Knaub, 'Report and Summary of the Horology in Art Symposium', held at the Museum of Fine Arts in Boston, 26–8 October 2017. Available at: http://www.horologyinart.com/report.html (accessed 3 December 2019).

Figure 215 William Hogarth, *Masquerade Ticket*, 1727. Engraving, British Museum, London inv. 1826,0313.8.

break. He rises from his desk to gird his sword and pass his troops in review.⁶²

Here the time displayed on the clock bears witness to Napoleon's dedication to the role of lawgiver, working through the night to enact his vision for the empire.

With its role in the regularization and regulation of the working day, the clock is especially prominent in industrial and capitalist contexts. An engraving in the Deutsches Museum showing an English razor factory depicts a clock mounted high on the factory wall, overseeing the various tasks being carried out beneath its gaze. In industrial society, obedience to the clock's dictates was instilled early in life: Turner notes the same prominence of a clock in a 1779 painting of a classroom, now owned by UNESCO World Heritage in Český Krumlov, Czech Republic. The children are being taught to read and being given religious instruction beneath a clock, the highest object in the room. Turner argues that the schoolroom was seen as the place to teach children obedience to clock time, preparing them for a life in industrial service.⁶³

Yet the clock was also enlisted by those who wanted to protest against the time-slavery enforced by the industrial way of life. It was particularly popular with writers and painters of everyday life and of the working classes. A 'workingman's clock' (as Steven Dillon has termed it) appears in Richard Redgrave's *The Sempstress* (1846), now in Tate Britain.⁶⁴ In a garret we observe a poor woman slaving over a shirt, the clock behind her showing the time 2:30am. But far from simply revealing the lateness of the hour, the clock is seen to supervise the seamstress's work, a 'tyrannical' symbol of 'discipline and control' that dictates her activities, and which perhaps alludes to her premature death as well.⁶⁵

While in earlier centuries interest from artists and writers derived from the clock's novelty and special value, its later reception seems to have depended on the opposite qualities of familiarity and ubiquity. As a now-commonplace object that was also fraught with symbolism, the clock became particularly popular with nineteenth-century writers and painters whose work aimed at social commentary, often pursued through a mode now described as 'Symbolic Realism'.⁶⁶ As Louise Cooling points out,

---

62 Cited in Eitner 2000, 200.
63 Turner 2018, 111–14.
64 See Dillon, 2002, 52–90, cited in Cooling 2018, 539.
65 Cooling 2018, 539–40.
66 Brooks 1984; *cf.* Cooling 2018, 528.

Figure 216 Jacques-Louis David, *Napoleon in his Study at the Tuileries*, 1812. Oil on canvas, National Gallery of Art, Washington DC. 1961.9.15.

painters in the Symbolic Realist tradition aimed to 'endow ... their realist paintings with an artistic and moral power via a pervasive symbolism attached to the exactly rendered details of everyday life'.[67]

Cooling suggests that many nineteenth-century examples took their inspiration from the popular eighteenth-century paintings and engravings of William Hogarth. For example, his 1732 engraving 'A Midnight Modern Conversation', showing revellers at a drinking club, makes use of a clock for both 'narrative and iconographic' reasons, revealing both 'the fashionable status of the club and the lateness of the hour'.[68] As an emblem of vanitas inspired by Steen's 'The World Turned Upside Down', Cooling suggests the clock may also provide a 'moralising undertone', strengthened by the presence of the guttering candle to the right.[69] Other of Hogarth's works also make use of the symbolic potential of clocks, for example 'The Lady's Last Stake' (c.1759), in which the clock marks a moment of crisis, when the lady must decide between financial and moral ruin: whether to play the man who has already won her fortune at cards, and either win back her husband's wealth or accept the card-player as a lover if she loses. Clocks are also found in the second and the sixth scenes of his series 'Marriage Á-la-Mode' (1743), in which a clock watches over

---

67 Cooling 2018, 528.
68 Cooling 2018, 531.

69 Cooling 2018, 531.

scenes of marriage breakdown, adultery, and finally suicide: incorporating the by now well-established connotations of temperance, right use of time, and mortality mediated by the clock.

Nineteenth-century artists continued to make symbolic use of the clock in genre paintings. According to George P. Landow, in William Holman Hunt's 'The Awakening Conscience' (1853) Hogarth's influence is seen in the choice of a 'contemporary moral subject' and use of 'pervasive integrated symbolism'.[70] Perhaps inspired by Hogarth's 'The Lady's Last Stake', Hunt has placed a clock on the mantelpiece, the dial of which reads five minutes to twelve. This indicates the approaching moment of crisis for the fallen woman, who half-starts from the lap of her illicit lover. Cooling also notes that the clock, with its French Rococo design, contributes to the feeling of 'vulgarity' in the gaudy interior.[71] Another moment of crisis is indicated by the clock in John Whitehead Walton's painting 'Anxious Moments: A Sick Child, Its Grieving Parents, a Nursemaid and a Medical Practitioner' (1894). Two timepieces, a longcase clock, and a pocket-watch in the hands of the doctor appear in the painting. The time on the longcase clock indicates that the scene is taking place in the early hours of the morning, while the doctor seems to use his watch to see when the medicine he has administered might be taking effect.[72]

The work of Charles Dickens is a good example of 'symbolic realism' in literature. As Stephen Franklin has observed, clocks 'lurk in the background of every Dickens novel'.[73] Though used to express a variety of ideas throughout his *oeuvre*, Franklin suggests that, for Dickens:

> ... [the] fascination rests on what they record, the flow of time and the action of time on existence, rather than on any particular form of time, such as history or the past [...] to Dickens, that the clock ticks on is of the utmost relevance; where its hands point means little.[74]

Franklin argues that Dickens uses clocks to draw out his characters' various approaches to time, within the framework of a Christian world-view which sees the acceptance of time's unstoppable onward rush as a virtue. Characters who refuse to acknowledge the impetuousness of time fall prey to 'self-delusion' and become instruments of 'social evil'.[75]

Take, for example, the characters in the short-lived periodical *Master Humphrey's Clock*, effectively a frame story in which a group of amateur writers meet to listen to each other's work, their manuscripts stored by Master Humphrey in the casing of a longcase clock. These characters describe themselves as 'alchemists who would extract the essence of perpetual youth from dust and ashes': an attitude which, Franklin notes, was out of step with both the symbol of Master Humphrey's clock and Dickens's own view that acceptance is the only proper, Christian, attitude towards the flow of time.[76] After a reading of *The Old Curiosity Shop*, Dickens has Master Humphrey meet with the realization that time is in fact relentless and that attempts to stop it are futile: all prompted by his recollection of a visit to the clock at St Paul's. After a reading of *Barnaby Rudge*, Master Humphrey notes the similarity between the sound made by the St Paul's clock and the ticking of his own clock, which prompts his awareness that time's onslaught can be 'the greatest kindness' as 'the only balm for grief and wounded peace of mind'.[77]

As Franklin notes, there are many other characters in Dickens's works whose refusal to acknowledge time's unstoppable nature marks them out as deluded, or even evil. Chief among them is Miss Havisham, whose clocks are all stopped at 9:12, the moment when her heart was broken many years previously. While she tries to impose stasis on her environment through the stopped clock and the trappings of her failed wedding day, the decay which surrounds her bears witness to the impossibility of her desire, and the evil that comes of it.[78] Meanwhile, Ebenezer Scrooge in *A Christmas Carol in Prose. Being A Ghost Story of Christmas* (1843) learns his moral about the right 'use of time'[79] through a spiritual clock-trick, the chiming of his clock eventually giving way to the realization that the spirits had 'done it all in one night', despite Marley's warning that Scrooge would be visited on three consecutive evenings. Here, in particular, Dickens brings to the fore the importance of memory for its ability to inform action in the present, not as a permanent state of retreat. In contrast, as Franklin suggests, the futility of Little Nell's attempt to retreat into a 'rural past' and escape the 'urban present' is also figured by the tolling bell of the church clock, which triggers her recognition that escaping the present moment is impossible, as she listens 'with solemn pleasure almost as a living voice'.[80]

This anthropomorphic image of the clock-as-person (rather than, as we saw earlier, person-as-clock) also features prominently in Dickens's non-fiction writing. In one piece he describes a clock 'importuning me in a highly vexatious manner to consult my watch, and see how I was off for Greenwich time'.[81] This clockwork anthropomorphism recurs in a variety of contexts, utilized

---

70 Landow 1979; Cooling, 2018, 537.
71 Cooling 2018, 537.
72 Cooling 2018, 542. Cooling reads this as a deathbed scene, in which case the clock would carry over its traditional symbolism of *memento mori* and mortality.
73 Franklin 1975, 18; *cf.* Bevis 2013, 47: 'there is always a clock in his novels'.
74 Franklin 1975, 2.
75 Franklin 1975, 3.

76 Franklin 1975, 4–5.
77 Franklin 1975, 6.
78 Franklin 1975, 27–8.
79 Franklin 1975, 15–16.
80 Charles Dickens, *The Old Curiosity Shop* (1840–1), quoted in Franklin 1975, 6–11.
81 Dickens, 'Out of the Season', *Household Words*, xiii (28 June 1856), 327; cited from Bevis 2013, 61.

both for its comic and tragic potential. A letter from him to a clock repairer survives, and reads as follows:

> My Dear Sir,
> Since my hall clock was sent to your establishment to be cleaned it has gone (as indeed it always had) perfectly well, but has struck the hours with great reluctance, and after enduring internal agonies of a most distressing nature, it has now ceased striking altogether. Though a happy release for the clock, this is not convenient to the household. If you can send down any confidential person with whom the clock can confer, I think it may have something on its works that it would be glad to make a clean breast of.
> Faithfully yours,
> Charles Dickens[82]

The humorous description of the failings of the clock contrasts with a later instance of horological anthropomorphism. Following Dickens's involvement in the Staplehurst railway accident, a tragic derailment in which ten people died, he described the horror of the accident through its effects on his watch, which subsequently had 'palpitations':

> Is it not curiously significant of the action of a great Railway accident on the nerves of human creatures, that my watch (a chronometer) got so fluttered in my pocket [. . .] that it has never since been itself to the extent of two or three minutes?[83]

The psychological persistence of the clock—its ability to intrude distressingly on human thought—is also treated in a number of well-known literary works. The first mention of the phrase 'tick-tock' in English literature does exactly that. 'They were both so silent that the ticktock of the Sacrifice of Iphigenia clock on the mantelpiece became quite rudely audible'.[84] In Edgar Allan Poe's short story 'The Tell-Tale Heart' (1843), the beat of the victim's heart which haunts the murderer and drives him to confess is compared to the ticking of a watch: 'It was *a low, dull, quick sound—much such a sound as a watch makes when enveloped in cotton*'.[85]

Returning briefly to Dickens, we also see the clock's persistence in *David Copperfield* (1850), as the title character waits for news of his fisherman friend Ham Peggotty in a storm, he describes how 'the steady ticking of the undisturbed clock on the wall tormented me'.[86] This scene is startlingly similar to that found in the 1888 painting 'The Hopeless Dawn' by Frank Bramley, an artist of the Newlyn school. Here a fisherman's wife forlornly awaits the return of her husband. A large clock in the background marks the coming of the dawn and tolls the death of the fisherman; meanwhile a candle on the windowsill has been extinguished. Cooling notes that both of these symbols, the clock and the candle, were not present in the preparatory oil sketch, but rather added by Bramley later on to reinforce the meaning of the final composition.[87]

This symbolic use of the clock is also found in the nineteenth-century genre paintings created in the United States. Ross Barrett has examined the use of two clocks, or a clock and a clock-case, in James Henry Beard's 'The Land Speculator' (1840). He suggests that the use of clocks in American genre paintings of this period incorporated many of the themes we have already seen: indicators of social standing, symbols of the moral order and its transgression, and the new 'tempo' of industrial capitalism. Barrett sees the clock in Beard's painting as alluding particularly to new regimented economic time, tied up with capitalist profit and production. The scene shows a family tempted by travelling salesmen to gamble their money on a scheme of land speculation. This risky activity was recognized at the time as having contributed to the latest period of recession, as a system of credit encouraged banks to lend beyond their reserves, creating a land bubble, which burst when smaller banks defaulted on their debts. Barrett identifies the painting as a moment of collision between the old notion of time, in which sustained work and domestic labour lead to slow accumulation of wealth, and the disingenuous, artificially accelerated land speculation that promised sudden fortune. In the background, a wall-mounted clock is paired with the empty cabinet of a longcase clock, alluding to the family's uncertain future as they consider embarking on the speculator's scheme.[88]

Barrett notes several other examples of clocks in nineteenth-century American genre paintings, many of which conform to the symbolism, already noted, of the 'workingman's clock' and the temporal pressure of wage-hunting and profit-making. For example, William Henry Burr, 'The Employment Office' (1849) is dominated by a Massachusetts clock, showing the agency's conformation to the rhythms of the working day.[89] The examples from the eighteenth and nineteenth centuries of clock symbolism discussed above were largely in keeping with this kind of objective, external time, the pressure of which was keenly felt in early industrial society. Such concerns, though part of a longer tradition, appear for example with particular vehemence in Charles Baudelaire's poem, *L'Horloge* (1861): 'The clock! A sinister, terrifying, inscrutable god, Whose finger threatens us and says "Remember!

---

82 Letter from Charles Dickens to John Bennett, 14 September 1863, cited from Bevis 2013, 49.
83 Bevis 2013, 69.
84 William Thackeray, *Vanity Fair* (1847), 194, see above, n. 2.
85 In some versions of the story this line is given as: 'It was a quick, low, soft sound, like the sound of a clock heard through a wall, a sound I knew well'.
86 Cited in Franklin 1975, 3.

---

87 Cooling 2018, 543.
88 Barrett 2017.
89 Barrett 2017.

– throbbing pains will soon be stabbing your cringing heart as if it were a target'".[90]

At the very end of the nineteenth century we start to sense the way these certainties are being demolished. Theodor Fontane's novel *Effi Briest* (1895) opens with the image of a sundial, set in a flowerbed, cast into shade by the shadow of a house. By rendering the sundial useless, the shadow symbolizes a move into subjective time, and the novel 'increasingly places Effi's perceptions at the centre of time's passage'. She is accused of 'never *holding on* to time', while her family and friends live strictly by the mechanical clock.[91] Finally, Effi is buried in the same flowerbed, her grave displacing the sundial which once stood there as her private, 'malleable' sense of time displaced the exactitude of objectively measured time.[92] Approaching the twentieth century, many of the seeming certainties of this earlier period continue to fall away and writers and artists seize on the return of subjective time, often also expressed through clock symbolism.

THE TWENTIETH CENTURY AND AFTER

Salvador Dalí's 'The Persistence of Memory' (1931) shows, literally, the melting away of old certainties about the human relationship to time. The limp, liquifying watches signify impermanence, the ants covering the orange watch in the bottom left a familiar Dalían emblem of decay and death. Even more significantly, the art historian Dawn Adès sees the 'soft watches' as an 'unconscious symbol of the relativity of space and time, a Surrealist meditation on the collapse of our notions of a fixed cosmic order'.[93] She suggests that this may have been inspired by Albert Einstein's theory of special relativity, which was gaining attention around this time: according to this theory time is no longer seen as universal and absolute, but rather dependent on spatial position and the observer's frame of reference. Dalí denied this when it was pointed out to him, saying that the image was instead inspired by the image of cheese melting in the sun, or as he put it, 'the camembert of time'.[94]

With the blossoming of 'stream of consciousness' narration, twentieth-century literature began to explore what Virginia Woolf called in *Orlando* (1928), 'the extraordinary discrepancy between time on the clock and time in the mind'. Woolf's novel *Mrs Dalloway* (1925) was originally going to be titled *The Hours*: its narrative takes place over the course of a single day and explores exactly the discrepancy Woolf described between formal or external time, artificially measured out in discrete units, and the more elastic and continuous sense of interior or mental time. Big Ben features repeatedly in the novel as a negative symbol of mortality, interrupting Clarissa Dalloway's thoughts, reminding her of the time she has already lost: 'a suspense [...] before Big Ben strikes. There! Out it boomed. First a warning, musical; then the hour, irrevocable. The leaden circles dissolved in the air'.[95] This symbol is to an extent counteracted by the bells of St Margaret's, a more positive reminder of the subtlety of passing time, the need to accept time and not fear its passing. The character Peter identifies St Margaret's chime with Clarissa herself:

Ah, said St Margaret's, like a hostess who comes into her drawing room on the very stroke of the hour and finds her guests there already. I am not late. No, it is precisely half past eleven, she says. Yet, though she is perfectly right, her voice, being the voice of the hostess, is reluctant to inflict its individuality. Some grief for the past holds it back; some concern for the present.[96]

The twentieth century also found other, more spectacular, ways to express the new-found precarity of time through the use of the clock symbol: Harold Lloyd's 1923 silent movie *Safety Last!* gave us the famous image of him dangling from the hands of an enormous clock, high above the Los Angeles traffic.[97] This iconic scene reverberated through twentieth-century film, with characters dangling from clocks in homage to Lloyd found in films as diverse as an episode of the British television series *Dad's Army* called 'Time on My Hands' (1972) and the blockbuster film *Back to the Future* (1985).[98] These homages have continued into our own century: in the film *Hugo* (2011), whose plot revolves heavily around the early days of silent film, we see a clip of Harold Lloyd's film: later the main character Hugo is also seen hanging from the hands of a large clock in a Parisian railway station in an attempt to escape capture by a railway guard.[99]

Despite this new awareness of the precariousness and subjectivity of time, much of the attention accorded to the clock in twentieth-century art and literature is precisely its ability to measure out exact parcels of time. In Aldous Huxley's *Brave New World*

---

90 Baudelaire 1986, i, 167. He also describes the clock's 'metal throat' which can 'speak all languages'.
91 Tucker 2007, 188.
92 Tucker 2007, 188.
93 Adès 1982, 179.
94 MOMA 2019. Dalí may have been inspired by a work by Italian metaphysical painter Giorgio de Chirico, 'The Enigma of the Hour' (1911). The painting shows an arcade with a covered walkway above it, on the exterior wall of which is mounted a clock showing the time six minutes to three.

95 Woolf 1925.
96 Woolf 1925. See Armentrout 1992.
97 Fred C. Newmeyer and Sam Taylor, dir., *Safety Last!* (Los Angeles: Pathé Exchange, 1923).
98 David Croft dir., 'Time on my Hands', *Dad's Army*, series 5, episode 13 (BBC Worldwide, 1972); Robert Zemeckis dir., *Back to the Future* (Universal Pictures, 1985).
99 Michael Scorsese dir., *Hugo* (Metropolitan Filmexport/Paramount Pictures, 2011), based on the book by Brian Selznick, *The Invention of Hugo Cabret* (New York 2007).

(1932), Chapter Ten opens with the line: 'The hands of all the four thousand electric clocks in all the Bloomsbury Centre's four thousand rooms marked twenty-seven minutes past two', emblematic of the exactness of time that dictates much of the dystopian society Huxley has created. We see this earlier in the book, too: when the clocks strike four the main day shift at the Central London Hatchery and Conditioning Centre hands over to the second day shift.[100] Meanwhile, in the 'ruthlessly industrialised world' represented in German expressionist science-fiction film *Metropolis* (1927), the 'Rulers' live on normal time while the 'Workers' live by clocks that measure ten hours, and are forced to perform two heavy-duty shifts of ten 'worker hours' every day.[101]

The potentially transformative effects of precise time measurement on those who are subjected to it can also be found in much twentieth-century literature. In Salman Rushdie's 'clock-ridden' novel *Midnight's Children*, the main character, Saleem Sinai, is born at the exact moment of Indian Independence, midnight on 15 August 1947, and as a result is granted telepathic powers and a superhuman sense of smell. This is later compounded by the discovery that all children born in India between midnight and 1 a.m. that day have also acquired magical abilities.[102] In Franz Kafka's *Metamorphosis* (1915) we meet the salesman Gregor Samsa, whose life and work as a travelling salesman is dictated by the new rhythms of industrial society. The failure of the alarm clock to sound becomes emblematic of Samsa's subjection to industrial time:

> He looked at the alarm clock ticking on the chest. Heavenly father! He thought. It was half-past six o'clock and the hands were quietly moving on, it was even past the half-hour, it was getting on toward a quarter to seven. Had the alarm clock not gone off? From the bed one could see that it had been properly set for four o'clock; of course it must have gone off. Yes, but was it possible to sleep quietly through that ear-splitting noise?[103]

The failure of the alarm clock corresponds with Samsa's monstrous transformation into an insect, rendering him incapable of continuing to perform the tasks of his employment. The monstrous body suggests Samsa's resistance to the relentless mechanization of his existence, the 're-turn-of-the-body' and the difficulty of regulating a biological entity by something as apparently artificial and external as clock time.[104]

We find resistance to clock time everywhere in twentieth-century literature: against the idea of the exactly measured modern moment is the ability of clocks to indicate a move beyond the normal frame of time. The opening of George Orwell's *Nineteen Eighty-Four* (1949), 'It was a bright day in April, and the clocks were striking thirteen', indicates the novel's position in an aberrant society, and highlights the need to call into question the so-called statements of truth in the society he has created. Similarly, the grandfather clock striking thirteen in *Tom's Midnight Garden* by Philippa Pearce (1958) signals the main character's timeslip from the present of the story into the Victorian era. The potential of the clock's liminal status, and its seeming ability to connect different times, is also reflected in the final chapter of Donna Tartt's *The Secret History* (1992), when the main character Richard encounters Henry in a ruined dreamscape, a place between life and death, in front of a mysterious clockwork machine: 'In the case was a machine revolving slowly on a turntable, a machine with metal parts that slid in and out and collapsed in upon themselves to form new images. An Inca temple ... click click click ... the Pyramids ... the Parthenon. History passing beneath my very eyes, changing every moment'.[105]

The tendency of clocks to be misaligned among themselves also inspired authors and artists: in Agatha Christie's *The Clocks* (1963) this tendency is used to nefarious ends, as a body is found surrounded by six clocks, four of which have been stopped at 4:13, and a fifth which reads 3 o'clock—all red-herrings. Meanwhile, "Untitled' (Perfect Lovers)', a 1991 work by the Cuban-born American artist Felix Gonzalez-Torres, consists of two identical battery-powered clocks, which will eventually fall out of sync or stop entirely. The artist spoke of the emotional import of the work, made shortly after his partner was diagnosed with AIDS: 'Time is something that scares me ... or used to. This piece I made with the two clocks was the scariest thing I have ever done. I wanted to face it. I wanted those two clocks right in front of me, ticking'.[106]

The potential for clocks to cause emotional disturbance is also a common theme in twentieth-century literature. In Adrian Conan Doyle's pastiche of his father's Sherlock Holmes series (1954), a story called *The Adventure of the Seven Clocks* features a character who is driven mad whenever he sees a clock—smashing, hiding, or burying them in his rages.[107] More disturbingly, in William Faulker's *The Sound and the Fury* (1929), we meet Quentin Compson, whose father had left him a pocket-watch, 'not that you may remember time, but that you might forget it now and then for

---

100 Aldous Huxley, *Brave New World* (1937), cited in Macey 1986, 102.
101 Fritz Lang dir., *Metropolis* (UFA, 1927); Wosk 2010, 403–8 at 403.
102 For a post-colonial reading of Rushdie's use of temporality, see Barrows 2011, *passim*.
103 Kafka 1983, 89–139, at 90, trans. Willa and Edwin Muir.
104 Shahar 2016, 253–62.

105 Tartt 1993, 628.
106 Anon., 'Felix Gonzalez-Torres 'Untitled' (Perfect Lovers), 1991', MoMA online catalogue. Available at: https://www.moma.org/collection/works/81074.
107 Doyle & Carr 1954.

Figure 217 Marc Chagall, *Le Christ à l'horloge* ('Christ as a Clock', sometimes called 'Christ in the Clock'), 1957. Lithograph. Photo: Artistshome e.K. Fine Art courtesy of Christine Blenner. © ADAGP, Paris and DACS, London 2020.

a moment'.[108] Contrasting with the free-flowing jumble of narrative time in other sections of the book, Quentin is obsessed with the measured passage of time. The character thinks constantly about time as he focuses on the pocket-watch's ticking: 'I could hear my watch ticking away in my pocket and after a while I had all the other sounds shut away, leaving only the watch in my pocket'.[109] The watch becomes a vehicle for an exploration of the nature of time, and the effect that measurement has on it: 'Because Father said clocks slay time. He said time is dead as long as it is being clocked off by little wheels; only when the clock stops does time come to life'.[110] Even after he tries to destroy the watch, he can't make the watch—or time—stop:

> I went to the dresser and took up my watch, with the face still down. I tapped the crystal on the corner of the dresser and caught the fragments of glass in my hand and put them into the ashtray and twisted the hands off and put them in the tray. The watch ticked on.[111]

Eventually its ticking drives the character to suicide, as he feels himself unable to escape his insignificance in the sweep of time and history.

Marc Chagall was an artist whose dreamlike ethereality penetrated everything he painted, not least the clocks that seem to have obsessed him. Clocks and sand-glasses appear in many of his paintings, often in conjunction with the image of the crucified Christ. Jewish himself by origin, Chagall chose the image of Christ to express the theme of the persecution of the Jewish people. In his lithograph 'Le Christ à l'Horloge (Christ as a Clock)' (1957, Figure 217) Chagall portrays Christ with a clock dial for a face, crucified against the body of a longcase clock.[112] In the upper left corner of the picture a purple horse reaches for a broken menorah, and beneath the horse's foreleg is a village, and a group of huddled figures, including several mothers with their children. The representation of Christ as a clock expresses the timelessness of Jewish suffering, but the presence of mothers with children suggests hope.

Another conjunction of clock with crucifixion occurs in Chagall's painting 'Self Portrait With Wall Clock' (1947). The artist leans towards the canvas of another crucifixion, where Christ is embraced by a figure in white, possibly the representation of Chagall's deceased and much-mourned wife, Bella. Meanwhile, next to the canvas, the artist, holding his palette, presses his face into the neck of a red donkey or cow; both observe us with a single, front-facing eye. Above their heads a pendulum clock seems to fly across the wall, sprouting a pair of blue arms or wings that seem both to hold it up and to carry it in flight. The time on the dial approaches three o'clock, the hour at which Christ is said to have expired on the cross. The clock's apparent flight is suggestive of the dove of the Holy Ghost, but also, perhaps, in the artist's grief, another emblem of hope: an ambivalent reference to the ability of time both to perpetuate memory of loss, and gradually to heal.

Clocks appear in numerous other paintings by Chagall: 'Clock' (1914); 'Time is a River Without Banks' (1930–9); 'Pendule au Ciel Embrasé (Clock in the Flaring Sky)' (1947); and 'L'Horloge (The Clock)' (1956). In 'Juggler' (1943), Chagall perhaps took some inspiration from Dalí, as the bird-juggler of the title is shown with a drooping pendulum clock slung over one arm, not unlike the melting watches of Dalí's dreamscapes.[113] Exactly what meanings Chagall invested in his clocks is difficult to discern, although the writer Yuri Trifonov described an encounter with Chagall when he showed the artist a reproduction of one of his paintings in which '[a]gainst a dark brown background an old grandfather clock in a wooden case stands slightly tilting to the side'. Trifonov describes how Chagall suddenly 'whispered barely audibly, not to us but to himself: "How miserable one must be to have painted this."...',[114] From Trifonov's description we might identify the picture with 'Clock' (1914), and the writer glosses Chagall's comment as follows:

> I thought: he had whispered off the very essence of it. To have been miserable in order to paint. Later you can be anything, but at first—miserable. The clock in a wood case tilts to the side. One must overcome the tilting time which tosses people all over: some it keeps in Vitebsk, others throws to Paris, and others yet to Maslovka Street [. . .].[115]

The old association between the ticking clock and human life continued in the twentieth century. A light-hearted example can be found in *The Wizard of Oz* (1939): towards the end of the film, when the Wizard gives the Tin Man the heart he has been seeking, it takes the form of a heart-shaped clock on a chain. 'It ticks!' he says delightedly as he passes it around, finally able to love again.[116] In other examples, the clock again stands for its old meanings of death and mortality. In one of Edvard Munch's final paintings, 'Self-Portrait Between the Clock and the Bed' (1940–43), the artist stands stiffly between two emblems of mortality: a grandfather clock and a bed. Is his stiffness defiance, determination to make the most of the remaining time? Or is he bracing himself for the final blow of death? In either case the clock brings out the themes of the passage of time and the limited span of human life: the artist died in 1944, the year after this portrait was finished. The poet W. H. Auden also turned to clocks in the face of death, for his poem 'Funeral Blues'. It first appeared in a play

---

108 Faulkner 1929, 73.
109 Faulkner 1929, 79.
110 Faulkner 1929, 81.
111 Faulkner 1929, 76
112 Gauss et al. 1988, 109, no. 196.
113 Guerman, Forestier, & Wigal 2007.
114 Trifonov & Shrayer 2005, 163–4.
115 Trifonov & Shrayer 2005, 164.
116 Victor Fleming and King Vidor (dir.), *The Wizard of Oz* (Loew's Incorporated, 1939).

cowritten with Christopher Isherwood called *The Ascent of F6* (1936) in a context of surreal political satire, but was reworked in 1940 as the lyrics for a cabaret song for the singer Hedli Anderson, which had been set to music by Benjamin Britten.[117] In these lines Auden expresses the enormity of the loss of a loved one, with the famous first couplet 'Stop all the clocks, cut off the telephone / Prevent the dog from barking with a juicy bone'.[118]

One of the most significant examples of clocks in contemporary art to appear in the twenty-first century is Christian Marclay's twenty-four-hour collage film installation 'The Clock' (2010). Using more than a thousand clips from films of a variety of genres, both famous and obscure, Marclay strings together a sequence which tells the time exactly and continuously for an entire day.

Each clock is embedded in its own narrative, with the setting and mood varying around the constant central emblem of passing time, as characters react with horror or surprise to an anticipated event or disaster.[119] By conflating the time of the clock within the artwork with the time outside it, Marclay brings together the fictive time of the represented clock—the clock as it is encountered in art and literature—and the functional aspect of the clocks, which mark the real time of the viewer's experience. In the centuries since it was invented, the clock has played an important symbolic, practical, and imaginative role in visual art and literature: Marclay's work and its critical acclaim are a testament to the permanent role of the clock in our lives.

---

117 Perry 2016, *passim*.
118 Roberts & Grigson 1938, 62.

119 Levinson 2015; Krauss 2011; Fowler, 2013.

CHAPTER TWENTY FIVE

THE LITERATURE OF HOROLOGY

Bernhard Huber

'Literature is the memory of humanity'.
Isaac Bashevis Singer

INTRODUCTION

Horological literature records the history of timekeeping, the diversity of ideas and inventions, as well as the biographies of the inventors and famous artists. Horological documents are an essential aid for the study of the history of timekeepers. They are our only source for understanding the development of horological technology and craft during many centuries. The amount of horological literature is vast. A survey is here offered of the primary horological literature. After reviewing dialling literature, the chapter explores mechanical timekeepers from the beginning of preserved records to modern times, although not simply in a chronological presentation. Over time, horological literature emerged in parallel in different countries for different audiences with different targets and the structure of the chapter is adapted to this. It highlights the various factors that have influenced the development of horological literature and provides typical examples for the countries involved. The survey is not, however, intended to replace existing bibliographies.[1]

DIALLING

Whatever its historical importance, dialling is a small subfield of elementary astronomy. Historically, elementary geometry, together with some spherical trigonometry for astronomical calculations, was sufficient for drawing dials and sections concerning them were included in textbooks thereof. Although the mathematics is straightforward, using only paper and pencil is annoying and tedious. Today calculation is done with computers so, even in recent times, new literature for dialling has been created.[2] Sundials were the main instrument for setting clocks until well into the mid-nineteenth century and manuals for their construction were many. In the late eighteenth century one commentator remarked, 'There are more books on gnomonics than one could ask for'.[3]

Already in the first century BCE, Vitruvius described the ancient sundials known to him. The astronomer Claudius Ptolemy (c. 100–c. 175 CE) wrote a comprehensive treatise (c. 140 CE) on mathematics and astronomy in thirteen books, known today as the *Almagest*.[4] This remained the standard work of astronomy in Europe until the end of the Middle Ages, while in his *Opera astronomica minora*[5] Ptolemy also discussed sundials. His text contains the graphical and computational construction of a plane sundial in any position. The *Almagest* presents the first known chart of the sine wave function (manually calculated!) and also an early method for calculation in what was later called spherical trigonometry. Particularly noteworthy is an explanation of the **equation of time** and its calculation. The text of Ptolemy was refined in the fifteenth

---

1 There is no complete bibliography of gnomonics, the fullest being that found in Houzeau & Lancaster 1964 with many works also listed in Tardy 1947/1980. For French works, see Faidit 2002, 119–38. For mechanical horology, see Baillie 1951; Baillie unpublished; Tardy 1947/1980; Watkins 2011; Graf 2011. For fuller details of works mentioned in the present chapter either by author and date only, or by an abridged title, see in the bibliography.

2 The author thanks Gerhard Aulenbacher for valuable information related to this section.
3 Klügel 1793.
4 Toomer 1984.
5 Heiberg 1907. An extensive analysis of Ptolemy's text is given by Drecker 1925, 4ff.

century by the medieval astronomer Regiomontanus (1436–76), who introduced more modern methods.[6] His fundamental work already reflects modern spherical trigonometry and is an early example of a scientific book in the first period of printing.

The earliest printed books on the construction of sundials were published during the first half of the sixteenth century. They also sometimes included instructions on the use of other instruments for astronomical purposes. Notable is the *Horologiographia* of Sebastian Münster (1488–1552), first published in 1531 in Latin, dedicated exclusively to sundials and nocturnals. A German translation appeared in 1537. Even today the reader is amazed at the enormous diversity of the dials described and the clear didactic structure.

The authors of the earliest dialling books could build on knowledge preserved from Antiquity. The arts of measuring and of dialling developed in the *artes liberales* tradition. Here the language of applied geometry, which could be used for the construction of sundials was well established. Textbooks summarized acquired knowledge. The large number of sundial books written from the sixteenth to the middle of the eighteenth century in France, Britain, Germany, Italy, and Spain shows that the sundial played an indispensable role in everyday life as an instrument for checking mechanical clocks, even into the nineteenth century. Possibly an additional aspect for the demand of these books is that the design of a sundial was well suited for a training in practical geometry and astronomy. Sundial books listed in the sale catalogue of Sotheran & Co., London for 1917, give an impression of the number of works available.[7]

Period	Before 1549	1550–99	1600–49	1650–99	1700–49	1750–99	1800–49	after 1850
Quantity	12	41	23	43	35	25	10	0

Language	Latin	Italian	German & Dutch	English	French
Quantity	49	47	35	30	28

The rapid spread of sundial books was also favoured because metric constructions could be easily described using established terms. These terms exist not only in Latin but also in the various national languages of Europe. This made it possible to translate Latin works into various languages.

From a modern viewpoint there are few significant differences in the content of these works; one gets the impression that each author has adopted much of his predecessors' material.[8] The concept of intellectual property did not exist. Works that should be mentioned include that by Christopher Clavius (1537–1612), probably the fullest and most pedantic work on dialling. Its seven hundred pages in folio format contain the first proof that the temporal hours on sundials are not straight lines. Despite its prolixity, it was appreciated by later generations of authors, while dialling was also discussed in comprehensive mathematical textbooks such as the *Cursus mathematicus* of Claude François Milliet de Châles (1621–78). In the Baroque period many popular treatises dedicated specifically to sundials were published. Well-known authors include Johann Friedrich Penther (1693–1749), Johann Peterson Stengel (*fl.* 1679), Dominique François Rivard (1697–1778), François Bedos de Celles (1700–79), and Charles Leadbetter (1681–1784). There is much overlap in their content and the *Gnomonica* of Eberhard Welper (1590–1664) can stand for all of them. The text of the last edition in 1708 benefited from additions and improvements by the Nuremberg mathematician Johann Christian Doppelmayr (1671–1750). The book is still valuable, decorated with thirty-four beautiful, precise copper engravings. Many exotic forms of sundials are explained.

After the Baroque era, sundial publications become more factual and less frequent. The work of Johann von Littrow (1811–77) falls into this category and is the very solid work of a renowned astronomer. Only plain standard sundials are treated. Here indeed 'Brevity is the soul of wit'. After a long break, a notable work was published by Joseph Drecker (1853–1931), in which the methods of Ptolemy are discussed with very good diagrams. The book is still indispensable today for study of the history of gnomonics. The author masters all the necessary mathematics and his work is of the highest level. He is hailed by the leading contemporary authority Denis Savoie for his innovative mathematical and historical approach to early dials.[9]

The availability of computers has inspired sundial enthusiasts. From about 1980 a considerable number of new sundial books have appeared. They are not only beautifully illustrated, but also contain construction manuals. During this time, national sundial societies were founded in France, Britain, The Netherlands, Belgium, Spain, and the USA. In a new age of gnomonics, the journals of these societies promote an understanding of sundials through a combination of historical and technical research. Two recent works, however, still need to be mentioned: firstly,

---

6 Regiomontanus 1496; Hughes 1967.
7 Private communication from Gerhard Aulenbacher. These figures should be compared with those established for France and Britain only in Turner 1987a, 346. See also graph 2 in Savoie 2014a, 15.
8 But compare Chapter 8.

9 Savoie 2014a.

that of Jörg Meyer, a physicist, who is particularly interested in embedding gnomonics not only in elementary astronomy, but also in the dynamic theory of the Earth's orbit. Thus, it is the only book that explains recent calculations of the equation of time. The circle is then closed by Denis Savoie, who offers a selection of interesting and challenging topics on gnomonics supported by computer programmes. One unusual topic which he treats is that of cornice or ledge dials.

MECHANICAL TIMEKEEPERS: THE EARLY PERIOD TO 1600

The beginnings of horological literature for mechanical timekeepers coincide approximately with the invention of printing around 1450. Somewhere in the later thirteenth century, weight-driven clocks were invented. Monasteries and abbeys in the Middle Ages were the great repositories of knowledge of arts and sciences at that time. They were a major vector for the rapid distribution of the new invention aided by Latin as the common language for communication and documentation throughout Europe. A few manuscripts with technical details are preserved from this early period. The oldest known description of a clock was written by Richard of Wallingford (c.1292–1336), mathematician and Abbot of St Albans Abbey, Hertfordshire. He described his clock in some detail.[10] Ramon Sanç documented in Catalan the construction of a clock for the royal palace at Perpignan in 1356 with detailed information about the costs.[11] Between 1348 and 1364 Giovanni de Dondi (1318–89) created a planetarium, for which he supplied comprehensive documentation with drawings and numbers of teeth for the wheels of this complex, weight-driven system. Several manuscripts accounts of it survive (Figure 98).[12] A major document for the early period of clockmaking is the notebook of Paulus Almanus, a German monk who was in Rome 1475–86.[13] His notebook, written in Latin, is highly practical and of special value because it describes thirty chamber clocks of differing types and origins with a wealth of technical details and drawings in their original size. It is the great merit of John Leopold that he published an annotated facsimile with a transcription and an English translation.[14]

BASIC OBSTACLES

At the time of the invention of the mechanical clock, in contrast with sundials, there were no established terms for the new technical device and its parts. It would require centuries for all the parts of a clock or watch to be designated by standard terms. Because of this linguistic problem, early horological books up to the end of the eighteenth century frequently contained an extensive glossary. The best illustration of these difficulties in describing the components of a clock is provided by the world's oldest printed book on the construction of mechanical clocks written by the Capuchin Giuseppe di Capriglia (mid-seventeenth century) in 1665.[15] The key problem for his work was the lack of common terms for the parts of a weight-driven clock. Capriglia therefore used analogies with well-known terms from daily life that are, today, often difficult to understand. For this reason, a German translation was published in 2018.[16] Unfortunately, Capriglia did not provide a glossary. As a result he often used different names for the same part—names that do not appear in any later horological work. Many terms demonstrate his creativity in comparing technical parts with everyday objects. The **verge** with its **pallets** and the **foliot** are referred as 'the two arms of time',[17] the escape wheel with its sharp teeth as a 'Catherine wheel' (following the legend that St Catherine was martyred on a wheel),[18] and the crown wheel as a wandering wheel.[19] One part is called a key, this being the lifting or trigger arm of the **striking** mechanism.[20] The term *pendicolo* used by him is misleading.[21] Since Capriglia did not know the pendulum clock, it only becomes clear from the context that he means the striking hammer, because it 'swings' back and forth when hitting the bell.

But it was not only the lack of suitable nomenclature for individual clock parts that was responsible for the hesitant development of horological literature. Another was the lack of knowledge and aspiration on the part of craftsmen. Conrad Dasypodius (1530–1600) complained about this problem. Dasypodius was responsible for calculating the second astronomical clock in Strasbourg cathedral (1572). He complains in his publication about it[22] that clockmakers and craftsmen lacked all basic principles of geometry and arithmetic, and that they had no ambition to acquire knowledge appropriate to their craft: 'They can do nothing but what their craft tells [them] and what they have learned from their masters.' Furthermore, Dasypodius laments that they can barely read or write and are completely inexperienced in the Latin language. Although c.1550 only some twenty per cent of

---

10 Texts edited by North 1976.
11 Beeson 1982.
12 Critical edition by Poulle 2003; facsimile of the earliest manuscript in Poulle 1998. There is a worthless English translation by Baillie *et al.* 1974, and a comparison by Dresti & Mosello 2020.
13 Staatsbibl. Augsburg, Ms 2° Cod. 209.
14 Leopold 1971.
15 Capriglia 1665. *Cf.* Robey & Linnard 2017.
16 Koch, 2018.
17 Capriglia 1665, 11.
18 Capriglia 1665, 29.
19 Capriglia 1665, 201.
20 Capriglia 1665, 32.
21 Capriglia 1665, 16.
22 Dasypodius 1580a; Dasypodius 1580b.

the entire population was literate, a slightly higher proportion could perhaps have been expected among skilled craftsmen such as clockmakers. If this was not the case, then it is not astonishing that they played no authorial role in the early days of horological literature. Even if the Reformation church supported literacy so that all classes should have access to printed religious works in the vernacular, the ability to write and express something precisely remained weak until well into the eighteenth century. In this respect, horological literature shows huge differences between the various authors.

Typical of this in early publications is the tendency to verbose phrases. An example can be taken from the works of John Harrison, who found writing difficult. In characterizing the 'Nature of a pendulum' he starts as follows:

> At first, or rather as here at the first [viz. as without taking any notice of the great or chief matter, viz. of what pertains to different Vibrations, or rather, as much as appropriate, of what Advantage pertains to, or accrues from the vibration of a vibration] the bare length of a pendulum can not otherwise be rightly considered or esteemed, but as only to what it bears, or may [as per the common application] bear in proportion to the length of the pallets.[23]

Analyzing the authors of horological books up to 1800, one finds a predominance of clerical writers usually able to express themselves succinctly and clearly. Examples are the Benedictine Richard of Wallingford (c.1291/2–1336), the Franciscan Sebastian Münster (1488–1552), the Jesuit Christopher Clavius, the Jesuit Caspar Schott (1608–66), the Jesuit Athanasius Kircher (1602–80), the Capuchin Giuseppe di Capriglia, the abbé Jean de Hautefeuille (1647–1724), the Anglican canon William Derham (1657–1736), the pastor Johann Georg Leutmann (1667–1736), the Benedictine Jacques Allexandre (1654–1734), the Benedictine François Bedos de Celle (1709–79), the Benedictine David a Sancto Cajetano (1726–96), the Jesuit Johannes Klein (1684–1762), the Augustinian Nicolaus Alexis Johann, the preacher Christof Wilhelm Forstmann (1736–83), and the pastor Philipp Matthäus Hahn (1739–90).

It was not until the early eighteenth century that lay horologists attempted to write. Among them may be noted John Smith (2nd half 17th century) in Britain, the 'father of Swiss watchmaking' JeanRichard (1665–1741), Henry Sully (c.1680–1728), John Harrison (1693–1776), Antoine Thiout (1692–1767), and Jean André Lepaute (1720–89). The last two, who were sons of locksmiths or toolmakers, achieved great success as clockmakers. They wrote extensive textbooks that became very well known. Slightly later, extremely complex clocks were constructed by Antide Janvier (1751–1835), who wrote copiously to validate his work.

Until the nineteenth century, horological books normally focused on either new inventions or on practice-proven horological techniques and rules. The authors report on an empirical basis without theoretical justification. France was a leader in this field. In the eighteenth century a clear, understandable, and exact language was developed for the description of technical parts, but only gradually in the nineteenth century was a more exact scientific approach established. It usually started, although marine chronometry offers some exceptions, with theoretical considerations from which practical conclusions were drawn. A very early example of this is Christiaan Huygens, who proposed his pendulum clock in 1657[24] after having already investigated the theory, although this would not be fully published for another seventeen years.

DISSEMINATING HOROLOGICAL KNOWLEDGE

The publications of the later seventeenth and eighteenth century were not primarily textbooks for practicing horologists but rather provided current knowledge about the calculation, design, and construction of watches and clocks. Illustrations were essential for an understanding of the text. Especially in France, capable illustrators and engravers were available and already by the middle of the eighteenth century clock parts were depicted in perspective drawings with exploded views. From 1750, the character of these horological books developed towards practice-oriented publications, which addressed active horologists or even laymen. An attempt was made to describe all work on a clock in an understandable way. In addition to these more or less comprehensive horological textbooks, special research results were published as separate contributions in scientific journals as early as the late seventeenth century.

The earliest book entirely devoted to the construction of mechanical clocks was that already mentioned by Giuseppe da Capriglia. He describes the construction of various clocks in full detail, his work being intended for practical use. In eighty pages and with numerous woodcuts, three types of clocks are treated: tower clocks (*rustico*), table clocks (*polito*), and watches (*lustro*). The clocks are of iron with verge **escapement** and foliot. The subtitle boasts that

> Measuring the time [is] taught with simple and safe rules to make new ones and to understand existing ones. So far never treated or published matter. Interesting and useful work not only for those who want to learn this art, but also for those who enjoy owning and handling watches of any kind. With this work, everyone can learn to repair, and preserve them.

It is interesting that even the clock owner is addressed. Capriglia created the only Italian textbook on mechanical clocks at a period when an extensive literature on sundials was being published. Figure 218 shows one of his woodcuts with the parts of a turret clock.

---

23 Harrison 1775, 8–9.

24 Whitestone 2012a, 2017.

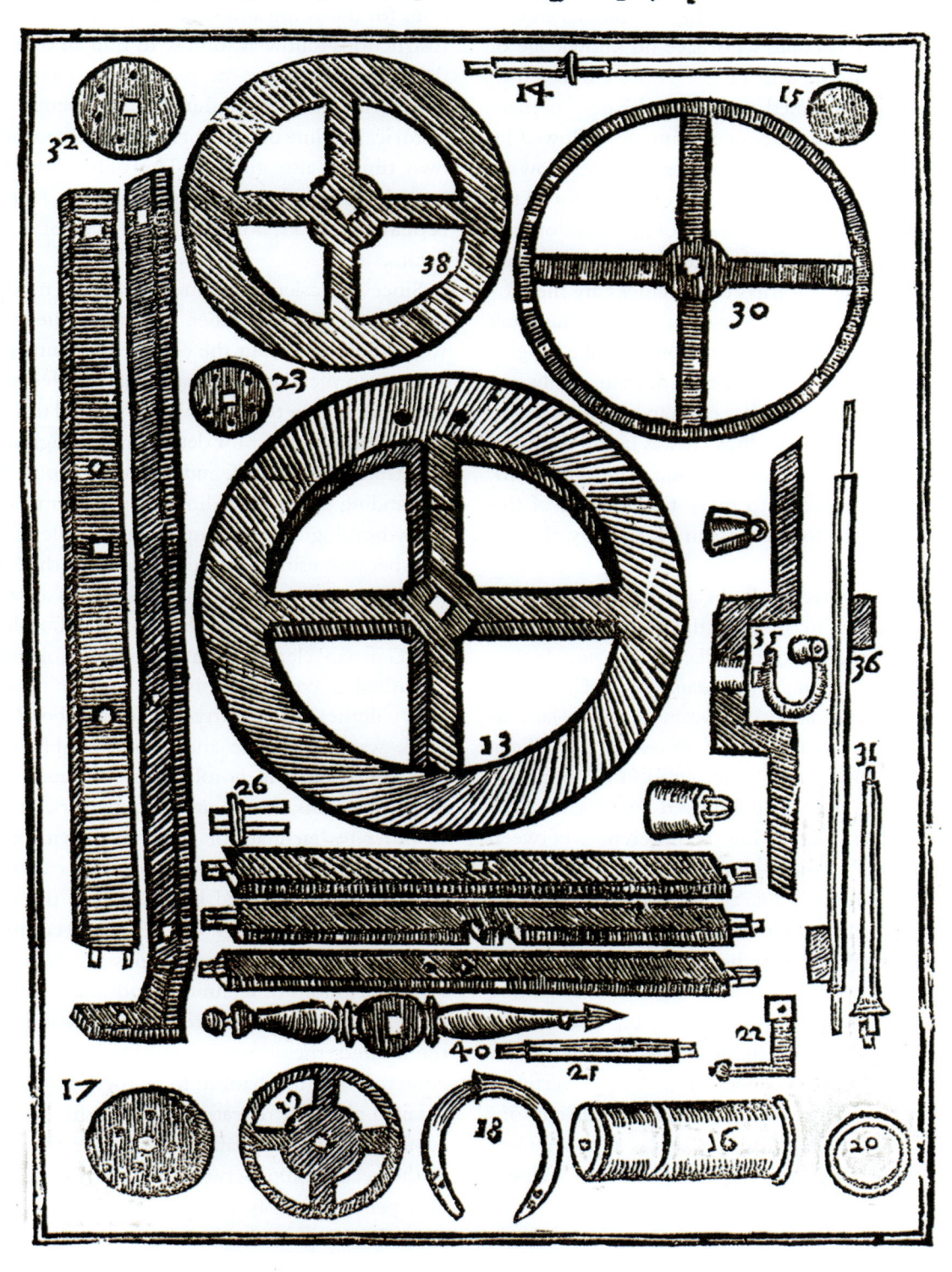

Figure 218 Parts of the going train of a turret clock from Capriglia, 1665.

In 1719, the Augustinian father Beuriot (d. 1739) wrote a small work of which the style and woodcut illustrations are comparable with those of Capriglia.[25] The publication is very rare today but had considerable consequences for the development of domestic clock production in the French countryside, particularly in Normandy. The book was entitled *Horlogeographie pratique ou la manière de faire les Horloges à poids . . .*, and Beuriot, who was not a clockmaker, explains in the minutest detail how to build a weight clock, 'for the satisfaction of people who have no knowledge of it'. It was the first book published in France entirely devoted to practical clockmaking. It was aimed at a lay, and barely numerate, audience. The book starts with basic arithmetic, followed by geometry. In the third part (pages 41–56), he indicates how to manufacture a dividing plate, including tables for its use for all the wheels needed. Finally, in the heart of the book, instructions are given for building a clock without striking but with an alarm, then for a clock with passing strike. All parts of the clocks are meticulously sketched and illustrated with twenty-four plates. The book ends with advice on how to weld metals and how to make weights and what shape to give them. The publication describes only one simple type of rustic clock with an iron cage and square section pillars, similar to the English lantern clock. It was useful for locksmiths or simple ironworkers in the countryside. The short work gave them the information needed to make timekeepers of this type with weights for local customers first in Normandy, then in other French regions.

The first book entirely devoted to clocks and watches in Britain had been published earlier, in 1675. Again the title is in the long, typical marketing, style of the period: *Horological dialogues. In three parts, shewing the Nature, Use, and right Managing of Clocks and Watches: with an Appendix Containing Mr Oughtred's Method for Calculating of Numbers. The whole being a work very necessary for all that make use of these kind of Movements.* The publication did not contain any details or drawings, but Smith followed it with two other works on the nature of time elucidating the differences between true solar time and the artificial mean-time of the clock.[26] In 1696, William Derham published *The Artificial Clockmaker*. In the preface to the fourth edition (1734), he retrospectively made clear for whom he had written it. With pride he states that 'it has done some, not inconsiderable, Good, in the World, not only among the Clock-Makers, and their poor Apprentices, but also among many Gentlemen and others, that delight in Mechanical Studies and Excercises: To whom it hath been an Innocent and vertuous Diversion'. It was an important work that enjoyed four editions in Derham's lifetime and a fifth edition as late as 1759. A German translation was published in 1708 as an appendix to Welper's *Gnomonica*,[27] while the French translation received two editions (1731 and 1746).[28]

Derham begins with a glossary of the essential parts of both the going and the striking **trains** of a clock. This makes it much easier to understand the description of the mechanism. It is interesting to see that most of the terms given by Derham have since remained in use unchanged. The second chapter deals with wheel calculations for the going and striking trains of both clocks and watches. Instructions follow as to how to find an optimal arrangement for the wheels between the plates. The next chapter discusses the correct length of pendulums, only recently introduced. A short historical outline of time measurement from antiquity to Derham's own time is also given.[29] The only illustration is of a pendulum clock, which is explained in detail. Derham's comments on **repeating** clocks, however, remain cursory without any technical details.

Since the calculation and making of models of the planetary orbits and those of other celestial bodies was popular as they would be throughout the eighteenth century, Derham described an astronomical gearing train that enabled the orbit of the four moons of Jupiter to be simulated. Behind this was the idea of using the moons of Jupiter for determining longitude. The use of equation tables is explained, and the publication ends with instructions for finding a meridian line. This information can be found in many horological books up to the middle of the nineteenth century because using a meridian transit of the Sun when visible was the easiest way to check the accuracy of a timekeeper.

Shortly before the earliest French horological text, in 1717 the German clergyman and scientist Johann Georg Leutmann published a comprehensive treatise on clocks and watches in two volumes, with thirty-one plates. For Leutmann, Derham's work was incomplete and imperfect. He aimed to enhance it and, understanding horology as a science, addressed himself to mathematicians, watch- and clockmakers, and other connoisseurs. In order to 'facilitate practical work with iron and brass' Leutmann passed on his own considerable experience. His textbook is based on a profound knowledge of the subject and treats all contemporary horology seamlessly and comprehensively.

Leutmann also starts with a glossary explaining the terms used. His descriptions of technical innovations are remarkable. To give some examples, he discusses the application of the pendulum and the cycloid, the construction of a seconds pendulum clock with constant force mechanism to be used as a time standard, and gives detailed plans for repeating mechanisms, including his own conceptions, and for equation clocks. He also discusses details such as the relationship between the driving weight and the weight of a pendulum bob. In addition, the calculation and construction of striking mechanisms is given. No author before Leutmann

---

25 Partially reprinted with an introduction in Sabrier 1988.
26 Smith 1686; Smith 1694.
27 Welper 1708.
28 According to a note on the title page of a copy in a private collection, the translator was one Alexander Mackensie, so far unidentified.
29 Discussed in Chapter 26, this volume.

published so many useful hints for practical work, including fire gilding, soldering, hardening drills, polishing, material properties, and the like. He also describes the making of a wheel-cutting engine and especially the making of a dividing plate.

Another topic is the equation of time and the difference between mean- and solar time. Leutmann's ambition is to produce a timekeeper that shows both times on one dial. He found a surprisingly simple solution. This is followed by detailed analysis of repeating watches for which he thought no satisfactory description existed. Highly uncommon are the chapters which describe the making of **mainsprings,** and that of a **fusee**-cutting machine—Leutmann's work was an up-to-the-minute description of horological technology at the time of its publication and remains today a milestone of horological literature. That it was not translated into any other language, however, underlines the predominance of Paris and London as the centres of horological science, art, and craft in the period.

The development of comprehensive horological textbooks in the eighteenth century was played out in France, where continuous development can be traced from Sully in 1717 to the comprehensive textbook of Claudius Saunier (1816–96) published in 1869. The latter offered clear theoretical considerations and their practical application. The result was a comprehensive and extensively illustrated textbook, still today a milestone of horological literature. It is impressive how fast the quality of the illustrations developed in some fifty years. With the help of large-format plates, authors visualized the various parts of clocks and watches.

A notable beginning was made by Jacques Allexandre in his *Traité general des Horloges* (1734). By the term *Traité* the author underlined his ambition to write a comprehensive horological treatise. He begins his extensive work of over 400 pages with the indispensable chapter on sundials, followed by a short description of water clocks, which includes an account of the successful compartmented cylinder model first described by Martinelli and relatively recently introduced into France.

At the heart of Allexandre's work is the construction of pendulum clocks. The descriptions contain specific details that prove the author's expertise. The same applies to watches. Compared to Derham or even Leutmann, the quality of the illustrations is considerably improved. The twenty-five copper-plate engravings offer detailed illustrations of cadratures, of striking mechanisms in various designs, or horological tools such as the wheel-cutting machine. The early bibliography of French and Latin clock literature of over 100 pages at the end of the book is also remarkable. It begins chronologically with the literature on sundials, water clocks, and hourglasses. This is followed by literature on mechanical clocks. The various publications receive extensive commentaries. Clearly Allexandre had access to the entire current (French) literature, since he comments on the translation of Derham published in 1731 just three years before his own work.

The next important textbook was published by Antoine Thiout in 1741. This *Traité de l'horlogerie méchanique et pratique* also has 400 pages. But it is not only extensive; it impresses above all by the ninety-one large-format plates, each providing several illustrations. Thiout also starts with a glossary. He then discusses in detail the various horological tools needed for practical work. For this he employs thirty-nine plates and provides an excellent survey of contemporary horological tools. The documentation and his comments on the various escapements proposed in 1741 are an impressive illustration of the inventivity of early eighteenth-century French makers.

The first volume contains an article by Enderlin on the irregularities of pendulums and two articles on the form of the teeth of wheels and **pinions** for improving clocks. The second volume shows a great variety of watch and clock calibres, striking trains and repeating mechanisms and equation movements. The value of the book lies mainly in the comprehensive collection of horological elements with detailed drawings. Only occasionally are critical remarks made on the advantages or defects of a particular device. This is typical for this early phase of technological development where, owing to the lack of a theoretical basis, trial and error were used and only after some time could an optimal solution emerged from a wide range of inventions. But the early practice-oriented text by Pierre Gaudron (c.1670–1745) on the investigation and detection of errors in a watch is noteworthy.[30]

A descriptive style also characterizes the work of Jean André Lepaute (1755), who described contemporary clocks and watches and reported improvements made since the treatise of Thiout. With 307 pages and seventeen plates, it was also a solid publication. It also gives interesting details about new inventions, such as a special pendulum compensation, Lepaute's **dead-beat** escapement for clocks, or his double virgule escapement for watches. Lepaute also includes equation clocks, two of which are to his own designs. The standard chapters on the form of the teeth of wheels and pinions, and the calculation of trains and the pendulum, were written by Lalande. One gets the impression that Lepaute used the publication as an advertising opportunity for his own timekeepers. The contents of this book are not didactically arranged.

Ferdinand Berthoud outperformed all previous authors with his *Essai sur L'Horlogerie* (1763). In ironic reference to his predecessors he did not call his work *Traité* (textbook), but simply an *essai* (trial). That was an understatement because, with over thousand pages and thirty-eight plates he delivered a highly impressive publication. It begins with a description of various clocks and watches. The drawings are admirable, many are in perspective and some show exploded views. Figure 219 shows a plate of 1763 with the design of his marine chronometer.

Berthoud's book has a textbook character. The actual operations of manufacturing clocks and watches, including springs, enamelling dials, and gilding are described in great detail. The accuracy of the descriptions reminds the reader that Berthoud also worked on the chapter on horology in the encyclopaedia of

---

30 Thiout 1741, 338–56.

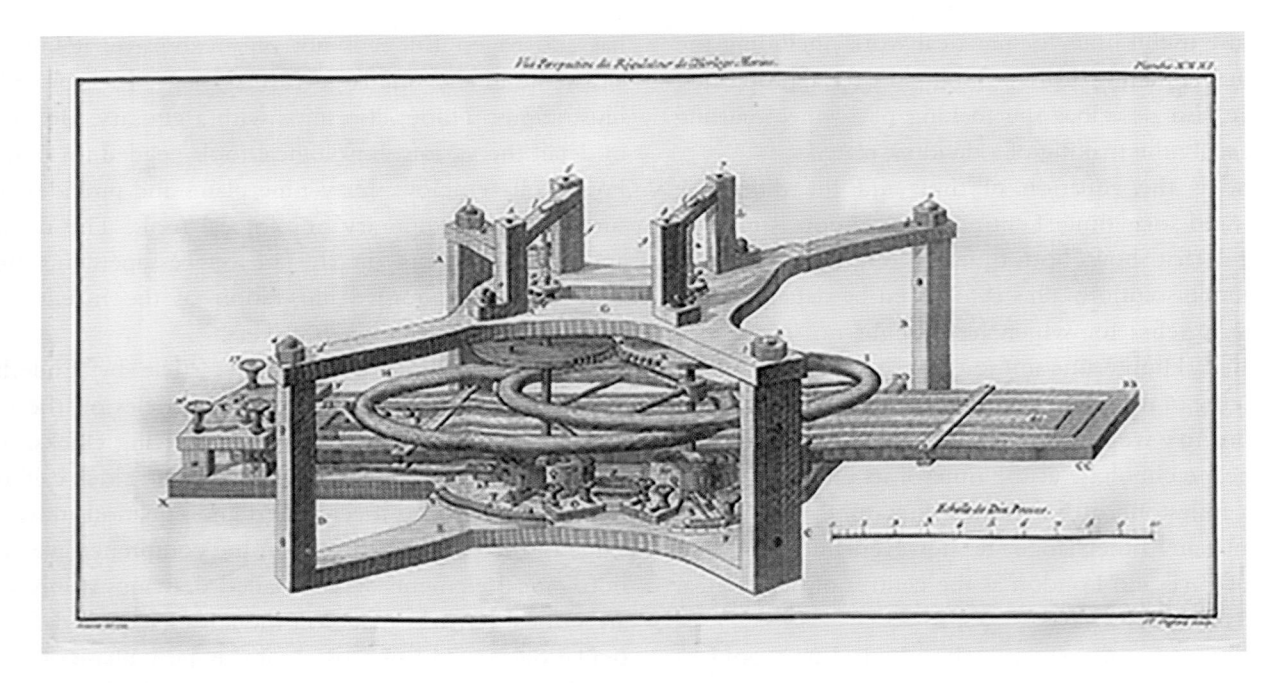

Figure 219 An illustration from Berthoud, 1786, pl. 31, Controlling mechanism for a marine timepiece.

Diderot and d'Alambert. Following a discussion of wheel teeth and the calculation of trains, a long series of experiments on the pendulum follows, including the effect of different escapements on its arc and isochronism and conclusions from the results achieved. After several chapters on temperature compensation of pendulums, a similar series of experiments is described on watch **balances**. Berthoud also describes an observatory clock and two designs of marine timekeepers, in which he uses curb pins with a gridiron compensator. In the supplement to this work, and his many others mainly concerned with his work on marine chronometers, Berthoud displays the transition from trial-and-error methods to solutions based on previous research work, although this was sometimes the unacknowledged work of others, such as Harrison and Pierre Le Roy. The verbosity of his explanations shows the continuing difficulty experienced by clockmakers in expressing themselves.

Even in the nineteenth century, French publications dominated the field. In 1827 an anonymous author ('sometime pupil of Breguet') published an extensive work of 620 pages with seventeen plates.[31] The thirty lessons ranged from the history of timekeeping to the construction of complicated clocks. Lessons twenty-nine and thirty are written for the owner of a watch and give advice on how to handle it and how to find and correct errors. In principle, this is a further evolution of the short explanations on the subject that Sully, Thiout, and Berthoud had already published. The book found favour and had four editions before 1845. It was translated into German in 1828 by G. Wolbrecht.[32]

An extensive manual in 1830 by Louis Sébastian Lenormand (1757–1837), in which he was probably helped by Antide Janvier (although he was only acknowledged from the second edition onwards[33]), was also successful, with a second edition as *Nouveau manuel ...* in 1837, followed by further editions in 1850, 1863, 1876, and 1896 in which Magnier and Stahl also collaborated. A Spanish translation was issued in 1849, and a German translation in 1851.

Louis Moinet's (1768–1853) *Nouveau traité géneral ... d'horlogerie pour les usages civils et astronomiques* published in Paris in 1848 was another comprehensive two-volume French textbook with fifty plates. An extended second edition followed in 1860. Moinet begins with a short history of time measurement, followed by a very detailed chapter of definitions. His work covers theory and practice in a balanced proportion. A third edition was published by Ledieu and Rodanet in 1877 and was enhanced by an appendix of 204 pages containing various contributions about electric clocks and new escapements.

Claudius Saunier (1816–96) wrote the most successful and comprehensive horological textbook of the nineteenth century. He attended the school for clockmakers founded in Mâcon in 1830. As the school was closed in 1836, he completed his apprenticeship in Switzerland. In 1841 he reopened the watchmaking school in Mâcon as its director. Owing to political events the school was closed again in 1848. Saunier then moved to Paris, where he worked until the end of his life. He knew from experience that for a clock- and watchmaker to be successful required the combination of practical skills and theoretical knowledge. His famous award-winning work is the culmination of comprehensive textbooks. The first edition of this *Traité d'Horlogerie moderne théoretique et pratique* was published in Paris in 1869 with over 800

---

31 Anon 1827.
32 Wolbrecht 1829.

33 Augarde & Ronfort 1998, 68.

pages and twenty-one plates. The content is comprehensive and well structured: escapements, gearing, the pendulum, the balance, the spiral spring, adjustments, and an appendix with a collection of various important contributions on contemporary topics for the theoretically interested horologist. Saunier treats all topics with unsurpassed expertise. Baillie comments positively that it was ' . . . the most complete work that exists on the construction of watches and clocks, other than turret clocks. It is particularly detailed in describing every escapement that has been used. It is a little weak in theory, but the practical side is very good. The drawings are beautiful.'[34] Two further editions, each enlarged, followed. In Germany, the work was translated by Moritz Grossmann and appeared in 1878 in three volumes, plus the illustrations. After a second enlarged edition, a third appeared in 1915. In Britain, this standard work was already translated in 1862 by Tripplin and Rigg. It would have nine editions, the last in 1952.

Eighteenth-century Britain produced nothing comparable to the extensive French publications. Only a very few works appeared during the century. This is remarkable because, at the time, Britain was the dominant commercial nation in Europe and the leader in horological development. However, there are practically no publications by the innovators apart from John Arnold, who printed between 1776–82 some patent specifications and small pamphlets, about ten pages each, written 'for the public' as propaganda texts for himself. In contrast with the impressive large-scale, fundamental works of the French authors, the preference in Britain was to publish new findings or investigations via short contributions in scientific journals. The records of the Board of Longitude between 1737–1828 provide good evidence of this.[35] Only a few English clockmakers wrote more extensive horological works.

Among those who did, a few are notable: Alexander Cumming (1731–1814),[36] whose book was translated into German by J. G. Geissler in 1789, and Thomas Hatton (fl. pre-1757–74), who wrote a comprehensive work.[37] They were followed by Thomas Reid in 1826.[38] The success of his book is evidenced by the appearance of a seventh edition in 1859. But Baillie is ambiguous:

> The Author was an eminent maker and his book shows a wide practical experience, but he introduces a good deal of theory in which he is very weak. Many clock and watch escapements are described, with excellent drawings. Compensated pendulums are treated at great length and several uncommon types using zinc and glass are described.[39]

In 1850, Edmund Beckett produced a successful work[40] justifying several editions, the eighth, in 1903, being considerably extended. It is an exceptionally good book on technical horology, dealing principally with clocks, and especially with turret clocks. The author, who was largely responsible for the design of the Westminster clock, shows a remarkable grasp of the practical side of clockmaking. A last, late, example of a general horological textbook with well-founded descriptions of the theoretical aspects of watchmaking was written by J. Eric Haswell in 1928, with further editions in 1928, 1929 (America), 1937, 1947, 1951, and 1975.

A landmark in nineteenth-century horological literature was written by the eminent horologist Urban Jürgensen, who first published his reflections on watches in Danish in 1804 and in French in 1805.[41] He gives a description of the best and most accurate way of measuring time and describes the best devices for temperature compensation together with an account of his observatory clock and a marine watch. His son Ludwig Urban edited a second enlarged edition that was issued in 1840 in German.[42] The appendix contains a description of the author's free escapement with separate wheels for locking and for impulse. It also contains an account of experiments on the effect of air density on observatory clock pendulums and marine chronometers. It ends with a useful bibliography.

As mentioned, Paris and London were the undisputed centres for the dissemination of horological knowledge in Europe during the eighteenth and early nineteenth century. By contrast, the situation for German countries was rather complex. By the end of the Thirty Years' War in 1648, the population had declined by about thirty per cent. The geographical area remained politically fragmented until the foundation of the 'Deutsches Reich' in 1871 and the region was completely impoverished. Economic stagnation prevailed until the middle of the nineteenth century and Germany lagged behind its neighbours in technical innovations. In consequence, a rich tradition of translations developed that was not limited to horology.[43] Many of the German horological textbooks of the eighteenth and also nineteenth century are translations from foreign publications making a large part of current horological knowledge available in German. The best example of this is the work of J.G. Geissler who, in ten parts from 1793–99, summarized much contemporary material. He claimed to have taken the best parts of French, English, and other horological writings and combined them with remarks and communications from German craftsmen. The contents derive mainly from Berthoud and Lepaute, but Mudge's constant force pendulum escapement is also described, showing that the author had access to the latest information. Many comments of Geissler show a thorough grasp of the subject and increase the value of this publication.

---

34 Baillie unpublished.
35 Baillie 1951, 192–8 offers a summary of some of these.
36 Cumming 1766.
37 Hatton 1773.
38 Reid 1826.
39 Baillie unpublished.

40 Beckett 1850.
41 Jürgensen 1805.
42 Jürgensen 1840.
43 Graf 2011.

EXPLORING: SCIENTIFIC HOROLOGY

The scientific exploration of the world was decisive for progress in time measurement. In parallel with horological textbooks primarily written for the practicing horologist, a new type of publication started in Britain and France in which inventions and new ideas were presented. Special questions were dealt with, taking up weak points in earlier designs and new solutions were proposed. Their experimental verification and the progress achieved were first empirically justified and became increasingly analytical later on. The material was not appropriate in book form, as the results of the investigations were presented concisely and the authors were also interested in rapid publication. Thus, scientific journals seemed a better alternative. The papers were intended for a rather small group of scholars or highly specialized horologists. In Britain and France contributions were often first published in the *Philosophical Transactions*, the *Mémoires de l'Académie Royale des Sciences*, and the *Journal des sçavans*. From the end of the eighteenth century, horological inventors increasingly used more recent journals, such as the *Observations sur la physique*, (1752–1823), *Nicholson's Journal of Natural Philosophy* (1797–1813), the *Bulletin de la Societe d'encouragement pour l'industrie nationale* (1801ff), the *Transactions of the Society of Arts* (1783–1843), or the *Philosophical Journal* (1798ff).

For example, Christiaan Huygens' idea of the balance spring was first published in the *Journal des Sçavans*.[44] An alternative was to issue a small, almost ephemeral, pamphlet such as he did with *Horologium*, a Latin description of his pendulum clock with a single illustration.[45] His full theory, however, would not be published until some fifteen years later.[46]

A significant example of early scientific reflection in the *Philosophical Transactions* is George Graham's 1726 contribution 'A contrivance to avoid irregularities in a clock's motion', in which he describes his experiments on the compensation of temperature effects on pendulums and suggests for the first time the use of mercury to negate them.[47] Another typical article in this journal was written by John Ellicott, also dealing with the expansion of metals by heat and its influence on the accuracy of clocks.[48] Many English contributions to the development of marine chronometers in the eighteenth century are also included in the reports of the Board of Longitude, where the dispute between the astronomer royal, Nevil Maskelyne, and John Harrison was documented. Baillie provides a good overview.[49]

The driving force in eighteenth-century horology was the search for the very accurate timekeepers needed to determine the longitude of ships at sea. In Britain, John Harrison, and his successors Kendall, Mudge, Arnold, and Earnshaw, achieved ground-breaking successes. In France, the leading watchmakers Ferdinand Berthoud and Pierre le Roy were competing with them. Their rival claims gave rise to a rich technical literature of a polemical nature, with some writing on the subject aimed at the general public. This was probably in order to strengthen the author's position. For example, Nevil Maskelyne published twenty-eight pages of critical remarks in 1767 in his *Account of the going of Mr Harrison's Watch, at the Royal Observatory* . . . . In the same year, Harrison replied with *Remarks on a pamphlet lately published by the Rev. Mr Maskelyne . . .*, and the detailed, though not lucid, description of his fourth timekeeper.[50] In 1799 Thomas Mudge Jr issued the extensive work *A Description with Plates of the Time-Keeper invented by the late Mr Thomas Mudge*. Another highlight of these scientific controversies is the publication in 1806 by the Board of Longitude called *Explanation of Time-keepers constructed by Mr Thomas Earnshaw and the late Mr John Arnold. Published by order of the Commissioners of Longitude* (Figure 121). In these papers Arnold's text confines itself to a simple description of the Arnold-type chronometer of that time, where Earnshaw's gives a general description of his chronometer, describes the escapement with spring detent, and includes a statement of the advantages of his escapement over that of Arnold. Two years later, in 1808, Earnshaw published his own lengthy account—*Longitude, An Appeal to the Public*—containing his complaints against the Board and his 'declared enemies' in the world of chronometry; it is a work full of vituperation but with much interesting 'inside' information.

In eighteenth-century France, the competitors Ferdinand Berthoud and Pierre le Roy in particular also wrote a series of treatises on sea chronometers. Important publications include:

- 1770: Pierre le Roy, *Mémoire sur la meilleure manière de mesurer le tems en mer* . . . (Figure 120).
- 1773: Ferdinand Berthoud, *Traité des horloges marines, contenant la théorie, la construction,* . . . . With 590 pages and 27 plates, it is a milestone in the literature of marine timekeepers and began the quarrel between Berthoud and le Roy.
- 1773: Ferdinand Berthoud, *Eclaircissemens sur l'invention, la theorie, la construction, et les épreuves des nouvelles machines proposées en France pour la determination des longitudes en mer par la mesure du temps*. This is a lengthy answer to Le Roy.
- 1774: Pierre le Roy, *Suite de précis sur les montres marines de France*. A reply to Berthoud's *Eclaircissemens*.

As knowledge advanced in the nineteenth century, scientific treatises on selected horological subjects increasingly adopted a strict international scientific methodology. An important example is given by Benjamin Lewis Vulliamy (1780–1854) in his 'On

---

44 25 February 1675.
45 Huygens 1658. But on this see Whitestone 2012a.
46 Huygens 1673.
47 Graham 1726.
48 Ellicott 1751.
49 Baillie 1951.

---

50 Harrison 1767. A French translation by Esprit Pezenas with commentary was published in the same year.

the theory of the dead escapement, and the reducing it to practice for clocks with seconds and longer pendulums' in 1824 with four folding plates. It contains a detailed discussion of Graham's dead-beat escapement. It provides reproductions of three designs by Berthoud, and points out their errors. Vulliamy believed that, for the first time, he gave the correct layout for a dead-beat escapement in this study. The fundamental observations made by Peter Barlow in 1821 on the disturbance effect of iron in ships also demonstrate the trend towards systematic research.

The in-depth state of horological research achieved in the middle of the nineteenth century can be inferred from Henri Robert's (1795–1874) book, *Etudes sur diverses questions d'horlogerie . . .*, 1852. Innovative results are presented for unsolved horological questions. Among other topics, it investigates isochronism in pendulums with flexible suspensions and free escapements for pendulum clocks. Another chapter is devoted to 'Practical considerations on the oil employed in horology'. This is apparently the first thorough treatment of oil and its behaviour. A further early example of scientific horology is the 1861 detailed essay by Isely on the influence of the suspension spring on the period of oscillation of a pendulum.

In the same year, Edouard Phillips (1821–89) published his important article 'Mémoire sur le spiral réglant des chronométres et des montres'. This is a fundamental work for the optimal layout of spiral springs in watches and shows again the scientific ambition of the nineteenth century striving for perfect solutions. Adrien Philippe (1815–94), meanwhile, had been developing his keyless winding system from 1842 onwards and from 1845 held a partnership in Patek et Cie, Geneva, which in 1851 would become Patek Philippe & Cie. In 1863 he published an account of his innovation in *Les montres sans clef* . . . . He gives a history of early attempts and presents his own superior construction.

During the second half of the nineteenth century, horological societies were founded all over Europe. They realized that a prize offered for research in special horological topics requiring improvement could lead to progress. A typical example is the British Horological Institute, which in 1865 proposed a prize for the best essay on the detached lever escapement. The winner was Moritz Grossmann with *Der freie Ankergang für Uhren. Preisschrift*, 1866. It was an essay on the construction of a simple and mechanically perfect watch. Baillie declared it was an admirable work, dealt with all parts of the watch from a practical standpoint, and was very clearly written.[51]

The theory of the adjustment of watches in various positions and the implication of this for terminal curves on balance springs was an important area of research in the last decade of the century. Several publications were made on the subject. The work of Louis Lossier (1847–93), *Étude sur la théorie du réglage des montres, suivi d'instructions et d'exemples pratiques*, is a representative example. Lossier was the director of the horological school in Besançon. His text was based on articles by Jules Grossmann (1829–1907), director of the horological school in Le Locle, published in the *Deutsche Uhrmacher Zeitung*.[52] Lossier's French translation simplified Grossmann's text by removing his complex mathematics. The approach was very successful and Lossier's work was translated back into German by L. Loeske.[53] Several German, English, and French editions were published soon after. Practical work for the adjustment of watches is discussed later in the section on 'Working at the bench'.

The best example of how systematic horological research can provide excellent results is perhaps that of Sigmund Riefler (1847–1912).[54] He was not a trained clockmaker but a brilliant engineer with strong analytical skills. Although the accuracy of precision pendulum clocks had already reached a high degree of maturity around 1880, Riefler's ambition was to perfect them further. He studied the still existing weak points and dramatically improved the accuracy of precision pendulum clocks within a single decade. This was only possible through radical new approaches to escapements, compensation pendulums, gearing, and winding. Around 1900, his clocks were considered to be the most accurate timepieces in the world and received numerous prizes. The study of Sigmund Riefler's publications is worthwhile, as it impressively demonstrates his analytical way of thinking.[55] He had familiarized himself with all the proposals for technical improvements in horology since the eighteenth century and successively found new concepts for the various parts of a precision timekeeper utilizing the latest innovations, such as Invar, which he adopted in 1897 for the pendulum rod only one year after Guillaume's publication of this new material. Thereafter, with theory and techniques at a high level, scientific work in the twentieth century concentrated on comparative analysis. A typical example of this is the work of Charles Gros, who made a comprehensive examination of the various escapements commonly used in clocks and watches, as well as escapements with constant force and with remontoirs.[56]

EDUCATING: HOROLOGICAL SCHOOLBOOKS

Towards the end of the eighteenth century, mechanical principles for the construction of clocks or watches were already quite mature: a necessary condition for the transition from manual work to nineteenth-century industrial mass production had thus been fulfilled. Mass production, however, still required skilled technicians as well as able repairers, suppliers of spare parts, dealers, and retailers. To meet the need, horological schools were founded in

---

51 Baillie unpublished.

52 Vols 6 (1882) & 7 (1883).
53 Loeske 1892.
54 See Chapter 19.
55 Riefler 1894; 1907; 1911.
56 Gros 1913.

Switzerland from 1824 onwards, followed by France, Germany, Britain, and the United States, with a special need for suitable textbooks setting out basic principles for apprentices, rather than the latest inventions. Such works provided the theoretical and practical knowledge needed for successful training. They were offered in a compact form as a catechism or professional handbook and compendium. The authors of this very specific literature were nearly always experienced directors of, or teachers in, technical schools and were dissatisfied with available publications. The textbooks also had to be continuously expanded and adapted to take account of changing work methods, new machines, and tools, as well as new technologies, such as the use of electricity in horology. They lasted until well into the middle of the twentieth century.

The large volume by Saunier (already mentioned) was the first of this new category of educational books. It was sufficiently mature when it appeared in 1869 that it became a standard, leaving little need for further new approaches in France. In other countries, however, various textbooks appeared after 1850. The publications were supported by the horological journals founded from 1858 (in Britain the *Horological Journal* was first published that year), and the horological societies. An early example of such works is the *Leitfaden für Uhrmacherlehrlinge* by Hermann Sievert (1845–98), which was awarded a prize in 1879. The publication filled an urgent need and is one of the most important of the German horological reference works. A total of fourteen editions were produced over six decades. Its success depended on the basic idea of giving instructions and explanations for the apprentice without previous training—something that earlier horological textbooks did not attempt. Thanks to the comprehensible language and the explanatory illustrations, the work rapidly became a recognized manual, well-suited for the apprentice to prepare for examinations with the help of 289 technical questions, which also included the correct answers. The active support of the German horological association certainly contributed to the success of the book. Pendulum clocks, pocket-watches, wristwatches, and even the first electric watches were all treated. All necessary maintenance and repair work, as well as adjustment settings, are discussed in detail. The work is supplemented by a short section on drawing, an introduction to trigonometric calculation, and tables of wheel and pinion sizes for pocket- and wristwatches, cylinder escapement, and pendulum lengths. Although the author died young, well-known specialists continued his work, which appeared in two revised French translations (1919 & 1924).

Another successful horological schoolbook was the *Katechismus der Uhrmacherkunst*. This was first published, with many illustrations, by H. Herrmann in Leipzig in 1863 with a second edition in 1874. The third edition was prepared by F. W. Rüffert in 1885, with a fourth in 1901. Baillie writes:[57] 'It forms an exceptionally good treatise on the subject. The chapter on toothed wheels, which forsakes the catechismal form, is one of the best ever written for horological work'. A fundamental textbook, *Die Uhrmacherlehre* by Julius Hanke, was published in 1911. Again the aim was to provide an apprentice with the necessary skills for his profession and to help him understand what he was doing. Only the repair and finishing of clocks and watches is taught since the realization of a single clock with all its parts no longer played a role. The whole content is given without any formulas or trigonometry, only with clear illustrations. Hanke's textbook can already be regarded as part of the practice of 'working at the bench'.

The discipline of technical drawing played a major role in the training of horological apprentices. The essential rules and dimensions to be observed in the design of the mechanical parts of a movement were taught step by step. For this purpose, a whole series of relevant textbooks with drawing instructions were published, for example, the works of Gustav Krumm, *Lehrgang für den Fachzeichen-Unterricht des Uhrmachers* in Leipzig in 1925, or Josef Linnartz, *Das Fachzeichnen des Uhrmachers* in Halle/Saale in 1938.

The British Horological Institute also supported educational publications. In 1885, its vice-president David Glasgow published *Watch and Clock Making*. This comprehensive treatise is a mixture of theory and (some) practice and provides the necessary basic knowledge on all important topics. These are the principles of trains, mainsprings, and fusees as well as the different types of escapements. Two short chapters on watch and chronometer adjusting, jewelling, and watch examination comprise the practical part of the book. Even after the Second World War new textbooks for clock- and watchmakers were created. The Chambre Suisse de l'Horlogerie, La Chaux-de-Fonds, published Defossez's *Théorie Générale de l'Horlogerie*. It is a comprehensive standard work. In a strictly analytical way all elements and functions of clocks and watches are treated. However, the textbook by Charles-Andre Reymondin and other authors published by the Federation des Écoles Techniques de Suisse, which is also available in German, English, and Spanish, is now used as the standard work for teaching in Swiss and German horological schools. With a modern layout, it offers a comprehensive presentation of the material in 368 pages and with some 940 coloured drawings and illustrations. It is also suitable for personal study.

This section of horological literature would be incomplete without mention of George Daniels' *Watchmaking* (1981), which has enjoyed several editions. This eminent maker, known for his high level of skill and knowledge, provides precise, practical information and excellent descriptions. Richard Watkins comments:

> This is a book of great value, being one of the very few which discusses watch making in detail, and the only significant English work since Glasgow, a hundred years earlier. After all, Daniels' [sic] is at the end of a long list of valuable books on watch construction and repair, and much of what he writes has been said before (although often not as well). Although written for the small band of people actually

---

57 Baillie unpublished.

wanting to hand make watches, it is of considerable value to repairers. Most of the 300 pages describe activities also undertaken by repairers.[58]

WORKING AT THE BENCH

Instructions for watch repair and error correction are an important addition to theory-oriented textbooks. Although handbooks for practical work with timekeepers have been published sporadically ever since the end of the eighteenth century, it was not until around 1880 that practice-oriented books were produced owing to the increased need for repair work on mass-produced clocks or watches. These books were written exclusively for watch and clock repairers and had to be easy to understand. In addition, they supplemented the courses taught at the horological schools. Similar publications appeared more or less simultaneously in all countries with a horological industry. It is typical that many of these publications were produced in several editions over a long period, showing the great demand for such literature. The object of these books was not to advocate any new or untried theory or to endorse any one opinion, but to tell the practical workman in plain words what the established practises were: the best ways known to the trade.

Although these books were mainly written to give advice for standard tasks at the bench, complex jobs also existed for which detailed instructions were necessary. The greatest challenge for the watchmaker at the bench in this respect was the adjustment of a watch in various positions and temperatures. Every high-precision watch needed such special treatment and the task required great skills and experience. For this delicate work, several high-quality texts were produced in Switzerland, Britain, the USA, and Germany, which complemented the basic instruction books. The chronological development of this category of horological literature is interesting.

The earliest such treatise for practitioners was written in Germany in 1779 by the pastor Christoph Wilhelm Forstmann (1736–83).[59] Although he was a passionate horologist it is his only book. He notes in the foreword that the existing literature lacked detailed information about pocket-watches and their problems Thanks to his extensive personal practice in watch repair and his talent for expressing things clearly, Forstmann's book is an outstanding early contribution to the repair of pocket-watches. He describes the tools and how to use them, ranging from the various types of files to the grinding of drills, work at the lathe, or the making of a dividing plate. His explanations of steel and brass and their treatment are also very practical and include soldering and chemical recipes for the treatment of surfaces. The central part of the book describes cleaning, repair work, and assembly, up to the adjustment of a watch in various positions. This section is supplemented by a cookbook type compilation for troubleshooting. The construction and repair of all possible types of repeating watches are also thoroughly explained.

Closely related to the horological work of Forstmann is that by Vigniaux, first published in 1788. For the first time, a practical manual was written both for apprentices and for the committed amateur. Thanks to its clear, understandable language and a good didactic structure, the aim was achieved. After an illustrated description of all the materials and tools needed for watchmaking, the production of a watch movement *en blanc* is described, including the escapement. The book concludes with notes on watch repair and also offers a glossary of terms used. A second edition was published in 1802.

Before watchmaking schools emerged in Europe, there was already a need for hands-on instruction specifically for the training of watchmakers in rural environments. Two examples are worth noting. In France in 1824, the watchmaker J. J. M. Ayasse from Angers (1774–1834) published his extensive *Manuel de l'apprenti horloger en Province*. The book, with 372 pages, consists of detailed workshop instructions, but there are no illustrations. It is noteworthy that the verge is the only escapement mentioned. For the same purpose, in Germany in 1827, Jacob Auch published a *Handbuch für Landuhrmacher*. Comprehensive and illustrated, it was successful and a third edition was published by Jürgen Meyer at Weimar in 1892. It is a handbook with step-by-step descriptions as to how to make a watch, take it apart and put it together, repair it, and regulate it. Primarily it was designed for apprentices and amateurs. The book contains very detailed workshop instructions for making every part of a watch. In this book, no escapement other than the verge is mentioned. Advanced watch escapement like the cylinder or the lever were still of no importance for the average watchmaker in the early nineteenth century.

In 1867, Erasmus Georgi produced the exhaustive and well-written treatise *Handbuch der Uhrmacherkunst*. It is a thorough guide to the manufacture and repair of all types of watches and includes instructions for making and repairing all types of timepieces. It still addresses the universal horologist capable of making a complete clock or watch with all its parts. Forty pages are devoted to turret clocks and their problems In Britain, Paul N. Hasluck (1854–1931) published his practice-oriented work *The Watch Jobber's Handybook* in 1887. The author wrote for the guidance of young beginners in their elementary practice of watchwork. A large portion of the work is devoted to an illustrated glossary of watchmakers' tools and machines in alphabetical order. It was very successful, with fifteen editions up to 1936. The author also published *The Clock Jobber's Handybook* in 1889. In his preface he suggests that 'a small amount of practice is sufficient to get the necessary manipulative skill. Thus clock jobbing offers an occupation easily acquired by those who have an aptitude for mechanical subjects and ... sufficient information is given to afford a guide to successful operations'. The advertising character sounds like a

---

58 Watkins 2011.
59 Forstmann 1779.

tempting promise to learn a new job in an easy way with the help of the book.

An impressive contrast to Hasluck is offered by the work of F. J. Britten, *On the Springing and Adjusting of Watches*, published in 1898. Britten does not deal with basic repair tasks—his book is focused on adjustment work—the greatest challenge for the watchmaker. The author honestly points out that his book is not suitable for beginners and the knowledge of many elementary facts is assumed. F. J. Garrard served an apprenticeship as a watchmaker and worked at the bench for the best years of his life. Later on he was adjuster of marine chronometers for the Admiralty. He wrote *Watch Repairing, Cleaning and Adjusting* (1903). In plain language and with 214 pages it covered the whole ground of watch repairing. The author addressed the workman, the apprentice, and the amateur. It was a very successful publication with a seventeenth edition in 1950. Another very successful English author was Donald de Carle who wrote a whole series of practical books. He published his first book, *The Watchmakers Lathe and how to use it*, in 1946. The inspiration for further publications came from Arthur Tremayne of the N.A.G. Press, who would publish all the later works of de Carle. Tremayne was convinced that only a book that did not disdain to explain even the simplest processes and operations could give the student full understanding. These principles were fully implemented in de Carle's *Practical Watch Repairing*, originally published as a series of articles in the *Horological Journal*. The first edition in book form appeared in 1946 and was immediately successful. Several English editions and also Italian and Spanish translations were produced. The processes of repairing and adjusting a modern watch are given in precise and meticulous detail. Taking nothing for granted, except the ability to read and to comprehend a simple description of a mechanical process, de Carle takes the reader through every stage and operation of watch repair. Twenty-three chapters with 553 accurate illustrations in 300 pages support the process.

Practical Clock Repairing was written by the same author and in the same style. It appeared in 1952 with a second edition in 1968. Again, Arthur Tremayne was godfather. He asked de Carle to write a book that dealt with clocks in exactly the same way—one that can be understood almost without reading it. For this reason it shows in every possible illustration the clockmaker's most important tools: his fingers! This is something completely new and especially interesting for laymen and hobby watchmakers. While the apprentice watchmaker usually learns from his master how to use the fingers correctly, for the untrained layman these illustrations were an excellent aid to correct work. De Carle continued the series of his books in 1956 with *Complicated Watches and Their Repair*, which discussed automatic watches, triple complicated watches, chronographs, calendar watches, and repeating watches. The action of each mechanism is briefly but clearly described, again with numerous illustrations. Finally, in 1964 de Carle published *Practical Watch Adjusting*. This was also welcomed by practitioners and a third edition was issued in 1977.

In the fast-rising horological industry of North America in the nineteenth century, instruction books were also needed for the many craftsmen who had not completed a thorough apprenticeship but were trained on the job. A whole series of books were written to meet the need in the second half of the century. A notable one among them, and probably the first work on technical horology to be written by a woman, Mary L. Booth's (1831–89) *New and Complete Clock and Watchmakers' Manual ...* was largely a translation of Magnier's edition of Lenormand & Janvier (1850) and was successful with several later editions.[60] In 1894 in Toronto, Canada, Charles Elgar Fritts published his comprehensive work *The Watch Adjuster's Manual: Being a Practical Guide for the Watch and Chronometer Adjuster in Making, Timing, Springing and Adjusting for Isochronism, Positions and Temperatures*, in which he shared his accumulated experiences and best practices. In 1904 the publisher of *The Keystone* in Philadelphia issued a third edition. The work, written with great care and thoroughness, was accurate and reliable. Theophilus Gribi wrote his compact but well-founded handbook *Practical Course in Adjusting* in 1891, which was published by *The Jewelers' Circular* in New York, and the *Journal Suisse d'Horlogerie* in La Chaux-de-Fonds. The English edition is based on lectures given to the American Horological Society in 1896–7 and presumably the French edition was derived from this. It is in three parts. The first deals with general principles for the adjustment of watches and chronometers, adjustment in positions, and compensation for temperature variation. The instructions are based on practical experiments conducted by Gribi. The second describes correct principles in the construction of watches and the practical work of adjusting. Important defects are also handled, as well as replacing balance springs, or forming terminal curves. Part three explains how to make a balance staff with modern tools, how to clean a watch properly, and is devoted to the lever escapement, its defects, and how to remedy them.

In addition to these publications, which were written for general use, some horological firms created specific manuals for the repair and servicing of their own movements. For example, in 1920, the Waltham Company in Massachusetts produced for its employees *Rules and Practice for Adjusting Watches* by Walter J. Kleinlein. The author had many years of experience in both factories and repair shops. For him, the time was past when an apprentice was to be taught the making of a complete watch. He even considered as overrated in 1920 the ability of a repairer to make new parts of a watch. The higher standard of timekeeping asked primarily for an understanding of the principles governing it and knowing the causes of variation. Therefore, his book presents the essential points of watch adjusting necessary for regulating and adjusting the better class of watch. Another excellent example of company-specific instructions is the 'Training units' issued by the

---

60 Booth 1860.

Bulova School of Watchmaking, founded in 1945, of which a thirteenth edition appeared in 2003. Richard Watkins commented on them that:

> These notes were written to support practical work done under supervision. Consequently many topics were not included because demonstrations to the students were sufficient. But what is included are the more difficult tasks of staff making and replacement, balance truing and poising, hairspringing, stem making, barrel repairs and escapement examination. The explanations are excellent with large, clear illustrations.[61]

In Germany, *Der Uhrmacher am Werktisch* by Wilhelm Schultz (1854–1921) achieved a success like no other publication before. It was first published in Berlin in 1902 by the *Deutsche Uhrmacher-Zeitung*. The fifth edition had already appeared by 1919 and the ninth in 1941. Schultz was the editor-in-chief of the *Deutsche Uhrmacher-Zeitung* in Berlin and had excellent theoretical and practical knowledge. He was asked to write a reference book for the watch repairer focusing on work at the bench. The idea was to treat the subject in a new way. For this reason, Schultz refrained completely from theoretical discussions, but described 1149 typical tasks that allow a repairer to amend a faulty pocket-watch. Until Schultz's book, there were no easily understandable publications available in Germany that supported the repairer without any theory. His book was not aimed at top technicians who carried out delicate precision work but was addressed to average watchmakers who were able to carry out standard work on watches quickly and inexpensively, but correctly. The various works are skilfully provided with useful drawings. The book was quickly translated into other languages: English, French (*L'Horloger à l'Etabli*), Danish, and Norwegian versions were available as early as 1919.

These manuals for practitioners, which became so popular from the end of the nineteenth century onwards, were constantly refined until around 1960. In 1924, Bruno Hillmann (1869–1928) presented his specialized work, *Die Reparatur komplizierter Uhren–Für den Selbstunterricht des Uhrmachers*. With 133 pages and 100 text figures, Hillmann offered a well-founded treatise for the repair of alarm clocks, calendar clocks, chronographs, and repeaters, ranging from verge watches to minute repeaters. French and English translations of this book also soon appeared. Because wristwatches had recently become fashionable, in 1925 Hillmann also wrote *Die Armbanduhr–Ihr Wesen und ihre Behandlung bei der Reparatur*.

The German watchmaker and instructor Hans Jendritzki (1907–96) was a successful author of numerous horological books and treatises in professional journals. He was distinguished by his great professional competence and his easy-to-understand presentation with text and illustrations. Some of his works were translated into ten different languages. His first success as an author in 1939 was his *Die Reparatur der Armbanduhr*, of which an eighth edition was already published in Halle/Saale by 1949. Very quickly this was followed in 1950 by the three parts of his practice-oriented *Werkstattwinke des Uhrmachers*. Even in the twenty-first century, English and French translations of his titles were produced.

In a similar manner, in Switzerland there was also a great demand for instruction manuals for the many watchmakers working at the bench in twentieth-century factories adjusting fine watches. For this purpose, in 1912 C. Billeter published *Le Réglage de Précision—Cours pratique et éléments théoretique*, Bienne, which had a second edition in 1921. Also the *Cours Élémentaire de Réglage destinée aux horlogers* by Eugène Jaquet and Leopold Defossez was used as the standard work for this delicate task. It was published in 1923 by the Technicum Neuchâtelois, Le Locle, and a fifth edition appeared in 1958. Meanwhile, the Technicum had published in 1928 a *Cours de réglage destinée aux régleuses . . .* .[62] Two late comprehensive books reflect the final phase in the development of the mechanical wristwatch. These are A. Helwig and Karl Giebel, *Die Feinstellung der Uhren. Ein Anleitungs- und Nachschlagewerk in zwei Teilen*, Berlin, 1952, and the instructive work of Emil Unterwagner published by the Neue Uhrmacher-Zeitung, *Die Feinstellung der Kleinuhren*, Ulm, 1949. Even more recently, some brilliant practical books have been written for watchmakers. A good example is *The modern watchmakers lathe and how to use it* by A. Perkins, published in 2003 by the American Watchmakers-Clockmakers Institute. The book provides detailed instructions for using the lathe to make parts, and covers pivots, balance staffs, stems, jewel bezels, repivoting, wheel and pinion cutters, and making wheels and pinions.

INSTRUCTING USERS

Many of the books on mechanical watches in the eighteenth and early nineteenth centuries were not written for horologists but address the watch owner, and Johannes Graf has done detailed research on this category of literature.[63] These mostly short and barely illustrated publications are intended to provide watch owners with what they need to know to handle a watch. They contain instructions for its proper care and regulation, the determination of time, and instructions for simple repairs. Because of their plain design and large circulation, such works were considerably cheaper than the comprehensive textbooks intended for professionals. This was certainly one of the reasons for their success. In addition, a watch was a sensitive and expensive item at that time, but of uncertain reliability. A small guide for the everyday use of timepieces was therefore almost mandatory for the owner of a watch. These publications often appeared in several editions and were translated into other languages. An important aspect of these brochures was also that they offered guidance before the purchase of a watch, whence amateurs could inform themselves

---

61 Watkins 2011.

62 Villeumier 1928, later editions 1944 and 1947.
63 Graf 2010.

about various quality grades of watches that were not immediately recognizable to them and thus understand price differences.

The first instructions for watch owners were published by Henry Sully in 1711 in a short pamphlet of only twenty-four pages.[64] It was clearly useful, and a substantially extended publication (with a different title) was published in 1714 in Vienna with 114 pages; further editions followed in 1717 and 1737. The great success of this work can also be seen in the German translations published in 1716, 1746, and 1754. Sully's close associate, Julien le Roy, published a similar pamphlet in 1719,[65] and Le Roy also verified what Allexandre wrote on this subject following Sully.[66] Of Julien le Roy's guide, two German translations of 1743 and 1759 exist. Even more successful was the guide written by Ferdinand Berthoud, 'for those who have no knowledge of horology to understand how to manage and regulate clocks and watches'. The first edition was published in Paris in 1759,[67] and reached a sixth edition in 1831, with translations published in Germany, Italy, Spain, and the Netherlands, each with several editions. Wilhelm Forstmann, however, went far beyond a short guide.[68] He wrote a complete treatise on ordinary and striking watches with more than 500 pages 'for those whose profession is the pen rather than the file'. The owner of a watch receives full instructions as to how to take it apart, clean it, and put it back together. An exhaustive list of defects that may occur is also included, with instructions for resolving them. This is one of the first publications dealing with these practical operations.

Even in the nineteenth century, there was still a constant need among clock and watch owners for easy-to-understand instructions and general information. Johann H. M. Poppe (1776–1854) was successful because of his clear style,[69] and an enlarged second edition of his work was published in 1822. A short, illustrated paper by Edward John Dent (1790–1853)[70] includes abstracts from two lectures on the construction and management of chronometers, watches, and clocks. The purpose was to spread knowledge needed for the purchase and subsequent use of watches. The paper gives a short account of the verge, cylinder, lever, duplex, and chronometer escapements. The author also compares Swiss and English watches for their quality, and a German translation was published by Littrow in Vienna in 1843.

In 1842 Gustav Jahn also published a work written for laymen.[71] The author recommends his book especially to all those who need a good knowledge of time such as landowners, clergymen, and country school teachers. Baillie notes:

This is an exceptionally good book on the determination of time by observation of the Sun and of the stars. . . . The use of both primitive and complete forms of sextant and of transit telescopes is explained. Every method is set out clearly with the help of examples. A valuable handbook for clockmakers and amateurs of astronomy.[72]

Also interesting is a small brochure of twenty-eight pages by J. H. Martens.[73] His instructions for 'the treatment of the pocket watch and how to regulate it' offered nothing new in terms of content, but the author included an advertisement for his own watches. This is an early example of the changing character of these brochures around 1900. They were a follow-on from the nineteenth-century practice among some retailers (e.g. the London retailers Kendal & Dent, and Adam Thompson) to produce small books on the history of clocks and watches to interest their customers. Now watch companies began offering their own brochures to inform customers about the differences between a cheap and a high-quality watch. The main motive was to promote their own products. A typical example of this type of information is the 1922 Waltham brochure *Your Watch and what you should know about it*. First, impressive figures are given to illustrate the daily workload of a pocket watch, followed by sections on buying a good watch—Waltham, inevitably, being the best choice. Only after this the essential parts of a watch are illustrated, such as the mainspring, balance-staff, jewels, and cleaning, magnetism, and adjustment are discussed.

The horological associations and national cooperatives also had a great interest in the education of their customers. Brochures were printed for the horological trade and distributed free of charge to customers. The success of this idea is shown by the guide published by the *Deutsche Uhrmacher-Zeitung*. It appeared for the first time in 1900 with a circulation of 100,000 copies.[74] This edition was sold out after only two years. In 1926 the tenth edition was printed with forty-seven pages. At this time translations already existed in ten languages, including Portuguese, Norwegian, and Russian. In this brochure the customer is first introduced to a watch as a masterpiece of mechanical art and it describes the differences in quality of watches. The main part informs the owner of a watch about how to wind and set it, about maintenance, humidity, mainsprings, warranty, and magnetism. It covers not only pocket-watches, but wristwatches, as well as various types of pendulum clocks and striking mechanisms. The descriptions were supported by illustrations. This guide was certainly a great success because the publisher was absolutely neutral–no manufacturer is mentioned in the whole brochure.

---

64 Sully 1711.
65 Le Roy 1719, second edition in 1741.
66 Allexandre 1734, 251–3.
67 Berthoud 1759.
68 Forstmann 1779.
69 Poppe 1818.
70 Dent 1841.
71 Jahn 1842.

72 Baillie unpublished.
73 Martens 1866.
74 Schultz 1900.

HOROLOGICAL JOURNALS

Why are horological journals so important? Primarily, because the content of these journals is an authentic reflection of the time in which the contributions were written. In any book of historical content the author presents his personal view about past events and persons and their achievements. A journal, on the other hand, provides vivid access to historical persons and their concerns at the very moment of its publication. It is an unvarnished primary source invaluable for any researcher.

Scientific and technical journals like the *Philosophical Transactions of the Royal Society* or the more general *Journal des Sçavans* date back to the later seventeenth century, appearing shortly after the first scientific societies such as the Royal Society in London (1660), and the Académie Royale des Sciences in Paris (1666 and 1699), on whose activities they drew for much of their content. The first German Academy, the Leopoldina, founded in 1652, also oriented towards natural sciences, started its own scientific journal *Nova Acta Leopoldina* in 1670. Technical and engineering studies appeared in such publications but the rapid development of industry from the mid-eighteenth century onwards led to the appearance of some journals more exclusively devoted to engineering and new commercial manufactures appearing. Similar to the *Transactions* of the English Society for the Encouragement of Arts, the *Journal for Factory, Manufacture, Act and Fashion* was published monthly in Leipzig from 1791 onwards and reported, among other things, on novel clocks and watches. It was aimed at the 'patriotic-thinking part of the bourgeois society as well as scholars, merchants and manufacturers'. The latter were asked to report on innovations in manufacturing industry and on new products. The articles submitted were an interesting mixture of technical information, advertisement, and price information.

In 1801, the Société d'Encouragement pour l'Industrie Nationale was founded in Paris by a high-ranking group of persons. In harmony with the Enlightenment, its goal was to aid economic progress through technical innovation and encouragement of the 'useful arts'. The monthly bulletin of the society ran from 1802 to 1943. In 1832 the *Annales de la Société Polytechnique* started in France with similar aims, and in 1809 the *Quarterly Journal of Science, Literature and the Arts* was established in London. To a large extent, the authors were associated with the Royal Institution of Great Britain, a London-based organization founded in 1799 and devoted to scientific education and research. Its purpose was 'diffusing the knowledge, and facilitating the general introduction of useful mechanical inventions and improvements; and for teaching, by courses of philosophical lectures and experiments, the application of science to the common purposes of life'.

Inspired by the success of the French *Encyclopédie*, the German chemist and manufacturer Johann Gottfried Dingler founded his *Polytechnische Journal* in 1820. It was the most important technical journal in Germany until 1855. Many of the contributions were taken from British periodicals. They were translated into German within a few weeks. But independent articles by various authors were also published. Dingler's aim was to create benefit through the 'spread of polytechnic knowledge' and to present the international results of technical research 'well-arranged for the instruction and practical use of all people'.

All these journals covered the entire field of mechanical engineering. They took into account mathematical and scientific fundamentals and addressed the knowledge and skills necessary for proper operation of the various trades and arts. Because of the comprehensive nature of these periodicals, they contained horological contributions from their very beginning. Their monopoly on technical periodical publication, however, gradually eroded around 1850 with the transition from the early phase of industrialization to the massive industrial growth of the second half of the nineteenth century. Specialization, expansion, and the increasing independence of the various technical disciplines led to the development of the first journals devoted to specific trades or professions.[75]

What was probably the first horological journal anywhere began in 1841 in Germany. The entrepreneurial Leipzig bookseller Carl Gottlob Schmidt published the *Zeitung fuer Uhrmacher*. He stated his aims in the preface to the first issue: 'To report everything new, which is of importance for horology. All new inventions, improvements etc., which have been made in Germany, Britain, France, Switzerland, America or elsewhere in the manufacture of watches, are communicated therein'. The second focus of the journal was competition: 'Nothing is more necessary now than to keep an eye on others who produce the same thing'. Finally, the practical use of the journal is emphasized: 'Our focus will be primarily on the fact that the newspaper gets practical value'. But in spite of this promising concept the third issue of the journal, published in 1844, already showed symptoms of fatigue, and thus the world's first attempt at a watchmakers' journal failed after three issues.

Both the timing and the basic concept of this journal seemed perfect. Germany had not much to offer in the horological area c.1840. Therefore, the extracting from foreign publications of interesting inventions and patents was a reasonable idea. One can see this as a continuation of the exploitation of foreign horological textbooks and their translation continuously practised in Germany since the eighteenth century. A horological journal could thus provide a cost-effective, convenient, and efficient insight into development and progress at home and abroad. It aimed at a broad audience and sought the connection of technical science and bourgeois readers. Why then did it fail? The reason was simply the early launching of this journal. No industrial centre for clock or watch production yet existed in Germany. Only in the Black Forest were clocks being produced on a large scale, but mainly in a small domestic production. By contrast, watches were already produced by French, Swiss, and English factories on an industrial scale. As a result, the local 'watchmaker' was degraded from a

---

75 For an illustration of this development in France in the third quarter of the century, see the technical journals listed by Hattin 1868, 592–5.

manufacturer to a mere repairer. This influenced the education of future clock- and watchmakers resulting in lower quality. A broad interest in horological progress made by theory and science was hard to find. Therefore, the average clock- or watchmaker was unsuitable as the target reader for a new demanding journal. He was fighting to survive commercially in a rapidly changing environment. He was unlikely to be motivated to read the high-quality contributions of a new horological journal.

After this journal failed, the Weimar publisher Bernhard Friedrich Voigt immediately started another in 1844, the *Zeitschrift fuer Groß- und Kleinuhrmacher jeder Gattung*. This journal had some success and was published continuously until 1855. The contributions were illustrated with numerous lithographic plates in folio format. Between two and three issues were published per year. It seems to be the earliest journal for horologists that was printed continuously over a reasonably long period. Again, the concept was convincing:

> to provide information for clock—and watchmakers regarding the latest inventions and improvements, machines and tools for horologists. This will be performed in a cost effective and efficient way, partly by publishing original contributions, partly by collecting the horological information that can be found in German, French and English periodicals of all kinds. Substantial financial funds are available for the latter part but the editor believes that the full advantage of the journal will only be reached when the many invitations sent out for the submission of original papers have been successful.[75a]

The editor of the journal succeeded in motivating renowned horologists to contribute, resulting in articles by Urban Jürgensen, Louis Breguet, P. Gaudron, Matthäus Hipp, Robert-Houdin, Adolph Poppe, Chr. Schwilgué, and J. Wagner. The index covers more than 380 entries dealing with clock and watch movements, marine chronometers, precision pendulum clocks, and tower clocks, as well as sundials and (of course) tools, techniques, and best practise for horologists. In addition to the original contributions, the journal regularly evaluated a broad spectrum of other technical periodicals and newspapers searching for news of all kinds. Apart from *Dingler's Polytechnisches Journal*, there are, for example, the *Bulletin de la Société d'Encouragement*, *The Practical Mechanics Journal*, *The Architect*, and *The Illustrated Newspaper*, among others. Primarily, it was financial problems that caused the failure of the journal, stemming from the low number of subscribers and the complete lack of advertisements. The importance of commercial advertising would be recognized as a major source of income in the next generation of horological journals.

The next attempt came from Paris in 1851, where Pierre Dubois started *La Tribune Chronométrique Scientifique et Biographique, à L'Usage des Membres dé la Corporation des Horlogers*. The publication was highly appreciated by Paris horologists and the first issue is full of welcome messages sent to the editor. But after only one year Dubois ceased publication from lack of support. This ended the short period of early horological journals issued by individual entrepreneurs without the support of established horological societies. Even so, despite the negative experience of Pierre Dubois and the failed German attempts, Claudius Saunier dared in 1855 to publish the first issue of the *Revue Chronométrique*.

Claudius Saunier was already distinguished by this time, a former director of the horological school in Mâcon and the secretary of the well-established French Société des Horlogers. The latter function was crucial as it led the dominating group of Parisian horologists to support the project. The national orientation was clearly pointed out in the preface of the first issue, emphasizing the threat posed by foreign competition (specifically the USA). With great emotion all French experts were asked to contribute actively to perfect the art of horology. The appeal worked and the *Revue Chronométrique* continued publication even after the death of Saunier (1896) until 1910. All issues contained many advertisements safeguarding the financial situation of the editor. The journal offered a good mix of interesting topics: technology, escapements, tools, practical procedures, metalworking, electrical timekeepers, improvements, biographies, exhibition reports, and news from the association.

Soon after the start of the *Revue Chronométrique*, the *Horological Journal* of the British Horological Institute in London was created in 1858 and became another successful journal. The British society had already decided, soon after its foundation in 1858, that 'a Journal devoted to the interests of horology should be established to appear monthly'. It should provide an arena 'for the discussion of disputed points, whose settlement may possibly lead to a fusion of those numerous classes, into which the horological trades are at present divided'. These considerations were the cornerstone for the success of the *Horological Journal*, which still exists today, being now the longest running horological journal in the world.

Compared with the first generation of horological journals there are three essential differences:

1. The publisher was usually a horological society with a non-profit orientation and not a private business.
2. A substantial number of advertisements complemented the text providing a reliable source of income which considerably stabilized the financial situation of the journal.
3. The members of the horological corporations were loyal supporters of their association and guaranteed a stable basis for the journal by their subscriptions.

Based on these principles after a gap of nearly twenty years, the number of horological periodicals exploded. The following list

---

75a Zeitschrift fuer Groß- und Kleinuhrmacher jeder Gattung, vol.1 (1845)/1

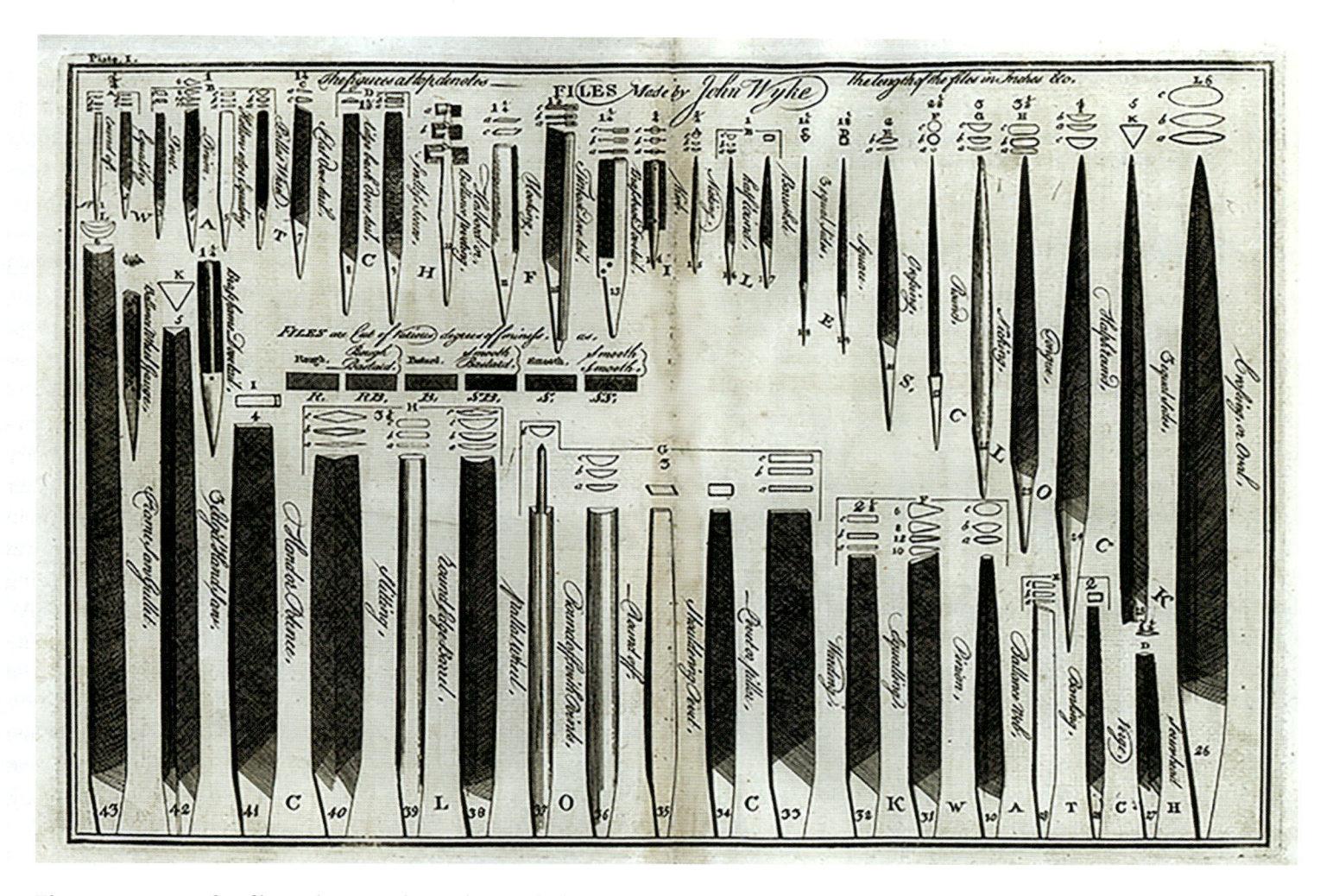

Figure 220 Forty-five files as shown in the catalogue of John Wyke, Prescot, c.1760, pl. 1.

summarizes major historical or still existing horological journals in chronological order:

Revue Chronométrique, (1855–1910), France

Horological Journal (1858–today), Britain

Journal Suisse d'Horlogerie et de Bijouterie (1876–1999), Switzerland

Allgemeines Journal der Uhrmacherkunst (1876–1943), Germany

Deutsche Uhrmacher-Zeitung (1877–1943), Germany

Schweizerische Uhrmacher-Zeitung (1878–1973), Switzerland

La suisse horlogère et revue internationale de l'horlogerie (1866–1983), Switzerland

Handels-Zeitung für die gesamte Uhren-Industrie/Leipziger Uhrmacher-Zeitung/Uhrmacherwoche, (1894–1943), Germany

Revue internationale de l'horlogerie et des branches qui s'y rattachent (1900–83), Switzerland

La France Horlogère. Revue universelle de l'Horlogerie, de la Bijouterie, de la Jouaillerie, de l'Orfévrerie, (1901–2008), France

Neue Uhrmacher-Zeitung (1947–93), Germany.[76]

When studying the nineteenth-century volumes of these journals the tremendous economic and social development that occurred between 1850 and 1900 in Europe becomes evident. Not only has a lengthy and rather diffuse way of expressing things been replaced by a prosaic and subject oriented style, but economic matters receive far more attention than previously and the horological societies as publishers act as strong political pressure groups to safeguard the interest of their members.

HOROLOGICAL TRADE CATALOGUES

Historical catalogues of horological manufacturers are often an indispensable aid for research. and so merit a discussion of their development.[77] John Wyke (1720–87), who was already established in 1746 as a manufacturer of watchmaking tools and watches in Prescot before he moved to Liverpool in 1761, played a major role in this development. He is regarded as the inventor of the modern trade catalogue. By 1758 at the latest, Wyke presented the

---

76 For a fuller, though summary, list of journals arranged by country see Tardy 1947/1980, 262–6.

77 For a fuller discussion see Crom 1989.

first copper engraved plates with exact illustrations of watchmaking tools. His catalogue, published around 1770, presented 450 tools shown with order numbers on 54 pages. Figure 220 shows an example.

Most of the tools in the catalogue are shown at full size. The price lists were not printed, but handwritten. This allowed a flexible distribution of selected sheets only and price adjustment for each customer or country. In addition, a retailer could show the catalogue to the end customer without him knowing the wholesale price. The advantages of such catalogues for the manufacturer were clear as they could be used universally for all business relationships: between tool manufacturer and wholesaler, between wholesaler and retailer, or between retailer and dealer. John Wyke visibly used his logo as a promotional element on every engraving in his catalogue. That was against the interests of a wholesaler or retailer because they wanted to prevent a direct contact between their customers and the manufacturer—a conflict still well known today.

The copper engravings necessary for these catalogues were expensive. Alan Smith suspects that John Wyke initially sold his catalogues only to selected English tool wholesalers and that a total of only about forty copies were printed.[78] Only three complete catalogues have survived.

The John Wyke catalogue probably set the model for future trade catalogues. William Whitmore already imitated it in 1778, producing his own tool catalogue. The catalogue of Peter Stubs appeared in 1801 and the copper plate tool illustrations of J. Smith & Sons, Sheffield in c.1825. In the second half of the nineteenth century, industrial exhibitions became popular. For these events, the exhibitors needed printed catalogues as advertising material, and this gave a strong impetus for the publication of catalogues.

In Germany, the oldest information about printed catalogues and brochures in the watch trade comes from the Black Forest. Hans-Heinrich Schmid[79] reports that printed sales information for clocks was sent to the USA in 1823. In 1849, Heine published a catalogue, with prices, of his company's products with watchmaking tools and accessories, ranging from the large lathe to small cylinder wheels as an appendix to his book.[80] It is probably the earliest German shop tool catalogue for watchmakers, but has no illustrations. Although full-size illustrations of clock designs had been produced for the use of French makers from the beginning of the nineteenth century, fully illustrated catalogues only gradually evolved after 1850. An example is the catalogue of the French tool shop of Auguste Moynet, Paris, c.1850. He is known since 1844 as a manufacturer of spare parts for verge and cylinder clocks. The first plate of the catalogue seems to be from this early period. With the growing product range more and more sheets were added.

The evolution of horological trade catalogues and the complete range of advertising alternatives can be followed closely in the brochures issued by the Lenzkirch company in the Black Forest. Founded in 1851, this industrial manufacturer developed successfully through the rest of the century and enjoyed a large export trade. Printed information with the full range of clocks offered in different countries was indispensable as a sales support, initially in France and later in Britain. The oldest catalogue in book form dates from 1855/57. It offers only sixteen clocks. All the illustrations were initially hand drawn and reproduced afterwards. A hand-written price sheet was loosely added. The standard technology for the illustrations in trade catalogues during the second half of the nineteenth century was wood-engraving, but photography-based techniques were soon used, becoming standard from about 1910 onwards.

During their first three decades Lenzkirch did not print annual catalogues, using instead the cheaper alternative of single sheets. For each clock model, Lenzkirch printed a small sheet on thin paper, which was left with the shop dealer as an order document. In the lower part of the sheet, the dealer could enter the required number of timepieces as well as the desired case design when placing his order. These individual sheets were also suited for the dealer to present his customers a selection of models. The salesmen were equipped with small Leporello (accordion) albums containing the latest and complete range of timepieces. Changes in the production range were simply communicated by pasting over existing entries. Thus the travelling salesman was always up to date. Other Black Forest manufacturers also used this system.

Large-format single-sheet prints were also a popular low-cost alternative to the bound catalogue. At a glance, they provided a comparative overview of the entire spectrum of a company. Until the 1890s, many horological manufacturers used such sheets as advantageous and cheap advertising material before fully fledged trade catalogues became standard about 1880. The size of these catalogues grew rather, partly issued even as hardcover editions with up to 400 pages. In parallel, the catalogue format continued to grow—for Junghans it had reached the size of a small folio by 1914.

After the Second World War, for cost reasons the leading horological companies stopped issuing catalogues with the full range of timepieces. Junghans published its last main catalogue in 1953 with 160 pages. Later on, only individual catalogues for the various types of timepieces (e.g. kitchen clocks, buffet clocks, alarm clocks, wristwatches, and so forth) were offered. These colour-printed catalogues on glossy paper could have between ten and seventy pages with texts in German, English, French, and Spanish. In addition, since the 1930s horological manufacturers printed modern flyers with selected products, which were given to retailers as advertising material for the potential customer.

At the latest with the release of a new catalogue, the previous one was superfluous and there was no reason to keep it. Therefore, outdated horological trade catalogues were usually thrown away and are rare to find today. On the other hand, there is a steady demand because illustrated trade catalogues are indispensable to determine the exact age of a specific timekeeper.

---

78 Smith 1978.
79 Schmid 2017.
80 Heine 1849.

Today, our digital world offers new possibilities for advertising with striking advantages over printed material. Online catalogues offer a photo quality that cannot be achieved in print. In addition, video sequences can be inserted without additional expenditure. Material, printing costs, and distribution expenses are saved. In addition, customers can be supplied promptly and comprehensively with press releases for every new timepiece launched. This trend is progressing rapidly and there are a number of horological manufacturers who no longer produce printed catalogues at all, but use only their websites for marketing.

HOROLOGICAL ALMANACS

Initially 'almanac' was used for astronomical ephemerides, which contained in chronological, or calendar-like, form predicted positions of the Sun, Moon, or planets associated with short comments. They retained their astronomical-astrological-calendar character well into the twentieth century. A horological example is by John Knapp, clock- and watchmaker in the city of Cork,[81] who in 1720 started to issue his annual *Almanack; or, Diary, astronomical, meteorological, and astrological*. It covered the eclipses of the luminaries and a table of sunrise and sunset for the city of Cork. The author also added 'some surprizing paradoxes, arithmetical, geometrical, and mechanical questions, enigma's &c. with many delightful particulars adapted for the use and diversion of both sexes. Also the method of the adjusting pendulum clocks and watches; with some remarks on our present astronomy'. This almanac was issued until at least 1733. A notable subsequent publication, entirely devoted to horological matters, was the *Etrennes chronométriques . . .*, published by Pierre Le Roy, which discussed time-measurement and its history, dialling, the way to choose and look after a watch, and accounts of some contemporary activity. Unfortunately, it had a limited success and ceased publication after only three issues (1759–61).[82] Nonetheless, it provided the basis for the almanac issued by Antide Janvier in the early nineteenth century as he attempted recover from bankruptcy.[83] His *Etrennes chronométrique* for 1811 appeared in two separate issues, and he reissued it under a new title in 1815 and 1821, the latter edition being significantly enlarged.

As with specialized journals and trade catalogues, it was only at the beginning of the nineteenth century that yearbooks specific to the subject began to emerge such as the *Almanach de la Fabrique de Paris . . . travaillant en Matiére d'Or, d'Argent et autres Meteaux* founded by Azur in Paris in 1803. In 1774, Johann E. Bode had founded the *Berlin Astronomical Yearbook*, the first almanac in Germany, to publish scientific contributions about time measurement. An example is the essay by Fr. von Zach 'On the accuracy of observations with English sea clocks', published in 1789.

Published in 1850, the first almanac fully intended for horologists was the initiative of a Parisian watchmaker Louis-Andre Borsendorff. The small-format booklet, with a caricature of a watchmaker on the cover, had the extravagant title of *La Loupe de L'Horloger–Almanac Astronomique, Historique, Scientifique, Comique, Prophetique, Critique, Pratique et Chantant*. It corresponded exactly to the taste of the time and provided a mixture of information and entertainment. The main contribution was a critical appraisal of the timepieces and tools on display in Paris at the French National Exhibition of 1849. Such contributions are interesting today because they express the opinion of the time and no other horological periodicals are available for this period. Only six issues of this almanac were published (1850–63). Already the third issue is fully focused on horology and reviews the horological products exhibited at the 1855 Paris World Fair. The almanac did not contain any advertising. Probably for this reason it had no economic success. Moreover, in 1859 Claudius Saunier had started a competing product, the *Annuaire Artistique et Historique des Horlogers*, also in a small format. This almanac had an impressive 144 pages, including an alphabetical list of Parisian watchmakers with a strong promotional character covering ten pages. It was published until 1869.

The next attempt was by the Swiss watchmaker Charles Gros with his *Almanach des Horlogers*. The first volume was published at St Imier in 1886. It appeared already with several pages of commercial advertisements, which certainly helped to secure its survival until 1913. With approximately 70–110 pages of technical articles, a bibliography, and some thirty pages of advertisements, these yearbooks provide detailed contemporary information. Gros outlines the motivation for his almanac in the preface of the first volume. He encourages his colleagues to concentrates their efforts on improving high-quality training for young people and the educational standard in the watchmaking schools. The aim of the *Almanach* was to disseminate information about new tools and efficient working methods. Gros wanted to gain an advantage over the semi-skilled factory workers in America. The background was the massive threat posed by the industrialized American watch companies to the still-traditional Swiss watch industry. It turned out at the 1876 World Fair in Philadelphia that America produced watches on large scale with maximum efficiency and sold them worldwide thanks to these attractive prices. This triggered a massive sales crisis in Switzerland.

In Germany, Moritz Grossmann in Glashütte published the first volume of his pocket-sized yearbook, *Note Calendar for Watchmakers*, in 1878. The focus was on high-quality articles for professionals. Grossmann's aim was 'to provide his colleagues with something useful, since such calendars had already proved themselves as reference works in many other trades, but for the watchmaker such a calendar was not yet available'. Thanks to high-quality, original contributions by recognized experts, the periodical was successful and appeared continuously until 1942. In parallel, Diebener in Leipzig started a calendar series in 1901 and continued this until 1997.

---

81 At 'The Dyal, St Peter's Church Lane' in 1724. Fennell 1963, 22.
82 Le Roy 1759–61.
83 Augarde & Ronfort 1998, 54. Janvier 1810, Janvier 1815, Janvier 1821.

In Switzerland, the *Annuaire de l'horlogerie Suisse* was a successful periodical published in Geneva between 1899 and 1977. It also covered jewellery and precision mechanics, so every issue comprised several hundreds of pages. Following constant inquiries from home and overseas, in Britain the British Horological Institute prepared a *Watch and Clock Year Book* in 1956. It offered some technical articles in every issue but is now mostly of interest for its advertisements and lists of suppliers. That it became a lasting success is uncertain, but it is at least traceable up to 1965. Another kind of yearbook, and an indispensable resource for horological historians, are the annual street directories. These offer comprehensive lists with the addresses of clock and watch manufacturers or suppliers of tools, machines, spare parts, shop supplies, etc. The oldest directory of this kind is *Azur*, founded in Paris in 1803. It was produced for over 200 years and covered the relevant trade in France, Monaco, and French overseas territories. In Switzerland, the oldest horological trade directory is the *Indicateur Davoine*. This publication, first produced in 1845 in Neuchâtel, serves as a reference work and contains an extensive directory of manufacturers from French-speaking Switzerland together with an extensive advertising section. Already the 1893 edition had 298 pages. The *Indicateur Suisse*, Bienne, is the third comprehensive directory, existing since 1913. In addition to the standard information, it publishes annually a huge list of all the horological trademarks registered in Switzerland. In 2018 the directory was acquired by the Swiss company Watch-Web. As for manufacturers catalogues, directories in the early twenty-first century have moved from print to a digital (virtual) form.

CHAPTER TWENTY SIX

COLLECTING AND WRITING THE HISTORY OF HOROLOGY

Anthony Turner

'History', the *Dictionary* of the French Academy tells us, is the 'narration of actions and things worthy of remembrance', although it can also designate 'All kinds of descriptions of natural things'.[1] In the hands of European antiquaries and naturalists describing the world around them in the seventeenth and eighteenth centuries, 'natural things' was extended to include the artefacts produced from them.[2] A sequential technical description of horological artefacts could therefore be considered as an history. So, too, could a chronological account of what was 'worthy of remembrance'. It was in this latter sense that Voltaire (1694–1778) approached the history of intellectual and material culture:

> Concerning the arts and sciences it is, I think, only needful, to trace the march of the human mind in philosophy, eloquence, poetry and criticism; to show the progress of painting, sculpture, and music; of jewelry, tapestry-making, glass-blowing, gold cloth-making and watch-making. On the way I want to depict only the geniuses who have excelled in these undertakings.[3]

This, a selective 'cherry-picking' procedure, does not recommend itself to modern approaches to the subject attempting to place the 'geniuses' in context, nor perhaps would it have been congenial to a contemporary like the Paris clockmaker Claude IV Raillard (1687–1762), whose aim in his treatise was to establish by whom, when, and where innovations in horology had been effected.

For doing this there was respectably an old, if somewhat empty, precedent. Already in his pioneering 1499 account of the origins of customs and things, Polydore Vergil (1470–1555) had devoted two short chapters to time.[4] The first treated of the divisions of the year and the months, and the second of hours, the dials, and clocks that divided them. Anaximenes is presented as the inventor of the sundial, while its use and lack of use in Rome is sketched mainly following Pliny; Ctesibios is seen as the inventor of water clocks that Scipio Nasica introduced to Rome. Afterwards, 'clocks made of metal were invented by ingenious minds, with toothed wheels and weights,...whose authors are as yet unknown'. Thin though his content is, Vergil's presentation of it set a literary pattern. Both Raillard and Jacques Allexandre (1653–1734) in the early eighteenth century would begin their accounts of horology, as would Saunier in the mid-nineteenth century,[5] by an account of the divisions of the year, the month, and the day, as Thomas Powell (1608–60) had also done in the mid-seventeenth century, although his account is more rambling, interspersed with moral reflections and an account of the *Cynocephalus*.[6]

With some seventy editions in Latin and over thirty-five in translation, Polydore Vergil's work was widely known. Like

---

1 *Dictionnaire*, 1771, i, 616.
2 For example, *naturalia* and *artificialia* were both described in the natural histories of Robert Plot (1640–96) and his successors in England.
3 Voltaire to Dubos, 30 October 1738.

4 Vergil 1499, bk ii, chs. iv, v. The work was expanded by a further five sections in 1521.
5 Saunier 1858.
6 The dog-headed ape, or baboon, associated with Thoth, was believed to urinate regularly, once every hour, during the day and night of the equinoxes. For this reason, the seated figure was placed over the outflow pipe of water clocks, the water issuing through his member. The story derives from the Greek *Hieroglyphica* of Horapollon the Younger (late fifth–early sixth century, a work that circulated widely in Latin translation during the Renaissance. *ODLA*, 2018, i, 741. Powell 1661, 1–3.

succeeding historical surveys of origins, such as those by Girolamo Cardano (1501–75) and Guido Pancirolli (1523–99)[7], it can be seen as a literary equivalent of the *Kunstkammer*, the 'Cabinet of Rarities', the 'Mirror of the World'.[8] Such collections had several social functions displaying curiosity, wealth, and virtue. Horological instruments could find a natural place within them. Certainly it so seemed to Henri Chesneau (? *c.*1610–post-60), when in the mid-seventeenth century he created a bogus inventory of the goods of the statesman Floribert Robertet (*c.*1460/65–1527), minister to Charles VIII, Louis XII, and François I. Chesneau's fake inventory, which was supposed to have been compiled by Robertet's widow, was part of a literary strategy to affirm and consolidate the social prestige and nobility of Charles, Marquis of Rostaing (1573–?), who was descended from Robertet through marriage in the female line and who, in 1633, had reacquired Robertet's château of Bury in the Loire valley.[9] Among the supposed possessions of Robertet, Chesneau included:

> Twelve watches (*monstres*),[10] seven of them striking, the other five mute in gold, silver and brass cases of different sizes; but amongst this number I [Robertet's widow] alone esteemed the large one made only of gilt copper (*cuivre*), that my spouse had had made, which showed all the heavenly bodies (*astres*), the signs [of the zodiac], and the celestial movements which he understood very well.[11]

Chesneau's purpose with this entry was to display Robertet's curiosity for a novelty, his culture in commissioning and understanding an astrological model, and his wealth. In inserting it, Chesneau displays a degree of historical understanding for watches in the first quarter of the sixteenth century in France that would indeed have been very rare, very expensive, and so exactly the kind of item to be included in a noble collection. Other horological items were included in verifiable collections of the period, a notable example being the mechanical nef (1585) in the collections of the Emperor Rudolph II, the complicated and decorative clocks in Kassel, and the outstanding pieces in the collections of the Dukes of Schleswig-Holstein in the Gottorp Palace.[12] These were all collections established in the later sixteenth century. They would survive, sometimes precariously, sometimes in part, occasionally augmented throughout the seventeenth century to become the subject of commentaries in the early eighteenth century. Then, as interest in the history of horology consolidated and attempts were made to survey it more fully, the timepieces contained in such collections would once again command attention.[13]

The first historical surveys of horology, however, remained in the earlier mould of a section in a wider account of the origins of things. More could now be said, thanks to the wider range of classical and post-Classical texts that the humanist endeavour of the Renaissance had made available. Thomas Powell in 1661 could cite from Macrobius, Cassiodorus, Vitruvius, and Plautus among the ancients, and from Busbecq, Kircher, Schott, even Gassendi among the moderns. His 'history' is meandering, slightly moralizing, but circumstantial. If Anaximenes is still the inventor of sundials, the rival claim of Berosus is advanced on the authority of Vitruvius; the lack of the division of the day into hours in Ancient Rome is likened to the similar lack among the modern Turks. Theodoric's desire of a clock to send to Gundibald of Burgundy in the fifth century is mentioned, as is Harūn al-Rashid's present of one to Charlemagne in the eighth, both being implicitly presented as weight- or spring-driven devices. These Powell then divides into striking and non-striking, admires their craftsmanship, especially those of miniature size, as well as that of the tools used to create them, 'no less admirable than the Engines themselves'. Several exceptional devices are then mentioned in an anecdotal ramble, timepiece-jewels, candle-lighting alarms, the Prague and Strasburg astronomical clocks, a saddle with moving eyes in the pommel and a clock behind driving them (Dresden), before he attempts to describe the cause of acceleration and retardation in clocks and watches. In the following chapter, Powell describes the spheres of Archimedes and of Sapor before turning to modern self-moving examples, in particular that of Cornelius Drebbel (1572–1634), described in some detail), and the works of Janello Torriano.[14]

Powell's inconsequential survey contrasts with the slightly earlier but more critical and pithy presentation of Sir Thomas Browne,[15] and even more markedly with that offered by William Derham (1657–1735) thirty-five years later. This distinguishes itself from any previous survey by being part of a practical, technical treatise entirely devoted to horology. For this, Derham had no models, for his two predecessors in the field of English books entirely devoted to horology by the clockmaker John Smith were primarily users' manuals devoid of any historical content. However, an example could have been offered by technical tracts on sundials, of which there is a longer tradition. Indeed, the very first of these in English, by Thomas Fale, began with a history of his subject.[16] This opened, as would Derham's history, with the dial of Achaz, 'the first dial that Histories remember . . . ', as Fale

---

7 Cardano 1557, ch. 47; Panciroli 1612, a Latin edition, prepared by his student Heinrich Salmuth, had appeared earlier in 1599.
8 On which, see Impey & McGregor 1985.
9 For Chesneau's forgery and its aims, see Herrara 2011.
10 Exactly what Chesneau intended by this term must remain doubtful, as he perhaps intended. *Monstre* could mean any kind of non-striking timepiece, but the indication of precious metals implies small, portable, items.
11 Translated from Grésy 1868. Although the authenticity of the text was questioned as early as 1869 (Herrara, 2011, 45, n. 18), Baillie 1951, 9, like some even later scholars, unfortunately accepted it as genuine.
12 For remarks on these and their context see Leopold 1995, 151–3.

13 See for example the account of items in the clock room in the Kunsthaus in Kassel given by John Carte in *c.*1710, Turner 2014a, 38–40.
14 Powell 1661, chs. 1 and 2. The book was successful with thirty-eight editions in the following three centuries.
15 Browne 1672, 301–2.
16 Fale 1593.

put it, or 'the earliest we read of', according to Derham. The latter however was expeditive: 'Some pretend to give a Description of this *Dial of Ahaz*: but it being meer guessing and little to my Purpose, I shall not trouble the Reader with the various Opinions about it'.[17]

This no-nonsense approach is typical of Derham. Two pages are allotted to sundials and water clocks in Antiquity on which he cites Favorinus, Censorinus, and Pliny, mentions the law-court water-timers,[18] the water clock of Scipio Nasica, and the earlier invention of sundials by Anaximenes. But the whole is rapidly dismissed: 'enough of these Ancient Time-Engines, which are not very much to my Purpose, being not pieces of Watch-work'. The distinction that would lead later writers to present sundials and water clocks as primitive forms of timepiece, fundamentally different from spring- and weight-driven pieces, is here already in place. Nonetheless, Derham has some doubts. He is unsure whether the machines of Dionysius and Sapor 'are pieces of Clock work or not', but partially contradicting what he has earlier said about water clocks, considers Ctesibios' machine as described by Vitruvius to be 'a piece of Watch Work, moved by an Equal influx of Water'.[19] It is notable that he here refers the reader to the recent French translation of it by Claude Perrault (1613–88) for an image.

Derham now moves towards his real subject, 'Watch and Clock-work', denying that it was invented in Germany less than 200 years earlier on the grounds that the sphere of Archimedes and that of Posidonius were both pieces of clockwork because they kept time with the heavens. After them, however, '*Barbarism* came on, and Arts and Sciences became neglected' until the sixteenth century, when 'Clock-work was revived, or wholly invented anew in *Germany*' (this because the oldest known pieces are German), 'but who was the Inventor, or in what Time, I cannot discover', although some think it was Boethius in 510. Derham's text here becomes confused and brief as he clearly has nothing to say. If it was not in Boethius' time, perhaps it was in Regiomontanus' time (said to be the end of the fourteenth century!), that clockwork was invented. Derham havers, but then proceeds to discuss the Hampton Court clock (1540), 'for its Antiquity and good Contrivance'.

As Derham moves into his own century, his account becomes more detailed. The possible origins of the pendulum are rehearsed. Astronomers had used them, but Huygens first applied them to clocks. The rival claims are aired but preference given to Huygens as a man of an integrity equal to his ingenuity. Development of the **anchor escapement** by either Clement or Hooke is recorded, some uses of pendulum clocks noted, and the proposal by Huygens of the circular pendulum with his claim duly disputed by Hooke. In the following chapter devoted to watches and the application of the **balance** spring, the technical matter is fuller to contrast clearly the claims of Hooke and Huygens, about which Derham makes no judgement. Thereafter, a brief, two-page, account of the origins of repeating clocks by Barlow and Quare (it being 'very frivolous to speak of the various Contrivances and Methods of Repeating work, and the Inventors of them'), brings Derham's history to a conclusion.

Derham's history is fuller than most that had gone before, but it still rests on a small group of well-known literary texts for what concerns Antiquity, has nothing to say about Medieval activity (the very existence of which is denied in these centuries of 'Barbarism'), and as the account gets more detailed in the modern period, it also becomes more Anglo-centric. Derham has done no original research except for the contemporary section, for which he had interrogated Robert Hooke and Tompion, 'the latter actually concerned in all or most of the late Inventions in Clock-work'. But the transformation of historical research and writing that had taken place during the later sixteenth and seventeenth centuries has passed him by.[20] The antiquaries' search for both objects and documents on which to base an account of the past, and the critical analysis of such materials and the assembling of them into a copious and truthful narrative, have no place in his approach, despite occasional sceptical reflexions and his wide acquaintance with ancient and contemporary natural philosophy. He is defensive about this in a conventional way, and the limitations imported by his nationalism also appear:

> I would have this little Treatise looked upon only as an Essay, which I hope will prompt some more able Undertaker to perform the Task better, especially in the Historical part. For since Watch-work oweth so much to our Age and Country, it is pity that it should not be remembered: especially when we cannot but lament the great Defect of History, about the Beginning and Improvements of this ingenious and useful Art.[21]

But for the clockmakers to whom Derham's book was partly addressed, no more needed to be said. When John Carte (c.1672–post-1715) settled to the writing of an extended technical and historical treatise on horology perhaps some ten years later, he simply copied out, with a few excisions, Derham's chapters 6–11, and his nationalism was equally strong.

Nevertheless, Carte did try to expand on Derham's account of the modern period, although in a disorganized way. He enters into precise technical details giving a clear description of the **stackfreed**, which he contrasts with the **fusee** and, relying on his own observations of old watches ('Besides the aforesaid I have met with another old Watch'), describes how to establish the age of a watch by its form. He then notes the improvement of watches in each country and in England, the latter getting

---

17 Derham 1696, 83. His view, however, did not trouble writers before or after him. For an introductory account of the subject with references, see Turner 1999.
18 See this volume, Chapter 1,
19 Derham 1734, 85; Vitruvius 1969, IX, viii. 4.

20 For which, in England, see Fussner 1962; for a European-wide survey Grafton 1997, chs. 5–6.
21 Derham 1696, a²v; Derham 1734 A⁶r.

twice as much space as the former.²² Carte, as already noted, was attentive to collections of timepieces. His description of that at Kassel shows that, in hereditary collections, old and new specimens were placed together, although in England Carte could have seen little. The 150-plus timepieces owned by Henry VIII had been dispersed, and those of his successors sold during the Commonwealth.²³ In Continental Europe, however, as at Kassel, some collections had remained nearly intact, while the systematic acquisition and preservation of exceptional clocks old or new had also begun. Olmi noted, when speaking of Manfredo Settala (1600–80), the Milanese collector and *amateur* of the sciences, that these 'were seen not so much as working instruments but as precious objets, to be appreciated more from an aesthetic than a practical point of view'.²⁴ Novel to many, they also represented up-to-date, fashionable and unusual objects stemming from modern technology.

Horological items constituted only one small section of Settala's vast collection of natural and technical objects. They had, therefore, to be striking. Much the same was the case for the collection of mechanical models created by Grollier de Servière in Lyons and by Brostrup de Schort in Kassel.²⁵ One of the main intentions of them all was to attract visitors for commercial and snobbish ends. All contained unusual clocks, although there was some overlap.²⁶ By the early eighteenth century, however, the emphasis on novelty and startling originality was giving way to a rational appreciation of ingeniously conceived and finely executed timepieces, masterpieces made following the 'rules of art'. As such, ten distinguished clocks would be found in the cabinet of Jean Baptiste Thomas, Marquis de Pange (1717–81).²⁷

As informed interest in contemporary advances in horology developed in the later seventeenth and eighteenth centuries, so, in parallel, it seems did interest in its earlier development. Derham, as noted, thought that a historical section would be of interest in the otherwise technical treatise that he intended for gentlemen and artisans, and Carte followed him. In France, there were similar developments. Marin Estienne (*fl.* 2ⁿᵈ half 17ᵗʰ century), about whom virtually nothing is known except that he was acquainted with the professor of mathematics at Caen University, Gilles-François Macé (1586–1637), and in 1668 and 1669 corresponded with Huygens,²⁸ wrote a short treatise 'Des machines de montres, & horloges. De leur invention et de leur progres et perfection où on les voit aujourd'hui', which he completed in 1693.²⁹ As he was also an engraver (a portrait of his co-citizen, the poet Jean Regnault de Segrais (1624–1701) is known),³⁰ he could show the layout of the new balance-spring watches with precise drawings of models by Martinot, Gribelin, and Thuret, all 'watches with a spiral [balance spring] improperly called pendulums' (fol. 42ʳ), as well as some rather more old-fashioned examples of his own (fols. 20ᵛ, 29ʳ, 30ʳ).

Estienne's is an early treatise displaying a modicum of artisan interest in the history of his subject and exceptional in being illustrated. In the following decades, interest in the subject developed on the part of both savants and artisans. The chief interest of savants lay in the divisions and measurement of time in antiquity since this posed interesting philological and critical problems; artisan explorers of the past of their subject wished more to show how present precision had been obtained. A notable learned study in the mid-seventeenth century was that of Leone Allacci (1586/7–1669), the erudite keeper of the Vatican Library. Among his successors was the equally erudite Claude Sallier (1685–1761), keeper of printed books in the Royal library in Paris in 1726–61, member of the Académie des Inscriptions et Belles Lettres in 1715, Professor of Hebrew in the Collège Royale in 1719, and elected to the Académie Française in 1729. He read a wide-ranging paper to the Académie des Inscriptions on 'the timepieces of the Ancients'. In it he surveys the extant literature, argues with his fellow scholars (particularly Saumaise) about the exact meaning of this or this Greek term, and gives full references. Sallier's is an essentially modern scholarly paper surveying and interpreting the whole known literature. What is lacking to him are vestiges—recovered examples of Greek and Roman dials and clocks. But the stream of archaeological discoveries that would transform the story of Ancient time-telling would not begin until some thirty or forty years later.³¹ At the end of the century, however, a notable, if internalist, study of Ancient dials would be offered by Delambre and the newly excavated examples subjected to close scrutiny.³²

There should be no rigid distinction between a learned tradition primarily concerned with Antiquity and developed by scholars (although Philippe de la Hire attempted some contemporary history)³³ and histories of horology produced by horologists themselves. Clockmakers could be cultivated men, and some were erudite. In a projected survey of horology that was primarily technical and intended to place his own innovations in an almost

---

22 Turner 2014a, 43

23 For Henry VIII's timepieces see Turner forthcoming; for what was dispersed during the Commonwealth, Gouk 1989, 392–7.

24 Olmi 1985, 12. For a general account of Settala's collection in scientific context see Tavernari 1976. In the contemporary catalogue of the collection by Scarabelli the timepieces occupy Ch VI. A summary of them is given by Baillie 1951, 74. For exceptional clocks in other European, mainly princely, collections see Impey & MacGregor 1985, 14, 230, 36, 42, 72, 82, 132, and 137.

25 For Grollier see Turner 2008; Turner 2016a. For de Schort, Monconys 1695, iv, 31–4.

26 Both Settala's and Grollier's cabinets, for example, contained examples of the **inclined plane clock.**

27 See Boileau 1781, lots 134–43.

28 *Œuvres*, letter numbers 1649, 1667, 1678, 1712, 1760, 1937, 1661, 1674, and 1759.

29 Musée des Beaux Arts, Caen, Collection Mancel MS 253. The text is signed on the final leaf, fol. 47ʳ on a small circular map of Caen.

30 Gombeux, 1936, 247.

31 Sallier 1716. For his career see Portes 2011.

32 Delambre 1814–27 and this volume Chapter 8.

33 Concerning the invention of the pendulum and the balance-spring. La Hire 1717, 80–2.

evolutionary sequence of improvement, Henry Sully (as Derham) included a section on the great names and inventions of the past. History for him, however, was largely descriptive, with no necessary chronological components and accounts of contemporary developments could be considered 'history'. It was the same for his friend and collaborator Julien Le Roy.[34] Their contemporary, Claude IV Raillard (1687–1762),[35] however, had both a wider view of his subject and an understanding of how to organize it. His intentions are best expressed by the long title of his unpublished treatise, which, in translation, reads:

> *An Historical and chronological treatise of the different divisions and distributions of night and day in, parts and in hours … ; of the manner in which different peoples begin the day and count the hours … : of the different kinds of timepieces, as much sciaterical as hydraulic, water-clocks and automata[36]: of the names of their inventors, of the time and the place where they were invented; and of their use and usefulness … .[37]*

Raillard's was a study of horology in its fullness from Antiquity to his own time. Parts of it had been delivered at four meetings of the Société des Arts, a body founded by Sully with the patronage of the Duc de Clermont to advance artisanal understanding of their crafts by uniting history and theory with their practice.[38] A member of this body from 1732 onwards, Raillard was, however, no admirer of its founder:

> If Mr Sully had been better historian than clockmaker, he would not have advanced this fact [that repeating clocks were invented in England towards the end of the reign of Charles II], but being neither the one nor the other, it is not astonishing that he is so often mistaken.[39]

Near contemporary events could already be subject to dispute—disputes coloured by a degree of nationalism.

Raillard was a cultivated man, acquainted with Greek and Latin, able to construct a literary work, and aware of the importance of the past. Alongside technical work, for example, his 1718 'Tables des Equations du mouvement du Soleil, et de L'acceleration des Etoiles fixes; avec leurs usages pour regler les montres ou pendules'

partially published by Thiout,[40] he also produced a fundamental calendar, *Extraits des principaux articles des statuts des maîtres horlogers de Paris* (1752), and beside this left a study 'Description des jettons des horlogers de Paris', as well as his History in manuscript.[41] Fundamentally, however, he was a compiler, not a critical historian. His account of the past of his subject is all-embracing:

Chapter 1: The different divisions of the day and might 'in parts and in hours' and the way different peoples divide it.
Chapter 2: The different constructions of sundials.
Chapter 3: Hydraulic clocks.
Chapter 4: Sand-glasses 'abusively called clepsydras'.
Chapter 5: 'Automatic clocks' (horloges automates), that is with springs or weights.
Chapter 6: 'The utility of clocks today', and of the superiority that the horological art should have over the other arts.

His account is also purely derivative as he avows in his preface: 'It will be very easy to see that there will be almost nothing of my own in this treatise'. He has cited the authors he quotes with precise references 'so as', he defensively adds, 'in this way to guarantee only my proper reflections'.[42] Unfortunately, Raillard reports without critical analysis what he finds in the texts he has read:

> Voelus, speaking of clocks in Chapter 1 of his book; the Venerable Bede, de diuisione temporum; and Clauis [Clavius], in the 1st book of his treatise of gnomonics place the invention of this art [dialling] before the time of the deluge, and believe that the glory should be given to the predecessors of Noah.[43]

If he notes that 'Pliny is very far from ascribing such high antiquity to sciaterical timepieces', it is only to use Achaz's dial to 'convict this historian of error'. This 'obliges us to believe … that it is probable that this invention is owed to one of the people of God, before the Deluge'. It thus predates knowledge of dials among the Babylonians, the Chaldaeans, and the Egyptians.

By moments, Raillard abandons all pretence at historical investigation. Advancing that sand-glasses are ancient, perhaps invented by Ctesibios, he adds 'whatever the case, as my idea is based only on conjecture, everyone is permitted to think as seems good to him, and to make the inventor of these clocks whomsoever he

---

34 See the 'historical' chapters in Leroy 1759–61.
35 The son of Pierre Labey and Madeleine Raillard, the daughter of Claude II Raillard, Rabey was apprenticed to his grandfather (Claude II Raillard) and adopted his patronym. He was received as a free master in the clockmakers' corporation 12 September 1700. Augarde 1996, 387.
36 As Raillard designated weight- and spring-driven time pieces.
37 Excerpt from a manuscript held in a private collection. The account that follows is based on a complete photostatic copy formerly in the library of Roberto Panicali and now in that of the author.
38 For the context and work of this group, see Bertucci 2017.
39 Raillard 1752, 206. An unpublished text from 1717, 'Des reflexions en forme de réponse sur le livre et la montre du Sr Sully', seems now to be lost. See also Augarde 1996, 308

40 Thiout 1741, 289–96.
41 The manuscript of the equation tables is now in the J.-C. Sabrier library in the offices of F. P. Journe, Geneva. The manuscript of the history is in a private collection in Italy. Both it and the 'Description des jettons … ' were once held in the collection of Jules Renée Olivier (1855–1933).
42 Raillard 1752, 8.
43 Raillard 1752, 60–1.

likes since no historian that I have read names him'.[44] Giving the name of the inventor, the place, and time preoccupy Raillard[45] and to find them he is as willing to cite modern, secondary (as they would be considered today) sources, as primary, contemporary testimony. Nor, as he advances into modern times, does he pay much attention to surviving clocks and watches as Carte had done, but continues to depend largely on published sources. Occasionally he cites contemporaries such as Isidore Guillaume Champion (1687–1765), who have actually seen old pieces, but he concludes his account of them lamely by saying that no more can be advanced 'because the historians say nothing'. The only exception is Cardan, but he is not very enlightening.[46]

Raillard's history is based solely on written, usually printed, sources, with no distinction made between primary and secondary accounts. It is also a progressive one. Clocks and watches especially in the century preceding him have been, as they still are, steadily advancing towards perfection. In tandem, his history becomes more detailed as it advances, but no more critical. The repeating clock is the finest, the most ingenious, and the most useful invention that has been made up to now, although, despite it being modern, its origins are unknown. They could not have been in England, as Sully affirms; rather, they most likely have a German origin and were introduced into France by Casimir after he had abdicated the throne of Poland in 1669. They were looked after by Simon Le Noir, according to his son Jean Baptiste Le Noir. Contemporary testimony can therefore play a part in Raillard's account, but this remains very partial, being anti-Huygens, anti-English, and anti-members of the Academy of Sciences. Huygens's cycloidal cheeks for Raillard were of no utility,[47] but while saluting Saurin's demonstration of this,[48] he is astonished that the members of the Academy had not realized it earlier. Clockmakers, on the contrary, 'without theory, only with the help of practice', realized this from the beginning. They never took any pains to make them with precision or according to the rules prescribed. If they put cycloidal cheeks in all their clocks it was only to content their customers.[49]

As he deals with the recent past and his own time, Raillard's text becomes increasingly a catalogue of recent inventions and a polemic. Like Carte, he has ideas about how to ensure the progress of horology, questions why there is no academy of horology as there is for the fine arts, and insists that neither Sully nor the Academy of sciences have advanced the subject, but rather it is by the clockmakers alone. They should, therefore, be supported by the state, and Raillard's account is intended to underpin this: 'I have begun by its antiquity, I continued with its utility, and I finish by the superiority that this art should have over all others'.[50] Raillard's history, the second, after Estienne to be entirely devoted to the history of the subject and to be written by a clockmaker, was one with a programme.

Although never published, Raillard's work was perhaps not unknown. If succeeding French clockmakers' treatises abandoned the historical narration in favour of the example of the latter part of Raillard's work, a simple semi-chronological presentation of new inventions,[51] it seems to have had some influence on Jacques Allexandre. He was an author from a different mould than Raillard. Allexandre was a learned Maurist,[52] a *savant* who spent forty years in the monastery of Bonne Nouvelle in Orléans where he wrote several works on mathematics, metallurgy, and the theory of the tides, although of these only the latter was published. Here also Allexandre thought and wrote about bell-making and horology, devoting himself in particular to devising and building an equation clock, although he also designed a large sundial for the garden of his Benedictine monastery, which was destroyed at the Revolution.[53] Cleric and *amateur* of horology, Allexandre resembled Derham in having a detailed knowledge of clockmaking, although not himself a professional of it. Like Derham also, a good deal of his book is concerned with the calculation of wheel-**trains** and finding the lengths of pendulums, but, like Derham again, Allexandre included a history of his subject. This began like that of Raillard with an account of the different divisions of the year, month, day, and hour among the ancients before surveying much of the same materials. Allexandre, however, was a closer, more critical, reader than Raillard. The clocks of Theodoric and Harūn al-Rashid he correctly designates as water clocks, he is doubtful that the spheres of Archimedes and Posidonius were weight-driven even if they employed wheelwork to simulate the celestial movements, and he was sceptical of claims that Pacificus in the mid-ninth century had invented an escapement, the honour of which he nonetheless ascribed to Gerbert (perhaps from pride in his order since Gerbert was also a Benedictine).

Allexandre's account of the Middle Ages is slightly fuller than that of Raillard, who leapt straight from Pacificus to Henri de

---

44 Raillard 1752, 104.
45 He is so obsessed by this that he specifically notes when he is unable to do it, e.g. 171, 176.
46 Raillard 1752, 169.
47 In this he was for once in agreement with Sully, who thought the same.
48 Saurin 1722.
49 Raillard 1752, 219–20. Earlier Raillard was equally dismissive of Varignon's work on the fusée 'because one can only have recourse, especially in small work, to the rules of mechanics.'.

50 Raillard 1752, 326.
51 Thiout 1741.
52 I.e. a Benedictine monk of the congregation of St Maur. The 'congregations' were groups of neighbouring monastic houses bound by a single code and under a single authority established as part of the reforms prescribed by the Council of Trent. The French Maurists had been established in 1612, the principal house being that of St Germain-des-Prés in Paris.
53 'M. Varignon a proposé de la part du P. Alexandre, une Pendule d'une construction nouvelle, qui suit le mouvement vrai du Soleil. On l'a trouvé fort ingénieuse'. *HARS*, ii, 341. *Cf.* Allexandre 1734, ch. iv. Cuissard 1897, 200.

Vic, in that he mentions Richard of Wallingford and the Courtrai clock removed to Dijon, these being 'the three oldest clocks that I find after Gerbert'.[54] Thereafter he passes, via the escapement, to Galileo and Huygens and modern clocks leading into the description of his own equation clock in order to have one that goes with the Sun, and so to the construction of clocks in general before ending with an annotated chronological bibliography that imparts further historical information. This is restricted to wheel clocks. For sundials and water clocks, Allexandre contents himself 'to give only the titles, because I put them in this treatise only to fulfill the idea of a complete treatise about timepieces'.[55] Although it is fuller, Allexandre's bibliography is very similar to Raillard's. Both follow a chronological order, and some entries are identical, Raillard's Latin note on Georg Hartmann being literally translated by Allexandre.[56] Unless Allexandre had knowledge of Raillard's work, they seem to be drawing on a common source. Like Raillard, Allexandre also has a high opinion of his subject: 'Horology which before was only treated as a mechanical art, will now be placed in the rank of the liberal arts, which it will not dishonour; and there it will not hold the least place, for Horology may be regarded as the master-piece of human invention'.[57]

The work of a well-informed and learned enthusiast, Allexandre's book is critical, carefully structured, detailed, and still of value. If the 'historical' sections of Julien Le Roy's edition of Sully are purely sequential accounts of contemporary work, in rewriting Sully's work in more structured and elegant form, Jean André Lepaute (1709–89) included a brief but full historical survey. In this he covers much the same ground as Allexandre but corrects his error concerning Gerbert, gives Richard of Wallingford's clock as the first 'of which history makes mention', mentions various well-known public clocks (Strasbourg, Anet, Lyon, among others), Passemant's sphere, and an automata clock at Versailles before moving on, in the wake of Sully, to supply the public with the information needed to look after and appreciate clocks.[58]

Although it thus emerged from an established style of writing, Lepaute's text also responded to the delight of fashionable society in the mid-eighteenth century for fine and novel horological products. The royal example of Louis XV in France and George III in England, both amateurs of the subject, filtered down through the court to nobles and wealthy bourgeois alike. Exceptional clocks, such as Passemant's sphere, could bring royal favour and courtly custom, outstanding pieces were the subject of speculation, displayed in commercial or private premises for a fee and/or for sale, and ambitious skilful craftsmen-entrepreneurs such as Charles Clay in London made extraordinary creations to tempt amateurs of horology, who were proud to acquire one or, like the Marquis de Pange, a whole, if small, collection of outstanding contemporary pieces.[59] Fascination as strong as this in polite society inevitably therefore led gentleman antiquaries to begin to examine the subject. A 'Discours on ancient and modern horology' was read by Borde to the Académie des Beaux Arts in Lyon on 4 May 1736.[60] An investigation notable for its range, learning, and completeness, as well as its errors, was that of Camille Falconet (1671–1762),[61] who tells us that the text of Jean Froissart's *l'horloge amoureuse* had been shown to Raillard, 'able clockmaker', who promises to give 'the explanation of all these old terms, comparing them with the new'.[62] He also cites the equal and unequal hour-showing clock made by Pierre Fardoil (d. post-1725) in the cabinet of Pajot d'Onsembray, the antiquity of horology here marching in step with the fascination for contemporary ingenuity.[63]

The fashion for horology in eighteenth-century Europe entailed that its development would be investigated in an age notable for the development of erudite, critical, document-based, history, which, at least in Falconet's view, should include the history of weapons, tools, the arts, and sciences.[64] Derivative accounts were given by Ferdinand Berthoud (1727–1807), based primarily on Allexandre and Derham,[65] and by Jaubert in his dictionary of the arts and crafts.[66] Daines Barrington (1727–1800) in England carried out more original textual research covering both England and continental Europe to arrive at the paradoxical conclusion that clocks were not 'excessively uncommon' nor yet 'generally used' in the Middle Ages. This he ascribes to the fact that they were at once imperfect, expensive, and of no great necessity.[67] While Barrington broke some new ground, a transformation in horological history was effected in Germany by

---

54 Allexandre 1734, 17.
55 Allexandre 1734, 26.
56 Raillard 1752, 18; Allexandre 1734, 278.
57 Allexandre 1734, 27.
58 Lepaute 1755, Preface.

59 Thus a 'singular and unique pendulum clock' built by Louis Charles Gallonde in 1740 that received an *approbation* from the Academy of Sciences was displayed by the procureur in Parlement le Febvre de Chantrainne at his house in the Cloître St Bénoit in 1758. *Description* 1758. For Clay see Jagger 1983, 77–81, Turner 2014b; for George III, Jagger 1983, 91–119, for Louis XV and the eighteenth-century French court, essays in Saule & Arminjon 2010. For the ten exceptional clocks in the cabinet of de Pange. See Boileau 1781, 37–42. One of them was the Phaeton longcase clock by Ferdinand Berthoud with case by Balthazar Lieutaud and bronzes by Philippe Caffieri now in Versailles. Augarde 1996, 266.
60 Académie des Sciences, Belles Lettres et Arts de Lyon, MS 182 ff 98, *et seq.*
61 Falconet 1745. It was perhaps this investigation that underlay the discourse on medieval horology included by La Curne de Sainte-Palaye in his unpublished Dictionnaire des antiquitez. Gossman 1968, 251.
62 Falconet 1745.
63 For Fardoil, see Augarde 1996, 313. The clock is now in the collections of the Observatoire de Paris
64 See the summary of Falconet's programme in Gossman 1968, 164–6.
65 Berthoud 1763, 1786, i, 34. His notice of the history of horology in the *Encyclopédie* of Diderot & Dalembert (vi, 1765, 299), however, is largely based on Falconet.
66 Jaubert 1773, i, 401–4.
67 Barrington 1778, 424–5.

Johann Beckmann (1739–1811) using the old mould of a general survey of the origins of inventions.[68] This was a densely documented survey that printed an unpublished paper by Hamberger read in Göttingen in 1758 augmented by the findings, with some corrections, of Daines Barrington and Beckmann's own contributions. Water clocks were dealt with in a separate chapter, but both reflect the critical methods now developing in German historiography generally, and are substantially accurate. Barrington's, Hamberger's, and Beckmann's work indeed form part of a wider movement of historical investigation of earlier, especially medieval, techniques,[69] although in the fourth English edition, prepared by Francis and Griffiths (1846), the account is brought forward anecdotally into the mid-nineteenth century.

Hamberger's and Beckmann's work placed the documentary history of horology on new and secure ground. They were followed by the young Johann Heinrich Poppe (1776–1854), whose short book set new standards of accuracy (although organized so traditionally that it mirrors that of Raillard).[70] Writing the history of their subject by clockmakers themselves reached an apotheosis in the vast, two-volume work of Berthoud in 1802. This, despite its greater scope, remained in the mould of Sully and Raillard in that, although offering a far more lucid and organized account of the technical development of clocks and watches, all this development was primarily seen as reaching its summit in Berthoud's own work. This is not to belittle the very substantial quantity of useful historical information contained in the book, but Berthoud wrote the history with a self-serving purpose.[71]

The writers of technical horological treatises would continue to incorporate brief introductory historical sections,[72] and objective (apparently) histories leading up to a celebration of the author's own work would continue to appear.[73] The development of collecting, however, would force a change in horological history. During the early nineteenth century, part of a rising interest in the arts of the Middle Ages and the Renaissance, everyday objects began to attract attention both in themselves and as possible models for contemporary craftsmen. Collections followed, among them collections of clocks and watches. Usually, these formed part of a far larger collection of everyday objects of all kinds, but their mechanical specificity gradually led to them being accorded a place apart. Already in the work of Pierre Dubois (1802–60),[74] a new ambition reveals itself—to write a chronological narrative technical history with historical problems discussed at the same level of sophistication as technical problems, the two interlocking to form a single whole and augmented by an examination of the corporative structures of the clockmakers and biographies of the major artists, this, as the publisher's presentation rightly noted, being a 'new and stimulating biography never before compiled'.[75]

Collecting *old* horology simply because it was old was developed by nineteenth-century *amateurs*. As specialized collections formed, details of the lives of the makers, and their social relations, became interesting, as did full, detailed descriptions of individual items. Very few individual watches or clocks had been described in pre-nineteenth-century literature, which had been primarily concerned with the generalities of their construction. Collecting changed this as collectors and antiquaries interacted. Octavius Morgan (1803–88) wrote to Sir Henry Ellis (1777–1869) in June 1848, 'On Thursday evening last, I took the liberty of laying on the table of the Society [of Antiquaries of London], for exhibition, a few ancient Watches which I had collected'.[76] This, he thought, would 'bring a new matter under the Society's consideration'. From this starting point Morgan developed a survey of the history of the watch. The same evening, W. H. Smyth (1788–1865) presented his detailed description of the astronomical clock by Jacob Zech (d. 1540) that belonged to the society.[77] Now such detailed studies would become more frequent, a model being provided by Dubois' catalogue of the watches in the collection of Prince Pierre Soltykoff (1804–89).

Typological and historical collecting of timepieces developed rapidly at the end of the nineteenth century and in the twentieth century. It derived in part from the application of historical evolutionist ideas as structuring frameworks for museums and collections,[78] in part from a growing interest in the details of historical timepieces on the part of horologists themselves who, as collectors increasingly demanded their services for the repair and maintenance of old specimens, sensed a new field for commercial operation which would range from repair and conservation to the creation of entirely new objects in 'antique' style (see Figure 159). Of this approach, the career of Matthieu Planchon (1842–1921) is exemplary, but such interest was not confined to mechanical timepieces. Although they now tended to be separated off from the former, sundials also benefitted from specialized attention, with some notable collections being formed by Lewis Evans (1853–1922), Max Elskamp (1862–1931), Joseph Drecker (1853–1931), Feliks (1872–1951), Tadeusz Przypkowski (1905–77), or Claude Basil Fry (1868–1942) within larger collections either of scientific or of horological instruments.[79]

---

68 Beckmann 1786–1805. First English edition 1797.
69 Of which a notable result was the publication of Theophilus' twelfth-century treatise on the practical arts. Dodwell 1961, 54–7.
70 Poppe 1799. Its chapters are itemized by Berthoud 1759/1805, ii, 429–30, who regretted that he had not been able to read it.
71 This, however, was common to historical writing of all kinds. See the discussion in Gossman 1968, 159–63.
72 Crespe 1804, but not well informed; Elève de Breguet 1827.
73 A notable example is Perron 1832.
74 Dubois 1849.

75 Lacroix 1849, iii.
76 Morgan 1848, 84. For Morgan as collector, see Thompson 2008.
77 Smyth 1848. *Cf.*, at the same period, Vulliamy's (1824) treatment of the movement of the Castlemaine Tompion, ch. 21, near n. 41.
78 On this see Skinner 1986.
79 For details of Elskamp and the context of his collecting see the essays in Frankinet *et al.* 2021.

Sundial collectors, however, were far outnumbered by clock and watch collectors. Whether professionals such as Planchon, fine connoisseurs like Courtenay A. Ilbert (1888–1956), or obsessive devotees like David Torrens (1897–1967),[80] their numbers increased rapidly.[81] The commercial interest of historical collections was also not neglected, that of the Denison Watch Case Company being exhibited at the Jewellers' Exhibition in London in 1913.[82] Inevitably, historical writing on horology adapted to these needs. The burgeoning numbers of professional horological journals included historical articles, and specialized studies began to appear. The notable work by F. J. Britten (1843–1913), that in its several editions became the *locus classicus* for the history of horology in the English-speaking world, began as anonymous articles contributed by Britten to the *Horological Journal*, of which he was editor, in response to the many queries that he received. These induced him 'to collate for publication, facts and information relating to the subject which I have been enabled to gather'.[83] A disorganized, rag-bag of a book, it nonetheless offered an exceptional list of early horologists with such scraps of information as could be found about them that provided a dating guide for generations of collectors alongside a general, but non-technical, survey of the whole sweep of horological history. Britten's book was perfectly adapted to the new market.

From the turn of the twentieth century, the history of horology exploded. To the antiquaries, savants, and clockmaker-historians who had provided most authors during the previous centuries, collectors themselves and a new breed of professional historians who distinguished themselves from 'mere antiquaries', were added. The first regional studies appeared,[84] an analytical study of the corporative structure of horological trade in a single city,[85] some general surveys,[86] and some specialized studies.[87] Catalogues of individual and institutional collections were produced, an especially notable example being Sir David Salomon's description of his own collection of Breguet,[88] while the Louvre dedicated a catalogue to the collection donated to it by Paul Garnier in 1916—a catalogue that was largely based on his own notes and descriptions. The foundations thus laid in the early part of the twentieth century have since carried an increasingly imposing edifice of detailed studies of an ever-increasing variety, while the grouping of themselves into societies of enthusiasts has also led to an efflorescence of research, as collectors (Figure 221), museum curators, clockmakers, and even professional historians have drawn closer together, mutually stimulating and informing each other.[89] At the same time, a technical literature specifically devoted to the repair and maintenance of early, historical items has developed, the lack of which was strongly felt earlier in the century when a learner's best resource was older clock repairers such as André Mongruel (1872–1960), who sought techniques in the earlier literature for the repairing of antique items, and exploited that literature commercially, by turning themselves into amateur historians.

Figure 221 The happy collector. An idealized image on the cover of an early magazine for collectors. Photo: Anthony Turner.

---

80 Torrens, an *amateur* in the best sense of the word, became a legendary figure for his knowledge in Britain in the first half of the twentieth century and although he wrote little, he had a strong influence on such diverse figures as Ilbert, Charles Allix (1921–2015), George Daniels (1926–2011), and Alan Smith (1925–2020). He was a pioneer of the collecting of horological tools. See Davies 1984, Froggat & Davies 1986.
81 See Chapuis 1942 who offers short biographies of some twenty collectors.
82 Dennison 1913. The collection included a late sixteenth-century Turkish watch by 'Nedjian', and two pieces by George Margetts.
83 Britten 1894, Preface.
84 Sandoz 1905; Develle 1917; Smith 1921; Vial & Côte 1927.
85 Babel 1916.
86 Saunier 1858; Havard n.d.; Franklin 1888b; Wood 1866; Benson 1875; Reverchon 1935.
87 Kendal 1892 and Baillie 1929 on watches; Perregaux & Perrot 1916 on automates; Ungerer 1926 on public clocks.
88 Salomons 1921.

---

89 The detailed knowledge accumulated in the journals of these societies such as *AFAHA*, *AH*, *Alte uhren*, *ANCAHA*, *Bulletin of the NAWCC*, *Chronometrophilia*, *Chronos*, *Clocks*, *DGC Yearbook*, *Klassik Uhren*, *La Voce di Hora*, and *Tijdschrift* probably far surpasses that presented in most general books and studies.

André Mongruel, who had been a working clock- and watch-maker in Paris, became increasingly specialized in antiquarian horology, and was charged with the collections of the Musée Carnavalet until his retirement in 1936. Thereafter, he lived in the provinces, at first in the Sarthe, later in an increasingly itinerant way as, after the death of his wife and tired of solitude, he rented lodgings with householders prepared to look after him in different parts of the country. During his early retirement he created for himself a new career as a writer and broadcaster on the history of horology, furnishing some 120 articles to professional reviews in France and elsewhere.[90] In these he was, as he said himself in a letter to Maurice Lesoive (d. 1966), merely a compiler:

> In all, I only re-edit what our fore-fathers formerly wrote—one invents nothing of the past—with documents one can make a compilation, all writers, novelists, historians, philosophers find their subjects only by the reference works which they possess or go to study in public libraries. One pulls it altogether and creates something new? already old![91]

Montgruel's attitude is the antithesis of that of more recent antiquaries and professional historians, who are constantly seeking new facts and new interpretations. Also new in the historiography of horology, as in many other subjects, is the increasingly important role of material evidence—that derived from the objects themselves. The study of surviving time artefacts from Antiquity has transformed what can be understood about that epoch, for which the documentary sources have been largely exhausted. If new documents for modern horology continue to be exhumed from hitherto little- or never-used sources such as newspaper files, ephemera,[92] literary correspondence, probate inventories, and other archival evidence,[93] the history that is written from them can never again be the same. Three-dimensional and two-dimensional evidence now both carry the same weight. The historiography of horology has been transformed, in part by the activity of collecting, and can never again be a purely literary exercise.

---

90 Most of these were written under his own name, but some were signed 'Sevigné' after the name of the street in Paris where he worked for forty years, and others were signed 'Gaillard', which was his mother's maiden name.

91 Translated, with permission, from a letter in a private collection.
92 On which, see Penney 2012.
93 For example, that of bankruptcy records, for which, in England, see Bryden 2018.

GLOSSARY

Many of the definitions given below have been derived, with permission, from Betts 2017. Others have been supplied by the authors, and some are adapted from Britten 1907, and Berner 1961, 1995.

Air brake: a device used to regulate the speed of a mechanical train. Usually referred to as the *fly*.

Alidade: a mobile sighting arm fitted with open or optical sights.

Altitude dials: sundials indicating time by the length of a shadow cast by the *gnomon*. This is a function of the height of the sun above the horizon. Compensation for the changing solar declination throughout the year has to be made when using them.

Anchor: an *escapement* with *pallets* resembling an anchor, though with the curved arms that carry the pallets turned downwards, rather than upwards; employed in various forms in combination with the pendulum from *c.*1670 onwards. In French horological terminology, however, an *ancre* also designates the watch escapement known in English as the *lever* escapement.

Android: an *automaton* made in the image of a human being and executing tasks which human beings commonly perform.

Analemmatic dials: a self-orienting instrument that combines a dial measuring time from the solar hour-arc (i.e. its position in the equinoctial) with an azimuth dial, in which the equinoctial circle is *orthographically* projected onto the plane of the horizon.

Aperture-gnomons or oculus: *gnomons* of a *sundial* in which instead of the edge or tip of the gnomon-shadow indicating the hour, a small hole in the centre of the shadow-caster allows time to be indicated by a luminous spot on the hour-scale.

Apsides: the points of greatest and least distance of a planetary body in its elliptical revolution around another body.

Arbors: the shaft or axle that carries rotating components such as wheels, *pinions* (often made in one piece with the arbor), or levers.

Aspect: the angular relation between the planets usually expressed as trine (30°), quadrature (45°), sextile (60°), quadrature (90°), and opposition (180°)—relations of importance in astrology.

Assortments: a complete set of parts that belong together as a *lever assortment* (*escape-wheel*, *pallets*, roller), or a *case assortment* (pendant, winding-button, bow).

Astrolabe: a model in two dimensions of the three-dimensional celestial sphere set in relation to the earth in which a movable volvelle can simulate the movement of the stars around the celestial pole and so supply by analogy a mechanical solution to various problems of calculation in astronomy and astrology. It is drawn in stereographic projection and is fitted with a sighting apparatus to enable it to be set and to effect basic measurements.

Astrolabe quadrant: a quarter of an astrolabe fitted with sighting apparatus primarily used for time determination.

Automaton: a self-operating mechanism, usually of animal or human form, that autonomously executes a pre-determined programme of actions.

Azimuth dials: see *direction dials*.

Babylonian hours: see Introduction.

Balance: an oscillator, usually in the form of a wheel (but see *Foliot*), that regulates the timekeeping of a clock or watch. See also *Balance wheel*.

Balance cock: a single-footed bracket that carries the top pivot of the balance.

Balance wheel: This term was used historically by watch and clockmakers when referring to the *escape wheel* of a balance-controlled clock or watch with a *verge escapement*.

Bassine: a form of watchcase that is rounded and completely smooth as if made of one piece.

Bellows: a bellows system for providing a source of air pressure, including feeders, an air reservoir, and valves to release the air and prevent it from leaking; used in mechanical organs and singing bird mechanisms

Bell top case: the top of a clock case with concave flaring sides. See also *inverted bell top*.

Bi-metal: a temperature sensitive combination of two different metals fixed together, of which the coefficient of expansion of the two is not the same resulting in a change of shape in different temperatures.

Blanc, blank, or ébauche: the basic, unfinished 'embryo' of the watch movement from which the finished movement is produced. The terms *blanc* and *ébauche* specifically refer to unfinished movements originating in continental Europe. The terms were not used by British makers who usually used the term *rough movement* to describe the equivalent partly finished movement.

Bob: see *Pendulum bob*.

Bombé: convex.

Brachiolus: an articulated arm, usually with two articulations.

Cadrature: under-dial striking or motion work.

Calibre: the specific design and size for a watch movement. In English watchmaking this was usually named 'caliper'.

Cam: a rotating disc whose profile is cut either on its periphery or face, to allow it to transmit a particular action, often part of a musical or mechanical programme.

Cartel: 1: an agreement between one or several companies for a specific operation, the companies retaining their judicial autonomy. A cartel is also created when it is agreed to establish an independent body to control competition between the parties to the agreement.[1] 2: term used to describe a form of French wall clock, usually in an ormolu Rococo-style case, popular from the mid-eighteenth century.

Cartelization: a voluntary process for limiting competition. It has three main features: shared governance, an influence on sales (through quota determination and price fixing), and geographical repartition or technical normalization. Cartels have a natural instability that shortens their existence.[2]

Cartonnier: an elaborate open filing cabinet for papers, dossiers, albums, and atlases, usually highly decorated.

Chalice dial: *sundials* in which the hour lines are drawn on the inside of a chalice or goblet.

Champlevé: a method of enamelling normally used on brass or copper in which the design to be enamelled was gouged out of the receiving surface and filled with the glass and oxides to be fused. The term also applies to metal watch dials of the late seventeenth and eighteenth century where the numerals are engraved in raised 'fields' (champlevé), the ground usually matted or chased.

Chamber master: a free master, but without a shop, who usually worked in his own premises as a sub-contractor for other clock or watchmakers.

Chatelaine: a chain of small, linked plaques depending from a larger plaque to which a flat hook was attached to pass behind the waist band. The watch was hung on a dependant swivel from the central chain, while shorter chains to the sides allowed keys, seals, or other trinkets to be attached. The term *chatelaine* seems not to have been used before 1828.[3]

Chiming: the audible feature in a clock movement by which the quarters of hours are sounded on bells or gongs. The simple sounding of the hour is properly always known as *striking*, not chiming. Chiming is usually achieved on a number of bells, the most familiar being the Westminster chimes (originally known as the Cambridge chimes, before being immortalized in the Great Clock at Westminster), which are sounded on four bells or gongs in a cycle of five four-note sequences. More elaborate forms employed six or eight bells playing a cycle of short tunes at each quarter. An earlier, simpler form, in which the quarters are sounded on two bells with a 'ting-tang' for each quarter being represented, is properly known as *quarter-striking*, rather than chiming.

Clepsydra: see *Water-clock*.

Clock: The term, deriving from its ancient origins, as a bell sounding device ('clocca', Latin for the sounding of a bell) was originally applied to mechanical devices that struck the hours (as opposed to the *Watch*, which did not).

Clock watch: a watch that strikes the hours in passing.

Comtoise: a low-cost, weight-driven long case clock typical of the Franche-Comté developed from the late seventeenth century onwards, ubiquitous in France and its colonies throughout the nineteenth and early twentieth centuries. They were also known as Morez and Morbier clocks.

Computus: the techniques deployed in the Christian churches for calculating the dates of the movable feasts, in particular that of Easter.

Contrate wheel: a wheel with teeth cut at right angles to its surface.

Count wheel: a circular plate with notches in its edge at distances corresponding with the number of hours to be struck. Sometimes referred to as a 'locking plate'.

Crown wheel: the *escape wheel* in a clock with verge escapement. In a watch or clock where the verge escapement is controlled by a balance, this wheel was termed the *balance wheel*.

Cubit: the length of the forearm from the point of the elbow to the fingertip, customarily standardized as eighteen inches.

Cylinder: a form of dead-beat escapement controlled by a balance. A hardened steel cylinder, forming the lower part of the balance staff, has an aperture through which the escape wheel teeth pass, delivering impulse as they do so.

Dead-beat escapement: In this the *escape wheel* remains stationary, in 'frictional rest', during the supplementary arc and only advances during impulse.

Decans: in origin, stars or groups of stars that rose at twelve regular intervals throughout the night and at ten-day intervals throughout the year. Used in Egypt to distinguish the

---

[1] Barjot 2010, 958; Levenstein & Suslow 2004, 43; Fear 2009, 269–70.
[2] Levenstein & Suslow 2006, 27, 57; Schröter 1994, 476; Suslow 2005, 730; Levenstein & Suslow 2011, 463.
[3] Cummins 2010, 21.

hours of the night. Absorbed into Egypto–Greek astrology; however, they were personified as rulers of each ten-degree division of the zodiac, and as such were transmitted to India.

Declination: the angular distance of a heavenly body north or south of the celestial Equator.

Deferent: a non-Ptolemaic medieval term to describe a circle that carries an epicycle in Ptolemaic astronomy.

Detent: a component that detains or locks another piece of a mechanism.

Dihedron: composed of two planes at angles to each other.

Direction dials: *sundials* in which time is indicated by the direction of the shadow of the *gnomon*.

Directoire: the period from 1795 during which France was governed by five 'directors', overthrown on 18 Brumaire year VIII [9 November 1799].

Dutch striking: in which the hours are fully sounded on a low-pitched bell and the half hours fully on a higher pitched bell, the number of strokes being that of the following hour.

Ecliptic: the great circle in the heavens that the sun appears to describe in the course of the year in consequence of the earth's orbit around it.

Engine-turned: see *Guilloché*.

Epicycle: a circle carried on the circumference of another circle, the deferent.

Epicyclic: a type of curve used in horological wheel tooth forms which, when used in conjunction with a *Hypocyclic* form of *pinion* leaf, produces constant relative velocities ('uniform lead') in the meshing between wheel and *pinion*.

Equal hours: the hours obtained by dividing the entire period of day and night into twenty-four invariable equal parts

Equant: the point from which angular motion of the centre of the epicycle is uniform. The equant is offset from the centre of the deferent by the same amount and in the same direction as the centre of the deferent is from the earth.

Equation of time: the difference between true solar time as indicated by a *sundial*, or derived from a solar observation, and the time shown by a mean-time timekeeper. It is caused firstly by the earth's orbit around the sun being an ellipse; secondly, by the earth's axis being tilted in relation to the celestial equator by some 23½ degrees. The equation of time has four days in the year when the mean-time will agree with true solar time. In the Gregorian calendar, these are 15 April, 13 June, 1 September, and 25 December. The difference can amount to the sun being fourteen minutes fast of mean time or just over fifteen minutes slow. Tables were published showing the changing difference throughout the year.[4]

Equinoxes: the two points of intersection of the ecliptic and the equator deriving their name from the fact that on these occasions day and night are equal in length throughout the world.

Escapement: the mechanism that communicates energy from the *train* to the oscillator.

Escape wheel: the last wheel in the *train*, which supplies energy to the oscillator.

Établir: the process of assembling a watch or clock so that it is ready (*établie*) for use. There is no equivalent term in English for the act of *Établissage*.

Établissage: a systematic delocalization of the production of individual parts between a multitude of dispersed workshops, then by the assembly of these, in centres of highly variable size, into semi– or entirely finished products.[5]

Fabrique: term used to describe everything connected with the gold and silversmiths' trades, jewellery, and watchmaking corporately organized in Geneva.

Filigree: ornamental work composed from fine, usually gold or silver, wire formed into a delicate tracery.

Finishing: the final work required to bring a clock, watch or chronometer from being a rough movement to its final state.

Fly: the term for a device in the form of an *air-brake*, used to regulate the speed of a mechanical *train*.

Foliot: a horizontal bar oscillator that serves as a *balance* in early clocks and some watches.

Form watches: watches in which the case is shaped in the form of another object, for example, a flower, an animal, a shell, an insect, a book, or a musical instrument.

Frictional rest escapement: escapements such as the verge and the cylinder in which the oscillator is constantly in frictional connection, even during the supplementary arc.

Fusee: a device for ensuring that a constant force is delivered from a coiled *mainspring*. The fusee has the form of a conical pulley that carries the great wheel and has a helical groove round it. Running in this groove is a cord or chain that wraps round a straight-sided barrel containing the mainspring. The cord pulls on the small diameter at the top of the fusee when the *mainspring* is fully wound up. As the clock or watch runs and the *mainspring* force gradually diminishes, the fusee and barrel slowly rotate with the cord coming off the fusee onto the barrel. As the fusee rotates, the cord is pulling at an increasing diameter, matching a diminishing force with a greater mechanical advantage, and equalizing the torque at the great wheel during the time the movement runs.

Galuchat: A covering for watchcases (and other items), named after the case maker Jean-Claude Galuchat (1689–1779), who developed a process for rendering the skin of ray fish and dogfish supple, smooth, and susceptible to being dyed. See *Shagreen*.

Ghati: see 'Introduction, Hour Systems'.

---

4 See further in Kitto 1999; Davis 2003, 139–40.

5 Loertscher-Rouge 1977.

Gimbals: a double-pivoted mounting that keeps the object carried horizontal regardless of angular displacement of its support in any direction.

Gnomon: The shadow-casting instrument that may be used for calendrical or hour-finding purposes. The shadow-casting element in a *sundial*.

Going train: the series of wheels and *pinions* that transfers energy from the driving force to the *escapement* in the timekeeping part of a horological movement.

Gravity escapement: an arrangement in which the clock *train* raises an arm of defined weight a short, but constant distance which, as it drops back to its original position, impulses the pendulum with a nominally constant force.

Great wheel: the first and usually the largest wheel in a *train* of wheels. In many watches and most chronometers, the great wheel forms part of the *fusee* assembly.

Grotesque, grottesche: in Renaissance art, wondrous, dream-like, disproportionate, and bizarre decoration of which the subjects are mischievous, humorous, and fantastic, for example, women transforming into columns, half-human/half-animal creatures, harpies, climbing plants sustaining impossible weights, and wide-mouthed masks (*mascheroni*).

Guilloché: ornamentation composed of lines or circles intersecting symmetrically but equidistant from each other. Carved or painted, it derives from Antiquity. In Renaissance and later horology, the term is applied to a ground thus decorated by hand using a burin, often destined to receive translucent enamel. With the development of specific attachments for lathes to do this work in the mid-eighteenth century, hence known as engine-turning in English, the technique was applied to the outer decoration of watchcases and to watch dials, with a smooth or a granular finish.

Hanging barrel: a going barrel, the arbor of which is fixed to the movement only at one end, usually beneath a bridge.

Hog's bristle regulator: a lever, registering against a scale marked on the back plate of a watch, is attached to one or two hog's (or other) bristles against which the arms of the *balance* bounce. Moving the bristles nearer the *balance* makes the watch run faster, further away slows it down.

Hour angle: the solar hour expressed in degrees.

Hour systems: see the section devoted to these in the Introduction.

Hour wheel: the wheel that carries the hour hand.

House (domicile): in astrology, divisions of the zodiac, not coinciding with the signs, that govern the different spheres of human life (mundane houses), or the one or two signs in which a particular planet is predominant, is the 'ruler' (domiciles).

Hunting case: a typical nineteenth-century watch case in which both the covers are solid, as in the earlier fully enamelled *bassine* watches, so that the dial, although not seen, is protected. In a half-hunter, a small glass is set at the centre with an hour scale enamelled around it on the outside of the cover. The cover is opened by a push piece set in the pendant.

Indian or Hindu circles: a way of finding a meridian in which a vertical *gnomon* is placed at the centre of a series of concentric circles drawn on a horizontal plane. The meeting points are marked with the apex of the shadow on one of the circles at a certain hour in the morning and at the corresponding hour in the afternoon. The two are joined and the mid-point of the line determined. A straight line drawn from it to the base of the *gnomon* then represents the meridian.

Invar: an iron nickel alloy with very low coefficient of thermal expansion developed by Charles-Edouard Guillaume and his associates c.1896, containing about thirty-six per cent nickel, the value of which is approximately a tenth of the dilatability of iron.

Inverted bell top: the dome on the top of a spring clock where it is formed of a lower section with convex sides, and a smaller upper section with concave sides.

Isochronous: the characteristic of a *balance* or pendulum that performs arcs, whether large or small, in equal time intervals.

Italian hours: see Introduction.

Keyframe: a row of levers in a musical clock that, when engaged by the pins of the barrel as it rotates, lift to admit air from the bellows to the organ pipes, or to activate the hammers striking carillon bells.

Lantern pinion: a composite *pinion* employing pins rather than leaves in a cage-like construction.

Latitude equivalent: the capacity of a planar dial of any orientation or inclination set up at a certain latitude to function as a horizontal dial at another 'equivalent' latitude.

Lever escapement: one in which a lever with the *pallets* fixed to it communicates the impulse between the *escape wheel* and the *balance* of a watch or clock.

Liberties: areas in incorporated towns or cities that were independent of civic and guild regulation usually as dependent upon a monastery, nunnery, or other ecclesiastical foundation, although some privileged sites, royal or legal precincts, were secular. The liberties in London included St Martin le Grand (north of St Paul's cathedral), St Bartholomew's, the Blackfriars, the Inns of Court, Chancery, the Tower of London, St Katherine's next to it, and outside the walls, the Whitefriars, Charterhouse, and Clerkenwell. In Paris outside the walls, the main *lieux privilégiés* were the Faubourg St Antoine, and the rue de l'Ursine, within the walls were the cloister and *parvis* of Notre Dame, the cloisters of St Bénoît, St Denis de la Chārtre, St Germain d'Auxerrois, the Abbey of St Germain-des-Prés, St Martin des Champs, the Hōpital de la Trinité and the enclaves of the Temple, l'Hōpital des Quinze Vingts of St Jean de Latran, and the cemetery of the Innocents.[6]

Locking plate: see *Count wheel*.

---

6 For London, see Rappaport 1989, 34–5; for Paris, see Augarde 1996, 46; for Europe in general, see Ogilvie 2019, 470–3.

GLOSSARY

Magnetic declination: the angle between the direction of geographical north (invariable in position) and that of magnetic north (variable in position).

Mainspring: a coiled ribbon of hardened and tempered steel that can act as the principal driving force for a watch or clock.

Maintaining power: a system used in timekeepers for keeping the movement running while being wound up

Masterpiece: The test-piece that an apprentice had to make as a partial qualification for being accepted as a master workman in a guild or corporation. The nature of the masterpiece was strictly defined and was not left to the apprentice to decide what he would make. In conservative corporations, this could lead to the test-piece becoming fossilized and standard drawings for its execution circulated. The importance of the masterpiece varied from trade to trade and city to city, with some, such as London, not requiring it at all.

Meridian: the great circle in the heavens or on the globe that passes through the zenith and the poles.

Moment of inertia: the property of a body that offers resistance to a change in its velocity of motion. The moment of inertia is equal to mk^2 where m = mass and k = the radius of gyration.

Motion work: the gearing for the hour, minute, and second hands, usually mounted between the front plate and the dial.

Nag's head striking: a system for unlocking the *striking train* common in Germany, the Low Countries, and other parts of Europe, but rare in England. In this system a pin on a slow-moving wheel first lifts a jointed spring-loaded tip on the end of the unlocking lever, and then lifts the lever itself unlocking the striking. As the *striking* begins, the unlocking lever is raised up and down again, the tip returning to its normal position, now behind the lifting pin.[7]

Nodes: the two points at which the orbit of a planetary body around the sun cuts the ecliptic.

Nuremberg hours: see the Introduction.

Nychthemeron: the entire twenty-four-hour cycle of day and night, i.e. from noon to noon or midnight to midnight.

Objets de vertu: generic term for all small, precious objects, usually produced by goldsmiths, that serve for personal decoration, such as small boxes for snuff, tobacco, perfume, pills, playing cards or sewing implements, fans, cane pommels, watches, chatelaines, and the like.

Oculus: see *Aperture-gnomon*.

Ornamentalist: a designer of decorative motifs for gold and silverware and other furniture.

Orthographic projection: projection of the celestial sphere taken from a point supposed to be at infinity onto the plane of the solstices.

Ouroboros: a snake eating its own tail; symbol of eternity.

Pallets: components that serve to lock and unlock the *escapement* or transfer impulse from the *escape wheel* to the *balance*.

---
[7] Details in Robey 2011.

Parachute or pare-chute: an early form of shock-proofing for the balance-staff pivots introduced by A. L. Breguet. One or both of the bearing jewels are fitted into sprung arms.

Patraque: a low-quality, even worn-out, watch.

Pegamoid: an imitation leather formed from a cotton base coated with nitrocellulose.

Pendulum bob: the mass at the bottom of a pendulum.

Periphrasis: figure of speech in which several words or a phrase are used to express the meaning of one; a roundabout expression.

Pinion: a gear with a small number of teeth usually integral to the steel *arbor*. In horology, pinions are generally driven but occasionally, as in the case of the minute-wheel pinion, act as the driving gear.

Polar gnomon: the *gnomon* of a *sundial* so placed that it is at right angles to the equator and parallel with the polar axis.

Polyhedral dials: geometric shapes of three or more faces, each carrying an hour-scale and *gnomon*.

Potence-plate: the frame plate carrying the potence between the plates that bears the balance staff

Profile cutter: tool for cutting a wheel or *pinion* with cutting teeth designed to produce the complete form of the tooth or leaf on the piece being machined.

Profiling: the act of giving the required form to a tooth or pinion being cut.

Putting-out system: the organization of manufacture such that it was effected in a worker's home without him owning the material or semi-finished goods on which he worked. The worker was paid a wage for his labour on them, usually in the form of a fixed rate (piece-work).

Rack-striking: a system of *striking* for domestic clocks developed in the 1670s in which the earlier *count wheel* system is replaced by a device with a rack which, being released by the going train, falls on a snail to select the required number of blows to be struck.

Réassujeté: a craftsman who, after completing his apprenticeship, voluntarily submits to a further period of low paid service with a master in order to acquire new skills or develop existing ones.

Recoil: momentary reversal of the direction of rotation of an escape wheel (and so the train wheels) as part of the action of such an escapement.

Reflex dial: a *sundial* operating from a reflected spot of light.

Regulator dial: a dial used for high-precision astronomical clocks, and some watches and chronometers, in which (usually) the minutes are marked around the circumference with subsidiary dials for the seconds at 12 o'clock and hour at 6 o'clock.

Remontoire: a device to ensure a uniform drive to the *escapement*; the main power source and the majority of the *going train* serving to wind a smaller spring or weight that drives the *escapement*.

Repeating: the audible feature in a clock or watch mechanism in which the previous hour can be sounded on a bell or gong.

Quarter repeating allows the last quarter to be discerned as well as the hour. In clocks that strike the hour, this feature was known as 'pull quarter repeating'. However, in clocks that had no striking (which were a *timepiece*) and thus did not previously strike the hour, the feature was termed 'silent pull quarter' (the theory being that you cannot *repeat* something that had not happened). In watches, however, no such distinction was made and, in spite of their not usually striking the hour, they were still referred to as 'repeating'.

Resille sur verre: lattice work enamelling.

Ronde bosse: work modelled in full relief, standing out from the background.

Secular years: end-of-century years such as 1700, 1800, 1900 of which the first two digits are not divisible by four and are therefore not counted as leap years.

Secular variation: the change across time of magnetic declination.

Shagreen: originally untanned leather or rawhide first from horse, mule, or donkey back but generally, in the seventeenth and eighteenth centuries, prepared from cured shark skin that was polished and dyed. See *Galuchat*.

Solstices: the points where the tropics meet the ecliptic in longitude 90° and 270°. The days thus marked are the longest (in summer) and the shortest (in winter), in the year.

Span: the distance on an outstretched hand between the tip of the thumb and the tip of the little finger customarily considered to be half a cubit and standardized at nine inches.

Speed governor: see *Air-brake* and *Fly*.

Spring barrel: cylindrical barrel containing the *mainspring*.

Stackfreed: a strong spring with a roller at its free end which acts on a snail-cam geared to the *mainspring arbor*, pushing against the *mainspring* to lessen its force when fully wound, and acting with it to augment its failing power as it unwinds. The gearing had stop-work that allowed the *cam* to rotate only once during the run of a watch.

Stem: The shaft connecting the winding crown to the movement in a keyless wound pocket- or wristwatch.

Stereographic projection: a projection used in astronomy and dialling in which the observer's eye is imagined to be placed on the surface of the celestial sphere at one of the poles whence visual rays are projected that locate the loci of the of the celestial sphere on to a plane, usually that of the equator.

Stereometry: the geometrical art of shaping stone and wood in architecture.

Strass or Rhinestone: a heavy leaded glass coloured with metallic oxides named after Georg Frederick Strass (1701–73), who developed it to imitate diamond from a crystal found in the Rhine.

Striking: the audible feature of a clock mechanism in which the hours of the day are struck on a bell or gong. See *Chiming*.

Striking train: the series of wheels and *pinions* that forms part of the *striking* mechanism of a horological movement.

Sub-style line: projection of the polar *gnomon* on the plane of the dial.

Sub-style line: the line on a sundial plate on which the gnomon is to be erected. Except in vertical declining dials, this usually coincides with the 12 o'clock line.

Sufflator: a vessel, often in the shape of a man's head, which is filled with water and heated. On boiling, steam is emitted through a small hole or holes worked in the vessel, which can thus be used as a blower for fires or for rotating a turbine.

Sundial: a device that simulates the apparent diurnal movement of the sun across the sky by means of the shadow of an indicator falling upon a specially constructed grid that indicates divisions of the daylight period and is permanently attached to the shadow indicator.[8]

Time-glass: Generic name for timekeepers composed of two glass ampoules set mouth-to-mouth with a liquid or granular substance flowing between them, such as sand-glasses, mercury-glasses, and log-glasses.

Timepiece: Generic term for any mechanical clock that does not strike the hours or chime the quarters.

Tortoiseshell: A decorative veneer used, inter alia, for clock and watchcases. The term is a misnomer since the material used was actually derived from the shell of the turtle. The best quality came from Indian or loggerhead sea turtles or the Caouane sea turtle, although Dampier, in his *Voyages* published in 1703, recommended Brazilian hawksbill turtle as the best.

Tourbillon: a revolving cage, invented by Breguet, that carries the *balance* and the *escapement* of a watch. During the period of rotation (usually a minute but other periods have been used), all the vertical positions, and so the errors they provoke, are passed through successively and thus average out.

Train: the toothed wheels in a watch or clock that connect the *barrel* or *fusee* with the escapement or which form part of the *striking* or *chiming* mechanism.

Trigon of signs: an instrument used to draw the day arcs on *sundials*.

Tropical period: the time taken for the earth in its revolution round the sun to leave and return to first point of Aries.

Tropical year: the year defined by the two passages of the Sun through the tropics or equinoctial points.

Unequal hours: hours obtained by dividing the period from sunrise to sunset into twelve equal parts. Since this period varies in length according to the season, the length of the twelfth part varies in length throughout the year. The night period, from sunset to sunrise was similarly divided, but except at the equinoxes, the lengths of day hours and night hours differed from each other.

Unlocking: the action of releasing a wheel or other mechanism to allow an action to occur.

Verge: the vertical rod carrying the pallets in a *verge escapement*. Historically the term also referred to the *pallet arbor* in clock *escapements*.

---

8 See further discussion in Turner 1989a, 303.

Vernis Martin: an imitation lacquer on a copal base largely utilized by the brothers Guillaume I, Julien, Etienne, Robert, and Guillaume II Martin in their (Paris-based?) varnishing workshops from about 1728 onwards.

Volvelle: a circular disc free to rotate within the circumference of a base or another disc with which it is concentric.

Watch: the act of keeping watch over something—a community at night, a flock of sheep, a ship at sea—and which evolved into a term to describe a timepiece specifically to indicate the time, without *striking* or *chiming*, and usually portable, to be carried on the person. Historically, however, it was used sometimes to describe any *timepiece*, whether portable or not, which did not strike.

Water-clock:
  Inflow: water from a reservoir flows into a vessel and fills it in a specific time.

Obliquity of the ecliptic: the inclination of the ecliptic to the equator at an angle that varies slightly over time.

Outflow: vessel filled with a quantity of water, which flows out
  in a specific time through a hole near the bottom of the vessel (Figure 24).

Sinking bowl: water entering through a hole in the bottom of
  a semi-circular bowl floated on a reservoir, causes it to sink in a pre-determined period (Figure 25).

Compartmented a cylinder is divided into compartments the divisions of which are pierced.

Clepsydra: to enable a liquid contained in them simultaneously
  to drive and break the movement of the cylinder down vertical cords or an inclined plane.

Worm: an endless screw.

Zenith: the point on the celestial sphere vertically above a place of observation.

BIBLIOGRAPHY & ABBREVIATIONS

The bibliography is specific to the present work. It is not a general bibliography of horology. It includes all works cited in abbreviated form in the notes. Works fully cited in the body of the text are not included unless they are elsewhere cited in abbreviated form.

AH	*Antiquarian Horology*
AVN	Archives de la Ville de Neuchâtel
BBSS	Bulletin of the British Sundial Society
BBT	Denis Beaudouin, Paolo Brenni, & Anthony Turner, *A bio-bibliographical Dictionary of precision Instrument-makers and related craftsmen in France and Switzerland, 1430–1960,* https://bibnum.explore.psl.eu/s/dictionarypim
BGE	Bibliothèque de Genève
BHI	British Horological Institute
BL	British Library
Brateau	Fiches Brateau
Cadran Info	Revue de la commission des cadrans solaires de la Société Astronomique de France
CRAI	Comptes rendu de l'Académie des Inscriptions.
DGC	*Deutsche Gesellschaft für Chronometrie Jahresschrift*
DSB	Charles Coulston Gillispie (Editor in chief), *Dictionary of Scientific Biography*, New York, 16 vols., 1970–80
EI 1	*The Encyclopedia of Islam*, 1913–38
EI 2	*The Encyclopedia of Islam*, 1954–2005
EIC	East India Company
GI	*Gnomonica Italiana*. Rivista di Storia, Arte, Cultura e tecnichi degli orology solari
HARS	*Histoire de l'Académie Royale des Sciences*
HCJ	House of Commons Journals
HJ	*The Horological Journal*, the journal of the British Horological Institute
ILN	*The Illustrated London News*
IOR	India Office Records (East India Company), held in the British Library, London
ISDN	*Illustrated Sporting and Dramatic News*
LFHS	*La Fédération Horlogère Suisse*, Swiss watch industry newspaper, La Chaux-de-Fonds
MPL	Jacques Paul Migne. *Patrologiae Cursus Completus*. Series Latina. 1844–68
NAWCC	National Association of Clock and Watch Collectors of America
OCCH	Société Hollandaise des Sciences, *Œuvres complètes de Christiaan Huygens*, 22 vols., Amsterdam 1886–1949, reissue Amsterdam 1967–1977
ODLA	Oliver Nicholson (ed.), *The Oxford Dictionary of Late Antiquity*, 2 vols., Oxford 2018
ODNB	*The Oxford Dictionary of National Biography*
OED	The Compact Edition of the *Oxford English Dictionary*, complete text reproduced micrographically, first edition (reprinted) 1972
PT	*Philosophical Transactions of the Royal Society*
PTL	East India Company: Private trade ledgers
T & R	Mario Arnaldi (ed.), *Tempus & Regula—orology solari medievali italiani*, 2011–19
TNA	The National Archives

WCB	NAWCC Watch & Clock Bulletin
WCM	The Watch and Clock Maker
WJS	The Watchmaker, Jeweller and Silversmith
Abbott 1893	Henry G. Abbott, *The American Watchmaker and Jeweller*, Chicago.
Abel 1893	Félix-Marie Abel, [no title], *Revue biblique*, 12, n.p.
Abeler 1968	Jürgen Abeler, *5000 Jahre Zeitmessung: Dargestelt im Wuppertaler Uhrenmuseum an der Privatsammlung der Urmacher-und Goldschmiedenfamilie Abeler*, Wuppertal.
Abeler 1977	Jürgen Abeler, *Meister der Uhrmacherkunst*, Wuppertal. Second significantly revised edition, published 2010.
Achelis 1937	Elizabeth Achelis, *The World Calendar*, New York.
Ackermann 1986	Christoph Ackermann, *Uhrmacher im alten Basel*, Basel.
Ackermann 1983	H. C. Ackermann, *Sammlung Carl und Lini Nathan-Rupp, Die Kutschenuhren*, Basel.
Ackermann 1991	Hans Christoph Ackermann, 'L'horlogerie dans les cantons de Bâle-Ville et de Bâle-Campagne', in Cardinal, Jequier, Barrelet, & Beyner 1991, 99–104.
Actes 2017	*Actes du cycle des conferences dans le massif du Jura. l'horlogerie, fille du temps*, Besançon.
Adam 2010	Jean-Luc Adam, 'Tianjin Sea-Gull Watch', *Europa Star* [online], Aug–Sept. https://www.europastar.com/magazine/highlights/1004082674-tianjin-sea-gull-watch.html.
Adam 2010	Shaul Adam, 'A Medieval Sundial in Jerusalem', *BBSS*, 13, 91.
Adams 1799	George Adams, *Lectures on Natural and Experimental Philosophy* . . ., 2nd edn., W. S. Jones (ed.), 5 vols., London.
Addomine 2007	Marisa Addomine, 'Ancora sull'orologio astronomico di Chiaravalle', *La Voce di Hora*, 23, 19–32.
Addomine 2016	Marisa Addomine, 'A Fourteenth-Century Italian Turret Clock', *AH*, 37, 213–22.
Addomine & Pons 2009	Marisa Addomine and Daniele Pons, 'Don Antonio Proviero e la suoneria senza ruote, *La Voce di Hora*, 26, 59–68.
Adelstein 2014	Tom Adelstein, 'Reissued Pilot Watches Shine a Light on Swiss Hamiltons and Four Chinese Watches', Asian-Watches.com [online], 28 August 2014. Available at: http://www.asian-watches.com/2014/08/reissued-pilot-watches-shine-light-on.html
Adès 1982	Dawn Adès, *Dalí and Surrealism*, New York.
Adler 1990	Mortimer J. Adler (ed.), *Gargantua and his Son Pantagruel* (c.1532–c.64), by Francois Rabelais, transl. Sir Thomas Urquhart and Peter Motteux (1693–14), Chicago.
Adler 2020	'Highlights of the Adler Planetarium: Scientific Instruments'. Available at: https://artsandculture.google.com/exhibit/highlights-of-the-adler-planetarium-scientific-instruments-adler-planetarium/7wKSiBDKz4X-LQ?hl=en
Aguillaume 2006	Cécile Aguillaume, 'The Swiss Watchmaking Industry Faced with Globalisation in the 1970s', *ICON*, 12, 190–217.
Aiken 1795	J. Aiken, *A Description of the Country from Thirty to Forty Miles around Manchester*, London.
Airy 1826	George Biddell Airy, 'On the Disturbances of Pendulums and Balances, and on the Theory of Escapements', *Transactions of the Cambridge Philosophical Society*, read 26 November 1826.
Aked 1973a	Charles K. Aked, 'Bewcastle Cross', *AH*, 8, 501–5.
Aked 1973b	Charles Aked, 'The First Free Pendulum Clock', *AH*, 8, 136–62.
Aked 1976	Charles K. Aked, *A Conspectus of Electrical Timekeeping*, Ticehurst.
Aked 1995a	Charles K. Aked, 'Bewcastle Cross (part I)', *BBSS*, 1, 2–8
Aked 1995b	Charles K. Aked, 'Bewcastle Cross (part II)', *BBSS*, 2, 10–18
Aked 1997	Charles K. Aked, 'Warren's Synchronous Clocks', Electrical Horology Group Technical Papers, 59, 1–11.
Aked & Rizzardi 1975	Charles K. Aked & P. Rizzardi, 'Dell' Orologio Applicato All Elettromotore Perpetuo', *AH*, 9, 524–39.
Aked & Severino 1997	Charles K. Aked & Nicola Severino, *Bibliografia della gnomonica*, 2nd edn, West Drayton.
Aked 1979	J. R. A. Aked, 'Sand-Glasses of the Historisches Uhren Museum Wuppertal, Germany', *AH*, 11, 293–8.
Alberi 2006	Auber Paolo Alberi, 'Gli orologi solari della torre dei venti a Atene e a Tinos', *Archeografo Triestino*, 66, 1–33.
Alberi 2011	Auber Paolo Alberi, 'L'obelisco di Augusto in Campo Marzio e la sua linea meridiana. Aggiornamenti e proposte', *Rendiconti Pontificia Accademia Romana di Archeologia*, 84, 447–580.
Alcouffe et al. 2004	Daniel Alcouffe, Anne Dion, & Gérard Mabille, *Les bronzes d'ameublement du Louvre*, Dijon.
Alexandre 1867	Félix Victor Alexandre. *Rapport adressé à la Commission d'Encouragement par la délégation des horlogers*, Paris.
Al-Faḍl 1948	Abū Al-faḍl, *Ā'īn-i Akbarī*, vol. 3, H. S. Jarrett (trans.), revised and further annotated by Jadu-Nath Sarkar, Calcutta.
Algarotti 1769	Francesco Algarotti, *Letters from Count Algarotti to Lord Hervey (etc)*, vol. 1, London.

Al-Hasan 1977	Ahmad Y. Al-Hasan, 'The Arabic Text of al-Jazarī's "A Compendium on the Theory and Practice of the Mechanical Arts"', *Journal for the History of Arabic Science*, 1, 47–64.
Al-Hassan 1976	Ahmad Y. Al-Hassan, *Taqi al-Din and Arabic Mechanical Engineering*, Aleppo.
Ali 1974	Maulana Muhammad Ali, *The Holy Qur'ân. Arabic text, English translation and commentary*, Chicago & Lahore.
Ali 1973	Mrs. Meer Hassan Ali & William Crooke (ed.), *Observations on the Mussulmauns of India*, Karachi.
Al-Khwârismî 1895	Abū Abdallah Al-Khworismî, *Liber Mafatih al-Olum*, G. Van Vloten (ed.), Leyden.
Allacci 1645	Leonis Allatii, *De mensura temporum Antiquorum et praecipue Graecorum exercitatio*, Rome.
Allexandre 1734	Jacques Allexandre, *Traité général des horloges*, Paris.
Allix 1981	Charles Allix, 'Mudge Milestones, Watch Dates', *AH*, 12, 627–34.
Allix & Bonnert 1974	Charles Allix and Peter Bonnert, *Carriage Clocks – Their History and Development*, Woodbridge.
Altick 1978	Richard Altick, *The Shows of London*, Harvard.
Ambronn 1899	Leopold Ambronn, *Handbuch der Astronomischen Instrumentenkunde*, vol. 2, Berlin.
Ancien élève 1827	Un Ancien Élévè de Breguet, *L'Art de L'horlogerie . . .*, Brussels; 2nd edn. 1828.
Anderson 2014	Benjamin Anderson, 'Public Clocks in Late Antique and Early Medieval Constantinople', *Jahrbuch des Österreichischen Byzantinistik*, 64, 223–32.
Anderson, Bennett, & Ryan 1993	Robert Anderson, J. A. Bennett & W. F. Ryan (eds.), *Making Instruments Count. Essays on Historical Scientific Instruments Presented to Gerard l'Estrange Turner*, Aldershot.
Andrewes 1996	William J. H. Andrewes (ed.), *The Quest for Longitude*, Cambridge (Ma).
Anon n.d.	Anonymous, *Moments of Eternity*: exhibition catalogue, Macao,
Anon 1760	Anonymous [?Laurence Sterne], *The Clockmakers Outcry Against the Author of The Life and Opinions of Tristram Shandy*, London.
Anon 1762	Anonymous, *Le pitture antiche di Ercolano e contorni incise con qualche spiegazione*, Naples.
Anon 1772	Anonymous [James Cox], *A Collection of Extracts . . . relative to the Museum in Spring Gardens*, London.
Anon 1809	Anonymous, 'Eclipses of the Satellites of Jupiter, observed by John Goldingham, Esq., F. R. S., and under his Superintendence, at Madras, in the East Indies,' *A Journal of Natural Philosophy, Chemistry, and the Arts*, 22, 153–6.
Anon 1817	Anonymous, *Report from the Committee on the Petitions of Watchmakers of Coventry*, Parliamentary Papers, London.
Anon 1827	Anonyme, *L'art de l'horlogerie enseigné en trente leçons; ou manuel complet de l'horloger et de l'amateur*, Paris.
Anon 1883	Anonymous, 'H. Grau's Elektrische Uhr', *Dingler's Polytechnisches Journal*, 247, 120–1.
Anon 1922	Anonymous, 'Heinrich Grau', *Die Uhrmacherkunst*, 8, 139.
Anon 1931	Anonymous, 'British-Made Clocks and Escapement', *The Watch and Clockmaker*, 4, 334–5.
Anon 1936	Anonymous, 'The Public Welcomes Synchronous Timekeeping', *Watch and Clock Maker*, 9, 279.
Anon 1967	Anonymous, Financial Times, 30 November.
Antiquorum 1994	Antiquorum Auctioneers, *The Art of Vacheron Constantin*, Geneva.
Antiquorum 2004	Antiquorum Auctioneers, *Artiste horloger*, 24 April, Geneva.
Antiquorum 2008	Antiquorum Auctioneers, *Imperial Treasures. A Private Collection of Pocket Watches made for the Chinese Market*, Geneva.
Antonini 1781-90	Carlo Antonini, *Manuale d varii ornamenti . . .*, Rome.
Apian 1532	Peter Apian, *Quadrans Apiani Astronomicus . . .*, Ingolstadt.
Apian 1533a	Petrus Apianus, *Horoscopion Apiani generale*, Ingolstadt.
Apian 1533b	Petrus Apianus, *Instrument Buch . . .*, Ingolsdadt.
Arberry 1955/1972	Arthur J. Arberry, *The Koran Interpreted*, London, reprinted.
Arcari & Costamagna 2002	Barbari Acari & Andrea Costamagna, *L'Abazia dell' Aquafredda*, Como.
Archinard 1988	Margarida Archinard, 'Les cadrans solaires rectilignes', *Nuncius*, 3, 149–81.
Archinard 1995	Margarida Archinard, 'Navicula de Venetis, une acquisition prestigieuse du Musée d'Histoire des sciences, Genève', *Genava*, 43, 87–94.
Archinard 2005	Margarida Archinard, 'Les cadrans solaires analemmatiques', *Annals of Science*, 42, 309–346.
Ardaillon 1900	Edouard Ardaillon, 'Horologium', in *Dictionnaire des antiquités grecques et romaines*, vol. 3, Paris, 256–64.
Argoud & Guillaumin 1997	Gilbert Argoud & Jean-Yves Guillaumin (eds.), *Les Pneumatiques d'Héron d'Alexandrie*, Saint-Etienne.
Armentrout 1992	Lisa Armentrout, 'The Significance of Time in Mrs. Dalloway', *Prized Writing* (1991–1992) [online]. Available at: https://prizedwriting.ucdavis.edu/significance-time-mrs-dalloway.
Armonville 1825	J.-R. Armonville, *La Clef de l'industrie et des sciences qui se rattachent aux arts industriels . . .*, 2 vols., Paris.
Arnaldi 1998	Mario Arnaldi, 'Da Buti, Danté and the Measure of Time', *BBSS*, 92, 14–16.

Arnaldi 1999	Mario Arnaldi, 'La frazioni dell' ora temporaria; dall' antichità al medioevo', *Gnomonica (Supplement to Astronomià)*, 5, 27–9.
Arnaldi 2000a	Mario Arnaldi, *The Ancient Sundials of Ireland*, Chingford.
Arnaldi 2000b	Mario Arnaldi, 'Medieval Monastic Sundials with Six Sectors: An Investigation into their Origin and Meaning', *BBSS*, 12, 109–15.
Arnaldi 2003a	Mario Arnaldi, 'Orologi solari medievali in provincial di Bari', *GI*, 1, 41–6.
Arnaldi 2003b	Mario Arnaldi, 'Orologi solari medievali a "tutto tondo" – origine et diffusione nei secoli XII-XV', *GI*, 2, 41–7.
Arnaldi 2005	Mario Arnaldi, 'Le ore "benedittine" e l'orologio solare medieval dell' abbazia dell' Aquafredda', *GI*, 3, 28–35.
Arnaldi 2006	Mario Arnaldi, 'Le ore italiane. Origine e decline di uno dei più importanti sistemiorarfi del passato (primo parte), *GI*, 1, 10–18.
Arnaldi 2007	Mario Arnaldi, 'Le ore italiane. Origine e decline di uno dei più importanti sistemi orari del passato (seconda parte), *GI*, 12, 2–10.
Arnaldi 2009	Mario Arnaldi, 'La regola di Erfurt in un codice appartenutgo a Frà Giocondo da Verona', *GI*, 6, 2–9.
Arnaldi 2011a	Mario Arnaldi, 'An Ancient Rule for Making Portable Sundials from an Unedited Medieval Text of the Tenth Century', *Journal of the History of Astronomy*, 43, 141–60.
Arnaldi 2011b	Mario Arnaldi, 'The Canterbury Pendant – Part 1: A New Insight from an Ancient Rule for Making Portable Altitude Dials', *BSSB*, 23, 2–7.
Arnaldi 2012a	Mario Arnaldi, 'The Canterbury Pendant – Part 2. Relationship with the Libellus Rule', *BSSB*, 24, 8–12.
Arnaldi 2012b	Mario Arnaldi, *De cursu solis: Medieval Azimuthal Sundials – from the Primitive Idea to the First Structured Prototype*, BSS Monograph Series, vol. 10, London.
Arnaldi 2019	Mario Arnaldi, 'L'orologi solare di Santa Maria della Strada a Taurisano: nuovi studi', in Ciurlia 2019, 171–99.
Arnaldi 2020	Mario Arnaldi, Gli 'Schemi delle ombre' nel Medio Evo latino - Addenda, *Orologi Solari*, 21, 17–34.
Arnaldi & Sanna 2015	Mario Arnaldi & Angelo Sanna, 'La Sardegna', T & R, 2, n.p.
Arnaldi & Schaldach 1997	Mario Arnaldi & Karlheinz Schaldach, 'A Roman Cylinder Dial: Witness to a Forgotten Tradition', *Journal of the History of Astronomy*', 28, 107–31.
Arnold 1780	John Arnold, *An Account kept during Thirteen Months in the Royal Observatory at Greenwich of the going of a Pocket Chronometer, Made on a new Construction . . .*, London.
Arnold 1791	John Arnold, *Certificates and circumstances relative to the going of Mr. Arnold's chronometers*, London.
Arnold & Dent 1833–1836	John Roger Arnold & Edward John Dent, 'Magnetic Experiments on 1833–1836 Chronometers', *The Nautical Magazine*, 2, 222–5; 353; 417–18 and 5, 705–17.
Arnold-Becker 2014	Alice Arnold-Becker, 'Friedberg—A Centre of Watch and Clock-Making in Seventeenth- and Eighteenth-Century Bavaria', Parts 1 and 2, *AH*, 35, 632–82, 783–95.
Arutyunov 1968	V. O. Arutyunov, A. M. Brodskii, & M. F. Fomichev, 'Leningrad Étalon Plant—The Experimental Base of the VNIIM', *Measurement Techniques*, 11 (November), 1446–50.
Āryabhaṭa 1976	*Āryabhaṭīya of Āryabhaṭa*, critically edited with Introduction, English Translation, Notes, Comments, and Indexes by Kripa Shankar Shukla, in collaboration with K. V. Sarma, New Delhi.
Aschoff 1980	Volker Aschoff, 'Über den byzantinischen Feurteleggraphen und Leon der Mathematiker', Deutsches Museum. Abhandlung und Berichte, 44, Munich & Düsseldorf.
Ashton 1939	T. S. Ashton, *An Eighteenth-Century Industrialist: Peter Stubbs of Warrington, 1756–1808*, Manchester.
Ashworth 1990	Peter Ashworth, 'Davis of Windsor, A Family Business', *Windelsora*, 9, 2–9.
Asprey et al. 1973	John Asprey, Sebastian Whitestone, & Anthony J. Turner, *The Clockwork of the Heavens*, London.
Asuka 1983	Asuka Historical Museum, The Water Clock in Asuka (Pictorial Records no. 11), Nara.
Atkins & Overall 1881	Samuel Elliott Atkins & William Henry Overall, *Some Account of the Worshipful Company of Clockmakers of London*, London.
Attali 1997	Jacques Attali, *Mémoire de sabliers: collections, mode d'emploi*, Paris.
Attinger 1973	Claude Attinger, 'Horlogerie', *Encyclopædia universalis*, 8, 561–5.
Aubert 1993	Daniel Aubert, *Montres et horlogers exceptionnels de la Vallée de Joux*, Neuchâtel.
Aubrey 1898	Andrew Clarke (ed.), *'Brief Lives', Chiefly of Contemporaries, Set Down by John Aubrey, between the Years 169 and 1696*, 2 vols., Oxford.
Aubriot 2004	Olivia Aubriot, *L'eau, miroir d'une société—Irrigation paysanne au Népal central*, Paris.
Auch 1827	Jacob Auch, *Handbuch für Landuhrmacher oder leichtfalsliche anleitung, wie man vom geringsten bis zumschwersten Stücke eine Taschenuhren bauen muss*, Ilmenau.
Augarde 1986	Jean-Dominique Augarde, 'Jean-Joseph de Saint-Germain (1719-1791). Bronzearbeiten zwischen Rocaille und Klassizismus', in Ottomeyer & Pröschel 1986, 521–53.

Augarde 1996	Jean-Dominique Augarde, *Les Ouvriers du temps, la pendule à Paris de Louis XIV à Napoléon Ier*, Geneva.
Augarde & Ronfort 1998	Jean-Dominique Augarde & Jean Nérée Ronfort, *Antide Janvier, mécanicien-astronome, horloger ordinaire du Roi*, Paris.
Avenier 1999	Cédric Avenier, 'Les cadrans solaires du Bas-Dauphiné à la période révolutionnaire', *Le Monde alpin et rhodanien*, 4, 7–22.
Ayasse 1824	J. J. M. Ayasse, *Manuel de l'apprenti horloger en Province, ouvrage éléme,naire à l'usage des amateurs et des apprenntis qui cultivent cet art*, Angers.
Azzarita 2005	Francesco Azzarita, 'Orologi considdeti canonic: consideraazioni su quello benedettino della chiesa di Ognissanti a Valenzano (BA)', Proceedings of the 13th National Seminar of Gnomonics, 8–10 April 2005, Lignano Sabbiadoro, 48–52.
B[euriot] 1719	B[euriot], *Horlogéographie pratique ou la manière de paire les horloges à poids . . .*, Rouen.
Baarsen et al. 1988	Reinier Baarsen, Gervase Jackson-Stops, Phillip M. Johnston, & Elaine Evans Dee, *Courts and Colonies: The William and Mary Style in Holland, England, and America*, Seattle/London
Babel 1916	Antony Babel, *Les Métiers dans l'ancienne Genève: Histoire corporative de l'horlogerie, de l'orfèvrerie et des industries annexes*, Geneva.
Babel 1938	Antony Babel, *La Fabrique genevoise*, Neuchâtel.
Bābur 2006	*Bābur-Nāma (Memoirs of Bābur)*, Annette Susannah Beveridge (trans.), Delhi.
Bach et al. 1992	Henri Bach, Jean-Pierre Rieb, & Robert Wilhelm. *Les trois horloges astronomiques de la cathédrale de Strasbourg*, Strasbourg,
Baden-Powell 1872	B. H. Baden-Powell, *Handbook of the Manufactures and Arts of the Panjab*, Lahore.
Bagheri 1998	Mohammad Bagheri, 'Sundials in Iran', *The Compendium*, 5, 24–5.
Bagheri 2002	Mohammad Bagheri, 'The First Analemmatic Sundial in Iran', *The Compendium*, 9, 24–5.
Bai & Li 1984	Bai Shangshu & Li Di, 'Zhou Shuxue's Contribution to Timepieces', *Studies in the History of Natural Sciences*, 3 (2), 138–144.
Bailey & Barker 1969	F. A. Bailey and T. C. Barker, 'The Seventeenth-Century Origins of Watchmaking in South-West Lancashire', in Harris 1969, 1–15.
Baillie unpublished	G. H. Baillie, '*Clocks and Watches, An Historical Bibliography*, volume II, n.d.
Baillie 1929	G. H. Baillie, *Watches, Their History, Decoration and Mechanism*, London.
Baillie 1951	G. H. Baillie, *Clocks and Watches, An Historical Bibliography*, London.
Baillie et al. 1974	G. H. Baillie, H. Alan Lloyd & F. A. B. Ward (eds.), *The Planetarium of Giovanni de Dondi, Citizen of Padua*, London.
Baillie et al. 1982	G. H. Baillie, C. Clutton, & C. A. Ilbert, *Britten's Old Clocks and Watches and their Makers*, London.
Bailly 2001	Sharon Bailly & Christian Bailly, *Oiseaux de bonheur, tabatières et automates/Flights of Fancy: Mechanical Singing Birds*, Geneva.
Bairoch 1996	Paul Bairoch, 'Les exportations d'articles manufacturés de la Suisse dans le contexte international (1840–1994)', in Körner & Walter 1996, 205–34.
Bairoch & Körner 1990	Paul Bairoch & Martin Körner (eds.), *La Suisse dans l'économie mondiale. Die Schweiz in der Weltwirtschaft*, Genève.
Baldi 2018	Rossella Baldi, 'La Suisse manufacturière au xviiie siècle: lectures croisées, numéro thématique de xviii.ch', in Rossella Baldi & Laurent Tissot (eds.), *Annales de la Société suisse pour l'étude du XVIIIe siècle*, vol. 9, Basel.
Baldini 1754	Gianfrancesco Baldini, 'Sopra un'antica piastra di bronze, che si suppono un orologio da sole', *Sgi di dessertazione accademiche pubbicamente lette nelle nobili Accademoa Etrfusca dell'antichissima città di Cortona*, 3, 184–94.
Ball 1930	A. Ball, 'My Experiences with Electric Clocks and Synchronisers', *the Horological Journal*, May, 175–8.
Ball 2006	Martin Ball, 'The Early Balance Spring Watch: A Brief History', *AH*, 29, 760–74.
Baloch 1979	N. A. Baloch, 'Measurement of Space and Time in the Lower Indus Valley of Sind', in Said 1979, Part I, 168–96.
Baran 1976	Neculai Baran, 'L'expression du temps et de la durée en latin', in R. Chevalier (ed.), *Aiôn. Le temps chez les Romains*, Paris, 1–21.
Bargès 1859	J. J. L. Bargès, *Tlemcen, ancienne capital du royaume de ce nom, sa topographie, son histoire, description de ses principaux monuments, anecdotes, legends et récits divers . . .*, Paris.
Barillet 1894	Jean Barillet, 'La Savoie horlogère et l'exposition de Cluses', *Revue chronométrique*, 11, 173–6.
Barjot 2010	Dominique Barjot, 'Cartels et ententes', in Daumas 2010, 958–64.
Barlow 1821	Peter Barlow, 'On the Effects Produced in the Rates of Chronometers by the Proximity of Masses of Iron', *PT*, 109, 361–89, & *The Edinburgh Philosophical Journal*, 5, 383–5.
Barnes 1940	Ralph M. Barnes, *Motion and Time Study*, 2nd edn., New York.
Barnfield 2011	Malcolm Barnfield, 'The Sundial Goes to War', *BBSS*, 23(2), 20–5; 23(3), 10–15.

Barnish 1992	S. J. B. Barnish, *Cassiodorus: Variae*, trans. with notes and introduction, Liverpool.
Barraclough 1990	Kenneth C. Barraclough, 'Swedish Iron and Sheffield Steel', *History of Technology*, 12, 1–39.
Barrelet 1987	Jean-Marc Barrelet, 'Les résistances à l'innovation dans l'industrie horlogère des montagnes neuchâteloises à la fin du XIXe siècle', *Revue suisse d'histoire*, 4, 394–411.
Barrelet & Cardinal 1999	Jean-Marc Barrelet & Catherine Cardinal (eds.), *Actes du colloque: apprendre la formation des horlogers, créer passé et l'avenir, transmettre*, La Chaux-de-Fonds.
Barret 2017	Ross Barrett, 'Sketching the Future: Time, Work and Art in James Henry Beard's "The Land Speculator"', Talk Given at the 'Horology in Art Symposium', held at the Museum of Fine Arts in Boston, 26–28 October 2017. Video available at: https://vimeo.com/showcase/5738130.
Barrington 1778	Daines Barrington, 'Observations on the Earliest Introduction of Clocks …', *Archæologia*, 5, 416–28.
Barrow 1804	John Barrow, *Travels in China*, London.
Barrows 2011	Adam Barrows, 'Time Without Partitions: *Midnight's Children* and Temporal Orientalism', *Ariel*, 42, 89–101.
Bartky 2000	I. Bartky, *Selling the True Time: Nineteenth-Century Timekeeping in America*, Stanford, CA.
Bartky 2007	I. Bartky, *One Time Fits All: The Campaigns for Global Uniformity*, Stanford, CA.
Basanta Campos 1972	José Luis Basanta Campos, *Relojeros de Espana, diccionario bio-bibliografico*, Pontevedra.
Bassermann-Jordan 1905	Ernst von Bassermann-Jordan, *Die Geschichte der Räderuhr unter besonderer Berücksichtigung der Uhren des Bayerischen Nationalmuseums*, Frankfurt.
Bassermann-Jordan 1927	Ernst von Bassermann-Jordan, *Die Standuhr Philipps des Guten von Burgund*, Leipzig.
Bastien 2012	Jean-Luc Bastien, 'Le temps civique et l'apparition des premières horloges à Rome', *Dossiers d'Archéologie*, 354, 1–15.
Bate 1634	John Bate, *The Mysteryes of Nature and Art*, London.
Bates 2018	Keith Bates, *Early Clock and Watchmakes of the Blacksmiths' Company*, Morpeth.
Baudelaire 1986	Charles Baudelaire, *The Complete Verse*, Francis Scarfe (ed. and trans.), London.
Baudrillard 1968	Jean Baudrillard, *Le Système des objets*, Paris.
Bayard 1999	Bayard, *Les réveils Bayard, plus d'un siècle de vue aliermontaise*, Rouen.
Bayard 2004	Françoise Bayard, 'Les horlogers à Lyon à l'époque moderne (XVIIe–XVIIIe siècles)', in Mauerhan 2004, 125–36.
Beaufoy 1822	Mark Beaufoy, *The Edinburgh Philosophical Journal*, 7, 189.
Beaujeu 1948	J. Beaujeu, 'La Littérature technique des Grecs et des Romains', in Actes du Congrès Guillaume Budé, 21–88.
Beaujour 1800	Félix Beaujour, *Tableau du commerce de la Grèce … depuis 1787 jusqu'en 1797*, 2 vols., Paris.
Bechmann 1991	Roland Bechmann, *Villard de Honnecourt, la pensée technique au XIIIe siècle et sa communication*, Paris.
Beck 1899	Th. Beck, *Beiträge zur Geschichte des Maschinenbaues*, Berlin.
Beckett 1850	Edmund Becket, *A Rudimentary Treatise on Clock and Watchmaking: With a Chapter on Church Clocks; and an Account of the Proceedings Respecting the Great Westminster Clock*, London.
Beckmann 1786–1805	Johann Beckmann, *Beitrage zur Geschichte der Erfindungen*, 5 vols., Leipzig.
Beckmann 1797/1817	John Beckmann, *A History of Inventions and Discoveries*, William Johnston (trans.), 3 vols., London.
Bedini 1956a	Silvio A. Bedini, 'Perpetuum Mobile: The Invention of Rolling Ball Clocks in the 17th Century', *NAWCC Watch & Clock Bulletin*, 8, 74–87.
Bedini 1956b	Silvio A. Bedini, 'Chinese Mechanical Clocks', *NAWCC Watch & Clock Bulletin*, 7, 211–22.
Bedini 1962	Silvio A. Bedini, 'The Compartmented Cylindrical Clepsydra', *Technology and Culture*, 3, 115–41.
Bedini 1965	Silvio A. Bedini, 'Adventures in Time VI: Time and Light', *La suisse horlogère*, 79, 29–38, & 80, 25–30.
Bedini 1975	Silvio A. Bedini, 'Oriental Concepts of the Measure of Time', in Fraser & Lawrence 1975, 451–84.
Bedini 1983	Silvio A. Bedini (ed.), *Giuseppe Campani, Discorso intorno a' suoi muti oriuoli, 1660 …*, Milan.
Bedini 1991	Silvio A. Bedini, *The Pulse of Time: Galileo Galilei, The Determination of Longitude and the Pendulum Clock*, Florence.
Bedini 1994a	Silvio A. Bedini, *The Trail of Time. Time-Measurement with Incense in East Asia, Shih-chien ti tsu-chi*, Cambridge.
Bedini 1994b	Silvio A. Bedini, 'In Pursuit of Provenance: The George Graham Proto-Orreries', in Hackmann & Turner 1994, 54–77.
Bedini & Maddison 1966	Silvio A. Bedini and Frances A. Maddison, 'Mechanical Universe, the Astrarium of Giovanni de' Dondi', *Transactions of the American Philosophical Society*, 56, 1–69.
Bedos de Celles 1760, 1774	Francois Bedos de Celles, *La Gnomonique Practique ou l'Art de tracer les cadrans solaire*, Paris.
Bedos de Celles 1766–78	François Bedos de Celles, *L'art du facteur d'orgues*, 3 vols., Paris.
Beeson 1963	C. F. C. Beeson, 'Some Tudor Clockowners', *AH*, 4, 86–8.
Beeson 1965	C. F. C. Beeson, 'A Clock-Watch by Michael Nouwen', *AH*, 4, 372.
Beeson 1971	C. F. C. Beeson, *English Church Clocks 1280–1850, History and Classification*, London.

Beeson 1982	C. F. C. Beeson, *Perpignan 1356. The Making of a Tower Clock and Bell for the King's Castle*, London.
Beeson 1989	C. F. C. Beeson, *Clockmaking in Oxfordshire 1400–1850*, 3rd edn., Oxford.
Beeson & Maddison 1969	C. F. C. Beeson & F. R. Maddison, 'An Early Anchor Escapement in a Turret Clock', *AH*, 6, 77–80.
Béguelin 1994	Sylvie Béguelin, 'Naissance et développement de la montre-bracelet: histoire d'une conquête (1880-1950)', *Chronometrophilia*, 37, 33–43.
Beillard 1895	Alfred Beillard, *Recherches sur l'horlogerie, ses inventions et ses célébrités, …*, Paris.
Belcher 1835	Edward Belcher, *A Treatise on Nautical Surveying*, London.
Beliard 1767	Beliard, *Reflexions sur l'horlogerie en general et sur les horlogers du Roi en particuier*, The Hague.
Bell 1763	John Bell, *Travels from St Petersburg in Russia to Diverse Parts of Asia*, Glasgow.
Bellettini 2007	Anna Bellettini, 'St Albans, John Whethamstead e il tratto di gnomonice di Robert Stickford (Amb. & 201*bis* sup.)' in Ferrari & Novoni 2007, 217–28.
Belmont 1975/1989	Henry L. Belmont, *La Bulle-Clock, horlogerie électrique, sonhistoire, sa fabrication et sa reparation*, Besançon (reprint London 1989).
Belot, Cotte, & Lamard 2000	Robert Belot, Michel Cotte, & Pierre Lamard (eds.), *La Technologie au risqué de l'histoire*, Belfort-Montbéliard/Paris.
Bender 1975	Gerd Bender, *Die Uhrenmacher des hohen Schwarzwaldes und ihre Werke*, 2 vols., Villingen.
Benedetti 1574	Battista Benedetti, *De Gnomonum umbrarumque solarium usu*, Turin.
Ben-Layish 1969–71	D. Ben-Layish, 'A Survey of Sundials in Israel', *Sefunim*, 3, 70–81.
Bennett 2002	James A. Bennett, 'Shopping for Instruments in Paris and London', in Smith & Findlen 2002, ch. 15.
Benschop & Draaima 2000	Ruth Benschop & Douwe Draaisma, 'In Pursuit of Precision: The Calibration of Minds and Machines in Late Nineteenth-Century Psychology', *Annals of Science*, 57, 1–25.
Benson 1875	James W. Benson, *Time and Timekeepers*, London.
Bent 1893	'Thomas Dallam, The Diary of Master Thomas Dallam, 1599–1600', in *Early Voyages and Travels in the Levant*, …, J. Theodore Bent (ed.), Part 1, London.
Bentley 1582	Thomas Bentley, *The Monument of Matrones*, London.
Béranger, Duffaut, Morlet, & Tier 1996	Gérard Béranger, François Duffau, Jean Morlet, & Jean-François Tier (eds.), *Cent ans après la découverte de l'Invar. Les alliages de fer et de nickel*, London.
Berg 2005	Maxine Berg, *Luxury and Pleasure in Eighteenth-Century Britain*, Oxford.
Berg & Nettell 1976	T. Berg & D. F. Nettell, 'Iron Marks on Turret Clocks', *AH*, 10, 78–81.
Bergeron 1995	Louis Bergeron, 'L'homme d'affaires', in Vovelle 1995, 146–51.
Berggren 1980	J. L. Berggren, 'A Comparison of Four Analemmas for Determining the Azimuth of the Qibla', *Journal of the History of Arabic Science*, 4, 69–80.
Berggren 1985	J. L. Berggren, 'The Origin of al-Bîrûnî's "Method of the Zijes" in the Theory of Sundials', *Centaurus*, 28, 1–16.
Berggren 2001	J. L. Berggren, 'Sundials in Medieval Islamic Science and Society', *The Compendium*, 8, 8–14.
Bergmann 2005	S. Bergmann, *Comtoise-Uhren*, Stolberg.
Beringen 2012	John Beringen, *Horloges van Nederlandse Uurwerkenmakers 1600–1800*, Schonhooven.
Berlinger-Konqui 1991a	Marianne Berlinger-Konqui, 'L'horlogerie dans le canton de Vaud', in Cardinal, Jequier, Barrelet & Beyner 1991, 179–86.
Berlinger-Konqui 1991b	Marianne Berlinger-Konqui, 'L'horlogerie dans le canton du Valais', in Cardinal, Jequier, Barrelet, & Beyner 1991, 173–8.
Bernard 1936	Henri Bernard, *Aux portes de la Chine: Les Missionnaires du XVIe Siècle* (Chinese edition), Beijing.
Bernardin 1855	Constant Flavien Bernardin, *Description de l'horloge astronomique de la cathédrale de Besançon*, Paris.
Berner 1932	G. A. Berner (ed.). *Quelques notes sur Pierre-Frédéric Ingold …, publié par la Société Suisse de Chronométrie …*, Berne.
Berner 1961/1995	Georges-André Berner, *Dictionnaire professionnel illustré de l'horlogerie and Supplément*, Biel/Bienne.
Berthier 1996	Christelle Berthier, *Les horlogers à Blois au XVIIe siècle*. Master's thesis, Université de Tours.
Berthoud 1759/1805	Ferdinand Berthoud, *L'Art de conduire et de régler les pendules et les montres*, Paris
Berthoud 1763, 1786	Ferdinand Berthoud, *Essai sur l'horlogerie; …*, 2 vols., Paris.
Berthoud 1787	Ferdinand Berthoud, *De la mesure du temps ou supplément au traité des horloges marines et a l'essai sur l'horlogerie*, Paris.
Berthoud 1802	Ferdinand Berthoud, *Histoire de la mesure du temps par les horloges*, 2 vols., Paris.
Bertucci 2017	Paola Bertucci, *Artisanal Enlightenment: Science and the Mechanical Arts in Old Regime France*, New Haven/London.
Bettray 1955	J. Bettray, *Die Akkomodationsmethode des P.M. Ricci S.J. in China*, Rome.
Betts 1982	Jonathan Betts, *Notes for Horological Conservators*, National Trust (internal document), London.
Betts 1985	Jonathan Betts, 'Problems in the Conservation of Clocks and Watches', *The Conservator*, 9, 36–44.

Betts 1989	Jonathan Betts, *A Report on the Precision Clocks at Armagh Observatory*, Old Royal Observatory, Greenwich.
Betts 1996	Jonathan Betts, 'Josiah Emery, Watchmaker of Charing Cross', *AH*, 12, 394–401, 510–23, & 23, 24–44.
Betts 1998	Jonathan Betts, 'Arnold and Earnshaw: The Practicable Solution', in Andrewes 1998, 311–30.
Betts 2003	Jonathan Betts, 'Jean Hyacinth Magellan (1722–1790), Horological and Scientific Agent', *AH*, 27, 509–17
Betts 2004	Jonathan Betts, 'Jean Hyacinth Magellan (1722–1790), Part 2: The Early Clocks', *AH*, 28, 174–83.
Betts 2007a	Jonathan Betts, 'Jean Hyacinth Magellan (1722–1790), Part 3: The Later Clocks and Watches', *AH*, 30, 25–44.
Betts 2007b	Jonathan Betts, 'Jean Hyacinth Magellan (1722–1790), Part 4: The Precision Pioneers', *AH*, 30, 365–79.
Betts 2017	Jonathan Betts, *Marine Chronometers at Greenwich*, Oxford.
Betts 2018	Jonathan Betts, 'Showtime at Oxnead: The Timekeepers Depicted in the *Paston Treasure*', *AH*, 39, 215–24.
Bevis 2013	Matthew Bevis, 'Dickens by the Clock', in Tyler 2013, 46–72.
BHI 1863	British Horological Institute, 'Trade Statistics', *HJ*, 6, n.p.
BHI 1879	British Horological Institute, 'Opening of the New Building', *HJ*, 22, 29–35.
BHI 1879	British Horological Institute, 'Trade Statistics', *HJ*, 22, 63.
BHI 1880	British Horological Institute, 'Trade Statistics', *HJ*, 23, 63.
BHI 1887	British Horological Institute, 'Trade Statistics', *HJ*, 30, 63.
BHI 1887	British Horological Institute, 'Jottings', *HJ*, 30, 151–4.
BHI 1906	British Horological Institute, 'The Tariff Commission', *HJ*, 48, 192.
BHI 1912	British Horological Institute, 'Clocks Without Works: Revolution Prophesied in the Clock Trade', *HJ*, 55, 26–7.
BHI 1916	British Horological Institute, 'H. Williamson Ltd.', *HJ*, LVIII, 97–99.
BHI 1951	British Horological Institute, 'Delhi's 900 'Slaves'', *HJ*, 93, 176.
BHI 1969	British Horological Institute, 'Ingersold Diamond Jubilee', *HJ*, 107, 27.
Biarne 1984	J. Biarne, 'Le temps du moine d'après les premières règles monastiques de l'Occident (VIe–VIe siècles)', in Leroux 1984, 99–128.
Bicknell 1996	Stephen Bicknell, *The History of the English Organ*, Cambridge.
Bickel & Gautschy 2014	Susanne Bickel & Rita Gautschy, 'Eine ramessidische Sonnenuhr im Tal 2014 der Könige', *Zeitschrift für Ägyptische Sprache und Altertumskunde*, 141, 3–14.
Bielefeld 2007	Claus-Ulrich Bielefeld, *Elektrik am Handgelenk. Geschichte und Technik der elektrischen Armbanduhr 1956 bis 2006. 50 Jahre elektrische und elektronische Armbanduhren*, Leipzig.
Bigourdan 1922	G. Bigourdan, *Gnomoniqe, traité théorique et pratique de la construction des cadrans solaires suivi de tables relatives aux cadrans et caledriers*, Paris.
Bilfinger 1886a	Gustav Bilfinger, *Die Zeitmesser der antiken Völker*, Stuttgart.
Bilfinger 1886b	Gustav Bilfinger, *Die Babilonische Doppelstunde*, Stuttgart.
Bilfinger 1888	Gustav Bilfinger, *Die antiken stundenangaben*, Stuttgart.
Bilfinger 1892	Gustav Bilfinger, *Die Mittelalterlichen Horen und die Modernen Stunden*, Stuttgart.
Bimbenet-Privat & Fuhring 2002	Michèle Bimbenet-Privat & Peter Fuhring, 'Le style 'cosses de pois': L'orfèvrerie et la gravure à Paris sous Louis XIII', *Gazette des Beaux-Arts*, 139, 1–224.
Bingham & Turner 2017	Tony Bingham & Anthony Turner, *Metronomes and Musical Time*, London.
Binns 1971	Malcolm Binns, 'Sun Navigation in the Viking Age and the Canterbury Portable Sundial', *Acta Archaeologia*, 43, 23–34.
Bion 1716/1725	Nicolas Bion, *Traité de la construction et des principaux usages des instrumens de mathematique*, Paris.
Biot 1811	Jean-Baptiste Biot, *Traité élémentaire d'astronomie physique*, Paris.
Bird 2012	D. Bird, 'Telephone Rentals and their Activities in Electric Clocks', Electrical Horology Group Technical Papers, 83.
Birth 2018	Kevin Birth, 'King Alfred's Candles and Anglo-Saxon Time-Reckoning', *Kronoscope*, 18, 117–37.
Bīrūnī 1910, 1964	Al-Bîrûnî, *Alberuni's India: An Account of the Religion, Philosophy, Literature, Geography, Chronology, Astronomy, Customs, Laws and Astrology of India about AD 1030*, Edward C. Sachau (trans.), 2 vols., London; reprint New Delhi.
Bīrūnī 1976	Al-Bîrûnî, *The Exhaustive Treatise of Shadows* (trans. & commentary by E. S. Kennedy, 2 vols., Aleppo.
Blackwell 1992	Dana Blackwell, 'Enquiries', *NAWCC Watch & Clock Bulletin*, 34, 606–7.
Blair, Blair, & Brownsword 1986	Claude Blair, John Blair, & R. Brownsword, 'An Oxford Brasiers' Dispute of the 1390s: Evidence for Brass-Making in Medieval England', *The Antiquaries Journal*, 64, 82–90.
Blakey 1780	William Blakey, *L'Art de faire les resorts de montres, suivi de la manière de faire les petits resorts de repetitions & les resorts spiraux*, Amsterdam.
Blancard 1806	Pierre Blancard, *Manuel du commerce des Indes Orientales et de la Chine* Paris.

Blanchard 2011	Philippe Blanchard, *L'établissage. Etude historique d'un système de production horloger en Suisse (1750–1950)*, Chézard-Saint-Martin.
Blanchard 1895	Raphaël Blanchard, *L'Art populaire dans le briançonnais – les cadrans solaires*, Paris.
Blaufox & Constable 2008	M. Donald Blaufox & Anthony R. Constable, 'The Sandglass and the Pulse', *Bulletin of the Scientific Instrument Society*, 46, 18–19.
Blondel et al. 1989	Christine Blondel, Françoise Parrot, Anthony Turner, & Mari Williams (eds.), *Studies in the History of Scientific Instruments Paper presented at the 7th Scientific Instruments Commission Paris 15–19 Septembert 1987*, London.
Blümmer 1875–86	H. Blümmer, *Technologie und Terminologie der Gewerbe und der Künste bei den Griechen und der Römerne*, 4 vols., Leipzig.
Blundeville 1594	Thomas Blundeveille, *His Exercises, Containing Six Treatises …*, London.
Blyelle 2011	Etienne Blyelle, 'Michel Joseph Ransonet, horloger-inventeur de génie, précurseur de la boîte à musique?: Ransonet à Nancy, un des pères de la boîte à musique', *Musiques mécaniques vivantes*, 80, 27–32.
Blyth 2014	T. Blyth (ed.), *Information Age: Six Networks That Changed Our World*, London.
Bobinger 1954	Maximilian Bobinger, *Christoph Schissler der Ältere und der Jüngere*, Augsburg.
Bobinger 1966	Maximilian Bobinger, *Alt-Ausburger Kompassmacher: Sonnen-, Mond- und Sternuhren, Astronomische und mathematikische Geräte Räderuhren*, Augsburg.
Bobynet 1663	Pierre Bobynet, *L'Horographie ingenieuse contenant des connaissances et des curiositez agreeable dans la composition des cadrans …*, Paris (1st edition, 1647).
Bodenmann 2011	Laurence Bodenmann (ed.), *Philadelphia 1876: Le défi américain en horlogerie. De l'unique à la série: l'interchangeabilité. Actes de colloque 19.10–20.10.2010*, La Chaux-de-Fonds.
Bohlhalter 2016	Bruno Bohlhalter, *Unruh: Die schweizerische Uhrenindustrie und ihre Krisen im 20. Jahrhundert*, Zürich.
Böhm 1866	Joseph G. Böhm, 'Beschreibung der alterthümlichen prager Rathhaus-Uhr', Abhandlungen der Königl. Böhmischen Gesellschaft der Wissenschaften *vom Jahre*, Berlin.
Bohnenstengel 1911	E. Bohnenstengel, *Elektrische Automaten Und Fernschalter*, Leipzig.
Boileau 1781	N. F. J. Boileau, *Catalogue des tableaux, pastels, gouaches, dessins, estampes, instrumens de musique et de physique, pendules, meubles vases précieux et porcélaines qui composoient le Cabinet de M. Thomas de Pange … dont la vente se fera en son Hôtel, rue des Saints-Peres, vis-à-vis la grande rue Taranne, le Lundi 5 Mars a 1781, & jours suivans, de relevée*, Paris.
Boillat 2010	Johann Boillat, 'Etat et industrie: l'exemple du cartel horloger suisse (1931–1951)', in Cortat 2010, 89–136.
Boillat 2012	Johann Boillat, 'Statut horloger', in Dictionnaire historique de la suisse [online]. Available at: https://hls-dhs-dss.ch/fr/articles/013790/2012-05-01.
Boillat 2013a	Johann Boillat, *Les véritables maîtres du Temps. Le cartel horloger suisse (1919–1941)*, Neuchâtel/La Chaux-de-Fonds.
Boillat 2013b	Johann Boillat, 'Les trois temps de l'Etat horloger (1934-1991)', *Chronométrophilia*, 73, 52–65.
Boillat 2016	Johann Boillat, 'L'or des horlogers. L'industrie neuchâteloise des métaux précieux (1846–1998): acteurs et réseaux', *Revue historique neuchâteloise*, 153, 23–45.
Boillat 2018	Johann Boillat, 'Birth of a Military Sector. The Case of the International Beryllium Industry (1919–1939)', in Garufo & Morerod 2018, 41–64.
Boillat 2019a	Johann Boillat, 'Des banquiers des horlogers aux horlogers des banquiers. Une analyse quantitative du patronat des cantons de Berne, Neuchâtel et Soleure (1900–1950)', in Flores *et al.* 2019, 177–99.
Boillat 2019b	Johann Boillat, 'La culture technique allemande de la chronométrie neuchâteloise: Jules Grossmann, l'Ecole d'horlogerie du Locle et l'Observatoire cantonal de Neuchâtel (1868-1907)', *Bulletin de la Société suisse de chronométrie*, 2, 41–50.
Boillat 2020	Johann Boillat, 'Le béryllium des horlogers. Transferts technologiques et district industriel dans l'Arc jurassien (XIXe–XXe siècles)', *Actes de la société jurassienne d'émulation*, 122, 115–40.
Boillat & Garufo 2012	Johann Boillat & Francesco Garufo, 'De la protection à la promotion: aux sources du Swiss made horloger (1924–1980)', in Decorzant *et al.* 2012, 209–26.
Boillat & Garufo 2013	Johann Boillat & Francesco Garufo, 'Au cœur du consensus helvétique: la commission consultative de l'horlogerie suisse (1946–1951)', in Fraboulet, Humair, & Vernus 2013, 55–67.
Bollen 1978	Ton Bollen, *Franse Lantaarnklokken*, Bussum.
Bolli 1957	Jean-Jacques Bolli, *L'aspect horloger des relations commerciales américano-suisses de 1929 à 1950*, La Chaux-de-Fonds.
Bond 1833	William Cranch Bond, 'Observations on the Comparative Rates of Marine Chronometers', *Memoirs of the American Academy of Arts and Sciences*, 1, 84–90.

Bönig 1993	Jürgen Bönig, *Die Einführung von Fließbandarbeit in Deutschland bis 1933. Zur Geschichte einer Sozialinnovation*, 1, Münster/Hamburg.
Bonnin 2011	Jérôme Bonin, 'Symbolic Meanings of Sundials in Antiquity', *Bulletin of the British Sundial Society*, 23, 6–10.
Bonnin 2012	Jérôme Bonin, Horologia Romana, Recherche sur les instruments de mesure du temps à l'époque romaine - Étude typologique, urbanistique et sociale. PhD thesis, l'Université de Lille. Available at: https://syrte.obspm.fr/astro/archeo/.
Bonnin 2015	Jérôme Bonin, *La mesure du temps dans l'Antiquité*, Paris.
Bonnant 1960	Georges Bonnant, 'The Introduction of Western Horology in China', La *suisse horlogère*, 1, 28–38.
Bonnant 1962	Georges Bonnant, 'Notes sur l'introduction de l'horlogerie occidentale en Extrême-Orient', *La suisse horlogère*, 1, 33–38.
Bonnant 1964	Georges Bonnant, 'Quelques aspects du commerce d'horlogerie en Chine à la fin du XVIIIe et au cours du XIXee siècles', *La suisse horlogère*, 3, 41–5.
Bonnet 2017	Michel Bonnet, 'Les fabricants d'outils d'horlogerie de Montécheroux', dans Actes 2017, 121–8.
Bonnin 2013	Jérôme Bonnin, 'Horologia et *memento mori*. Les hommes, la mort et le temps dans l'antiquité gréco-romaine', *Latomus*, 72, 468–91.
Boorstin 1983	Daniel J. Boorstin, *The Discoverers*, New York.
Booth 1860	Mary L. Booth, *New and Complete Clock and Watchmakers' Manual. Comprising Descriptions of the Various Gearings, Escapements, and Compensations Now in Use in French, Swiss, and English Clocks and Watches, …*, New York (editions 1869, 1870, 1872, 1877, 1882, 1889).
Boquen 2019	Manon Boquen, '1925: Les radium girls reclament la justice', Causette.fr, 1 October [online]. Available at: https://www.causette.fr/societe/a-l-etranger/1925-les-radium-girls-reclament-justice.
Borchardt 1920	Ludwig Borchardt, 'Die Ältagyptische Zeitmessung', in E. Von Bassermann-Jordan (ed.), *Die Geschichte der Zeitmessung und der Uhren*, vol. 1, Lg. B, Leipzig/Berlin.
Borrel 1901	Georges Borrel, *Exposition Universelle international de 1900. Rapports du Jury International, classe 96 – Horlogerie*, Paris.
Borst 1998	Arno Borst, *Die karolingische Kalenderreform*, Stuttgart.
Bosard 1989	J. Bosard, 'Le cadran solaire décimal de l'ère républicaine', *Le Ciel*, 51, 291–2.
Bosticco 1957	Sergio Bosticco, 'Due frammenti di orologi solari egiziani', in *Studi in onore di Aristide Calderini e Roberto Paribeni I*, Milan, 33–49.
Bouchard 2015	André Bouchard, *A la recherche de la beauté dans un cadrans solaire, esthétique et gnomonique*, Outremont.
Boulle c.1724	André Charles Boulle, *Nouveaux desseins de meubles et ouvrages de bronze et de marqueterie. Inventés et gravés par André-Charles Boulle*, Paris.
Boursier 1936	Charles Boursier, *800 devises de cadrans solaires*, Boulogne.
Bouthier 2011	Alain Bouthier, 'L'acier dans le Nivernais et l'expérimentation à Cosne', in Philippe Dillmann, Liliane Hilaire-Perez, & Catherine Verna (eds.), *L'Acier en Europe avant Bessemer*, Toulouse, 315–40.
Bouwens & Dankers 2013	Bram Bouwens & Joost Dankers, 'Competition and Coordination: 2013 Reconsidering Economic Cooperation in Dutch Business, 1900–2000', *Revue économique*, 64, 1105–24.
Bowen 2007	Huw Bowen, 'Privilege and Profit: Commanders of East Indiamen as Private Traders, Entrepreneurs and Smugglers 1760–1813', *International Journal of Maritime History*, 19, 43–88.
Bradley 2015	Howard Bradley, 'Foliot Revisited – The Origins of the Word', *AH*, 36, 91–6.
Braga 1967	J. M. Braga, 'A Seller of "Sing-Songs" - A Chapter in the Foreign Trade of China and Macao', *Journal of Oriental Studies*, 6, 61–108.
Brand 1999	Stewart Brand, *The Clock of the Long Now: Time and Responsibility*, London.
Brandi 1955	C. Brandi, 'Chiaramenti sul "Buon Governo" di Ambrogio Lorenzetti' *Bolletino d'Arte*, 40, 119–23.
Braun & McCarthy 2000	Theodore E. D. Braun & John Aloysius McCarthy (eds.), *Disrupted Patterns: On Chaos and Order in the Enlightenment*, Amsterdam.
Bredekamp 1995	Horst Bredekamp, *The Lure of Antiquity and the Cult of the Machine, The Kunstkammer and the Evolution of Nature, Art and Technology*, Princeton.
Bredekamp 2000	Horst Bredekamp, *Antikensehnsucht und Maschinenglauben*, (new edn.), Berlin.
Breguet 1997	Emmanuel Breguet, *Breguet, Watchmakers since 1775: The Life and Legacy of Abraham-Louis Breguet (1747–1823)*, Paris.
Bret 2019	Patrice Bret, 'Instruments of Knowledge and Power in a Colonial Context: Scientific Istruments during the French Occupationof Egypt, 1798–1801', in Brown *et al.* 2019, 206–28.
Brett 1954	Gerard Brett, 'The Automata in the Byzantine "Throne of Solomon"', *Speculum*, 29, 477–87.
Brevet & Chapiro 1981	Léonide Brevet & Adolphe Chapiro, 'Un ingénieux dispositif à renvoi pour améliorer le rendement du travail sur tout à pivoter inventé dans la région de Cluses', *Horlogerie ancienne*, 31, 29–37.

Brice et al. 1976	Willlliam Brice, Colin Imber, & Richard Lorch, *The Dâ'ire-yi Mu'addel of Sedî 'Alî Re'îs* (Seminar on Early Islamic Science Monograph 1), Manchester.
Brieux & Maddison (2021)	Alain Brieux & Francis Maddison, *Répertoire des facteurs d'astrolabes et de leurs œuvres en terre d'Islam*, Turnhout.
Briselance 2015	Claude Briselance, *Les écoles d'horlogerie de Besançon. Une contribution décisive au développement industriel local et régional (1793–1974)*. Unpublished thesis, Université de Lyon II.
British Clock Mfs Ass 1933	'The British Clock Manufacturers', in *Special Supplement to* The Watch and Clock Maker …
Britten 1894	F. J. Britten, *Former Clock & Watchmakers and Their Work*, …, London.
Britten 1907	F. J. Britten, *The Watch and Clock Makers' Handbook, Dictionary and Guide*, 11th edn., London.
Britten 1929	F. W. Britten, *Horological Hints and Helps*, London.
Brix 1981	H. Brix, 'Another Standard-Time Sundial', *Journal of the British Astronomical Association*, 92, 16–21.
Brockhaus 1955–60	Otto Lehman Brockhaus, *Lateinische SWchriftquellen zur Kunst in England, Wales, und Schottland vom Jahr 901 bis zum jahr 1307*, 5 vols., Munich.
Brockliss & Jones 1997	Laurence Brockliss & Colin Jones, *The Medical World of Early Modern France*, Oxford.
Bromley 1982	A. G. Bromley, 'Charles Babbage and the Invention of Workmen's Time Recorders', *AH*, 13, 444.
Bromley 1977	John Bromley, *The Clockmakers' Library. The catalogue of the books and manuscripts in the library of the … Clockmakers Company*, London.
Brooks 1984	Chris Brooks, *Signs for the Times: Symbolic Realism in the Mid-Victorian World* London.
Brooks 2000	Hero Boothroyd Brooks, *A Short History of IIC*, London.
Brown 2000	David Brown, 'The Cuneiform Conception of Celestial Space and Time', *CAJ*, 10, 103–22.
Brown et al. 2019	Neil Brown, Silke Ackermann, & Feza Günergun (eds.), *Scientific Instruments between East and West*, Scientific Instruments and Collections, vol. 7, Leiden.
Brown, Fermor, & Walker 2000	David Brown, John Fermor, & Christopher Walker, 'The Water Clock in Mesopotamia', *AOF*, 446, 130–48.
Brown 2010	Nancy Marie Brown, *The Abacus and the Cross*, New York.
Brown 1982	Olivia Brown, *The Whipple Museum of the History of Science: Catalogue 1, Surveying*, Cambridge.
Brown 1971	Peter Brown, *The World of Late Antiquity AD 150–750*, London.
Brown 2015	R. Brown, 'Time Travelers: A History of Electric Clocks for the American Automobile', *NAWCC Watch & Clock Bulletin*, 57, 307–22.
Browne 1672	Thomas Brown, *Pseudoxia Epidemica: or, Enquiries into Very Many Received Tenents and Commonly Presumed Truths*, 6th edn., London.
Bruce 1912	J. Douglas Bruce, 'Human Automata in Classical Tradition and Medieval Romance', *Modern Philology*, 10, 511–26.
Brusa 1978	Giuseppe Brusa, *L'arte dell'orologeria in Europa. Sette Secoli di orologi meccanici*, Milan.
Brusa 1980	Giuseppe Brusa, 'La navicelle orarie di Venezia', *Annali dell'Istituto e Museo di Storia della Scienza*, 5, 51–9.
Brusa 1981	Giuseppe Brusa, 'Orologi Meccanici', in Carlo Pirovano (ed.), *Museo Poldi Pezzoli, Orologi, Oreficerie*, Milan.
Brusa 1990	Giuseppe Brusa, 'Early Mechanical Horology in Italy', *AH*, 19, 485–513.
Brusa 1994	Giuseppe Brusa, 'L'orologio dei pianete di Lorenzo della Volpaia', *Nuncius, Annali di Storia della Scienza*, 9, 645–69.
Brusa 2005	Giuseppe Brusa, *La Misura del Tempo: l'antico splendore dell'orolgeria italiana dal XV al XVIII secolo*, Trento.
Brusa & Allix 2006	Giuseppe Brusa and Charles Allix, 'The Eminent Pierre Le Roy in the Art of Timekeeping', *AH*, 29, 645–62, 775–89.
Brusa & Leopold 1999	Giuseppe Brusa & J. H. Leopold, The Vertical Stackfreed: Mechanical Evidence and Early Eye-Witnesses, *AH*, 25, 166–79.
Bruton 2002	Eric Bruton, *The History of Clocks and Watches*, London.
Bryden 2018	D. J. Bryden, 'The Bankruptcy of W & T Gilbert, Optical and Mathematical Instruments Makers', *London's Industrial Archaeology*, 16, n.p.
Bucciantini, Camerota, & Roux 2007	Massimo Bucciantini, Michele Camerota, & Sophie Roux, *Mechanics and Cosmology in the Early Modern Period*, Florence.
Buchanan-Brown 1972	John Buchanan-Brown (ed.), *John Aubrey: Three Prose Works, Miscellanies, Remaines of Gentilisme and Judaisme, Observations*, Fontwell.
Buchner 1992	Alexander Buchner, *Les Instruments de musique mécanique … adaptation française de Philippe Rouillé*, Paris.
Buchner n.d.	Alexander Buchner, *Mechanical Musical Instruments*, Iris Unwin (trans.), London.
Buchner 1976	Edmund Buchner, 'Solarium Augusti und Ara Pacis', *Römischen Mittheilungen*, 83, 319–65.
Buchner 1982	Edmund Buchner, *Die Sonnenuhr des Augustus*, Mainz.

Buchon 1828	J. A. Buchon (ed.), 'Discours de Nicetas Chroniates sur les monuments détruits ou mutilés par les croisés en 1204', in *Chroniques de la prise de Constantinople par les Francs écrite pa Geoffroy de Ville-Hardouin*, Collection des Chroniques Nationales Françaises III, Paris, 323–38.
Bud & Warner 1998	Robert Bud & Deborah Jean Warner, *Instruments of Science, An Historical Encyclopedia*, New York.
Buffard 2017	François Buffard, 'L'Horlogerie à Morbier de 1684 à aujourd'hui', in *Actes 2017*, 31–8.
Buffard et al 2019	François Buffard et al., *L'horloge comtoise et ses horlogers les paysans-horlogers, les négociants, les émailleurs, les maîtres de forges*, Morez.
Buffard, Dumain & Renard 2013	François Buffard, Michel Dumain, & Marie-Paul Renard, *Petite histoire des horloges d'édifice: les fabricants du Haut-Jura*, Morez.
Bugati 1570	Gasparo Bugati. 'Historia universale', in *Vinegia: Appresso Gabriel Giolito de Ferrari*, Italy.
Bull 1978	Simon P. Bull, *Geneva Watches 1630–1720*, Basle.
Bullo 2003	Aldo Bullo (ed.), *Tractatus astrarii di Giovanni Dondi dall' Orlogio*, Conselve.
Burchall 1985	Adrian Burchall, 'The Noctuary or Watchman's Clock: Its Introduction and Development', *AH*, 15, 231–51.
Burke 1969	Peter Burke, *The Renaissance Sense of the Past*, London.
Burnett et al. 2004	Charles Burnett et al. (eds.), *Studies in the History of the Exact Sciences in Honour of David Pingree*, Boston.
Burton 1994	John Burton, *An Introduction to the Hadîth*, Edinburgh.
Bush 1880	H. Bush, 'Pneumatic Clocks', *HJ*, 23, 61–3.
Butler 1978	Alfred J. Butler, *The Arab Conquest of Egypt and the Last Thirty Years of Roman Dominion*, 2nd edn., Oxford.
Butler 1949	H. E. Butler (ed.), *The Chronicle of Jocelin of Brakeland*, London.
Byrd 1992	William A. Byrd (ed.), *Chinese Industrial Firms Under Reform*, Oxford.
Byrd & Tidrick 1992	William A. Byrd & Gene Tidrick, 'The Chongqing Clock and Watch Company', in Byrd 1992, 58–119.
Cabanelas 1958	Dario Cabanelas, 'Relojes del sol hispano-musulman', *Al-Andalus*, 25, 391–406.
Calmet 1728	Augustine Calmet, *Histoire ecclésiastique et civile de Lorraine*, Nancy.
Camerer Cuss 1976	Terence Camerer Cuss, *The Camerer Cuss Book of Antique Watches*, Woodbridge.
Camerer Cuss 2009	Terence Camerer Cuss, *The English Watch 1585–1970*, Woodbridge.
Camerota 2000	Filippo Camerota, 'La meridiana "tetracycla" del Quirinal', in Francesco Borromini (ed.), *Proceedings of the International Conference of Studies, Rome, Palazzo Barberini - Hertziana Library, January 13–15, 2000*, Milan, 233–41.
Campbell 1747	R. Campbell, *The London Tradesman being a Compendious View of all the Trades, Professions, Art . . . Now Practised in the Cities of London and Westminster . . .*, London.
Capart 1938	Jean Capart, 'Horloges Egyptiennes', *Bulletin des Musées Royales de l'Art et l'Histoire*, 3, n.p.
Capital Museum 2015	Capital Museum (ed.), *Geneva at the Heart of Time: The Origin of Swiss Watchmaking Culture*, Beijing.
Caplan 2012	James Caplan, 'Following the Stars: Clockwork for Telescopes in the Nineteenth Century', in Alison Morrison-Low, Sven Dupré, Stephen Johnston, & Giorgio Strano (eds.), *From Earth-Bound to Satellite. Telescope. Skills and Network*, Leiden/Boston, 155–76.
Carafa 1689	Carolo Maria Carafa, *Exemplar Horologium Solarium*, Mazzarino.
Carandell 1984	Joan Carandell, 'An Analemma for the Determination of the Azimuth of the Qibla in the *Risala fî' Ilm al-Zilâl* of Ibn Raqqâm', *Zeitschrift für Geschichte der Arabisch-Islamischen Wissenschafften*, 1, 61–72.
Carandell 1989	Joan Carandell, 'Dos cuadrantes solares andalusies de Medina Azahara', *Al-Qantara*, 10, 329–42.
Carboni 2007	Stefano Carboni, *Venice and the Islamic World, 828–1797*, New York.
Carcopino 1939	Jérôme Carcopino, *Rome à l'apogée de l'Empire, Ier siècle après J.-C.*, Paris, 183–93.
Cardano 1557	Girolamo Cardano, *De varietate rerum, libri xxii*, Basle.
Cardinal 1983	Catherine Cardinal, *L'horlogerie dans l'histoire, les arts et les sciences. Chefs d'œuvres du Musée International d'horlogerie de La Chaux-de-Fonds*, Lausanne.
Cardinal 1984	Catherine Cardinal, *Les montres et horloges de table du Musée du Louvre. La collection Olivier*, Paris.
Cardinal 1985	Catherine Cardinal, *The Watch, from its Origins to the XIX Century*, Lausanne.
Cardinal 1989	Catherine Cardinal (ed.), *La Révolution dans la mesure du temps: calendrier républicain, heure decimal 1793–1805*, La Chaux-de-Fonds.
Cardinal 1997	Catherine Cardinal (ed.), *Abraham-Louis Breguet 1747–1827. L'art de mesurer le temps*, La Chaux-de-Fonds.
Cardinal 1998	Catherine Cardinal, *Le metre et la seconde: Charles-Edouard Guillaume (1861–1938), prix Nobel de physique*, La Chaux-de-Fonds.
Cardinal 1999a	Catherine Cardinal (ed.), *Apprendre, créer, transmettre. La formation des horlogers, passé et avenir. Actes du colloque*, La Chaux-de-Fonds.
Cardinal 1999b	Catherine Cardinal, *Splendeurs de l'émail. Montres et horloges du XVIe au XXe siècle*, La Chaux-de-Fonds.
Cardinal 2000	Catherine Cardinal, *Les montres et horloges de table du musée du Louvre*, vol. 2, Paris.
Cardinal 2002	Catherine Cardinal, 'Les émailleurs et la décoration des montres', *Les arts décoratifs sous Louis XIII*, Dijon.

Cardinal 2010	Catherine Cardinal, 'Les années 1760 à 1820 en France: un temps de perfection pour la montre', in *Montres et Merveilles. Collection Musée du Temps Besançon*, Besançon.
Cardinal 2018a	Catherine Cardinal, 'New Light on French Enamel Painting (1630–1660): *Tondi* as Models for the Decoration of Watches', *AH*, 39, 69–79.
Cardinal 2018b	Catherine Cardinal, 'Jacques Goullons (*c*.1600–1671), Master Clockmaker on the Île de la Cité, Paris', *AH*, 39, 185–201.
Cardinal & Galli 2017	Catherine Cardinal & Lavinia Galli, 'Watches', in *Art Deserves More Space; New Collections for the Poldi Pezzoli Museum*, Milan.
Cardinal et al. 1991	Catherine Cardinal, François Jequier, Jean-Marc Barrelet, & André Beyner (eds.), *L'homme et le temps en Suisse 1291–1991*, La Chaux-de-Fonds.
Cardinal & Mercier 1993	Catherine Cardinal & François Mercier, *Musées d'horlogerie, La Chaux-de-Fonds, Le Locle*, Zurich.
Cardinal & Piguet 1999	Catherine Cardinal & Jean-Michel Piguet, *Catalogue d'oeuvres choisies du Musée international d'horlogerie*, La Chaux-de-Fonds.
Cardinal & Sabrier 1987	Catherine Cardinal & Jean-Claude Sabrier, *La dynastie des Le Roy, horlogers du roi*, Tours.
Cardinal & Vingtain 1998	Catherine Cardinal & Dominique Vingtain (eds.), *Trésors d'horlogerie: le temps et sa mesure du Moyen Age à la Renaissance*, Avignon.
Carrel 1936	Laurent Carrel, *Normalisierung in der Schweizerischen Uhrenindustrie*, Bern.
Carrera 1980	R. Carrera, *Les heures de l'amour*, Lausanne.
Carrington 1978	R. W. & R. F. Carrington, 'Pierre Frederic Ingold and the British Watch and Clockmaking Company', *AH*, 11, 698–714.
Casanova 1923	Paul Casanova, 'La Montre de Sultan Nôur ad-Dîn', *Syria. Revue d'art oriental et d'archéologie*, 4, 283–99.
Cassini 1732	Jacques Cassini, 'De la méridienne de l'Observatoire', *HARS* 1732, 452–470.
Cassis & Tanner 1993	Youssef Cassis & Jakob Tanner (eds.), *Banken und Kredit in der Schweiz. Banques et crédit en Suisse:* (1850–1930), Zürich.
Casulleras 1993	Josep Casulleras, 'Descripciones de un cuadranta solar atipico en el occidente Musulman', *Al-Qantara. Revista des estudos Arabes*, 14, 65–87.
Catamo 2008	Mario Catamo, *L'Evoluzione della misura orario del temmpo e le meridian di Civita Castellano . . .*, Castellana.
Caudine 1992	Alain Caudine, *La grande horloge, la Comtoise au XIXe siècle*, Paris.
Cavin 1983	Arnold Cavin, 'L'outillage des penduliers', in Chapuis 1983, ch. 10.
Chabert 1978	Philippe-Gérard Chabert, *La Pendule au 'Nègre'*, Saint-Omer.
Chaldecott 1987	John A. Chaldecott, 'Platinum and Palladium in Astronomy and Navigation', *Platinum Metals Review*, 31, 91–100.
Chamberlain 1941	Paul Chamberlain, *It's About Time*, New York.
Chambre . . . Doubs 2019	Chambre de commerce et d'industrie du Doubs, Table ronde, Horlogerie, 2019. La montre ne sert plus à lire l'heure, et alors? Situation, enjeu et perspective.
Chan 2007a	Joel Chan, 'Fist (*sic*) made Chinese Watch - Shanghai Watch Factory', Micmicmor [blog], 6 March 2007. Available at: http://micmicmor.blogspot.com/2007/03/fist-made-chinese-watch-shanghai-watch.html.
Chan 2007a	Joel Chan, '70s & 80s Shanghai watches purchasing coupon', Micmicmor [blog], 25 April 2007. Available at: http://micmicmor.blogspot.com/2007/04/70s-80s-shanghai-watches-purchasing.html.
Chan 2007c	Joel Chan, 'Shanghai Zuanshi (Diamond) Watch Info', Micmicmor [blog], 20 October 2007. Available at: http://micmicmor.blogspot.com/2007/10/shanghai-zuanshi-diamond-watch-info.html.
Chan 2007c	Joel Chan, '上海牌 A581 手表 China', Micmicmor [blog], 8 April 2009. Available at: http://micmicmor.blogspot.com/2009/04/article-wrote-on-may-2008-for-magazine.html.
Chan 2008b	Joel Chan, 'Additional informaton for National Standard Movement SZ1 (Tongji)', Micmicmor [blog], 24 April 2008. http://micmicmor.blogspot.com/2008/04/additional-informaton-for-national.html.
Chandler & Vincent 1980a	Bruce Chandler & Claire Vincent, 'To Finance a Clock: An Example of Patronage in the 16th Century', in Maurice & Mayr 1980, 103–13.
Chandler & Vincent 1980b	Bruce Chandler & Clare Vincent, 'A Rock Crystal Watch with a Cross-Beat Escapement', *The Metropolitan Museum Journal*, 15, 193–201.
Chandra 1949	Moti Chandra, *Jain Miniature Painting from Western India*, Ahmedabad.
Chang 2016	David Chang, *Bovet 1822: The Legend*, Plan-les-Ouates.
Chang 2020	David Chang, 'Macao: A Turning Point in the "Horological Road"', Sina [blog], 9 March 2020. Available at: http://blog.sina.com.cn/s/blog_4c7248850102yqtj.html.
Chang & Bai 2009	David Chang & Bai Yingze, China and Horologe, Shanghai.
Chapiro 1988	Adolphe Chapiro, *Jean-Antoine Lépine, horloger (1720–1814) Histoire du développement de l'horlogerie en France de 1760 à l'Empire*, Paris.
Chapiro 1991	Adolphe Chapiro, *La montre française*, Paris.

Chapiro et al. 1989	Adolphe Chapiro, Chantal Meslin-Perrier, & Anthony Turner, *Catalogue de l'horlogerie et des instruments de précision du début du XVIe au milieu du XVIIe siècle*, Paris.
Chaponnière 1924	H. Chaponnière, *Le Chronographe et ses applications*, Bienne/Besançon.
Chapuis 1917	Alfred Chapuis, *Histoire de la pendulerie Neuchâteloise (horlogerie de gros et de petit volume)*, Paris/Neuchâtel.
Chapuis 1919	Alfred Chapuis, *La montre 'chinoise'*, Neuchâtel.
Chapuis 1931	Alfred Chapuis, *Pendules neuchâteloise: documents nouveaux*, Neuchâtel.
Chapuis 1942	Alfred Chapuis, *A travers les collections d'horlogerie*, Neuchâtel.
Chapuis 1944	Alfred Chapuis, *Montres et Émaux de Genève: Louis XIV, Louis XV, Louis XVI et Empire*, Lausanne.
Chapuis 1953	Alfred Chapuis, *A. L. Breguet pendant la Révolution française, à Paris, en Angleterre et en Suisse*, Neuchâtel.
Chapuis 1954	Alfred Chapuis, *De horologies in Arte, l'horloge et la montre à travers les ages d'après les documents du temps*, Lausanne.
Chapuis 1955	Alfred Chapuis, *Histoire de la Boîte à musique*, Lausanne.
Chapuis 1957	Alfred Chapuis, 'L'identification d'un automate Jaquet-Droz', *Musée neuchâtelois*, 44, 54–7.
Chapuis & Droz 1949	Alfred Chapuis & Edmond Droz, *Les Automates. Figures artificielles d'hommes et d'animaux*, Neuchâtel.
Chapuis & Gélis 1928	Alfred Chapuis and Edmond Gélis, *Le Monde des automates*, 2 vols., Paris.
Chardin 1711	Jean Chardin, *Voyages de Mr le Chevalier Chardin en Perse, et aux autres lieux d'Orient*, Paris.
Charette 2003	François Charette, *Mathematical Instrumentation in 14th-Century Egypt and Syria*, Leiden.
Chasles 1837	Michel Chasles, *Aperçu historique sur l'origine et le développement des méthodes en géométrie*, Brussels.
Chavigny 1991	Richard Chavigny, *Les Brocot, une dynastie d'horlogers*, Neuchâtel.
Chavigny 1997	Richard Chavigny, 'Pierre-Honoré-César Pons, pionnier de l'horlogerie industrielle', *Bulletin d'ANCAHA*, 80, 45–56.
Chavigny 2007	Richard Chavigny, *La pendule de Paris et sa cadette la pendulette de voyage*, Paris
Chayette & Cheval 2019	Chayette & Cheval, *Horlogerie . . . 2 décembere 2010*, Paris.
Chayette & Cheval 2019	Chayette & Cheval, *Sciences, techniques, Horlogerie 21 juin 2019*, Paris 2019.
Cheetham 2002	John Cheetham, 'More Mudge Milestones', *AH*, 27, 90–3.
Chen 1983	Chen Jiujin, 'Study on Chinese Ancient Time System and its Conversion', *Studies in the History of Natural Sciences*, 2, 118–32.
Chen 1981	Chen Kaige, 'Clockmaking in Suzhou during the Qing Dynasty', *Palace Museum Journal*, 4, 90–4.
Chen 2014a	Chen Mang, 'British Horologe in China', *National Humanity History*, 14, 120–3.
Chen 2014b	Chen Mang, 'Fascinating in the Far East: Wonderful James Cox's Horological World', *National Humanity History*, 18, 116–19.
Chen & Hua 2011	Chen Meidong & Hua Tongxu (eds.), *General History of Chinese Timing Instrument (Ancient Volume)*, Hefei.
Chen et al. 2018	Yuyu Chen, Mitsuru Igami, Masayuki Sawada, & Mo Xiao, 'Privatization and Productivity in China', *VoxChina*. 31 January 2018. Available at: http://voxchina.org/show-3-64.html.
Chen 2016	Chen Zungui, *History of Chinese Astronomy*, Shanghai.
Chen 1987	Chen Zuwei, 'Clockmaker Xu Chaojun in the Qing Dynasty and an Illustrated Treatise of Horology', *China Historical Materials of Science and Technology*, 8, 43–5, 58.
Chenakal 1972	Valentin L. Chenakal, *Watchmakers and Clockmakers in Rusia, 1400 to 1800*, W. F. Ryan (trans.), Ramsgate.
Cheong 1965	W. E. Cheong, 'Trade and Finance in China: 1784–1834", *Business History*, 7, 34–56.
Cheong 1979	W. E. Cheong, *Mandarins and Merchants*, Curzon.
Cheong 1997	W. E. Cheong, *The Hong Merchants of Canton*, Curzon.
Cherbel 1971	C. D. Cherbel, 'Some Features of 16th-, 17th- and 18th-Century Italian Clocks', *AH*, 7, 198–204.
Cheruel 1844	Adolphe Chéruel, *Histoire de Rouen pendant l'époque communale –1382*, Rouen.
Chevallier 1991	Bernard Chevallier, *La mesure du temps dans les collections du musée de Malmaison*, Paris.
China Daily 2015	'Time to Celebrate a Century of Clock-Making', *China Daily*, 14 August 2015. Available at: https://www.chinadaily.com.cn/culture/art/2015-07/14/content_21271840.htm.
Chipps-Smith 2013	Jeffrey Chipps-Smith, 'Peter Flötner and the Theater of the World', in Kaschek, Muller, & Schauerte 2013, 175–95.
Choisselet & Vernet 1989	D. Choisselet & P. Vernet, *Les Ecclesiastica Officia Cisterciens du XIIème siècle*, Turnhout.
Christie, Manson, & Woods Ltd. 1980	Christie, Manson, & Woods Ltd., *The Codex Leicester by Leonardo da Vinci*, London.
Christie's 1792	Christie's, *Catalogue of various musical clocks, pieces of mechanism [. . .] late the property of James Cox [etc]*, 16 February, London.
Christie's 2007	Christie's, *Art of the Islamic and Indian Worlds*, 17 April 2007, London.
Ciotti 1991	Ugo Ciotti, 'Testimonianze di magistrature e sacerdozi a Mevania', in A. E. Feruglio, L. Bonomi Ponzi, & D. Manconi (eds.), *Mevania: Da centro umbro a municipio romano*, Perugia, 81–5.
Cipolla 1967/2003	Carlo M. Cipolla, *Clocks and Culture 1300–1700*, London.

Ciurlia 2019	Antonio Ciurlia (ed.), *Nuovi Studi sulla chiesa di Santa Maria della Strada di Taurisano. Atti di Convegno du studi – 2000*, 2001, 2004, Taurisano.
Clagett 1995	Marshall Clagett, *Ancient Egyptian Science II: Calenders, Clocks, and Astronomy*, American Philosophical Society Memoirs 214, Philadelphia.
Clark 1882	Latimer Clark, *A Treatise on the Transit Instrument as Applied to the Determination of Time*, London.
Clark 1888	Latimer Clark, *Transit Tables for 1888 . . .*, London
Classet 1880	Jean Classet, *Histoire de l'Eglise au Japon in Dajo-kan*, trans. sec., *Japanese Christian History* (日本西教史), 1.
Clavius 1581	Christopher Clavius, *Gnomonices libri octo in quibus non solum horologiorum solarium*, Rome.
Clavius 1586	Christopher Clavius, *Fabrica et usus instrumenti ad horologium*, Rome.
Clavius 1599	Christopher Clavius, *Horologium nova descriptio*, Rome.
Clavius 1605	Christopher Clavius, *Tabulae Astronomicae Nonnullae ad Horologium*, Rome.
Clerizo 2013	Michael Clerizo, *George Daniels: A Master Watchmaker & His Art*, London.
Cliborne 1981	A. F. Cliborne, 'Mudge Milestones, Addenda', *AH*, 13, 144–5.
Clifton 1995	Gloria Clifton *Directory of British Scientific Instrument Makers 1550–1851*, London.
Cloutman & Linnard 2003	Edward Cloutman & William Linnard, *Henry Williams Lancarvan: 'A Clock and Watchmake and a Great Farmer' in the Vale of Glamorgan*, Cardiff.
Clutton & Daniels 1975	Cecil Clutton & George Daniels, *Clocks and Watches in the Collection of the Worshipful Company of Clockmakers*, London.
Clutton & Daniels 1979	Cecil Clutton & George Daniels, *Watches: A Complete History of the Technical and Decorative Development of the Watch*, London.
Cluzon, Delpont, & Mouliérac 1994	Sophie Cluzon, Eric Delpont, & Jeanne Mouliérac, *Syrie. Mémoire et Mouliérac Civilisation*, Paris.
Cohen & Drabkin 1966	Morris R. Cohen & I. E. Drabkin (eds.), *A Source Book of Greek Science*, Cambridge, MA.
Coleman 1960	D. C. Coleman, *The Domestic System in Industry*, Historical Association: Aids for Teachers Series N° 6, London (reprinted 1964).
Collin 2003	Dominique Collin, 'Les cadrans solaires verticaux à deux gnomons rectilignes quelconques [Généralisation des cadrans bifilaires de Michnik]', *Observations & Travaux*, 55, 12–31.
Collin 2013	Dominique Collin, 'L'astrolabe d'Oughtred', *Cadran Info*, 27, 21–38.
Cologni et al. 1994	Franco Cologni, Giampero Negretti, & Franco Nencini, *Montres et Merveilles de Piaget 1874–1994*, Lausanne.
Colon de Carjaval 1987	J. Ramon Colon De Carjaval, *Catalogo del Patrimonia Nacional*, Madrid.
Commandino 1562	Federico Commandino, *Claudii Ptomemæi, Liber de Anelemmate*, Rome.
Constable 1975	Giles Constable, 'Horologium stellare monasticum', *Consuetudines Benedictinæ variæ*, 6, 2–11.
Conybeare 1969	F. C. Conybeare, *Philostratus. The Life of Apollonius of Tyana . . .*, 2 vols., London.
Coole & Neumann 1972	P. G. Coole & E. Neumann, *The Orpheus Clocks*, London.
Cooling 2018	Louise Cooling, 'Timepieces in Victorian Narrative Painting', *AH*, 39, 528–44.
Cools & Liska 2016	Arthur Cools and Vivian Liska (eds.), *Kafka and the Universal*, Berlin.
Coppard 1969	George Alfred Coppard, *With a Machine Gun to Cambrai*, London.
Coquery 2011	Natacha Coquery, *Tenir boutique à Paris au XVIIIe siècle. Luxe et demi-luxe*, Paris.
Cordier 1922	Henri Cordier, *Les Correspondants de Bertin, secretaire d'etat au XVIIIe siècle*, Leiden.
Cornec & Segalen 2010	Jean-Paul Cornec & Pierre Labat-Segalen, *Cadrans solaires de Bretagne*, Skol Vreizh.
Cortat 1999	Alain Cortat, 'Vivre chez son maître, vivre avec son maître. Les conditions de l'apprentissage dans l'horlogerie. Montagnes Neuchâteloises (1740–1810)', in Barrelet & Cardinal 1999, 7–16.
Cortat 2010	Alain Cortat (ed.), *Contribution à une histoire des cartels en Suisse*, Neuchâtel.
Cournarie 2011	Emanuelle Cournarie, *La mécanique du geste: trois siècles d'histoire horlogère et mécanique à Saint Nicolas d'Aliermont en Normandie*, Rouen.
Cousins 1969	Frank W. Cousins, *Sundials*, London.
Cowham 2007	Mike Cowham, 'Early French "Shell" Dials', *BBSS*, 29, 19–23.
Cox 1772a	James Cox, *A Descriptive Catalogue of the Several Superb and Magnificent Pieces of Mechanism and Jewellery . . .*, London.
Cox 1772b	[James Cox], *Descriptive Catalogue . . . of the Spring Gardens Museum*, London.
Cox 1775	James Cox, *The Arrangement, Allotment and Particular Description of the Several Prizes in the Museum Lottery*, London.
Cranmer-Byng 1962	J. L. Cranmer-Byng, *An Embassy to China: Being the Journal Kept by Lord Macartney during His Embassy to the Emperor Ch'ien-lung, 1793–1794*, London.
Craven 2015	Maxwell Craven, *John Whitehurst of Derby, Clockmaker and Scientist 1713–86*, Mayfield.
Crawford 1929	S. J. Crawford (ed.), *Byrthfert's Manual*, London (reprinted 1966).
Crespe 1804	François Crespe, *Essai sur les montres à repetition*, Geneva.

Crom 1970	Theodore R. Crom, *Horological Wheel Cutting Engines*, Gainsville.
Crom 1980	Theodore R. Crom, *Horological Shop Tools, 1700–1900*, Melrose.
Crom 1987	Theodore R. Crom, *Horological and Other Shop Tools, 1700–1900*, Melrose.
Crom 1989	Theodore R. Crom, *Trade Catalogues 1542 to 1842*, Melrose.
Cros 2018	Lucie Cros, *Les ouvrières et le mouvement social: retour sur la portée subversive des luttes de chez Lip à l'épreuve du genre*. Unpublished thesis, Université de Franche-Comté.
Cuissard 1897	Charles Cuissard, *Etude sur le commerce et l'industrie à Orléans avant 1789*, Orléans.
Culme 1987	John Culme, *The Directory of Gold and Silversmiths, Jewellers and Allied Traders, 1838–1914*, Woodbridge.
Cumhail 1991	P. W. Cumhail, *Investing in Clocks and Watches*, London.
Cumming 1766	Alexander Cumming, *The Elements of Clock and Watch-work, Adapted to Practice. In two essays*, London.
Cummings & Simpson 2010	Brian Cummings & James Simpson (eds.), *Cultural Reformations: Medieval and Renaissance in Literary History*, Oxford.
Cummins 1994	Genevieve Cummins, *Chatelaines: Utility to Glorious Extravagance*, Woodbridge.
Cummins 2006	Genevieve Cummins, *Antique Boxes - Inside and Out for Eating, Drinking and Being Merry, Work, Play and the Boudoir*, Woodbridge.
Cummins 2010	Genevieve Cummins, *How the Watch was Worn, A Fashion for Five Hundred Years*, Woodbridge.
Cummins & Ò Gràda 2019	Neil Cummins & Cormac Ò Gràda, *Artisanal Skills, Watchmaking, and the Industrial Revolution: Prescott and Beyond*, University College Dublin Centre for Economic Research, Working paper series 19/24, Dublin.
Cuneo et al. 1988	Paolo Cuneo *et al.*, *Architettura armena. Dal quarto al diciannovesimo secolo*, 2 vols., Rome.
Cuoq 1979	Joseph Cuoq (trans.), Abd al-Rahmân al-Jabartî, *Journal d'un notable du Caire durant l'expédition française 1798–1801*, Paris.
Cutmore 1991	N. Cutmore, *Pin Lever Watches*, Crapstone.
Czajkowski 2019	Michael Czajkowski, 'Interpretation of the Mogila Monastery Clock Dial, Kraków, Poland and Suggestions for the Associated Clock Mechanisms', Available at: https://www.academia.edu/32769208/Interpretation_of_the_Mogi%C5%82a_ Monastery_Clock_Dial_Krak%C3%B3w_Poland_and_Suggestions_for_the_Associated_Clock_Mechanisms.
Daclin 1968	M. Daclin, 'La crise des années 30 à Besançon', *Annales littéraires de l'université de Besançon*, 96, 136.
Dai 1988	Dai Nianzu, *History of Chinese Mechanics*, ShijiazhuZang.
Dai 2010	Dai Nianzu, *A General Survey of Chinese Physics and Mechanics*, Shanghai.
Dain 1933	Alphonse Dain, *La tradition de texte d'Héron de Byzance*, Paris.
Dallas n.d.	K. Dallas. Available at: https://www.stolenhistory.org/.
D'Ancona 1889	Alexandra d'Ancona, *Journal du voyage de Michel de Montaigne en Italie...*, Città di Castello.
Dane 1973	E. Surrey Dane, *Peter Stubbs and the Lancashire Hand Tool Industry*, Altrincham.
Daniel 2005	Christopher St J. H. Daniel, 'The Equation of Time: The Invention of the Analemma, A Brief History of the Subject', *BBSS* 17, 91–100, 142–54.
Daniell 1975	J. Daniell, *Leicestershire Clockmakers*, Leicester.
Daniels 1975	George Daniels, *The Art of Breguet*, London.
Daniels 1981	George Daniels, 'Thomas Mudge, The Complete Horologist', *AH*, 13, 150–74.
Daniels & Markarian 1980	George Daniels & Ohannes Markarian, *Watches & Clocks in the Sir David Salomons Collection...*, London & Jerusalem.
Danisan 2020	Gaye Danisan, 'Cylinder Dials in the History of Ottoman Astronomy', *BBSS*, 32, 10–15.
Danjon 1980	André Danjon, *Astronomie Générale*, Paris.
Darius & Thomas 1989	Jon Darius & P. K. Thomas, 'French Innovation in Clockwork Telescope Drives', in Blondel *et al.* 1989, 145–54.
Darken 2006	Jeff Darken, *Time & Place, English Country Clocks 1600–1840*, London.
Darken & Hooper 1997	Jeff Darken & J. Hooper, *English 30 Hour Clocks*, Great Yarmouth.
Darnell & Nielsen 2013	Poul Darnell & Frank Nielsen, *Ole Roemer's Eclipsareon and Planetarium*, Denmark.
Dante 2011	Robert M. Durling (ed. and trans.), *Dante Alighieri: The Divine Comedy: vol. 3 Paradiso*, Oxford.
Dasypodius 1580a	Conrad Dasypodius, *Warhafftige Außlegung und Beschreybung des Astronomischen Uhrwercks zu Straßburg*, Strasbourg.
Dasypodius 1580b/2008	Conrad Dasypodius, *Heron Mechanicus: seu de mechanicus artibus, atque disciplinis...*, Bernard Aratowsky (trans.), Introduction and Commentary by Günther Oestmann, Augsburg.
Daumas 2004	Jean-Claude Daumas (ed.), *Les systèmes productifs dans l'Arc jurassien. Acteurs, pratiques et territoires, XIXe–XXe siècles*, Besançon.
Daumas 2010	Jean-Claude Daumas (ed.), *Dictionnaire historique des patrons français*, Paris.
Daumas 2018	Jean-Claude Daumas, *La révolution matérielle au fil de l'histoire*, Paris.

David 1877	Jacques David, *Rapport à la Société intercommunale des industries du Jura sur la fabrication de l'horlogerie aux États-Unis*, reprint St Imier 1992.
Davies 1996	A. Davies, 'Technical Expertise and Public Decisions: British Watchmaking 1842–43: A Case Study', *1996 International Symposium on Technology and Society Technical Expertise and Public Decisions, Princeton, NJ, 21–22 June 1996*, New York, 438–47.
Davies 1984	Alun C. Davies, 'David Smyth Torrens, 1897–1967, An Horological Memoir', *AH*, 14, 564–84.
Davies 1985	Alun C. Davies, 'Rural Clockmaking in Eighteenth-Century Wales: Samuel Roberts of Llanfair Caereinion, 1755–1774', *The Business History Review*, 59, 49–75.
Davies 2008	Alun C. Davies, 'An Invasion in Time: American Horology and the British Market', *AH*, 30, 829–44.
Davies 2009	Alun C. Davies, 'The Ingold Episode Revisited: English Watchmaking's Pyrrhic Victory', *AH*, 31, 637–54.
Davies 2011	Alun C. Davies, 'English Horology and the Great War', *AH*, 32, 636–50.
Davies 2012	Alun C. Davies, 'Horology at International Exhibitions', *AH*, 33, 591–608.
Davies 2016	Alun C. Davies, 'Swiss Watches, Tariffs and Smuggling with Dogs', *AH*, 37, 377–83.
Davies & Fouracre 2010	Wendy Davies & Paul Fouracre (eds.), *The Languages of Gift in the Early Middle Ages*, Cambridge.
Davis 1966	Dorothy Davis, *Fairs, Shops, and Supermarkets: A History of English Shopping*, Toronto.
Davis 2003	John Davis, 'The Equation of Time as Shown on Sundials,' *BBSS*, 16, 135–43.
Davis 2011	John Davis, 'Robert Stickford's "De umbris versis et extensis"', *BBSS*, 23, 24–28.
Davis & Lowne 2009	John Davis & Michael Lowne, *The Double Horizontal Dial*, British Sundial Society Monograph 5, London.
Davis 1818	T. S. Davis, 'An Enquiry into the Geometrical Character of the Hour Lines upon the Antique Sundials', *Transactions of the Royal Society of Edinburgh*, 8, 77–122.
Davison 1993.	G. Davison, *The Unforgiving Minute: How Australia Learned to Tell the Time*, Melbourne.
Dawe 1974	Donovan Dawe, 'The Mysterious Pyke, Organ Builder', *The Musical Times*, 115, 68–70.
Dawson, Drover, & Parkes 1982	Percy G. Dawson, C. B. Drover, & D. W. Parkes, *Early English Clocks, A Discussion of Domestic Clocks up to the Beginning of the Eighteenth Century*, Woodbridge.
Day 1578	Richard Day, *A Book of Christian Prayers*, London.
DeB Beaver 1970	Donald deB Beaver, 'Bernhard Walther: Innovator in Astronomical Observation', *Journal for the History of Astronomy*, 1, 39–43.
De Beaurepaire 1892	Charles de Beaurepaire, *Dernier recueil de notes historiques et archéologiques concernant … la Seine-Inférieure et … la ville de Rouen*, Rouen.
De Beaurepaire 1905	Charles de Beaurepaire, 'Notes sur d'anciennes fabriques de Rouen d'après des declarations d'exportation', *Bulletin de la commission des antiquités de la Seine-Maritime*, 15, 420ff.
De Camus 1735	Charles Etienne Louis de Camus, 'Sur le Figure des dents des roues …', *HARS pour 1733*, Paris, 117–40.
De Cangey 1517	Mathurin de Cangey, *Liber Usum Cisterciensis*, Paris.
De Carle 1947	Donald de Carle, *British Time*, London.
De Carle 1975	Donald de Carle, *Watch and Clock Encyclopedia*, London.
De Châles 1690	Milliet de Châles, *Cursus mathematicus*, 3 vols., Lyon.
De Charmasse 1888	Anatole de Charmasse, *Journal de la société archéologique d'Autun* (La Société éduenne), i.
Decorzant et al., 2012	Yann Decorzant, Alix Heiniger, Serge Reubi, & Anne Vernat (eds.), *Made in Switzerland: Mythen, Funktionen, Realitäten*, Basel.
Décret	Convention Nationale, *Décret … 1793 du 4ᵉ jour de Frimaire, an second de la République Française … sur l'ère, le commencement & l'organisation de l'année, …*, Paris. Reprinted in Droz & Flores 1989, 91–109.
De Floutières 1619	Pierre de Floutrières, *Traitté d'horlogeographie … avec un moyen et invention nouvelle …*, Paris.
Defossez 1946	Léopold Defossez, *Les Savants du XVII Siecle et La Mesure du Temps*, Lausanne.
Defossez 1950	Léopold Defossez, *Théorie Générale de l'Horlogerie*, 2 vols., La Chaux-de-Fond.
Defossez 1956	Léopold Defossez, 'A propos de l'origine de la montre', *Journal suisse de l'horlogerie*, 81, 175–80.
DGC 2019	Deutsche Gesellschaft für Chronometrie (DGC), *Jahresschrift, Time Made in Germany, 12–15 September 2019*, Nürnberg, 58, Germany.
De Granada 1599	Luis de Granada, *An Excellent Treatise of Consideration and Prayer* (trans. Anonymous, of *Libro de la oracion y meditacion*, 1573), Salamanca.
De Guignes 1808	C. L. J. De Guignes, *Voyages à Peking …*, Paris.
De Hautefeuille 1694	Jean de Hautefeuille, *Sentiment … sur le different du R. P. Malebranche … et de M. Regis touchant l'apparence de la lune vüe à l'horizon & au meridian, avec quelques particularitez concernant l'horlogerie …*, Paris.
De Hautefeuille 1718	Jean de Hautefeuille, *Deux problèmes d'horlogerie proposez pour resoudre*, Paris.
Dejardin 2019	Camille Dejardin, *Madame Blakey*, Rennes.
De Jussieu 1719	Antoine de Jussieu, 'Observations sur ce qui se pratique aux mines d'Almaden en Espagne pour en tirer de mercure', *Mémoires de mathématique et de physique … de l'Académie Royale des Sciences*, 349–62.

Dekker 1986	Dekker *The Leiden Sphere: An Exceptional Seventeenth-Century Planetarium*, Leiden.
De Laborde 1851	Le Comte de Laborde, *Les ducs de Bourgogne, études sur les lettres, les arts et l'industrie pendant le XVe siècle*, 2 vols., Paris.
Delahar 1993	Peter Delehar, 'Illustrations of Scientific Instruments in the Gentleman's Magazine, 1746–1796', in Anderson, Bennett, & Ryan 1993, 383–94.
De Lalande 1803	Jérôme de Lalande, *Bibliographie astronomique avec l'histoire de l'astronomie depuis 1781 jusqu'à 1802*, Paris.
Delalande, Delalande, & Rocca 2020	Dominique Delalande, Eric Delalande, & Patrick Rocca, *Astrolabes*, Paris.
Delambre 1814	J. B. J. Delambre, 'D'un cadran trouvé à Délos, et par occasion de la gnomonique des anciens', *Analyse des travaux de la classe des sciences mathématiques et physique de l'Institut Royal de France*, n.p.
Delambre 1814–27	Jean-Baptiste Delambre, *Histoire de l'astronomie . . .*, 4 vols., Paris.
De la 'Peyronie 1793	Gauthier de la Peyronie (ed.), *Voyages de M. P. S. Pallas en differentes provinces de l'empire de Russie*, 4, Paris.
De la Verpilliere 2018	Lorraine de la Verpilliere, *Interactive and Sculptural Printmaking in the Renaissance*, Boston.
De Liman 1854	L. Raguet de Brancion de Liman, *Nouveau cours d'horlogerie . . . à l'usage des fabriquants et des rhabilleurs . . .*, Paris/Besançon.
De Loarte 1579	Gaspar de Loarte, *The Exercise of a Christian Life*, I. S. (trans.), London.
Dembowski 1986	Peter F. Dembowski (ed.), *Jean Froissart: le paradis d'amor, l'orloge amoureus*, Geneva.
De Mercey et al. 2016	Olivier de Mercey, Jean-Paul Crabbe, & Christian Mangé, *L'heure de Vérité, horloge astronomique de la cathédrale de Beauvais*, Saint-Rémy-en-l'Eau.
De Mornay 1587	Philippe de Mornay, *A Woorke Concerning the Trewnesse of the Christian Religion*, Philip Sidney and Arthur Golding (trans.), London.
Deng 1832	Deng Meisheng, *Elementary Treatise on Heaven, Earth, and Human* (reprinted 1991), Yangzhou.
Denison 1854	E. B. Denison, 'Clock and Watch Work', *Encyclopaedia Britannica*, 8th edn., 7, Edinburgh, 25–7.
Denison 1860	E. B. Denison, *A Rudimentary Treatise on Clocks, Watches, and Bells*, London.
Dennison 1913	Dennison Watch Case Company, *Historic Horology, Being a Catalogue of a Collection of Antique Watches belonging to the Franklin Dennison Collection and Exhibited by the Dennison Watch Case Company at the Jewellers' Exhibition, 1913*, London.
Dent 1841	John Edward Dent, *An Abstract from Two Lectures on the Construction and Managment of Chronometers, Watches and Clocks*, London.
Deparcieux 1741	Antoine Deparcieux, *Nouveaux traités de trigonométrie rectiligne et sphérique avec un traité de gnomonique*, Paris.
Département 1912	Département de l'instruction publique (ed.), *L'Observatoire cantonal neuchâtelois, 1858–1912: souvenir de son cinquantenaire et de l'inauguration du Pavillon Hirsch*, Neuchâtel.
Dequidt 2008	Marie-Agnès Dequidt, 'Implantation, transport et finances: l'expérience d'un négociant horloger parisien en 1780 vue au travers de sa correspondance', *Histoire urbaine*, 23,169–84.
Dequidt 2013	Marie-Agnès Dequidt, 'L'horlogerie parisienne entre art et industrie (1750–1850)', in Pierre Lamard (ed.), *Art & Industrie, XVIIIe–XXIe siècle*, Paris, 95–106.
Dequidt 2014	Marie-Agnès Dequidt, *Horlogers des Lumières; temps et société à Paris au XVIIIe siècle*, Paris.
Déré, Duffaut, & de Liège 1996	Anne-Claire Déré, François Duffaut, & Gérard de Liège, 'Cent ans dede Liège. Science et d'industrie', in Béranger, Duffaut, & Tiers 1996, 3–23.
De Regiomontanus 1490	Johannes de Regiomontanus, *Tabulæ directionum profectionumque*, Augsburg.
De Regiomontanus 1496	Johannes de Regiomontanus, *Almagesti Epytoma in almagestum ptolomei*, Venice.
De Regiomontnus 1533	Johannes de Regiomontanus, *De triangulis omnimodis libri quinque: . . .*, Nuremberg.
De Rey-Pailhade 1894	Joseph de Rey-Pailhade, 'Le temps décimal: avantages et procédés pratiques avec un projet d'unification des heures des colonies françaises', *Bulletin de la société de géographie de Toulouse*, 1–2, 534–41.
De Rey-Pailhade 1899a	Joseph de Rey-Pailhade, *Décimalisation du jour et du cercle. Table à neuf chiffres pour la transformation des angles et des degrés en fractions décimales de jour et du cercle*, Toulouse.
De Rey-Pailhade 1899b	Joseph de Rey-Pailhade, *Extension du système métrique à la mesure du temps et des angles . . . suivi d'un essai de bibliographie annotée . . .*, Paris.
De Rey-Pailhade 1909a	Joseph de Rey-Pailhade, 'La décimalisation du temps et des angles. Le système C. G. S. et la montre décimale, *L'Horloger*, 3, 1–8.
De Rey-Pailhade 1909b	Joseph de Rey-Pailhade, *La Montre décimale à l'usage des ingénieurs, médecins et sportsmen*, Toulouse.
De Rey-Pailhade 1910	Joseph de Rey-Pailhade, *Traité pratique de la montre décimale*, Toulouse.
Derham 1696, 1734	William Derham, *The Artificial Clockmaker. A Treatise on Watch, and Clock-work . . . Also the History of Clock-Work Both Ancient and Modern. . . .*, 1st edn; 4th edn, London.
Derham 1731, 1746	William Derham, *Traité d'horlogerie pour les montres et les pendules . . . l'histoire ancienne et modern de l'horlogerie*, Paris.
Dermigny 1964a	Louis Dermigny, *La Chine et l'Occident. Le commerce à Canton au XVIIIe siècle 1719–1833*, 3 vols., Paris.

Dermigny 1964b	Louis Dermigny, *Les Mémoires de Charles de Constant sur le commerce à la Chine*, Paris.
De Rivières 1877–1885	Baron Edmond de Rivières, 'Inscriptions et devises horaires', *Bulletin monumental ou collection de mémoires*, 43, 44, 47, 49, 50, 51.
De Rochas [1884]	Albert de Rochas, *La Science dans l'antiquité. Les origines de la science et ses premières applications*, Paris.
De Roulet 2018	Daniel de Roulet, *Dix petites anarchistes*, Paris.
De Saavedra Fajardo 1640	Diego de Saavedra Fajardo, *Idea de un principe politico christiano*, Amsterdam.
Desaguliers 1744	J. T. Desaguliers, *Course of Experimental Philosophy*, 3rd edn., London.
de Sainte-Croix 1810	Félix Renouard de Sainte-Croix, *Voyage commercial et politique aux Indes Orientales*, ..., 3 vols., Paris.
De Sainte Marie Magdelaine 1641	Pierre de Sainte Marie Magdelaine, *Traité d'Horlogiographie*, ..., Paris.
Des Bruslons 1750	Savary des Bruslons, *Dictionnaire du Commerce*, Paris.
Descartes 1974	Réné Descartes, *Œuvres de Descartes publiées par Charles Adam & Paul Tannery ... nouvelle presentation* [par Josephe Beaude & Pierre Costabel, 1969], second issue, 11 vols., Paris.
Deschales 1674	Claude-François Milliet Deschales, *Cursus seu Mundus Mathematicus*, Lyon.
Description 1758	*Description d'une pendule singulière & unique dans son espece, qui a eu l'approbation de l'Académie des Sciences, & des conooisseurs ende genre: ... par M. Gallonde*, Paris.
de Séjour 1761	Pierre Dionis du Séjour, *Recherches sur la gnomonique*, Paris.
De Solla Price 1975	D. J. de Solla Price, 'Clockwork before the Clock and Timekeepers before Timekeeping', in Frazer & Lawrence 1975, 367–80.
De Solla Price 1955	Derek de Solla Price, *The Equatorie of the Planetis*, Cambridge.
De Solla Price 1980	Derek de Solla Price, 'Philosophical Mechanism and Mechanical Philosophy: Some Notes Toward a Philosophy of Scientific Instruments', *Annali dell'Istituto e Museo di Storia della Scienz di Firenze*, 5, 75–85.
De Solla Price 1959	Derek J. de Solla Price, 'On the Origins of Clockwork, Perpetual Motion Devices and the Compass', *United States National Museum Bulletin*, 218, 81–112.
De Solla Price 1962a	Derek J. de Solla Price, 'Mechanical Water Clocks of the 14th Century in Fez, Morocco', *Actes du Congrès International d'Histoire des Sciences, Ithaca, 26 August) 2 September 1959*, Paris, 599–602.
De Solla Price 1962b	Derek J. de Solla Price, 'Portable Sundials in Antiquity, Including an Account of a New Example from Aphrodisias', *Centaurus*, 14, 242–66.
[d'Espagne] 1757	[d'Espagne], *Regles des horloges et explication des deux meridiennes du tems-vrai et tems-moyen*, Blois.
de Thaon 1873	Philippe de Thaon, *Li Cumpoe ...*, E. Mall (ed.), Strasburg.
de Thaon 1984	Philippe de Thaon, *Comput (Ms Bl. Cotton Nero A.V.)*, Ian Short (ed.), London.
De Vallemont 1707	Pierre le Lorrain de Vallemont, *La Sphère du monde selon l'hypothèse de Copernic ...*, Paris.
Develle 1917	E. Develle, *Les horlogers Blésois au XVIe et au XVIIe siècle*, 2nd edn., Blois.
De Vries 1998	Fer J. de Vries, 'Universal Card Dial with Nomograms for Babylonian, Italian, and Antique Hours', *The Compendium*, 5, 13–19.
De Vries 1999	Fer J. de Vries, 'A 'Universal' Capuchin Dial (or The Sailing Wooden Shoe)', *The Compendium*, 6, 1999, 4–8.
De Zach 1819	Baron de Zach, *Correspondance astronomique, géographique, hydrographique et statistique*, Genoa.
D'Haenens 1980	Albert d'Haenens, 'La quotidienneté monastique au Moyen Āge', in *Klösterliche sachkultur des spatmitelalters*, Vienna, 31–42.
D'Hollander 1999	Raymond D'Hollander, *L'Astrolabe, histoire, théorie, pratique*, Paris.
Di Capriglia 1665	Guiseppe Di Capriglia, *Misura del Tempo*, Padua.
Dick 1925	Otto Dick, *Die Feile und ihre Entstehungsgeschichte*, Berlin.
Dick & Hamel 2019	Wolfram R. Dick & Jürgen Hamel (eds.), *Beiträge zur Astronomiegeschichte*, Band 1, Acta Historica Astronomiae series, 66, Leipzig.
Dictionnaire 1772	Académie Française, *Dictionnaire de l'Académie françoise*, 4th edn., 2 vols., Paris.
Didier 2010	Mélanie Didier, 'Pairs of "Chinese" Watches: The Mirror of Seduction', in Tellier & Didier 2010, 27–9.
Dieboldt & Zylberberg-Hocquard 1984	Evelyne Dieboldt & Marie-Helène Zylerberg-Hocquard (eds.), *Aline Valette, Marcelle Capy, femmes au travail au XIXe siècle*, Paris.
Diels 1917a	Hermann Diels, *Über die Prokop beschreibene Kunstfuhr von Gaza*, Berlin.
Diels 1917b	Hermann Diels, *Antike Technik*, Leipzig/Berlin.
Dienstag 1910	Paul Dienstag, *Die deutsche Uhrenindustrie. Eine Darstellung der technischen Entwicklung in ihrer volkswirtschaftlichen Bedeutung*, Leipzig.
Dijksterhuis 1955	E. J. Dijsterhuis, *The Principal Works of Simon Stevin*, Amsterdam.
DIK 1997	DIK, 'Die Zeitmessung hinter dem Eisernen Vorhang', *Uhren Juwelen Schmuck*, 7, 38–42.

Dikeç & Galviz 2016	M. Dikeç & C. Lopez Galviz '"The Modern Atlas": Compressed Air and Cities *c*.1850–1930', *Journal of Historical Geography*, 53, 11–27.
Dillon 2002	Steven Dillon, 'Illustrations of Time, Watches, Dials and Clocks in Victorian Pictures', in Richard Maxwell (ed.), *The Victorian Illustrated Book*, ch. 2, Charlottesville.
Ditisheim et al. 1940	Paul Ditisheim, Roger Lallier, L. Reverchon & [Jean-Baptiste] Vivielle, *Pierre Le Roy et la chronométrie*, Paris.
Dizer 1977	Muammer Dizer, 'The Dâ'irat al Mu'addal in the Kandili Observatory and Some Remarks on the Earliest Recorded Islamic Values of the Magnetic Deviation', *Journal for the History of Arabic Science*, 1, 257–62.
Dizer 1980	Muammer Dizer (ed.), *International Symposium on the Observatories in Islam,* Istanbul.
Dodge 1970	Bayard Dodge (ed. & trans.), *The Fihrist of al-Nadîm: A Tenth Century Survey of Muslim Culture*, 2 vols., New York.
Dodwell 1961	C. R. Dodewell (ed. & trans.), *Theophilus, de Diuersis artibus*, London.
Doensen 1994	Pieter Doensen, *Watch: History of the Modern Wrist Watch*, Ghent.
Dohrn-van Rossum 1987	Gerhard Dohrn-van Rossum, 'The Diffusion of the Public Clocks in the Cities of Late Medieval Europe, 1300–1500', in Lepetit & Hoock 1987, 29–43.
Dohrn-van Rossum 1996	Gerhard Dohrn-van Rossum, *History of the Hour: Clocks and Modern Temporal Orders*, Chicago.
Dohrn-van Rossum 1998	Gerhard Dohrn-van Rossum, 'Le temps et les horloges à la fin du Moyen Āge', in Cardinal & Vingtain 1998, 11–17.
Dominique & Delalande 2014	Anna, Dominique, & Eric Delalande, *Cadrans solaires – sundials, catalogue d'exposition*, Paris.
Dominique & Delalande 2015	Anna, Dominique, & Eric Delalande, *Sabliers d'autrefois*, Paris.
Dong & Wang 2009	Dong Yunfeng & Wang Tintin '"China's First Watch" The Old Times of China and the New Long March', *NetEase*. 21 August 2009. Available at: http://money.163.com/09/0821/18/5H8T0OBG00253JP4.html. 董云峰 & 王婷婷. '"中华第一表"的旧时光与新长征.
D'Onsembray 1735	Léon Louis Pajot, Comte d'Onsembray, 'Description et usage d'un metrometre ou machine pour battre les mesures & les temps de toutes sortes d'airs', in *Mémoires de mathématiques et de physique, tirées des registres de l'Académie Royale des Sciences de l'anée MDCCXXXII*, Paris.
Donzé 2004	Pierre-Yves Donzé, 'Les industriels horlogers du Locle (1850-1920). Un cas représentatif de la diversité du patronat dans l'Arc jurassien', in Daumas 2004, 61–82.
Donzé 2006	Pierre-Yves Donzé, *Histoire d'un syndicat patronal horloger. L'association cantonale bernoise des fabricants d'horlogerie (ACBFH)/Association patronale de l'horlogerie et de la microtechnique (APHM), 1916–2006*, Neuchâtel.
Donzé 2007	Pierre-Yves Donzé, *Les patrons horlogers de La Chaux-de-Fonds: dynamique sociale d'une élite industrielle (1840–1920)*, Neuchâtel.
Donzé 2008	Pierre-Yves Donzé, 'De l'excellence à l'utilitarisme. Culture technique et enseignement professionnel dans les écoles d'horlogerie suisses (1850–1920)', *Histoire de l'éducation*, 119, 5–28.
Donzé 2012, 2015	Pierre-Yves Donzé, *History of the Swiss Watch Industry: From Jacques David to Nicolas Hayek*, Bern.
Donzé 2014	Pierre-Yves Donzé, *'Rattraper et dépasser la Suisse': Histoire de l'industrie horlogère japonaise de 1850 à nos jours*, Neuchâtel.
Donzé 2015a	Pierre-Yves Donzé, 'Global Value Chains and the Lost Competitiveness of the Japanese Watch Industry: An Applied Business History of Seiko since 1990', *Asia Pacific Business Review*, 3, 295–310.
Donzé 2015b	Pierre-Yves Donzé, *Histoire du Swatch Group*, Neuchâtel.
Donzé 2017a	Pierre-Yves Donzé, *L'Invention du luxe, histoire de l'industrie horlogère à Genève de 1815 à nos jours*, Neuchâtel.
Donzé 2017b	Pierre-Yves Donzé, 'Fashion Watches: The Emergence of Accessory Makers as Intermediaries in the Fashion System' *International Journal of Fashion Studies*, 4, 69–85.
Donzé 2020	Pierre-Yves Donzé, *Des nations, des firmes et des montres: histoire globale de l'industrie horlogère de 1850 à nos jours*, Neuchâtel.
Dorikens 2002	Maurice Dorikens (ed.), *Scientific Instruments and Museums: Proceedings of the XXth International Congress of History of Science, 20–26 July 1997, Liège*, 14, Turnhout.
Douglass & Gedye 1911	William Tregarthen Douglass & Nicolas G. Gedye, 'Lighthouse', *The Encyclopædia Britannica*, 11th edn., 14, 627–51.
Downey 1947	Glanville Downey, 'Pappus of Alexandria on Architectural Studies', *Isis*, 28, 197–200.
Doyle & Carr 1954	Adrian Conan Doyle & John Dickson Carr, *The Exploits of Sherlock Holmes*, London.
Doyon & Liaigre 1966	André Doyon & Lucien Liaigre, *Jacques Vaucanson, mécanicien de genie*, Paris.
Drachmann 1948	A. G. Drachmann, *Ktesibios, Philon Heron. A Study of Ancient Pneumatics*, Copenhagen.
Drake 1974	Stillman Drake, *Galileo Galilei, Two New Sciences Including Centers of Gravity & Force of Percussion*, Madison.
Drecker 1925	Josef Drecker, *Theorie der Sonnenuhren*, Berlin.

Dresti & Mosello 2020	Guido Dresti & Rosario Mosello, 'A Comparative Analysis of Twelve Manuscripts of Giovanni Dondi's Astrarium', *AH*, 41, 473–504.
Dreyer 1913–29	J. L. E. Dreyer (ed.), *Tycho Brahe, Opera Omnia 1586*, Kopenhagen.
Drover 1954	C. B. Drover, 'A Medieval Monastic Water-Clock', *AH*, 1, 54–8.
Drover 1962	C. B. Drover, 'The Brussels Miniature: An Early Fusee and a Monastic Alarm', *AH*, 3, 357–61.
Drover 1980	C. B. Drover, 'The 13th-Century "King Hezekiah" Water Clock', *AH*, 12, 160–9.
Drover et al. 1960	C. B. Drover, P. A. Sabine, C. Tyler, & P. G. Coole, 'Sand-Glass "Sand", Historical, Analytical, Practical', *AH*, 3, 62–72.
Droz 2012	Yves Droz, *Les clefs du temps: le grand livre des clefs de montre*, Villers-le-Lac,
Droz 2018	Yves Droz, *Les horlogers du Val de Morteau de 1700 à nos jours, catalogue raisonné*, édition de la Communauté de communes du Val de Morteau.
Droz & Flores 1989	Yves Droz & Joseph Flores (eds.), *Les heures révolutionnaires*, Besançon.
Dubois 1849	Pierre Dubois, *Histoire de l'horlogerie depuis son origine jusqu'à nos jours . . .*, Paris.
Dubois 1858	Pierre Dubois, *Collection archéologique de Prince Pierre Soltykoff: horlogerie . . .*, Paris.
Duby & Perrot 1991	Georges Duby & Michelle Perrot (eds.), *Histoire des femmes*, vol. 3, *XVIe—XVIIIe siècle*, Paris.
Du Cange 1885	Charles du Fresne, sieur du Cange, *Glossarium mediæ et infimæ latinitatis*, Niort.
Duesberg & Audiau 2004	François Duesberg & Stéphane Audiau, *Musée François Duesberg, arts décoratiofs 1775–1825*, Mons.
Dujro 1879	Cesário Fernández Duro, *Los Ojos en el Cielo*, Madrid.
Du Moncel 1872–78	Théodore Du Moncel, *Exposé des applications de l'électricité*, 3rd edn., 4 vols., Paris.
Dunlop 1971	D. M. Dunlop, *Arab Civilization to A.D. 1500*, London/Beirut.
Dupré 2003	Sven Dupré, 'The Dioptric of Refractive Dials in the Sixteenth Century', *Nuncius*, 28, 39–67.
Dupuy-Baylet 2006	Marie-France Dupuy-Baylet, *Pendules du mobilier national, 1800–1870*, Dijon.
Dürer 1525	Albrecht Dürer, *Underweysung der Messung, mit dem Zirckel und Richtscheyt, in Linien, Ebenen und gantzen corporen*, Nüremberg.
Dussard 1928	René Dussaud, 'Un gnomon syriaque', *Syria. Revue d'art oriental et d'archéologie*, 9(1), 80.
Dzik 2019	Sunny Dzik, *Engraving on English Table Clocks: Art on a Canvas of Brass 1660–1800*, Oxford.
Eagleton 2006	Catherine Eagleton, 'Medieval Sundials and Manuscript Sources: The Transmission of Information about the Navicula and the Organum Ptolomei in Fifteenth-Century Europe', in Kusukawa & MacLean 2006, 41–72.
Eagleton 2009	Catherine Eagleton, 'Oronce Finé's Sundials: The Sources and Influence of *De solaribus horologiis*', in Marr 2009, 83–99.
Eagleton 2010	Catherine Eagleton, *Monks, Manuscripts and Sundials: The Navicula in Medieval England*, Leiden/Boston.
Eamon 1983	William Eamon, *Technology as Magic in the Late Middle Ages and Renaissance*, Cambridge.
Earle 1902	Alice Morse Earle, *Sundials and Roses of Yesterday*, London.
Eden & Lloyd 1900	H. F. K. Eden & Eleanor Lloyd, *The Book of Sun-dials Originally Compiled by the Late Mrs Alfred Gatty*, London.
Edey 1982	Winthrop Edey, *French Clocks in North American Collections*, New York.
Edgcumbe 2000	Richard Edgcumbe, *The Art of the Gold Chaser in the Eighteenth Century*, London.
Edidin 1992	Michael Edidin, 'English Watches for the American Market', *NAWCC Watch & Clock Bulletin*, 34, 515–44; 659–93.
Edidin 2000	Michael Edidin, 'Chinese Clock Label', *NAWCC Watch & Clock Bulletin*, 42, 804.
Edwardes 1965	Ernest L. Edwardes, *Weight-Driven Chamber Clocks of the Middle Ages and Renaissance*, Altrincham.
Edwardes 1977	Ernest L. Edwardes, *The Story of the Pendulum Clock*, Ashbourne.
Edwardes 1996	Ernest L. Edwardes, *Weight-Driven Dutch Clocks and Their Japanese Connections*, Ashbourne.
Eglash 2007	Ron Eglash, 'Broken Metaphor: The Master-Slave Analogy in Technical Literature', *Technology and Culture*, 48, 360–9.
Eichholz-Bochum 2010	Klaus Eichholz-Bochum, 'Die polyeder-Sonnenuhr des Ludwig Hohenfeld von 1596', *Jahresschrift der Deutschen Gessellschaft für Chronometrie*, 49, 169–86.
Eiseman 1990	Fred B. Eiseman, Jr, *Bali: Sekala and Nishkala. Volume II: Essays on Society, Tradition and Craft*, Berkeley/Singapore.
Eisler 2014/15	Wiliam Eisler, 'A Calvinist Republican at the Court of His Catholic Majesty: Jacques François Deluc, The *fabrique de Genève* and Genevan-Spanish Relations during the Eighteenth Century,' *Bulletin de la société d'histoire et d'archéologie de Genève*, 44, 20–43.
Eisler 2016/18	William Eisler, 'A Calvinist Republican at the Court of His Catholic Majesty: Jacques François Deluc, the *fabrique de Genève* and Genevan-Spanish Relations during the Eighteenth Century, Part II: The *Fabrique* Intervenes in Spain', *Bulletin de la société d'histoire et d'archéologie de Genève*, 45, 3–23.

Eitel 2018	Florian Eitel, 'Anarchistische Uhrmacher in der Schweiz', *Mikrohistorische Global Geschichte zu den Anfängen der anarchistischen Bewegung im 19. Jahrhundert*, DeGruyter.
Eitzner 2000	Lorenz Eitner, *French Paintings of the Nineteenth Century, Part I: Before Impressionism*, Washington DC.
Elève de Breguet 1827	Elève de Breguet, *L'Art d'horlogerie enseigné en trente leçons . . .*, Paris.
Elkhadem & Bracke 2004	Hossam Elkhadem & Wouter Bracke, *Simon Stevin, 1548–1620, L'émergence de la nouvelle science*, Turnhout.
Ellicott 1751	John Ellicot, 'A Description of Two Methods, by which the Irregularity of the Motion of a Clock, Arising from the Influence of Heat and Cold upon the Rod of the Pendulum may be Prevented', *PT*, 47, 479–94.
Elliott 1726	Henry Elliott, *The Clockmaker's Assistant: . . .*, London.
Elliot & Dowson 1871	Henry M. Elliot & John Dowson, *The History of India, as Told by its Own Historians: The Muhammadan Period, Edited from the Posthumous Papers of the Late Sir H. M. Elliot, . . . by Professor John Dowson, . . .*, vol. 3, London.
Elton 2002	Chris Elton, 'Obituary of Dan Parkes', *AH*, 27, 28–9.
Emerson 1769	William Emerson, *A System of Astronomy*, London.
Emmison 1964	F. G. Emmison, *Tudor Food and Pastimes*, London.
Encyclopédie 1751–80	Denis Diderot & Jean le Rond d'Alembert, *L'Encyclopédie ou dictionnaire raisonné des sciences, des arts et des metiers*, 45 vols., Paris.
Engelbert 1994	André Engelbert, 'Technologie', in Speiser 1994, 179–239.
Engishiki 1929	Engishiki kōtei, The Code of the Engi Period, revised, Volume 1, *Engishiki* (延喜式), Institute for the Study of Japanese Classics and National Association of Shinto Priests Revision, Ōokayama Shoten.
Epstein & Prak 2008	S. R. Epstein & Maarten Prak (eds.), *Guilds, Innovation and the European Economy, 1400–1800*, Cambridge.
Ereira 2001	Alan Ereira, 'The Voyages of H1', *The Mariner's Mirror*, 87, 144–9.
Escuder 2005	Olivier Escuder (ed.), *Paroles de Soleil*, 2 vols., Paris.
Eser 2014	Thomas Eser, *Die Älteste Taschenuhr der Welt? – der Henlein-Uhrenstreit*, Nürnberg.
Eser 2016	Thomas Eser, 'The Henlein Exhibition at the Germanisches Nationalmuseum: A Look Back, A Look Forward and New Discoveries', *AH*, 37, 199–212.
Essen 1977	Louis Essen, 'Quartz Clocks and Atomic Time', *Electrical Horology Group Technical Papers*, 19, 1–8.
Etienne 1810	J. P. Etienne, *Notions sur l'horlogerie, pour l'instruction des personnes qui font usage des Montres*, Nancy.
Evans 2010	David Evans, 'Peter Litherland, Liverpool and The Rack Lever', *AH*, 32, n.p.
Evans 1998	James Evans, *The History and Practice of Ancient Astronomy*, New York/Oxford.
Evans 2017	James Evans, 'Images of Time and Cosmic Connection', in Jones 2017, 143–69.
Evans & Berggren 2006	James Evans & Lennart Berggren, *Geminus' Introduction to the Phenomena: Translation and Study of a Hellenistic Survey of Astronomy*, Princeton/Oxford.
Evans 1999	Jeremy Evans, 'Horological Wheel-Cutting by Mechanical Means: An Early Reference and Some Further Notes', *AH*, 24, 551–5.
Evans 2000	Jeremy Evans, 'Scallop-Shell Marked Turret Clocks: Leonard Tennet Turret-Clock Maker and his School, Part III', *AH*, 25, 389–406.
Evans 2002	Jeremy Evans, 'Mainspring Makers of London and Liverpool – Some Observations and Lists', *AH*, 27, 63–89.
Evans 2006	Jeremy Evans, *Thomas Tompion at the Dial and Three Crowns*, Ticehurst.
Evans, Carter, & Wright 2013	Jeremy Evans, Jonathan Carter, & Ben Wright, *Thomas Tompion, 300 Years. A Celebration of the Life and Work of Thomas Tompion*, Stroud.
Evans & McBroom 2020	Jeremy Evans & Anne McBroom, 'The Chandos Delander', *AH*, 41, 521–30.
Evans 1901	Lewis Evans, 'On a Portable Sundal of Gilt Brass Made for Cardinal Wolsey', *Archæologia*, 57, 331–3.
Ew 1983	Ew, 'Drahtseilakt zwischen hoher Exportabhängigkeit und fehlendem Inlandsmarkt', *Uhren Juwelen Schmuck*, 19, 61–2.
Eyraud 2004	Charles-Henri Eyraud, *Horloges astronomiques au tournant du XVIIIe siècle : de l'à-peu-près à la précision*. Unpublished PhD thesis, University of Lyon. Available at: https://theses.univ-lyon2.fr/documents/lyon2/2004/eyraud_ch#p=0&a=top.
Fabian 1983	Dietrich Fabian, *Kinzing und Roentgen Uhren aus Neuwied*, Bad Neustadt.
Fabian 1977	Larry L. Fabian, 'Could It Have Been Wren?', *AH*, 10, 550–70.
Faidit 2002	Jean-Michel Faidit, 'Le Cadran astronomique, géographique et lunaire du Père Emanuel de Viviers (1737) suivi d'une bibliographie gnomonique de langue française', in Dorikens 2002, 109–41.
Falck et al. 1996	Per Falck, Kersti Holmquist, & Maj Nodermann, *Tidsfodral*, Stockholm.
Falconer 1987	Kenneth John Falconer, 'Digital Sundials, Paradoxical Sets, and Vitushkin's Conjecture', *The Mathematical Intelligencer*, 9, 24–7.

Falconet 1745	Camille Falconet, 'Dissertation sur Jacques de Dondis, auteur d'une horloge singulière, et à cette occasion sur les anciens horloges', *Mémoires de literature de l'Académie des Belles Lettres*, 34, 217–49.
Fale 1593	Thomas Fale, *Horologiographia. The Art of Dialling: Teaching an Easie and Perfect Way to Make all Kinds of Dials . . .*, London.
Falk 2016	Sebastian Falk, 'Improving Instruments: Equatoria, Astrolabes, and the Practices of Monastic Astronomy in Late Medieval England'. Unpublished PhD thesis, Cambridge University.
Fallet 1991	Estelle Fallet, 'L'horlogerie dans le canton de Fribourg', in Cardinal, Jequier, Barrelet, & Beyner 1991, 117–22.
Fallet 1995	Estelle Fallet, *La mesure du temps en mer et les horlogers suisses*, La Chaux-de-Fonds.
Fallet 1999	Estelle Fallet, 'L'apprenti à l'établi: du b a ba à la maîtrise. Reflexions sur le contenu des apprentissages de l'horlogerie au XVIIIe siècle', in Barrelet & Cardinal 1999, 17–32.
Fallet 2010	Estelle Fallet, 'Nicolas G. Hayek', in *Dictionnaire historique de la suisse* [online]. Available at: https://hls-dhs-dss.ch/fr/articles/035280/2010-06-30.
Fallet 2012	Estelle Fallet, *L'horlogerie à Genève. Magie des métiers. Trésors d'or et d'émail*, Genève.
Fallet 2015	Estelle Fallet, 'Master Enameller', in Capital Museum 2015, 260–3.
Fallet & Cortat 2001	Estelle Fallet & Alain Cortat, *Aprendre l'horlogerie dans les montagnes neuchâteloise, 1740–1810*, La Chaux-de-Fonds.
Fallet & Simonin 2010	Estelle Fallet & Antoine Simonin (eds.), *Dix écoles d'horlogerie suisses. Chefs-d'œuvre de savoir-faire*, Neuchâtel.
Fallet 1912	Marius Fallet, *Le travail à domicile dans l'horlogerie suisse et ses industries annexes . . .*, Berne.
Fallet 1948	Marius Fallet, 'Le Rayonnement séculaire de l'horlogerie Suisse, une synthèse historique', in Chapuis 1948, 9–66.
Fantoni 1988	Giromamo Fantoni, *Orologi solari: trattato complete di gnomonica*, Rome.
Faraday 2019a	C. J. Faraday, 'The Concept of Liveliness in English Visual Culture, c.1560–c.1630'. Unpublished PhD thesis, University of Cambridge.
Faraday 2019b	C. J. Faraday, 'Tudor Time Machines: Clocks and Watches in English Portraits c.1530–c.1630', *Renaissance Studies*, 33, 239–66.
Farré-Olivé 1989	Eduard Farré-Olivé, 'A Medieval Catalan Clepsydra and Carrillon', *AH*, 18, 371–80.
Farrer 1903	K. E. Farrer (ed.), *Letters of Joseph Wedgwood, I, 1762–1772*, London.
Faulkner 1929	William Faulkner, *The Sound and the Fury*, London.
Favre 1991	Maurice Favre, *Daniel JeanRichard, 1665–1741*, Le Locle.
Favre 1992	Maurice Favre, 'Daniel JeanRichard, premier horloger des montagnes neuchâteloise et personage de légende', *Musée Neuchâtelois*, 2, 45–56.
Fawcett 2016	Julia H. Fawcett, *Spectacular Disappearances: Celebrity and Privacy, 1696–1801*, Ann Arbor.
Faye 1854	Hervé Faye, *Leçons de cosmographie*, Paris.
Fear 2009	Jeffrey Fear, 'Cartels', in Jones & Zeitlin 2009, 268–92.
Feinstein 1995	George Feinstein, 'F. M. Fedchenko and his Pendulum Astronomical Clocks', *NAWCC Watch & Clock Bulletin*, 37, 169–84.
Feldhaus 1930	F. M. Feldhaus, 'Die Uhren des Königs Alfons X von Spanien', *Deutsche Uhrmacher-Zeitung*, 54, 608–12.
Fennell 1963	Geraldine Fennell, *A List of Irish Watch and Clock Makers*, Dublin.
Ferguson 1760	James Ferguson, *Lectures on Select Subjects* in *Mechanics, Hydtostatic, Pneumatics, Optics* and *Astronomy*, London.
Fermor & Steele 2000	John Fermor & John M. Steele, 'The Design of Babylonian Water Clocks: Astronomical and Experimental Evidence', *Centaurus*, 42–43, 210–22.
Fernandez & Fernandez 1987	Maria P. Fernandez & Pedro C. Fernandez, 'Davis Mell, Musician and Clockmaker and an Analysis of Clockmaking in 17th-Century London', *AH*, 16, 602–17.
Ferrari 1998	Gianni Ferrari, 'Relazioni e formule per lo studio delle meridian piane', in *Atti del IX Seminario Nazionale di Gnomonica*, San Felice del Benaco, 217–29.
Ferrari 1999	Gianni Ferrari, *Alcune meridian basate sulla coordinate tolemaiche*, private publication.
Ferrari 2000	Gianni Ferrari, 'Alcune note su un ororologio murale sulla cattedrale di Brunswick', *Gnomonica*, 7, 58–74.
Ferrari 2000	Gianni Ferrari, 'Uno studio sull'orologio romano consciuto come "Prosciutto di portici"', *GI*, 15, 2–12.
Ferrari 2009	Gianni Ferrari, 'Gnomonici ottomani', *GI*, 18 (July).
Ferrari 2012	Gianni Ferrari, *Le Meridiani dell' Antico Islam. Il tempo nella civiltà Islamica. Caratteristiche, descrizione e calcolo dei quadranti e degli orologi solari islamici*, 2nd revised edn., Modena.
Ferrari & Navone 2007	M. Ferrari & M. Navone, *Nuove ricerche sui codici in Scrittura Latino dell'Ambrosiana*, Milan.
Ferreira 2004, 2005	Alain Ferreira, 'Les horloges solaires d'Ernest et d'Amédée Bollée', *L'Astronomie*, 23, 3.
Field 1990	J. V. Field, 'Some Roman and Byzantine Portable Sundials and the London Sundial-Calendar', *History of Technology*, 12, 103–35.
Field 1997	J. V. Field, *The Invention of Infinity: Mathematics and Arts in the Renaissance*, Oxford.

Field & Gray 1987	J. V. Field & J. J. Gray, *The Geometrical Work of Girard Desargues*, New York.
Filippetti 2000	S. Filippetti, 'La meridiana di Bevagna. Una nuova proposta di cronologia e di analisi', *Studi classici, Annali della Facoltà di lettere e filosofia, Università degli studi di Perugia*, 33, 49–64.
Fima 2010	Hélène Fima, 'Robert Robin, un horloger de la cour de Louis XVI à l'aube de la Révolution Française', *Horlogerie ancienne, revue de l'AFAHA*, 67, 7–35.
Fima-Leonardi 2019	Hélène Fima-Leonardi, *Le Magicien des Maillardet. L'aventure d'un automate hors du commun*, La Croix-sur-Lutry.
Finch, Finch, & Finch 2017	Adrian A. Finch, Valerie J. Finch, & Anthony W. Finch, 'Edward East (1602–c.1695) Part 1 – Early Stuart Period and Commonwealth', *AH*, 38, 343–64; 'Part 2 – The Restoration and Latter Years of the East Business', 478–90.
Finch, Finch, & Finch 2019	Adrian A. Finch, Valerie J. Finch, & Anthony W. Finch, 'David Ramsay, c.1580–1659', *AH*, 40, 177–99.
Fine 1532	Oronce Fine, *Protomathesis*, Paris.
Fine 1534	Oronce Fine, *Quadrans astrolabicus*, Paris.
Fine 1560	Oronce Fine, *De solaribus horologiis et quadrantibus libri 4*, Paris.
Fisher 1961	F. J. Fisher (ed.), *Essays in the Economic and Social History of Tudor and Stuart England in Honour of R. H. Tawney*, Cambridge.
Fisher 1820	George Fisher, 'On the Errors in Longitude as Determined by chronometers at Sea Arising from the Action of the Iron in the Ships, upon the Chronometers', *PT*, 110, 196–208.
Fleet 1915	J. F. Fleet, 'The Ancient Indian Water-Clock,' *Journal of the Royal Asiatic Society*, 47, 213–30.
Fleury 1994	Philippe Fleury (ed.), *Vitruve, De l'architecture, livre I*, Paris.
Flores et al. 2007	Joseph Flores, Denis Kleinknecht, & Marc Augereau, *Le comput ecclésiastique de Frédéric Klinghammer*, Besançon.
Flores et al. 2019	Juan Flores Zendejas, Gisela Hürlimann, Luigi Lorenzetti, & Hans-Ueli Schiedt (eds.), *Texte und Zahlen. Der Platz quantitativer Ansätze in der Wirtschafts- und Sozialgeschichte. Des textes et des chiffres. La place des approches quantitatives dans l'histoire économique et sociale*, Zürich.
Flores & Mundschau 2017	Joseph Flores and Heinz Mundschau, 'Histoire et étude du "tambourin", le chaînon entre l'horloge et la montre', *Horlogerie ancienne*, 82, 7–30.
Fludd 1617	Robert Fludd, *Utriusque cosmi historia, liber septimus*, Frankfurt-am-Main.
Folkerts & Lorch 2000	Menso Folkerts & Richard Lorch (eds.), *Sic itur ad astra. Studien zur Geschichte der Mathematik und Naturwissenschaften. Festschrift für de Arabisten Paul Kunitzsch zum 70. Gebusrstag*, Wiesbaden.
Folta 1997	Jaroslav Folta, 'Clockmaking in Medieval Prague', *AH*, 23, 405–17.
Forcellini 1871	Aegidii Forcellini, *Lexicon totius latinitatis*, 4 vols., Padua.
Forrer, Le Coultre, Beyner, & Oguey 2002	Max Forrer, René Le Coultre, André Beyner, & Henri Oguey (eds.), *L'aventure de la montre à quartz. Mutation technologique initiée par le Centre électronique horloger Neuchâtel*, Neuchâtel.
Forstmann 1779	Christian Wilhelm Forstmann, *Ausführlicher Unterrricht von zeigenden und schlagenden Taschenuhren*, Halle/Saale.
Forte 1988	Antonino Forte, *Mingtang and Buddhist Utopias in the History of the Astronomical Clock. The Tower, Statue, and Armillary Sphere Constructed by Empress Wu*, Rome & Paris.
Foster 2008	Martin Foster, 'Chinese Persistence with Exquisite High-End Watches is Paying Serious Dividends - But Can They Ever Catch the Swiss?', *Europa Star*, November 2008. Available at: https://www.europastar.com/magazine/features/1003878817-chinese-persistence-with-exquisite-high-end.html.
Foster 1638	Samuel Foster, *The Art of Dialling*, London.
Foster 1654	Samuel Foster, *Elliptical or Azimuthal Horologiography*, London.
Foucher 1850	Paul Foucher, *Manuel d'horlogerie, . . . contenant l'Art de connaitre et de faire l'echappement a cylindre, du repassage des montres qui portent cet echappement, . . .*, Bourges.
Fournier 1878	Edouard Fournier (ed), *Le Livre commode des adresses de Paris pour 1692 par Abraham de Pradel*, 2 vols, Paris.
Fourrier 2000	Thibaud Fourrier, *Dictionnaire des horlogers de Blois*, Mosnes.
Fourrier 2001–2005	Thibaud Fourrier, *Les Maîtres Horlogers de Blois*, 3 vols., Mosnes.
Fowler 2013	Catherine Fowler, 'The Clock: Gesture and Cinematic Replaying' *Framework: The Journal of Cinema and Media*, 54, 226–42.
Fraboulet et al. 2013	Danièle Fraboulet, Cédric Humair, & Pierre Vernus (eds.), *Coopérer, négocier, s'affronter: les organisations patronales et leurs relations avec les autres organisations collectives*, Rennes.
Fracheboud 2016	Virginie Fracheboud, 'L'horlogerie et les autorités fédérales suisses face aux Américains lors de la "Guerre des montres": entre performances et revers (1953–1956)', *Revue suisse d'histoire*, 66, 381–400.
Fraiture 2009	Eddy Fraiture, *Belgishe Uurwerken en Hun Makers AZ*, Leuven.

Fraiture & Van Rompay 2011	Eddy Fraiture and Paul Van Rompay, 'Clock and Watchmaking in Belgium 1300–1830', *AH*, 33, 27–45.
Frankfort 1933	Henri Frankfort, *The Cenotaph of Seti I at Abydos*, 2 vols., London.
Frankinet et al. 2021	Baptiste Frankinet, Anthony Turner et Philippe Tomsin, *Cadrans solaires. Collections du Musée de la Vie Wallonne,* Liège.
Franklin 1888a	Alfred Franklin, *La Vie privée d'autrefois . . . La cuisine*, Paris.
Franklin 1888b	Alfred Franklin, *La Vie privée d'autrefois . . . La Mesure du temps*, Paris.
Franklin 1975	Stephen L. Franklin, 'Dickens and Time: The Clock without Hands', *Dickens Studies Annual*, 4, 1–35 & 167–71.
Fraser, Friis-Hansen, & Kline 1988	J. T. Fraser, Dana Friis-Hansen, & Katy Kline *Clockwork. Timepieces by Artists, Architects and Industrial Designers*, Cambridge, MA.
Fraser & Lawrence 1975	J. T. Frazer & N. Lawrence (eds.), *The Study of Time II*, Berlin.
Fraser, Lawrence, & Haber 1986	J. T. Fraser, N. Lawrence, & F. C. Haber (eds.), *Time, Science, and Society in China and the West*, Amherst.
Frémontier-Murphy 2002	Camille Frémontier-Murphy, *Les instruments de mathématiques XVIe–XVIIIe siècle*, Paris.
French & Wright 1985	Anne French and Michael Wright, *John Joseph Merlin: The Ingenious Mechanic*, London.
Friess 1999	Peter Friess, 'Rediscovering Josef Weidenheimer (1758–1795), and Clockmaking in the German-Speaking Countries', *AH*, 24, 523–38.
Frischer 2017	B. Frischer, 'Edmund Buchner's *Solarium Augusti*: New Observations and Empirical Studies', *Rendiconti della Pontificia Accademia Romana di Archeologia*, 89, 1–92.
Fritsch 1999	Julia Fritsch (ed.), *Ces cucrieux navires, trois automates de la Renaissance*, Paris.
Fritsch 2010	Peter Fritsch, *Wiener 'Reiseuhren'/Viennese Travelling Clocks*, Vienna.
Froggatt & Davies 1986	Peter Froggatt & Alun C. Davies, 'David Smyth Torrens, Physiologist and Horologist', *Hermathanea*, 140, 11–31.
Froidevaux 2000a	Yves Froidevaux, 'Banque publique régionale et industrie: Les engagements industriels de la Banque cantonale neuchâteloise dans l'entre-deux-guerres', in Marguerat, Tissot, & Froidevaux 2000, 251–70.
Froidevaux 2000b	Yves Froidevaux, 'State Intervention in Regional Economy during the Interwar Years in Switzerland: The Example of Two Public Regional Banks', in Kuijlaars, Prudon, & Visser 2000, 65–78.
Frois 2000	Luis Frois, Ki-ichi Matsuda, & Momota Kawasaki (tr.), *Complete Translation of Frois's Japanese History* (完訳フロイス日本史), Tokyo.
Fung 2004	Fung K. W., 'The Construction of Western Sundials and Transmission of Related Books in Late Ming China: With Special Reference to Christopher Clavius' Gnomonices (1581)', in Rong & Li 2004, 337–65.
Fussner 1962	Théodore Ungerer & André Gloria, 'L'astrologue au cadran solaire de la cathédrale de Strasbourg (1493)', *Archives alsaciennes d'histoire de l'art*, 12, 73–107.
Gabourd 1885	Amadée Gabourd, *Histoire de Paris*, 4 vols., Paris.
Gagnaire 1999	Paul Gagnaire, *Cadrans solaires en Savoie*, Chambéry.
Gagnaire 2000	Paul Gagnaire, 'L'équerre et l'oiseau, l'art et la manière de Zarbula', *Cadran Info,* i, n.p.
Gagnaire 2004	Paul Gagnaire, 'Cadran solaire: les cadrans de temps moyen . . ., le cadran à l'équation de l'abbé Guyoux', *L'Astronomie*, 117, 121ff.
Gagnebin-Diacon 2006	Christine Gagnebin-Diacon, *La fabrique et le village: la Tavannes Watch Co., 1890–1918*, Porrentruy.
Gahtan & Thomas 2001	Maia Wellington Gahtan & George Thomas, 'Philip Melanchthon's Watch Dated 1530', *AH*, 24, 377–88.
Gallon 1735	Jean-Gaffin Gallon, *Machines et inventions approuvées par l'Académie Royale des sciences depuis son établissement jusqu'au present . . .*, 6 vols., Paris
Galluci 1590	Paolo Galluci, *Della Fabrica & Uso del Novo Horologio*, Venice.
Galvez-Behar 2007	Gabriel Galvez-Behar, 'Brevet d'invention', in Stanziani 2007, 35–47.
Galvez-Behar 2019	Gabriel Galvez-Behar, 'L'impossible institutionnalisation de la propriété scientifique, 1919–1939', in Le Bot, Commaille, & Able 2019, 327–36.
Gapaillard 2011	Jacques Gapaillard, *Histoire de l'heure en France*, Paris.
Garbers 1936	Karl Garbers, *Ein Werk Thabit B. Qurr'a uber ebene Sonnenuhren*, Berlin.
Garel 2015	Gilles Garel, 'Lessons in Creativity from the Innovative Design of the Swatch', *Technology Innovation Management Review*, 5, 34–40.
Garnier 1774	Joseph Blaise Garnier, *Gnomonique mise à la portée de tout le monde ou méthode simple et aisée pour tracer les cadrans solaires*, Marseille.
Garnier & Hollis 2018	R. Garnier & L. Hollis, *Innovation & Collaboration*, Santon.
Garnier & Carter 2015	Richard Garnier & J. Carter, *The Golden Age of English Horology*, Dorchester.
Garufo 2015	Francesco Garufo, *L'emploi du temps. L'industrie horlogère suisse et l'immigration (1930–1980)*, Lausanne.
Garufo & Morerod 2018	Francesco Garufo & Jean-Daniel Morerod (eds.), *Laurent Tissot, une passion loin des sentiers battus,* Neuchâtel.

Gatto 2010	Martin Gatto, *The Tavern Clock*, Bath.
Gaulke 2007	Karsten Gaulke (ed.), *Der Ptolemäus von Kassel: Landgraf Wilhelm IV von Hessen-Kassel und die Astronomie*, Kataloge der Museumslandschaft Hessen Kassel 38, Kassel.
Gaulke 2010	Karsten Gaulke, 'Wilhelm IV von Hessen-Kassel. Der Nutzen der Astronomie für einen Fürstenhof des 16. Jahrhunderts', in Gaulke & Hamel 2010, 47–66.
Gaulke & Beck 2019	Karsten Gaulke & Michael Beck, 'On Minutes and Seconds: The Significance of Timepieces for Determining the Positions of Celestial Bodies at the Observatory of Landgrave Wilhelm IV of Hesse-Kassel, from 1560 to 1589', *DGC*, 58, 111–31.
Gaulke & Hamel 2010	Karsten Gaulke & Jürgen Hamel (eds.), *Kepler, Galilei, das Fernrohr und die Folgen*, Acta Historica Astronomiae, 40, Frankfurt am Main.
Gaulke & Korey 2007	Karsten Gaulke & Michael Korey, 'Alltag Uff der Aldaun: die Vermessung des Fixsternhimmels', in Gaulke 2007, 43–60.
Gaup 1708	Johannes Gaup, *Tabulae gnomonicae oder Tafeln zur mechanischen Sonnen-Uhr-Kunst*, Lindau.
Gaup 1720	Johannes Gaup, *Gnomonica Mechanica Universalis*, Leipzig & Frankfurt.
Gausse et al. 1988	Ulrike Gauss (ed.), & Christofer Conrad, Henri Deschamps, Ulrike Gauss, *et al.*, *Marc Chagall: The Lithographs*, Stuttgart.
Gazeley 1956	W. J. Gazeley, *Clock and Watch Escapements*, London.
Geisberg 1974	Max Geisberg, *The German Single-Leaf Woodcut: 1500–1550*, 4 vols., New York.
Geissler 1793–99	J. G. Geissler, *Der Uhrmacher oder Lehrbegriff der Uhrmacherkunst aus den besten englischen, französischen und anderen Schriften darüber zusammen getragen, . . .*, Leipzig.
Gélis 1949	Edouard Gélis, *L'Horlogerie ancienne, histoire, décor et technique*, Paris.
Gensheimer 2005	Joseph M. Gensheimer, 'Even More Questions and Answers on Time Recorders and Master Clocks', *NAWCC Watch & Clock Bulletin*, 47, 51–61.
Gent & Co. 1911	Gent & Co. *'Pulsynetic' Electric Impulse Clocks: Catalogue*, Herne Bay.
Gent & Co. 1929	Gent & Co., *Catalogue*, Herne Bay.
Gent & Co. 1932	Gent & Co., *Catalogue*, Herne Bay.
Gent & Co. 1938	Gent & Co., *Catalogue*, Herne Bay.
Gent & Co. 1939	Gent & Co., *Catalogue*, Herne Bay.
Georgeon 2012	François Georgeon, 'Temps de la réforme, réforme du temps. Les avatars de l'heure et du calendrier à la fin de 'Empire Ottoman', in Georgeon & Hitzel 2012, 241–80.
Georgeon & Hitzel 2012	François Georgeon & Frédéric Hitzel (eds.), *Les Ottomans et le temps*, Leiden/Boston.
Georgi 1867	Erasmus Georgi, *Handbuch der Uhrmacherkunst eine grundliche Anleitung zur Anfertigung und Reparartur aller Arten von Uhren*, Altona.
Gerland 1906	Ernst Gerland, *Leibnizens nachgelassene Schriften physikalischen, mechanischen und technischen Inhalts*, Leipzig.
Gersaint 1744	E. F. Gersaint, *Catalogue raisonné d'une collection . . . de divers curiosités . . . contenues dans le Cabinet de feu Monsieur Bonnier de la Mosson . . .*, Paris.
Gessner 2010	Samuel Gessner, 'The Use of Printed Images for Instrument-Making at the Arsenius Workshop', in Jardine & Fay 2014, 124–52.
Gessner, Korey, & Gaulke 2020	Samuel Gessner, Michael Korey, & Karsten Gaulke, 'The Anomalous Sun. Variant Mechanical Realizations of Solar Theory on Planetary Automata of the Renaissance', *Nuncius*, 25, 191–234.
Gibbs 1999	R. Gibbs, 'In Search of Ambrogio Lorenzetti's *Allegory of Justice* in the Good Commune', *Apollo*, 149, 11–16.
Gibbs 1976	Sharon Gibbs, *Greek and Roman Sundials*, New Haven/London.
Gifford 1793	John Gifford, *The History of France, from the Earliest Times to the Accession of Louis the Sixteenth*, vol. 1, London.
Gilchrist 1795	John Gilchrist, 'Account of the Hindustanee Horometry,' *Asiatick Researches*, 5, 81–9.
Gille 1980	Bertrand Gille, *Les mécaniciens grecs. La naissance de la technologie*, Paris.
Gilomen et al. 2001	Hans-Jörg Gilomen, Rudolf Jaun, Margrit Müller, & Béatrice Veyrassat (eds.), *Innovationen. Voraussetzungen und Folgen-Antriebskräfte und Widerstände. Innovations. Incitations et résistances – des sources de l'innovation à ses effets*, Zurich.
Gingerich 1996	Owen Gingerich, 'Cranks and Opportunists: "Nutty" Solutions to the Longitude Problem', in Andrewes 1996, 133–47.
Ginzel 1914	Friedrich Karl Ginzel, *Handbuch des mathematischen und technischen Chronologie*, vol. 3, Leipzig.
Giovio 1552	Paolo Giovio, *Historiae sui temporis*, vol. 2, Florence.
Girardier 2012	Sandrine Girardier, 'Les Jaquet Droz et Leschot, Virtuosité mécanique, commerce international et utilité publique', in Junier & Künzi 2012, 20–30.
Girardier 2017	Sandrine Girardier, 'Réseaux réels et rêvés; l'exemple des Jaquet-Droz et Leschot', in *La Neuchâteloise, Histoire et technique de la pendule neuchâteloise, XVIIIe–XXIe siècle*, Neuchâtel.

Girardier 2020	Sandrine Girardier, *L'entreprise Jaquet-Droz. Entre merveilles de spectacle, mécaniques luxueuses et machines utiles 1758–1811*, Neuchâtel.
Glaser 1990	Günther Glaser, *Handbuch der Chronometrie und Uhrentechnik*, 2 vols., Stuttgart.
Glasmeier 1991	Amy Glasmeier, 'Technological Discontinuities and Flexible Production Networks: The Case of Switzerland and the World Watch Industry', *Research Policy*, 20, 469–85.
Glasmeier 2000	Amy Glasmeier, *Manufacturing Time. Global Competition in the Watch Industry 1795–2000*, New York.
Glennie & Thrift 2009	Nigel Glennie & Paul Thrift, *Shaping the Day: A History of Timekeeping in England and Wales 1300-1800*, Oxford.
Glick 1969	Thomas Glick, 'Medieval Irrigation Clocks', *Technology and Culture*, 10, 424–8.
Glucksmann 1990	M. Glucksman, *Women Assemble*, London.
Goldsmiths 1901	Goldsmiths and Silversmiths Company, *Watch and Clock Catalogue 1901*.
Goldstein 2012	Catherine Goldstein, 'Les fractions decimals: un art d'ingénieur?', HAL Archive Ouvertes. Available at: https://hal.archives-ouvertes.fr/hal-00734932v1.
Golvers 1999	Noël Golvers, *François de Rougemont, S.J., Missionary in Ch'ang-shu (Chiang Nan): A Study of the Account Book (1674–1676) and the Elogium*, Leuven.
Gombeux 1936	Edmond Gombeux, *Le graveur Michel Lasne*, Caen.
Good, Gregory, & Bosworth 1819	John Mason Good, Olinthus Gregory, & Newton Bosworth, *Pantologia. A New Cabinet Encyclopedia Comprehending a Complete Series of Essays, Treatises and Systems . . . with a General Dictionary of Arts, Sciences and Words*, 12 vols., London.
Good 1978	Richard Good, *Watches in Colour*, Poole.
Good 1981	Richard Good, 'A Watch by Thomas Mudge, London, No. 574 with Perpetual Calendar Mechanism', *AH*, 13, 178–87.
Goodison 2002	Nicholas Goodison, *Matthew Boulton: Ormolu*, London.
Goold 1977	G. P. Goold (ed. & trans.), *Manilius, Astronomica*, London.
Gordon-Smith 1938	Allan Gordon-Smith, 'Clocks Just as British as Greenwich', *The Daily Telegraph Supplement*, 7 November 1938, 9.
Gorgé 2013	Viktor Gorgé, 'Charles-Edouard Guillaume', in *Dictionnaire historique de la suisse* [online]. Available at: https://hls-dhs-dss.ch/fr/articles/028841/2013-12-12
Gossman 1968	Lionel Gossman, *Medievalism and the Ideologies of the Enlightenment: The World and Work of La Curne de Sainte-Palaye*, Baltimore.
Gotoh 2003	Akio Gotoh, 'Orologi solari in Giappone (period Edo 1600–1867), *GI*, 4, 9–13.
Gotteland 1988	Andrée Gotteland, 'Pourquoi un gong-méridien au jardin du roi en 1787?', *Horlogerie ancienne*, 23, 101–17.
Gotteland 1993	Andrée Gotteland, *Cadrans solaires de Paris*, Paris
Gotteland 2002	Andrée Gotteland, *Les cadrans solaires et méridienes disparus de Paris*, Paris.
Gotteland 2008	Andrée Gotteland, *Les méridiennes du monde et leur histoire*, 2 vols., Paris.
Gouk 1988	Penelope Gouk, *The Ivory Sundials of Nuremberg, 1500–1700*, Cambridge.
Gouk 1989	Penelope Gouk, 'Horological, Mathematical and Musical Instruments. Science and Music at the Court of Charles I', in MacGregor 1989, 387–402.
Gould 2013	Rupert Gould, *The Marine Chronometer*, 3rd edn., Woodbridge.
Gounaris 1980	Georgios Gounaris, 'Anneau astronomique portative antique découvert à Philippes', *Annali dell'Istituto e Museo di Storia della scienze di Firenze*, 5, 1–18.
Gow & Page 1965	A. S. L. Gow & D. L. Page, *The Greek Anthology. Hellenistic Epigrams*, 2 vols., Cambridge.
Gowing 1997	Ronald Gowing, 'Pierre Varignon and the Measurement of Time', *Revue d'histoire des sciences*, l, 361–8.
Graf 2008a	Johannes Graf, 'Herausforderung Quarzuhr. Die deutsche Uhrenindustrie in den 1970er Jahren', in Graf 2008b, 62–75.
Graf 2008b	Johannes Graf (ed.), *Die Quarzrevolution. 75 Jahre Quarzuhr in Deutschland 1932-2007. Vorträge anlässlich der Tagung im Deutschen Uhrenmuseum Furtwangen am 20. und 21. August 2007*, Furtwangen.
Graf 2010	Johannes Graf, *Der kunststreiche Uhrmacher. Kostbarkelten aus der Bibliothek des Deutsches Uhrenmuseums*, Furtwangen.
Graf 2011	Johannes Graf, 'Von Null auf Hundert in 40 Jahren', *DGC*, l, 241–62.
Graf 2013	Johannes Graf, 'Uhren Im Gleichtakt. Wilhelm Foerster Und Die Zeitsynchronisation in Deutschland', in C. Kassung & T. Macho (eds.), *Kulturtechniken der Synchronisation*, Munich, 161–87.
Graf 2015	Johannes Graf, 'Quarzuhren Bestehen Nicht Aus Quarz: serienmässige Quarzuhren Der Zwischenkriegszeit', *DGC*, 54, 67–90.
Graf 2019	Johannes Graf, 'Uhrenindustrie im 20. Jahrhundert', *DGC*, 58, 241–259.
Graf 2020	Johannes Graf, 'Time on the Dashboard. Car Clocks from Germany', *AH*, 41, 357–72.
Grafton 1997	Anthony Grafton, *The Footnote, A Curious History*, Cambridge, MA.

Graham 1726	George Graham, 'Contrivance to Avoid the Irregularities occasioned by Heat & Cold Upon the Rod of the Pendulum', *PT*, 39, 40–4.
Grandjean de Fouchy 1770	Grandjean de Fouchy, 'Eloge de M. de Parcieux', in *HARS*, Paris.
Grant 1971	Edward Grant (ed.), *Nicolas Oresme, De Commensurabilitate sive Incommensurabilitate Motuum Celi*, Madison, WI.
Greenberg 1951	Michael Greenberg, *British Trade and the Opening of China 1800–42*, Cambridge.
Greenwood 1606	Henry Greenwood, *A Treatise of the Great and Generall Daye of Judgement Necessarie for Everie Christian that Wisheth Good Successe to his Soule, at that Great and Terrible Day*, London.
Grésy 1868	Eugène Grésy (ed.), 'Inventaire des objets d'art ... trouvé en 1532 au Château de Bury ... l'héritage de Messire Flormond Robertet, minister et seul secrétaire des finances de François premier ... dressé de la main meme de sa veuve madame Michèle Gaillard de Longjumeau ... 4 Aoust 1532', *Mémoires de la Société nationale des antiquaries de France*, 3rd series, 10, 1–66.
Grierson 1964	Herbert Grierson, *The Poems of John Donne*, Oxford.
Griffith 1889	Ralph T. H. Griffith (trans.), *The Hymns of the Rigveda*, with a popular commentary, vol. 1, Benares.
Griffiths 1985	John Griffiths, *Clock and Watchmaking today ...*, Prescot.
Griffiths 1994	John Griffiths, 'The Rise and Fall of Toolmaking in Lancashire', *Journal of the North West Society for Industrial Archaeology*, 3–8.
Griffiths 2002	R. J. Griffiths, 'The Early Watchmakers of Toxteth Park Near Liverpool and the Origin of the Industry, The Aspinwalls with Some Notes on their Successors', *AH*, 27, 163–78.
Grimbergen et al. 1996	C.A. Grimbergen, D. Lazoe, L. H. Van Der Tweel, & C. J. Wijnberg, *Willem Barentsz en zijn Uurwerk*, Zaanse Schans.
Grisard 1974	Jean Grisard, 'François Viète: un homme du XVIe siècle, mathématicien du XVIIe siècle?', *Actes des journées internationales d'étude du Baroque*, 7, 1–9.
Grisel & Bouquet 1643	Hercule Grisel & Valentin Bouquet, *Les fastes de Rouen*, Paris.
Groiss 1980a	Eva Groiss, 'The Augsburg Clockmakers' Craft', in Maurice & Mayr 1980, 57–86.
Groiss 1980b	Eva Groiss, 'Automatic Music: The Bidermann-Langenbucher Lawsuit', in Maurice & Mayr 1980, 125–30.
Gros 1913	Charles Gros, *Échappements d'Horloges et de Montres*, 2nd edn., Paris.
Grossinger 2002	Christa Grossinger, *Humour and Folly in Secular and Profane Prints of Northern Europe, 1430–1540*, London.
Grossmann 1866	Moritz Grossmann, *Der freie Ankergang für Uhren. Preisschrift*, Glasshutte.
Grossmann 2004	Peter Z. Grossmann (ed.), *How Cartels Endure and How They Fail. Studies of Industrial Collusion*, Cheltenham.
Grunebaum 1970	G. E. von Grunebaum, *Classical Islam, A History 600–1258*, London.
Guan 2000	Guan Xueling, 'The Reform of Clocks and Watches during the Emperor Qianlong's Period', *Palace Museum Journal*, 2, 85–91.
Guan 2011	Guan Xueling, 'Watches and Clocks Carried by the Emperor of the Qing Dynasty', *Trends Time*, 6, 140–2.
Guerman, Forestier, & Wigal, 2007	Mikhail Guerman, Sylvie Forestier, & Donald Wigal, *Chagall: Vitebsk–Paris–New York*, London.
Guilbert 1966	Madeleine Guilbert, *Les fonctions des femmes dans l'industrie*, 8, 249–50.
Guillaume 1897	Charles Edouard Guillaume, 'Recherches sur les aciers au nickel', *Comptes rendus de l'Académie des sciences*, 125, 125–8.
Guillaume 1921	Charles Edouard Guillaume, 'Esquisse de ma vie', L'Université de Paris, May, 1–39.
Guillaume 1922	Charles-Edouard Guillaume, *L'invar et l'elinvar. Conférence Nobel*, Neuchâtel.
Guilmard 1881	D. Guilmard, *Les maîtres ornemanistes, dessinateurs, peintres, architectes, sculpteurs et graveurs, ...*, Paris.
Gümbel 1924	Albert Gümbel, 'Peter Henlein der Erfinder der Taschenuhr', *Verband der Deutschen Uhrmacher*, 5, n.p.
Gunalla 2019	Alessandro Gunella, 'Giovanni Battista Benedetti: de *Gnomonum umbrarumque solarium usu*', *Orologi Solari*, 19, 56–7, annexes.
Gunella 2006	Allessandro Gunella, 'Un metodo grafico approssimativo, che si trova nel primo libro gnomonica in italiano, il "Vimercato"', *GI*, 11, 36–8.
Gunter 1623	Edmund Gunter, *De Sectore et Radio*, London.
Gunter 1624	Edmund Gunter, *The Description and Use of the Sector*, London.
Gunter 1673	Edmund Gunter, *The Works of Edmund Gunter*, 5th edn., London.
Gunther 1932	Robert T. Gunther, *The Astrolabes of the World ...*, 2 vols., Oxford.
Guo 1988	Guo Shengchi, *The Science of Timing in Ancient China*, Beijing.
Guo 2011	Guo Shengchi, 'Mechanical Timepieces, Incense Stick, and Others', in Chen & Hua 2011, 439–546.
Guye 1955	Samuel Guye, 'Les Unités horlogères et leur adaptation au système métrique', *La Suisse Horlogère*, 3, 3–6.
Guyou 1902	Emile Guyou, 'Application de chronomètres décimaux à la pratique de la navigation', in *Congrès Internationale de Chronométrie. Comptes-rendus des travaux, procès-verbaux, rapports et mémoires*, Paris, 116–21.

Guyou 1903	Emile Guyou, 'De l'extension du système décimal à la mesure de la circonférence. Ephémérides et tables numériques enpartgies décimales de quart de cercle préparées pour une application à la navigatioin', *Bulletin du Bureau des Longitudes*, 6, C3–89.
Habib 1977	Irfan Habib, 'Cartography in Mughal India,' *Medieval India: A Miscellany*, 4 122–34.
Hachenberg 1991	Karl Hachenberg, 'Brass in Central European instrument-making from the 16th through the 18th centuries', *Historic Brass Society Journal*, 4, 229–47.
Hachette 1828	Jean Nicolas Pierre Hachette, *Traité de géométrie descriptive*, Paris.
Hackmann & Turner 1994	W. D. Hackmann & A. J. Turner, *Learning, Language and Invention: Essays Presented to Francis Maddison*, Aldershot/Paris.
Hafter 1998	Daryl M. Hafter, 'Les veuves dans les corporations de Rouen sous l'ancien régime', in *Veufs, veuves et veuvage dans la France d'ancien régime, colloque à Poitiers, juin 1998*, 121–33.
Hahn 1892	Theodore Hahn, *Montres-controle portative et fixes patentees Hahn*, Stuttgart.
Haigh 1879	Haigh, Daniel Henry, 'Memoirs on Yorkshire Dials', *Yorkshire Archeological and Topographical Journal*, 5, 134–222.
Hall 1997	John Hall, *Anglo Saxon Sundials in Ryedale*, Leeds.
Hall & Hall 1970	A. Rupert Hall & Marie Boas Hall, *The Correspondence of Henry Oldenburg, vol vii, 1670–1671*, Madison.
Hallinger 1983	Kassius Hallinger (ed.), *Consuetudines Cluniacensium Antiquiores cum Redactionibus Derivitatis*, Siegburg.
Hallo 1930	Rudolf Hallo, 'Von alten Uhren im Hessischen Landesmuseum und von der Uhrmacherkunst in Kassel', *Die Uhrmacherkunst*, 32, 657–66.
Hamann 1980	Günther Hamann (ed.), *Regiomontanus-Studien*, Sitzungsberichte Österreichische Akademie der Wissenschaften, Philosophisch-Historische Klasse, 364, Vienna.
Hamel 1998	Jürgen Hamel, *Die Astronomischen Forschungen in Kassel unter Wilhelm IV*, Acta Historica Astronomiae Series, vol. 2, Frankfurt am Main.
Hamel & Müsch 2018	Jürgen Hamel & Irmgard Müsch, *Die Sonnenuhren des Landesmuseums Württemberg Stuttgart*, Leipzig.
Hamilton 1926	Henry Hamilton, *The English Brass & Copper Industries to 1800*, London.
Hammerstein 1986	R. Hammerstein, *Macht und Klang. Tönende Automaten als Realität und Fiktion in der alten und mittelalterlichen Welt*, Wiesbaden.
Hanke 1911	Julius Hanke, *Die Uhrmacherlehre*, Leipzig.
Hanloser 2015	Bettina Hanloser, *Der Uhrenpatron und das Ende einer Ära. Rudolf Schild-Comtesse, Eterna und die schweizerische Uhrenindustrie*, Zürich.
Hannah 1979	Leslie Hannah, *Electricity before Nationalisation: A Study of the Electricity Supply Industry in Britain to 1948*, London.
Hannah 1982	Leslie Hannah (ed.), *From Family Firm to Professional Management: Structure and Performance of Business Enterprise*, Budapest.
Hannah 2008	Robert Hannah, 'Timekeeping', in Oleson 2008, 1031–59.
Hannah 2009	Robert Hannah, *Time in Antiquity*, London.
Harcourt-Smith 1933	Simon Harcourt-Smith, *A Catalogue of Various Clocks, Watches, Automata and Other Miscellaneous Objects of European Workmanship Dating from the XVIIIth and the Early XIXth Centuries in the Palace Museum and the Wu Ying Tien Peiping*, Beijing.
Harris 1988	Denis Harris, 'Wrist Watches 1910–1920', *AH*, 17, 357–66.
Harris 1732–1734	Joseph Harris, *The Description and Use of the Globes, and the Orrery*, London
Harris 1969	J. R. Harris, *Liverpool and Merseyside: Essays in the Economic and Social History of the Port and its Binterland*, London.
Harrison 1767	John Harrison, *The Principles of Mr. Harrison's Time-Keeper, with Plates of the Same. Publised by Order of the Commissioners of Longitude*, London.
Harrison 1775	John Harrison, *A Description Concerning such Mechanism as will Afford a Nice, or True Mensuration time; Together with some Account of the Attempts for the Discovery of the Longitude of the Moon; and also an Account of the Discovery of the Scale of Music*, London.
Harrison 1848	William Harrison, *The Tongue of Time*, 5th edn., London.
Harrold 1984	M. C. Harrold, 'American Watchmaking: A Technical History of the American Watch Industry 1850–1950', *NAWCC Watch & Clock Bulletin* Suppl, 14, 1–444.
Harrold 2007	M. C. Harrold, 'An Economic Look at the American Watch Industry, The Pocket Watch Era 1860–1930', *NAWCC Watch & Clock Bulletin*, 49, 425–41.
Harsdörfer 1651	Georg Philip Harsdörffer, *Delitiæ Mathematicæ et Physicæ*, Nürnberg.
Hart 1962	A. Tindal Hart, *Country Counting Hosuse. The Story of two Eighteenth-Century Clerical Account Books*, London.
Hart 1962	Ivor B. Hart, *The World of Leonardo da Vinci, Man of Science, Engineer and Dreamer of Flight*, New York.

Harte 1973	N. B. Harte, 'Rees's Clocks, Watches Chronometers and Naval Architecture: A Note', *Maritime History*, 3, 92–5.
Hartmann 1827	Hartmann, *Le temps vrai et le temps moyen*, Paris.
Hasim 2012	Ahmet Hasim, 'Le Temps musulman', in Georgeon & Hitzel 2012, 371–3.
Hasluck 1887	Paul N. Hasluck, *The Watch Jobber's Handybook*, London.
Haspels 1987	Jan Jaap Haspels, *Automatic Musical Instruments, Their Mechanics and Their Music 1580–1820*, Utrecht.
Haspels 2006	Jan Jaap Haspels (ed.), *Royal Music Machines, National Museum from Musical Clock to Street Organ*, Utrecht.
Haselberger 2011	Lothar Haselberger, 'A Debate on the Horologium of Augustus: Controversy and Clarifications', Journal of Roman Archaeology series 24, 47–73.
Haselberger 2014	L. Haselberger (ed.), *The Horologium of Augustus: Debate and Context*, Journal of Roman Archaeology 99, 1–206.
Haswell 1928	J. Eric Haswell, *Horology - The Science of Time Measurement and the Construction of Clocks, Watches and Chronometers*, London/New York.
Hatin 1868	Eugène Hatin, *Bibliographie historique et critique de la presse périodique française*, Paris.
Hatton 1773	Thomas Hatton, *An Introduction to the Mechanical Part of Clock and Watch Work in Two Parts: ...*, London.
Havard 1887–90	Henri Havard, *Dictionnaire d'ameublement et de la decoration, depuis le XIIIe siècle jusqu'à nos jours*, 4 vols., Paris.
Havard n.d.	Henri Havard, *L'Horlogerie*, Paris.
Henry Bédat 2006	Jacqueline Henry Bédat, *Témoignages. Histoire d'un syndicat patronal horloger 1916–2006: l'Association cantonale bernoise des fabricants d'horlogerie (ACBFH)/Association patronale de l'horlogerie et de la microtechnique (APHM)*, Neuchâtel.
Hayard 2004	Michel Hayard, *Chefs d'oeuvre de l'horlogerie ancienne. Collection du musée Paul-Dupuy de Toulouse*, Toulouse.
Hayard 2011	Michel Hayard, *Antide Janvier 1751–1835*, Paris.
Haye 1716, 1726, 1731	Thomas Haye, *Règle horaire universelle*, Paris.
Hayward 1979	J. F. Hayward, *English Watches*, London.
Heiberg 1907	J. L. Heiberg (ed.), *Claudii Ptolemaei opera quae extant omnia, ii. Opera astronomica minora*, Leipzig.
Heilbron 1999	J. L. Heilbron, *The Sun in the Church: Cathedrals as Solar Observatories*, Cambridge/London.
Heine 1849	Xavier Heine, *Allgemeine Grundsätze Über die Uhrenmacherei ... oder Handbuch für die Schwarzwälder Uhrenmacher und Uhrenhändler*, Vöhrenbach/Villingen.
Heller 1991	Otto Heller, 'L'horlogerie dans le canton de Schaffhouse', in Cardinal, Jequier, Barrelet, & Beyner 1991, 155–60.
Hellman 1898	G. Hellman (ed.), *Rara Magnetica*, Berlin.
Hellyer & Hellyer 1971	Brian Hellyer & Heather Hellyer, 'The Astronomical Clock at Hampton Court Palace', *Journal of the British Astronomical Society*, 81, 215–19.
Hellyer & Hellyer 1973	Brian Hellyer & Heather Hellyer, *The Astronomical Clock Hampton Court Palace*, London.
Herkner 1988	Kurt Herkner (ed.), *Glashütte und seine Uhren*, 2nd revised and expanded edition, Dormagen.
Herkner 1994/95	Kurt Herkner, *Glashütte und seine Uhren: Glashütter Armbanduhren, von der ersten Fertigung bis zur Gegenwart, die Weiterentwicklung der Unternehmen nach 1945*, Dormagen.
Heslin 2007	Peter Heslin, 'Augustus, Domitian and the So-Called *Horologium Augusti*', *Journal of Roman Studies*, 97, 1–21.
Heslin 2011	Peter Heslin, 'The Augustus Code: A Response to L. Haselberger', *Journal of Roman Studies*, 24, 74–7.
Hevelius 1673	Johannes Hevelius (Johann Hevel), *Machina Coelestis*, Danzig.
Hewitt 1992	P. A. Hewitt, 'Clock and Watchmaking in Leicestershire and Rutland 1680–1900', *AH*, 20, 31–55.
Higton 1995	Hester Higton, 'Dating Oughtred's Design for the Equinoctial Ring Dial', *Bulletin of the Scientific Instrument Society*, 44, 25.
Higton 2001	Hester Higton, *Sundials, An Illustrated History of Portable Dials*, London.
Higton 1996	Hester Katharine Higton, Elias Allen and the Role of Instruments in Shaping the Mathematical Culture of Seventeenth-Century England. PhD thesis, University of Cambridge.
Higton et al. 2002	Hester Higton, Silke Ackermann, Richard Dunn, Kiyoshi Takada, & Anthony Turner, *Sundials at Greenwich. A Catalogue of the Sundials, Nocturnals, and Horary Quadrants in the National Maritime Museum, Greenwich*, Oxford.
Hilaire-Perez et al. 2012	Liliane Hilaire-Perez, Anne-Laure Carré, M.-S. Corcy, & Christiane Demeulenaere-Douyère (eds.), *Les expositions universelles en France au XIXe siècle, techniques, publics, patrimoines*, Paris.
Hill 1994	David K. Hill, 'Pendulum and Planes: What Galileo Didn't Publish', *Nuncius*, 9, 499–515.
Hill 1974	Donald R. Hill (ed. & trans.), *The Book of Knowledge of Ingenious Mechanical Devices (Kitāb fī ma 'rifat al-hiyal al-handassiyya*, Dordrecht/Boston.
Hill 1979	Donald R. Hill (trans. & annotated), *The Book of Ingenious Devices (Kitāb al-Hiyal) by the Banu Musa (Sons of) Mūsà bin Shākir*, Dordrecht.
Hill 1981	Donald R. Hill, *Arabic Water-Clocks*, Aleppo.
Hill 1984	Donald R. Hill, *A History of Engineering in Classical and Medieval Times*, London/New York.

Hill 1998	Donald R. Hill, *Studies in Medieval Islamic Technology*, Aldershot.
Hill 1976	D. R. Hill (ed. & trans.), *On the Construction of Water Clocks: kitāb Arshimidas fi 'amal al-binkamat*, London.
Hillard & Poulle 1971	Denise Hillard, Emmanuel Poulle, 'Oronce Fine et l'horloge planétaire de la Bibliothèque Sainte-Geneviève', *Bibliothèque d'Humanisme et Renaissance*, 33, 311–51.
Hindle 1964	Brooke Hindle, *David Rittenhouse*, Princeton.
Hinüber 1978	Oskar von Hinüber, 'Probleme der Technikgeschichte im alten Indien', *Saeculum*, 29, 215–30.
Hiraoka 2020	Ryuji Hiraoka, 'Jesuits and Western Clock in Japan's "Christian Century" (1549-c.1650)', *Journal of Jesuit Studies*, 7, 2094–2220.
Hirsch 1883	A. Hirsch, 'La Pendule Électrique de Précision de M. Hipp', *Bulletin de la Société des Sciences Naturelles de Neuchâtel*, 14, 3–18.
History of Science Museum 2020	History of Science Museum. Available by searching the Object Collection Database at: http://www.hsm.ox.ac.uk/database, and searching on inventory number 39896.
Hitzel 2012	Frédéric Hitzel, 'De la clepsydre à l'horloge. L'art de mesurer le temps dans l'Empire Ottoman', in Georgeon & Hitzel 2012, 13–38.
HKSM 2018	Hong Kong Science Museum (ed.), *Treasures of Time*, Hong Kong.
HMGSIA 1972	HMGSIA: The Historical Metallurgy Group of the Swedish Ironmasters' Association, *Iron and Steel on the European Market in the 17th Century*, Stockholm.
Hoag 2006	C. Hoag, 'The Atlantic Telegraph Cable and Capital Market Information Flows', *The Journal of Economic History*, 66, 342–53.
Hockey 1975	S. F. Hockey, *The Account Books of Beaulieu Abbey*, London.
Hoët-Van Cauwenberghe 2010	Christine Hoët-Van Cauwenberghe with Éric Binet, 'Un cadran solaire portatif sur os decouvert à Amiens', *Bulletin de la Société nationale des Antiquaires de France*, 2009, 309–17.
Hoët-Van Cauwenberghe 2012a	Christine Hoët-Van Cauwenberghe, 'Le disque de Berteacourt-les Dames (cite des Ambiens), et les listes gravés sur cadrans solaires portatifs pour voyageurs dans le monde romain', *Archéologie de la Picardie et de Nord de la France, Revue du Nord*, 94, 97–114.
Hoët-Van Cauwenberghe 2012b	Christine Hoët-Van Cauwenberghe, 'Les cadrans solaires portatifs à Amiens et dans le monde romain', *Dossiers d'Archéologie*, 354, 40–7.
Hoët-Van Cauwenberghe 2012c	Christine Hoët-Van Cauwenberghe, 'Cadrans solaires portatifs antiques: un exemplaire inédit provenant des Balkans', *Archäologisches Korrespondenzblatt*, 42, 555–71.
Hoët-Van Cauwenberghe & Binet 2012	Christine Hoët-Van Cauwenberghe & Éric Binet, 'Le temps et l'espace dans l'empire romain', *Les Dossiers d'Archéologie*, 354, 66–9.
Hoeve & Thompson 2014	Johan ten Hoeve & David Thompson, 'A Flemish Clock at the Shogun's Shrine', *AH*, 35, 1063–77.
Hof et al. 1900	Emile Hof, Henri-Aug. Delachaux, Ami Rosset, & Ch. Bonifas, *La Boîte de montre, fabrication et decoration*, Genève.
Hofbauer & Solombrino 2009	Karl G. Hofbauer & Patrizia Solombrino, *Zeit im Buch. Die Sonnenuhren des Johannes Gaupp*, Basle.
Hoff & Geddes 1960	Hebbel E. Hoff & L. A. Geddes, 'The Technological Background of Physiological Discovery: Ballistics and the Graphic Method', *Journal of the History of Medicine and Allied Sciences*, 15, 345–63.
Hoffmann 2014	Oliver Hoffmann, *Innovation neu denken. Histozentrierte Analyse der Innovationsmechanismen der Uhrenindustrie*, Wiesbaden.
Hohmann 2009	Frank L. Hohmann III, *Timeless: Masterpiece American Brass Dial Clocks*, New York.
Hoke 1991	Donald Robert Hoke, *The Time Museum Historical Catalogue of American Pocket Watches*, Rockford, IL.
Holland 2019	Julian Holland, 'John Cuff (1707?–after 1772): "The Best Workman of His Trade in London" Part I', *Bulletin of the Scientific Instrument Society*, 141, 2–18.
Honig 1980	Peter S. Honig, 'History and Mathematical Analysis of the Fusee', in Maurice & Mayr 1980, ch. 10.
Hooper 1950	D. G. Hooper; 'Letter, 20 February to P. B. Hunt, Board of Trade (Distribution of Industry and Regional Division)'. BT 177/910: S. Smith and Sons Ltd (Smiths' Clocks), Cerfin and Wishaw. The National Archives Catalogue, Kew.
Hope-Jones 1900	Frank Hope-Jones 'Electrical Time Service', *HJ*, 42, 61–9.
Hope-Jones 1913	Frank Hope-Jones, 'Wireless Time Signals from the Eiffel Tower', *HJ*, 55, 158–61.
Hope-Jones 1949,1951	Frank Hope-Jones, *Electrical Timekeeping*, London.
Hopper 1864	Clarence Hopper, 'On Clocks and Watches Belonging to Queen Elizabeth', *Journal of the British Archaeological Association*, 20, 348–52.
Hordijk 2018	Ben Hordijk, *The Life and Work of Nicolas Hanet*, Blaricum.

Hordijk 2020	Ben Hordijk, 'Nicolas Hanet – A New Discovery', *AH*, 41, 539–45.
Hornsby MS	Thomas Hornsby, Bodleian Library, MS Radcliffe Trust Manuscripts, d.9.
Horský 1964	Zdeněk Horský and Emanuel Procházka, 'Pražký orloj', *Acta historiae rerum naturalium nec non technicarum*, 9, 83–146.
Horton 2015	Chris Horton, 'Watches: Made in China', *The New York Times,* 29 September 2015. Available at: https://www.nytimes.com/2015/09/30/fashion/watches-made-in-china.html.
Horton & Marrison 1928	J. W. Horton & W. A. Marrison, 'Precision Determination of Frequency', *Proceedings of the Institute of Radio Engineers*, 16, 137–54.
Hou 2009	Hou Hau-Chih, 'Superb Craftsmanship Excelling Nature: Private Gadgetry in the Prosperous Ch'ing Dynasty', *Journal of the Historical Studies*, 24, 87–118.
Houzeau & Lancaster 1964	J. C. Houzeau & A. Lancaster, *Bibliographie générale de l'astronomie jusqu'en 1880*, 2 vols in 3, new edition by D. W. Dewhirst, London.
Hovey 1986	R. A. Hovey, 'The Phenomenon of the South Korean Clock Industry, 1964–1984', *NAWCC Watch & Clock Bulletin*, 28, 275–90.
Howse 1970a	Derek Howse, 'The Tompion Clocks at Greenwich and the Dead-Beat Escapement', *AH*, 7, 18–34.
Howse 1970b	Derek Howse, 'The Tompion Clocks at Greenwich and the Dead-Beat Escapement, Part 2', *AH*, 7, 114–33.
Howse 1971	Derek Howse, *Greenwich Time and the Longitude*, London.
Høyrup 1998	Jens Høyrup, 'A Note on Water Clocks and the Authority of Texts', *AOF*, 44–45, 192–4.
Hua 1991	Hua Tongxu, *Chinese Clepsydra*, Hefei, Anhui Science & Technology Publishing House.
Huang 2006	Huang Chunyan, *The Spread and Manufacture of Western Clockwork in China at the Turn of the Ming and Qing Dynasties*. Master's thesis, Jinan University, Guangzhou.
Huang 2013	Huang Qingchang, *A Brief Account on Clocks Made in Guangzhou in the Qing Dynasty*, Guangzhou.
Huber 2019	Bernhard Huber, 'The Engineer of Precision Time: pendulum Clocks by Sigmund Riefler', in *Time Made in Germany: 700 Years of German Horology*, Nuremberg, 260–83.
Huber & Banbery 1993	Martin Huber & Alan Banbery, *Patek Philippe Genève*, Genève.
Hufton 1991	Olwen Hufton, 'Le travail et la famille', in Duby & Perrot 1991, 27–57
Hughes 1967	Barnabas Hughes (trans.), *Regiomontanus on Triangles*, Madison.
Hughes 1994	Peter Hughes, *French Eighteenth-Century Clocks and Barometers in the Wallace Collection*, London.
Huguenin, Piguet, & Baldi 2017	Régis Huguenin, Jean-Michel Piguet, & Rossella Baldi (eds.), *La neuchâteloise: histoire et technique de la pendule neuchâteloise, XVIIIe–XXIe siècle*, Neuchâtel.
Hukada 1832	Masatsune Hukada (ed.), 'Tokei (自鳴磬) Section', *Owari-shi*, vol. 4.
Hume 1639, 1640	James Hume, *Méthode Universelle . . . , pour faire . . . toutes sortes de quadrans & d'horloges*, Paris.
Hummel 1968	Charles F. Hummel, *With Hammer in Hand: The Dominy Craftsmen of East Hampton, New York*, Charlottesville.
Hunter & Schaffer 1989	Michael Hunter & Simon Schaffer, *Robert Hooke, New Studies*, Woodbridge.
Huo 2019	Huo Feile, *The Life of a Horologist*, Huo Feile Collection.
Hurrion 1993	Christopher Hurrion, 'Paul Garnier's Engine Counter', *AH*, 20, 541–46.
Hutchinson 1992	Beresford Hutchinson, 'A Visit to Vulliamy's Premises 7 September, 1786', *AH*, 20, 66–8.
Huygens 1658	Christiaan Huygens, *Horologium*, The Hague.
Huygens 1673	Christiaan Huygens, *Horologium oscillatorium sive de motu pendulorum ad horologia aptato demonstrations geometriae*, Paris.
Huygens 1675	Christiaan Huygens, 'Extrait d'une lettre de Mr Hugens à l'Auteur du Journal, touchant une nouvelle invention d'horloges tres-justes & portatives', *Journal des Sçavans*, 25 February, 68–70.
Impey & MacGregor 1985	Oliver Impey & Arthur MacGregor (eds.), *The Origins of Museums: The Cabinet of Curiosities in Sixteenth- and Seventeenth-Century Europe*, Oxford.
'Invention' 1971	Various. 'The Invention of the Anchor Escapement', *AH*, 7, 225–8.
Isambert-Jamati 1955	Viviane Isambert-Jamati, *L'industrie horlogère dans la région de Besançon*, Paris.
Isely 1861	Iseley, 'Influence du Ressort de Suspension sur la Dureé des Oscillations du Pendule', *Bulletin de la Société des Sciences Naturelles*, 5, 648–74.
I-Tsing 1896, 1966	I-Tsing, *A Record of the Buddhist Religion as practised in India and the Malay Archipelago (A.D. 671–695)*, J. Takakusu (trans.), London; reprint Delhi.
Jackson 1929	J. Jackson, 'Shortt Clocks and the Earth's Rotation', *Monthly Notices of the Royal Astronomical Society*, 88, 239–50.
Jacob 1985	André Jacob, 'Le cadran solaire "Byzantin" de Taurisano en terre d'Otrante', *Mélanges de l'Ecole française de Rome; Moyen Age, temps modern*, 97, 7–22.
Jacobi 1920	Hermann Jacobi, 'Einteilung des Tages und Zeitbestimmung im alten Indien,' *Zeitschrift der Deutschen Morgenländischen Gesellschaft*, 74, 247–63.

Jacquemard et al. 2007	Catherine Jacquemard, Olivier Desbordes, & Alain Hairie, 'Du quadrant *vetustior* à l'*horologium viatorum* d'Hermann de Reichenau: étude du manuscrit Vaticano, BAV Ott. Lat. 1631, f. 16-17^v', *Kentron*, 23, 79–124.
Jacquinot 1545	Domique Jacquinot, *L'Usaige de l'astrolabe avec un traité de la sphere*, Paris.
Jagger 1983	Cedric Jagger, *Royal Clocks, the British Monarchy and its Timekeepers 1300–1900*, London.
Jagger 1988	Cedric Jagger, *The Artistry of the English Watch*, Newton Abbot.
Jahn 1842	G. A. Jahn, *Anleitung zur genauen Bestimmung des Ganges und Standes der Uhren*, Leipzig.
Jamieson 1883	G. Jamieson, 'The Tributary Nations of China', *China Review*, 12, 94–109.
Janin 1972a	Louis Janin, 'Le cadran solaire de la Mosquée Umayyade à Damas', *Centaurus*, 14, 285–98. Reprinted in Kennedy & Ghanem 1976, 102–21.
Janin 1972b	Louis Janin, 'Les méridiennes du château de Versailles', in *Revue de l'histoire de Versailles*, 5–10.
Janin 1974	Louis Janin, *Le cadran analemmatique – histoire et développement*, Besançon.
Janin 1979	Louis Janin, 'Astrolabe et cadran solaire en projection stéréographique horizontale', *Centaurus*, 2, 298–314.
Janin & King 1977	Louis Janin & D. A King, 'Ibn al-Shāṭir's *Sandūq aal-Yawāqīt*: An astronomical Compendium', *Journal for the History of Arabic Science*, 1, 187–256.
Janin & King 1978	Louis Janin & D. A King, 'Le cadran solaire de la Mosquée d'Ibn Ṭūlūn au Caire', *Journal for the History of Arabic Science*, 2, 331–57. Reprinted in King 1987, ch. 16.
Janvier 1810	Antide Janvier, *Etrennes Chronométriques pour l'an 1811 ou précis de ce qui concerne le tems, ses divisions, ses mesures, leurs usages . . .*, Paris.
Janvier 1811	Antide Janvier, *Essai sur les horloges publiques pour les communes de la campagne*, Paris.
Janvier 1812	Antide Janvier, *Des Révolutions des Corps Célestes par le Mécanisme des Rouages*, Paris.
Janvier 1815	Antide Janvier, *Manuel Chronométrique ou précis de ce qui concerne le temps, ses divisions, ses mesures, leurs usages . . .*, Paris.
Janvier 1821a	Antide Janvier, *Manuel Chronométrique ou précis de ce qui concerne le temps, ses divisions, ses mesures, leurs usages . . .*, Paris.
Janvier 1821b	Antide Janvier, *Recueil de Machines*, Paris.
Janvier 2019	Antide Janvier, *Description d'une sphère mouvante*, transcription of Janvier's manuscript by Michel Hayard, privately printed.
Jaquet 1943	Eugène Jaquet, 'Les montres en forme d'animaux, contribution à l'histoire des horlogers genevois due XVIIe siècle', *Journal suisse d'horlogerie et de bijouterie*, 68, 97–105.
Jaquet & Chapuis 1970	Eugène Jaquet & Alfred Chapuis, *Technique and History of the Swiss Watch*, new edition with added material, London.
Jardine 2008	Lisa Jardine, *Going Dutch*, London.
Jardine & Fay 2014	Nicholas Jardine & Isla Fay (eds.), *Observing the World Through Images*, Leiden.
Jars 1774	Gabriel Jars, *Voyages métallurgiques ou recherches et observations sur les mines . . .*, 3 vols., Paris.
Jaubert 1773	Pierre Jaubert, *Dictionnaire raisonné universel des arts et metiers contenant l'histoire, la description, la police des fabriques et manufactures de France & des pays etrangers . . .*, new edn., 3 vols., Paris.
Jawad 1960	Naji Jawad, *The Story of Time*, n.p.
Jeffries 1884	Richard Jeffries, *Red Deer*, London.
Jeon 1974	Sang-Woon Jeon, *Science and Technology in Korea: Traditional Instruments and Techniques*, Cambridge, MA.
Jequier 1972	François Jequier, *Une entreprise horlogère du Val-de-Travers: Fleurier Watch Co. SA. De l'atelier familial du XIXe siècle aux concentrations du XXe siècle*, Neuchâtel.
Jequier 1983	François Jequier, *De la forge à la manufacture horlogère (XVIIIe–XXe siècles): cinq générations d'entrepreneurs de la vallée de Joux au cœur d'une mutation industrielle*, Lausanne.
Jequier 1998	François Jequier, 'Essai d'analyse comparé de la gestion de deux entreprises horlogères suisses de 1914 à 1925', in Michèle Merger, Dominique Barjot (eds.), *Les entreprises et leurs réseaux: hommes, capitaux, techniques et pouvoirs*, Paris, 557–70.
Jequier & Landes 1982	François Jequier & David Saul Landes, 'Swiss Watch Supremacy Under Challenge: A Case Study in Entrepreneurial Response', in Hannah 1982, 60–75.
Jerome 1860	C. Jerome, *History of the American Clock Business for the Past Sixty Years and Life of Chauncey Jerome*, New Haven.
Jervis-Smith 1911	Frederick John Jervis-Smith, 'Chronograph', in Hugh Chisholm (ed.), *Encyclopaedia Britannica*, vol. 6, Cambridge, 301–5.
Jiang & Niu 1998	Jiang Xiaoyuan & Niu Weixing, *A General Survey of Ancient Chinese Astronomy*, Shanghai.
Jin 2013	Jin Bo, 'Seeking Old Brands: Seagull Watch Soars for a Half Century', *People's Daily*, 22 May 2013. Available at: http://cpc.people.com.cn/n/2013/0522/c83083-21569191.html. 靳博.'觅迹老品牌：海鸥表飞翔半世纪'

Jin & Wu 2007	Jin Guoping & Wu Zhiliang, *Essay of Early Macau History*, Guangzhou.
Jones 2014	Alexander Jones, 'Some Greek Sundial Meridians', in Nathan Sidoli & Glen Van Brummelen (eds.), *From Alexandria, through Baghdad. Surveys and Studies in the Ancient Greek and Medieval Islamic Mathematical Sciences in Honor of J. L. Berggren*, Heidelberg, 175–88.
Jones 2017a	Alexander Jones (ed.), *Time and Cosmos in Greco-Roman Antiquity*, Princeton.
Jones 2017b	Alexander Jones, *A Portable Cosmos: Revealing the Antikythera Mechanism, Scientific Wonder of the Ancient World*, New York.
Jones & Zeitlin 2008	Geoffrey Jones & Jonathan Zeitlin (eds.), *The Oxford Handbook of Business History*, New York.
Jones 2000	T. Jones, *Splitting the Second: The Story of Atomic Time*, Bristol.
Jordan & King 1988	D. Jordan & David A. King, *Überlegungen zur Angelsächsischen Sonnenuhr von Canterbury/Reflections on the Canterbury Sundial*, Johann Wolfgang Goethe Universitat: Institute für Geschichte der Naturwissenschaften, Preprint Series 9, Frankfurt am Main.
Jubinal 1874	Achille Jubinal, *Œuvres completes de Rutbœuf, trouvère du XIIIe siècle*, Paris.
Judet 2004	Pierre Judet, *Horlogeries et horlogers du Faucigny (1849–1934), les métamorphoses d'une identité sociale et politique*, Grenoble.
Junier & Künzi 2012	Caroline Junier & Claude-Alain Künzi (eds.), *Automates et merveilles, Les Jaquet Droz et Leschot*, Neuchâtel.
Jürgensen 1805	Louis Urban Jürgensen, *Principes généraux de l'exacte mesure du temps par les horloges; Ouvrage contenant les principes élémentaires de l'art de la mesure du temps par les horloges, la description de plusieurs échappements et deux nouveaux proposés aux artistes par l'Auteur, les meilleurs moyens de compensation des effets de la temperature . . . la description d'une pendulum astronomique et d'une montre marine projetées par l'Auteur*, Copenhagen.
Jürgensen 1840	Louis Urban Jürgensan, *Allgemeine Grundsätze der genauen Zeitmessung durch Uhren, oder Zusammenfassung der Grundsätze des Uhrenbaues zur sorgfältigsten Zeitmessung, mit einem Anhange versehen, enthaltend zwei Abhandlungen Über die Uhrmacherkunst. Nach der zweiten durch Ludwig Urban Jürgensen besorgten und vermehrten Ausgabe deutsch bearbeitet*, Leipzig.
Jüttemann 1991	Herbert Jüttemann, *Schwarzwälder Uhren/Black Forest Clocks/Pendules de la Forêt Noir*, Karlsruhe.
Kafka 1983	Franz Kafka, *The Complete Stories*, Nahum N. Glatzer (ed.), New York.
Kahlert 2007	Helmut Kahlert, *300 Jahre Schwarzwälder Uhrenindustrie*, 2nd revised edn., Gernsbach.
Kahlert, Mühe, & Brunner 1990	Helmut Kahlert, Richard Mühe, & Gisbert L. Brunner, *Armbanduhren. 100 Jahre Entwicklungsgeschichte*, Munich.
Kaltenbök/Schwank 1983	Kaltenbök/Schwank, *Watch-keys, three centuries of history and development*, Münchberg.
Kamp 2019	Artur Kamp, 'Preiswerte Uhren aus Ruhla', in DGC, 218–39.
Kamp et al. 2015	Artur Kamp, Rainer Paust, & Klaus Mleinek, *150 Jahre Gebrüder Thiel – uhrenwerke Ruhla – gardé – decke MAHO Seebach. Eine Zeitreise mit Fotos und Dokumenten aus der Ruhlaer Uhren- und Maschinenproduktion anlässlich der Jubiläen im Jahr 2012*, 2nd edn., Ruhla.
Kaneko & Tachibana 1955	Motoomi Kaneko & Munetoshi Tachibana, *Interpreted Makuranososhi,* revised edn. (改稿枕草子通解), Tokyo.
Kangxi 1994	Kangxi Xuanye, *The Motto in the Imperial Palace*, Zhengzhou.
Karr Schmidt 2017	Suzanne Karr Schmidt, *Interactive and Sculptural Printmaking in the Renaissance* Leiden.
Kaschek, Muller, & Schauert 2013	Bertram Kaschek, Jurgen Muller, & Thomas Schauerte (eds.), *Von der Freiheit der Bilder: Spott, Kritik und Subversion in der Kunst der Durerzeit*, Petersberg.
Katzir 2003	Shaul Katzir, 'The Discovery of the Piezoelectric Effect', *Archive for History of Exact Sciences*, 57, 61–91.
Kauṭilya 2010	Kauṭilya, *The Kauṭilīya Arthaśāstra, Part II: translation with Critical and Explanatory Notes*, K. P. Kangle (ed & trans.), 7th reprint, Delhi.
Kawamoto 2013	Nobuo Kawamoto, 'Consideration on the Start Time of Mechanical Clock Production in Japan', *Oryō shigaku*, 39, 149–73.
Keith 1883	W. H. Keith, *A Family Tale*. Unpublished MS. Available at: http://www.watkinsr.id.au/Keith.html.
Keller 1985	Alexander Keller, 'Mechanics and the Origins of the Culture of Mechanical Invention', *Minerva*, 23, 348–61.
Kemp 1981	Robert Kemp, *The Fusee Lever Watch*, Altrincham.
Kemp 1982	Robert Kemp, 'The Massey Watch Escapement', *AH*, 13, 558–64.
Kendal 1892	James Francis Kendal, *A History of Watches and Other Timekeepers*, London.
Kennedy 1959	E. S. Kennedy, 'Bīrūnī's Graphical Determination of the Local Meridian', *Scripta Mathematica*, 24, 251–5.
Kennedy 1960	E. S. Kenedy, *The Planetary Equatorium of Jamshîd Ghiyâth al-dín al-Kashî (d. 1429)*, Princeton.
Kennedy 1985	E. S. Kennedy, 'Al-Bīrūnī on the Muslim Times of Prayer', in Peter J. Chelkowski (ed.), *The Scholar and the Saint: Studies in Commemoration of Abū l-Rayhān al-Bīrūnī and Jalāl al-Dīn al-Rūmī*, New York, 83–94.
Kennedy & Ghanem 1976	E. S. Kennedy & Imad Ghanem, *The Life and Work of Ibn al-Shātir: An Arab Astronomer of the Fourteenth Century*, Aleppo.

Kennedy & Ukashah 1969	E. S. Kennedy & Walid Ukashah, 'The Chandelier Clock of Ibn Yūnis', *Isis*, 60, 543–5.
Kenney 1979	George C. Kenney, 'Daniel Quare Keyhole Clocks', *Horological Dialogues*, 1, 39–48.
Kenney 2016	George C. Kenney, 'Daniel Quare's Numbered Clocks', *AH*, 37, 37–54.
Kepler/Frisch 1858–1871	Johannes Kepler, *Opera omnia*, C. Frisch (ed.), Frankfurt.
Kepler 1938	Johannes Kepler, *Gesammelte Werke*, Munich.
Kessler 2016	Marlene Kessler *et al.* (eds.), *The European Canton Trade 1723*, Oldenburg.
Kieckhefer 1989	Richard Kieckhefer, *Magic in the Middle Ages*, Cambridge.
Kienast 2014	H. J. Kienast, *Der Turm der Winde in Athen, mit Beiträgen von Pavlina Karanastasi zu den Reliefdarstellungen der Winde und Karlheinz Schaldach zu den Sonnenuhren*, Wiesbaden.
King 1974	David A. King, 'An Analog Computer for Solving Problems of Spherical Astronomy: The *Shakkāzīya* Quadrant of Jamāl al-Dīn al-Māridīnī', *Achives Internationales d'Histoire des Sciences*, 24, 219–42. Reprinted in King 1987, ch. 10.
King 1975a	David A. King, 'Medieval Mechanical Devices. A Review of D. R. Hill, *The Book of Knowledge of Ingenious Mechanical Devices*', *History of Science*, 13, 1975, 284–9. Reprinted in King 1987, ch. 20.
King 1975b	David A. King, 'Ibn al-Shatir', in DSB, vol. 12, 357–64.
King 1977	David A. King, 'A Fourteenth-Century Tunisian Sundial for Regulating the Times of Muslim Prayer', in Y. Maeyama & W. G. Saltzer (eds.), *Prismata. Naturwissenschaftesgeschichten Studien. Festschrift für Willy Hartner*, Wiesbaden, 187–202. Reprinted in King 1987, ch. 18.
King 1978	David A. King, 'Three Sundials from Islamic Andalusia', *Journal for the History of Arabic Science*, 2, 358–92. Reprinted in King 1987, ch. 15.
King 1979	David A. King, 'An Islamic Astronomical Instrument', *Journal for the History of Astronomy*, 10, 51–3. Reprinted in King 1987, ch. 13.
King 1980	David A. King, 'Astronomical Timekeeping in Ottoman Turkey', in Dizer 1980, 245–69.
King 1983	David A. King, *Al-Khwārizmī and New Trends in Mathematical Astronomy in the Ninth Century*, Occasional Papers on the Near East, no. 2, New York.
King 1985	David A. King, 'The Medieval Yemeni Astrolabe in the Metropolitan Museum of Art in New York City', *Zeitschrift für Geschichte der Arabisch-Islamischen Wissenschaften*, 1, 99–122. Reprinted in King 1987, ch. 2.
King 1987	David A. King, *Islamic Astronomical Instruments*, London.
King 1990	David A. King, 'A Survey of Medieval Shadow-Schemes for Simple Time-Reckoning'. *Oriens*, 32, 191–249.
King 1992	David A. King, 'Los Cuadrantes Solares Andalusies', in Juan Vernet & Julio Samsó (eds.), *El Legado Cientifico Andalusi*, Madrid, 89–102.
King 1993a	David A. King, '*Mīqāt*: Astronomical Timekeeping', in *EI 2*, 7, 27–32. Reprinted in King 1993b, ch. 5.
King 1993b	David A. King, *Astronomy in the Service of Islam*, Aldershot.
King 1994	David A. King, Contributions to Cluzon, Delpont, & Mouliérac 1994.
King 1997	David A. King, 'Astronomie et société musulmane: "qibla", gnomonique, "mīqāt"', in Ragheb & Morelon 1997, vol. 1, 173–216.
King 1999	David A. King, *World Maps for Finding the Distance and Direction of Mecca; Innovation and Tradition in Islamic Science*, Leiden.
King 2002	David A. King, 'A Vetustissimus Arabic Treatise on the *Quadrans vetus*', *Journal for the History of Astronomy*, 33, 237–55.
King 2003	David A. King, '14th-Century England or 9th-Century Bagdad? New Insight on the Elusive Astronomical Instrument called Navicula de Venetiis', *Centaurus*, 45, 204–26.
King 2004–5	David A. King, *In Synchrony with the Heavens: Studies in Astronomical Timekeeping and Instrumentation in Medieval Islamic Civilisation: I: The Call of the Muezzin; II: Instruments of Mass Calculation*, Leiden.
King 1912	L. W. King, *Cuneiform Texts from Babylonian Tablets . . ., in the British Museum*, vol. 33, London.
King & Millburn 1978	Henry C. King in collaboration with John R. Millburn, *Geared to the Stars: The Evolution of Planetariums, Orreries and Astronomical Clocks*, Bristol.
Kintz 1982	J. P. Kintz, 'Jean Baptiste Sosime Schwilgué', in Nouveau *dictionnaire de biographie Alsacienne*, vol. 34, Strasbourg
Kircher 1646	Athanasius Kircher, *Ars magna Lucis et Umbrae*, Rome.
Kitto 1999	Tony Kitto, 'John Flamsteed, Richard Towneley and the Equation of Time', *AH*, 25, 180–4.
Kiu 2006	Kiu Tai Yu, *Antique Chinese Calibre Pocket Watches: Collections of Kiu Tai Yu Museum*, Hong Kong.
Kjellberg 1997	Pierre Kjellberg, *Encyclopédie de la pendule française, du Moyen Âge au XXe siècle*, Paris
Kleutghen 2014	Kristina Kleutghen, 'Chinese Occidenterie: The Diversity of "Western" Objects in Eighteenth-Century China', *Eighteenth-Century Studies*, 47, 117–35.
Klügel 1793	Georg Simon Klügel, *Anfangsgründe der Astronomie . . . und Gnomonik*, Berlin.

Knap 1984	Johan Knap, 'A Danish Maker of Clocks and Watches', *AH*, 8, 617–18.
Knaub 2017	Katie Knaub, 'Report and Summary of the Horology in Art Symposium', Museum of Fine Arts in Boston, 26–28 October 2017. Available at: http://www.horologyinart.com/report.html.
Knorr 1997	Wilbur R. Knorr, 'The Latin Sources of *Quadrans vetus* and what they Imply for its Authorship and Date', in E. Sylla and M. McVaugh (eds.), *Texts and Contexts in Ancient and Medieval Science: Studies on the Occasion of John E. Murdoch's Seventieth Birthday*, Leiden, 23–67.
Knox 1984	Ellis Lee Knox, *The Guilds of Early Modern Augsburg: A Study in Urban Institutions*. PhD thesis. Available at: https://scholarworks.umass.edu/dissertations_1/1136.
Knox 1681, 1995	Robert Knox, *An Historical Relation of the Island Ceylon in the East Indies, London*, with an Introduction and Afterword by Dr H. A. I. Goonetileke, 3rd facsimile reprint, New Delhi.
Koch 2018	Ekkehard Koch, *Das Messen der Zeit - Misura del Tempo*, Georgsmarienhütte.
Kochmann 1974	Karl Kochmann, *Clockmaking in Europe – the Gustav Becker Story*, Concord.
Kochmann 1976	Karl Kochmann, *The Junghans Story*, Concord.
Kochmann 1978	Karl Kochmann, *Industrialized Clockmaking in Europe – 25 Year Anniversary 1899–1924 United Freiburg Clock Factories, Former Gustav Becker Works*, Concord.
Koeppe 2012	Wolfram Koeppe (ed.), *Extravagant Inventions. The Princely Furniture of the Roentgens*, New York.
Koestler & Ceccarell 2012	T. Koestier & M. Ceccarelli (eds.), *Explorations in the History of Machines and Mechanisms*. History of Mechanism and Machine Science Series, vol. 25, Dordrecht.
Koller 2003	Christophe Koller, *L'industrialisation et l'État au pays de l'horlogerie, contribution à l'histoire économique d'une région suisse*, Courrendlin.
Korey 2007a	Michael Korey, 'Gantz und gar entzunden von Wilhelms Instrumenten: august von Sachsen, seine in Hessen hergestellte Planetenuhr und die Funktion der Astronomie am kursächsischen Hof', in Gaulke 2007, 93–106.
Korey 2007b	Michael Korey, *The Geometry of Power, the Power of Geometry: Mathematical Instruments and Princely Mechanical Devices from around 1600 in the Mathematisch-Physikalischer Salon*, Munich.
Körner & Walter 1996	Martin Körner & François Walter (eds.), *Quand la Montagne aussi a une Histoire, Mélanges offerts à Jean-François Bergier*, Berne/Stuttgart/Vienna.
Kragten 1989	Jan Kragten, *The Little Ship of Venice*, Eindhoven.
Krämer 2019	R. Krämer, 'The Cottage Industry in the Black Forest Region', in *Time Made in Germany*, Nuremberg.
Krauss 2011	Rosalind E. Krauss, 'Clock Time '. *October*, 136, 213–17.
Kreiser 2012	Klaus Kreiser, 'Les tours d'horloges ottomans: inventaire préliminaire et remarques générales', in Georgeon & Hitzel 2012, 61–76.
Kremer 2016	Richard L. Kremer, 'Playing with Geometrical Tools: Johannes Stabius's *Astrolabium imperatorium* (1515) and its Successors', *Centaurus*, 58, 104–34.
Krenn 1977	Claudia Krenn, 'The Traveller's Dial in the Middle Ages: The Chilinder', *Technology & Culture*, 18, 419–35.
Krishnan 2013	Shekhar Krishnan, Empire's Metropolis: Money, Time & Space in Colonial Bombay, *1870–1930*. PhD thesis, MIT. Available at: http://hdl.handle.net/1721.1/86283.
Krombholz 1984	L. Krombholz, *Frühe Hausuhren mit Gewichtsantrich. Der Beginn der mechanischen Zeitmessung*, Munich.
Kubitschek 1928	W. Kubitschek, *Grundriss der Antiken Zeitrechnung*, Munich.
Kuenzl 2003	Ernst Kuenzl, 'Ein römischer Himmelsglobus der mittleren Kaiserzeit. Studien zur römischen Astralikonographie', in Jahrbuch der Römisch-Germanischen Zentralmuseums, 47, 496–594.
Kugel 2016	Alexis Kugel, *Un bestiaire mécanique. Horloges à automates de la Renaissance 1580-1640*, Paris.
Kugel et al. 2002	Alexis Kugel, Kœnraad Van Cleempoel, & Jean-Claude Sabrier, *Spheres, The Art of the Celestial Mechanic*, Paris.
Kuijlaars et al. 2000	Anne-Marie Kuijlaars, Kim Prudon, & Joop Visser (eds.), *Business and Society, Entrepreneurs, Politics and Networks in a Historical Perspective*, Rotterdam.
Kullberg 1887	Victor Kullberg, 'Centrifugal Force and Isochronism' *The Horological Journal*, 30, September, 6–7.
Kurz 1975	Otto Kurz, *European Clocks and Watches in the Near East*, London.
Kusukawa & MacLean 2006	Sachiko Kusukawa & Ian MacLean (eds.), *Transmitting Knowledge. Words, Images and Instruments in Early Modern Europe*, Oxford.
Labarte 1847	Jules Labarte, *Description des objets qui compose la collection Debruge Dumenil . . . d'art*, Paris.
Labarte 1878	Jules Labarte, *Inventaire du Mobilier de Charles V, Roi de France*, Paris.
Lachat 2014	Stéphanie Lachat, *Les pionnières du temps, Vies professionnelles et familiales des ouvrières de l'industrie horlogère suisse (1870–1970)*, Neuchâtel.
Lachat 2017	Stéphanie Lachat, *Longines through Time, The Story of the Watch*, Neuchâtel.
Lacroix 1878	E. Lacroix, *Etudes sur l'exposition de 1878*, Paris.
Lacroix [1849]	Pierre Lacroix, *Prospectus pour Pierre Dubois, Histoire de l'horlogerie depuis son origine jusqu'à nos jours, . . .*, Paris.

Lagadha 1985	Lagadha *Vedāṅga-jyotiṣa Vedāṅga Jyotiṣa of Lagadha, in its Ṛk and Yajus Recensions*, T. S. Kuppanna Sastry (trans. & notes), K. V. Sarma (ed.), New Delhi.
La Hire 1717	Philippe de la Hire, 'Recherche des dates de l'invention du micromètre, des horloges à pendule, & des lunettes d'approche', *HARS* Paris, 78–87.
Lake 1916	B. C. Lake, *Knowledge for War: Every Officer's Handbook for the Front*, London.
Lalande 1769	Jérôme Lalande, *Voyage d'un français en Italie fait dans les années 1765 & 1766*, 2 vols., Paris.
Lalla 1981	Lalla, *Siṣyadhīvṛddhida Tantra of Lalla, with the Commentary of Mallikārjuna Sūri, Critical Edition with Introduction, English Translation, Mathematical Notes and Indices by Bina Chatterjee: Part I: critical Edition with Commentary; Part II: Translation and Mathematical Notes*, New Delhi.
Lamalle 1940	Edmond Lamalle, 'La propagande du P. Nicolas Trigault en faveur des missions de Chine (1616)', *Archivum Historicum Societatis Jesu*, 9, 49–120.
Lamard 1984/1988	Pierre Lamard, *Histoire d'un capital familial au XIXe siècle: le capital Japy (1777–1910)*, Belfort.
Lamard 1985	Pierre Lamard, 'Japy et ses ouvriers au XIXe siècle', *Société d' Emulation de Montbéliard*, 80, 103–33.
Lamard 1999	Pierre Lamard, *Frédéric Japy et son heritage*, Belfort.
Lamard 2004	Pierre Lamard, 'Contraintes économiques, transferts technologiques, attitudes techniques: regard sur l'horlogerie en France et en Suisse dans la seconde moitié du XIXe siècle', in Jean-François Belhoste et al. (eds.), *Autour de l'Industrie. Histoire et patrimoine. Mélanges offerts à Denis Woronoff*, Paris, 569–87.
Lambret & Saindrenan 1996	Eric Lambret & Guy Saindrenan, 'The Discovery of Invar and the Metallurgical Works of Charles-Edouard Guillaume', in Wittenauer 1996, 9–47.
Lamprey 2002	John Lamprey (ed. & trans.), *Hartmann's Practika: A Manual for Making Sundials and Astrolabes with the Compass and Rule*, Bellvue.
Lan 2016	Lan Xiang, 'World Expo Gold Medal Screen Clock in 1915'. 25 April 2010. http://xmwb.xinmin.cn/history/xmwb/html/201004/25/content_496666.htm 一九一五年世博金奖插屏钟
Landau 1967	Rom Landau, *Morocco*, London.
Landes 1979	David S. Landes, 'Watchmaking: A Case Study in Enterprise and Change', *Business History Review*, 53, 1–39.
Landes 1983/2000	David S. Landes, *Revolution in Time: Clocks and the Making of the Modern World*, Cambridge, MA.
Landes 1987	David S. Landes, *L'Heure qu'il est. Les horloges, la mesure du temps et la formation du monde moderne*, Paris.
Landes 1990	David S. Landes, 'Swatch! Ou l'horlogerie suisse dans le contexte mondial', in Bairoch & Körner 1990, 227–36.
Landow 1979	George P. Landow, *Replete with Meaning: William Holman Hunt and Typological Symbolism*, New Haven, CT.
Lang 2004	Hans Lang, *Die Hans-Lang-Uhr, Eine astronomische Kunstuhr der Superlative*, Munich.
Laplace 1796	P. S. Laplace, *Exposition du système du monde*, Paris.
Laresche 1828	M. Laresche 'Procédé pour la preparation de l'huile d'olive à l'usage d'horlogerie', *Bulletin de la Société de l'Encouragement pour l'Industrie Nationale*, 284, 60–1.
Latzel 2009	Robert W. Latzel, *Die Entwicklung der Taschenuhr für jederman in Deutschland*, Norderstedt.
Law 1891	Ernest Law, *The History of Hampton Court Palace*, 2 vols., London.
Leadbetter 1737	Charles Leadbetter, *Mechanick Dialling . . .*, London.
Lebrère 2015	Marylène Lebrère, 'L'artialisation des sons de la nature dans les sanctuaires à automates d'Alexandrie, du IIIe s. av J.-C. au Ier s. apr. J.-C.', *Pallas, Revue d'études antiques*, 98, 31–53.
Le Bot et al. 2019	Florent Le Bot, Jacques Commaille, & Virginie Albe, *L'échelle des régulations politiques, XVIIIe-XXIe siècles. L'histoire et les sciences sociales aux prises avec les normes, les acteurs et les institutions*, Villeneuve d'Ascq.
Le Cerf 1932	G. Le Cerf & E.-R. Labande (eds. & trans.), *Les traités d'Henri Arnaut de Zwolle et de divers anonymes (MS. B. N. Latin 7295)*, Paris.
Leclerq & Cabrol 1907–53	Henri Leclerq & Fernand Cabrol, *Dictionnaire de l'Archéologie chrétienne et de 1907-liturgie*, 15 vols., Paris.
Lefranc 1925	Etablissements Lefranc, *Aide mémoire*, Paris.
Le Goff 1960	Jacques le Goff, 'Au Moyen Age: temps de l'église et temps de marchand', *Annales E. S. C.*, 417–53. Reprinted in Le Goff 1977, 46–65.
Le Goff 1977	Jacques le Goff, *Pour un autre Moyen Age, temps, travail et culture en occident: 18 essais*, Paris.
Le Goff 1980	Jacques le Goff, *Time, Work and Culture in the Middle Ages*, Chicago.
Lehoux 2007	Daryn Lehoux, *Astronomy, Weather, and Calendars in the Ancient World: Parapegmata and Related Texts in Classical and Near-Eastern Societies*, Cambridge.
Lehr 1981	André Lehr, *De geschiedenis van het Astronomisch Kunstuurwerk: Zijn techniek en muziek*, The Hague.
Lejbowicz 1992	Max Lejbowicz, 'Computus. Le nombre et le temps altimédiévaux', in Ribémont 1992, 151–95.
Le Moyne 1666	Pierre Le Moyne, *De l'art des devises*, Paris.
Lenfeld 1984	J. Lenfeld, *Slunecni hodiny ze shirek Upm v Praze*, Prague.
Leng 2012	Leng Dong, 'The Horologe Manufacturing of Thirteen Hongs in the mid-Qing Dynasty', *Lingnan Culture and History*, 1, 49–51.

Lennox-Boyd 2005	Mark Lennox-Boyd, *Sundials, History, Art, People, Science*, London.
Le Normand 1830	Sébastien Le Normand, *Manuel de l'horloger, ou guide des ouvriers qui s'occupent de la construction des machines propres à mesurer le temps*, Paris.
Le Normand et al. 1896	S. Le Normand, Janvier, D. Magnier, & LST, *Nouveau manuel complet de l'horloger*, Paris.
Lenotre 1986	Jean Lenotre, 'L'Horloge de Fécamp', *Horlogerie ancienne*, 19, 49–54.
Leopold 1971	J. H. Leopold, The Almanus Manuscript: Staats-und Stadtbibliothek Augsburg, *Codex in 2°, No. 209, Rome circa 1475–circa 1485*, London.
Leopold 1979	J. H. Leopold, 'Christiaan Huygens and his Instrument Makers', *Studies on Christiaan Huygens: Papers from the Symposium . . . 22–25 August 1979*, Amsterdam, 221–33.
Leopold 1986	J. H. Leopold, *Astronomen, Sterne, Geräte: Landgraf Wilhelm IV und seine sich selbst bewegenden Globen*, Lucerne.
Leopold 1995	J. H. Leopold, 'Collecting Instruments in Protestant Europe before 1800', *Journal of the History of Collections*, 7, 151–7.
Leopold 2003	J. H. Leopold, 'Almanus Re-Examined', *AH*, 27, 665–72.
Leopold 2005	J. H. Leopold, 'Some More Notes on the Coster-Fromanteel Contract', *AH*, 28, 568–70.
Leopold & Smith 2016	John Leopold & Roger Smith (eds.) *The Life and Travels of James Upjohn*, London.
Lepaute 1755	J. A. Lepaute, *Traité d'horlogerie contenant tout ce qui est nécessaire pour bien connoitre et pour regler les pendules et les montres, . . .*, Paris.
Lepaute 1767	J. A. Lepaute, *Traité d'Horlogerie* (new edition). Paris.
Lepetit & Hoock 1987	B. Lepetit & J. Hock, *La Ville et l'innovation en Europe, 14e–19e siècles*, Paris.
Leroux 1984	Jean-Marie Leroux (ed.), *Le temps Chrétien de la fin de l'Antiquité au Moyen Age, IIe – XIIIe siècles*, Paris.
Lerouxel 1981	Gérard Lerouxel, *Les horloges de la Basse-Normandie*, Bayeux.
Le Roy 1719	Julien Le Roy, *Avis contenant les vrais moyens de régler les montres tant simples qu'à repetition*, Paris.
Le Roy 1737	Julien Le Roy (ed.), *Regle artificielle du temps, . . . de Mr. Henry Sully, . . .*, new edn., Paris.
Leroy 1759–61	Pierre Leroy, *Etrennes chronométriques ou calendrier pour l'année . . . Contenant ce qu'on sçait de plus intéressant sur le temps, ses divisions, ses mesures, ses usages, etc . . .*, Paris.
Leroy 1766	Pierre Le Roy, 'Memoire sur la Meilleure Maniere de Mesurer le Tems en Mer…1766' in [J. D.] Cassini, *Voyage Fait par Ordre du Roi en 1768 pour éprouver les montres marines inventées par M. le Roy*, Paris.
Lescure 2006a	Michel Lescure, 'Introduction générale. Le territoire comme organisation et comme institution', in Lescure 2006b, 1–7.
Lescure 2006b	Michel Lescure (ed.), *La mobilisation du territoire. Les districts industriels en Europe occidentale du XVIIe au XXe siècle*, Paris.
Le Strange 1900/1972	G. Le Strange, *Baghdad during the Abbasid Caliphate*, Oxford/London.
Leupold 1724	Jacob Leupold, *Theatrum machinarum generale. Schau-Platz des Grundes mechanscher Wissenschaften. das ist deutliche Anweisung zur Mechanic oder Bewegungs Kunst . . .*, Leipzig.
Leutmann 1722	Johann Georg Leutmann, *Vollständige Nachricht von den Uhren, Erste Continuation oder, zweiter Theil . . .*, Halle.
Levenson & Massing 1991	Jay A. Levenson & Jean Michel Massing, *Circa 1492 Art in the Age of Exploration*, Washington, DC.
Levenstein & Suslow 2004	Margaret C. Levenstein & Valerie Y. Suslow, 'Studies of Cartel Stability: A Comparison of Methodological Approaches', in Grossmann 2004, 9–52.
Levenstein & Suslow 2006	Margaret C. Levenstein & Valerie Y. Suslow, 'What Determines Cartel Success?', *Journal of Economic Literature*, 44, 43–95.
Levenstein & Suslow 2011	Margaret C. Levenstein & Valerie Y. Suslow, 'Breaking Up Is Hard to Do: Determinants of Cartel Duration', *The Journal of Law & Economics*, 54, 455–92.
Levinson 2015	Julie Levinson, 'Time and Time Again: Temporality, Narrativity, and Spectatorship in Christian Marclay's "The Clock"', *Cinema Journal*, 54, 88–109.
Levy-Leboyer 1975	Maurice Levy-Leboyer, *Le patronat de la seconde industrialisation*, Paris.
Lewis 2009	Michael Lewis, 'Theoretical Hydraulics, Automata, and Water Clocks', in Ö. Wikander (ed.), *Handbook of Ancient Water Technology*, Leiden, 343–69.
Leybourn 1682/1700	William Leybourn, *Dialling*, London.
Li 2014	Li Li, 'Western Clocks on Images in the Qing Dynasty', *Rong Bao Zhai Magazine*, 4, 206–15.
Li 2015	Li Tong Tong, 'Polaris Beijixing Wristwatch Brand Introduction', *Xbiao*, 14 April 2015. http://www.xbiao.com/20150414/30985.html. 李童童. Polaris 北极星手表品牌介绍.
Li 2012	Li Yu-ju, *Clocks, Clock Towers, and Standard Time: Western-style Timepieces and their Interaction with Chinese Society (1582–1949)*, Department of History, National Chengchi University.
Li 2014a	Li Zhichao, *History of Chinese Water Clock*, Hefei.
Liao 2002	Liao Pin (compiler), *Clocks and Watches of the Qing Dynasty, from the Collection in the Forbidden City*, Beijing.

Liengme Bessire, & Barrelet 1996	Marie-Jeanne Liengme Bessire & Jean-Marc Barrelet, 'L'évolution des structures de la production dans l'industrie horlogère des Montagnes jurassiennes à la fin du XIXe siècle. Une mutation escortée par l'histoire', in Pfister, Studer, & Tanner 1996, 49–64.
Lightbown 1989	Ronald Lightbown, *Sandro Botticelli: Life and Work*, London.
Lin 2013	Lin Fangyin (ed.), *Yuan Ming Yuan: Qing Emperor's Splendid Gardens*, Eurographics Association.
Linder 2007	Patrick Linder, *At the Heart of an Industrial Vocation, Longines Watch Movements (1832–2009)*, St. Imier.
Linnard 2015a	William Linnard, 'The Word "Foliot" ', *AH*, 36, 100–1.
Linnard 2015b	William Linnard, 'The Measurement of Time in 1665: Giuseppe di Capriglia's *Misura del tempo*', *The Horological Journal*, Nov., 505–5.
Linnard & Owen 2012	William Linnard & Anne Parry Owen, 'Horological Requests in Early Welsh Poems', *AH*, 33, 631–6.
Lippincott 1999	Kristen Lippincott (ed.), *The Story of Time*, London,.
Lipson 1948	E. Lipson, *The Economic History of England*, 5th edn., 3 vols., London.
Lis & Soly 2008	Catharina Lis & Hugo Soly, 'Subcontracting in Guild-Based Export Trades: Thirteenth – Eighteenth Centuries', in Epstein & Prak 2008, 81–113.
Liu 1962	Liu Xianzhou, *History of Chinese Engineering Inventions*, Part 1, Beijing.
Livingston 1992	John Livingston 'The Mukhula: An Islamic Conical Sundial', *Centaurus*, 26, 299–308.
Lixfeld 2011	Gisela Lixfeld, 'Amerikaneruhren – made in Germany: einflüsse der amerikanischen Uhrenindustrie auf Gestaltung und Vermarktung in Deutschland produzierter Uhren für den Massenbedarf', in Bodenmann 2011, 181–91.
Lloyd 1958	H. Alan Lloyd, *Some Outstanding Clocks Over Seven Hundred Years*, London.
Lloyd 1959	H. Alan Lloyd, 'The Burgundy Clock', *AH*, 2, 235.
Lloyd & Drover 1955	H. Alan Lloyd & C. B. Drover, 'Nicholas Vallin (c. 1565–1603)', in *The Connoisseur Yearbook*, 110–16.
Lloyd et al. 1992	Steven A. Lloyd with Penelope Gouk & A. J. Turner, *Ivory Diptych Dials 1570–1750*, Cambridge, MA.
Lochner 1875	G. W. K. Lochner, *Des Johann Neudörfer Schreib- und Rechenmeisters zu Nürnberg Nachrichten von Künstlern und Werkleuten daselbst aus dem Jahre 1547, nebst der Fortsetzung des Andreas Gulden. Nach den Handschriften und mit Anmerkungen herausgegeben*, Vienna.
Loertscher-Rouge 1977	Françoise Loertscher-Rouge, 'La politique de la FOMH dans l'horlogerie lors de la crise des années trente (1930-1937)', *Revue européenne des sciences sociales*, 15, 143–99.
Loeske 1892	M. Loeske, *Das Regulieren der Uhren in den Lagen, in Theorie und Praxis*, Bautzen.
Lombardi & Gebus 2015/16	Marianne Lombardi & Eric Gebus, 'La plus ancienne horloge Saint Nicolas', *Bulletin d'ANCAHA*, cxxx 53–58.
London Chamber 1937	'The Watch Trade', London Chamber of Commerce, *The Chamber of Commerce Journal*, 68, June.
Long 1860	S. Long, 'On the Preparation of Oil for Horological Purposes', *HJ*, 99–100.
Loomes 1974	Brian Loomes, *The White Dial Clock*, Newton Abbot.
Loomes 1976	Brian Loomes, *Country Clocks and their London Origins*, Newton Abbot.
Loomes 1981a	Brian Loomes, *The Early Clockmakers of Great Britain*, London.
Loomes 1981b	Brian Loomes, *White Dial Clocks*, London.
Loomes 1994	Brian Loomes, *Painted Dial Clocks*, Woodbridge.
Loomes 2006	Brian Loomes, *Watchmakers and Clockmakers of the World . . .*, London.
Loomes 2008	Brian Loomes, *Lantern Clocks & Their Makers*, Ashbourne.
Loomes 2014	Brian Loomes, *Clockmakers of Britain 1286–1700*, Ashbourne.
Lorain 1845	P. Lorain, *Histoire de l'abbaye de Cluny depuis sa fondation jusqu'à sa destruction à l'époque de la Révolution française*, Paris.
Lorch 1980	Richard Lorch, 'Al-Khâzinî's Sphere that Rotates by Itself', *Journal for the History of Arabic Science*, 4, 287–329.
Lorch 1981	Richard Lorch, 'A Note on the Horary Quadrant', *Journal for the History of Arabic Science*, 5, 115–20.
Lory 1781	Michael Lory, *Gnomonik*, Salzbourg,
Löschner 1906	Hans Löschner, *Über sonnenuhren*, Gratz.
Losito 1989	Maria Losito, 'Il IX Libro del *De Architectura* di Vitruvio nei Commentari di Daniele Barbaro (1556–1567)', *Nuncius. Annali di Storia della Scienza*, 5, 3–42.
Lossier 1890	Louis Lossier, *Étude sur la théorie du réglage des montres, suivi d'instructions et d'exemples pratiques*, Geneva.
Lovell & Kluger 1995	J. Lovell and J. Kluger, *Apollo 13*, London.
Low 1998	Morris F. Low (ed.), 'Beyond Joseph Needham; Science, Technology, and Medicine in East and Southeast Asia', *Osiris*:13.
Lowne 2001	C. M. Lowne, 'The Design and Characteristics of the Double-Horizontal Sundial', *BBSS*, 13, 138–46.

Lu 2017	Lu Tian-ze, 'Research on the Technological Origin of the Mechanical Clock', *Studies in Dialectics of Nature*, 33, 65–70.
Luckey 1927	Paul Luckey, 'Das Analemma von Ptolemäus', *Astronomische Nachrichten*, 230, 18–46; C. Segard & D. Collin (trans.), 'The Analemma of Ptolemy', *Cadran Info*, 18, 20–35.
Luckey 1937–8	Paul Luckey, 'Thabit b. Qurra's Buch uber die ebenen Sonnenuhren', Quellen und Studien zur Geschichte des Mathematik, Astronomie und Physik, B.4, 95–148.
Ludlam 1769	William Ludlam, *Astronomical Observations made in St John's College, Cambridge in the years 1767 and 1768 . . .*, Cambridge, 1769.
Lunardi 1974	Heinrich Lunardi, *900 Jahre Nürnberg - 600 Jahre Nürnberger Uhren*, Vienna, 99–113.
Lush 2011	Julian Lush with M. Arnaldi, 'Sundial in Armenia', *T & R*, 1, 333–54.
Luyken 1694	Jan Luyken, *Het menselijke bedrijf*, Amsterdam.
Ma 2009b	Yingjian Ma, 'Tianjin Clockmaking's Number One Person: Chen Yuwu'. Sina Blog. 7 December 2009. Available at: http://blog.sina.com.cn/s/blog_62a2dc9b0100fs93.html. 马樱健. '天津制表第一人——陈玉吾'.
Maccagni 1969–71	Carlo Maccagni, 'The Florentine Clock and Instrument-Makers of the della Volpaia Family, *Der Globusfreunde*, 18–20, 921–9.
Maccagnolo 1980	Enzo Maccagnolo, *Il Divino o il Megacosmo: testi folosofici e scientifici della scuola di Chartres*, Milan.
Macey 1986	S. L. Macey, 'Literary Images of Progress: The Fate of an Idea', in Fraser, Lawrence, & Haber 1986, 93–103.
Macgowan 1852	Daniel J. Macgowan, 'On Chinese Horology, with Suggestions on the Form of Clocks Adapted for the Chinese Market', in United States Patent Office, *Report of the Commissioner of Patents for the Year 1851, Part I: Arts and Manufactures*, Washington, 335–42.
MacGregor 1989	Arthur MacGregor (ed.), *The Late King's Goods. Collections, Possessions and Patronage of Charles I in the Light of the Commonwealth Sale Inventories*, Oxford.
Macmillan 1898	Hugh MacMillan, *The Clock of Nature*, London.
Madden 1855	Frederick Madden, 'Agreement between the Dean and Chapter of St. Paul's, London, and Walter the Orgoner ... Relating to a Clock ... November 22 1344', *The Archæological Journal*, 12, 173–7.
Maddison 1969	Francis Maddison, *Medieval Scientific Instruments and the Development of Navigational Instruments in the XVth and XVIth Centuries*, vol. 30, Lisbon.
Maddison 1987	Francis Maddison, 'Al-Jazarī's Combination Lock: Two Contemporary Examples', Oxford Studies in Islamic Art, vol. 1, 141–57.
Maddison 1994	Francis Maddison, 'Masculine Clocks in French', *AH*, 21, 554.
Maddison 1997	Francis Maddison, 'The dā'irat in Maddison and Savage-Smith 1997, n.p.
Maddison 1997	Francis Maddison & Emilie Savage-Smith, *Science, Tools and Magic, Part One. Body and Spirit, Mapping the Universe, The Nasser D. Khalili Collection of Islamic Art*, vol. 12, London.
Maddison, Scott, & Kent 1962	Francis Maddison, Bryan Scott, & Alan Kent, 'An Early Medieval Water Clock', *AH*, 3, 348–51.
Maddison & Turner 1999	Francis Maddison & Anthony Turner, 'The Names and Faces of the Hours', in Loti Nauta & Arie Johan Vanderjagt (eds.), *Between Demonstration and Invention. Essays in the History of Science and Philosophy Presented to John D. North*, Leiden, 124–56.
Maguire 1997	Henry Maguire (ed.), *Byzantine Court Culture from 829–1204*, Cambridge, MA.
Mahrer 2012	Stephanie Mahrer, *Handwerk der Moderne: jüdische Uhrmacher und Uhrenunternehmer im Neuenburger Jura 1800–1914*, Cologne.
Maignan 1648	Emmanuel Maignan, *Perspectiva horaria sive de horographia gnomonica tum theoretica, tum practica libri quatuor*, Rome.
Maignien 1887	Edmond Maignien, *Les artistes grenoblois, . . .*, Grenoble.
Maitzner & Moreau 1985	Francis Maitzner & Jean Moreau, *La Comtoise (la Morbier, la Morez), histoire-technique*, Dreux.
Mâle 1931	Émile Mâle, *L'Art Religieux de la Fin du Moyen Âge en France*, 4th edn., Paris.
Mann 1992	Vivian B. Mann, Thomas F. Glick, & Jerrilynn D. Dodds (eds.), *Convivencia. Jews, Muslims and Christians in Medieval Spain*, New York.
Manuel 1827	*L'Art de l' horlogerie enseigné en trente leçons ou manuel complet de l'horloger et de l'amateur*, Paris.
Marçais & Marçais 1903	William Marçais & Georges Marçais, 'El-Mansourah', in *Les Monuments Arabes De Tlemcen*, Paris, 192–22.
Marder 1929	A. E. Marder, 'Conical Sundial and Ikon Inscription from the Kastellion Monastery on Khorbet el-Merd in the Wilderness of Juda', *The Journal of the Palestine Exploration Society*, 9, 122–35.
Marey 1979	Bernard Marey, *Les grands magasins des origines à 1939*, Paris.

Marguerat et al. 2000	Philippe Marguerat, Laurent Tissot, & Yves Froidevaux (eds.), *Banques et entreprises en Europe de l'Ouest, XIXe–XXe siècles: aspects nationaux et régionaux*, Neuchâtel.
Marguet 1931	F. Marguet, *Histoire générale de la navigation du XVe au XXe siècle*, Paris.
Markusen 1996	Ann Markusen, 'Sticky Places in Slippery Spaces: A Typology of Industrial Districts', *Economic Geography*, 72, 293–313.
Marquet 1999	Nicole Marquet, 'Six cadrans solaires décimaux', *Observations et Travaux*, 51, 26–7.
Marr 2009	Alexander Marr (ed.), *The Worlds of Oronce Fin: Mathematics, Instruments and Print in Renaisssance France*, Donington.
Marrison 1948a	W. A. Marrison, 'The Evolution of the Quartz Crystal Clock, Part 2', *HJ*, 90, 342–5.
Marrison 1948b	W. A. Marrison, 'The Evolution of the Quartz Crystal Clock, Part 3', *HJ*, 90, 402–7.
Marrison 1948c	W. A. Marrison, 'The Evolution of the Quartz Crystal Clock, Part 4', *HJ*, 90, 460–6.
Marrison 1948d	W. A. Marrison, 'The Evolution of the Quartz Crystal Clock, Part 6', *HJ*, 90, 588–92.
Martene 1736/7	Edmond Martene, *De antiquis ecclesiae ritibus. Tractatum de antiqua ecclesiae*, Antwerp.
Martens 1866	J. H. Martens, *Rathgeber für den Uhrenbesitzer. Belehrung Über die Behandlung der Taschenuhr und das Reguliren derselben*, Furtwangen.
Marti 2001	Laurence Marti, 'Entre la ferme et l'usine. Essai d'histoire orale', in *Pour une histoire des femmes dans le Jura*, Porrentruy, 131–16.
Marti 2016	Laurence Marti, *Le renouveau horloger contribution à une histoire récente de l'horlogerie suisse (1980–2015)*, Neuchâtel.
Martial 1990	D. R. Shackleton-Bailey (ed.), *Martial: Epigrams*, London.
Martinelli 1669	Domenico Martinelli, *Horologi elementary divisi in qvattro parti*, Venice.
Martini 1777	Georg Heinrich Martini, *Abhandlung von des Sonnenuhren der Alten*, Leipzig.
Martyn 1701	Henry Martyn, *Considerations on the East India Trade . . .*, London.
Maslot 1718	Jean Maslot, *Les loix universelles en nombres, poids et mesures*, Troyes & Paris.
Massé 2008	Yvon Massé, 'De la resolution du triangle sphérique de position par l'analemme à différents cadrans de hauteur (3e partie)', *Le Gnomoniste*, 15, 4–13.
Massé 2009	Yvon Massé, *De l'analemme aux cadrans de hauteur*, Pontoise.
Mateos 1964	Juan Mateos, 'Un horologion inédit de Saint-Sabas, le codex sinaitique grec 863 (IXe) siècle', in *Mélanges Eugène Tisserant*, vol. 3, Studi e testi Series 233, Vatican City.
Matsuda et al. 1988	Francisco Pasio, Ki-ichi Matsuda, & Toshimitsu Ie-iri (trans.), 'Japanese Matters in 1601–1602 (1601–1602年の日本の諸事)', *Reports of Jesuit Society in the 16th and 17th Century*, ph. 1, vol. 4, Dohosha.
Matthes 2015	Dietrich Matthes, 'A Watch by Peter Henlein in London?', *AH*, 36, 183–94.
Mattthes 2017	Dietrich Matthes, 'Tools from Henlein's Workshop Discovered?', *AH*, 38, 326–7.
Matthes 2018	Dietrich Matthes, *Zeit haben, Tragbare Uhren vor 1550*, Dover, DE.
Matthes 2019	Dietrich Matthes, 'Les Horloges gothiques de Mehun sur Yèvre', *Horlogerie Ancienne*, 101–17.
Matthes 2020	Dietrich Matthes, *Spring-Driven Horology before 1510*, Dover, DE.
Matthes & Sánchez-Barrìos 2017	Dietrich Matthes & R. Sánchez-Barrìos, 'Mechanical Clocks and the Advent of Scientific Astronomy', *AH*, 38, 328–42.
Matthes & Sánchez-Barrios 2019	Dietrich Matthes & R. Sánchez-Barrios, 'From Gold to Iron', *Clock and Watch Bulletin*, 439, 221–8.
Mauerhan 2004	Joëlle Mauerhan (ed.), *Pratiques et mesure du temps*, Actes du colloque des sociétés historiques et scientifiques [édition électronique], Besançon.
Mauerhan 2018	Joëlle Mauerhan, *Horlogers et horlogères à Besançon, 1793–1908, un passé prêt à revivre*, Paris.
Maurice 1967	Klaus Maurice, *Die Französische Pendule des 18 Jahrhunderts, Ein Bertrag zur Ihrer Ikonologie*, Berlin.
Maurice 1976	Klaus Maurice, *Die deutsche Räderuhr: zur Kunst und Technik des mechanischen Zeitmessers im deutschen Sprachraum*, 2 vols., Munich.
Maurice 1980a	Klaus Maurice, 'Jost Bürgi, or on Innovation', in Maurice & Mayr 1980, ch. 8.
Maurice 1980b	Klaus Maurice, 'Propagatio fidei per scientias: Jesuit Gifts to the Chinese Court', in Maurice & Mayr 1980, ch. 4.
Maurice & Mayr 1980	Klaus Maurice & Otto Mayr (eds.), *Die Welt als Uhr*, München (in English: *The Clockwork Universe: German Clocks and Automata 1550–1650*), Washington DC.
Maxe-Werly, 1887	Léon Maxe-Werly, 'Note sur des objets antiques découverts à Gondrecourt et à Grand', in *Mémoire de la Société nationale des antiquaires*, Paris, 170–8.
Mayall 1982	R. Newton Mayall, 'A Bit of Porcelain', *Sky & Telescope*, 63, 16–17.
Mayaud 1991	Jean-Luc Mayaud, *Les patrons du Second Empire, Franche-Comté*, Paris
Mayaud 1994	Jean-Luc Mayaud, *Besançon horloger*, Besançon.
Mayaud & Henry 1995	Jean-Luc Mayaud & Philippe Henry (eds.), *Horlogeries, le temps de l'histoire*, Besançon.

Mayer & Bentley-Cranch 1994	Claude-A Mayer & Dana Bentley-Cranch, *Florimond Robertet (?–1527) homme d'état français*, Paris.
Mayer 1956	L. A. Mayer, *Islamic Astrolabists and Their Works*, Geneva.
Mayette 1889	J. Mayette, 'De la mesure du temps et du réglage des montres et horloges', *Annales de la Société d'Agriculture de Lyon*, 2, 213–74.
Mayr 1980	Otto Mayr, 'A Mechanical Symbol for an Authoritarian World', in Maurice & Mayr 1980, ch. 1.
Mayr 1986	Otto Mayr, *Authority, Liberty and Automatic Machinery in Early Modern Europe*, Baltimore.
Mayson 2000	Geoffrey T. Mayson, *Mechanical Singing-Bird Tabatières*, London.
Mazard 2011	Chantal Mazard, *Les cadrans solaires en Isère*, Grenoble.
Mazur 2008	Agathe Mazsur, *Ō Temps! Suspends ton vol. Catalogue des pendules et horloges du Musée des Arts décoratifs de Lyon*, Lyon.
McCamp & Armstrong 1979	John McCamp & Joe E. Armstrong, 'Notes on a Water Clock in the Athenian Agora', *Hesperia*, 46, 147–61.
McCann 1976	Justin McCann (trans.), *The Rule of St. Benedict*, London.
McCleod & Millburn 1998–1999	W. R. and V. McLeod and John R. Millburn, *Horological Advertisements in the Reign of Queen Anne (1702–1714)*, privately published, Aylesbury.
McCluskey 1990	Stephen C. McCluskey, 'Gregory of Tours, Monastic Timekeeping and early Christian Attitudes towards Astronomy', *Isis*, 81, 9–22.
McConnell n.d.	Anita McConnell, 'David Ramsay (c.1575–1660)', in *ODNB*. Available at: https://doi.org/10.1093/ref:odnb/23080.
McDonald & Hunt 1982	Donald McDonald & Leslie Hunt, *A History of Platinum and its Allied Metals*, London.
McEvoy. 2011	Rory McEvoy 'Admiral Lord Nelson and the Commemorative Domestic Clock', *AH*, 32, 665–88.
McEvoy & Betts 2020	Rory McEvoy & Jonathan Betts (eds.), *Harrison Decoded: Towards a Perfect Pendulum Clock*, Oxford.
McKay 2010	Chris McKay, *Big Ben: The Great Clock and the Bells of the Palace of Westminster*, Oxford.
McKay 2013	Chris McKay, *The Turret Clock Keeper's Handbook*, new revised edn., Wimborne Minster.
Meerwalt 1921	J.-D. Meerwalt, 'De Trimalchionis, Ctesibii, Platonis Automatis', *Mnemosyne*, 49, 406–26.
Mégnin 1909	Georges Mégnin, *Naissance, développement et situation actuelle de l'industrie horlogère à Besançon*, Besançon.
Meis 1986	Reinhardt Meis, *Das Tourbillon. Faszination der Uhrentechnik*, Munich.
Meis 1987	Reinhardt Meis, *Pocket Watches*, Westchester (Pa.).
Meis 2011	Reinhardt Meis, *A. Lange & Söhne. Eine Uhrmacherdynastie aus Dresden*, 2 vols., Munich.
Mendel 1914	G. Mendel, *Musées Impériaux ottomans: catalogue des sculptures grecques, romaines et byzantines*, vol. 2, Constantinople.
Menz & Sturm 2003	Cäsar Menz & Fabienne Xavière Sturm, 'Musée de l'horlogerie et de l'émaillerie Genève: catalogue des pieces dérobées le 24 novembre 2002, *Genava*, 51, 1–134.
Mercer 1962	R. Vaudrey Mercer, 'Peter Litherland & Co.', *AH*, 11, 316–23.
Mercer 1972, 1975	R. Vaudrey Mercer, *John Arnold and Son, Chronometer Makers 1762–1843*, & *Supplement*, London.
Mercer 1977	R. Vaudrey Mercer, *The Life and Letters of Edward John Dent, Chronometer Maker, and Some Account of his Successors*, London.
Mercier 2014	Eric Mercier, 'Cadrans portatifs de Dieppe (XVIIe)', *Cadran Info*, 30, 45–65.
Mercier 2015	Eric Mercier, 'Cadrans portatifs et déclinaison magnétique (XVIe–XVIIIe siècles)', *Cadran Info*, 32, 61–76.
Mersmann 2015	Jasmin Mersmann, 'Moving Shadows, Moving Sun, Early Modern Sundials Restaging Miracles', *Nuncius*, 30, 96–123.
Mertens, 2009	Joost Mertens, 'Éclairer les arts: Eugène Julia de Fontenelle (1780–1842), ses manuels Roret et la pénétration des sciences appliquées dans les arts et manufactures', *Documents pour l'histoire des techniques* [online]. Available at: http://journals.openedition.org/dht/339.
Mertens 2011	Joost Mertens, 'Le déclin de la technologie générale: Léon Lalanne et l'ascendance de la science des machines', *Documents pour l'histoire des techniques*, [online]. Available at: http://journals.openedition.org/dht/1749.
Mesnage 1949	Pierre Mesnage, 'Un chef-d'oeuvre de Jost Bürgi au Conservatoire des Arts et Metiers de Paris', *Journal suisse d'horlogerie et de bijouterie*, May–June, 198–207.
Messerli 1995	Jacob Messerli, 'Gleichmässig, Pünktlich, Schnell: zeiteinteilung und Zeitgebrauch in Der Schweiz Im 19. Jahrhundert', Zurich.
Messerli 2005	Jakob Messerli, 'Präzision und Massenfertigung. Der amerikanische Einfluss auf die europäische Uhrenindustrie im 19. Jahrhundert', in Schmittgen 2005, 178–94.
Mesturini 2015	Giorgio Mesturini, 'Cadrans de guerre d'Albertis', *Cadran Info*, 32, 89–92.
Meyer 2019	Alexander Meyer, 'The Vindolanda Calendrical Clepsydra: Time-Keeping and Healing Waters', *Britannia*, l, 185–202.

Meyer 2008	Jörg Meyer, *Die Sonnenuhr und ihre Theorie*, Frankfurt am Main.
Meyer 2013	Tobias Meyer (ed.), *Beiträge zur Wissenschaft des 18. Jahrhunderts im Lichte neuerer Untersuchungen*, Leipzig.
Meyer 1980	Wolfgang Meyer, 'Sundials of the Osmanic Era in Istanbul', in Dizer 1980, 193–202.
Michel 1954	Henri Michel, 'Les tubes optiques avant le telescope', *Ciel et terre*, 70, 175–84.
Michel 1960	Henri Michel, '*L'Horloge de sapience* et l'histoire de l'horlogerie,' *Physis*, 2, 291–8.
Michel 1962	Henri Michel, 'Some New Documents in the History of Horology', *AH*, 3, 288–91.
Michel & Ben-Eli 1965	Henri Michel & Arie Ben-Eli, 'Un cadran solaire remarquable', *Ciel et terre*, 81, 214–16.
Michel-Nozières 2000	C. Michel-Nozières, 'Second Millennium Babylonian Water Clocks: A Physical Study', *Centaurus*, 42, 180–209.
Michnik 1914	Hermann Michnik, *Beiträge zur Theorie der Sonnenuhren*, Leipzig.
Michnik 1923	Hermann Michnik, 'Theorie einer Bifilar-Sonnenuhr', *Astronomische Nachrichten*, 217, 81–90.
Miclet 1976	Bernard Miclet, 'Le prototype Vérité; Bilboquet électrique (1853)', *Revue de l'Ancaha*, 17, 47–8.
Miclet 1977	Bernard Miclet, 'Un horloger du XIXe siècle: Auguste-Lucien Vérité', *Revue de l'Ancaha*, 18–19, 27–61; 5–44.
Midleton 2007	Alan Midleton, 'The History of the BHI: The Early Years', *HJ*, 149, 312–15.
Migeon 1917	Gaston Migeon, *Musée du Louvre: Collection Paul Garnier . . .*, Paris.
Migne 1863	Jacques Paul Migne, *Patrologia Græce*, 142, Paris.
Miles 2011	Robert Miles, *Synchronome: Masters of Electrical Timekeeping*, Ticehurst.
Millas-Vallicrosa 1931	J. M. Millas-Vallicrosa, *Assaig d'Historia da les idees fisiques i matematiques a la Catalunya medieval*, Barcelona.
Millas-Vallicrosa 1932	J.-M. Millas Vallicrosa, 'La introduccion del cuadrante con cursor en Europa', *Isis*, 17, 218–58.
Millas-Vallicrosa 1950	J. M. Millas-Vallicrosa, *Estudios sobre Azarquiel*, Madrid.
Millburn 1973	John R. Millburn, 'Benjamin Martin and the Development of the Orrery', *British Journal for the History of Science*, 6, 378–99.
Millburn 1976	John R. Millburn, *Benjamin Martin. Author, Instrument-Maker, and 'Country Showman'*, Leyden.
Millburn 2000	John R. Millburn, *Adams of Fleet Street, Instrument Makers to King George III*, Aldershot.
Millburn & King 1988	John R. Millburn with Henry C. King, *Wheelwright of the Heavens. The Life and Work of James Ferguson, FRS*, London.
Miller 1908	W. Miller, *The Latins in the Levant*, London.
Mills 1982	A. A. Mills, 'Newton's Water Clock and the Fluid Mechanics of Clepsydrae', *Notes & Records of the Royal Society*, 37, 35–61.
Mills 1985	A. A. Mills, 'Heliostats, Siderostats, and Coelostats: A Review of Practical Instruments for Astronomical Applications', *Journal of the British Astronomical Association*, 95, 89–99.
Mills 1986	A. A. Mills, 'Portable Heliostats (Solar Illuminators)', *Annals of Science*, 43, 369–406.
Mills 1988	A. A. Mills, 'The Mercury Clock of the Libros del Saber', *Annals of Science*, 45, 329–44.
Mills 1995	Allan A. Mills, 'The "Dial of Ahaz", and Refractive Sundials in General', *Bulletin of the Scientific Instrument Society*, 44, 21–4; 45, 25–7.
Minns & Wallis 2013	Chris Minns & Patrick Wallis, 'The Price of Human Capital in a Pre-Industrial Economy: Premiums and Apprenticeship Contracts in 18th-Century England', *Explorations in Economic History*, l, 335–50.
Mirot 1900	Léon Mirot, 'Le procès de Maître Jean Fusoris', *Mémoires de la Société de l'Histoire de Paris et de l'Ile de France*, 27, 137–287.
Mirror 2010	Patek Philippe Museum, *The Mirror of seduction, prestigious pairs of "Chinese" watches*, Geneva.
Moevs 1999	Christian Moevs, 'Miraculous Syllogisms: Clocks, Faith and Reason in Paradiso 10 and 24', *Dante Studies*, 117, 59–84.
MOMA 2019	MOMA, ' Salvador Dalí's "The Persistence of Memory" (1931)'. Exhibition catalogue entry [online]. Available at: https://www.moma.org/collection/works/79018. Accessed 24/12/19; excerpt from MoMA Highlights: 375 Works from The Museum of Modern Art, New York.
Monconys 1695	Balthazar de Monconys, *Voyages de M. de Monconys . . .*, 5 vols., Paris.
Monks 1990	Peter Rolfe Monks, *The Brussels Horloge de Sapience: Iconography and Text of Brussels, Bibliothèque Royale, MS IV 111*. Leiden.
Monnier 2000	Raymond Monnier, *La Radioactivité et l'industrie horlogère*.
Monnin 2015/16	Reuben Monnin, 'Démarrage de la fabrication des pendules de voyage chez L'Epée', *Horlogerie ancienne. Bulletin de l'Association nationale des collectionneurs et amateurs d'horlogerie ancienne et d'art*', cxxx, 31–46.
Montañes 1968	L. Montañes, *Relojes Espagnoles*, Madrid.
Montucla 1798–1802	J. E. Montucla, *Histoire des mathématiques dans laquelle on rend compte de leurs progress depuis leur origine jusqu'à nos jours, . . .*, new edn., 4 vols., Paris.
Moore 1945	C. W. Moore, *Timing a Century*, Cambridge.

Moore 2003	Denis Moore, *British Clockmakers & Watchmakers Apprenticeship Records, 1710–1810*, Ashbourne.
Morales 1575	Ambrosio de Morales, *Las antiguedades de las ciudades de España que van nombradas en la Coronica, con la aueriguacion de sus sitios, y nōbres antiguos*, n.p.
Moran 1977	Bruce T. Moran, 'Princes, Machines and the Valuation of Precision in the 16th Century', *Suddhoffs Archiv*, 59, 209–28.
Moreau 1914	Moreau, 'Note sur un cadran solaire vertical à style vertical', *Nouvelles annales de la construction*, 7ᵉ série, 1, 115–21.
Morelon 1987	Régis Morelon (ed. & trans.), *Thābit ibn Qurra, œuvres d'astronomie*, Paris.
Morgan 1960	Morris Hickey Morgan, *Vitruvius: The Ten Books of Architecture translated by . . .*, New York.
Morgan 1849	Octavius Morgan, 'Observations on the History and Progress of Watchmaking, from the Earliest Period to Modern Times, . . .', *Archæologia*, 33, 84–100 & 293–307.
Morgan 1652	Sylvanus Morgan, *Horologiographia optica*, London.
Morini 1999	Carla Morini, 'Horologium e Daegmæl nei manosccritti angloassoni del computo', *Aevum. Rassegna di scienze storiche, linguistiche e filosofiche . . .*, 73, 273–93.
Morpurgo 1950	Enrico Morpurgo, *Dizionario degli Orologiai Italiani*, Roma.
Morpurgo 1954	Enrico Morpurgo, *L'origine dell'orologio tascabile*, Rome.
Morpurgo 1958	Enrico Morpurgo, 'The Clock and the Pendulum', *AH*, 2, 138–42, 151.
Morpurgo 1970	Enrico Morpurgo, *Nederlandse klokken en horlogemakers vanaf 1300*, Amsterdam.
Morse 1926–1929	H. B. Morse, *Chronicles of the East India Company Trading to China 1635–1834*, 5 vols., Oxford.
Morrison-Low et al. 2012	Alison Morrison-Low, Sven Dupré, Stephen Johnston, & Giorgio Strano (eds.), *From Earth-Bound to Satellite Telescope. Skills and Network*, Leiden/Boston.
Morrison-Low et al. 2016	A. D. Morrison-Low, Sara J. Schechner, & Paolo Brenni, *How Scientific Instruments Have Changed Hands*, Leiden.
Morsman 2006	Marieke Morsman, *Quicquid rarum, occultum et subtile: Augsburg Musical Automata around 1600*, Faculty of Humanities thesis, Utrecht University.
Mortensen 1957	Otto Mortensen, *Jens Olsen's Clock*, Copenhagen.
Morton & Wess 1993	Alan Q. Morton & Jane A Wess, *Public & Private Science: The King George III Collection*, Oxford.
Mosley 2019	Adam Mosley, 'Sundials and Other Cosmographical Instruments': Historical Categories and Historians' Categories in the Study of Mathematical Instruments and Disciplines', in Nall, Taub, & Willmoth 2019, 55–81.
Mottu-Weber 1970	Liliane Mottu-Weber, 'Apprentissages et economie genèvoise au début du XVIIIe siècle', *Revue suisse d'histoire*, 20, 341.
Mottu-Weber 2004	Liliane Mottu-Weber, 'Inventeurs genevois aux prises avec la maladie des doreuses et doreurs en horlogerie (fin XVIIIe–début XIXe siècle)', *Cahiers d'histoire et de philosophie des sciences*, 52, 283–96.
Moxon 1677/78	Joseph Moxon, *Mechanick exercises: Or the Doctrine of Handy Works . . .*, London.
Mraz 1980	Gottfried Mraz, 'The Role of Clocks in the Imperial Honoraria for the Turks', in Maurice & Mayr 1980, ch. 5.
Mudge 1799	Thomas Mudge [Jr], *A Description with Plates of the Time-Keeper Invented by the Late Mr Thomas Mudge. To which is prefixed a narrative . . .*, London.
Mühe & Vogel 1976	Richard Mühe and Horand M. Vogel, *Alte Uhren, ein handbuch europäischer Tischuhren, Wanduhren und Bodenstanduhren*, Munich.
Müller 2012	Reto Müller, 'Reinhard Straumann', in *Dictionnaire historique de la suisse* [online]. Available at: https://hls-dhs-dss.ch/fr/articles/016400/2012-06-29.
Muller & Simonin 2006	Bernard Muller & Antoine Simonin, *Les métiers de l'horlogerie*, Chézard-St Martin.
Mumford 1934	Lewis Mumford, *Technics and Civilization*, New York.
Munck 2005	Thomas Munck, *Seventeenth Century Europe. State, Conflict and the Social Order in Europe, 1598–1700*, 2nd edn., Basingstoke.
Münster 1531	Sebastian Münster, *Compositio horologiorum, in plano, muro, truncis, anulo, concavo, cylindro & variis quadrantibus*, Basel.
Münster 1533	Sebastian Münster, *Horologiographia*, Basel.
Munz 2016	Hervé Munz, *La transmission en jeu. Apprendre, pratiquer, patrimonialiser l'horlogerie en Suisse*, Neuchâtel.
Murdoch 1985	Tessa Murdoch, *The Quiet Conquest*, London.
Murdoch 2008	Tessa Murdoch, 'Daniel Marot (1661–1752)', in ODNB. Available at: https://doi.org/10.1093/ref:odnb/39328.
Murdoch 2013	Tessa Murdoch, 'Time's Melody', *Apollo*, November, 78–85.
Nadelhoffer 1984	Hans Nadelhoffer, *Cartier: Jewelers Extraordinary*, London.

Naffah 1989	Christiane Naffah, 'Un cadran cylindrique ottoman du XVIIIe siècle', in Turner 1989b, 37–52.
Nagel & Wood 2005	Alexander Nagel & Christopher S. Wood, 'Towards a New Model of Renaissance Anachronism', *The Art Bulletin*, 87, 403–15.
Nagel & Wood 2010	Alexander Nagel & Christopher S. Wood, *Anachronic Renaissance*, New York.
Nall, Taub & Willmoth 2019	Joshua Nall, Liba Taub, & Frances Willmoth (eds.), *The Whipple Museum of the History of Science*, Cambridge.
Nallino 1899–1907	C. A. Nallino, *Al-Battani sive Albateni opus astronomicus*, (Publicazione dei Reale Osservatorio di Brera in Milano, xl), 3 vols., Milan & Rome.
National Bureau of Standards 1978	National Bureau of Standards. 'The Atomic Clock: An Atomic Standard of Frequency and Time', *The Journal: Coast and Geodetic Survey*, 105–9.
National Bureau of Standards 1978	National Bureau of Standards. *Atomic Clocks of the National Bureau of Standards*, Gaithersburg.
Needham 1959	Joseph Needham with Wang Ling, *Science and Civilisation in China, 4, Physics and Physical Technology, Part II: Mechanical Engineering*, Cambridge.
Needham 1965	Joseph Needham & Wang Ling, *Science and Civilisation in China, 3: Mathematics and the Sciences of the Heavens and the Earth*, Cambridge.
Needham et al. 1986	Joseph Needham, Wang Ling, & Derek J. De Solla Price, *Heavenly Clockwork*, 2nd edn, Cambridge.
Needham, Gwei-Djen, Combridge, et al. 1986	Joseph Needham, Lu Gwei-Djen, John H. Combridge, & John S. Major, *The Hall of Heavenly Records: Korean Astronomical Instruments and Clocks 1380–1780*, Cambridge.
Nelthropp 1873	H. L. Nelthropp, *A Treatise on Watch-Work, Past and Present*, London.
Neret 1992	Gilles Néret, *Boucheron, le joaillier du temps*, Lausanne.
Nemrava 1975	Steve Z. Nemrava, *The Morbier 1680–1900 . . .*, Portland, Or.
Neuburger 1919	A. Neuburger, *Die Technik des Alterfums*, Leipzig.
Neugebauer 1947	O. Neugebauer, 'Studies in Ancient Astronomy VIII: The Water Clock in Babylonian Astronomy', *Isis*, 37, 37–43.
Neugebauer 1975	Otto Neugebauer, *A History of Ancient Mathematical Astronomy*, 3 vols., New York/Berlin.
Neugebauer & Parker 1960	Otto Neugebauer & Richard Parker, *Egyptian Astronomical Texts I*, Providence.
Neumann 1958	E. Neumann, 'Christopher Margraf and the Invention of the Rolling Ball Clock', *La Suisse horlogère*, International edition, 83ff.
Neumann 1967	Erwin Neumann, *Der königliche Uhrmacher Moritz Behaim und seine Tischuhr von 1559*, Luzern.
Nève 2016	Jacques Nève, *Les Pendules d'Hubert Sarton, 1748–1828, horloger, mécanicien, inventur*, St Ouen/Paris.
Newman 2010	Sue Newman, *The Christchurch Fusee Chain Gang. Who they were; what they did; how they lived*, Stroud.
Newman 2020	R. Newman 'Early American Watches', https://www.colonialwatches.com.
Nicholson 1803	William Nicholson (ed.), *Journal of Natural Philosophy*, (second series) 5, 158–9.
Niehüser 1999	Elke Niehüser, *French Bronze Clocks, 1700–1830, A Study of the Figural Images*, Atlgen, PA.
Niklès van Osselt 2013	Estelle Niklès Van Osselt, 'From Swiss Watches to Chinese Antiques: The Story of the Loup Family', *Arts of Asia*, 43, 76–84.
Nisbet 1901	John Nisbet, *Burma under British Rule–and Before*, Westminster.
Noble & Price 1968	Joseph V. Noble & Derek J. de Solla Price, 'The Water Clock in the Tower of the Winds', *American Journal of Archaeology*, 72, 345–55.
Nordon 1990	Marcel Nordon, 'Sur un objet en bronze découvert à Grand en Lorraine, en 1886', *Bulletin de l'Association nationale des collectionneurs et amateurs d'horlogerie ancienne*, 57, 27–42.
Nordon 1991	Marcel Nordon, *Histoire de l'hydraulique. L'eau conquise. Les origins et le monde antique*, Paris.
North 1976	J. D. North (trans. & English commentary), *Richard of Wallingford. An Edition of his Writings with Introductions*, 3 vols., Oxford.
North 1978	J. D. North, 'Nicolas Kratzer – The King's Astronomer', *Studia copernicana*, 16, 205–34.
North 1966	John D. North, 'Opus quorundam rotarum mirabilium', *Physis: Rivista Internazionale di Storia della Scienza*, 8, 337–372.
North 2005	John D. North, *God's Clockmaker: Richard of Wallingford and the Invention of Time*, London.
Nye 2014	James Nye, *A Long Time in Making: The History of Smiths*, Oxford.
Nye & Bird 2001	James Nye & Derek Bird, 'Electrical Horology Group: Tour on 8–9 June', *AH*, 24, 302.
Nye & Rooney 2007	James Nye & David Rooney, '"Such Great Inventors as the Late Mr Lund": An Introduction to the Standard Time Company, 1870-1970', *AH*, 30, 501–23.
Obrist 1997	Barbara Obrist, 'Wind Diagrams and Medival Cosmology', *Speculum*, 72, 33–84.

Obrist 2000	Barbara Obrist, 'The Astronomical Sundial in Saint Willibrord's Calendar and its Early Medieval Context', *Archives d'Histoire doctrinale et litteraire du Moyen Age*, 67, 71–118.
Ò Croinin 1983	Daibhi Ò Croinin, 'The Irish Provenance of Bede's Computus', *Perita*, 2, 229–47.
Oddi 1638	Mutio Oddi, *De gli horologi solari nelle superficie piane*, Venicse.
Oechslin 1996	Ludwig Oechslin, *Astronomische Uhren und Weltmodelle der Priestermechaniker im 18. Jahrhundert*, Neuchâtel.
Oestmann 2012	Gunther Oestmann, *Auf Dem Weg Zum 'Deutschen Chronometer'*, Bremen.
Oestmann 2013	Günther Oestmann, 'Gemma Frisius und die Verwendung von Uhren zur Bestimmung der geographischen Länge', in Meyer 2013, 137–45.
Oestmann 2018	Günther Oestmann, 'Early Watches – The Argument over Priority in Italy and Germany', *AH*, 39, 92–7.
Oestmann 2019	Günther Oestmann, 'Die "planetenuhr" und "Bergkristalluhr": zwei Hauptwerke Jost Bürgis im Kunsthistorischen Museum zu Wien, *Acta Astronomicae*, 66, 37–58.
Oestmann 2020	Günther Oestmann, *The Astronomical Clock of Strasbourg Cathedral*, Leiden.
Ogilvie 2019	Sheilagh Ogilvie, *The European Guilds, An Economic Analysis*, Princeton.
Ogle 2015	Vanessa Ogle, *The Global Transformation of Time*, Cambridge, MA.
Ōhashi 1993	Yukio Ōhashi, 'Development of Astronomical Observation in Vedic and Post-Vedic India,' *Indian Journal of History of Science*, 28(3), 185–251.
Ōhashi 1994	Yukio Ōhashi, 'Astronomical Instruments in Classical Siddhāntas,' *Indian Journal of History of Science*, 29(2), 155–314.
Ōhashi 1998	Yukio Ōhashi, 'The Cylindrical Sundial in India,' *Indian Journal of History of Science*, 33, S147–205.
Oleson 2008	J. P. Oleson (ed.), *Handbook of Engineering and Technology in the Classical World*, Oxford.
Olivier 1847	Théodore Olivier, *Applications de la géométrie descriptive*, Paris.
Olivier 2000	Jean-Marc Olivier, 'Trois cycles technologiques à Morez, Haut Jura', in Belot, Cotte, & Lamard 2000, 25–42.
Olivier 2002	Jean-Marc Olivier, *Une industrie à la campagne, le canton de Morez entre 1780 et 1914*, Salins-les-Bains.
Olivier 2004	Jean-Marc Olivier, *Des clous, des horloges et des lunettes. Les campagnards moréziens en industrie, 1750–1914*, Paris.
Ollion 1912	H. Ollion, *Lettres inédites de John Locke à ses amis Nicolas Thoynard, Philippe von Limborch et Edward Clarke . . .*, The Hague.
Olmi 1985	Giuseppe Olmi, 'Science-Honour-Metaphor: Italian Cabinets of the Sixteenth and Seventeenth Centuries', in Impey & MacGregor 1985, 5–16.
Olsen 2013	Sherri Olson, *Daily Life in a Medieval Monastery*, Oxford.
Olszawski 2012	Marek Titian Olszawski, 'Les cadrans solaires dans les mosaïques antiques', *Dossiers d'Archéologie*, 354, 6–17.
Omnès 1997	Catherine Omnès, *Ouvrières parisiennes*, Paris.
O-okayama Shoten	O-okayama Shoten, National Diet Library, Digital Library. https://dl.ndl.go.jp/, bibliography ID:000000739583, keyword: 延喜式.
Opizzo 1998	Yves Opizzo, *Les ombres des temps. Histoire et devenir du cadran solaire*, Vannes.
Ord-Hume 1982	Arthur W. J. G. Ord-Hume, *Joseph Haydn and the Mechanical Organ*, Cardiff.
Ord-Hume 1995	Arthur W. J. G. Ord-Hume, *The Musical Clock*, Ashbourne.
Oresme 1377	Nicolas Oresme, *Le livre du ciel et du monde*, A D Menut & A J Deverny (eds.), A. D. Menut (trans.), Madison.
Orton, Wood, & Lees 2007	Fred Orton & Ian Wood, with Clare A. Lees, *Fragments of History: Rethinking the Ruthwell and Bewcastle Monuments*, Manchester.
Osborne 1966	C. A. Osborne, 'The Birmingham Clock Dial Makers', *AH*, 5, 94–95.
OSC 2003	Office of Shanghai Chronicles, 'Complementary Collaboration', Shanghai Light Industry, Part I, Chapter 8, Section 4. 26 December 2003. Available at: http://www.shtong.gov.cn/dfz_web/DFZ/Info?idnode=68972&tableName=userobject1a&id=66700. 协作配套
Ottomeyer & Pröschel 1986	Hans Ottomeyer & Peter Pröschel, *Vergoldete Bronzen. Die Bronzearbeiten des Spätbarock un Klassizismus*, 2 vols., Munich.
Oughtred 1636, 1652	William Oughtred, *The Description and Use of the Double Horizontal Dyall*, London.
Ozanam 1685	Jacques Ozanam, *Méthode générale pour tracer des cadrans sur toutes sortes de plans*, Paris.
Ozanam 1694, 1741, 1774	Jacques Ozanam, *Recreations mathematiques et physiques . . . avec un traité nouveau des horloges elementaires*, Paris.
Ozanam 1697	Jacques Ozanam, *Cours de mathématiques*, Paris.
Pagani 2001	Catherine Pagani, *Eastern Magnificence and European Ingenuity*, Ann Arbor.
Pagliari 1989	P. N. Pagliari, 'Della Volpaia', in Pavan Massimiliano (ed.), *Dizionario biografico degli Italiani*, vol. 37, Rome, 789–802.
Panciroli 1612	Guido Panciroli, *Raccolta breue d'alcune cose piu segvnalate c'hebberogli antichita e d'alcune altre trouata da moderni*, Venice.
Pardies 1673	Ignace Gaston Pardies, *Description et explication de deux machines propres à faire les cadrans avec très grande facilité*, Paris.

Panicali 1988	Roberto Panicali, *Orologi e orologiai del Rinascimento Italiano – la Scuola Urbinate/Sixteenth Century Italian Chamber Clocks and the Urbino School*, Urbino.
Parker 2000	Jo Alyson Parker, 'The Clockmaker's Outcry: *Tristram Shandy* and the Complexification of Time', in Theodore E. D. Braun and John Aloysius McCarthy (eds.), *Disrupted Patterns: On Chaos and Order in the Enlightenment*, Amsterdam, 147–60.
Parker 2019	Jon Parker, 'Was There High-Quality, Wholesale Movement Manufacture in Seventeenth-Century London?', *AH*, 40, 469–86.
Parmenter 1625	John Parmenter, *Heliotropes or New Posies for Sundials Written in an Old Book . . . and Expounded . . .*, Percy Landon (ed.), 1904, London.
Parsons 2016	Geoff Parsons, 'The Development and Use of the Pilkington and Gibbs Heliochronometer and Sol Horometer', in Elisa Felicitas Arias, Ludwig Combrinck, Pavel Gabor, Catherine Hohenkerk, & P. Kenneth Seidelmann (eds.), *The Science of Time. Time in Astronomy & Society, Past, Present and Future*, 47–48.
Paspates 1893	A. G. Paspates, *The Great Palace of Constaninople*, London.
Pasquier 1900	Ern. Pasquier, 'De la décimalisation du temps et de la circonférence', *Annales de la Société Scientifique de Bruxelles*, 24, 59–104.
Pasquier 2005	Hélène Pasquier, 'Die technischen Fachkräfte in der Schweizer Uhrenindustrie bis zur Mitte der 1950er Jahre', *Technikgeschichte*, 4, 313–32.
Pasquier 2008a	Hélène Pasquier, *La 'Recherche et Développement' en horlogerie. Acteurs, stratégies et choix technologiques dans l'Arc jurassien suisse (1900–1970)*, Neuchâtel.
Pasquier 2008b	Hélène Pasquier, 'Uhren, Kompasse und elektrische Zähler. Longines, 1910–1925', in Rossfeld & Straumann 2008, 151–69.
Pasquier 2008c	Hélène Pasquier, 'Acteurs, stratégies et lieux de "Recherche et Développement" dans l'industrie horlogère suisse, 1900–1970', *Entreprises et histoire*, 52, 76–84.
Pasquier 2018	Hélène Pasquier, 'Swatch Group', in *Dictionnaire historique de la suisse* [online]. Available at: https://hls-dhs-dss.ch/fr/articles/041990/2018-01-23.
Pastre 1721	S. Pastre, *Description d'une grande & merveilleuse Horloge portative en pandule, le plus admirable & surprennante qui aïe jamais parū au monde . . .*, n.p. [?Amsterdam].
Patek 2010	Patek Philippe Museum, *The Mirror of Seduction. Prestigious Pairs of Chinese Watches*, Geneva.
Patrizzi 1980a	Osvaldo Patrizzi, 'The Watch Market in China: The First Period', *Arts of Asia*, 10, 65–75.
Patrizzi 1980b	Osvaldo Patrizzi, 'The Watch Market in China: The Second Period', *Arts of Asia*, 10, 100–11.
Patrizzi 1998	Osvaldo Patrizzi, *Dictionnaire des horlogers genevois*, Geneva.
Patrizzi & Sturm 1979	Oswaldo Patrizzi & Fabienne Sturm, *Montres de fantaisie*, Genève.
Pattenden 1979a	Philip Pattenden, *Sundials at an Oxford College*, Oxford.
Pattenden 1979b	Philip Pattenden, 'Sundials in Cetus Faventinus', *The Classical Quarterly*, 29, 203–12.
Pattenden 1983	Philip Pattenden, 'The Byzantine Early Warning System', *Byzantion*, 53, 258–99.
Paulson 1965	Ronald Paulson, *Hogarth's Graphic Works*, vol. 1, New Haven.
Pedersen & de Clercq 2010	Kurt Moller Pedersen and Peter de Clercq, *An Observer of Observatories*, Aarhus.
Pedretti 1957	Carlo Pedretti, *Studi Vinciani: documenti, analis e inediti leonardeschi*, Geneva.
Pelliot 1920	Paul Pelliot, 'Bulletin Critique: Alfred Chapuis, La Montre "Chinoise"', *T'oung Pao*, 20, 61–8.
Penney 2006-7	David Penney, 'Evidence from the transient: the Importance of Ephemera for a proper understanding of the Watch and Clock Trades: Part 1', *AH* xxix, 790–803; 'Part II', *AH* xxx, 45–65; 'Part III', *AH* xxx, 177–195.
Penney 2012	David Penney, 'Watches and Watchmaking: A Short Appraisal', *AH*, 33, 389–92.
Penther 1768	Johann Friedrich Penther, *Gnomonica Fundamentalis & Mechanica*, Augsburg.
Pérez Álvarez 2012	Victor Pérez Álvarez, 'Tiempo, Agua y Vida Artificial: clepsidras y Autómatas de Tradición Helenística en la Edad Media, in Isabel Del Val Valdivieso & Antonio Bonachio Hernando (eds.), *Agua y Sociedad en la Edad Media Hispana*, Grenada, 174–207.
Pérez Álvarez 2013	Victor Pérez Álvarez, 'Mechanical Clocks in the Medieval Castilian Royal Court', *AH*, 34, 489–502.
Pérez Álvarez 2015a	Victor Pérez Álvarez, 'The Role of the Mechanical Clock in Medieval Science', *Endeavour*, 39, 63–68.
Pérez Álvarez 2015b	Victor Pérez Álvarez, 'From Burgundy to Castile. Retracing and Reconstructing a Fifteenth-Century Golden Clock', *AH*, 36, 249–254.
Pérez Álvarez 2018	Victor Pérez Álvarez, *Técnica y fe: el reloj medieval de la catedral de Toledo*, Madrid.
Pérez Álvarez 2020	Victor Pérez Álvarez, 'Life & Works of Martin Altman, Engineer to the Hapsburgs', *Bulletin of the Scientific Instrument Society*, 147, 11–19.
Perregaux & Perrot 1916	Charles Perregaux & F.-Louis Perrot, *Les Jaquet-Droz et Leschot*, Neuchâtel.
Perrenoud 1993	Marc Perrenoud, 'Crises horlogères et interventions étatiques: le cas de la Banque cantonale neuchâteloise pendant l'entre-deux-guerres', in Cassis & Tanner 1993, 209–40.

Perret, Beyner, Debély, & Tissot 2000	Thomas Perret, André Beyner, Pierre-Alain Debély, & Laurent Tissot (eds.), *Microtechniques et mutations horlogères: clairvoyance et ténacité dans l'Arc jurassien: un siècle de recherche communautaire à Neuchâtel*, Hauterive.
Perrin 1902	Narcisse Perrin, *L'Horlogerie Savoisienne et l'Ecole Nationale d'Horlogerie de Cluses*, Thonon-les-Bains (facsimile reprint 2004).
Perrin 1991	Noel Perrin, *Giving Up the Gun* (鉄砲を捨てた日本人), Heita Kawakatsu (trans.), Tokyo.
Perron 1832	L. Perron, *Essai sur l'histoire abrégée de l'horlogerie . . .*, Paris/Besançon.
Perrot 1978	Michelle Perrot, 'De la nourrice à l'employée, travaux de femmes dans la France du XIXe siècle', *Le mouvement social*, 105, 3–10.
Perry 2016	Seamus Perry, 'An Introduction to "Stop all the Clocks"', British Library [online] Discovering Literature: 20th Century (25 May 2016). Available at: https://www.bl.uk/20th-century-literature/articles/an-introduction-to-stop-all-the-clocks.
Persegol 1895	J.-E. Persegol, *Nouveau Manuel complet de l'horloger rhabilleur*, Paris.
Petroski 1996	Henry Petroski, *Invention by Design: How Engineers Get from Thought to Thing*, Cambridge, MA.
Pettifer 2020	Martyn Petifer, 'Another Diamond', *AH*, 41, 209–13.
Pezenas 1767	Esprit Pezenas (trans.), *Principes de la montre de Mr. Harrison, avec les planches relative à la meme montre, . . .*, Avignon.
Pfleghart 1908	Adolf Pfleghart, *Die schweizerische Uhrenindustrie: ihre geschichtliche Entwicklung und Organisation*, Leipzig.
Pfister et al. 1996	Ulrich Pfister, Brigitte Studer, & Jakob Tanner (eds.), *Arbeit im Wandel: Deutung, Organisation und Herrschaft vom Mittelalter bis zur Gegenwart. Le travail en mutation: interprétation, organisation et pouvoir, du Moyen Age à nos jours*, Zurich.
Philippe 1863	Adrien Philippe, *Les montres sans clef, ou se montant et se mettant à l'heure sans clef*, Paris.
Phillips 1861	Edouard Phillips, 'Mémoire sur le spiral réglant des chronométres et des montres', *Annales des Mines* Series 5, vol. 20, 1–107.
Phillips 1864	Edouard Phillips, *Memoire sur le Spiral Réglant*, Paris.
Picard 1892	A. Picard (ed.), *Exposition universelle de 1889. Rapports du Jury international*, Paris.
Picard 1921	Etienne Picard, 'Le Jaquemart de l'église Notre-Dame de Dijon', *La revue de Bourgogne*, 15 March, 77–82.
Picard 1693	Jean Picard, 'De la pratique des grands cadrans par le calcul', in *Mémoires de l'Académie Royale des Sciences*, 6, Paris (1730), 481–531.
Picard 1993	Jacques Picard, 'Swiss Made oder Jüdische Uhrenfabrikanten im Räderwerk von Politik und technischem Fortschritt. Einige Notizen Über einen zeit- und grenzgeschichtlichen Forschungsgegenstand', in *Allmende. Alemannisches Judentum. Versuche einer Wiederannäherung*, vol. 13, 85–105.
Pickford 1991	Chris Pickford, *Bedfordshire Clock and Watchmakers 1352–1880*, Bedford.
Picolet 1987	Guy Picolet (ed.), *Jean Picard et les débuts de l'astronomie de precision en France au XVIIe siècle*, Paris.
Pieper 1992	Wolfgang Pieper, *Geschichte der Pforzheimer Uhrenindustrie 1767–1992*, Pforzheim.
Pillet 1921	Jules Pillet, *Traité de géométrie descriptive*, Paris.
Pinches, Sachs, & Strassmaier 1955	T. Pinches, A. Sachs, & J. Strassmaier, *Late Babylonian Astronomical and Related Texts*, Providence.
Pini 1598	Valentino Pini, *Fabrica de gli horologi solari*, Venice.
Pipping 2000	Gunnar Pipping, 'A Review of Clockmaking in Sweden before 1900', in *AH: the Proceedings of the Oxford 2000 Convention. . . A Supplement to Antiquarian Horology*, 25, 20–7.
Pipping et al. 1995	Gunnar Pipping, Elis Sidenbladh, & Eik Elfström, *Urmakere och Klockor i Sverige och Finland*, Upsala.
Pingré 1764	Alexandre-Guy Pingré, *Mémoire sur la colonne de la Halle aux Blés*, Paris.
Pingree 2009	David Pingree, *Eastern Astrolabes*, Historic Scientific Instruments of the Adler Planetarium and Astronomy Museum Series, vol. 2, Chicago.
Pitiscus 1600	Bartholomaeus Pitiscus, *Trigonometriae, sive de Dimensione triangulorum libri quinque . . .*, Augsbourg.
Piton 1973	Monique Piton, *C'est possible. Le récit de ce que j'ai éprouvé durant cette lutte de Lip*, Besançon.
Planchon 1900	Mathieu Planchon, *Musée Rétrospectif de la Classe 96: horlogerie à l'Exposition universelle internationale de 1900 à Paris*, Paris.
Plaßmeyer & Gluch 2015	Peter Plaßmeyer & Sibylle Gluch (eds.), *Einfach – vollkommen. Sachsens Weg in die internationale Uhrenwelt. Ferdinand Adolph Lange zum 200. Geburtstag*, Berlin/Munich.
Plommer 1973	Hugh Plommer, *Vitruvius and Later Building Manuals*, Cambridge.
Plomp 1979	Reinier Plomp, *Spring-Driven Dutch Pendulum Clocks*, Schiedam.
Plomp 1999	Reinier Plomp, 'A Longitude Timekeeper by Isaac Thuret with the Balance Spring Invented by Christiaan Huygens', *Annals of Science*, 56, 379–94.
Plomp 2009	Reinier Plomp, *Early French Pendulum Clocks, 1658–1700, Known as 'Pendules Religieuses'*, Schiedam.
Pogo 1935	Alexander Pogo, 'Gemma Frisius, His Method of Determining Differences of Longitude by Transporting Timepieces (1530) and his Treatise on Triangulation (1533)', *Isis*, 22, 469–85.

Pointon 1999	Marcia Pointon, 'Dealer in Magic: James Cox's Jewelry Museum and the Economics of Luxurious Spectacle in Late-Eighteenth-Century London', *History of Political Economy*, 31, 423–51.
Polak 1986	B. Polak, *Staropprasske slunecni hodiny*, Prague.
Pollet 2001	Christophe Pollet, *Les gravures d'Etienne Delaune (1518–1583)*, 2 vols., Villeneuve d'Ascq.
Pomel 2012	Fabienne Pomel, 'Pour une approche littéraire des cloches et horloges médiévales: réflexions méthodologiques et essai de synthèse', in Pomel 2012, 9–36.
Pomel 2012	Fabienne Pomel, *Cloches et horloges dans les textes médiévaux: mesurer et maîtriser le temps*, Rennes.
Pomian 1987	Krzyzstof Pomian, *Collectionneurs, amateurs et curieux, Paris, Venise: XVIe–XVIIIe siècles*, Paris.
Pommier 1978	Charles Pommier, 'Cadran solaire à équation', *L'Astronomie*, 92, 283–6.
Poniz 2021	Philip Poniz, 'Adolphe Nicole and the Chronograph Watch', *NAWCC Watch & Clock Bulletin*, 63, n.p.
Ponsford 1985	Clive N. Ponsford, *Devon Clocks and Clockmakers*, London.
Ponsford 2008	Clive N. Ponsford, 'Wills of Clock and Watchmakers On-Line from the National Archives: Part Iv, Case Makers and Other Specialist Tradesmen (London area)', *AH*, 31, 31–42.
Poole 1915	Rachel Poole, 'A Monastic Timetable of the Eleventh Century', *Journal of Theological Studies*, 16, 98–104.
Popplow 2007	Marcus Popplow, 'Setting the World Machine in Motion: The Meaning of *machina mundi* in the Middle Ages and Early Modern Period', in Bucciantini, Camerota, & Roux 2007, 45–70.
Portes 2011	Laurent Portes, 'Claude Sallier (1685–1761), dans la République des Lettres', *Revue de la bibliothèque national de France*, 38, 57–63.
Portuondo 2009	M. Portuondo, 'Lunar Eclipses, Longitude and the New World', *Journal for the History of Astronomy*, 40, 249–76.
Post & Turner 1973	J. B. Post & A. J. Turner, 'An Account for Repairs to the Westminster Palace Clock', *The Archaeological Journal*, 130, 217–20.
Poulle 1963	Emmanuel Poulle, *Un constructeur d'instruments astronomiques au XVe siècle. Jean Fusoris*, Paris.
Poulle 1964	Emmanuel Poulle, 'Le Quadrant nouveau médiéval, i', *Journal des savants*, 2, 147–67.
Poulle 1972	Emmanuel Poulle, 'Les instruments astronomiques de l'Occident latin aux XI^è et XII^è siècles', *Cahiers de civilisation médiévale*, 57, 27–40.
Poulle 1974	Emmanuel Poulle, 'Les mécanisations de l'astronomie des épicycles: l'horloge d'Oronce Finé', *Comptes rendus des séances de l'Académie des Inscriptions et Belles-Lettres*, 118, 59–79.
Poulle 1980a	Emmanuel Poulle, *Les instruments de la théorie des planètes selon Ptolémée: Équatoires et horlogerie planétaire du XIIIe au XIVe siècle*, 2 vols., Geneva.
Poulle 1980b	Emmanuel Poulle, 'L' horloge planétaire de Regiomontanus', in Hamann 1980, 335–41.
Poulle 1983	Emmanuel Poulle, *Les instruments astronomiques du Moyen Age*, Astrolabica 3, Paris.
Poulle 1985	Emmanuel Poulle, 'L'astronomie de Gerbert', in Giacomo Barabino (ed.), *Gerberto: scienza, storia a mito: atti del Gerberti symposium, Bobbio, Italy, 25–17 July 1983*, 397–617.
Poulle 1998	Emmanuel Poulle, *Horologicum Amicorum: L'Astrarium de Giovanni Dondi*, Paris
Poulle 1999	Emmanuel Poulle, 'L'horloge a-t-elle tué les heures inégales?', *Bibliothèque de l'Ecole de Chartes*, 157, 137–56.
Poulle 2003	Emmanuel Poulle (ed.), *Giovanni Dondi dall' Orologio: Tractatus astrarii. Edition critique et traduction de la version A*, Geneva.
Poulle, Sändig, Schardin, & Hasselmeyer 2008	Emmanuel Poulle, Helmut Sändig, Joachim Schardin, & Lothar Hasselmayer, *Jahresschriften der Deutschen Gesellschaft für Chronometrie: Die Planetenlaufuhr ein Meisterwerk der Astronomie und Technik der Renaissance, geschaffen von Eberhard Baldewein, 1563–1568*, 47, Stuttgart.
Powell 1661	Thomas Powell, *Humane Industry, or, a History of most Manual Arts Deducing the Original Progress, and Improvement of Them, ...*, London.
Pradère 2003	Alexandre Pradère, *Charles Cressent sculpteur, ébéniste du Régent*, Dijon.
Prager 1968	Frank D. Prager, 'Brunelleschi's Clock?', *Physis*, 10, 203–16.
Prasad 2015	R. Prasad, *Tracks of Change: Railways and Everyday Life in Colonial India*, Cambridge.
Prenot-Guinard 2019	Martine Prenot-Guinard (ed.), *De la mine à l'horloge: Des horlogers dans le triangle d'or du Haut-Doubs forestier*, Pontarlier.
Price & Noble 1968	Derek J. Price & Joseph V. Noble, 'The Water-Clock in the Tower of the Winds', *American Journal of Archaeology*, 72, 345–55.
Priestley 2000	Philip T. Priestley, *Early Watch Case Makers of England*, Columbia, PA.
Priestley 2009	Philip T. Priestley, *Aaron Lufkin Dennison, An Industrial Pioneer and His Legacy*, Columbia, PA.
Priestley 2018	Philip T. Priestley, *British Watchcase Gold & Silver Marks, 1670–1970. A History of Watchcase Makers and Registers of Their Marks, ...*, Columbia, PA.
Principe 2014	Lawrence M. Principe, *Alchemy and Chemistry: Breaking Up and Making Up (Again and Again)*, Washington DC.

Pritchard 1936	Earl H. Pritchard, *The Crucial Years of Early Anglo-Chinese Relations 1750–1800*, Pullman. Reprinted 1970, New York.
Pritchard 1997	Kathleen H. Pritchard, *Swiss Timepiece Makers, 1775–1975*, 2 vols., West Kennebunk.
Pritchett 1963	W. Kendrick Pritchett, *Ancient Athenian Calendars on Stone*, University of California Publications in Classical Archaeology IV, 4, Berkeley, 267–402.
Prosper 1727	Le Comte Prosper, 'Horloge à sable', in Gallon, vol. 5, Paris, 23–9.
Pryce & Davies 1985	W. T. R. Pryce & T. Alun Davies, *Samuel Roberts Clock Maker: An Eighteenth-Century Craftsman in a Welsh Rural Community*, St Fagan's.
Przypkowski 1967	Tadeusz Przypkowski, 'The Art of Sundials in Poland from the Thirteenth to the Nineteenth Century', *Vistas in Astronomy*, 9, 13–23.
Puraye 1982	Jean Puraye, *Les porte-montre des XVIII et XIXe siècles*, La Chaux-de-Fonds.
Pyenson 1998	Lewis Pyenson, 'Assimilation and Innovation in Indonesian Science', *Osiris*, 13, 34–47.
Qaisar 1982	Ahsan Jan Qaisar, *The Indian Response to European Technology and Culture (AD 1498–1707)*, Delhi.
Qian 1997	Qian Yong, *Notes of Lu Yuan*, Zhonghua.
Qing 2016	Qing Long, *Time and Technology ii. Collection and Appreciation of Antique Pocket Watches*, Guangming Daily Press.
Quan 2013	Quan Hejun, 'Gnomon and Sundial', in Wu & Quan 2013, 364–87.
Quill 1960	H. Quill, 'Adjusting a Raingo Orrery-Clock', *HJ*, 102, 774–780.
Quincy 1847	Josiah Quincy (ed.), *Journals of Major Samuel Shaw . . .*, Boston.
Quinlan-McGrath 1995	Mary Quinlan-McGrath, 'The Villa Farnesina, Time-Telling Conventions and Renaissance Astrological Practice', *Journal of the Warburg & Courtauld Institutes*, 58, 52–71.
Rabb 1962	T. K. Rabb, 'The Effects of the Thirty Years' War on the German Economy', *The Journal of Modern History*, 34, 40–51.
Radi 1665	Archangelo Maria Radi, *Nuovo scienza di horologi a poluere: che mostrano e' suonano distintamente tutti l'hore*, Rome.
Raffaelli 2019	Ryan Raffaelli, 'Technology Reemergence: Creating New Value for Old Technologies in Swiss Mechanical Watchmaking, 1970–2008', *Administrative Science Quarterly*, 64, 1–43.
Ragep & Kennedy 1981	Jamil Ragep & E. S. Kennedy, 'A Description of Zâhariyya (Damascus) MS 4871: A Philosophical and Scientific Collection', *Journal for the History of Arabic Science*, 5(1–2), 85–108.
Raillard, n.d. [1720s]	Claude IV Raillard, *Traité historique et chronologique des differentes divisions et distributions du jour . . . de toutes les differentes sortes d'horloges . . . .* [Paris], MS.
Raillard 1752	Claude IV Raillard, *Extraits des principaux articles des statuts des Mes. Horlogers de Paris, des années 1544, 1583, 1646, 1707 et 1719 . . .*, Paris.
Rājaśekhara 1934	*Kāvyamīmāṃsā of Rājaśekhara*, C. D. Dalal & R. A. Sastry (eds.), revised and enlarged by K. S. Ramaswami Sastri, Baroda.
Rāmacandra 1886–92	Rāmacandra Vājapeyin, *Yantraprakāśa.*, together with an auto-commentary, MS 975 of the Bhandarkar Oriental Research Institute, Pune.
Ramasubramanian et al. 2019	K. Ramasubramanian et al. (eds.), *Bhāskara-prabhā*, New Delhi.
Rambaux 1984	C. Rambaux, 'Conquête et fuite de temps dans l'enseignement spirituel de Tertullien' in Leroux 1984, 168–74.
Ramus 1567	Petrus Ramus, *Prooemium Mathematicum ad Catharinam Mediceam, Reginam, Matrem Regis*, Paris.
Rancière 1975	Jacques Rancière, 'En allant à l'Expo: l'ouvrier, sa femme et les machines', *Les Revoltes logiques*, 1, 5–22.
Randall 1988	A. G. Randall, 'Thomas Earnshaw's Numbering Sequence', *AH*, 17, 367–71.
Randall & Good 1990	A. G. Randall & Richard Good, *Catalogue of Watches in the British Museum, Vol. VI: Pocket Chronometers, Marine Chronometers and Other Portable Precision Timekeepers*, London.
Rappaport 1989	Steve Rappaport, *Worlds Within Worlds: Structures of Life in Sixteenth-Century London*, Cambridge.
Rapport 1844	*Rapport du Jury central de l'Expositinno des produits de l'Industrie française en 1844*, Paris.
Rapport 1850	*Rapport du jury central sur les produits de l'agriculture et de l'industrie exposés en 1849, Vol. 2: instruments de précision, section horlogerie*, Paris, 478–512.
Rapport 1856	*Exposition universelle de 1855. Rapports du jury mixte international publiés sous la dir. de S.A.I. le Prince Napoléon, président de la commission impériale*, Paris, 406–24.
Rapport 1876	'Délégation ouvrière française à l'Exposition Universelle de Vienne, 1873', in A. Morel (ed.), *Rapport d'ensemble*, Madison, WI.
Rashed 1996–2008	Roshdi Rashed, *Les mathématiques infinitésimales du IXe au XIe siècle*, 6 vols., London.
Rashed & Morelon 1997	Roshdi Rashed & Régis Morelon (eds.), *Histoire des sciences arabes*, 3 vols., Paris.

Rau 1962	Reinhard Rau, 'Die Kunstuhr des Philipp Imser', *Tübinger Blätter*, 49, 25–33.
Rau 1999a	Herbert Rau, 'Bedeutung der sektorartigen Kratzungen bei mittelalterlichen Sonnenuhren in Europa? Mittelalterliche Vormittags Sonnenuhren?', *DGC*, 80, 44–45.
Rau 1999b	Herbert Rau, 'Sonderformen mittelterlicher sonnenuhren', *DGC*, 81, 38–40.
Rau 2000	Herbert Rau, *über Mittelalteriche vertikale Sonnenuhren in Mecklenurg-vopommern in 1899-1995. Und die Entdeckung der mittelalterlichen Sonnenuhr an der St Georgeskapelle in Neubrandenburg*, Neubbrandenburg, 33–45.
Read 2005	David Read, 'Pneumatic Clocks', *AH*, 28, 754–64.
Read 2015	David Read, 'The Marine Chronometer in the Age of Electricity', *AH*, 36, 343–60.
Redslob 1933	Robert Redslob: *L'horloge astronomique de la cathédrale de Messine, œuvre d'un maître*, Strasbourg.
Rees 1819–20	Abraham Rees, *The Cyclopaedia; or Universal Dictionary of Arts, Sciences and Literature*, London.
Reglements 1763	*Reglements des maîtres horlogers de la ville de Lyon . . .*, Delaroche, Lyon.
Rehm & Weiss 1903	Albert Rehm & Edmund Weiss, 'Zur Salzburger Bronzescheibe mit Sternbildern', *Jahreshefte des Österreichischen Archäologischen Institut*, 6, 35–49.
Reich & Wiet 1939–40	S. Reich & G. Wiet, 'Un astrolabe syrien du XIVe siècle, *Bulletin de l'Institut français d'archéologie orientale*, 39, 195–202. Reprinted in Kennedy & Ghanem 1976, 36–44.
Reid 1826	Thomas Reid, *A Treatise on Clock and Watchmaking, Theoretical and Practical*, Edinburgh.
Remelec 1900	Remelec ' Leaflet about the Remelec Electric Clock', IET Archives: UK0108 SPT/P/I/065/58.
Reti 1974	Ladislao Reti (ed.), *The Unknown Leonardo*, London.
Reverchon 1935	Leopold Reverchon, *Petite histoire de l'horlogerie*, Besançon.
Reymondin 1998	Charles-André Reymondin, *Théorie d'horlogerie*, Neuchâtel.
Reynolds 2005	Colin F. Reynolds, *The Gent Who Really Started Something*, privately printed.
Rheticus 1540/1943	Georg Joachim Rheticus, *Erster Bericht Über die 6 Bücher des Kopernikus von den Kreisbewegungen der Himmelsbahnen*, Karl Zeller (trans.), München.
Rheticus 1551	Georg Joachim Rheticus, *Canon doctrinae triangulorum*, Leipzig.
Ribémont 1992	Bernard Ribémont (ed.), *Le temps et sa mesure et sa perception au Moyen Âge. Actes du Colloque d'Orléans 12–13 avril 1991*, Caen.
Rice 1914	William Gorham Rice, *Carillon Music and Singing Towers of the Old World and the New*, New York.
Richard 1856	Louis Richard, *Rapport présenté au Comité du canton de Neuchâtel pour l'Exposition universelle de 1855 à Paris*, Neuchâtel.
Richon 2007	Marco Richon, *Omega: A Journey Through Time*, Bienne.
Richon 2014	M. Richon, 'En voiture, Simon!', *Chronometrophilia*, 75, 11–31.
Rico y Sinobas 1863–6	M. Rico y Sinobas (ed.), *Libros del Saber de Astronomia*, 5 vols., Madrid.
Riefler 1894	Sigmund Riefler, *Die Präcisions-Uhren mit vollkommen freien Echappement und neuem Quecksilber-Compensationspendel sowie die Regulirung und Behandlung derselben*, Munich.
Riefler 1907	Sigmund Riefler, *Präzisions-Penduluhren und Nickelstahl-Kompensationspendel*, Munich.
Riefler 1911	Sigmund Riefler, *Einfluß der Pendellänge, der Gravitation und der Luftdichte auf das schwingende Pendel*. [Munich] MS.
Riegel 2006	Pete Riegel, 'Roasting Jack Operation Revealed!', *NAWCC Watch & Clock Bulletin*, 48, 28.
Rienstra 1986	M. Howard Rienstra (ed. & trans.), *Jesuit Letters from China 1583–84*, Minneapolis.
Rieu 2014	Jean Rieu, *Les cadrans solaires de l'abbée Guyoux*, n.p.
Rifkin 1973	Benjamin A. Rifkin, *The Book of Trades [Ständebuch] Jost Amman & Hans Sachs*, New York.
Rigot 2017	Georges Rigot, *Montres de souscription et à tact de Breguet*, Lyon.
Rihaoui 1961–2	Abdul Kader Rihaoui, 'Découverte de deux inscriptions arabes', *Les annales archéologiques de Syrie*, 11–12. Reprinted in Kennedy & Ghanem 1976, 69–71.
Riley 2019	W. J. Riley, 'A History of the Rubidium Frequency Standard', IEEE UFFC-S History *2019*, 1–35. Available at: http://ieee-uffc.org/aboutus/history/a-history-of-the-rubidium-frequency-standard.pdf.
Riolini-Unger 1993	Adelheld Riolini-Unger, *Friedberger uhren*, Friedberg.
Rivard 1742	Dominique François Rivard, *La Gnomonique, ou l'art de faire de cadrans*, Paris.
Robert 1852	Henri Robert, *Etudes sur Diverses Questions d'horlogerie*, Paris.
Robert 1878	Henri Robert fils, *Exposition universelle de 1878: les recompenses de la Classe 26 (Horlogerie)*, Paris.
Roberts 1989	Derek Roberts, *Continental and American Skeleton Clocks*, West Chester.
Roberts 1993	Derek Roberts, *Carriage and Other Travelling Clocks*, Atglen.
Roberts 2003a	Derek Roberts, *Precision Pendulum Clocks: The Quest for Accurate Timekeeping*, Atglen
Roberts 2003b	Derek Roberts, *English Precision Pendulum Clocks*, Atglen.
Roberts 1970	Ken Roberts, 'Some Observations Concerning Connecticut Clockmaking', *NAWCC Watch & Clock Bulletin*, Suppl 6, 1–44.

Roberts 1976	Kenneth D. Roberts, *Tools for the Trades and Crafts: An Eighteenth-Century Pattern Book, R. Timmins & Sons, Birmingham*, Fitzwilliam.
Roberts & Grigson 1938	Denys Kilham Roberts & Geoffrey Grigson (compilers), *The Year's Poetry*, London.
Robertson 1931	J. Drummond Robertson, *The Evolution of Clockwork with a Special Section on the Clocks of Japan . . .*, London.
Robey 2001	John A. Robey, *The Longcase Clock*, Ashbourne.
Robey 2005	John A. Robey, 'Who Invented Rack-and-Snail Striking? The Early Development of Repeating and Rack Striking', *AH*, 28, 584–601.
Robey 2011	John A. Robey, 'Nag's Head Striking', *HJ*, November, 494–7.
Robey 2012a	John A. Robey, 'Moorfields and Clock-Brass Founders, Part 1: The London Horological Trades in Moorfields', *AH*, 33, 479–86.
Robey 2012b	John A. Robey, 'Moorfields and Clock-Brass Founders, Part 2: The Mayor Family and Other Founders', *AH*, 33, 609–23.
Robey 2015	John A. Robey, 'The English Use of Foliot and Balance', *AH*, 36, 239–43.
Robey 2016	John A. Robey, 'The Origin of the English Lantern Clock Part 1: Comparison with European Gothic Clocks', *AH*, 37, 511–21.
Robey 2017	John A. Robey, 'The Origin of the English Lantern Clock Part 2: The Earliest Lantern Clocks', *AH*, 38, 35–50.
Robey 2019	John A. Robey, 'William Whitaker and William Shreeve, Dialmakers of Halifax, West Yorkshire', *AH*, 40, 360–76.
Robey 2020	John A. Robey, 'Two More Diamond Dials', *AH*, 41, 400–1.
Robey & Gillibrand 2013	John Robey & Leighton Gillibrand, 'The Porrvis Clock of 1567 – The Earliest Surviving Domestic Clock Made in England', *AH*, 34, 503–18.
Robey & Linnard 2017	John A. Robey & William Linnard, 'Early English Horological Terms', *AH*, 38, 191–201.
Robinson 1957	F. N. Robinson (ed.), *The Works of Geoffrey Chaucer*, 2nd edn., London.
Robinson 1981	Tom Robinson, *The Longcase Clock*, Woodbridge.
Robinson & Adams 1935	Henry W. Robinson & Walter Adams (eds.), *The Diary of Robert Hooke . . . 1672–1680*, London.
Roegel 2010a	Denis Roegel, 'A reconstruction of the tables of Rheticus' Canon doctrinae triangulorum (1551)', [Research Report]. Available at: https://hal.inria.fr/inria-00543931/PDF/rheticus1551doc.pdf.
Roegel 2010b	Denis Roegel, 'A reconstruction of the tables of Briggs' Arithmetica logarithmica (1624)', [Research Report]. Available at: https://hal.archives-ouvertes.fr/inria-00543939.
Roegel 2011	Denis Roegel, 'A Reconstruction of the Tables of Pitiscus' Thesaurus mathematicus (1613)', [Research Report]. Available at: https://hal.inria.fr/inria-00543933/en.
Roguet 1912	Daniel Roguet, 'Le cadran solaire de l'observatoire Flammarion de Juvisy', *L'Astronomie*, 26, 445–6.
Rohou 2019	Julie Rohou (ed.), *Graver la Renaissance. Etienne Delaune et les arts décoratifs*, Paris.
Rohr 1986	René R. J. Rohr, *Les cadrans solaires, histoire, théorie, pratique . . .*, revised & enlarged edition, Strasburg.
Rohr 1988	René R. J. Rohr, 'Le Da'ire-yi mu'addil', *Horlogerie ancienne*, 23, 67–74.
Rolex 1946	Rolex Jubilee, *Vade Mecum*, 4 vols., Geneva.
Romer 2010	Daniel Romer, 'Tragbare Nachtwachter-Kontrolluhren und ihr Hersteller', *DGC*, 49, 31–50.
Romer 2011	Daniel Romer, 'Burks "Original"', *DGC*, l, 179–98.
Ronfort 2009	Jean Nérée Ronfort, *André-Charles Boulle (1642–1732). Un nouveau style pour l'Europe*, Paris.
Rong & Li 2004	Rong Xingjiang & Li Xiaocong (eds.), *History of Sino-Foreign Relations: New Historical Materials and New Issues*, Beijing.
Rooney 2006	David Rooney, 'Maria and Ruth Belville: Competition for Greenwich Time Supply', *AH*, 29, 614–28.
Rooney 2008	David Rooney, *Ruth Belville: The Greenwich Time Lady*, London.
Rooney & Nye 2009	David Rooney & James Nye, '"Greenwich Observatory Time for the Public Benefit": Standard Time and Victorian Networks of Regulation', *British Journal for the History of Science*, 42, 5–30.
Rose 1994	Ronald E. Rose, *English Dial Clocks*, Woodbridge.
Rosen 1959	Edward Rosen (ed.), *Three Copernican Treatises*, New York.
Rosenberg 1963	C. Rosenberg, 'Pitkin Fact and Fiction', *NAWCC Watch & Clock Bulletin*, 10, 582–90.
Roslund 2005	Curt Roslund, 'The Intriguing Case of the Braunschweig Sundial', *BBSS*, 17, 116–19.
Rossfeld & Straumann 2008	Roman Rossfeld & Tobias Straumann (eds.), *Der vergessene Wirtschaftskrieg. Schweizer Unternehmen im Ersten Weltkrieg*, Zürich.
Rossi 1593	Theodosius Rubeus Privernas, *Tabulae XII*, Rome.
Rothmann 1589	Miguel A. Granada, Jürgen Hamel, & Ludolf von Mackensen (eds.), 'Christoph Rothmanns Handbuch der Astronomie von 1589', in *Acta Historica Astronomiae*, vol. 19, Frankfurt am Main.
Rozier 1793	Francois Rozier, *Observations sur la Physique, sur l'histoire naturelle et sur les arts*, Paris.
Ruan 1935	Ruan Yuan, *Biographies of Mathematicians and Scientists*, vol. 1, Shanghai.

Ruellet 2016	Aurélian Ruellet, *La Maison de Salomon: histoire du patronage scientifique et technique en France et Angleterre au XVIIe siècle*, Rennes.
Runciman 1951–54	Stephen Runciman, *History of the Crusades*, 3 vols., Cambridge.
Ryckmans & Moreau 1926	G. Ryckmans & F. Moreau, 'Un gnomon arabe du XIIe siècle', *Le Museon Revue d'études orientales*, 39, 33–40.
Ryder 1939	William Matthews (ed.), *The Diary of Dudley Ryder, 1715–1716*, London.
Sabel & Zeitlin 1997	Charles F. Sabel & Jonathan Zeitlin (eds.), *World of Possibilities: Flexibility and Mass Production in Western Industrialization*, New York.
Sabra 1996	A. I. Sabra, 'Situating Arabic Science: Locality Versus Essence', *Isis*, 87, 654–70.
Sabrier 1988	Jean-Claude Sabrier, *Les horloges lanternes francaises: les trois dernières parties de l'ouvrage du Père Beuriot Horlogéographie pratique, réeditée avec une introducion*, London/Le Mesnil-le-Roi.
Sabrier 1993	Jean-Claude Sabrier, *La Longitude en mer à l'heure de Louis Berthoud et Henri Motel*, Geneva.
Sabrier Rigot, 2005	Jean-Claude Sabrier, Georges Rigot, *Steel Time*, Ed. F.-P. Journe, Geneva.
Sachs 1955	Abraham Sachs, in Pinches, Sachs, & Strassmaier 1955.
Sachse 1895	Julius F. Sachse, 'Horologium Achaz (Christophorus Schissler, Artifex)', *Proceedings of the American Philosophical Society*, 34, 21–30, n. 147.
Sadler 1995	Philip M. Sadler, 'An Ancient Time Machine: The Dial of Ahaz', *American Journal of Physics*, 63, 211–16.
Safadi 1978	Y. H. Safadi, *Islamic Calligraphy*, London.
Saff 2019	Donald Saff, *From Celestial to Terrestrial Timekeeping. Clockmaking in the Bond Family*, London.
Sagot 1988	Robert Sagot, 'Le test des tangentes', *Observations & Travaux*, 13, 20–26.
Sāʿīd 1991	Sāʿīd Al-andalusī, *Science in the Medieval World: 'Book of the Categories of Nations'*, Semʿan I. Salem & Ajok Kumar (ed. & trans.), Austin.
Said 1979	Hakim Mohammad Said (ed.), *History and Philosophy of Science: Proceedings of the International Congress of the History and Philosophy of Science, Islamabad, 8–13 December 1979*, Karachi.
Saint-Léon 1922	Etienne Martin Saint-Léon, *Histoire des corporations de metiers depuis leur origins jusqu'à leur suppression en 1791 …*, 3rd revised & enlarged edn., Paris.
Saito 1995	Kuniji Saito, *Ancient Time System of Japan, China, and Korea—Verification by Ancient Astronomy* (古代日本、中国、朝鮮の時刻制度), Yuzankaku.
Salamé 2016	R. Salamé, *Clocks and Empire: An Indian Case Study*. Brown University 2014 Undergraduate Research Prize for Excellence in Library Research.
Sallier 1716	Claude Sallier, 'Recherches sur les horloges des anciens', *Mémoires de Littérature tirez des régistres de l'Académie des Inscriptions et Belles Lettres*, 5, 142–60.
Salodius 1617, 1626	Hyppolyte Salodius, *Tabulae gnomonicae*, Brescia.
Salomons 1921	David Lionel Salomons, *Breguet (1747–1823)*, London.
Saluz 1996	Eduard Saluz, *Klangkunst, 200 Jahre Musikdosen*, Zürich.
Saluz 2015	Eduard Saluz, 'Nicht nur in Sachsen–zur Fabrikation von Taschenuhren in Deutschland in der zweiten Hälfte des 19. Jahrhunderts', in Plaßmeyer & Gluch 2015, 117–36.
Samsó 1994a	Julio Samsó, 'Andalusian Astronomy: Its Main Characteristics and Influence in the Latin West', in Julio Samsó, *Islamic Astronomy and Medieval Spain*, Aldershot.
Samsó 1994b	Julio Samsó, *Islamic Astronomy and Medieval Spain*, Aldershot.
Sandoz 1905	Ch. Sandoz, *Les horloges et les maîtres horlogers à Besançon, du XVe siècle à la Révolution Française*, Besançon.
Sarazin 1630	Jean Sarazin, *Horographum Catholicum sev vniversale …*, Paris.
Sarma 1948	M. Somasekhara Sarma, *History of the Reddi Kingdoms (circa 1325 A.D. to circa 1448 A.D.)*, Andhra University Series No. 38, Waltair.
Sarma 1986–87a	Sreeramula Rajeswara Sarma, 'Astronomical Instruments in Brahmagupta's Brāhma-sphutasiddhānta,' *Indian Historical Review*, 13, 63–74. Reprinted in Sarma 2008, 47–63.
Sarma 1986–87b	Jagannātha Samrāṭ, *Yantraprakāra of Sawai Jai Singh*, Sreeramula Rajeswara Sarma (ed. & trans.), *Supplement to Studies in History of Medicine and Science*, 10–11 (1986–87).
Sarma 1992	Sreeramula Rajeswara Sarma, 'Astronomical Instruments in Mughal Miniatures,' *Studien zur Indologie und Iranistik*, 16–7, 235–76. Reprinted in Sarma 2008, 76–121.
Sarma 1994	S. R. Sarma, 'The Bowl that Sinks and Tells Time', *India Magazine, of her People and Culture*, 14, 31–6. Reprinted in Sarma 2008, 125–35.
Sarma 2004	Sreeramula Rajeswara Sarma, 'Setting up the Water Clock for Telling the Time of Marriage', in Burnett 2004, 302–30. Reprinted in Sarma 2008, 147–75.
Sarma 2008	Sreeramula Rajeswara Sarma, *The Archaic and the Exotic: Studies in the History of Indian Astronomical Instruments*, New Delhi.

Sarma 2014	S. R. Sarma, 'Astronomical Instruments presented by the Maharaja of Benares to the Prince of Wales,' *Bulletin of the Scientific Instrument Society*, 122, 12–15.
Sarma 2019a	Sreeramula Rajeswara Sarma, *A Descriptive Catalogue of Indian Astronomical Instruments*, 2nd revised edition, Düsseldorf. Available at: http://srsarma.in/catalogue.php, and at CrossAsia-Repository: https://crossasia-repository.ub.uni-heidelberg.de.
Sarma 2019b	Sreeramula Rajeswara Sarma, 'Astronomical Instruments in Bhāskarācārya's *Siddhāntaśiromaṇi*', in Ramasubramanian 2019, 320–58.
Sarton 1935	George Sarton, 'The First Explanation of Decimal Fractions and Measures (1585). Together with a History of the Decimal Idea and a Facsimile of Stevin's *Disme*', *Isis*, 23, 153–244.
Sasaki et al. 1989	Katsuhiro Sasaki *et al.*, 'On the History of the Large Pillar Clock with Detailed Hour Scale Exhibited at the National Science Museum', *Bulletin of the National Museum of Nature and Science Series E*, 12, 47–58.
Sasaki 1996a	Katsuhiro Sasaki, 'The Scent of Time—The Story of Incense Clocks (の香り - 香のはなし)', *World Wrist-Watch*, 26, 113–16.
Sasaki 1996b	Katsuhiro Sasaki & Makoto Watanabe, 'The 13 Dividing Hour System Used in the Old Fief of Kaga in Edo Period', *Bulletin of the National Museum of Nature and Science Series E*, 19, 17–42.
Sasaki 2002	Katsuhiro Sasaki, 'A Japanese Hanging Clock with Single Foliot Made by Risuke Katsu', *Bulletin of the National Museum of Nature and Science Series E*, 25, 23–34.
Sasaki et al. 2005	Katsuhiro Sasaki, *et al.* 'The Mechanism of Automatic Display for the Temporal Hour in the Japanese Clocks', *Bulletin of the National Museum of Nature and Science Series E*, 28, 33–6.
Sasaki et al. 2010	Katsuhiro Sasaki *et al.*, 'The Structure and Characteristics of Double Foliot Mechanism with "Fuji-Guruma" Escape-Wheel', *Bulletin of the National Museum of Nature and Science Series E*, 33, 9–19.
Sasaki et al. 2015	Katsuhiro Sasaki *et al.*, 'The Single Foliot Lantern Clock with Astronomical Display Made by Msashi Hirayama', *Bulletin of the National Museum of Nature and Science Series E*, 38, 9–22.
Sasaki & Kondo 2008	Katsuhiro Sasaki & Katsuyuki Kondo, 'The Japanese Clocks Made by Sukezaemon Tsuda and Their Characteristics', *Bulletin of the National Museum of Nature and Science Series E*, 31, 1–14.
Sasaki & Kondo 2009	Katsuhiro Sasaki & Katsuyuki Kondo, 'A Wall Clock by Arimasa in 1834 with the Pie Chart Style Dial and the Automatically Expanding-Contracting Hand', *Bulletin of the National Museum of Nature and Science Series E*, 28, 33–36.
Sasaki & Saito 2016	Katsuhiro Sasaki & You Saito, 'The Clockwork of the Table Clock Signed by Hans de Evalo in 1581 which is Stored in the Kunozan Toshogu Shrine, and its History', *Bulletin of the National Museum of Nature and Science Series E*, 39, 9–22.
Sasch 2010	Stephen Sasch, 'The Marean-Kielhorn Director', *The Compendium*, 17, 36.
Saule & Arminjon 2010	Béatrix Saule & Catherine Arminjon (eds.), *Science & Curiosité à la cour de Versailles*, Paris/Versailles.
Saunier 1858	Claudius Saunier, *Le temps, ses divisions principales, ses mesures et leurs usages aux époques anciennes et modernes*, Paris.
Saunier 1881	Claudius Saunier, *The Watchmakers' Hand-book* [1873/4], English translation with additional material by Tripplin & Rigg, London.
Saurin 1722	Joseph Saurin, 'Remarques sur les horloges à pendule', *Histoire de l'Académie Royale des Sciences année MDCCXX avec les Mémoires de mathématique et physique . . .*, Paris, 208–30.
Sautebin 1982	M. T. Sautebin, 'Femmes dans l'horlogerie, quelle avenir', *Femmes suisses et movement féministe*, 70, 12–14.
Savian 2009	Fabio Savian, 'Le ore antiche, il percorso travagliato del primo sistema orario', *Gnomonica Italiana*, 6, 39–47.
Savoie 1998	Denis Savoie, 'L'ancien cadran solaire de la colonne Catherine Médicis à Paris', *L'Astronomie*, 112, 38–43.
Savoie 2001	Denis Savoie, *La Gnomonique*, Paris.
Savoie 2003	Denis Savoie, 'Les cadrans solaires du château de Denainvilliers', *Bulletin de l'ANCAHA*, 97, 79–82.
Savoie 2007	Denis Savoie, 'Le cadran solaire grec d'Aï Khanoum: la question de l'exactitude des cadrans antiques', *CRAI*, 151-2, 1161–90.
Savoie 2008	Denis Savoie, 'L'aspect gnomonique de l'oeuvre de Fouchy: la méridienne de temps moyen', *Revue d'Histoire des Sciences*, 61, 41–62.
Savoie 2012a	Denis Savoie, 'Les cadrans solaires de hauteur', *Dossiers d'Archéologie*, 354, 48–51.
Savoie 2012b	Denis Savoie, 'Pourquoi "cadran de berger"?', *Cadran Info*, 25, 79–81.
Savoie 2012c	Denis Savoie, 'Cadran de hauteur Volpaia', *Cadran Info*, 26, 99–105.
Savoie 2014a	Denis Savoie, *Recherches sur les cadrans solaires*, Des Diversis artibus: Collection de travaux de l'Académie Internationale d'Histoire des Sciences, 96, Turnout.
Savoie 2014b	Denis Savoie, '*Quadrans vetus*: cadran portable médiéval', *Cadran Info*, 30, 93–6.
Savoie 2017a	Denis Savoie, 'Cadrans solaires et châteaux', in Turner 2017, 60–6.
Savoie 2017b	Denis Savoie, 'Le cadran solaire de hauteur de Wenzel Jamnitzer de l'Observatoire de Paris', *Cadran Info*, 34, 120–35.

Savoie & Goutaudier 2012	Denis Savoie & Marc Goutaudier, 'Les disques de Berteaucourt-les-Dames et de Mérida: méridiennes portatives ou indicateurs de latitude?', *Revue du Nord*, 44, 115–19.
Savoie & Goutaudier 2016	Denis Savoie & Marc Goutaudier, 'Le cadran solaire inversé de la Cité des Sciences et de l'Industrie', *Cadran Info*, 35, 100–5.
Savoie & Turner 2014	Denis Savoie & Anthony Turner, 'An Exceptional Sundial', *The BSS Bulletin*, 26, n.p.
Sawyer 1978	Fred Sawyer, 'Bifilar Gnomonics', *Journal of the British Astronomical Association*, 88, 334–51.
Sawyer 1995	Fred Sawyer, 'Serle's Dialing Scales', *The Compendium*, 2, 5–9.
Sawyer 1997	Fred Sawyer, 'Towards a General Theory of Dialing Scales', *The Compendium*, 4, 14–20.
Sawyer 2003	Frederick W. Sawyer III, *The Analemmatic Sundial Source Book*, Glastonbury, CT.
Sawyer, Schilke, & Severino 2009	Fred Sawyer, John Schilke, & Nicola Severino, 'Andreas Schöner's Stereographic Sundial Design', *The Compendium*, 16, 15–18.
Scarabelli 1666	P. F. Scarabelli, *Museo e galleria adunata dal sapere, e dallo studio del Signore canonico Manfredo Settala*, Tortona.
Schaarf 1988	B. Schaaf, *Schwardswalderuhren*, Freiburg im Breisgau.
Schäfer 1914	K. H. Schäfer, *Die Ausgaben der apostolischen Kammer unter Benedikt XX, Klemens VI und Innocenz VI (1335–1362)*, 3 vols., Paderborn.
Schaefer 1976	Scott Jay Schaefer, *The Studiolo of Francesco I de Medici in the Palazzo Vecchio in Florence*. PhD thesis, Bryn Mawr College, Ann Arbor.
Shahar 2016	Galili Shahar, 'The Alarm Clock: The Times of Gregor Samsa', in Cools & Liska, 2016, 257–69.
Schaldach 1996	Karlheinz Schaldach, 'Vertical Sundials of the 5th–15th Centuries', *BBSS*, 46, 32–8.
Schaldach 2002	Karlheinz Schaldach, 'Di due manoscritti di Rostock e la regola di Erfurt', *GI*, 2, 28–31.
Schaldach 2004	Karlheinz Schaldach, 'The Arachne of the Amphaiaerion and the Origins of Gnomonics in Greece', *Journal for the History of Astronomy*, 35, 435–45.
Schaldach 2006	Karlheinz Schaldach, *Die antiken Sonnenuhren Griechenlands*, Frankfurt-am-Main.
Schaldach 2008	K. Schaldach, 'Gli schemi delle ombre' nel Medio Evo lathinoi', *GI*, 16, 9–16.
Schaldach 2017	Karlheinz Schaldach, 'Measuring the Hours: Sundials, Water-Clocks and Portable Sundials', in Jones 2017, 63–93.
Schechner 2008	Sara Schechner, 'Astrolabes and Medieval Travel', in *The Art, Science, and Technology of Medieval Travel*, Aldershot, 181–210.
Schechner 2016	Sara J. Schechner, 'European Pocket Sundials for Colonial Use in American Territories', in A. D. Morrison-Low, Sara J. Schechner, & Paolo Brenni (eds.), *How Scientific Instruments Have Changed Hands*, Leiden, 119–70.
Schechner 2019	Sara J. Schechner, *Time of Our Lives: Sundials of the Adler Planetarium*, Chicago.
Scheibe & Adelsberger 1932	Adolph Scheibe & Udo Adelsberger, 'Eine Quarzuhr Für Zeit- Und Frequenzmessung Sehr Hoher Genauigkeit', *Physikalische Zeitschrift*, 33, 835–41.
Scheibe & Adelsberger 1936	Adolph Scheibe & Udo Adelsberger, 'Schwankungen der astronomischen Tageslänge und der astronomischen Zeitbestimmung nach den Quarzuhren der Physikalisch-Technischen Reichsanstalt', *Physikalische Zeitschrift*, 37, 185–203.
Scheicher 1985	E. Scheicher, *The Collection of Archduke Ferdinand II at Schloss Ambras: Its Purpose, Composition and Evoluhtion*, in Impey & MacGregor 1985, 28–38.
Schelhorn 1731	Johann Georg Schelhorn, 'De vita et meritis Joannes Homelii', *Amoenitates literariae*, 14, 403–467.
SH 1915	Schweizerisches Handelsamtsblatt. Berne, December 23, 1733–4.
Schenk & Woltey 2006	Adolf Schenk & Georg Von Woltey, *Die Uhrmacherfamilie Liechti von Winterthur und ihre Werke*, Winterthur.
Scheurenbrand et al. 2005	Hans Scheurenbrand, Michael Schwarz, & Hermann-Michael Hahn, *Festo Harmonices Mundi-Astrolabium, Kalenderuhr und Glockenspiel. Konstruktion-Funktion-präzision*, Stuttgart.
Scheurer 1995a	Hugues Scheurer, 'Paysans-horlogers: mythe ou réalité?', in Mayaud & Henry 1995, 45–53.
Scheurer 1995b	Hugues Scheurer, 'Une entreprise familiale entre La Cibourg et Lisbonne (fin XVIIIe–début XIXe siècle) in Mayaud & Henry 1995, 157–68.
Schiavon 2013	Martina Schiavon, *Itinéraires de la précision; géodésiens, artilleurs, savants et fabricants d'instruments de précision en France, 1870–1930*, Nancy.
Schilling 1943	Dorotheus Schilling, with Takeo Yanagiya (trans.), 'Education of Jesuits in the 16th and 17th Centuries (16·17世紀に於けるゼスス会士教育事業)', in Schülwesen der Jesuiten in Japan 1551–1614, Munster.
Schinto 2005	Jeanne Schinto, 'The Clockwork Roasting Jack, or How Technology Entered the Kitchen', *NAWCC Watch & Clock Bulletin*, 47, 25–31.
Schmid 2002	Werner Schmid, 'Nachtwachter-Kontrolluhren aus Stuttgart', *DGC*, 41, 167–76.
Schmid 2006a	Hans-Heinrich Schmid, 'Die Gründerväter der Uhrenindustrie in Deutschland', *DGC*, 45, 69–78.
Schmid 2006b	Werner Schmid, 'Tragbare Nachtwächter-Kontrolluhren und ihre Hersteller', *DGC*, 45, 79–96.
Schmid 2007	Werner Schmid, 'Tragbare Nachtwächter-Kontrolluhren und ihre Hersteller ii', *DGC*, 46, 19–32.

Schmid 2009	Werner Schmid, 'Tragbare Nachtwächter-Kontrolluhren und ihre Hersteller iii', *DGC*, 48, 7–14.
Schmid 2011a	Hans-Heinrich Schmid, 'Der Einfluss der amerikanischen Uhrenindustrie auf die Fertigung von Uhren in Deutschland', in Bodenmann 2011, 196–202.
Schmid 2011b	Hans-Heinrich Schmid, 'Die Serienherstellung von Uhren und der Einfluss der amerikanischen Uhrenindustrie auf die Fertigung von Uhren in Deutschland', *DGC*, l, 199–222.
Schmid 2013	Hans-Heinrich Schmid, 'Die Uhrenfabrik Lenzkirch. Ihre Entwicklung im Spiegel zeitgeschichtlicher Umstände und ihrer Besitzverhältnisse', *DGC*, 52, 97–114.
Schmid 2017a	Hans-Heinrich Schmid, *Lexikon der Deutschen Uhrenindustrie 1850–1980, vol. 1:* Firmenadressen, Fertigungsprogramm, Firmenzeichen, Markennamen, Literaturverzeichnis, 3rd expanded edn., Nürnberg/Berlin.
Schmid 2017b	Hans-Heinrich Schmid, *Lexikon der Deutschen Uhrenindustrie 1850–1980, vol. 2:* Firmenbeschreibungen, 3rd expanded edn., Nürnberg/Berlin.
Schmid 2017c	Hans-Heinrich Schmid, *Lexikon der Deutschen Uhrenindustrie 1850–1980*, Nürnberg.
Schmidgen 2005	Hennin Schmidgen, 'Physics, Ballistics and Psychology: A History of the Chronoscope in/as Context 1845–1890', *History of Psychology*, 8, 46–78.
Schmidt 1987	G. Schmidt, *Die Comtoise-Uhr*, Villingen.
Schmittgen 2005	Henning Schmittgen (ed.), *Lebendige Zeit: Wissenskulturen im Werden*, Berlin.
Schneider 2007	Denis Schneider, 'Cadrans canoniaux', *BCCS*, 15, 81–86.
Schneider 2017	Denis Schneider, 'Goethe, ses éditeurs et l'heure italienne …', *Cadran Info*, 36, 152–165.
Schönberger 1622	Georg Schönberger, *Demonstratio et constructio horologiorum nouorum radio recto, refracto in aqua, reflexo in speculo, solo magnete horas astronomicas, Italicas, Babylonicas indicantium*, Fribourg en Brisgau.
Scoresby 1823	William Scoresby, 'Observations on the Errors in the Sea-Rates of Chronometers Arising from the Magnetism of their Balances; With Suggestions for Removing this Source of Error', *Transactions of the Royal Society of Edinburgh*, 9, 353–64.
Schott 1664	Gaspar Schott, *Technica curiosa, sive mirabilia artis, Liber IX*, Nuremberg.
Schoy 1923	Karl Schoy, *Gnomonik der Araber, Die Geschichte der Zeitmessung und der Uhren*, E. Bassermann-Jordan (ed.), part I.F,Berlin/Leipzig.
Schraven 2003	Thomas Schraven, 'The Hipp Chronoscope'. Meeting of the Electrical Group of the Antiquarian HorologicalSociety at the Science Museum, London, on November 23. Available at: http://vlp.uni-regensburg.de/documents/schraven_art13.pdf.
Schreck & Zheng 1628	Johannes Schreck & Wang Zheng, *Die wunderbaren Maschinen des fernen Westens in Wort und Bild*, Beijing.
Schreiber 2017	Jürgen Schreiber, *Uhren–werkzeugmaschinen–rüstungsgüter. Das Familienunternehmen Gebrüder Thiel aus Ruhla 1862–1972*, Köln.
Schröter 1994	Harm G. Schröter, 'Kartellierung und Dekartellierung 1890–1990', *Vierteljahrschrift für Sozial- und Wirtschaftsgeschichte*, 81, 457–93.
Schukowski 2006	M. Schukowski, *Wunderuhren*, Schwerin.
Schukowski 2009	Manfred Schukowski, 'Uhren in Kirchen aus hansische Zeit', *DGC*, 48, 69–83.
Schultz 1900	Wilhelm Schultz, *Unsere Zeitmesser und ihre Behandlung. Anleitung zur sachgemäßen Behandlung der Taschen- und Zimmeruhren*, Berlin.
Schulz 1999	Berndt Schulz, *Swatch oder die Erfolgsgeschichte des Nicolas Hayek*, Düsseldorf.
Schumann 1716	Johann Friedrich Schuman, *Dissertatio Horographica sistens Horologium Universale Munsterianum,*
Schwilgué 1857	C. Schwilgué, *Notice sur la vie, les travaux et les ouvrages de mon père J. B. Schwilgué, ingénieur-mécanicien, officier de la Légion d'honneur, créateur de l'horloge astronomique de la Cathédrale de Strasbourg, etc.* Strasbourg.
Science Museum 2020	Science Museum Group, Wooden pulley Orrery by James Ferguson, London to illustrate the motions of the Moon and Earth around the Sun. 1755–56. Available at: https://collection.sciencemuseumgroup.org.uk/objects/co8006072/mahogany-pulley-orrery-components-demonstration-models-orreries-planetaria-models.
Scobie-Youngs 2018	Keith Scobie-Youngs, 'Salisbury, Wells and Rye–The Great Clocks Revisited', *AH*, 39, 327–41.
Thomas Scot 1616	Thomas Scot, *Philomythie, or, Philomythologie Wherin outlandish Birds, Beasts, and Fishes, are Taught to Speake True English plainely*, London.
Scott 1816	John Scott, *A Visit to Paris in 1814*, London
Scott & Cowham 2010	David Scott & Mike Cowham, *Time Reckoning in the Medieval world. A Study of Anglo-Saxon and Early Norman Sundials*, London.
Sédillot 1835	J. J. Sédillot (trans.), *Traité des instruments astronomiques des arabes compose au treizième siècle par Aboul Hasssan Ali de Maroc …*, 2 vols., Paris.
Sédillot 1844	Louis Amélie Sédillot, 'Mémoire sur les instruments astronomiques des arabes', *Mémoires presents par divers savannts à l'Académie Royale des Inscriptions et Belles-Lettres*, 1, 1–229.

Seller 1669	John Seller, *Practical Navigation . . .*, London.
Semedo 1655	Álvaro Semedo, *The History of That Great and Renowned Monarchy of China*, London.
Semphill 1635	Hugh Sempill, *De Mathematicis Disciplinis Libri Duodecim*, Anvers.
Sen 1984	Geeti Sen, *Paintings from the Akbarnama: A Visual Chronicle of Mughal India*, Calcutta.
Sen 2015	Joydeep Sen, *Astronomy in India, 1784–1876*, London.
Serge & Pins 2014	Nicolas Serge & Delphine Pins, 'La Loi de Piéron et les premiers instruments de mesure des temps de réaction', *Bulletin de Psychologie*, 47, 385–407.
Serle 1657	George Serle, *Dialling Universal*, London.
Service Hydro-Graphique 1859–95	Service Hydrographique de la Marine, *Recherches sur les chronomètres*, 16 vols., Paris.
Settle 1961	Thomas Settle, 'An Experiment in the History of Science', *Science*, 113, 19–23.
Severino 1995	Nicola Severino, *Antologia di storia della gnomonica*, Roccasecca.
Severino 1997	Nicola Severino, 'The Portici Ham', *The Compendium*, 4(2), 23–5.
Severino 2000	Nicola Severino, 'Sulla successione cronologica degli orologi solari d'Altezza Rettininei', *Gnomonica*, 5, 38–44.
Severino 2007	Nicola Severino, *La Gnomonica in Alcuni libri di prospettiva e geometria descrittiva*, Roccasecca.
Severino 2011	Nicola Severino, *Saggi di Storia della gnomonica*, vol. 2, Roccasecca.
Severino & Colombo 2009	Nicola Severino, 'Storia dell'orologio solare a rifrazione'. Available at: http://www.nicolaseverino.it [online,now defunct]; and response from Lino Colombo, 'Gli orologi solari a rifrazione in Italia'. Available at: https://www.ta-dip.de/fileadmin/user_upload/bilder3/258d73f93813070bf75e3788c1062d57_Gli_orologi_solari_a_rifrazione_in_Italia.pdf[online].
Sezgin 1974; 1978	Fuat Sezgin, *Geschichte des arabischen Schriftums*, vols. 5 & 6, Leiden.
S'Gravesande 1711	Willem Jacob s'Gravesande, *Essai de perspective*, La Haye.
Shank 2007	Michael H. Shank, 'Mechanical Thinking in European Astronomy (13th–15th centuries)', in Bucciantini, Camerota, & Roux 2007, 3–27.
Sharma 2016	Virendra Nath Sharma, *Sawai Jai Singh and His Astronomy*, Delhi; 2nd revised edn, Delhi.
Shaw 2011	Matthew Shaw, *Time and the French Revolution: The Republican Calendar 1789–Year XIV*, Woodbridge.
Shen 2011	Shen Ming, 'History of Huacheng Watches', *China Business Daily*. 5 August. Available at: http://www.gucn.com/Info_KnowLedgeList_Show.asp?Id=8162.历史上的华成表.
Shenton 1994	Alan Shenton, 'Who was Charles Shepherd?', *AH*, 21, 438–45.
Shenton 1985	Alan Shenton & Rita Shenton, *The Price Guide to Collectable Clocks, 1840–1940*, Woodbridge.
Sheridan 1896	Paul Sheridan, 'Les inscriptions sur ardoise de l'abbaye de Villers', *Annales de la Société d'Archéologie de Bruxelles*, 10, 203–15, 404–51.
Shukla 1967	Kripa Shankar Shukla, 'Āryabhata I's Astronomy with Midnight Day- Reckoning,' *Gaṇita*, 18(1), 83–105.
Siddiqi 1927	A. Siddiqi, 'Construction of Clocks and Islamic Civilization', *Islamic Culture*, 1, 245–51.
Siegenthaler 1996	Hansjörg Siegenthaler (ed.), *Historische Statistik der Schweiz. Statistique historique de la Suisse. Historical statistics of Switzerland*, Zürich.
Sievert 1879	Hermann Sievert, *Leitfaden für Uhrmacherlehrlinge. Handbuch für Lehrmeister und Lehrbuch für Lehrlinge, sowie zur Vorbereitung auf die theorischen Fach prüfunge. . . .*, Berlin.
Sievert 1910	Hermann Sievert, *Guide manuel de l'apprenti horloger avec apendice et supplement*, L. Fresnard (trans.), Paris.
Sievert 1921	Hermann Sievert, *L'apprenti horloger. Manuel pour maîtres d'apprentissage et cours d'études à l'usage des apprentis . . .*, 2nd edn., revised by H. Pignet, Paris.
Silent Electric Clock Co. 1911	*Catalogue*, London.
Simon-Perret 2020	Martine Simon-Perret, *Dieudonné Sarton (1730–1801), de Liège à Lyon*, Lyon.
Simoni 1954	Antonio Simoni, 'A New Document and Some Views about Early Spring Driven Clocks', *The Horological Journal*, n.v., 590.
Simoni 1965	A. Simoni, *Orologi Italiani dal "500–800"*, Milan.
Simoni 1968	Antonio Simoni, 'Le sfere italiane e la trasmissione ad angolo retto', *La Clessidra*, 24, n.p.
Simoni 1971	Antonio Simoni, 'La sfera dei pianeti', *La Clessidra*, 27, 17–25.
Singer 1992	Aubrey Singer, *The Lion and the Dragon*, London.
Siraisi 1990	Nancy Siraisi, *Medieval and Early Renaissance Medicine: An Introduction to Knowledge and Practice*, Chicago.
Sleeswyk 1979	André Wegner Sleeswyck, 'The 13th Century 'King Hezekiah' Water Clock', *AH*, 11, 488–494.
Skinner 1986	Ghislaine M. Skinner, 'Sir Henry Wellcome's Museum for the Science of History', *Medical History*, 30, 383–418.
Sloley 1931	R. W. Sloley, 'Primitive Methods of Measuring Time, with Special Reference to Egypt', *Journal of Egyptian Archaeology*, 17, 166–78.

Smeur 1965	Simon Stevin, *De Thiende*, Facsimile with an introduction by A. J. E. M. Smeur, Nieuwkoop.
Smith 1978	Alan Smith, (ed.), *A Catalogue of Tools for Watch and Clock Makers by John Wyke of Liverpool*, Charlottesville.
Smith 1985	Alan Smith, 'An Early 18th-Century Watchmaker's Notebook: Richard Wright of Crompton and the Lancashire-London Connection', *AH*, 15, 605–25.
Smith 1968	Cyril Stanley Smith, *Sources for the History of the Science of Steel, 1532–1786*, Cambridge, MA.
Smith 1675	J[ohn] S[mith], *Horological Dialogues. In Three Parts, Shewing the Nature, Use, and Right Managing of Clocks and Watches: With an Appendix Containing Mr. Oughtred's Method for Calculating of Numbers. The Whole Being a Work Very Necessary for All that Make Use of These Kind of Movements*, London.
Smith 1686	John Smith, *Of the Unequality of Natural Time with its Reason and Causes. Together with a Table of the True Aequation of Natural Dayes . . .*, London.
Smith 1694	John Smith, *Horological Disquisitions Concerning the Nature of Time, and the Reasons Why All Days, From Noon to Noon, Are Not Alike Twenty Four Hours Long . . .*, London.
Smith 1921	John Smith, *Old Scottish Clockmakers from 1453 to 1850*, Edinburgh.
Smith 1738	Robert Smith, *A Compleat System of Opticks . . .* , 2 vols., Cambridge,
Smith 1999	Roger Smith, 'The Devil Tavern Group, an Eighteenth-Century Horological Trade Association', *AH*, 24, 427–31.
Smith 2000	Roger Smith, 'James Cox (*c.*1723–1800): A Revised Biography', *The Burlington Magazine*, 142, 353–61.
Smith 2006	Roger Smith, 'Some Mid-Eighteenth-Century Craftsmen: Gray and Vulliamy Outworkers and Suppliers, *c.* 1760', *AH*, 29, 348–58.
Smith 2008	Roger Smith, 'The Sing-Song Trade: Exporting Clocks to China in the Eighteenth Century', *AH*, 31, 629–58.
Smith 2013	Roger Smith, 'The Export of Clocks and Similar Luxury Goods from Britain to China in the Eighteenth Century', in Lin 2013, 19–27.
Smith 2016	Roger Smith, 'James Cox's Silver Swan. An Eighteenth-Century Automaton in the Bowes Museum', *Artefact. Techniques, Histoire et Sciences Humaines*, 4, 361–5.
Smith 2017	Roger Smith, 'Penduliers neuchâtelois et connexions Londoniennes' in Huguenin, Piguet & Baldi 2017, 179–92.
Smith 2018	Roger Smith, 'Les artisans étrangers au service de James Cox . . . ', in Baldi & Tissot 2018, 73–99.
Smith Fussner 1962	F. Smith Fussner, *The Historical Revolution: English Historical Writing and Thought, 1580–1640*, London.
Smith & Findlen 2002	Pamela H. Smith & Paula Findlen (eds.), *Merchants & Marvels. Commerce, Science and Art in Early Modern Europe*, New York/London.
Smyth 1848	W. H. Smyth, 'Description of an Astrological Clock Belonging to the Society of Antiquaries of London . . . ', *Archæologia*, 33, 8–35.
Smyth 2002	Alfred P. Smyth, *The Medieval Life of King Alfred the Great: A Translation and Commentary on the Text Attributed to Asser*, Basingstoke.
Solente 1936	Suzanne Solente (ed.), *Le Livre de fais et bonnes meurs du bon sage roy Charles V, par Christine de Pisan* (Société de l'Histoire de France, 256), 2 vols., Paris.
Soler 2009	Rafael Soler, 'A Double Catenary Bifilar Sundial for the Balearic Islands University Campus', *BBSS*, 21, 34–7.
Somerville 1990	Andrew Somerville, *The Ancient Sundials of Scotland*, London.
Sommervogel 1890–1911	Carlos Sommervogel, *Bibliothèque de la Compagnie de Jésus*, new edition, 12 vols., Brussels & Paris.
Sonderegger 2013	Helmut Sonderegger, 'A Universal Sundial Presented by Johannes Gaupp', *The Compendium*, 20, 26–34.
Song 1960	Song Boyin, 'A Survey of the Clockmaking Industry of Nanjing and Suzhou in the Late Qing Dynasty', *Cultural Relics Magazine*, 1, 18–21.
Song 2014	Song Lian, *Collected Works of Song Lian*, No. 7, Hangzhou.
Sotheby's 1994	Sotheby's, *Clocks, Watches, Wristwatches, Barometers, Mechanical Musical Instruments and Instruments of Science and Technology, 3 and 4 March 1994*, London.
Sotheby's 2004	Sotheby's, *Masterpieces from the Time Museum, Part Four, Volume II, Chronometers and Scientific Instruments*, New York.
Sotheby's & Bobinet 2006	Sotheby's & Bobinet, *George Daniels Retrospective Exhibition*, London.
Sotheran 1927	*Sotheran's Price Current of Literature No. 804. Annotated and classified catalogue of Rare and Standard works on Astronomy, including chronology, geodesy, horology, dialling and octher collateral subjects . . .*, London.
Soubiran 1969	J. Soubiran, *Vitruve: de Architectura*, vol. 9, Paris.
Souchier 2018	Côme Souchier, *Maîtriser le Temps*, Vulaines-sur-Seine.
Sougy 2007	Nadège Sougy, 'Liberté, légalité, qualité: le luxe des produits d'or et d'argent à Genève au XIXe siècle', *Entreprises et Histoire*, 46, 71–84.

Sougy 2018	Nadège Sougy, 'Les montres de Genève au XIXe siècle. La fabrique des qualités', *Revue d'histoire moderne et contemporaine*, 45, 7–28.
Spätling & Dinter 1985	L. G. Spätling and P. Dinter (eds.), *Consuetudines Fructuarienses –Sanblasiana*, Siegburg.
Speel 2008	Erica Speel, *Painted Enamels, An Illustrated Survey 1500–1920*, London.
Spence 1984	Jonathan Spence, *The Memory Palace of Matteo Ricci*, London.
Spencer 1973	Thomas Spencer, *The Sundicator*, Santa Barbara.
Spenser 1963	Eleanor P. Spenser, '*L'Horloge de Sapience:* Bruxelles, Bibliothèque Royale, MS IV, III,' *Scriptorium*, 17, 277–99.
Speiser 1994	David Speiser (ed.), *Die Werke von Daniel Bernouilli . . .*, 7, Basel.
Spierdijk 1965	C. Spierdijk, *Klokken en Klokkenmakers: zes eeuwen uurwerk*, 2nd edn., Amsterdam.
Sposato 1993	Kenneth A. Sposato, *The Dictionary of American Clock & Watchmakers*, White Plains, NY.
rīdhara 1959	S Śrīdhara, *The Patiganita of Sridharacarya*, Kripa Shankar Shukla (ed. & trans.), Lucknow.
Staeger 1997	Hans Staeger, *One Hundred Years of Precision Timekeepers: From John Arnold to Arnold and Frodsham*, Stuttgart.
Stanziani 2007	Alessandro Stanziani, *Dictionnaire historique de l'économie-droit XVIIIs–XXe siècles*, Paris.
Starmer 1904–5	William Wooding Starmer, 'Carillons', in Proceedings of the Musical Association, 31st Session, Manchester, 43–61.
Starmer 1907–8	William Wooding Starmer, 'Chimes', in Proceedings of the Musical Association, 34th Session, Manchester, 1–24.
Starmer c.1908	William Wooding Starmer, *Quarter-Chimes and Chime Tunes*, London.
Starsy 1985	P. Starsy, 'Ein Brief von Kaspar von Schöneich an Peter Henlein', *Uhren und Schmuck*, 22, 23
Staunton 1797	Sir George Staunton, *An Authentic Account of an Embassy . . . to the Emperor of China*, 3 vols., London.
Stebbins 1961	Frederick A. Stebbins, 'A Medieval Portable Dial', *Journal of the Royal Astronomical Society of Canada*, 55, 49–56.
Steblin 2013	Rita Steblin, 'Maelzel's Early Career to 1813: New Archival Research in Regensburg and Vienna', *Regensburger Studien zur Musikgeschichte*, 10, 161–210.
Steiner 1991	Reinhard Steiner, *Prometheus*, Grafrath.
Steinhaus & Beugnon 2008	Hans Steinhaus & Guilhem Beugnon, *Dom Bedos de Celles, Entre orgues et cadrans solaires: vie et travaux d'un Bénédictin du Languedoc (1709–1779)*, Béziers.
Stengel 1679	Johann P. Stengel, *Gnomonica universalis, sive praxis amplissima geometrice describendi horologia solaria . . .*, Ulm.
Stephen 1948	W. B. Stephens, 'Hermann Wenzel and His Air Clock', *California Historical Society Quarterly*, 27 (March), 1–8.
Sternheim 1842	Hermann Sternheim, *Populäre Gnomonik*, Weimar.
Stevens 1962	John C. Stevens, 'The Will of Isaac Pluvier Containing an Inventory of a London Clockmaker's Stock Just Prior to the Great Fire', *AH*, 1, 18–21.
Stirling-Middleton 2018	Emma Stirling-Middleton, 'Clockmaking Industry in England', in HKSM 2018, 30–63.
St-Louis 2020	Robert St-Louis, 'Pierre-François LeRoy: The Lesser-Known Brother of Julien LeRoy', *NAWCC Watch & Clock Bulletin*, November–December, 389–406.
Stolberg 1993	Lukas Stollberg, *Die Kutschenuhr*, Munich.
Stöffler 1518	Johanes Stöffler, *Calendarium Romanum Magnum*, Oppenheim.
Stone 1723	Edmund Stone, *The Construction and Principal Uses of Mathematical Instruments . . .*, London.
Strano 2005	Giorgio Strano, 'Orologi astronomici e planetari', in Brusa 2005, 128–39.
Strauch 2002	Ingo Strauch, *Lekhapaddhati-Lekhapañcāśikā: Briefe und Urkunden im mittelalterlichen Gujarat*, Text, Übersetzung, Kommentar, Glossar, Berlin.
Strauss 2012	Johann Strauss, '*Kurūn-I vista*: la découverte du "Moyen Āge" par les Ottomans', in Georgeon & Hitzel 2012, 205–40.
Strubel 1992	A. Strubel, *Lettres gothiques*, Paris.
Stutzinger 2001	Dagmar Stutzinger, *Eine romische Wasserauslaufuhr (Patrimonia)*, 195, Berlin.
Strzygowski 1894	J. Strzygowski, 'Inedita der Architektur und Plastik auf der Zeit Basilios I (867–886), *Byzantinische Zeitschrift*, 3, 1–16.
Subrahmanian 1966	N. Subrahmanian, *Saṅgam Polity: The Administration and Social Life of Saṅgam Tamils*, Bombay.
Sully 1711	Henry Sully, *Abrége De quelques Régles pour faire un bon usage des Montres, Avec des Réflexions utiles . . .*, Paris.
Sully 1737	Henry Sully, *Regle Artificielle du Temps*, Paris, 3rd edn., (Julien Leroy).
Sun 1925	Sun Meitang, *May War Lee & Co: Fiftieth Anniversary Souvenir Book*, May War Lee.
Suslow 2005	Valerie Y. Suslow, 'Cartel Contract Duration: Empirical Evidence from Inter-War International Cartels', *Industrial and Corporate Change*, 14, 705–44.
Su Song 1090	Su Song, *Xinyi Xiangfayao (Essentials of a New Method for Mechanising the Rotation of an Armillary Sphere and a Celestial Globe)*, n.p.

Swift 1726	Jonathan Swift, *Gulliver's Travels*, London.
Swift 1940	Emerson Haviland Swift, *Hagia Sophia*, New York.
Symonds 1951	R. W. Symonds, *Thomas Tompion, His Life and Work*, London.
Symons 1999	Sarah Symons, *Ancient Egyptian Astronomy: Timekeeping and Cosmography in the New Kingdom*. Unpublished PhD Thesis, University of Leicester. Available at: https://leicester.figshare.com/articles/thesis/Ancient_Egyptian_astronomy_timekeeping_and_cosmography_in_the_new_kingdom/10097984.
Symons 2002	Sarah Symons, 'Egyptian Shadow Clocks', in Dorikens 2002, 11–20.
Symons & Khurana 2016	Sarah Symons & Himanshi Khurana, 'A Catalogue of Ancient Egyptian Sundials, *Journal for the History of Astronomy*, 47, 375–85.
Synchronome 1899	The Synchronome Syndicate Limited, 'Report and Statement of Accounts', *IET Archives: UK0108 SPT/P/I/065/42*.
Szabo & Maula 1986	Árpád Szabo & Erkka Maula, *Les Débuts de l'astronomie, de la géographie et de la trigonométrie chez les Grecs*, M. Federspiel (trans. German), Paris.
Tadić 1988	Milutin Tadić, *Katalog antikih i srednjevjekovnish sunanika u Jugoslavij*, Sarajevo.
Tadić 1997	Milutin Tadić, 'Dalmatinische Halbe holkalotten-sonnenuhren (ungenaue Nachbildungen nach Berosos)', Rundschreiben, 14, 6–7.
Tadić 2002	Milutin Tadić, *Sunani asovnici, Zarod za ud benike i nastavnasredstva*, Belgrade.
Tailliez 1999	Bernard Tailliez, 'Heures Italiques et Italiennes, Babyloniques ou Bohemiennes', *Observations & Travaux*, 51, 20–1.
Tait 1983	Hugh Tait, *Clocks and Watches*, London.
Tait 1986	Hugh Tait, *Catalogue of the Waddesdon Bequest in the British Museum: I Jewels*, London.
Tait & Coole 1987	Hugh Tait & P. G. Coole, *The Stackfreed (Catalogue of Watches in the British Museum I)*, London.
Talbert 2017	Richard J. A. Talbert, *Roman Portable Sundials, The Empire in your Hand*, New York.
Talbert 2019	Richard J. A. Talbert, 'A Lost Sundial Found, and the Role of the Hour in Roman Daily Life', *Indo-European Linguistics and Classical Philology*, 23, 971–88.
Talbot 1991	Alice-Mary Talbot, 'Horologion', in *The Oxford Dictionary of Byzantium*, Alexander P. Kazhdan (ed.), Oxford, n.p.
Talbot 2016	Michael Talbot, 'Gifts of Time: Watches and Clocks in Ottoman-British Diplomacy, 1693-1803', *Jahrbuch für Europäische Geschichte*, 17, 55–79.
Tang 2012	Tang Kaijian, *A History of Catholicism in China during the Ming and Qing Dynasties, Volume 1: Starting Out from Macau*, Macau.
Tang 2017	Tang Kaijian, *Chinese Literature Data Collection and Explanation about Matteo Ricci during the Ming and Qing Dynasties*, Shanghai.
Tannery 1897	Paul Tannery, 'Le Quadrant de Maître Robert Anglès, (Montpellier, XIIIe siècle)', *Notices et extraits des manuscrits de la Bibliothèque Nationale*, 35, 2e partie, 561–640.
Tardy 1947/1980	Tardy, *General Bibliography of Time Measurement*, 1st & 2nd edns., Paris.
Tardy 1972	Tardy, *Dictionnaire des horlogers français*, 2 vols., Paris.
Tardy 1981	Tardy, *French Clocks the World Over (La Pendule Francaise dans le monde)*, Paris.
Tartt 1993	Donna Tartt, *The Secret History*, London.
Taton 1981	René Taton, *L'Œuvre mathématique de G. Desargues*, 2nd edn., Paris.
Taton 2000	René Taton, 'La mathématisation des techniques graphiques. Les grandes étapes des origines à Dürer, à Desargues et à Monge', in *Etudes d'histoire des sciences*, Turnhout, 305–23.
Tavernari 1976	Carla Tavernari, 'Manfredo Settala, collezionista e scienzato Milanese del '600', *Annali dell'Istituto e Museo di storia della scienza di Firenze*, 1, 43–61.
Taylor 2010	John C. Taylor, 'The Coster-Fromanteel Contract: John Fromanteels' Brass and Steel', *AH*, 32, 336–42.
Tekeli 1962	Sevim Tekeli, 'Equatorial Armilla of Iz ad-Dîn b. Muhammad al-Wafai and Torquetum', *Ankara Universitesi dil ve tarih, cogrgya Fakultesi Dergivisi*, 18, 227–59.
Tekeli 1963	Sevim Tekeli, *Meçhul Bir Yazarın İstanbul Rasathanesinin Āletlerinin Tasvirini* Veren 'âlât-ı Rasadiye li Zîc-i Şehinşâhiye', Adlı Makalesi, *Araştırma*, 1, 71–122.
Tekeli 1966	Sevim Tekeli, *The Clocks in Ottoman Empire in 16th-century and Taqi al-Din's 'The Brightest Stars for the Construction of the Mechanical Clocks'*, Ankara.
Tellier 2010	Arnaud Tellier, 'Development of Horological Trade with China', in Tellier & Didier 2010, 13–23.
Tellier & Didier 2010	Arnaud Tellier & Mélanie Didier, *The Mirror of Seduction: Prestigious Pairs of 'Chinese' Watches*, Geneva.
Teng 2015	Teng Xiaobo. "Lasting: Chinese Clock and Watch Design", Zhuangshi, 9, Available at: http://www.izhsh.com.cn/doc/320/3337.html. 滕晓铂. '历久弥新：中国钟表设计', 装饰.
Tennant 2009	M. F. Tennant, *The Art of the Painted Clock Dial*, Mayfield.

Ternant 2004	Evelyne Ternant, *La dynamique longue d'un système productif localisé: l'industrie de la montre en Franche-Comté*. Unpublished PhD thesis, Université de Franche-Comté.
Thacker 1714	Jeremy Thacker, *The Longitudes Examin'd*, London.
Theodossiou, Katsiotis, Manimanis, & Mantarakis 2010	Efstratios Theodossiou, Markos Katsiotis, Vassilios N. Manimanis, & Petros Mantarakis, 'The Large Built Water Clock of Amphiaraeion, *Mediterranean Archaeology and Archaeometry*, 10, 59–167.
Thiout 1741	Antoine Thiout, *Traité de l'horlogerie méchanique et pratique . . .*, Paris.
Thirsk 1961	Joan Thirsk, 'Industries in the Countryside', in Fisher 1961, 70–88.
Thirsk 1978	Joan Thirsk, *Economic Policy and Projects. The Development of a Consumer Society in Early Modern England*, Oxford.
Thomas 1904	Jules Thomas, *Epigraphie de l'église Notre Dame de Dijon*, Paris.
Thomas 1979	A. A. L. Thomas, 'The Nineteenth-Century Watchman's Clock', *AH*, 11, 496–503.
Thomas 1998	Alfred Thomas, 'Watchman's Clocks', *AH*, 24, 141–6.
Thompson 1996	David Thompson, 'Huguenot Watchmakers in England: With Examples from the British Museum Horological Collections'.
Thompson 1997	David Thompson, 'The Watches of Ellicott of London', *AH*, 23, 306–21; 429–42.
Thompson 2004	David Thompson, *The British Museum, Clocks*, London.
Thompson 2007	David Thompson, *Watches in the Ashmolean Museum*, Oxford.
Thompson 2008	David Thompson, *The British Museum, Watches*, London.
Thompson 2019	David Thompson, 'The Art and Mystery of Watchmaking. A Detailed Account of the Making of a Watch c.1650', *AH*, 40, 560.
Thompson & Sasaki 2012	David Thompson & Katsuhiro Sasaki (trans.), 'A Gilded-Brass Spring-Driven Table Clock, Madrid, Date 1581', in *Report on Observation and Consideration from the Survey by the British Musem Curator*, 4–47, Shizuoka.
Thompson & Wycherley 1972	Homer A. Thompson & Richard-Ernest Wycherley, *The Athenian Agora XIV*, Princeton.
Thorndike 1929	Lynn Thorndike, 'Of the Cylinder Called the Horologe of Travellers', *Isis*, 13, 51–2.
Thorndike 1941	Lynn Thorndike, 'Invention of the Mechanical Clock about 1271', *Speculum*, 16, 242–3.
Thorndike 1949	Lynn Thorndike, *The Sphere of Sacrobosco and its Commentators*, Chicago.
Thorndike 1964	Lynn Thorndike, 'Relations between Byzantine and Western Science and Pseudo-Science before 1300', *Janus*, 51, 1–48.
Thornton 2000a	W. J. Thornton, 'Samuel Deacon: An Account of Myself Part I', *AH*, 25, 324–32.
Thornton 2000b	W. J. Thornton, 'Samuel Deacon: An Account of Myself Part II–The First Notebook', *AH*, 25, 532–40.
Thornton 2001	W. J. Thornton, 'Samuel Deacon: An Account of Myself Part III–The First Notebook', *AH*, 26, 263–73.
Thornton 2002	W. J. Thornton, 'Samuel Deacon: An Account of Myself Part IV–The First Notebook', *AH*, 27, 52–62.
Thorsten n.d.	K. Thorsen, The Bidstrup & Anker Families [website]. Available in Danish at: http://www.bidstrup.cc/slaegt/.
Thurston 1907, 1975	Edgar Thurston, 'Steel-Yards, Clepsydras, Knuckle-Dusters, Cock-Spurs, Tallies, Dry Cupping', in *Ethnographical Notes in Southern India*, Part II, reprint, Delhi, 560–6.
Tihon 2000	Anne Tihon, 'Un texte byzantin sur une horloge persane', in Folkerts & Lorch 2000, 523–35.
Tissot 1982	André Tissot, *Voyage de Pierre Jaquet-Droz à la cour du roi d'Espagne 1758-1759, d'après le journal d'Abraham Louis Sandoz son beau-père*, Neuchâtel.
Tissot & Daumas 2004	Laurent Tissot & Jean-Claude Daumas (eds.), *L'arc jurassien, histoire d'un espace transfrontalier*, Vesoul/Yens-sur-Morges.
Todorović & Žitnik 2019	N. Todorović, & I. Milić Žitnik, 'The Astronomical Observatory in Belgrade: Then and Now', *Romanian Astronomical Journal*, 29, 167–76.
Toesca & Manetti 1927	E. Toesca & Antonio Manetti, *Vita di Filippo di ser Brunellescho*, Florence.
Tölle 1969	Renate Tölle, 'Eine spätantike Reisuhrr', *Archaeologische . . .*, 3, 309–17.
Tonnerre 2019	Quentin Tonnerre, 'Une question de prestige dans le domaine international de l'industrie horlogère. Diplomatie suisse et chronométrage sportif (1964-1970)', *Relations internationals*, 177, 129–44.
Toomer 1984	G. J. Toomer (ed. & trans.), *Ptolemy's Almagest*, London.
Travail 1856	'L'exposition de 1855', in *Le travail universel: revue complète des oeuvres de l'art et de l'industrie exposées à Paris en 1855*, Paris, 387–992.
Travaux 1854–1873	'Horlogerie civile, dite de commerce', in *Travaux de la commission française sur l'industrie des nations*, vol. 3, part 2, group II, Jury X, 23–5.
Travaux 1857	Arthur Morin, Claude Arnoux, & Jean-Victor Poncelet, *Travaux de la Commission française sur l'industrie des nations*, vol. 3, Part 1, Section 1. IIe groupe. Ve, VIe jurys, Paris.

Treherne 1977	Alan Treherne, *The Massey Family: Clock, Watch, Chronometer and Nautical Instruments Makers*, Newcastle-under-Lyme.
Treherne 2009	Alan Treherne, 'The Contribution of South-West Lancashire to Horology Part 1: Watch and Chronometer Movement Making and Finishing' *AH*, 32, 457–76.
Trifonov & Schrayer 2005	Yuri Trifonov, 'A Visit with Marc Chagall', Maxim D. Shrayer (trans.), *Agni*, 61, 156–65.
Trigge 1602	Francis Trigge, *The True Catholique Formed According to the Truth of the Scriptures, and the Shape of the Ancient Fathers, and Best Sort of the Latter Catholiques, which Seeme to Favour the Church of Rome*, London.
Trincano 1940/1990–1	Louis Trincano, 'Histoire de l'industrie horlogère', *Revue de l'ANCAHA*, 57, 43–52; 58, 46–54; 59, 65–9; & 60, 57–67.
Trueb 2005	Lucien F. Trueb, *The World of Watches. History, Technology, Industry*, New York.
Trueb 2006	Lucien F. Trueb, *Zeitzeugen der Quarzrevolution*, La Chaux-de-Fonds.
Trueb 2008	Lucien F. Trueb, *Kinder der Quarzrevolution*, La Chaux-de-Fonds.
Trueb, Ramm, & Wenzig 2011	Lucien F. Trueb, Günther Ramm, & Peter Wenzig, *Die Elektrifizierung der Armbanduhr*, Ulm.
Tsukada 1960	Taizaburo Tsukada, 'Dials', in Japanese Clocks, Toho Shoin.
Tsunoyama 1984	Sakae Tsunoyama, *The Social History of Watches and Clocks* (時計の社会史), Tokyo.
Tuck 1989	Paul Tuck, 'Fine Examples of Horology Seen during the Scottish Tour, 1988', *AH*, 18, 181–6.
Tucker 2007	Brian Tucker, 'Performing Boredom in Effi Briest: On the Effects of Narrative Speed', *The German Quarterly*, 80, 185–200.
Tumanian 1974	Benik E. Tumanian, 'Measurement of Time in Ancient and Medieval Armenia', *Journal for the History of Astronomy*, 5, 91–98.
Tupper 1895	Frederick Tupper, 'Anglo-Saxon Dæg-mæl', *Journal of the Modern Languages Association of America*, 10, 111–341.
Turicchia 2018	E. Turicchia, *Nuovo Dizionario degli Orologiai Italiani*, n.p.
Türler et al. 2013	Franz Türler, Jörg Spöring, & Ludwig Oechslin, *Das Unikat, Die Türler-Uhr, Modell des Kosmos*, Zürich.
Turner 1972	A. J. Turner, 'The Introduction of the Dead-Beat Escapement: A New Document', *AH*, 7, 12.
Turner 1973a	Anthony Turner, *The Clockwork of the Heavens, An Exhibition of Astronomical Clocks, Watches and Allied Instruments . . .*, London.
Turner 1973b	Anthony Turner, 'Mathematical Instruments and the Education of Gentlemen', *Annals of Science*, 30, 51–88.
Turner 1977	A. J. Turner, *Science and Music in 18th-Century Bath*, Bath.
Turner 1982a	A.J. Turner, 'William Oughtred, Richard Delamain and the Horizontal Instrument in 17th-Century England', *Annali dell'Istituto e Museo di Storia della Scienza di Firenze*, 6, 99–125. Reprinted in Turner 1993, ch. 7.
Turner 1982b	A. J. Turner, '"The Accomplishment of Many Years": Three Notes Towards a History of the Sand-Glass', *Annals of Science*, 39, 161–72. Reprinted in Turner 1993, ch. 5.
Turner 1984a	A. J. Turner, 'Anglo-Saxon Sundials and the "Tidal" or "Octival" System of Time Measurement', *AH*, 15, 76–7.
Turner 1984b	A. J. Turner, *Water-Clocks, Sand-Glasses, Fire-Clocks* (The Time Museum, Catalogue of the Collection I: Time-measuring Instruments part 3), Rockford.
Turner 1985	A. J. Turner, *Astrolabes: Astrolabe-Related Instruments* (The Time Museum, Catalogue of the Collection I: Time-measuring Instruments part 1), Rockford.
Turner 1987a	Anthony J. Turner, 'La Gnomonique en France à l'époqe de Jean Picard', in Picolet 1987, 345–62.
Turner 1987b	Anthony Turner, 'Dialling in the Time of Giovan Battista Benedetti', in *Cultura, scienza e techniche . . . Ati del Convegno internazionale di studio Giovann Battista benedetti . . .*, Venice, 311–20.
Turner 1987c	Anthony Turner, *Early Scientific Instruments. Europe 1400–1800*, London.
Turner 1989a	A. J. Turner, 'Sundials: History and Classification', *History of Science*, 27, 303–18. Reprinted in Turner 1993, ch. 2.
Turner 1989b	A. J. Turner (ed.), *Etudes 1987–1989* (Astrolabica 5), Paris.
Turner 1990	A. J. Turner (ed.), *Time*, The Hague.
Turner 1993	A. J. Turner, *Of Time and Measurement: Studies in the History of Horology and Fine Technology*, Aldershot.
Turner 1994	A. J. Turner, *Mathematical Instruments in Antiquity and the Middle Ages*, London, 1994.
Turner 1995	A. J. Turner, 'Yūnus the Candle-Clock Maker and Babylonian Functions', *Nuncius*, 10, 321–3.
Turner 1996	A. J. Turner, 'In the Wake of the Act, But Mainly Before', in Andrewes 1996, 115–32.
Turner 1999	A. J. Turner, 'A Biblical Miracle in a Renaissance Sun-Dial', *Bulletin of the Scientific Instrument Society*, 41, 11–14.
Turner 2000	A. J. Turner, 'The Anaphoric Clock in the Light of Recent Research', in Folkerts & Lorch 2000, 536–47.
Turner 2002a	A. J. Turner, 'Donald Hill and Arabic Water Clocks', *AH*, 27, 206–13.

Turner 2002b	Anthony Turner, 'Antoine d'Abbadie et son observatoire decimal à Hendaye', in Turner & Poirier 2002, 7–87.
Turner 2004	Anthony J. Turner, 'A Use for the Sun in the early Middle Ages: The Sun-Dial as Symbol and Instrument' *Micrologus*, 12, 27–42.
Turner 2007	Anthony J. Turner, *Istituto e Museo di Storia della Scienza: Catalogue of Sun-Dials, Nocturnals and Related Instruments*, Florence.
Turner 2008	Anthony Turner, 'Grollier de Servière, The Brothers Monconys: Curiosity and Collecting in Seventeenth-Century Lyon', *Journal of the History of Collections*, 20, 205–15.
Turner 2009	Anthony Turner, 'Dropped Out of Sight: Oronce Finé and the Water-Clock in the Sixteenth and Seventeenth Centuries', in Marr 2009, 191–205.
Turner 2010	Anthony Turner, 'Watch Pads and Watch Papers', *AH*, 32, 417–23.
Turner 2011	Anthony Turner, 'The Tribulations of Jean Helot', *AH*, 32, 779–804.
Turner 2014a	Anthony Turner, *John Carte on Horology and Cosmology: A Transcription with Introduction and Notes of Bodleian Library MS Carte 264, ff. 18r–57r*, Ticehurst.
Turner 2014b	Anthony Turner, 'Charles Clay: Fashioning Timely Music', *AH*, 35, 929–48.
Turner 2014c	Anthony Turner, 'From Sun and Water to Weights: Public Time Devices from Late Antiquity to the Mid-Seventeenth Century', *AH*, 35, 649–62.
Turner 2015a	Anthony Turner, 'An Exemplary Clock-Maker', *AH*, 36, 97–9.
Turner 2015b	Anthony Turner, 'The Eclipse of the Sun: Sundials, Clocks and Natural Time in the Late 17th Century, *Early Science and Medicine*, 20, 169–86.
Turner 2016a	Anthony Turner, 'Concerning Some Curious Clocks in the Cabinet of Grollier de Servières', *AH*, 37, 349–65.
Turner 2016b	Anthony Turner, 'The Origins and Diffusion of Watches in the Renaissance . . ., France', in Zanetti 2016, 141–6.
Turner 2017	Anthony Turner (ed.), *Les Sciences à l'âge des Lumières*, actes du colloque tenu au Château de la La Roche-Guyon, 14 juin 2014, La Roche-Guyon.
Turner 2019	Anthony Turner, 'A Mingling of Traditions: Aspects of Dialling in Islam', in Brown *et al.*, *Scientific Instruments Between East and West*, Leiden, 108–21.
Turner 2021	Anthony Turner, 'Collectionner les sciences à la fin du XIXe siècle: une contexte pour la Collection Max Elskamp' in press.
Turner forthcoming	Anthony Turner, 'Time Measurement and Practical Mathematics in the Inventory of Henry VIII', in David Starkey & Maria Hayward (eds.), *The Inventory of Henry VIII: Essays and Commentary*, vol 4, Turnhout.
Turner et al. 2018	Anthony Turner, with Silke Ackermann & Taha Yasin Arslan, *Mathematical Instruments in the Collections of the Bibliothèque Nationale de France*, Paris/New York.
Turner & Crisford 1977	A. J. Turner & A. C. H. Crisford, 'Documents Illustrative of the History of English Horology, I: Two Letters Addressed to Thomas Mudge', *AH*, 10, 580–2.
Turner & Poirier 2002	Anthony Turner & Jean-Paul Poirier, *Antoine d'Abbadie*, Paris.
Turner 2018	Jonathan Turner, *Telling Timepieces: Representations of the Timepiece within Literature and Visual Culture of the Eighteenth Century*. Unpublished PhD thesis, University of Roehampton, London.
Tuve 1963	Rosemond Tuve, 'Notes on the Virtues and Vices', *Journal of the Warburg and Courtauld Institutes*, 26, 264–303.
Tyler 2013	Daniel Tyler (ed.), *Dickens's Style*, Cambridge.
Tyler 1976	E. J. Tyler, 'Rye Church Clock', *AH*, 10, 41–54.
Tyler 1977	E. J. Tyler, *Black Forest Clocks*, London.
Tyler 1991	E. J. Tyler, 'The Clocks of St Andrew's Church, Hornchurch, Essex', *AH*, 19, 617–19.
Uchida 1985	Hoshimi Uchida, 'The Establishment of Seikosha', in *The Development of Watch and Clock Industry*, Tokyo.
Uchida 2002	Hoshimi Uchida 'The Spread of Timepieces in the Meiji Period', *Japan Review*, 14, 173–92.
UJS 1973	'In der Weltproduktion an 2. Stelle', Uhren, Juwelen, Schmuck, 1, 7.
Ullmann 2015	Mathias Ullmann, 'Der Weg von Dresden nach Glashütte', in Plaßmeyer & Gluch 2015, 102–15.
Ungerer 1922	Alfred Ungerer & Théodore Ungerer, *L'horloge astronomique de la Cathédrale de Strasbourg*, Strasbourg.
Ungerer 1926	Alfred Ungerer, *Les horloges d'édifice, leur construction, leur montage, leur entretien, . . .*, Paris.
Ungerer 1931	Alfred Ungerer, *Les horloges astronomiques et monumentales les plus remarquables de l'Antiquité jusqu'à nos jours*, Strasbourg.
Ungerer 1934	Théodore Ungerer, 'L'Horloge astronomique de la cathédrale de Messine', *Annales françaises de chronometrie*, 4, 63–98.
Unver 1954	A. Suheyl Unver, 'Sur les cadrans solaires horizontaux et verticaux en Turquie', *Archives internationales d'histoire des sciences*, 7, 254–66.

Urai 2014	Sachiko Urai, 'The Life of the Edo People and Time Bells (江戸庶民の生活と時の鐘), *Special Feature: Time, Quarterly 'Yukyu (悠久)*', 138, 85–91.
Urquizar-Herrara 2011	Antonio Urquizar-Herrara, 'La Mémoire des choses passées: Florimond Robertet, Charles de Rostaing, Henri Chesneau and the Place of Social Narratives in French Early Modern Noble Collections', *Journal of the History of Collections*, 23, 29–47.
Utrecht n.d.	Nationaal Museum van Speelklok tot Pierment, *De Koekoek en de nachtegaal: Elf vroege cylinderorgels*, compact disc STP 002, Utrecht.
Uttinger & Papera 1965	Hans W. Uttinger & Robert D. Papera, 'Threats to the Swiss Watch Cartel', *The Western Economic Journal*, 3, 200–16.
Valdés Carracedo 1997	Manuel Valdes Carracedo, *Horologios, tablas de pies*, Madrid.
Van Beeck Calkoen 1797	Jan Frederik Van Beeck Calkoen, *Dissertatio mathematico-Antiquaria*, Amsterdam.
Van Boxmeer 1995–98	Henri van Boxmeer 'Les méridiennes de Quetelet: Celle de Malines', *Ciel et terre*, 111, 22–4; 112–14; 112, 15–17, & 79–82; 113, 205–7; 114, 33.
Van Cleempoel 2002	Koenraad Van Cleempoel, *A Catalogue Raisonné of Scientific Instruments from the Louvain School, 1530 to 1600*, Turnhout.
Van den Ende et al. 2004	H. Van den Ende, F. Van Kersen, M. F. Van Kersen-Halbertsma, J. C. Taylor, & N. R. Taylor, *Huygens Legacy: The Golden Age of the Pendulum Clock*, Castletown.
Van Dyke 2005	Paul A. Van Dyke, *The Canton Trade: Life and Enterprise on the China Coast, 1700–1845*, Hong Kong.
Van Helden 1996	Albert van Helden, 'Longitude and the Satellites of Jupiter', in Andrewes 1996, 86–100.
Van Kersen 2005	Fritz Van Kersen, 'The Coster-Fromanteel Contract Re-Examined', *AH*, 28, 561–7.
Van Wely 2010	Bob van Wely (ed.) *Singsong. Schatten uit de Verboden Stad*, Utrecht.
Varāhamihira 1968	Varāhamihira, *The Pañcasiddhāntikā of Varāhamihira*,. G. Thibaut and Sudhakara Dvivedi (ed. & trans.), 2nd edn., Varanasi.
Väterlein 1989	Christian Väterlein (ed.), *Philipp Matthäus Hahn 1739–1790: Pfarrer, Astronom, Ingenieur, Unternehmer, Ausstellungen des Württembergischen Landesmuseums Stuttgart und der Städte Ostfildern, Albstadt, Kornwestheim, Leinfelden-Echterdingen*, Part 2, Stuttgart.
Vaucanson 1738/1985	Jacques Vaucanson, *Le mécanisme du flûteur automate, présenté à messieurs de l'Académie Royale des Sciences*. 1st edn. reprint with a preface by Catherine Cardinal, J. T. Desaguliers trans., 1742, Paris.
Vaucher 2003	'Les Vaucher: horlogers originaires de Fleurier'. Available at: http://www.iro.umontreal.ca/~vaucher/Genealogy/Documents/Horlogerie/Vaucher_horlogers.html.
Vaughan 1994	Denys Vaughan 'Charles Shepherd's Electric Clocks', *AH*, 21, 519–30.
Vaulezard 1640	Vaulezard, *Traicté ou usage du quadrant analematique*, Paris.
Vaulezard 1644	Vaulezard, *Traitté de l'origine, demonstration, construction & usage du quadrant analematique*, Paris.
Vaupel 2015	Elisabeth Vaupel, 'Edelsteine aus der Fabrik. Produktion und Nutzung synthetischer Rubine und Saphire im Deutschen Reich (1906–1925)', *Technikgeschichte*, 82, 273–302.
Vehmeyer 2004	H. M. Vehmeyer, *Clocks: Their Origin and Development 1320–1880*, 2 vols., Gent.
Venkataramanaiah 1974	N. Venkataramanaiah (ed.), *Inscriptions of Andhra Pradesh, Warangal District*, Hyderabad.
Vergil 1499	Polydore Vergil, *De rerum inventoribus, libri octo,* n.p.
Verheul 1981	Ambroise Veheul, 'La prière monastique chorale avant S. Benoît: son influence sur le culte en Occident', *Questions liturgiques*, 62, 227–42.
Verdet 1985	Jean-Pierre Verdet 'Sur Nicolas Kratzer (vers 1487–après 1550)', *L'Astronomie*, 99, 343–5.
Verhoeven 2010–2012	Peter Maria J. Verhoeven, *Die Monumentaluhr von Daniel Vachey*, 4 vols., La Chaux-de-Fonds.
Vérité 1860	Auguste-Lucien Vérité, *Notice descriptive de l'horloge astronomique de l'église cathédrale de Besançon*, Besançon.
Verlet 1987	Pierre Verlet, *Les bronzes dorés français du XVIIIe siècle*, Paris.
Vermot et al. 2014	Michel Vermot, Philippe Bovay, Damien Prongué, & Pascal Winkler *Traité de construction horlogèr*, Lausanne
Vever 1906–1908	Henri Vever, *La bijouterie française au XIXe siècle 1800–1900*, 3 vols., Paris.
Veyne 2005	P. Veyne, 'L'identité grecque contre et avec Rome', in *L'Empire gréco-romain*, Paris.
Veyrassat 1997	Béatrice Veyrassat, 'Manufacturing Flexibility in Nineteenth-Century Switzerland: Social and Institutional Foundations of Decline and Revival in Calico-Printing and Watchmaking', in Sabel & Zeitlin 1997, 188–237.
Veyrassat 2000	Béatrice Veyrassat 'Aux sources de l'invention dans l'Arc jurassien. Une approche par les brevets', in Belot, Cotte, & Lamard 2000, 69–76.
Veyrassat 2001	Béatrice Veyrassat, 'De la production de l'inventeur à l'industrialisation de l'invention. Le cas de l'horlogerie suisse, fin XIXe siècle–seconde guerre mondiale', in Gilomen, Jaun, Müller, & Veyrassat 2001, Zurich, 367–83.
Vial & Côte 1927	Eugène Vial & Claudius Côte, *Les Horlogers Lyonnais de 1550 à 1650*, Lyon.
Vida 1550	Marco Girolamo Vida, *Cremonensium Orationes III adversus Papienses in Controversia Principatus*, Cremona.
Vigniaux 1788	P. Vigniaux, *Horlogerie pratique, a l'usage des apprentis et des amateurs*, Toulouse (2nd edn. 1802).

Viladrich Grau 2002	Maria Mercè Viladrich Grau 'Medieval Islamic Quadrants for Specific Latitudes: Their Influence on the European Tradition', in Dorikens 2002, 73–108.
Vilhjálmsson 1991	Thorstein Vilhjálmsson, 'Time Reckoning in Iceland Before Literacy', in L. N. Ruggles (ed.), *Archaeoastronomy in the 1990s*, Loughborough, 69–76.
Vilhjálmsson 1997	Thorstein Vilhjálmsson 'Time and Travel in Old Norse Society', *Disputation*, 2, 89–114.
Villeumier 1928	Alfred Villeumier, *Cours de réglage destiné aux régleuses*, La Chaux-de-Fonds.
Vincent 1858	A. J. H. Vincent, 'Extraits des manuscripts relatifs à la géométrie pratique des Grecs', *Notices et extraits des manuscrits de la Bibliothèque Impériale: Académie des Inscriptions et Belles Lettres*, 19, pt 2, 157–431.
Vincent 2007	Clare Vincent 'Some Seventeenth-Century Limoges Painted Enamel Watch-Cases and Their Movements', *AH*, 30, 317–46.
Vincent, Leopold, & Sullivan 2015	Clare Vincent, J. H. Leopold, & Elizabeth Sullivan, *European Clocks and Watches in the Metropolitan Museum of Art*, New Haven.
Vitruvius 1969	*Vitruve, De l'architecture, livre ix*, Jean Soubiran (ed., trans., & commentary), Paris.
Vogt 1942	Susanne Vogt, Ebert Baldewein, der Baumeister Landgraf Ludwigs IV von Hessen-Marburg 1567–1592. Unpublished PhD thesis, Marburg.
Voltaire 1734/1937	Voltaire, *Treaté de Métaphysique 1734*, H. Temple Patterson (ed.), Manchester.
Von Archenholtz 1785	Johann W. Von Archenholtz, *England und Italien*, vol. 1, pt. 1, Leipzig.
Von Bertele 1953	H. Von Bertele, 'Precision Timekeeping in the Pre-Huygens Era', *The Horological Journal*, 96, 794–816.
Von Bertele 1954	H. Von Bertele, 'Early Clocks in Denmark', *The Horological Journal*, 96, 784–96.
Von Bertele 1955	H. Von Bertele, 'Jost Bürgis Beitrag zur Formentwicklung der Uhren', *Jahrbuch der Kunsthistorischen Sammlungen in Wien*, 51, 169–88.
Von Bertele 1961	H. Von Bertele, *Globes and Spheres*, Lausanne.
Von Bertele & Neumann 1963	H. Von Bertele & E. Neumann, 'Der Kaiserliche Kammeruhrmacher Christoph Margraf und die Erfindung der Kugellaufuhr', *Jahrbuch der Kunsthistorischen Sammlungen in Wien*, 10, 56, 74ff.
Von Castillon 1784	Johann von Castillon, 'Sur la gnomonique', *Nouveaux Mémoires de l'Académie Royale des Sciences et Belles-Lettres*, Berlin, 259–96.
Von Felten 1991	Luca von Felten, 'L'horlogerie dans le canton du Tessin', in Cardinal, Jequier, Barrelet, & Beyner 1991, 167–72.
Von Kues 1982	Nikolaus Von Kues, *Philosphisch-Theologische Schriften*, Vienna.
Von Littrow 1831, 1838	Johann von Littrow, *Gnomonik oder Anleitung zur Verfertigung aller Arten von Sonnenuhren*, Wien.
Von Mackensen 1991	Ludolf von Mackensen, *Die Naturwissenschaftlich-technische Sammlung. Geschichte, Bedeutung und Austellung in de Kasseler Orangerie*, Kassel.
Von Poppe 1799	Johann H. M. von Poppe, *Theoretisch-praktisches Wörterbuch der Uhrmacherkunst*, vol. 1, Leipzig.
Von Poppe 1822	Johann H. M. von Poppe, *Die Wand-, Stand-, und Taschenuhren. Der Mechanismus, die Erhaltung, Reparatur und Stellung derselben*, Frankfurt.
Von Stetten 1779	Paul von Stetten, *Kunst-, Gewerb- und Handwerks-Geschichte der Reichs-Stadt Augsburg*, Augsburg.
Voore 2005	Tim Voore, 'The Quartz Revolution', Electrical Horology Group Technical Papers, 71, n.p.
Vovelle 1995	Michel Vovelle (ed.), *L'homme des lumières*, Paris.
Vulliamy 1824	Benjamin Lewis Vullijamy, 'On the Theory of the Dead Escapement, and the Reducing it to Practice for Clocks with Seconds and Longer Pendulums', *Journal of Science, Literature and the Arts*, 16, 1–24.
Waddington 1998	Trevor Waddington, 'Archibald Miller Clockmaker and the Hammermen of Glasgow', *AH*, 24, 229–31.
Wadsworth 1965-66	Francis Wadsworth, 'A History of Repeating Watches', *AH*, 4, 142–76; 5, 194–201.
Wåhlin 1923	Theodor Wåhlin, *Horologium mirabile Lundense: det astronomiska uret i Lunds Domkyrka*, Lund.
Wåhlin 1932	Theodor Wåhlin, 'Astrolabe Clocks and Some Thoughts Regarding the Age, and Development of the Astrolabe', in Gunther 1932, 540–9.
Walker 1810	Ezekiel Walker, 'On Purifying Olive Oil for the Pivots of Chronometers', *The Philosophical Magazine*, 36, 81–5.
Walker 2012	Alicia Walker, *The Emperor and the World: Exotic Elements and the Imaging of Middle Byzantine Imperial Power, Ninth to Thirteenth Centuries C.E.*, Cambridge.
Walker & Barber 1895	G. Walker & W. N. Barber, 'The Theory of Timing', *The Horological Journal*, 37, n.p.
Waltham 1921	Waltham Watch Co., *The Story of the Waltham Watch*, Waltham, MA.
Wang 1986	Wang Lixing, 'Research on the Ancient System of the Chronometrical Nomenclatures Used in Historical Records of Astronomical Events', in Wang et al. 1986, 1–47.
Wang 1989	Wang Zhenduo, *Essay of Science and Technology Archaeology*, Beijing.
Wang et al. 1986	Wang Lixing et al. (eds.), *Anthology of Chinese Astronomy History*, No. 4, Beijing, Science Press.

Wang & Qi 2017	Wang Jin & Qi Haonan, *I am Repairing Horologes in the Forbidden City: The British Horologe*, Beijing.
Ward 1980	F. A. B. Ward, 'A Fifteenth-Century Italian Clockmakers Workshop', *AH*, 12, 172–4.
Watkins 2011	Richard Watkins, *Mechanical Watches: An Annotated Bibliography of Publications*. Extended edition, Parts 1 and 2, Tasmania.
Wartmann 1873	Elie-François Wartmann, 'Notice historique sur les inventions et les perfectionnements faits à Genève dans les champs de l'industrie et dans celui de la médecine', *Bulletin de la Classe d'industrie et de commerce de la Société des arts de Genève*, 101, n.p.
Way 1868	Albert Way, Ancient Sundials, Especially Irish Examples Illustrated by the late George du Noyer', *The Archaeological Journal*, 25, 207–23.
Wayman 2000	M. Wayman (ed.), *The Ferrous Metallurgy of Early Clocks and Watches – Studies in Post Medieval Steel*, British Museum Occasional Paper 136, London.
Weaver 1982	J. D. Weaver, *Electrical & Electronic Clocks & Watches*, London.
Webb 2017	P. A. Webb, *The Tower of the Winds in Athens: Greeks, Romans, Christians, and Muslims: Two Millennia of Continual Use*, Memoirs of the American Philosophical Society, vol. 270, Philadelphia.
Webber Jones 1946	Leslie Webber Jones (trans.), *An Introduction to Divine and Human Readings by Cassiodorus, Senator*, New York.
Webster 1975	Charles Webster, *The Great Instauration, Science, Medicine and Reform 1626–1660*, London.
Weidner 1924	E. F. Weidner, 'Ein babylonisches Kompendium der Himmelskunde', *American Journal of Semitic Languages and Literatures*, 40, 186–208.
Weiss 1982	L. Weiss, *Watchmaking in England 1720–1820*, London.
Weitzmann 1977	Kurt Weitzmann, *Late Antique and Early Christian Book Illustration*, New York
Welch 1978	Stuart Cary Welch, *Imperial Mughal Painting*, London.
Welper 1708	Johann Gabriel Doppelmayr (ed.), *Neu vermehrte, oder gruendlicher Unterricht und Beschreibung, Wie man alle reguläre Sonnen-Uhren auf ebenen Orten leichtlich aufreissen*, 3rd edn., Nuremberg.
Wenham 1853	F. H. Wenham, 'On a Method of Constructing Glass Balance Springs and their Application to Timekeepers', *Journal of the Society of Arts*, June 1853, 325–7.
Wensinck 1913–38	A. Wensinck, 'Miqât', in EI 1 1913–38, 3, 559–60.
Wenzel 1883	The Wenzel Pneumatic Clock Company, *The Wenzel Pneumatic Clock*, Washington, DC. Available at: http://www.survivorlibrary.com/library/the_wenzel_pneumatic_clock_1883.pdf.
Wetherhold 2012	Sherley Wetherhold, 'The Bicycle as Symbol of China's Transformation', The Atlantic, 30 June. Available at: https://www.theatlantic.com/international/archive/2012/06/the-bicycle-as-symbol-of-chinas-transformation/259177/.
White 1962	Lynn White Jr, *Medieval Technology and Social Change*, Oxford.
White 1969	Lynn White Jr, 'The Iconography of *Temperantia* and the Virtuousness of Technology', in Theodore K. Rabb & Jerrold E. Seigel (eds.), *Action and Conviction in Early Modern Europe: Essays in Memory of E. H. Harbison*, Princeton, 197–219. Reprinted in White 1978, 181–204.
White 1975	Lynn White Jr, 'Medical Astrologers and Late Medieval Technology', *Viator*, 6, 295–308.
White 1978	Lynn White Jr (ed.), *Medival Technology and Religion: Collected Essays*, Berkeley.
White 1989	George White, *English Lantern Clocks*, Woodbridge.
White 2009	George White, 'Not a Bad Timekeeper ... ', *AH*, 31, 621–6.
White 2012	Ian White, *English Clocks for the Eastern Markets. English Clockmakers Trading in China & the Ottoman Empire, 1580–1815*, Ticehurst.
White 2019	Ian White, *The Majesty of the Chinese-Market Watch. The Life and Collection of Gustave Loup ...*, London.
White 2017	Graham White, *Early Epicyclic Gears*, 3 vols., London.
Whitestone 1993	Sebastian Whitestone, 'A Minute Repeating Watch circa 1715: Friedberg's Ingenuity in a Biased Market', *AH*, 20, 145–57.
Whitestone 2008	Sebastian Whitestone, The Identification and Attribution of Christiaan Huygens' First Pendulum Clock, *AH*, 31, 201–22.
Whitestone 2010	Sebastian Whitestone, 'Minute Repeating in Tompion's Lifetime', *AH*, 32, 525–31.
Whitestone 2012a	Sebastian Whitestone, 'Christiaan Huygens' Lost and Forgotten Pamphlet of his Pendulum Invention', *Annals of Science*, 69, 91–104.
Whitestone 2012b	Sebastian Whitestone, 'The Chimerical English Pre-Huygens Pendulum Clock', *AH*, 33, 347–58.
Whitestone 2017	Sebastian Whitestone, 'Galileo, Huygens, and the Invention of the Pendulum Clock', *AH*, 38, 365–84.
Whitestone 2019	Sebastian Whitestone, 'Time before the Oscillator: Horology in the 13th-Century Manuscripts of the *Libros del Saber*', *AH*, 40, 487–500.

Whitestone 2020	Sebastian Whitestone, 'Revelation in Revision, How Alterations to a Woodcut Block Change the History of Huygens' Pendulum Clock Invention', *AH*, 41, 197–208.
Whitestone 2022	Sebastian Whitestone, 'Huygens' pendulum clock invention, a conclusive proof of its first printed image', *AH* March (forthcoming).
Whiston & Ditton 1714	William Whiston & Humphrey Ditton, *A New Method for Discovering the Longitude Both at Sea and Land, Humbly Proposed to the Consideration of the Publick*, London.
Whyte	Nicolas Whyte, 'The Astronomical Clock of Richard of Wallingford'. Available at: http://nicholaswhyte.info/row.htm
Wickersheimer 1979	Ernest Wickersheimer, *Dictionnaire biographique des médecins en France au Moyen Age*, 2 vols., and *Supplément* by Danielle Jacquart, Geneva.
Wiedemann 1909	E. Wiedemann, 'Uber Musikautomaten bei den Arabern', in *Centenario della Nascita di Michele Amari*, vol. 2, 64–185.
Wiedemann & Würschmidt 1916	E. Wiedemann & J. Wurschmidt, 'Über eine arabische kegelformige Sonnenuhren', *Archiv für die Geschichte der Naturwissenschaften und der Techniks*, 7, 359–76.
Wietrzyński 2018	Rafał Wietrzyński, 'Die Vereinigten Freiburger Uhrenfabriken (VFU)', *DGC*, 57, 59–78.
Wikander 2010	Johann Anton Wikander, 'Norwegian Medieval Sundials', in *T & R*, i, 307–12.
Wilkins 1648	John Wilkins, *Mathematical Magic; Or the Wonders that May be Performed by Mechanical Geometry*, London (reprint edn. London 1802/1970).
Wilkinson 1987	W. R. T. Wilkinson, *The Makers of Indian Colonial Silver*, London.
Willan 1970	T. S. Willan, *An Eighteenth-Century Shopkeeper: Abraham Dent of Kirkby Stephen*, Manchester.
Willan 1976	T. S. Willan, *The Inland Trade. Studies in English Internal Trade in the Sixteenth and Seventeenth Centuries*, Manchester.
Willard 1962	Charity Cannon Willard, 'Christine de Pisan's "Clock of Temperance"', *L'Esprit Créateur*, 2/3, special issue: 'Didactic Literature of the Middle Ages', 149–54.
Williams 2005	Chris H. K. Williams, 'Seventeenth- and Eighteenth-Century Clock Demand. Production and Survival, an Economic and Statistical Analysis', *AH*, 28, 571–83.
Williams 1933	Clare Williams, *Sophie in London, 1786: being the diary of Sophie v. la Roche*, J. Cape, London.
Williamson 1912	George Charles Williamson, *Catalogue of Watches, The Property of J. Pierpont Morgan*, London.
Williamson 1975	A. R. Williamson, *Eastern Traders*, London.
Willmoth 1993	Frances Willmoth, *Sir Jonas Moore, Practical Mathematics and Restoration Science*, Woodbridge.
Wilson 1622	John Wilson, *The Treasury of Devotion Contayning Divers Pious Prayers, & Exercises both Practicall, and Speculative*, Saint-Omer.
Wilson 1976	Gillian Wilson, *French Eighteenth-Century Clocks in the J. Paul Getty Museum*, Malibu.
Wilson et al. 1996	Gillian Wilson, David Harris Cohen, Jean Nerée Ronfort, Jean-Dominique Augarde, & Peter Friess, *European Clocks in the J. Paul Getty Museum*, Los Angeles.
Winter 1980	R. Winter, *Paulus Fabricius – ein Wiener Universitätsprofessor des 16. Jahrhunderts*, Vienna (unpublished).
Winter 1964	H. J. J. Winter, 'A Shepherd's Time-Stick, Nāgarī Inscribed,' *Physis: Rivista Internazionale di Storia dello Scienze*, 4, 377–84.
Wittenauer 1996	Jerry Wittenauer (ed.), *The Invar Effect. A Centennial Symposium*, Warrendale, PA.
Wolbrecht 1829	Georg Wolbrecht, *Die Uhrmacherkunst, vorgetragen in 30 Vorlesungen*, Leipzig.
Wood 1866	Edward J. Wood, *Curiosities of Clocks and Watches from the Earliest Times*, London
Woodbury 1958	Robert S. Woodbury, *History of The Gear-Cutting Machine . . .*, Cambridge, MA.
Woodward 2011	Philip Woodward, 'Middle Temperature Error', *The Horological Journal*, April, 156–8.
Woolf 2005	Daniel R. Woolf, 'From Hystories to the Historical: Five Transitions in Thinking about the Past 1500–1700', *Huntington Library Quarterly*, 68, 33–70.
Woolf 1925	Virginia Woolf, *Mrs Dalloway*, London.
Wordsworth 1897	Christopher Wordsworth, *Statutes of Lincoln Cathedral*, Cambridge.
Wormald n.d.	Patrick Wormald, 'Asser', in ODNB. Available at: https://doi.org/10.1093/ref:odnb/810.
Wosk 2010	Julie Wosk, 'Metropolis', *Technology and Culture*, 51, 403–8.
Wright 1989	Michael Wright, 'Robert Hooke's Longitude Timekeeper', in Hunter & Schaffer 1989, 63–118.
Wright 2000a	M. T. Wright, 'Greek and Roman Portable Sundials. An Ancient Essay in Approximation', *Archives for the History of the Exact Sciences*, 55, 177–87.
Wright 2000b	M. T. Wright, 'Moxon's Mechanick Exercises: Or Every Man His Own Clock-Smith', *AH*, 25, 524–31.
Wright 2007	Michelle Wright, *Time, Consciousness and Narrative Play in Late Medieval Secular Dream Poetry and Framed Narratives*. PhD Thesis, University of Glamorgan.
Wright 2019	M. T. Wright, 'An Early Use of Zinc in the Compensation Pedulum', *AH*, 40, 343–59.

Wu 1984	Wu Youchang, 'A Survey of the Clockmaking Industry of Yangzhou in the Late Qing Dynasty', *Nanjing Museum*, 7, 106–15.
Wu & Quan 2013	Wu Shouxian & Quan Hejun (eds.), *Ancient Chinese Astrometry and Astronomical Instruments*, Beijing.
Wulff 1966	Hans E. Wulff, *Traditional Crafts of Persia: Their Development, Technology and Influence on Eastern and Western Civilizations*, Cambridge, MA.
Wulff 1968	H. E. Wulff, 'The Qanats of Iran,' *Scientific American*, 218 (April), 94–105.
Wynne 1682	Henry Wynne, *The General Horological-Ring or Universal Ring-Dial*, London.
Xu 1963	Xu Guangqi, *Collected Works of Xu Guangqi*, Beijing.
Xu & Gao 2019	Xu Kun & Gao Bin, *Masterpieces of Antique Watches and Clocks*, Kulangsu Gallery of Foreign Artefacts.
Xue 2003	Xue Jixuan, *Collected Works of Xue Jixuan*, Shanghai.
Yagou 2019	Artemis Yagou, 'Novel and Desirable Technology: Pocket Watches for the Ottoman Market (Late 18th–Mid-19th c.)', *Icon*, 24, 78–107.
Yaguï-Betrán & Adam 2002	Adam Yagüi-Beltrán & Laura Adam, 'The Imprisonment of David Kinloch, 1588–1594 … ', *The Innes Review*, 53, 1–39.
Yan 2007	Yan Hong-Sen, *Reconstruction Designs of Lost Ancient Chinese Machinery*, Dordrecht.
Yan 2015	Yan Zonglin, *History of Communication between China and the West*, Taiyuan.
Yang 1987	Yang Boda, *Tributes from Guangdong to the Qing Court*, Hong Kong.
Yang 2007	Kerning Yang, *Entrepreneurship in China*, Aldershot/Burlington.
Ye 2008	Ye Nong, 'Guangzhou and the Western Clock's Trade in the Ming and Qing Dynasties', *Social Sciences in Guangdong*, 2, 128–35.
Ye 2019	Ye Nong (ed.), *Escritos de Diego de Pantoja, S. J.*, Guangzhou/Guangdong.
Young 1939	Suzane Young, 'An Athenian Clepsydra', *Hesperia*, 8, 274–84.
Yun 2008	Yun Limei, 'British Timepieces', in *Timepieces in the Imperial Palace*, Beijing.
Zanetti 2014	Cristiano Zanetti, *The Microcosm: Technological Innovation and Transfer of Mechanical Knowledge in Sixteenth-Century Habsburg Empir*, The Medici Archive Project, New York.
Zanetti 2016	Cristiano Zanetti, *Janello Torriano, A Renaissance Genius*, Cremona.
Zanetti 2017	Cristiano Zanetti, *Janello Torriani and the Spanish Empire: A Vitruvian Artisan at the Dawn of the Scientific Revolution*, Leiden.
ZBYJSa	ZBYJS: Xi'an Horological Research Institute of Light Industry Corporation, Ltd. 'Company Profile'. Available at: http://www.zbyjs.cn/About.asp?.Types=2轻工业钟表研究所.'企业简介'.
ZBYJSb	ZBYJS: Xi'an Horological Research Institute of Light Industry Corporation, Ltd. 'Memorabilia'. Available at: http://www.zbyjs.cn/About.asp?Types=217轻工业钟表研究所.'大事记'
Zech 2019	Heike Zech, 'Ein neuer Blick auf die Burgunderuhr. Forschungsgeschichte und perspektiven', *DGC*, 58, ch. 5.
Zeeman 1978	J. Zeeman, *De Nederlandse Stoelklok*, Assen/Amsterdam.
Zeeman 1996	J. Zeeman, *De Nederlandse Staande Klok*, Zwolle.
Zeitlin 2006	Jonathan Zeitlin, 'Districts industriels et flexibilité de la production hier, aujourd'hui et demain', in Lescure 2006b, 447–72.
Zeitlin 2008	Jonathan Zeitlin, 'Industrial Districts and Regional Clusters', in Jones & Zeitlin 2008, 219–43.
Zek & Smith 2005	Yuna Zek & Roger Smith, 'The Hermitage Peacock. How an Eighteenth-Century Automaton Reached St Petersburg', *AH*, 29, 699–715.
Zerella y Ycoaga 1791	Don Manuel de Zerella y Ycoaga, *Tratado General y Matematico de Reloxeria*, Madrid.
Zerubavel 1981	Eviatar Zeubavel, *Hidden Rythms: Schedules and Calendars in Social Life*, Chicago.
Zervos 1935	Christian Zervos, *L'art en Grèce des temps préhistoriques au début du XVIIIe siècle*, 2nd edn., Paris.
Zhan 2010	Zhan Xiaobai, The Study of the Changing Time System and Concept of Time in Modern China. PhD thesis, Renmin University of China, Beijing.
Zhang 2012	Baichun Zhang, 'The Transmission of European Clock-Making Technology into China in the 17th–18th Centuries', in Koetsier & Ceccarelli 2012, 565–77.
Zhang G. 2019	Zhang Guogang, *A General History of Sino-Western Cultural Relations*, Peking
Zhang 2014	Zhang Li, '"Three-Five" Brand Clock: Legend of Chinese Clock Industry', *Eastday* [online]. Available at: http://history.eastday.com/h/shlpp/u1a7874954.html.张励. '"三五"牌时钟：中国钟表行业的传奇'.
Zhang 2016	Zhang Xialing, *General History of Chinese Timing Instruments, Modern and Contemporary Volumes*, Hefei. 中国计时仪器通史 近现代卷
Zhang Y. 2019	Zhang Yantian, *Study of Ancient Chinese Timing Recording System*, Shanghai.
Zhang & Zhang 2019	Zhang Baichun & Zhang Jiuchun, 'A Review of the Reconstruction of the Astronomical Clock-Tower', *Journal of Dialectics of Nature*, 41, 47–55.

Zheng 2001	Zheng Xiyuan, *China in The New York Times*, Beijing.
Zhu et al. 1999	Zhu Huisen *et al.* (eds.), *Annotation of the Draft of History of Qing Dynasty*, No. 2, Taipei.
Zhu 2001	Zhu Weizheng (ed.), *Chinese Works of Matteo Ricci*, Shanghai.
Zhu 2018	Zhu Yancun, 'A Brief Explanation of the Evolution and Definition of Guang's Clock in the Qing Dynasty', *Horology of Guangdong*, 4, 45–6.
Ziegeltrum 2018	Francis Ziegeltrum, 'Une histoire illustrée des échelles gnomoniques', *Cadran Info*, 37, 141–54.
Ziner 1930	Ernst Zinner, '*Horologium viatorum*', *Isis*, 14, 385–97.
Zinner 1939	Ernst Zinner, 'Die Ältesten Rädenuhren und modernen sonnenuhren. Forschungen den Ursprung der modernen Wissenschaft', *Gericht der Naturforschenden Gesellschaft in Bamberg*, 28, n.p.
Zinner 1954	Ernst Zinner, *Aus der Frühzeit der Räderuhr*, München.
Zinner 1957	Ernst Zinner, 'Die Planetenuhren von Dondi und Regiomontan', *Die Uhr*, 21, 18–21.
Zinner 1964	Ernst Zinner, *Alte Sonnenuhren an Europäischen Gebäuden*, Wiesbaden.
Zinner 1976	Ernst Zinner, *Deutsche und Niederländische astronomische Instrumente des 11.–18. Jahrhunderts*, Munich.
Zinner 1979	*Deutsche und Niederländische astronomische Instrumente des 11.–18. Jahrhunderts*, Munich.
Zinner 1990	Ernst Zinner, *Regiomontanus: His Life and Work*, Ezra Brown (trans.), Amsterdam.
Zou 2009	Zou Boqi, *Posthumous Manuscript of Zou Boqi: Commemorating the 190th Anniversary of Zou Boqi's Birthday*, Guangdong.
Zubal 1989	Ulrike Zubal, 'Die Werkstatt Philipp Matthäus Hahns und die Aufhebung sder Uhrmacherzunft in Württemberg', in Väterlein 1989, 391–402.
Zunino 1989	Florence Zunino, *Les sociétés horlogères de l'arrondissement de Besançon, 1883–1914*. Unpublished MA thesis, Université de Franche-Comté.
Zuzzeri 1746	Giovani Luca Zuzzeri, *D'una antica villa scoperta sul doso del Tusculo, e d'un antico oroloio a sole tra lerovine ritrovato*, Venice.

INDEX

A General History of horology – index entries suffixed *t* refer to tables, *f* to figures and n to notes

A

Aachen, clocks 177
Abbas, Shah of Persia 37
Abel, Clarke 51n
Abî al Farāj 'Īsā 28, 86
Abū Bakr al-Ṭabarī 83
Abū Ḥammū 98
Académie Royale des Sciences (Paris) 637
 Mémoires de 630
accessories 555, 557–8
 bells and gongs 558
 chains 560–1
 chatelaines and necessaries 559
 dust caps 199, 558–9
 imitation watches 560
 watchcases 556–7, 562–7
 watch stands 561–2
Account of the going of Mr Harrison's Watch (Maskelyne) 630
Accutron wristwatches 492, 526
açenna 126
acrylic crystals 486
Adam, Urbain 246, 365
al-Ādamī, Abū 'Alī al-Ḥusayn ibn Muḥammad 83
Adams, George 572, 573–4
Adamson, John 426
 watch 427f
Adelsberger, Udo 507
Adès, Dawn 615
adjustable twin-pallet escapements 181
Adler Planetarium and Astronomical Museum, Chicago 29n, 115n
The Adventure of the Seven Clocks (Conan Doyle) 616
advertising 227
Aegler, Hermann (Aegler Bienne) 489
Aelianus Asclepiodotus 19
æolopiles 594
aesthetics 421

after-sales 541, 543
Aghia Triada, sundial 107, 108
Aghios Laurentios, sundial 108
Aghjots Vank, sundial 108
Agora (Athens) 7
Ahaz dial 239, 240, 644–5
Aḥmad Mukhtar Pasha 94
Aḥmad Pasha 92
Ahmet Ziya Akbulut 94
Ahya'i, M. A. 94
air brakes 144, 289, 653
air-density compensation 403
Airy, George Biddell 407, 409
 Airy's bar 414
 detached detent escapement 408f
 electric clock system 497
Akbar 35
Aked, Charles 415, 525
Aktiengesellschaft für Uhren-fabrikation, Lenzkirch 379, 398
alarm clocks
 England 355
 France 363, 399
 Germany 378, 388
 guard-tower clocks 153
 United States 358
 water clocks 122
alarm mechanisms 193
Albert chain 474, 561
Alciat, André 251
Alfonsine tables 258, 269f
Alfred the Great 135
Algarotti, Count 456
alidades 249, 653
All British Escapement Company (ABEC) 350
Allacci, Leone 646
The Allegory of Good Government (Lorenzetti) 600
Allegory of Vanity (de Pereda) 605, 605f

Allen, Thomas 599n
Allexandre, Jacques 330, 624, 627, 643, 648–9
Allgemeine Electicitäts-Gesellschaft (AEG) 388
Allgemeine Journal der Uhrmacherkunst 551
Almadén 133
Allix, Charles 651n
Almagest (Ptolemy) 253, 261, 621
al-Ma'mūn 83
Almanach Dauphine 595n
Almanach de la Fabrique de Paris 641
Almanach des Horlogers (Gros) 641
Almanack; or, Diary, astronomical, meteorological, and astrological (Knapp) 641
almanacs 641–2
Almanus, Paulus 167, 542, 623
 Manuscript 311
almucantar-quadrants 91, 92, 93f
Alonso de Santa Cruz 161
Alpirsbach Abbey sundial 238
altitude dials 118f, 231, 232, 233f, 653
 portable 232–7
Amant, Jean-Louis 150
Amant, Louis 181
American clockmaking 355–7
American Horological Society 634
American lever watches 359
American Watch Company 359, 360, 360f
 waterproof wristwatches 487
American Watchmakers-Clockmakers Institute 635
American watchmaking 358–60
American-style movements 379
Amfissa, sundial 107, 108
Amigoni, Jacopo 296
Amman, Jost 218
ammonia-molecule clock 505f

Amory, Edward 296
Amsterdam
 fritillary form watch 190
 Huguenots 428
 longcase clocks 176–7, 183, 429
 metronomes 582
 Rijksmuseum 177n, 189n, 426n, 431n
 watchmaking 210–11
anachronism 606n
Analemma (Ptolemy) 77n, 84
analemmatic dials 82, 88, 231, 242, 243f, 653
anaphoric clock 11–12, 12f
Anaximander of Miletus 5–6, 5n
Anaximenes 5, 643, 644
anchor escapements 142, 173n, 407, 584, 645, 653
 superseding verge 150, 171, 173, 318, 463, 549
al-Andalus, sundial in 84–5
Andreas Haller company 388, 388n, 390
androids 299–300, 653
Andronikos Cyrrestes 21, 23
Anglo-American Watch Company 361
Anglo-Celtic Watch Company 352–3
angular values, Erfurt rule 120t
animal clocks 434f
animated figures
 cathedral clocks 292
 musical automaton clocks 294
 water clocks 290
Annales de la Société Polytechnique 637
Annuaire Artistique et Historique des Horlogers (Saunier) 641
Annuaire de l'horlogerie Suisse 642
Anquetin, Modeste II 344
anti-flexure balance rims 410
Antikythera Mechanism 254n
Antioch, *horologion* 94

INDEX

Antiquarian Horological Society (UK) 553
antiquarian horology, appreciation of 549, 552, 553, 652
antique clocks 552
Antiquity, time measurement 1, 24–5, 645
   Ancient Greece 5–7
   Ancient Rome 7–9
   calenders 20–1
   horologium 9–17
   makers and methods 17–20
   Meridian of Augustus 23–4
   oriental origins, time measurement 1–5
   Tower of the Winds 21–3
Antiquity scenes, neoclassicism 436–7
anthropomorphism 613–14
Antwerp, Musée des Beaux Arts 157n
Anxious Moments: A Sick Child, Its Grieving Parents, a Nursemaid and a Medical Practitioner (Walton) 613
aperture-gnomons 13, 14f, 19, 244, 653
Apian, Peter 116, 234n, 237, 237n, 255
Apollonius of Perge, mathematician 84
Apollonius of Tyana 79, 79n, 95
Apollonius the carpenter, the geometrician 79, 95
apostles 110, 276, 293
apostolic clocks 282
apparent time, mechanism 276
apprenticeships 219, 219n
apsides 282, 284, 653
Aquileia, sundials 19
al-ʿArabī, Muḥammad 96
Arabic sundials 81–94
Arago, François 342, 578
Aragon, Louis 367
arbors 126, 143, 193, 197, 589, 653
   roasting-jacks 594
   spring-driven clocks 156–7, 160
   vertical clocks 166, 166n, 167n, 168
arched dials, longcase clocks 176
Archimedes 11
   On Floating Bodies 94
Archinard, Margarida 116
Arctos Uhrenfabrik Philipp Weber KG 388
argument from design 607n
Argumentum Horologii 117
Arimasa Tatsunosuke Iyo 72, 73f
aristocracy, musical and automaton clocks 294–5
Aristophanes 7, 20n
Armagh Observatory 403, 407
armaments industry 382
armillary instruments 44, 570, 574
armillary sphere 45f
Armstrong, Joe E. 7
Army Trade Pattern (ATP) wristwatches 490

Arnault, Henri de Zwolle 154–6, 156n, 157n, 160, 266
Arnold, John 205, 209, 333–4, 335
   bi-metallic balances 410
   double S balance 202
   finger ring watches 474
   marine chronometer 337f, 405, 630
   publications 629–30
   watch No. 1/36 336
Arnold, John Roger 205, 405, 406, 410
L'Art de l'horlogerie enseigné en trente leçons 399
Art Deco
   and Art Nouveau 440–1
   digital readout wristwatches 490
   watch stands 562
   watches 560, 561
L'art du facteur d'orgues (Bedos) 290
L'Arte dell' orologeria in Europa, sette secoli di orologi meccanici (Brusa) xii
Art Nouveau, 440
   watch stands 562
art objects 555
artefacts from Antiquity 94
Arthaśāstra of Kautilya 27
The Artificial Clockmaker (Derham) 543, 626
artificial jewel-stones 376
artistic and literary interpretations 599
   13th–15th centuries 602
   16th & 17th centuries 604–9
   18th & 19th centuries 609–15
   20th & 21st centuries 615–19
Arts and Crafts movement 552
Arundel, Alethea, Countess of 189
Āryabhaṭa 32
al-Aṣbāḥī 81
The Ascent of F6 (stage play) 619
Asclepiodotus, Aelianus 19
Ascoli Piceno, cathedral sundial 103
Ashmolean Museum, Oxford 186
   book clocks 423
   form watches 423n
   table clock 182n
   watch cases 425
   watches 187n, 188n, 190, 191n, 194n, 196, 198, 425, 430n
al-Ashraf Shaʿbān, Sultan 89
aspect relations 267, 653
Aspinwall, Samuel 221
Aspinwall, Thomas 219n
Aspreys, jewellers 474, 559
assay offices, Switzerland 373
Asser, *Life of Alfred* 136
assortiment 367
assortments 653
astral sciences 254, 266n
astrarium 129, 132, 263, 266, 285
Astro-Chron clocks 388, 525
astrolabe-quadrants 91, 92, 93f, 234, 653
astrolabes 234, 267n, 653

early public clocks 148
   Europe 127, 132
   Festo astronomical clock 285
   Islam 82, 83
   planetary clocks 258
astrology 148, 267
Astron, Seiko watches 74, 75, 526
astronauts' wristwatches 491
astronomical clocks 130–2, 141–4, 148–50, 177, 365
   Beauvais 278, 365
   Besançon 278, 365
   Bourges 148–9, 281
   Chartres Cathedral 126, 148
   Chiaravalle 143, 146
   China 55
   Clusone, Lombardy 471
   exhibition clocks 281
   Hampton Court 281
   India 38
   Japan 68
   Lübeck 281
   Lund 281
   Lyon, St Jean Cathedral 148
   Macerata 281
   Mantova 281
   Messina, Sicily 469
   Münster 281
   Norwich Cathedral 129, 130, 139
   Olomouc (Olmütz) 281
   Padua 141
   Prague 126, 149–50, 281
   public clocks 273–80
   reconstructions 280–1
   repairs 543
   smaller clocks 281–5
   St Alban's Abbey/Cathedral 130, 132, 144
   Strasbourg 263n, 469, 623
   Venice 281
astronomical hours 243
astronomical precision measurements 311–12
astronomical regulator 281, 335, 410, 414, 499, 501, 503, 504f
astronomical ring dials 232, 237
astronomical timepieces
   China 43–4
   Japan 68
astronomical timers 578
astronomy
   China 41
   decimal time 341–5
   Egypt 89
   Islam 83, 89
   sundials 231
ASUAG 376
Athenaeus 11, 20n
Athens 7
Atkinson, Robert d'Escourt 280
ATO electric clocks 522–3
atomic clocks 419, 508–9

atomic frequency standard (AFS) 508
Aubrey, John 599n
Aubriot, Olivia 36
Auch, Jacob 633
Auden, W. H. 618
Augsburg
   centre of trade 228
   clockmaking 168, 218
   direction dials 237
   horological masters 225
   musical automaton clocks 294–6, 302
August, Elector of Saxony 254, 266
Augustus Caesar
   Meridian of Augustus 23–4
   Roman calendar 24
Aulenbacher, Gerhard 622n
Aumerle, Marguerite Elisabeth 532
Aurelius a San Daniele (Michael Fras) 273
Australia xii, 249, 281, 348, 349, 352, 356, 360, 513, 521, 522
automata/automaton 147–8, 289, 653; *see also* musical and automaton clocks *and* planetary clocks
   Beauvais astronomical clock 278
   Byzantium 94
   China 445
   dulcimer player 300
   exports to East Indies 451
   flute player 299
   harpsichord player 299
   religious themes 293
   Strasbourg clock 292–3
   watches 307–8
automatic light switching 527
automatic winding, watches 209, 489, 492
automatic wristwatches 352
Automatical Museum, London 301
automaton clocks *see* musical and automaton clocks
automaton watches 307–8
The Awakening Conscience (Hunt) 613
al-Awzāʿī 81
Ayasse, J-J-M 549–50, 633
azimuth dials 231, 241–2, 653
   Islam 85
   medieval Latin world 110
Azur 641

B

Bābur 35
Babylon, time measurement in 4–5
Babylonian hours 105, 119
Bach, Henri 280
Bacher, A. 474
Baden, Black Forest 378–80, 382
Bagheri, Muhammad 94
Baghdad
   automata 122

INDEX

Baghdad (continued)
   Mustansiyya college water clock 96
   new city plan 82
Baillie, Granville Hugh 314, 629, 630, 631, 632, 636
Bailly-Comte company 464
Bain, Alexander 414, 503, 514, 519
balance cocks 653
   spring-driven clocks 154, 160
balance spring 209, 210, 316–23, 324, 430, 543
   English watchmaking 196–7
   isochronous 331, 332–3
balances 129, 132, 513
   American clocks 356
   astronomical clocks 274
   brass 405
   compensation 323–9, 332, 334, 407–10, 481
   electrical contacts 520, 523
   integral balances 334, 415, 417
   Japanese clocks 66
   non-magnetic 374n
   spring-driven clocks 154–7
   travelling clocks 182
   watches 197, 350, 627
balanciers 395
balata 85
Baldewein, Eberhard 254, 257–63, 264t, 266, 267, 270
Baldewein I planetary clock 254, 257, 260, 264t, 266–7, 269, 270, 270n
Baldewein II planetary clock 254, 257, 258, 261f, 260, 264t, 266–7, 270, 271
Balestri, Domenico 314
ball-dropping candle lamp 136
ball-dropping water clocks 95, 96
ballistics 575
Baltazar, Charles 47, 184f
Baltimore, Walters Art Gallery 186n, 423n, 426n
Bandel (Bengal), clock tower at 38
Banks, Sir Joseph 335
Banū Mūsà ibn Shākir 135, 291
Banū Mūsà 83
Baoshi Clock Factory 53
Barapullah Flyover, New Delhi sundial near 40
Barbaro, Danielo 231, 239
barillet movement 308–9
Barlow, Edward 174n, 198
Barlow, Peter 631
Baronneau, Joseph 430
baroque style 176, 425–31, 562
Barraud, 405, 550, 591
barrels, automata 289
Barrett, Rose 614
Barrier, René 584
Barrington, Daines 649, 650
bars, watchcases 488

Bartholomew, clockmaker at St Paul's 139, 542
Barwise, John 474
Basel/Basle 3, 146
   Fair 353, 489, 525
   Historical Museum 305
bassine 557, 653
Bate, John 127
al-Battānī 84
battery-driven watches 362
Baudelaire, Charles 614
Baudrillard, Jean 555n
Baugrand, Gustave 438
Bauhaus school 441
Baullier & Fils 438
Baume & Co 486
Bayard company 364, 399
Bayt al-Ḥikma 83
BBC, six-pips broadcast 499
beadwork pouches 564
Beale & Co 459
Beard, James Henry 614
Beaulieu Abbey clock 129
Beaumarchais *see* Caron, Pierre Auguste
Beauvais, Antoine 421–2
Beauvais astronomical clock 278, 279f, 287, 292, 365
Bechmann, Roland 123n
Becker, Gustav 379
Beckett, Edmund 519, 629
Beckmann, Johann 650
Bede 77, 99, 105
Bedford Museum 143
Bedos de Celles, François 238n, 244, 246–7, 290, 622, 624
Beg, Ulugh 312
Begg, Samuel 500f
Beijing Watch Factory 55
Beliard, François II 225n
Bell, Benjamin 221
bell founders 215
bell music 292
bell ringing 145
bell top case 179, 653
bell towers 144
Bellanger, François Joseph 435
bellows 296, 300, 304, 653
bells 145, 296; *see also* Big Ben
   watches and clocks 558
Belville, John, and family 497
Benares College, hours/ghatis 38
Benedetti, Giovan Baptista 240
Benedetto de Blachis 146
Bengg, Paulus 195
Bennett, Edward Bright 596
Bennett, Edward Holmes 596
Bennett, John 348, 410
Benoît, Achille 550
Bérain, Jean 430
Bergauer, Michael 249
Berggren, J. L. 77, 82

Berlin, Ägyptisches Museum 2n
Berlin, network infrastructures 511, 516f
Berlin, Staatliche Museen, 436n
Berlin Astronomical Yearbook 641
Berlioz, J. 438
Bernard, Augusto 466
Bernardin, Constant Flavien 278
Bernoulli, Daniel 134n
Berosus 644
Berthoud, Ferdinand xii, 178, 181, 207, 274, 325, 329, 404, 435, 627–8, 649
   decimal time 342
   Encyclopédie 391
   equation of time 212
   historical survey 650
   How to Manage and regulate Clocks and Watches 545–6
   longcase clocks 435, 649n
   marine chronometer 628f
   marine timekeepers 330, 333, 337
   publications 630
   specialized trades 556
Berthoud, Jean Henry 207
Berthoud, Louis 337–8
   decimal time 342
Besançon 368
   astronomical clock 278, 287, 365
   female workers 534, 536–7, 537f, 539
   horological school 631
   Musée du Temps 538n
   watchmaking 363
Bettwiller (Alsace), sundial 19
Beuriot 626
Beurnier company 366
Beuve, Alexandre 595
Bevagna, sundial 9f, 9
Bharatpur 35
Bhāskarācārya 32
Bidermann, Samuel 296
Bienaimé (Fournier) 583
bi-filar dials 251
Big Ben 403, 464, 465f, 468–9
   electric telegraph time signals 496, 497
   Mrs Dalloway (Woolf) 615
   standardized time 496, 514
Billeter, C. 635
Billiard, Lewis 141
bi-metal 324, 326, 334, 407, 654
   wristwatches 481
Bion, Nicholas 241
Birch & Gaydon 474
Bird, John 329
Bird, Rufus 38n
birdcage frames 142, 143, 463, 466
al-Bīrūnī 34, 35, 77, 81, 133
Black Forest, clockmaking 184, 226, 378–9, 386–7
   trade catalogues 640

black marble clocks 393f, 394f, 396, 398
blacksmiths 215
Blaeu, Willem Janszoon 570
Blake, Maurice 404
Blakey, William 532
Blanc, Henry 441
Blancpain Fifty Fathoms 491
blanc-roulants 391, 391n, 395, 397, 399
blancs 366, 391, 654
   German watchmaking 387
   marine chronometers 363
   Swiss manufacturers 372
blessing automata 48
Bletsoe triangular frame clock 143
blocked watch wristlets 475
Blois
   clockmaking 164
   painting on enamel 225, 426
   watches 198, 423, 424f
   watchmaking 188–9, 228
Bloxam, James 403, 469
Blundeville, Thomas xii–xiii
Board of Longitude 203, 327–3, 337, 403, 405–6, 541
   disbanding 406
   Explanation of Time-keepers constructed by Mr Thomas Earnshaw and the late Mr John Arnold 630
   reports 629, 630
bob *see* pendulum bob
Bobynet, Pierre 136, 244
Bockelt, Matthys 191
Bockelts, Jan Janssen the elder 190, 190f
Bode, Johann E. 641
Bodet, Paul 366
Bodet company 366
Bodo of Hardessen 141
Boeotia, sundial 108
Boer War, wristwatches 478–9
Boethius 645
Bogaert, Gaston frontispiece
Bohemus, astrolabe clock 423
Böhm, Herman 440
bolt-and-shutter 173, 175f
Bombay 39, 522
bombé 654
Bond, William Cranch 406
Bond of Boston 414, 499
Bonnant, Georges 49n
Bontems family 310
Book of Christian Prayers (Day) 600n
Book of Knowledge of Ingenious Mechanical Devices (al-Jazarī) 291
The Book of the Balance of Wisdom (al-Khāzinī) 133–4
Book of the Water Clock (Rico y Sinobas) 125
Book on the Description of the Instrument which Sounds by Itself (Banū Mūsà) 291

INDEX

book-clocks 423
Booth, Mary L. 634
Boquel, Joseph 399
Borda, Chevalier 342, 342n, 344
Bordes, Philippe 610
Bordier, Jacques 427
Borel-Jaquet, Abram 222
Borer, Emile 489
Borgel, Francois 484
Borgel screw case 484, 485f, 487
Borghesi, Francesco 273
Borletti 588
Boromée Delépine & Lanchy 396
Borrell, Henry 302–3
Borrell, Jean-Henry 448, 462n
  musical clock 450f
Borrelli, Giovanni Alfonso 576
Borsendorff, Louis-Andre 641
Boston Museum of Fine Arts 35n, 610n
Boston Watch Company 359, 467
Botticelli, Sandro 605
bottle-jacks 596
Boucher, François 431, 432, 610
Boucheron, Frederick 440
Bouguet, David 189, 194
Boulle, André Charles 176, 430–1
  floor-standing clock 432f
Boulton, Matthew 448, 591
Bourdon, Pierre 429
Bourdon, Sebastian 426
Bourg-en-Bresse, analemmatic dial 242
Bourges Cathedral, astronomical clock 148–9, 281
Bouquet, David *see* Bouguet, David
Boursier, Charles 251
Bovet, Edouard 49, 50f, 456n
Bowell, George Bennett 518–20
Bowes Museum, Barnard Castle 194n, 451
Bowrey, Thomas 37
Bowyer, William 169, 169f, 172
bracelets, leather 566
brachiolus 235, 654
bracket clocks, Japan 66
Bradley, Langley 149
Bradshawe, Ellis 221
Brahe, Tycho 127, 312
Brahmagupta 28
Bramer, Paulus 177
Bramley, Frank 614
Brandt, Félicité 533, 534f
brass 166
  balances 405
  clockmaking/watchmaking 168, 168n, 188, 220, 222–3, 227
Braunschweig cathedral, sundials 120n
Brave New World (Huxley) 615
Breguet, Abraham-Louis 182, 202, 208–10, 229, 338, 342, 436, 437–8
  bells, in watches 558
  chronographs 578
  decimal time 342
  forgeries 214
  glass balances 405
  keys 566
  lubrication 551
  Marie-Antoinette watch 565
  metronome 582
  precision regulators 404
  shock resistant wristwatches 488
  twin-barrel chronometers 404
  watch 437f
  wristwatch 473
Breguet, Antoine Louis 342
Breguet, pupil of 628
Breitling 588
Brentel, Georg 231
Bresseo, Villa Cavalli clock 144
Breviloquium de Virtutibus (John of Wales) 602
Brewster, David 506
bridle wheel 132
Briggs, Henry 239
Brisset family, casemakers 197
Brillié Frères 517
bristles, escapement development 162
Britain; *see also* England
  18th century publications 629
  19th century horological industry 347–9
  20th century horological industry 347–53
  decline of horological industry 353–4
  exports to China 445–7
Britannia Act 1697 198n
British Army
  wristlets 475f, 474
  wristwatches 482, 484f, 490
British Clock and Watchmaking Company 348
British Clockmakers Association 351
British East India Company *see* East India Company
British Horological Institute (BHI) 348, 550–1
  essay prize 631
  formation 410
  Horological Journal 638
  publications 632
  Watch and Clock Year Book 642
British Jaeger 588
British Library 35n
British Museum
  astronomical clock 282
  automaton watches 308n
  coach watches 206n
  dual time watches 210n
  forgeries 213
  James Cox watch 213n
  lantern clocks 66
  lever escapements 201n, 202
  minute-repeating watches 199
  musical and automaton clocks 294n, 295n
  organ clock 303
  pendulum clocks 429n
  perpetual calendar watches 204
  pocket chronometer 205
  sinking bowl water-clocks 4n
  souscription watches 209n, 214
  table clocks 157n, 158n, 159n, 167n
  watchcases 426
  watches 189n, 191n, 193n, 194n, 195n, 196, 212n, 424n, 425n
British Sundial Society 251n
British Watch and Clock Making Company 348
British Watch and Clockmakers' Association 351
British Watch Company 358n
Britten, F. J. xii, 551, 634, 651
Brockbanks, John 337
Brocot, Achille 549
Brocot family 391–3, 399
  black marble clock 394f
  escapement/suspension 391–3, 394f, 399, 549, 552
broken arch pediments 180, 180f
brooch watches 560f
Brookes, Edward 174
Brown, David 4n
Browne, Sir Thomas xiii, 644
Bruce, Alexander 315
Bruguier family 310
Brunelleschi, Filippo 143, 154, 157n, 167
Brusa, Giuseppe xii, 281, 285
Brussels, carillon 291
Bucharest, National Archaeological Museum 19
Bucher, Hans 260, 261f
Buchner, Alexander 289
Buchner, Edmund 24
buckles 561
Buddhagosa 35
Buddhist temples 60
Bugge, Thomas 335
Bulcke, Jacques 191
Bull, Edmund 194
Bull, John 548
Bull, Randolf 191
Bullant, Jean 231
Bulle electric clocks 523
Bulletin de la Société d'Encouragement pour l'Industrie Nationale 630
Bulova 375
  Accutron wristwatches 492, 526
Bulova School of Watchmaking 635
Bundi 35
Buonsignori, Stefano 239
buoys, time-switches 527
Burchall, Adrian 591
Bureau International de L'Heure (BIH), Paris 499
Buren factory (Williamsons) 588
Burghley House 591
Bürgi, Jost 159, 159–60n, 162, 271, 318
  Vitruvian artisan 266
Burgundy
  clockmaking, 16th & 17th century 164
  state, formation of 157n
Bürk, Johannes 379
Burnap, Daniel 223, 355
Burr, William Henry 614
Bury St Edmunds Abbey, *horologium* 123
Bury St Edmunds, John Gershom Parkington Memorial Collection 30n
Busbecq 644
Buschmann, Hans II 423
bushing 545
Bushman, John 198
Butterfield, Michael 238
buttonhole watches 561
al-Buzjānī, Abū al-Wafā' 84
BWCMA 351
Byzantium
  sundials 77–80
  water clocks 94–5

C

cabinet of curiosities/*cabinets de curiosités* 163, 294
Cadenzia metronome 584
cadratures 224, 627
Cady, Walter Guyton 507
Caen, Musée des Beaux Arts 646n
caesium atomic clocks 419, 508–9
caesium fountain clocks 419
Caffieri, Jacques 431, 435
caged roller-bearing 324
Caillart, Jacques 426
Caillatte, Catherine 557
Cairo 88, 89
  Museum of Egyptian Antiquities 1n
Cajetano, David a Sancto (David Ruetschmann) 273, 624
calculating programs 251
Calcutta 39, 278, 455n, 522
calendar, Egyptian decadal 341
calendar change, Japan 72
calendar clocks 286
calendar dials 254, 277, 279
calendar keys 566
calendar watches 190f, 193
Calendarium Romanum (Stöffler) 238
calendars 20–1, 286
  ancient China 41n
  Chinese mechanism 43
  Egyptian 341
  Roman 24
  Roman errors 8

INDEX

cam 654
Cambridge, Whipple Museum 341n
Cambridge chimes 469
Cameel, Caspar 424
camembert of time (Dali) 615
Campani brothers 133, 180
Campani, Giuseppe 163
campanile 144
Campanus de Novara 254
Campbell, R. 221, 223, 224
cams, automata 289, 300, 304
Camus, Charles 411
Camus, François Joseph de 321
candle clocks 96, 135–6
Cannes, Nautical Club of 344
canonical hours 102, 102n
    sundial 110f
Canonico family 466
Canterbury Cathedral, clock 129, 139
Canterbury pendant 115, 115f
Canton (Guangzhou) 444–7, 451, 454–5, 456ff, 522
Capriglia, Guiseppe di 623, 624
    turret clock 625f
Capt, Henry Daniel 308, 437
Capt & Piguet 437
capuchin 116, 236
capucines 182
car clocks 350, 586
    electromechanical clocks 588–9
    future developments 590
    mechanical clocks 586–8
    quartz clocks 589
Carafa 243
Cardan, Jerome 136
Cardano, Girolamo 644
carillon clocks 145
    baroque style 429
carillons 150, 291–2, 296
carillons, water clocks 123
Carl Werner company 586
Carlin, Martin 435
Caron, Pierre Auguste 207
Carpenter, Alexander 293
Carpenter, William 302, 305
Carracci, Annibale 604
carriage clocks
    Armand Couaillet 399
    St Nicolas d'Aliermont 363
carrying cases, clocks 563
Carte, John 222, 645–6
Carte, Kohn 644n
cartel d'alcove 184, 184f
cartelization 374, 377, 654
cartels 181, 184, 217f, 654
Carteron, Stéphane 425
Cartier 441, 479, 488, 491
cartonniers 433, 654
Case, Dr John 607, 608f
case making trade 556
case manufacturers 356
Casimir 648

Casio, electronic wristwatches 492
Cassian, John 117
Cassini, Jacques 244
Cassini, Jean Dominique 313
Cassiodorus 10, 10n, 121, 137, 644
cast iron 147
Castle Combe, St Andrew's Church 144
Castlemaine Tompion 549
Catania, sundial 8
cathedral decoration 438
cathedrals, musical and automaton clocks 292–3
Cattelin brothers 310
Celebi, Evliya 134n
celestial globes/spheres 43, 44, 45f, 133, 253, 271
    Antide Janvier 274
    Hans Lang 284f
    Jean-Baptiste Schwilgué 276
    planetary clocks 257
celluloid watch crystals 486
Censorinus 10, 99, 645
central heating, time switches 527
Centre Electronique Horloger (CEH) 492, 526
centre seconds 50
Centre Suisse d'Électronique et de Microtechnique (CSEM) 493
Centurion 323
cercles tournantes 181
CETEHOR (Technical Centre for Horology) 369
Cetius Faventinus 17, 19, 21, 77
Chagall, Marc 617f, 618
L. Cailly aîné 396, 397
chains, spring-driven clocks 158
chains, watch 560–1
chair/double frames 464
chalice dials 231, 654
Chalier, Valentin 47, 302, 445n
Cham (craftsman in Suzhou) 51n
Chamaret Cathedral, sundial 119
Chamberlaine, Thomas 194
chamber-masters 224
Champion, Isidore Guillaume 648
champlevé 199, 424, 425, 654
    Art Nouveau 440
    watch dials 557
Chapuis, Alfred 49n
Charlemagne, water clock 290–1, 644
Charles, Jacques Alexandre 342
Charles Frodsham & Co 354, 586
Charles of Lorraine, Cardinal 254
Charles the Bold, Duke of Burgundy 422, 594
Charles V of France 153
Charles V, Emperor 167, 170, 254
Charlton, John 194
Chartres Cathedral 146, 148
    astronomical clock 126, 148
    clock dial 147f

chasing 425
Chateaubriand 438
chatelaines 306f, 433, 438, 440, 474, 555–6, 558–61, 565, 654
Chaucer, Geoffrey xiii, xiiin, 293n
Chaumet 473
Chaux-de-Fonds *see* La Chaux-de-Fonds
Chelsea Clock Co. 589
Chesneau, Henri 644, 644n
Chester, William 609
Chevalier, Michel 364
Chiaravalle, astronomical clock 143, 146
Chicago. Adler Planetarium and Astronomical Museum 29n, 115n
child labour 535
chiming 654
    Big Ben 468
    clocks 129
    mercury clocks 133
    public clocks 145, 150
    trains 142, 150
    watches 49
    water clocks 125
China
    ancient astronomy 41
    astronomical timepieces 43–5
    electronic wristwatches 492
    European clockmakers 456
    European clocks 45–7, 444–5
    hour systems xiv
    imperial clocks 47–8
    lantern pinions 594
    modern era 53–8
    pocket watches 48–51
    popular clockmaking 51–2
    sinking bowl water-clocks 37
    trade with Europe 48–51
    traditional timepieces 41–3
    watchcases 564
China Clock Factory 53
Chinese calibre watches 49–50, 50f
Chinese market watch 49
Chinese scenes, rococo clocks 433
Chinese Temple automata clock 451, 451n, 454
chinoiserie 176, 451
Chioggia, St Andrew's Church, public clock 139, 140
Chitqua 454
Le Christ à l'Horloge (Chagall) 617f, 618
Christchurch, Hampshire 535
Christianity
    concept of day 99
    time perception 101
Christianity, early 79–80
Christie, Agatha 616
A Christmas Carol (Dickens) 613
chronograph function, wristwatches 489

chronographs 344, 576–8, 577f, 579f, 584
chronometer watches 335, 417
chronometers
    English *vs.* French approaches 411
chronomètres 582, 583f
chronometrical thermometers 409
Chronos clocks 471
chronoscopes 344, 575–6
church clocks 137, 195, 511, 542n, 543, 606n
    astrolabes 126
    Charles Dickens 613
    disuse 471
    unreliability 207, 511
Church Hanborough, pendulum clock 142
cigarette lighters 561
Cipolla, Carlo M. 47n
circular plane vertical declining dials 16, 17f
Cistercian order, rules 123, 141
Citizen 375
    electronic wristwatches 492
civil day, ancient Rome 99
CK (watchmaker) 187f
Clarke, Christopher 174, 210
Clarke, Henry 347
Classet, Jean 61
Clavius, Christopher 232, 235, 239, 240, 241, 243, 622, 624, 647
Clay, Charles 296, 649
clay oil lamps 135
cleaning, clock repairs 546–7
cleaning solutions 551
Clement, William 173
clepsammia xiii
clepsydrae xiii, 6–7f, 122–3, 127, 647, 654, 659; *see also* water clocks
    China 43f
    Greece 10–11
    India 34
    Islam 96
    Renaissance 127
Clerkenwell, clock- and watchmaking 404–5, 410
Cleveland Museum of Art 170n
Les Cloches de Bâle (Aragon) 367
clock, etymology 654
Clock and Watch Research Institue, Xi'an 55–6
Clock Cleaning and Repairing (Jones) 552
Clock Jobber's Handybook, The (Hasluck) 552, 633
The Clock of Love (Froissart) 600
Clock of the Long Now 285
clock repairing *see* repairing clocks and watches
Clock Repairing and Making (Garrard) 552
The Clock (film installation) 618–19

INDEX

clock watches 193
clock-as-person, metaphor 613–14
clocking-in clocks 364f, 512, 527
clock-keeping 141–2, 541
Clockmakers' Company, Worshipful
    Company of 204, 205, 213, 348,
    429, 460
  BHI formation 410
  Huguenot members 197–8
  incorporation under Royal
    Charter 194
  library 328n, 544n, 547n
  opposition to Nicolas Fatio de
    Duillier 200
Clockmakers Handbook (Georgi) 551
The Clockmakers' Outcry (anon) 610
Clockmaker's Soup (cleaning
    solution) 551
clockmakers/clockmaking
  Britain 347–54
  Byzantium 94–5
  early public clocks 142–4
  France 363–9, 391–401
  Germany 378–90
  Japan 65–72, 68f
  Liège 215–16
  mass-production 410
  Paris 216
  Switzerland 370–7
  United States 355–62
clocks
  collections 651
  as furniture 431
  as metaphors 599, 602, 606–7, 609n
  as musical instruments 599n
The Clocks (Christie) 616
clock towers 38, 300, 522
clock-winding 471
  automated 471
clockwise rotation 146
clockwork applications 569
  car clocks 586–90
  metronomes 582–5
  planetary models 570–4
  roasting jacks 594–7
  timing and driving systems 575–81
  watchmen's clocks 586–90
clockwork metaphor, cosmos 266
Closon, Peter 172
cloutage 557
Clowes, James, longcase clock 175f
club-tooth lever escapements 203
Cluny Abbey 292
Cluses, Haute Savoie region 366–8,
    399
Cluses, horological school 400
Clusone, Lombardy, astronomical
    clock 471
coach watches 196, 563
coastal time signals 497
Cochin, church of St Francis and
    clock 38

Cochin, Daniel 211
Cochstedt, sundial 109
cock crow 276, 293
cockfights 37
Cocleus, Johannes 185
Codex Atlanticus 132, 595n
coelostats 578–81
coil springs *see* helical springs
Cole, Benjamin 572
collecting activity 163, 552–4, 555,
    644, 646, 650, 651f, 652
Collin, Armand Francois 344, 365,
    592, 597
Grand Palais clock 281
Cologne, clockmakers 129
column dials 28–30, 30f
comb and cylinder movements, musical
    and automaton clocks 308–9
Combret, Pierre 189
Commandini, Frederico 239, 241
commercial travelling 227
common hours 105
communal use, clocks 138, 144
compartmented water-clocks 659
compass dials 220, 237–8
compensated pendulum 178, 181
compensation balances 323–9, 334,
    407
complexity, early public clocks 147
Complicated Watches and their Repair (de
    Carle) 634
compline 102
Composite Signals Organisation
    (CSO) listening stations 418
Compositio Horologiorum (Münster) 116,
    232
compound circular motions 286
compressed air 509–11
computatrix (star) 117
computers, personal 251
computus 654
  astronomical clocks 273
  Beauvais 278
  calendrical data 286
  Copenhagen 282
  Daniel Vachey 283
  Easter, determination 287
  medieval Latin world 100
  Philippe de Thaon 105–10
  Strasbourg 286
Comte, Victorine Lucie Marie
    Bailly 533
Comtoise clocks 177, 363, 366, 536,
    654
Conan Doyle, Adrian 616
concave dials, Byzantium 79
Congreve, William 414
conical dials 14, 15f, 77, 86
Connecticut clock industry 356, 358
conservation movement 553
Conservation of Clocks and Watches
    (Wills) 554

*Considerations Pratiques sur L'huile
    Employee en Horlogerie*
    (Robert) 550
constant force escapements 403, 414
constant pressure pendulum clocks 502
Constanta, sundial 19
conticium (early part of night) 99
contrate wheels 130, 192, 654
controlling clocks 511, 514
Convention Nationale, decimal
    time 341, 342, 342n
Conversano, sundial 107, 108n
conversions, pocket- to
    wristwatches 486
Cooke, William 497
Cooling, Louise 611, 612, 613
Copenhagen City Hall, astronomical
    clock 282, 283f
Copenhagen watchmakers 411
Copernicus 311n, 574
copper, clockmaking 222
corona freni 144
Cortébert Watch Company 588
Corum watches 442
Cosmographia Pomponii Melae
    (Cocleus) 185
cosmological models 271
cosmos, clockwork metaphor 266
Coster, Salomon 142, 171–2, 314, 316
Cotehele House, Saltash,
    Cornwall 143
cottage industry 224, 226
Couaillet, Armand 399
Coudray, Julien 164
Coullery, Henri 584
count wheels 142, 144, 174, 464, 520,
    654
  domestic wall clocks 154
  pendule de Paris 391
coup perdu (lost beat) escapements 181
*Cours Élémentaire de Règlage destiné aux
    horlogers* (Jaquet & Defossez) 635
Coustou, Nicolas 430
Coventry, clock- and
    watchmaking 218, 460
Cox, James 179, 300, 301, 447, 451
  bankruptcy 459
  cabinet musical clock 449
  chronoscope 451, 454, 453f
  exports to China 49, 455, 458, 460
  exports to India 456
  goat clock 452f
  watchcases 558
  workshop 448
Cox, John Henry 49, 455–6
Cox, Robert Harvey 535
craftsmen 17–20, 224, 228
  guilds 215–18
Crayle, Richard 194
credit 218–9, 227n, 356, 357, 378, 448,
    450, 455, 462, 527, 588, 614
crepusculum (twilight) 99

Cressent, Charles 431
Cretin-L'Ange, A. 464
  flat-bed tower clock 466f
Croft, David 615n
cross-beat escapements 162
Croutte, Charles-Antoine 229n
Croutte & Cie 396, 397
crown brakes 144
crown wheels 46, 130, 132
  escapement development 161
crucible steel 222
crucifix watches 424
crutch, lever escapements 201
crystal chronometers *see* quartz
    watches
Crystalline planetary clock
    (Torriani) 266
crystals, unbreakable 486
CSG system 342
Ctesibios 11, 94, 290, 643, 647
cubits 28, 654
cuckoo clocks 184
Cūḍā-yantra *see* ring dials
Cuff, Peter 224
Culpeper, Edmund 242
Cumming, Alexander xii, 338, 469,
    629
Cuper family 188, 423
Curci, Alfonso 466
curfew bells 145
cursors 234
Cursus mathematicus (de Châles) 622
Curtis, Samuel 359
Cusin, Jean 424
Cusin, Noël I 188, 188n
Cutts, Harry 489
CW *see* Werner, Caspar
cuvette, 50, 308, 559
cyclists, wristwatches 476, 477
cylinder dials 28–30, 114, 232
cylinder escapements 325–6, 367, 367f
  wristwatches 481
cylinders, automata 289
cylindrical springs *see* helical springs
Cynocephalus 643

D

d'Abbadie, Antoine 344
Dacre, Mary Neville Baroness 163
Dad's Army (tv series) 615
dā'irat al-muʿaddil 91
Dalavan, Geoffrey 542
d'Albertis, Enrico 250
Dali, Salvador 615, 618
Dallam, Thomas 301
Dalrymple, Alexander 334, 335
Damascus, sundial 30, 81, 89–90, 93–7
Damian, Peter 117
Danck, Johannes 119
dangle spits 596
Daniels, George 214n, 354, 651n
  Watchmaking 632

735

Dante Alighieri 599–600
   Divine Comedy 106
   medieval hour systems 103, 104, 104f, 105
   Paradiso 129
Dardel, M. 564–5
Darier, David 566
Darier, Hughes 566
Darjeeling 30
d'Arsonval, Arsène 576
Das, Debasish 38n
dashboard clocks 565, 587
Dasypodius, Conrad 141, 149, 263n, 293, 623
date indications, wristwatches 489
Dauville, Noel 421
David, Jacques 534–5
David, Jacques-Louis 610, 612f
David Copperfield (Dickens) 614
Davies, Alun 348
da Vinci, Leonardo
   clock drawings 132, 146
   roasting-jacks 594, 595
   stackfreed 159
   strob escapement 144
   time-glass 133
   water timer 127
Davis, John I 595
Day, Richard 600n
Day, Samuel 591
day divisions
   Babylonian 5
   Indian 27
   Roman 100t
day-and-night indicator 254
de Bernon, Yves 424
de Béthune, Marie-Henri 181
de Blachis, Benedetto 146
de Boissières, Claude 240
de Bry, Theodore 423
de Camus, François Joseph 321, 430
de Carle, Donald 552, 634
de Caus, Salonon 231, 290
de Celles, François Bedos 246–7, 622
de Cerceau, Androuet 421
de Châles, Claude François Milliet 622
de Charmes, Simon 179
de Chirico, Giorgio 615n
de Corville, Raoul 542
de Coulomb, Charles-Augustin 506
de Courtecuisse, Jehan 602
de' Dondi, Giovanni 129, 132, 141
   astrarium 263
   planetary clock 253, 258f, 260, 263, 264t, 265t, 266, 623
   Vitruvian artisan 266
de' Dondi, Jacopo 141, 148, 266
de Duillier, Nicolas Fatio 200
de Evalo, Hans 63, 64f, 170
de Flamanville, Painel 326
de Floutrières 241
de Groot, Jan Cornets 162

de Hautefeuille, Jean 316, 325, 624
de Heck, Gérard 193–4, 225n
de la Cour, William 433
de la Garde, Jacques 423
   spherical watch 424f
de la Hire, Philippe 247, 646
de la Hyre, Laurent 426, 428
de la Joue, Jacques 431
de La Mettrie, Julien 299
de la Roche, Jean 220
de Magalhães, Gabriel 47
Delépine and Canchy 397
de Medici, Francesco I 163
de Metz, Petrus 228n
de Meung, Jean 129, 153
de Mornay, Philippe 604, 607
de Pantoja, Diego 41, 46
de Pereda, Antonio 605, 605f
de Pisan, Christine 600, 602, 603
de Pisan, Thomas 603–4
De Quatuor Virtutibus 602
De Saavedra Fajardo, Diego 163
de Sacrabosco, Johannes 128
de Sainte-Croix, Renouard 459, 460
de Santa Cruz, Alonso 161
de Schort, Brostrup 646
de Séjour, Dionis 244
de Servière, Grollier 163, 646
De Sphera Mundi (de Sacroboso) 128
de St Germain, Jean Joseph 431, 433
de Steur, Jacobus 341n
de Stokes, Roger and Laurence 130, 228n
de Thaon, Philippe, computus 105–10
de Troestenberch, Nicholas 65
de Vallemont, Abbé 570
de Varignon, Pierre 158n
de Vaulezard, J.L. 242
de Ventavon, Jean-Mathieu 47
de Vita, Domenico 466–7
Deacon, Samuel 223, 223n
dead-beat escapements 178, 181, 200, 654
   Benjamin Vulliamy 549, 630
   Graham 178
   Greenwich clocks 318
   London clockmakers 321, 326, 335, 338, 403, 414
   Paris clockmakers 393, 627
   pinwheel type 404
   scientific applications 403, 414
Debaufre, Pierre and Jacob 200
decadal calendar, Ancient Egypt 341
decans 28, 654
Decharmes, Simon 197
De Choudens 436
decimal time 341–5
decimal watches 203, 210, 343f, 344
declination curves xii–xiii, 13, 655
declining dials 239
declining plane vertical dials 16, 16f, 245f

decorative arts 421
   art nouveau and art deco 440–1
   baroque style 425–31
   historicism 438–40
   modern movement 430
   neoclassicism 433–8
   postmodernism 441–2
   renaissance and mannerist periods 421–5
   rococo 431–3
Décret 342n
deferent 260, 261–2, 655
Defoe, Daniel 222
Defossez, Leopold 632, 635
Dehind brothers 218n, 221
Delambre, Jean-Baptiste 247, 646
Delander, Daniel 197, 204, 330
Delander, Nathaniel 198
Delaune, Etienne 421, 422, 423
Delépine & Conchy 396, 397
Delhi, water clocks 35
Delhi Observatory 28
Della Rovere, Eleonara Gonzaga 163
Della Volpaia family 141, 234
Della Volpaia, Girolamo 118
Della Volpaia, Lorenzo 234
   planetary clock 262, 264t, 265t
   Vitruvian artisan 266
Delos, sundials 19
Delvaux frontispiece
Demainbray, Stephen 572
demures 106n, 106
Deng Xiaoping 56
Denison, Edmund Beckett 403, 468
   Cambridge chimes 469
Denizet, J.F. and René Barrier, watchmakers 584
Denmark
   chronometer making 411
   longcase clocks 177
Denis, Gustave 399
Dennison, Aaron 359
Dennison Watch Case Company 359, 485, 566
Dent, Abraham 227
Dent, E. J. 406, 407, 412, 551–2, 592, 636
   Big Ben 465f, 468
   decimal regulator 344
   regulator 413f
   watchmen's clocks 592–3
Deparcieux, Antoine 246, 247
department stores 400
Depollier wristwatches 487
Derham, William 198, 543, 544, 624, 626, 627, 644–5
Desaguliers, John Theophilus 572
Desargues, Girard 231, 248
Descartes, René 266, 299, 607
Description and Use of the Globes, and the Orrery (Harris) 572
descriptive geometry 248

Desmares 300
d'Espagne, père 244n
destruction, public clocks 470–1
detached detent escapement 408f
detached escapements 330–1, 334, 403, 407
detached lever escapements 331
detents 330, 403, 407, 655
Detouche, Constantin 281
Deutsche Uhrmacher-Zeitung 635, 636
Deutsches Gesellschafte für Chronometrie (Germany) 553
Deutsches Museum, Munich 251n
Deutsches Uhrenmuseum, Furtwangen 283, 284f, 421n, 423n, 646n
dial of Ahaz 239, 240, 644–5
dialling *see* sundials
dialling scale 243
dialling tables 242
diamond-shaped wall clocks 172
Dickens, Charles 561, 613–14
Dictograph 512
Diego de Pantoja 41
Diepel, Hermann 260, 261f, 270
digesting duck 299
digital readout wristwatches 490
digital sundials 252
dihedrons 13, 655
di Jacopo, Mariano 144
Dijksterhuis, E. J. 162n
Dijon, Notre-Dame, Jaquemarts 293–4
diluculum (dawn) 99
Dimier Brothers & Co., wristwatches 479–80
al-Dīn, Taqī 89n, 98, 312, 594
Dingler, Johann Gottfried 637
Diodorus Siculus 77, 82
Diogenes Laertius 5, 10n
diptych dials 238
direction dials 231, 232, 655
   Egypt 3
   portable 237–41
Directoire 176, 655
distribution, manufacture and trade, 16th–18th century 226–7
Ditton, Humphrey 318–19
divers' wristwatches 490–1
diversification 354, 368–9
dividing plates 221
division of labour 223–4, 226
   Black Forest, clockmaking 378–9
   German clockmaking 381
   planetary clocks 263
   Swiss watchmaking 410
   women 533, 534, 536, 538
Dodderidge, Lady 606
Dodillet company 366
Dohrn-van Rossum 102, 216n
Dôle, time announcer 292n
dollar watches 361, 492

INDEX

domed top case 178, 179f
domestic clocks, 13th–18th century
   escapement development 161–3
   geared clocks 153–4
   introduction of pendulum 171–3
   longcase clocks 173–8
   luxury, power and knowledge 163–4
   regional clockmaking centres (Europe) 164–70
   spring-driven clocks 154–60, 178–82
   travelling clocks 182
   wall clocks 182–4
domestic clocks, electric clocks 523
Dominy family 549
Dondi, Giovanni de' 129, 132
Dondi, Jacopo 141, 148
Dongfeng watch movement 55, 56
Donne, John 606, 606n
Donzé, Pierre-Yves 527
d'Onzembray, Comte 582
door frame clocks 143
Doppelmayr, Johann Christian 622
d'Orléans, Philippe 557
Dorn, Hans 237n
double dials 242
double foliot system (Japan) 66
double hours 105
   China 41, 42f
double S balance 202
double three-legged gravity escapement, Big Ben 469
double-ampoule sand-glass 134
double-foliot mechanisms 67f, 66, 67
Doubs 222, 367–9, 535
Doxa, dashboard clocks 587
Doyet, Marie 538
Dr John Case (portrait, unknown) 608f
dragon boat clocks 42, 42f
Drebbel, Cornelius 644
Drecker, Joseph 251n, 622
Dresden
   16th century planetary clock 254
   clock and watchmakers 335
   planetary clock 254
driving systems *see* timing and driving systems
Droz, Jean-Pierre 395, 395n
drum chronographs 499, 576
drum movements 391, 395, 397, 398
   watchmen's clocks 591
Dubois, Louis 418
Dubois, Pierre 638, 650
Duboule, Jean Baptiste 196, 424
Duchesne, Claudius 179, 296
Duchesne, Pierre 208f, 430
Ducommun, George 587
Dueber, John 360
du Guernier, Louis 427
Duguore, Jean Démosthène 435
Duhamel, Pierre 431
Duhamel, Pierre II 196

dumb-repeaters 198
Dumesnil, Debruge 214
Duncan, Jonathan 38
Dundee Medical History Museum 606n
Dunod, Claude 249
Dunstable Priory
   clock repairs 542
   horologium 129, 138–9
Duplessis, Jean Claude 431
duplex escapements 203–4, 330
duplex watches for China 460, 461f
Duquesnoy, François 431
Durand, William 99, 129
Dürer, Albrecht 239, 605
Durham Cathedral, astronomical clock 149
dust caps 199–200, 558–9
Dutch dials 210
Dutch East India Company 443
Dutch forgeries 214, 347
Dutch striking 177, 655
Dutertre, Jean-Baptiste 204
Dutton, William 178, 204
Duval, John 447
Duverdrey, Paul 399
Dvchesne, P 208f *see also* Duchesne, Pierre
Dye, David W. 507

E

Eagleton, Catherine 115n, 116n, 235n
Earle, Alice Morse 251
Earnshaw, Thomas 204–5, 336–8, 404, 406, 541
   dead-beat escapement 403
   Longitude, An Appeal to the Public 630
   marine chronometer 337f
Earnshaw, Thomas junior 403
East, Edward 173, 194
East Germany 384–5
East India Company (EIC) 37, 38, 46, 48, 443
   exports to China 445–7, 455, 457–9, 458t, 461
   marine chronometers 405
   mechanical clocks 37–8
   musical and automaton clocks 302
   planetary model 572
   precision timekeeping 334–7
East Indies 18th century trade 443–4, 458t, 461–2
   automata 451–4
   Canton 446–7, 456–7
   clock trade with China 444–5, 457–9
   design 451–4
   export and sale 455–6
   manufacture 447–51
   watches 460
Easter, determination 273, 274, 279, 283, 287

Beauvais astronomical clock 278
   Copenhagen astronomical clock 282
   Strasbourg astronomical clock 277
Eayre, Joseph 543
ébauches 212, 391, 533, 584, 654
Ebauches SA 369, 442
Ebsworth, John 183f
Echternach, sundial 111
eclipses 4, 267, 573, 641
   astronomical clocks 276–7, 279–80, 285
   Jupiter satellites 38
   lunar 130, 148, 258, 311, 312
ecliptic 253, 262f, 274, 655
Ecouen, Musée national de la Renaissance 295n, 422n, 424n, 425n
Edinburgh
   electric clock systems 517
   floral clock 469
   Observatory 503
Edmond Jaeger company 587
Edward III, King of England 147
Effi Briest (Fontane) 615
Effi Briest, guard-tower clocks 153
Egypt, time measurement in 1–4
Egyptian decadal calendar 341
Egyptian obelisks, Rome 23
Egyptian time measurement 1–5
Eichens, William (Paris) 580f
Eiffe, John Sweetman 407, 408–9
Eiffel Tower, time-signal transmissions 498–9, 500f
eight-sector dials, medieval hour systems 103–5, 106f
Eisinga, Eise 574
elasticity, balance springs 408
electric chronographs 344
electric clocks 351, 414, 511–23
electric motor-driven clocks 357, 365
electric telegraph time signals 497
electric winding systems 465, 467
electrical contacts 519–20
electrical/electronic wristwatches 491–2
electricity supply systems 524
electromagnetic pendulums 514, 517
electromechanical clocks 588–9
electromechanical/electronic public clocks 471–2
Electronic Horology Centre, Neuchâtel 375
electronic technology 387, 418–19; *see also* quartz watches
elephant water clock 97f
eleven-sector dials 107–8, 111f
Elgiloy, alloy 492
Elgin National Watch Company 360–2, 482, 492, 587
Elinvar (alloy) 374, 415, 492, 506

Elizabeth I, Queen of England 37, 168, 191, 301, 423, 470, 473
Elizabeth II, Queen of England 470
el-Levi, Samuel 136
Ellicott, John 199, 321, 326–7
   in *Philosophical Transactions* 630
Ellicott family 199, 201, 204, 429, 437, 547
Elliott, Henry 542
Ellis, William 414
Elskamp, Max 251, 650
Eltric company 589
Ely Abbey, clock 129, 139
Embe-Fahrzeuguhren GmbH 589
Emerson, William 341
Emery, Josiah 202, 335, 337
Emile Martin 396, 397
Emmoser, Gerhard 259f, 263
The Employment Office (Burr) 614
enamel painting 426, 427f, 428, 428f
enamelling 425
   Art Deco 440–1
   Art Nouveau 440
   watchcases 188–9, 212, 558, 559f
Encomion Sanitatis (Fabricius) 312
Encyclopédie 341, 391
Encyclopédie méthodique: mathématiques (Bossut) 237n
Enderlin 627
endplates, spring-driven clocks 160
Engel, David-Guillaume 299
Enfield Clock Company 352–3
Engeringh, Cornelius 297
Engilbert, Jean 153
engine-turned *see* guilloché
England
   astronomical clocks 149
   clockmaking, 16th & 17th century 168–9
   lantern clocks 168–9, 182, 183f
   longcase clocks 173–5
   musical clocks 296
   public clocks 138–9
   sinking bowl water-clocks 121
   turret frames 464
   watchcases 557n
   watchmaking, 16th–18th century 191
English clock model 429
English levers 203
Engramelle, Marie-Dominique-Joseph. 289–90, 297
engravers/engraving 425, 556
The Enigma of the Hour (de Chirico) 615n
Enlightenment 300
epacts 278, 280, 282, 286–7
l'Epée company 366
ephemerides 240, 267, 268f
epicycle 259–62, 655
Epistre d'Othéa (de Pisan) 600, 602

INDEX

Epperlein, Helmut 388
Eppner, Eduard 380–1
equal hours xiv, 655
    Egypt 2–3
    Europe 118, 119
    fire clocks 135
    Greece 6
    India 37
    Islam 84, 89
    medieval Latin world 99, 110
    polar gnomons 232, 238
    rectilinear dials 115
equant 261–2, 655
equation clocks 627
equation kidney 178
equation of time 129, 212–13, 626–7, 655
    correction system 249
    pendulum clocks 178, 317–18
    Ptolemy 621
equatorial sundials 41
equinoctial hours *see* equal hours
equinoctial sundials 238, 240
    China 41
    Islam 84, 90
equinoxes xiii, 29, 258n, 503, 643n
    astronomical clocks 274, 276, 280, 284
    azimuth dials 110, 111
    calculated angular values 119t
    meridians 244
    Navicula de Venetiis 115–16, 235
    ring dials 237
    sundials 5, 6, 19, 24, 30, 41n, 83, 88, 102, 108, 234n, 249
    water clocks 12, 122
Erfurt rule 119–20, 119t, 120t
Erghum, Bishop Ralph 140
erotic watches 308
escape wheels 130, 192n, 655
    equalizing tools 221
    guard-tower clocks 153
escapement conversions 171, 472, 549, 552
escapement error 407
escapements 655
    19th century public clocks 468
    Ancient China 44
    China 46
    conversions 549
    domestic clocks, 13th–18th century 161–3
    duplex 203–4
    early public clocks 142
    early watches 186
    Europe 127–32
    lever 201–3
    specialist makers 224
    spring-driven clocks 160
    virgule 207
Eschle company 310
Esquivillon, Louis 436

Essai sur L'Horlogerie (Berthoud) 546, 627
Essen, Louis 419, 507, 508
Ester, Henri 424
Estienne, Henri 595
Estienne, Marin 646
ETA 376
établir 655
établissage 370–7, 371f, 536, 538, 655
établisseurs 372
Étalon factory 503
Etienne, J. P. 541
Etrennes chronométriques 641
Etudes sur diverses questions d'horlogerie (Roberts) 631
Etymologiae (Isidore) 99
Eureka electric balance clocks 523
Europe
    domestic clocks, 13th-18th century 153–70
    early oscillator escapements 127–32
    early public clocks 139t
    exports to China 45–6
    exports to Japan 63, 64f
    hour systems xiv–xv
    Islamic water clocks 121–2
    longcase clocks 174
    planetary clocks, 14th–16th century 263–72
    public clocks 13th–18th century 137–51
    trade with China 48–51, 444–7
    watches, 16th-18th century 185–214
    water clocks 121–7
Evans, Lewis 251
Evelyn, John 595n
EverBright(Ebohr) quartz watches 58
Exeter Cathedral
    clock 129, 144, 149, 542
    horologium 139
    Lady Dodderidge monument 606
Exhaustive Treatise on Shadows (al-Bīrūnī) 81
exhibition clocks 281
exhibitions, industrial 411, 640; *see also* international exhibitions
An Explanation of my Watch (Harrison) 328
Explanations of the Clock (Schreck) 46
exports
    automata to East Indies 451
    Britain to Japan 445–7
    Chinese market 213
    East India Company to China 445–7, 455, 457–9, 458t, 461
    Europe 225
    Europe to China 45–6
    Europe to Japan 63, 64f
    Gent & Co 522
    Germany to Britain 349

    James Cox to China 49, 455, 456, 458, 460
    luxury objects to China 454, 462
    Russia to Britain 353
    Swiss mechanical and electronic watches 376f
    Swiss watches 373f, 375f
    Switzerland to Britain 348, 352
    Switzerland to China 460
    Synchronome Company 38, 522
    Turkish market 213
    USA to Asia 357–8
    USA to Britain 349
eykt (eighth) 103

F

Fabricius, Paulus 312
fabrique 226, 655
Facini, Bernardo 273
Fahrenheit, Daniel G. 579
Fajardo, Diego de Saavedra 163
Falconer, Kenneth John 252
Falconet, Camille 649
Falconet, Etienne Maurice 435
Fale, Thomas xiii, 240, 244, 644
Falize, Alexis 438
Falize, Lucien 440
fantasy watches 561
Fantoni, Girolamo 116
Far East and India, musical and automaton clocks 302
Far East, history of horology 35–8
Fardoil, Peter (Pierre) 198, 649
al-Farghānī (Alfraganus), Aḥmad ibn Muḥammad ibn Kathīr 83
fashion accessories, watches 476, 555
Father Time 293, 436, 610
Fatio, Nicolas *see* de Duillier, Nicolas Fatio
Fatton, Louis Fréderic 578
Faulkner, William 618n
Favarger, Albert 576, 577f
Faventinus, Cetius 17, 19, 21, 77
Favorinus 645
Favre, Antoine 308
Favre-Brandt & Co 73
al-Fazārī, Ibrāhīm ibn Ḥabīb 82
Fécamp, pendulum clock 142
Fedchenko, Feodosii Mikhailovich 419, 503, 504f
Federation des Écoles Techniques de Suisse 632
Federation of the Swiss Watch Industry 373
Die Feinstellung der Kleinuhren (Unterwargner) 635
Die Feinstellung der Uhren. Ein Anleitungs- und Nachschlagewerk in zwei Teilen (Helwig & Giebel) 635
Feldhaus, F. M. 126

Fell, Andrew 352
female workers 534–5, 535f, 536n, 537, 537f, 588
Ferguson, James 572, 573
Ferracina, Bartolomeo 142, 148
Ferrerius, Joannes 240
Ferrié, Gustave-Auguste 498
Ferry, Maurice Bernard 424
Féry, Charles 517
Festo astronomical clock 285
Fez
    Būʿanāniyya mosque 98
    Qarawiyyin mosque water clock 96
Fiamma, Galvanus 145
field-gate frames 143
Fiester, John 282
Filassieri, P. 166f
filigree 558, 655
Fillary, Thomas 592
financial sector, standardized time 496
Fine, Oronce 127, 235, 236f, 238, 239, 242
    Navicula de Venetiis 236f
    planetary clock 254, 258, 258f, 260, 262, 264t
Fine planetary clock 254, 258, 260, 262, 263n, 264t
finger ring watches 205, 473, 474
finishing and profiling 224
fire clocks 60–1, 135–6; *see also* incense sticks
fire-gilt 187, 187n
First World War, wristwatches 482–4
Fischer, Martin 511, 521
Fisher, Rev George 405
Fitch, Ezra C. 488
fixed vertical sundials 110
Fiyta quartz watches 58
Flajoulot, E. 310
Flamsteed, John 178, 243, 317–8
Flanders, watchmaking, 16th-18th century 189–91
flat springs, spring-driven clocks 154
flatbed frames 142, 150f, 150–1, 463, 464, 466f
Fléchet, Victor 249
Fleury, Victor, astronomical clock 281
flip displays 471
floral clocks 469–70, 470f
Florence
    Museo Galileo 235n
    Santa Maria del Fiore clock 146
    Santa Maria del Fiore dial 247
    Santa Maria Novella Railway Station 471
flötenuhren (organ clocks) 177
Flötner, Peter 421, 609
Fludd, Robert 290
fly *see* air brake
fob watches 559 *see also* pocket watches
foliot 129, 130, 655
    China 46

INDEX

escapement development 161, 162
spring-driven clocks 160
Fondulo, Giorgio 266
Fonnereaux, Zacharias 424
Fontaine, Pierre François 436
Fontana, Cesare 466
Fontane, Theodor 615
foot board clocks 565
force equalization, spring-driven
 clocks 157–60
forgeries 206, 213–14, 347
form watches 190, 423, 425f, 655
 Art Nouveau 440
 England 194
 Netherlands 190–1
 Switzerland 196
Förster, Wilhelm 515
Forstmann, Christoph Wilhelm 624
Fortin, Auguste 216
Fortis
 Aquatic 487
 Spacematic 491
Foster, Samuel 242, 243
Foucault, Léon 578, 579
 siderostat 580f
Foucher, Paul 550
Foullet, André 431
foundry work 225
fountain feed lamp 136
Fouquet, Alphonse 440
Four Ages of Man 196, 276, 293, 294
The Four Times of Day—Morning
 (Hogarth) 610
Fourier, Joseph 286
Fournier, Bienaimé 583
four-sector dials, medieval hour
 systems 103
France
 18th century publications 627
 Besançon, clock- and
  watchmaking 368
 clockmaking 164–6, 166f, 391–401
 decimal time 341–5
 exhibitions 364, 368, 373, 395–7,
  401, 440, 441, 533
 female workers 537–8
 flatbed frames 464
 horological literature, 19th
  century 411
 industrialization 365–7
 The Jura, clock- and
  watchmaking 367
 lantern clocks 182, 183f
 longcase clocks 176–7
 marine chronometry 337–9
 musical clocks 297
 Paris and St Nicolas d'Aliermont,
  clock- and watchmaking 363–4
 public clocks 365
 rococo watches 433
 sundials 106, 108f
 table clocks 180, 181, 421–2, 422f

watchmaking 188–9, 206–10, 208f,
 228n–9n
La France Horlogère (journal) 400
Frankfordiense emporium 595n
Frankfurt, DCF77 Radio Signal 472,
 526
Frankfurt, book fair 595
Frankfurt Archaeological Museum 12
Franklin, Stephen 613
Fras, Michael 273
Frassoni family/company 466, 467
free pendulum 414–15, 502–3
free trade 354
Freeman, J. G. 231, 252
Freiburg, Silesia, clockmaking 378,
 379, 382
Freischwinger escapements 468
French marine chronometry 337–9
French Revolution 182, 208, 341–2,
 391, 393
 working rights for women 533
Frers, Johann Eggerich 341n
frictional contact 330
frictional rest escapements 200, 330
Fried, Henry B. 552
Friedberg
 clockmaking 225, 229, 378
 watchmaking 198–9, 200n, 206, 225
Friedrich Krupp AG 382
Frisard, Jacob 301, 304, 305
Frisch & Shauerte 404
Frisius, Gemma 161, 313, 314
fritillary form watch 190
Fritts, Charles Elgar 634
Frocourt, astronomical clock 278
Frodsham, W. J. 406
Frodsham, Charles 354, 412, 586
Frois, Luis 63
Froissart, Jean 600, 603, 649
Fromanteel, Ahasuerus 172, 174
 hooded clock 173f
 longcase clock 429
 pendulum clocks 172, 172f, 316,
  318, 429n
Fromanteel, John 172, 314, 316
Fromanteel & Clarke, Amsterdam 174,
 210
Fruttuaria Abbey, water clock 122
functionalism, modern movement 441
Funeral Blues (Auden) 618
furniture incorporating clocks 431
Furtwangen
 clockmaking school 378
 German Clock Museum 283, 284f,
  421n, 423n, 646n
fusees 645, 655
 Antide Janvier 274
 cutting tools 221
 English watchmaking 191–3
 French watchmaking 188
 metronomes 583
 roasting-jacks 596

specialist makers 224
spring-driven clocks 156, 158, 159
Fusoris, Jean 148–9, 156, 164n, 233n,
 238n, 239, 281
Fussner, F. Smith 645n

G

G10 strap 480
Gabriel de Magalhães 47
Gagneux, Stephan 404
Gainsborough, Humphrey 331
Galileo, Galilei 127, 162, 649
 longitude determination 313–5
 pendulum 171, 314
 water clock 127
Galileo, Vincenzo 314
Galli, Antonio & Luigi 467
gallicinium (ninth hour of night) 99
Gallonde, Louis Charles 649n
Galluci 241
galuchat 557, 655
galvanometers 575
Gambey, Henri-Prudence 579
Gamot, Grégoire 426
Gardner, William 516
Gargantua and Pantagruel (Rabelais) 609
Garnier, Jean-Paul 514–15
Garnier, Joseph Blaise 242, 247
Garnier, Paul 242, 365, 398, 438, 651
Garon, Peter 197
Garrard, F. J. 552, 634
Garstin, A, & Co 475
Gassendi 644
Gatty, Margaret 114, 250–1
Gaudron, Antoine 174, 174n, 180,
 430, 431
Gaudron, Pierre 627
Gaupp, Johannes 243, 246, 247
Gavril, Jacques 227
Gaza 3, 94, 290
Gazeley, William J. 552
gear wheels 321, 411
 astronomical clocks 273–4, 276,
  285, 287, 311
 calculation 140, 269f
 cutting 221, 224
 domestic clocks 153–4, 161
 epicyclic 201
 metronomes 582
 nocturnals 118
 planetary models 254, 258, 258n,
  262–3, 264–5t, 573
 public clocks 142
 roasting-jacks 594
 watch movements 185, 186, 188,
  192, 192n, 193, 197, 208
 water clocks 125, 137, 291
geared portable dials, Byzantium 78f
Gebrüder Junghans *see* Junghans
 company
Gebrüder Staiger *see* Staiger company
Gebrüder Thiel 381, 382, 587

Geiger, Thomas I 595
geisha houses 61
Geissler, J. G. 629
General Electric company 589
General Society of the Swiss
 Horological Industry SA 374
Geneva
 Bibliothèque 304n
 centre of trade 228
 forgeries 214
 Musée d'Art et d'Histoire 305,
  556n, 558f, 559f, 560f, 562f, 563f,
  564f, 566f, 567f
 Musée de l'horlogerie et de
  l'émaillerie 196
 Musée d'Histoire des Sciences 235n,
  236n
 musical and automaton
  watches 307–8
 Patek Philippe Museum 159n, 166f,
  196, 196n, 306f, 307f, 308n, 426
 rules for women 531
 Société des Arts 305, 308–10, 533,
  647
 specialized trades 556, 557
 watchmaking 195, 211
Geng Xun 43
Genoa, first clock in 141
Gent, J. T. 593f
Gent & Co 511, 512, 520, 522, 525,
 593
Georgi, Erasmus 551, 633
Gerbert of Aurillac 122, 649
Germain, Thomas 431
German Clock Museum, Furtwangen 283, 284f, 421n, 423n,
 646n
German Hydrographical Institute 412
Germanisches Nationalmuseum,
 Nuremberg 153n, 156f, 156n,
 157n, 312n
German-style dials 146
Germany
 18th century publications 629
 chronometer making 411–12
 clockmaking, 16th & 17th
  century 166f, 167–8, 169f, 423
 clockmaking, 19th century 378–80
 clockmaking and watchmaking,
  20th century 381–90
 exports to Britain 349
 German Democratic Republic
  (GDR) 384–5
 longcase clocks 177
 musical clocks 297
 standardized time 507
 textbooks 632
 timecode transmissions 499
 time-signal transmissions 498
 turret frames 464–5
 watch movements 186
 watchcases 187–8

739

INDEX

Germany (*continued*)
  watchmaking, 16th–18th century 185–8, 206
  watchmaking, 19th–20th century 380–2
Gerry, James H. 360
Getty Museum 174n, 430n, 433n
ghaṭikā-yantra 30, 32–3, 32f, 34
ghatis 27, 28, 30, 38
  fractions 33
Gib family 211
Gibbs, George James 249
Gibbs, Joshua 429
Giebel, Karl 635
Gifford, John 290–1
gifts of state, musical and automaton clocks 301–2
gilding 533, 538
Giles Wales & Co 360
Gillett & Bland 149
Gillett & Johnston 520
Gilliszoon de Wissekerke 263, 266
gimbals 317, 656
Giocondo of Verona 120n
Girard, Jaquet 491
Girard-Perregaux company 475
Girardon, François 430
girls, minor trades 533
Gisze, Georg 421
Glasgow, David 632
Glashütte 384–5
glass balance springs 405
glass domes 563
Glick 36
global positioning systems (GPS) 419, 492, 529, 590
Glyn Roberts, Robert David 584
Glynne, Richard 570
gnomon-shadow 27
A Gnomon for the Determination of Noon (al-Fazārī) 82
gnomonic declination 232n
Gnomonica (Welper) 622
Gnomonices (Clavius) 240
gnomonics *see* sundials
Gnomonique pratique (Bedos de Celles) 246–7
gnomonists 19
gnomons 12, 13–15, 653, 656
  bibliography 621n
  China 41
  Egypt 2
  fixed/mobile 242
  Greece 5–6
  India 27–30
  virgula ferrea 110
Goat clock (Cox) 451, 452f
God as clockmaker 602, 607
Godet, Noël 557
Goediej, Nicholas 192
going train 656
gold balances 327, 405

gold watchcases 197, 557
golden number 278, 280, 282, 286
Goldingham, John 38
gongs 558
Gonzalez-Torres, Felix 616
Good, Richard 352
Gordon-Smith, Allan 351
Gorla, Alberto 281
Goryu Asada 68
gothic iron clock 154
Göttweig, Benedictine Abbey 122
Gould, Rupert T. 323, 331–3, 417
Goullons, Jacques 189, 426, 427
Goumois, Frédéric 456n
Gounouilhou, Pierre Simon 308
Gouthière, Pierre 435
Gowland, James 403, 469
GPS *see* global positioning systems
grades (day subdivisions) 344
Grafton, A. 645n
Graham, George 178, 200–1, 324, 433
  cylinder escapements 325–6
  dead-beat escapement 178, 403, 631
  equation clock 429
  in *Philosophical Transactions* 630
  planetary models 570–2
  regulator compensated pendulum 321, 322f
  watch 430
Graizely, Adrien 539
Granaglia & Co 466
Grand (Vosges, France), anaphoric clock 12
Grand Ducal Baden Clock and Watchmaking School 380
grand orrery 572, 574
grand style, France 428, 430
grandfather clocks 429n
Grandjean de Fouchy 246
Grant, John 202
graph-type dial clock 73f
grasshopper escapements 321
Grau, Heinrich 518
Gravesande, Willem Jacob, 's 579
gravity escapements 282, 403, 656
  Big Ben 469
  electrically reset 511, 514, 520
  turret clocks 468
Graz, Austria 146
Great Exhibition (1851) 348, 364, 395, 397, 401, 411, 414, 474
  planetary models 574
great wheel 656
Greece
  sundials 5–6, 107–8
  water clocks 6–7
Greek hours xv
Greenleaf, William 589
Greenwich clocks 318
Greenwich Mean Time (GMT) 406, 514
Greenwich meridian 249

Greenwich observatory 314, 319, 514
  chronometer testing 407, 409f
  marine chronometers 406
  standardized time 414, 497
  time ball 406, 497
  time-signal transmissions 499
Greenwood, Henry 604
Gregorian calendar 72, 273
Gregory of Tours 117
Gregson, John 474, 532
Gribelin family 188, 422, 426, 429, 430, 646
Gribi, Theophilus 634
gridiron pendulums 178, 278, 404
  John Harrison 321
Griesbaum family 310
Griessbach, Richard 418
Grimthorpe, Lord 403, 514
Grinkin, Robert junior 194
Grinkin, Robert senior 192f
Gros, Charles 631, 641
Großenwieden, sundial 109
Grossmann, Jules 631
Grossmann, Moritz 629
  Note Calendar for Watchmakers 641
grotesque (grottesche) 423, 656
Gruber, Hans 166f, 196
Grundzüge der physiologischen Psychologie (Wundt) 575
Grünwald, Simon 440
Guang clocks 51–2, 51n
Guangxu, Emperor 51
Guangzhou
  clockmaking in 46–50, 51, 52n, 55, 444
  Watch Factory 58
guard-tower clocks 153, 155f
Gudin, Jacques Jérôme 431
Guerier, Marin 160
Gugumus brothers 365
guibiaos 59
Le guide-Manuel de l'horloge (Saunier) 400
guilds, craftsmen 215–18, 378n
Guillaume, Charles-Edouard 374, 374n, 408, 415, 482, 492, 631
guilloché 656
Gujarat 35
Gulliver's Travels (Swift) 609–10
Gundobad 137
Gunmaking in Japan 65
Gunter, Edmond 234
Guo Shoujing 44
gut line, spring-driven clocks 158
Gutkaes, Friedrich 335
Guyou, Emile 344
Guyoux, Jean Marie Victor 249

H

Ḥabash ibn ʿAbd Allāh al-Marwazī al-Ḥāib 83

al-Ḥabbāq, al-Tilimsānī, Muḥammad ibn 96
Habrecht, Isaac 149, 282, 293, 294
Habrecht, Josias 149, 293
Habrecht family, clock repairs 542
Hachette, Jean Nicolas Pierre 248
haggling 218–19
Haghartsin, sundial 108
Hagia Sophia, *horologion* 95
Hague clocks 171, 172f, 174, 178, 180, 180f
Haguenau (France), astronomical clock 280
Hahl, Augustus 510
Hahn, Philipp Matthäus 206, 249, 273, 380n, 574, 624
Hahn, Theodore 592–3
  watchmen's clocks 593f
Haigh, Daniel 103, 109
Haldiman, Beat 404
Hales, Stephen 570
half-hunter wristwatches 487
Hall, John 109
hall marking 198
Haller, Thomas 379, 381
Halley, Edmund 319, 321
hallmarking, British 349
ham dial, Herculaneum 114
Hambledon 12
Hamberger, George Erhard 650
Hamblet, J. Jr. 592–3
Hamburg American Clock Company 586
Hamburg Chronometer Works 418
Hamburg-Amerikanische Uhrenfabrik 380, 382
Hamilton company, USA 362, 375, 388
  electric wristwatches 491, 526
  Model 21 417, 418
Hamiyyat Allah al-Astūrlabī 115
Hampden Watch Company 360
Hampton Court, astronomical clock 149, 281, 645
Han Gonglian 44
hand finished watches 354
handbag watches 561
handbooks, practical work 633–5
Handbuch der Uhrmacherkunst (Georgi) 633
Handbuch für Landuhrmacher (Auch) 633
Hanet, Nicolas 172
hanging barrel 49
hanging lantern clocks, Japan 66
Hanke, Julius 632
Hanover
  Historisches Museum 153n, 155f
  tower-guardian clock 153n, 155f
Hanwell, pendulum clock 142
Hansa region 144, 148
Harder, Dr Oscar E. 492

INDEX

Hardy, William 338, 502
Harland, Admiral Robert 334
Harrān 84
Harris, Dennis 484
Harris, Joseph 572
Harrison, James III 403, 469
Harrison, James IV 403
Harrison, John 178, 319–24, 320f, 339, 624, 630
   air-density compensation 403
   H1–H3 timekeepers 321–7
   H4 timekeeper 201, 324, 327–30
   H5 timekeeper 329
   longitude determination 204
   pendulum clocks 319–21, 624
   temperature compensation 321, 332
Harrison, William 327–8
Harūn al-Rashīd, 121, 290, 644, 648
Harsdörfer, Georg Philip 247
Hartford, Devon, St John's Chapel clock 143
Hartmann, Georg 239, 648
Hartnup, John, senior 406
Harvard College Observatory 414, 499
Harvey, Robert 169
Harwood
   John 489
   Self-Winding Watch Company 489
Hasan ibn 'Alī 86n
Hasan Kashif Jarkis 92
Hasius, Jacob 174, 429
Hasluck, Jacob 210
Hasluck, Paul 552, 633
Hassler, Hans Leo 296
Haswell, J. Eric 629
Hatot, Etablissement Leon 418, 523
Hatton, Thomas 629, xii
Hattori-Seiko 73–5, 375
Hauser, Eduard 379
Havard, Henri 562n, 597
Haydn, Joseph 296
Haye, Thomas 243, 570
Hayek, Nicolas 527
al-Haytham 85
Head and Tail of the Dragon (moon phases) 258
Heath and Wing 574
Heavenly Pagoda clock 51
Heidegger, John James 610
Heine, trade catalogue 640
Hele, Peter 185–6 *see also* Henlein, Peter
helical springs 157n, 334
heliocentric system 271
heliostats 579–81, 580f
heliotrope 5, 23
Helot, Jean 545
Helwig, A. 635
hemihedral axes 506
hemispherical dials 14, 14f, 86
Hendaye 344
Henlein, Andreas 186

Henlein, Peter 158, 162, 186, 186n, 218n, 221n
Henrion, Didier 239
Henry of Langenstein 603n
Henry VIII 168, 559
Heraeus-Vacuumschmelze company 374
Herat library water clock 98
Herculaneum 114, 248
Hercules clock
   Gaza 94, 95
   Rennaissance 163, 165f
Hérigone, Pierre 244
Herlin, Christian 293
Hermann the Lame 234
hermetic watches 561
Hermitage Museum, St Peterburg 182, 448, 451, 459
Hero of Alexandria 10n, 11, 94–6, 127, 290
   automata 290
   mechanical theatre 133f
   self-feeding lamps 135
Hero of Byzantium 79
Herodotus 5–6
Héroy, Noël 227
Herrmann, H. 632
Herschel, William 273
Hertz, Heinrich 498
Hesdin Palace clock 129
Hesdin Palace water features 122
Hetzel, Max 492, 526
Heuer Watch Company 344, 490, 584, 587
Hevelius, Johannes 314
Hewitt, Thomas P. 349
Hideyoshi Toyotomi 63
Hieroglyphica 643n
high-end watchmaking 354
high-frequency balances 325
Hight, Robert 588
Hilberg, Wolfgang 388
Hildemar de Corbie 121
Hilderson, John 315
Hill, Donald 79, 96–6, 127, 134–6
Hillmann, Bruno 635
Hindu circles 119, 656
Hindu temples, water clocks used in 35, 39
Hipp, Matthäus 502, 518–19
   chronograph 577f
   chronoscopes 344, 575–6
Hipp toggles 523
Hippias, water clock 10
Hirsch, Adolphe 519, 575
Hisashige Tanaka 72
historical surveys
   20th century developments 651–2
   Claude IV Raillard 647–8
   Ferdinand Berthoud 650
   Henri Chesneau 644
   Jacques Allexandre 648–9

Jean André Lepaute 158n, 627
Johann Beckmann 649–50
John Carte 645–6
Marin Estienne 646
Raillard, Claude IV 647–9
Thomas Powell 644
Vergil, Polydore 643–4
William Derham, 644–5
historicism 438–40
   watch 439f
Historisches Museum Hanover 153n, 155f
History of Science Museum, Oxford 29, 30f, 85n–6n, 114, 114n, 121, 574n
Hogarth, William 610, 611f, 612
hog's bristle regulator 186, 656
Hohwu, Andreas 412
Holbein, Hans 609
Holland
   longcase clocks 176
   musical clocks 297
   stoel clocks 183
   watchmaking, 16th–18th century 189–91, 210–11
Holmden, John George 405
Holmes, Robert 315
Homer 5, 436
Hong Kong Observatory, time ball 498f
hooded clocks 183
Hooke, Robert 127, 157n, 173n, 182, 196, 316, 318, 544, 645
Hope-Jones, Frank 502, 513, 517, 520
The Hopeless Dawn (Bramley) 614
hora completa 145
hora incipiens 146
Hora Italica 145, 146
Horapollon the Younger 643n
horarium (timetable), medieval Latin world 101
horary quadrants 91, 234
horizon, visual/virtual 93
horizontal dials 30, 31f
   fixed latitude 83
   Islam 82
L'Horloge (Baudelaire) 614
Horlogeographie pratique ou la manière de faire les Horloges à poids (Beuriot) 626
horloges d'edifice 144
horn, watchcases 197
Hornchurch, pendulum clock 142
Hornsby, Thomas 574
horizon
   visual 93
   virtual 93
horologia 9–10
   medieval Latin world 121–3
   precision 17
   public clocks 138
   shadow-length schemes 113

St Willibrord 112f
sundials 12–17
water clocks 10–12
horologia hiberna (winter clock) 11
horologion
   Antioch 94
   Byzantium 94–5
Horological Conservation and Restoration conference 554
Horological Dialogues (Smith) xiii, 543, 626
Horological Disquisitions (Smith) 544
horological guilds 216t, 216
horological hat 602
Horological Journal 348, 411, 414, 550–1, 632, 638, 651
horological oil 550
Horological Road, West to East 46
horological schools 538, 631
horological societies 631, 632
Horologiographia (Münster) 116, 622
Horologium (Huygens) 314, 630
Horologium Floræ (Linnaeus) 469
Horologium Oscillatorium (Huygens) 315
Horologium Sapientiae (Suso) 602, 603f
horologium viatorum 113
horology, definitions and meanings xii–xiii
horoscopion (sundial) 116
Horton & Morrison company 419
Horwitt, Nathan George 441
Hörz, Philipp 280
Houblin, Eustache François 216, 219n
   cartel 217f
Houdon, Jean Antoine 435
hour, subdivisions, medieval Latin world 101t
hour angle 234n, 656
hour hand 146
hour lines, sundials 13, 13n
hour systems xiii–xv, 37, 656
   Early Modern period 232
   medieval Latin world 102–5
hour wheels 153, 656
hours
   ab occasu (Italian) 118
   ab ortu (Babylonian) 118
hours, Greek 6
hours of time/hours of prayer 102
houses (domiciles) 267, 656
Houzeau & Lancaster 621n
How to Manage and regulate Clocks and Watches (Berthoud) 545–6
E. Howard and Company 360
Howard, Edward 359, 467
Howard Clock Company 467–8, 592
Howard Miller Clock Company 357
Huacheng Watch Case Factory 53
Huang Luzhuang 51n
Huaud family 212, 428
Huber, Bernhard 157n
Huber, Johann Jacob 336, 338

INDEX

Hugo (film) 615
Huguenots
  leaving France 206
  in London 178, 193, 197–8
  in Netherlands 211
  in United States 356
  watchmaking 188, 189, 191
Huguet, Jacques 430
Huguet, N. 341n
Hyu-Yuan 37
The Human Sundial (Flötner) 609
Hunt, William Holman 613
Hunter, Ernest R. 584
hunter movements 481, 482
hunter wristwatches 486–7
hunting cases 559, 656
Huntsman, Benjamin 222
Hurtu, Jacques 425
Huxley, Aldous 615
Huygens, Christiaan 178n, 272, 314–7
  balance spring 182, 316–17, 630
  correction table 178
  cycloidal cheeks 178, 178n, 648
  endless chain (winding system) 471
  equation of time 178
  pendulum clock 142, 171, 314–15, 315f, 624, 645
  time determination 544
Huygens, Constantyn 544
Huygens, Steven 174
Hynam, Robert 179

I

Ibn ʿAbd al-Barr 81
Ibn al-Akfānī 89
Ibn al-Haytham, Abū ʿAlī al-Ḥasan ibn al-Ḥasan (Alhazen) 84, 85
Ibn al-Mahallabī 89
Ibn al-Nadīm 82–4, 96
Ibn al-Raqqām 88
Ibn al-Shāṭir 30, 89–90, 238
Ibn al-Zarqellu 96
Ibn Bāṣo 88
Ibn Tulum mosque, Cairo 88
Ibn Yūnus 84
Ibrāhīm ibn Sinān 77, 82, 84
Ieyasu, Tokugawa 63
Ilay, Early of 321
Ilbert, Courtenay A. 651, 651n
Ilbery, James 49, 456n
Ilbery, John 49
Ilbery, William 49, 213, 456n, 460, 461f
ilm al-mīqāt 81
imitation watches 560
immigrant watchmakers, England 191
Imperial Clockmaking Workshop, China 444, 445n, 454, 456–7
imperial clocks, China 47–8, 48f
import quotas/duties, Britain 351, 354
Imser, Philip 254, 257–63, 264t, 267n

Imser planetary clock 254, 258, 258f, 260–3, 264t, 267n
Incabloc 488–9
incense sticks 41–2; *see also* fire clocks
incense-clocks 135
inclined cylindrical dials 15, 15f
inclining dials 239
inclining–declining planar dials 244
India
  column dials 28–30
  gnomons 27–8
  horizontal dials 30
  Indian Standard Time 38–9
  musical and automaton clocks 302
  ring dials 28
  standardized time 522
  sundials as sculpture 39–40
  units, time measurement xiv, 27
  vocabulary 32–5
  water clocks 30–5, 39
Indian circles 119, 656
Indian Standard Time 38–9
Indicateur Davoine 642
Indicateur Suisse 642
Indonesia, sinking bowl water-clocks 35, 37
industrialization 184
  England 176
  France 365–6, 391, 400–1
  German clockmaking 378–9, 381
  interchangeable parts 463–4
  network infrastructures 509
  new tempo 614
  night watchmen 591
  standardized time 495, 496
inflow water-clocks 659
  China 43
  Japan 59
Ingatestone Hall, roasting-jacks 595
Ingersoll, Robert 349, 492
Ingold, Pierre-Frédéric 348, 358, 410, 474
Ingraham, Edward 357
Ingraham, Elias 356–7
Ino Tadataka Museum 67
Institut du Monde Arabe, Paris 29n
Institute of Conservation (ICON) 553, 554
integral balances 415, 417
intempestum (central hours of the night) 99
International Centre for the Study of the Preservation and Restoration of Cultural Property (ICCROM) 553
International Exhibitions 364
  Besançon 368
  Brussels 1958 441
  London 1851 348, 364, 395, 397, 401, 411, 414, 474
  Paris 1849 396–7

  Paris 1855 364, 368, 373, 395, 401, 533
  Paris 1878 440
  Paris 1937 441
  Philadelphia 1876 364, 368, 372, 373, 398, 641
  Sidney 1879 465
International Institute for Conservation (IIC) 553
International Meridian Conference, Washington 249
International Watch Company of Schaffhausen 361
Invar (alloy) 374, 415, 482, 502, 656
invariable hours *see* equal hours
'Iran, the philosopher' 126
iron
  clockmaking 154, 164, 222
  recycling of 139, 223
irrigation clock, sinking bowl 36
Isaac b. Sid 133
al-Isfahānī, Muḥammad ibn Ḥāmid ibn Maḥmūd 89
Isidore of Seville 99
Islam
  sundials 81–94
  water clocks 95–8
isochronism 162, 628, 631, 656
  balance spring 196, 327, 331, 332–3
  oscillator 317
Istanbul observatory 29n, 312
Italian hours xv, 118, 120, 232, 243
Italy
  clockmaking, 16th & 17th century 166f, 167
  lantern clocks 183
  public clocks 466
  table clocks 180
IWC 482, 567

J

al-Jabartī, ʿAbd al-Rhaman ibn Ḥasan 92
al-Jabartī, Ḥasan ibn Ibrāhīm ibn Ḥasan al-Zaylaʿī 92
Jaccard, Hélène 538
jacks and jaquemarts 293–4
Jacob, Jean-Aime 412
  movements 413f
Jacobus de Novaria 141
Jacot-Guillarmod brothers 226, 227
Jacques, William 198
Jacques Depollier & Son 488
Jacquinot, Dominique xiii
Jaeger LeCoultre 441, 565, 587, 588
Jagellonian globes 570
Jahāngīr, Emperor 37
Jahn, Gustav 636
Jaipur
  Museum of Indology 28
  observatory 28, 29n, 30
Jakob, Hans 184

James Ritchie & Son 470
Jamnitzer, Wentzel 231, 234
  altitude dial 233f
Janvier, Antide 181, 246, 274, 275f, 276, 278, 281, 284, 286, 342, 404, 436, 570, 624, 628, 634
  astronomical clock 275f
  decimal time 342
  Etrennes chronométriques 641–2
  meridians 246
  precision regulators 404
Japan
  clock- and watchmaking 357
  clockmaking 65–72
  clock types 66, 68f
  electronic wristwatches 492
  European clocks 61–5
  factories 73–4
  fire clocks and incense sticks 60–1
  hour systems xiii
  Myriad year clock 285
  names of clockmakers in 65
  sundials and water clocks 59–60
  watch and clock industry 72–5
Japanese scenes, watches 438
japanning 176
Japy, Frédéric 229, 358, 393, 395n
Japy family/company 358, 363
  clockmaking 395–6, 397–9, 400
  machine tools 366
  metronomes 584
  watchmaking 393
Japy Frères & Cie 399, 440
Jaquard, Sntoine 423
Jaquemart watches 307
Jaques, Auguste 488
Jaquet, Eugène 635
Jaquet-Droz, Henri-Louis 299, 448
Jaquet-Droz, Pierre 177, 227, 297, 299, 431, 565
Jaquet-Droz 49, 213, 227, 354, 448n, 460, 473
  automata 299–300
  musical clock 431
  painting on enamel 435, 437
  workshops 303–5
Jaubert, Pierre xiii, 221, 224, 649
Jauch & Schmid 589
al-Jazarī, Ibn al-Razzāz 96, 136, 291
Jean de Hautefeuille 158
JeanRichard 212, 624
Jeffries, Richard 597
Jefferson, Elizabeth 532
Jefferys watch 204, 327
Jehan de Lyckbourg 157n
Jemina family 464, 466
Jendritzki, Hans 635
Jennings, Henry Constantine 133
Jerger, Wilhelm 379
Jerome, Chauncey 356
Jerome, Noble 356
Jerome, St 605

742

INDEX

jewelled lever escapements 492
jewelled watches 361
jewelling 200, 204, 330
Jiaqing, Emperor 48, 51, 53, 459, 461
Ji Tanran 51
Jin Xing Industrial Association 53, 55
jobbing work, clock repairs 546
Jocelyn de Brakelond 123
Johann, Nicolaus Alexis 624
Johannsen & Co 411, 418
John Gershom Parkington Memorial Collection, Bury St Edmunds 30n
John of Saxony 119
John of Wales 602
John Smedley textile mill 512–13
John Smith and Sons 591
jokobans (fire clocks) 60–1, 62f
Jolly, Josias 424
Joly, Jacques 196
Jones, Bernard 552
Jones, Henry 173
Jones, Robert Lewis 517
Jones, William 574
Josef Kaiser GmbH 387
Joseph, Father Hermann 125
Jöstel, Melchior 267
Journal des Sçavans 630, 637
Journal for Factory, Manufacture, Act and Fashion 637
journals 400, 550–1, 637–9, 651, 651n
Juggler (Chagall) 618
Junghans company 375, 379, 379n, 380f, 381, 382–4, 390, 525, 586
  alarm clock 'gymnast' 384f
  British copies 352
  competition for France 398
  quartz clocks 388
  trade catalogue 640
Junghans Microtech GmbH 384
Jupiter, satellites 272, 273, 274, 313, 626
Jura region 367
  clockmaking 177, 184, 464
  clockmaking schools 400
  female workers 536, 538
  horological industry 226, 297, 365–7, 368–9
  manufacturing hinterland 229
  watchmaking 211, 415, 460
Jürgensen, Urban 629
Jürgensen family 411
Jury, Frederick 448

K

Kaeser, Moilliet & Co., Schweizerische Celluloidwaren Fabrik 486
Kafka, Franz 616
Kaifeng, China, water clock 291
Kalendarium (Muller) 116
Kalimpong 30
Kandili Observatory, Istanbul 29n
Kangxi, Emperor 47
Karner family, Nuremberg 224
Karr-Schmidt, Suzanne 609
Kasā-yantra *see* column dials
Kassel 159
  centre of astronomy 270, 312
  clocks 644, 644n, 646
  planetary clock 254, 266
Katechismus der Uhrmacherkunst (Herrmann) 632
Kater, Henry 403, 469
Kautilya 27
keihyos (sundials) 59
Keith, William 359
Kelvin, Lord 283, 414, 415
Kemlo, Francis 551
Kemlo's Watch Repairer's Guide 551
Kendall, Larcum 630
  K1 timekeeper 329
  K2 timekeeper 333
Kepler, Johannes 163, 266, 570
  Mysterium Cosmographicum 271
  planetary clocks 271–2
Kerela 38
Kessels, Heinrich Johann 412
keyframe 656
keyless winding 566
key-making industry 565–6
Khalīl ibn Ramtash 89
al-Khamāʾirī, Muḥammad ibn Fattūḥ 89
al-Khāzinī 133
al-Khwārizmī, Muḥammad ibn Mūsā 83–4, 133
Kienzle company 587, 589; *see also* Schlenker & Kienzle
al-Kindī, Abū Yūsuf Yaʿqūb ibn Isḥāq 84
King, David A. 83–4, 87–92, 94, 96, 98, 115–6
Kinloch, Dr David 606
Kintaro Hattori 73
Kinzing, Peter 177, 297, 300, 335
Kirby Beard company 482, 483f
Kircher, Athanasius 247, 290, 624, 644
Kitāb al-Hiyal (Book of Ingenious Devices) 135
Kitakyushu museum 66
Kittel, Adolph August 412
Kleemayer, Christian Ernst 297
Klein, Johannes 273, 624
Klingenberg, Hans Ulrich 491
Klinghammer, Frédéric 280
Klock, Pieter 174
Knapp, John 641
Knibb, Joseph 142, 173, 173n, 181, 317, 318
Knibb, Samuel 173, 429
Knittl, Franz Anton 249
Knoblich, Theodor 412
Knox, Robert 36
Koch, Hans 187, 239, 423
Korea, clockmaking 357–8
Kota 35
Kraków, time announcer 292
Kratzer, Nicolas 231, 239
Krause, Jean Christian 436
Krille, Friedrich Moritz 412
Krotz-Vogel, Werner 252
Krumm, Gustav 632
kufic script 88, 88n
Kuhlo, Ernst 527
Kulibin, Ivan Petrovich 182
Kullberg, Victor 410, 412, 417, 418
Kundo company 388, 388n, 390
Kunozan Toshogu Museum, Shizuoka 64f
Kunozan Toshogu Shrine, clock 63
Kunsthistorisches Museum, Vienna 159n, 160n, 312, 423n, 605
Kunstkammer 269, 294
al-Kurdī, Ibrāhīm al-Faradī 91
Kvarnberg, Ivan Ivanovich 503

L

La Chaux-de-Fonds, Switzerland; *see also* Musée international d'horlogerie
  clockmaking 177
  horological industry 367
  metronomes 584
  musical clocks 297
  watchmaking 211, 489
La France Horlogère (journal) 400
Labarte, Jules 153n
Lacher & Co 388
Laco-Uhrenfabrik, Pforzheim 382
ladies watches 476–7
The Lady's Last Stake (Hogarth) 612
Lagrange, Joseph Louis 342
Laidrich, Edouard 50
Lalande, Jérôme 237n, 247
Lalla 32
Lambert, Jean-Henri 242
Lambert clocking-in clocks 364, 364f
lamellas 575, 576
lamp-clocks 135, 135f
al-Lamtī 87
Lancashire, watchmaking 225, 229
Lancashire Watch Company 349
The Land Speculator (Beard) 614
Landes, David xii, 9, 9n, 591
Landow, George P. 613
Lang, Fritz 616n
Lang, Hans, astronomical clock 283–4, 284f
Lange, Ferdinand Adolph 381
Lange & Söhne, A 382, 384
Lange company 412, 418
Langenbucher, Veit 296
Langevin, Paul 506
Lansberg, Philip 240
lantern clocks 173f, 182, 183f
  England 168, 169f
  Japan 66–7
lantern pinions 594, 656
Laplace, Pierre-Simon 273, 342
Latham, John 227
latitude
  dialling tables 242–3
  sundials 231, 232n
latitude-equivalent 244
latitude-independent dials 252
Latz, Jean Pierre 431
Laurence of Stoke 130
Laurent, Félix 456
Lavet, Marius 418, 517, 523
LCD (liquid crystal display) watches 492
Leadbetter, Charles 622
leap years 273
  Antide Janvier 274
  Beauvais astronomical clock 279
  Besançon astronomical clock 278
  calendar clocks 286
  Copenhagen astronomical clock 282
  Messina astronomical clock 280
leather
  bracelets 566
  cases for clocks 563
  Norwegian belts 559
  watchcases 197, 557, 558
  workers 215, 224
  wrist straps 479
  wristlets 475f
le Blon, Michel 423
Lebru, Philippe 366
Leclanché, George 514
LeCoultre company 587
Le Count, Daniel 430
Ledieu 628
Lefebvre, François 426
Lefranc, Etablissements 135
Légaré, Gilles 426
Le Goff, Jacques 138
Lekceh 198
Leicester, Society of Watch and Clockmakers 546
Le Locle
  clockmaking 177
  Musée d'horlogerie 304, 304n, 307, 308n, 429n, 430n
  watchmaking 211, 415
Le Noir, Etienne 433
Le Noir, Jean-Baptiste 648
Le Noir, Simon 648
Le Play, Frederick 364
Le Roy, Charles, 435
Le Roy et fils 476–7, 489
  ladies watches 477f
Le Roy, Julien 142, 207, 220n, 238n, 330, 432f, 433, 545, 636, 647
Le Roy, Pierre 204, 207, 323, 325, 330, 331–4, 630, 641

INDEX

Le Roy, Pierre (Continued)
   montre marine 331–2, 332f, 333
   publications 630
   Règle Artificielle du Temps 330, 545
Le Verrier, Urbain 366
Leeds, Temple Newsam House 297
legitimate day 99
Leiden Sphere 574
Leitfaden für Uhrmacherlehrlinge (Sievert) 632
Lemaindre, Nicolas 424
Lemaire, Jean 180
Lemercier, Balthazar 425
Lenoir, Nicolas 216
Lenoir, Robert 350
Lenormand, Louis Sébastien 550, 628
Lenzkirch company 379, 382, 640
Leopold, John 623
Leopoldina (German Academy) 637
Lepaute, Henri 365
Lepaute, Jean André 181, 207, 223n, 435, 624, 627, 649
   decimal time 342
   flat-bed frames 464
   lubrication 546
Lepautre, Jean 428
Lépée, Auguste 395
L'Epée company 366
Lépine, Jean-Antoine 207, 338, 404, 438
   watch 209f
Lépine calibre 49
Lépine watches 480–1, 481f
Leroux, John 202
Leroy & Cie 344, 415, 502, 503, 517, 522
   Chronostat models 418, 524
Leroy-Beaulieu, Paul 535
Leroy, Léon 439f
Leschot, Jean-Frédéric 297, 308, 448, 460, 473
Lestourgeon, David 197, 198
Letter, Jean Baptiste 195
Leutmann, Johann Georg 206, 624, 626–7
lever escapements 182, 201–3, 331, 417, 481, 656
Levy frères 438
Leybourn, William 243
Leyland, Thomas 405
Li Dongshan 53
Li Lan 43
Li Zhizao 46
Liang Lingzan 43
Liaocheng Zhongtai Watch Co. Ltd 58
liberties 215, 656
Libro del Saber de Astrologia (el-Levi) 133, 136
Libros del Saber 125–6, 128
Liechti family, Winterthur 154, 184
Liège, clockmakers/clockmaking 215–16

Lierre (Belgium), astronomical clock 280
Lieutaud, Balthazar 435
Life of Alfred (Asser) 136
lighthouses, time-switches 527
lignum vitae 321
Liman, Louis-Marie-Josserand de Raguet de Brancion de 550
Lincoln, John
   clock repairs 542
Lindley, James 456n
Linnaeus, Carl
   floral clock 469
Linnartz, Jodef 632
Linwood, John 596
Liobard 250
Lip, Fred 368, 539
Lip company 362, 539
   advertisement 369f
   electric wristwatches 491, 526
   trades unions 369
Lipmann, Ernest 363, 368, 537
Lippman, Garriel 506
liquid crystal display (LCD) watches 362, 492
liquor licensing, standardized time 496
Lissajous, Jules 506
literature of horology 621
   17th–18th centuries 624–9
   almanacs 641–2
   early mechanical timekeepers 623–4
   journals 637–9
   practical guides 633–5
   scientific approaches 630–1
   sundials 621–3
   textbooks 631–3
   trade catalogues 639–41
   user manuals 635–6
literary interpretations *see* artistic and literary interpretations
Litherland, Davies & Co 406
Litherland, Peter 202
   watch 203f
Liutprand, Bishop of Cremona 122
Liverpool
   Bidston Observatory 406, 414
   chronometers 406
Livre du Ciel et du Monde (Oresme) 603
Lloyd, Alan 126, 589
Lloyd, Harold 615
local mean-time 249
locations, public clocks 144
locking plate 656
locksmiths 215, 225
Lockton, sundial 109
locust's leg pillar dial 86
Loeske, L. 631
logarithms 239, 266
London
   Blacksmith's Company 191, 222n
   centre of trade 228
   clock- and watchmaking 404–5

   Clockmakers' Company *see* Clockmakers' Company, Worshipful Company of
   Goldsmiths' Company 191, 198
   horological industry 302
   horologium, St Paul's Cathedral 139
   network infrastructures 511
   watchmaking 200
longcase clocks 173–5
   18th century 176–7
   baroque style 429
   precision 178
Longines 376, 441, 482, 486, 490, 526, 534, 536n, 561
   chronograph 579f
Longitude, An Appeal to the Public (Earnshaw) 630
Longitude, Board of *see* Board of Longitude
longitude, dialling tables 242–3
Longitude Act 1714 319
longitude determination 161, 203, 249, 313–16, 406, 626
Lorenzetti, Ambrogio 600, 601f
Lorenzo de' Medici 254
Lorenzo della Volpaia, planetary clock 262
Lorgoner, Wauter 141, 223n, 542
Lossier, Louis 411, 631
Loup, Pierre 50
Louis XI 136, 163
Louis XV style 176, 181
Louis XVI style 176, 182, 184
Loulié, Etienne 582
   chronomètre 583f
Louteau, Pierre 189
Louvre Museum
   book clocks 423n
   clocks 431n
   form watches 423n
   musical and automaton clocks 297n
   musical watch 305
   Olivier collection 433n
   painting on enamel 437n
   rococo decorations 431n, 433
   sundial 247
   watches 192n, 424n, 425n, 426, 428n
Low Countries, carillons 292
Lowndes, Jonathan 544
   domed spring clock 179f
Loyseau, J. 428f
Lu Cai 43
Lübeck
   astronomical clock 281
   St Mary church clock 126
lubrication 140, 200, 542, 546
   horological oil 550, 630
Lucian 10
Ludlam, William 329
Luis de Granada 606
Lully, Jean-Baptiste 584
luminous dials, wristwatches 485–6

lunar distance method 313, 319
lunar eclipses 311, 312
lunar information, astronomical clocks 148, 274, 282, 284
Lunar Society, Birmingham 591
lunar sphere 276
lunarium 573
Lund, astronomical clock 126, 281
Lund, John Alexander 515
Lund Cathedral clock 126
luxury goods 163–4, 189, 215
   haggling 218–19
luxury market 354
luxury objects 391, 401, 421, 436, 604
   exports to China 454, 462
   gold clocks 422
   planetary clocks 268
   watches 424
   wristwatches 473, 493, 567–8
Lyon
   astronomical clock, St Jean Cathedral 148
   Musée des Beaux Arts 431n
   watchmaking 189
Lyons, Harold 419, 508–9
Lyra 20

M

Maar of Zurich 588
Macao 45
macaroni chains 560
Macartney, Lord 48, 445, 447, 456
Macerata, astronomical clock 281
Macgowan, Daniel J. 52, 52n
machine tools 366–7
Mackensie, Alexander 626n
Mâcon, clockmakers school 400n, 628
Macrobius 644
Madras Time 38
Madrid codex 132, 158
Maelzel, Johann Nepomuk 301, 582–4
   metronome 584f
Maestlin, Michael 272
Magellan, John Hyacinth 335
Magneta system 521
magnetic compass 90–1, 114, 237
magnetic declination 237, 657
magnetic forces 405
Magniac, Charles 49
Magniac, Francis 447, 455, 459
Magnier 628
Magritte frontispiece
Maḥmūd ibn al-Ḥasan al-Nīshī 92
mahogany cases 176
Maignan 241
Maillardet, Henry 301, 448, 451
Maillardet, Jean David 301
Maillard-Salins 366
maillechort 567f
Maimonides 85
mainsprings 627, 657

744

INDEX

China 51
  early watches 186
  English watchmaking 192
  spring-driven clocks 157–60
maintaining power 657
maintenance, contract 545
maintenance, early public clocks 141–2
Maison Chaumet 473
makers and methods, Antiquity 17–20
Mâle, Émile 602
Malta, water clock 122
mane (morning) 99
Manetti, Antonio 154
Mangot, Pierre 422
Manley, Henry 201, 433
Mannendokei 72
Mansura 87
mantel clocks 181, 433–5, 436
Mantio Ito 63
Mantova, astronomical clock 281
Manuel de L'apprenti horloger (Ayasse) 549–50, 633
Manuels Roret 399–400
manufacturing hinterlands 229
manufacture and trade, 16th–18th century 215
  centres of trade 228–9
  distribution 226–7
  guilds and regulation 215–18
  shops workshops and tools 218–21
  specialization 221–6
Manufrance company 400
Mao Zedong 55
Mappin & Webb, wristwatches 478
Mappin Brothers, Sheffield 477
  wristwatches 478f
al-Maqsī, Shihāb al-Dīn 89
Marcel, J. J. 88
Marclay, Christian 619
Marconi, Guglielmo 417, 498
Marcus Manilius 105
Margetts, George 337
Margotin 595
Margotin, Pierre 182
Margraf, Christoph 162
Marguerat, Charles-Armand 310
Marguerite, Mathieu 180, 182, 430
The Marine Chronometer, its History and Development (Gould) 417
marine chronometers 337f, 373, 628f
  demise 417–19
  England 404, 405, 410, 411
  Ferdinand Berthoud 627
  France 404
  maintenance 512
  quartz 418
  repairs 548
  standardized time 495
  Switzerland 415–17
marine clocks 313
marine timekeeper H1 321, 323
marine timekeeper H2 323–4, 334

marine timekeeper H3 324, 327
marine timekeeper H4 324, 327–31
marine timekeeper H5 329
markets, financial, standardized time 496
Markham, Robert 303
Markwick, Markham, & Perigal 213
Markwick Markham company 179
Marlys wristwatches 489
Marot, Daniel 429
marquetry 430, 436
Marqüch 198
Marradi, abbey sundial 103, 104f
al-Marrākushī, Abū ʿAlī Ḥasan ibn ʿAlī 28, 82, 85, 89n
  sundials 86–8, 91
Marriage À-la-Mode (Hogarth) 612
marriage alliances 532
Marrison, Warren 419, 502, 506–7, 511
Marston Magna 144
Martens, J. H. 636
Marti, Fritz 399, 489
Marti, Laurence 536
Marti-Roux & Cie 366, 397
Martial 101
Martin, Abraham 430
Martin, Benjamin 572
  table orrery 573f
Martincourt, Etienne 435
Martineau, Joseph 305
Martinelli, Domenico 134
Martinot, Balthazar 206
Martinot, Gilles (Paris) 316
Martinot, Gilles (Rouen) 432f
Martinot, Jérôme 570, 571f, 646
Martinot family 206, 430
Martyn, Henry 221
Masatsugu Fukada 65
Maskelyne, Nevil 328, 329, 333, 335, 336, 630
Masquerade Ticket (Hogarth) 610, 611f
mass culture 364
Massé, Yvon 116, 116n
Massey, Edward 202–3
Massey, Nicholas 179
mass-production 410, 474
Massy, Henry 197
Master Humphrey's Clock (Dickens) 613
masterpieces 216, 219, 220, 657
Masterson, Richard 194
materials
  clockmaking/watchmaking 220–1
  specialist suppliers 222
mathematics
  clockmaker's knowledge 263n, 266
  Islam 83–4
  planetary clocks 253, 263
  sundials 231
Mathieu, Claude 342
matins 102
Matthews, William 329

Matthieu, Césaire, Cardinal – archbishop of Besançon 278
matutinum (pre dawn) 99
Maurolico, Francesco 240
Mauthe, Friedrich 379, 387n
Mauthe Uhren 387
Mavelot, Charles 557n
May War Lee company 52
Mayet family 177n
Mayr, Otto 603n, 607
Mayrhofer, Carl Albert 510
Mazurier, Antoine 426
McCamp, John 7
McHattie, John 469
McKenna duties 349
mean solar time 248–9
mean-time meridian 247
Meccano parts, astronomical clocks 281
mechanics, Byzantium 94–5
Mechanik Lange & Söhne VEB 384
mechanism of apparent time 276
mechanistic philosophy 607
mechanization *see* industrialization
medicine, astral sciences 266n
medieval hour systems 104f, 104n
medieval Latin world 99–101
  computus 105–9
  early sundials 110–13
  hour systems 102–5
  monastic time-telling 101–2
  portable altitude dials 113–20
Mégevand, Laurence 368, 533
Mei, Pietro 467
Meihuali Watch & Clock Co, Ltd 53
Meissonnier, Juste-Aurèle 431
Melanchthon, Philipp 186, 267
Mell, Davis 224
Melland, Jean 557
memento mori imagery 599, 604
Mémoire sur la meilleure Manière de mesurer le Temps en Mer (Le Roy) 331
Mémoire sur le spiral réglant des chronométres et des montres (Phillips) 631
men's wristwatches 482
Mercer, Thomas 418
Mercury, observations 311
mercury clocks 125–6, 126f, 133
mercury gilding 533
Meridian of Augustus 23–4
meridians 244–6, 626, 657
meridies (midday) 99
Mérindol-Les-Oliviers, six-sector dial 106, 108f
Merlin, John Joseph 302, 448, 454, 596
Merton College, Oxford, clock 129, 139
Merz, Georg 578
mesh guards, wristwatches 486

Messina, Sicily, astronomical clock 280, 469
metalworking, early public clocks 140
metalworking guilds 216
Metamorphosis (Kafka) 616
metric system 342, 345
métromètres 582, 583f
metronomes 582–5
Metropolis (film) 616
Metropolitan Museum of Art, New York
  astrolabes 423n
  automata 423n
  clocks 435n, 436n
  forgeries 214
  musical and automaton clocks 295n, 296n
  rococo decorations 431n, 433n
  watches 189, 194n, 210n, 211n, 424n, 425n, 426n, 428n
Metzger, Jeremias 168, 423
Meung, Jean de 129, 154, 599n
Meuron & Cie 308
Meyer, Jörg 623
Meyerstein, Moritz 579
Meylan, Philippe Samuel 308, 309, 437
Michell, John 329
Michnik, Hermann 251
Microcosm planetary clock (Torriano) 263, 266
microtechnology 369
middle-temperature error (MTE) 405, 407–8, 414, 415
Midnight Modern Conversation (Hogarth) 612
Midnight's Children (Rushdie) 616
Milan
  early public clocks 145, 146
  monastery clock 129
  Museo Poldi Pezzoli 426n
Milham, Willis I. xii
military developments, wristwatches 477–8, 490
military timepieces 382
Mills, A. A. 126, 128n
mill clocks 138
miniature watches 205
miniaturized clocks 160–1, 161f
minute hand 146
minute-repeating watches 198–9
miqati (astronomers) 88
Miroglio 466
al-Miṣrī, Najm al-Dīn 89
missionaries to China 444
al-Mizzī, Muḥammad ibn Aḥmad 88, 88n, 89, 89n, 92, 96
mobile phones 493
modern movement 441
The modern watchmakers lathe and how to use it (Perkins) 635
Moebius, oil manufacturer 550

745

INDEX

Moinet, Louis 578, 628
Möllinger, Christian 297
Molyneux, Robert 409
moment of inertia 186, 657
monasteries
  early public clocks 140, 141
  sundials 80
  time-telling 101–2, 137
Mondanomalienuhr planetary clock (Bürgi) 266
Mondon, Thomas 431
Monge, Gaspard 246, 248
Mongruel, André 651–2
'monkey up the rope' (clock winding) 471
monograms 557
Montaigne 595, 596n
Montbéliard 363
Montecosaro, abbey sundial 103
montre marine 331–2, 332f, 333
moon, phases 258
moon ball 148
Montucla 236, 248
Moray, Sir Robert 128
Morbier clocks 177n
More, William 168
Morel, Nicolas, watch 428f
Morez clocks 177n
Morez-Morbier clocks 365, 464
Morgan, Octavius 650
Morin, Gabriel 280
Morin, Jean-Baptiste 313
mortality, clock as symbol 605, 607
Morteau 367
Moseley, Charles 359, 360
Moser, George Michael 433, 437
Mossman, James 470
Mossoke, Thomas 221
Motel, Jean-François Henri 404
motion work 470, 657
Mouret 281
Movado 441, 588
movement, engraved 50
movement manufacturers 356
movements en blanc 633
moving sphere 570, 571f
Moxon, Joseph 570, 595
Moynet, Auguste 640
Mrs Dalloway (Woolf) 615
Mudge, Thomas 178, 199, 201–2, 204, 325, 405, 433, 469, 629–30
  A Description with Plates of the TimeKeeper 630
  detached lever escapement 182, 331, 335–6
  equation of time 212
  expert panel 329
  gravity escapements 403
  remontoire 338
  temperature compensation 326
Mudge & Dutton 204
Mughals, water clocks 35

Mugnier 438
Muḥammad ibn al-Ṣabbāḥ 83
Muḥammad ibn Ismāʿīl Nifarawī 92
muhurtas 27
Mukhula dial 87, 115
Mul-Apin clay tablets 4–5
Mulchand, B. 38
Müller, Johannes 116, 235
multiple face dials 16, 17f
multiple hour dials, early public clocks 146
Munch, Edvard 618
Munster, astrolabe dial 148, 281
Münster, Sebastian 116, 232, 235, 236, 238, 239, 240, 242, 622, 624
Munz, Hervé 539
muqawwar, equinoctial dial 91
al-Murādī, Ibn Khalaf 133
Musashi Hirayama 66, 69f
Musée Carnavalet, Paris 398f
Musée Condé, Chantilly 421n
Musée d'Art et d'Histoire, Geneva 556n, 558f, 559f, 560f, 562f, 563f, 564f, 566f, 567f
Musée d'art et d'histoire, Neuchâtel 300n
Musée de l'horlogerie et de l'émaillerie, Geneva 196
Musée de Vannes 242n
Musée des Antiquités Nationales, St Germain-en-Laye 12n
Musée des Arts Décoratifs, Lyon 430n
Musée des Arts Décoratifs, Paris 430n, 435n, 438n, 440n
Musée des Arts et Métiers, Paris 135n, 178n, 331, 431n, 436n
Musée des Beaux Arts, Caen 646n
Musée des Beaux Arts, Lyon 431n
Musée d'Histoire des Sciences de Genève 235n, 236n
Musée d'horlogerie de Saint-Nicolas d'Aliermont 364f, 392f, 396f
Musée d'horlogerie, Le Locle 430n
  automaton watches 308n
  pendulum clocks 429n
  singing birds 304, 304n, 307
Musée du Petit-Palais, Paris 422n, 424n, 428n
Musée du Temps, Besançon 538n
Musée Internationale d'Horlogerie, La Chaux-de-Fonds
  Art Deco 441n
  astronomical clock 283
  early watches 191n
  form watches 440n
  historicism 438n
  musical clock 431
  painting on enamel 427f, 437n
  post modernism 442n
  table clock 422n
Musée national de la Renaissance, Ecouen 295n, 422n, 424n, 425n

Musée national des techniques, Paris 212n
Museo Galileo, Florence 235n
Museo Poldi Pezzoli, Milan 426n
museum collections 553–4
Museum of Egyptian Antiquities, Cairo 1n
Museum of Fine Arts, Boston 35n, 610n
Museum of London 191n, 213n
Museum of Modern Art, New York 441
Museum of the Stoa of Attalos 12n
musical and automaton clocks 289–90, 310, 449f, 450f
  14th–16th century developments 291–6, 295f
  17th–18th century developments 296–309
  19th–20th century developments 309–10
  decorative features 423
  early developments 290–1
  exports to East Indies 451
musical chronometers *see* metronomes
musical instruments 169f
musical public clocks 150
musical watches 302, 303f, 305–8, 306f, 307f, 309
Muslim day 82
Mustansiyya college water clock, Baghdad 96
Musurgia Universalis (Kircher) 290
muwaqqits (religious timekeepers) 88, 89
Myanmar, sinking bowl water-clocks 36
Myriad year clock 72, 72f, 285
Mysterium Cosmographicum (Kepler) 271

N

Nādikā-yantra 30–2
Nagasaki, clockmaking 65
nag's head striking 168, 657
Najm al-Dīn 89
nālikas 27
Nanjing clocks 45, 51, 52, 53, 54f, 55
  watches 56, 58
Napier, John 239
Naples, National Museum 114n
Napoleon in his Study at the Tuileries (David) 610, 612f
Nardin, Ulysse 417
nasaba sāʿat 81
Nashua Watch Company 359
National Army Museum 474, 475f
National Association of Watch and Clock Collectors (USA) 553
National Bureau of Standards (NBS) 508
National College of Horology 354

National Museum of Nature and Science, Tokyo 59, 61, 72
National Watch Company of Elgin, Illinois 360
Nativity Clock 302
NATO strap 480
natural day 99, 103
Natural History (Pliny) 8, 9
naturalia 163
nature, idealization 181
Nautical Almanac 319
Navicula de Venetiis sundial 115–16, 232, 235, 236f
navigation at sea 313
Nayi 36–7
al-Nayrīzī, al-Faḍl ibn Ḥātim 84n
Naze, Jean 422
necklace chains 560
Needham, Joseph 37, 42n, 43n, 44n, 46n, 51n, 59n, 291, 594
nefs (miniatures) 294
Nelthropp, Rev H. L. 551
neoclassicism 433–8, 454
neo-gothic 438
Nepal, sinking bowl water-clocks 36
Neptune, discovery 273
Neßtfellm, Johann Georg 273
Netherlands
  longcase clocks 176
  musical clocks 297
  stoel clocks 183
  watchmaking, 16th–18th century 189–91, 210–11
Netherlands Sundial Society 251n
network infrastructures 509, 590
network timing 529
networked computers 529
networks of precision times 509–23
Neuchâtel
  clockmaking 177
  Electronic Horology Centre 375
  Musée d'art et d'histoire 300n
  observatory 373, 415, 575
Neuchâteloise clocks 181, 184
Neudorfer, Johann 186
New and Complete Clock and Watchmakers' Manual (Booth) 634
New Haven Clock Company 356, 588
New York
  Metropolitan Museum of Art *see* Metropolitan Museum of Art, New York
  Museum of Modern Art 441
New Zealand xii, 348, 352
Newark Watch Company 359
Newsam, Bartholomew 168
Newton, Isaac 127, 317, 319
Newton, William 574
Newtonian mechanics, planets 574
Nice, Nautical Club of 344
Nicholas of Cusa 127
Nicholson, William 403

INDEX

Nicholson's Journal of Natural Philosophy 630
nickel based alloys 374
Nicole, Adolphe 576
Nicole Nielsen company 350, 586
Nicolson, Alexander McLean 507
Niemecz, Joseph 296
night clocks 180
night watches, China 41, 47
night-time tables 117
Nihonshoki 59
Nimrud, sinking bowl water-clocks 4n
Nineteen Eighty-Four (Orwell) 616
Nisbet, John 36
Nitot 473
Nivarox balance spring 374, 376, 492
Noakes, Derek 404
Nobel Prize, Charles-Edouard Guillaume 415
Nobunaga Oda 63
noctuaries *see* watchmen's clocks
nocturnals (astronomical instruments) 117–18, 117n, 118f, 121
nocturne 102
nodes 284, 285, 657
Nolin, Pierre 425
nomenclature, clock parts 623
nones 102, 103, 105
non-isochronic behaviour 162
non-uniform motions, planets 285
noon-guns 246
Norden, John 163n
Normal-Zeit GmbH 515
   electric clock systems 516f
Normand 183f
Normandy, clockmaking 365, 395–6, 399, 401
Norris, Joseph 174
North, John 130–2
North American Sundial Society 251n
north pole projection 126
Norton, Eardley 547
Norwegian belts 559
Norwich Cathedral, astronomical clock 129, 130, 139
Note Calendar for Watchmakers 641
Nouet 92
Nouveau Manuel Complet de l'horloger (Roret) 399
Nouveau manuel (Lenormand) 550
Nouveau traité géneral (Moinet) 628
Nouwen, Francis 190–1, 199
Nouwen, Michael 190–1, 223n, 225
Nova Acta Leopoldina 637
Nova reperta 146
Novgorod, Nizhny 182
Nūr al-Dīn Maḥmūd ibn Zangī 86
Nuremberg
   centre of trade 228
   closed trades 216
   development of watches 186
   Germanisches Nationalmuseum 156n
   guard-tower clock 153n
Nuremberg hours xv, 118, 157n, 657
nychthemeron xiv, 657
   China 41
   decimal time 344
   medieval hour systems 103
   medieval Latin world 99
   sixteen divisions 109

O

oak cases 177
objets de vertu 657
obliquity of the eliptic 244
Octo 587, 588
oculus *see* aperture-gnomons
Oddi, Muzio 240
Odobey family/company 365, 366, 464
Odyssey (Homer) 5
Oechslin, Ludwig 284
Ogilvie, Sheilagh 215n, 218n
'oignon' watch 206–7, 208f, 211, 214, 430, 557
oiling *see* lubrication
Old Clocks and watches and their makers (Britten) xii
The Old Curiosity Shop (Dickens) 613
Olivier, F & fils 344
Olivier, Théodore 248
Olomouc, astronomical clock 281
Olsen, Jens, 282
   astronomical clock 282, 283f, 284, 286
Omega 353
   dashboard clocks 587
   Marine divers' watches 490
   Museum 588
   Speedmaster Professional 491
   tuning fork timekeepers 526
   wristwatches 482, 483f
On the Disturbances of Pendulums and Balances, and on the Theory of Escapements (Airy) 407
On the Springing and Adjusting of Watches (Britten) 634
Oosterwyck, Severyn 180, 315, 544
Opera astronomica minora (Ptolemy) 621
Oppenordt, Gilles-Marie 176, 430
   floor-standing clock 432f
optical lattice clocks 509
Opticks (Smith) 544
Orchomenos, sundial 108
Oresme, Nicholas 602–3, 603n
Orford 323
organ clocks 177, 296–7
Organi, Giovanni degli 141
organum Ptolomei sundial 115–16
oriental origins, time measurement 1–5
Orlando (Woolf) 615
Orly-West airport (Paris), astronomical clock 280
ornamentistes 421, 428, 657
Oropos 7
orreries 570, 278–85, 570, 572–4
orthographic projection 242, 657
Orwell, George 616
Oshakan, sundial 108
Osmond, Robert 435
Ottery St Mary, astronomocial clock in 149
Ottheinrich, Elector of the Palatinate 254
Ottoman empire 91–2
Ottoman trade 229, 301–2, 443, 558
   musical and automaton clocks 301, 302–4
Oudin, Charles 438
Oughtred, William xiii, 237, 242, 626
ouroborus 657
Oursian, Nicolas 149, 542
outflow water-clocks 659
   Babylon 4
   China 42
   Egypt 3–4
   India 30–2
outworking 223n, 224n
   American watchmakiers 358
   German clockmakers 379
   Japanese clockmakers 357
   London clockmakers 405, 448
Oxford, Ashmolean Museum 186
   book clocks 423n
   form watches 423n
   table clocks 179n
   watch cases 425n
   watches 187n, 188n, 190, 191n, 194n, 196, 198, 425n, 430n
Oxford, History of Science Museum 29, 30f, 86n, 114, 114n, 121, 574n
Oxford, Wadham College 142, 173n
Oyster wristwatch 488
Ozanam, Jacques 134n, 136n, 234n, 236, 246, 247

P

Pacificus, Archdeacon of Verona 121, 648
Paddy, Sir William 606–7
Padua, astronomical clock 141
Padua Cathedral, clock dial 148f
Pagani, Catherine 47n
Pahozin 36
Paillard, Charles-Auguste 412
painted clock dials 154, 176, 184, 355
painting on enamel 426, 427f, 428, 428f, 558
pair-cases 197, 430, 557, 558f
pairs of watches 49, 303f, 309, 437, 451, 460, 560
Pajot, Léon Louis 582
métromètre 583f
Pajou, Augustin 435
Palabhā-yantra horizontal dial 30
Palace Museum, Beijing 49n, 451, 454
Palermo, water clock 122
Palladio, Andrea 433
palladium alloys, balance springs 412
Palladius, Rutilius, shadow scheme 113, 113t
Pallas, Peter Simon 456
pallets 130, 657
   escapement development 162
   spring-driven clocks 160
Pancirolli, Guido 644
Panier, Samuel 174
pantograph 214
Pappus 94
Paquet, Ph-H 397
parachute (pare-chute) 209, 657
Paradiso (Dante) 599–600
parallelogram (rectiliinear dial) 116
parapegmata (calendars) 20
Pardies, Gaston 241
Paris
   16th century planetary clock 254
   axis with St Nicolas d'Aliermont 363–4
   clockmakers/clockmaking 164, 216, 228
   exhibitions 364, 368, 373, 395–7, 401, 440, 441, 533
   food market dial 247
   Institut du Monde Arabe 29n
   Musée Carnavalet 398f
   Musée des Arts Décoratifs 300n, 430n, 435n, 438n, 440n
   Musée des Arts et Métiers 135n, 178n, 331, 431n, 436n
   Musée du Petit-Palais 422n, 424n, 428n
   Musée national des techniques 212n
   Observatory 313, 517
   planetary clock 254
   School of Horology 363, 368
   speaking clock 251, 497
   specialized trades 556
   vertical declining dial, Sorbonne 245f
Paris, Charles 47
Paris, Charles Baltazar 184f
Paris time 249
Parker, William 348
Parkes, Dan 547
Parry, Jack 419
Parsons, I. H. 593
Pascal, Blaise 316
Pascal, Claude 171
Pascalis, Hyacinthe 250
Pasha, Ahmad 92, 94
Passemant, Claude-Siméon 273, 649
Patek Philippe company 440, 473, 584, 631

INDEX

Patek Philippe Museum, Geneva 159n, 196, 196n, 308n, 426
   French table clock 166f
   musical watch 306f, 307f
patraque 368, 657
Paulson, Ronald 610
Pavia, *astrarium* 132, 266
pay rates, discriminatory 535
pea pod design 425
Peacock clocks (Cox) 448, 451, 454, 459
Pearce, Philippa 616
E. J. Pearson and Sons, leather wrist straps 480
Pearson, William xii, 221
peasant clockmakers 226n
pegamoid 565, 657
peg-and-slot mechanism 262, 263
pegging out 547, 550, 551
Pellegrin, Francisque 421
Pelliot, Paul 47n
pendant watches 560, 561
Pendleton, Richard 202
pendule religieuse 180
pendules 391, 396–7, 398f, 400, 401
   marble clocks 392f
pendules nègres 181
pendulum clocks 172f, 543, 645, 657
   domestic clocks, 18th century 171–3
   public clocks 142, 150
   Japan 68
   longcase clocks 173–80
   metronomes 582
   precision 404, 419
   transformative effect 150
pendulum regulators 501–6
pendulum swing counter 314
Penther, Johann Friedrich 241, 622
Percier, Charles 436
Pérez Álvarez, Víctor 125
periodic motions 286
periphrasis 657
Perkins, A. 635
perpetual calendar watches 204
Perpignan, public clock 139, 623
Perrault, Claude 10n, 645
Perret, David 418
Perret, Paul 482, 492
Perrot, Michelle 531
The Persistence of Memory (Dali) 615
Peshawar 34
Petit, Nicolas 435
Petitot, Jean 427
Petitpierre, Charles Henri 456n
Petronius 20n
Peyer, Favarger & Cie 577f
Pforzheim, watch industry 380, 380n, 381, 386, 387, 388
Pforzheimer Uhren-Rohwerke GmbH (PUW) 388
Pherecydes 5
Philip II of Burgundy 291

Philip II of Spain 313
Philip the Good of Burgundy 156f, 157, 157n, 158, 163, 254
Philippe, Adrien 566, 631
Philippe de Mornay 163n
Phillipps, Edouard 411, 631
Philippus, Quintus Marcius 8
Philo of Byzantium 11, 290
Philosophical Transactions of the Royal Society 630, 637
Phinney Walker 588
physics laboratories 508–9
Piaget watches 442
Picard, Jean 244, 245f
Pictet, Marc-Auguste 308
piezoelectricity 492, 506–7
Pigeon, Jean 570
Piguet & Meylan 303f, 309, 437
Piguet, Isaac Daniel 308, 309, 437, 565
pila 125
pillar clocks, Japan 66, 70f
pillar sundials 86–7, 87f, 114
pillar-plates, English watchmaking 192
pilot's watches 382, 490
Pinchbeck, Christopher 300
Pinel & Pinel 565
Pingré, Alexandre-Guy 247
pinions 126, 140, 366, 657
pinmaking 224
pinned cylinders 292, 296
pin-pallet escapements 352, 481, 492
pinwheel escapements 150, 181, 404, 463, 468
Pipelard, Pierre 153n
Pistor & Martins (Berlin) 581f
Pitiscus, Bartholomew 244
Pitkin brothers 358–9
Pitt Rivers Museum of Ethnology, Oxford
   water clocks 33f, 35, 36
plain horizontal dials 15, 16f
planar dials 244
   Arab-Islamic 82
   Islam 86
Planchon, Matthieu 650
plane dial, inclined 16
plane vertical dihedral dials 16, 18f
planetarium 21, 573, 574, 623
planetary clocks 253–8, 258f, 259f, 260f, 261f, 285; *see also* automata
   early known examples 255–6t
   mechanisms 258–63
   symbols of power 269–72
   uses 266–8
   well-documented examples 264–5t
planetary models 570–4
planets, non-uniform motions 285
planisphere
   Beauvais astronomical clock 280
   Copenhagen astronomical clock 282
Plantart, Nicolas 422

plate and spacer frames 464, 465
platina, balances 405
Platnauer, Louis 477
Plato 11
Plautus 8, 644
Pleiades 20
Plenderleith, Dr Harold 553
Pliny the Elder 5n, 8, 9, 20, 23–4, 110, 123n, 643, 645, 647
Ploërmel (France), astronomical clock 280
Plutarch 5
Pluto
   Messina astronomical clock 280
   Türler astronomical clock 284
Pluvier, Isaac 223n
pneumatic clock networks 509–11, 510f
Pneumatics (Hero) 290
pocket chronometers 205, 411
   Switzerland 415
pocket watches 195, 417
   China 48–51
   Japan 73f
   precursors to wristwatches 474, 480–1
   Switzerland 361
   temperature compensation 326
Poe, Edgar Allan 614
Poerson, Charles 426
Poissy Abbey clock 129
Poitevin, Isaac 423
polar gnomons 120, 231, 238, 243, 657
   Damascus sundial 89
   Egypt 3
Polaris wristwatches 55
polarity reversing 518
polarized dial movements, electric clock systems 518
pole star, position 117–18
Politzer, Ludwig 440
polos 5–6
Polwarth clock 182, 201
polwender 518
polyhedral dials 231, 239, 657
Polytechnische Journal 637, 638
Pompei 19, 248, 433, 436
Pond, Chester 516
Pond, John 406
Pons, Aristide 436
Pons, Honoré 363, 396, 396f
pop culture 441
Popp, Viktor Antoine 510
Poppe, Johann Heinrich M. 636, 650
popularity, public clocks 147
Porlock, Somerset, clock 143
Porrvis, James 168
portable clocks 160–1, 161f
   wristwatches 473
portable dials 16–17, 232–41, 251
   British Isles 114f
   Byzantium 78, 78f

   Islam 90
   Japan 59, 60f
   medieval Latin world 113–20
   primitive 114
portraits 163
Post Office
   electric telegraph time signals 497
   speaking clock 497–8
posted frames 464
Postlip Hall 600n
postmodernism 441–2
potence-plate 192
pouches 564, 564f
Pouillet, Claude 575
Poulle, E. 262, 623n
Pouzait, Jean Moïse 212
Powell, Thomas xiii, 35, 643, 644, 644n
power, symbols 163
power station turbines 524
Practical Clock Repairing (de Carle) 552, 634
Practical Course in Adjusting (Gribi) 634
practical handbooks 633–5
Practical Watch Adjusting (de Carle) 634
Practical Watch Repairing (de Carle) 552, 634
Prager, Frank D. 157n
Prague, astronomical clock 126, 149–50, 281
prahara 35
Pratt, Derek 354
prayer times
   Christian 81, 121
   Islam 81, 82, 84–6, 88–90
prayer-wheels 594
precision timekeeping 205, 311–13
   19th & 20th century 403–19
   balance springs 316–23
   compensation balances 323–9
   East India Company 334–7
   French marine chronometry 337–9
   John Arnold 333–4
   John Harrison 329–31, 339, 624
   longcase clocks 178
   longitude determination 313–16
   Paris 363
   Pierre Le Roy 331–3
   sundials and water clocks 17
   Switzerland 373
   watches 330
Premium Trials, Greenwich 406
Prescot, Lancashire, watchmaking 225, 405, 406, 639
preservation/restoration, public clocks 472
Prêtre, Théophile 365
Priapus 610
Prieur, Jean Louis 435
prime 102, 103
primitive portable dials 114
primum mobile 132

INDEX

The Principles of Mr Harrison's Timekeeper (Harrison) 329, 330, 333
printing, invention 623
Prior, Edward 303
Prior, George 179, 213, 303, 456
Procopius 94, 95, 290
Produktionsgemeinschaft Precis 384
Professional Accreditation of Conservator-Restorers (PACR) 554
profile cutting 221, 224, 657
programmable systems 472
programme, musical 289, 292
programming, automata 310
Prost company 464
protectionism 374
psychological persistence 614
Ptolemy, Claudius
   Almagest 253, 621
   Analemma 77, 82, 84, 116, 239–42, 248
   hour subdivisions 101
   observation errors 312
   Opera astronomica minora 621
   planetary models 262f, 263
   planetary system 570
public clocks *see also* turret clocks *and* tower clocks
   13th–18th century Europe 137–51
   19th and 20th century 463–72
   astronomical clocks 273–80
   India 39
   maintenance 511
   standardized time 496
   unreliability 517
publications, modern 554
publications, scientific 411
publications, early 543
Pulsar quartz watch 362
pulsometers 578
puncta 146
Purcell, Henry 582
puritan watches 194
Purmann, Marcus 231
putting-out system 215, 223, 226, 657
Pyenson, Lewis 37
Pyke, John & George 296–7
pyramid lantern clocks, Japan 66
pyrotechnology 506

Q

qanāt system, irrigation 36
Qarawiyyin mosque water clock, Fez 96
Qarmatian script 88n
al-Qaṭṭān, Qāsim ibn Mutarfiff 85
Qianlong, Emperor 47–9, 50, 445, 451, 454
Qian Yong 51
qibla 82, 82n, 85, 86, 88, 91, 92
qiblanumā 91
quadrans vetus 234, 234n
quadrants 91
Quare, Daniel 173, 176, 198, 199, 200, 224
   forgeries 213–14
   longcase clock 429
   watches 430
quarter hours 146
Quarterly Journal of Science, Literature and the Arts 637
quarter-striking 465
quarters system, China 41, 44, 50
quarter-spherical dials 14, 14f
quartz
   clocks 388, 419, 506, 525, 589
   crystal timekeepers 506–7
   marine chronometers 418
   movements 389f
   watches 353, 362, 492, 527
   watches, China 56, 58
   watches, France 368–9
   watches, Germany 388
   watches, Japan 74–5
   watches, Switzerland 375, 376f, 377f
Quasar watch 353
Queen's Watch, the 335
Quetelet, Adophe 246
Quirinal Garden dial 240
al-Qūshjī, ʿAlā al-Dīn ʿAlī ibn Muḥammad 91
Qusṭā ibn Lūqā 96

R

Rabelais, François 609
Rabiçag 126
rack clocks 184
rack lever escapements 202
rack-and-snail striking 174, 464
rack-striking 144, 198, 657
radar equipment 417
Radi, Archangelo Maria 134
radio communications 417
radio signals *see* time-signal transmissions
radioactivity exposure 538
radio-controlled timepieces 388, 526
radium, painting 485–6, 538
Rado 442
Raillard, Claude IV 643, 647–8
railway time 249, 474
railways
   clocks 249, 414, 497
   electric telegraph time signals 497
   flip displays 471
   standardized time 496
Raineri family 148
Raingo, Zacharie-Nicholas-Amé-Joseph, orrery clocks 281
Les raisons des forces mouvantes (de Caus) 290
Ramnagar 35
Rampet 36
Ramsay, David 193, 194, 225, 225n
Ramus, Petrus 270
Ransonet, Michael-Joseph 308
Raoul de Corville 542
al-Rashīd, Harūn 121, 290, 644, 648
Rasht 94
ratchet-tooth levers 203
rate, clock 321
rate, watch 328
rationalism, modern movement 441
rattrapante 578
Ratzerdorfer, Hermann 440
Ravenna, water clock 122
raw materials, clockmaking 222
Raytheon transistors 526
ray-tracing 251
Read, David 524
réassujetés 219
Recherches sur les Chronomètres (French Hydrographic Service) 411, 411n
reciprocating armatures 518
recoil escapements 130, 162
Récréations mathématiques (Ozanam) 236
rectilinear dials 115–16, 232
Redgrave, Richard 611
reflex dials 231, 240, 657
Regiomontanus (Johannes Müller) 116, 235, 239n, 255, 263, 266, 621, 622, 645
Regiomontanus dial 116, 232, 235, 236, 237
regional clockmaking schools, 16th & 17th centuries 164
   Burgundy 164
   England 168–9
   France 164–6
   Germany 167–8
   Italy 167
   Spain 169–70
regional concentrations and specializations, Switzerland 371, 371f
Reggio Emilia 143
Le Réglage de Précision—Cours pratique et éléments théoretique (Billeter) 635
Règle Artificielle du Temps (Le Roy/Sully) 330, 545
regulations, guilds 216–18, 533
regulator, astronomical 281, 335, 410, 414, 499, 501, 503, 504f
regulator dials 202, 657
regulator pendulums 321, 322f, 338
   decimal time 342
   longcase clocks 178
   neoclassicism 436
Rehe, Samuel 448
Reid, Thomas 338, 502, 629
religious literature 604
religious themes, automata 293
religious timekeeping 4, 84, 137
   Europe 121
   Islam 89, 94
   mechanical clocks 129
   vs. secular timekeeping 138
remontoire 278, 338, 468, 657
   balance spring 317
   clock winding 471
   spring-driven clocks 159
Renaissance 163, 231
renaissance and mannerist periods 421–5
repairing clocks and watches 541–555, 651
   handbooks 633–5
   watch repairs 548–9
Die Reparatur der Armbanduhr (Jendtitzki) 635
Die Reparatur komplizierter Uhren–Für Selbstunterricht des Uhrmachers (Hillmann) 635
repeating strikes 657–8
repeating watches 198–9
Report and Summary of the Horology in Art Symposium (Knaub) 610n
repoussé work 425, 430, 433, 558
Resch, Gebrüder 510
research, recent developments xii
restoring force 314
retailers, Switzerland 372
retrofitting 543
Reuge company 310
Revolution in time, clocks and the making of the modern World (Landes) xii
Revue Chronométriques 411, 550–1, 638
Reymondin, Charles-Andre 632
Rey-Pailhade, Joseph-Charles-François de 344, 344n
RGM company, USA 362
Rgveda (Indian text) 27
Rheticus, Joachim 163, 239n, 607
Rhind, W. 546
rhinestone 658
Rhinoceros clocks (Cox) 454
Ricci, Matteo 45–6, 47, 302, 444
Rice, Charles 467
Rice, William Gorham 292
Richard of Wallingford 119, 130, 139, 140, 144, 624, 649
   clock repairs 542
   earliest description of clock 623
   planetary clock 254, 262, 264t, 265t
Rico y Sinobas 125n, 126f, 133n
Riḍwān 96
Riefler, Sigmund 415, 631
Riefler family 501–2, 503, 522
Riepold, Mathias 426
Rieussec, Nicolas Mathieu 578
Rigault family 563
Rigg, Edward 629
Rijksmuseum, Amsterdam 177n, 189n, 426n, 431n
Rijksmuseum Boerhaave, Leiden 172f
Rimbault, Stephen 179n
rim-wound clocks 588

INDEX

ring dials 28, 29f
ring watches 561
rise-and-fall regulation 178
Risuke Katsu 73
Ritchie, Frederick James 517
Ritchie, James & Son 470
Rittenhouse, David 574
Rivard, François 246
Rivières, Baron Edmon de 251
roasting jacks 594–7
Robbins, Royal E. 359
Robert, Henri 550, 631
Robert, Josué & fils 229
Robert family, La Chaux-de-Fonds 299
Robert II Count of Artois 122
Robert the clockmaker, seal ii
Robert the Englishman (Robertus Anglicus) 128–9, 234
Robertet, Floribert 644
Roberts, Samuel 223, 223n, 224n
Roberts, Simon A. 492
Robin, Robert 178, 181, 202, 210, 338, 342, 404, 435, 436, 438
Robinson, T. R. 140
Robinson, Thomas 403, 407
Rochat family 309–10
rock-crystals, case in 195, 424
Rockford Time Museum 29, 426
Rockwatch (Tissot Le Locle) 442f
rococo 431–3
Rococo style 176, 179, 181, 562
Rodanet, Auguste 368, 628
Rodella, Giambattista 144
Roe, Sir Thomas 37
Roemer, Ole 570
Roentgen, David 177, 289, 297, 300, 436
Roger de Stoke 139–41
Rogers, Isaac 455, 456
Rohde & Swartz 507
Rohm and Haas Company 486
Roiz, Pedro 240
Roizin 595
rokokus (water clocks) 59–60, 61f
Rolex Watch Company 441, 488, 491
Rollenhagen, Gabriel 251
rolling ball clocks 41, 42, 162–3
Rolls Royce, car clocks 586
Roman de la Rose (de Meung) 129, 153, 154n
Roman time measurement 1, 7–9, 24–5
Rome, sundials 19
Romme, Gilbert 341
roof dials 247
Ropford, Roger de 542
Roret, Manuels 399
Roskell, Robert 203, 406
Roskopf pin-pallet escapements 481
Rossi, Teodosio 240
double face dial 241f

Rossini, quartz watches 58
rotary dials, electric clock systems 518
Rotherham & Sons, Coventry 349, 479
Rothmann, Christoph 266, 267, 268, 270, 271
Rouen
centre of trade 228
clock repairs 542
musical clock, Abbey of Sainte Catherine 292
pendulum clock 142
rough movements 337
Rousseau, noon-gun maker 246
Rousseau, Jean 195, 196, 246, 424
Roux, M. 366, 395
Rowbotham, J. 592
Rowley, John 570, 572
Roy, Samuel 299
Royal Air Force, pilot's watches 490
Royal Institution of Great Britain 637
Royal Mail, standardized time 497
Royal Navy, marine chronometers 405, 406, 417–18
Royal Observatory *see* Greenwich observatory
Royal Society (London) 637
Ruan Yuan 42n
Rubeus, Theodore 240
double face dial 241f
rubidium atomic clocks 509, 529
Rudd, Robert James 415, 502
Ruetschmann, David 273
Rugendas, Niklaus 188
Ruggieri, Michele 45
Ruhla 384–5
ruler-type dials 2, 3f
Rules and Practice for Adjusting Watches (Kleinlein) 634
Rushdie, Salman 616
Russia
English clocks 179
exports to Britain 353
travelling clocks 182
Ruteboef (French poet) 129
Rutilius Palladius, shadow scheme 113, 113t
Rye church clock, Sussex 140, 141

S

's Gravesande, Willem Jacob 579
al-Sāʿātī, ʿAlī ibn Taghlib ibn Abī al-Ḍiyā 96
Ṣandūq al-yawāqīt 89, 90
al-Ṣabbāḥ, Muḥammad ibn 83
Sacrobosco, Johannes de 234
Safety Last! (film) 615
al-Ṣaffār, Aḥmad ibn 84n, 85n
al-Saʿīd, ʿAbbās 87
St Alban's Abbey/Cathedral astronomical clock 130, 132

clockmaking skills 140–1
early public clock 144
fire clocks 136
St Benedict 80
St Gall Abbey, sundial 109
St Germain-en-Laye, Musée des Antiquités Nationales 12n
St John of Acre, mosque 91
St Nicolas d'Aliermont 229, 363–4, 396, 412
Saint-Omer clock 148
St Paul's Cathedral clock 129, 139, 542, 613
St Petersburg State Hermitage Museum 182, 448, 451, 459
Saint-Rigaud, François de 234n, 236
Sainte Marie Magdeleine 241
St Thomas Aquinas 99
St Willibrord 111n
horologium 111, 112f
Salisbury Cathedral clock 139–40
Sallier, Claude 646
Salmon, Antoine 310
Salodius 242
Salomon, Sir David 651
Salzburg, anaphoric clock 12, 12f
Samos 7
Sampurnanand Sanskrit University 38
Samso, Julio 85n
Sanç, Ramon 623
sand 44–5
Sandberg collection, watches 192n
sand-clocks/glasses 133–5, 134f, 600, 601f, 604
maritime use 134n
Sanderson, George 566
Sandoz 588
Sandringham Palace 38
Ṣandūq al-yawāqīt (portable sundial) 90, 91
Sanskrit texts 28
Sapor 644
Sarazin 241
Saram, Gokula Natha 30, 31f
Sarton, Hubert 181
Satapatha-Brahmana 27
satellite-based GPS 419 *see also* global positioning systems
Saturn, satellites 271, 273
Saunderson, Nicholas 544
Saunier, Claudius 398, 399, 627, 628, 632, 643
Annuaire Artistique et Historique des Horlogers 641
Revue Chronométriques 411, 550–1, 638
technical schools 400n
Saurin 648
Sautebin, M. T. 538
sautoir chains 560
Savarin, Jean-Bruno 547
Savoie, Denis 6n, 8, 623

Savoie region 399
savonette watches 481, 481f
Sawai Jai Singh of Jaipur 28, 29f, 30
al-Ṣawwār, Aḥmad ibn 84, 85
SBS movement 58
Scandinavia, longcase clocks 177
scaphe 6
scaphium inversum 11
Scappi, Bartolomeo, spring-driven jack 596f
Scarperia, Brunelleschi clock 143
Schaldach, Karkheinz 107, 108, 119n
Scharstein, Daniel 252
Scharstein, Hans 252
Scheibe, Adolf 507
Scheufele, Karl I 566
Scheurenbrand, Hans 285
Scheveningen, pendulum clock 142
Schickard, Wilhelm 272
Schierwater, Charles Adolf 487
Schissler, Christoph 267, 267n
Schlenker & Kienzle 381;
see also Kienzle
Schlottheim, Hans 294, 301–2
Schmidt, Carl Gottlob 637
Schmidt, Joseph 584
Schneider company 365
Schnier, Hans 423
Schönebeck, sundial 109
Schöner, Andreas 235, 237, 239, 241n
Schöpperle, Ignaz 379
Schott, Gaspar 292, 624, 644
Schramberg (Germany), astronomical clock 280
Schreck, Johann Terrenz 46
Schultz, Wilhelm 635
Schwilgué, Jean-Baptiste 274, 276, 278, 286, 465
astronomical clock 149, 277f
turret clocks 469
watchmen's clocks 592
sciatherre 241
Science Museum 573n
scientific publications 411, 630–1
scilla 122
Scobie-Youngs, Keith 140
Scoresby, William 406
Scorsese, Michael 615n
Scot, Thomas 606n
scrapping, clocks 139
screw-down crowns 488
script, kufic 88, 88n
script, Qarmatian 88n
scrupuli (hour subdivisions) 101
Scrymgeour, James 405
Seaforth Highlanders 474, 475f
seasonal hours *see* unequal hours
secondary dials, early public clocks 146
seconds hands 312
The Secret History (Tartt) 616
Sectronic movements 353
secular figures, automata 293–4

secular timekeeping *vs.* religious timekeeping 138
secular variation 237, 658
secular years 273, 658
Seignior, Robert 198
Seiko 73–5, 353
   Astron watches 74–5, 492, 526
   automatic winding watches 490
   electronic wristwatches 492
   Museum 67, 74
   post-war development 74–5
   quartz watches 74, 526
   sports timers 523
Seikosha clock and watchmakers 73–5, 74f, 357
Sei-shonagon 60
Self Portrait Between the Clock and the Bed (Munch) 618
Self Portrait With Wall Clock (Chagall) 618
self-feeding lamps 135
self-help guides 545
self-regulation, metaphor 599, 600
self-sufficiency, clockmaking 223
Self-Winding Clock Company (SWCC) 516
self-winding wristwatches 489
Semedo, Álvaro 51
semiconductors 362, 529
semihorae (half hours) 100
The Sempstress (Redgrave) 611
Seneca 1n, 17
senkodokei (incense-stick clock) 61, 64f
serinette 304
Serle, George 243
Sermand, Jacques 195, 196
Settala, Manfredo 646
seventeen-jewel movement, China 56
Severino, Nicola 116
sext 102, 103
Seydi Ali Reis 91
Seyffert, Johann Heinrich 335
shadow clocks 2f, 20–1
shadow-length schemes and tables 20–1, 27, 113
shagreen 197, 213, 557, 658
Shakespeare, William 169, 604
shakkaziya quadrant 91–2
Shams-i Sirāz 'Afīf 36
Shanghai Clock and Watch Company 56
Shanghai wristwatches 53–5
Shank, Michael 263
al-Shaṭawī, Abū 'Abd Allāh Muḥammad ibn Ḥasan ibn Akhrī Hishām 83, 95
Sheffield, Mappin Bothers 477
Shelton, John 178, 321
Shen Kuo 43
Shepherd, Charles 414, 497, 511, 514, 519
'shepherd's dial' 233

Sherwood, Napoleon P. 359–60
Shibaura Seisakaju Co. Ltd 72
Shigetomi Hazama 68
Shinminato Museum 68, 70f
Shinola company, USA 362
Shirley, Sir Robert 37
shock resistant wristwatches 488
shojibans (pillar clocks) 68, 70f
shops, clock- and watch-makers 218–21, 219f
Shortt clock 415, 416f, 419, 499, 502–3, 522
Shortt, William Hamilton 502
Shui Yun Ti Xiang Tai (astronomical timepiece) 44
sidereal train, astronomical clock 276
siderostats 579, 580f
Siegel & Augustin 565
Siemens 388
Siemens & Halske 518, 521
Sievert, Hermann 632
al-Sijzī 83, 84
Silbermann, Jean-Thiébault 579
silent escapements 180
silicon parts, wristwatches 493
Silver Swan (Cox) 451
silver watchcases 197, 557
silver watches 194
Simon, Jules 534
Simplex wristwatch 480
sine quadrants 91
sines, tables 239
singing birds
   musical and automaton clocks 304–8
   watches 305, 309–10
single-foliot lantern clocks Japan 69f
single-foliot lantern clocks, Japan 66–7
sing-songs 48, 445, 447
al-Ṣinhāghī, Muḥammad 96
sinking bowl water-clocks 659
   Babylon 4
   England 121
   India 30, 32, 36
   Islam 95
   origin and diffusion 35–7
Sinnington, sundial 109
al-Siqilī, Ibn Yaḥyā 87, 88
six-pips time-signal transmissions 499, 501f
six-sector dials 105, 106, 109f
sixteen-sector dials 109
Skirmisher wristwatch 477, 478f
skulls 604
Slano convent, sundial 119
smart phones 493
smart watches 492
Smeaton, John 321
Smith, J. L. & W., of Clerkenwell 592
Smith & Sons, J, Sheffield 640
Smith, John 543–4, 624, 644

Horological dialogues 543, 626
Smith, John and Sons 591
Smith, Michael (Miguel) 199, 204
Smith, Richard Bartholomew 282
Smith, Robert 544
Smith, Roger (historian) xv
Smith, Roger (watchmaker) 354
Smith & Sons, S. (clock- and watch-makers) 349–53, 525, 587f; *see also* Smiths Industries
   dashboard clocks 586, 588
   shop 350f
   showroom 587f
Smiths Industries 525, 586, 589
smoke-jacks 594, 596
Smyth, W. H. 650
Smyth, Walter 168
Snellen, Willem 330
snuffboxes 304
Société de Commentry-Fourchambault & Decazeville 374
Société d'Encouragement pour l'Industrie Nationale 637
Société des Arts, Geneva 305, 308, 310, 533, 647
Société des Horlogers 411, 638
Society of Watch and Clockmakers of Leicester 546
soft watches, Dalían 615
solar altitudes 232
solar chronometer 249
solar system, models 273
Solari, Fratelli 466, 471
solaria 10
Soler, Rafael 252n
Solis, Virgil 421, 423
Solomon 602
solstice diurnal arcs 111
solstices 5, 13, 244
Somersall, Mandeville 227
Sotheran & Co., London 622
The Sound and the Fury (Faulkner) 616–18
souscription watches 208–9, 210, 214, 437f
south and east Asia *see* East Indies
Soviet Union 503
Spain, clockmaking, 16th & 17th century 169–70
span (distance) 28, 658
spare parts, Swiss manufacturers 372
Sparrevogn, Anders Christian 584
speaking clocks 251, 497–8
specialization 228, 370–1
   manufacture, 16th - 18th century 221–6
speed governors 289, 658
speedometers 350, 586
Spezimatic, watch 385, 386f
sphère mouvante 274, 570, 571f
spherical dials 14, 14f, 15f, 19, 77

spherical trigonometry 621–2
spherical watch, Jacques de la Garde 424f
Spiegel, Joseph 206
Spilhaus, Athelstan, Space Clock 281
spit-jacks *see* roasting-jacks
split-piece dials 66, 67f, 70, 71f, 72
split-seconds hand 578
Spöring, Jörg 284
spring barrels 596, 658
spring bars, wristwatches 488
spring detent escapements 205, 336
Spring Gardens Museum (James Cox) 451n, 454, 454n
spring-driven clocks 157–60, 178–82
   Burgundy 164
   China 45, 46
   development of watches 185
   earliest example 156f
   France 164, 166
   mainsprings 157–60
   miniaturization 160–1
   steel 222
   United States 356–7
spring-driven jacks 596, 596f
spring pallets 338, 415, 502
springs, steel 223, 405
squilla 122
Sri Lanka, sinking bowl water-clocks 36
SSIH 376
staartklok ('tail clock'). 183
Staatliche Museen, Berlin 436n
Stabius, Johannes 236
stackfreeds 645, 658
   early watches 186, 187f
   horizontal 159
   spring-driven clocks 158, 159
   vertical 158
Stadlin, Frantz 47
Stahl 628
Staiger company 388, 388n, 389f, 390
   Chrometron CQ 2000 525
   quartz movements 389f
Stampioen, Nicolaas 574
standardization, clock parts 398–9
standardized time 495
   demands for standardization 495–6
   domestic timekeepers 523–9
   electric clock systems 511–23
   observatories 499–501
   pendulum regulators 501–6
   physics laboratories 508–9
   pneumatic clock networks 509–11
   quartz crystal timekeepers 506–7
   telephone speaking clocks 497–8
   time signals 496–7, 552
   time signals by radio 498–9
   tuning fork timekeepers 506
Starley, John Kemp 477
stars observation 117–18

INDEX

State Hermitage Museum, St Petersburg 182, 448, 451, 459
status symbols 604
   clocks 163–4
   watches 186
   wristwatches 493
Staunton, Sir George 456
Staveley, A. W. 593
Stebbins, Frederick A. 116
steel
   clockmaking 222
   springs 223, 405
Steele, Richard 531
steelyard clepsydra 43, 44
Steen, Jan 609, 612
Steenwyck, Harmen 604
Steinfeld, Abbey of 125
Stella, Jacques 426
stem 658
stem sealing glands 488
Stengel, Johann Peterson 622
Stephenson, Luther 467
stereographic projection 234, 242, 658
stereometry 248, 658
Sterne, Laurence 610
Stevens, Henry 169
Stevin, Simon 162, 341
Stewart Museum, Montreal 166n
Stewart Warner Corporation 589
Stikford, Robert 105, 119
still-life paintings 604
stoel clocks 183
Stöffler, Johannes 234, 235, 238
Stogden, Matthew 198
Stollenwerck, Michel 297, 431
'stop all the clocks' (Auden) 619
'stop the water' 11
stopped clock, Miss Havisham's (Dickens) 613
stopping holes 545
stopwatches 55, 344, 576, 578
stop-work 584
stop-work fingers 192
Stradanus, Johannes 146
Stralsund St Nicholas church clock 126
Strand, per 27, 298f
Strasbourg
   astronomical clock 126, 141, 147, 149, 263n, 274–7, 277f, 285, 287, 469, 623
   astronomical clock replicas 281–2
   automaton 147, 292–3
   clock repairs 542
   musical clock 292
   replicas 282
   sundial 238
strass 558, 658
Strasser, Ludwig 415
Strasser & Rohde 415, 503
Stratton, Nelson P. 359
Straumann, Reinhard 374, 374n, 492
stress corrosion cracking (SCC) 551

strike patterns, public clocks 145
striking 658
striking mechanisms, English watchmaking 193–5
striking trains 157n, 658
   domestic clocks 161f, 166f
   domestic wall clocks 157
   early public clocks 142, 144
   spring-driven clocks 156–7
   travelling clocks 182
stringed instrument clocks 296
strob escapement 130, 131f, 132, 144
strontium atomic clocks 509
Strubel, A. 153n
Strutt, William 591
Struve, Friedrich Georg 578
Stubs, Peter 640
St-Yrieix-le-Perche (Vienne), six-sector dial 106
Su clocks 51–2
Su Song 44, 44n, 45f, 291
subcontracting 223–4
subdivisions, minutes and seconds 344
The Sublime Methods of Spiritual Machines 594
Submarine wristwatch 487f
sub-style line 244, 658
sufflators 594, 658
al-Ṣūfī, Shams al-Dīn Muḥammad ibn Abī al-Fatḥ 90
sugar tongs temperature compensation 205
suiyokyu-gi (pendulum bob) 66, 70f
Sukezaemon Tsuda 65, 67
Sullivan, Raymond H. 589
Sully, Henry 244, 319, 330, 411, 545, 589, 624, 627, 647
   lubrication 546
   Règle Artificielle du Temps 330, 545
   watch owners' handbook 636
Sun Meitang 52, 53
Sun Ruli 51n
Sun Tinguan 52
Sun Yuanhua 41
sun-and-moon dials 210
sundials 12–17, 231–2, 658
   17th century developments 241–4
   18th century developments 246–8
   19th century developments 248–50
   20th century developments 250–2
   Ancient Greece 5–6
   Babylon 4–5
   Bevagna 9f
   Byzantium 77–80
   canonical hours 104f
   Catania 8
   checking mechanical clocks 622
   China 41, 41n
   collections 650
   craftsmen 220
   decimal time 342
   Greece 5–6, 10

   India 28–30, 40
   Islam 81–94
   Japan 59
   literature 621–3
   medieval Latin world 102, 110–13
   meridians 244–6
   mottoes on 250
   parts 13f
   portable altitude dials 232–7
   portable direction dials 237–41
   urban sculpture 39–40
suprema (hours before sunset) 99
sur plateau movement 309
Suso, Heinrich 602, 603f
suspension springs 501
Suwa Seikosha 75
Suzhou clock-making centre 52
Swatch Group 376
   plastic watches 376, 442
Swift, Jonathan 610
Swiss Chamber of Horology 373, 374
Swiss Company of Microelectronics and Horology 376
Swiss made 376, 377
Swiss Society of Horological Industry SA 374
Switzerland
   clockmaking 226
   exports to Britain 348, 353
   exports to China 460
   female workers 537, 539
   longcase clocks 177
   musical and automaton clocks 297–8
   product innovation 372–5
   production post 1970 375–7
   production system structure 370–2
   rivalry with France 368–9
   wall clocks 184
   watches, 16th–18th century 195–6
   watchmaking, 16th–18th century 211–12
symbolic realism 611–12, 613
Symons, Sarah 1n
sympathetic pendulums 517
synchronized grid systems 525, 525f
synchronizing systems 513
Synchronome Company 38, 502–3, 512, 513
   Benares College clock 38, 39f
   exports 522
   switch 519–20
synchronous motors 351, 513, 517
Synthetic Jewels Ltd 352, 353
synthetic quartz 376
Syracuse 5
Syros 5

T

tabatières 305, 310
table clocks 624
   baroque style 429

   China 51
   France 421–2, 422f
   Germany 168
table orrery 573
tableau animé 300
tables, dialling 242
tachymeters 578
Tadayuki Iwano 70, 71f, 72f
Takanori Endo 70
Tak On Clock Factory 53
Takanori Endo 70
Tamilnadu 36
Tanaka, Hisashige 72, 285
Tanaka-seizoujo 72
tangents, tables 239
tank clock/regulator (air-tight pressure container) 415, 501, 502, 511, 519, 521, 522
al-Ṭanṭāwī, 'Abd al-Qādir ibn Muḥammad 93
tape chronographs 576
Taqī al-Dīn 89n, 98, 312, 594
Tard, Antoine 438
Tardy 621n
Tartt, Donna 616
Tartu Observatory (Estonia) 578
Taurisano, sundial 106, 110f
Tavannes Watch Co 487
   Captive model 561
   car clocks 588
   Submarine model 487f
Tavernier, Etienne 566
taximeters 586
Taylor, C. & Co. 592
Taylor, Frederick W. 578
Technica Curiosa (Schott) 292
Technical Centre for Horology (CETEHOR) 369
technical drawing 632
technical schools
   France 363, 367–8, 380, 400, 400n
   Germany 378, 371
   Switzerland 373, 377
Technicum Neuchâtelois, Le Locle 635
technology, astronomical clocks 285–7
Telč, sundial 109, 112f
telegraph systems, pneumatic 510
Telephone Rentals 512
telephone speaking clocks 497–8
telescope drives 578–9
telleruhr ('dish clock') 184
The Tell-Tale Heart (Poe) 614
tell-tale watches 379
tellurians 281, 285, 570, 573
temperature compensation 178, 205, 209, 321, 323, 326–7, 331, 332, 407, 409, 411, 492
   Big Ben 469
   wristwatches 481–2
Temperance, allegorical figures 600–2, 601f

INDEX

Temple Newsam House, Leeds 297
temporal hours *see* unequal hours
ten-sector dials 108, 112f
Tenji, Emperor 59–60
Tensho Boy Mission 63
terce 102, 103, 105
terminal curves 336
Terrile brothers 466
territorial expansion, early public clocks 147
Terry, Eli 355–6
Tertullian 79–80
têtes-de-poupées ('dolls' heads') 180, 182
textbooks 631–3
 Claudius Saunier 628
 Ferdinand Berthoud 627
Thāqeb Astronomical Society 94
Thābit ibn Qurra 83–4, 86, 244
Thacker, Jeremy 502, 544n
Thackeray, William 614n
Thailand, sinking bowl water-clocks 37
The Hague Museum 582n, 609
Theatre of the Muses 300
Thebes 5
Theodoric 137, 644, 648
Theorica planetarum (Campanus de Novara) 254
Théorie Générale de l'Horlogerie (Defossez) 632
Thièble brothers 399, 399n
Thiel brothers, watchmaking 381–3, 587
Thiele, Johann Georg 411
Thiout, Antoine 330, 624, 627, 647
thirteen, clock striking 616
thirteen-division unequal hours 70
thirteen-equal-division 70
thirteen-sector dials 108–10
Thomas, François 194
Thomas More and his Family (Holbein) 609
Thomire, Pierre-Philippe 435, 436
Thompson, David 63
Thompson Electric Clock Co. 589
Thomson, William (Lord Kelvin) 414, 502
Thoughts on the Means of Improving Watches (Mudge) 201, 331n
Thresher & Glenny wristwatches 484f
Thuret, Isaac 171, 172, 180, 182, 314–16, 646
Thuret, Jacques 430
Thuret, Jacques III 431
Thurston 36
Thwaites, A & J 405
Thwaites, Aynsworth 547
Thwaites and Reed, clockmakers 302, 303, 591
 repairs 547
Tianjin Sea-Gull 56, 58
Tianjin wristwatches 53–5, 55f

'tick tock', first use 599n
tid(e), eighth portion of day 103
tides
 astronomical clocks 286
 Beauvais astronomical clock 280
tierces 582
Tiffereau, Théodore 135
Tillier, Paul 189
time and motion studies 578
Time and timekeepers (Milham) xii
time ball
 Greenwich observatory 406, 497
 Hong Kong Observatory 498f
 Washington Observatory 516
 Western Union, New York 516
time distribution systems 365–6, 496–7
 electric clocks 499, 517, 521–2, 524–5
 pneumatic clocks 509–11
 radio communications 417
 telegraph lines 529
Time Museum, Rockford 29, 426
time-switches 527–9, 528f
timecode transmissions 499
time-glasses 133, 658
timepieces 658
'times out of joint' (Shakespeare) 609
time-signal transmissions 417, 495, 496–7
 Eiffel Tower 498–9, 500f
 Frankfurt DCF77 472, 526
 international 499
 radio 498–9
Timex 349, 361, 375
 Dundee factory 354
 wristwatches 492
time zones 39, 283
timing and driving systems 575
 chronographs and stopwatches 576–8
 chronoscopes 575–6
 heliostats, siderostats and coelostats 579–81
 telescope drives 578–9
tipsy key 566
Tissot watches 442, 487, 561, 567
Titian 604
Tlemcen, mosque 87
Tobias, Wenzel 312
toddy lifter 7
Tokugawa Ieyasu 63
Tokyo
 National Museum of Nature and Science 59, 61, 72
 Olympic Games, 1964 74, 75, 523
 Seiko Museum 67, 74f
Tokyo Electric Company 72
Toledo 96
 Cathedral musical clock 292
Tompion, Thomas 173, 176, 178, 196, 200, 318, 325, 548, 645
 Castlemaine clock 549

 clock repairs 544–5
 dual control clocks 182
 forgeries 213–14
 longcase clocks 429
 pendulum clocks 173, 318
 planetary models 571–2
 'regulator' 197
 shop 218
 tools 221
 table clock 429
 watch 196, 198, 201f, 430
 workshop 222n, 544–5
Tompion and Quare 176
Tompion regulator 197
Tom's Midnight Garden (Pearce) 616
tondi 426
tongji movement 56, 56f
La Tonotechnie ou l'Art de noter les cylindres (Engramelle) 289–90
toolmaking/toolmakers 222
tools, clockmaking/watchmaking 220–1
Topkapi Museum, Istanbul 456n
Tordesillas, treaty of 161, 312
Torrens, David 651, 651n
Torriani, Giovanni/Turriani, Juanello 164, 167, 170, 221n, 644
Torriani, Janello 164, 170, 221n, 644
 Microcosm 263
 Vitruvian artisan 266
Torricelli 127
torsion spits 596
tortoiseshell 197, 658
Toscanelli, Paolo 244
Toshiba 72
tourbillon 58, 209, 658
Toutin, Henri 189, 426
Toutin, Jean 189, 425
Toutin, Jean II 426
tower clocks 144, 168
Tower of the Winds, Athens 21–3, 22f, 94
 vertical plane dial 15
town hall clocks 137, 145
Towneley, Richard 178, 318
Tozaburo Toda 67
Tractatus astrarii (de' Dondi) 254, 257f, 623
Tractatus horologii astronomici (Richard of Wallingford) 130–2, 144
Tracy, Stephen (London) 174
Tracy, Steven, (Rotterdam) 574
trade catalogues 639–41
trade development, China and Europe 48–51
trades unions 539
traditional timepieces, China 41–3
train 658
training 538, 552
Traité de l'horlogerie méchanique et pratique (Thiout) 330, 627

Traité d'Horlogerie moderne théoretique et pratique (Saunier) 399, 628
Traité General des Horloges (Allexandre) 330, 627
Transactions of the Society for the Encouragement of Arts 630, 637
transducers 506
transistors 526
traveller's cases 565
travelling clocks 182
Trebino company 472
Trebino family 466
Treffler, Johann Philipp 180, 314
Tremayne, Arthur 634
Tremont Watch Company 359, 361
trench watches 479, 484–5
triangular frame clock 143
triangular pendulum 316–17
La Tribune Chronométrique 411, 638
Trifonov, Yuri 618
Trigault, Nicolas 302
trigon 240
trigon of signs 235, 658
trigonometry 243, 248
 Islam 83, 91
 spherical 239
Trippli, Julien S. 629
Triptych Portrait of Henry and Dorothy Holme (unknown) 605
Tristram Shandy (Sterne) 610
tropical period 244, 658
tropomètres 344
Troughton, Edward 412
Tucker, W. E. 479
Tughluqs, water clocks 35
tulip form watches 190–1
tumbling-pallet escapements 326
tuning fork timekeepers 506, 526
Turkish market watches 213
Türler, Franz 284
Türler astronomical clock 284
Turner, Jonathan 610, 611
turret clocks 144, 624
 birdcage frames 463
 flat-bed frames 463
 France 365
 gravity escapements 403
Tuve, Rosamond 602
twelve hours xiv
twelve-sector dials 102–3
twenty-four hours xiv
twin-barrel chronometers 404
twin-pallet escapements, adjustable 181
Tyrer, Thomas 204

U

Uccello, Paolo 146
Udaipur 35
Uhrenfabrik Villingen 387
Uhrenfabriken Gebrüder Junghans AG 382

Uhrenmuseum Winterthur 169f
Uhrenwerk-Ersingen 388
Der Uhrmacher am Werktisch (Schultz) 635
Die Uhrmacherlehre (Hanke) 632
Ulm (Germany), astronomical clock 280
Ulrich, J. G. 405
Ulrich & Weule company 465–6
Ulysse Nardin company 415–17
Umar ibn ʿAbd al-ʿAzīz 81
Umayyad mosque 90
unequal hours xiv, 232, 658
  chandelier clock 136
  concordance with equal hours 6t
  Europe 118, 119, 121
  Greece 6
  India 38
  Islam 82
  Japan 59, 66, 68, 70, 72
  medieval Latin world 99, 111
  monastic day 102
  planetary clocks 267
Ungerer, Théodore 122n, 280
Ungerer brothers/company 246, 278, 280, 281, 366, 465, 469
United States
  clockmaking 355–7
  competition with Black Forest clocks 378
  exports to Asia 357–8
  exports to Britain 349
  horological industry, decline and fall 361–2
  National Bureau of Standards 508
  turret clocks 467–8
  watchmaking 358–60
United States Watch Company 360
Universal Exhibitions *see* International Exhibition
unlocking 157, 658
unpaid work, women 531
Unterwagner, Emil 635
'Untitled' (*Perfect Lovers*) (Gonzalez-Torres) 616
Upjohn, James 205, 227, 456
Urania-Uhren company 515
Uranus, discovery 273
Ursus, Nicolaus Reimarus 266, 271
user manuals 635–6
Utinam company 366
Utrecht Cathedral clock 171
Utrecht sundial 238
Uzerche, six-sector dial 106, 107f

V

Vacheron Constantin business 227
Vachey, Daniel 283
vacuum, clocks in (air-tight pressure tank) 415, 501, 502, 511, 519, 521, 522
vacuum watches 491
Valeran, Florant 422
  table clock 422f
Valesco, Rodrigo de Vivero y 63
Valette, Aline 535, 536
Vallier, Jean 189, 424
Vallignano, Alessandro 63
Vallin, Nicholas 191, 294, 424
van Ceulen, Johannes 180, 316
van de Cloese, Barent 317
van der Plancke system 519
van der Straet, Jan (Stradanus) 146, 220f
van der Woerd, Charles 359
van Gheele, Ghylis 191
van Pilcom, Daniel 191
Vanderpool 7
vanitas imagery 599, 604, 606, 607
Vanity Fair (Thackeray) 614n
variable hours *see* unequal hours
Varignon, Pierre de 158n, 648n
Varro 8, 21, 23
Vaucanson, Jacques 299, 509
Vaudoux family 565
Vauquer, Jacques 426
Vauquer, Robert 426
Vautier, Louis 189
VDO quartz clocks 589
VEB Feingerätewerk Weimar 385
VEB Glashütter Uhrenbetriebe (GUB) 384–5
  wristwatch 386f
VEB Kombinat Mikroelektronik Erfurt 385
VEB Mechanik Glashütte Uhrenbetriebe 384
VEB Uhren- und Maschinenfabrik Ruhla 385
  wristwatch 386f
VEB Uhrenwerke Ruhla 385
Vedanga-jyotisa 27
Vedic texts 27
Veglia Borletti 588
Velazquez, Diego 604
Venice
  astronomical clock 281
  early public clocks 142, 148
  marble dial, Arsenal 250
  St Marks, jacks 293
Verbiest, Ferdinand 68
Vereinigte Freiburger Uhrenfabriken 382, 383f
verge and foliot escapements 143, 171, 624
  guard-tower clocks 153, 156
  spring-driven clocks 160
verge escapements 367
  conversion to anchor 549
  early watches 186
verge watches 208f, 207
verges 130, 658
  spring-driven clocks 156–7
Vergil, Polydore 643, 643n
Vérité, Auguste-Lucien 278, 280, 286, 365
vernis Martin 557, 659
Verrien, Nicolas 557n
Versailles, meridian 244
vertical clocks 166
vertical declining dial 245f
vertical plane dials 15, 16f
vertical stackfreed 159, 159n
vesper (sunset) 99
vespers 102, 103, 105, 106
Venus, transit of 499
Vickery wristwatch 485f
Victor wristwatch 480
Victoria & Albert Museum 422n
  Bull clock 451n
  clocks 436n
  painting on enamel 437n
  paintings 605
  rococo style 438
  sundials 89
  verge watches 206n
  watchcases 426
  watches 193n, 194n, 425n, 427n, 430n
Vienna
  16th century planetary clock 254
  enamel 438
  Kunsthistorisches Museum 159n, 160n, 312, 423n, 605, 605f
  planetary clock 254
Viet, Charlemagne 221
Viète, François 239
Viger, François 297
vigil 102
Vigniaux, Pierre 223, 541, 547, 633
Villarceau, Antoine-Yvon 578
Villard de Honnecourt 123
Villeneuve, Adelaïde 534
Villers Abbey, water clock 117n, 127
Villon, Albert 399
Vimercato, Giovanni Battista 240
Vincenti, Jean 366, 395, 397
Vindolanda 12
Vion, François 435
Vipiteno, public clock 142
virgula ferrea (metal gnomon) 110
virgule escapement 207–8
virgulis aeneis 11
Visbach, Pieter 180, 180f
Visconti, Gian Galeazzo 132
Vitruvian artisan 266
Vitruvius 9, 10, 644
  anaphoric clocks 11–12, 20
  clepsydrae 122
  portable dials 113
  sundials 13, 17, 77, 231, 239, 621
  Tower of the Winds, Athens 21
  water clocks 10n, 11, 17n, 137
Viviani, Vincenzo 127, 180, 314
Vivot, Alexandre 425
vocabulary, Indian 32–5

Voellus 647
Voigt, Bernhard Friedrich 638
Voltaire 643
volvelles 187, 659
Von Bertele, H. 162
von Bruhl, Hans Moritz 335
von Fraunhofer, Joseph 578
von Humboldt, Alexander 578
von Littrow, Johann 622
von Peyer, Alfred 576
von Steinheil, Carl August 514
voorslag 292
Vosseler, Michael 379
Vouet, Simon 426, 428
Vrard, Ludovic 50
Vroezen, Adriaen 574
Vuillemin company 366
vulgar hours 105
Vulliamy, Benjamin Lewis 468, 471, 549, 630
  and Hampton Court Clock 149, 223n
  On the theory of the dead escapement 630–1
Vulliamy, Justin 548
Vulliamy and Dutton 592
Vulliamy family 303, 339, 436, repairs 548–9

W

WA (initials) 189
Wadham College, Oxford, pendulum clock 142
al-Wafāʾī al-Mīqātī, ʿAbd al-ʿAzīz ibn Muḥammad 90, 91
Wager, Sir Charles 323
Wagner, Carl Theodor 518, 521
Wagner, J. Bernard Henry 365, 597
Wagner family 365, 365n, 583–4
waistcoat chains 560–1
Wales, Albert Edward, Prince of 38
Walker, Ezekiel 550
wall clocks 182–4
Wallace Collection, London 431n, 433n, 435, 436n
Wallingford *see* Richard of Wallingford
Walter, David 404
Walters Art Gallery, Baltimore 186n, 423n, 426n
Waltham watches 359–61, 482, 588, 634, 636
Walther, Bernhard 311
Walton, John Whitehead 613
Wang Zheng 46
Wanli (Zhu Yijun) 46–7, 302
Warren, Henry E. 524
Washington, DC, time-signal transmissions 498
The Watch Adjuster's Manual (Fritts) 634
Watch and Clock Making and Repairing (Gazeley) 552
Watch and Clock Making (Glasgow) 632

INDEX

Watch and Clock Year Book 642
watch chains 555–6, 560–1
  as metaphors 610
watch dials 199, 210, 224, 557
The Watch Jobber's Handybook
  (Hasluck) 552, 633
watch keys 565–6, 566f
watch No. 1/36 (Arnold) 336
The Watch Repairer's Manual (Fried) 552
Watch Repairing, Cleaning and Adjusting
  (Garrard) 552, 634
watchcases 187, 556–7, 562–7, 563f
  Besançon 368
  decorative features 423–4
  enamelling 188–9, 425–6, 427f,
    428f, 558
  pouches 564f
  rock-crystals 195
watches and watchmaking 185 *et
  seq*, 624, 659; *see also* wristwatches
  adjustments 631
  Britain 191, 196–205, 347–54
  China 50–5
  collections 651
  decorative features 423
  dials 199
  East Indies 18th century trade 460
  exports and forgeries 213–14
  factories 372
  France 188–9, 206–10, 363–9,
    400–11
  fusee cutting engine 627
  fusees 191–3
  Germany 185–8, 206, 378–90
  high-frequency balances 325
  Japan 66–73, 68f
  Liège 215–16
  machinery 358, 361
  maritime use 202–3
  mass-production 410
  Netherlands and Flanders 189–91,
    210–11
  Paris 216
  pouches 564, 564f
  repairing *see* repairing clocks and
    watches
  stands 561–2, 562f
  striking and alarm
    mechanisms 193–5
  Switzerland 195–6, 211–12, 370–7
  United States 355–62
  waist bags 559
  watchmaking schools 380
The Watchmakers Lathe and how to use it
  (de Carle) 634
Watchmaking (Daniels) 632
watchmen's clocks 591–3
water clocks 10–12, 659
  accuracy 127
  Ancient Greece 6–7
  automata and sounds 290–1
  Babylon 4

Byzantium 94–5
China 42–3
Egypt 3–4
Europe 121–7, 124f, 125f
Far East, history of horology 33f
Greece 10
India 30–2
Islam 95–8, 97f
Japan 59–60
linkwork mechanisms 44n
places of worship 39, 117
public clocks 137
scientific use 127
Waterbury Clock Company 356, 361,
  492
waterproof wristwatches 487–8
Watkins, Richard 632, 635
Watson, Ernest Ansley 351
Watson, Samuel 222, 544
Way, Albert 114
Webbe, John 168
Weber, Albert 441
Weckherlin, Elias 426
Wedgewood, Josiah 591
Weeks, Thomas 301
Weems, Phillip Van Horn 490
weight-driven clocks 156–7
  China 45, 46
weight-driven jacks 595, 596
Weiss company, USA 362
Wells Cathedral
  clock double dial 149f
  public clock 140
Welper, Eberhard 622, 626
Weltmaschine 574
Wempe company 418
Wenham, F. H. 405
Wenzel, Hermann 510
Werkstattwinke des Uhrmachers
  (Jendtitzki) 635
Werner, Caspar 186
Werner, Johann 313
Westminster chimes 469
Wetzel, Marie-Françoise 532–3, 533f
Weule, Johann Friedrich 465
Weule company, Bockenem 464, 465
  turret clock 467f
Wheatstone, Charles 497, 575
wheel-lock chain 159
Whichcote, Samuel 545
Whipple Museum, Cambridge 341n
whistles, singing birds 304
Whiston, William 318–19
Whitaker, Leah and Sarah 532
White, Charles A. 584
White, Lynn Jr 602
Whitehurst of Derby 591
Whiteside, Thomas 202
Whitestone, Sebastian 314
Whitmore, William 640
wholesalers, foreign, Switzerland 372
widows, working 532

Wiener planetenuhr 160n
Wilhelm IV, Landgrave of Hesse-
  Kassel 254, 260, 261f, 262, 270–1,
  269f
Wilkins, John 594–5
Willan, John 223
Willard, Simon 355
Willebrand, Johann Mathias 222, 222n
William, Abbot of Hirschau 121
William of Conches 99
Williams, Henry 224n
Williamson, Joseph 318
Williamson, Timothy 302, 305, 447,
  451
H. Williamson Ltd, Coventry 349,
  479, 484, 588
Willoughby, John 198
Wills, Peter 554
Wilsdorf, Hans 488
Wimborne Minster, astronomical
  clock 149
Winde, William 545
Winkel, Nicolas Dietrich 582–3, 582n
Winnerl, Joseph Thaddeus 412
winter clock (*horologia hiberna*) 11
Winterhalder and Hoffmeyer 592
wireless telegraphy 417
wisdom, allegorical figures 602
With the Watchmaker at the Bench (de
  Carle) 552
The Wizard of Oz (film) 618
Wolf, Andreas 312
Wolfe, Catherine 538
Wollaston, William Hyde 412
women in horology
  16th & 17th century 225, 531–2
  18th century developments 532–3
  19th century developments 534–5
  20th & 21st century
    developments 535–9
women's strike, Switzerland 539
women's wristwatches 474, 477, 479,
  483f
Woods, J. F. 592
Woolf, Virginia 615
workers rights, standardized time 496
workingman's clock 611, 614
workplace clocks 512–13
workshops 19, 215, 218–21, 220f
The World Turned Upside Down
  (Steen) 609, 612
worm gears/wheels 197, 260, 589, 659
  planetary models 573
  roasting-jacks 594, 595
  telescope drives 578
Worshipful Company of Clockmakers 194, 198, 197n, 200,
  348
Wright, Richard 225, 227, 546
Wright, Thomas 337, 572
wristlets 474, 475f, 476f, 478
wristwatches 193, 473, 556

  16th–19th century 473–5
  19th century developments 475–9
  20th century developments 479–93
  21st century developments 493
  automatic 352
  clip-joints 567
  East Germany 384–5
  electric 526–7
  postmodernism 441–2, 442f
  production in China 51–5
  rationalism 441
  replacing pocket watches 374
wrought iron 147
Wullf 36
Wunderkammern 163
Wundt, Wilhelm 575–6
Wuppertaler Uhrenmuseum 169f,
  186n
Württemberg, clockmaking 378–9,
  382
Wuxing (prototype wristwatch) 53
Wybrandi, Wybe 190
Wycliffe, John xiii
Wyke, John 222, 639–40
  trade catalogue 222n, 639f

X

Xavier, St Francis 61, 444
Xu Chaojun 46
Xu Guangqi 47
Xue Jixuan 41

Y

yāma 35
Yamaha Clickstation 585
Yan Su 43
Yang Tingyun 46
Yangzhou, clockmaking 52
Yantai Baoshi Clock Factory (Yantai
  Polaris) 53, 55
  anniversary clock 57f
Yantraprakara 30
Yi Jing (I-Tsing) 34
Yi Xing 43
Yongzheng 50
York Minster, astronomical clock 280
Yoshitoko Takahashi 68
ytterbium atomic clocks 509
Yūnus al-Husayn al-Asturlābī 136

Z

Zaandam clocks 183
Zamboni, Giuseppe 514
zappler (German clock) 180, 184
al-Zarqellu, Ibn 96
Zech, Jakob 158n, 162, 650
*Zeitschrift fuer Groß- und Kleinuhrmacher
  jeder Gattung* 638
Zeitung fuer Uhrmacher 637
zenith 659
Zenith company 490, 587, 588
Zephyr 20

INDEX

Zerbola (Zarbula), Giovanni Francesco 249–50
Zhan Xiyuan 44
Zhang Heng 43
Zhang Shuosen 51n
Zhang Sixun 43
Zheng, Wang 46
Zhou Hui 51n
Zhou Shuxue 45

Zimmer, Louis 280
Zinner, Ernst 120, 251
zodiacal dials 96, 234
zodiacal signs 23, 237n, 279–85, 644
   anaphoric clock 11
   Arab-Islamic 86, 88, 92, 96
   Chinese xiv
   English 212–13, 237
   European 106, 125, 146, 148, 177, 193, 196, 232, 235
   Indian 28, 39
   Meridian of Augustus 23
   musical and automaton clocks 291, 299
   planetary clocks 253, 254, 255t, 257–60, 263, 267–74
   planetary models 570, 572
   position of sun 20
Zou Boqi 46
Zuanshi (Diamond) brand name 53
Zug watchmaking district 195
Zurlauben, Hans Jacob 195
Zvarnots, sundial 108